Studies in History and Philosophy of Science

Volume 51

More information about this series at http://www.springer.com/series/5671

Andrea Strazzoni

Burchard de Volder and the Age of the Scientific Revolution

Springer

Andrea Strazzoni
Gotha Research Centre of the University of Erfurt
Gotha, Germany

ISSN 0929-6425 ISSN 2215-1958 (electronic)
Studies in History and Philosophy of Science
ISBN 978-3-030-19877-0 ISBN 978-3-030-19878-7 (eBook)
https://doi.org/10.1007/978-3-030-19878-7

This Springer imprint is published by the registered company Springer Nature Switzerland AG.
The registered company address is: Gewerbestrasse 11, 6330 Cham, Switzerland

Acknowledgments

The preparation of this book has been made possible by a Christoph-Martin-Wieland Postdoctoral Fellowship, which allowed me to focus on my research on Burchard de Volder at the Gotha Research Centre of the University of Erfurt over the years 2016–2018. My deepest thanks go to the directors of the center: Martin Mulsow, Markus Meumann, and Iris Schröder, who supported and enabled my research in Gotha. Very special thanks go to my colleagues Dirk Jansen (who crucially helped me with translations of Dutch texts) and Lucinda Martin (who helped me with transcriptions from English manuscripts). The completion of the book has equally been facilitated by the revision of the English text by Mark Rogers. Last but not least, I wish to thank, for their readiness and openness, the editorial body of the Springer series Studies in History and Philosophy of Science: Stephen Gaukroger, Lucy Fleet, and Prasad Gurunadham.

Contents

List of Figures

List of Tables

Chapter 1
Introduction

Abstract In this chapter, I briefly introduce the figure of Burchard de Volder (1643–1705) in his historical significance, as well as the state of the art of the studies on him. Moreover, I consider how my study relates to the current historiography of philosophy and science, and I provide a summary of the contents of the book.

Amongst the various judgments on him made by the contemporaries of Burchard de Volder (1643–1705), four are particularly illuminating on his intellectual path. First, once Jakob Bernoulli, who had witnessed De Volder's experimental lectures at Leiden in the early 1680s, visited the professor of philosophy Johannes de Raey at Amsterdam, and expounded to him the results of De Volder's teaching of what we know as the Boyle-Mariotte law, De Raey (according to Bernoulli's *Methodus ratiocinandi, sive Usus logicae in praeclaro aliquo phaenomeno physico enodando*, 1686),

> he immediately looked at me with a harsh face, then he asked who I was, [and] after [this] he was indignant, irritated, inveighed against experimental philosophy, and labelled it as histrionic, and he almost [...] threw me out of his house by force. This was, actually, the solution of that famous Cartesian. We, on the other hand, in order to address this topic less militarily, and more philosophically, establish before anything else some principles [...].[1]

In turn, the brother of Jakob Bernoulli, Johann, in his letter of 31 October 1722 to Willem Jacob 's Gravesande – who was going to assume the professorial chair of natural philosophy left vacant by De Volder – labelled 's Gravesande's predecessor a "rigid Cartesian," who would be convinced of the idea of living force (i.e. that the force of motion of a body is proportional to the square of its speed), only after lengthy efforts by Leibniz and Johann Bernoulli himself.[2] Leibniz, in turn, in

[1] Translations, unless explicitly indicated or unless referring to a modern, translated edition, are mine. Original text from Bernoulli 1686a, 257–258. For the full quotation and more details, see Sect. 5.6.3.3, *De Volder's assumption of Boyle's law in the works of his students*.

[2] The letter is included in the online source *Briefwechsel von Johann I Bernoulli. Korrespondenz zwischen Bernoulli, Johann I und Volder, Burchard de* (http://www.ub.unibas.ch/bernoulli/index.

A. Strazzoni, *Burchard de Volder and the Age of the Scientific Revolution*, Studies in History and Philosophy of Science 51,
https://doi.org/10.1007/978-3-030-19878-7_1

commenting in a letter to Pierre-Daniel Huet of April 1695 upon De Volder's *Exercitationes* (1690–1693) against Huet's *Censura philosophiae Cartesianae* (1689), noted that "it appeared to me that he [i.e. De Volder] not rarely deserts Descartes while he seems to defend [him]."[3] Eventually, Joachim Lange (theologian at Halle) labelled De Volder as the "little flower plucked from Spinoza's garden" in the second edition of his *De causa Dei* (1727, first edition 1723).[4]

These judgments and statements provides us with a picture of De Volder as a complex figure: in whom Cartesianism, Spinozism, an experimental approach, as well as anti-Cartesian instances merge and cannot be easily captured. As Tammy Nyden has recently put it, "what makes de Volder an interesting and valuable object of study is precisely that he does not fit neatly into the categories of Cartesian, Newtonian, and Spinozist, or perhaps we should say, his case indicates how untidy these categories actually were in the seventeenth century."[5] In fact, at a level of generality De Volder can be considered a witness to a change between two overall, identifiable natural-philosophical models: namely, to the passage from the natural philosophy of René Descartes (which in the 1670s had become a standard in the Dutch academic framework) to the emergence of Newtonianism, which De Volder himself – *teste*, amongst others, Jean Le Clerc[6] – contributed to making known in Continental Europe. A passage which was enriched – especially in the Dutch academic framework – by the emergence of experimentalism as a crucial component of natural philosophy, and of Spinozism as (at least according to some interpretations) the extreme derailment of Cartesianism in metaphysics. De Volder, who witnessed all these four components, is thus per se worthy of a dedicated study, as revelatory of what can be roughly defined as a change in paradigm, i.e. of its historical and theoretical underpinnings. Namely, he was a witness of the actual ways in which such two models were confronted in printed texts and academic disputations, quarrels, teaching practices, in epistolary exchanges, as well as of the scientific and philosophical problems (or: natural-philosophical and logico-metaphysical problems respectively) at stake in such intellectual exchanges. More particularly, De Volder's significance in the history of philosophy and science stands in his witnessing the final evolution of Cartesianism in natural philosophy: namely, in providing us with a picture of a Cartesianism more lively than that offered to us by the proponents of the 'rationalism vs. empiricism' dichotomy, which had dominated until recently the histories of early modern philosophy and science.[7] In fact, only the erosion of the rationalism-empiricism divide (in which Descartes and his followers were placed in the first category) has allowed historiogra-

php/Kategorie:Volder_Burchard_de-Bernoulli_Johann_I (accessed 13 January 2019)). For the full quotation and more details, see Sect. 2.4.2, *De Volder's retirement, death and legacy*.

[3] GP III, 19. For the full quotation and more details, see Sect. 2.2.3, *The mid-1670s clash at Leiden and the foundation of the experimental theatre*.

[4] Lange 1727, 47. I owe this reference and translation to Israeli 2001, 547. For the full quotation and more details, see Sect. 3.1.4.2, *The Wittich-Driessen quarrel – and beyond*.

[5] Nyden 2013, 228–229.

[6] See Sect. 2.3.3, *Mathematician and scientific expert*.

[7] For a thorough discussion, see Dobre and Nyden 2013, *Introduction*.

phy to capture figures like De Volder, who do not clearly fit into either of the two groups, and which are potentially more easily matched by new categories such as those of 'experimental' and 'speculative' philosophy.[8]

To date, our view of De Volder's philosophical and scientific thought is, however, incomplete. In fact, his views have been for long considered in the light of the judgment given by Le Clerc, according to which he came to be disenchanted with Cartesianism through the theories of Englishmen such as Isaac Newton, but was unwilling to develop a new system of his own.[9] De Volder's significance, indeed, has for long been recognized mostly in his role as founder of the Leiden experimental theatre (1674–1675),[10] as a 'discontent' Cartesian who converted to Newtonianism,[11] and as a correspondent of Leibniz.[12] Only from the 1980s has his thought been considered worthy of dedicated studies. Wim Klever has focused on De Volder's supposed 'crypto-Spinozism',[13] while more recently Gerhard Wiesenfeldt, Tammy Nyden and Wiep van Bunge have contextualized De Volder's teaching activities through experiments in the Leiden didactic and philosophical tradition,[14] and have considered their epistemic premises. Nyden has highlighted how "*scientia* and experiment are essential components of [De Volder's] natural philosophy,"[15] while Wiesenfeldt has provided recognition of some handwritten notes describing De Volder's experimental lectures, titled *Experimenta philosophica naturalia* (1676–1677) – a transcription of which is appended to this book[16] – comparing them with De Volder's coeval disputations *De aëris gravitate* (1676–1678).[17]

And yet, our knowledge of De Volder's philosophical and scientific thought remains incomplete. More precisely, we lack a unitary view of how De Volder's positions on the method of natural philosophy relate to his actual natural-philosophical theories – which have been barely surveyed in themselves – and to his broader metaphysical views (intended as a theory of mind and knowledge, and rational theology). Besides his not clearly fitting into the above-mentioned historiographic categories, the lack of such a view is due to the fact that the failure of

[8] As to these categories, see Anstey 2005; Anstey and Vanzo 2016.

[9] Le Clerc 1709, 382 and 398. See Sect. 3.2.3.2, *De Volder's methodological liberalism – and its continuity over time*.

[10] De Pater 1975.

[11] Sassen 1970; Ruestow 1973, 89–112. Against such interpretation, see Krop 2003d.

[12] Russell 1927–1928; Hall 1982; MacDonald Ross 1987; Rey 2009a, b, 2016; Lodge 2013.

[13] Klever 1988, 1989, 1997, chapter 11, 2000. Klever's positions have been criticized in Lodge 2005.

[14] Wiesenfeldt 2002, 54–64 and 99–132, 2003; Nyden 2013, 2014. On De Volder and the Leiden experimental tradition, see, moreover, Gori 1972; Otterspeer 2001; Mijers 2012, chapter 2; Van Bunge 2013, 2017.

[15] Nyden 2013, 244.

[16] Wiesenfeldt 2002, chapter 3. These notes were first brought to the fore and commented on in the doctoral dissertation of Adriaan de Hoog: De Hoog 1974.

[17] De Volder 1676–1677, 1676–1678. Both the lectures and the disputations were public: on the various forms of academic teaching in Dutch universities, see Wiesenfeldt 2002, 118–123; Van Miert 2009, part 2.

Cartesian physics in the second half of the seventeenth century, that is, its speculative claims, together with its lack of a mathematical formulation of natural laws, and the ambiguous use of experience and experiment have brought about an image of Cartesianism as a dead branch in the history of early modern philosophy and science. Accordingly, scant attention has been paid to the interrelation of physics, metaphysics, and other branches of philosophy apart from the case of Descartes. As a consequence, Descartes's followers have mostly been labelled uninteresting figures in early modern philosophy and science, whereas their master brought about an actual novelty in the history of philosophy. Eventually, a third factor hindering the appreciation of De Volder's thought is that he never made his theories public in a unitary form, nor did he allow them to circulate outside the University: his works which appeared as self-standing monographs, indeed, were not authorized by him, and were the result of the collation of his series of academic disputations.[18] Moreover, besides his series of disputations and his academic orations,[19] De Volder's theories are expounded in handwritten sources (i.e. the aforementioned *Experimenta* and his academic *dictata* on Descartes and Jacques Rohault),[20] in his vast correspondence,[21] and in the disputations he presided over, but not included in the aforementioned, unauthorized editions.[22] Moreover, many of the disputations he presided over (especially those taking place after 1680), and in which he figures amongst the dedicatees – so that we can legitimately assume that he had a hand in their composition – were officially written by the defenders, who figure as their authors.[23] In other cases, disputations dedicated to him were *pro gradu*, so that in both cases the actual influence of De Volder has to be assessed from the ground.[24]

---

[18] These are his *Disputationes de rerum naturalium principiis* (ed. 1674–1676 as disputations printed by the academic publishers, 1681 as a monograph), *Disputationes de aëris gravitate* (as in the previous case, 1676–1678 and 1681), *Disputationes contra atheos* (1680–1681 and 1685), and his *Exercitationes* (1690–1693 and 1695) against Huet's *Censura philosophiae Cartesianae*. For more details, see the bibliography.

[19] De Volder 1682, 1698. On the 'academic rituals' in the early modern Dutch context, see Wiesenfeldt 2016.

[20] I extensively deal with De Volder's handwritten sources in Sects. 2.2.2, *The academic* dictata, and 2.2.4, *The sources for De Volder's experimental lectures*.

[21] I extensively deal with De Volder's correspondence in Chap. 2.

[22] See, amongst others, De Volder 1671a, b, 1672, 1673a, b, 1675. For complete bibliography, see the section of bibliography.

[23] De Volder and Enyedi 1671; De Volder and Van Alphen 1672; De Volder and De la Bassecour 1672; De Volder and Lufneu 1676; De Volder and Helvetius 1677; De Volder and Szathmári 1681; De Volder and Köleséri 1681; De Volder and Nemethi 1682; De Volder and Derecskei 1682; De Volder and Van Alphen 1684; De Volder and Vander Codde 1684; De Volder and Von der Lahr 1684; De Volder and Van Bronchorst 1685; De Volder and Huneken 1685; De Volder and Van den Honert 1686; De Volder and Serrurier 1688; De Volder and Mandeville 1689, De Volder and Vander Tak 1689, De Volder and Voogd 1691; De Volder and Huszthi 1693; De Volder and Casembroot 1694; De Volder and Robberghtsten 1694; De Volder and Rouxel 1695; De Volder and Westhovius 1698; De Volder and Van Bashuysen 1698; De Volder and De Beaumont 1698a, b, c, De Volder and Copius 1699.

[24] De la Bassecour 1672; Rattrey 1679, Köleséri 1681; De Sadeler 1684; Wariensis 1685; Schuyl 1688a, b; Boerhaave 1690; Serrurier 1690; Oosterdijk Schacht 1693; Casembroot 1696; Gale

Accordingly, the study of De Volder's thought poses two intriguing challenges: first, (1) his theories need to be studied from the ground, namely, by considering a large bulk of primary sources largely neglected until now, either handwritten or printed. These texts need to be put in the historical context of their publication and, in the case of those whose attribution to De Volder is ambiguous (for instance, those whose defender was also the author), one needs to ascertain the role that De Volder played in their preparation. Furthermore, (2) the study of De Volder's thought can be the testbed for two, novel approaches to the history of philosophy and science. First of all (2.a), De Volder's theories can be better mapped other than by the aforementioned traditional categories, those of 'experimental' and 'speculative philosophy'. Indeed, in De Volder's theories Cartesian notions are embedded in an experimental approach consisting, for instance, in the repetition of the experiments of Robert Boyle. Also, De Volder criticized Descartes's account of the heart beat on an experimental i.e. chemical basis as early as his *De natura* (1664), well before his supposed farewell to Cartesianism in his *Oratio de rationis viribus* (1698), where, in turn, he attacked the use of the notions of pores and particles in medicine as not based on sound experiments. De Volder's overall approach, therefore, is potentially better captured through the categories of 'experimental' and 'speculative philosophy' rather than through those of 'Cartesianism' or 'rationalism'. Second (2.b), the broadness of De Volder's interests and commitments requires a further, new approach, which I have identified as the 'integrated history and philosophy of science' (&HPS).[25] With this approach, De Volder's (a) declared epistemic-methodological tenets, and (b) his actual scientific practices (both in discovery and exposition or teaching) are studied by considering

1. their mutual relations: this aspect turns out to be particularly important in De Volder's case, as he did not adhere to a univocal methodology, therefore his method (both purported and actual) has still to be clarified;
2. their historical context and development, depending on the appearance of concurrent theories, debates, and quarrels, concerning natural philosophy, metaphysics, theology;
3. their more systematic aspects. These are, for instance, the explanatory role of De Volder's scientific theories, their conditions of certainty, their hypothetical status, and the role of experiments in establishing or confirming a theory.

It should be noted that this integrated approach also concerns De Volder's metaphysical theories, mainly because in the seventeenth century what we now label as the scientific and philosophical or metaphysical aspects could not be clearly disentangled. The study of these theories, concerning the nature and functioning of mind,

1699; Zoutmann 1702; Vander Ghiessen 1703. The complete list of De Volder's students who graduated under him has been reconstructed and discussed in Wiesenfeldt 2002, 247–257. Wiesenfeldt has based himself, amongst others, on the manuscript source ASF (Archief Senaat en Faculteiten) of the Leiden University, 348 and 349, extant at the Leiden University Library.

[25] For a more detailed account of contemporary perspectives on and methods of the historiography of early modern philosophy, see Laerke et al. 2013; Lenz and Waldow 2013.

and rational theology, is necessary in order to ascertain the positions of De Volder on the theory of knowledge, and on the use of metaphysics with respect to physics. Moreover, given the fact that he directly intervened in the seventeenth century philosophical debates, his own natural-philosophical thought was historically and conceptually driven by coeval discussions in metaphysics. In order to re-assess De Volder's role in the history of early modern philosophy and science, I will follow a &HPS approach in proposing here the first study of the thought of De Volder that takes account of, on the one hand the texts printed as monographs and his orations, and on the other his academic disputations not included in his unauthorized monographs, and the handwritten sources on his teaching, i.e. the *Experimenta philosophica naturalia* and his various academic *dictata*, or *collegedictaten* on Descartes's *Principia philosophiae* and *Meditationes metaphysicae*. This book is thus intended to be a contribution to the current state of studies on Dutch and European Cartesianism and on its interrelation with the seventeenth and early eighteenth century history of science.

The scholarly tradition on the dissemination and reception of Descartes's philosophy and science – for which I address the reader to the introduction by Mihnea Dobre and Tammy Nyden to the collective volume *Cartesian Empiricisms* (2013) for an up-to-date, thorough account – is nowadays substantiated by a huge body of literature which had its initiators in the classic studies of Ast, Damiron, Bouillier, Cousin, Ueberweg, Monchamp, Bohatec, Brulez, Brunet, Mouy, Thijssen-Schoute and Dibon.[26] Cartesian studies have mostly focused on those geographical contexts in which Cartesianism inspired the curricula of universities – as in the Dutch and German areas – and of learned circles and scientific academies – as in France.[27] As to the Dutch context, Cartesian studies have been largely dominated by an approach focusing on the institutional clashes between philosophers and theologians, and on the interrelations of Cartesianism and Spinozism.[28] In Germany, Cartesianism has been studied especially with regard to its uses in medicine,[29] to its influence on seventeenth and eighteenth century *Schulphilosophie*,[30] and to its contribution to the German Radical Enlightenment.[31] In France, Cartesian studies have focused mostly on the uses of Cartesianism in natural philosophy,[32] theories of matter and mind, and

[26] Ast 1807; Damiron 1846; Bouillier 1854; Cousin 1866; Ueberweg 1866; Monchamp 1886; Bohatec 1912; Brulez 1926; Brunet 1926; Mouy 1934; Dibon 1954; Thijssen-Schoute 1954.

[27] On the history of French scientific institutions, see Brown 1934; Brockliss 1981, 1987; Hahn 1971; Hirschfield 1981.

[28] Brunet 1926; Dibon 1954, 1990; Thijssen-Schoute 1954; Ruestow 1973; De Hoog 1974; McGahagan 1976; Verbeek 1992a; Van Ruler 1995; Van Bunge 2001. See also Dijksterhuis 1950; Vanpaemel 1985; Frijhoff and Spies 2004; Schmaltz 2005a, 2016.

[29] Rothschuh 1953, 1968; Trevisani 1992; Smith 2013; Omodeo 2017; Theis et al. 2009. See also Angyal 1941.

[30] Althaus 1914; Wundt 1939, Freedman 1984; Wollgast 1988; Mulsow 2009. See also Blum 1998.

[31] Mulsow 2002, 2015; Suitner 2016.

[32] Mouy 1934; McClaughlin 1977, 1996, 2000; Vanpaemel 1984; Clarke 1989, Brockliss 1995; Roux 1998, 2006, 2013a, b; Des Chene 2002; Goldstein 2008; Shank 2008; Dobre 2010, 2013c, 2017; Borghero 2011; Milani 2010, 2013, 2014a, 2015; Ariew 2014; Spink 2018.

the polemics these provoked.[33] Furthermore, those contexts in which Cartesianism did not constitute a dominant paradigm (as the emergence of new paradigms was embodied in more experimentally-oriented bodies of knowledge) – such as in England[34] and Italy[35] – have enjoyed a remarkable focus of attention. The view of the transnational development of Cartesian philosophy, finally, is at a mature stage of development.[36] This study, actually, aims at bridging the results achieved in such a body of literature with the efforts achieved in the history of science – especially in those studies paying attention to the philosophical underpinnings of scientific theories. As far as the Dutch context is concerned, I refer in particular to the discussions of the natural-philosophical ideas of Simon Stevin, Willebrord Snel, Isaac Beeckman, Descartes himself, Christiaan Huygens,[37] as well as of Newtonians like Bernard Nieuwentijt and 's Gravesande,[38] and – in medicine – those of the Cartesian school (such as Henricus Regius, Florentius Schuyl, Cornelis van Hogelande, Jan Swammerdam, Theodoor Craanen), as well as Franciscus Sylvius, Archibald Pitcairne, Frederik Ruysch, Niels Steensen, and Hermann Boerhaave.[39] As far as the French context is concerned, this study builds upon the recent scholarship on Blaise Pascal, Gilles Personne de Roberval, Marin Mersenne, Pierre Gassendi, Edme Mariotte, as well as on Cartesians like Rohault, Fontenelle, and Pierre-Sylvain

[33] Watson 1966, 1998; McClaughlin 1979, Lennon 1993; Schmaltz 2002, 2005b; Shelford 2007; Lennon 2008; Nadler 2011b; Ariew 2013; Milani 2014b.

[34] Webster 1969, Gabbey 1982; Hutton 1990; Fouke 1997; Atherton 2005; Jesseph 2005; Dessì and Lotti 2011; Jalobeanu 2011; Hatfield 2013.

[35] Belgioioso 1985, 1992, 1999, 2005; Armogathe 2005.

[36] For a comparative treatment, see Schmaltz 2016, and the collections of essays of Lennon, Sorell, Schmaltz, Borghero, Del Prete, Garber, Roux, Antoine-Mahut, Gaukroger: Lennon et al. 1982; Sorell 1993; Lennon 2003; Schmaltz 2005a; Borghero and Del Prete 2011; Garber and Roux 2013; Kolesnik-Antoine 2013a; Antoine-Mahut and Gaukroger 2016; Schmaltz 2016.

[37] Aiton 1955a, 1957, 1958a, b, 1972; Blackwell 1966, 1977; Dijksterhuis 1970; Sassen 1970; Elzinga 1971; Westfall 1971; Hatfield 1979; Guéroult 1980; Mahoney 1980, 1995; Gaukroger 1980, 2000a, b, 2002; Grabow 1985; Snelders 1989; Mormino 1990a, b, Garber 1992; Schliesser and Smith 1996, 2000; Schliesser 1997, 2000; Garber 2001, 2002; Trevisani 1992; Voss 1993; Erlichson 1994; Chareix 1996, 2004; Slowik 1996, 1997; Yoeder 1998; Gaukroger et al. 2000; Fuchs 2001; Gaukroger and Schuster 2002; Jalobeanu 2002; Slowik 2002; Dijksterhuis 2004; Andriesse 2005; Lüthy 2006; Arthur 2007; Osler 2007; Calero 2008; Ducheyne 2008; Howard 2008; Manchak 2009; Anstey and Jalobeanu 2011; Pasnau 2011; Schuster 2013; Van Berkel 2013; Petrescu 2013; Yoder 2013; Van der Schoot 2014; De Grijs 2015; Chalmers 2017; Dobre 2017; Antoine-Mahut and Roux 2018.

[38] Gori 1972; Lunsingh Scheurleer and Posthumus Meyjes 1975; Vermij 1991, 2002; Van Berkel et al. 1999; Jorink 1999; Jorink and Maas 2012; Van Besouw 2016; Mojet 2016, 2017; Van Bunge 2018.

[39] Baumann 1949; Underwood 1972; Lindeboom 1975; Luyendijk-Elshout 1975; Rothschuh 1968; Sassen 1970; Beukers 1980, 1999; Guerrini 1987; Gariepy 1990; Trevisani 1992; Bitbol-Hespériès 1993; Cook 1993; Lech 1993; Kardel 1994; Hansen 1996; Cobb 2000, 2006; Verbeek 2000; Debus 2001; Fuchs 2001; Knoeff 2002; Krop 2003d, Schmaltz 2005a; Ragland 2008; Miniati 2009; Jorink 2010; Bellis 2013; Kolesnik-Antoine 2013b; Jorink 2016a, b; Schmaltz 2016; Bertoloni Meli 2017; Omodeo 2017; Andrault and Laerke 2018; Strazzoni 2018b.

Régis,[40] while, as to the English environment, on the studies on Boyle, Robert Hooke, John Wallis, the activities of the Royal Society, and of course Newton.[41] An analysis of De Volder's ideas, eventually, cannot ignore the scientific debates in Italy – viz. the ideas of Galileo Galilei, Evangelista Torricelli, Giovanni Alfonso Borelli, and the Accademia del Cimento[42] – and the German and Swiss areas: Leibniz, Otto von Guericke, Johann Christoph Sturm, Gaspar Schott, Johann and Jakob Bernoulli.[43] The ideas of all these authors, actually, constitutes the background to the development and dissemination of those of De Volder, who came to discuss or overtly confront himself with most of them.

This abridgement between different bodies of ideas and literature (both primary and secondary) will guide this study of De Volder's figure and role in early modern philosophy and science, guided by the following theoretical and historical themes, treated in four main chapters. Thus, after the reconstruction (in Chap. 2) of De Volder's intellectual formation, career and network, I will focus (in Chap. 3) on his ideas in metaphysics, which in the seventeenth century was understood as rational theology, as the discipline concerning the functioning human mind, as ontology, as well as a discipline provided – according to Descartes – with a foundational role

[40]Thirion 1907–1909; Hoskin 1961; McClaughlin 1977, 1996; Vanpaemel 1984; Gabbey 1986; Joy 1987; Lennon 1993; Brockliss 1995; Roux 1998, 2006, 2012a, b, 2013b; McClaughlin 2000; Jones 2001; Schüller 2001; Palmerino 2001; Shea 2003; Des Chene 2002; LoLordo 2005, 2007; Goldstein 2008; Milani 2010, 2013, 2014a, b, 2015; Torero-Ibad 2011; Grosslight 2013; Dobre 2013a, b; Schmaltz 2016; Spink 2018.

[41]Snow 1924; Armitage 1950; Aiton 1955b; Rosenfeld 1969; Home 1968; Westfall 1971, 1981; Gabbey 1980; Shapin and Schaffer 1985; Cook 1986; Débarbat 1986; Snelders 1989; Pourciau 1991; Weinstock 1992; Bertoloni Meli 1993, 2006; Mahoney 1993; Arthur 1994; Erlichson 1994; Wilson 1994; Slowik 1997; Clericuzio 1998, 2000; Guicciardini 1999, 2005; Schüller 2001; Anstey 2005; Eaton 2005; Nauenberg 2005; Newman and Principe 2002; Newman 2006; Shank 2008; Borghero 2011; Ducheyne 2011; Murray et al. 2011; Jalobeanu 2011; Hyslop 2015; Anstey and Vanzo 2016; Kochiras 2017; Tho 2017.

[42]Gliozzi 1930; Knowles Middleton 1963, 1971; Koyré 1973; Baldini 1976; Torrini 1977; Le Grand 1978; Galluzzi 1979; Naylor 1980; Prager 1981; Zouckermann 1981; Redondi 1983, 1985; Miniati 1987; Nonnoi 1988; Shea 1989; Renn 1992; Gorman 1994; Conti 1997; Thurston 1999; Palmerino 2000, 2001; Fredette 2001; Wohlwill 2001; Shea 2004; Des Chene 2005; Pope 2005; Palmieri 2005; Bertoloni Meli 2006; Boschiero 2007; Beretta et al. 2009; Pisano and Capecchi 2010; Valleriani 2010; Maffioli 2011; Shank 2012; West 2013.

[43]Slaby 1907; Russell 1927–1928; Guéroult 1934; Aiton 1955b, 1989; Grieser 1969; Fox 1970; Iltis 1971, 1974; Costabel 1973; Gale 1973, 1988; Fleckenstein 1977; Heimann 1977; Bernstein 1981; Bosl 1983; Sayre-McCord 1984; Bell 1986, chapter 8; Dunham 1987; MacDonald Ross 1987; Hartz and Cover 1988; Khamara 1988; Palaia 1989; Khamara 1993; Lariviere 1989; Mugnai 1990, 1992; Bodeüs 1991; Van Helden 1991; Sierksma 1992; Bazerman 1993; Bertoloni Meli 1993; Van Maanen 1993; Adams 1994; Arthur 1994, 2013; Gorman 1994; Guicciardini 1995; Rutherford 1995; Shafer 1996; Fichant 1998; Cover et al. 1999; Look 1999; Mercer 1999; Sierksma and Sierksma 1999; Mercer 2001; Israel 2001, chapter 23; Ahnert 2002; Freudenthal 2002; Herrmann and Platz 2003; Lodge 2003; Roberts 2003; Gaab et al. 2004; Phemister 2005; Hackmann 1979; Hartz 2006; Knobloch 2006; De Risi 2007; Harsch 2007; Antognazza 2008; Futch 2008; Jauernig 2008; Garber 2009; Rey 2009a, b; Conlon 2011; Smith and Nachtomy 2011; Laerke 2013; Lodge 2013; Schneider 2013; Secundant 2016; Guicciardini 2017; Tho 2017; Adomaitis 2019; Garber and Tho 2018.

with respect to natural philosophy.[44] In particular, I will focus on De Volder's ideas on certainty and on the method of natural philosophy by considering them with respect to Descartes's own ideas, and to their elaboration in the Dutch context. In this chapter – which also has the broader function of a historical introduction – I will also consider De Volder's (alleged) Spinozism, for the reason that his dealing with the theory of knowledge, and with the function of metaphysics for natural philosophy led him to confront Spinoza's ideas on substance, God and necessitarianism. In the chapter, I will argue that De Volder stood for a sort of 'methodological liberalism' according to which experience can offer a kind of certainty which is ultimately indubitable, and which allowed him to include in natural philosophy the conclusions reached through observation and experiment. Moreover, I will show that De Volder pursued a detachment of physics from metaphysics which was historically rooted in Dutch Cartesianism, and which was theoretically functional to a principle of economy in explanation. Of course, the detachment of physics from metaphysics, and the acquisition of an 'experimental' kind of certainty in physics are related, as the project of an absolutely or metaphysically certain natural philosophy pursued by Descartes came to be weakened in De Volder's hands. In turn, I will show that De Volder's dealing with the (metaphysical) ideas of substance, attributes and properties, as well as his theory of certainty as indubitability led him to embrace a form of physical and mental necessitarianism which led him to consider favourably some of Spinoza's ideas (even if he cannot be labelled a thorough-going Spinozist).

This assessment of De Volder's positions on methodology and metaphysics will constitute the background for the treatment of his ideas in natural philosophy. So, in Chap. 4 I will consider De Volder's treatment of the idea of material substance and of the problem of its source of activity, i.e. of the cause and laws of motion. These issues dominated not only his correspondence with Leibniz, but also his natural-philosophical disputations and handwritten texts, in which he aimed at not reverting to the (metaphysical) idea of God in accounting for the movement of bodies. This notwithstanding, De Volder never arrived at a solution to the problem of activity, as he did not accept Leibniz's views on substance, or Nicolas Malebranche's occasionalism, while at the same time he accepted the idea that the quantity of movement is not necessarily conserved in the world. His detachment of physics from metaphysics, moreover, is observed in his treatment of the problem of the cohesion of bodies – traced back by Descartes to the metaphysical idea of rest, and which De Volder solved by experimental means (viz. by considering the pressure of air, which is the proper subject of Chap. 5). In turn, it was through experiments that De Volder came to accept the correct laws of impact for elastic and non-elastic bodies, already formulated by Huygens, Mariotte, Wallis, and Christopher Wren.

I will then focus – in Chap. 5 – on De Volder's treatment of hydrostatics, which was at the core of his teaching activities at the Leiden experimental theatre. The chapter is devoted to the discussion of De Volder's attempt to combine two main models in hydrostatics. First, the Cartesian one – which Descartes expounded by

[44] As to Descartes's foundationalism and its reception, see in particular Garber 1992, 2006; Dobre 2017; Strazzoni 2018a.

providing his theory of gravity – according to which bodies (either solid or fluid) provided with the same specific weight constantly nullify their reciprocal pressure, so that, for instance, there is no increasing pressure in deep water. Second, the Archimedean model (assumed over the same years by Boyle) according to which the conditions of the floatation of bodies are determined by the different pressures exerted within a fluid. In the chapter, I will show that these two models are ultimately inconsistent with each other, so that in refraining from publishing his disputations, De Volder might have been partially motivated by this issue. Moreover, I will consider De Volder's application of such a hydrostatics model to the explanation of the effects of the pressure of air, and its partial overcoming by his assumption, as a key factor in pneumatics, of the idea of the elasticity of air. He applied this concept also to physiology, and it further bears witness to his appropriation of the Boyle-Mariotte law.

Finally, in Chap. 6 I will consider De Volder's approach to cosmology, which most probably led him to the relinquishment of Cartesianism in the early 1700s. I will argue that the appearance of Newton's *Philosophiae naturalis principia mathematica* (which De Volder received as early as 1687) – containing a striking criticism to Descartes's vortex theory of planetary motions – and the failure of Huygens to provide experimental confirmation of his own alternative model, with which he corrected Descartes's model of gravity and opposed Newton's idea of universal gravitation, eventually led De Volder to renounce teaching natural philosophy, as he could accept neither Newton's nor the Cartesian model. In particular, I analyse De Volder's role in evaluating the results obtained by Huygens in two distinct trials to calculate the longitude at sea by means of pendulum clocks, whose shortening according to latitude was calculated by Huygens by taking into account his model of gravity. The outcome of these trials led De Volder not accept Huygens's theory, and ultimately to dismiss Descartes's model. Moreover, I consider De Volder's attempt to combine Newton's and Descartes's ideas in the explanation of the movement of the Moon, a case in which Descartes's vortex theory was immune to Newton's criticism, but which could not be extended to the explanation of planetary movements.

Since the last chapter testifies to De Volder's eventual dismissal of the Cartesian model, which he had already come to criticize since his treatment of the laws of motion, and of his missed acceptance of the models of Newton and Huygens, it also serves as the conclusion to the book. Instead of a concluding chapter; instead, I provide an appendix, namely a transcription (with annotations) of what is probably his most valuable text, the handwritten report (*Experimenta philosophica naturalia*, in English) of the experimental lectures he carried out at Leiden, which is extensively introduced and commented upon in the course of the book.

Chapter 2
The Life and the Works of Burchard de Volder

Abstract In this chapter I reconstruct De Volder's education and intellectual career at Amsterdam, Utrecht and Leiden, by taking into account the biographical elements extant on him – including those on his family, which are important as they offer insights into his Menonnite background, and by extent into De Volder's ideas on religion – as well as by offering a recognition of the sources for him (especially handwritten ones), of his activities as the founder of the Leiden experimental theatre, and of his epistolary network.

Most of the biographical information on Burchard de Volder comes from two mains sources: the *Laudatio funebris* that Jakob Gronovius (1645–1716), professor of history and Greek at Leiden,[1] read at the University on 20 April 1709, and the *Éloge de feu Mr. De Volder Professeur en Philosophie et aux Mathematiques, dans l'Academie de Leide*, written by Jean Le Clerc (1657–1736) a few weeks after Gronovius's *Laudatio*[2] and published in the issue 18 of the *Bibliothèque choisie* (1709). Both Gronovius and Le Clerc knew De Volder in person, and relied on information not extant in any other printed source.[3] All the other biographies of De Volder – which appeared in various journals and encyclopedias in the eighteenth and

[1] On him, see Graupe 2012, section 3.

[2] "Mr. Gronovius, Professeur aux Belles-Lettres, a fait son éloge à Leide, il y a peu de semaines, par ordre du Sénat Academique," Le Clerc 1709, 347. Whenever necessary, I have partially modernized quotations from unedited primary printed sources.

[3] Le Clerc used to meet De Volder in Amsterdam, which De Volder visited twice a year; they had known each other since 1689 circa: see Le Clerc 1709, 347 and 367: "Mr. De Volder, que j'aî eu l'honneur de connoître, depuis plusieurs années, & qui ne venoit guere en cette ville, où il faisoit un petit voyage une, ou deux fois l'année, sans que j'eusse l'avantage de l'entretenir quelque heures, sur toutes sortes de sujets. [...] Mr.de Volder, au moins depuis une vintaine d'années en çà, que j'ai eu l'honneur de le connoître [...]." Only one letter is extant from their correspondence, viz. a cover letter for the sending of Le Clerc's *Ars critica* (1696) to De Volder, whose minute is dated 1697: it is transcribed in Le Clerc 1987–1997, volume 2, 257 (letter 281).

A. Strazzoni, *Burchard de Volder and the Age of the Scientific Revolution*, Studies in History and Philosophy of Science 51,
https://doi.org/10.1007/978-3-030-19878-7_2

the nineteenth century – relied on these two main sources.[4] Moreover, essential biographical and historical sources on De Volder are the *Bronnen tot de geschiedenis van de Universiteit Leiden* (1913–1924) published by Philip Christiaan Molhuysen, the *Album studiosorum* of the universities of Utrecht and Leiden, the city archives of Amsterdam and Leiden, the funerary orations De Volder wrote for his friends and colleagues Sibert Coeman (1679) and Lucas Schacht (1689), his *Oratio de novis et antiquis* (1699, posthumously published by Hermann Boerhaave (1668–1738) in 1709), and his *Oratio qua sese laboribus academicis abdicavit* (1705). Another important source is the *Reisejournal* of Gottlieb Stolle and his fellow Georg Theophil Hallmann, who had a colloquium with De Volder at Leiden on 18 July 1703.[5] In what follows, I will reconstruct De Volder's life by relying on these sources, as well as on his correspondence and on the relevant information provided by his other texts, correspondence, and other primary sources referring to him.

2.1 Early Life and Studies

Burchard de Volder was born in Amsterdam on 26 July 1643, son of Joost de Volder (1607/1608–1669) and his second wife Margareta Burgers Lievelt, whom he had married on 17 December 1639. Joost de Volder was a landscape painter, active in Haarlem until July 1643, when he moved to Amsterdam with his wife. In Haarlem, Joost was a prominent figure in the Guild of Saint Luke, of which he had been member since 1632, being its vice-secretary in 1640, and secretary in 1642. Once in Amsterdam, Joost stopped his activities as a painter and become a successful mediator and accountant for merchants. In his spare time, or *otia*, Joost translated into Dutch the *De veritate religionis Christianae* (1627) of Hugo Grotius – which was to become a key text in Remonstrant theology[6] – published in Amsterdam in 1653 as *Van de waerheyt der Christelijke Godts-dienst wt het Latijn vertaalt, door I. D. Volder.*[7]

[4] See, amongst others, Niceron 1733; Van Hoogstraten and Schuer 1733; Zedler 1746; Jöcher 1751; Dunkel 1755; Chauffepié 1756; Paquot 1765; Wagenaar 1767, 1790; Eloy 1778; Kok and Fokke 1793; Water 1802, 196–197; Jourdan 1825; Van der Vijver 1844; Kobus and De Rivecourt 1870; Van der Aa 1876. See also Van Berkel 1999a. A review of Gronovius's *Laudatio* appeared in the issue of May 1709 of the *Nouvelles de la République des Lettres*, by Jacques Bernard.

[5] The original copies of their *Reisejournal* are preserved at the Warsaw University Library (ms. R 766 and Cod. IV oct. 49) and at the National Library of Russia, Saint Petersburg (ms. Nem. IV F 158). Parts of the *Reisejournal* (from the manuscripts now extant in Warsaw) are transcribed in Guhrauer 1847 and Freudenthal 1899. An edition by Martin Mulsow and Olaf Simons is forthcoming.

[6] See Kaplan 1982, 279; Heering 1994; Stanglin 2016.

[7] "Domum certe negari nequit honestam fuisse, ex qua matrem, eius Mariam van Liesveld antea in viduitate contubernalem sororis, in extrema aetate aliquamdiu domesticam filio vidimus et sepelivimus, frugi et modestam; quae nupta Iusto de Volder, inter aliam subolem, nec numero arctatam, nec sexu uniformem, sed utroque satis prosperam hunc Burcherum quarto coniugii anno peperit sexto et vigesimo die mensis Iulii anno tertio et quadragesimo praeteriti proximi saeculi, qui erat

He died on 12 December 1669, while they were living on the Prinsengracht.[8] Less information is available on De Volder's mother, from Haarlem, who after the death of her husband lived with her sister (who then made Burchard her heir)[9] and then, in her last years, with her son. She died on 21 November 1689, with the Leiden professors attending her funeral and burial at the Hooglandse Kerk (or Pancraskerk) around 24 November. Burchard was not an only child: according to Gronovius, he had at least one sister, who survived the infancy, but died before her brother.[10]

Dominicus noctis hora sesquiprima. Quod ita acriter et perscrutari mihi libuit et palam prodere, non quo tempora discernam instar astrologi et ex genitali hora dimensus spatia siderum vaticinari velim, quales annos diesque foret noster, sed ut laetae relationis auditu illam iubeamus macte esse praeclarae frugis ubertate. Neque est, quod ulteriores avias revellam. Risisse ipsum in ulnis parentum et circa eorundem genua lusisse, etiam provolvente aetate per literularum exordia teneritudinem eius productam fuisse, ultro intelligitis. Non tamen progredi oportet, retinente hanc narrationem prima quoque indolis festivae in acriculo nostro Burchero exundantia. Nam quum pater primos annos impendisset pingendo, tam feliciter, ut plane existimaretur habilis; postea tamen mutavit vitam adversariis et computis ratiociniorum inter mercatores comparandis; ita ut animo et corpore involutus numeris, hunc nostrum exactissimum numerorum non in pecuniariis sordibus, sed coeli terraeque retrusis inveniendis ratiocinatorem genuerit. Eam vero artem sive operam ita sancte, ita perite dexterrimus vir exercuit, ut quantam, vellet pecuniam offerrent primarii in foro Amstelodamensi; immo flerent, siquos ob quascunque causas putaret deserendos. Inter haec tamen quasi forensia otium domesticum utiliter collocans, in quorumvis hominum, sed a Latinis minus paratorum gratiam, libros VI. Hugonis Grotii de veritate religionis Christianae in sermonem vernaculum transtulit; ut inde pateat, quo se tulerit viri illius ingenium, cuius quidem libri adiutorio sicut usus fuerat in firmanda notitia Dei, ita eidem vocanti obtemperans decessit anno LXIX. Ipso, nempe, qui antecessit vocationem filii Leidensem," Gronovius 1709, 10–12.

[8] The life and the significance of Joost de Volder for the history of art are reconstructed in Van der Willigen 1870, 319; Renckens 1966; Stechow 1966, 3–4; Beck 1972–1991, volume 4, 422–427; Fechner 1980; Briels 1997, 402; Weller 2009, 216–219; Lambour 2012. These sources omit, however, to consider the latter phase of his life (namely his Amsterdam years), which can be reconstructed through Gronovius's *Laudatio*. Joost's father was named Joost as well, and he had come from Eekelsbeek in Flanders, and on 14 June 1603 he married at Amsterdam Judith Jansdr. (from Antwerp as well).

[9] "Decessit nullo testamento; non qui nesciret vulgarem modum (fecerat enim olim, quod rupit in morte materterae, quam haeredem instituerat), [...]" Gronovius 1709, 33.

[10] See *supra*, n. 7. Gronovius 1709, 10–11; cf. Molhuysen 1913–1924, volume 4, 64–65: "Nov. 23. Senatui, in horam 10 indicto, significavit Magn. Rector matrem Clarissimi collegae D. de Volder 21. huius mensis mortalitatem exuisse; quaesitumque an Celeberrimo collegae per Rectorem Magnificum et facultatum Seniores illo nomine contestandus esset dolor; et an Professores togati funus essent prosequuturi." The burial certificate of what appears to be De Volder's mother (as Margareta Burgers Lievelt) is extant at the Leiden City and Regional Archives (Erfgoed Leiden en Omstreken): Inventaris van het stadsarchief van Leiden: 1313–1343, Registers van te Leiden begravenen, register 1327: 1691 augustus 4–1698 december 27, 326. In the Church registers of Amsterdam Mennonites (to whom De Volder's parents belonged) – extant at the Amsterdam City Archives (Stadsarchief Amsterdam) – it is reported that on 18 February 1680 a certain Geertruijd de Volder became a member of the Church, with her mother Griettie Borgers and her maternal aunt Hester Borgers as witnesses: Lidmaten Doopsgezinden, 18-02-1680, Archief 1120, inv. nr. 213, 78. A certain Geertruijdt de Volder, who lived on Oudezijds Achterburgwal, is reported as buried at the Amsterdam Westerker on 15 October 1681: see Begraafregistrer, 15-10-1681, Wester Kerk, DTB 1102, 102. No information with regard to the burial date or place of Hester Borgers has been found. No sisters are mentioned in De Volder's *Collaterale successie*, extant at the Leiden city archive (which I consider in Sect. 2.4.2, *De Volder's retirement, death and legacy*).

According to Gronovius – who most probably relied on the memories of De Volder's mother, as he seems to have been familiar with her – De Volder was laughing while his parents took him in their arms for the first time; in underlining his vivacity further, Gronovius notes how he broke everything he touched, with the exception of papers and books. Noting this, his father used to give him papers and books as presents, up to a mature age (when from time to time he came to visit his parents) – prompting the hilarity of his mother.[11] In fact, Burchard's father started to teach him the alphabet in his infancy. Later, in his boyhood, he enrolled him at the Latin School of Cornelius Sladus on the Oude Zijde (one of the two Latin schools of Amsterdam),[12] where Burchard studied Latin and Greek with success: being thin and weak (Stolle and Hallmann would indeed report that De Volder was short),[13] De Volder could not bring home the works of Seneca that Sladus gave him as a present in 1657, when he finished at school.[14] Freed from this duty, De Volder started his pre-university education, attending the lectures of two famous professors at the Amsterdam Athenaeum Illustre (at that time an institution of high education, but not authorized to confer university degrees),[15] namely Alexander de Bie (1620–1690, professor of mathematics) and Arnold Senguerd (1610–1668, professor of philosophy and father of Wolferd, De Volder's prospective colleague at Leiden), both praised by De Volder in his *Oratio funebris in obitum Siberti Coeman* (1679). This *Oratio* offers some detail on De Volder's higher education, which he shared with Coeman, born in Amsterdam on 15 July 1643, who had attended the Nieuwe Zijde Latin School of Adrianus Junius, before hearing the lectures of De Bie and Senguerd with De Volder, for three years (i.e. between 1657 and 1660), when the two, on the suggestion of an unmentioned fellow who then practised medicine at Amsterdam (and died in 1679), moved to Utrecht in order to obtain a degree.[16] At Amsterdam, De Volder reports studying

[11] Gronovius 1709, 12–13.

[12] See Van Miert 2009, chapter 3, 2018a; Frijhoff 2016.

[13] See *infra*, n. 342.

[14] "Latinis et Graecis inspiciendis ad ludos literarios filium duxit, repulsis aliis vitae ornamentis, quibus destinari poterat. Et sic cernimus Volderum nostrum transscribi in cellam sudatoriam, cui pueritiam et adolescentiam assimulabat Secundus. Regebat tunc eos ludos Cornelius Sladus, vir artificii istius laudatus, cuius quoque gratam sibi esse recordationem nobis declaravit filio eius Matthaeo, Medico et Graecis literis mirabiliter dedito, nuper Oxonii mortuo et sepulto, in hospitium come admisso. Enimvero patefacta hac area effusius exseruit, quid in ipso lateret, nuspiam tergiversatus aut remisse agens, quin percurreret istarum Scholarum sacra; et sicut exacte annos mensesque annotari non refert, ita in offenso impetu vehens per medias et magistrorum et aequalium laudes ac praemiorum testimonia, et plerisque quae ibi initiatus, anno septimo et quinquagesimo supremam scholam laudatus deseruit, testante id Annaeo Seneca, cuius operibus eum tunc iussu Scholarcharum donavit Sladus, quasi abhinc philosophica commendantium. Toto autem isto tempore ita exiguae staturae itaque, macilentus fuit, noster, sit illius praemii lator domum esse posset, capsario identidem eum adiuvante," Gronovius 1709, 14–15. Gronvius reported De Volder's interests for Greek and Latin literature: Gronovius 1709, 28.

[15] On the history of the Amsterdam Athenaeum, see Van Miert 2009.

[16] "His in studiis, cum ultra triennium elaborasset sedulo cupido hominem incessit, philosophica laurea condecorari se ut gestiret. Quam in sententiam et me et alterum sodalem nostrum eodem in

mostly Scholastic philosophy: De Bie, sympathiser with Cartesianism, at that time was professor of mathematics, while Senguerd was an Aristotelian.[17] Under their guidance, however, De Volder and Coeman linked the study of mathematics and philosophy – a link which De Volder extensively praises in this and in other academic orations.[18]

At Amsterdam, De Volder acted as *respondens* in two of De Bie's disputations, pertaining to mixed mathematics – and thus concerning physical topics – namely the *Disputatio mathematica de profunditate maris* (12 October 1658) and the *Disputatio mathematica de linea, quam globus per aërem describit missus e tormento* (January 1659). The first disputation is devoted to the explanation of different methods to measure the depth of the sea, including those of Marino Ghetaldi, Leon Battista Alberti, Giuseppe Biancani, and arguing that water constitutes 1/1,000 of the mass of the Earth.[19] The second disputation concerns the definition of the line described by a projectile, which is established as being a curved line, due to the projectile's 'moving faculties' (the nature of which is not considered by De Bie), one bringing it downwards, and one horizontally, and it is aimed against Galileo's theory of parabolic motion, given in the fourth *Giornata* of his *Discorsi e dimostra-*

---

stadio currentem, qui nuper Amstelodami praxin feliciter exercens medicam diem suum obiit, suo et exemplo et hortatu adduxit. Nos igitur terni hac de causa Ultraiectinam adire Academiam," De Volder 1679, 10–11 (unnumbered). See also Molhuysen 1913–1924, volume 3, 343; Le Clerc 1709, 374–375. Coeman later pursued studies in law.

[17] On the teachings of De Bie and Senguerd at Amsterdam, see Van Miert 2009, chapters 2, 8 and 10. Lucas Schacht had also attended the lectures of Senguerd and De Bie, as well as the Latin School of Sladus (albeit some years before De Volder): see De Volder 1689, 6–7.

[18] "Versabamur siquidem, iisdem in studiis, utebamur iisdem praeceptoribus Viris Clarissimis Arnoldo Senguerdio Collegae mei Senguerdii parente, philosophiae Amstelodami Professore Primario [...], et Alexandro de Bie Mathesin ibidem tunc temporis, nunc et Philosophiam profitente. Horum ductu mathematica disciplinas cum Philosophica copulavit, neque separandas eas scientias censuit, quas et ratio, et rerum coniunxit natura. Philosophorum sola disiunxit ignorantia. Praeterquam enim quod is esset, qui nullam sibi studiorum negligendam partem censeret, facile sibi persuaderi passus est, Mathematicas scientias, licet primo aspectu non tantum videantur spondere fructum, habere, tamen in recessu quamplurima ad vitae usum, ad rationis rectae regimen, ad veritatis inquisitionem utilissima. Non enumerabo, quod facile possem, mathematicarum rerum plurimas omnino non contemnendas utilitates: praetermittam absque harum usu eximiam rerum Physicarum notitiam habiturum neminem. Nec adducam harum disciplinarum neglectu in tenebris et squallore multos annos Philosophiam iacuisse. Id unum addam, quod nemo diffitebitur, nihil quibuscunque etiam in negotiis iudicii robore esse praestantius. Nihil certo veri et falsi indicio pretiosius. Nihil denique mentis in dignoscenda veritate exercitatione omnes ad res praeclarius. Ad quae nobis vel comparanda vel corroboranda certe, plurimum momenti afferre artium harum inquisitionem, vel perfunctorie eas contemplanti, et certas earundem demonstrationes cum aliarum, nisi geometrico ritu tractentur, quod nostro demum coeptum est saeculo, incertis plerumque coniecturis conferenti, planum erit. Glorientur vulgo de dialectices praestantia, audiat illa unica veri falsive disceptatrix et iudex; certe mathesis quae re praestat, quod verbo dialectica promittit, suam sibi gloriam apud rerum peritos praeripi non patietur," De Volder 1679, 8–9 (unnumbered). As to the other orations, see Sect. 3.2.3.2, *De Volder's methodological liberalism – and its continuity over time*. Cf. also *infra*, n. 24.

[19] De Bie 1658, theses 6–8 and 24.

zioni matematiche intorno a due nuove scienze (1638).[20] Moreover, at Amsterdam De Volder most probably attended the lectures in anatomy of Gerard Blasius, mentioned as a dedicatee, and labelled as a mentor in his anatomical studies, in De Volder's 1660 inaugural disputation at Utrecht (to which I shall return in a moment). Notably, the first disputation held by De Bie was dedicated by De Volder to Galenus Abrahamsz. de Haan (1622–1706), at that time physician and minister at the Mennonite Church congregation at the Amsterdam church Bij 't Lam (i.e. Singelkerk) – attended also by De Volder's father[21] – and who had strong links with the Collegiants. As reconstructed by Andrew Cooper Fix, De Haan had a markedly 'liberal' approach to the articles of faith, deeming as necessary only those aiming at salvation, and prompting the attacks of more conservative Mennonites.[22]

At the time of his move to Utrecht,[23] De Volder was imbued with Aristotelian theories, even if – as pointed out in his *Oratio* for Coeman – he had started to read the works of Galileo and Gassendi. In fact, he was still not acquainted with Descartes's ideas, into which he was initiated by Johannes de Bruyn (1620–1675),[24]

[20] As reconstructed by De Bie, Galileo's theory is based on two premises: (1) that a projectile covers equal horizontal distances in equal times, (2) that the two moving faculties of the projectile (viz. the horizontal and perpendicular) move it as if they were acting independently of each other. De Bie criticizes such premises, by claiming that the faculty moving the projectile horizontally is violent, external to it, and decreases in time (such as the experiment with the pendulum reveals), so that in equal times unequal spaces are covered (see De Bie 1659, theses 23–29). The second premise is criticized on the grounds that a projectile in motion does not fall in the same time as if only its 'weight' would have brought it to the ground (theses 31–34). Therefore, De Bie claims that projectiles do not move through parabolic but through curved lines (thesis 35). On this issue, see Naylor 1980; Renn 1992; Wohlwill 2001.

[21] See Meihuizen 1952; Lambour 2012.

[22] See Fix 1990. Notably, in the pamphlet *Het compromis tusschen Dr. Galenus Abramsz, nevens sijne medestanders, en Tobias Govertz van den Wyngaert* (1665), Joost de Volder is cited as siding with Galen: De Haan 1665, 7 and 13.

[23] The date of De Volder's enrolment at Utrecht is not reported in the *Album studiosorum Academiae Rheno-Traiectinae MDCXXXVI–MDCCCLXXXVI* (1886). For his graduation at Utrecht, see the *Album promotorum qui inde ab anno MDCXXXVI usque ad annum MDCCCXV in Academia Rheno-Trajectina gradum doctoratus adepti sunt* (1936), 16.

[24] "Erat noster, eramus omnes peripateticae philosophiae imbuti principiis, nisi quod in physicis, ex Galilaeo, Gassendo et id genus scriptis nonnihil verioris et mechanicae rationis degustaramus. Summum illud saeculi nostri lumen, illustrem mechanicae philosophiae propagatorem, Renatum Cartesium videramus nondum, nisi id vidisse est, quae non intelligas perlegisse. Huius viri, circa ea praesertim quae Deum et mentem spectant, scrutandis dogmatis totum se dedit, amica hac in parte usus opera Clar. de Bruyn Physices et Mathesios dignissimi ibidem Professoris. Cui eo nomine plurimum debet noster, et me debere profitebor, dum spiritus hos reget artus," De Volder 1679, 12 (unnumbered). Cf. Gronovius 1709, 15–16: "[a]b isto etiam adductiore disciplinae modo liber et iam magis sui arbitrii, nihil moratus ad maiora se in patria urbe contulit. Nam quum Amstelodamo tunc ita bene esset provisum, ut Philosophicas varietates liceret accipere simul mathematicis, credidit se tam robusta et capacis esse mentis, ut quum duo incliti Doctores separatim eas traderent, sub utroque posset crescere. Horum nomina ipse ne sine gratissimae voluntatis professione et iustae laudis commemoratione indicavit in utraque, quam dixi, Laudatione, quum Coemannum et annis aequalem et eiusdem Musae haberet aemulum, sive ab ipsa tirociniorum origine additum, etsi hic elegisset scholam Iunianam, sive ab consortio philosophiae, et consensu

who was the first dedicatee of his *Disputatio philosophica inauguralis de simplicitate Dei, echo et republica* (17 October 1660), together with Daniel Berckringer (professor of eloquence and practical philosophy at Utrecht), as well as his previous teachers Senguerd, De Bie and Blasius. The disputation – as noted by Gerhard Wiesenfeldt – was thoroughly Cartesian.[25] The section *De simplicitate Dei* De Volder defended God's simplicity, viz. the absence of any composition, which depends only on the weakness of our reason in conceiving Him.[26] In the section *De echo*, De Volder analyses the transmission of sound with the same method used in the study of light, namely by considering its reflection and refraction (that is, with a mixed-mathematical approach), but without entering into much detail either on its nature, understood as 'crumbled air' (*aër contritus*, thesis 2), or on that of light. Certainly, this section was Cartesian, but was also compatible with contemporary explanations of sound, and even with late-Scholastic approaches.[27] Finally, in the section *De republica* (which I analyse below), De Volder proposed an account of the formation of society relying on the jusnaturalism of Grotius.[28]

Once he obtained his degree as doctor of philosophy and master in the liberal arts, De Volder came back to Amsterdam, cultivating now mechanical philosophy, and rejecting the ideas of substantial forms and occult qualities.[29] Both as the result

---

animi electam. Cui industriae, ne existimetur bonus eventus defuisse, enimvero ita res cecidit, ut quum per mediocre tempus illic exercitiis adfuisset, magnus iam et plane optabilis incrementi noster praebuerit se spectandum Ultraiectinis, ita ut post disputationem de Simplicitate Dei, de Echo, de Republica coronatus fuerit in Philosophia et Liberalium Artium Magisterio die Octobris octavo decimo anni sexagesimi."

[25] "Der Teil de echo entsprach dem Vorbild seiner Amsterdamer Disputationen, indem er sich mit der Ausbreitung des Echos und dabei auftretenden Phänomenen beschäftigte, die er mit geometrischen Hilfsmitteln beschrieb. De Volder verwies hier ausdrücklich auf das Vorbild der geometrischen Optik. Im metaphysischen Teil bejahte er die Möglichkeit unmittelbarer Gotteserkenntnis, nicht unbedingt eine klassische cartesianische Position; Descartes wird dagegen nur mit seiner Theorie des Magneten erwähnt, die verschiedene Wirkungen erklären könne," Wiesenfeldt 2002, 55.

[26] God's simplicity is intended as the absence of any composition and as entailed by His perfection (De Volder 1660, *De simplicitate Dei*, theses 2–3). According to De Volder, such a simplicity does not contradict the freedom of God's decrees, for the reason that one has to distinguish, in God, between His essence or vital act, and His determination of this or that object, which is free (thesis 5). In turn, the distinction between divine essence and divine determinations or decrees depends only on the weakness on our reason, and does not pose a real distinction in God (thesis 6). As to the issue of the nature of God, see Sect. 3.1.4.2, *The Wittich-Driessen quarrel – and beyond.*

[27] On the idea of the nature of sound in late Scholastic and Cartesian sources, see Van Ruler 1995, chapter 4.

[28] See Sect. 2.3.5, *De Volder's views in politics and religion.*

[29] "Hoc stadio ita fauste et properanter decurso ipse quidem regressus est domum, sed substitit Ultraiecti Coemannus nescio qua existimans indagandum et acerbius inspiciendum, quod semiclausis oculis e longinquo fulgere obiter, monstraverat: Clarissimus Professor Philosophiae de Bruyn; tanti aestimatum semper nostro, ut propterea exclamarit se plurimum ei debere professurum, dum spiritus hos regeret artus, doctrinam quippe se sectatos perceperant spinarum matrem, frumenti novercam, et in metaphysicis simulacra acceperant, corpus, et mera veritas aberat. Hinc illae densae literae de mutuis profectibus inter nostrum et Coemannum; tunc ferbuit ista sapiens ac meditabunda lectio Cartesii; testante palam nostro tunc se animadvertisse, quam perfunderetur

of his family's wishes, who wanted him to become a physician in the Mennonite community, and through his own desire of discovering scientific approaches alternative to Aristotelianism,[30] De Volder then moved to the University of Leiden, the oldest of the Dutch universities,[31] where he enrolled on 22 February 1661,[32] and lived in the house of his father's friend Laurentius Borman. At Leiden, De Volder studied medicine under Franciscus Sylvius (1614–1672), graduating three years later with a *Disputatio medica inauguralis, de natura* (3 July 1664),[33] dedicated to

sensim maiori veritatis luce, experiretur mentis dispelli tenebras, et habere nos in nobis ipsis, modo dotibus a Deo concessis rite utamur, sufficientem circa has res veritatis scaturiginem. Id quod veritatem appellat, dulcedine sua excellere et animo humano bene constituto solum placere posse et debere, etiam ubi forsan radit malignas aures. Adeo patet nequivisse pati nostrum, ut post fruges repertas glandibus uteretur deinceps. Atque ita praeparata molliter et expansa intus sede cuiuslibet boni accipiendi perita, non continuit noster, quin se concuteret vehementius, ut ulteriora obtineret," Gronovius 1709, 16–17; cf. De Volder 1679, 12–13 (unnumbered): "[q]uam viam ingressi, et per litteras, aberam enim ego, de progressibus nostris communicantes, sensim animadvertimus maiori, quam hactenus exerti essemus, nos veritatis luce perfundi, mentis dispelli tenebras, et habere nos in nobismet ipsis, modo dotibus a Deo concessis rite utamur sufficientem circa has res veritatis scaturiginem. In rebus Physicis, non eam invenimus difficultatem, utpote ad quas recentiorum scripta nobis perlecta aditum iam fecissent. Iamdudum enim formis substantialibus, qualitatibus realibus et id genus figmentis nuncium miseramus, quo promtius mechanicam philosophandi rationem amplecteremur. Quam sane in physicis, ut et priorem ex intimis mentis haustam penetralibus in et metaphysicis unicam esse viam, qua veritas detegi potest, asseverare nullus dubito. Quae et sua perspicuitate sibi viam fecit apud Anglos, Gallosque, et faciet porro, ubicunque liberior spirat aura."

[30] "Burcher de Volder étoit né le 26 de Juillet MDCXLIII à Amsterdam. Son Pere se nommoit Juste de Volder, & sa Mere Marie de Liesveld. Ils éleverent ce fils, de qui ils sembloient avoir conçu de fort bonnes esperances, avec autant de soin, que leur condition & leur fortune, qui n'étoit pas considerable, purent le permettre, & il répondit très-bien à leurs esperances. Ils étoient de ceux qu'on nomme Mennonites en Hollande, & ailleurs Anabaptistes; parmi lesquels il y a beaucoup de gens, dont la modestie & la regularité sont de bon exemple, mais très-rarement des gens de Lettres. Ses Parens avoient des sein de le faire étudier en Médecine, & aparemment de l'employer en suite à prêcher dans quelques unes de leurs Assemblées, selon leur coutume; car leurs Ministres exercent très-souvent la Médecine, en même tems. Après avoir étudié quelque tems en Philosophie sous Arnaud Senguerd & en Mathematique sous Alexandre de Bie Professeurs à Amsterdam; ou, pour parler plus exactement, en son particulier, car ces Mrs. ne pouvoient conduire personne sort loin dans ces sciences; il se fit recevoir Maître aux Arts à Utrecht le 18. d'Octobre MDCLX. II avoit d'abord étudié la Philosophie, comme on faisoit encore communément en ce tems-là; c'est à dire, qu'il s'étoit appliqué à celle de l'Ecole, plus propre à gâter l'esprit qu'à l'éclaircir & à le conduire, dans la recherche de la Verité. Cependant il avoit l'esprit si bon, que peu de tems après, plûtòt de son propre choix, que par le conseil de personne, il tourna, comme on le verra, ses études d'un autre côté," Le Clerc 1709, 348–349.

[31] On the history of Leiden University, see Dibon 1954; Ruestow 1973; Lunsingh Scheurleer and Posthumus Meyjes 1975; Grafton 1988; Bergh 2002; Krop 2003a.

[32] See *Album studiosorum Academiae Lugduno-Batavae MDLXXV–MDCCCLXXV*, column 486, where the wrong age of 20 is given for De Volder.

[33] "In his quum ad medicarum rerum notitiam vergeret, credidit eas castius tractari in hac Academia, quum novo tunc et placitissimo docendi et exercendi modo in infinita audientium corona istic eniteret Franc. Sylvius. Sed adolescens noster iam octodecim fere annorum, etsi philosophiae quidem sacratus et mente excelsus, habebat tamen aliquid imbecillum, quod pater vix sustinebat convictui hominum ignotorum, domui extrariae, tanto peregrinorum concursui credere, praesertim

Sylvius himself and to Johannes Hudde (1628–1704), mathematician, mayor of Amsterdam, a director of the Dutch East India Company (Vereenigde Oostindische Compagnie, or VOC), correspondent of Spinoza, Christiaan Huygens, Johann Bernoulli, Newton and Leibniz, and who was to recommend De Volder for a position at Leiden.[34] As I am going to analyse in more detail in Sect. 3.2.3.3, *The role of experience in De Volder's natural philosophy*, De Volder's *De natura* mostly represents the iatrochemical views of Sylvius on fermentation – the most important physiological process – but it also includes attacks on Jean Fernel and Aristotle (as emphasised by Le Clerc),[35] and a correction of the positions of Descartes and the Cartesian physician Cornelis van Hogelande on fermentation itself.[36]

After having obtained his degree, De Volder came back to the house of his parents in Amsterdam, where he become physician to the Remonstrant Church, and where he continued his 'private' meditations (especially on Cartesian philosophy), and participated in De Bie's lectures as polemical opponent (i.e. *antisophistas*), making his name widely known.[37] At that time, as indicated by Pieter Baert (correspondent of Christiaan Huygens), he was probably acquainted, besides Hudde, with Huygens and Spinoza, as Baert reported him to have had contacts with all them from 1665 onwards.[38] Moreover, he was acquainted with Constantijn Huygens, father of Christiaan.[39] Eventually, it was on the recommendation of Hudde, and after having been proposed by Frederik van Dorp Jr., Lord of Maasdam and Curator of the University, that De Volder was called (*vocatus*) to a chair as ordinary professor in philosophy at Leiden in 1670, as testified to by the *Resoluties* of the University

quum hic novellus philosophus libros adfuisset aperire, non patinas. Itaque extorsit Laurentio Borman amico suo et civi huius urbis, ut tecto et mensa sua recipere filium vellet, ita procedente mansionis huius ingenio, ut post annos quatuor sibi curaret induci quoque nomen medici post Disputationem de ipsa natura die III Iulii anno quarto et sexagesimo," Gronovius 1709, 17. No information on Laurentius Borman could be found.

[34] On Hudde, see De Waard 1911; Vermij and Atzema 1995; Klever 1997, chapter 3; Van Berkel 1999b; De Jong and Zuidervaart 2018.

[35] "Il alla étudier la Médecine à Leide, où il fréquenta les leçons de François de Leboé Sylvius. Il y soûtint des Theses de la Nature, très-opposées aux idées Péripateticiennes, le 3. de Juillet MDCLXIV & fut reçu Docteur en Médecine," Le Clerc 1709, 349–350.

[36] See De Volder 1664, theses 2–3 and 9.

[37] "Sic gemina adorea insignitus revertit ad suos, neutram Musam omittens, quin assiduitate omni expoliret, seu privatim meditationibus in haerendo, seu publice valetudines regendo, cui magnam fenestram illi aperuit decreta praefectura valetudinarii Remonstrantium. Sed praecipuus amator philosophiae diligenter se conferebat, et quidem cum frequente comitatu, ad diatribas, Doctoris Biaei, non auditor, sed antisophista, saepe vi sua praesidem turbans, ut hinc efflorescentis etiam rumor spargi coeperit inter eruditos magnae urbis de nota optima," Gronovius 1709, 17–18; cf. Le Clerc 1709, 350: "[i]l s'applica à la pratique de cette science, pendant quelques années, qu'il fut le Médecin des pauvres de l'Eglise des Rémontrants d'Amsterdam; sans néanmoins quitter l'étude de la Philosophie & des Mathematiques. Comme la Philosophie de Descartes faisoit grand bruit en ce tems-là, & qu'elle commençoit à prendreledes fus, en Hollande; il s'y attacha avec beaucoup d'application, & y fît les progrès, qui parurent dans la suite."

[38] See Sect. 3.1.1, *Some direct evidence on De Volder's views and relation with Spinoza*.

[39] See Sect. 2.2.3, *The mid-1670s clash at Leiden and the foundation of the experimental theatre*.

Curators of 15 April 1670. He was called, apparently, as the professor of philosophy, Henricus Bornius, had suspended his teaching activities as a consequence of his duties as bailiff to Prince William of Orange.[40] Another chair in philosophy was granted in the same year to Theodoor Craanen (1633–1688), Cartesian and former professor at Nijmegen, who was going to assume a chair in medicine in 1673.[41] Before assuming the post – as he was going to do on 18 October of the same year (where he gave an inaugural oration, whose text is now lost)[42] – De Volder – who certainly had Remonstrant sympathies, as testified to by his activities as a physician in Amsterdam and close contacts with Philipp van Limborch (1633–1712)[43] and Le Clerc (respectively professors of theology and philosophy at the Amsterdam Remonstrant Seminary from 1668 and 1684) – entered the Walloon Church. According to Gronovius, the Curators had been informed that De Volder was a Mennonite, and was not baptized. De Volder came to be acquainted with this, and informed the Curators that he had already decided to enter the Walloon Church, and

---

[40] "Op 't versouck van D. Henricus Bornius, Professor Philosophiae, omme tot bevorderinge van syne domestique saeken met de effective exercitie ofte reassumtie van syne professie te mogen supersederen tot nae de expiratie van de groote vacantie in de hontsdagen, is nae gehoudene deliberatie goedgevonden ende verstaen ende vervolgens hem aengeseyt dat hy tot verrightinge van syne particuliere affaires soodanigh een tyd sal können besteeden als hem goed- duncken sal: des dat denselven tyt geescouleert wesende hy alvorens de voors: professie te exerceren gehouden sal syn te versoeken speciale admissie, als wanneer dan op 't selve versoek met een vrye deliberatie nae gelegentheyt van saeke sal werden gedisponeert. Op 't geproponeerde van den heere van Maesdam, off D. Burcherus de Volder, gedesigneert Professor ordinarius Philosophiae in dese Universiteyt, tot nogh toe door den doop niet ingelijft synde in de Gereformeerde kercke, alvorens denselven in siju beroep te installeren niet behoorde gedisponeert te werden omme hem door den doop in deselve kerke te laeten inlyven, is goedgevonden ende verstaen, dat soo lange het voors. Sacrament by hem niet sal wesen gecelebreert, hy tot het doen van de inaugurale oratie niet sal wesen admissibel, ende is hem 't selve door den secretaris uyt den naeme van de H. C. ende B. aenstonts aengeseyt met versoek dat hy sigh hoe eer hoe beeter daer toe soude believen te laeten bewegen," Molhuysen 1913–1924, volume 3, 241–242. See Gronovius 1709, 18–19: "[n]eque tamen ista moenia signare et includere poterant limitem famae tam constantis et illustris. Egressa est in omnes appositas plagas. Sed nullo tempore crediderim famam minus mentitam, esse, quam quum Illustri Curatorum Academiae et Consulum Collegio Volderum tradidit. Quum enim in hac ipsa locus esset vacuefactus discessu alterius, ac non minori cautela et inquisitione mereretur suppleri, quam qua vel in Senatum Laconicum aut in Atticum Areopagum substitutus olim fuisse legitur, miro consensu praeconceperunt omnes virtuti per tot experimenta probatae in adolescente in dolis quidem certae, sed nondum sedis, prorsus responsuram fortunam claro loco adfixam nomineque et munere insignem, ipsiusque adeo philosophiae vel magis veritatis discipulum agnitum iri, cuius arcem fortissime nosset tueri. Sic urgente fato in thronum hunc Philosophicum invectus est, quum et ipse non dubitaret huc accedere, duntaxat ut denique desineret vel sibi soli inhaerere, vel privatae et ignotae meditationi servire; et pro more ad inaugurandum ex hoc ipso loco die octo decimo mensis Octobris anno septuagesimo, quo penes patrem meum erant fasces, habuit concionem." See also Le Clerc 1709, 350: "[e]n ce tems-là, une Chaire en Philosophie vint à vaquer à Leide, & ayant été recommandé, par le moyen de Mr. Hudde, depuis Bourgmestre à Amsterdam, & grand Mathematicien, à l'un des Curateurs de l'Academie, il y fut appellé & fit la Harangue Inaugurale le 18. d'Octobre en MDCLXX." On Bornius, see Krop 2003b.

[41] On Craanen, see Sect. 3.2.3.1, *The case of explanations in physiology*.

[42] See Molhuysen 1913–1924, volume 3, 238.

[43] See Sect. 3.1.3.1, *The relation between material and immaterial substances*.

that he could renounce the post (so that they would think that he was converting on purpose). After this statement, and after his baptism, they allowed him to assume the post on 30 September.[44] As noted by Le Clerc, De Volder joined the new Church without condemning his former faith, as he maintained that the differences between the two churches concerned only inessential things.[45] As seen above, this was the

---

[44] In fact, as testified to by the letter of De Volder to Van Limborch of 18 July 1687, the academic authorities were not very strict on asking for a declaration of faith from professors – at least in philosophy. It is worth quoting here the full text of the letter (these are the conventions adopted in the transcription: 1) the text deleted in the manuscript has been put between brackets thus < >; whenever possible, I have provided the deleted text; otherwise, I have used dots instead of the illegible letters; 2) the text in the margin or between the lines is put between the symbols \ /; 3) doubtful text is put between brackets { }; whenever possible, I have provided the text, otherwise, I have used dots instead of the illegible letters; 4) my additions are put between brackets [ ];): "Vir Amplissime, / Recte quidem existimasti subscriptionem, de qua agitur non exigi a Philosophiae aut Graecae Linguae professoribus, nec credo illum exigi ab eo qui Hebraeam Linguam profitetur. Certo tamen asserere non possum. Quamdiu enim hic fui, nullum Haebraicae Linguae habuimus Professorem praeter Uchtmannum, qui me aliquot mensibus praecessit. Cur tamen non credam causae hae sunt; In statutis Collegii ordinum expresse id exigitur ut Proregens et Confessioni et Catechismo, et Canonibus Dodracenis subscribat, quod tamen D. Crane nunquam fecisse saepius ex ipso audiv{i} {<...>}. Nec memini ullius subscriptionis nequidem a Theologo factae, quanquam novi Doctores Theologiae ubi creantur, ad illam subscriptionem ante promotionem compelli; Verum hac in re ne quid forte {q} esset quod me quia inter Theologos res perageretur fugeret, adii D. Le Moyne; et inter alios sermones quid huius rei esset quaesivi. Respondet se non putare quempiam subscribere illis sive Confessioni et Catechismo, sive Canonibus praeter Theologiae Professores, imo se jam per quadriennium Professoratu suo functum, antequam subscripsisset, nec tunc temporis subscripturum fuisse nisi forte in conventu Facultatis Theologicae, non recordabatur autem qua de causa <insti> {conrevissent}, ille liber in quo haec sacra continentur incidisset in manis Spanhemii. Illeq[ue] non inveniens manum D. Le Moyne monuisset de subscriptione facienda, quam ille lubens agnovisset. / Vides quam negligenter haec orthodoxiae conservandae fundamenta nunc tractentur, Sic solet ubi aestus animorum deferbuit. Putem itaq[ue] illi erudito viro nihil periculi fore, si in aliis Academiis harum provinciarum res eodem ac apud nos peragatur modo; Accurate enim qu{...} apud vicinos fieri solet non novi. Vale et ama. / Tuum / B. De Volder. / Lugd. Batav. 18 Iulii, 1687," Amsterdam University Library, OTM hs. J 83 a. For a commentary, see Strazzoni 2019.

[45] "Avant qu'il reçût ses Patentes, il y eut quelques difficultez, sur sa vocation, fondées sur ce qu'il avoit fréquenté jusqu'à lors les Assemblées des Mennonites. Mais comme il fut que les Curateurs de l'Université déliberoient là-dessus, il leur fit dire qu'il avoit dessein de se joindre à l'Eglise Wallonne Réformée de Leide; mais qu'au reste il ne pretendoit pas acheter par-là la Chaire de Professeur, où on l'avoit appelle, ni qu'on lui reprochât à l'avenir d'avoir rien fait contre sa conscience; qu'autrement il aimeroit mieux renoncer à cette vocation. Cette déclaration leva entierement la difficulté, & il fut en état de commencer l'exercice de sa Charge. Il ne faut pas s'imaginer là-dessus que Mr. De Volder, en passant dans une autre Assèmblée Chrétienne, condamnât celle qu'il venoit de quitter; comme s'il avoit cru, qu'il y eût du danger pour le salut d'y demeurer, & qu'il regardât avec aucune aversion ceux avec qui il avoit été élevé, ou qu'il eût plus mauvaise opinion de leur doctrine qu'auparavant. Il n'a jamais été du nombre de ceux, qui condamnent leur prochain, pour des opinions, comme celles-là; et il n'avoit pas ce zele amer, qui a donné la naissance à tant de Sectes, & qui est encore cause des divisions de la Chrétienté. Il crut apparemment être plus utile, non seulement à lui même, mais encore au Public, en entrant dans une autre Societé, où il trouvoit les mêmes principes de Religion, à quelques differences près, qui ne sont pas essentielles, & dans lesquelles la profession l'exemptoit en quelque sorte d'entrer. D'autres ne le voudroient peut être pas faire, mais c'est à châcun à se conduire, par ses propres lumieres, & non par l'exemple d'autrui," Le Clerc 1709, 350–352; cf. Gronovius 1709, 19–21; Molhuysen 1913–1924, volume 3, 242 and 244.

attitude of De Haan, who deemed as necessary only those articles concerning salvation. Or, as Erik Jorink has put it in his study on the Dutch connections of the Danish naturalist Niels Steensen (who studied at Leiden in 1663–1664, graduating *in absentia* under Sylvius, and had contacts with Blasius, as well as with two other acquaintances of De Volder, namely Jan Swammerdam (1637–1680, graduating in 1667) and Frederik Ruysch (1638–1731, graduating in 1664, and later physician in Amsterdam, like De Volder)),[46]

> on the suggestion of the increasingly important Hudde, he [i.e. De Volder] was appointed as professor [...] at Leiden University, on condition that he convert to the Reformed church. De Volder, who had collegiant affinities and was sympathetic to the concept of the invisible church, did this without hesitation. Like so many of his peer-group [...] De Volder was convinced that religious denominations and outward appearances were of far less importance than inner devotion. Although we have no direct evidence of this, Steno must certainly have known De Volder. Steno's laments about the Dutch "political way of living" could refer to incidents like De Volder's apparently completely unproblematic change of religion for the sake of a professorship: such a move appeared insincere and cynical to the outsider but was perfectly sensible to a "Christian without church."[47]

2.2 De Volder's Career Up to the Establishment of the Experimental Theatre

2.2.1 The Early Years at Leiden

In his first years as a professor, De Volder taught logic – using the textbooks of Franco Burgersdijk,[48] of which he probably left some, now lost *dictata*[49] – and, upon his request (reported on 8 November 1670), natural philosophy, each discipline twice a week.[50] Before the opening of the physical theatre or auditorium in 1675, he was teaching, in the summer term of 1671, the logic of Burgersdijk on Mondays and Tuesdays (at the *hora prima* in the afternoon), and on Thursdays and Fridays the 'causes of meteors' – a topic probably allowing to him to teach the whole of natural philosophy. In the winter term of the same year, he taught metaphysics on Monday and Tuesday, and general physics on Tuesday and Friday (so that he most probably

[46] On Ruysch, see Dohmen 1982; Hansen 1996; Kooijmans 2004; Kooijmans 2011; Knoeff 2015. On Swammerdam, see *infra*, n. 187. On De Volder's contacts with Swammerdam and Ruysch, see Sect. 2.3.1, *The correspondence with Thévenot.*

[47] Jorink 2018, 26.

[48] On Burgersdijk, see Kuiper 1958; Bos and Krop 1993; Krop 2003c.

[49] See *Bibliotheca Triglandiana*, 201: "Cl. Viri Burcheri de Volder Dictata ad logicam."

[50] See Molhuysen 1913–1924, volume 3, 245; according to Le Clerc, De Volder taught the whole logic of Burgersdijk in few lessons, and then he switched to natural philosophy and to metaphysics: see Le Clerc 352: "[i]l débuta dans sa Profession, par l'explication de la Logique de Burgersdice, qui s'enseignoit en cette Academie; mais il la finit en peu de leçons, & l'on s'apperçut bien tôt qu'il n'étoit rien moins que Péripateticien. Dans la suite, il fit rouler ses leçons publiques sur divers sujets de Physique & de Metaphyisique, qu'il choisissoit, selon son goût."

finished teaching Burgersdijk's logic in one term, from January to July). On 23 January 1672, the Senate conceded to De Volder teaching in the morning at the *hora nona*. In December 1673, to De Volder was conceded the *hora decima*.[51] According to Le Clerc, his lectures (both public and private)[52] attracted numerous students, thanks to his cleverness in explaining.[53] Indeed, as Gerard Noodt was to note in a letter of 15 July 1709 to Johann Bernoulli, De Volder had circa 80 private students in philosophy, and often more than 100.[54]

From January 1671, De Volder started to preside over his first disputations. Those which have been recovered are his *Disputatio physica de corpore* (28 January), *De motu prima* (18 April), *De anima humana prima–secunda* (May–June), *De Deo* (31 October), while in 1672 he continued with the *De ideis* (16 March, authored by the *respondens*), *Disputatio philosophica, in qua quid de principiis philosophiae vulgaris sit censendum disquiritur* (19 March, authored by the *respondens*), *De motu secunda* (14 December). In 1673, he presided over his *De unitate Dei* (25 March, *respondens* Bernard Nieuwentijt), *De criterio veritatis prima* (28 June). Moreover, on 31 January 1674 he started to preside over a long series of 15 disputations, namely his *De rerum naturalium principiis*, which continued until 1676. In 1674 he did not preside over any other disputations. In conformity with his academic duties, all these disputations (the contents of which will be dealt with in the rest of the book) concern physics and metaphysics. In turn, logic is avoided, in accordance with his stress on distancing himself from logic as a suitable instrument of discovery and of exercising the mind).[55] As revealed in Le Clerc's eulogy, De Volder used the same strategy pursued by the Cartesian professor Johannes de Raey (1620–1702) (whose lectures De Volder could have attended at Leiden) in order to support Descartes's ideas in natural philosophy.[56] Namely, De Volder often read to his students passages from Aristotle – purposely omitting his name – which seemed to be taken from Descartes's, especially on the use of doubt in philosophy, on which De Volder quoted from book B, chapter 1 of Aristotle's *Metaphysica*.[57] Analogously, as

[51] See Molhuysen 1913–1924, volume 3, 257–258, 268, 334, 235* and 237*. His very first lectures were temporarily given at the *hora tertia*: Molhuysen 1913–1924, volume 3, 238. No details on De Volder's further lectures are available before the ones of 1681.

[52] As to the differentiation between public and private lectures (which granted substantial honorariums to the professors), see Wiesenfeldt 2002, chapter 3.

[53] "Mr. de Volder avoit un concours extraordinaire d'Auditeurs, soit dans ses leçons publiques, soit dans les particulieres, où il expliquoit la Physique & la Métaphysique de Descartes. Cela ne venoit pas seulement de l'attachement, que l'on avoit alors pour la Nouvelle Philosophie; mais aussi de la maniere claire, nette & ingenieuse, dont il l'expliquoit. C'est un talent particulier qu'il avoit, dont on peut voir des preuves dans les Theses de Philosophie, qu'il a fait soûtenir en divers tems, & dont je parlerai dans la suite. Ceux qui l'ont ouï faire ses leçons lui donnent géneralement cette louange," Le Clerc 1709, 353–354.

[54] See Sect. 2.4.2, *De Volder's retirement, death and legacy.*

[55] See Sect. 3.2.1, *De Volder on clarity and distinctness.*

[56] See Strazzoni 2011, and Sect. 3.2.2.1.4, *De Raey's concealment and defence of Descartes's metaphysics.*

[57] "Comme il falloit encore ménager les esprits de certaines gens, qui etoient prévenus contre la Philosophie de Descartes, sans savoir pourquoi; il fait souvent voir que des Dogmes de Descartes,

I am going to show,[58] De Raey taught Descartes's physics by commenting upon those passages, in the Aristotelian corpus, which were (somehow) consistent with Descartes's positions, and presenting his views in his *Clavis philosophiae naturalis, Aristotelico-Cartesiana* (first edition 1654, second edition 1677).[59]

2.2.2 *The Academic* Dictata

In these first years at Leiden, De Volder could have completed, and used in his lectures one or both of his two extant commentaries or *dictata* on Descartes's *Principia philosophiae naturalis* (1644).[60] What I presume is the first of De Volder's *dictata* is extant, in a complete form, in an undated manuscript preserved at the Hamburg State and University Library Carl von Ossietzky (and to which I will hereafter refer as 'Hamburg 273'),[61] and in a partial form in the first column of a copy of his *dictata* preserved at the Dutch National Library in The Hague, which is dated 1 June–August 1690,[62] and – according to the catalogue – written by the student Adriaan Pieter de Hinojosa (later president of the Hof van Holland, viz. the High Court of the provinces of Holland). The other series of De Volder's *dictata*, in turn, is preserved in more copies: at Hamburg (henceforth 'Hamburg 274'),[63] in the second column of the The Hague manuscript (albeit partially), at the Leiden University Library (and concerning only the fourth part of Descartes's *Principia*),[64] at the

contre la nouveauté desquels on s'échauffoit, se trouvent dans les Anciens, & même dans Aristote. L'on m'a assuré qu'il avoit fait porter quelque fois lès Ouvrages dans la Chaire & qu'il en avoit lu des passages à ses Auditeurs; pour faire voir que bien des choses, que l'on accusoit de nouveauté, se trouvoient dans cet Auteur. On a accusé, par exemple, Descartes d'avoir voulu établir le Pyrrhonisme, pour avoir dit qu'il falloir commencer, par douter de tout, si l'on vouloit parvenir à une verité assurée. Cependant cela se trouve dans Aristote, Livre II de sa Métaphysique chap. I où il entreprend de montrer l'usage du doute, & les choses, dont il faut d'abord douter," Le Clerc 1709, 352–353; cf. *Metaphysica*, 995a24–995b4.

[58] See Sect. 3.2.2.1.4, *De Raey's concealment and defence of Descartes's metaphysics.*

[59] On the attempts to develop 'novantique' philosophical theories, see Bodeüs 1991; Greene 1993; Strazzoni 2011.

[60] Besides, of course, his now lost *dictata* on logic, and, possibly, the third, now lost *dictata* on the *Principia.*

[61] As *Notulae quaedam in nobilissimi doctissimique viri Dni Renati Descartes Principiorum philosophiae* [...] *scriptae ex ore clar. doctissimique viri Burcheri de Volder medicinae ac philosophiae doctoris, et in Academia Lugduno Batava professoris* (Cod. philos. 273).

[62] As *Dictata Cl. Viri Dn. B. de Volder Professor. Publici in Celeberrim. Acad. Lugdun. Batava ad Principi. philosoph. Cartesi* (ms. 72 A 7).

[63] As *D. D. Burcheri De Volder dictata in Carthesii Principia philosophica* (Cod. philos. 274). The digitized text of both the Hamburg series of *dictata* is available at the following addresses: https://resolver.sub.uni-hamburg.de/kitodo/PPN1014827108 and https://resolver.sub.uni-hamburg.de/kitodo/PPN1014826934 (accessed 5 February 2019).

[64] As *B. de Volder Med. et Phil. Doct. eiusque professoris Dictata in Principiorum philosophiae partem quartam de terra* (ms. BPL 2841).

Table 2.1 De Volder's *dictata* on Descartes's *Principia philosophiae*

Descartes's *Principia*	Hamburg cod. 273 copy of De Volder's *dictata*	Hamburg cod. 274 *dictata*	The Hague *dictata* Column A	The Hague *dictata* Column B	Leiden *dictata*	British Library *dictata*	Warsaw *dictata*	Pretoria *dictata*
Part 1	V	X	X	*Text not appearing in any other copy*	Absent	X	X	X
Part 2	V	X	V	X (partial)	Absent	X	Absent	X
Part 3	V	X	V	X (partial)	Absent	X	Absent	X
Part 4	V	X	V	X (partial)	X	X	Absent	X

British Library,[65] at the National Library of South Africa (Pretoria) – dated 15 December 1687–26 November 1688[66] – and at the National Library of Poland (Biblioteka Narodowa, Warsaw), where only the commentary on the first part of the *Principia* is reported, as well as the only copy of De Volder's commentary on Descartes's *Meditationes*[67] (which he might have been using since his first years at Leiden as well).[68] Notably, the manuscript preserved in Pretoria contains some marginalia, in a different hand from that of the main text. The marginalia consists of clarifications of the arguments given in the main text, references to other authors (such as Galileo), as well as some passages from the original of the French version of Descartes's *Principia*, which De Volder translated into Latin and included in his commentary. The The Hague copy, in the second column, includes some of these Latin translations, as given in the Hamburg 274 manuscript. The reader can have a synopsis of the contents of the different manuscripts in Table 2.1, *De Volder's* dictata *on Descartes's* Principia philosophiae, where I indicate by 'X' the text of Hamburg 273, and by 'Y' the text of Hamburg 274.

The hands of these *dictata* are all different, and none of them is De Volder's. A further series of *dictata* on Descartes's *Principia* might have been used by De Volder, as a further manuscript preserved at Hamburg, and concerning the second part of the *Principia* was lost during the Second World War,[69] and column B of the

[65] As *Dictata viri clarissimi Burcheri de Volder in Principia Cartesii* (ms. Sloane 1216).

[66] As *Principia Renati des Cartes dictata a doctissimo celeberrimoque domino Burchero de Volder* (MSD27).

[67] As *B. de Volder Annotationes in Meditationes Renati Descartes. Accedunt quoque notae quaedam Burcheri de Volder in primum librum Principiorum Cartesii De cognitione humana* (ms. BN Rps 3365 II).

[68] In the course of the book, I will refer to the Hamburg copies of De Volder's *dictata*, which are the most complete versions. Whenever necessary, I will indicate the variants with respect to the other copies.

[69] At Hamburg three *codices* including De Volder's *dictata* were extant, according to the historical catalogue of the Library: Cod. philol. 273 "Codex chartaceus CCCLX pagg. quo continetur Notulae quaedam in Renati desCartes principiorum Philosophiae partem primam scriptae ex ore Burchardi de Wolder. Ex bibliothekca Uffenbachiana"; Cod. philos. 274: "Codex chartaceus CXXI

The Hague manuscript includes a commentary on some articles of the first part of Descartes's *Principia* which cannot be traced back to any other copy. Moreover, copies of De Volder's *dictata* on Descartes's *Principia* and *Meditationes* are listed in the auction catalogues of the private libraries of Johan de Witt Jr. (1696 and 1701), Cornelius Nicolai (1698), Henricus Christianus Henninius (1705), Abraham Wasteau (1719), Johannes Theodoor Schalbruch (1722), Joan Duncan (1754), Emo Lucius Vriemoet (1761), Nicolas Barkey (1787), P. de Mey (1796), and an anonymous one (1764).[70] The Hamburg copies, as reported in the *ex libris*, were acquired by the German book collector Zacharias Conrad von Uffenbach – probably during his visit at Leiden in 1711 (to which I shall return below) – and then passed to Johann Christoph or Johann Christian Wolf. In fact, two copies of De Volder's *dictata* (one as *notae*, and the other as *annotationes breviores*, most probably corresponding to the Hamburg 274 version, which is shorter) are listed in the *Bibliotheca Uffenbachiana manuscripta* (1720).[71]

I presume that De Volder used, in his first years of teaching, the Hamburg 273 version of his *dictata*, as in this version – as I am going to show in Sect. 6.2.1.2, *Huygens's quantification of centrifugal force* – De Volder provided a wrong quantification of centrifugal force. In the Hamburg 274, he provided a correct quantification, based on the idea of accelerated motion of falling bodies, such as that one can find in Huygens's *Horologium oscillatorium*, appeared in 1673. Moreover, amongst the items of his experimental theatre, listed in the inventory De Volder prepared in 1705, is listed a "wooden circle rotating on an axis, to demonstrate the centrifugal force,"[72] so that De Volder probably ascertained the right quantification of centrifugal force after 1673 or 1675 (when the theatre was opened). In any case, the first version of his *dictata* (Hamburg 273) was still copied or dictated as late as in 1690, as testified to by The Hague manuscript. Also, there are traces of a lost series of *dictata* of De Volder on the *Traité de physique* (1671) of Jacques Rohault (1618–1672) – to which, according to Le Clerc, De Volder shifted in his last years of teaching – dated 1698 and 1699, listed in the auction catalogue of a private library of

---

pagg. continens Eiusdem idem scriptum. Ex bibliotheca Uffenbachiana"; Cod. philos. 275: "Codex chartaceus CXCIV pagg. continens Eiusdem annotationes in Ren. d. Cartes principiorum Philosophiae partem secundam. Ex bibliotheca Uffenbachiana." I thank Monika Müller, librarian at the Staats- und Universitätsbibliothek Hamburg Carl von Ossietzky, for having provided me with this information, and for having assisted me in locating De Volder's *dictata*. The Cod. philos. 275 was lost during the war.

[70] *Catalogus De Witt* (1696), 335; *Catalogus De Witt* (1701), 62; *Bibliotheca Nicolaiana*, 148; *Catalogus* [...] *Henninius*, 120; *Bibliotheca Wasteau*, 53; *Bibliotheca Schalbruchiana*, 122; *Bibliotheca Duncanniana*, 67; *Bibliotheca Vriemoetiana*, 114; *Catalogus J. A. C.*, 44; *Bibliotheca Barkeyana*, 412; *Catalogue De Mey*, 241. See also *Bibliotheca Musschenbroekiana* (of Petrus van Musschenbroek, 1764–1823), 183, where a dictation of De Raey is wrongly attributed to De Volder.

[71] Uffenbach 1720, column 671. Also in Uffenbach 1729–1731, volume 3, 201–202. This catalogue, moreover, includes a reference to De Volder's *Animadversiones* (now lost) on Descartes's *Principia*: Uffenbach 1729–1731, volume 3, 463.

[72] "Een houte circul drajende op een ax ad demonstrandam vim centrifugam," Molhuysen 1913–1924, volume 4, 105∗. See Sect. 6.2.1.2, *Huygens's quantification of centrifugal force*.

1752.[73] Notably, De Volder had already referred to Rohault before these now lost *dictata*, namely, while commenting upon Descartes's *Principia* I.71 in his Hamburg 273 series, concerning the origin of philosophical prejudices in sensory perception: as in the case, for instance, of the lack of feeling of pressure underwater (a topic which I will deal with extensively in Sect. 5.5, *De Volder's Cartesian hydrostatics: a mixed approach*), or the case of the bigger appearance of the Moon at the horizon, explained by De Volder by re-directing the reader to Rohault's *Traité*, part 1, chapter 32, § 22.[74] This reveals that De Volder completed his Hamburg 273 *dictata* after the appearance of Rohault's *Traité* in January 1671. Moreover, since Rohault's *Traité* appeared in French, it might even be that De Volder wrote his *dictata* after the appearance of the Latin translation of Rohault's *Traité* by Théophile Bonet in 1674 (as *Tractatus physicus*).

2.2.3 The Mid-1670s Clash at Leiden and the Foundation of the Experimental Theatre

The year 1674, actually, constituted a fundamental turning point in De Volder's career. In that year, a quarrel over Cartesianism exploded at Leiden, leading De Volder – together with the Cartesio-Cocciean theologians Abraham Heidanus (1597–1678) and Christoph Wittich (1625–1587) – to defend the religious and political orthodoxy of Cartesian philosophy. Moreover, during the summer holiday of 1674 De Volder travelled to England (where he met Newton in Cambridge), and at the end of the year he requested the academic Curators and the Leiden city authorities the permission and funds to establish a place to perform experiments at the University, which was granted the following year. As far as the 1674–1676 quarrel at Leiden is concerned, I will deal with it in Sect. 3.1.2.2, *Cartesianism under attack at Leiden in the mid-1670s*. Here, it is worth recalling that it all started when students, allegedly stirred up by Cartesian sympathisers like Cornelis Bontekoe (former pupil of Arnold Geulincx and editor of some of his posthumous works), Johannes Swartenhengst (lecturer at Leiden, and expelled from the University in 1675), and Cornelis van der Linden repeatedly interrupted the lectures of Aristotelian professors. On 3 March 1674, Gerard de Vries (1648–1705), professor of philosophy, could not complete his lecture and complained to the academic authorities. As reported by Le Clerc and in the academic proceedings, the Curators admonished the

---

[73] Namely, "Annotata in Jacobi Rohaulti, Tractatum physicum, a clariss. et nobil. Professore Burchero de Volder, 1698. MS.," and "Annotata in Rohaultii, tertiam Partem de rebus terrestribus a clar. et nobili Professore Buchero de Volder, A° 1699," *Bibliotheca Martiniana*, 116 and 427. Le Clerc states that De Volder shifted to Rohault's physics with not great satisfaction: "il étoit ennuyé d'expliquer [...] la Physique de Rohaut, qu'il avoit expliquée dans les derniers tems; parce qu'il voyoît combien peu il y avoit de choses assurées, dans ces Livres," Le Clerc 1709, 398. Also Ruardus Andala reports De Volder's use of Rohault: Andala 1718, 124.

[74] Hamburg 273, 69.

professors not to mix theology and philosophy (in accordance with a regulation of 1656) and to avoid further quarrelling.[75] On 28 June 1674, De Volder himself, together with Wittich and Heidanus, spoke with the Grand Pensionary of Holland Gaspar Fagel, attempting to convince him that Cartesian philosophy did not pose any danger either to religion or to the state. Apparently, Fagel was positively impressed by De Volder's words (which are reported by Le Clerc).[76] Yet, the quarrel continued, as the enemies of Cartesianism made another appeal to the Academic Curators (according to Le Clerc, because they could not rebut Cartesianism by sound arguments),[77] proposing to them in January 1676 a list of propositions concerning philosophy and theology, and addressing, besides the ideas of Descartes, those Lodewijk Meijer and Spinoza, allegedly taught by De Volder. On 16 January, the Curators forbade discussion of a reworked list of propositions.[78] In the same year, Wittich, Heidanus and De Volder defended the Cartesian positions by publishing their *Consideratien, over eenige saecken onlanghs voorgevallen in de Universiteyt binnen Leyden*, in which the theological propositions were discussed by Wittich, those in philosophy by De Volder, while Heidanus wrote a preface, and officially authored the whole book. For this reason, on 5 May 1676 he was sum-

---

[75] "Comme il étoit encore jeune, & qu'il ne s'étoit pas encore apperçu des défauts de cette Philosophie, comme il l'a fait depuis; il s'interessoit davantage dans ces sortes d'exercices & soûtenoit ce qu'il disoit avec beau coup de feu, quoi qu'il fût d'ailleurs très-moderé à l'égard des personnes. Il avoit alors pour collegue feu Mr. de Vries, depuis Professeur en Philosophie à Utrecht, qui l'attaquoit souvent dans ses Theses & dans ses Corollaires; & dont les fauteurs parloient mal de lui, sans qu'il leur en eût donné sujet. Les Etudians se partagerent entre eux, & les Cartesiens, comme il semble, se trouverent superieurs de beaucoup. Cela fit que dans une Dispute de Mr. de Vries, tenue le 3. de Mars 1674. ils firent quelques desordres dans l'Auditoire de Philosophie, qui obligerent ce Professeur desortir de Chaire, avant que l'heure fût écoulée. Châcun porta ses plaintes au Sénat Academique. Mrs. les Curateurs de l'Academie firent dire aux Professeurs en Philosophie, qu'il ne vouloient pas leur ôter la liberté de philosopher, mais qu'ils trouvoitnt à propos premierement, quel'on ne mêlât pas, ainsi quel'on faisoit, des questions Théologiques dans la Philosophie; comme cela avoit été déja défendu par un reglement de l'an 1656 & en second lieu, que châcun proposât ses opinions, soit Cartesiennes, soit Péripateticiennes avec modestie, & sans s'emporter les uns contre les autres," Le Clerc 1709, 354–355; cf. Molhuysen 1913–1924, volume 3, 280–281, 290–294 and 313–314.

[76] "Comme en ce tems là les Cocceïens & les Cartesiens s'étoient unis ensemble, non qu'il y eût aucune liaison entre leurs sentimens particuliers; mais parce que les Voetiens s'opposoient également aux uns & aux autres, Mrs. Heidanus, Wittichius & de Volder, furent voir Mr. Fagel, Conseiller Pensionaire de la Province de Hollande, pour le desabuser des préventions, qu'on avoit voulu lui donner contre eux," Le Clerc 1709, 356. For a full quotation, see Sect. 3.1.2.2, *Cartesianism under attack at Leiden in the mid-1670s*.

[77] "Ne sachant pas combattre leurs Adversaires, par la Raison, ils vouloient employer- l'autorité de l'Etat, pour accabler le Cartesianisme. Pour cela, il falloit faire accroire qu'il etoit opposé à la Théologie des Réformez, & même au Gouvernement d'alors; qui étoit celui d'un Stadthouder, dont l'autorité, presque collaterale à celle des Etats, régloit tout. Il n'y avoit rien, dans cette Philosophie, qui eût du rapport avec aucune maxime contraire au Gouvernement, ou à la Théologie. Mais ceux, qui étoient opposez à Descartes, firent en sorte que le Gouvernement fit quelques dé marches contre les Cartesiens," Le Clerc 1709, 360–361.

[78] Le Clerc 1709, 368–370; cf. Molhuysen 1913–1924, volume 3, 317–321. For full quotations, see Sect. 3.1.2.2, *Cartesianism under attack at Leiden in the mid-1670s*.

moned by the Curators and lost his post at Leiden.[79] As to De Volder, he lost the opportunity to become a rector in 1674, when the University Curator Hieronymus van Beverningh proposed his name to the Stadtholder, William of Orange, for the position of rector. According to Gronovius and Le Clerc, a letter by the Prince recommending the nomination of De Volder was sent to one of the mayors of Leiden, Jan Rippertsz. van Groenendijk. However, Van Groenendijk was absent from the town, so he could not communicate the Prince's order to the Academic Senate. The messenger, who knew the content of the letter, had in the meantime spread the news. Hence, a second letter from the Prince was sent to another major, Herman van der Meer, recommending the nomination of the professor of law Johann Friedrich Böckelmann, who became rector in 1675. According to Le Clerc, this was due to the fact that the Prince and the Curators, ignorant of philosophy, were often pushed by the enemies of Cartesianism to hinder its dissemination: in the case of De Volder's rectorate, it had been suggested to the Prince that the Cartesians were against the House of Orange and the government of the Stadtholder.[80]

---

[79] Le Clerc 1709, 370–373; cf. Molhuysen 1913–1924, volume 3, 324–327. For full quotations, see Sect. 3.1.2.2, *Cartesianism under attack at Leiden in the mid-1670s.*

[80] "Ipsi autem acciderat, ut literas ab Gubernatore subscriptas et in hanc urbem iam missas, quibus destinabatur Rector Magnificus, quorumcunque artificio exaratae aliae, quibus avertebatur ab isto decore Volderus, postridie exceperint. Statim nosse sortem suam potuit noster, quod in rebus tetricis magnum est et gloriosum et plenum solatii. Hinc ad despiciendos et illos, per quos venerant, atque et iam, id ipsum, propter quod venerant, concitatus, magnisque et sanctis, munimentis ab hac parte se circumsepiens, inprimisque hinc semper interius flectens, fortunae inde venienti laqueos et ungues paravit. Volder non erat possidere aliquid precibus expressam, animoque servili praemia ineptiae reputare. Volderus semper, quasi ignoraret sese, modestus, et credens plura peccari ab eo, qui magnates adulatione demereretur; quam si offenderet, silentio et tunc et semper maluit se involvere tanquam libertatis seu spretionis satis magnifico generi," Gronovius 1709, 34–35; "[c]es Leçons lui firent beaucoup d'honneur, & furent très-avantageusês à l'Academie, à laquelle sa réputation attira beaucoup du monde. Feu Mr. de Beverning, qui étoit Curateur de l'Academie, prit une amitié particuliere pour lui, sans qu'il l'eût autrement recherchée; & fit en sorte, auprès du Prince d'Orange, qu'il le nomma Recteur cette année-là. La Patente même en fut expediée, & en voyée par un Messager à Leide à l'un des Bourgmestres [Mr. Groenendyck], selon la coûtume. Ce Bourgmestre se trouva absent & ne revint en ville qu'assez tard, de sorte qu'il n'étoit pas tems de convoquer le Sénat Academique, pour y déclarer l'élection du Recteur. Ce pendant le Messager ayant ouï lire la lettre, dont il étoit chargé, ne manqua pas de publier que Mr. de Volder avoit été fait Recteur. Mais il arriva un Courrier, la même nuit, adressé à un autre Bourgmestre [Mr. van der Meer], avec une seconde Lettre, où le Prince revoquoit l'élection précedente & choisissoit pour Recteur Mr. Boekelman. Quelques uns des ennemis, non de la personne de Mr. de Volder, qui n'étoit pas homme à s'en attirer par aucune mauvaise conduite, mais du Cartesianisme, ayant su sa nomination, qui étoit un effet, comme jl'ai dit, des bons offices de Mr. de Beverning, firent entendre au Prince que c'étoit un Cartesien, peu affectionné à la Maison d'Orange, & au Gouvernement du Stadthouder, & obtinrent de lui, dans l'absence de Mr. de Beverning, que son élection seroit revoquée. C'est ainsi qu'un grand Prince, par complaisance pour des gens, qui ne lui pouvoient guere rendre de service, & encore moins à l'Etat, faisoit quelque fois des mécontens, sans aucune nécessité; contre les regles de la bonne Politique, qui demande qu'on ne mécontente ja mais personne, sans qu'il l'ait mérité, par une conduite contraire aux interêts du Public. Ils croit bien à souhaiter que les nuages de la passion & des interêts des particuliers ne puisent jamais parvenir jusqu'à ceux, que leur haute dignité doit élever au-dessus de toutes les considerations, qui n'ont pas pour leur sin le bien de l'Etat. Mais on ne voit que trop souvent le contraire, & les grandes

Before expending more words on De Volder's own political and religious views (as I will do below), is worth noting that, as reported by Le Clerc, De Volder after this quarrel over Cartesianism at Leiden was much more cautious in teaching Descartes's metaphysics, waiting until the dust of the quarrel settled.[81] This is confirmed by his subsequent public disputations. In 1674–1676 and after, he presided over his disputations *De rerum naturalium principiis* – in which any metaphysical consideration is purportedly avoided[82] – his *De motu tertia* (13 November 1675), *De materiae divisibilitate in infinitum* (author and *respondens* Hermann Lufneu, 27 January 1676), his *De aëris gravitate* (14 October 1676–29 June 1678) – also touching upon the contents of his experimental lectures – the *De magnete* (10 July 1677, author and *respondens* Johannes Balthasar Helvetius), and the *Disputatio philosophica inauguralis exhibens vitri thermometrici historicam simul et acroamaticam delineationem* (30 May 1679) of Jacobus Rattrey, in which De Volder acted only as a nominal promoter.[83] Moreover, in 1679 he gave the funeral oration for his friend Coeman, who was just appointed professor of law at Leiden, but who could not assume his post. Thus far, De Volder waited until 1680 to preside over disputations in metaphysics. Namely, with the series of four disputations *Contra atheos* (1680–1681), in which two Cartesian proofs for the existence of God are set over against the 'atheists', thereby showing the orthodoxy of Cartesianism.[84] Moreover, De Volder would repeatedly refuse to authorize the publication, as separate monographs, of his series of disputations: namely, of his *De rerum naturalium principiis* and *De aëris gravitate* (which had different and collated editions in 1681), *Contra atheos* (1685), and of his *Exercitationes philosophicae adversum Censuram*, a series of 28 disputations presided over between 1690 and 1693, and aimed at rebuking the criticisms of Descartes's philosophy moved in the *Censura philosophiae Cartesianae* (1689) of Pierre-Daniel Huet. We know that De Volder purportedly avoided publishing any text aimed at circulation outside academia from his colloquium with Stolle and Hallmann – where he declared that he would rather reduce

qualitez du Prince, dont je viens de parler, nel'ont pas toûjours pu garentir de cette sorte de surprises. On les doit néanmoins attribuer plûtôt à ceux qui l'environnoient qu'à lui même, qui n'avoit souvent aucun interêt, dans ce qu'on lui faisoit faire," Le Clerc 1709, 364–366 and 373.

[81] "Mr. de Volder, à qui l'on fit entendre que ce n'étoit pas à lui, à qui l'on en vouloit, continua d'enseigner son Cartesianisme, mais avec un peu plus de précaution. Dans la suite, quand la chaleur de ces disputes fut passée, il explica en public non seulement la Physique, mais encore la Métaphysique de Descartes," Le Clerc 1709, 373–374.

[82] See Sect. 3.2.2.2, *De Volder on the 'truth' of Cartesian principles.*

[83] Rattrey, from Sweden, never mentions De Volder in the disputation; moreover, he enrolled at Leiden only on 17 May 1679 (see *Album studiosorum Academiae Lugduno-Batavae MDLXXV–MDCCCLXXV*, column 628). In the disputation he refers to his teacher at Uppsala Petrus Hoffwenius, and to De Raey, who had been teacher of Hoffwenius at Leiden. See Sect. 5.6.3.3, *De Volder's assumption of Boyle's law in the works of his students.*

[84] I discuss this topic in Sect. 3.1.2.3, *The demonstrations of the existence of God.*

the number of books[85] – from Le Clerc's eulogy,[86] and from the preface of Boerhaave to De Volder's posthumous *Oratio de novis et antiquis* (1709), originally held on 3 July 1699 on the occasion of the honorary graduation of John Gale at Leiden.[87] Boerhaave was, at first, in charge of discovering any testament or will of De Volder at his house just after his death. Hence, while looking further at his papers,[88] Boerhaave found that De Volder had left many things unfinished, and this *Oratio* was the only publishable work. According to Boerhaave, De Volder, like Socrates, indeed preferred not to publish anything.[89] As far as Huet's *Censura* is concerned, Le Clerc reports that De Volder told his friends that he refused to publish his *Exercitationes* as a monograph in order to avoid quarrels – because this work had chiefly a didactic function (as I am going to show in a moment).[90] In fact, his *De*

[85] "Er hat ein groß ingenium, und weis artig zu schertzen; Sein Judicium scheint nicht geringer zu sein. Wenn er was schreiben sollte, so würde er ohnezweifel einen Satyricum agiren, denn er ist ein spitziger Gast. Allein er hat nicht in Willens die unzehliche Menge der Bücher zuvermehren, die man vielmehr vergeringern sollte," Warsaw University Library, Cod. IV oct. 49, 594. Transcription kindly provided by Martin Mulsow.

[86] "Comme il n'a donné au Public, que quelques Harangues (car les Theses, que l'on a publiées de lui, ont été publiées sans son aveu,) je croi qu'on ne sera pas fâché que je parle de châcune, en son tems [...]," Le Clerc 1709, 374.

[87] On Gale, see Gordon 2004.

[88] See Sect. 2.4.2, *De Volder's retirement, death and legacy*. Boerhaave looked again at De Volder's scientific papers after his first check, as he had to put seals on De Volder's belongings after having looked for his testament on 28 March 1709. Boerhaave's preface to De Volder's oration is dated 18 May 1709.

[89] "Tanti viri dotes nosse qui desiderant, nae habebunt illi, unde votis satisfaciunt suis, in perfectissima eximii Gronovii oratione, digna Volderi meritis, qua, quae aeternitatem meruit, defuncti memoriam mori vetuit. Ego acerbissimo mihi fato amici, qui nulli flebilior quam mihi occidit, natam iacturam sarcire studens, animo solatium meo, et publico utilitatem quaesivi in manuscriptis, quae vir optimus reliquerat. Quam speravi inventum iri, quod illi laudem, omnibus usum, nobis gaudia, praeberet! Sed spe magna decidi! Inveni quippe multum laboris, digesti nihil. Quid mirum? Illi ad altiora semper nitenti, iam perspecta ut ordinaret, nec otium superfuit, nec amor: quumque omni vita Socratem expressisset, et ipsa hacce re imitatus eundem fuit, ut editum quidpiam a se noluerit. Solummodo haec se obtulit oratio, hanc expolivit, hanc, efflagitantibus amicis, nec reperto, quod vetaret, mandato auctoris, edendam curavi," De Volder 1709, *Lecturo*, 2 (unnumbered). See also Le Clerc 1709, 390–392.

[90] "Mr de Volder est obligé, dans ces Theses, d'expliquer souvent les sentimens de Descartes contre ses Adversaires, qui ne les entendoient pas assez bien. Cela l'avoit d'abord à degoûté de cet Ouvrage: 'qu'importe, disoít-il, de rechercher si Descartes a cru ceci, ou cela? Lors qu'on est assuré de son sentiment, il est alors de l'interêt des Sciences de savoir s'il est veritable'. Mais l'utilité, qu'en pouvoit retirer la Jeunessè l'obligea de passer par dessus cette consideration. C'est pour ce la, qu'il s'est plus étendu qu'il n'auroit fait, dans un Ouvrage d'une autre nature, & qu'il a redit des choses déja expliquées par d'autres, qu'il auroit entierement omises. Aussi ne voulut-il jamais achever cette défense de Descartes, ni la publier lui même; & un Libraire l'ayant fait imprimer à Amsterdam, en MDCXCV in 8, il desavoüa cette Edition, & ne put néanmoins se résoudre à en donner une meilleure. Quoi qu'il en soit, il n'y a point de livre, que je sâche, où les sentimens de Descartes sur les matieres qu'il traite, soient expliquez plus nettement & défendus avec plus de méthode & de force. Il y a encore une chose, qu'on ne peut trop louer, c'est que Mr. de Volder y traite par tout son Adversaire, avec toute la civilité possible, & ne s'attache qu'aux choses, sans rien mêler de personnel. Cet ouvrage, revu avec soin, lui pouvoit assurément faire honneur. Mais l'Auteur étoit résolu à ne rien publier de semblable, quelque talent qu'il eût à se bien

rerum naturalium principiis, *De aëris gravitate* and *Contra atheos* had already been published without his consent, as is clear from the short review by Pierre Bayle (a direct acquaintance of De Volder)[91] of the 1685 edition of De Volder's *Contra atheos*, which appeared in the February 1685 issue of the *Nouvelles de la République des Lettres*, in which their purely didactic function is remarked upon.[92] Moreover, De Volder did not like – as reported by Prosper Marchand – that the Middelburg edition of his *De rerum naturalium principiis* had been profanely dedicated by the printer Remigius Schrever "to God One and Three, and to [...] Burchard de Volder,"[93] and in a letter to Melchisédech Thévenot dated 28 January 1683 (on which I will be back in Sect. 2.3.1, *The correspondence with Thévenot*) he complained that he was obliged to make his 1682 *Oratio de coniungendis philosophicis et mathematicis disciplinis, cum philosophicae professioni adiunctam mathematicam rite*

expliquer. Ne dire que ce que d'autres avoient dit ne lui paroissoit pas une occupation digne de lui; & avancer quelque chose de nouveau étoit, comme il croyoit, s'exposer à des contradictions, qui finissoient souvent par des querelles. C'est comme il parloit quelque fois à ses Amis," Le Clerc 1709, 383–384. Cf. De Volder 1695, *De dubitatione universali*, thesis 2.

[91] De Volder acted as intermediary between Bayle, Leibniz and Johann Bernoulli: see the letter of Leibniz to Bayle of 19 August 1702, of Bernoulli to Leibniz of 16 September 1702, of Leibniz to Bayle of 3 October 1702, of Bernoulli to Leibniz of 18 November 1702.

[92] "Il y a une fausseté dans le titre de ce Livre, car on y assure que toutes les Theses que Monsieur de Volder a faites contre les Athées, & il n'est pas vrai qu'elles soyent toutes ici. Il avoit expressément marqué dans celles qu'il a fait soûtenirà ses Disciples, sur cette importante question de l'existence de Dieu, qu'il lui en restoit d'autres à faire, & cela paroît assez. Comme il n'avoit pas dessein qu'elles servissent à d'autres usages qu'à une Dispute Académique, il n'a point voulu les achever, croyant par là faire en sorte qu'aucun Libraire n'en entreprit l'édition. Mais se soins ont été inutiles, car on a publié malgré lui ce qui en avoit parû sur les bancs de son Auditoire, & on a retranché l'endroit où il avoit averti, que l'Ouvrage n'étoit pas encore à sa fin. Comme donc il n'avoit pas eu dessein de travailler pour le Public, en faisant ces Theses pour ses Ecoliers, & que d'ailleurs cette édition a été faite sans qu'il en ait été averti, qu'elle est pleine de fautes d'impression, & qu'il est faché qu'on l'ait entreprise, il déclare qu'il ne reconnoît point pour sa produttion cet Ouvrage-ci. Il le desavouë entierement. Il contient néanmoins de très-bonnes choses, qui peuvent servir d'un bon Commentaire à quelques endroits difficiles des Méditations de Monsieur Descartes. On avoit déja joüé une semblable piece à Monsieur de Volder, à l'égard d'une Dissertation sur les principes du Corps, & sur la pesanteur de l'air, qu'on publia à son insçû l'an 1681. C'est une bonne marque, & qui doit le consoler des supercheries que les Libraires lui font. Il y a bien des gens qui voudroient être en la même peine; ceux par exemple qui ne trouvent personne qui se veuille charger de leurs Ouvrages, quoi qu'ils sollicitent de porte en porte tous les Libraires à les imprimer. Que. plaisir ne seroit-ce pas pour eux, si on les imprimoit sans aucune sollicitation, & sans attendre leur consentement? Monsieur de Volder est Professeur en Philosophie & en Mathématique à Leide, & s'est acquis une grande réputation," Bayle 1685, 225–227. A short review of De Volder's *Contra atheos* appeared also in the issue 31 of the *Journal des Sçavans* (19 November 1685).

[93] See Marchand's *Dictionnaire historique* (1758–1759), article *Albizi*, where Marchand remarks how the Protestant meddled profane and sacred things, and noted how De Volder was displeased by such a dedication: "Professeur Burcher de Volder [...], qui a souffert qu'on lui adressàt cette étrange Dédicace: Deo tri-uni, et Clarissimo Domino Burchero de Volder," Marchand 1758–1759, volume 1, 8. Marchand's source on the anecdote is unknown. A short review of the collated edition of De Volder's *De rerum naturalium principiis* and *De aëris gravitate* (1681) appeared on the May 1682 issue of the *Acta Eruditorum*.

auspicaretur printed.[94] In the case of his *Exercitationes*, De Volder publicly distanced himself from this unauthorized publication by Arnold van Ravestein in Amsterdam (which might have been prompted by the publication of the second edition of Huet's *Censura* in 1694),[95] by a public letter which appeared in the *Histoire des Ouvrages des Sçavans* (directed by Henri Basnage de Beauval) in May 1695, and which is worth quoting in full:

> I (M. de Volder) cannot stop myself from complaining in public of the greed of the booksellers, who undertake[,] without any consideration[,] to print whatever they think fit to bring them some profit. In the functions of my profession, I had composed some Academic Exercises in which I examined the *Censura philosophiae Cartesianae* by Mr. Huet: and this only for the use of my students. But as I did not intend them for print, I did not work on them with the same application, or the same exactness as if I had other aims. Besides[,] having no other plan than the instruction of my students, you know that it is necessary to explain more fully, and more extensively some principles, that one poses in a public writing, without stopping to prove them, because it is supposed that they are already known to the reader. Moreover[,] I have not pretended to explain my own feelings: I have only proposed to report the opinions of Descartes, and to defend them against the objections of the Bishop of Avranches. But I do not take a position, and I do not express my particular judgment. As to the rest[,] this edition, which is done without my participation[,] is so full of mistakes, that there are many things that can be misunderstood, and whose meaning is very doubtful and ambiguous. So that in order that one does not ascribe to neither the faults, nor the opinions of others, I pray you to disavow for me these Exercises on which the Booksellers have put my name without my consent.[96]

De Volder, thus, did not aim his *Exercitationes* at wide circulation, as he intended them only at the instruction of his students, and, moreover, he claims not to have endorsed their contents, but only to have aimed them at a scholarly defence of Descartes. In turn, in the first theses of the first disputation of the series, *De*

[94] The correspondence is extant at the Staats- und Universitätsbibliothek Göttingen, Cod. Ms. Hist. nat. 102, fasc. 25, ff. 44–69. A complete edition of this correspondence is currently in processing by the author. See 63r: "[...] {une} harangue de ma facon, qu'il m'a falu publier contre mon gré par l'authorité de ceux qui gouvernent nostre Academie."

[95] On Huet's *Censura* and its reception, see Rapetti 1999, 2003, 2018; Shelford 2007; Lennon 2008.

[96] "Je (Mr. de Volder) ne puis m'empêcher de me plaindre au public de l'avidité des Libraires, qui entreprennent sans aucuns égards d'imprimer tout ce qu'ils jugent propre à leur apporter quelque profit. Dans les fonctions de ma profession, j'avois composé quelques Exercitations Academiques où j'examinois la Censura Philosophiae Cartesianae de Mr. Huët: & cela uniquement pur l'usage de mes auditeurs. Or comme je ne les destinois point à l'impression, je ne les ai pas travaillées avec la même application, ni la même exactitude que si j'avois eu d'autres vuës. D'ailleurs n'ayant d'autre dessein que l'instruction de mes disciples, vous savez qu'il faut expliquer plus amplement, & avec plus d'étenduë certains principes, que l'on pose dans un écrit public, sans s'arrêter à les prouver, parce que l'on suppose qu'ils sont dêjà connus du lecteur. De plus je n'ai point pretendu expliquer mes propres sentimens: je me suis uniquement proposé de rapporter les opinions de Descartes, & de les defendre contre les objections de Mr. l'Evêque d'Avranches. Mais je ne prens point de party, & je n'allegue point mon jugement particulier. Au reste cette édition qui se fait sans ma participation est si pleine de fautes, qu'il y a beaucoup de choses qui peuvent être mal entenduës, & dont le sens est fort douteux & fort ambigu. Ainsi afin que l'on ne m'impute ni les fautes, ni les sentimens d'autrui, je vous prie de desavouër pour moi ces Exercitations ausquelles les Libraires ont affiché mon nom sans mon consentement," Basnage de Beauval 1695, 421–422. On this letter, see Wiesenfeldt 2016.

dubitatione universali, he maintained that he was asked by his students to prepare a defence of Descartes's ideas against Huet, whose criticisms De Volder's students considered unpersuasive, in the form of academic disputations allowing them to exercise themselves. De Volder himself hesitated to accept his students's invitation, as he would have needed to repeat many things already stated by many thinkers; moreover, Huet often just misinterpreted Descartes's words, so that a reply was going to involve a clarification of Descartes's own positions, rather than a genuine discussion of ideas. Eventually, he accepted in order to allow his students to exercise themselves in public disputations, as well as to defend some general ideas of Descartes (as remarked by Le Clerc), whose positions he did not entirely follow.[97] This is consistent with the fact that many of the positions defended by De Volder in the *Exercitationes* – for instance, his idea of substance, which was perfectly Cartesian, or his demonstrations of the existence of God, which were just inspired by Descartes, and which cost him accusations of Spinozism by Ruardus Andala (1665–1727)[98] – recur also in his other texts, where he however did not endorse all the positions of Descartes. As Leibniz pointed out to Huet in a letter of April 1695, "it appeared to me [while reading De Volder's *Exercitationes*] that he deserts Descartes not rarely while he seems to defend [him]."[99] In turn, Le Clerc is certainly

[97] "Placuit nuper Illustri Suessoniensi, Episcopo Cartesianae, quam vocat Philosophiae edere Censuram, qua non omnis, sed praecipua et capitalia eius vitia persequitur. Nec defuere inter eos qui illam Philosophandi rationem amant minus, qui in hoc reconditae sane eruditionis viro magnum suae causae patrocinium ponerent, non tam, si quid video, rationibus ab ipso allatis, quam viri summi merito suo nominis auctoritate ad id adducti. Qua re factum est, ut ex meis discipulis nonnulli egregii adolescentes, cum animadverterent me et Reverendi huius Praesulis et aliorum multorum argumentationibus, quae contra Cartesii Philosophandi viam, eiusque dogmata passim producerentur, parum commoveri, nec in illis existimare tantum esse virium, quantum praedicarent multi, me rogaverint, ut quo illis sese exercendi detur occasio, Academicarum Disputationum forma exponere velim, quidnam aut de Cartesii sententiae veritate, aut de aliorum in Cartesium allatis rationibus sentirem. [...] Cui petitioni an obsequerer diu dubius haesi. Facile enim videbam multa dicenda esse, quae ab aliis dicta non sine nausea possent repeti. Deinde et illud apparebat longe aliam mihi in nonnullis casibus esse sententiam de Cartesii mente, quam eius adversariis, adeoque quaerendum nonnunquam fore, quaenam fuerit mens Cartesii. Quae disputationis ratio, ut ad personarum, ubi de iis agitur, defensionem multum potest, ita ad veritatis perquisitionem nihil confert. Sive enim hoc senserit Cartesius, sive illud, quid attinet inquirere? An verum sit quod senserit, ubi de sensu constat, hoc demum scientiarum et veritatis interest. [...] Ex altera tamen parte multa videbantur suadere, ne illorum voluntati refragarer; vix enim, quamdiu in hoc publice docendi munere versor, videbar honeste denegare posse id, quod et moris est Academici, et quod in sui commodum redundare non posse immerito fortasse arbitrabantur. Accedebat et illud, quod hac occasionis sperabam, me eam philosophandi rationem, quam licet nequaquam in omnibus, in generalibus tamen sequendam existimo, ab innumeris liberaturum cavillationibus, et praecauturum hac via, ne Tyronum, quibus haec solummodo destinantur, animi per illas avocentur a rei veritate exacte inquirenda et agnoscenda," De Volder 1695, *De dubitatione universali*, theses 1–3. Cf. Le Clerc 1709, 381–382: "[l]'Auteur déclare, dès le commencement, qu'il croyois que l'on doit suivre les principes géneraux de la Philosophie de Descartes, mais qu'il ne la faut pas suivre en tout." Cf. *supra*, n. 90.

[98] I will consider such demonstrations in Chap. 3.

[99] "Forte nuper amicus ad me attulit quae Burcherus de Volder clarissimus apud Leidenses professor pro Cartesio in tuam censuram novissime scripsit, publice edita. Viri fama, ipsiusque amici desiderium fecere, ut perlegerem solito attentius; visusque mihi est non raro deserere Cartesium cum defendere videtur," GP III, 19.

exaggerating when he claims that De Volder was "by no means tempted by the philosophy he taught."[100] Even if, as reported by Le Clerc himself, in the last phase of his career De Volder was decidedly discontented with Descartes's physics and metaphysics – especially after having read Newton and the books of other "Englishmen" (such as John Locke, as I am going to show in Sects. 2.3.3, *Mathematician and scientific expert*, and 3.2.3.2, *De Volder's methodological liberalism – and its continuity over time*)[101] – De Volder certainly endorsed many of Descartes's ideas throughout his career. It might be, thus, that he refrained from allowing his *Exercitationes* to circulate outside academia in order to avoid quarrels (as suggested, again, by Le Clerc).[102] The separate publication of his *Exercitationes* as disputations (usually printed in 100 copies)[103] certainly would not prevent De Volder from being involved in an academic quarrel within the Netherlands: however, it could have restricted the dissemination of them. Indeed, Huet himself received only handwritten copies of some of these disputations, from Madame de Tilly.[104]

Thus far, De Volder assumed, after the clash over Cartesianism at Leiden in 1674–1676, a decidedly cautious attitude towards philosophical quarrelling. We might therefore interpret, at least partially, the experimental activities he carried out since 1675–1676 as a way to teach without inserting himself in academic debates by means of printed texts. Or, rather, that the academic authorities, as suggested by Wiesenfeldt, allowed the establishment of his academic auditorium or laboratory in an attempt at preserving the *libertas philosophandi* and settle further quarrelling.[105] Of course, De Volder already had genuine interests in experimental philosophy. For instance, he presents and discusses the ideas of Boyle as early as in his Hamburg 273 *dictata* (as I show in Sect. 5.5.3, *De Volder's Archimedeo-Cartesian hydrostatics: an impossible synthesis?*). Moreover, he travelled to England in the summer of

[100] "Il n'étoit nullement entêté de la Philosophie, qu'il enseignoit," Le Clerc 1709, 398.

[101] "Je ne doute pas que le Livre de Mr. Newton ne lui eût ouvert les yeux, sur la doctrine des tourbillons de Descartes; que ce Livre a entierement renversée, à cela près, qu'il établit le Systeme de Copernic, comme lui; en reconnoissant le Soleil, comme le centre commun des mouvemens de la Terre & des autres Planetes. Cette partie de la Physique de Descartes avoit néanmoins paru à bien des gens la mieux imaginée. On voit en cette rencontre, comme en plusieurs autres, l'exemple d'un grand Mathematicien, qui ne raisonnoit point mathematiquement, ni même conséquemment. [...] Il n'étoit nullement entêté de la Philosophie, qu'il enseignoit; sur la fin de ses jours, & même quelques années auparavant, il avoit reconnu le foible du Cartesianisme; autant apparemment, par sa propre méditation, que par le secours des habiles Anglois, qui ont établi d'autres principes. Je l'ai ouï se moquer, plus d'une fois, d'une bonne partie des Méditations de Descartes, quoi qu'il les eût expliquées pendant long-tems. S'il est permis d'ajoûter cette raison à celles, qu'il a rendues de l'envie qu'il avoit de vivre en repos; je croirois qu'il ne vouloit pas composer un nouveau Systême, & qu'il étoit ennuyé d'expliquer ces Méditations, aussi bien que la Physique de Rohaut, qu'il avoit expliquée dans les derniers tems; parce qu'il voyoît combien peu il y avoit de choses assurées, dans ces Livres. Il lui échappoit de tems en tems des plaintes, en s'entretenant avec ses amis, du peu de progrès que l'on avoir fait, dans la connoissance de la Verité," Le Clerc 1709, 382 and 398–399.

[102] See *supra*, n. 90.

[103] See Van Miert 2009, chapter 5.

[104] See Rapetti 1999, 113–117 and 140–143. A handwritten copy of the first disputation is extant at the Bibliothèque Nationale de France, amongst Huet's papers: Ms. F. Lat. 11451, 2r–8r.

[105] See Wiesenfeldt 2000, 2002, chapter 2, 2008.

1674, just after his colloquium with Fagel, and well before the firing of Heidanus. So that his subsequent request to establish an experimental theatre was most probably independent from the growing disagreements over Cartesianism, which might however have favoured the establishment of such a laboratory, as a means to settle academic quarrelling and opening new ways of teaching.

As far as De Volder's journey to England is concerned, Wiesenfeldt – to whose *Leerer Raum in Minervas Haus* I redirect the reader as to the intellectual and historical context of De Volder's setting up of an experimental theatre, as well as for the subsequent experimental activities at Leiden, chiefly in natural philosophy, chemistry and medicine – has dated it to between 13 August (when De Volder acknowledged his salary receipt) and 1 October 1674 (when he was present at a public disputation at Leiden).[106] Moreover, since the summer holiday at Leiden usually started around mid-July and ended at the beginning of September,[107] it might be that De Volder travelled to England between 16 July and 13 August.[108] The only certainties are that he travelled with George Konrad Crusius (professor of law) and Christiaan Melder (professor of mathematics), and they requested letters of presentation to Van Limborch[109] (correspondent of De Volder at least since 1676),[110] who at that time was in contact with Henry More and Ralph Cudworth.[111] Moreover, we know that De Volder met Newton at Cambridge: according to a letter De Volder wrote to Newton on 24 November 1684, and which served as a recommendation for Johann Christoph Zimmerman, nephew of Christoph Wittich and heading to

[106] See Wiesenfeldt 2002, 61, referring to manuscripts AC (Archief van Curatoren) 87, and ASF (Archief Senaat en Faculteiten) 296, 51v, extant at Leiden University Library.

[107] See Molhuysen 1913–1924, volume 2, 37, 108, 109, 142, 297–298 and 267*.

[108] No clearer information is given by Gronovius and Le Clerc, who only mention De Volder's travel to England: see Gronovius 1709, 23–24; Le Clerc 1709, 362–363. According to Le Clerc, De Volder stayed in England a few weeks; he knew some English, enough to read it.

[109] I owe this information to Abraham des Amorie van der Hoeven's *De Philippo a Limborch, theologo, dissertatio historico-theologica* (1843): "[t]anti Limborchium faciebant viri docti in Anglia ut viri clarissimi Crusius Juris, Volderus Philosophiae, Melderus Mathematices in Academia nostra primaria Lugduno-Batava professores, cum anno 1674 per ferias caniculares in Angliam excurrere statuissent, ab illo literas commendatitias sibi efflagitandas censuerint," Des Amorie van der Hoeven 1843, *De Philippo a Limborch, theologo, dissertatio historico–theologica*, 39. Des Amorie van der Hoeven does not declare his source: he refers to "[m]anuscripta varii generis quae servantur in Bibliothecis Remonstrantium Amstelodamensi et Roterodamensi," Des Amorie van der Hoeven 1843, *De Philippo a Limborch, theologo, dissertatio historico–theologica*, 39. These manuscripts are now extant at the Amsterdam University Library and at the municipal library of Rotterdam. On Van Limborch's correspondence, see Simonutti 1984; Simonutti 1990; Van Rooden and Wesselius 1987; Walsh 2018.

[110] Two letters by De Volder to Van Limborch of 1676 and 1688 are reported in a auction catalogue of 1862, organized by Martinus Nijhoff: "Volder, B. De, Prof. de philos. à Leide, cartésien célèbre. 2 Ltt. aut. sig. en Holl. à S. v. Limborch. 1676, 88. Avec cachet. Lettres spirituelles. 'Hoe souw ik int goudt gaan als men met dadelycke proeven kon toonen dat onse leerwijse soude dienst doen aan de vrienden'," Nijhoff 1862, 85. These two letters have not been recovered. Four letters, in Latin, between De Volder and Van Limborch, dating back to 1687–1699, are extant at the Amsterdam University Library.

[111] See Simonutti 1990.

England to study mathematics, De Volder had met Newton in Cambridge ten years before, but no detail is provided.[112] Wiesenfeldt had advanced the hypothesis that De Volder could have been put in contact with John Wallis (active in Oxford) by Hudde and Frans van Schooten, and with More (active in Cambridge) by Van Limborch.[113] Probably, De Volder had some contact with Robert Hooke (correspondent of Constantijn Huygens) and Boyle in London, where they lived in 1674.[114] In fact, the air pump built for De Volder by Samuel van Musschenbroek (1640–1681) was based on the second air pump built by Hooke for Boyle (as I am going to show). Once he came back to Leiden, De Volder, according to the proceedings of the University Curators of 3 December 1674,

> Mr. Burchard de Volder, Professor of Philosophy in this University, has expounded to the Curators and to the Mayors, both verbally and in writing, the usefulness and the great advantages which could be brought to this University, if – following the example of other foreign academies and illustrious schools – here in this University could be taught and demonstrated by experiments, the truth and the certainty of those theses and doctrines, taught to the students in theoretical physics [*physica theoretica*]; professing that there was nothing he would like to see more, than that Curators and Mayors would be pleased to admit

---

[112] See the full text of the letter, preserved at the Uppsala University Library (Waller Ms. benl-00770): "Viro Amplissimo Newton. / Mathesios Professori / S. P. / B. De Volder. / Quanquam decennium est, a quo te Cantabrigiae vidi, me tamen nequaquam {oblitum} summae, quam tunc expertus sum, humanitatis tuae ex hisce cognoscere. Ea enim mihi tanta visa fuit, ut, licet eam promereri hactenus non potuerim, commendare tamen tibi {sust} meam hunc Ioh. Christ. Tsimmermans, qui litteras hasce ad te deferet. Iuvenis est, honesto loco natus, Reverendi hac nostra in Academia Theologiae Professoris Wittichi ex sorore nepos. huic, postquam per triennium circiter rebus ex philosophicis et theologicis diligentem {navassit} operam apud vos commorari nunc est animus, et se inter alia mathematicis etiam dare. Optime autem secum actum iri cum iure existimet, si tua ope et consilio hosce in conatibus suis adiuvetur, iustissima eius petitioni denegare non potui, quo minus te rogarem, ut pro summa tua in exteros humanitate et pro eximia in rebus mathematicis peritia egregii huius iuvenis, qui suam tibi sat scio, probaturus est diligentiam, studia adiuvare non dedigneris[,] quo ipso summopere tibi eos qui ipsum proprius contingunt honestissimos viros, devincies, et meo ad omnia vicissim officia praestanda habebis paratissimum[.] / Vale. / Lugd. Batav. a.d. VIII Calend. Decembr. / CIƆIƆCLXXXIV." English translation in Hall 1982, 11. Later, in June 1702, Newton recommended Frans Burman Jr. to be admitted to the meetings of the Royal Society as he had studied under De Volder: "Sr. / The Gentleman who brings you this, is on [sic] of the chaplains to the Dutch Embassadors. I beg the favour that by the leave of the R. Society you would introduce him to see one of their meetings. He has heard Monsr. Volders Lectures & has a curiosity about Mathematicae et Philosophicae things. If he brings a friend with him, I beg the favour that you will treat them with respect. I am / Your humble servant / Js. Neuwton / Jerounstreet, / June 2. 1702," Capadose 1828, 10. Burman would be received by Newton, on 13 June 1702, because he was a student of De Volder: "D. 13 p. m. Invisi ad Celeberrimum D. Isaacum Neuwtonum, qui me humanissime excepit vel solius Volderi causa, cuius me discipulum ferebam. Actum praecipue de systemate universi," Capadose 1828, 9. Burman did not defend disputations under De Volder.

[113] See Wiesenfeldt 2002, 61. See also De Hoog 1974, 144; Van Helden 1991, 163.

[114] Thomas Burnet might also have had direct contacts with De Volder. Indeed, he was present at the inaugural oration of Jacques Bernard, who succeeded De Volder at Leiden: see his letter to Leibniz of 29 December 1705: "[j]e quita la Haye pour 3 ou 4 jours pour allér à Leyde entendre le Harangue Inaugurale de Monsr Bernard (à cette heure lectur en philosophie dans la place de Monsr Volder et aussi ministre des Eglise fransoise dans cette ville)," A XXV1, 460.

> him to the exercise of the said experimental physics [*physica experimentalis*], and to institute a post, together with such instruments and further necessities, as the said demonstration would require; the said De Volder being absolutely convinced that, apart from the utility and the delight of the said experiment to be undertaken, many students from other academies and schools would be attracted here, and the study of physics could be greatly facilitated. After due deliberation, it has been decided and resolved that this proposal should be examined in greater detail, and that it should be considered which and what places could best be used, and in how far the University should be charged with this, and with the purchase and maintenance of the necessary instruments and preparations; and, this having been done, [the proposal] should be decided according to what shall be found.[115]

Thus far, the auditorium had, for De Volder, the function of showing and confirming the results of theoretical physics by experiments. As I am going to show in Sect. 3.2.3.3, *The role of experience in De Volder's natural philosophy*, De Volder did not consider experience as a source of the formulation of scientific principles, namely as a source of the ideas at the basis of physics (viz. matter and motion). However, he ascribed to experience an essential role in choosing how to apply such theoretical principles in the explanation of phenomena themselves. Moreover, for him experience serves to confirm or to demonstrate the truth of such principles, because their applicability is made evident by experiments. As I am going to show, this was even to supplant, in De Volder's perspective, the metaphysical demonstration of their being actually at work behind phenomena. As Le Clerc noted in commenting on the opening of the auditorium (the idea of which, according to him, was inspired in De Volder by the very example of the Royal Society),[116] De Volder was often heard to claim that in physics experience is more important than reason, while in metaphysics one has to follow more reason that experience.[117] In turn, from a broader, insti-

---

[115] "D. Burchardus de Volder, Professor Philosophiae in dese Universiteyt, heeft aen de H. C. ende B. soo mondelingh als by geschrifte gerepresenteert de nuttigheyt ende de groote avantages, die dese Universiteyt soude konnen werden toegebraght, indien nae het exempel van andere uytlandsche academien en illustre scholen alhier in dese Universiteyt by experimenten moghten werden gedoceert en aengewesen de waerheyt ende seekerheyt van die stellingen ende leeren, die in Physica theoretica de studenten werden voorgehouden, met presentatie dat hy niets soo lieff soude sien dan dat C. ende B. voorn. hem tot de exercitie praefatae Physicae experimentalis soude believen te admitteren ende daer toe te subministreren soodanigh een plaets, mitsgaders alsulke instrumenta ende verdere nootsaekelycheden, als de voors. demonstratie soude komen te vereyschen; sijnde de voorn. de Volder absolutelyck gepersuadeert dat, behalve de nuttigheyt ende het vermaek van de voors. te doene demonstratie, door deselve veele studenten van andere academien ende scholen herwaarts aengeloct ende het studium Physices seer gefaciliteert soude konnen werden. Waerop gedelibereert sijnde is goedgevouden ende geresolveert dat dese saeke wat nader sal werden geexamineert, ende overwoogen welke ende hoedanige plaetsen dairtoe best soude konnen werden geappliceert en hoeverre dese Universiteyt daer mede als oock met den incoop ende onderhoud van de nodige instrumenten ende preparatien soude werden belast, om 't selve gedaen sijnde, alsdan soodanigh te werden geresolveert als bevonden sal werden te behooren," Molhuysen 1913–1924, volume 3, 298. See also Nyden 2013, 233–234.

[116] Le Clerc 1709, 362–363. For a discussion of the differences in function of the experiments carried out at the Royal Society and at Leiden, see Wiesenfeldt 2002, 111–113.

[117] "Ses Amis & ses Disciples se souviennent de lui avoir ouï faire alors un discours, où faisoit voir qu'en matieres de Physique, l'Experience sert plus que le Raisonnement. Une autrefois il en fit un, où il montra que, dans les choses Métaphysiques & de pure spéculation, le Raisonnement alloit

tutional perspective, the auditorium served to attract students and to increase the prestige of the University of Leiden, on the model of some foreign institution (as I am going to show in). On 26 January 1675, the Curators eventually agreed to establish the theatre in order to allow De Volder to teach 'experimental physics'. This did not come without conflict. Indeed, Carel de Maets, professor of chemistry, on 3 December 1674 had asked to be allowed to teach experimental physics himself, in his laboratory, where he already had instruments and funding. A conflict arose, and after the failed mediation of one of the mayors of Leiden, Herman van der Meer, the Curators allowed (on 26 January 1675) both De Maets and De Volder to teach experimental physics. In fact, De Volder got the most part of the funding, while De Maets could perform only limited physical experiments. In turn, De Volder performed limited experiments in chemistry in his theatre or auditorium.[118]

For the establishment of the auditorium, a house was bought in Nonnesteeg for 2,500 guilders (refurbished in 1743, and no longer extant; see Fig. 2.1, Bylaert 1743, in which it is depicted on the left, while on the right there is the house of the professor of botanics), round the corner from the University, while De Volder was granted 400 guilders per year to buy the apparatuses for the theatre.[119] As documented by Wiesenfeldt, in the course of 1675 the sum used to buy equipment – built by the Van Musschenbroek workshop in Leiden, in 1675 headed by Samuel van

---

plus loin que l'Experience. Soit qu'il n'eût pas écrit ces discours, ou qu'il les ait négligé depuis, ils ne se sont pas trouvez parmi ses Papiers," Le Clerc 1709, 364.

[118] On De Volder's experiments in chemistry, see Sect. 5.6.4, *The application of Boyle's pneumatics to physiology*. On the conflict between De Maets and De Volder, see Wiesenfeldt 2002, 199–201. In his reconstruction, to which I re-direct the reader for more details, Wiesenfeldt quotes and refers to Molhuysen 1913–1924, volume 3, 298, 301, 269∗ and 273∗, and to the manuscripts, preserved at Leiden, AC 108, AC 16, ff. 50–56, AC 26, ff. 673–680. In 1705, De Volder publicly thanked the Curators for the opening of the theatre: see De Volder 1705, 33.

[119] "1675. Jan. 26. By resumtie gedelibereert wesende op het versoek van den Professor Burchardus de Volder, breder onder de notulen van den 3en Decemb. jonxstl. geextendeert, is goedgevonden ende geresolveert denselve Professor de Volder tot de institutie van de Physica experimentalis te admitteren, gelijck deselve daertoe geadmitteert werd mitsdesen: dienvolgende dat de huysinge van den rentmeester van der Meer, staende bewesten de woniuge van den Professor Botanices in de Nonnesteegh by de Universiteyt, sal werden aengenomcn tot soodanigh een prijs, als hetselve met relatie tot de waarde, die daer is geweest voor den laetste ongeluckige oorlogh, by neutrale mannen sal werden getauxeert, ende dat ter concurrenter somme van deselve tauxatie aen den voorn, rente- meester van der Meer in plaetse van kooppenningen sal werden ter hand gestelt een obligatie van de H. C. ende B. met stipulatie van alsulke intresten, als in de jonxste gedane negotiatie aen andere des Universiteyts crediteuren siju uytgelooft; dat voorts deselve huysinge soodanig sal werden geapproprieert, als de gelegentheyt van de te houdene exercitiën mitsgaders de te doene demonstratien in dicta Physîca experimentali sal komen te vereyschen: dat eyndelyck aen den voorn. D. de Volder sal werden ter hand gestelt een somme van 400 guldens omme beheert te werden tot incoop van alsulke instrumenten ende praeparatien, als tot den aenvanck ende uytvoeringe van de voors. exercitiën en demonstratien onvermydelyck sullen werden vereyst, des dat hy van de uytgifte der voors. Somme sal hebben te houden pertinente notitie ende daervan des versoght sijnde in handen van de H. C. ende B. overleveren specifycque declaratien. En sal aen den voorn. ü. de Volder mitsgaders aen den Rector ende de Senatus Academiae toegesonden werden een extract authentycq uyt dese resolutie tot hare naerigtiuge," Molhuysen 1913–1924, volume 3, 301.

Fig. 2.1 The external facade of the Leiden experimental theatre, in Nonnesteeg, around 1743. The theatre was in the building on the left. (Source: Bylaert 1743. Stadsarchief Leiden, Erfgoed Leiden en omstreken, signature: PV13541, https://www.erfgoedleiden.nl/collecties/beeldmateriaal/zoeken-in-beeldmateriaal/detail/197775d6-26bd-11e3-a46c-3cd92befe4f8/media/2d8cadce-1094-d816-d508-8895abc78292. Public domain mark)

Musschenbroek[120] – increased to more than 937 guilders, while the total sum used by De Volder up to his retirement in 1705 amounted to 4,313 guilders and 13 stuivers.[121] On 15 August 1675, projects for the internal refurbishment of the house,

---

[120] For a thorough reconstruction of the history of the Van Musschenbroek workshop, see De Clercq 1988, 1989, 1991, 1997a, b.

[121] See Wiesenfeldt 2002, 63. Wiesenfeldt refers to the manuscripts AC 88–91, especially AC 88, f. 39; to AC 27, ff. 15v and 46. According to Le Clerc, De Volder spent more than 400 guilders and he used his own money for the additional expenses. In the following years, he spent less money; he was anyway granted additional sums: see Le Clerc 1709, 362–363: "[i]l le leur persuada & ces Mrs. acheterent un endroit près de l'Academie, qu'ils firent disposer en forme de théatre, afin que les Etudians puisent voir commodément les Experiences de Physique & de Méchanique, que Mr. de Volder feroit devant eux. Ils lui assignerent pour cela quatre cents francs par an, qu'ils lui donnerent la liberté de dépenser, pour acheter les instrumens nécessaires & pour faire les Experiences qu'il trouveroit à propos. La premiere année, il dépensa bien davantage; mais il ne voulut mettre à

making it fit to host students in an auditorium, were presented to the Curators and Leiden town authorities.[122] Eventually, the auditorium was constructed as an anatomical theatre,[123] as testified to by Uffenbach in his travel journal of 1711, where he provides a description of the bad conditions of the auditorium after the death of De Volder, when it was directed by Wolferd Senguerd (1646–1724):

> January 19 [1711] [...] So we looked at the academic physical laboratory by the garden behind the gallery. It consists of a not so big room on the ground [floor], which is not really clean[,] especially for Holland. It had benches elevated around it, like an anatomical theatre. In the middle stood also a raised table. On this [there] was a quite large pneumatic pump, still of the old invention of Samuel Musschenbroek, inclined with a trunk, to put water in it. On the side [there] was another pump, also of the first inventions, but vertical on a tripod, albeit very bad, as in that case the cylinder was only one inch thick. Behind this room were moreover a few bad instruments in a lodge, mostly recipients for the pump, but so unclean and lying under each other, also broken for the most part. A thermometer was also in pieces. A glass tube about ten feet long, and [of] two inches in diameter, but we could not see what it was for. A little aeolipile. A small sextant of wood, to make the movement of pendulums with. As one assures us, so the public lectures of Mister Senguerd take place here four times in the week; Mister de Volder[,] however[,] had more listeners, he had also been more curious in his experiments. Whether it looks pretty bad in this auditorium or laboratory, it would be desirable if all the universities were the same, and the experiments were performed in public.[124]

---

compte à l'Université, que la somme qu'on lui avoit assignée. Il paya le surplus, du sien; mais les années suivantes, les dépenses n'égalerent pas cette somme, & il avoit soin de fournir des comptes exacts de ce qu'il avoit dépensé, sans profiter jamais de la liberalité de Mrs. les Curateurs, pour son avantage particulier; quoi qu'il l'eût pu faire, sans qu'on y eût trouvé à redire. Aussi faut-il rendre ce témoignage à Mr. de Volder, que c'étoit l'homme du monde le moins interessé. Jamais il ne fit le moindre mouvement, pour faire augmenter ses gages, qui le furent néanmoins plusieurs fois, par la génerosité de Mrs. les Curateurs; car d'abord ils avoient été fort petits." See also De Clercq 1997a, 137.

[122] "Aan Curateuren ende Burgermeesteren verthoont wesende verscheyde figurative projecten, volgens de welke sonde können werden gedresseert de sitplaetse ende het theatrum in het Auditorium Philosophiae experimentalis, sijn Burgermeesteren ende Regierders der stad Leyden versoght ende geauthoriseert om uyt deselve projecten soodanigh een te verkiesen, ende vervolgens de voors. Sitplaetseu ende het theatrum soodanigh te disponeren, als ten meeste dienste van de voors. exercitiën, ende tot commoditeyt van de auditores bevonden sal werden te behooren," Molhuysen 1913–1924, volume 3, 312.

[123] See Wiesenfeldt 2002, 118–124.

[124] "Den 19. Jan. [1711] [...] Wir besahen also bey dem Garten hinter der Gallerie das Laboratorium physicum Acad. Es bestehet in einem nicht gar grossen Zimmer auf der Erde, das sonderlich vor Holland nicht gar sauber ist. Es hatte rings herum erhöhte Bäncke, wie ein Theatrum anatomicum. In der Mitte stund auch ein erhöhter Tisch. Auf diesem war eine ziemliche grosse antlia pneumatica, noch von der alten Invention des Samuel Musschenbroeks, incliniert mit einer cista, Wasser darein zu thun. An der Seite stunde noch eine Antlia, auch von den ersten Erfindungen, aber vertical auf einem Dreifuß, wiewohl sehr schlecht, wie dann der Cilinder nur einen Zoll stark war. Hinter diesem Zimmer waren in einem Kämmergen noch einige schlechte Instrumente, meist Recipienten zur Antlia, so aber unsauber und unter einander lagen, auch guten theils zerbrochen waren. Ein Thermometrum war auch in Stücken. Ein gläserner Tubus bey zehen Schuh lang, und zwey Zoll im Diameter, da wir aber nicht sehen konnten, wozu er diente. Eine kleine Aeolipila. Ein kleiner Sextans von Holz, den motum pendulorum damit zu machen. Wie man uns versichert, so geschehen allhier viermal in der Wochen Lectiones publicae von Herrn Senguerd; Herr de Volder

The theatre opened between September and December 1675. In a letter of 12/22 August 1675 to Henry Oldenburg, Henri Justel wrote that he heard from Caspar Bartholin that "a philosophical laboratory, where all sorts of experiments will be made, has been established there by the States [of Holland]. Someone named Mr. Voldred [*sic*] is in charge; he is clever, and able to use it properly."[125] Hence, in a letter to Oldenburg dated 21/31 January 1676, Constantijn Huygens wrote that

> your latest *Transactions* were welcome to me, and first Mr. Volder[,] Professor of Philosophy at Leiden, who lately made us some fine and attractive demonstrations of the weight and static force of the air, came to borrow them from me. I have encouraged him to produce also his experiments on his own, so that it would not be that they are just heard [*ne semper sit auditos tantum*], and I assure myself that he will listen to my remonstrations.[126]

Certainly, the theatre opened before March 1676, as the first lecture reported in the notebook in which the most part of De Volder's lectures are extant (to which I shall return below) is dated 12 March 1676 (most probably, the date of transcription), and mentions "some other experiments before now."[127] Usually, De Volder performed lectures on Mondays at the *hora undecima* before noon, as testified to by Le Clerc[128] and by the *series lectionum* of the University for the summer and winter terms of 1681 – while on the other days of the week he taught upon "famous problems in philosophy."[129] According to a resolution of the University Curators of 9 November 1676, the caretaker of the theatre was a certain Hieronymus Meyer (who was also

---

aber habe mehr Zuhörer gehabt, er seye auch in seinen Experimenten curiöser gewesen. Ob es gleich in diesem Auditorio oder Laboratorio ziemlich schlecht aussahe, so wäre doch zu wünschen daß auf allen Universitäten dergleichen wären, und die Experimente öffentlich gemacht würden," Uffenbach 1753–1754, volume 3, 420 and 425–426. Please note that the inclined air-pump saw by Uffenbach was that of Senguerd. On the other instruments, see *infra*. References to the auditorium are given also in *Les délices de Leide* (1712): "[i]l y a pur cela un beau Laboratoire dans le Nonnesteeg, fourni de tous les Instrumens necessaires pour ce sujet, où le Professeur conduit des Disciples & les Curieux, raisonne sur les experiences qu'il fait, & en démontre l'evidence & la justesse," Van der Aa 1712, 82. Wiesenfeldt cites also an anonymous handwritten *Journal of my Travels* (24 June – 20 August 1691), extant at the British Library (Ms. Add. 20705, ff. 1–46): "[...] fine physick garden and Elaboratory with many fine rarities," f. 32r. I quote from Wiesenfeldt 2002, 68.

[125] Oldenburg 1965–1986, volume 11, 454. Cf. the original text: "[o]n luy écrit de leyde que Messrs. les Estats y ont establi un laboratoire philosophique ou on fera toutes sortes d'experiences. Cest un nommé Mr. Voldredus qui en a soin qui est habile et capable de s'en bien servir," Oldenburg 1965–1986, volume 11, 453.

[126] "Vos dernieres transactions m'ont esté les bienvenues, et d'abord le Sr. Volder Professeur en Philosophie à Leiden, qui dernièrement nous a faict de belles demonstrations attractives du poids et de la force statique de l'Air, me les est venu empruntés. Je l'aij animé à produire aussi ses experiences de son costé, ne semper sit auditos tantum, et je m'asseure qu'il escouttera mes semonces," Oldenburg 1965–1986, volume 12, 145. I slightly modified the translation provided in this edition: Oldenburg 1965–1986, volume 12, 147.

[127] De Volder 1676–1677, 78r. As to the opening of the theatre in 1675, see also *infra*, n. 139.

[128] Le Clerc 1709, 363–364.

[129] "D. Burcherus de Volder singulis Lunae diebus in Theatro Physico in hunc usum constructo experimenta Physica cuiuscunque generis publice demonstrabit; ceteris vero diebus celebriora Philosophiae problemata exponet," Molhuysen 1913–1924, volume 3, 269∗; see also 272∗.

the University head gardener), who received an annual salary of 50 guilders for this task of keeping the instruments in order.[130] As a sort of special assistant or technician, the Leiden instrument maker Antonius Hoevenaer enrolled at Leiden on 25 March 1683 as "*amanuensis* of Prof. De Volder[,] in what concerns mathematical instruments."[131]

As testified to by Uffenbach, after De Volder's death the auditorium, used by Senguerd, was in a bad condition. In De Volder's time, on the other hand, it attracted numerous students and travellers.[132] As far as the instruments of the laboratory are concerned, the reader can find an exhaustive and complete treatment in the studies of Wiesenfeldt, Peter de Clercq, Cornelis de Pater, Hans Hooijmaijers, Ad Maas, Anne C. van Helden and Maurice Daumas.[133] Here, it is worth offering a comparative account of the items of the auditorium (see Table 2.2, *De Volder's natural-philosophical instruments*), by considering 1) the inventory received from De Volder by Senguerd when he took over responsibility for the auditorium from De Volder, and compiled on 14 November 1705,[134] 2) the evidence from the various reports on

[130] "Extract uyt de Resolutien van de Curateuren der Universiteyt binnen Leyden, en Burgermeesteren der selver Stadt, dato den 9 November 1676. Op de Requeste van Hieronimus Meyer, Dienaer des Universiteyts Kruythof, om geloont te werden voor den dienst van het schoon den van Instrumentenen anderen, die hy in 't Auditorium Philosophiae experimentalis, op ordre van den Heere Professor de Volder, eenigen tydt herwaerts hadde gedaen, midtsgadersom voor toekomende op een sortabel Tractement in den voorszdienst gecontinueertte mogen werden. Waer op gedelibereert zynde, is goedtgevonden en verstaen, dat den voornoemden Hieronimus Meyer in den voorsz dienst sal werden gecontinueert, en dat op een Tractement vanvijftigh guldens jaerlijcks; te rekenen van dien tydt af dat hy sich tot den voorsz dienst effectivelijck heeft laten employeren, ende dat daer mede alle syne pretensien van voorgaende diensten sullen komen op te houden ende te cesseren," Rosenboom 1697, part 1, 89–90. This resolution is not reported by Molhuysen. See Wiesenfeldt 2002, 63–64, referring to ms. AC 27, 83; see also Rosenboom 1697, part 1, 14–15, and part 2, 23.

[131] "Antonius Hoevenaer, D^{i} Prof^{is} Volderi amanuensis quod ad instrumenta mathem.," *Album studiosorum Academiae Lugduno-Batavae MDLXXV–MDCCCLXXV*, column 659.

[132] On students's attendance at lectures in Leiden, see Wiesenfeldt 2002, 100–105. No information about the number of people attending De Volder's experimental lectures is extant. On the attendance to De Volder's private lectures, see his letter to Johannes Braun (Cocceian professor of theology and Hebrew at Groningen) of 9 May 1689: see *infra*, n. 267.

[133] Daumas 1972, part 1; De Pater 1975; De Clercq 1988, 1989, 1991, 1997a, b; Van Helden 1991; Wiesenfeldt 2002, chapter 3; Hooijmaijers and Maas 2013. See also the *Catalogue of the Special Loan Collection of Scientific Apparatus at the South Kensington Museum* (1876), 135–137.

[134] It is worth quoting it here in full; for a translation, see Table 2.2, *De Volder's natural-philosophical instruments*: "1705 Nov. 14. Lyste en register van de machinen en instrumenten behoorende tot het Theatrum Experimentale Physicum deser Universiteit, door de heer Professor de Volder den 14en Novemb. A^{o} 1705 aan my overgelevert, en waarvan ik aan syn Ed. quytantie gepasseert hebbe. Een kopere antlia pneumatica in een kopere kas, staande in een houte back met loot beslagen met verscheyde sleutels daar toe behoorende. Twee kopere platen om de recipiënten in te stellen. Twee kopere pijpen tot die platen. Een copere omgeboge pijp mede tot de antlia behoorende. Verscheyde recipiënten tot die selvige behoorende, waar van sommige boven open en sommige boven toe sijn. Verscheyde decksels tot de recipienten. Een recipient om een klockje in te stellen met sijn decksel en het klockje etc. Een recipient met een kopere pijp om de lugt boven uyt te halen tot vis etc. Een dexsel met een hevel, daer door verscheyde glase hevels, als oock glase in ver-

Table 2.2 De Volder's natural-philosophical instruments

Instruments mentioned in the inventory of November 1705 (Molhuysen 1913–1924, volume 4, 104*–106*)	Instruments mentioned in the *Experimenta philosophica naturalia*, 1676–1677	Instruments mentioned in other primary sources on De Volder's experiments	Secondary sources treating such instruments
A copper pneumatic pump in a copper container, standing in a wooden box covered in lead with various keys belonging to it.	Used in experiments 3–14, 16 and 27.	Mentioned in De Volder 1676–1678, disputation 5; Senguerd 1681, 65; Senguerd 1685, 88; Senguerd 1715, 3–4.	Daumas 1972, 84–85; De Pater 1975, 314–319; Van Helden 1991, 173; De Clercq 1997a, 137–139; De Clercq 1997b, 67–68 (item 60); Hooijmaijers and Maas 2013, 33–35.
Two copper plates on which to place the recipients.			
Two copper pipes belonging to these plates.			
A curved copper pipe that also belongs to the pump.	Probably used in experiment 11.		
Several recipients belonging to the same, some of which are open above, and some closed.	Probably used in experiments 4–7, 9, 11–14, 16, 19–20, 22–24 and 26.	In Bernoulli 1683, 108–110, it is described a glass container including two smaller containers connected by a syphon. Various glass containers and pipes are depicted in Van Musschenbroek 1700, plate 2. Uffenbach mentions various recipients for the air pump: Uffenbach 1753–1754, volume 3, 425.	
Various covers for the recipients.			
A recipient in which to place a little bell, with its cover, and the little bell etc.	Used in experiment 10.	This instrument is depicted in Van Musschenbroek 1700, plate 2.	
A recipient with a copper pipe to extract the air at the top [above] the fish etc.	Used in experiment 8.		

A cover with a syphon, with several glass syphons belonging to it, as well as glasses blown in various forms.	Probably used in experiments 4–5.		
Various glass pipes, some closed on one side, some open on both [sides], some with bulbs, some without bulbs.	Probably used in experiments 17 and 23–26.	Various pipes are mentioned in De Volder 1676–1678. Uffenbach mentions a glass tube of 10 feet, with a diameter of 2 inches: Uffenbach 1753–1754, volume 3, 425	
Two copper hemispheres with a wooden stand for this.	Probably used in experiments 3 and 27.	De Volder 1676–1678, disputation 5; Dalrymple 1686, 542. This apparatus is depicted in Van Musschenbroek 1700, plate 3.	De Clercq 1997b, 77 (item 70; uncertain) and 78 (item 71).
Two marble cylinders polished to fit onto one another.	Used in experiments 1–2.	Dalrymple 1686, 539 (reporting a diameter of almost 3 inches). This instrument is depicted in Van Musschenbroek 1700, plate 4.	
Two glasses with pipes to draw up quicksilver.	Probably used in experiments 24 and 26.		
Two tin boxes.			
In addition, another [one], with a separation between both parts, and with holes at the bottom.			
Two glass pipes, one closed without air, the other open, each containing little ball [of wool?] to show the difference in [the speed of] falling.			
A large wooden box with long and short glass pipes.			
A part of a hollow copper cone.			

(continued)

Table 2.2 (continued)

Instruments mentioned in the inventory of November 1705 (Molhuysen 1913–1924, volume 4, 104*–106*)	Instruments mentioned in the *Experimenta philosophica naturalia*, 1676–1677	Instruments mentioned in other primary sources on De Volder's experiments	Secondary sources treating such instruments
A glass for the rarefaction of air by heat.			
A big balance with wooden trays.	Used in experiments 1–2. Probably used in experiment 27.		De Clercq 1997b, 174 (item 276; dated 1677, built by Johannes Andries Groengraft, Amsterdam instruments maker).
Between 1,200 and 1,300 pounds weight of weighing masses	Probably used in experiments 1–3 and 27.		
Besides some copper weights.	Probably used in experiments 1–3, 5, 15, 17–22 and 27.		
Also a box with a small weight.			
Some lead cylinders, eight of 10 pounds and four of 5 pounds.			
A smaller balance with two pairs of trays.	Used in experiment 21. Probably used in experiments 17–20 and 27.	De Volder 1676–1678, disputation 5.	De Clercq 1997b, 178 (item 248; uncertain).
A screw that is fixed.	Probably used in experiments 9, 13 and 27.		
A big [writing] slate.	Probably used in experiment 18.		
Two aeolipiles.			De Clercq 1997b, 138 (item 222; uncertain).
A wooden ball with two triple counter screws.			
A small copper syringe.	Probably used in experiment 4.		De Clercq 1997b, 76 (item 68; uncertain).

A wooden circle rotating on an axis, to demonstrate the centrifugal force.			
A box with four pounds of quicksilver.			
Another jug of quicksilver.			
A glass plumb bob.			
A glass with the four elements.			
A lodestone with an anchor attached to it.			De Clercq 1997b, 161 (item 261; uncertain).
A box with some magnetic needles, as well as a small unarmed lodestone.			De Clercq 1997b, 144 (items 232–234; uncertain).
A microscope with two glasses.			De Clercq 1997b, 126 (item 197, 198; uncertain).
Various microscopes with one glass.			De Clercq 1997b, 125–127 (items 195–200; uncertain).
A metallic perspective mirror with its painting[s].			De Clercq 1997b, 114 (item 145; uncertain).
A barometer.			
A box with two glasses belonging to two double barometers of Mr. Huygens, together with the frame.			
A thermoscope.		In Bernoulli 1683, 214, two thermoscopes are mentioned, one closed and one opened. Uffenbach mentions a broken thermometer: Uffenbach 1753–1754, volume 3, 425.	

(continued)

Table 2.2 (continued)

Instruments mentioned in the inventory of November 1705 (Molhuysen 1913–1924, volume 4, 104*–106*)	Instruments mentioned in the *Experimenta philosophica naturalia*, 1676–1677	Instruments mentioned in other primary sources on De Volder's experiments	Secondary sources treating such instruments
A hygroscope.			
A copper oven with a leaden box [tray?].			
A wooden instrument standing on its pedestal, with below two pieces of copper, circles divided and marked in grades to demonstrate the rules of motion.		Described in various letters of Johannes Joosten van Musschenbroek sent to Marburg (see Sect. 4.2.3.4.2, *De Volder on elastic impacts*), and based on the model of Mariotte (described and depicted in Régis 1691, volume 1, 366–367). Uffenbach mentions a little sextant of wood, by which one can show the movement of pendulums; he could have been referring to the same instrument: Uffenbach 1753–1754, volume 3, 426. Later, 's Gravesande used a similar apparatus (see De Clercq 1989, 27–31).	De Clercq 1991, 100–101 and 120; De Clercq 1989, 27–31.
For this purpose 16 ivory and 3 wooden spheres.			
Another little wooden gutter for the same purpose.			
A copper machine for boiling bones [until they are] soft.			
A Württemberg syphon.		Mentioned in Leupold 1724, 11.	
A copper fountain with various pipes belonging to it.			
Another fountain of glass with two copper pipes.			

A glass with an opening in the middle to demonstrate the lateral pressure of the air.			
A box with balls of copper, of lead, and of a mixture of both, serving to weigh these metals, as well as to determine the alloy of these two [metals].	Used in experiments 19–20.		
A hollow copper cube of half a foot, intended to determine the weight of a foot of water.	Used in experiment 15.	In De Volder 1676–1678, disputation 5, Dalrymple 1686, 255, 604, and Van Musschenbroek 1748, 383, there is mentioned the weight of a cubic foot of water measured by De Volder, most probably with this cube. Analogous measurements are also reported in the letter of Henri Justel to Locke of 24 July/3 August 1681 (letter 651), and in Van Nierop 1680, 156. The cube was sold by the Van Musschenbroek's workshop: its use is described in a letter sent to Marburg by Johannes Joosten van Musschenbroek (see Sect. 5.2, *The conditions of hydrostatic equilibrium in De Volder's experimental lectures*).	
A copper tube marked in feet to weigh and measure the perpendicular pressure of the water as well as its speed.			

(continued)

Table 2.2 (continued)

Instruments mentioned in the inventory of November 1705 (Molhuysen 1913–1924, volume 4, 104∗–106∗)	Instruments mentioned in the *Experimenta philosophica naturalia*, 1676–1677	Instruments mentioned in other primary sources on De Volder's experiments	Secondary sources treating such instruments
A copper cylinder with a plunger of ½ foot diameter inserted, with several copper pipes and a chain going from the mobile plunger through the pipe up to its end.	An idealized version of this apparatus is mentioned in experiment 28.	An inverted T-shaped container, composed by two cylindrical vessels, is labelled as *Volderi cylinder* and described in various letters of Johannes Joosten van Musschenbroek sent to Marburg (see Sect. 5.4.2.3, *…and De Volder's refinement of it*). It is also mentioned in the sales catalogue of the Van Musschenbroek workshop published in Valentini 1714, 104 (see also 43). The upper cylinder described by Johannes Joosten van Musschenbroek has various taps, and the lower cylinder has a diameter of 6 inches (i.e. ½ foot), and a height of 1 foot. The upper cylinder is 6 feet high. The bottom of the lower cylinder is made of brass. The upper cover of this cylinder is fixed by four screws. A marginal annotation in this catalogue refers to an almost identical machine presented in Valentini 1709, 59–60, whose lower cylinder has a diameter of 6 inches (i.e. ½ foot), and its bottom is connected to a balance by a chain. This machine basically the same model presented in Lufneu 1685 and Lufneu 1687, which has slightly different dimensions and features (the bottom, is connected by a rope, and the whole apparatus is 6 feet high instead of 7): see below in this column. Nieuwentijt 1747, table 20, presents an idealized version of this instrument, which was devised for the first time by Simon Stevin. This kind of apparatus is also described in Jacob Leupold's *Antlia pneumatica illustrata* (1712): Leupold 1712, volume 2, 28–29.	De Clercq 1991, 101–102 and 104.

Two copper boxes fitting into one another, with a wooden stand for them.			
Some glass pipes welded together.			
A large wooden container with a wooden divider in which there are several openings, with lids for them.			
Also several copper boxes of one and a half foot, with lids, for use in the wooden container, to measure the horizontal pressure of water, and at any angle.		De Volder's experiments on the oblique pressure of water are mentioned in Van Nierop 1680, 153.	
A big glass with a copper lid, with a pipe [attached] at the side, to sustain the weight of the water in this sidepipe standing [rising] above the weight.			
A magic lantern with its paintings.			
Two foot measures, a Dutch and a Parisian foot.			
	Experiment 3: "a brass plate."		
	Experiment 5: "a pretty big bladder the which by the uppermost end was tied by a string to the upper part of said glass. And at the lower end was tied to the bladder a weight of 5 lb which was surrounded by some cancells of brass."		

(continued)

Table 2.2 (continued)

Instruments mentioned in the inventory of November 1705 (Molhuysen 1913–1924, volume 4, 104*–106*)	Instruments mentioned in the *Experimenta philosophica naturalia*, 1676–1677	Instruments mentioned in other primary sources on De Volder's experiments	Secondary sources treating such instruments
	Experiment 13: "a bladder having within it a round brass circle, within the pretty big glass and over the bladder did hang perpendicular a cupping glass." Experiment 14: "Two several bladders with cupping-glasses sticking to them."		
	Experiment 18: "a cord of hair of a horse's tail."		
	Experiment 21: "a crystal globe, of the bigness of a small apple." Experiment 22: "a round glass, or ball, of crystal, which seemed to be hermetically sealed. The crystal glass, which was of the bigness of an ordinary apple, or somewhat less, had a neck, unto which any thing might be tied, or hanged."		

	An idealized version of this apparatus is mentioned in experiment 28.	Lufneu 1685, 384–386: inverted T-shaped container whose lower cylinder has a height of 1 foot, whose upper cover is fixed with four screws, while the bottom is blocked from falling by a border of 6 lines. The upper cylinder has a height of 5 feet, and a diameter of 6 lines. The bottom of the cylinder is connected, by a rope passing through the smaller cylinder, to a balance (probably the big balance mentioned in the 1705 inventory).	
		Lufneu 1685, 389: a not described apparatus aimed at demonstrating a version of the hydrostatic paradox, consisting of a simple container. I discuss this apparatus in Sect. 5.4.2.1, *Early discussions of the paradox: Stevin, Galileo and Schott*, as *libra hydrostatica mirabilis*.	
	An idealized version of this apparatus is mentioned in experiment 28.	Lufneu 1687, 242–243: an apparatus similar to the one described in experiment 28, namely an inverted T-shaped container, whose lower part was made by 4 panels of wood.	

De Volder's lectures, and 3) the holdings of the Museum Boerhaave in Leiden. Moreover, De Volder had some instruments at his house, both physical and astronomical, and reported in the selling catalogue of his book collections, viz. the *Bibliotheca Volderina* (1709), as 'mathematical instruments' (*mathematise instrumenten*). Amongst the instruments for physics one can find two *Grand thermometre de Florence*, three "curious magnets," and a glass pipe.[135]

In fact, only a few of all the instruments listed in the 1705 and 1709 inventories survive today. Already in 1705, moreover, some of them were broken, as reported by Senguerd after Johannes Joosten van Musschenbroek (1660–1707) inspected the collection.[136] The most notable surviving object is the air pump used by De Volder

scheyde forme geblasen. Verscheyde glase pypen, sommige aan een kant toe, sommige aan byde open, sommige met bollen, sommige sonder bollen. Twee kopere hemisphaeria met een houte stellingh daar toe. Twee marmere cylinders op elkander gesleepen. Twee glasen met pijpen om quicksilver op te trecken. Twee blicke dosen. Daarenboven nogh een met een schijdingh tussen byde en onder met gaten. Twee glasen pypen, een toe sonder lugt, de ander open, byde met een wolletge daerin om het onderscheyd in het valle te toonen. Een groote houte back met lange en korte glase pypen. Een stucx van een holle copere conus. Een glas tot de rareficatie van de lugt door de warmte. Een groote bilance met houte schalen. Tussen de 1200 en 1300 pond swaar gewicht. Nevens eenigh koper gewigt. Als ook een doosje met klyn gewigt. Eenige lode cylinders, agt van 10 pond en vier van 5 pond. Een klynder bilance met twee paar schaalen. Een schroef die vast gestelt is. Een groote ley. Twee aeolipilae. Een houte bol met twee driedubbelde contrarie schroeven. Een koper speuytge. Een houte circul drajende op een ax ad demonstrandam vim centrifugam. Een doos met vier pont quicksilver. Nogh een kannetje met quicksilver. Een glase pasloot. Een glas met de vier elementen. Een zylsteen met een anker daar aen hangende. Een doosje met eenige acus magneticae, nevens een kleyne zijlsteen ongewapent. Een microscopium met twee glasen. Verscheyde microscopia met een glas. Een metale perspectief spiegel met sijn schilderye. Een barometrum. Een doos met twee glasen tot twee dubbelde barometra van de Hr. Huygens, nevens het stelsel. Een thermoscopium. Een hygroscopium. Een kopere oven met een lode back. Een hout instrument staande op syn kruck, onder met twee stucken van kopere afgedeelde circuls ad regulas motus. Hier toe 16 ivore en 3 houte bollen. Nogh een hout gootje tot het selfde gebruik. Een kopere machine om bene sagt te koken. Een sipho Wirtenburgensis. Een fonteyn van koper met verscheyde pypen daertoe behoorende. Nog een fonteyn van glas met twee kopere pypen. Een glas met een openingh in het midden ad demonstrandam aëris pressionem lateralem. Een doos met copere, lode, en uyt beide gemengde bollen, dienende tot het wegen van die metalen, als ook te kennen het mengsel van die beyde. Een holle copere cubus van een half voet, sijnde tot het wegen van de swaarte van een voet waters. Een copere buys verdeelt in voeten om de perpendiculare pressie van 't water als ook de snelheyt van die te wegen en te meten. Een copere cylinder met een ingesette boom van ½ voet diameter, met verscheyde kopere pypen en een kettingh gaande van de losse boom door de pyp naar boven tot de selfde eynde. Twee kopere backen de een in de ander passende, nevens een houte stelsel daertoe. Eenige glazen pijpen aan een gesoldeert. Een groote houte back met een houte schydingh waar in verscheyde openingen zyn, met dexsels daertoe. Alsmede verscheyde kopere backen van een en van een half voet met decksels om daardoor in de houte back te meten pressionem aquae horizontalem et ad quemlibet angulum. Een groot glas met een koper decksel met een pyp ter zyde, om gewigt te sustineren door water in de sypyp staande boven het gewigt. Een toverlantaarn met syn schilderytjes. Twee voetmaten, een Hollandsche en een Parijsche voet," Molhuysen 1913–1924, volume 4, 104∗. See also 222.

135 "Le Grand Thermometre de Florence, 1 ——— dito, 3 Curieuse Magneetsteenen, [...] 1 Glaase Pijp," *Bibliotheca Volderina*, 95.

136 In particular, the barometer of Huygens (or the 'double thermometer'), the hygroscope, a *pijlgewigt* (viz. a set of hollow weights fitting inside each other – probably the hollow cone mentioned

in his lectures, which deserves some mention. Built in 1675 by Samuel van Musschenbroek, owner of the family-owned Van Musschenbroek workshop at Leiden (originally makers of lamps), De Volder's model was based on the second model of an air pump designed by Hooke, namely a single-barrelled vertical pump (see Fig. 2.2, Boyle 1669b, volume 3, 180, plate 1),[137] in which both the container of the barrel, as well as the basis of the pump with the pipes connecting the barrel with the container were filled with water to seal the system.[138]

De Volder's pump (which was slightly modified by Willem Jacob 's Gravesande (1688–1742) in the early eighteenth century, who made it simpler to use),[139] extant at the Museum Boerhaave (see Fig. 2.3, *Luchtpomp van Burchardus de Volder*), was first described and depicted in the *Philosophia naturalis* (first edition 1681, second edition 1685) of Senguerd (see Fig. 2.4, Senguerd 1681, 65), as well as in his *Rationis atque experientiae connubium* (1715), where Senguerd claimed that it needed two men to operate it, so that in 1679 he asked Samuel van Musschenbroek

---

by De Volder), and the large glass with a copper cover. Cf. Molhuysen 1913–1924, volume 4, 106∗: "[d]en 14en Novemb. A°. 1705 heeft de Hr. de Volder my de boven gespecificeerde goederen overgelevert, en heb ik den 18en daar aan volgende aan sijn Ed. deswege quitantie gepasseert. Dogh ten selven dage nevens Musschenbroek het selvige goet nader oversiende, heb bevonde dat verscheyde van de boven gespecificeerde goederen gebrocken en aan stucke waren als onder anderen. Als de barometrum van de Hr. Huygens, waar van een ledige lyst daer toe was en de glase niet gevult. Het hygroscopium aan stucken, synde alle de snaartjes gebrooken. Het koper pijlgewigt gebrooken, het hantvat aan stucken. Alsmede het groot glas met een koper dexsel met een pyp ter zyde om gewigt te sustineren, in welck glas een groot gat gebrooken is, en geheel onbequaem gebruyckt te worden." See also 185∗–186∗.

[137] Cf. Lufneu 1679, thesis 12: "[...] antliam pneumaticam (cuius iconem si quis desiderat, adeat Boylaei Tractatum de vi aëris elastica, ubi parum diversam reperturus est figuram)." According to a recollection of the Van Musschenbroek family tracing back to the 1760s, and written by Jan Willem van Musschenbroek (1729–1807) (*Geslachtlyst van Van Musschenbroek*), the pump was built by Samuel van Musschenbroek under the supervision of De Volder: see De Clercq 1997a, 254. The recollection is extant at the Van Musschenbroek Foundation (The Hague). Other papers of the Van Musschenbroeks are extant at Leiden University Library: mss. BPL 240 (viz. the papers of Pieter van Musschenbroek, 1692–1761), BPL 246 (including four letters of Jan van Musschenbroek, 1687–1748), BOERH g a 138 d and BOERH a 251 (family documents). For more details, see De Clercq 1997a, 307–308.

[138] See Sect. 5.1.4, *De Volder's criticism of the* metus vacui.

[139] "Inventor Antliae Pneumaticae, circa elapsi seculi medium, in Germania fuit Otto Guirikius, qui pulcherrima cum ea fecit pericula. Haec impulerunt Nob. Boyleum, adiutum opera [...] Hookii, et Papini, ad similem antliam in Britannia construendam, quacum plurima instituendo experimenta philosophiam naturalem summopere promovit, hinc machina, vel antlia Boyleana, saepe vocatur. Eadem tempestate Leydae a Cl. Voldero alia inventa fuit, quacum an. 1675 Laboratorio physico experimenta instituta Leydae fuerunt. Hanc machinam, nostra tempestate Cl. s' Gravesandius ad magnam perfectione et simplicitatem reduxit, ita ut levi opera, et brevissimo tempore, ex recipientibus maxima aëris copia exantliari possit," Van Musschenbroek 1734, 381. 's Gravesande also removed the basin of the pump, and added a pedestal, as is evident by comparing Senguerd's depiction of the pump, and the pump as it is now.

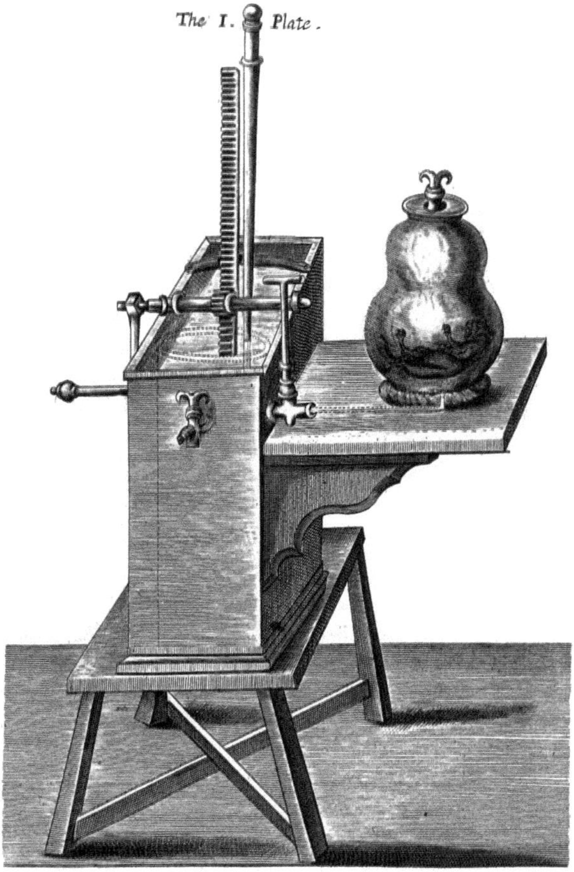

Fig. 2.2 The second model of the air-pump built by Robert Hooke. (Source: Boyle 1669b, volume 3, 180, plate 1. London, Wellcome Collection, shelfmark: 14991/B, https://wellcomecollection.org/works/hnjqcvh6. Attribution 4.0 International (CC BY 4.0))

Fig. 2.3 The air-pump built by Samuel van Musschenbroek for De Volder in 1675. (Source: *Luchtpompen* [*sic*] *van Burchardus de Volder*. Leiden, Museum Boerhaave, inventory number: 2.24.10.02, http://proxy.handle.net/10648/af5524ce-d0b4-102d-bcf8-003048976d84. Public domain (CC0))

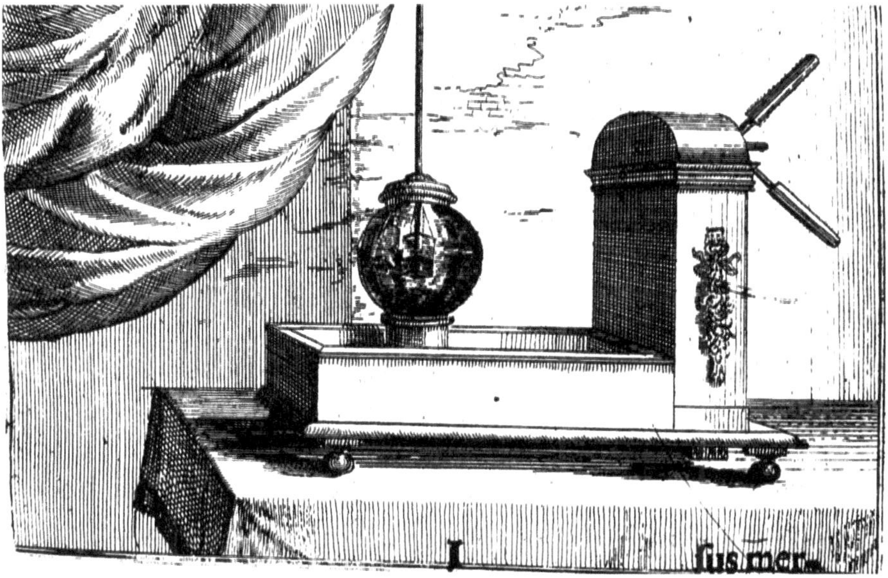

Fig. 2.4 The model of the air-pump built by Samuel van Musschenbroek for De Volder in 1675. (Source: Senguerd 1681, 65. Smithsonian Libraries, Special Collections (Dibner), call number: QC17.S47. Public domain)

to build another,[140] viz. one with a diagonal or inclined barrel, which is also extant at the Museum, and which is mentioned by Uffenbach, as seen above.[141]

A detailed scheme of De Volder's air-pump is given in *Les instruments scientifiques aux XVIIe et XVIIIe siècles* of Maurice Daumas (1953), (see Fig. 2.5, Daumas 1972, 86). As far as its functioning is concerned, a man moved the piston up and

[140] "[C]irca annum 1675 consummatae magis structurae, experimentis minori cum molimine, et periculo peragendis, antlia, ut et apparatus ad illa requisiti, in lucem prodiret. Antliae tubus perpendiculariter erectus erat, et immobilis; inferiori eius parti ad angulum rectum affixus erat minor tubulus, cui recipientia, campanulae, tubi, quaeque porro evacuanda veniebant, adaptabantur. Huius formae antliam illo tempore, ad experimenta molienda, adhibui. Verum, cum et illius: structurae, situs, ac constitutionis antliam multis obnoxiam difficultatibus, atque incommodis; exquisitissimae aëris evacuationi minus convenire experirer, (duplicis enim, quo gaudebat, verticilli usus, plus involvit molestiae, et vitio inficitur facilius, quam unius: binos ad experimenta tentanda requirebat operarios, alterum ad antliam agitandam, et aërem subducendum; qui versurae epistomii vacaret alterum) antliam, quae minorem exigeret apparatum, exquisitiori aëris exhaustui, validiori eius constipationi, pluribus experimentis perficiendis, quibus prior impar erat, inservire aptam, minusque vitiis obnoxiam, excogitavi, atque ut mihi construeretur, anno 1679 cum artifice conveni," Senguerd 1715, 3–4.

[141] Notably, at Leiden there was present, as a kind of precedent for this pump, the single-barrelled vertical model used by Boyle, which is extant at the Museum and is described by Uffenbach as "vertical on a tripod, albeit very bad, as in that case the cylinder was only one inch thick" (whereas the model used by De Volder had a cylinder of a diameter of 9 centimetres, viz. more than 3 inches). Such a pump is still preserved at the Museum Boerhaave. It is unclear where the vertical pump mentioned by Uffenbach comes from, as it does not have the Van Musschenbroek's trademark (viz. a lamp), while De Volder's and Senguerd's pumps bear one: for more information, see De Clercq 1997b, 66.

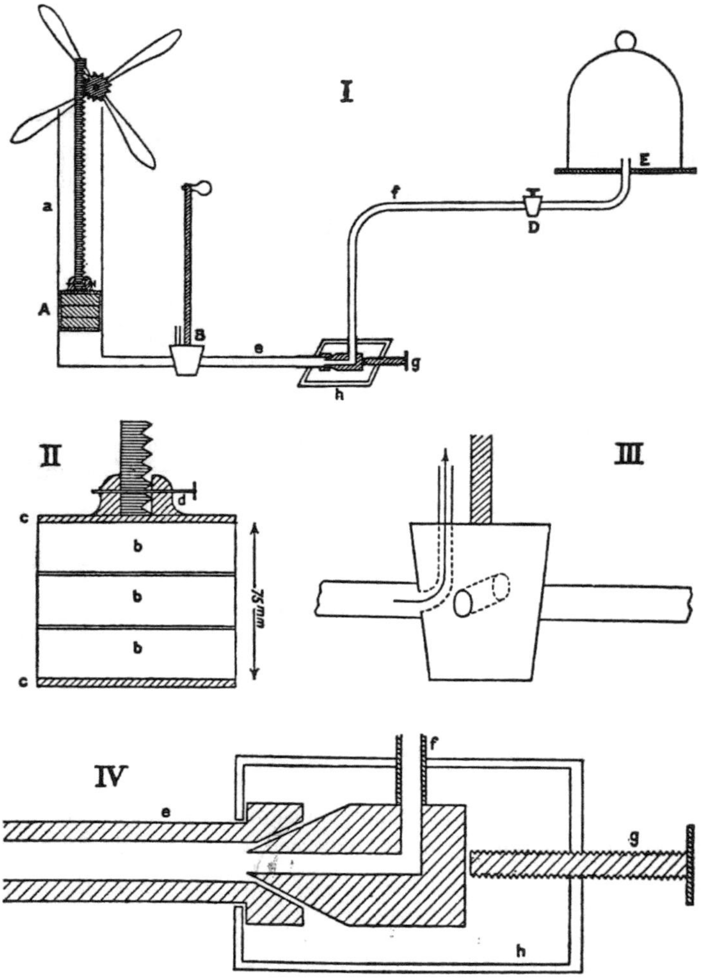

Fig. 2.5 The internal structure of De Volder's air-pump. (Source: Daumas 1972, 86. © Pavilion Books Group Limited)

down, while another opened and closed the stopper B, with a key, putting the piston in communication alternatively with the container or with the external air.[142]

[142]Cf. Daumas's explanation: "[t]he air-pump of Samuel van Musschenbroek. Figure I is a schematic representation of the air-pump. The single cylinder (a) is of brass. The piston A (Fig. II), of almost 9 cm diameter, is of cork, covered with three strips of leather, and held between two brass plates (c). The piston rod consists of an iron rack attached by means of a pin to a small cylinder fixed to the upper of the two brass plates of the piston, there are no valves. After each movement of the piston the stop-cock B is opened or closed (Fig. III). In the position illustrated the pump is in communication with the air; in the reversed position, with the receiver. The tube (f) between the pump and the plate E is of copper. Near the receiver is a tap D. The junction of the tube (f) with the

2.2.4 The Sources for De Volder's Experimental Lectures

As to the sources for De Volder's experimental lectures, the first and foremost to be considered is the handwritten *Experimenta philosophica naturalia auctore M[a]gis[tro] De Valdo*, namely the description of 28 lectures or *experimenta* performed by De Volder, each one dated and ranging from 12 March 1676 to 25 March 1677.[143] Unlike in his lectures on Descartes's *Principia*, such experimental lectures were not dictated by De Volder, even if the ordering and numbering of the theses suggests that De Volder presented them in a highly systematic manner, with a "proposal at the beginning [of] the experiment,"[144] a description of the experiment itself, and the presentation of its conclusions, including some digressions. Moreover, most of the text is in English, with some propositions in Latin (usually the more theoretical contents of the lectures, such as definitions or formulation of physical principles), these being the only parts probably reported *ex ore* by the writer. As far as the writer is concerned, he can be identified as Christopher Love Morley (1645/1646–1702), an Englishman, student at Leiden from 18 February 1676 (when he was 30 years old),[145] and who was going to graduate in medicine in March 1679 with a disputation *De rachitide*, his foremost teacher being Lucas Schacht (1634–1689, from 1670 professor of medicine).[146] Morley would deserve a dedicated study: in fact, the *Experimenta philosophica naturalia* is just a small portion of an impressive series of handwritten transcriptions of lectures he himself took note of, or that he recollected and transcribed from other students at Leiden, and now extant at the British Library amongst the Sloane manuscripts. In circa 36 volumes, Morley personally transcribed students's notes and academic *dictata* at Leiden, bringing them with him when he returned to England in 1679. A recognition of his commonplace book preserved at the British Library[147] has allowed to ascertain his hand behind the *Experimenta*, which is further confirmed by the fact that another item present in the volume in which the *Experimenta* are transcribed is titled "[o]tras cosas de el mismo autor Drelincurtio recogidas par el Dr. D. Carlos Vinson,"[148] namely a report, whose main text is Latin, of anatomical lectures of the professor Charles Drelincourt, selected by another English student at Leiden, viz. Charles Vinson (to which the *Experimenta* are usually attributed). Spanish was, together with English, a language well mastered by Morley, as his commonplace book is written mostly in Spanish: so that the writer of the *Experimenta* was certainly Morley himself. The aforementioned Vinson (from Plymouth), in turn, had been a student at Leiden from 25

---

tube (e) of the pump is shown in Fig. IV. The interior of the end of the tube (e) is ground away in conical form to receive the conical end of the tube (f). The latter is held in place by means of a screw passing through the collar (h)," Daumas 1972, 86.

[143] Commentaries on this document are given in De Hoog 1974, 194–239, and Wiesenfeldt 2002, 108–130.

[144] De Volder 1676–1677, 134r.

[145] *Album studiosorum Academiae Lugduno-Batavae MDLXXV–MDCCCLXXV*, column 602.

[146] See Sect. 3.2.3.1, *The case of explanations in physiology*.

[147] Ms. Sloane 1281.

[148] Ms. Sloane 1292, 39r.

August 1671 (when he was 20 years old), and then from 22 May 1674,[149] before graduating with a medical disputation *De partu difficili* on 25 September 1675 (undedicated). Some information on Morley's activities as a copyist is revealed by the preface of the typographer of one of the only two printed texts based on Morley's collection of manuscripts, namely the *Collectanea chymica Leidensia, id est, Maetsiana, Margraviana, Le Mortiana. Opus quingentis, et amplius processibus adornatam, omnibus et medicis, et chimicis, et pharmacopaeis, imprimis utile collegit, digessit, edidit Christophorus Love Morley*. The book is a collection, as the title says, of more than 500 chemical processes, which are explained by giving the relevant accounts of the Leiden professors Carel de Maets, Christiaan Marggraf and Jacob le Mort, in a comparative way for each process. The preface of the printer provides precious information about Morley. In it, the printer clarifies that a friend of his saw, in London, the "treasury" (*gazophylacium*) of Morley, composed by circa 40 handwritten volumes (roughly corresponding to the number of manuscripts now present at the British Library), of which the printer (an Englishman, Henry Drummond) hopes to publish something more, and which Morley collected by attending lectures (public and private), by taking notes and dictations on a 'palimpsest', and then by copying them in notebooks.[150] Moreover, Morley certainly

---

[149] *Album studiosorum Academiae Lugduno-Batavae MDLXXV–MDCCCLXXV*, columns 570 and 589.

[150] "Commodaverat hoc opus, sua manu scriptum, praestantissmus Author, Christophorus Love Morley, Medicinae Doctor, Anglus, amico, hac familiari, sito hic in Batavia, commoranti, ea de causa, ut inde, quidquid vellet, excerperet, ac in propriam penum transferret. Is ad me iam brevi peregre abiturus ad Indos, mutuum volumen defert; ac, ut praelo committam, orat. Esse enim, quod lucem afficeret, perquam sane dignum; quodque omnium, qui Chemiae vel student, vel hanc artem profitentur, perpetuo in manibus versetur, ac ore. Quid enim? Multa hic reperiri Chemicae facultatis documenta oppido non vulgaria; insignem in rebus singulis operandi varietatem demonstrari; trium hoc opus Auctorum complecti institutiones admodum accuratas; proindeque triplex esse volumen uno comprehensum. [...] Is enim, quemquidem huius editionis Aucthorem nominavi, Christoph. Love Morley (ut iam tandem de toto hoc opere, eiusque authoribus, aliquid edisseram) magna diligentia, indefesso labore, ac studio, tum haec, tum alia scripta quamplurima, collegisse fertur hic, ante annos aliquot, in Hollandia. Huc et enim, id est, Lugdunum Batavorum, se contulit, post varias peregrinationes terra, marique obitas; postquam Academias non paucas, alibi terrarum, invisisset. Medicinam, praestantissimam artem, de novo, uti ille perhibebat, addiscendi causa. [...] Verum enimvero tanta incesserat hominem addiscendi cupiditas, tam uberem hic disciplinae Asclepiadea et copiam, et proventum, reperit; tot, talesque viros ista martem profitentes, magnum in primis Schachtium, praestantissimum virum, cuius quidem assiduus auditor erat, et comes, Drelincurtium, Craanium, Syënum, Maëtsium, Margravium, alios, ut ab illis incessanter audiendis, illorum lectionibus, collegis, consortiis frequentandis, per tot annos, avelli non potuerit. Quo vero accuratius faceret, in promptuque omnia haberet, cum memoriam non nisi summe labilem esse cognosceret, quidquid in Collegiis, Lectionibus, Theatris Anatomicis, Nosocomii, privatisve Laribus, sive oculis, sive auribus percepisset, protinus in palimpsestum referebat. Indeque, cum per otium suppetebat, in adversaria admodum curiose confecta, ordine quaeque suo digerere, ac denuo transcribere, consuevit. Ac fidem propemodum superat, quam ingentem utique molem scriptorum omnis generis, ad Medicinam, pertinentium, diligentissimus simul, ac studiosissmus vir, hinc inde corraserit. Ut Bibliothecam paene integram, ac plenam, propria manu conscriptam, sibi confecisse; certe supra quadraginta in quarto, uti vocant, volumina, bene crassa, ac ponderosa, quemadmodum amicus eius mihi retulit, qui isthaec non solum vidit, verum, etiam numeravit; ex hac Academia,

collected materials from other students, as amongst his papers we do find lectures tracing back before his enrolment at Leiden, such as those of Vinson (who graduated before Morley enrolled).[151] The whole, transcribed text of the *Experimenta* is

---

secum asportasse perhibeatur. Tandem in Angliam, iucundissimam patriam, tanquam ex postliminio reversus, Londini, uti audio, celeberrima in Urbe, ac Emporio, ad Medicinam faciendam domicilium sibi fixit; ubi vivus, valensque, etiamnum degit. Caeterum, primo post reditum anno, conscripsit ad Sodales Collegii Medicorum Londinensium florentissimi, in quorum numerum, ac ordinem Honorarium, per eos dies cooptatus fuerat, Narrationem, de morbo Epidemico, tam illius, quam superioris anni. Cui item adiunxit Lucae Schachtii, Praeceptoris sui, ac Medicinae Professoris praestantissimi, de eodem morbo ad ipsummet epistolam, ad modum accurate conscriptam. [...] Certo tamen scio, pro eo, quo pollet, ingenio, et omnigenae litteraturae, in primisque rei Medicae, amore, pollet autem ingenti, et maximo, oportune sperari posse, ut ex eadem pharetra, qua haecce Collectanea, et illa Historiola, multo plura in diem, et his maiora; siqua istis maiora, utiliorave esse queant, proditura sint. Certe, quod haec clanculum ex ipsius gazophylacio, illoque inscio, tam a me, quam ab ipso, qui haec mihi dono dedit, id est, a duobus sibi amicissimis hominibus, typis nunc vulgata, idque etiam sub illius nomine, sint, id ipsum illi, pro ea, qua praeditus est, humanitate, et morum suavitate insigni, incitamento non levi, cum ut alia nobis edenda communicet, tum ut haec ipsa in bonam partem accipiat, fore confido," Morley 1684, *Typographus lectori*, 1–5 (unnumbered). Cf. Morley and Schacht 1680. The printer thus provides some insight into the interesting story of Morley and of the *Collectanea*. Indeed, he declares he received the manuscript (prepared by Morley) circa the beginning of 1683, while Morley was travelling to the Indies. In fact, the genesis of this book is interesting, because it reveals a broader intellectual network. In the Morley corpus, indeed, there is a letter (ms. Sloane 1270, 218r–219r) in Spanish sent to him on 24 July 1681 by a certain Alonso de Buena Maison, or Bonne Maison, physician in Amsterdam and student at Leiden in the same years as Morley. In it, Buena Maison writes that he was awaiting the originals of the *Collectanea* from the (unmentioned) printers, which were to be given either to a certain Don Gabriel de Toledo, living in Rotterdam, or to a certain Moses Baruch. Probably, Morley was eager to get back the manuscript. However, something then happened. Buena Maison was known as a translator: he translated, from Dutch to Spanish, the famous book on pirates by Alexandre Olivier Exquemelin, namely his *De Americaensche Zee-Roovers* (1678), as *Piratas de la America* (1681). Moreover, Buena Maison revised the Spanish translation (made by Melchor de Novar) of the *De bello Belgico* (1632–1647) of Famiano Strada (published in 1682) and, in 1681, he apparently wrote a book called *Bárbaras tirantas cometidas en Indias por los españoles contra Dios y conciencia*, the copies of which were all bought by a Spanish nobleman, Don Gabriel Fernández de Villalobos, who complained to the Amsterdam city authorities about the book, which he deemed offensive to Spain. In the same year, both Buena Maison and Exquemelin (they both lived in the same house in Amsterdam), fled to Jamaica: for a full reconstruction, see Den Boer 2008. Apparently, the publication of the *Collectanea chymica* was retarded until 1684 for this reason. The *Collectanea chymica* had seven editions between the seventeenth and eighteenth century. On Morley, see Payne 1894, volume 39, 73.

[151]Another text which was put into print can be found in the Sloane manuscript containing De Volder's *Experimenta*: namely the *Experimenta anatomica, ex vivorum sectionibus petita* of Charles Drelincourt, published in 1681 by Ernst Gottfried Heyse (professor at Danzig and student at Leiden in 1680), then reprinted in 1684 and 1727. I am quite sure that Morley was the source of the manuscript used by Heyse, for two reasons: first, in the preface of the book, Heyse states that he published it "visis Angliae luminibus": see Heyse's preface in Drelincourt 1681, 5 (unnumbered). Second, the manuscript of Morley is more complete, namely it reports, for each "canicidium" – as the dissections of dogs are labelled – a description of the dog is absent from the printed text. For instance, at the beginning of the first *canicidium* there is reported a description of the dog, and the fact that the dog was given by a *praefectum* of the town of Leiden (ms. Sloane 1292, 1r).

given as an appendix to this book, while its contents are discussed in the next sections. Here, it is worth mentioning that the dates of each experiment do not correspond either to the Julian or to the Gregorian calendar to Mondays (the day on which De Volder usually performed his public lectures, at least from 1681),[152] and are moreover scattered over the week, so that it might be that these dates refer to the day of transcription rather than to the day of the lecture (as in the case of the dates reported in The Hague and Pretoria *dictata*, written over spans of time not compatible with each other). Moreover, as – for instance – in experiment 13, dated 29 September 1676, it is reported that De Volder had performed the same experiment two days before,[153] it might be that De Volder, before 1681, performed experiments not just on Mondays. In any case, it is unprobably that much time had passed between the lecture and the date reported in the transcription: indeed, the first lecture is dated 12 March 1676, less than one month after Morley's enrolment.

Further evidence for De Volder's experimental lectures is provided by another student of De Volder, Hermann Lufneu (1657–1744, who would later become physician at Rotterdam),[154] who enrolled at Leiden on 19 January 1675, and graduated in medicine in June 1679 with a *Disputatio chemico-medica inauguralis de fermentatione* (not dedicated to any professor in medicine or philosophy).[155] Already in his *Disputatio* Lufneu provides an evidence of a chemical experiment of De Volder, consisting of mixing acids in a vacuum or in open air, and demonstrating that air is necessary for their fermentation. The same experiment was then to be recalled by Craanen to claim that respiration serves the fermentation of blood, in his posthumous *Tractatus physico-medicus de homine* (1689).[156] Moreover, Lufneu offers a description and explanation of De Volder's demonstration of the hydrostatic paradox (which I deal with in Sects. 5.4.2.3, *...and De Volder's refinement of it*, and 5.4.2.4, *De Volder's Archimedean approach to the paradox*) in two communications published in the *Nouvelles de la République de Lettres*, in April 1685 and March 1687, namely his *Memoire communiqué sur une expérience curieuse d'hydrostatique* and his *Réponse à la difficulté que Mr Pujolas luy a faite dans les Nouvelles du mois de Janvier dernier*, refuting the criticisms of Moïse Pujolas (fellow of the Royal Society from 1695) on the validity of the paradox. Lufneu's communications are particularly interesting because the last experiment reported by Morley (whose report is dated 25 March 1677) consists of the presentation of the paradox, which is then fully described in De Volder's experimental demonstrations by Lufneu, who reports, in his first communication, to have witnessed them seven or eight years

[152] There is no evidence that De Volder performed experiments with an air-pump at home.

[153] See De Volder 1676–1677, 111v.

[154] On him, see Thijssen-Schoute 1960.

[155] *Album studiosorum Academiae Lugduno-Batavae MDLXXV–MDCCCLXXV*, column 594. He was also author and *respondens* of the disputation *De materiae divisibilitate in infinitum*, presided over by De Volder on 27 January 1676, and *respondens* of the last disputation of the series *De rerum naturalium principiis* (4 June 1676).

[156] I deal with this topic in Sect. 3.2.3.1, *The case of explanations in physiology*.

before, namely, around 1677–1678.[157] So that Lufneu's reports ideally complete those of Morley.

Another witness of De Volder's experiments was Jakob Bernoulli (1654–1705, older brother of Johann, 1667–1748), who in his *Dissertatio de gravitate aetheris* (1683) describes De Volder's experiment on the elasticity of air by showing the principle of communicating vessels within a glass container which is emptied of air, as well as its demonstrations with closed and open thermometers (and probably taking place around 1681).[158] Moreover, James Dalrymple (1619–1695, who lived in Leiden from October 1684 up to 1689), who in his *Physiologia nova experimentalis* (1686) reports De Volder's demonstrations with cohering plates of marble and with hemispheres, as well as De Volder's measurement of a cubic foot of water, also reported (from an unmentioned source) in the *Institutiones physicae* (1748) of Pieter van Musschenbroek (1692–1761), and in the *Tweede deel op de Wiskonstige rekening* (1680 – concerning a half-cubic foot) of Dirk Rembrandtsz. van Nierop (1610–1682), who also reports De Volder's measurement of the pressure of water on a oblique surface.[159] Also, a mention of De Volder's experiment with a 'Wittenberg syphon' (also listed in the laboratory catalogue) and taking place in 1686 is given by Jacob Leupold (1674–1727) in his *Theatrum machinarum hydralicarum* (1724).[160]

Last but not least, De Volder himself presented some of his experimental practices in his *Disputationes de aëris gravitate*, held between 14 October 1676 and 29 June 1678, which is worth briefly summarizing here. De Volder presents in the first disputation some evidence for the weight of the air (such as Evangelista Torricelli's experiment with mercury, or the case of the Magdeburg hemispheres – which he recalls performing at Leiden). Hence, he explains the conditions of the floatation of bodies in water, as well as the phenomena of the cohesion of bodies by means of Archimedes's principles (disputation 2), which constitute, together with a corpuscular theory of matter, the natural-philosophical backbone of his experiments. Moreover, he criticizes some objections to the idea of the weight of the air (disputation 3) as well as the idea of the fear of a vacuum and action at a distance (disputation 4). Only in disputation 5 De Volder presents, in detail, his measurement of the weight of a separate volume of air, consisting of weighing, on a balance, a container emptied of air (also described in the *Experimenta philosophica naturalia*, and mentioned by Dalrymple). Thus, as Wiesenfeldt has put it,

> while in the lectures the immediate experimental experience was in the foreground, [...] in the disputations the experiment was more firmly rooted in natural-philosophical debates. In the lecture the experiment always preceded the theoretical interpretation, in his disputation series at least one experiment followed at the end, to confirm the previous remarks. In this difference, the differentiation between the two forms of teaching becomes clear. The lec-

---

[157] See Sect. 5.4.2, *The hydrostatic paradox*.

[158] I deal with this in Sect. 5.6.3.3, *De Volder's assumption of Boyle's law in the works of his students*.

[159] I deal with this in Sect. 5.4, *The explanation of the hydrostatic paradox*.

[160] I deal with this in Sect. 5.1.3, *De Volder's 'didactic' ascertainment of the pressure of the atmosphere – and related experiments*.

tures served to impart relatively elementary knowledge to a broad audience whose professional interest was limited.[161]

As I am going to show, the natural-philosophical topics dealt with in the *Experimenta* can be found also in the printed disputations of De Volder, as well as in his handwritten *dictata* on Descartes's *Principia*. Certainly, however, the *Experimenta* represented another, and a new form of teaching with respect to disputations and academic *dictata*. In fact, such experimental lectures fit one of the purposes declared by De Volder for the establishment of the theatre. Namely, the confirmation of the principles of theoretical physics by means of experiments, or, in other words, of mechanical principles (consisting of Archimedes's model of floatation, based in turn on the model of the lever, which is the founding idea of Descartes's dynamics) are used to explain the behaviour of air and water both in the *Disputationes* and the *Experimenta*. In turn, the idea that phenomena are ruled by mechanical principles (*in nuce*, by matter and motion) is set out in its most systematic manner by De Volder in the preceding series of disputations *De rerum naturalium principiis*, held from 31 January 1674 to 4 June 1676. We might understand the *Disputationes*, finally, as the fulfilment of Constantijn Huygens's suggestion quoted above, if we interpret it as an invitation to De Volder to publish the results of his experimental activities ("so that it would not be that they are just heard") also as an invitation "to produce also his experiments on his own." Indeed, as I am going to show in Chap. 5, De Volder hardly invented completely new experiments in hydrostatics: however, he certainly provided improvements and refined versions of others experiments, for instance, of the weighing of a separate volume of air, which had not be attempted by anyone before him.

2.2.5 *The Leiden Didactic Tradition*

De Volder's experimentalism, in fact, did not come from nothing. I will deal more extensively with the role of experience in Dutch Cartesianism, and in its influences on De Volder in Sect. 3.2.2.1, *Physics and metaphysics in the history of Cartesianism*. Moreover, the reader can find a thorough account of the Leiden philosophical tradition in the works of Van Bunge, Nyden, De Pater, Ruestow and Wiesenfeldt.[162] Here, it is worth mentioning that, as noted by Cornelis de Pater, Willebrord Snel, who in

[161] "Während also in den Vorlesungen die unmittelbare experimentelle Erfahrung im Vordergrund stand, die von spektakulären Effekten wie von Hinweisen auf Moral und Nützlichkeit eingerahmt war, stand in den Disputationen das Experiment fester im Rahmen naturphilosophischer Debatten. In der Vorlesung ging das Experiment der theoretischen Deutung stets voraus, in seiner Disputationsreihe folgte wenigstens das eigene Experiment erst am Ende, zur Bestätigung der vorherigen Ausführungen. In dieser Differenz wird der Unterschied zwischen den beiden Lehrformen deutlich. Die Vorlesungen dienten der Vermittlung relativ elementaren Wissens an ein breit gestreutes Publikum, dessen fachliches Interesse begrenzt war," Wiesenfeldt 2002, 129.

[162] See Sect. 2.2.5, *The Leiden didactic tradition*.

1613 assumed a chair in mathematics at Leiden, under the influence of Ramism had already encouraged students to "investigate subjects that would be of use in everyday life," and carried out experiments with refraction since 1620s (albeit he did not perform experiments for his students), later followed by his successor, Jacob Golius.[163] In the early 1640s, Adriaan Heereboord first taught some of Descartes's ideas at Leiden, also showing consistent appreciation for Baconianism.[164] With a more consistent introduction of Cartesianism at Leiden by De Raey (from 1648) – who himself had praised Francis Bacon, and who proposed a discussion of Torricelli's barometer in his *Clavis philosophiae naturalis*[165] – as well as with the teachings in medicine of Sylvius, the idea that experience and experimentation play an essential role in philosophy (at least as the *explanandum*) became a standard at Leiden.[166] Thus far, the very cultural milieu of the University encouraged alternative forms of teaching – with respect to the presentation of abridged forms of Aristotle's texts, such as those embodied by the textbooks of Burgersdijk.[167] De Volder, thus, was already well imbued with an experimental tradition, or rather with a philosophical tradition open to the use of experience, before travelling to England and asking for the establishment of an experimental theatre. Moreover, as early as in 1672 Johann Christoph Sturm (1635–1703, himself a student at Leiden under De Raey in 1660) began to give a *collegium experimentale*, i.e. a series of lectures consisting of the performing of experiments at the University of Altdorf, later presented in his *Collegium experimentale sive curiosum* (first part 1676, second part 1685).[168] Sturm certainly played some role in prompting De Volder to open the experimental theatre at Leiden, moreover, Sturm probably acquired a copy of De Volder's disputations *De aëris gravitate* – even if he was later to attack De Volder's criticism of eclecticism.[169] Also, in 1660s and 1670s other figures were performing experiments open to small circles across Europe: this was the case, for instance, with the experiments taking place during *Mercredis* of Rohault at the Académie Montmor in Paris,[170] or the experimental lectures of Jean-Robert Chouet at the Academy of Geneva,[171] not to mention the activities of the Royal Society and the Accademia del Cimento in

---

[163] De Pater 1975, 309. Willebrord's father, Rudolph Snel, professor at Leiden from 1581 to 1613, was also a foremost Ramist: see Verbeek 2001.

[164] See Strazzoni 2012.

[165] See Sects. 5.1.4, *De Volder's criticism of the* metus vacui, and 5.6.3.3, *De Volder's assumption of Boyle's law in the works of his students*.

[166] De Raey, actually, was to become from the late 1660s a fierce opponent of the idea that experience could be the source of scientific knowledge: see Strazzoni 2015, 2018a, chapter 5; Del Prete 2018.

[167] See Bos and Krop 1993.

[168] On Sturm, see Bosl 1983; Herrmann and Platz 2003; Gaab et al. 2004; Wiesenfeldt 2004.

[169] Wiesenfeldt 2002, 159–160. See Sect. 3.2.3.2, *De Volder's methodological liberalism – and its continuity over time*.

[170] On Rohault, see Hoskin 1961; McClaughlin 1977, 1996, 2000; Clair 1978; Vanpaemel 1984; Schüller 2001; Des Chene 2002; Roux 2012b, 2013a, b; Dobre 2013a, b; Milani 2015; Spink 2018.

[171] See Heyd 1983.

Italy.[172] In fact, as De Volder himself put it in his request in late 1674, his theatre was to be established after the example of foreign academies. Moreover, as analysed in the works of Wiesenfeldt and Nyden, De Volder's own theatre prompted the creation of analogous institutions in the Netherlands and abroad. For instance, in 1697 Johann Bernoulli was requested to perform experimental lectures at Groningen, "after the example of De Volder of Leiden."[173] In 1706, De Volder's former student Joseph Serrurier was appointed professor at Utrecht, were he gave public experiments and who had already published a *Physicae experimentis innixae compendiosa tractatio, in usum scholarum domesticarum* (1700).[174]

Thus far, the opening of the experimental theatre was rooted in a long-standing philosophical and scientific tradition. As noted by Wiesenfeldt, moreover, the theatre also had the function (besides attracting students and increase the prestige of the University) of settling the disputes over Cartesianism, granting a free space in which a less dogmatic philosophy could be taught. As reported by Le Clerc, De Volder told Fagel that Cartesianism had already inspired an institution like the Royal Society, so that the theatre could allow the teaching of an experimentally oriented form of Cartesianism which could, at least partially, get rid of the philosophical quarrels and promote an advance in learning.[175] In fact, the opening of the theatre in 1675, as well as the appointment of De Volder in 1670 have to be considered in the light of the effort, by the University Curators, to grant a free space in which the new philosophy could be taught, along with the old ones. This effort could be noted as early as 1651, when De Raey was put in charge, by the Curators, of teaching Cartesian physics by showing its agreement with the thought of Aristotle (as De Raey himself declared in the dedicatory letter of his *Clavis philosophiae naturalis*).[176] Later, the teaching of Cartesian philosophy was granted by the Curators by the appointment of Geulincx (encouraged by Heidanus). In turn, after 1669 Geulincx had died during the Leiden plague, and De Raey moved to Amsterdam Athenaeum Illustre, the Curators appointed (in 1670) Craanen and De Volder as professors in philosophy. Moreover,

---

[172] See Knowles Middleton 1971; Miniati 1987; Beretta et al. 2009; Boschiero 2007.

[173] Cf. his letter to Leibniz of 8 January 1698: "[o]rdines nostri novam mihi imposuerunt docendi provinciam, atque in eum finem certam decreverunt summam ad emenda instrumenta experimentalia, ut exemplo Volderi Lugdunensis Studiosos nostros etiam experimentis Mathematico-physicis exerceam et delectem," A III7, 702. For a discussion, see Wiesenfeldt 2002, 291.

[174] See Sects. 5.6.3.3, *De Volder's assumption of Boyle's law in the works of his students*, and 2.4.2, *De Volder's retirement, death and legacy*.

[175] Le Clerc claimed that the example of the Royal Society inspired De Volder: Le Clerc 1709, 362: "[c]e fut apparemment la Societé Royale d'Angleterre, toute occupée à rechercher la Verité, par la voie des Experiences, & peut être l'exemple de l'illustre Mr. Boyle, qui en étoit comme l'ame, & qui faisoit une grande dépense en son particulier pour cela, qui firent naître à Mr. de Volder la pensée de proposer à Mrs. les Curateurs de l'Academie de faire quelque dépense, pour la méme chose."

[176] "Vos estis, qui me ex doctore privato publicum professorem creastis, et ut eam philosophandi rationem, quam pluris a me fieri atque etiam Aristoteli valde adversam ab aliquibus censeri notum erat, cum Aristotele componerem, haud obscure imposuistis," De Raey 1654, 24 (unnumbered). See Strazzoni 2011.

they granted – after De Vries's dramatic move to Utrecht – the teaching of more traditional ideas by appointing Wolferd Senguerd (already a lecturer at Leiden) and Wilhelm Wilhelmi (who died in 1677) as ordinary professors of 'Peripatetic philosophy' in January 1676.[177] In fact, Senguerd can hardly be labelled an Aristotelian. Even if he from time to time criticized Descartes in his *Philosophia naturalis*, he carried out an experimental programme which was, to some extent, provided with a more remarked heuristic value than that of De Volder, for whom experiments had a chiefly didactic function. Namely, De Volder did not undertake a programme of observations such as that followed by Senguerd, testified to by his *Inquisitiones experimentales* (first edition 1690, second edition 1699), namely the presentation of a series of experiments carried out with the air pump on living bodies and bodily organs, and the *Rationis atque experientiae connubium* (1715). Senguerd performed his experiments at home – where he owned his own air pump, built in 1679 by Samuel van Musschenbroek – but also at the University (as the *series lectionum* of 1681, 1694, 1696–1700, 1703 and 1704 report that he performed public experiments on Tuesdays, even if the place is not specified, while De Volder steadily performed experiments in the physical theatre).[178] In 1705, on De Volder's retirement, he was allowed to teach at the *hora undecima* (viz. De Volder's usual lecturing time), he became director of the theatre, and from 1706 he started to perform experiments in it.[179] Notably, Senguerd himself, like De Volder, used Descartes's *Principia* to teach natural philosophy, providing two *dictata* on them.[180]

Last but not least, the experimental theatre of De Volder was not the only space in which professors performed public experiments at Leiden. As reconstructed by Wiesenfeldt, the University of Leiden already had several places in which sciences were cultivated outside the *aulae*, namely, the library (set up in 1585), the botanical garden (1590), the anatomical theatre (1593), the astronomical observatory (1632), and the chemical laboratory (1669)[181] – in which De Maets himself performed experiments in natural philosophy, while De Volder, at his own theatre, performed experiments in chemistry, as seen above.

---

[177] Molhuysen 1913–1924, volume 3, 319.

[178] Molhuysen 1913–1924, volume 3, 268*–269* and 271*–273*; volume 4, 45*–46*, 55*–58*, 65*–67*, 70*–71*, 80*–81*, 93*–94* and 99*–102*. Moreover, in Senguerd's *Inquisitiones experimentales* experiments are described as taking place in public, while in his *Connubium* he reports that he used, for some time before 1679, De Volder's air pump: see *supra*, n. 140.

[179] Molhuysen 1913–1924, volume 4, 216, 221 and 107*.

[180] Both extant at The Hague: *Dictata Clarissimi Doctissimi Viri D. D. Wolferdii Senguerdii Philosophiae Professoris in Renati Des-Cartes Principia Anno 1690* (ms. 72 A 8), and *Nobillissimi, Doctissimi Wolferdi Senguerdi notae ad Cartesii principia Philosophiae ad dua priores partes seu libros* (ms. 133 M 71). See Wiesenfeldt 2002, 91.

[181] I owe this list to Wiesenfeldt 2002, 67. See also Lunsingh Scheurleer and Posthumus Meyjes 1975.

2.3 De Volder as Scientific *Savant* in the 1680s and 1690s

As seen above, after 1675 De Volder did not immediately preside over disputations in metaphysics. Only in 1680–1681 he presided over his *Contra atheos*, in which Cartesian arguments for the existence of God are presented. In 1681–1682, moreover, he presided over (but not authored) a *Disputatio philosophica de aeternitate Dei* (5 March 1681) of Johannes Pap Szathmári, as well as an *Exercitium philosophico-metaphysicum de perenni duratione mentis* (March 1682) of Samuel Nemethi, both dedicated to De Volder and to Wittich. In natural-philosophical domains, he presided over a *Disputatio mathematico-physica de lumine* (12 and 15 March 1681) of Samuel Köleséri, who would also graduate under De Volder with his *Disputatio philosophica inauguralis de systemate mundi* (30 June 1681), and the *Exercitatio philosophica, coelorum, siderumque lucidorum originem, et phaenomena, methodo synthetico-mathematica demonstrans* (March and May 1682) of Pál Derecskei. In fact, leaving aside the major series of disputations mentioned above, after the opening of the theatre De Volder mostly presided over disputations not officially authored by himself: an indirect sign of his progressive 'detachment' from academic debates. This, of course, does not mean that he stopped his activities as a professor and as a *savant*, especially in natural philosophical and mathematical fields, viz. what we may understand today as 'scientific' domains. The years 1681–1682 indeed represented another turning point in De Volder's career.

2.3.1 The Correspondence with Thévenot

In summer 1681 De Volder travelled to Paris in order to acquire further scientific instruments for the theatre, while others were acquired in 1685.[182] Most probably, he travelled there just after the beginning of the summer holiday, as his presence is reported in a letter by Henri Justel to Locke of 24 July/3 August 1681, according to which he wished to talk with the members of the Académie Royale des Sciences (and in which De Volder's experimental weighing of different kinds of water is briefly mentioned).[183] As I am going to show in Sects. 5.4.2.4, *De Volder's*

[182] See Le Clerc 1709, 376. According to Le Clerc, De Volder spoke French well, as he had had occasions to practise it in Holland. According to Gronovius, De Volder was asked by the Curators to travel to France: Gronovius 1709, 23–34. As reported in Rosenboom 1697, part 2, 40 and 52, De Volder was reimbursed 695 guilders and 5 stuivers for having bought scientific instruments in France and elsewhere, in 1681 and 1685. In particular, in November 1681 he was reimbursed 442 guilders and 19 stuivers, and in November 1685 he was reimbursed 252 guilders and 6 stuivers. See also Wiesenfeldt 2002, 91–92, referring to the Leiden manuscript AC 44.

[183] "Monsieur Warder [*sic*] qui est professeur en physique et qui a le soin du Theatre physique de Leyde est ici. Il m'est venu voir et ma dit qu[']il avoit dessein de conferer avec Messieurs de l'Academie. Je ne croy pas qu[']il apprenne beaucoup de choses deux. il a faict plusieurs experiences sur l[']eau et a trouvé que toutes les eaux estoyent dune égale pesanteur, au moins que la difference en estoit peu considerable," Locke 1976–1989, volume 2, 426–427. This is letter 651 of Locke's correspondence.

Archimedean approach to the paradox, and 4.2.3.4, *Collision in De Volder's late thought*, in Paris De Volder was probably inspired by the activities of Edme Mariotte (1620–1684), as De Volder devised a collision apparatus for his theatre (described by Johannes Joosten van Musschenbroek in some of the letters he sent to the Marburg professor Johannes Dorstenius, and later mentioned by Uffenbach, as a "wooden sextant") based on the model already used by Mariotte. In any case, De Volder certainly met some members of the Académie, as – besides Huygens and Justel – he was in contact with Melchisédech Thévenot (circa 1620–1692),[184] founding member of the Académie, with whom he corresponded at least until August 1683. This is testified to by a series of 14 letters by De Volder to him ranging from 15 October 1680 (but which however can probably be traced to 15 October 1682)[185] to 27 August 1683, and including two undated letters. The correspondence – which is quantitatively inferior only to De Volder's correspondence with Johann Bernoulli and Leibniz (starting from 1698) – did not touch upon philosophical topics – the only exception being some short remarks of De Volder on his attempts at the production of phosphorus (thus further testifying to his interests in chemistry).[186] In fact, De Volder's letters mostly focus on the *Nachlass* of Swammerdam (correspondent of Thévenot, whom Swammerdam met in Paris in 1664–1665, together with Steensen),[187] amongst whose papers the letters by De Volder are now extant at Göttingen.[188] As reconstructed by Boerhaave in his biography of Swammerdam – published as a preface to the 1737 Latin edition of Swammerdam's *Bybel der natuure of Historie der insecten* – before dying on 17 February 1680, Swammerdam gave his unpublished Dutch manuscripts to a certain Herman Wingendorp in Leiden, in order for them to be translated into Latin. However, Wingendorp did not want to give the manuscripts to Thévenot – to whom Swammerdam had left them as inheritance – unless he were paid a large sum of

[184] See *infra*, n. 196.

[185] See *infra*, n. 195.

[186] In his letter of 31 April 1682 De Volder claims to have been unable to do so: "[d]escriptio phosphori, nescio an meo an descriptionis vitro mihi tentanti non successit. Si commodior via phosphoros efficiendi inventia sit est quod gratulemur promotae rerum naturalium cognitioni, praesertim si eo res deducta sit ut lucernas sive veras sive fictas sepulchrales veterum aut imitari, aut nunc primum conficere possimus. Quia de re si quid porro inaudieris gratissimum erit si communicare digneris," Staats- und Universitätsbibliothek Göttingen, Cod. Ms. Hist. nat. 102, fasc. 25, 50r. In his letter of 29 October 1682, De Volder mentions that he ignored that phosphorus could melt metals, and that Paul Hermann, professor of botany at Leiden, had brought from England two specimens of phosphorus (studied by Boyle in his *The Aerial Noctiluca, or, Some New Phenomena and a Process of a Factitious Self-Shining Substance: Imparted in a Letter to a Friend Living in the Country*, 1680, listed in the *Bibliotheca Volderina*, 8): "[p]hosphoro metalla fluidi res apud nos est plane inaudita. D. Hermans secum attulit ex Anglia duos Boyleanos alterum liquidum alterum <consistentem> durum, quorum priorem aeram noctilcuam posteriorem, glacialem vocat ipse Boyle," Cod. Ms. Hist. nat. 102, fasc. 25, 60r.

[187] See Jorink 2003, 2010, chapter 5, 2016a, 2018. See also Nordström 1955; Kardel 1994; Cobb 2000, 2006, chapter 2; Ruestow 1996, chapter 5; Miniati 2009; Kooijmans 2011.

[188] See *supra*, n. 94.

money, or unless he were ordered to do so by a court. In turn, Thévenot asked Daniel de Hoest (husband of the main heir of Swammerdam, Margarita Volckers), if he could get the manuscripts from Wingendorp, in order to prepare their translation. However, it was only with the mediation of De Volder, and after an order from a court of April–May 1682 the papers could be eventually given to De Volder himself in November of the same year, and then forwarded to Thévenot in January 1683. However, Thévenot did not made them translated (Swammerdam's *Biblia naturae*, in fact, was published in Latin by Boerhaave only in 1737).[189]

Besides the *affaire* of Swammerdam's papers, the correspondence is important because it reveals other details on De Volder's intellectual network. First, it reveals a very probably direct acquaintance with Swammerdam himself. Swammerdam, as anticipated above, had studied at Leiden under Sylvius between 1663 and 1667, and his inaugural disputation *De respiratione* was going to be taken into account by De

[189] "Ubi omnia ingravescebant, testamentum scripsit XV Ianuarii, anno MDCLXXX, in quo legavit Melchisedeco Thevenoto, Legato quondam Regis Galliarum ad Rempublicam Genuensium, omnia sua manuscripta originalia, quae pertinent ad historiam Apum, Papilionumque, Naturalem, et Anatomicam, una cum tabulis eo spectantibus lii. Quae omnia erant eo tempore in aedibus Hermanni Wingendorpi, Leydae. Iussit, tradenda haec omnia legatario intra annum a morte testatoris. Rogavitque serio, ut tractatus de Apibus ederetur etiam sermone belgico: quia sapientia, et omnipotentia, Dei tam apodictice in eo demonstrantur. Pauculum superstitis porro vitae Deo adorando, et amando, unice dedit. XVII Februarii eodem anno moriebatur. Haeredem scripsit Margaritam Volckers, uxorem Danielis de Hoest, doctoris medicinae. Exsecutores testamenti constituit Christophorum van Wyland, et Margaritam Volckers, quae, Wylando haud diu ab his mortuo, sola in hoc munere permansit. Statim ab exsequiis Dominus Ort, rogatus ab exsecutoribus testamenti, scripsit Thevenoto, XXIX Martii sequentis, de legato apud Wingendorpium recondito. Swammerdammius parum versatus in Latinis, volens tamen opera sua lingua latina evulgare, dederat ea omnia huic Wingendorpio, ut latine verteret: hic enim ipsi opem praestiterat hanc, in sermonem romanum vertendo eius scripta de Utero. Urgebat Thevenotus de Hoestium, ut curaret tradendum sibi legatum. Wingendorpius pauper, sustinebat vitam vertendo in varias linguas, et quaerebat, omni astu credita sibi retinere, frivola millena praetexens, tandem aperte negabat, se redditurum, nisi sententia iudicis coactus. Itum ergo in ius; neque ante maium anni MDCLXXXII, post taediosas lites, decreto iudicum, omnia simul tradita sunt Viro Longe celeberrimo, Burchero de Volder, Mathesios, et Philosophicae, Professori Clarissimo. Hunc Thevenotus rogaverat, ut ius suum tueretur amici amicus. Et, abfuisset fidelis sapientia Volderi, vix Thevenoti ad manus pervenissent. Postquam iam possidebat, voluit ea belgico sermone edere ilico. Verum mutavit sententiam, sibique in Gallias mittenda curavit," Swammerdam 1737–1738, volume 1, *Vita auctoris*, 34–36 (unnumbered). Cf. the letter of De Volder to Thévenot of 31 April 1682, in which he communicated to Thévenot the positive conclusion of the quarrel: "hucusq[ue] responsum ad tuas distuli quod putarem me paucorum dierum moram nuntio <pa>transacti tandem negotii de M.S. Swammerd. facile compensaturum. De qua re sic habe. Eo adductos esse adversarios, ut contentionem deposuerint, manuscripta tibi deberi fateantur," Cod. Ms. Hist. nat. 102, fasc. 25, 49r. See also his letter of 19 November 1682 (in which De Volder communicates to Thévenot that he took the manuscripts from the court: "j'ay retiré les ecris du Greffe," Cod. Ms. Hist. nat. 102, fasc. 25, 61r), and the letter of of De Volder to Thévenot of 28 January 1683 (in which De Volder communicates to Thévenot that he had sent the manuscripts to Paris probably by Hendrik van Bleiswijk (one of the mayors of Delft), who was following the delegation of Anthonie Heinsius to France: "Monsr. Bleyswijck [...] s'en partira {pet} peut estre qu'il est desia parti pour Paris en la suite de Monsr. <{...} > Heinsius qui s'en va en Ambassade vers le Roy de France. Ie luy ay confié ces papiers, ce que j'espere qu'il ne vous sera pas desagreable," Cod. Ms. Hist. nat. 102, fasc. 25, 63r). On Heinsius, see De Jongste and Veenendaal 2002.

Volder in his dealing with respiration in the *Experimenta*.[190] As revealed by a letter to Thévenot of 30 October 1670, in Amsterdam Swammerdam was part of an informal circle including, amongst others Hudde, Matthaeus Sladus (son of Cornelius, and physician), the cartographer Nicolaas Witsen, and De Volder.[191] At Leiden, Swammerdam had contacts with his fellow students Steensen and Ruysch.[192] The latter, in turn, was going to mention De Volder – a man "who was not used to asserting something immediately" – as an attentive witness of his public dissections in Amsterdam, and as endorsing his ideas on the structure of the brain cortex against Govert Bidloo (from 1694 professor of medicine at Leiden).[193] In fact, De Volder – while not himself a professor of medicine – was to praise, in his 1698 *Oratio de*

---

[190] See Sect. 5.1.4, *De Volder's criticism of the* metus vacui.

[191] "De andere vrienden als quina, utenbogert, Hudden, Sladus, Witsen, de volder varen seer wel," Lindeboom 1975, 53. For a commentary, see Jorink 2016b.

[192] See the studies of Jorink mentioned *supra*, n. 187.

[193] In his *Adversariorum anatomico-medico-chirurgicorum decas tertia* (1717), Ruysch mentions De Volder while reporting how the features of the skin of the body were usually neglected by physicians. De Volder, however, paid much attention to it: "[s]ed etiam Chirurgi norunt quam difficile semper sit, per medicamenta epulotica, quae frigefacientia debent esse, et exsiccantia, epidermidem rursum superducere cuti; quin et ubi per haec non procedit cicatricis suae productio, cogimur aliquando confugere ad ea emollientia, quae simul humectantia sunt, ut ad Digestivum aureum, et similia. Parco hac occasione et illos memorare, quibus ab actione incensi pulveris pyrii cutis decoriata fuit. Certe deinde hac ex causa tanta quotidie puris copia gignitur, ut exhausti viribus, idoneoque nutrimento privati mortui sint. Qui casus aliquoties nobis obvenere. Videtur itaque non decorum, de hisce integumentis vel contemptim loqui! Vel negligenter de iis tractare: neque sane oblitus rei illius sum, quae multos ante annos nihil tale merenti mihi accidit. In publica quidem tunc demonstratione anatomica, coram frequentissima honoratorum spectatorum corona, meo more communia corporis extimi integumenta haec, atque in his solis tantum demonstrandum habebam, ut iis sesqui horam integram impenderem. Inter erat huic demonstrationi Medicus Amstelaedamensis, practicus famigeratissimus, una cum Burchero de Volder, Philosophiae et Mathesios in illustri Academia Lugduno-Batava Professore Eximio. Hic ille medicus omnibus aeque in demonstrationem intentis per contemptum, finita demonstratione, in haec verba erupit: An igitur vilia haec merentur, ut iis cognoscendis sesqui horam perdamus? An quidem decet Anatomicum nostrum talibus tractandis nos tanto tempore detinere? Miratus hominis dictum Egregius Volderus regessit, Itane tu de hoc negotio sentis? Ego numquam tam subtilia, atque vera tamen, in corporis humani partibus scivi, aut vidi. Atque sic ostendebat, severum talium spectatorem fuisse," Ruysch 1717, 25. It is unclear who was the other physician mentioned by Ruysch. Moreover, in his *Responsio ad Dissertationem epistolicam Jo. Christoph. Bohlii quam de usu novarum cavae propaginum in systemate chylopoeo, nec non de cortice cerebri, conscripsit* (1725), Ruysch recalls how De Volder could not observe the existence of round glands in the cortex of the brain, as maintained by Bidloo: "[q]uod attinet corpora illa dicta olivaria, haec sane est veritas quaedam perversa sine ulla entitate, de quibus memini, quid mihi accidebat, cum Profess. Bidloo pro sua defensione contra me volebat demonstrare, quod cortex cerebri corpora olivaria essent, et hoc quidem in publica anatomia. Tenebat nempe corticem cerebri in manu sua et auditoribus interrogabat, num non viderent, quod cortex cerebri tales essent glandulae, ac in suo opere dixisset, cum vero ad Professorem Volderum venisset, (qui non ex nunc solebat aliquid asserere) et eum interrogabat, respondebat: Non, mi Domine. Et ita adhuc nunc non nulli dantur, et hoc propter micas! Si nunc et ex Te V. Doct. peterem, ut mihi haec praetensa corpora olivaria demonstrares, nonne idem tibi contingeret? Milites quando hostium adventum audiunt, clamant: ad arma! ad arma! Sic ego dico hic: ad visum! Ad visum!" Ruysch 1725, 11–12. See Kooijmans 2011, 253.

rationis viribus, a physiology based on experience only, rather than on the speculative ascertainment of invisible features of the body.

Second, the correspondence reveals De Volder's contacts with Théodore Barin and Adrien Auzout – whose expected coming to Leiden, in order for them to deal with the issue of Swammerdam's papers – is mentioned in a letter dated 15 October 1680 (but probably tracing to 15 October 1682).[194] As it emerges from the subsequent letters, De Volder met Barin, while it is unclear whether he could also meet Auzout in Holland.[195] It is unclear whether De Volder met them in Paris in 1681, though he most likely met Thévenot.[196] As I am going to show below, moreover, De Volder

[194] "I'attens avec une grande impatience l'arrivé de Monsr. Barin ou de Monsr. Auzout pour me gouverner selon leurs conseils," Cod. Ms. Hist. nat. 102, fasc. 25, f. 44r. The letter continues with a reference to a medal done and sent to the chemist and instruments builder Jacques Borelly: "[e]nfin, Monsieur, la medaille pour Monsr. Borelli est achevée et je l'ay donné par Monsr. Clignet a Monsr. Merault marchand demeurant chez Messr. De Bernjers rue St Honoré a Paris, qui se est chargé pour le luy faire rendre. / Pardonez moy que je vous envoye cette lettre pour luiy. Ie ne doute pas qu'il {ne ...'ait deja} receue \la medaille/, en quel cas je vous prie de luy recommander, de notifier par une lettre a Messr. Les Curateurs de l'Academie et Bourguemaistres de cette ville qu'il l'a receu, car cela servira a me decharger. S'il ne l'a pas receu, de la prie d'en faire l'enqueste chez ce Monsr. Merault," Cod. Ms. Hist. nat. 102, fasc. 25, 44r–v. As I show in the next footnote, this reveals that the date of this letter, as written by De Volder, was most probably wrong, and was written after 17 August 1682. This would place all the correspondence after De Volder's trip to Paris in 1681. As to Borelli, or Borelly, see below.

[195] In his letter of 15 July 1682 De Volder claims that he still had not met either Barin or Auzout, and – notably – that the medal mentioned in the previous footnote had not yet been sent to Borelly: "je n'ay pas encor eu l'honneur de voir Monsr. Barin, ny Monsr. Auzout, je m'estimerois fort heureux de leur pouvoir faire quelq[ue] service en ce pais. Enfin la medaille pour Monsr. Borelli est preste pour luy estre envoyée a la premier occasion," Cod. Ms. Hist. nat. 102, fasc. 25, 53r. It might be that the first letter of the correspondence, unmistakeably dated '1680', was wrongly dated by De Volder, and has to be dated between 17 August and 19 November 1682. In the letter of 17 August 1682, indeed, De Volder informs Thévenoth that the medal was ready and could be sent to him: see *infra*, n. 219, while in the letter of 19 November 1682 he shows his happiness at knowing that Borelly had appreciated the gift of Leiden University: "[i]e suis ben aise que Monsr. Borelli ait eu pour agreable le present de Messr. Les Curateurs," Cod. Ms. Hist. nat. 102, fasc. 25, 61r. In the letter of 17 August 1682, De Volder complains that he could not meet again Barin once again: "j'ay esté si malheureux de ne voir pas Monsr. Barin estant a l'Université quand il estoit chez moy, et n'ayant pas l'occasion de le revoir {par} son soudain depart. [J]'espere que je seray plus heureux a son retour," Cod. Ms. Hist. nat. 102, fasc. 25, unnumbered folium. In turn, the letter of 29 October 1682 suggests that Barin and Auzout were expected to come back to Holland: "[a] quo ultimas tuas accepi, nihil expectavi ardentius quam aut evidens eorum quae fieri a me velles, mandatum, aut adventum sive D. Auzotii sive D. Barin, quo illorum virorum prudenter {...} consilius in tua causa uti possim. [...] Interea temporis [...] D. Barin aut Auzout huc accedant illis omnem rem aperiam, [...] etsi forte mecum non sentirent, ita agam," Cod. Ms. Hist. nat. 102, fasc. 25, 59r and 60v. Also in the letter of 19 November 1682, De Volder notes that Auzout was still to come: see *infra*, n. 225. Finally, in the letter of 27 August 1683, it is mentioned that Auzout visited Amsterdam, but the attempts of De Volder and Hudde to meet him were in vain: "[i]e suis fort faché de n'avoir pas peu rencontrer Monsr. Auzout quoy que je l'aye cherché avec Monsr. Hudde a Amsterdam puisque au mesme temps il a passé par cette ville," Cod. Ms. Hist. nat. 102, fasc. 25, 64v.

[196] See the letter of De Volder to Thévenot of 24 September 1682: "[...] et Dieu vueille, que je pousse avoir occasion de vous pouvoir montrer en un autre cas, avec plus de succes combien je<{...}> \me {sens}/ redevable a vous pour toutes les bontés que vous avez eu pour moy a

probably instructed the chemist and instrument builder Jacques Borelly (or Borelli) to provide a lens with a focal distance of 18 feet for the University. This lens, however, was not the only astronomical instrument considered by De Volder in his correspondence with Thévenot. In his letter of 31 April 1682, De Volder expresses an interest in a micrometer (namely an instrument to measure the angular distance between the stars, which I consider below as well), as well as for a pendulum, mentioned by Thévenot in a previous letter (now lost).[197] The pendulum, actually, is also mentioned in the the letter of 28 May 1682,[198] as well as in the first undated letter (which can be tentatively dated between 19 January and 31 April 1682),[199] in which De Volder thanks Thévenot for his care for it (but no details are provided).[200]

Third, the correspondence reveals De Volder's connections with De Raey, namely, with the Professor at Amsterdam, and with his son Johannes de Raey *de jonge* (who died before 1700), author of a *Dictionarium geographicum ofte Schaten woordt-boeck des aerdt-rycks* (1680, with further editions in 1700, 1709 and 1710), for whom De Volder acted as intermediary and dealt with geographical topics.[201] Notwithstanding their divergent views on natural philosophy (as De Raey

Paris, de<{...}> \les quelles/ {il vous a pleu} de continuer jusques icy," Cod. Ms. Hist. nat. 102, fasc. 25, 57r.

[197] "Quae de pendulo et micrometro scribis, scias me ea quidem avidissime desiderare," Cod. Ms. Hist. nat. 102, fasc. 25, 49r.

[198] "[J]e vous remercie tres humblement des {...} qu'il vous a {pleu} de prendre de mon pendule," Cod. Ms. Hist. nat. 102, fasc. 25, 52r.

[199] In the letter there is mentioned a letter by Thévenot of 19 January, and it is declared that the Swammerdam affaire had not been solved yet (so that it has to be dated before 31 April 1682, when De Volder communicated to Thévenot the conclusion of the quarrel (cf. *supra*, n. 189)): "Il me fa<c>sche fort que je ne \vous/ puis encore ecrire, que l'affaire de Mons. Swammerdam est achevée," Cod. Ms. Hist. nat. 102, fasc. 25, 66r.

[200] "[J]e vous remercie tres humblement de la {peine que} vous prenez de mon pendule et certes j'ay me {...} de l'attendre plus long temps et de l'avoir fort ex[missing text]ct que de l'avoir bien tost, s'il manquoit quelque chose a l'exactitude, ce qui sans doute {...} [missing text]out," Cod. Ms. Hist. nat. 102, fasc. 25, 67r.

[201] See the letter of 31 April 1682: "[p]rioribus tuis acceptis litteris scripsi continuo ad D. de Raey, a quo ante {trid...} responsum demum accipi, quod hisce inclusum vides. Distulit ille scriptionem reditum frustra expectans eius ex quo sperabat se plura de terra {jecmensi} auditurum {Re.}, ut ad me{,} scribit, in {mora} erit, quo simulae quodpiam explicatus fuerit, id tecum communicet. Prodiit de novo praeter libellos, quos D. De Raey enumerat, iter Nieuhof, quem simul cum illis quos illa notat libellis si eos nondum {haereas} prima quavis occasione data, ut habeas officiam si per quempiam ex amicis id commode fieri possit, utar ea via sui {minuis}, utar {mani}," Cod. Ms. Hist. nat. 102, fasc. 25, 49v. De Volder refers to Johan Nieuhof's *Zee- en Lant-Reise door verscheide Gewesten van Oostindien, behelzende veele zeldzaame en wonderlijke voorvallen en geschiedenissen. Beneffens een beschrijving van lantschappen, dieren, gewassen, draghten, zeden en godsdienst der inwoonders: En inzonderheit een wijtloopig verhael der Stad Batavia*, or to his *Gedenkweerdige Brasiliaense Zee- en Lant-Reise und Zee- en Lant-Reize door verscheide Gewesten van Oostindien*, both published in 1682. In the letter of 28 May 1682, De Volder mentions a visit that a certain "Monsieur Ray" paid to him: "Mr. Ray m'est venu trouver il ya quelques jours. Il m'a dit clairement que des terres ny australes ny septentrionales il ne scavoit rien plus que ce que il vous avoit ecrit. Car qu'on n'avoit d'icy fait aucun voyage vers {ces terres}, <sinon> excepté ceux dont il avoit fait mention dans ses memoires que je vous ay envoyéz," Cod. Ms. Hist.

senior was to inveigh against experimental philosophy in a private colloquium with Jakob Bernoulli in 1682),[202] and, as reported by Stolle, De Volder once heard De Raey wagering five ducatons for each error anyone could show him in Descartes's texts,[203] they seem to have maintained a close relation.

Thus far, this correspondence reveals how De Volder was well inserted into a network ranging from Paris to Amsterdam, and which had its ramification in the Amsterdam circle of anatomists – amongst whom one can count Swammerdam, Ruysch and Blasius,[204] – as well as with the Utrecht *Collegie der Sçavanten*, in which Swammerdam figured too, together with De Bruyn, Johannes Georgius Graevius, Ludwig Wolzogen, Frans Burman, Regnier van Mansvelt and Lambert

nat. 102, fasc. 25, 52r. Cf. also the first undated letter of De Volder (i.e. that one probably tracing back between 19 January and 31 April 1682, see *supra*, n. 199): "j'ay escrit a monsr. De Raey qui m'a repondu comme vous verrez dans ce memoire[.] Monsr. son pere {adjouste} que son fils a {desia} tant de marquez de ce monsr. {Jelmer}, qu'il est presque impossible de me le trouver pas, et qu'il me fera scavoir, en que son fils pourra connoistre de luy," Cod. Ms. Hist. nat. 102, fasc. 25, 67r, and the second one: "la lettre que vous m'{ariez} envoyé pour un amy de Monsr. Raey est envoyé a luy a Amsterdam [missing text]tatim ab acceptis litteris," Cod. Ms. Hist. nat. 102, fasc. 25, 69r. De Raey senior was in contact with Swammerdam, and is mentioned in Swammerdam's correspondence with Thévenot: and Swammerdam: see Lindeboom 1975, 58 and 64. Notably, Jan Baptist van Lamsweerde attacked, in 1674, both Swammerdam's theory of respiration and De Raey's theory of logic in his *Respirationis Swammerdammianae exspiratio, una cum Anatomia neologices Joannis de Raei*. Geography recurs again in the correspondence: for instance, in the letters of 31 April, 28 May, 15 July, 24 September 1683, and in the first undated letter, De Volder provides discussions of the voyages and discoveries of Paul Hermann, professor of botany at Leiden. A handwritten *Descriptio Moscoviae*, in possession of the heirs of the Dutch ambassador Nicholas Heinsius (who died in 1681), is mentioned in the letter of 31 April 1682. In the letter of 17 August 1682, De Volder writes to Thévenot that this manuscript could not be found: see Cod. Ms. Hist. nat. 102, fasc. 25, 50r and 55. Amongst the items in De Volder's private collection of mathematical instruments, there is mentioned a *Landkaart van Moscovia* of Nicolaas Witsen (which could be Witsen's *Nieuwe lantkaarte van het noorder en ooster deel van Asia en Europe strekkende van Nova Zemla tot China* (1687), *Noord en oost Tartarye* (1692), or the *Imperii Russici sive Moscoviae, status generalis, in eius regna, ducatus, provincias, populosq*[*ue*] *subiacentes divisus, et ex tabula spectat. mi domini N. Witsen* [...] *pro maiori parte excerptus, per F. de Witt* (circa 1690): *Bibliotheca Volderina*, 96. The *Descriptio Moscoviae* is not mentioned in the *Bibliotheca Heinsiana* (1682).

202 See Sect. 5.6.3.3, *De Volder's assumption of Boyle's law in the works of his students*.

203 "Raë, der anfangs hier profitiret, und dem nach Amsterdam kommen, habe sich obligiret, einem vor jeden errorem, den er ihm in Cartesii Schrifften weisen würde, 5 Ducaten zugeben," Warsaw University Library, Cod. IV oct. 49, 592. Transcription kindly provided by Martin Mulsow. The anectode is reported also in Stolle's *Introductio in historiam litteraria*: Stolle 1728, 676 (cited also in Koecher 1738, 121). De Volder did not read, however, De Raey's *Cogitata de interpretatione* (1692), but committed himself to the quality of the book: "[d]ieser Raë habe einen tractat de Interpretatione geschrieben, den ihm (Voldern) einst ein gewisser Rathherr in Amsterdam tefflich gelobt. Als er (Volder) ihn nun gebethen, er möchte doch so gut sein und ihm die excellentia loca mit der feder zeichnen, weil er nicht wohl Zeit habe, das Buch zu durchlesen, so habe er auf den Tisch geschlagen und geschrien: omnia esse medullam omnia excellentissima," Cod. IV oct. 49, 592–593. Transcription kindly provided by Martin Mulsow.

204 See Ruestow 1996; Jorink 2010, chapter 5; Cunningham 2010, chapter 4; Bertoloni Meli 2017.

van Velthuysen, and which was modelled upon the Académie of Thévenot.[205] Moreover, as anticipated, it provides insights on his activities as director of the Leiden academic observatory, which he was going to direct from 1682.

2.3.2 The Assumption of the Chair of Mathematics and the Directorship of the Observatory

With the death of Christiaan Melder, professor of mathematics at Leiden, on 25 April 1682 De Volder was appointed as his successor, with a salary of 300 guilders (raised to 400 on 12 May), and with teaching duties on Wednesdays and Saturdays.[206] For the position, he gave on 15 June 1682 his inaugural *Oratio de coniungendis philosophicis et mathematicis disciplinis, cum philosophicae professioni adiunctam mathematicam rite auspicaretur*. According to Le Clerc, it was the University Curator Hieronymus van Beverningh who asked De Volder – who was in fact already giving private lectures on algebra at least since 1677,[207] and who had already provided mathematical advice for the treatise in ship-building *Aeloude en heden-*

---

[205] As Eric Jorink has put it, "[t]he Utrecht Cartesians were united in the *Collegie der Scavanten*, a learned society not unlike Thévenot's *Academie*. (It should be noted that Graevius and Thévenot maintained a lively correspondence, and often met in person). At weekly meetings new books, opinions and discoveries were discussed, corpses were dissected, and the members made fun of scholastic philosophers and their views on the order of nature. In true Cartesian fashion, a mechanical world-view was propagated and, what is equally important, war was declared on the belief in magic, spirits and hidden meanings in nature, as for example Van Velthuysen's much reprinted *Tractaet van de Afgoderye en Superstitie* ('Treatise on Idolatry and Superstition', 1669) is testimony. Graevius' correspondence reveals the highly significant fact that Jan Swammerdam was also a frequent visitor at the meetings held in Utrecht. This not only supports my idea that Swammerdam was a much more sociable scientist than is usually assumed, it also puts him firmly in an intellectual movement. The *Collegie* was abhorred by orthodox Calvinists, and repeatedly attacked by them. Swammerdam can therefore without any reservation be counted among Holland's leading rationalists. Not only did he belong to the inner circle of prudent Cartesians such as Hudde and De Raey, as his letters to Thévenot reveal, he was also on friendly terms with more daring intellectuals such as Graevius, Van Velthuysen and Wolzogen, who all had outspoken ideas on 'superstition' and the place of rationalism in science and scholarship," Jorink 2003, 95–96. See also Hartog 1876; Gootjes 2019.

[206] Molhuysen 1913–1924, volume 4, 13. On 19 June 1682, it was conceded to him to teach at the *hora undecima*: Molhuysen 1913–1924, volume 4, 4.

[207] In a letter dated 15 October 1677 of Albert Jansz. van Dam, nephew of Dirck Rembrantsz. van Nierop, to Klaas Dirksz. Zalm, at that time student at Leiden, Van Dam asks his friend the "order" (we may read this as a teaching programme or task) used by De Volder in teaching algebra: "[i]k [...] verzoek ook van UE dat gy my die order eens overzend die myn Hr. de Volder gebruykt om iemant de Algebra te leeren, of met wat order gy 't zelve by hem geleert hebt," Rijks 2012, 248. According to Gronovius, this appointment was the outcome of the excellent results of the teaching of De Volder in natural philosophy and metaphysics: "[h]aec et multa ex illis quaere censui, agnoscitis varia esse, quae tunc demum explicare incepit, postquam Illustres Curatores et Consules videntes Philosophiam suam splendidissime collocatam, sane et mathesin illi sociandam esse censuerunt anno praeteriti saeculi secundo et octogesimo," Gronovius 1709, 25–26.

daegsche scheeps-bouw en bestier (1671) of Nicolaas Witsen[208] – to suggest a candidate to him. De Volder mentioned someone (whose name cannot be found), but eventually the chair was offered to him.[209] A decision which seems to have pleased De Volder a lot, as it appears from his letter to Thévenot of 17 August 1682.[210] Together with the teaching of mathematics, De Volder was also put in charge of the direction of the astronomical observatory, at that time located in a small wooden tower on the roof of the main building of the University (see Fig. 2.6, Hagen 1670), representing the main University building and the academic garden around 1670).

He also taught astronomy itself: according to the *series lectionum* of the winter and summer terms 1694, besides giving experimental lectures on Monday De Volder was teaching astronomy on Tuesday and Wednesday, metaphysics on Thursday and Friday (always at the *hora undecima* in the morning).[211] In the following years up to his retirement in 1705, he continued to give experimental lectures on Mondays, while on the other days he taught physics, metaphysics, as well as mixed disciplines like geography and cosmology for the rest of the week.[212]

His activities as director of the observatory and as professor of mathematics were no less important than those at the experimental theatre. As it has been reconstructed by Huib J. Zuidervaart, when his predecessor, Christiaan Melder, died, his widow refused to leave to the University his collections of instruments. The observatory, at that time, certainly hosted the famous azimuthal quadrant of Willebrord Snel, built around 1610.[213] After having been requested to check what was needed for the observatory, De Volder ordered the Leiden instrument maker Antonius Hoevenaer – who was going to enrol at Leiden on 25 March 1683 as his *amanuensis*[214] – some

[208] Witsen 1671, *Appendix*, 29. Discussed in Rank 1984, chapter 8. See also Jorink 2018.

[209] Le Clerc 1709, 376–378.

[210] "Je vous suis fort obligé, Monsieur, de vostre bonté que vous temoignés en me felicitant sur ma nouvelle profession. Il a {pleu} a nos Messr de m'en {pouvoir}, et quoy que j'eusse {desia} assez de affaires je l'ay accepteé, parce que j'ay toujours eu un grand penchant pour les mathematiques," Cod. Ms. Hist. nat. 102, fasc. 25, unnumbered.

[211] Molhuysen 1913–1924, volume 4, 45*.

[212] In the autumn and winter term of 1696 and in the summer term of 1697, he taught geography on Tuesday and Wednesday, physics on Thursday and Friday (besides providing his experimental lectures on Monday, and always at the *hora undecima* in the morning). In the summer and winter terms of 1698 he switched from geography to cosmology (the two fields being anyway somewhat akin). In the autumn and winter term of the same year, he lectured – besides providing experiments – on 'meteors' on all the other days of the week – apparently dropping the teaching of metaphysics. In the winter and summer terms of 1699, 1700, 1703, 1704 he taught, besides experiments on Mondays, metaphysics, physics and mathematics on the other days of the week. See Molhuysen 1913–1924, volume 4, 56*–58*, 65*–67*, 71*, 80*, 93*, 99* and 101*. A summary of De Volder's fields of lecturing is given in Gronovius 1709, 24, adding mechanics, hydrostatics and optics to the list of these topics. Gronovius also adds a summary of De Volder's metaphysics, which I discuss in Sect. 3.2.1.2, *Clarity, distinctness and indubitability of simple and complex ideas*.

[213] On the Leiden observatory, see De Sitter 1933; Zuidervaart 2007, 2012a, b; Wiesenfeldt 2008; Baneke 2010; Hooijmaijers and Maas 2013.

[214] See *supra*, n. 131.

Fig. 2.6 The Leiden University building and the Leiden botanical garden around 1670. Only one turret is present at this date on the roof of the University building. (Source: Hagen 1670. Erfgoed Leiden en omstreken, signature: PV12512, https://www.erfgoedleiden.nl/collecties/beeldmateriaal/zoeken-in-beeldmateriaal/detail/aa4cfdb6-26bc-11e3-b6dc-3cd92befe4f8/media/230f31fd-707d-f78e-3a9e-709a65c67323. Public domain mark)

modifications to the quadrant already in possession of the observatory, acquired a mast and a stand for installing large telescopes, and acquired a helioscope, according to the resolutions of the Curators of 1682.[215] De Volder's other activities as a director can in fact be reconstructed from his correspondence with Thévenot. In his letter dated 15 October 1680 (but which can probably be traced to 15 October 1682) De Volder mentions that a medal for "Monsr Borelli" – certainly the the chemist and instrument builder Jacques Borelly (1623–1689), member of the Académie Royale[216] – had been

[215] "[...] daarop de groote verrekijkers soude moghen rusten," "[...] een pool hoog 40 a 50 voeten met zijnen blikken passer en houte goot voor 't gebruk van 't groote telescopium," "[...] instrument am des sonsvlackte te bestuderen" Zuidervaart 2007, 162, 164 and 168; see also 12 and 24; Zuidervaart refers to the manuscript, preserved at the Leiden University Library, AC I-7, 186, reporting the Curators's resolutions of 8 August 1682.

[216] On him see Chabbert 1970.

made and delivered to him by a certain Mr. Clignet through Mr. Merault (a merchant resident in Paris).[217] Most probably, the medal was granted to him as a sign of appreciation for his qualities as a lens maker, even if his relations with De Volder and the University seem not to have been entirely positive. In the letter of 31 April 1682, for instance, he writes that Borelly "did not accept so far[,] after six months[,] the remuneration established for him, even if in this matter there is no fault by me," and refers to some negligence by Borelly – even if he then positively refers to a "second glass" by which Borelly was going to gladden him.[218] It might be, thus, that De Volder had already acquired an instrument for the University from Borelly in Paris in 1681. In a posterior letter (17 August 1682), De Volder acknowledged to have received from a certain Mr. Vander Burgh "the glass of Mr. Borelly,"[219] while in the letter of 19 November 1682 he communicates that the University authorities were satisfied with Borelly's offer for a further lens, with a longer reach, even if they were doubtful whether this lens could be used in the observatory, in which there was no space enough to host a telescope carrying such a lens:

> our academy is very much obliged to him for the offer he makes of another [lens] of a longer range. Even if it is doubted a lot whether one can make use [of it] in the place aimed at for it, and whether it would not be better for us, especially in this beginning of activities, to have one more of 18 feet, and for convenience one of 7 or 8. But it is not for us to prescribe him how to direct us.[220]

Thus, Borelly built for the Leiden observatory a lens of 18 feet, plus a further lens for De Volder himself (of 12 feet),[221] and offered to the University a bigger lens. Indeed, in the second undated letter of the correspondence – most probably dating back between 19 January and 31 April 1682[222] – De Volder reports how one of the Curators had expressed an interest in the building of a place in which a telescope (*lunette*) of 50 or 60 feet could be hosted; this, however, seems not to have pleased De Volder, who preferred to furnish the observatory (which already had one of 25

---

[217] See *supra*, n. 194.

[218] "Pudore profecto suffundor cum de Borellio cogito, qui hactenus jam ante semestre ipsi decretum redhostimentum nondum accepit, quanquam ea in re nulla mea culpa est. Bene equidem habet, quod de negligentia eius cui ea res commissa est, ad te scripsi. Neque enim licet sexcenties {commoruerim} hominem impellere hactenus potui, hae in re ut officium praestaret {sinue}, sive id {sollerui} ejus {ereriat} incuria sive peculiari quadam, cujus aliunde indicia habeo sat evidentia, in me malevolentia. \{Borellius}/ equidem est quod gratias habeam quod altero me beare voluerit vitro," Cod. Ms. Hist. nat. 102, fasc. 25, 49r–v.

[219] "J'ay receu de Monsr. Vander Burgh le verre de Monsr. Borelli, dont je suis fort obligé et a vous et a luy. [J]e differeray neantmoins d'ecrire a Monsr. Borelli, pour le remercier de sa bonté jusques a ce, qu'on envoyera la medaille de l'academie, qui est enfin achevée," Cod. Ms. Hist. nat. 102, fasc. 25, unnumbered.

[220] "[N]ostre Academie est a luy fort obligé pour l'offerte qu'il fait d'un autre d'une plus longue portée. Quoy que se doûte fort si on si en pourra servir dans le lieu a cela destiné, et si il ne scauroit pas plus propre pour noûs, principalement dans ce commencement des affaires, d'en avoir encor un des 18 pieds, et pour la commodité un de 7 ou 8. Mais ce n'est pas a nous de luy perscrire comment nous regoler," Cod. Ms. Hist. nat. 102, fasc. 25, 61r.

[221] See *infra*, n. 224.

[222] See *supra*, n. 199.

feet – and whose provenance is unknown)[223] with smaller objects, and who seemed to show appreciation of the lens makers in Delft.[224] Furthermore, De Volder attempted to aquire from Auzout a micrometer (as mentioned above) for the observatory. It was Thévenot who supervised the order and the construction – an "enterprise [...] really worthy of the King of France." However, it seems that the instrument never reached Leiden, even after repeated requests from the De Volder, noting in the letter of 29 October 1682 that he would be happy "if you send it to me at the first occasion, while our [Curators] [...] prepare [...] the small tower for the observers, which up to now has lain forgotten." In fact, in this and in other letters De Volder repeatedly asked Thévenot about the instrument,[225] which is however not present in De Volder's inventory of the observatory (7 February 1706):

---

[223] Zuidervaart mentions, in his catalogue, a telescope of 25 feet with a tube made of iron plate on a wooden stand, mentioned in a catalogue of 1742: Zuidervaart 2007, 48 and 64. Zuidervaart also mentions two telescopes, of 16 and 22 feet, which could have entered the observatory in 1682: Zuidervaart 2007, 164.

[224] "[J]'ai communiqué a un des principaux de nos Curateurs de l'Academie <d>l'intention de Monsr. Borelli qui lui estoit fort agreable. Mais sa responce {me} fait changer de sentiment. Car il t<o>emoignoit <f>d'avoir asses de curiosité po<{r}>ur preparer un lieu pour se pouvoir servir d'une lunette de 50 ou 60 pieds. Pourquoy je vous prie, fines ad huc est in integro, de faire tant auprez monsr. Borelli, qu'il demeure dans son premier sentiment. Asseurement Mons{r.} Borelli m'accablé de ses bontés en m'offrant encor{e} une pour mon usage. et I'a<j>vois dessein de luy en temoigner ma reconnoissance, mais je suis si tard retourné de la Haye, que l<a>e temps me presse trop pour le pouvoir faire a cette occasion. Cependant je vous diray brievement, qu'il me semble, que de luy prescrire la longueur de la lunette qu'il me fait {bstrir}, s<{otd}>eroit d'assez mauvaise grace. I<s>l se pourra asseurer, que tout ce que v<c>iendra de luy me sera fort agréable. Mais si j'ose vous dire cela en confidence, puis qu'il a eu la bonté de me regaler d'un verre de 12 pieds et que celle de l'Academie de 2{5} pieds est aussy a mon usage, on peut a[missing text] voir qu'une longueur \entre ces deus/ moyenne {<luij>} seroit foi[missing text] Mais je vous prie derechef d'en rien dire a [missing text] peut estre, qu'il temoigne luy mesme d'[missing text] dessein. On a icy quelques messrs a Delft [missing text] {d}e pouvoir faire des verres d'une figur[missing text]," Cod. Ms. Hist. nat. 102, fasc. 25, 68v.

[225] Cf. the letter of 17 August 1682: "[q]uant au micrometre dont vous parlez dans la lettre de Monsr. vander Burgh, je vous prie tres humblement de me le faire avoir le plus prontement que se pourra, car vous voyez bien que j'en ay maintenant plus de besoin que jamais," Cod. Ms. Hist. nat. 102, fasc. 25, unnumbered; and the letter of 29 October 1682; "[e]st profecto quod tibi summas agam gratias pro ea cura qua confectioni micrometri invigilasti. Gaudebo equidem, si, dum nostri suam iam turriculam quae hactenus neglecta iacuit, serio ad observationes praeparant, id prima quavis occasione ad me transmittas," Cod. Ms. Hist. nat. 102, fasc. 25, 60r; the letter of 19 November 1682: "[p]uisque Mr Auzout ne viendra pas si tost dans ce pais, je vous prie de me envoyer le micrometre par la premiere occasion," Cod. Ms. Hist. nat. 102, fasc. 25, 61r; the letter of 28 January 1683: "[missing text] vous prie Monsr. de vous [missing text]enir du micrometre," Cod. Ms. Hist. nat. 102, fasc. 25, 63r, and the letter of 27 August 1683, in which De Volder writes to Thévenot that a certain Mr. van Heul, namely an acquaintance of De Volder who was carrying the letter for Thevenot – as well as a letter for Johan de Witt Jr. – would probably visit Thévenot to take it: "[i]e me console en partie du ritardement du micrometre, parce que cel est arrivé a cause d'une enterprise que est vrayement digne du Roy de France. Mais j'espere que il sera enfin achevé. Monsr. Van Heul, qui vous donnera cette lettre, est particulièrement des mes amis. Ie l'ay prié de se charger encore une autre a Monsr. de Witt le fils de feu Monsr. Le Pensionaire d'Hollande, qui s'en est allé il y a quelque temps a Paris, pour s'en retourner, dans le mois de Septembre. Si le micrometre est achevé je ne doute pas que il ne {sien} charger volontiers, et qu'il ne prenne la

An azimuthal quadrant
A sextant
A quadrant
Two clocks measuring seconds
A metal mirror
A pair of compasses
An armillary sphere
Two old globes
An objective lens of 50 feet
Another one of 18 feet
A little copper quadrant
A telescope of 12 feet
Various other eyeglasses, a few telescopes etc.[226]

The inventory includes a further lens of 50 feet, which was built in 1688 by the optician and natural philosopher Nicolaas Hartsoeker (1656–1725), in 1675 a student of De Volder, and then active in Paris.[227] His lens was praised by De Volder in his correspondence with Huygens and Leibniz,[228] and was personally shown by De Volder to Huygens at his house in March 1689, from which, however, they could only observe the other side of the canal.[229] In turn, the telescope of 12 feet was built

peine de vous venir trouver pour cet effet. Ie vous suis infiniment obligé du soin qu'il vous a {pleu} de prendre en la construction de cette machine, et de la generosité dont {vous vais servez envoie moy}. Mais je vous prie encore une fois de vous en servir un peu moins en cette occasion, et pour ne me {combler} tant a fait de {...}, et parce que la machine appartient a nostre université," Cod. Ms. Hist. nat. 102, fasc. 25, 64r–v. Johan de Witt Jr. owned some of De Volder's *dictata*: see *supra*, n. 70.

[226] "1706 Febr. 7. Inventaris van het Observatorium. Een quadrans azimuthalis. Een sextans. Een quadrans. Twee horologien wysende seconden. Een metael spiegel. Een spangpasser. Een armilla. Twee oude globi. Een objectif glas van 50 voet. Nogh een van 18 voet. Een kleyne kopere quadrant. Een kycker van 12 voet. Nevens verscheyde oogglasen, minder kyckers etc.," Molhuysen 1913–1924, volume 4, 107*–108*.

[227] On him, see Hill 1985; Dijksterhuis 2007; Abou-Nemeh 2012.

[228] See the letter of De Volder to Huygens of 26 April 1689 (which is the first extant from their correspondence): "[i]ck sal oock, soo UEd. geen volkomen genoegen had in de proef op de solder, maar liever het glas aan de maan selfs probeerde, niet manqueren op het aldereerst advys, UEdt. de blickke ring met de steert toe te stueren sij is wel heel ruw gemaackt, maar sal voor een proefje meen ik, genoech verstrecken," Huygens 1888–1950, volume 9, 315 (letter 2537). Cf. the letter by Leibniz to De Volder of 6 July 1701 (GP II, 224–228), the letter by De Volder to Leibniz of 7 October 1701 (GP II, 228–232), the letter by Leibniz to De Volder of 27 December 1701 (GP II, 232–235), the letter by Bernoulli to Leibniz of 14 January 1702 (A III9, 6–9).

[229] See the letter of Christiaan Huygens to his brother Constantijn of 22 March 1689: "[j]e fus voir a Leyden le professeur de Volder qui me montra un verre de 50 pieds que Hartsoecker a fait pour l'Academie, selon l'inscription qu'il a mise dessus. Il avoit un bord de maroquin rouge, avec quelque dorure de la largeur de 2 doigts. Il est de matiere tres belle a ce que j'ay pu voir et bien espaisse, mais quant a la bontè je n'en suis pas bien certain, n'y aiant pas de place chez de Volder pour l'essaier si non une allée de sa maison, ou je m'en enloignay tant que je vis les pierres des maisons de l'autre costè du canal renversées, mais seulement avec les yeux sans oculaire, parce que pour cela il auroit falu de l'obscuritè tout au tour," Huygens 1888–1950, volume 9, 313 (letter 2534). On the lens, see Zuidervaart 2007, 52, 166. See also Crommelin 1929.

by Huygens in 1683, and probably entered the observatory in 1690,[230] but it might even be that it is the lens provided by Borelly to De Volder, mentioned above. Moreover, in 1685 De Volder purchased a big, brass Tychonic sextant allowing the measurement of the angular distance between the stars, which he had built by a craftsman in Amsterdam, Coenraet Metz, and for which De Volder had asked, according to the Resolutions of the Curators of 8 November 1685, 1,200 guilders, which eventually cost, according to the University budget of the same year, 1,050 guilders. In 1689, the installing of this instrument finally required the construction of a second tower, which is depicted in engravings – together with a support for installing a telescope, viz. one of the tools acquired by De Volder in 1682 – appeared on the front page of Boerhaave's *Index plantarum* (1710), as well as in Pieter van der Aa's *Les delices de Leide* (1712) (see Fig. 2.7, Van der Aa 1712, plate 5), and are duly described and depicted by Uffenbach, together with a pedestal aimed at installing telescopes (see Figs. 2.8, Uffenbach 1753–1754, volume 3, plate 14, and 2.9, Uffenbach 1753–1754, volume 3, plate 16).[231]

---

[230] Zuidervaart 2007, 164. As indicated by Zuidervaart, the "inventories of the observatory [...] suggest that at the turn of the century there were at least eight telescopes present. These inventories also reveal the habit of storing the large telescopes in pieces: the long tubes were stored in the upper attic of the former convent and the lenses were kept in a cupboard elsewhere. So apparently a large telescope was assembled on the spot, each time an observation was made," Zuidervaart 2007, 12–13.

[231] "Den 14. Jan. waren wir erstlich bey unserm Kauffmann Herrn Brunk. Hernach besahen wir die sogenannte Academie oder Universitäts-Collegium, und die darinn befindliche Auditoria. [...] Das vornehmste und beste ist wohl das Observatorium, oben auf dem Collegio, als welches sehr geräumig und gut ist. Man kan sich sehr wohl auf der Gallerie über die Stadt umsehen. Auf bey den Seiten sind zwey Häusgen, in deren einem ist ein besonders schöner und grosser Sextans, welchen der Mathematicus Mez in Amsterdam vor hundert und fünfzehen [*sic*] Gulden gemacht. Er siehet sehr wohl aus, und soll auch gar accurat seyn. In dem andern Häusgen gegen über ist noch ein Quadrant, welches derjenige seyn wird, den Vossius de scientiis mathematicis C. XXXVI. §.47. sq. p.199. sq. so sehr rühmet, und welchen der berühmte Bleau gemacht haben soll, wiewohl wir nirgends keinen Namen darauf finden können, ausser daß auf dem Fuß auf eine Platte von Messing gestochen war: Antonius Hoevenaer fecit Leidae. Wir fragten also, ob etwan noch ein Quadrant hie wäre, allein der Pedell, so uns herum führte, wolte nichts davon wissen. Es scheinet also, daß ihn dieser Mann verfertiget, der berühmte Bleau aber ihn angegeben, und die Direction dabey gehabt, daher Vossius ihme selbigen zuschreibt. Es soll aber dieser Hoevenaer ein guter Uhren macherge wesen seyn. Das Dach von beyden Häusgen war artig gemacht, daß man es ringsherum drehen könnte, wie der Abriß Fig. XIV. und XV. zeiget. Dieser Riß stellet nemlich das in Leiden auf dem Collegio erbaute Observatorium vor, welches sehr gut und bequem ausgesonnen ist. Es bestehet nemlich aus einem oben auf dem Dach stehenden runden Thürngen, dessen gleichfalls rundes und spizig zugehendes Dach sich herum drehen lässet, gleichwie man an den Schiffkranen sehen kan. Diese Bewegung war vermittelst eines Triebes und eines einwärts zahnigten Rades gar leicht ins Werk zu richten, denn wie man in Fig. XIV. und noch besser in Fig XV. siehet, so hat das Dach unten herum einwärts einen runden eisernen Cirkel mit starken Zähnen, (aa) und unter demselben sind in einer Höhlung viele kleine aber doch dicke und starke Walzen oder Rollen bevestiget, darauf sich das Dach desto leichter und bequemer rücken lässet. Diese Rollen (b) sind auswendig mit einem überhängenden Brett bedeckt, damit kein Regen noch Tag dazwischen in das Häusgen fallen kan. An einer Seite des Thürngens ist oben ein starkes Eisen, (c) wie ein Galgen formirt, vest, mit welchem, wie auch mit den andern kleinen Banden (dd) der Trieb (ee) bevestiget ist. Dieser Trieb (e) greiffet oben in die Zähne des Cirkels (aa) an dem Dach, und wenn er vermittelst des andern untersten kleinen Triebs (f) herum gedrehet wird, so beweget er allgemach das ganze

Fig. 2.7 The Leiden University building and the Leiden botanical garden around 1712. Two turrets and a pedestal for telescopes are present at this date on the roof of the University building. (Source: Van der Aa 1712, plate 5. Leiden, Museum Boerhaave, object number: RP-P-AO-10-21A-3, http://hdl.handle.net/10934/RM0001.COLLECT.393538. Public domain mark 1.0)

Dach oben herum, so daß das Fenster, (gg) wohin man es haben will, sich stellet. Fig. b. ist der grosse Trieb (ee) apart, fig. c. aber der kleinere, so in Fig. XVI. mit f. bezeichnet ist, und fig. d. ist die eiserne Hand habe, so unten den kleinen Trieb f. hält, und darinn er sich beweget. In einem andern Zimmer stunden noch einige Instrumente, nebst etlichen alten Tubis, wie z. E. aus dem Abriß Fig. XVI. zu ersehen. Unter andern kostbaren und grossen Instrumenten haben wir auch diesen kleinen Fuß gefunden, darauf man die Tubos mit dem hintersten Theil leget, und unvermerkt hin und her bewegen kan. Es hat derselbe unten ein Creuz, und einen dicken Stock darauf, in welchem ein zahnigtes Holz gleich in einer Winde stecket, und das man auch auf gleiche Art mit einem kleinen Trieb und Handhabe (aa) in die Höhe und hinunter stellen kan. Auf diesem Holze war ein von eisernen Linealen gemachter Triangel, (b) welcher an der obersten Seite zwey runde

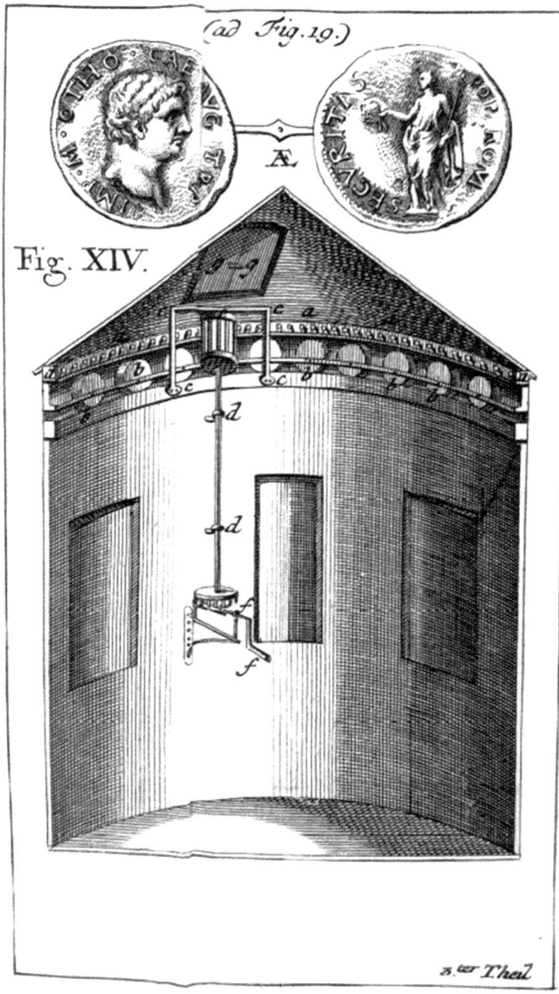

Fig. 2.8 The internal structure of the turrets of the Leiden astronomical observatory. (Source: Uffenbach 1753–1754, volume 3, plate 14. Zentralbibliothek Zürich, signature: NR 619, https://www.e-rara.ch/zuz/doi/10.3931/e-rara-55535. Public domain mark 1.0)

aufrecht stehende Blättgen (cc) hatte, darinnen sich eine lange Schraube (e) vermittelst der Handhabe (d) drehete. An dieser Schraube (e) war eine kleine eiserne Maschine, (f) wie sie hier zu sehen, auf geschraubet. Diese bestunde aus zwey Blättern, (g) die mit einem halben eisernen Bogen oder Cirkel (h) zusammen vest gemacht waren. Auf diesen halben Bogen (h) legte man den Tubum, und band ihn mit den zweyen daran hangenden Bändern vest. Wenn man den hernach die Handhabe (d) umdrehete, so schraubete sich die Maschine (f) mit dem Tubo entweder hinauf oder herunter, und das mit einem so gleichgehenden und sanften Gang, daß man das Object immer vor dem Tubo behalten, und wenn sich selbiges beweget, mit dem Tubo auch nachrücken könnte," Uffenbach 1753–1754, volume 3, 395–398.

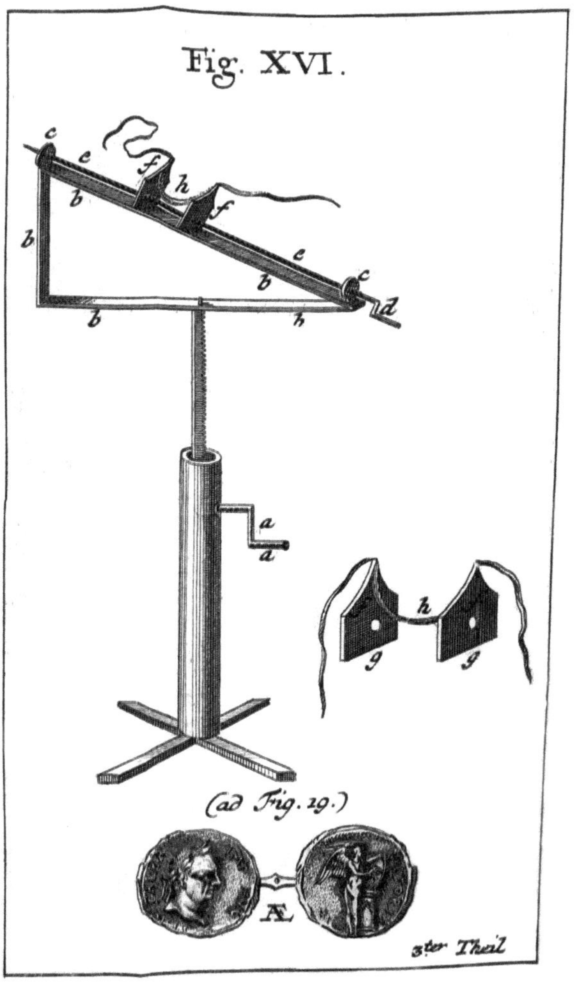

Fig. 2.9 A pedestal to install telescopes present at the Leiden astronomical observatory. (Source: Uffenbach 1753–1754, volume 3, plate 16. Zentralbibliothek Zürich, signature: NR 619, https://www.e-rara.ch/zuz/doi/10.3931/e-rara-55535. Public domain mark 1.0)

Moreover, in 1706 De Volder ordered from Metz a brass quadrant with a radius of 3 feet, later acquired and modified by 's Gravesande.[232]

[232] See the aforementioned entries given *supra*, n. 213; cf. Molhuysen 1913–1924, volume 4, 38–39 and 107*–108*; Rosenboom 1697, part 1, 120; part 2, 40 and 54. For the renewal of the observatory, according to the academic budget 1,800 guilders were spent at the beginning of 1688, and 2,031 guilders were spent in 1690: see Rosenboom 1697, part 1, 105; part 2, 40 and 52. In particular, 1,796 guilders were paid to Jan Bastiannsz. Lopik for carpentry work on the observatory towers, and 235 were paid to Adriaan Kramer for iron work in the towers: Rosenboom 1697, part 2, 52. Hieronymus Meyer took care of the cleaning of the observatory, while Metz took care of the maintenance of the instruments: see Molhuysen 1913–1924, volume 4, 225–226; Rosenboom 1697, part 2, 108. With De Volder's permission, the observatory was used also by Lotharius

De Volder also had a personal collection of astronomical (and geographical) instruments. According to the *Bibliotheca Volderina*, these included:

Two curious astrolabes
1 copper sundial
1 telescope for the observation of the Moon
1 small binocular [or telescope]
1 pedestal to install a big telescope for the observation of the Moon
1 large wooden quadrant with copper fittings standing on a pedestal.
The *Astrolabium catholicum* of Voogt [in fact, the book by Nicolaus Jansz. Voogt, 1680]
1 perpetual lunar calendar in a black frame
1 big sphere
1 smaller [of the] same
1 surveryor's chain
1 terrestrial and celestial globe
1 tape [cloth] measure to measure a flat globe or a map.[233]

Moreover, several atlases and maps are listed in the *Bibliotheca*.[234]

2.3.3 Mathematician and Scientific Expert

All this testifies to De Volder's interests in astronomy, geography and of course mathematics. These had their outcome, not only in De Volder's activities as a teacher and director of the observatory at Leiden, but also in his being a scientific expert for the VOC, on whose board of directors sat Hudde himself, and for the States of Holland. In 1689 and 1693, indeed, De Volder acted for the VOC as reviewer of the proposal of Christiaan Huygens on a method for calculating longitude at sea with pendulum clocks on board ships (trials which were actually carried out by Huygens). As I am going to show in more detail in Sect. 6.2.2, *Newton's critique of Descartes's vortex theory and its reception by Huygens and De Volder*, De Volder did not

Zumbach de Koesfeld, student at Leiden from 1688 to 1693, and private teacher of mathematics and astronomy. He left Leiden in 1708, after having been for a short time director of the observatory: see Molhuysen 1913–1924, volume 4, 222, 225–226 and 234. On him, see Zuidervaart 2007. As professor of mathematics, on 5 June 1698 De Volder allowed a certain Stavinskii to give private lectures, in Leiden, on military and civil architecture, and on mechanics and perspective: see Molhuysen 1913–1924, volume 4, 162.

233 "Twee Curieuse Astrolabia [...]. 1 Kopere Sonnenwijser. 1 Maankijker. 1 Kleyne Verrekijker. 1 Pedestaal om een groote maankijker op te legge [...]. 1 Groote houte Quadrant met koper beslag staande op een voet. Astrolabium Catholicum, van Voogt. 1 Eeuwigdurende Maanwijser in een swarte lijst. 1 Groote Sphera. 1 Dito kleynder. [...] 1 Landmeeters ketting. 1 Globus Celestis en Terrestris. 1 Doekmeter om een platte Globus of Landkaart te meten," *Bibliotheca Volderina*, 95–96.

234 *Bibliotheca Volderina*, 24–25.

ultimately accept Huygens's conclusions, which were based on a Cartesian idea of gravity, and served him in attacking Newton's idea of universal gravitation. De Volder's judgment on Huygens's trials, as I will argue, probably undermined his own Cartesianism. Also, De Volder acted as expert for the States of Holland in a further case in which a method for calculating longitude at sea was proposed for patenting (which I also consider in more detail in Sect. 6.2.2, *Newton's critique of Descartes's vortex theory and its reception by Huygens and De Volder*). In spring 1689 he was involved as an examiner, together with Huygens, of the method proposed by Lieuwe Willemsz. Graaf, which revealed itself to be completely unsound. In 1697, eventually, De Volder rejected – with a written report which has survived – a method proposed by Antonius de Saint Julien-Pottier to the VOC, based on the observation of the position of the Moon from ships.

Moreover, De Volder was renowned in the Netherlands as a 'pure' mathematician, and as one of the very first to study Newton's use of mathematics in physics. As reported by Le Clerc and Gronovius, De Volder applied himself to the study of integral and differential calculus, the mathematics of the ancients (especially Euclid and Diophantus), as well as that of the analytical geometry of the moderns: viz. that of Descartes and of Cartesian mathematicians such as Hudde, Johan de Witt, Hendrik van Heuraet, Florimond de Beaune, René François Walther de Sluze (Slusius),[235] and Huygens himself. Moreover – as stressed by Gronovius – De Volder was one of the first to study the infinitesimal calculus, developed by Isaac Barrow, Newton and Leibniz,[236] also explaining it to Huygens.[237] As a 'pure' mathematician,

[235] Frans van Schooten included tracts and letters by Hudde, De Witt, Van Heuraet and De Beaune in his 1659 Latin edition of Descartes's *Géométrie*. On them, and on the reception of Descartes's geometry, see Boutroux 1914; Grootendorst and Van Maanen 1982; Schoneveld 1983, chapter 2; Brigaglia 1995; Van Berkel et al. 1999, part 1.

[236] "Mathesin didicerat primo vulgarem, quam Syntheticam vocant; paulo post excoluit subtiliorem, quae Analytica appellatur, qua veteres usi ad optima inventa; sed quam caute texerunt, exceptis Euclide et Diophanto, qui aliquas edidere scintillas. Hic statim immensum profecit; Cartesii, Huddii, Heuratii, deBeaunii, Wittii, Slusii, Leibnitzii, Hugenii, Newtoni inventa excussit, addidicit. Ita ut tandem Logisticam infinite parvorum sciverit inter paucos optime. Illa mirifica humanae sciendi in Mathematicis facultatis meta: quam Barrovius, Leibnitzius, Newtonus, Bernouilleus ita in sui seculi celebritatem excoluere," Gronovius 1709, 26–27. De Volder was praised as skilled in mathematical analysis by the German astronomer Johann Gabriel Doppelmayr in his letter to Leibniz from Leiden of April 1701: "[k]ein pars in der Mathesi specialiori gehet hier mehr in Schwang, als die fortification; was die Mathesin universalem anlanget, so ist die Analysis allhier noch in sehr gutem Stand, und haben wir hier auch ziemlich viel gute Analysten, des Hl de Volders Collegia Mathematica sind meistens von der Analysi," Gottfried Wilhelm Leibniz Bibliothek Hannover, ms. LBr 971, 60r, transcription from *Briefwechsel Johann Gabriel Doppelmayr*, https://www.astronomie-nuernberg.de/index.php?category=doppelmayr&page=1701-04-00-Allgoewer (accessed 13 January 2019). As an evidence of De Volder's teachings, the auction catalogue (1739) of the private library of Boerhaave, reveals that in 1689 Boheraave wrote some *Problemata arithmetica et geometrica sub auspicio B. de Volder methodo Cartesii soluta* (now lost): *Bibliotheca Boerhaaviana*, 27.

[237] "Pendant que Mr. de Volder s'appliquoit à étudier & à enseigner les Mathematiques, de tréshabiles gens découvrirent ces nouvelles Méthodes aux quelles on a donné les noms de Calcul differentiel & de Calcul integral, & faisoient de tems en tems paroître quelque chose en public, qu'ils

indeed, De Volder was in contact with Ehrenfried Walther von Tschirnhaus, whom he had met in Leiden in September 1682.[238] In 1687, De Volder helped him in solving the problem of the construction of the tangent of a curve, with the use of three fixed points within the curve. Tschirnhaus had proposed a method in his *Medicina mentis* (1687), then criticized by Nicolas Fatio de Duillier in the 1687 April issue of the *Bibliothèque universelle et historique*, with his *Reflexions sur une méthode de trouver les tangentes de certaines lignes courbes, laquelle vient d'être publiée dans un livre intitulé Medicina mentis*. Tschirnhaus, aiming at replying, sent an *Apologia* to Pieter van Gent, in order for it to be published in the *Bibliothèque*. In turn, Van Gent forwarded this reply to Huygens, to ask his advice, and to De Volder[239] (who, as he was to report to Stolle and Hallmann in 1703, was decidedly not impressed by Tschirnhaus's *Medicina mentis*).[240] Eventually, Van Gent reported the judgment he had received from De Volder by copying it in his letter to Huygens of 26 July 1687, pointing out Tschirnhaus's errors.[241] As reported in the letter of Van Gent to Huygens of 11 September 1687, De Volder then eagerly asked Van Gent whether he had received any reply from Tschirnhaus,[242] which only came with a letter dated 23 August, in which Tschirnhaus acknowledged his errors and thanked De Volder.[243] More importantly, De Volder was a member of what Rienk Vermij has

---

avoient découvert, par ces Méthodes. Nôtre Philosophe s'applica aussi à ces nouvelles manieres, & dès que le livre de l'Illustre Mr. Newton, des Principes Mathematiques de la Philosophie Naturelle, eut paru en MDCLXXXVII. Il s'attacha fortement à cette lecture, pour découvrir les principes sur les quels l'Auteur s'étoit fondé; & l'on a encore trouvé parmi ses papiers les calculs de ses démonstrations. On sait que ce livre suppose une connoissance des Mathematiques, qui n'est pas commune. Je me souviens même d'avoir ouï dire à Mr. de Volder, que peu de tems après qu'il parut, feu Mr. Huygens, qui étoit un grand Mathematicien, mais à qui les nouvelles méthodes, dont j'ai parlé, étoient inconnuës, le vint voir à Leide & le mit sur le sujet du livre de Mr. Newton. Il avoüa à Mr. de Volder, qu'il trouvoit ce livre extrémement obscur, & lui demanda ce qu'il en pensoit. Nôtre Philosophe lui répondit qu'il n'étoit pas en effet facile de pénetrer les principes des démonstrations de l'Auteur, mais qu'il avoit trouvées veritables celles qu'il avoit examinées. On ne peut pas néanmoins douter que Mr. Huygens ne fût un excellent Mathematicien, & que la Physique & les Mathematiques ne lui soient très-redevables," Le Clerc 1709, 379–380.

238 "In einem kurzen Schreiben, datirt Amsterdam d. 11 Sept. 1682 giebt Tschirnhaus Nachricht von neu erschienen Büchern, und meldet 'Hrn. Hugens habe in Hage gesprochen; er gehet gleich nach Paris. Zu Leiden habe Hrn. Voldern gesprochen'," Leibniz 1899, volume 1, 445.

239 See Van Gent's letter to Huygens of 26 June 1687: "[e]andem Epistolam, seu Apologiam ad D. De Volder mittere animus est, ut exauditam sententiam ex illo expiscarer," Huygens 1888–1950, volume 9, 175 (letter 2467).

240 "Daß Tschirnhaus was Spinosistisches in seiner Medicina mentis habe, oder auch selbsten mit Spinosa bekandt gewesen, wüste er auch nicht. Er sagte, er habe die Medicinam Mentis gelesen, habe aber nichts dergleichen angetroffen, soviel er sich zu entsinnen wüßte," Freudenthal 1899, 229.

241 "Excidit sane N.V. num hanc D. de Volder Epistolam tecum communicarim [...]," Huygens 1888–1950, volume 9, 194 (letter 2475).

242 "Convenit me ante aliquot septimanas D. prof. de Volder, sciscitans numne quicquam literarum a D. Tschirnh. receptum a me esset," Huygens 1888–1950, volume 9, 214 (letter 2485).

243 "D. Voldero obligatus sum, nam hic perfecte non solum meum errorem detexit sed et aliqua ex parte unde meus lapsus originem traxerit," Huygens 1888–1950, volume 9, 215 (letter 2486).

labelled the "Amsterdam mathematical amateurs," as revealed by his contacts with David Gregory, from 1691 professor of astronomy at Oxford. Gregory had been a student at Leiden, and in 1693 he visited Holland. According to one of his memoranda, he was aiming at establishing a correspondence, not only with the publishers Reinier Leers and Pieter vander Aa, but also with Adriaen Verwer and with De Volder himself.[244] Verwer was a leading figure of the Amsterdam mathematical amateurs, a circle formed by Lambert ten Kate, Abraham de Graaf, Johannes Makreel and Bernard Nieuwentijt (1654–1718), who had been a student of De Volder at Leiden. In turn, 'external' members were Gregory, De Volder, as well as Archibald Pitcairne (1652–1713), who was for short time professor of medicine at Leiden in 1692–1693,[245] and who befriended De Volder, according to Gronovius, as they both shared mathematical interests.[246] Common friends at Amsterdam, in fact, are mentioned by De Volder in his letter to Gregory of 31 December 1695.[247] Moreover, as early as 1691 Gregory reported to Newton (in a letter of 27 August) to have received from Amsterdam some letters on mathematics including "a demonstration and improvement (as the author says) of Slusis Methodus tangentium by de Volder at Leyden, wherin in my opinion he hath not shewn great skill in Algebra."[248] In fact, amongst Gregory's papers there is extant a transcription by Makreel, labelled by Verwer as a "disciple" of De Volder, of a five-page manuscript by De Volder on Slusius, as well as a further manuscript on the matter, including some marginalia by Gregory himself.[249] The problem of tracing tangents was thus at the centre of the mathematical interests of De Volder, since in his letter to Guillaume de L'Hôpital of 16 June 1694, Christiaan Huygens reports to have received, from someone whose identity is not disclosed, a solution given by De Volder to the problem of the quadrature of the *folium* of Descartes, namely, the problem of tracing a tangent on a leaf-

[244] The memorandum is dated 29 June 1693: "[t]o settle a correspondence with Volder, Leers, Verwer, vander Aa." I quote from Vermij 2003, 186. Originally in Ms. 247, f. 81, of the Royal Society, London.

[245] See Vermij 2003; Mojet 2016, 2017.

[246] "Quanta erat coniunctio nostri cum brevi gaudio nostro Archibaldo Pitkarnio?" Gronovius 1709, 27. On Pitcairne's influence on De Volder, see Sect. 3.2.3.1, *The case of explanations in physiology*. See also the letter by De Volder to Gregory of 31 December 1695, quoted *infra*, n. 261, in which is evident De Volder's closeness with Pitcairne.

[247] See *infra*, n. 261.

[248] Newton 1959–1977, volume 3, 166.

[249] For the first manuscript: Edinburgh University Library, ms. Coll-33/Quarto A [39]), titled (according to the catalogue) *Methodus Tangentium demonstrata...*, circa 1691. I owe this information to Newton 1959–1977, volume 3, 167. For the second manuscript: Edinburgh University Library, ms. Coll-33/Folio C [81], titled *Methodus tangentium Slusii illustrata a Burchero de Volder*, 1691. Cf. the letter by Verwer to Gregory of 23 August 1691: "Slusianae Tangentium Methodi exegesis a D. burghero de Volder expedita, ac discipulo cuidam suo tradita," which I quote from Mojet 2016, 86. See also Mojet 2017.

shaped curved.[250] Indeed, amongst Huygens's papers a handwritten solution to this problem, written by the hand of De Volder, is extant.[251]

Leaving aside Gregory's remarks on De Volder's skills in algebra, De Volder discussed infinitesimal calculus with his former student Nieuwentijt and with Johann Bernoulli. As early as 1695 Bernoulli reported to Leibniz, in a letter of 8/18 October, that according to De Volder Leibniz's method of infinitesimals traced back to that of Slusius.[252] In turn, in his answer (20/30 October 1695) Leibniz noted that

---

[250] See Paradís et al. 2004; Knobloch 2006.

[251] "L'on m'a donné depuis peu une solution du probleme de la quadrature de la Feuille de Des Cartes par les appliquées à l'axe, qui pourtant sera differente, comme je crois, de celle que vous m'aviez promise, par ce qu'elle va par de grands détours et par la comparaison des termes des equations à la maniere de Des Cartes. Ces solutions se trouvent, lors qu'on en a desia d'autres, mais je ne laisse pas de l'estimer. J'ay veu que Mr. de Volder, Professeur à Leyde, en est l'autheur," Huygens 1888–1950, volume 10, 623. Cf. the words of the editors of Huygens's *Ouevres*: "[n]ous possédons dans la collection Huygens deux solutions du problème de la quadrature du folium de Descartes auxquelles les qualifications du texte sont plus ou moins applicables. Toutes deux ont passé sous les yeux de Huygens puisque sur chacune d'elle on trouve une petite annotation de sa main. L'une d'elle, notre pièce No. 2861, qui va par de plus 'grands détours' que l'autre, est rédigée en langue hollandaise. Elle n'est certainement pas de l'écriture de de Volder, écriture que nous croyons reconnaître avec sûreté dans la seconde, notre No. 2862, qui est rédigée en Latin. Elles se distinguent l'une de l'autre principalement parce que dans la première la methode de différentiation de de Sluse et dans la seconde celle de Leibniz a été suivie. Nous sommes inclinés à supposer qu'elles sont toutes deux de de Volder qui aurait fait copier la première par un de ses disciples. Dans ce cas la seconde constitue une rédaction améliorée de la première [...]," Huygens 1888–1950, volume 10, 624. In these fragments, De Volder used Barrow's theorem of calculus in his solution. Notably, Newton's copy of Barrow's *Lectiones geometricae* (1669) reports De Volder's name on the front, written by Newton: Newton Project website, http://www.newtonproject.ox.ac.uk/his-library/books-in-newtons-library (accessed 13 January 2019). They might have discussed Barrow's ideas in person. De Volder might even have met Barrow in 1674. In his letter to L'Hôpital of 4 October 1694, in turn, Huygens noted the difference between his own and De Volder's solutions: "[j]e crois que ma quadrature de la feüille de Descartes par les appliquées à l'axe sera fort differente de celle de Mr. de Volder; car elle est uniquement fondée, comme je vous ai deja mandé sur quelques regles que j'ai pour prendre les sommes et dont je vous ferai part lorsque j'aurai un peu de loisir," Huygens 1888–1950, volume 10, 687 (letter 2879). Huygens refers to his letter of L'Hospital of 18 January 1694.

[252] "Lugduno Bat. transii, ubi Volderum Mat. P. adii, quem breve post colloquium reliqui, praesertim ubi illum non ita bene de nostra methodo sentire audirem, quam totam ex Slusiana deductam dicebat," A III6, 518. Moreover, infinitesimal calculus is dealt with in the letter of Bernoulli to De Volder of 27 June 1698: "[e]cce mitto ut promiseram enodationem difficultatis paulo ante discessum meum a Te motae contra infinitorum methodum." The letters between De Volder and Bernoulli (besides those pertaining to their correspondence with Leibniz) are preserved at the Basel University Library, and have been digitized and transcribed in the *Briefwechsel von Johann I Bernoulli. Korrespondenz zwischen Bernoulli, Johann I und Volder, Burchard de* (http://www.ub.unibas.ch/bernoulli/index.php/Kategorie:Volder_Burchard_de-Bernoulli_Johann_I (accessed 13 January 2019)), to which I owe this and the other quotations from their correspondence. As to the original of this letter: Basel University Library, ms. L Ia 675, 161–162. No other letters exchanged between De Volder and Bernoulli (with the exception of the letter by De Volder to Bernoulli for Leibniz, 21 November 1698, in A II 3B, 476–482) could be found.

De Volder did not pay much attention to his method of infinitesimals.[253] Eventually, the topic resurfaces in the Leibniz-Bernoulli-De Volder correspondence. In the first letter of the correspondence (Bernoulli to Leibniz, 5/15 July 1698). Bernoulli tells Leibniz that De Volder had proposed to him a difficulty in infinitesimal calculus, namely that of the squaring of a hyperbola, which Nieuwentijt was also not able to solve.[254] Bernoulli, who had proposed his own solution to De Volder in a letter to him dated 27 June 1698, then forwarded this letter to Leibniz himself.[255] In his letter to Bernoulli of 29 July/8 August 1698, Leibniz approved Bernoulli's argument against De Volder's "very clever and elegant objection to infinitesimal calculus," which does not resurface in De Volder's correspondence.[256]

Thus, De Volder came to be known as a mathematician from the 1680s. His expertise in mathematics – even if did not result in any published theory – according to Gronovius led Huygens himself to note with pride, in a private paper, to have been mentioned by De Volder in his 1682 inaugural *Oratio*. Moreover, it led De Volder to be offered a chair in Padua – with the guarantee of religious freedom (an offer which has no other evidence but Gronovius's *Laudatio funebris*), and, as I am

---

[253] "Volderi, viri licet egregie docti, judicium de nostris Methodis non est quod nos magnopere moveat, videtur enim in hanc Analyseos partem minus inspexisse. At Hugenius ipse, quo nemo melius ista dijudicare poterat, et cui Slusiana et multo ampliora erant perspectissima, de praestantia nostrae Methodi magnifice sentiebat," A III6, 528.

[254] "Volderus sub discessum meum proponebat mihi difficultatem contra calculum infinitorum, quam sibi se eximere non posse neque a Nieuwentiitio cui dudum eandem proposuerat hactenus enodationem accepisse rebatur, rogans ut ego sibi hac super re satisfacerem, quod etiam libenter promisi me scilicet satisfacturum statim ac huc redux factus essem, et revera nudius tertius solutionem ipsi misi," A III7, 812. In his letter to Bernoulli of 12/22 July 1698, Leibniz showed no wonder that Nieuwentijt could not solve De Volder's problem: "[m]irum non est Dn. Nieuwentiit ipsi non satisfecisse circa calculum differentialem quem ab [...] ipso non satis profunde penetratum constat," GM III, 515. In the words of the editors of the Leibniz *Akademie Ausgabe*, the question was whether two unequal parts of the hyperbola, which are contained in one another, can have the same content, namely, if this is infinite: "[v]on de Volders Schwierigkeiten mit Leibniz' Infinitesimalkalkül berichtete Joh. Bernoulli, der ihn auf einer Reise durch die Niederlande kennengelernt hatte [...]. Im Gegensatz zu Nieuwentijt und Clüver hatte de Volder kein alternatives Konzept. Er hatte sich aus Interesse mit dem Infinitesimalkalkül beschäftigt und war bei der Betrachtung der Quadratur einer Hyperbel auf einen (vermeintlichen) Widerspruch gestoßen. [...] Dabei ging es darum, ob zwei ineinander enthaltene nicht gleiche Teilflächen der Hyperbel gleichen Flächeninhalt haben können, nämlich wenn dieser unendlich ist, und allgemein um die Frage, ob die Axiome des Endlichen, z. B. dass das Ganze größer als der Teil ist, im Unendlichen weiter gelten. De Volders mathematische Schwierigkeiten waren nach Joh. Bernoullis Erläuterungen [...] wohl ausgeräumt," A III7A, *Einleitung*, XL.

[255] See the letter of Bernoulli to De Volder of 27 June 1698: "[e]cce mitto ut promiseram enodationem difficultatis paulo ante discessum meum a Te motae contra infinitorum methodum," transcription from *Briefwechsel von Johann I Bernoulli* (Basel University Library, ms. L Ia 675, 161–162). The solution was attached to his letter to Leibniz of 23 July/2 August 1698, in GM III, 516–518.

[256] "Rectissime etiam solvisse mihi videris Viri Clarissimi Objectionem sane peringeniosam et elegantem contra calculum infinitesimalem," A III7, 855. Moreover, Leibniz and De Volder touched upon the problem of the hyperbole: see, for instance, the letter by De Volder to Bernoulli of 21 November 1698, and the letter by Leibniz to De Volder of 27 December 1698: GP II, 148–152 and 153–163.

going to show in more detail, to be charged by Huygens with publishing his posthumous works.[257] Huygens, moreover, consulted De Volder as an expert, when Newton's *Philosophiae naturalis principia mathematica* came to the fore in 1687. Indeed, De Volder was one of the first in Continental Europe to read the *Principia* and understand them. He had received a copy directly from Newton through Zimmerman, as testified to by a second letter by De Volder to Newton (dated 14 August 1686 – but in fact traced to 1687):

> I thought it my duty to thank you for the most splendid offspring of your intellect, with which you have blessed me. Absence from my study at home has prevented me going over it carefully, but a glance through it shows it to contain recondite mysteries of great value to natural philosophy [...]. Soon I will enjoy it in greater detail.[258]

As reported by Le Clerc, De Volder said that Huygens thereafter paid a visit to him in Leiden, in order to get clarifications on the contents of the *Principia* – whose demonstrations, according to Le Clerc, De Volder found to be *veritable*. Moreover, amongst De Volder's papers were found repetitions of Newton's demonstrations.[259] However, it seems that De Volder was not so impressed by Newton's *Principia*.

---

[257] "Hinc fieri debuit, ut latissime didita Volderi fama maximam isti nomini conciliaverit auctoritatem. [...] Quid loquar vel inter liturarios Nobilissimi Hugenii repertam chartulam, in qua is accinuit sibi ab Voldero in oratione Professionis Mathematum factam esse sui honoratam mentionem, quasi oculis suis et voluptati non satisfieret, si nosset in ipsa oratione id legi, nisi chirographo suo scriptum aliquoties etiam videret? Ab eodem ultimis mandatis cautum est, ut pars postuma scriptorum ab Voldero ederetur, honorato tamen mille florenis. Hinc Amplissimus Huddius pari liberalitate inter legatarios prosequutus fuit. Hinc in Academiam Patavinam vocatus libera opinione in rebus sacris promissa," Gronovius 1709, 27. In a not perfectly clear statement, in this passage Gronovius notes that De Volder was also mentioned in Hudde's testament (please note that Hudde was not mentioned in Huygens's testament: see Huygens 1888–1950, volume 22, 773–778): see Sect. 2.4.2, *De Volder's retirement, death and legacy*. Notably, in 1699 a Latin edition of Galileo's *Dialogo sopra i due massimi sistemi del mondo* (1632), viz. *Systema cosmicum* was dedicated to De Volder by the publishers, Friedrich Haaring and David Severin of Leiden: see Galileo 1699. Johann Bernoulli congratulated De Volder for this in his letter of 31 May 1699.

[258] Hall 1982, 11. Cf. the original text of the whole letter (to which I shall return below): "Viro Amplissimo D. Is. Newton. / S. P. B. De Volder. / Cum D. Zimmerman, ad vos reditum pararet mei officii esse {putavi}, tibi hisce gratias {agere} maximas pro luculentissimo ingenii tui {factu}, quo me beare voluisti. \Nam,/ quam <{...}> propter meam a domo, Musaeoq[ue] absentiam {integrum} accurate pervolvere non licuerit, vel obiter inspicienti satis {patiunt} contineri in illo reconditae eruditionis, et summae ad res Physicas utilitatis mysteria. Sed brevi mihi dabitur illis impensius delectari. D. Zimmerman numere apud vos {Ecclesiastae} fungi gratissimum {fuit}, eiq[ue] praeterea tuam amicitiam quam ille non immerito maximi facit singulariter gratulatus sum; doluiq[ue] certe mihi deesse rationem qua ulteriorem quam <{...}> quaerere videtur, apud vos commendationem ipsi procurare possim. {Interum} maxime {gavisus} sum cum intelligerem ex ipso tibi omnia esse recte. Iam enim tibi metueram, cum ex novellis, quas vocant, intellexissem, re etiam periculosissimo hocce tempore libertatis et privilegiorum Acad. patrocinium generosissime suscepisse. Sed bene habet, quod vanus fuerit hicce meus metus. Utinam ita semper eveniat, ubi libertatis, et Religionis defensoribus, qui hisce temporibus oppido quam pauci sunt, aliquid mali {imaninere} videtur. Tibi certe ex animo voco ut in Acad. gloriam, in Reip. letterariae emolumentum perpetua felicitate frui contingat. / Vale. / Dabam Amstelod. / 14 Augusti. 1686," Uppsala University Library, Waller Ms. benl-00771.

[259] See *supra*, n. 237.

Even if Le Clerc repeatedly claimed that Newton and some "Englishmen" opened De Volder's eyes on Descartes's vortex theory and on Cartesianism in general[260] – which might have been true, as far as the criticisms to Cartesianism are concerned – and notwithstanding De Volder's praise for Newton in two of his academic orations (as I am going to discuss in Sect. 2.3.3, *Mathematician and scientific expert*), in the letter he addressed to Gregory on 31 December 1695, De Volder wrote that "you already saw the book of Newton[,] as I understood from people from Amsterdam, and in it, as I deem [*ut opinor*] [you did not find] anything extraordinary."[261] This is indirectly confirmed by two letters by Johann Bernoulli to De Volder of May and November 1706, in which Bernoulli – commenting upon the *Introductio ad veram physicam seu Lectiones physicae* (1702) of John Keill, namely, the first adaptation of Newton's natural philosophy to a student audience[262] – remarks how the book hardly contained anything original, as it was based on the ideas of Newton, and moreover contained outdated criticisms of Descartes's vortex theory, not taking into account Huygens's improvement of it.[263] According to De Volder, who had asked for Bernoulli's opinion on it, the book was appreciated by Newton, Gregory, and by

[260] Cf. the text quoted *supra*, n. 101. See also Le Clerc 1709, 374: "[...] il explica en public non seulement la Physique, mais encore la Métaphysique de Descartes. Il en étoit en ce tems-là beaucoup plus grand admirateur, qu'il ne l'a été sur la fin de sa vie, qu'il en avoit reconnu le foible; comme il l'a avoué plusieurs fois à ses Amis, & ainsi que je lui ai ouï dire moi-même."

[261] Cf. the full text of the letter: "Viro Amplissimo. Iacobo Gregorio, Matheseos Saviliano Professori S. P. B. De Volder. / Clarissimus vir Godfridus Bidloo, Anatomes et Medicinae, apud nos Professor, iussu regio Angliam petit. Hic meis hisce litteris tibi innotescere voluit, quanquam nihil, ut opinor, necesse est. Ipsorum opere Anatomico, quod ante multos annos edidit, tibi vestratibusq[ue] notus erit satis superq[ue]. Quod se tanem ex mea commendatio adjicere huic rei vel {tantillima} queat, eam hisce lubentissime adjungo. / Deinde non possum, {quam} tibi hac occasione gratias agam pro Optices libello, quo me donare dignatus {es}. Caeterum apud nos. Vic quidquam in re litteraria {boni} geritur. Librum Newtoni ut ex Amstelodamensibus intellexi jam vidisti, nec in eo, ut opinor {quidquam} extraordinari. Bernoulli <{...}> vicinia {degit} apud Groninganos vocatus illic Matheseos professor. Hugenius diem suum, ut proculdubio jam nosti, obiit. Manuscripta sua legavit Acad. nostre; ea tamen lege, ut mecum D. Fullenuius Matheseos apud Franekeranos Professor, ea perlegeret, et si {quae praelo} idonea forent, praelo {commuteret}. Verum hactenus apud haeredes latent, {b...eroi} tamen, nisi fallor, si valetudo {suaerit} ea inspiciendi dabitur facultas. De Pitcarnio omnis spes nostra evanuit. Huic meis verbis multam salutem dicito, et me propediem ipsi responsurum, ubi occasio erit, si placet indicabo. Vale. / Lugd. Batav. / Prid. Calend. Ianuar. / anni CIƆIƆCXCVI, qui tibi ut ineat et exeat felix, Deum precer. Iterum vale," Bodleian Library, ms. Tanner 24, f. 189. De Volder probably received Gregory's *Catoptricae et dioptricae sphaericae elementa* (1695). Bernard Fullenius was professor of mathematics at Franeker. In 1705, Gregory even suspected that De Volder (who inspected Huygens's papers after his death, and published with Fullenius his *Opuscula postuma*) could have organized the printing of Huygens's remarks on Newton's *Principia*, appeared as an appendix to the *Historia cycloeidis* of Johann Groenig in 1701: see Newton 1959–1977, volume 3, 169. This claim could not be substantiated.

[262] Or, as Bernard Cohen has put it, "Keill could be assumed to speak with an authentic Newtonian voice," Cohen 1997, 125. On Keill, see also Thijssen 1992; Guicciardini 1995; Wilson 2009, chapter 1; Carpenter 2011, chapter 1; Palmerino 2013.

[263] I explore this topic in Sect. 6.2.1, *Huygens confronting Descartes's theory of weight*.

Thomas Herbert, Earl of Pembroke, just because Keill showed appreciation for them. In his letter to De Volder of 13 November 1706, Bernoulli reiterates his judgment.[264] In his letter to Jakob Hermann of 3 May 1710, Bernoulli reports that De Volder had asked for his judgment on Keill's book, and which he fully approved after having received it: in fact, De Volder was so discontent with Keill's book, that he vetoed the possible hiring of Keill to replace him.[265]

---

[264] Cf. the text of the two letters: "[n]on aegre adducor ut credam quod scribis Johannem Keill Scotum Vestris vehementer commendatum fuisse a Comite Pembrokiae, Newtono, Gregorio, hos enim omnes singulari modo obstrinxit; nam Pembrokio dedicavit suam introductionem physicam, in qua Newtonum singulis fere paginis novis elogiis condecorat, Gregorium vero popularem suum arrepta levi qualibet occasione extollit quandoque etiam ultra meritum. Quantum ad ipsum hoc opusculum, quod Tibi visum puto; profecto illud non tanti deprehendi, quam primo aspectu sperabam; particulares enim corporum affectiones eorumque phaenomena physica plane non explicat; in praefatione sugillat quidem modum communem explicandi gravitatis causam per pressionem materiae subtilis, et absurdissimas diuque obsoletas objectiones contra eum movet, quo ostendit se non legisse Dissertationem incomparabilis Hugenii De causa gravitatis; interim quod mireris per totum libellum melioris explicationis nec volam nec vestigium reperies; multa reprehendit alia, quibus tamen mediocri in contextu substituere (ut quidem ab eo expectabam) postea non meminit: Habet quidem in sequentibus nonnulla physico-mathematica, de viribus et motu gravium quae sunt alicujus momenti et quibus indicat, se aliquousque ad haec interiora penetrasse, sed si verum dicere fas est, nihil novi reperio quod non compilaverit ex Newtonianis aut ex Hugenianis aut ex Nostris etiam; quamvis peregrinos plerumque tanti non aestimet ut quid ex illis hauserit profiteatur, more scilicet multorum Britannorum, adeo ut nihil novi ex sua penu nobis dederit." "Quid in Keylii libello sit quod eum commendet tanquam totius Europae post Newtonum, summum Geometram, meorum certe oculorum aciem subterfugit; nisi quis velit eum ideo tanto elogio dignum, quod fuerit Newtoni fidelis exscriptor et suorum plagiorum cautus celator; interim miror si parem non habet excepto Newtono, Keylius, quod vili adeo precio intra 1000 flor. se venum offerat, siquidem vinum vendibile quod ajunt non indiget haedera suspensa," transcription from *Briefwechsel von Johann I Bernoulli* (Basel University Library, ms. L Ia 675, 191).

[265] "[...] recordor, ipsum jam dum viveret Volderum ad me scripsisse et petiisse meum judicium de ejus Introductione de veram physicam; quod cum dedissem, rescripsit se omnino mecum sentire, neque opusculum illud, (quod mera esset rapsodia, continens varia a saniorum Philosophorum praesertim Volderi ipsius placitis dissentientia) tanti ponderis deprehendere, ut ideo cathedram suam ad quam nonnulli eum commendaverant promereretur; ex quo et ex aliis circumstantiis colligere potui, Volderi animum ab isto Keylio omnino fuisse abalienatum; et ideo quod Curatores multum deferre solebant Voldero parum spei relictum esse Keylio, equidem nescio quomodo nunc post fata Volderi Lugdunenses, sint animati, quod dies docebit," transcription from *Briefwechsel von Johann I Bernoulli* (Basel University Library, ms. L Ia 659, number 6). See also the letter of Jakob Hermann to Bernoulli of 12 July 1710: "[h]aec Curatorum perseverantia satis evidens praebet indicium, quantum Curae ipsis sit ut statio quam Cel. Volderus decessu suo vacantem reddidit, Eximio Viro instauretur, simulque satis aperte monstrant Academiae Leidensis decori non posse felicius consuli, quam si Bernoullio potiantur; adeo ut hinc nullus dubitem quin efficacissima quaevis media adhibituri sint, ut ad se trahant; sed in honorem Universitatis Patriae et confirmationem Valetudinis Tuae quae nova soli mutatione detrimentum capere posset, optarem ut eorum conatus irriti fiant. Si Volderus adhuc in vivis esset, haud dubie Keilio spes omnis Provinciam obtinendi quam Tu detrectasses, praecisa esset, sed Celebri Viro nunc fatis functo, non penitus incredibile duco, ut Anglus iste scopum suum assequatur," transcription from *Briefwechsel von Johann I Bernoulli* (Basel University Library, ms. L Ia 659, number 19∗).

2.3.4 De Volder's Academic Activities After 1682

After 1682, thus, De Volder acquired international fame: most of his extant correspondence (viz. with Newton, Huygens, Bernoulli and Leibniz), indeed, dates back after this year. As far as his academic activities are concerned, De Volder continued to preside over disputations the greater part of which was not officially authored by him. Amongst others, it is worth indicating here the two-fold series of disputations encompassing the *Disputatio philosophica de motu* and *De absoluta quiete* (November–December 1684), moreover, the *De vera gravitatis causa* (1685), the anti-Spinozist *De mundi in tempore productione* (1686), and Hermann Schuyl's *Disputatio philosophica inauguralis de vi corporum elastica* (1688) – in which De Volder's rules of impact are expounded, together with a Huygenian explanation of gravity (as I am going to show in Sects. 4.2.3.4, *Collision in De Volder's late thought*, and 6.2.2.3.2, *De Volder's Huygenian theory of weight (and its relation to the idea of cohesion)*). In these years, famous students of De Volder graduated under him as promoter, or under his guidance: for instance, Bernard de Mandeville (with a Cartesian *Disputatio de brutorum operationibus*, 1689), Boerhaave, with his *Disputatio de distinctione mentis a corpore* (1690)[266] – attacking both Spinoza's and Thomas Hobbes's ideas of substance – and Joseph Serrurier, with his *Disputatio de gravitate aëris* (defended on the same day as Boerhaave's). Before starting his aforementioned *Exercitationes* against Huet's *Censura* in 1690, moreover, De Volder gave the *Oratio in obitum Lucae Schacht* (1689) – in which he anticipates some of the methodological positions which he would develop later. By the end of the 1680s, moreover, the situation was finally stable at Leiden. As De Volder wrote in a letter of 9 May 1689 to the Cocceian theologian Johannes Braun, professor at Groningen, "[for] what it concern us, everything is here at rest and in peace. Everyone is doing what he wants; and it seems here that it will last, until the troubles from outside come to blow about here."[267]

---

[266] As reconstructed by Wiesenfeldt (Wiesenfeldt 2002, 234), Boerhaave did not officially graduate with De Volder as promoter, and had also been a student of Senguerd. However, Boheraave from time to time praised De Volder as his teacher.

[267] In the letter, De Volder thanked Braun for having received a copy of his *Doctrina foederum sive Systema theologiae didacticae et elencticae* (1688): another copy was for Hieronymus van Beverningh, protector of De Volder at Leiden – and sympathiser with Cartesianism and Cocceianism. As to the rest, the letter contains some news on other colleagues of De Volder, and on the attendance at lectures at Leiden. For a commentary, see Wiesenfeldt 2002, chapter 2. Cf. the full text: "Mijn Heer, / UEdts aangename van den 28ste Maart is mij eerst den 25ste April, ter handt gekomen. Ick dank UEdt harterlyck voor het overgesonden exemplaar van UEdts Systema Theologiae. Het ander heb ick aan de Hr. Bewerningh doen behandingen. {Lis mij lief}, dat de treves tusschen UEdt, en UEdts Collegaas UEdt tydt geeft om uw gedachten op beter en {voorst} gemeen nutter dingen te leggen als op twist schriften, uijt {welcten} seer selden, soo {wyt} eenigh woordere te verwachten is. Wat ons aangaat, alles is hier in rust en Vrede. Elck doedt wat hij wil; en 't schijnt hier wel te sullen duuren, ten waar de onlusten van buijten hier quamen over te waayen. Maar dan dach en raadt. Men doet her sijn uijterste best om de wereld te doen geloven dat de Theologische Faculteijt nu immers so wel floreert als te vooren. De Hr. Trigland {...} over 3 a 4 maanden, dat in een Collegie 60 studenten had die reedts haar naam hadde geteyckent, buijten die

In turn, the 1690s represented another crucial period in De Volder's intellectual path. In these years, De Volder became – as indicated by Le Clerc, and as confirmed by his own works – more and more 'discontent' with respect to Cartesianism. Moreover, as seen above, he entered into correspondence with some of the foremost scientific figures of his age. For instance, the extant correspondence with Johann Bernoulli spans from 1698 to 1706 (14 extant letters of Bernoulli, and one of De Volder), that with Huygens from 1689 to 1693 (nine extant letters by the two), the correspondence with Leibniz (with the intermediation of Bernoulli) from 1698 to 1706 (67 letters of the three philosophers). In particular, the correspondence with Leibniz – which has been exhaustively dealt with in the studies of Paul Lodge and Anne-Lise Rey[268] – arose when Bernoulli, aware of De Volder's growing 'discontent' with some aspect of Descartes's philosophy (chiefly, Descartes's theory of the

---

geen die sonder naam te teyckenen quamen. De Hr. Spanheym door imant van de Magistraat naa 't getal der studenten gevraaght sijnde, bekende wel dat 'er minder studenten waren als voorheen, maar {weghde} 'er terstont bij, dat die geen die vertrocten waren, geen Theologenten waren, maar Iuristen, die omdat de officier de studenten soo hard handelde, hier van {daan} waren geretireert. Maar soo UEdt mij vraaghd, wat ick 'er van gevoet; so sal ick UEdt seggen, dat 'er saeckerlyck vrij minder studenten syn als voorheen; {hoewel} ick 's {oeck} dat moet bij seggen, dat 'er noch meerder sijn; als ick verwacht had. Maar dat men dese minderingh op de Iuristen soeckt te schuijven is maar {een praatijes}. De Hr. Spanheym heeft wel Collegie begonnen, waar van ick het getal der studenten heb hooren begrooten op 90; maar door onpasselyckheyt ingeen 3 maanden iets gedaan. {2.} 60 van de Hr. Trigland sijn {al} rijckelyck op de {helft gesanost}. So men het getal van die uijt het Collegie uijtkomen geloven mach. / Ick {beet} mij {ewermelin}, dat de Hr. Trigland soo langh als ons Academie soo als nu bestaat, hier het Catje vande baan sijn sal. Want de Hr. Spanheijn is te luij; ende Hr. Le Moyne {altes} geleert. De Hr. Gaillard heeft al eenige maanden {h' te} huijs gehouden, en veeltjdts bedleegerigh geweest, sonder hoope van restitutie. Ick geloof echter niet, dat men, so hij quam te overlyden, een ander {voepen souw}. Want een Voetiaan heeft men niet van doen, en een Coccejaan {wilmen} niet, in 'tis oock {sra} veler oordeel 'buijten twijfel beter dat Franeker van studiosi Theologiae opgepropt is, als dat men hier {weer den droessem} (soo spreecken sommige, {siende} op de Hr. Wittichius) dienen nu quijt is, soude inhalen. En synd' er wat minder studenten, de Theologie is hier weer noijt soo suijver geheet, als te deser {standt}. De Hr. Voet beroepen sijnde tot Utrecht, en ten minsten minen maackende van te willen gaan, heeft verkregen den tytel van Professor Iuris Practici, merers 400 gl. jaarlycks, sodat hij nu met den Hr. Noot gelijck staat; 't geen <{...}> de Hrn. Matthaeus en Vitrarius \seer/ gespeten heeft. De Hren Noot en Voet vechten hier om de Roosenkraus dubio ad huc dum Marte, alhoewel dit jaar de schaal wat {schijnt} aan de kant vande Hr. Noot {overteslaan}. In de Medicijnen hebben wij een nieuw Anatomicus gekregen, ende Hr. Drelincourt is vande {last} vande Anatomie geexcuseert, behoudende evenwel sijn tractement. Maar dit sal UEdt buijten twijsel al langh bekent sijn. De Hr. Van Es is over 2 a 3 maanden Doctor Theologiae \van ons/ gemaackt, welcke tijtel hem door die van Haerderwijck gepraesenteert was. Veele meenen dat dit siet op de successie van de Hr. Gaillard. Maar ick geloof niet, dat de Hr. Van Es het sonder tractement soude begeren; en oock niet dat de Hren hem het op een ander wijs sullen geven. 't welck, soo 't evenwel geschiet, mirabor, quod hisce temporibus saepius feci. De Hr. Spanheim heeft voor 8 a 10 dagen een bastert pleuris gehadt niet sonder peryckel. Maar is tegenwoordigh weer {hoet} aande beterhaut. Verders weet niet dat hier iets meer nieuws is; soo dat ick sal eijndigen met UEdt te verseekeren dat ick waarlyck ben / Mijn Heer / UEdts ootmoedige Dienaar / B. De Volder / Leyden, den 9 maij, 1689," British Library, Ms. Add. 24712, 76r–77r. Another curator protector of Cartesianism at Leiden was Cornelis van Beveren: see De Volder 1689, 1.

[268] See Chap. 4.

conservation of the quantity of motion, the idea of material causality and the theory of the body-mind relation) put De Volder in contact with Leibniz after having paid a visit to De Volder himself. Bernoulli, actually, vainly hoped that De Volder would teach Leibniz's ideas at Leiden.[269] Leibniz, moreover, already knew De Volder's experimental lectures,[270] he knew that De Volder was charged with editing Huygens's posthumous works in 1695,[271] and had partially appreciated De Volder's *Exercitationes* against Huet's *Censura*, deeming the *Exercitationes* worthy of being re-published together with a new edition of the *Censura* and with his own *Animadversiones* on Descartes's philosophy.[272]

Along with and after the *Exercitationes*, De Volder was then to preside over disputations in physiology and medicine (matters which he was not in charge of teaching), authored by his students, such as the *Exercitium medico-physico aphoristicum de morborum causis* (1691), the four-fold *Dissertatio physiologica* (1693), the *Dissertatio medico-physica de circulatione sanguinis* and *Dissertatio physica de circulatione sanguinis in foetu* (1698). At the same time, he presided over or acted as promoter in disputations in cosmology, namely a *Disputatio philosophica quae est de mundi systemate* (1694), Gysbertus Casembroot's *Disputatio philosophica inauguralis de aestu marino* (1696), and a disputation *De Telluris motu* (1698). It is worth remarking these disputations because they were in fields – namely, physiology and cosmology – which were at the core of De Volder's disenchantment with

[269] See the letter of Bernoulli to Leibniz of 6 December 1698: "non dubito quin systema si quod componeres, felicissimum successum habiturum esset, sunt enim in Batavis nonnulli egregii Viri qui hunc Tuum ingenii foetum fortiter foverent et defenderent, interque illos ipse Dn. Volderus qui cum Cartesiana principia tanquam insufficientia et plurimum falsa ut ipse mihi fassus est iam a longo tempore deseruerit, si Tua semel probe percepisset et imbibisset, dubium non est quin ea gnaviter propagaturus suisque Discipulis quorum semper insignem numerum habet adeo esset inculcaturus, ut forte brevi dominium haberet supra Cartesianam et Aristotelicam, Veteresque cum modernis quasi reconciliaret. Rogo Te etiam atque etiam velis de hoc cogitare, deque Orbe philosophico bene mereri; consule quaeso nobis et posteritati, Tuoque nomini erige monumentum; quid Te Cartesio inferiorem putas?" A III7, 960.

[270] See, for instance, the letter of Leibniz to Johann Jacob Spener of 3 January 1691: "[c]eleberrimus Hugenius, qui Te Hanovera transiisse credebat in Epistola quadam sua inter alia et de Te quaesivit, meminitque nescio cujus experimenti a Te insinuati, quod nondum sibi Volderoque successerit Mercurium per Siphonem attrahere volentibus, credo mente tua non satis percepta," A III5, 16. See also his letter to Cornelius Dietrich Koch of 29 May 1702: "Volderus philosophiae experimentalis cognitione praestat, et licet Cartesianus sit habitus, in non paucis tamen recedit a Magistro, et quaedam etiam mea assensu suo comprobavit," A I21, 283.

[271] See the letter of Henri Basnage de Beauval to Leibniz of 12 September 1695 (GP III, 119–120) and of Leibniz to Basnage de Beauval of the second half of September 1695 (GP III, 120–123).

[272] See the letter of Leibniz to Simon Foucher of 16 April 1695: "[u]n professeur celebre à Leide, nommé Mons. Volder ayant publié sur la fin de l'année passée une réponse à la Critique de Mons. l'Evêque d'Avranches; un amy qui me la porta, me pria de luy en dire mon sentiment; En la lisant je fis des remarques, car il me sembloit, qu'il ne satisfaisoit pas assés. Un jour cela se pourra joindre à d'autres animadversions sur la philosophie de M. des Cartes, que j'ay faites, sur tout si l'on songeoit encor à une nouvelle edition de la *Censure* de M. d'Avranches," GP I, 420–421. See also the letter of Leibniz of Huet of April 1695, mentioned in Sect. 2.2.3, *The mid-1670s clash at Leiden and the foundation of the experimental theatre*, and the letter of Leibniz to Joachim Bouvet of 15 February 1701: see A I19, 411. See Rapetti 1999, chapter 2; Laerke 2013.

Cartesianism towards the end of the seventeenth century. If De Volder's correspondence with Bernoulli and Leibniz (as well as his academic *dictata*) reveal De Volder's issues with Descartes's justification of the law of the conservation of the quantity of motion and with the mind-body relation,[273] Le Clerc's eulogy reveals a discontent with Descartes's cosmology, and in his 1698 *Oratio de rationis viribus, et usu in scientiis* contains a criticism of the application of Cartesian ideas to physiology, as done in Holland by his colleague Theodoor Craanen, whose idea of *oeconomia animalis* was harshly attacked by De Volder in his *Oratio*.[274]

2.3.5 De Volder's Views on Politics and Religion

De Volder's 1698 *Oratio* was given by him on 8 February 1698,[275] as the conclusion of his service as rector of the University in 1697. After his missed appointment as rector in 1674, and after having been proposed by the Senate as a possible rector to William of Orange, now King of England, in 1693 and 1695, in January 1697 he was eventually nominated rector, assuming the post on 8 February.[276] As reported by Le Clerc, even those who opposed his previous designation recognized that William III – no longer fearing Cartesianism – wanted to do justice to De Volder.[277] In turn, De Volder seems to have had no great an opinion of William himself. As seen above, he attempted to convince Fagel of the a-political nature of Cartesianism. On the other hand, as it emerges from the colloquium of De Volder with Stolle and Hallmann of 1703, De Volder did not deem William as particularly protective of Cartesianism,

---

[273] De Volder's discontent with Descartes's metaphysics is also testified to by Le Clerc: "[j]e l'ai ouï se moquer, plus d'une fois, d'une bonne partie des Méditations de Descartes, quoi qu'il les eût expliquées pendant long-tems [...] il étoit ennuyé d'expliquer ces Méditations," Le Clerc 1709, 398.

[274] See Sect. 3.2.3.1, *The case of explanations in physiology*.

[275] See Molhuysen 1913–1924, volume 4, 151.

[276] Molhuysen 1913–1924, volume 4, 100, 116 and 139–140. Before assuming this post, De Volder had been secretary of the Academic Senate in 1682 and 1691 (Molhuysen 1913–1924, volume 3, 363 and 371; volume 4, 12, 13, 70 and 87) and he was assessor of the Senate in 1685, 1692, 1693, 1700 (Molhuysen 1913–1924, volume 4, 61, 90, 100 and 173). He was *quaestor* in 1705 (Molhuysen 1913–1924, volume 4, 211) He acted as pro-rector later in 1698 (Molhuysen 1913–1924, volume 4, 172–173).

[277] "L'an MDCXCVII. il fut nommé pour être Recteur de l'Academie, & choisi par le Roi d'Angleterre; sans qu'il eût fait aucune démarche, pour s'attirer cet honneur, ni cessé de parler avec la même liberté, dont il avoit accoûtumé de se servir. Mais en ce tems-là, la haine du Cartesianisme, & les autres préventions, que l'on avoit autre fois données à ce Prince, s'étoient presque entierement évanouies. Il étoit parvenu à un degré de grandeur & d'autorité, qui le mettoit entierement au-dessus de toutes ces menues considerations. Aussi plusieurs de ceux, qui s'étoient opposez douze ans auparavant à son election, reconnurent que le Roi d'Angleterre lui avoit rendu justice; car enfin il avoit été très-utile à l'Academie, par le grand nombre d'Etudians, qu'il y avoit attirez; & personne n'avoit sujet de se plaindre, en quoi que ce fût, de sa conduite," Le Clerc 1709, 386. See also Gronovius 1709, 35.

as "the King of England [...] favoured the Voetians in Holland and the Episcopalians in England."[278] De Volder, moreover, had complained of the dominance of the 'Voetians' in Holland as late as 1694 and 1696. As revealed in the letter by Huygens to Leibniz of 8 June 1694, where is reported De Volder's recommendation of Johannes Teyler for a position as professor of mathematics at the Ritterakademie of Wolfenbüttel (on request of Leibniz),[279] as late as 1694 the struggle between 'Cartesians' and 'Voetians' was lively in the Dutch universities, as Teyler – who had been a student and then successor of Craanen at Nijmegen in 1670 – was not hired at Utrecht because he was a Cartesian.[280] Moreover, as testified to by his letter to Abraham van den Ende (an Amsterdam lawyer, and common acquaintance of De Volder and Van Limborch)[281] of 26 December 1696, "our theologians seek [to impose] tyranny on students," and providing, as an example, the case of a student accused of Socinianism.[282] In any case, as confirmed by Verwer in a letter to Gregory

---

[278] "In aulis non curari religionem, sed utile. Ita Regem Angliae nuper demortuum, in Hollandiâ favisse Voetianis et simul in Angliâ Episcopalibus," Warsaw University Library, Cod. IV oct. 49, 591–592. Transcription kindly provided by Martin Mulsow.

[279] See *infra*, n. 330.

[280] "J'ay parlé du depuis à Mr de Volder pour m'informer touchant ce que je vous avois mandé, qui m'a nommé encore quelques personnes qu'on pourroit proposer pour l'employ dans l'Academie inconnüe, mais m'a assuré en mesme temps qu'il n'en connoissoit pas de plus capable que le Sr Teiller dont vous m'aviez escrit. Il m'en a dit aussi touchant ses bonnes qualitez des choses que je ne scavois pas, et entre autres qu'il avoit voiagé en Italie[,] en Sicile, et jusqu'au Caire, et qu'il avoit dessiné en tous ces pais une infinité d'antiquitez et de belles vues. Au reste que sa solicitation ou celle de ses amis pour la profession de Mathematique à Utrecht n'avoit point reussi, seulement par ce qu'il avoit esté disciple de Mr Cranen, car ces partialitez du Cartesianisme et du Voetianisme s'etendent jusques mesme les professions où il n'est pas question de Theologie," Huygens to Leibniz, 8 June 1694, GM I, 177–178.

[281] On him, see Sects. 2.4.2, *De Volder's retirement, death and legacy*, and 3.1.3.1, *The relation between material and immaterial substances*.

[282] "Ex Literis B. d. V ad A. v. E. / In Academia nostra nihil fit novi, nisi quod indies magis magisq[ue] innotescat, Theologos nostros tyrannidem quaerere in studiosos. Huius exemplum hoc sit. Ante sex septemve hebdomadas cum {examinaventur} studiosi, qui in Collegio Ordinum degunt, fuit, qui respondere Martio, quaerenti, cur Cristus diceretur aeternus Dei Filius? Causam huius esse, quod ab aeterno destinatus esset a Patre ad munus mediatoris. Quod cum bis terve repeteret, reconduit Martius, et accusarit hominem Socinianismi. Ita quidem tum discessum est. Circa vesperam adfuit studiosus hicce proregenti, excusavit factum, nescire se quid responderit, fuisse se animo perturbato, nunquam aut Roëlii libros aut Socinianorum legisse, sed in Burmanno, alioq[ue] eijusdem commatis, cuijus nomen non memini, omnem operam posuisse. De Roelii opinionem se nihil unquam intellexisse, nisi ex sermonibus inter studiosos habitis. Postmodum vocatus est ad Professores Theologos, sed ea lege, ut statim cogeretur illos adire, cum bedello, qui vetabat, ne cum ullo antea colloquium haberet. Conqueritur se ab illis acerbe receptum, praesertim a tuo \Triglandio/ illo cliente, quo neminem ait fuisse acerbiorem, ut Spahnemio meliorem neminem. Illos excusationem suam non accepisse: dimissum esse hisce cum minis, scire sese quid sibi hac in re faciendum foret. Proxime habito Curatorum conventu adiit eos Martius, rogans ut velint ipsis auxiliari, ne novae hae opiniones in nostra Academia propagentur: Curatores ea de re monuerunt Regentes Collegii, rogarunt, qualisnam hic juvenis foret; et cum uterq[ue] omnia bona de ipso testarentur, jusserunt, ut invigilarent huic negotio: talia ipsos ferre non posse: addideruntq[ue], se postmodum quid ipsis hac de re animi foret Regentibus indicaturos. Ex quibus mihi persuadeo rem dilatam in adventum Opdamii: ille enim aberat. Vides facile quidnam nostri moliantur, ut sc.

of 1691,[283] De Volder never revealed his actual opinions on politics or religion. We may partially reconstruct his political views by considering Le Clerc's biography and De Volder's inaugural disputation (1660), his *Oratio* for Coeman and *Oratio de novis et antiquis*. According to Le Clerc, De Volder considered a republic – viz. the form of republicanism envisaged by Aristotle – as the best form of government for Holland and capable of granting freedom. A freedom intended – as revealed by De Volder's *De novis et antiquis*, commented upon by Le Clerc – as liberty of expression.[284] Such views, actually, are consistent with De Volder's (probable) closeness with Johan de Witt – with whose son Johan Jr. (1662–1701) he was a correspondent in the 1680s[285] – and with his judgment on William III. Moreover – in accordance with his Mennonite roots – De Volder might be labelled a pacifist. In fact, he came to disagree with Le Clerc on what is known today as the War of the Spanish Succession, whose end in May 1709 – Le Clerc noted – could not have been acknowledged by De Volder.[286] In particular, his political views seem to have been

---

reverentiam, quam putant sibi deberi, et quam meritis adipisci se posse desperant, metu obtineant, aut certe externum obsequium. / Lugd. Batav. a.d. 26 Decemb. 1691," Amsterdam University Library, OTM hs. Gs 24.

[283] "Academiarum moderatores sunt apud vos Presbyteriani, apud nos Episcopalibus propiores, dogmatis sc. rigore, non institutis. Hi autem licet alteri ab alteris e diametro diversi, tamen utrobique in eo tertio tam mirifice consentiunt, quod stupiditati plerique, ceu Deae, immolent, (saniori cuiquam praeiudicatum nolo) quod puriori [sic, pro purioris] philosophiae et certitudinum mathematicarum sint osores et hostes capitales, ideo quia ignorantes: et si res ex eorum cederit voto, tractionem earum facultatum si non exilui plecterent, saltem tam angustis circumscriberent cancellis, ut parum esset reliqui. Habet Lugdunum Batavorum suum Fredericum Spanhemium, Theologum, qui clavo assidet, censoria potestate. Haec et similia in causa esse ferme audio, quod D. Burghero de Volder minus sit animi quaedam in publicum protudere, quamvis ego etiam metuam ne assiduo isto silentio, velut rubigine, torpeant ingenia," transcription from Mojet 2016, 85. See also Mojet 2017.

[284] "Quas vero ad se non pertinere Magistratuum censuit prudentia, acribus quidem exarsisse initiis, sed incurioso sine sponte sua, vel rixarum non sine labore et absque emolumento exercitarum taedio si non compositas, ad silentium certe redactas. Neque enim existiment Magistratus, se suis edictis animis et opinionibus civium imperare posse. Quasi vero ut in Reip. regimine multa fiunt arbitrarie, quanquam haec ipsa eo meliori loco constituta est, quo maior in legibus, minor in Magistratuum arbitrio vis ac potestas est, itidem sese res haberet in scientiis. Quae vera sunt, sive velint, sive nolint, qui rerum potiuntur, erunt vera, nec ab horum sive favore sive odio dependent scientiae," De Volder 1709, 13–14. Cf. Le Clerc 1709, 391.

[285] See *supra*, n. 225.

[286] "Il est vrai que Mr. de Volder, au moins depuis une vintaine d'années en çà, que j'ai eu l'honneur de le connoître, m'a toujours paru extrêmement porté pour la Liberté & pour la forme du Gouvernement, qui est la plus conforme aux Loix de ces Provinces en géneral & de la Hollande en particulier. Mais il n'y avoit rien que de louable, dans cette disposition, & il étoit très-éloigné de tout ce que l'on peut appeller esprit de brouillerie & de tumulte. Tout le desordre venoit de ceux, qui vouloient parvenir à leurs fins, par des voies extraordinaires, & non par la Justice & par la Raison. Au reste, on ne doit pas faire honneur de cela au Cartesianisme, ou à un esprit de nouveauté; Descartes, comme nôtre Philosophe le disoit à Mr. Fagel, n'a jamais touché ces matieres; mais Aristote a donné en ses livres de Politique, tous les principes, que l'on nomme Républicains, dans la plus grande étendue qu'ils puissent avoir. Dans les Républiques de la Grece & de l'Italie, en son tems, c'étoit un honneur, que d'être bon Républicain: au lieu que nous avons vû, da nôtre, regarder comme une espece d'héresie des sentimens; que l'on devroit faire enseigner

inspired by the theory of natural law of Grotius – whose *De veritate religionis Christianae* had been translated into Dutch by his father – as appears from the last part of his inaugural disputation of 1660, namely the *De republica*. In it, De Volder argues that men formed societies in order to avoid the constant fear of an imminent death, by renouncing their freedom and, in the case that power is conceded to a collegial institution rather than to a monarch, constituting a republic.[287] In turn, the power conceded to rulers could be limited by some "fundamental laws," (as in the case of Christian societies) or unlimited, with the exception of those natural and divine laws, which hold in any kind of society. In any case, once conceded to the rulers, the power or *imperium* could be taken back by the people, in the form of a right of resistance.[288] In his *Oratio funebris* for Coeman, in turn, De Volder showed

publiquement dans tous les Etats bien reglez, en donnant de bons gages à ceux qui s'en acquiteroient le mieux. [...] A l'égard des matieres d'Etat, il étoit extrêmement zelé pour la République & pour la Liberté, & il ne dissimuloit point ses sentimens là-dessus. La crainte même, où il étoit qu'il n'arrivât, pendant la guerre, quelque desastre, qui nuisît à la liberté de l'Etat, lui faisoit souhaiter passionnément la paix. Il ne s'échauffoit jamais tant, avec ses amis, que sur cet article & je me souviensque nous avons souvent parlé de cette guerre, avec un peu de chaleur; sans blesser néanmoins le moins du monde les égards, que les honêtes gens doivent avoir les uns pour les autres. J'aurois bien souhaité, qu'il eût pu voir la fin glorieuse de cette guerre. Il auroit eu meilleure opinion de ceux, qui y ont le plus contribué," Le Clerc 1709, 367–368 and 399–340. Moreover, in his *Oratio de novis et antiquis*, De Volder praised the freedom granted by the States of Holland: "[q]uam ad rem nunquam sat magnis laudibus extolli poterit promulgatum ante paucos annos Illustrissimorum Hollandiae Procerum decretum, quod ut sententiarum admittit libertatem in iis, quae in controversia sunt, ita acerbitati tollit, moderationi praemia ponit quam iustissime," De Volder 1709, 17. Le Clerc and De Volder met for the last time in Summer 1708: see Le Clerc 1709, 397.

287 "Quis vero dubitet talem statum et ad humani generis, et ad cuiuslibet hominis conservationem esse inidoneum? Quis non concedat tale vitae periculum omnibus quibus fieri potest modi propulsandum et fugiendum esse? Quis non praeferendum existimabit perpetuo iamiam imminentis mortis metu amissionem libertatis? Hoc igitur communi mortis periculo ad societates constituendas adducti sunt mortales, cum intelligerent, sese in tali statu diu vivere non posse seque nullis pactis ad mutuum auxilium obstrictos, a vi externa tutos esse non posse, ob eamque causam accedente forsan instinctu cuiusdam, qui inter ipsos prudentior esse videbatur, societates constituerunt, seseque ad mutua auxilia obstrinxerunt, summumque imperium, sine quo societatem istam consistere non posse perspiciebant, vel uni, vel cuidam collegio concesserunt, quae societas respublica nominata est," De Volder 1660, *De republica*, thesis 3.

288 "Huic autem vel monarchae, vel collegio, quibus summum concesserant imperium, vel illimitatam dedere potestatem, quale imperium despoticum apud barbaros plerosque viget, quarum principes nullis legibus, praeterquam naturalibus et divinis, sese obligari patiuntur, vel quibusdam legibus, quas fundamentales vocant, limitatam quas ipsi principi reliquis legibus civilibus soluto non observare illicitum, qualia regna apud Christianos in usu sunt. [...] Actus vero principis sive eius qui summum habet imperium a nemine mortalium irriti redi possunt, cum nullius iuri subsint. Unde pernitiosissima illa opinio falsitatis convincitur, quae docet, penes populum summam semper residere potestatem, eumque omnes omnino reges, si imperio abutantur coercere et punire posse. Quae opinio quam facile flagitiosis hominibus ansam tradere queat ad principum exitium conspirandi quis non videt? Quis talem opinionem, nulli firmae rationi subnixam quae tot malis occasionem dare possit ex hominum animis plane extirpandam non iudicet? Praesertim cum validissimae rationes contrarium astruant. Quidni enim populus regendi sui ius in principem transferre posset, nulla eius parte retenta, cum privatus sese tali modo alterius potestati subiicere queat? Quodque privato conceditur, cur populo sui iuris id denegabimus? Verum tamen est, si princeps ita

appreciation of the views of his friend on the origin of society. According to De Volder (as no texts of Coeman are extant besides his untitled inaugural oration of 28 June 1679) Coeman acknowledged the origin of society not in fear (*metus*) – as maintained by Hobbes – but in a rational union of men following divine will – which perfectly matches natural or rational law: prescribing the fundamental norms – formulated by Ulpian – of living honestly, not to harm others, and to give everyone his due.[289] Once again, such a position seems to conform to pre-Hobbesian forms of natural law theories.

De Volder's political views are reflected in his ideas on religion. We have seen above that De Volder deemed as necessary to be held only those articles of faith concerning salvation, thus ideally belonging to the 'invisible church'. Thus far, he was a supporter of religious freedom – with the exception of Catholics, as appears from his second letter to Newton, in which De Volder congratulates him for having opposed the appointment by James II of a Catholic fellow, Alban Francis, at Cambridge in 1687.[290] Moreover, De Volder certainly had sympathy for the Remonstrants, as testified to by his activities as a physician of the Amsterdam Remonstrant Church and by his close contacts with Le Clerc and with Van Limborch, with whom he discussed theological questions (and agreeing with the latter's theory that the miracles of Moses were impostures).[291] As far as his approach to Scripture

---

se gerat, ut in totius populi perniciem conspirasse videatur, a populo ei iure resisti posse, rei iudicium et resistentia non cuilibet privatis competit, sed toti populo," De Volder 1660, *De republica*, theses 4–5. In the *corollaria*, moreover, De Volder claims that some good is antecedent to God's will (corollary 3) and that the pacts contracted by fear are anyway valid (corollary 8).

289 "[...] summo argumentorum pondere, refutantem Hobbii principia, quibus ad omnia scelera Natura factum hominem, eademque, quae meditatur, ab aliis sibi metuentem fingit. Ostendisset contra iuris, aequi bonique principia, non ex externo a sceleribus patratis deducenda metu, sed ex rationis rectae repetenda fontibus. Hanc velle docuisset ad perfectissimum, quantum in nobis est, Dei exemplar, ad sanctissimam Divinae voluntatis normam, componenda esse nostra omnia. Facile autem Divina attributa contemplanti liquere, eam eius voluntatem esse, quae benignissimae, iustissimaeque Naturae respondeat. Ad quam si advertamus mentem, cognituros nos, non sceleribus, non caedibus, non cuiusvis generis maleficiis, Dei nos Naturam moribus exprimere, sed honesta vita, neminem laedendo, et suum cuique tribuendo, quae prima sunt iuris fundamenta," De Volder 1679, 32 (unnumbered). See State 1991; Zagorin 2007, 2009; McConnell 2008; Greene 2010.

290 Cf. the text quoted *supra*, n. 258. In his colloquium with Stolle, De Volder remarked on how James II was in fact Catholic: "[c]redere se Jacobum ex animo fuisse pontificium, alias se (propter religionem) non passus esse regno eiici," University Library, ms. R 766, 454.

291 De Volder commented upon Van Limborch's *De veritate religionis Christianae amica collatio cum erudito* (1687) – viz. Orobio de Castro – in his letter to Philipp van Limborch of 3 November 1687. In the letter, De Volder does not show sympathy for the Cocceians, who stood for the so-called 'prophetic theology' (which was the case, at the time of the letter, of Salomon van Til and Campegius Vitringa; see Daugirdas 2017). Notably, De Volder shows appreciation of Van Limborch's comparison between Moses (whose miracles, according to Van Limborch – probably under the influence of Spinoza: see Landucci 2015 – were impostures) and Christ (who performed actual miracles; see Van Limborch 1687, 172–175). Moreover, De Volder polemically refers to the Cartesian theology of the Franeker theologians, viz. Hermann Alexander Roëll and and his nephew Gysbert Wessel Duker (see Bordoli 2009). Cf. the full text: "Amplissime Vir, / Quod jamdudum factum oportuit, id nunc demum ago, gratias quas debeo, tibi ut agam maximas pro libro, quem ad

is concerned, as I am going to show in Sect. 3.1.2.2, *Cartesianism under attack at Leiden in the mid-1670s*, De Volder was a foremost supporter of the 'separation thesis', i.e. of the idea that theology and philosophy have different purposes and different methods. This was a view he shared with Cartesio-Cocceian faction in the

---

me misisti. Fuit profecto ille mihi et dignitate, quam {tractas}, rei, et quod te Auctorem agnoscit gratissimis. Illa {qmdem}, quam instituis comparatio inter causas propter quas Iudaei Mosi, nos Christo credamus, mihi admodum placuit. Neq[ue] enim alia ratione arctius constringi {Iudaeis} potuit; Nisi forte Franequerani nostri aliam meliorem viam ex sola ratione invenerint. Sed quidquid huius rei {sit, nullius} dubito, {quni} ingenua tua confessione nullum talem, qualem Iudaeus ille quaerebat locum extare, offenderis publicae Religionis Theologos, et praesertim Coccejanos, quos vocant, qui, ut nosti, quam clarissim[missing text] se ubiq[ue] in veteri testamento Christum invenire depraedicant. In qua opinione confirmor ex eo, qud ante biduum duos ex eo genere, quorum tamen neutri prelectus erat libertinus, ratiocinantes inter se audivi de locis quibusdam e Psalmis et Ieremia, ex quibus evidentissime Iudaeos convinci posse asserebant. Sed hactenus de tuo libro nihil, quod {et}in causa est, cur te hisce tam tarde compellarim. Sperabam enim aliquid fore, quo nimiam {hancae} meam moram compensarem. Nunc cum spes haec frustranea fuerit, ad tuam ut confugiam aequitatem necesse est, quae facile {huic} meae negligentiae ignoscet. Vale et ama / Tuum ex asse, / B. De Volder. / Lugd. Batav. / a.d. III Nonas Novembr. / CIƆIƆCLXXXVII," Amsterdam University Library, OTM hs. J 83 b. The identity of the two Cocceian theologians mentioned by De Volder cannot be ascertained. For a commentary, see Strazzoni 2019. Moreover, De Volder is reported to have commented upon Van Limborch's *Historia Inquisitionis* (1692), judging it positively against Jacobus Trigland II (grandson of Jacobus Trigland, a contra-Remostrant): see letter 1640 of Locke's correspondence (Van Limborch to Locke, 16 June 1693): "D. Voldero, non diu post editam meam historiam aderat cum aliis Triglandius Professor. Vidit historiam meam, eaque leviter inspecta rogat eum Volderus, an eam viderit ac legerit. Alter negat; additque: homines hi credunt se anno huius saeculi decimo octavo laesos, ideoque istud tempestive intempestive inculcare ac repetere solent. Volderus respondit, nihil horum, uno ne unum quidem verbum, exstare in tota hac historia; nisi fortasse existimet, lectores, qui hic processus Inquisitionis graphice depictos conspiciunt, credituros gesta anni decimi octavi hisce per omnia esse similia, atque ea ratione gesta haec libro condemnari," Locke 1976–1989, volume 4, 695. Van Limborch refers to the Synod of Dordrecht (1618). See Simonutti 2002. See also Sect. 3.1.3.1, *The relation between material and immaterial substances*, where I analyze De Volder's role in the Locke-Hudde dialogue (mediated by Van Limborch) on the uniqueness of God. On Jacobus Trigland II, see also De Volder's letter to Braun: *supra*, n. 267. Also, in 1694, as member of the Academic Senate, De Volder helped Van Limborch in obtaining a copy of the academic *testimonium* given by Leiden University to the widow of Jacobus Arminius in 1611 (see Molhuysen 1913–1924, volume 2, 18). De Volder himself noted a difference between the *testimonium* for Arminius, and the academic *testimonium* for Franciscus Gomarus (conceded to him in the same year, when he was going to leave his post after Conrad Vorstius, an Arminian, was appointed in replacement of Arminius himself). This was going to be remarked on by Van Limborch in his posthumous *Relatio historica de origine et progressu controversiarum in Foederato Belgio de praedestinatione*. See the full text of the letter by De Volder to Van Limborch of 16 November 1694: "Vir Reverende, / Commodum accepi tuas eo ipso tempore, quo mihi adeundus esset Senatus Acad. Quare statim quaesivi ex eo, qui Senatui est a Secretis hunc illumve Actorum librum, quorum jam binos frustra pervolveram, cum incidit, repertum in forte hoc testimonium inter acta quae diu dispersa jacuerunt, et nunc demum ante paucos annos sine ulla temporis ratione in unum volumen erant compacta; Adii itaq[ue] et illud, et cum jam de successu desperarem <{...}> obtuilit sese tandem hoc ut opinor, quod quaerebam, cujus apographum hic ad te mitto. Aderat simul et testimonium ab eodeom Senatu Gomaro datum adhuc reverti, et stationem suam relinquere meditanti. In hoc loco ejus, quod in Arm. testimonio est, (nam de controversiis non judicamus) habetur (controversias enim aliis relinquimus). Quod eam ob causam addo, ne forte existimes hanc clausulam Arminio peculia-

Dutch universities, which at Leiden had its foremost supporters (in 1670s) in Heidanus and Christoph Wittich. In fact, the suspicions of Stolle and Hallmann that De Volder was a secret atheist[292] are exaggerated. Indeed, as reported by Gronovius De Volder immersed himself in a daily study of the Bible – reported also by his servants – in order to ascertain those passages aimed at salvation, in a way consistent with the Cartesio-Cocceian approach to theology,[293] with his own Mennonite roots, and also with the Remonstrant theology of Van Limborch and Le Clerc, both sharing the idea that (in the words of John Marshall), "the Christian religion consisted principally of obeying the precepts of the gospel, that there were many doctrines in the Scripture that were not absolutely necessary to salvation, and that there were only a few essential or 'fundamental' articles which were clearly revealed."[294] Amongst De Volder's papers, moreover, Gronovius found a *Collectio eorum quae aperto sensu in SS. commendantur ut omnino necessaria scitu vel factu ad salutem, sequentibus numeris ad denotanda illa loca*, as well as a paper in which the Greek text of the Letters of Paul and the Apocalypse was compared (with extensive annotations) with the Vulgate.[295] This testifies to a philological interest in Scripture, which

---

rem esse. Sigillum erat munus Acad. quo in litteris et testimoniis utitur Senatus. / Hodie novus hic, absente licet Obdamio, creatus est Medicinae Professor Dekkers, Auctum simul est salarium D. Bidloo ad 600 flor. Annuos, Ex quibus facile vides, aliis ex oris jam apud nos spirare ventum, quam solebat. Quid inde mutationis futurum sit dies docebit. Nil mediocriter exspecto sive boni, sive mali. Vale, et si quid porro sit, in qui tibi inservire queam, impera / Tuo / B. de Volder. / Lugd. Batav. / a.d. XVI Novemb. 1694," Amsterdam University Library, OTM hs. J 83 c. For a commentary, see Strazzoni 2019. Cf. Van Limborch's *Relatio historica*: "Gomarus professionem suam resignavit Academiae Curatoribus, et in eius locum vocatus est Joannes Polyander. Testimonium Gomaro perhibitum est a Curatoribus, perinde atque Arminio defuncto; nisi quod, cum in Arminii Testimonio legeretur, nam de controversiis non iudicamus, idem in Gomari verbis paululum immutatis expressum sit: controversias enim aliis relinquimus," Van Limborch 1730, volume 1, 16. The *testimonium* for Arminius was then printed, in its full form, in the 1704 edition of the *Epistolae praestantium ac eruditorum virorum, ecclesiasticae et theologicae varii argumenti* (as epistle 5), edited by Van Limborch (first edition 1660).

[292] "[...] weis ich wohl, daß man von ihm nicht, wie von einem andern lieben Manne sagen kann: nihil est intus, denn das Gemüthe dieses holländischen Philosophi ist eine solche Tieffe, darinne viel sonderbahre Gedancken, und darunter allem Ansehn nach auch einige Atheistische verborgen liegen," Warsaw University Library, Cod. IV oct. 49, 595–596. Transcription kindly provided by Martin Mulsow.

[293] See Sect. 3.1.2.2, *Cartesianism under attack at Leiden in the mid-1670s*.

[294] Marshall 2006, 481.

[295] "[F]requentissimus in lectione verbi Divini non tantum certis horis, sed fere quoties opera cessabat. Hoc inspicere solebat non decurrens, verum ad singulos articulos meditabundus, eosdemque non semel, crebro repetens; et siquid aut non satis arriperet, aut existimaret propius ad se pertinere, surgentem et velut in ruminatione acri, dum ambulat, susurrantem ac volutantem domestici saepe viderunt, sive infigere vellet memoriae, sive interiorem eius causam scrutaretur. Venit in manus meas charta, in quam coniecerat omnia rarius occurrentia Graeca vocabula Epistolarum Pauli Apostoli et reliquorum, etiam Apocalypseos, quae cum vetere versione contulerat, saepe ad scripto quod ipsi videretur; ut pateat ea ab illo emeditata fuisse. Extat altera, cui inscriptum est: Collectio eorum quae aperto sensu in SS commendantur ut omnino necessaria scitu vel factu ad salutem, sequentibus numeris ad denotanda illa loca. Et quam vim pietatis censeret, demonstravit affatim disciplina et forma vitae quam tenuit, velut ad sanctissimum aliquod praescriptum gubernata. Hinc

was however quite different from the hermeneutical approach adopted by Spinoza – aimed at ascertaining the human and historical character of the Biblical message,[296] with whom De Volder was from time to time associated. In any case, as I am going to show in Chap. 3, De Volder cannot be meaningfully labelled a Spinozist, even if he was committed to a peculiar form of necessitarianism (highlighted by Gronovius as well). His approach to Revelation, in fact, was more inspired by the ideal of the 'invisible church', on the one hand, and on the separation thesis on the other, according to both of which Revelation is aimed only at the salvation of man, and Holy Writ is to be interpreted in the light of itself. This approach De Volder also appreciated in his friend Coeman,[297] and in Lucas Schacht.[298]

2.4 The Last Phase of De Volder's Career

The last phase of De Volder's career, which we may tentatively date as starting in 1698, was characterized by what Le Clerc has labelled as De Volder's acknowledgment of *le foible du Cartesianisme*.[299] In this phase, De Volder dramatically reduced his activities as *praeses* of disputations – even if, as seen above, he probably continued to teach Descartes's philosophy, through *dictata* upon Rohault's *Traité*. After 1698, he formally acted as promoter of John Gale (who was to become a foremost Baptist theologian),[300] graduating *more maiorum* (viz. with an honorary degree) and presenting an *Inquisitio philosophica inauguralis de lapide Solis* (2 July 1699, dedicated to Senguerd), which De Volder's *Oratio de novis et antiquis* followed on the next day.[301] Further disputations were the standard Cartesian *Disputatio philosophica de mentis humanae et divinae existentia et reale a corpore distinctione* (1699), authored by the *respondens*, as well as inaugural disputations in which De Volder

potuit extrema Coemannii et Schachtii tam pie depingere, scilicet quod sibi in antecessum optaret similia evenire," Gronovius 1709, 28–29. De Volder owned several editions and commentaries of the Bible: see *Bibliotheca Volderina*, 38–53.

[296] On Spinoza's biblical hermenutics and its Dutch context, see Frampton 2007; Douglas 2015; Van Miert et al. 2017; Morrow 2017; Van Miert 2018b; Touber 2018.

[297] According to De Volder's oration for Coeman, his friend had shown sympathies for Cocceianism, and entered into theological discussions (for instance, with Wittich) only on salvation: see De Volder 1679, 20–22 (unnumbered).

[298] In his *Oratio*, De Volder praises 1) Schacht's approach to theology by considering Holy Writ only, 2) by comparing the Old and the New Testament, and 3) by using as the sole norm of Biblical interpretation Holy Writ itself. Such an approach had been inspired in Schacht by Heidanus, Johannes Cocceius, and Jacobus Trigland: see De Volder 1689, 9. See also Molhuysen 1913–1924, volume 4, 61.

[299] Le Clerc 1709, 398.

[300] See *supra*, n. 87.

[301] The main tenet of De Volder's *Oratio*, as remarked also Le Clerc, is that it is absurd to favour a priori the ancients or the moderns, as both endorsed true and false opinions: see De Volder 1709, 2–5; Le Clerc 1709, 390–392. The *Oratio* is concluded by a commendation of Gale: see De Volder 1709, 18–19. See also Molhuysen 1913–1924, volume 4, 76*–77*.

acted as promoter, in physiology[302] and philosophy, the last one being the *Disputatio philosophica inauguralis de mundo* of Jacob vander Ghiessen (20 September 1703) – which does not have a markedly cosmological character as it is aimed at demonstrating the non-eternity of the world. In the disputation, moreover, his name never appears.[303] Finally, the De Volder's last text was his *Oratio qua sese laboribus academicis abdicavit*, given on 19 October 1705 on the occasion of his retirement (to which I shall return in a moment).[304]

2.4.1 De Volder as Editor of Huygens's Posthumous Works

Between the two centuries, whereas the academic activities of De Volder slightly declined, his scientific activities reached their peak with his correspondence with Leibniz, as well as with the editorial preparation of Huygens's unpublished treatises. In accordance with his testament of 23 March 1695, all the papers of Huygens (who died on 8 July 1695) were left to the Leiden University Library. In turn, De Volder and the professor of mathematics at Franeker Bernard Fullenius (1640–1707) demanded that his unpublished though completed treatises be published, namely, his *Dioptrica*, a treatise on the fabrication of glasses (in Dutch – then translated by Boerhaave),[305] his *Dissertatio de coronis et parheliis*, and his *De motu corporum ex percussione*,[306] for which De Volder and Fullenius were each endowed with 1,000 guilders.[307] In fact, De Volder and Fullenius did not publish only the treatises indicated by Huygens in his testament. As I am going to show in Sect. 6.2.1.2, *Huygens's quantification of centrifugal force*, they also published Huygens's *De vi centrifuga* (re-ordering its contents) and *Descriptio automati planetarii*; moreover, they provided some propositions of Huygens's *Dioptrica* with new demonstrations – as duly explained in the preface (written, according to Johann Bernoulli, mostly by De Volder)[308] opening Huygens's *Opuscula postuma*, which

[302] Viz. the *Disputatio medico-philosophica inauguralis de imaginationis maternae viribus in foetum* and the *Disputatio medica inauguralis physico-anatomica de spirituum animalium vera existentia et operatione* of Johannes Zoutmann, 1702.

[303] See Sect. 3.1.2.5, *De Volder's necessitarianism reassessed.*

[304] See Molhuysen 1913–1924, volume 4, 216 and 220.

[305] See Huygens 1703, *Praefatio*, 8 (unnumbered).

[306] On Fullenius, see Dijkstra 2007; Dijkstra 2012.

[307] See Huygens 1888–1950, volume 22, 773–778. An extract is provided in Molhuysen 1913–1924, volume 4, 58*–59*. The testament was read at the University Senate on 8 February 1697: see Molhuysen 1913–1924, volume 4, 151–152. See also 183, 207–208, 310–311 and 118*–119*; Gronovius 1709, 27; Le Clerc 1709, 289. See, moreover, the letter of De Volder to Gregory of 31 December 1695, quoted *supra*, n. 261.

[308] "Perlustravi haec opuscula, et nihil non dignum reperi Auctoris sui existimatione. Legi quoque Vestram eruditam praefationem, quae licet etiam Cl. Fullenii nomen in fronte praeferat, a Te tamen conceptam conscriptamque esse sunt quaedam indicia quae me dubitare non sinunt," transcription from *Briefwechsel von Johann I Bernoulli* (Basel University Library, ms. L Ia 675, 176–181).

was eventually published in 1703. Notably, in working on the *Dioptrica*, De Volder and Fullenius were helped by Newton, who communicated to them the results of his experiments on optical aberrations.[309] Moreover, De Volder took care of the last phase of the publication of Huygens's *Cosmotheoros* (1698 – to which he wrote a short preface),[310] the publication of which had been initially charged to Christiaan's brother Constantijn Jr., but which had been interrupted when Constantijn died on 12 November 1697.[311] In this case, however, De Volder seems not to have provided any change to the text itself.

2.4.2 *De Volder's Retirement, Death and Legacy*

De Volder's edition of Huygens's works was his last, main contribution in the history of philosophy and science. In 1704 his last, great patron – Johannes Hudde – had died, leaving to him 1,500 guilders.[312] In 1705, finally, De Volder resigned from

[309] "Nec possumus quin hac occasione referamus V. Cl. Newtonum nobiscum communicasse, recentioribus se experimentis accuratius didicisse, [...]" Huygens 1703, *Praefatio*, 9 (unnumbered). No letter of Newton to De Volder or Fullenius is extant as to this point. Leibniz showed much interest in De Volder's edition of Huygens's posthumous works: see, for instance, the letters of Bernoulli to Leibniz of 17/27 December 1695 and 21/31 July 1696, and the letter of Leibniz to Bernoulli of 20/30 October 1695: A III6, 586–591, GM III, 295–302, GM III, 801–810. See also the letter of Leibniz to Antonio Magliabechi of 3 October 1702: A I21, 543–546. I will deal with Leibniz's interests for De Volder's editorial work on Huygens's posthumous works also in Sect. 4.2.3.4.1, *De Volder on non-elastic impacts*. Notably, in his letter to Leibniz of April 1701, Johann Gabriel Doppelmayr reports to Leibniz that he would see Huygen's manuscripts before their publication: "[n]och was neues muß ich vermelden, daß der Hl. Professor de Volder ein Tractat von dem Her. Hugenio von seinem Automate planetarum bald ediren werde. Bei Hl. Volder ist sein Automaton zu sehen, so sehr courieus. Sonsten hat Hugenius Hl Volder unterschiedl: Scripta zum ediren hinterlaßen, die aber alle affecta sint," Gottfried Wilhelm Leibniz Bibliothek Hannover, ms. LBr 971, 60v, transcription from *Briefwechsel Johann Gabriel Doppelmayr*. This is confirmed by the letter by Johann Georg Eckhart to Leibniz of 23 June 1702: "H. Volder zu Leyden soll vom Hugenio eine Dyoptricam (wo mir recht ist) unter der Preße haben. Es hat besagter Monsieur die übrigen schedas Hugenianas bey ihm gesehen. Der nahme dieses frembden ist Joh. Gabr. Doppelmayer und ist er von Nürnberg bürtig: ich hoffe noch viel von ihm zuhören, wenn ich übermorgen als auf St. Johannis mit ihm nach Heringhausen fahren werde," A I21, 33.

[310] "Il n'y avoit pas long-tems, que Mr. Huygens étoit mort en ce tems-ci, car son Cosmotheoros, qui avoit été commencé à imprimer pendant sa vie, parut l'an MDCXCVIII avec un petit avertissement de Mr. de Volder, mais sans nom," Le Clerc 1709, 389.

[311] On the *Cosmotheoros*, see Ait-Touati 2011, chapter 4; Van der Schoot 2011, 2014.

[312] "L'année suivante MDCCIV Mr. Hudde, Bourgmestre d'Amsterdam, mourut le 15 d'Avril, & pour montrer l'estime, qu'il faisoit de Mr. de Volder, il lui fit un légat de quinze cents florins. Il donna une semblable marque d'amitié à Mr. Hubers d'Utrecht, leur ami commun, dont il connoissoit aussi le mérite depuis longues années. C'étoit un grand Magistrat, & à la mémoire duquel la ville d'Amsterdam fera éternellement redevable," Le Clerc 1709, 393. See *supra*, n. 257. The identity of Mr. Hubert, from Utrecht, could not be ascertained. He is mentioned in the correspondence of Locke, as a friend of Mattheus Sladus: see Locke 1976–1989, volume 2, 656 (letter 794);

his academic positions. Various reasons have been put forward for De Volder's decision. Gronovius, Le Clerc and De Volder himself have mentioned De Volder's problems with health. For instance, Gronovius reports that, after De Volder's corporeal constitution improved in his adulthood, in his mature age he was frequently subject to illness.[313] He suffered jaundice, and when this had passed (after a treatment prescribed by his physician and friend, a certain Hoogmadius, who died before De Volder), he diagnosed himself with cancer, while urination was painful and showed blood(also mentioned by Le Clerc).[314] A further reason for his retirement was offered by De Volder as his chief motivation in his 1705 *Oratio*, namely, the search

---

volume 4, 64 (letter 1286). He is also mentioned in the letter of Swammerdam to Thévenot of 30 October 1670: "Doctor Huyberts versta ik met gesontheid te venetien is gekoomen," Lindeboom 1975, 53. Hudde's posthumous papers are discussed by De Volder in his correspondence with Leibniz: see the letter of De Volder to Leibniz of 5 January 1706 (GP II, 279–280). In 1705, De Volder remarked that all his colleagues of 1670 had died in the meantime: De Volder 1705, 28–29.

[313] For instance, in the documents at Leiden University it is reported that on 20 April 1693 he could not travel to The Hague to meet the delegates of Dutch cities as a consequence of a colic: Molhuysen 1913–1924, volume 4, 110. De Volder's problems of health are mentioned by De Volder in his letters Leibniz of 18 February 1699 and 5 January 1706, by Bernoulli in his letter to De Volder of 8 April 1699, and by Leibniz (who had heard about them from Bernoulli) in his letter to De Volder of 30 June 1704. As to Gronovius and Le Clerc, see the next note.

[314] "Undiquaque enim intentus amoenitati suae Volderus quum caveret acriter ne quem diem perderet, nec aliquod officii intermitteret munus, atque omnia desiderat habere evigilata, allevabatur quidem lenociniis istis et gaudebat animus. Corpus, quod ab ista primae aetatis macie inceperat iniustam aequalitatem ac congruentiam temperari et convalescere, satis firmum et par exantlando quod ab eo exigi debebat, denique vires suas obtundi seque hebetari interdum ostendit. Aetas quoque illum iam ad cellam frigidariam (senectutem intelligitis) admoverat. Corpus ergo velut destitui se querens et minans proditionem, nisi studiosius amare vellet, non omisit aliquoties admonere, ut parcius alteri indulgeret, non absumendo artus externos, sed interiore motu concitato, ac denique per vias a natura perforatas indicando, quantis miseriis premeretur. Saepius in morbos satis saevos ac truces coniectum vidimus; sed in eis tandem regnaverunt indicia icteri, non cute modo ad deformitatem flavae bilis tormento infecta, sed urina quoque fuscam nigredinem attrahente. Neque sic semel convellisse satis fuit; recurrebat identidem velut ad constitutum, quoties nihil moratus, solennia laboriosae vitae repeteret assidue Volderus; praesertim postquam sibi persuadens certum praesensque remedium ab Hoogmadio, Medico et longi temporis Amico, excogitatum esse, eo etiam post mortem istius utens experiretur sibi conducere. Sed qualiscunque tandem fuerit morbi istius sive remotio sive remissio, non fuit adeo coecus Volderus, ut non sentiret adesse aliquid, quo labefactaretur, et hostem in ulnis suis foveri. Quod ubi non tenui suspicione, sed rebus et documentis deprehendit, prudentiam deferens nunquam, in hac etiam parte iudicavit sanum esse oportere, praesertim quum exterreret iam cruentans se urina non sine cruciatu," Gronovius 1709, 29–31. No precise information could be found on Hoogmadius. A certain J. V. Hoogmade, medical doctor in Leiden, is mentioned in the medical *Opera omnia* (1703) of Johannes Dolaeus as having appreciated his works in 1689. Moreover, Dolaeus provides the text of two letters of Hoogmade to him, of 1678 and 1687: see Dolaeus 1703, 17 (unnumbered), 28, 94. De Volder's jaundice is mentioned in the letter of Leibniz to Bernoulli of 6 June 1704: see A III9, 373–375. See also Le Clerc 1709, 393: "[e]nfin l'an MDCCV. après avoir été incommodé quelque tems, en sorte que pour peu de mouvement qu'il se donnât, il rendoit de l'urine sanglante; comme il ne pouvoit plus faire ses leçons publiques & particulieres, sans s'incommoder davantage, il demanda sa démission à Mrs. les Curateurs de l'Academie."

for a literary *otium* and the avoidance of further academic quarrels.[315] Behind this we can discern the quarrels which started once more to agitate the Dutch academies: for instance, in 1699 and 1702 Johann Bernoulli – as I am going to show in Sect. 3.1.4.1, *The Bernoulli-Hulsius quarrel* – was involved in a theological quarrel at Groningen (of which De Volder was duly kept informed). Moreover, in 1717–1720 Jacob Wittich (1677–1739, nephew of Christoph), who had been a student of De Volder, was involved in the polemic on the nature of God which extended to Jena, and caused by some (allegedly) Spinozist positions he held in his 1711 disputation *De natura Dei*, some of which were directly inspired by De Volder, as Andala was going to note in 1710s. Andala himself, actually, in 1705 had sent to De Volder some critical remarks on his metaphysics, and causing – as Andala maintained – De Volder's resignation after the summer break.[316] This might also have been a cause of his resignation, so that even if earlier in his career he was apt to reject others's criticisms, and at the same time avoid overt quarrels and attacks, as emphasised by his biographers,[317] later in his career he was no more likely to do so. Probably for the reason that, as reported by Le Clerc, he was less and less 'tempted' by the philoso-

---

[315] "Et sane si ulli, hi certe, qui in Academiis docent, seclusi quasi a reliquis sibi tranquillissime viverent, nisi suam animi inquietudinem, sua vitia in hanc scenam afferrent, et quibus carerent haud difficulter, ipsi sibi turbas et molestias pararent. Hoc enim nisi foret, unde, quaeso, tot inter scientiarum cultores rixae, tot iurgia, tot simultates summis saepe cum inimicitiis exercitae? [...] Quod cum omnibus extra partes constitutis, et rem recto iudicio ponderantibus perridiculum necessario apparere debeat [...]. Quid ergo? Egone Otium tantopere appetendum censeo? Otium, quod omnium vitiorum audit mater; Otium, quod homines stupidos, inertes, sibique et aliis inutiles, et infructuosos facit. [...] Ego vero Otium summopere laudo, sed non illud [...] quod vera hominis vivi sepultura est, sed illud, quod studiis impenditur; quod in veri inquisitione, cognitioneque versatur; quod in mente perficienda elaborat; quodque nullam ob causam Otium dicitur, quam quia secretum a rebus huius vitae soli contemplationi vacat. [...] At ego vitae meae genus mutabo nequaquam, sed faciam, nisi fallor, liberius, faciam iucundius. Neque enim id molior, ut me alii aut Disciplinae, aut Professioni addicam, sed ut in iisdem, quibus assuevi, studiis, placide consenescam; [...] Potestne Academia ulla ratione conqueri, quasi a me praemature derelinqueretur, cum rari admodum sint, qui suam operam in una Academia tot, quot ego in hac feci, annos collocent? Si illis, qui ab una Academia ad aliam transeunt, id ipsum honori ducitur, non vitio vertitur, quanquam eos, qui id faciunt, sui emolumenti gratia id facere nemo ambigat, mihi ne crimini dabitur, quod non quaestus, sed quietis et tranquillitatis studio a laboribus Academicis discedam? Et discedam quidem post tanti temporis in ipsa moram, ubi ingruentis aetatis incommoda, quorum initia iam persentisco, effectura sint, ut minus porro ipsi prodesse queam," De Volder 1705, 7–10, 21 and 26. See also Le Clerc 1709, 394–395.

[316] See Sect. 3.1.3, *De Volder on substance monism.*

[317] For instance, Gronovius reports how a young scholar from Brabant once came to Leiden – under pressure from others – to attack De Volder's metaphysics, with himself posing as a defender of religion. After De Volder's answers to him, this scholar – whose identity cannot be ascertained – agreed with De Volder. Moreover, Gronovius reports that during a journey by sea, a famous man from another town attacked De Volder's ideas in the presence of De Volder himself, without knowing who he was. De Volder, not revealing his identity, managed to convince this man who only afterwards discovered that he had been talking with De Volder himself: see Gronovius 1709, 27–28. According to Le Clerc, when De Volder did not want other people to know his ideas, he used to give generic answers: Le Clerc 1709, 397–399. Le Clerc also underlines his respectful manners, for instance, in his criticisms to Huet's *Censura*: Le Clerc 1709, 381.

phy he taught, and annoyed to continue teaching Descartes's physics and metaphysics, which he saw as offering no possibilities of progress in knowledge – without however embracing any new system.[318] Indeed, in his 1705 *Oratio* De Volder himself claimed that the "system of the sciences" is suitable to be used for teaching if it is perfect and complete, otherwise, it causes quarrels, such as those provoked by the use of doubt in philosophy, leading to accusations of atheism and scepticism – viz. those he fought in the 1680s.[319] However, it seems that he was not eager to defend his Cartesian positions any longer.

In fact, from the autumn 1705 De Volder retired from academic teaching – and from the related direction of the astronomical observatory – but he still maintained a role at the University, together with a decreased salary of 1,000 guilders per year, conceded to him on 6 October 1705. At the end of his career, as reported by Gerard Noodt in a letter of 15 July 1709 to Johann Bernoulli – who had been offered the succession to De Volder as professor,[320] but who did not ultimately accept the posi-

[318] See *supra*, n. 101.

[319] "[N]ecesse est, ut scientiarum systemata tractemus, quae ad docendum ut sunt accommodatissima, et si ad perfectionis culmen pervenissent scientiae, omne punctim ferrent; ita hoc in rerum statu, quo a perfectione absumus quam longissime, scientiis promovendis multo magis nocent, quam prosunt. His, ut omittam, quae de rerum contentionibus modo attigi, accedunt innumerabiles logomachiae, quae sedem quasi suam ita in Academiis fixisse videntur, ut illis, prorsus abstinere omnino non liceat. Quid est, quod non tantum Academia, sed, et Ecclesiae cathedras maiori contentionis implevit strepitu, quam illa in principio philosophiae universalis de omnibus dubitatio? Ab his tanquam summopere necessaria propugnata, ab illis reiecta non modo, sed cum dubitationem Dei involvere videretur, scepticismi, atheismi, et nescio cuius non impietatis insimulata, cum tota haec res tantis utrinque animis agitata, si rite perpendatur, in mera consistat verbi controversia," De Volder 1705, 18–19.

[320] In his letter to Johann Bernoulli of 21 May 1709, Noodt announced De Volder's death and manifested his hopes that Bernoulli would succeed him. Moreover, Noodt praised De Volder's ideas in physics and metaphysics, and his activities as director of the observatory: "[n]escio, Vir Clarissime, an allatus ad te sit nuncius de morte Viri admirabilis, Burcheri de Volder, Collegae mei, et communis nostri amici. Abiit ille ad plures a. d. 28. Martii: mihique et omnibus qui ejus virtutes norant, incredibile sui reliquit desiderium. Jacturam hanc sentit literatus orbis; sed maxime nostra Academia: cui doctrinae, ingenii, et sapientiae, veneratione ingens erat ornamentum. Quod restat, cogitant nunc, ad quos ea res pertinet, de substituendo ei successore qui possit tanti hominis vices supplere. Commendantur autem complures: inter quos aliqui neque absurdi, et doctrina spectati. Ego vero et amici mei quem tecum comparemus, nullum novimus: itaque quem lubentius accipiamus Collegam praeter te non desideramus alium. Quid futurum sit, nescimus: quamquam speramus, fore: ut in te possit cadere consensus procerum qui veram amant doctrinam. Nil tamen videntur tentaturi, quamdiu incerti sunt; utrum patriam Academiam nostrae sis praelaturus: cum praesertim timeant, spe sua iterum excidere cum aliqua Academicae dignitatis imminutione. Accedit, quod nesciunt, quibus conditionibus moveri possis ad suscipiendam apud nos Philosophiae et Matheseos professionem, exemplo Clarissimi Volderi, in omnibus suis partibus exercendam. Eget Academia nostra, praeter Physicam, in primis Metaphysices doctrina qua excelluit amicus noster. Sunt praeterea in turri Astronomica insignia Astronomiae instrumenta quorum negligi usum, minime convenit. Quis dubitet, tua opera, eruditione, fama, magnum ex omni Philosophiae et Matheseos disciplina ad Academiam nostram perventurum fructum?" transcription from *Briefwechsel von Johann I Bernoulli* (Basel University Library, ms. L Ia 713, 31–32). On De Volder's death, see also the letter of Bernoulli to Noodt of 5 June 1709 (in *Briefwechsel von Johann I Bernoulli*).

tion as, for him, it was underpaid – De Volder earned 1855 guilders per year, almost the amount then offered to Bernoulli (1,800 guilders). It is worth quoting Noodt's letter, because it offers a significative overview of De Volder's career:

> for what it concerns the annual salary: I will speak as with a friend, simply. The most renowned De Volder had annually 1,800 guilders of our currency. To these 55 guilders [were] added for the honour of the gown, and for buying back the right to hunt. These sums[,] together[,] make annually 1,855 guilders. And however that egregious man did not always enjoy such a great salary. At the beginning he stayed within 1,000 guilders. Then the teaching of mathematics added, if I am not mistaken, 400. Afterwards the physical experiments 200. [When he was] already old[,] the liberality of the Curators, for a few years[,] honoured [him] further another with 200 guilders annually. So that the maximum salary which [he] had, was 1,855 guilders. And, before De Volder, it was not as much for any philosopher or mathematician. [...] I will not be silent (indeed it is of a certain importance) that we are all exempted from the taxes on wine, beer, soap, coffee and tea. Moreover the dignity of [being] Rector Magnificus moves annually through the disciplines, and the function of cashier of the Senate: both the profits [are] in fact large enough. [...] There is no need to add [something] about the frequent promotions of doctors [...]. I also reminded [in a previous letter] about private lessons: I will anyway say, that in these at that time 80 philosophers listened De Volder, often 100, and more: from each as honorarium [...] was given 10 [...] ducatons. For algebra[,] on the other hand[,] at whose explanation he had generally 4 students, he did not want more, was given by each 100 thalers.[321]

---

[321] "Quod vero attinet ad annuum salarium: loquar ut apud amicum, simpliciter. Habuit clarissimus Volderus quotannis mille et octingentos florenos nostrae monetae. His accedebant quinque et quinquaginta floreni pro honore togae, et pro redimendo jure venationis. Hae summae junctae efficiunt quotannis mille octingentos quinque et quinquaginta florenos. Nec tamen egregius ille vir semper fuit tanto fruitus stipendio. Primum stetit intra mille florenos. Adjecit deinde Matheseos doctrina, nisi fallor, quadringentos. Mox Physica experimenta ducentos. Jam senem Curatorum liberalitas, ante paucos annos ultro honoravit aliis ducentis annuis. Ita maximum quod habuit salarium, fuit mille octingentorum quinque et quinquaginta florenorum. Nec, ante Volderum, ullius Philosophi aut Mathematici fuit tantum. Nescio quoque, an totum impetraturus sis, ob aerarii quam dixi, inopiam. Eo tamen aut paulo minore si possis esse contentus: videbimus, quid procurare tibi valeamus per amicos qui amant famam eruditionis et nominis tui. Plus certe nihil impetrabis. Etsi, si Deus faveat fortunae Academicae, et tuae, fieri possit; ut aliquis deinde accedat cumulus: sic enim aliquando contigit, atque ipsi Voldero. Verum hoc, ut futura omnia, sperare potes, polliceri tibi non potes. Illud non silebo; (est enim alicujus momenti) quod omnes sumus immunes a vectigalibus vini, cerevisiae, saponis, coffae et theae. Ambulat praeterea annuatim per disciplinas Magnifici Rectoris dignitas, et munus Actuarii Senatus: utriusque vero emolumenta satis ampla. Alterum florenos reddit plerumque mille et quod excurrit: alterum sexcentos circiter aut septingentos. Nil opus addi de frequentibus Doctorum promotionibus: de his enim jam superiore dixi epistola. Etiam monui de domesticis scholis: dicam tamen, quod in his audiebant olim Volderum octuaginta Philosophi, saepe centum, et plures: a singulis autem pro honorario dabantur deni Philippici, sive Dukatoni. Pro Algebra vero in cujus explicatione habebat passim quatuor Auditores, plures nolebat, solvebantur a singulis imperiales centeni," transcription from *Briefwechsel von Johann I Bernoulli* (Basel University Library, ms. L Ia 713, 33–34). The *Bronnen* does not report De Volder's very first salary. However, on 15 August 1671 it was raised to 1,000 guilders per year: see Molhuysen 1913–1924, volume 3, 252. In 1681, it was raised to 1,200 guilders: see Molhuysen 1913–1924, volume 3, 369. In 1694, 1695 and 1696, he had a yearly salary of 1,600 guilders (including 400 guilders as a professor of mathematics): see Rosenboom 1697, part 1, 45, 49, 52 and 53; part 2, 60; see also part 1, 50 and 101, as well as Molhuysen 1913–1924, volume 4, 114 and 197 (where occasional, additional sums given to De Volder are mentioned). 1 Dutch ducaton amounted to 3 guilders; 1 thaler to 2.5 guilders. Anyway, Bernoulli complained about the condi-

As reported by Le Clerc, the Academic Curators allowed him to retire from teaching on the condition that he would continue to participate in some academic activity, such as examining students.[322] Moreover, De Volder continued to act as assessor of the Academic Senate in the years 1707 and 1708, and was also nominated for the year 1709.[323] His duties as director of the experimental theatre were then undertaken by Senguerd, as seen above. Moreover, as director of the observatory there was appointed Lotharius Zumbach de Koesfeld (who was already using it), with De Volder's approval.[324] In philosophy and mathematics, there was appointed as *lector*, with a salary of 400 guilders (viz. the same salary De Volder had received as professor of

tions of employment in his letter to Noodt of 4 August 1709: "[q]uod addis Cl. Volderum postremo stipendio suo non semper fruitum fuisse, id lubens crediderim; ille enim qui fuit Belga aut Belgio vicinus, juvenis, Coelebs, et nullo alio in munere antea constitutus, ubi quantumvis parum sibi oblatum pro lucro habuisset, quid mirum, si primam quamvis conditionem amplexus est? Aliam vero meam sortem nunc esse; et longe diversam," transcription from *Briefwechsel von Johann I Bernoulli* (original at Berlin, Staatsbibliothek, Sammlung Darmstaedter, unknown signature). See also Bernoulli's letter to William Burnet of 9 April 1710: "[v]ous dites Monsieur, qu'on n'a jamais offert à aucun professeur en Philosophie ces 1800 fl. et que Mr. de Volder ne les a pas obtenu qu'aprez 25 ans de service; je le veux croire, Mr. Noodt m'a dit la meme chose, mais il sçait ce que je luy ay repondu là dessus: Mr. De Volder n'avoit ni femme ni enfans, il n'avoit ni charge ni profession à quitter, sa Patrie n'étoit pas loin de Leyde, si j'étois dans ce cas là, je ne balancerois pas longtemps, j'acceptrois les offres à mains ouvertes, mais je suis dans un cas extraordinaire qui merite aussy quelque consideration extraordinaire," transcription from *Briefwechsel von Johann I Bernoulli* (Basel University Library, ms. L Ia 654, number 5). Discussions on employment conditions recur frequently in Bernoulli's letters. According to Le Clerc, De Volder never asked for an increase in salary: see *supra*, n. 121.

[322] "Il auroit pu continuer à tirer ses gages, sans rien faire, à cause de son in commodité, sans qu'on l'eût trouvé mauvais; car enfin après trente-quatre ans de Professorat, pendant les quels il s'étoit toujours aquité très-régulierement de sa Charge, personne n'auroit été surpris, qu'il eût cherché un repos, qui lui étoit nécessaire, au dépends de l'Academie, qu'il avoit servie si long-tems. Aucun Professeur n'avoit encore eu de semblable scrupule; & il y a bien de l'apparence qu'il n'y en aura guere, qui l'imitent. Comme il n'avoit jamais été marié, & qu'il vivoit sans aucun faste, quoi qu'il n'y eût rien de sordide dans ses manieres, & qu'il reçût & régalât ses Amis dans les occasions, aussi bien que ceux de son Ordre puissent le faire; il lui étoit assez resté de bien, pour pouvoir se passer des émolumens de l'Academie, sur tout à l'entrée de la Paix, qui se négotioit alors. Mais Mrs. les Curateurs ne voulurent pas lui donner son congé, de la maniere dont il le demandoit. Ils lui continuerent une bonne partie de son gage, & le prierent de demeurer dans l'Academie, sans l'obliger à faire aucunes leçons ni publiques, ni particulieres, & sans l'exclure des autres emplois, ou émolumens qu'on y peut avoir; seulement à condition qu'il ne refuseroit pas ses conseils à ceux, qui l'iroient consulter sur leurs études. Il accepta leurs offres, mais il ne voulut avoir aucune sorte de charge. Il s'employoit en cette occasion, avec beaucoup de zele & d'adresse, pour ses Amis, à qui il a souvent rendu de très-bons offices, en ces sortes de choses," Le Clerc 1709, 394–395; cf. Molhuysen 1913–1924, volume 4, 220. See also Gronovius 1709, 31–32: "[e]nimvero tunc sibi consulendum ratus, non si omnia morarentur, Academica munera ulterius obire statuit, et e vestigio se iis abdicavit fere ante quadriennium, honoris ergo indulgens Ill. Curatoribus et Consulibus, ut tamen Academiae adesse et in hac urbe consistere perseveraret." See Wiesenfeldt 2002, 92–93, referring to ms. ASF 299, ff. 75 and 91, preserved at Leiden University Library.

[323] Molhuysen 1913–1924, volume 4, 224, 231 and 236.

[324] See *supra*, n. 232.

mathematics)[325] Jacques Bernard, the French theologian who was directing the *Nouvelles de la Republique des Lettres.*[326] In fact, as revealed by his letter to Leibniz of 5 January 1706, De Volder attempted to have Johann Bernoulli appointed as his successor: however, in 1705 Bernoulli had preferred to move to Basel – probably because the vacant position was going to be filled by a *lector* rather than by a professor, as I am going to show in a moment – to take the position of his younger brother Jacob (who had been a student of De Volder), who had died in August 1705. In turn, Johann Bernoulli had recommended for De Volder's vacant position a student of Jakob Bernoulli and *protégé* of Leibniz, namely Jakob Hermann (1678–1733, from 1707 professor of mathematics at Padua), who however was not deemed sufficiently competent in metaphysics by De Volder.[327] De Volder, actually, had already tried to have appointed at Groningen a student of his, Hermann Schuyl (1665–1748, who became doctor of philosophy with perhaps the most advanced inaugural disputation

[325] See *supra*, n. 321.

[326] Molhuysen 1913–1924, volume 4, 221. On him, see Juillard 1999; Palladini 2011, chapter 8.

[327] "Forsitan rumor ad te detulit, me laboribus Academicis valedixisse, et in id, quantum poteram, incubuisse, mihi ut succederet amicus noster communis Bernoullius, sed frustra cum ille se suis Basileensibus addixerit. Nostri interea Do. Bernard, qui scribit les Nouvelles de la Rep. des Lettres, facultatem dederunt docendi Philosophiam et Mathesin, quem tamen Professoris titulo non donarunt, visuri, opinor, rei successum. Clariss. Bernoullius mihi commendavit D. Hermannum, qui me dum ante paucos annos has oras inviseret, me compellavit, et cuius egregia quaedam iisdem in Novellis, specimina in Mathematicis vidi. Provocat noster ad tuum de eodem Hermanno testimonium, sed nihil adiungit, quidnam ille praestare queat in Philosophicis, Metaphysicisque. Qua de re si, quod nosti, perscribere non dedigneris, possit hoc aliquando in eius utilitatem cedere," GP II, 280. Hermann had published in 1704 a *Méthode géométrique et générale de déterminer le diamètre de l'Arc-en-ciel*, in the *Nouvelles de la Republique des Lettres*. Cf. Leibniz's answer: "P. S. Dom. Hermannus apud Basileenses vir doctus, de quo sententiam meam quaeris; in Mathesi, ut scis, egregius est. Quid in aliis praestiterit, equidem satis exploratum non habeo, cum neque collocutus sim unquam, neque nisi de Mathematicis cum eo per literas communicaverim: Non dubito tamen, qui possis majus, posse minus, et cum Theologiae se dederit, philosophiam excoluisse," GP II, 283. Hermann, actually, had already been recommended for a chair in philosophy mathematics at Utrecht by Bernoulli to De Volder: see the letter of Bernoulli to De Volder of 5 January 1705. Hermann thanked Bernoulli with a letter of 3 December 1704. In turn, De Volder indicated the name of Hermann to Pieter Burman with a letter dated 27 January 1705, to which De Volder's attached his copy of a part of Bernoulli's letter. In his own letter, De Volder noted that "[i] n rebus philosophicis quid praestiterit ignoro prorsus." On Hermann's missed appointment at Utrecht, see also the letter of Johann Bernoulli to him of 28 June 1704, in which it was traced back to the internal quarrels at the academy: "[v]idebis Lycaeum ad quod Te pertrahere tentavi esse Ultrajectinum, theatrum sane, si ullum aliud, in Orbe splendidissimum quod Famae Tuae augendae foret accomodatissimum, ubi meam quam ob sonticas rationes quamvis aegre recusavi sedem, debuisses adimplere; difficultatem quam nonnulli studiis melioribus parum faventes de juvenilibus Tuis annis et fama nondum satis magna excitata movent, amovere conabor: putem tamen veram causam frigiditatis illius, de qua Volderus, esse, motus illos et tumultus civiles, qui in Provincia Ultrajectina ut et in aliis circumjacentibus ab aliquo tempore grassantur et Procerum animos ab academia restituenda hactenus avocant; quibus autem sedatis spes forte melior affulgebit," transcription from *Briefwechsel von Johann I Bernoulli* (Basel University Library, ms. L Ia 659, number 3). Bernoulli might have been referring to Burman. Hermann had personally met De Volder at Leiden in May 1701, according to Hermann's *Stammbuch* (beloning to a private collection in Basel), as well as his letter to Johann Jakob Scheuchzer of 20 June 1703 (in *Briefwechsel von Johann I Bernoulli*).

held at Leiden in the late seventeenth century, as I am going to show in Sects. 4.2.3.4, *Collision in De Volder's late thought*, and 6.2.2.3.2, *De Volder's Huygenian theory of weight (and its relation to the idea of cohesion)*), as De Volder declares in a letter to Pieter Burman (professor at Utrecht) of 19 November 1705. The letter by Burman to which De Volder answered with one of his is lost; however, as is evident from De Volder's letter, Burman had asked him his opinion on some possible candidates to fill the chair in philosophy left vacant by Gerard de Vries (who was to die on 1 December 1705). In his letter, De Volder considers some of his former students: Boerhaave and Hermann Oosterdijk Schacht – who was stepson of Lucas Schacht, and had graduated under De Volder at Leiden[328] (neither got the chair as they were busy with other occupations), Schuyl – whom he had recommended both to Groningen – where Johann Bernoulli was preferred to him, being favoured by Huygens's recommendation – and to Leiden (albeit in vain). At last, De Volder mentions Joseph Serrurier, with whom he had lost contact after his graduation at Leiden in 1690,[329] and who was in fact to be

[328] See his *Disputatio philosophica inauguralis de sensibus internis memoria et imaginatione*, 6 July 1693.

[329] See the full text of the letter: "Celeberrimo viro / Petro Burmanno / S. P. / B. de Volder / Quod tuo iudicio, id quod egi, probes, in eo mihi admodum gratulor. Verum est, quid ais, id nos tarde demum consequi posse, sed quid agas? Praestat pauculo tempore otio frui, quam nullo. Quae porro adiungis de iis, quos proceribus nostris commendavi, ea docent te satis fido usum amico, quae et accurate et vere tibi, quae acta sunt, exposuit, modo illud addiderit, licet eius mentionem nullam {facias}, me eos commendasse, tanquam tales, qui in Academia res suas egregie peregerant, in Philosophicis et Mathematicis prae caeteris elaborarant, et de quibus spes esset, eos, si iisdem studiis post discessum ab Academia porro operam dedissent, quod me nescire {probitebar}, et ad professionis munus evocati, iisdem {pentas incumberent}, aliquid egregi {praestituros}. Eadem spe me vocatum, vocatum Senguerdium, et, quod unico addo, vocatum vestrum Vriesium. / De Boerhavio tecum sentio, illum a vestris {...} non posse. Idem de Oosterdyckio sentire, nisi ex amicis ejus nonnulli mihi in aurem insusurrassent, illum frequentissimae et molestissimae praxeos medicae onere oppressum, non abhorrere a professione, si illa tolerabilius conditionibus ipsi offeratur, etiamsi magnam {...} partem eo amissurus esset. Quae si vere sentit, ut mihi ferme persuasum est, nae ille haud mecum sentit. {Sed} hoc nihil ad rem, cum ipsius negotium agatur. Credo tamen ejus rationem maiori cum emolumento et suo, et Academiae haberi posse, si forte professio Medica vocaverit. / Schuylius Medicinae Doctor est Delftis. Illum ante aliquod annos commendaram Groninganis, sed ei ex Hugenii commendatione merito praelatus est Bernoullius. {Commendavi} quoque nostris, sed Bernoullii verbis, quem et vobis antehac commendarat, D. Hermanni (nisi quod adderem, me quaedam eius in Mathematicis egregia specimina vidisse) de quo plura ex Celeber. Viro Leibnitio sciri posse, Bernouillius adjunxerat. / De Serrurerio nihil habeo, quod addam. Ex dictis enim constat, quid de eo existimem. Unicum illud scrupulum mihi facit, quod nesciam, qui vitam suam transegerit, postquam a nobis discessit, num his studiis nostris valedixerit, an vero iisdem porro se oblectaverit. Priori enim in casu non ausim hominem vobis commendare, in posteriori vero minus forem difficilis. / Haec de iis, quae petis, tibi dicere habeo, quibus ut in commodum Acad. vestrae utaris, ut non intercedo, sic a prudentia et aequitate tua expecto, ut caveas, me, quem libere et ex animi sententia tibi dixi, mihi {nullam} apud vos, quorum interest, invidiam {...}. Vale / Lugd. Batav / a.d. XIX Novembr. / CIƆIƆCCV." Leiden University Library, ms. BUR Q 27. See also the letter of Huygens to Wicher Wichers of 15 June 1694: Huygens 1888–1950, volume 10, 620–621 (letter 2858). In fact, Bernoulli was recommended by Huygens after the efforts of L'Hôpital in securing him the position: see Sierksma 1992; Van Maanen 1993. See also Wiesenfeldt 2002, 243–245 and 256. On Bernoulli, see also Fleckenstein 1977; Heimann 1977; Bell 1986, chapter 8; Dunham 1987; Aiton 1989; Sierksma 1992; Shafer 1996; Sierksma and Sierksma 1999; Guicciardini 2017.

appointed as professor on 28 December 1705. Serrurier was going to introduce experimental philosophy at Utrecht, and had already published a textbook resulting from his activities as a private teacher (i.e. his *Physicae experimentis innixae compendiosa tractatio, in usum scholarum domesticarum*, 1700). In turn, in his letter to De Volder of 8 May 1706, Johann Bernoulli asked De Volder whether his former position could be filled by a professor, viz. whether they could offer him good conditions of employment, and complaining about those offered to him at Utrecht in 1703, when Burman had offered him a chair in mathematics, which was then going to be filled in 1705 by Serrurier, the "disciple" of De Volder.[330] In turn, in his letter to Wicher Wichers of August 1711, Bernoulli reports that he had been offered a professorial chair of

[330] "Tibi interim Vir Cl. hanc meam mentem volui aperire, sed tamen adhuc silentio committendam; ut cum nullus hucusque successor Ordinarius Tibi suffectus sit, mihi quantocyus scribere digneris, an ex Tua commendatione aliqua mihi vocationis spes supersit, et quales conditiones sperandae; nosti vero haud dubie illas quas Trajectini postremo mihi obtulerunt; Non crediderim Trajectinos nunc alium desiderare Professorem mathes. et philos. postquam Cl. Serrurierium Tuum quondam discipulum vocarunt, res haud dubie satis feliciter et ex voto gerentem; quare illos mei jam oblitos esse facile colligo, eoque magis quod Cl. Burmannus jam a longo tempore ad ultimas meas ad se datas nihil respondit," transcription from *Briefwechsel von Johann I Bernoulli* (Basel University Library, ms. L Ia 675, 189–190). In his letter, Bernoulli also commends De Volder's *otium*. De Volder provided recommendations also on behalf of Huygens and Leibniz. In his letter to Leibniz of 29 May 1694, Huygens wrote to Leibniz that he was going to ask De Volder whether he knew of any suitable candidate for a position as professor of mathematics at the Ritterakademie of Wolfenbüttel (about which Leibniz had told him in his letter of 26 April/6 May 1694). Huygens refers to Johannes Teyler – a colour print fabric producer, and former professor of mathematics at Nijmegen between 1670 and 1676, who was known to Leibniz: see the letter of Huygens to Leibniz of 29 May 1694: Huygens 1888–1950, volume 10, 609–612 (letter 2854). In his letter to Leibniz of 8 June 1694, Huygens reports that De Volder had recommended Teyler (who he seems to know well) for the position, above all the other possible candidates. However, he remarks that his recommendation of Teyler at Utrecht had been unsuccessful, because Teyler had been a student of Craanen, at a time in which the struggle between the Cartesians and Voetians was still lively. Indeed, Teyler succeded at Nijmegen to Craanen himself in 1670, and left his position in 1676. See *supra*, n. 280. The position at Wolfenbüttel was not filled by Teyler, who apparently did not accept it. See also Huygens's *sommaire* of the letter: Huygens 1888–1950, volume 10, 617–618 (letter 2856). For the whole story, see Grieser 1969. Moreover, on 10 March 1709, Johannes Scheuchzer reported to Bernoulli the text of an undated letter from Pieter Valkenier, dealing with the discussion on 18 February about the assumption of a professorial post of botany and medicine by Boerhaave (up to that time a lector at Leiden). Apparently, Bernoulli wrote to De Volder asking about the position (vacant after the death of Petrus Houttuyn), probably in favour of Johann Jakob Scheuchzer, brother of Johannes: his letter (now lost), forwarded by Valkenier, was returned to him as the post was taken by Boerhaave: "[l]e 18. de ce Mois [viz. February] il y eut un grand debat entre Messrs les Curateurs à Leyde sur nôtre affaire de succession de tous les Pretendents personne ne Venoit en consideration que Vous et Monsr Bourhave, mais celuy-cy l'emporta à la fin, par le moyen de tant d'Amis qu'il a. On alleguoit, qu'en le prenant on menageroit les 600 Florins, qu'il tiroit deja en Lecteur, et qu'on n'auroit à faire quelque depense pour son transport, ce qui seroit pourtant necessaire, quand on prendroit un etranger, outre ce que Mr Bourhave avoit deja promesse de la premiere place Vacante dans la faculté, et qu'il avoit aussi avancé, qu'en cas qu'on luy prefereroit un autre, il vouloit quitter son Lectorat, et se retirer ailleurs avec bonne partie d'Estudiants, principalement des Anglois, Ecossois et Irlandois, lesquels il possede presque tous. – J'ay donné adresse à la Lettre de Monsr Bernoulli pour Monsr Volder, la quelle me fut rendue un jour apres, qu'on eut disposé du Professorat," transcription from *Briefwechsel von Johann I Bernoulli* (Basel University Library, ms. L Ia 668, number 29∗). See also Molhuysen 1913–1924, volume 4, 242.

mathematics at Leiden after De Volder's death, but that he did not accept it because the salary was too low.[331] This chair, in 1719, was eventually to be filled by 's Gravesande, who enrolled at Leiden on 6 June 1704 and graduated under Boerhaave, and who probably witnessed De Volder's last experimental lectures in 1704–1705.[332]

Eventually, De Volder's health deteriorated during the harsh 1708–1709 winter. While in summer 1708 his condition was good, as witnessed by Le Clerc (who met him in Amsterdam), and his urine became normal, during the autumn his condition deteriorated, as he started to vomit, could no longer eat meat, could only drink milk, and fell into a progressive languor. Fearing the possibility of contagion, he refused to meet people in the last months of his life.[333] As reported by Gronovius, in his last hours he progressively felt a sensation of cold, "appearing to enter into a second winter," and he lay in bed until the morning of 28 March, murmuring unintelligibly. At the *hora quinta* (viz. at 10–11 a.m.), he asked to be moved to another bed, were he was had lain the day before. While the bed was being prepared, and one of his servants was putting a pillow under his head, he died in the presence of his friends, the lawyers Abraham and Johan van den Ende (father and son).[334] De Volder never married, and he had no children: according to

---

[331] "Il y a quelque temps que je voyois quelque apparence pour y retourner, en ce qu'on m'avoit offert la chaire de Mathematique à Leyde vacante par la mort de Mr. de Volder, mais les conditions n'etant pas comme on me les fit esperer auparavant, et meme moindres encore que celles qu'on m'a presentées autrefois à Utrecht, je n'ay pas encore accepté cett'offre," transcription from *Briefwechsel von Johann I Bernoulli* (Basel University Library, ms. L Ia 674, 108). Indeed, Bernoulli had been offered a chair in mathematics at Utrecht in 1703. In his letter to Bernoulli of 30 August 1709, William Burnet reported to have heard that Bernoulli was De Volder's successor at Leiden (in *Briefwechsel von Johann I Bernoulli*).

[332] See *Album studiosorum Academiae Lugduno-Batavae MDLXXV–MDCCCLXXV*, column 780. On 's Gravesande's education, see Gori 1972, chapter 2; Van Besouw 2016.

[333] "Il sembloit, pendant quelque tems, que Mr. de Volder se trouvât mieux, & même il parut tout à fait délivré de l'incommodité dont j'ai parlé. Nous le vimes à Amsterdam l'Eté passé, en assez bonne santé. Mais vers l'automne il tomba peu à peu dans un dégoût si grand, qu'en fin il ne pouvoit souffrir aucune chair, & qu'il ne se soûtenoit qu'en buvant du lait. Cela le jetta dans une si grande langueur, qu'il fallut enfin qu'il gardât le lit, o ù il mourut tranquillement le 28. de Mars au matin, cette année MDCCIX," Le Clerc 1709, 397; cf. Gronovius 1709, 31–32: "[q]uanta erat illa tarditas in ingressu? In rebus plane omnibus nausea, aut immensa indifferentia? Ut quasi exul ab vita, quam carpebat, suas ipse per singulos passus exequias duceret. Perinde et ipse opinatus fuit, ita ut more Scythico solum lac admittens amicis interdum clare fatentur nolle se ulterius morbum alere; tanquam vivere molestum foret ultra quam posset esse utilis, et neque ipsi quidquam posset esse utile nisi iam plane migrare ad propiorem usum eius, partim cogitando, partim Divini scriptis legendis hic tam calide perceperat."

[334] "Denique quum in sedente, stante, incedente omnis calor se frangeret, ut ipse mihi ante dies confitebatur, ab nimio frigore in lectum prostratus traxit usque ad horam sesquisextam matutinam diei octavi et vigesimi mensis Martii. Qui enim truculentos hiemis gelidissimae rigores toleraverat et integer transierat, illius noctis paucarum horarum frigore sic dilaceratus, sic ustus fuit, ut aliam hiemem intravisse appareret. Itaque assiduum tunc intra labra, sed inarticulatum nec intelligendum murmur in eo observatum fuit, sive id ipsum querente, sive Dei implorante auxilium. Tandem ad horam quintam mente instantem migrationem divinante quum clamavisset transferri se debere in alium lectum, qui in eodem cubiculo die praecedente stratus erat, dum ad id omnia parantur, videns famulus capite nimis depresso et humili incommode cubare, rogansque an pluribus cervicalibus fulciri mallet, ubi visus fuit assentire, insiliens in lectum et divaricatis pedibus levans aegrum, ut pulvini commode stiparentur, repente comperit eum dilabi, anhelitum pressiorem esse et fauces

Gronovius, because he did not want to increase the number of 'subjects of tribulation', and preferring to have his students as his family members. Moreover, he did

---

arctari, ut vix advocatis praestantissimis Viris et huius morientis longe suavissimis Amicis et tunc hospitibus Vanden-Endiis Patre et filio, in eorum conspectu defecerit, cum admirabili aequanimitate exantlatis tot truculentis senectutis et mortis oppugnationibus, migrante anima ex illo lecto, quod per corpus praedixerat," Gronovius 1709, 32–33. In his *Oratio qua repurgatae medicinae facilis asseritur simplicitas*, held on 20 March 1709, Boerhaave lamented De Volder's absence: see Boerhaave 1721, 31. De Volder wrote an attestation of studies for Boerhaave: see *Testimonium voor Herman Boerhaave* (1668–1738), August 1683, The Hague, Koninklijke Bibliotheek, ms. 121 E 4. Two letters by De Volder to Abraham vanden Ende, traced to 1686 and 1691, are extant at Amsterdam. In the first letter, dated 6 May 1686, De Volder informs about and comments upon the Walloon Synod taking place at Rotterdam of April 1686, concerning the declaration of faith to be subscribed by Huguenot pastors (see Cerny 1987, 63–64; Van Eijnatten 2003): "[e]x Literis B. d. V ad A. v. E. / Interfuit Synodo Roterodamensi Spanhemius, Le Moyne, Gaillardus: quorum ille casu, ut ait, paucis ante inchoatam Synodum diebus Roterodamum venit; Le Moyne a suo Synedrio erat deputatus: Gaillardus vero accessit Synodum licet non deputatus, more, ut fertur, apud Gallos satis solemni. Formulam cui subscribere ex Gallia profugi, antequam ad munus Ecclesiasticum hisce in regionibus vocari possunt, concinnandum dedit Synodus quatuor Professoribus, adjungns tribus illis modo nominatis D. Jurieun, unà cum quatuor pastoribus. Formula autem continet tria capita. I. Ut suscribant Synodo Dordracenae, Confessioni, Formulis consenus, seque subijciant disciplinae Ecclesiasticae, quae sic obtinet. 2. Ut promittant se illas opiniones, quae ante aliquot annos Galliae Ecclesias turbantur, nec publice nec privatim docturos. 3. Ut eos qui Pelagianismum in Ecclesiam per Gratiam Mediatam introducere tentant se non toleraturos spondeant, nec cum iis facturos. Primum caput solemne est: hoc enim ut agant quotquot sive Gallicae sive Belgicae Ecclesiae fiunt pastores, necesse est. Hoc et egit noster Le Moyne cum ad hanc Ecclesiam vocaretur. Alterum caput Amyraldistas, Tertium, ut nosti, Pajonistas spectat. Laudant nostri inde reduces moderationem Spanhemianam, quanquam nonnulli addunt, illum, secus hi egisset, nihil profecturum fuisse. Huic jam formulae centum octoginta subscripserunt. Verum duo, qui de Pajonismo maxime suspecti sunt, vocati se non stitere, sive quod itineris molestia et sumptus, Groningae enim commorantur, ipsos abstenuerit, sive quod incerti quidnam illis subscribendum foret maluerint exitum rei absentes exspectare. Praeter hos duos neminem esse audio, quem haec formula feriat. Ita quidem agunt placidissime confecta et transacta esse omnia. / Lugd. Bat. pridie nonas majas / CIƆIƆCLXXXVI," Amsterdam University Library, OTM hs. Gs 24. As to the second letter, see *supra*, n. 282. Besides the correspondents of De Volder mentioned and discussed in this section, it is worth mentioning that De Volder corresponded with Albert Jansz. van Dam, nephew of Dirck Rembrantsz. van Nierop. Albert asked for De Volder's opinion on Henry More's criticism of Descartes in a letter dated 8 November 1677 (extant in Oostwoud 1754–1759, volume 6, 402–403; Rijks 2012, 249; I briefly discuss it in Sect. 6.2.2.3.4.1, *De Volder's review of Huygens's report on the first trial at sea*). No answer from De Volder is extant. A previous letter of Van Dam to De Volder is mentioned in a letter (dated 16 October 1677) by Van Dam to Klaas Dirksz. Zalm, student at Leiden, who had to forward it to De Volder: see Oostwoud 1754–1759, volume 6, 316–317; Rijks 2012, 247–248. In this letter to Zalm, Van Dam also mentions De Volder's teaching of algebra: see *supra*, n. 207. Moreover, De Volder corresponded with Theodorus Janssonius van Almeloveen, physician and professor of Greek and history at Harderwijk, to whom De Volder sent two copies of his 1698 *Oratio* on 19 March 1698: one for him, and one for Johannes Meyer, professor of theology. Two minutes of a letter of thanks by Almeloveen to De Volder, dated 22 October 1698, are extant at Utrecht: Utrecht University Library, Hs. 995, 163–164, and Hs. 996, 458–460. Finally, a letter by De Volder to the Rotterdam printer Reinier Leers, dated 14 May 1695, is extant at Leiden. In it, De Volder points out errors he has found in a book he received from Jakob Perizonius: namely, the *Mémoires de mathématique et de physique tirés des registres de l'Académie Royale des Science* (1693), of which De Volder owned a defective copy: Leiden University Library, ms. PAP 15.

not prepare a testament, preferring that his belongings be distributed with equity, according to the law.[335] Indeed, as testified to by the proceedings of the Leiden Academic Senate of 1705, on that same day of his death the Senate was informed, and Boerhaave and Noodt (professor of law) were charged with organizing the funeral, and to visit De Volder's house in order to discover any papers reporting his testaments (before sealing everything). On 1 April, after a short speech by the rector (Johannes Voet), Boerhaave and Noodt testified that no testament had been found, witnessed by Abraham van den Ende and Petrus la Court (apparently, a member of the De la Court family). The day after, some collateral relatives of De Volder (also mentioned in De Volder's private papers) were heard by the Senate, and they organized the burial for the afternoon of the following Wednesday, viz. the day after, 3 April 1709. Gronovius was asked to give a funeral speech, given on 15 April. The burial, which took place at the Hooglandse Kerk, in an individual tomb (not extant),[336] was attended by the Academic Senate, in a procession from De Volder's, located in Houtstraat.[337] His belongings, including his house, were inherited by collateral succession, while his library (the catalogue of which of circa 1,860 books, plus his 'mathematical instruments', is extant as the *Bibliotheca Volderina*, which is a precious document for De Volder's sources) was sold at an auction taking place on 14–18 October 1709 at the printing workshop of Johannes vander Linden Jr. According to De Volder's *collaterale successie* (dated 25 April (*solvit*)), De Volder left a house in Houtstraat valued 1,200 guilders (viz. on the value of which taxes of 20% were applied); moreover, he left life annuities (most probably as dowries) to nine girls of the families Van Roijen, De Bije, Van den Bergh, Geerling and Alberthoma for 4,275 guilders, and bonds for 31,700 guilders (before taxes): two of them in the name of Abraham vanden Ende and Jacob van der Meer, Lord of Alphen, and one (dated 1644) in the name of Engel Clementsz., a ship-maker. De Volder's heirs were Burger van Singel and Abraham van Zingel – of whom no consistent traces can be found.[338] Apparently, De Volder's papers were dispersed after his death: his books were sold, – together with his 'mathematical instruments', including instruments for physics,

[335] De Volder 1705, 30; Gronovius 1709, 33–35.

[336] Gronovius 1709, 7.

[337] Molhuysen 1913–1924, volume 4, 238–239 and 178*. Apparently, De Volder lived on the corner between Houtstraat and Rapenburg, as in a letter of Christiaan Huygens to his brother Constantijn it is reported that Christiaan and De Volder could observe the canal (which is on Rapenburg) with a lens: see *supra*, n. 229. On De Volder's home and burial place, see Inventaris van het stadsarchief van Leiden: 1313–1343, Registers van te Leiden begravenen, register 1329: 1704 maart 29–1709 juni 1. 1704–1709, 265.

[338] As to De Volder's *Collaterale successie*, see Inventaris van het stadsarchief van Leiden, Memoriaalboek, inventarisnummer 210+6, 71v–75r. Further money was given to De Volder's heirs for his academic duties in February and December 1710: see Molhuysen 1913–1924, volume 4, 241 and 246.

astronomy,[339] as well as a small collection of musical instruments and other items[340] – while all the letters directed to him are extant as minute copies. Boerhaave, who inspected his papers in April–May 1709 (as testified to in his preface to De Volder's *Oratio de novis et antiquis*), did not find any complete, unpublished treatises. Moreover, once Johann Bernoulli expounded to 's Gravesande, on 31 October 1722, the argument he used to convince De Volder, *rigide cartesien*, that the force of motion is calculated by the formula $m \cdot s^2$ rather than $m \cdot s$ (as I am going to show in Sect. 4.2.3.3.1, *The quantification of motive force*), he invited 's Gravesande to ask De Volder's heirs for his papers, amongst which Bernoulli's original letter (of 13 February 1700) was extant.[341] Apparently, however, 's Gravesande never succeeded in doing so. It is worth closing this presentation of De Volder's life with the words of Stolle and Hallmann:

> [...] Professor [de] Volder is a small person, underpowered in his body, and without danger over 50 years old. Even if he is a mathematician, that's not how one looks at him. He is polite, without much ceremony. His philosophy is eclectic but has a lot of Cartesian in itself. He is one of the most famous professors in Holland. The Latin, which he speaks well, flows perfectly from his mouth. He has great intelligence, and knows how to joke. His judgment does not seem to be inferior. If he should write anything, he would undoubtedly conceive a Satyricon, for he is a sharp guest. But he does not have the will to multiply the countless number of books, which one should rather have reduced. In conversation he is a great diplomat, at least he has showed himself as such before us. Because he speaks little, and thinks much, but what he speaks, he expresses nicely and briefly, and asserts it with great emphasis. He is more serious than friendly, but [this] does not make him so annoying,

339 See *supra*, nn. 135 and 233.

340 "3 Winkelhaaken. 1 Koper Paralel Liniaal. 2 Kopere Trekpennen [...] 1 Cartesiaanse Studeer-Lamp met een blikke Tregter. 1 Toetsteen. 1 Stuk Moscovis glas. 1 Landkaart van Moscovia door Nic. Witsen. 1 Curieus Verkeer-bort met de Schijven en twe Blikke kandelaaren. 1 Seer Curieus Notebome Schaakbort met des selfs curieuse Yvoire en swartes Ebbenhoute stelsel. 1 ——— Dito. Alle door Haka gemaakt.[:] 1 Palmhoute Heaubois. 1 ——— Dwars-Fluyt. 2 ——— dito kleynder. 1 Ebbenhoute Dwars-Fluyt. 1 ——— dito kleynder. 1 ——— Flute doux. 7 Boeken-kassen getekent N. 1. 2. 3. 1 Staande dubbelde Studeer-lessenaar met twee Kasten daar onder boven met Rusleer bekleet en met goede slodten voorsien. 1 Lessenaar op deselve wijs bekleet en een goed slot daar aan," *Bibliotheca Volderina*, 95–96.

341 "Puisque je me suis mis en train de Vous ecrire une longue lettre, je veux bien vous la communiquer, j'espere qu'elle vous fera plaisir, d'autant plus que c'est par cette meme demonstration que j'eus le bonheur, il y a environ 23 ans, de convertir feu Mr. de Volder Vôtre Predecesseur, rigide Cartesien, s'il en fut jamais, apres que Mr. Leibnitz employa inutilement tous ses arguments (dans un long commerce de lettre qu'il y avoit entre eux deux et qui passoient toujours par mes mains) pour le convaincre de la verité. Il seroit à souhaiter que les heritiers de Mr. de Volder voulussent vous communiquer ses papiers, vous y trouveriez une de mes lettres, datée je crois dans l'année 1700 qui contient la demonstration dont je Vous parle et dont voici le contenu [...]," transcription from *Briefwechsel von Johann I Bernoulli* (Basel University Library, ms. L Ia 674, 50–61).

because [...] he also sweetens his mouth a lot. For he does not easily begin to speak for himself, even if there were silence for a quarter of an hour; and if you put him directly to discussion, he says almost nothing more, than what he has been asked. The reason for this is hidden from me, but I know well that one can not say of him, as may be said of another famous man: there is nothing inside [*nihil est intus*], for the mind of this Dutch philosopher is such a deep thing, [that] in it are a lot of peculiar thoughts, and under all appearances also some atheist [thoughts] lie hidden.[342]

[342] "Der Herr Professor Volder ist eine kleine Person, dabey untersetzt von Leibe, und ohngefahr über 50 Jahre alt. Ob er wohl ein Mathematicus ist, so siehet man es ihm doch nicht an. Er ist höfflich, ohne viel caeremonien. Seine Philosophie ist Ecclectica hat aber viel Cartesianisches an sich. Er ist hierinn einer von den berühmtesten Professoribus in Holland. Das Latein, so er wohl spricht, gehet ihm fertig vom Munde. Er hat ein groß ingenium, und weis artig zu schertzen; Sein Judicium scheint nicht geringer zu sein. Wenn er was schreiben sollte, so würde er ohnezweifel einen Satyricum agiren, denn er ist ein spitziger Gast. Allein er hat nicht in Willens die unzehliche Menge der Bücher zuvermehren, die man vielmehr vergeringern sollte. In der Conversation ist er ein großer Politicus, wenigstens hat er sich gegen uns so erwiesen. Denn er redet wenig, und denckt viel, was er aber redet, bringet er nett und kurtz vor, und behauptet es mit ungemeinem Nachdruck. Er ist mehr ernsthafft als freundlich, doch macht ihn dieses nicht so verdrüßlich, als daß er seinen Mund so gar sehr menagiret. Denn er fängt nicht leicht selbst zureden an, und wenn auch eine 4tel Stunde silentium sein sollte; und wenn man ihn gleich auf einen discours bringt, so sagt er doch fast nichts mehr, als was er gefragt worden. Die Ursache hiervon ist mir verborgen, das aber weis ich wohl, daß man von ihm nicht, wie von einem andern lieben Manne sagen kann: nihil est intus, denn das Gemüthe dieses holländischen Philosophi ist eine solche Tieffe, darinne viel sonderbahre Gedancken, und darunter allem Ansehn nach auch einige Atheistische verborgen liegen," Warsaw University Library, Cod. IV oct. 49, 593–596. Transcription kindly provided by Martin Mulsow. The reference is to the common saying 'nihil est intus / dixit Carolus Quintus': see, for instance, Oertel 1842, 35; Lenau 1993, volume 7, 295.

Chapter 3
De Volder's Views on Metaphysics and on the Method of Natural Philosophy

Abstract In this chapter I focus on De Volder's ideas on metaphysics, namely, on De Volder's alleged Spinozism, as well as on his ideas on certainty and on the method of natural philosophy. The two points are deeply interconnected as De Volder's dealing with the theory of knowledge and with the function of metaphysics for natural philosophy led him to confront Spinoza's ideas on substance, God and necessitarianism. I argue that De Volder stood for a sort of 'methodological liberalism', according to which experience can offer a kind of certainty which is ultimately indubitable, and which allowed him to include in natural philosophy the conclusions reached by observations and experiments. Moreover, I show that De Volder pursued a detachment of physics from metaphysics which was historically rooted in Dutch Cartesianism, and which was theoretically functional as a principle of economy in explanation. Also, I show that De Volder's dealing with the (metaphysical) ideas of substance, attributes and properties, as well as his theory of certainty as indubitability led him to embrace a form of physical and mental necessitarianism which led him to consider favourably some of Spinoza's ideas, even if he cannot be labelled an out-and-out Spinozist.

3.1 De Volder's Spinozism: A Reassessment

As a matter of fact, the positions of De Volder in metaphysics attracted more attention than those he held on natural philosophy. Notwithstanding his attempts to limit the circulation of his ideas in the form of disputations – such as in the case of his *Exercitationes* against Pierre-Daniel Huet's *Censura philosophiae Cartesianae* (1689) – his teaching prompted two different quarrels: between Johann Bernoulli and Paulus Hulsius at Groningen (1702),[1] and between Jacob Wittich and Antonius Driessen, whose quarrel originated in Groningen but spread beyond the Netherlands (1717–1720).[2] Generally speaking, De Volder's ideas on metaphysics – notwithstanding his genuine religious

[1] See Steenbakkers 1997; Klever 2000; Israel 2001, chapter 23.

[2] See Israel 2001, chapter 23; Goudriaan 2003, 2006, chapter 2.

A. Strazzoni, *Burchard de Volder and the Age of the Scientific Revolution*, Studies in History and Philosophy of Science 51,
https://doi.org/10.1007/978-3-030-19878-7_3

beliefs and interests, as seen in Sect. 2.3.5, *De Volder's views on politics and religion* – were accused of being Spinozist or as favouring Spinozism, and had their foremost opponent in the Franeker Cartesian professor of philosophy and theology Ruardus Andala, who attacked De Volder's views mostly in texts which appeared after his death.[3] Besides by Andala, De Volder was suspected or accused of being a Spinozist, and an atheist, by Gottlieb Stolle and Georg Theophil Hallmann – who interviewed him on 18 July 1703 – Pierre Poiret, Johannes Regius, Jacob Leydekker and Joachim Lange. De Volder's (supposed) 'crypto-Spinozism' has in recent decades attracted the attention of Wim Klever, who has argued, in his article *Burchard de Volder (1643–1709). A Crypto-Spinozist on a Leiden Cathedra* (1988) that De Volder upheld the following Spinozist ideas:

1. epistemological naturalism,
2. necessitarianism,
3. mind-body parallelism,
4. substance monism.

In 2005, Paul Lodge has convincingly criticized the evidence adduced by Klever to argue for De Volder's crypto-Spinozism, admitting that "there is compelling evidence only for the first of these,"[4] namely (as Klever put it) the idea that "it is impossible to have a clear understanding of something and not to adhere to it."[5] In what follows, I will assume De Volder's supposed Spinozism as the starting point for a survey of all his positions on metaphysics – and on the method of natural philosophy. By re-considering his positions, I will ascertain whether De Volder purportedly taught Spinoza's ideas at Leiden, whether he came to the same conclusions as Spinoza by starting from Cartesian premises, or whether he proposed positions distant from both these intellectual benchmarks. In doing so, I will inquire into the motives prompting the aforementioned quarrels by considering his influence and intellectual heritage. Building upon this survey, I will then consider De Volder's views on the method and limits of philosophical knowledge, which had a strict relation to what Klever has labelled his 'epistemological naturalism'.

3.1.1 Some Direct Evidence on De Volder's Views and Relation with Spinoza

As I am going to show in more detail in Chap. 6, the only overt positive mention of Spinoza by De Volder is given in the Hamburg 273 series of his *dictata* on Descartes's *Principia*. In commenting upon article 33 of the second part of the *Principia* (II.33), De Volder re-directs the reader to Spinoza's *Principia philosophiae Cartesianae*

[3] On Andala, see Galama 1954; Bordoli 2009; Miranda 2012; Caroti 2014.

[4] Lodge 2005, 131.

[5] Klever 1988, 196; see Lodge 2005, 120.

(1663), axiom 14, in order to illustrate the principle – adopted by Descartes – according to which a fluid, passing from a large to a narrow tube, increases its speed. Apparently, this positive use of Spinoza's exposition of Descartes's philosophy by De Volder went unnoticed by his contemporaries. In fact, it concerns a hydrodynamical principle which traces back at least to Benedetto Castelli,[6] and which cannot be considered as Spinozist or as having metaphysical consequences. Yet, the use of Spinoza in academic *dictata* certainly points to an appreciation by De Volder of his thought, and to its free circulation and use by the students themselves, even after the appearance of Spinoza's *Tractatus theologico-politicus* in 1670, the year in which De Volder started to lecture at Leiden. Second, there could have been direct contacts between De Volder and Spinoza. As indicated by Klever, Pieter Baert (from 1676 professor of hydrography at Dunkirk and then at Toulon) wrote on 5 February 1676 to Christiaan Huygens that

> [...] since the year 1665, when I went to live in Amsterdam, where during many wonderful meetings, I have often heard your eminence spoken highly of in the arts of philosophy and mathematics, in particular by Mr. Johannes Hudde, Benedictus de Spinoza and Dr. De Volder, professor of philosophy at the University at Leiden.[7]

Even if this does not prove of a *direct* meeting between Spinoza and De Volder, their common acquaintance with Hudde, Huygens and Baert himself makes this likely.

Third, we find a direct opinion expressed by De Volder on Spinoza's ideas in the travel journal of Stolle and Hallmann, as they visited Leiden and met De Volder on 18 July 1703. As appears from the colloquium, the two were already suspecting De Volder of atheism and Spinozism. Moreover, Hallmann – after having spoken with Jan Rieuwerts Jr., son of the late publisher of Spinoza's *Principia* – noted that that "it is [...] very probable, that [...] there are many people, who even [if] they do not profess themselves together, or [do not] behave like fellows of a famous society, or who [do not] worship the person of Spinoza, but nevertheless [they] esteem the doctrines of Spinoza in silence."[8] Amongst the answers of De Volder to Stolle and Hallmann on Spinoza and (his) atheism, it is worth drawing attention to the following:

1. He found Pierre Bayle's exposition of Spinoza's theories very clear, but his criticisms of him as missing the point;
2. The most relevant errors of Spinoza were (a) his absolute necessitarianism, (b) his confusion of body and mind.

---

[6] See Sect. 6.1.2, *Descartes's cosmogony and cosmology*.

[7] "[...] sedert den jaere 1665 als wanneer ick tot Amsterdam ginck woonen, alwaer ick in veel heerlijcke vergaderingen, als ook int bysonder, van dhr. Johannes Hudde, Benedictus de Spinosa ende dr. Voldere professeur der hoogeschole in philos. tot leyden; menigmael loffelyck hebbe ooren spreken, van uE overtreffelijckheyt in de konsten der philosophie en mathematique," Baert to Huygens (letter 2085), 5 February 1676, in Huygens 1888–1950, volume 18, 3–4.

[8] "[...] es ist doch sehr wahrscheinlich, daß es [...] viel Leuthe giebt, die Sich zwar nicht zusammen bekennen, oder sich alß Collegiaten von der Bekanten Societät aufführen, oder die Person Spinosae anbeten, aber doch in der Stille Spinosae Lehrsätze amplectiren," Freudenthal 1899, 225–226.

3. De Volder could not say that Spinoza was an atheist, since Spinoza maintained that God exists by Himself (*a se*);
4. Spinoza wrongly maintained that the attribute of extension pertains to God, and that God is an immanent cause.[9]
5. Spinoza did not think highly of the Christian religion, as he compared the doctrine of the Incarnation to the appearance of a triangular circular: however, Spinoza was not a Christian, but a Jew.[10]
6. Finally, that "the books of Spinoza are read by few [people], are understood by very few [people], even by those, who nevertheless want to be considered followers of Spinoza."[11]

The existence of a sort of Spinozist 'sect', thus far, is corroborated by the words of De Volder, who seems to give the impression of having understood Spinoza's words better than his supposed followers – with whom, however, he does not overtly associate himself.

As to atheism as such, De Volder (answering to Stolle and Hallmann's request for a judgment on Christian Thomasius) declared that no learned man could be an athe-

[9] "Bayle habe in seinem Dictionaire des Spinozae Meinung mit solchen Worten vorgestellt mit welchen sie vielleicht Spinoza selbst würde exprimiret haben, wenn er sich hätte erklähren wollen. Allein die argumenta, so er contra Spinozam vorbringe, wären Spinozae Meinung gar nicht directè contrair, sondern bewiesen was anders als sie beweisen sollten. Des Spinozae Haupt-Irrthümer wären 1. Absoluta fatalitas, 2. confusio mentis cum corpore. Ob er ein Atheus sey, könne er nicht sagen. Denn er concedire Deum esse à se, er sage auch: daß er creatorem à creaturâ diversum statuire, nur daß thue er, daß er Deo corporis attributa zueigne und Deum pro causâ immanente et non transeunte halte," Warsaw University Library, Cod. IV oct. 49, 589–590. Transcription kindly provided by Martin Mulsow. Cf. proposition 18 of Spinoza's *Ethica*, and Bayle 1697. Moreover, De Volder claimed that in his letters Spinoza expounded himself better than in his *Tractatus theologico-politicus*: "[i]n seinen Briefen gebe sich Spinosa noch mehr bloß, als im Tractatu Theologico politico," Freudenthal 1899, 228–229. Also, De Volder reported that he did not know anything about Spinoza's supposedly lost treatise on rainbow (about which Stolle and Hallmann conferred with Rieuwerts); as well as about Tschirnhaus's Spinozism. Also, that he did not hear that Arnold Geulincx appreciated Spinoza, that there existed unpublished letters of Spinoza, that Spinoza was a teacher at the school of Franciscus van den Enden (opposer of Louis XIV the Great): "[v]on Spinosae tractatu de Iride habe er nichts gehört, und als wir sagten, es stände in praefation operum posthumorum, wollte er es nicht wohl glauben. Daß Tschirnhaus was Spinosistisches in seiner Mediana mentis habe, oder auch Selbsten mit Spinosa bekandt gewesen, wüste er auch nicht. Er sagte, er habe die Medicinam Mentis gelesen, habe aber nichts dergleichen angetroffen, soviel er sich zu entsinnen wüßte [...] Se non audivisse quod Geuligius Spinosa aestimaverit. eius ethicam obscure et nimis concise forte etiam Scriptum, ut occultare hoc modo quaedam potuerit. Concedebat, Spinosam forte plures Scripsisse epistolas quam eius amici publicare ausi sunt. Haec se audisse, Spinosam fuisse Submagistrum in Schola, quam Eudius habuit Latinam. Hunc Eudium fuisse acerrimum Regis Galliae Hostem," Freudenthal 1899, 229.

[10] "[D]aß er nicht viel von der Christlichen Religion halte, e. g. wann Er sage, daß es ihm ebenso unmöglich scheine, Deum factum esse Hominem als triangulum esse Circulum [...] Das sey gewiß, Spinosam nunquam fuisse Christianum, sed Judaeum," Freudenthal 1899, 229. Cf. Spinoza's epistle 73, to Henry Oldenburg.

[11] "Spinosae libros a paucis legi, a paucissimis intelligi etiam ab iis, qui pro Spinosiastis tamen haberi volunt, ut sepius ipsi iam compertum sit," Freudenthal 1899, 229.

ist, unless he has no unquestionable arguments for the existence of God.[12] This, of course, is perfectly consistent with his judgment on Spinoza's atheism. This judgment seems not to have been accepted by the two travellers: indeed, even if De Volder had already devoted several disputations to the demonstration of the existence of God, he was nevertheless suspected of atheism by Stolle and Hallmann, who close their report on De Volder by stating that "the mind of this Dutch philosopher is such a deep thing, in which many strange thoughts, and under all the appearance also [...] some atheist [thoughts] lay[,] hidden."[13]

Thus far, what emerges from the *Reisejournal* of Stolle and Hallmann is De Volder's acquaintance with Spinoza's ideas (which is quite obvious). Moreover, despite the philosophical errors he acknowledged in Spinoza, his judgment is certainly more positive than that of many of his contemporaries: first and foremost, he denies that Spinoza was an atheist. At least from a speculative or philosophical standpoint, according to De Volder Spinoza had a convincing argument for the existence of God.[14] Moreover, it seems that for De Volder Spinoza's negative judgment on the doctrine of the Incarnation was mostly due to his being a Jew.[15] This, however, does not amount to saying that De Volder purportedly followed some of Spinoza's opinions.

Eventually, a fourth association of De Volder with Spinoza derives from the *Laudatio funebris* of Jakob Gronovius for De Volder (1709), in which Gronovius provides a summary, in ten propositions, of De Volder's positions on metaphysics as they had been taught to his students. Amongst others, Gronovius mentions the proposition that "there is no free decision [*liberum arbitrium*] in the mind, [and] no chance [*casum*] in the body," reasoned on the ground that things exist as God conceives and wants them to be with His most perfect understanding and will.[16] As

---

[12] "Auf die Frage: ob auch Thomasius hier bekannt sey, antwortete er: o, ja iste vir ingeniosas scribit satyras. Und als wir ihm erzehlten: daß Thomasius in öffentlichen Schrifften zum Atheo gemacht worden, sagte er: Qui virum ingenio pollentem pro Atheo habet, is aut ipse argumenta certa de existentiâ Dei non habet et ipse est Atheus, aut certè nescit, quid dicat. Umb Atheos practicos, derer es unter dem Pövel viel gäbe, kümmere man sich nicht; aber Atheos speculativos wolle man immer unter klugen Leuthen finden," Cod. IV oct. 49, 590–591. Transcription kindly provided by Martin Mulsow. As seen in Sect. 2.3.3, *Mathematician and scientific expert*, Stolle and Hallmann also asked De Volder for a judgment on Tschirnhaus's alleged Spinozism, to which De Volder could not give an answer: "[d]aß Tschirnhaus was Spinosistisches in seiner Medicina mentis habe, oder auch selbsten mit Spinosa bekandt gewesen, wüste er auch nicht," Freudenthal 1899, 229.

[13] "[...] das Gemüthe dieses holländischen Philosophi ist eine solche Tieffe, darinne viel sonderbahre Gedancken, und darunter allem Ansehn nach auch einige Atheistische verborgen liegen," Cod. IV oct. 49, 596. Transcription kindly provided by Martin Mulsow.

[14] Please note that for De Volder practical atheists where usually to be found in unlearned people: see *supra*, n. 12.

[15] As seen in Sect. 2.3.5, *De Volder's views on politics and religion*, in his letter to Van Limborch of 3 November 1687, De Volder shows appreciation for Van Limborch's comparison between Moses (whose miracles, according to Van Limborch – probably under the influence of Spinoza –, could be explained as priestly impostures) and Christ (who performed real miracles; see Van Limborch 1687, 172–175; Landucci 2015).

[16] "[S]ic in metaphysicis illis, quae de primis humanae scientiae fundamentis tractantia in omnibus aliis ut firma adsumi solent, procedente Volderi ingenio, ut colligeret I. Omnium rerum esse Deum

I am going to show, Gronovius's report was to be widely cited as evidence of De Volder's Spinozism.

In amongst this evidence, the most promising for the investigation on De Volder's (supposed) metaphysical Spinozism is that on Spinoza's atheism, and the problem of necessitarianism, i.e. on the relation between cause and effect, and between God and creature. In the light of these reports, indeed, De Volder seems to favour Spinoza by (1) rejecting his alleged atheism, and (2) upholding necessitarianism in the physical and spiritual domains.

3.1.2 De Volder on Atheism and on the Relation of Cause and Effect

De Volder's views on the relation of cause and effect were expounded for the sake of his demonstrations of the existence of God. The foremost place in which he deals with this topic is a series of four disputations authored and presided over by him in 1680–1681, namely, his *Disputationes contra atheos*. These were all defended by the Hungarian student Samuel Köleséri as *respondens*, with the exception of the third, defended by Conrad Philip Limmer. Later, this series was re-printed as a monograph, in 1685 (in Middelburg) and 1719 (in Utrecht – and at the peak of the Driessen-Wittich controversy). Moreover – as anticipated in Sect. 2.4.2, *De Volder's retirement, death and legacy* – in his 1705 *Oratio qua sese laboribus academicis abdicavit* De Volder was to mention the previous arguments he had deployed against the 'atheists' and the 'sceptics', viz. the demonstrations of the existence of God, and the overcoming of doubt through an analysis of ideas only.

The actual function of this series of disputations and of its unauthorized edition is not entirely clear. In 1684 Noël Aubert de Versé published in Amsterdam his *Impie convaincu, ou Dissertation contre Spinosa*, the subtitle of which is in itself revelatory of the attempt to link Spinozism, atheism, and Cartesianism: *Dans laquelle l'on refute les fondemens de son atheisme. L'on trouvera dans cét ouvrage non seulement la refutation des maximes impies de Spinosa, mais aussi celle des principales hypotheses du cartesianisme, que l'on fait voir être l'origine du spinosisme.*[17] The publication of this work might have prompted the 1685 unauthor-

unam causam. II. Hunc vero esse unum cogitans sibi se totum debens, nec ulli omnino obnoxium causae extra se. III. Hunc maxime distinctum non modo a corporea materie, sed et ab omnibus aliis finitis cogitantibus. IV. Minime ergo Dei cogitationem esse conflatam ex congerie omnium cogitationum finitarum. V. Contra vero uno simplici omnium simul contuitu cuncta complecti. VI. Hanc scientiam Dei nullarum habere in Deo causam praeter ipsam Dei perfectionem infinitam. VIII. Reliqua quae sunt, ideo esse, quia perfectissimus Dei intellectus ea ita concepit, et perfectissima eius voluntas ita voluit. IX. Non ergo haec a Deo intelligi, quia sunt, sed esse, quia a Deo sic cogitantur. X. Nullum igitur liberum arbitrium in mente, nullum casum in corpore. Quid hac causa simplicitate sincerius? Quid oppositius ad vulgatas illas desipientium strophas elidendas? Quod confertius tegmen contra quascunque Petitiones necti potest?" Gronovius 1709, 24–25.

[17] See Benigni 2017, 2018; see also De Versé 2015.

ized edition of De Volder's *Contra atheos* – which had a far wider circulation than the disputations themselves. Thus far, one might interpret the appearance of this edition of De Volder's *Contra atheos* as a commercial attempt at exploiting the current debates over atheism, Cartesianism, and Spinozism. In themselves, De Volder's disputations expounds two demonstrations of the existence of God based on Descartes's ideas, so that De Volder might have been attempting, in 1680–1681, to vindicate Descartes's orthodoxy without, however, touching upon or criticizing Spinoza's own positions. In order to have a clearer grasp of the significance of De Volder's disputations, however, it is necessary to look at the Dutch debates over Spinozism in 1660s–1670s, which came to be entangled with those on Lodewijk Meijer.

3.1.2.1 The Early Reactions to Spinoza and Meijer

During Spinoza's lifetime, accusations of atheism had been levelled at him by the Utrecht physician Lambert van Velthuysen – himself a sympathizer with Descartes's and Hobbes's ideas.[18] In his letter to Jacob Ostens of 24 January 1671, Van Velthuysen admitted that Spinoza recognized the existence of God as creator of the world, but as he "envisages such a God as can not move men to reverence by his divinity, since he himself is subject to fate; no room is left for divine governance and providence, and the assignment of punishment and reward is entirely abolished," he can be labelled as "teaching sheer atheism with furtive and disguised arguments."[19] Later, Van Velthuysen would publish three treatises against Spinoza: namely, his *Tractatus moralis de naturali pudore et dignitate hominis* (1676), *De articulis fidei fundamentalibus* and *De cultu naturali, oppositus Tractatui theologico-politico et Operi posthumo Benedicti de Spinoza* (both published in his *Opera omnia*, 1680).[20] These are just a small number of a plethora of treatises that appeared or were widely disseminated in the Netherlands, aimed against Spinoza's *Principia*, *Tractatus* and later against his *Ethica* (1677): Jean Baptiste Stouppe's *La religion des Hollandois* (1673), Regnerus van Mansvelt's *Adversus anonymum Theologo-politicum liber singularis* (1674; written in 1671), Johannes Batalerius's *Vindiciae miraculorum* (1674), Willem van Blijenbergh's *De kennisse Godts en Godts-Dienst veweert tegen d'uytvluchten der atheisten* (1671 (first edition, now lost, 1663): concerning Spinoza's exposition of Descartes's *Principia*), *De waerheyt van de Christelijcke Godts-Dienst en de authoriteit der H. Schriften of een Wederlegginge van dat Godtlasterlijcke boeck, genoemt Tractatus theologico-politicus* (1674), *Wederlegging van de Ethica of Zede-kunst, van Benedictus de Spinosa* (1682), Johannes Bredenburg's *Enervatio Tractatus theologico-politici* (1675) – Bredenburg being later attacked as a Spinozist himself, viz. on the appearance of his *Wiskunstige*

[18] On him, see Klever 1991; Van Bunge 1995, 2001, chapters 3–4; Bordoli 1997, chapter 4; Krop 1999, 2004. See also Van Velthuysen 2013.

[19] Spinoza 1995, 236.

[20] For a discussion, see Van Bunge 2001, chapter 4.

demonstratie, dat alle verstandelijke werking noodzaakelijk is in 1684 (written circa 1676),[21] Frans Kuyper's *Arcana atheismi revelata* (1676), Huet's *Demonstratio Evangelica* (1679), Adriaen Verwer's *'t Mom-aensicht der atheistery afgerukt* (1683). In the meantime, Spinoza's *Tractatus* had been condemned by the States of Holland (1674).[22]

Amongst these works, however, only a few explicitly address the relations of Cartesianism with Spinozism and atheism. For instance, Van Velthuysen – himself a foremost Cartesian – attempted to distance his own positions from Spinoza's.[23] Van Mansvelt, in turn, claimed that Spinoza had presented his own principles as if they were Cartesian: accordingly, he must have constructed his atheism in spite of Descartes's arguments against it.[24] Van Blijenbergh, who was a correspondent of Spinoza himself, in fact feared the similarities between Spinoza's and Descartes's positions, when he deemed Spinoza to be making God accountable for evil in his correspondence with Spinoza.[25] Even an anti-Cartesian like Huet, in turn, does not relate Descartes's ideas to Spinoza's in his *Demonstratio Evangelica*.[26] In short, even if most of these authors were to some extent sympathizers with Descartes, little effort was devoted to show the distance between Descartes's and Spinoza's ideas.

The attacks on Spinoza came to be entangled with those levelled against another, key figure who applied a sort of Cartesian standard to Biblical interpretation, namely, Lodewijk Meijer, who had published in 1666 his *Philosophia Sacrae Scripturae interpres, exercitatio paradoxa*. The book had prompted a wave of criticism – fully analysed by Roberto Bordoli[27] – analogous to that levelled against Spinoza's *Tractatus*. In this case, however, Cartesian scholars were much clearer in detaching Meijer's ideas from those of Descartes. For instance, in 1668 Van Velthuysen had argued in his *Dissertatio de usu rationis in rebus theologicis* that Meijer applied Descartes's criterion of clarity and distinctness to matters where the principles of rationality do not pertain, such as articles of faith. In the same year, this defensive strategy was adopted by Ludwig Wolzogen[28] in his *De scripturarum*

[21] For a thorough discussion of Bredenburg's ideas and intellectual path, see Van Bunge 1990; Israel 2001, chapter 19.

[22] For a history of the early Dutch reception of Spinoza's *Tractatus* and *Ethica*, see the entries cited *supra*, n. 17, as well as Van Bunge 1989; Israel 1996, 2010; Nadler 2011a; Douglas 2015, chapters 4–5; Mignini 2015; Touber 2018, chapter 2.

[23] See especially Krop 1999; Van Bunge 2001, chapter 3; Douglas 2015, chapter 4.

[24] See Van Mansvelt 1674, 4. For a discussion, see Israel 2010.

[25] See Klever 1993; Douglas 2015, chapter 3.

[26] See Laerke 2016; Shelford 2002.

[27] See Bordoli 1997, chapter 4.

[28] Both Van Velthuysen and Wolzogen were members of the informal Collegie der Sçavanten in Utrecht, active in the mid-1650s and led by Van Velthuysen himself. It included the theologian Frans Burman, Johannes de Bruyn (professor of natural philosophy, and defender of Descartes's metaphysics in his *Defensio doctrinae Cartesianae de dubitatione et dubitandi modo* (1670)), the professor of history Johannes Georgius Graevius. The circle had connections with Heidanus and Christoph Wittich: on it, see Hartog 1876; Gootjes 2019.

interprete,[29] in turn overtly taken up by Samuel Desmarets (who had already provided a refutation of Meijer) in the following year.[30] In 1669, Abraham Heidanus went further with this strategy, as he published an *Advijs* to the theological faculty of Leiden, rejecting the idea that Meijer's *Interpres* was drawn from Cartesian principles, as this text was written by a rogue.[31]

3.1.2.2 Cartesianism Under Attack at Leiden in the Mid-1670s

In the midst of all this, Cartesian philosophy came under attack at Leiden. On the one hand, it was prompted by what had become a habit at the University: namely, the interruption and disruption of lectures and disputations by the Aristotelians by students.[32] The foremost case was that of the lectures of Gerard de Vries, former student of Gijsbert Voet (or Voetius, 1589–1676), and sub-regent of the theological college of the University of Leiden (Statencollege). His lecture of 3 March 1674 was interrupted by the students, and he left for Utrecht in May 1674, after having complained to the Academic Senate.[33] As noted by Jean Le Clerc, the Curators then admonished professors not to mingle philosophy and theology (as already stated in a regulation of 1656, as I shall show in a moment), and to avoid quarrelling.[34] De Volder submitted to this recommendation, but noted (apparently, in a private form) that De Vries himself had accused the Cartesians of truly doubting the existence of God, while disputing original sin.[35] In any case, those deemed responsible for these

[29] For a discussion of such strategies, see Krop 1999; Van Bunge 2001, 97–100 and 111–113.

[30] Desmarets criticized the application of a Cartesian philosophical standard in his 1667 *Disputationes theologicae refutatoriae*: see Bordoli 1997, 235–245.

[31] Discussed in Van Bunge 2001, 100. Besides these foremost cases, reactions to Meijer came from Pieter Beelthouwer, Petrus Serrarius, Johannes Amos Comenius, Christianus Schotanus, Reinier Vogelsang, Nicolaus Arnoldus: see Bordoli 1997, chapter 4; Sdzuj 2005; Goudriaan 2013.

[32] As to these interruptions, see Bertrand 2015, 52–60.

[33] See Molhuysen 1913–1924, volume 3, 280–281 and 290–294.

[34] See the proceedings of the Academy Senate of 9 March 1674, in Molhuysen 1913–1924, volume 3, 280–281 (addressing the interruption of lectures), as well as a letter of the University Curator Hieronymus van Beverningh of 18 June 1674 (addressing the mingling of theology and philosophy, and De Vries's resignation): Molhuysen 1913–1924, volume 3, 290–294, and the resolutions of the Curators of 6 December 1675, addressing the regulation of 1656: Molhuysen 1913–1924, volume 3, 313–314.

[35] "Les Etudians se partagerent entre eux, & les Cartesiens, comme il semble, se trouverent superieurs de beaucoup. Cela fit que dans une Dispute de Mr. de Vries, tenue le 3. de Mars 1674. ils firent quelques desordres dans l'Auditoire de Philosophie, qui obligerent ce Professeur de sortir de Chaire, avant que l'heure fût écoulée. Châcun porta ses plaintes au Sénat Academique. Mrs. les Curateurs de l'Academie firent dire aux Professeurs en Philosophie, qu'il ne vouloient pas leur ôter la liberté de philosopher, mais qu'ils trouvoitnt à propos premierement, que l'on ne mêlât pas, ainsi quel'on faisoit, des questions Théologiques dans la Philosophie; comme cela avoit été déja défendu par un reglement de l'an 1656 & en second lieu, que châcun proposât ses opinions, soit Cartesiennes, soit Péripateticiennes avec modestie, & sans s'emporter les uns contre les autres. Mr. de Volder se soumit très-volontiers à leurs sentimens, mais il se plaignit que son adversaire avoit fait tout le contraire; puis qu'il avoit traité du Peché Originel, qu'il avoit

interruptions were Cornelis Bontekoe (former pupil of Arnold Geulincx and editor of some of his posthumous works), Johannes Swartenhengst (lecturer at Leiden, and expelled from the University in 1675), and Cornelis van der Linden, as testified to by the proceedings of the University of July 1674, January and December 1675.[36] However, what stoked the atmosphere were not only these 'anti-Aristotelian' uproars, but, more generally, the mingling of Cartesian philosophy with (revealed) theology and politics, such as that practiced by Meijer and Spinoza. Indeed, this mingling came to be discussed on 28 June 1674 by a delegation composed by Heidanus, Christoph Wittich and De Volder, who held a colloquium with Gaspar Fagel, Grand Pensionary of Holland.[37] Heidanus, Wittich and De Volder constituted the so-called Cocceio-Cartesian faction at the University, notoriously opposed to the Voetian faction (usually favoured by the political authorities, according to De Volder himself).[38]

The main tenet of the so-called 'Cartesio-Cocceians' faction in Dutch universities, which included Wittich, Heidanus, Johannes de Raey, but also Balthasar Bekker, Salomon van Til, Petrus Allinga, Campegius Vitringa, and Frans Burman, was the distinction between philosophy and theology.[39] In particular, Wittich and Heidanus – although theologians – are to be regarded as shaping the development of

prouvé par des passages de l'Ecriture, & invectivé contre les sentimens de son Collegue, com me s'ils tendoient à rendre l'existence de Dieu douteuse," Le Clerc 1709, 354–355. In the *Bronnen*, no trace of De Volder's accusation of De Vries could be found. De Vries, in his *Narrator confutatus, sive Animadversiones in Narrationem de controversiis nuperius in Academia Ultraiectina* (1679), reports how in a series of disputations *De authoritatis in philosophia usu et abusu*, held under the auspices of the Leiden Statencollege, the very prohibition of the Curators of mingling philosophy and theology had been discussed, as well as the topic of original sin, and the use of doubt in theology, addressed, in accordance with Le Clerc's report, by commenting upon Biblical passages: see De Vries 1679, 27–28. No trace of such disputations could be retrieved.

[36] Molhuysen 1913–1924, volume 3, 283, 300–301 and 313–315.

[37] "Comme en ce tems là les Coccеïens & les Cartesiens s'étoient unis ensemble, non qu'il y eût aucune liaison entre leurs sentimens particuliers; mais parce que les Voetiens s'opposoient également aux uns & aux autres," Le Clerc 1709, 356.

[38] See Sect. 2.3.5, *De Volder's views on politics and religion*. Cf. also Le Clerc's own judgment: "ce qui m'étonne le plus c'est que le Prince d'Orange & les Curateurs, qui apparemment n'avoient jamais étudié ces matieres, & dépendoient purement des raports qu'on leur en faisoit, voulussent bien se prêter à la pasion de si mal-habiles gens, & déclarer punissables ceux qui enseigneroient des doctrines, de la fausseté desquelles ils n'étoient nullement instruits, & qui n'avoient aucun rapport avec l'administration des affaires de l'Etat. Que leur importoit-il que l'on fût Péripateticien, ou Cartesien? Ne leur importoit-il pas au contraire de ne faire point de mécontens, sans nécessité? Sans doute l'importunité des Anti-Cartesiens, & les suggestions secretes, contre les personnes, l'emporterent sur la prudence & sur la modération du Prince & de ces Messieurs. Si ceux qui gouvernent les Royaumes & les Républiques se proposoient toûjours pour regle de leur conduire cette belle maxime, de faire facilement du bien à tout le monde, mais de faire du mal avec beaucoup de peine; on verroit rarement des exemples de semblables partialitez, qui sacrifient, sans fruit, les accusez aux accusateurs," Le Clerc 1709, 371–372. Please note that De Volder did not always agree with the positions of the Cocceians: see Sect. 2.3.5, *De Volder's views on politics and religion*.

[39] For a thorough discussion, see Van der Wall 1996; Douglas 2013, 2015, chapter 2; Goudriaan 2016; Strazzoni 2018a, chapter 4.

Cartesian philosophy itself, as it was taught in the faculty of the arts. If Wittich set the standard for the separation of theology and philosophy – to the extent that this does not serve as a hermeneutic criterion – Heidanus, minister and then professor of theology in Leiden from 1648, used Cartesianism in the early 1640s as a source of arguments against Remonstrant theologians such as Simon Episcopius, at whom he directed his *De causa Dei* (1645), and Voetius, who aimed to develop a 'neo-Scholasticism' in support of Reformed theology.[40] In fact, Heidanus defended some principles of Cartesianism in order to rebuke Voetius for his project of using philosophy as the handmaid of theology. Under the pseudonym of Irenaeus Philalethius, Heidanus quarrelled with the Voetius faction in two pamphlets over Cartesianism which he wrote with the cooperation of De Raey in 1656, namely the *Bedenkingen op den Staat des geschils over de Cartesiaensche philosophie en op de Nader openinghe over eenige stucken de theologie raeckende*, and *De overtuigde quaetwilligheidt van Svetonius Tranquillus*. These were replies to Svetonius Tranquillus – whose identity has not been ascertained, but was certainly a Remonstrant theologian – who had published various anti-Cartesian pamphlets in the same year, attacking Descartes's philosophy as irreconcilable with Christian Faith, forcing Cartesian theologians to adopt the accommodation principle.[41] Eventually, the quarrel ended on 30 September 1656, with a resolution by the States of Holland sanctioning the principle of the separation of theological and philosophical discussions. Such a resolution was favourable to the Cartesian faction, as it allowed the teaching of Cartesianism at the faculty of arts.[42]

As another 'crisis' came to the fore in the 1670s at Leiden, a delegation composed of De Volder, Heidanus and Wittich conferred with the Grand Pensionary himself. According to Le Clerc, who based his account on a now lost manuscript by De Volder, the latter defended Cartesian philosophy by claiming that (1) it did not concern politics or affairs of the state; (2) that if someone allegedly professing Cartesian ideas had meddled in such matters should be punished – but Cartesian ideas should not be; (3) that Descartes's philosophy had flourished under different kinds of government, and inspired learned institutions such as the Royal Society.[43]

---

[40] See Duker 1897–1915; Van Asselt and Dekker 1995; Van Ruler 1995; Beck 2001, 2007; Goudriaan 2006; Woo 2013.

[41] See Svetonius Tranquillus 1656a, b, c. See also Vermij 2002, 305–306.

[42] See Van Bunge 2012, chapter 9.

[43] "Je ne m'arréterai pas à ce que les Théologiens lui dirent, par ce qu'il ne s'agit pas d'eux, en cettê occasion, & que les matieres Philosophiques & Théologiques, agitées de ce tems là, n'avoient aucune liaison naturelle ensemble. Je dirai seulement que nôtre Professeur en Philosophie représenta à Mr. le Pensionaire que 'tout ce dont on chargeoit les Cartesiens, se pouvoit réduire à deux chefs: L'un étoit qu'ils se mêloient de choses, qui concernoient l'Etat & qu'ils n'étoient pas bien affectionnez pour le Gouvernement présent: L'autre étoit qu'il traitoient de choses Théologiques & qui regardoient l'Eglisè: Que l'une & l'autre accusation étoit si mal fondée, que ce ne seroit pas la peine de les réfuter, si l'on n'avoit pris soin de les répandre partout: Que l'on ne trouveroit pas un seul Dogme, concernant la Politique, dans Descartes, puis qu'il n'avoit écrit que de Physique & de Métaphysique: Que s'il y avoit quelcun, qui fût du sentiment de Descartes, & qui se mêlât de parler d'affaires d'Etat, sans y être appellé, par son employ, ce n'étoit pas la faute de la Philosophie, qui n'y a aucun rapport, & qu'on pourroit dire à cette occasion; puniantur personae,

Moreover, De Volder (4) re-instated the separation thesis between revealed theology and Cartesian philosophy (based on different principles, and finalized with different aims), (5) claimed that Descartes's ideas conformed to religion as he maintained the distinction of body and soul, (6) claimed that doubting the existence of God was nothing but fictitious, and that (7) the Cartesians set forth demonstrations of the existence of God against the atheists, demonstrations with which the adversaries of Cartesianism could not but agree. De Volder concluded his petition by asking simply for freedom of philosophy – the only condition for the advancement of the sciences. In fact, the subsequent opening of the Leiden experimental theatre in 1675 testifies to the success of the petition of De Volder – of whom Fagel had a definitely positive judgment.[44] The appointment of Wolferd Senguerd in the same year, moreover, granted the *libertas philosophandi* sought by De Volder, as Senguerd was supposed to teach more traditional ideas at the University.[45] However, the Cartesians were put under further pressure.

As reported by Le Clerc, at the beginning of 1676 the opponents of the Cocceians and the Cartesians presented to the Curators of the University a list of propositions

non res: Que la Philosophie de Descartes étoit estimée, en toutes sortes de Gouvernemens, comme en France & en Angleterre, où la Societé Royale étoit établie sur les fondemens du Cartesianisme; & qúe cette Philosophie avoit même fleuri dans les Provinces Unies, ce qui étoit une marque que l'on pouvoit très-bien souffrir les Cartesiens sous toutes sortes de Gouvernemens, & qu'ils suivoient cette maxime, boni civis esse praesentem Reipublicae statum tueri: Que le Gouvernement appartenoit à ceux, que Dieu avoit établi pour cela, à que, si Mr. le Pensionnaire vouloit examiner la chose, il trouveroit, que ceux qui s'ettoient voulu mêler des affaires de l'Etat, depuis quelque tems, étoient tous du parti opposé'," Le Clerc 1709, 356–358. On the 'Cartesian' character of the Royal Society, see Armitage 1950; Jalobeanu 2011; Henry 2013.

[44] "[...] 'Qu'à l'égard de la seconde accusation, elle étoit si mal fondée, que les Cartesiens étoient persuadez que la Philosophie étoit appuyée sur ses propres sondemens, tout differens de ceux de laThéologie, & que leurs ennemis mêmes le leur reprochoient: Que les sentimens de Descartes, touchant Dieu & l'Ame humaine, ne choquoient nullement la Théologie, puis qu'il enseignoit qu'il y a un Dieu que l'Ame est distincte du corps, ce qui assurément ne devoit scandalizer personne: Que d'autres choses, dont on avoit parlé, n'étoient que des bagatelles, & des disputes de mots, comme ce qu'on lui objectoit touchant le doute: Que si quelcun disoit que l'on doit douter sérieusement s'il y a un Dieu, tous les Cartesiens rejetteroient cette proposition, aussi bien que les autres; mais que si l'on disoit que l'on doit rechercher, contre les Athées, les argumens qui démontrent certainement qu'il y a un Dieu, comme les Cartesiens le faisoient, ceux du parti opposé ne pourroient pas s'empêcher d'enconvenir: Qu'on trouveroit qu'il en étoit de même de toutes les questions & que si ceux, qui etoient opposez aux Cartesiens, vouloient avoir la moitié de la moderation, que les Cartesiens se croyoient obligez de garder, toutes les disputes s'en iroient en fumée. Mr. de Volder conclut delà que, puisque cette Philosophie ne renfermoit aucun Dogme, qui fût contraire aux maximes de l'Etat & de l'Eglise; en l'enseignant on ne faisoit rien d'opposé à l'intention Mrs. les Etats, qui vouloient que l'on travaillât à l'avancement des Sciences dans leur Academie; ce qui ne se pouvoit faire, sans quelque liberté de philosopher, & que c'étoit tout ce que les Cartesiens demandoient. Mr. Fagel, après avoir fait quelques difficultez, parut assez satisfait du discours de Mr. de Volder, & dit qu'il faudrait chercher quelque temperament, qui satisfît les deux partis & que si ceux qui étoient opposez au Cartesianisme faisoient quelques propositions làdessùs, on les leur communiqueroit. Mr. Fagel parla depuis de Mr. de Volder, comme d'un homme d'esprit & qui savoit très-bien se défendre'," Le Clerc 1709, 358–360.

[45] For a thorough discussion, see Wiesenfeldt 2000, 2002, chapter 2.

concerning philosophical and theological positions, accusing Wittich and Heidanus to teach those in theology, and De Volder those in philosophy. Moreover, they allegedly pointed out the places, in the works of these thinkers, in which these positions could be found.[46] On 16 January 1676, the Curators forbade the discussion of a list of 21 propositions, addressing – amongst others – the use of clarity and distinctness as a norm of the truth in theology (proposition 7), the idea that "Scripture speaks according to the erroneous prejudices of the common people" (proposition 8) the idea that will is "absolutely free and undetermined and equally infinite according to its objects as it is the will of Good" (proposition 16) and that "error is only in will" (proposition 18), that "is to be doubted of everything, even of the existence of God" (proposition 19), "philosophy is the interpreter of Holy Writ" (proposition 21).[47] These propositions concern both Descartes's philosophy as well as its (supposed) use by Meijer and Spinoza (as far as Biblical interpretation is concerned). Plainly, the more Cartesian of these propositions were still discussed after the prohibition,

[46] "L'année suivante MDCLXXVI. il arriva une chose qui fit du bruit à Leide, & dans toute la Province; mais dont je ne dirai que ce qui a quelque rapport à nôtre Philosophe. Les ennemis des Cocceïens & des Cartesiens, que l'on regardoit comme unis, parce qu'ils avaient également sujet d'être peu contents de la maniere dont on les traitoit, présenterent vint propositions aux Curateurs de l'Academie & aux Bourgmestres de Leide, comme peu orthodoxes, pour obtenir qu'ils défendissent aux Professeurs de les enseigner. Les unes étoient Théologiques, & lés autres concernoient la Philosophie. [...] On accusoit Mrs. Heidanus & Wittichius, Professeurs en Théologie, d'avoir enseigné les premieres, & Mr. de Volder les secondes [...]. Ce sont les Propositions de Philosophie, que l'on accusoit d'hétérodoxie, & que l'on publia avec les lieux des Auteurs, d'où l'on prétendoit les avoir tirées. Les Curateurs défendirent aux Professeurs d'enseigner ces propositions ni en public, ni en en particulier, & compritent dans la même défence la Métaphysique de Descartes, dont ils défendirent même de tirer aucunes Theses; le tout sous peine aux contrevenans d'être déposez & chassez de l'Université. C'est ce que porte la résolution des Curateurs du 16. de Janvier 1676," 368–370. No reference to their alleged teaching by Heidanus, Wittich and De Volder is made in the *Bronnen*.

[47] "1. Patribus Veteris Testamenti ante ultima tempora non fuisse collata vera et permanentia bona ipsamque salutem non fuisse ipsis reapse patefactam. 2. Habuisse spiritum servitutis, versatos semper in trepidatione et metu, quod pro peccatis ipsorum nondum esset satisfactum. 3. Non potuisse habere conscientiam tranquillam. 4. Spiritum Sanctum in ipsis non produxisse effectus Deo ut patre dignos nec ipsis ut filiis dignos. 5. Obnoxios per omnem vitam fuisse imperio Diaboli et metui mortis. 6. Solum Decalogum fuisse foedus gratiae in Veteri Testamento, inscriptionem vero legis in cordibus fidelium non esse inter bona Veteris Testamenti. 7. In rebus fidei normam et mensuram veritatis esse claram et distinctam perceptionem. 8. Scripturam loqui secundum erronea vulgi praeiudicia. 9. Omnipraesentiam Dei esse efficacissimam Dei voluntatem, qua omnia sustentat ac gubernat, explicandam ab operatione, qua extra se aliquid producit. 10. Angelos posse esse praesentes in spatiis dissitis in iisque simul operari. 11. Omnem philosophiam esse religionis expertem, summumque hominis bonum esse animum sua sorte contentum. 12. Mundum esse ortum ex seminibus. 13. Eum extensione infinitum esse ita ut impossibile sit dari plures mundos. 14. Animam hominis nil nisi cogitationem esse eaque sublata posse hominem vivere et moveri. 15. Nihil ad humanam naturam facere praeter cogitationem. 16. Voluntatem hominis absolute liberam esse et indeterminatam ac aeque infinitam esse ratione obiectorum ac est Dei voluntas. 17. Deum posse fallere, si velit. 18. Habere nos facultatem, qua cavere possumus ne unquam erremus; errorem vero tantum esse in voluntate. 19. De omnibus rebus esse dubitandum, etiam de Dei existentia, et ita dubitandum ut habeantur pro falsis. 20. Homines habere ideam adaequatam de Deo. 21. Philosophiam esse S. Scripturae Interpretem," Molhuysen 1913–1924, volume 3, 320.

even if – for instance, as in the case of the scope of human will and the use of doubt as the foundation of knowledge – these would be tempered by De Volder in his teaching.[48] In the same year, a book authored by Heidanus, Wittich and De Volder (but bearing the name of Heidanus only), viz. the *Consideratien, over eenige saecken onlanghs voorgevallen in de Universiteyt binnen Leyden* appeared in three Dutch editions, after which a Latin one followed in 1678. It was intended by the Cocceio-Cartesian faction at Leiden as a defence of Cartesianism. In it, as reported by Le Clerc, Wittich wrote the commentary on the propositions concerning theology, while De Volder that on those concerning philosophy. Once again, he aimed at showing that the condemned propositions were either alien to Descartes's philosophy, or they were misinterpreted by the Curators. In turn, Heidanus wrote a preface.[49] In the preface, notably, this prohibition was compared to the 1209, 1251, and 1366 bans on Aristotelian books at Paris,[50] and ultimately caused the firing of Heidanus on 5 May 1676.[51]

---

[48] See Sect. 3.2.1.2, *Clarity, distinctness and indubitability of simple and complex ideas.*

[49] "Ceux que cette défense regardoit résolurent de donner un Ecrit au Public, pour se défendre contre des accusations, qui pouvoient leur nuire; car quoi qu'il n'y eût personne denommé, on savoit assez à qui la défense s'addressoit. Mrs. Wittichius & De Volder se partagerent l'examen des propositions. Le premier se chargea d'expliquer celles qui regardoient la Théologíe, & le second les Philosophiques. Pour Mr. Heidanus, il convint de faire l'Avant-propos, & voulut bien que tout l'Ouvrage parût sous son nom. Mr. de Volder, pour ne parler que de lui, fit voir que presque toutes les Propositions, que l'on objectent aux Cartesiens, étoient mal conçues, équivoques, ou tournées d'une maniére odieuse, que quelques unes étoient entierement fausses selon leurs principes, & que l'on ne pouvoit pas les leur attribuer, & d'aunes tout à fàit orthodoxes, à moins que de les prendre à contre-sens. Ceux qui ont quelque connoissance de la Philosophie Cartesienne en tomberont facilement d'accord, à présent que la querelle est finie & que les esprits sont revenus de leur premiere chaleur. Il seroit surprenant que les Anti-Cartesiens eussent si mal attaqué les sentimiens de Descartes, & eussent si peu sûprendre leurs avantages contre lui; s'ils n'avoient pas fait profession de renoncer à leur propre Raison, comme à une lumiere trompeuse, car après cela on ne peut plus rien dire de raisonnable, que par hazard," Le Clerc 1709, 370–371. The propositions more related to theology are 6–7 and 10–20.

[50] "Hier sien wy dan D. des Cartes nu vo or de tweedemaal in dese Academie deselve fortuyne loopen, die eertijts Aristoteles in de Universiteyt van Parys tot verscheydemalen geloopen heeft: gelijck ons de Historie daar van seer punctuelijck is beschreven in een seecker boeck, in den Hage Anno 1656 gedruckt, de Varia Aristotelis fortuna in Academia Parisiensi, auctore Ioanne de Lannoy, Theologo Parisiensi. In 't welcke wy sien, dat Anno 1209 devoornaamste Boecken van Aristoteles ten vyere gedoemt zijn, en het lesen der selve verboden. Daarnae Anno 1215 sijne Dialectica toegelaten, maar sijne Metaphysica en Philosophia Naturalis verboden. Maar Anno 1366 is hy weder wat genadeliker gehandelt, wanneer verscheyde sijne Boecken te voren geinterdiceert, weder werden toegelaten, doch niet alle," Heidanus et al. 1676, 26.

[51] "L'Ouvrage de ces trois Messieurs ayant paru, sous le nom de Mr. Heidanus, il fut appellé devant les Curateurs; & l'ayant reconnu en leur présence, pour sien, sans accuser ses Collegues d'y avoir eu part, il fut déposé de sa Profession en Théologie, & demeura seulement Ministre de Leide. Il sauva ainsi ses deux Collegues, qui auroient été obligez de quitter l'Academie, sans cela. On ne peut pas manquer de louer fa génerosité, en cette occasion; mais on assure que ce qui avoit le plus choqué, c'étoit l'Avant-propos, qui, comme je l'ai dit, étoit de sa façon," Le Clerc 1709, 373. See Molhuysen 1913–1924, volume 3, 261∗–262∗.

3.1.2.3 The Demonstrations of the Existence of God

In the light of all this, De Volder's *Disputationes contra atheos* can be interpreted in three different ways: (1) as a defence of Descartes's demonstrations of the existence of God, aligned with the contents of De Volder's colloquium with Fagel; (2) as a criticism of the idea that Spinoza was an atheist (a criticism that De Volder was actually to make in his colloquium with Stolle and Hallmann), i.e. that he did not provide arguments for the existence of God – regardless of his approach to revealed religion; (3) a criticism of Spinoza's atheism and a vindication of Descartes's demonstrations of the existence of God. Since the specifics of Spinoza's atheism – for instance, if we consider the allegation of atheism by Van Velthuysen, the refusal of divine freedom, providence, and of punishments and rewards – does not concern the 'bare' demonstration of the existence of God, we need to exclude hypothesis 3. In turn, if we look at Bayle's exposition and criticism of Spinoza's ideas which was to appear in Bayle's *Dictionnaire* (which De Volder owned in its second edition, 1702), where Bayle's atheism is labelled a form of immanentism, and is traced back to a misuse of Descartes's ideas,[52] something which De Volder was going to criticize as missing the real meaning of Spinoza's words, as reported in Stolle-Hallmann's journal,[53] we can assume that De Volder was (just) showing that Descartes's demonstrations of the existence of God had nothing to do with any allegations of atheism: namely, that he wrote his *Contra atheos* in accordance with hypothesis (1). This is somehow confirmed by De Volder's 1705 *Oratio*, where his earlier demonstrations of the existence of God are recalled as an example of how he had to fight to rebuke the criticisms of those attacking the use of doubt in philosophy.[54] However, this

---

[52] On Bayle's article *Spinoza*, see Brykman 1987–1988. See also Bayle 1983.

[53] See *supra*, n. 9.

[54] As he puts it, the quarrels arising from this use were the sign of the imperfection of the system he was teaching, by which polemics or *logomachiae* could not be avoided. This confirms his late 'discontentment' with Cartesianism: "necesse est, ut scientiarum systemata tractemus, quae ad docendum ut sunt accommodatissima, et si ad perfectionis culmen pervenissent scientiae, omne punctim ferrent; ita hoc in rerum statu, quo a perfectione absumus quam longissime, scientiis promovendis multo magis nocent, quam prosunt. His, ut omittam, quae de rerum contentionibus modo attigi, accedunt innumerabiles logomachiae, quae sedem quasi suam ita in Academiis fixisse videntur, ut illis, prorsus abstinere omnino non liceat. Quid est, quod non tantum Academia, sed, et Ecclesiae cathedras maiori contentionis implevit strepitu, quam illa in principio philosophiae universalis de omnibus dubitatio? Ab his tanquam summopere necessaria propugnata, ab illis reiecta non modo, sed cum dubitationem Dei involvere videretur, scepticismi, atheismi, et nescio cuius non impietatis insimulata, cum tota haec res tantis utrinque animis agitata, si rite perpendatur, in mera consistat verbi controversia. Neque enim qui hanc dubitationem impugnant argumentis demonstrandam Divini numinis existentiam negabunt, et quidem quam certissimis, quibus rigidissime licet excussis, ita ut rigidius a pertinacissime dubitante excuti nequeant, sua tamen intemerata constet fides. Neque aliud volunt huius propugnatores; qui sicut omnium, quae ad scientias pertinent, ita et Divini Numinis demonstrationes conficiendas urgent solidissimas, quae ne scepticorum quidem longe petitis ratiunculis vel minimum infringantur, adeoque non ex externis, quae scepticus inter dubia recenset, sed solis ex cogitationibus nostris hauriantur. Harum enim in nobis conscientia omnem, ne diffitentibus quidem scepticis, dubitationem excludit. Hae solae per se cognoscuntur, reliqua omnia per illas. Qua via ut procedat, qui a se impetrare potest, ille profecto

hypothesis is also consistent with hypothesis (2), namely, by vindicating Descartes's orthodoxy he could also have been arguing that Spinoza, as a supposed follower of Descartes, was not an atheist.

The contents of De Volder's *Contra atheos* are inspired by Descartes, even if Descartes is not overtly mentioned. Rather, De Volder relies first on the indubitable existence of thought as maintained in Augustine's *Soliloquia*, as the foundation of all knowledge.[55] Having stated in this way the existence of thought and ideas, he sets forth two demonstrations for the existence of God, in order to criticize those whom he labels sceptics, rather than atheists – with quotations from Cicero and Sextus Empiricus.[56] The first proof (which De Volder had already provided, in a different form, in his 1671 *Disputatio philosophica de Deo*)[57] is based on two assumptions:

---

omni hac scepticarum dubitationum farragine, quae tantas tragoedias excitavit, nequaquam indiget. Haec et id genus infinita, cum nullius licet momenti, maxime tamen serio in Academiis agitari soleant, plurimum necesse est taedii, plurimum ulteriori scientiarum profectui nocumenti afferant," De Volder 1705, 18–20.

[55] "Verum omnis difficultas in eo est, taliane principia inveniri possint. Qui enim omnia negantibus, aliquid extorqueri potest quod non negent? [...] Quantumcunque enim dubitationi indulgeat, non tamen hoc efficiet, ut se cogitare nesciat. Quod ipsum es quod Augustinum impulit, ut, in inquisitione, quam instituit de Deo et mente, cogitationem pro fundamento poneret. Soliloq. I. 2," De Volder 1685, 9–10. Cf. Augustine's *Soliloquiorum libri duo,* II, I, 1: "R. Tu qui vis te nosse, scis esse te? A. Scio. [...] R. Cogitare te scis? A. Scio. R. Ergo verum est cogitare te. A. Verum," Augustine 1986, 45.

[56] De Volder 1685, theses 2–10. Andala was later to note that De Volder was attacking the sceptics rather than the atheists: see Andala 1716, 29.

[57] The disputation was presided over and authored by De Volder in 1671. In it, he claims that we do have an innate idea of God, demonstrated by the fact that we assign to the word 'Deus' a meaning (theses 1 and 3–4), that this represents the most perfect being (thesis 2). Hence (theses 9–13), he proposes an ontological argument, based on the premises that (1) all our knowledge comes from ideas, (2) "whatever is contained in the idea of a thing, or whatever follows from the idea of something, the same can be truly affirmed of the thing of which it is the idea" – a principle justified on the ground of its evidence; (3) the fact that a concept represents a given essence, and not another one: "IX. Quibus ita constitutis restat porro inquirendum, an ea quam nobis adesse ostendimus, idea Dei talis sit, ut nos doceat obiecti sui exsistentiam. Etenim quidquid cognoscimus per ideas quas in nobismet ipsis reperimus, cognoscimus. Nec datur ulla de rebus extra nos existentibus cognitio nisi mediantibus ideis, quarum ope rerum ipsarum novimus attributa. Ex quibus equidem evidentissime sequitur, aut nullam dari ne quidem de evidentissimis cognitionem, nullam scientiam, aut si detur, verum esse, quidquid continetur in idea rei, sive quidquid ex idea rei alicuius sequitur, id ipsum de re cuius est idea, vere affirmari posse. Neque enim ullam habeo de rebus scientiam, nisi per ideas. Ideae autem nihil docent de rebus, praeter quae ex iis sequuntur, sive quod eodem redit, quae in iis continentur. Quae si falsa esse queant, simul etiam omnia poterunt esse falsa, quaecunque etiam sunt evidentissima. Certus sum totum esse maius sua parte, sed quam ob causam? Ullamne aliam, nisi quia concipio hanc proprietatem in idea totius contineri? Hoc sane non demonstro ex eo, quod extra meos conceptus est, sed ex ea totius idea, quae menti meae inhaeret. Unde fieri si possit, ut id quod in idea totius contentur, de toto verum non sit, fieri etiam poterit, ut falsum sit totum sua parte maius esse. Quod nemo facile dixerit. Quin etiam ex triangulo extra me posito non deduco eius angulos duobus rectis aequales esse, sed ex eo tantum, quod advertam hanc proprietatem in ipsa trianguli idea contineri. Quae unica ratio est cur hoc de triangulo ab aeterno verum esse, nec aliter sese habere posse asseverem. In quibus si falli hoc ipso indicio non possum, quid certius est, quam ea quae in idea rei continentur de ipsa re vere posse affirmari? [...]

1. what necessarily follows from the nature of something is an attribute of that thing,
2. what necessarily follows from the idea of something necessarily follows from the nature of that thing.

The second assumption is grounded in the very nature of ideas as these are objective beings (or entities). Since ideas are natures (or essences) qua represented, one can find the same connection between the nature of things and their attributes, and between ideas and what ideas entail.[58] In other words, De Volder assumes that essences exist as they are represented in ideas, so that the order of ideas matches the order of things represented by the ideas themselves, and which can exist independently from ideas themselves. A keystone in De Volder's metaphysics is, in fact, that ideas are our only means of knowing reality, and in order to form any judgment on reality we have to assume that extra-mental, or extra-ideal entities can exist as they are represented by ideas. The move is Cartesian: namely, like Descartes De Volder also starts his metaphysics by considering ideas only. Yet, as I am going to show, De Volder would make a more limited recourse to the idea of God in justifying whether our ideas can match extra-mental entities. This is part of De Volder's aforementioned epistemic naturalism, viz. the autonomous foundation of the powers of reason.

The same tenet is upheld with more clarity in the section *De Deo* of his *Exercitationes* against Huet's *Censura*: if we doubt that ideas are able to represent natures, or if we doubt that the consequences between ideas match those in natures or *res*, we would have no means of knowing extra-mental realities.[59] This approach,

XI. Nec te audio, si dicas cogitationem meam nullam imponere rebus necessitatem, atque adeo ex eo quod rem ita cogitem non sequirem ita esse. Tantum enim abest ut credam meos conceptus aliquam imponere rebus necessitatem, ut e contrario existimem ipsas res imponere necessitatem meis conceptibus. Neque vero res ita se habet, quia ego talem percipio, sed talem eam percipio, quia ita se habet, quia ego talem percipio, sed talem eam percipio, quia ita se habet; unde nec fundamentum nostrae assertionis est ipse conceptus quatenus conceptus est, sed quatenus hanc mihi repraesentat essentiam et non aliam," De Volder 1671c, theses 9 and 11. The demonstration itself is given in thesis 13. In this disputation, De Volder does not provide an a posteriori argument, and does not give insights on his idea of causality, as he was to do in his *Contra atheos*.

[58] "Quae itaque paucis ut ob oculo ponatur, assumo, ea omnia quae per necessariam, et certissimam consequentiam ex rei cuiuscunque natura deducuntur, ea esse rei istius attributa, rei isti certo competere, nec absque iis rem illam aut existere aut concipi posse. [...] Assumo deinde, ea omnia, quae in idea rei alicuius continentur, sive quae ex idea istius rei necessario sequuntur, necessario quoque sequi ex ipsa natura. Nam quid aliud sunt quam naturae ipsarum rerum repraesentationes? Ex quibus quidquid sequitur, sequitur ex natura rerum quas repraesentat. Quis enim non videt, inter naturam rei, quatenus ab idea offertur menti, et attributa, quae ex eadem necessario fluere mens percivit, eandem omnino connexionem esse, quae est inter rei naturam extra nos existentem, et attributa quae producit? Ita quidem ut sicut omnino nequeo habere ideam rei sine illo attributo quod ex idea sequitur, sic res illa omnino nequeat existere absque eodem illo attributo. Nullam enim aliam ob causam ex idea id sequitur, quam quod cum ipsa rei natura necessario copulatur. [...] Manifesto siquidem eadem analogia et connexio est inter ideam rei et ea quae in idea continentur, quae est inter naturam rei, et ea quae ex illa fluunt," De Volder 1685, theses 26–27.

[59] "Nisi enim ex ideis de rebus iudicare possum, nullum mihi aliud iudicium adest. [...] Fierine potest, ut hic nexus in ideis sit, in re ipsa non sit? [...] Sin vero possit, iam idea mea non respondet

however, is not merely 'pragmatic': namely, De Volder does not merely hypothesize that ideas can match reality because otherwise we would not even be able to conduct our lives. Or: he does not make appeal to a moral certainty with regard to the reliability of ideas in representing things. This notwithstanding, his justification of such reliability comes from a consideration of the limits of our reason. First of all, he claims that we have no creative power over simple ideas: namely, all the ideas may come either from the senses, or they are innate to us.[60] That's the case, for instance, with our conceiving a third nature or attribute of body and thought: this is just impossible.[61] In other words, ideas are natures – viz. objective natures – which are

---

triangulo, quod manifestissime falsum est; iam aut falso, aut saltem incerto assuerunt mathematici, triangulo habere angulos aequales duobus rectis; conceptus trianguli hoc quidem involvit, sed si ex conceptu ad rem ipsam argumentari non licet, dubitari potest, an idem hoc, quod ex conceptu trianguli sequitur, sequatur ex trianguli natura. Quod si dubium esse queat, iam profecit nulla datur via, per quam in rerum cognitionem deveniam," De Volder 1695, *De Deo*, thesis 14. Notably, in his 1698 *Oratio de rationis viribus*, De Volder claims that ideas and things coincide, for the reason that things are defined through ideas, and not vice versa: see De Volder 1698, 5–6: "[n]ullae igitur certae emanabunt conclusiones, praeterquam ex ideis evidentissime distinctissimeque perceptis. Quarum hoc praecipuum est, ut ex illis eaedem profluant in cogitationibus sequelae, quae ex ipsis sequuntur rebus. Nam cum ideae rebus, quarum sunt ideae, conveniant exactissime (non enim ex rebus ideae, sed ex ideis res definiendae sunt) inter quarumque rerum ideas idem consequenter nexus erit, quem ipsae inter se res habent. Quod attendenti ita perspicuum est, nihil ut clarius percipi queat. Ex his itaque, et per has ideas omnis comparatur scientia. Hisce suas vires, suam in ratiocinando efficaciam debet mens nostra." In the *Disputatio philosophica de essentia et existentia Dei* (1718), presided over by Jacob Wittich, and written and defended by Johann van den Honert, son of Taco Hajo van den Honert (whose relation with De Volder I discuss below), De Volder is mentioned amongst those maintaining that knowledge of things depends on ideas only. This is a premise for the demonstration of the existence of God from His idea: see Wittich and Van den Honert 1718, 2.

[60] "Non [...] quaeri, an possimus ideas, quas a rebus externis accepimus variis inter se modis operatione mentis nostrae componere, verum an possimus plane novam effingere, sic Cartesius facillime huic allato exemplo oggeret, non quaeri, an nos qui habemus idea extensionis et motus [...] possimus illas variis inter se modis disponere pro arbitrio intellectus nostri, et si nihil extra nos sit, quod illi compositioni simile sit; sed id agi, an illas ideas extensionis et motus habere possimus, si nihil foret, quod extensioni et motui respondeat. Quae duo admodum a se differre, vel hoc ipsum Chimerae, et c. exemplum evincit. Chimaeram enim, et Pegasum ut non possem effingere, nisi animalia, ex quorum compositione illa effingo, sensibus percepissent, ita nec possem mundum illum alterum effingere, nisi haberem ideam extensionis et motus, ex quorum in se actione omnia haec mea figmenta oriuntur. Ex eo itaque quod possim ideas, quas habeo extensionis et motus pro arbitrio meo varie inter se componere, et ex hac compositione alterum quendam mundum effingere ab hoc nostro modorum respectu diversum, concludere velle, me posse pro meo quoque arbitrio, illas ipsas extensionis et motus idea efformare [...] ac si qui concludere vellet, ex eo quod colores, quos visu [...] percepit, varie inter se miscere possit pro arbitrio suo, se quoque, utut ab ineunte aetate caecus fuisset, adeoque eos visu nunquam percepisset, efformare tamen in se pro arbitrio suo has ipsas colorum ideas posse," De Volder 1695, *De Deo*, thesis 11.

[61] "Nos mentem habemus, quae duo rerum percipit genera, cogitantium et extensarum; tertium genus non percipimus. Neque tamen ullus est, opinor, qui ex eo concludere valeat, tertium genus rerum non posse dari. Ponamus itaque tertium aliquod rerum genus, et praeterea dari substantiam cogitantem finitam, quae praeter illa duo rerum genera, quae et nos percipimus, etiam percipiant hocce tertium. Haec tertii generis rerum idea in illa substantia cogitante, erit mera modificatio, et tamen mens nostra, quae est substantia, et cogitans substantia non habet in potestate sua, ut illam

independent from us, and such independence allows for their being representative.[62] With this, De Volder moves within the boundaries of Cartesianism. As Descartes put it in his *Meditatio quinta*,

> I discover in myself an infinitude of ideas of certain things which cannot be esteemed as pure negations, although they may possibly have no existence outside of my thought, and which are not framed by me, although it is within my power either to think or not to think them, but which possess natures which are true and immutable. For example, when I imagine a triangle, although there may nowhere in the world be such a figure outside my thought, or ever have been, there is nevertheless in this figure a certain determinate nature, form, or essence, which is immutable and eternal, which I have not invented, and which in no wise depends on my mind, as appears from the fact that diverse properties of that triangle can be demonstrated, viz. that its three angles are equal to two right angles, that the greatest side is subtended by the greatest angle, and the like, which now, whether I wish it or do not wish it, I recognize very clearly as pertaining to it, although I never thought of the matter at all when I imagined a triangle for the first time, and which therefore cannot be said to have been invented by me.[63]

Like Descartes, De Volder assumes this as a criterion to allow for their being capable of representing something other than themselves, before his demonstration of the existence of God. On this ground, in fact, De Volder argues that as the idea of existence is entailed by the idea of God, existence necessarily belongs to the nature of God – this being his first proof.[64] The proof is kindred to Descartes's ontological argument, expounded in the fifth of his *Meditationes*, even if De Volder builds upon different premises. Once again: that the objective contents of ideas can correspond to or represent the essence of extra-mental entities, and that the series of essences and their attributes in ideas is identical to the series of essences and their attributes as extra-mental natures. This twofold tenet is granted by considering the nature of ideas only, as representative beings, and by considering their independence from the powers of reason. On the other hand, Descartes relies on doubt and on the *cogito*. Having established that clarity and distinctness, granted by the *cogito*, are the mark of the truth of present perceptions, our clear and distinct perception that the idea of God includes that of His existence is enough to demonstrate that God exists outside His idea.[65] As I am going to show, De Volder does not reject clarity and distinctness

---

alterius ideae realitatem obiectivam [...] in se producat, cum tamen illa, qualiscunque etiam sit cogitatio videatur a re cogitante eodem modo procedere, ac eae quae sunt in nobis videntur procedere a mente nostra," De Volder 1695, *De Deo*, thesis 13.

[62]"Proculdubio enim ideae respondent obiectis, sive rebus repraesentatis; cumque hic nexus, sequatur non ex communi hoc, quod ideae sint, sed ex repraesentamine idearum, sive ex naturis repreaesentatis," De Volder 1695, *De Deo*, thesis 14.

[63]Descartes 1996, 91.

[64]"Profecto si quis ea quae [...] de connexione inter ideas, et rerum quae repraesentent naturas diximus, attente applicet ideae divinitatis videbit evidenter, istius naturae existentiam necessarium esse attributum, ideoque de ea non posse non affirmari," De Volder 1685, thesis 29.

[65]"[N]ihil absurdius est quam ideam hanc statuere effictam. Nam quid quaeso aliud effícta idea est quam quae ex pluribus sola mentis operatione pro arbitrio est conflata. Quod certe de hac nostra dici non potest, cui nihil omnino addi, a qua nihil potest demi, quin eo ipso tota Divinitatis destruatur notio. Si enim ab hac ipsa Dei Idea quidpiam separatum velimus, nonne eo ipso tollitur Natura entis

as marks of the truth. Quite the contrary, his repeated references to such features of ideas as criteria to recognize their truth constitute another aspect of his epistemic naturalism. Moreover, also in his *Contra atheos* De Volder repeatedly refers to clarity and distinctness as marking the very contents of the idea of God, viz. the necessary connection between God as the most perfect being and His existence, viz. his being an *ens a se*.[66] Yet, the path De Volder follows is not that of Descartes, as,

absolute infiniti? Unde sequitur necessariam omnino connexionem et inseparabilem nexum inter omnes perfectiones dari, aut potius unam tantum in ipsa Natura Dei omnia comprehendentem contineri perfectionem. Quid itaque a fictione distat magis, quid ab arbitrio nostro dependet minus, quam Divinitatis idea? Qua ne minimam quidem sine totius rei destructione admittere potest separationem Ubi igitur nullae continentur diversae partes, Ubi omnia referuntur ad unum, illic nulla est compositio, nulla effictio, illic simplicissima est idea, et minime arbitraria. [...] Verum, inquiet forte quispiam, ex cogitationibus nostris, quae multa comprehendunt, quae in rerum non natura reperiuntur, non licet concludere rerum existentiam. Imo vere existentiam concludere nisi ex cogitationibus nullo modo licet. Res quidem non minus existent, sive ego de iis cogitem, sive minus, sed tamen ex solis cogitationibus meis patebit mihi rem existere aut non existere. Quaecunque enim extra me fiunt, nisi mentem meam afficiant, nisi certas cogitationes in ea excitent, certe cognitionem meam iuvabunt nihil. Quid quid igitur sive existentiae sive quarumcunque proprietatum de rebus concludo, id solis pendet ex conceptibus. Nemo itaque mirabitur Divinam quoque existentiam nos ex ideis deducere, cum ea eo evidentius deducatur, ex idea sui, quod ad illam demonstrationem nulla alia requiratur idea, quod in nulla alia re obtinet. Solem ex. gr. (idem enim est quacunque de re loquar,) ut existere concludamus, ad id non faciet ipsa natura Solis in coelo, sed ipse conceptus meus quem habeo de sole, ex quo solo cum non sequatur Solis existentia, necesse est ut eiusdem dependent iam ob aliarum rerum, quarum existentiam aliunde novi, ideis animadvertam, sicque demum, non tamen aliunde quam per cogitationes meas, Solis concludam existentiam. Ex quo liquet omnium rerum existentiam ex solis conceptibus concludi, sive illa sequatur ex ipsius rei conceptu, ut hoc in sola Dei idea obtinet, sive ex dependentia ab aliarum rerum, ipsiusque demum Dei conceptibus," De Volder 1685, theses 34–35. Cf. Descartes 1996a, 91–92: "[b]ut now, if just because I can draw the idea of something from my thought, it follows that all which I know clearly and distinctly as pertaining to this object does really belong to it, may I not derive from this an argument demonstrating the existence of God? It is certain that I no less find the idea of God, that is to say, the idea of a supremely perfect Being, in me, than that of any figure or number whatever it is; and I do not know any less clearly and distinctly that an [actual and] eternal existence pertains to this nature than I know that all that which I am able to demonstrate of some figure or number truly pertains to the nature of this figure or number, and therefore, although all that I concluded in the preceding Meditations were found to be false, the existence of God would pass with me as at least as certain as I have ever held the truths of mathematics (which concern only numbers and figures) to be."

[66] "Tam aperta et illustria sui indicia animis nostris omni ex parte impressit Deus O. M. ut eius existentiam vel Mathematica evidentia monstrare volenti, non tam laborandum videatur, argumenta ut quaerat, quam ut ex iis, quae undique attendenti occurrunt, ea seligat, quae obviam magis et facilem, certam tamen et inconcussam habeant demonstrationem; sive enim ipsam Dei Naturam contemplemur, manifesto convincimur eam illam esse, quae in se sufficiens habeat existentiae suae, eiusque necessariae et aeternae fundamentum; sive eius effecta respiciamus, et ea quidem quorum notitia intime mentibus nostris insculpta nullum omnino dubitationi locum relinquit, illa dilucide demonstrant, causam summe perfectam, summe infinitam; sive denique quidpiam arripiamus vel quod existat revera, vel quod existere fingamus illa sive vera sive ficta exsistentia clare patefacit, existere aliquod ens a se. Quae quidem ut cognoscantur evidentius, singula paulo accuratius prosequenda erunt. In illa autem existentiae divinae demonstratione, quae ex natura Dei fluit, morabor minus, quia memini me alias hanc rem deduxisse latius," De Volder 1685, thesis 25. On the other hand, in his *Exercitationes*, section *De Deo*, De Volder was to be more constant to Descartes's version of the argument, on which he comments extensively: having re-stated his con-

having granted the existence of mind by an Augustinian, quasi-Cartesian argument, he demonstrates the existence of God without any consideration of the reliability of our mental faculties besides that one granted by their very limits.

The second proof set by De Volder consists in the application of the principle of causality to ideas, and corresponds to Descartes's a posteriori proof.[67] Also in this case, De Volder provides a demonstration based on ideas only, namely, besides considering the idea of God he also justifies the validity of the principle of causality on the basis of the very notion of idea. First, he rejects any proof based on a commonsensical persuasion that every idea has a cause provided with those features it represents.[68] As seen above, he does not rely on any 'moral certainty' in the reliability of ideas. Second, he focuses again on the nature of ideas: since these are, *ut obiectum*, 'natures', they differ from each other according to their representative contents. In the case that the things ideas represent exist, therefore, the differences between things would correspond to those between ideas, insofar as these represent natures. As a consequence, the very connection between the natures of cause and effect can be found between the ideas of cause and effect,[69] and that the difference

siderations on ideas as the only way to know external things (De Volder 1695, *De Deo*, thesis 14), De Volder does not built upon the premises he set in his *Contra atheos*, but he rather uses the notion of the necessary, viz. clear and distinct connection between ideas to argue for the entailment of the idea of existence in that of God (De Volder 1695, *De Deo*, thesis 19). See also De Volder's *dictata* upon Descartes's ontological argument. In commenting upon *Principia* I.14, De Volder is more constant to Descartes's 'short' version of the argument: namely, he considers 1) the immediate awareness we have of the idea of God, and 2) His essence as *ens summe perfectum*: Hamburg 273, 13–14; Hamburg 274, 9–10. The same account is followed in De Volder's *dictata* on *Meditatio quinta*, articles 9–11: cf. the *dictata* extant at Warsaw: 84–86. De Volder's a priori argument for the existence of God is recalled as exemplifying the Cartesian theories in Justus Wessel Rump's *Institutiones theologicae* (1721): Rump 1721, 8–9.

[67] "Post eam existentiae divini numinis demonstrationem, quae suam efficaciam debet ipsi naturae Dei [...] proximum est, ut videamus quidnam sequatur ex eadem illa idea, eam si consideremus, non in sua natura, sed tanquam causae alicuius effectum. Cui rei non inutile erit praemittere quaedam generalia ipsam naturam causae et effecti," De Volder 1685, thesis 37. Notably, Le Clerc claims that De Volder used, in his *Contra atheos*, not only Descartes's demonstration of the existence of God (probably referring to the a priori one), but also an argument taken "ab effectibus": see Le Clerc 1709, 385–386.

[68] "Cur enim quaeso dicimus terram, coelum, sidera, idem autem est de quibuslibet rebus existere? Nonne quia ea videmus aut aliis quibusdam sensibus percipimus? Visio autem haec nobis ne quidem persuadere, multo minus nos certos reddere posset, de rei alicuius existentia, nisi mentem afficeret. Quid enim evidentius, quam si corpus afficeretur, mens non afficeretur, nunquam nos visuros, nec ex ea corporis affectione si eius conscii non fiamus, nos nobis unquam persuasuris dari aliquid extra nos. Sola igitur idea est, quae persuadet [...]. Qua autem, quaeso, ratione hoc potest idea, nisi persuasissimi essemus ea causa requirere extra nos existentem, et talem quidem, quae illius repraesentatis perfectionibus respondeat? Quae non eo adduco, quasi ex hac persuasione argumentum petere velim, sed solummodo ut ostendam eos qui, ubi de idea Dei agitur existentia, hanc de causis idearum veritatem in dubium trahunt, eandem illam, ubi de rerum sensilium existentia agitur, extra omnem dubitationis aleam ponere, sive sibimet ipsis pugnantia loqui," De Volder 1685, thesis 59.

[69] "Quod si idearum nostrarum naturam vel obiter contemplemur, facile liquebit inter eas respectu rerum quas repraesentant eandem omnino diversitatem esse, quae foret inter ipsas res quarum sunt

between causes matches that between effects.[70] In this case, De Volder also uses a demonstration *ab absurdo*: namely, it is absurd to suppose that the conclusions drawn from two identical data can be different, and that the same conclusion can be drawn from different data.[71] The validity of the principle of causality, therefore, is grounded in the nature of ideas as objective beings or entities and our processes of thoughts: then, this is applied to ideas as these are effects of something else. This has a further consequence, however. As he assumes geometrical inferences as the paradigm to exemplify the relation between data and conclusion, and between cause and effect, he relies on an absolutely necessitarian account of causality. First, he relies on the traditional idea of proximate cause, according to which it is defined as what 'if it is in place, the effect is in place, and if it is removed, the effect is removed'. Second, he emphasizes the necessity of such a relation as "not dissolvable by any reason," and he compares it to the very relation between attributes and their properties (for instance, between the definition of a triangle and the sum of its internal angles):

> in a word[,] here we call cause only just that one which is vulgarly said proximate, which [once is] in place the effect is in place, and which [once is] removed [the effect] is removed. [...] However[,] of this cause it appears very evidently, that [such cause] [...], has a necessary connection with the effect[,] not dissolvable by any reason. Indeed[,] what stricter union can be conceived, than that, which has been just described, and which entails [...] that the one without the other can neither be in place nor be removed?

and then:

> that [...] weakness of human understanding is [of that sort], that it cannot apprehend and understand by one concept the nature of any thing, and all [things] which are in it [...], and so it proposes to itself the nature of one and the same thing [...] by many concepts, which even if they are in reality different concepts, they however represent the one and the same nature of the thing, [so that] it is allowed[,] on the ground of the diversity of concepts according to the vulgar way of speaking, that attributes of the thing, or properties, are labelled [as] different. In this way I contemplate the nature of the triangle by completely different concepts; either indeed I conceive [it] like a figure of three sides, or like a figure, whose biggest side is opposed to biggest angle, or [...] like the figure whose angles are equal to two straight [angles]. It appears that these different concepts are [...] from that,

ideas, si eae forte existerent [...] Ex quibus nec difficile erit advertere eundem hunc nexum, qui est inter naturam operationemque causae, et effectum quod producit, esse quoque eadem omnino necessitate, inter causae, eiusque operationis ideam, et ideam effecti," De Volder 1685, theses 42–43.

[70] "Quibus omnibus consentaneum est, quantum inter diversas causas varietatis est tantundem necessario diversitatis inter earum effecta fore, et viceversa, quantum est inter effecta variarum causarum discriminis, tantundem quoque inter ipsas causas reperiri differentiae," De Volder 1685, thesis 44.

[71] "Ut enim in ratiocinio, fieri nequit, ut ex iisdem plane datis diversis concludantur eadem, sic nec fieri poterit, ut ex communibus iisdemque in utraque causa proprietatibus, operationes diversae sequantur, neque ut ex diversis proprietatibus, operationes diversae sequantur, neque ut ex diversis proprietatibus sequantur eaedem. Hoc etenim si fieret, sequeretur diversam plane esse inter causam et effecta connexionem ab ea quae est inter ideam causae, eiusque effecti; quod absurdum esse [...] evicimus. Merito igitur licebit concludere, eandem omnino causarum, quae effectuum, et viceversa varietatem poni," De Volder 1685, thesis 45.

because we can know one from these attributes concerning the triangle, the others being unknown. However[,] one cannot negate that [these] are concepts only just of one thing[,] because they fit only to the triangle, from whose nature they follow necessarily, and they reciprocate to each other as much as with the same [nature], as much as with themselves. [...] However between these attributes, between which, if we consider the thing as such, there is no diversity, our mind poses some diversity by its way of conceiving. Even if indeed those [attributes] pertain to the thing in the same way, our mind however is used to conceive one of these [...] like the primary one, either because it comes first to the mind, or because it allows an easier deduction of the others[.] Which [one] calls the very nature of the thing, and from which the other things, which [one] conceives like attributes, [one] demonstrates that they follow necessarily, not because these constitute the nature of the thing, in a way different from that [one], but because such is the order of our [...] knowledge of things. From this [...] follows, [that] that primary [attribute] which is the foundation of all our knowledge and reasoning about thing[s][,] is conceived by us as the efficient cause of those properties, which we deduce, and conclude, from that [one]. [...] And indeed if rightly, with nobody dissenting, we say that the attributes are effected and produced by the nature of the things, from which however they do not differ, why it is not allowed to say the same about the [attribute of] existence in that case, [...] in which there is the attribute of the thing, from whose nature it follows necessarily[?] Why cannot we conceive the one and the same, both as it is the giver, and as it is the receiver? Why can't we conceive in this way the one and the same thing [...] with two different concepts, which are said to constitute different attributes of the thing, as I conceive that the one gives its being to the other one, and the other one receives it from the other, that is, the one is the cause of the other one? In reality[,] it is granted that between these two attributes, as far as they are in the very thing, there is no difference. And[,] in fact[,] there is [no need] that they claim that the cause has to be preceding the effect in time. Indeed[,] [what] only lacks in order for this to be true, [is that] to the contemplator of the necessary connection which is between the cause and the effect, and without which the cause would not be a cause, becomes clear that the case[,] to the extent that it is a cause, cannot be by any means preceding the effect in time. What indeed is more evident than that [thing with which] all people agree, to put the cause in place is to put the effect in place, and [once] taken away, [the effect] is removed[:] such a reciprocation soundly excludes any differentiation in time. Being [all] this proper to the cause, and nevertheless fitting no less to all the attributes which are the same with the thing, of which they are attributes, what is, please, [that reason] for which we negate that the same [thing] can exist by reason of itself, which is a cause by reason of the effect?[72]

[72] "Uno verbo eam hic solummodo causam vocamus quae vulgo proxima dicitur, qua posita effectum ponitur, et qua sublata tollitur. [...] De hac autem causa evidentissime liquet, eam omnem cum effecto, necessariam nec ulla ratione solubilem habere connexionem. Quae enim arctior copulatio concipi potest, quam ea, quae modo descripta est quaecunque id continet, ut alterum absque altero nec poni nec tolli queat? [...] Ea siquidem imbecillitas humani intellectus est, ut nequeat cuiuscunque etiam rei naturam, et omnes quae in ea sunt, unico conceptu apprehendere intelligereque. Quod autem illa unico non valet, id pluribus praestare et exequi iure merito tentat, sicque unam eandemque rei etiam individuae naturam pluribus sibi proponit conceptibus, qui utut diversi revera sunt conceptus, repraesentant tamen unam eandemque rei naturam, licet propter diversitatem conceptuum diversa vulgo, nec male, dicantur eiusdem rei attributa, aut proprietates. Sic trianguli naturam diversis omnino contemplor conceptibus; vel enim concipio tanquam figuram trium laterum, vel tanquam figuram, cuius maximum latus opponitur maximo angulo, vel denique ne plura enumerem, tanquam figuram cuius anguli duobus rectis sunt aequales. Diversos hosce conceptus esse vel ex eo liquet, quia unum ex hisce attributis de triangulo possum noscere non cognitis reliquis. Unius autem solummodo rei conceptus esse negari nequit quia soli triangulo conveniunt, ex eius natura necessario sequuntur, et tum cum eadem, tum secum invicem reciprocantur. [...] Inter haec autem attributa, inter quae ipsam rem si spectemus, nulla est diversitas, mens nostra suo con-

As I am going to show in a moment, this account of causality – which is also defended in his *dictata* on Descartes's *Principia*[73] – will cost De Volder the accusa-

cipiendi modo diversitate quandam facit. Quanquam enim illa eodem modo conveniunt rei, solet tamen mens nostra unum ex hisce concipere, tanquam primarium, vel quia primo menti occurrit, vel quia faciliorem reliquorum deductionem praebet, quod ipsam rei naturam vocat, et ex quo reliqua, quae instar attributorum concipit, necessario sequi demonstrat, non quod haec non aeque ac illud, rei naturam constituant, sed quod talis ordo sit nostrae quam habemus de rebus cognitionis. Ex quo et illud fluit, illud primarium quod est nostrae cognitionis et ratiocinii omnis de re fundamentum a nobis concipi tanquam causam efficientem illarum proprietatum, quas ex ea deducimus, concludimusque; quod ipsum sufficit ad id quod nobis nunc est prae manibus. [...] Etenim si recte, nemine dissentiente, attributa a rerum natura, a qua tamen non differunt, effici producique dicimus, quidni liceat idem dicere de existentia eo quidem in casu, quo est rei, ex cuius natura necessario fluit, attributum. Quidni unum et idem possum concipere, et tanquam esse dans et tanquam accipiens? Quidni unam eandemque rem reipsa duobus possum conceptibus, qui diversa constituere dicantur rei attributa, ita concipere, ut concipiam alterum esse suum dari alteri, et alterum ab altero accipere, hoc est, alterum esse alterius causam? Licit revera inter haec duo attributa, quatenus ipsa in re sunt, nulla sit differentia. [...] Neque vero est quod urgeant causam effecto tempore priorem esse debere. Tantum enim abest hoc ut verum sit, ut contemplanti necessarium qui est inter causam et effectum nexum, et sine quo causa non esset causa, planum fiat, causam quatenus causa est, tempore effecto priorem nullo modo esse posse. Quid enim evidentius quam id quod omnes confitentur, posita causa poni effectum, et sublata tolli, quae sane reciprocatio omne excludit temporis discrimen. Haec autem cum causae propria sit, nec minus tamen conveniat omnibus attributis quae eadem sunt cum re cuius sunt attributa, quid est quaeso cur negemus idem esse posse ratione sui ipsius, quod est causa ratione effecti?" De Volder 1685, theses 37–38 and 66–69. In his *Exercitationes* the reference to the attribute-property disappears, even if De Volder still reduces any causation to efficient one, labelled as necessary: De Volder 1695, *De Deo*, thesis 3: "[r]elatio autem causae ad effectum talis est, ut posita causa ponatur effectum, adeoque ex causae sive natura, sive operationibus effectum sequatur necessario. Quo ipso facile liquet sermonem hic esse de ea causa quam vulgo proximam vocant, neque causae vocem hic, ut in vulgari usu loquendi saepe fieri solet, ita laxe sumi, ut designet omnia, quae in re sunt, quae causa dicitur, sed praecise tantum illud quod in re illa est, quod causat, id est, aliquid efficit, nulla habita ratione reliquorum sive attributorum, sive operationum, quae dum ad hoc effectum non concurrunt, nec huius effecti causae dici queunt. Hunc autem necessarium nexum inter causam et effectum dari, vel ex eo liquet, quod pro sufficienti cuiuscunque etiam effecti causa sibi obtrudi patietur nemo id, quod licet existat et operetur, effectum tamen possit non producere: semper enim ulterius quaeret, cum possit illud, quod pro causa adducitur, operari, ita tamen ut effectum, cuius causa dicitur, ex ea operatione non sequatur, quidnam accedere necessarium sit, ut effectum sequatur, quod ipsum aut solum aut priori illi operationi iunctum demum pro vera agnoscet causa. Quod ipsum si necesse foret innumeris in exemplis facile ostendi posset." The sample of fire is used also in De Volder's *Contra atheos*: "[q]uae exemplo ut illustrentur supponamus, ignem aquam calefacere. Quod, nisi necessario sequatur ex natura ignis agentis in aquam, poterit fieri, ut ignis in aquam agat, et eam tamen non calefaciat. Haec enim si seiungi non posse statuas; eadem opera fatearis oportet esse necessarium quem urgeo nexum inter ipsius ignis in aquam actionem, et aquae calefactionem. Quem nexum si neges, et nihilominus ignis actionem solitariam esse caloris introducti velis causam, plane loqueris repugnantia. Ponamus enim nexum hunc si fieri potest dissolutum, ita ut ignis agere queat in aquam et eam non calefacere, concipiamusque duo vasa, eiusdem figurae et magnitudinis, eadem aquae quantitate plena, eidemque igni eodem ritu apposita in quorum altero si fieri queat, aqua contenta incalescat; in altero maneat frigida. Quibus ita positis, si quis nunc de te causam quaerat cur altera aqua calefiat, risui profecto te expones, si dicas id fieri propter actionem ignis," De Volder 1685, thesis 38.

[73] Cf. his commentary on I.18, in Hamburg 273, 17–18: "[i]n altero versamur argumento in quo per ipsam ideam q[ua]m habemus, concludimus a posteriori Deum existere i.e. tanquam ex effectis idea sc. {q[uae]} causa indiget, q[uo]d argumentum ut intelligatur, praemonenda sunt

tion of being a Spinozist, viz. as reducing any causal relation to geometrical causality.

On the ground of his justification of the principle of causality (as well as on the assumption that ideas can match extra-mental entities),[74] De Volder re-installs Descartes's a posteriori proof, by considering the idea of God and its difference from the ideas of other things.[75] In this case, De Volder does not diverge much from Descartes's causal argument for the existence of God, set forth in the third of his *Meditationes*:

> in order for a given idea to contain such and such objective reality, it must surely derive it from some cause which contains at least as much formal reality as there is objective reality in the idea. [...] If the objective reality of any of my ideas turns out to be so great that I am sure that the same reality does not reside in me, either formally or eminently, and hence that I myself cannot be its cause, it will necessarily follow that I am not alone in the world, but that some other thing which is the cause of this idea also exists.[76]

q[uae]dam de relatione inter causam et effectum, et {q[uo]d} primo ex causa quatenus causa sequi effectum: dico quatenus causa, i.e. quia habentur {q[uae]dam} pro causa actionum, v.g. loquela, motus, etc.: q[uae] solis a musculis et {...} fiunt: dein quando hic agitur de causa, agitur de parte, {q[uae] q[uo]d} efficit ut cum v.g. {o[mnin]o} potens tollere 100 ll tollit 50, non omnes concurrunt vires, sed de iis tunc t[an]tum loquimur, q[uae] concurrunt: ex hac causa sequitur effectus necessarius, qui non potest non sequi: ut e.g.: ut si quis 2 vasa eiusdem magnitudinis et figura apponat igni aquae repleta, si aquam in uno vase calefiat, in altero non, certe ignis non potest dici eius calori causa, si vero ambo calefiant causa dici debet, quae positae effectus ponit, et qua sublata tollitur: si nunc diversae causae omnibus viribus agentes producunt effectum aliq[uo]d, quo profecto causa est perfectior, una alia, eo et unus effectus alio erit melior eodem modo, quo ex diversis effectis unum altero melius est, meliusque concludere licet, eo unius causam altera meliorem esse. Deniq[ue] notissimum est effectum non posse plura continere quam causam [...]."

[74] De Volder stresses again that in the case where we are not convinced that ideas require a cause external to themselves that matches their contents, we could not assess the existence of anything. Actually, we have no means besides mere ideas to grasp reality: "[v]erum, inquiet forte quispiam, ex cogitationibus nostris, quae multa comprehendunt, quae in rerum natura non reperiuntur, non licet concludere rerum existentiam. Imo vero existentiam concludere nisi ex cogitationibus nullo modo licet," De Volder 1685, thesis 35. See also thesis 41: "nam si omnes meae, quas habeo, ideae tales sint, ut nullam extra me causam agnoscant, de nullius quoque rei a me diversae existentiae certus fieri potero." Accordingly, he assumes that we conceive things as they are represented by ideas, in the case that such things exist outside mind: "[a] quo argumento antequam discedam, non inutile forte erit ex iisdem fundamentis [...] demonstrare conclusionem. [...] Quod ut fiat, illud primum considerari velim, omne id quod sub perceptionem nostram cadit, si forte non existat, existere tamen ex sua natura posse," De Volder 1685, theses 48–49.

[75] "Quae quidem omina non difficulter evincent Dei existentiam, cum iam constet, ideam, quam habeo divinitatis, tantopere differre ab ideis aliarum rerum, quantopere ipse Deus sic existere ab illis quoque rebus existentibus diversus foret. Habeo ego ideam Dei, illa certe causam requirit. Nihil enim aut dari aut concipi potest quod existentiae suae essentiaeve causam non habebit, sive a semet ipso, sive ab alio. A qua igitur causa illa est? Vel certe ab ea quae est infinita et omnino perfecta, vel a finita: illud si quis dixerit, eo ipso fatebitur rem infinitam, hoc est Deum dari," De Volder 1685, thesis 46.

[76] Descartes 1996b, 28–29.

De Volder reworks this argument, both in his *Contra atheos* and in his *Exercitationes*,[77] without touching its substance, viz. without questioning the differentiation between formal and objective being.

3.1.2.4 De Volder's 'Spinozism' Under Attack by Andala

His additions to Descartes's arguments, however, did not go unnoticed by De Volder's contemporaries. If in the 1680s and 1690s no attacks on his ideas on metaphysics – viz. to his *Contra atheos* – were forthcoming, starting from the eighteenth century his figure came to be overtly associated with Spinozism. We find a first trace of this first in the 1703 report of Stolle and Hallmann. It was, however, with the analysis to which De Volder's ideas on metaphysics were subjected by a Cartesian of the last generation, namely, Ruardus Andala (from 1701 professor of philosophy and from 1713 professor of theology at Franeker), that De Volder's name came to be (posthumously) associated with Spinozism. Namely, it was within the Cartesian circle that De Volder, who during his life made considerable efforts to secure the *libertas philosophandi* mostly through his attractive experimental lectures, was first accused of heterodoxy.

In a sort of *crescendo*, we find circumstantiated criticisms of De Volder's *Contra atheos* in all the major works of Andala. In a nutshell, Andala accused De Volder of having corrupted Descartes's demonstrations by introducing a necessitarian idea of causality, and by relying on a *petitio principii*, viz. on the unjustified assumption that ideas represent extra-mental realities. In particular, in his *Exercitationes academicae in philosophiam primam et naturalem* (1708), *Exercitatio quarta, de cognitione existentiae Dei a posteriore, ex idea Dei*,[78] Andala attacks De Volder's idea of necessary causation as it is applied to the divine creation of the world. As De Volder reduced any kind of causality to efficient causation, and equated this with the geometrical relation between natures and attributes, he ultimately imposed necessity upon God.[79] By extension, an absolutely necessitarian or efficient account of

[77] The same argument is given in his *Exercitationes*, section *De Deo*, thesis 3 (where he reinstate the idea of cause as given in his *Contra atheos*), theses 4–6 (as to the differences between causes, matching the differences between effects), thesis 7 (providing the demonstration itself). In his *dictata*, as it was for the a priori proof, De Volder is more constant to Descartes's version of the argument: see his commentary on *Principia* I.16, in Hamburg 273, 17; Hamburg 274, 10. Cf. his commentary on Descartes's *Meditatio tertia* (to which De Volder refers in his commentary on the *Principia*), articles 13–16, in Warsaw *dictata*, 54–57.

[78] The original text of the disputation is not recoverable.

[79] "B. de Volder in quadam [...] Disputationibus contra atheos [...] quae supponit, mihi videntur minus esse solida, obscura, ambigua. Supponit tantundem esse discriminis inter effecta, quantum est inter eorum causas. Et viceversa, tantundem discriminis esse inter causas diversorum effectuum, quantum est inter diversa illa effecta Si Vir. Clar. intelligeret veras causas externas efficientes, quae revera agunt et operantur, et vera causalitate, actione et efficientia aliquem effectum producunt, falsa mihi videretur sententia, falsum axioma. Deus sane diversa effecta, corpora et mentes, vere produxit. Ita una et eadem causa diversa effecta producit. Uti et diversae causae eundem effectum producere possunt. Corpus enim corpus, et mens corpus movere potest. Sed aliud

causality, coupled with De Volder's premise that the differences in ideas are due to the difference in their causes, entails the consequence that God cannot have but infinite effects. So that the only idea that God could have created is the very idea of God, whose objective being is proportionate to His essence. Ideas of finite beings or entities, for instance, the idea of a triangle, cannot but have as cause finite beings.[80] This, of course, is absurd from a Cartesian standpoint. To this attack on the principle of causality, an accusation of *petitio principii* is added in Andala's *Dissertationum philosophicarum heptas* (1711) – viz. in the *Dissertatio philosophica prima, in qua ostenditur existentiam Dei non tantum a posteriore, sive per effectus, sed et a priore, sive per ipsam eius naturam, validissime demonstrari* (originally held in June 1705). Besides re-stating his accusation of necessitarianism (which was to become a leitmotiv for Andala),[81] Andala notes that De Volder's first argument for the exis-

per causam efficientem Vir. Clar. intelligit, hinc negat diversa effecta ab eadem, vel a diversis causis idem effectum sequi posse. Scilicet id ex quo aliud quid necessario sequitur haberet relationem causae efficientis, et id quod ita fluit et necessario sequitur, haberet relationem effecti. Ita natura rei esset causa efficiens, et proprietates, quae ex ea fluunt, essent effecta. Quaecunque attributa, ex quibus alia fluunt, essent quoque causae efficientes, sed ea, quae inde fluunt, essent effecta. Et ita ex eadem natura non possent sequi diversa effecta, sive attributa, nec ex diversis causis idem effectum sive attributum. Hoc sensu procederet quidem axioma" Andala 1708, 36–37.

[80] "Nec felicius pergit Vir. Clar. applicans praecedens axioma ideis et earum causis. Ut ostendat quantum discrimen sit inter diversas ideas seu diversa effecta, ponit [...] inter ideas diversas aut potius idearum repraesentamina tantundem esse discriminis, quantum foret inter res ipsas, quae per illas repraesentantur menti, si forte illae existerent. E. gr. ideam trianguli et circuli respectu repraesentaminis tantundem differre, ac triangulum et circulus, si existerent. Hinc et inter ideam rei finitae et rei infinitae ideam tanta deberet esse differentia, quanta esset inter rem finitam et infinitam, si forte existerent. Cum vero tantundem discriminis debeat esse inter harum idearum seu effectorum causas, quanta est inter ideas seu effecta, sequitur id discrimen esse infinitum. Cumque rei finitae ideae causa proculdubio sit finita, uti triangulum est causa ideae trianguli, circulus causa ideae circuli, ergo ideae rei infinitae causa debet esse infinita, id est Deus. Ergo Deus existit. Sed ad haec Viri Subtilissimi intricata ratiocinia haec noto. 1. Non debere ita confundi obiectum et repraesentamen rei. Quod passim fit potissimum [...] repraesentamen rei est idea ipsa, quae obiectum menti repraesentat. Obiectum est ipsa res, quae cogitatur a mente; est actu extra mentem, sed dicitur esse obiective in mente. Hinc 2. inter obiectum finitum et obiectum infinitum discrimen est infinitum. 3. Sed id discrimen minime est inter ideas eorum obiectorum. Omnes ideae sunt modi cogitandi et finiti. Ideam Dei non esse infinitam et summe perfectam, Vir. Clar. [...] agnoscit [...]. Idea Dei, quae finita est, repraesentat nobis Deum, ens infinitum, summe perfectum. Hoc ens, hoc obiectum est infinitum, non vero idea, non illius entis species, sive repraesentamen. Ergo 4. quia non quaeritur causa obiecti, entis illius, sed ideae, qua ens infinitum nobis repraesentare possumus, non habemus effectum infinitum, sed finitum. Sed 5. vellem intelligere ex Viro Clar. quomodo statuat ideae rei finitae esse causam finitam, et ideae rei infinitae esse causam infinitam? An idea rei finitae necessario fluit ex re et causa finita? E. gr. idea trianguli et circuli ex ipso triangulo et circulo? An ita idea rei infinitae fluit et sequitur necessario ex re infinita? Non enim aliam causae efficientis et effecti notionem agnoscere videtur Vir Celeberrimus. Patet itaque, quod hic non habeamus effectum infinitum, sed finitum; cuius tamen causa exemplaris efficiens nulla alia nisi Deus ipse esse potest, quia obiectum infinitum menti repraesentat. Adeoque longe aliter hoc argumentum proponendum est et explicandum, ac a Viro Clarissimo est factum," Andala 1708, 37–38.

[81] "Quod autem Vir. Clar. hic dicat, naturam rei sua attributa producere, et ita dici causam eorum efficientem, [...] nollem tanto Viro excidisset. Prudentius hodie loquendum esse certum est. Praestat utique attributa et proprietates ab effectibus distinguere; sic et attributum primarium, quod

tence of God relies upon the assumption – as seen above – that the objective contents of ideas represent something which can exist outside mind itself, as stated by De Volder both in his *Contra atheos* and in his subsequent *Exercitationes* against Huet. This makes De Volder's argument a sophism, due to the fact that De Volder was too involved in mathematical demonstrations, so that for him it was enough to presuppose that extra-mental entities can exist as they are represented by ideas, without further justifications.[82] Eventually, an overt accusation of Spinozism came in Andala's *Dissertationum philosophicarum pentas* (1712), viz. in the anti-Leibnitian *Dissertatio philosophica prima, in qua vera substantiae notio vindicatur, simulque confutatur vis activa, a G. G. L. cuilibet substantiae, tam corporeae, quam spirituali, in Specimine dynamico attributa* (held in 1711, and defended by Campegius Vitringa Jr.), where Andala remarks that De Volder perverted Descartes's a posteriori demonstration, coming to "favor Spinoza," as the idea that any causation which amounts to an efficient or geometrical one makes God an immanent cause, as everything follows from Him by necessity as the cause of everything.[83]

---

proprietates necessario consequuntur, quodque sit quasi fons, unde fluunt et manant omnes, a causa efficiente. Cui usui haec confusio?" Andala 1711, 38.

[82] "Hoc argumentum est merum sophisma, prout ab alio Viro Clar. est propositum. [...] argumentum Cartesii a priore [...] vindicari minime potest, prout illud proposuit, et defendere voluit contra Illustrem Huetium Vir Subtilissimum de Volder; qui id tractat Dissert. de Deo thes. 14–24. Clar. de Volder statim thes. 14 a recto tramite aberrat, quando disputat et urget, nos ad res quae sunt extra nos quidquam pertinere scire non posse, nisi ex ideis, quas habemus de iis. [...] Respondeo, haec quidem esse verissima, et nos teneri ex illis ideis de rebus ipsis iudicare, ubi nobis constat, quod revera tales res extra mentem exsistant, quarum ideae sint in mente. E. gr. si atheus concedat dari revera triangulum, omnino tenetur illi attribuere proprietates, quas idea eius necessario continet. Sic si quis Dei exsistentiam agnoscat, tenetur ex eius idea de Deo cogitare, eique concedere omnes illas perfectiones, ipsam quoque necessariam exsistentiam et aeternitatem. Ceteroquin enim non existeret exsistentiam, quod tamen supponebatur. Sed si quis neget rei alicuius exsistentiam, uti atheus Dei exsistentiam revera negat, ne hilum proficiet Vir Subtilissimus, urgendo et disputando, ex ideis de rebus, quae sunt extra nos, iudicandum esse. Ceteroquin actum esse de omni cognitione nostra. Quin ridebit atheus et iure postulabit, ut primo sibi demonstretur existere Deum. Non ergo hic laborandum est et quaerendum de nexu inter ideam Dei et Deum ipsum extra intellectum existentem: sed id quaeritur, an hoc obiectum, quod idea menti repraesentat, et dicitur obiective esse in intellectu, reapse quoque sit extra intellectum. Hoc non licere praesupponere, miror non animadvertisse Virum Acutissimum, mathematicis demonstrationibus assuetum," Andala 1711, 34–35

[83] "Clar. de Volder in suis Disputationibus contra Atheos manifestasse novam vocis causae et effectus, sic et actionis, notionem. Dicit § 37 id praecise causam, cuius ex attributis viribusque effectum sequitur. Et § 43 si alterum ex altero necessario sequatur, eorum alterum causae, alterum effecti vicem supplet. Ubi vero ea connexio necessaria non est, ibi non est relatio causae et effecti; ita ut nexus debeat esse inter naturam causae [...] et effectum, quod producit, id est quod fluit ex natura rei. Quod ipsum repetit § 62. Ut porro hoc mysterium capias, scribit § 62 licet inter quaedam rerum attributa nulla sit diversitas, mens tamen nostra inter illa suo concipiendi modo quandam diversitatem facit. Scilicet unum concipimus ut primarium, sive naturam rei; reliqua instar proprietatum sive attributorum. Hinc primarium dicit a nobis concipi ut causam efficientem illarum proprietatum, quae ex eo sequuntur, sive quas ex eo deducimus. Uti ait § 60 ex trianguli natura deducimus, quod tres eius anguli sint duobus rectis aequales. Et ne quid dubites, tandem scapham scapham dicit § 68, recte nemine dissentiente (ego vero puto nemine orthodoxo consentiente,) attributa a rerum natura effici producique dici. Unde intelligi potest aenigma, quod habet § 44. quantum inter diversas causas varietatis est, tantundem necessario diversitatis inter earum effecta fore, et vicev-

This idea, as seen above, was precisely De Volder's criticism of Spinoza in his colloquium to Stolle and Hallmann – where De Volder also attacked Spinoza's necessitarianism. In fact, it seems that De Volder himself was aware of the potential, Spinozist outcome of his treatment of causality, especially that one given in the *Contra atheos*, as in his *Exercitationes* he was no longer to reduce efficient to geometrical causality, even if he still maintained the idea of necessary causation as the ground of his demonstration.[84]

3.1.2.5 De Volder's Necessitarianism Reassessed

This necessitarianism, in fact, came to be mitigated by De Volder. After the appearance of the unauthorized edition of the disputations *Contra atheos* (1685), in July 1686 De Volder presided over a disputation defended by Taco Hajo van den Honert (1666–1740), who was to become professor of theology at Leiden in 1714, and who would defend Jacob Wittich (a student of De Volder) against the accusations of Spinozism in the 1710s (as I am going to show below). His *Disputatio philosophica de mundi in tempore productione* was overtly aimed at criticizing Spinoza's idea that the world is *ab aeterno*, as a consequence of the absolute necessity by which the intellect and the will of God follow from His nature or essence. By "following the path of [...] [his] most renowned preceptors," Van den Honert aims at showing that the world is not *ab aeterno* in the way Spinoza, the "famed one of our century" characterized it, and that God was free in creating the world, which does not necessarily follow from His essence.[85] By assuming the definition of 'world' as that

ersa. Scilicet ex diversis naturis fluere non possunt eaedem proprietates, nec ex iisdem diversae. Uti id explicat § 45 et § 44 exemplo circuli et trianguli illustrat. Eadem habet Diss. de Deo contra Huetium §. 3. 4. 5. 6. etc. atque ita pervertit etiam argumentum Cartesii a posteriore ex idea Dei. [...] Quomodo haec notio actionis, causae et effectus, faveat Spinosae, statuenti omnia ex necessitate Divinae naturae sequi, nec tamen ab ipsa distingui; et ideo dicenti Deum omnium rerum causam immanentem, non transeuntem, cuivis oculato iam satis constabit. Imo haec subtilis notio causae efficientis totius illius Spinosismi erit fundamentum," Andala 1712, 39–41

[84] See *supra*, n. 71.

[85] The author refers to *Ethica*, part 1, proposition 32 – stating that divine will is not a free cause – and quotes from its second corollary: "[a]b aeterno autem, ut reliquos taceam, mundum statuere videtur famigeratus ille nostri seculi Spinosa: qui cum omnia educat per absolutam necessitatem ex natura divina, omnia licet non aeterna, ab aeterno tamen dicere cogitur: hoc est mundum quidem statuit dependentem a Deo. Sed fluentem eadem necessitate ex natura divina, omnia licet non aeterna, ab aeterno tamen dicere cogitur: hoc est mundum quidem statuit dependentem a Deo, sed fluentem eadem necessitate ex natura Dei, qua Dei voluntas, ac intellectus ex eadem fluit. Quod patet proposit. 31 et 32 partis I suae Ethicae, et in primis ex corollario huius 2 [...]. Haec pauca annotasse, sufficere potest, cum mei instituti non sit catalogum variarum circa hanc rem, sententiarum exhibere, sed quid vestigia clarissimorum praeceptorum premendo ex ipsa natura rei investigare queam, ostendere," De Volder and Van den Honert 1686, thesis 3. Cf. the corollary: "will no more appertains to God than does anything else in nature, but stands in the same relation to him as motion, rest, and the like, which we have shown to follow from the necessity of the divine nature, and to be conditioned by it to exist and act in a particular manner," Spinoza 1912, volume 2, 70.

which is other than God, and of 'eternal' as that whose essence entails existence,[86] Van den Honert argues, first, that the idea of the world (composed by extended and thinking substance) does not entail the idea of existence.[87] Hence, he shows that

1. will and intellect can be clearly and distinctly conceived as two different attributes,
2. that in God they are "one and the same" thing, even if we cannot conceive them together as we cannot conceive God through "one act."
3. that in God, will precedes the intellect, viz. "God perceives to the extent that He is aware of His will," so that divine will is not bounded but by itself.

Being divine will the efficient cause (*causa effectrix*) of the world, this was created freely by God.[88] Notably, Van den Honert specifies that he was proposing his own elaboration of the ideas apprehended by his preceptors – according the list of the dedicatees of the second part of the disputation states – Christoph Wittich, "my only one preceptor in theology, and benevolent adviser in the other parts of [my] studies," and De Volder, "[my] preceptor in mathematics and philosophy" (besides Gronovius, professor of Greek at Leiden).[89] Between the two, De Volder seems to be a better

[86] "Per mundum itaque intelligo omne id, quod est praeter Deum, secundo notari velim, me id vocare aeternum, cuius essentia necessariam involvit existentiam adeoque quod nunquam non fuit, nec unquam non erit. Essentiae enim rerum omneque illud, quod ex illorum essentia fluit, sunt aeternae," De Volder and Van den Honert 1686, thesis 4.

[87] De Volder and Van den Honert 1686, theses 6–8.

[88] "Sed forte quaeret quis, si mundus non est a se ipso, nec ex Dei natura fluit, unde ergo originem habet? Priusquam huic respondeam, necessum est, ut Dei, quem unico actu non possum concipere, voluntatem et intellectum separatum considerem. Non, quod putem, illos esse ab invicem distinctos ut in nobis; novi enim haec duo attributa unum et idem esse in Deo. Sed tamen haec distinctio non parum claros et distinctos nobis exhibebit conceptus. Quantum itaque ad intellectum, ille nihil aliud potest, quam rerum ideas percipere, et quicquid est vel in ipso Deo, vel extra Deum cognoscere. Voluntatis autem natura consistit in assensu, aut dissensu rerum, sicque intellectus sese habet mere passive proprie loquendo; voluntas active, eodem modo ac in nobis, nisi quod nostra voluntas sequatur intellectum, Dei voluntas praecedat: noster enim intellectus exhibet et proponit voluntati ideas, e contrario vero Deus res percipit, quatenus conscius est suae voluntatis, videtque quo pacto hanc vel illam rem voluerit. [...] [E]ius [i.e. of God] voluntas quodammodo prior est intellectu, adeoque non potest ab intellectu ita distrahi et detineri. Sed id liberum voco, quod a nullo determinatur ad agendum, necessarium vero quod ita determinatur. Iam ut respondeatur, rursus distinctio instituenda inter Dei intellectum et voluntatem, ut sic manifestum fiat, num causa mundi, hoc est voluntas Dei egerit libere, vel necessario. Sed illa non fuit determinata ad producendum mundum a sua natura, prout illa voluntati eius contradistinguitur. Hoc enim satis patet [...] nec ab ipso mundo, utpote qui nondum fuit, ergo voluntas Dei produxit mundum, ita, ut a nulla alia re ad id determinaretur. Quid quaeso magis liberum? Quam ut quis suae ipsius obsequatur voluntati, E. summa libertate mundum produxit," De Volder and Van den Honert 1686, theses 12 and 15. On Descartes's ideas on divine understanding and will – given especially in his *Responsiones secundae* – see Menn 1998; Carriero 2016.

[89] "Christophoro Wittichio, [...] in theologicis praeceptori meo unico, et in reliquis studiorum partibus consultori benevolo [...]. Burchero de Volder, [...] in mathesi et philosophia praeceptori, et hoc tempore praesidi meo plurimum venerando," De Volder and Van den Honert 1686, part 2, unnumbered page. The disputation is also dedicated to Carolus Schaaf, professor of oriental languages.

candidate as the direct inspirer – if not the writer – of the contents of the disputation. Indeed, if in Wittich's *Anti-Spinoza* (which was to appear in 1690) we find the idea that world is not eternal as justified on the Cartesian ground of the idea of matter as indefinite extension (as in Van den Honert's disputation),[90] we find also that Wittich overtly rejected the Cartesian idea that, in God, will precedes intellect.[91]

This disputation distances De Volder from Spinozism, in a way that can also be appreciated in the very last disputation in which De Volder acted as *promotor*, namely the *Disputatio philosophica inauguralis de mundo* of Jacob vander Ghiessen (September 1703) – where De Volder's name never appears, and which exhibits a dedicatory poem by Jacobus Trigland II (professor of theology) – in which the idea that there is only one substance is criticized by taking into account the essential differentiation between attributes, and the idea of the eternity of the world is criticized by considering that matter and thought are not *a se*, but must have been created by an external cause.[92] This notwithstanding, the positions given in these disputations

[90] Wittich 1690, 402–403.

[91] "Deus debet habere ideas rerum ex semetipso venientes. At Deus debet habere ideas rerum ex semetipso. Quaeritur, an eas habeat ex intellectu suo, an ex voluntate? Per voluntatem intelligimus id quo iudicat, i.e. quo affirmat vel negat. Dein etiam id, quo probat vel aversatur. Non potest Deus iudicare, non potest affirmare, corpus esse extensum, mentem esse cogitantem, triangulum esse figuram tribus lateribus comprehensam, nisi hoc intelligat. Ergo prius concipiendus est a nobis divinus intellectus, quo intelligit corpus, mentem, triangulum, non potest Deus etiam probare vel improbare id, quod non cognoscit, non intelligit. Ergo etiam huius volitionis respectu praecedit intellectus," Wittich 1690, 343. Please note that the idea that understanding precedes will, in God, is not overtly discussed in Wittich's *Annotationes* on Descartes's *Meditationes*, printed and published in 1688; in it, however, he does not criticizes Descartes's positions as to this point. Following Descartes, Wittich recognizes in following what is clear and distinct in perception the freedom of man; the freedom of God, on the other hand, is posed in its spontaneity (see Wittich 1688, 102). Indeed, "[q]uia Deus [...] verus et bonus iam determinavit, et constituit. Ergo hominis libertas in eo tantum consistit, ut illud verum et bonum pro tali agnoscat," Wittich 1688, 104–105.

[92] "Caput III. Agitur hic de universi existentia a se [...]. Spinosae virus detegitur et denique demonstratur et materiam et formam esse a Deo solo [...]. At vero ridebunt [...] hodierni athei [...] qui non materiam uti superiores qua talem a se esse, sed in quantum est unum attributorum infinitorum substantiae quae a se est, quae summe est perfecta ac realis asserant. Deum enim illi definiunt: ens absolute infinitum id est substantiam constantem infinitis attributis, quorum unumquodque aeternam et infinitam essentia exprimit. Verum ne ita gaudeant, res non adeo est confecta; non ii sumus nos, qui patimur nobis obtrudi inanes et sine mente sonos pro mysteriis haud vulgaris sapientiae. Quid quaeso volunt per ea infinita attributa, quae praeter extensionem et cogitationem [...] in suo concipiunt Deo; quidquid nos concipimus, pertinet ad extensionem vel cogitatione, si illi, non dicam infinita alia, sed vel unicum concipiant aliud, cur hactenus id non protulerunt ad retundendam hanc fortem adeo obiectionem; dicunt ergo quod ipsi ignorant: nec hoc tantum, sed et quam nam quaeso concipiunt substantiam cuius duo attributa possent esse cogitatio et extensio? Omnis siquidem substantiae natura est simplicissima, omniaque attributa tum cum se invicem, tum cum substantia ipsa, identificantur [...]. Ut ergo sciamus num duo vel plura quae nos concipimus, sint attributa eiusdem substantiae, videndum unice num conceptus illi inseparabiliter ita ac reciproce cohaereant, uti ex. gr. angulum maximum oppositum lateri maximo, ac tres habere angulos aequales 2 rectis. Iam vero ideam cogitationis cum idea extensionis non tantum non cohaerere, sed se mutuo ita fugere, ut quidquid de una dicatur, de altera eo ipso negetur, quemcunque docebit propria, ad quam in talibus provocandum est, conscientia. Neutiquam ergo sunt eiusdem rei attributa, sed plane diversae res nil habentes commercii inter se. Unde eadem opera evertitur, quod

are still consistent with Gronovius's summary of De Volder's metaphysics, according to which there is neither chance nor free will, in the physical and mental domains, as the direct consequence of the tenets for which "VIII. [...] [the things] which exist, thus exist, because the most perfect intellect of God conceives them as they are, and the most perfect will of Him wants them as they are. IX. Therefore these [things] are not understood by God, because they exist, but they exist, because they are so thought by God."[93] Even if rejecting the idea that God is an immanent cause, in the disputation it is still maintained that divine will and intellect are one thing, and our conceiving them as different depends only on our limited understanding. A position which, in fact, De Volder had already touched upon in his 1660 *Disputatio inauguralis de simplicitate Dei, echo et republica*, stating the perfect coincidence of divine essence (which is necessary) and divine decrees.[94]

---

dicunt duo attributa esse rei quae est a se, nam dum talis rei attributa necessario continentur in eius natura, quae est a se, cohaerent quoque reciproco cum se invicem nexu, nisi enim ita cohaereant alterum ab altero separari potest, unde fluet ut vel utrumque eius rei non fuerit attributum, vel alterutrum saltem, at vero separabilia esse extensionem et cogitationem modo ostendimus, substantia ergo illorum, sive ens infinitum male dicitur esse a se. Sunt ergo per demonstrata cogitatio et extensio duo separata entia, quae cum utraque non possint esse a se talis enim unica tantum dari potest, nam quia natura quae est a se in uno individuo plene ac cum omnibus suis attributis potest existere, non requiret ea natura, ut plura existant quam unum ens a se, si ergo plura existerent ea non existerent vi eius naturae quae est a se ac proinde non essent a se sed ab alia causa. Videant autem illi, contra quis hic ago, quid inde porro fluat. Cum vero omnia entia praeter illud quod est a se, non sint causae suae existentiae multo minus poterunt esse cur altera res existat, omnia ergo ea reliqua entia existentiae suae causam habebunt ab unico illo ente, quod est a se. Id vero cum non possit esse extensio a rationes supra datas, necessario erit cogitatio, quae proinde erit a se, independens, infinita, perfectissima, necessarioque existens, hoc est illud ipsum ens quod nobis est Deus, a quo proinde pendebit extensio, et si quae plura praeter hanc putant esse entia, id est, Deus erit creator, materia erit creatura Dei; Q.E.D.," Vander Ghiessen 1703, 14 and 24–25. In the disputation, moreover, the author refers to the cosmogonies of Descartes and Thomas Burnet as offering physical accounts of the history of the world (which is dealt with in the last part of the disputation): "[...] Genesin mundi [...] inter eas quas recentiores excogitarunt, celebres imprimis sunt Cartesii: quam ut pote satis notam hic omitto et Burneti, quae summatim haec est, [...]. Quae et similes plures etsi in se non sint improbabiles, pro veris tamen propter rationem datam admitti neutiquam possunt," Vander Ghiessen 1703, 45. Also, a standard, Cartesian demonstration of the actual existence of the external world, viz. of extended substance, based on the awareness that I cannot be the creator of my sensations, and that God is not a deceiver, is provided: Vander Ghiessen 1703, 12–13. The standard, Cartesian contents of this disputation, paradoxically, are consistent with De Volder's eventual unconcern with Cartesianism: namely, the disputation could be the result of De Volder's teaching Descartes's philosophy, without overtly questioning it anymore. However, the very hand of De Volder can be ascertained neither by dedications nor by internal references.

[93] "VIII. Reliqua quae sunt, ideo esse, quia perfectissimus Dei intellectus ea ita concepit, et perfectissima eius voluntas ita voluit. IX. Non ergo haec a Deo intelligi, quia sunt, sed esse, quia a Deo sic cogitantur," Gronovius 1709, 25.

[94] In his *Disputatio inauguralis de simplicitate Dei, echo et republica*, defended at Utrecht in 1660 and dedicated, amongst others, to the Utrecht professor De Bruyn, as well as to the Amsterdam professors Arnold Senguerd (father of Wolferd and Aristotelian), Alexander de Bie and Gerard Blasius (professor of medicine), De Volder had argued that God is the most simple being as being composed entails imperfection (thesis 3), so that He is not composed by matter and form (thesis 4), and His essence consists of His very decrees, which are different only according to our ways of

This, actually, could have reinforced the idea, in Gronovius (dedicatee of the disputation), that De Volder inclined towards necessitarianism, even if not towards immanentism. Later, in his *Vera et cognita omnium prima, sive Disquisitio theologico-philosophica, in qua Spinosismus et Socinianismus tuto praevertuntur* (1715), Pierre Poiret was to explicitly relate Gronovius's statement on De Volder's necessitarianism to De Volder's theory of eternal truths, given in his *Exercitationes* (section *De mente, quod existat*).[95] Indeed, De Volder's *Exercitationes* had been positively quoted by Abraham Pungeler (professor of theology at Herborn) in his 1712 *Dissertatio de rerum possibilium ideis in Deo, opposita Cl. Poireto* (aimed against Poiret's *Cogitationes*). Pungeler quotes from the *Exercitationes* a statement by De Volder according to which eternal truths are not eternal in themselves, but insofar as they "flow from the eternal and immutable intellect of God." In this, as De Volder himself pointed out, he distanced himself from Descartes's theory of eternal truths, according to which these are the result of His will and decrees (and which was followed in Van den Honert's disputation, where the precedence of will over intellect is maintained).[96] According to Poiret, this idea, in the hands of De Volder, led to necessitarianism: as he put it,

considering them (thesis 5). Also His attributes are differentiated only according to our ways of considering divine essence (thesis 6). De Volder addresses neither the nature of such attributes, nor that of the freedom of God. He maintains, however, that the freedom and variety of His decrees do not hinder his simplicity, insofar as we consider them as directed to this or that object: "[q]uae autem de libertate, pluralitate, varietate, et contrarietate decretorum affirmant, aut falsa sunt, si nempe de […] essentia divina intelligantur, quod ipsum evincit argumentum, vel ad Dei simplicitatem destruendam nihil ponderis habent, si de determinatione ad diversa obiecta intellecta velint," De Volder 1660, *De Deo*, thesis 5. As to the treatment of understanding and will by Spinoza, see in particular Della Rocca 1996; Nyden 2007; Hübner 2018.

[95] See the judgment of De Volder on Poiret: "Poiretum pollere ingenio sed iam magis imaginatione uti. In cogitationibus ration. eum multa praeclara demonstrasse atque hunc librum testem esse magni ipsius ingenii. Allein nachdem er seine Oeconomie gelesen, darinne er die Hauptstücke der Religion aus der Vernunfft demonstriren wollen, so lese er kein Buch mehr von Poiret. Denn er setze allda sehr viel voraus, so er beweisen sollte, und indulgirte trefflich seiner imagination und Phantasia. Indessen lebe Poiret gar wohl nam eum, qui sciat Prophetam agere, nunquam eo destitui, unde vivat," Warsaw University Library, Cod. IV oct. 49, 588–589. Transcription kindly provided by Martin Mulsow. De Volder refers to Poiret's *Cogitationum rationalium de Deo, anima, et malo libri IV. In quibus quid Cartesius eiusque sequaces, de hisce senserint, omnisque philosophiae certiora fundamenta, atque in primis tota metaphysica verior continentur* (1677), and *L'oeconomie divine ou Système universel et démontré des oeuvres et des desseins de Dieu envers les hommes* (1687, translated as *Oeconomia divina, in quibus Dei erga homines proposita, agendi rationes, atque opera demonstrantur* (1705)).

[96] "[C]um Cartesio respondere possem, Deum si voluisset, ut duo et duo non facerent quatuor, simul etiam nobis exhibiturum suis e mentem, quae duo et duo facere quatuor evidenter non perciperet. Hoc sc. naturam Dei ab omni deceptione alienam necessario requirere. Mutata nimirum ipsa rei veritate, mutari etiam conceptus nostros. Qui cum tales sint, quales eos nunc experior, indicio mihi sunt, noluisse Deum, ut repugnantia effata simul vera esse possint. Sed agam ingenue magis. Dabo V. R. haec duo pugnare secum invicem. Quid inde? Alterutrum igitur erit falsum. Procul omni dubio. Non tamen falsum erit me existere, qui cogito. Huius enim rei me certissimum esse experior. Falsum ergo erit Deum posse efficere, ut repugnantia effata simul sint vera. Nullus dubito. Fateor enim me hac in parte Cartesio assentiri non posse; non quod velim aeternas illas

> those [...] words of De Volder, do not help Pungeler after I have demonstrated that nothing flow necessarily and essentially or naturally from the eternal intellect of God, besides God himself and [those things] which are are essential to His nature and to the attributes of [His] nature: [and that] the other things are not necessary to Him [...]. Truly, all [those] who are acquainted with his [i.e. De Volder's] principles, know that that man (I mean De Volder) stated that [there is no] freedom neither in God, nor in man, and so that he[,] forced by his own principles, could not admit anything else.[97]

In other words, for Poiret De Volder embraced a form of necessitarianism according to which all that follows from God (such as eternal truths) does so by an absolute necessity. This does not make De Volder a Spinozist, but a necessitarian *sui generis*.

Further evidence for De Volder's necessitarianism comes from his correspondence with Leibniz, viz. from his presentation to Leibniz of the contents of Bredenburg's *Wiskunstige demonstratie, dat alle verstandelijke werking noodzakelijk is* (written around 1676, and published without Bredenburg's consent in 1684 by Frans Kuyper).[98] As De Volder summarizes Bredenburg's argument in his letter of 18 October 1700,

> his demonstration [...] contains four propositions. The first of these says that there is a necessarily existing being; the second, that all the attributes of that being exist necessarily; the third, that [...] genuinely and necessarily acting force, is an attribute of this being; and the last that the effects of this being necessarily exist, and exist just the way they do. The axioms from which he deduces these things are: 1) Nothing is produced from nothing; and 2) That which is, or is something, never passes away into nothing, since there is an infinite distance between being and nothingness; and as a result, 3) No more is contained in the effect than in the cause.

---

veritates non pendere a solo Deo, sed vel ab humano intellectu, vel ab aliis rebus existentibus, quod argumentum Cartesii est, sed quod existimem has aeternas veritates dici, non quod ex se aeternae sint, hoc enim soli Deo competit, sed quod fluant ab aeterno et immutabili intellectu Dei, adeoque hoc sensu et de aeternitate, et simul de immutabilitate participent. Verum hanc rem latius deducere non est huius loci," De Volder 1695, *De mente, quod existat*, thesis 8. Cf. Pungeler 1712, 122. As to Descartes's theory of eternal truths, see his letters to Marin Mersenne of 15 April 1630 (AT I, 144–146), 27 May 1630 (AT I, 151–153), and his replies to the *Obiectiones quintae*, of Gassendi (AT VII, 380): for a thorough treatment, see Pavelich 1997; Cunning 2003.

[97] "[...] ista, inquam D. Volderi verba, nihil iuvant D. Pungelerum post quam ostendi ex intellectu Dei aeterno nihil necessario et essentialiter sive naturaliter fluere praeter ipsummet Deum et quae in eius natura naturaeque attributis sunt essentialia: cetera non esse ei necessaria; nec ex eo se qui nisi quia placuit divino intellectui nonnulla exercitii liberrimi ac indifferentissimi gratiae excogitare atque fingere. Verum, libertatem neque in Deo, neque in homine statuis eum virum (D. Volderum dico), adeoque eum principiis sibi propriis coactum, non potuisse aliter dicere, sciunt omnes qui eiusdem principia cognoverunt," Poiret 1715, 262–263. See also Poiret 1715, 282–283: "[v]ivos non nomino: inter defunctos [...] erat Celeberrimus Volderus, paulo ante a nostro laudatus, qui ex eo quod ideae necessario et immutabiliter ex intellectu divino, ut sentiebat, sequerentur, eo devenerat, ut omnem et Deo et hominibus libertatem aperte tolleret, quod et in Laudatione eius funebri minime diffitetur Cl. Gronovius, quando aliquot eiusdem docti viri principia metaphysica ibidem enarrans, hoc inter alia recenset decimum numero: nullum igitur liberum arbitrium in mente, nullum casum in corpore." Later, this passage would be quoted by Christiaan Ebherard Weismann in his *Commentatio theologica de praeiudicio* (1722), reported, in turn, in Joachim Lange's *Modesta disquisitio novi philosophiae systematis de Deo, mundo et homine* (1723): see Weismann 1722, 23; Lange 1723b, 56–57.

[98] The whole story was given, as reported by De Volder himself in his correspondence with Leibniz, in Bayle's *Dictionnaire*, entry *Spinoza*.

Such ideas cost Bredenburg harsh accusations of atheism, "on the grounds that, as even Bredenburg admitted, it [i.e. religion] could not be made consistent with God's acting necessarily" – as De Volder himself put it, implicitly referring to the foremost reason for the accusation of atheism against Spinoza.[99] In presenting this account, De Volder does not express any judgment on it. After Leibniz's letter of 31 December 1700, according to which "Bredenburg's demonstration [...] suffers from the common fallacy that the determination or infallibility of what will happen is confused with necessity,"[100] in his subsequent letter (13 February 1701) De Volder remarks that "I am not at all surprised that Bredenburg's demonstration did not please you," and adds that "the notions of cause and effect seem to incline very strongly toward necessity. Certainly it seems that nothing happens without a cause, and every cause seems to produce its effect necessarily."[101] It is hard not to think that De Volder's positions on causality, as expressed in his *Contra atheos* and then in this letter – de facto endorsing Bredenburg's views – lead to an absolute necessitarianism, just partly avoided in Van den Honert's disputation.

3.1.3 De Volder on Substance Monism

Thus far, there is strong evidence that De Volder could not avoid necessitarianism. In turn, it is hard to claim that he was consistently inclined towards Spinozism. As pointed out by Lodge, "it is reasonable to take the adoption of one particular thesis as a necessary condition for any Spinozist, namely substance monism."[102] In the case of De Volder, on the contrary, we find a strict dualism between extended and thinking substance. Paradoxically, it was exactly his dualism that led his followers, in particular Jacob Wittich, to be accused of holding Spinozist ideas as the only way to account for the interacting of the two kinds of substances.

We find a first, evident trace of De Volder's commitment to such a dualism in his dictata on Descartes's *Principia*, especially in commenting upon *Principia* I.53, stating that "each substance has one principal attribute, thought, for example, being that of mind, and extension that of body."[103] As De Volder put it in the Hamburg 273 series of *dictata*, this topic was often a matter of error – probably referring to Henricus Regius (1598–1679) and Spinoza, both assuming that a substance can

[99] Lodge 2013, 191–192.

[100] Lodge 2013, 201.

[101] Lodge 2013, 205.

[102] Lodge 2005, 131.

[103] Descartes 1982, 23. This translated edition is based on the Latin version of Descartes's *Principia*, but it also includes the parts added in the French edition (1647). The additions present in the French edition are indicated by the following parentheses: { } (which as to all the other texts quoted in this book are used to denote dubious words). Whenever a more direct grasp or comparison with the Latin or French version is required, I will quote the Adam-Tannery edition of Descartes's texts.

have more than one essential attribute.[104] On the other hand, De Volder infers the substantiality of any thing provided with an attribute, and rejects the idea that a substance can have more than one attribute, viz. that two different natures or essences can co-exist in the same substance, as contradictory. In fact, for De Volder a substance can have more attributes only insofar as we conceive it under different concepts (as seen above), while in fact what we call the different 'attributes' of a substance are one and the same thing. In the case of matter, for instance, extension is usually regarded as its primary attribute, while those of of divisibility and impenetrability are regarded as derived attributes or properties: in reality, however, they are one and the same thing, just as a triangle can be equally defined as a figure of three sides and a figure whose angles are 180°. Eventually, the criterion used by De Volder to infer the substantiality of an attribute is its epistemic independency: for instance, extension – to which De Volder refers as the attribute of matter – does not require any other concept to be conceived but by itself, and cannot be reduced to any other nature or attribute.[105] In turn, in commenting upon I.51–52 (where Descartes treats the problem of the equivocacy of the term 'substance', which I will discuss in more detail in Sect. 4.1.2, *De Volder's monist view of material substance*), he points out further requirements to argue for the substantiality of an attribute. First, in accordance with the epistemic independence of attributes, a substance can be conceived as a simple concept: the concept of body, for instance, involves only a "simple and not divided nature." The second criterion is the ontological or causal independence of a substance from any other created thing: matter, for instance, does not depend for its existence but upon God.[106] The features of the concept of substance[107] can thus be reduced to (1) its causal and conceptual independency (or its being conceivable separately from its cause (viz. God) and from any other concept), and (2) its conceptual simplicity, would then be more clearly expressed by De Volder in his correspondence with Leibniz: namely, "from which nothing of what is represented I can take away, without the whole would perish."[108]

[104] For discussions of Regius's idea of substance, see Verbeek 1992a, b; Wilson 2000; Strazzoni 2014.

[105] For full quotations, see Sects. 4.1.1, *De Volder's Cartesian standpoint on the ontology of physics*, and 4.1.2, *De Volder's monist view of material substance*. In the series Hamburg 274, on I.53 and I.63, De Volder points out, more briefly, that attributes can be conceived independently from any other concept: Hamburg 274, 20–21 and 24. Also in his (very long) commentaries upon I.60 (concerning real distinction) in both the series, De Volder points out that a substance is conceived independently from anything else, and that it is created by God: Hamburg 273, 54–61; Hamburg 274, 22–23. As he discusses the idea of substance in commenting on the *Meditationes*, De Volder re-directs the reader to the *Principia*: see Warsaw *dictata*, 36v.

[106] See Sect. 4.1.2, *De Volder's monist view of material substance*.

[107] For a thorough discussion, see Lodge 1998, 2001.

[108] "Si ergo conceptus meos contempler, in iis mihi invenire videor hanc diversitatem. Vel enim conceptus mihi unum quid repraesentat, a quo nihil repraesentaminis possim tollere, quin pereat totum; vel duo mihi repraesentat, quorum alterum seorsim concipere queam, alterum nequeam. Si primum contingat, dico conceptum hunc esse conceptum rei, sive substantiae, et obiectum ei respondens voco rem sive substantiam. Sin posterius, hos conceptus voco conceptus modorum, et obiecta horum conceptuum voco modos," De Volder to Leibniz, 18 October 1700, GP II, 215.

Thus far, De Volder argued for substantiality from the simplicity and independence of attributes. He does so, besides in his *dictata* and correspondence with Leibniz, in his *Disputationes de rerum naturalium principiis* – where he claims that ascribing different attributes to the same substance would lead to a confusion of natures[109] – and in his *Exercitationes* against Huet, viz. in the section *De corpore*. In it, De Volder demonstrates that matter is a substance which cannot have any other attribute than extension: which would be either something following from extension (i.e. extended as well), or something external (i.e. an accident), or something completely (viz. *toto genere*) different, so that substance would be something composed by more substances.[110] As to the essence of mind, in his *dictata* he from time to time

Leibniz found this passage rather obscure (see his letter to De Volder of 31 December 1700). De Volder provided them first in his letter to Leibniz of 18 October 1700, quoted in the next note. In his letter to Leibniz of 13 February 1701, thus, De Volder would explain himself as it follows: "[s]ubstantiae notionem quod attinet; hunc mihi substantiae conceptum dixi, qui ita unum repraesentat, ut nihil ab eo repraesentaminis possit tolli, quin pereat totum. De huius obscuritate quereris, nec putas eam coincidere cum altera descriptione, qua dixi substantiam esse id, quod seorsim concipi potest. Ego autem non satis capio, quid hic tibi obscuritatis sit, conabor tamen mentem meam, quantum pote, illustrare. Ubi unum quid concipio, manifestum mihi videtur, illud aut totum, ut ita loquar, mihi repraesentari, aut nihil eius. Non enim habet, ex quibus componitur. Si rem concipiam, quae aut partes nullas habeat, aut, si quidem habeat, eae omnes sint unius, parilisque naturae, nonne verum est, eam rem, aut unam illam parilemque omnium partium naturam, aut totam concipi, aut nihil eius? Unde cum hic conceptus nihil aliud repraesentet praeter hoc unum, quid manifestius, quam hoc unum seorsim a reliquis omnibus concipi posse? Quod etiam substantiae ascripseram," GP II, 222. In a further letter, he would clarify that the idea of substance does not include the idea of its cause (viz. of God): see the letter of De Volder to Leibniz of 13 February 1701: "[o]biicis [...], praeter primam substantiam nullam per se posse concipi, quia sc. habent causam. Ego vero existimo, ad substantiae concipiendam existentiam necessario requiri causam, non vero ad essentiam, de qua hic solummodo agitur," GP II, 223–224.

[109] See De Volder 1681, thesis 173. See also theses 15, 60 (where he clarifies that impenetrability is a way of considering extension by a negation) and 166.

[110] "Verum, dicet forte quispiam, hoc ratiocinium non concludere in sola extensione consistere naturam corporis, quod tamen asserit Cartesius, sed solummodo nos praeter extensionem nihil de Corpore nosse. Hoc autem, licet verum esse concedatur, non impedire, quin fieri possit, ut aliud quid nobis incognitum simul cum extensione demum perficiat naturam corporis. Qui sane scrupulus, cum sponte videatur enasci ex ratiocinio, quo hactenus usi sumus, nec careat omni specie veri, meretur considerari paululum propius. Illud ergo, quod praeter extensionem adesse forte posset corpori, et simul cum extensione corporis constitueret naturam, aut necessario sequetur ex natura extensionis, aut erit quidpiam, quod ab externa causa extensioni tribui potest, aut erit denique quidpiam ab ipsa extensione toto genere diversum. Prius si sit, non differet ab extensione, utpote ex cuius natura necessario sequitur, et cum qua nexum idcirco habet indissolubilem. Sin a causa externa extensioni tribui potest, erit accidens corporis, non ipsa substantia corporea, de qua tamen solum quaeritur. Si denique quidpiam sit ab extensione toto genere diversum, illud sive substantia fuerit sive accidens, vel erit ipsum substantia ab extensione distincta, vel inerit certe tali substantiae, eiusque erit modus. Quod si sit, corpus non erit una substantia; sed compositum quid ex duabus pluribusve substantiis toto genere a se invicem distinctis; quod non facile admittet, ni si fallor, aut V. Ill. aut quisnam etiam alius; hanc enim ideam plurium substantiarum cum voce corporis, quod sciam, coniunxit nemo. [...] Sed agamus liberaliter, ponamus corpus revera esse quid compositum ex pluribus substantiis; confitendum tamen erit, illarum diversarum substantiarum, quae secundum hanc hypothesin componunt corpus nullam a nobis concipi praeter extensionem solam, eamque idcirco, si forte non absolvat omnem naturam corporis, certe absolvere omne id,

referred to the mind as a power or faculty, rather than as a continuous action of thinking, whereas extension is perpetually extended.[111] Moreover – with an anti-Spinozist stance – that to distinguish minds by their thoughts means to mistake substances with their modes.[112] Accordingly, De Volder strongly supported a Cartesian dualism, insofar as mind and body have nothing in common.

Despite his efforts, his positions were from time to time labelled Spinozist, or at least as monist. In replying to De Volder's letter of 7 October 1701 – where De Volder claimed that one cannot include in the notion of a substance the idea of its cause, for the reason that this cause is God, i.e. it is not explicable[113] – Leibniz maintained that De Volder's idea of substance leads to the consideration that only God Himself is a substance, as an uncreated being. Namely (according to Leibniz's letter 27 December 1701), that

> [i]f the notion of substance in general does not apply to anything except the most simple or primary substance, then this alone will be a substance. I admit that you are within your right to understand the word 'substance' so that God alone is a substance, and the other things are called something else.[114]

---

quod a nobis de corpore sciri potest. Quibus consequens est, non male Cartesium, pro fundamento suis rebus Physicis ratiocinii, omnis suae inquisitionis in phaenomena natura posuisse solam extensionem; cum manifestum sit, utut etiam fingere quis velit, plura talia coniunctim ad corporis naturam requiri, illud corporis, quod a nobis cognosci potest; illud, quod causa est omnium phaenomenum, quorum causa a nobis unquam detegi potest, unice esse extensionem. Si enim vel maxime fingamus multa phaenomena esse, quae ab illis coniunctis substantiis pro ducantur, cum illorum causae essent illae ab extensione diversae substantiae, quarum notionem habemus nullam, nec istarum causarum ullam nunquam possumus adipisci notitiam. Quare, quidquid huius sit, aut corpus erit una substantia, et consistet in sola extensione? Aut erit ex variis composita, et tamen nihil de eo sciri poterit praeter extensionem, in nullas eius proprietates, effecta inquiri poterit, nisi sint proprietates et effecta extensionis. Quae itaque non male sola a Cartesio consideratur, tanquam basis vel omnium rerum physicarum, vel certe omnis nostrae, quae haberi potest, de rebus physicis notitiae," De Volder 1695, *De corpore*, theses 9–10.

[111] See his commentary on *Principia* I.53 and I.63 in Hamburg 273: Hamburg 273, 62–65. See also Sects. 4.1.1, *De Volder's Cartesian standpoint on the ontology of physics*, and 4.1.2, *De Volder's monist view of material substance*. In the *Exercitationum duodecima, continens theses selectas de partibus generationi* (1703) of Günther Christoph Schelhammer, the idea that mind always thinks is criticized by appealing to Huet, and referring to De Volder's attempt to refute him; in any case, De Volder does not touch upon this topic in his *Exercitationes*: see Schelhammer 1703, thesis 13.

[112] In Hamburg 273, on I.64: see Sect. 4.1.2, *De Volder's monist view of material substance*.

[113] "Quod de causa substantiae dicis, existimem si nullius substantiae essentia concipiatur absque causa eius possibili, nullius essentiam concipi. Modorum siquidem causas possibiles concipere mihi videor, sed substantiarum nullas. Hinc vulgo eas dicimus creari, hoc est, habere causam a qua producantur, sed modo nescio quo," GP II, 229. In fact, for Leibniz the idea of a substance has to include that of its cause, in order for it to be conceived as a complete being, i.e. an individual substance different from all the other ones: see the letter of Leibniz to De Volder of 6 July 1701: "cum ais substantiam esse quae per se concipitur, cui sententiae opposui, effectum non melius concipi quam per causam, omnes autem substantias prima demta causam habere. [...] [A]d concipiendam substantiae existentiam requiri causam, non vero ad concipiendam eius essentiam. Ego vero duplico: ad concipiendam eius essentiam requiri conceptum causae possibilis, [...] iam olim apud me referri debere," GP II, 255.

[114] Lodge 2013, 223.

This amounts almost to an accusation of Spinozism (which was then actually levelled by Leibniz at De Volder, but just for the reason that Leibniz mistook De Volder's monism on material substance for monism as such, as I show in Sect. 4.1.2, *De Volder's monist view of material substance*).

An overt accusation of monism can actually be found in the *Cartesius verus Spinozismi architectus* (1719) of Johannes Regius (1656–1738), who reports a colloquium he had with an unnamed minister, "addicted to [...] Descartes" who had been a student of De Volder. According to Regius, this minister reported that "De Volder was suspected of Spinozism by many," and, after Regius asked him "tell me in good faith whether [...] De Volder did not teach to you[,] from Descartes[,] that [there] is only one substance?" he answered, "with a certain vehemence," "what else would [he] have taught? There is and can be only one substance, and [De Volder] asserted that with great confidence as something certain, evident and known." In fact, there is no trace of this teaching in De Volder's printed and manuscript texts, nor in his correspondence with Leibniz. The attempt by Johannes Regius was to link monism to Descartes's ideas, as Descartes – as De Volder was to do – maintained that substance is what does not need anything else to subsist (in fact, Descartes meant, as De Volder was to do, only natural causes for subsistence).[115] It might even be that this unnamed witness was Andala himself, whose positions are extensively criticized in Regius's *Architectus* (this being a reply to Andala's *Apologia pro vera et saniore philosophia* (1718), aimed, amongst others, against Regius's *Kartezius Spinoza voorlichtend*, 1714).[116] Andala, indeed, briefly studied in Leiden (in 1687), was a minister of the Reformed Church, and whose words on De Volder, as given in his *Examen Ethicae Geulingii* (1716), are textually echoed in Regius's *Architectus*.[117] In his *Examen*, Andala re-stated his former criticisms of De Volder, claiming that he had perverted Descartes's a priori demonstration, reducing it to a sophism and to a

[115] "Tum ego, dic mihi bona fide docuitne te Cl. de Volder e Cartesio, unicam tantum esse substantiam? Regessit ille quadam cum vehementia, quid aliud docuisset? Unica tantum est substantia et esse potest, illudque tanquam certum aliquod, evidens, et notum magna cum fiducia asseruit. Hoc antehac quam luculentissime demonstraveram, et etiamnum contra omnes Viri Cl. exceptiones, interpretationes, et distinctiones clarissime demonstro: nihil hic moror, quae hic ob ipsius rei conditionem admitti nequit, inanem emphasin, multorum ad incitas redactorum hodie refugium. Concludo, me summa iniuria vocari impudentem calumniatorem, perfidum, dolosum, fraudulentum, ob hunc syllogismum: substantia nulla re indiget ad existendum. Res creatae re quadam indigent ad existendum. Ergo res creatae non sunt substantiae," Regius 1719, 126–127. Regius refers here to Descartes's *Principia*, I.51–52, where Descartes claimed that created substances are still substances, as they do not need any other created substance to exist, but only the concurrence of God. De Volder followed this tenet in justifying the numerical unicity of material substance; moreover, he objected to Leibniz that this tenet does not imply that created substance are not truly substances, because we can still conceive them per se, without taking into account God as their cause, for the reason that God is a supernatural cause and is anyway inscrutable: see Sect. 4.1.2, *De Volder's monist view of material substance*.

[116] See Israel 2001, chapter 25; Van Ruler 2003; Bordoli 2009, part 2, chapters 6–7.

[117] See *infra*, n. 120.

petitio principii,[118] and embracing a necessitarian view of causality.[119] Notably, in this text Andala recalls the disputation, mentioned above, he held in June 1705, whose text was then reprinted in his *Heptas*, viz. his *Disputatio philosophica, in qua ostenditur existentiam Dei non tantum a posteriore, sive per effectus, sed et a priore, sive per ipsam eius naturam, validissime demonstrari*. Andala recalls how this disputation – as reported in some letters sent to Andala from De Volder himself, now lost – caused the "wrath and indignation" of De Volder, who, having received its text before the summer holiday of 1705, after this holiday abruptly resigned from his academic position.[120] Andala, thus, seems to argue that it was his criticism – which

---

[118] "Perverse admodum explicat, pervertit et corrumpit, intricat, et subtilitatibus multis obscurat argumenta duo Cartesii, qua dicuntur a priore, sive per ipsam Dei essentiam, et a posteriore ex idea; quae tamen vult videri defendere contra Huetium. Prius plane in sophisma convertit, et in meram petitionem principii; praesupponens, de rebus quae sunt extra nos iudicandum est ex earum rerum ideis; adeoque nexum, et convenientiam inter res ipsa, et rerum ideas," Andala 1716, 16.

[119] "Sed ut pateat, Clar. de Volder non sensim deseruisse Cartesii philosophiam, sed alia arcana in pectore clausa et tecta aluisse, etiam illis annis, quando Cartesii philosophiam quam maxime docuit, citandae erunt subtilitates intricatae, difficiles nugae, imo sententiae paradoxae et heterodoxae, quas Disputationibus contra atheos, et contra Huetium immiscuit. Definit causam Disp. C. A. § 37 id praecise, cuius ex attributis viribusque effectum sequitur. Verba habes Lector, sed sensum nondum assequeris. Ubi scribit § 38 causa existente et operante effectum esse, notat de vocabulo operante per parenthesin, quod vel omitti posset, cum sine operatione causam non sit causa. Ergo causa existente effectum necessario erit. Necdum restenetur. § 43 ait, quaecumque ita se habent, ut alterum ex altero necessario sequatur, eorum alterum causae, alterum effecti vicem supplere, non est quod dubitet. Sed ubi ea connexio non est, de iis alterum causa non est, nec alterum effectum. Parum luminis affulget § 44, quando tandem quaerit, fieri ne potest, ut a diversa trianguli et circuli natura, quatenus diversa est, eadem fluant attributa? Videtur enim Vir Subtilissimus trianguli naturam habere pro causa efficiente, et eius proprietates, quae ex eius natura necessario sequuntur, pro effectis. Imo ita res se habet. Haec est clavis, qua multa aenigmata reserantur; et ne quis dubitet de vero sensu, in § 68. sine circuitione aperte ita scribit. Ex quo et illud fluit, illud primarium, quod est nostrae cognitionis et ratiocinii omnis de re fundamento, a nobis concipi tamquam causam efficientem illarum proprietatum, quas ex ea deducimus, concludimusque. Addit § 68. Recte, nemine dissentiente, at tributa a rerum natura, a qua tamen non differunt, effici, producique dici. Hinc deducit ibid. quidni unum et idem possum concipere et tamquam esse dans, et tanquam accipiens? Haec eadem immiscuit Disp. de Deo contra Huetium § 3. Ex causae sive natura, sive operationibus, effectum sequi necessario; § 4 hinc deducit, quantum diversitatis est inter causas; tantundem esse inter effecta et viceversa. Recordare Lector, naturam esse causam efficientem, et eius proprietates esse effecta illius causae. Has omnes et longe plures subtilitates percipi debent, antequam id unum argumentum pro exsistentia Dei ex idea ut effectu quis perciperet. Sed cui usui insolita, paradoxa et subtilis, ista notio vel explicatio causae efficientis et effecti? Ego hactenus nullum alium invenire potui, quam ut B. De Spinosa secundum has subtilitates Deum possit dicere causam rerum omnium immanentem, non transeuntem; statuit enim omnia ex necessitate Divinae naturae necessario fluere. Si ergo omnia illa sint in Deo, et sequantur ex eius natura necessario, haec erunt effecta, Deus causa et creator erit. Ita mentem Spinosae sequi possumus mediante hac clavi," Andala 1716, 16–18.

[120] "Habeo in memoria, quantopere incurrerim in iram et indignationem Clar. De Volder, quando in Dissertatione A.D. 17 Jun. A. 1705, ventilata non tantum vindicavi argumentum Cartesii a priore, sive per ipsam eius naturam, a depravatione et sophisticatione Clar. de Volder, in sophisma id convertentis; sed et potissimum notavi quaedam periculosa dogmata de notione causa efficientis, quorsum talia tenderent, non obscure innuens. Quam aegre id tulerit, quantopere mihi succensuerit, cognovi ex literis ad me datis, postquam opera fidelis cuiusdam amici et discipuli, cupientis de hac

would later become an overt accusation of Spinozism – that caused the resignation of De Volder (who anyway does not mention this issue in his 1705 *Oratio*). In turn, the opinion of the aforementioned, unnamed witness of De Volder's lectures, as reported by Regius, textually echoes Andala's own description of De Volder's wrath as given in the *Examen*.[121] In his subsequent *Cartesius verus Spinozismi eversor*, in turn, Andala attacked Johannes Regius as improperly reducing Descartes's idea of substance to Spinoza's, and does not mention De Volder as to this point.[122] In fact, in his *Heptas* Andala had positively referred to De Volder's *Exercitationes* – viz. to those theses of the section *De corpore* quoted above – in order to argue, like De Volder, for the substantiality of extension.[123] In his answer, thus, Andala aimed at separating the original thought of Descartes from that of Spinoza, without however labelling De Volder a monist – but only attacking, once again, De Volder's idea of causality as an example of how many of those inspired by Descartes misrepresented his thought.[124]

To sum up, De Volder's idea of substance, either through arguments (as those discussed in his correspondence with Leibniz), or through rumours (as in Regius's vitriolic attack) was from time to time associated with Spinoza's. This notwithstanding, it is indubitable that De Volder made a significant effort to develop an idea of substance – on a Cartesian basis – entailing a substantial or real distinction between attributes. In fact, Hermann Boerhaave – who would later repeatedly praise De Volder as his mentor[125] – defended a Cartesian dualism in his inaugural disputation *De distinctione mentis a corpore* (1690)[126] expressly against Spinoza's monism of substance, as well as against Hobbes's and Gassendi's theories of the

re cum magni Nominis Viro colloqui, meam Dissertationem ad ipsum transmiseram et cum ipso communicaveram. Quam primum ad has literas serio, solide, perspicue responderam, statim post ferias, a munere docendi publico habita Oratione se abdicavit," Andala 1716, 23. The same themes recurs in Andala's *Apologia pro vera et saniore philosophia* (1718), where he recalls again the "wrath and indignation" of De Volder, and the absolute necessitarianism entailed by his idea of causality: Andala 1718, 137, 1719b, 23, 28 and 189.

[121] "Fuit mihi praeterita aestate sermo cum quodam Reverendo V. D. Ministro, Viro erudito, partibus Cartesii addicto, qui olim Cl. de Volder usus erat, Praeceptore. Inter alia dixi, Cl. de Volder multis de Spinozismo fuisse suspectum, quod admodum aegre tulit Vir venerandus, mihique succensuit," Regius 1719, 126. Cf. the quotation given *supra*, n. 119. The wrath of this Minister, in fact, was not caused by Regius himself, but was rather the result of De Volder's alleged endorsement of Spinozism: see *supra*, n. 114.

[122] See, for instance, Andala 1719a, 37–38.

[123] See Andala 1711, 59 and 82; see also Andala 1708, 184, 197, 212, 232 and 235, where a Cartesian view of matter is defended by referring to De Volder's works. It might be, actually, that Andala – if we suppose that he spoke with Regius – was referring to the unicity of material substance only. Even Leibniz, indeed, mistook De Volder's ideas on the unicity of material substance with those on the unicity of substance as such: see Sect. 4.1.2, *De Volder's monist view of material substance*.

[124] Andala 1719a, 28 and 35.

[125] See Sects. 2.4.2, *De Volder's retirement, death and legacy*, and 3.2.3.1, *The case of explanations in physiology*.

[126] It is discussed in Boerhaave 1983, 20 and 24.

soul.[127] What, actually, would cost De Volder a further accusation of Spinozism was the way in which he addressed the problem of the union between the two kinds of (created) substances, viz. soul and body.

3.1.3.1 The Relation Between Material and Immaterial Substances

As far as the union of soul and body is concerned, De Volder did not provide a unitary exposition or discussion of the issue. If we look at his commentary upon Descartes's *Meditatio sexta*, viz. on the proposition "these sensations of hunger, thirst, pain, etc. are in truth none other than certain confused modes of thought which are produced by the union and apparent intermingling of mind and body,"[128] he does not show any commitment to the modalities or the reality of their interaction, but only that there is some sort of union between them: as "colours, smells, sounds etc. cannot arise [*oriri*] in the soul unless through the body, so that [they] do not indicate anything else, than that union which is between the soul and the body."[129] As to the articles of Descartes's *Principia* where union is considered (I.48, II.3 and IV.189), De Volder claims that body and soul constitutes a "compositum,"[130] that soul and body have a "close union," so that the perception of food "follows or accompanies" the movement in a body in which hunger consists,[131] that sensations "arise [...] from the conjunction i.e. relation [of soul and body] to each other."[132] Eventually, De Volder rejects the idea that the soul 'informs' the body (as the meaning of "informare" is far from being clear), and maintains that the connection between soul and body consists in the fact that their mutations "follow" each other,[133] and that "from the the movement of the brain the

[127] On Spinoza, see especially chapter 5, where Boerhaave reduces the substantial difference between two different attributes, on the one hand to their being conceivable independently from each other, and on the other hand to the evidence of their being different: "[c]oncedit enim attributa duo complete, clare et distinet intellecta, ita se habere ut unius conceptus commode formari possit dum nihil in eo ingreditur quod ad alterius ullo modo conceptum pertinet, et viceversa: aio ergo, dum clare et distincte et plene ex hypothesi concipio illa attributa esse diversa, concipio eorum substantias esse diversissimas, ipsa namque rerum attributa aliud nil sunt quam ipsa substantiae natura atque vera essentia (def. 4 et prop. X p. I Eth.) quamobrem in quod de attributis rerum, id substantiae denegari nihilo iure potest, proindeque colligere nequeo, quam diversas esse diversorum attributorum substantias. Atqui inflat de attributi cuiusque essentia est per se concipi," Boerhaave 1690, chapter 5, 12 (unnumbered).

[128] Descartes 1993, 93.

[129] "[...] colores odores soni etc. non possunt oriri in anima nisi per corpus, unde etiam nihil aliud denotant, quam eam quae est corpus inter et animam unionem," Warsaw *dictata*, 46v.

[130] Hamburg 273, 37 (on I.48).

[131] See Hamburg 274, 19 (on I.48): "[...] sic in appetitu famis fit quaedam mutatio in ventriculo, quae actio hactenus mere corporea est, illam mutationem ex arcta unione quam habet animam cum corpore, insequitur vel comitatur perceptio quaedam in animae et desiderium quoddam cibi."

[132] See Hamburg 273, 82 (on II.3): "[s]ensuum perceptiones non oriuntur ex sola mente, vel solo corpore, sed ex coniunctione i.e. relatione ad invicem. Sensus non docent {q[uo]d} sit in corpore, sed {q[uo]d} possit nobis prode[ss]e vel nocere." The same passage, slightly changed, is present in Hamburg 274, 33.

[133] In Hamburg 273, 344 (on IV.189): "[e]xplicui. An[im]am inquit informare totum corpus, quo

thoughts arise in the soul."[134] Thus far, De Volder did not commit himself to any view as to the relation of soul and body. Nor does he do so in his *Exercitationes* – where he claims that perceptions 'accompany' bodily movements.[135] Notably, in the *Disputatio philosophica de mundi in tempore productione*, authored by Van den Honert, an occasionalist account of the relation of body and mind is given. Provided that sensations could not be arbitrarily caused by mind itself (as we are not aware of how to produce them), and that it is absurd to conceive an interaction between soul and body, for the author God Himself causes sensations, on the occasion of corporeal movements.[136] This idea cannot be found in Christoph Wittich's *Annotationes* on Descartes's *Meditationes*,[137] so that it could have been just the result of Van den Honert's own elaboration, albeit certainly approved by De Volder, at least as a possible solution to the problem of the union. Eventually, in his 1698 *Oratio de rationis viribus*, De Volder claimed that "the whole way is precluded to reason" in understanding how to proceed "from thought to movement" and vice versa, as we are only aware that movements and thought "go along together" (*pari passu*).[138] No explana-

obscuro vocabulo vulgo putant se indicare coniunctionem corporis cum anima, verum si roges {q[uo]d} sit informare? Respondebunt efficere, ut sit [h]o[m]o et non brutum. Author [n]a[m] vult eam dari connexionem inter corpus et a[n]i[m]am, ut mutationes cogitationum in mente sequantur mutationes et actiones q[uas]dam c[or]poris, et viceversa. C[um] vero dicitur e[ss]e in toto corpore, falsum dicunt, cum a[n]i[m]ae nec locus, nec situs adscribi po[ss]unt. Author itaque vult id {q[uo]d dixere} modo: etsi ergo corpus informet, i.e. ex {...} corporis, {...} mutationibus sequantur cogitationes, et viceversa, vult tamen praecipuam mentis sedem e[ss]e in cerebro."

[134] In Hamburg 274, 119 (on IV.189): "[s]edem suam habere in cerebro. Sedem dici anima habere in cerebro hanc unicam ob causam, eo q[uo]d in cerebro exercet praecipuas suas operationes, quatenus ex motu cerebri oriuntur cogitationes q[uas]dam in animam."

[135] See De Volder 1695, *De mente*, thesis 15. Other disputations presided over by De Volder, and authored by his students, do not offer many more insights on the essence of mind and its union with the body: Johannes Enyedi's *Disputatio philosophica de anima humana prima–secunda* (1671), Samuel Nemethi's *Exercitii philosophico-metaphysici de perenni duratione mentis pars prima–tertia* (1682), Joseph Serrurier's *Disputatio philosophica de mente humana secunda* (1688), Stephan Huszthi's *Dissertatio physiologica secunda, quae est de natura mentis eiusque functionum* (1693), Theophilus Copius's *Disputatio philosophica de mentis humanae et divinae existentia et reale a corpore distinctione* (1699).

[136] "Adsum ego e.g. igni, statim sentio calore, huius ideae inquiro causam. 1. Ipse ignis eius causa esse non potest, tum quod nihil in corpore meo producere valeat, quam motum quendam solito vehementiorem, qui ope nervorum communicatur cerebro. Cum hactenus hic nihil aliud sit quam motus corporeus, mens autem mea sit incorporea. [...] 2. Nec ipse sum causa istius ideae, possem enim alias illam excitare, sive adessem igni, sive non, quod tamen est falsum, [...] imo ipse deberem esse conscius, me posse illam producere, quae omnia cum non sint in me, nec ipse eius ideae causa sum. Cum ex his evidenter pateat, nec ignem, nec me ipsum esse causam istius ideae, necessum est, ut aliam eius assignem causam, quae occasione istius ignis in me producat ideam caloris, et illa causa, cum agat in subiectum cogitans, non potest nisi per efficaciam suae voluntatis agere, adeoque non potest non esse Deus," De Volder and Van den Honert 1686, thesis 13.

[137] Wittich, like De Volder in his own *dictata*, admits that there is a strong connection between soul and body, but he does not enter into detail: see Wittich 1688, 150 and 165.

[138] "Inter ea, quae in rerum natura posuit Divinum Numen, maximam omnium admirabilitatem facit homo. Pessime [...] ille, qui hominem dixit, id quod omnes norunt. Verius dixisset homine, quod ignorant omnes. Etenim, quid sit humanum corpus, quid mens, si vel maxime nobis putaremus perspecta, quod a rei veritate abest quam longissime, nemo tamen est, nemo, qui horum inter

tion, thus far, is provided: rather, De Volder commits himself to the impossibility of establishing rationally any link between the two. The topic then resurfaced in his correspondence with Leibniz. Indeed, the issue of the soul and body relation (besides that of the activity of material substance, which I will deal with in Sect. 4.1.4, *The search for physical unity and activity: De Volder vis-à-vis Leibniz*), was one of the problems which De Volder deemed to be unsolved by the Cartesians. For instance, he found the hypothesis of interaction – in his letter to Leibniz of 12 November 1699 – as simply "absurd": this being a foremost example of those tenets of Descartes which De Volder did not avail himself of, and which he warned not to be attributed to him.[139]

In fact, De Volder seemed to incline to a theory of pre-established harmony as a solution to this problem. He had found Leibniz's solution of pre-established harmony – as he declared first in his letter to Leibniz of 18 February 1699 – as "wonderfully pleasing" to him. However, De Volder reinterpreted it in his own way, providing Leibniz with his own explanation. De Volder accounts for it by hypothesizing two machines: a real machine, and an adequate idea of this machine and of all its parts. According to De Volder, the series of movements in the machine, and the series of adequate ideas of the changes of the machine go along together. The same holds for the adequate ideas of what happens to a part of the machine: from this idea, the mind will know everything following from the force of this part. In turn, the mind will have an inadequate idea – namely, it will have a sensorial perception – of what happens to the machine, or to a part of it, if this is due to an external agent.[140] Or, in De Volder's own words,

---

se nexum, in quo proprie natura hominis consistit, mente sua assequatur. Sunt enim duo haec, mens et corpus, res toto genere diversae, et ita quidem diversae, ut, quod nullis aliis in rebus usu venit, et qua maior discrepantia ne cogitari quidem potest, nihil in unius conceptu contineatur, quod alterius notio non excludat. Quare, dum inter haec nihil percipimus commune, rationi omnis interclusa via est, qua progrediatur a proprietatibus unius, ad proprietates alterius, sive a cogitatione ad motum, vel a motu ad cogitationem. Id unum novimus, non quidem ex ipsarum rerum natura, quae perfectissima demum scientia est, sed experientiae solius ope, ipsa nos hoc docente conscientia, arctissimam, quae inter haec est, societatem ea lege coiisse, ut corporearum mutationum nonnullae, cum mutationibus cogitationum, et hae vicissim cum illis pari ambulent passu," De Volder 1698, 1–2.

[139] "Iam ex prioribus tuis, Vir Illustris, intellexi, quod postremae hae confirmant admodum, suspicari te, ita me deditum esse Cartesio, ut eius auctoritati plane assentiar, et in eo putem me intelligere quae in aliis me intelligere negem. Cuius rei nullam faciam Apologiam, utpote quae, cum eam facerent etiam illi, qui ipsi mordicus adhaerent, nihil merito proficeret. Id tamen addam, me multos et magnos etiam circa principia tam in Physicis, quam in Metaphysicis agnoscere eius errores, licet in multis generalibus praesertim ipsi assentiar. In exemplum adducam illud, quod ad rem nostram facit, me neutiquam intelligere, quid sibi velint Cartesius et Cartesiani, quando de anima humana corpori conjuncta, et in corpus agente, et vicissim a corpore patiente loquuntur; verba haec sunt, nullum meo judicio sensum habentia, nisi ostenderint, quo pacto ex cogitationibus sequatur motus, et ex motu cogitationes, quod tantum abest ut facere illos putem, ut e contrario nihil absurdius esse existimem," GP II, 198. See also the letter of De Volder to Leibniz of 30 October 1703.

[140] "Unum tamen silere non possum, mirabiliter mihi arridere tuum illud inter mentem et corpus harmoniae, ut vocas commercium, cuius quem conceptum formem, inde exemplo conabor explicare, rectene an secus, tui erit iudicii. Supponamus machinam multis ex partibus conflatam, suisque

> [i]magine a machine put together from many parts [...]. Imagine a mind, in which there is an adequate idea of this machine and of all its parts [...]; nothing will be represented in it which is not in the machine, and whatever change obtains among the parts of that machine, a representation of this will follow necessarily from the idea. And vice versa [...]. If we now imagine that there is a mind which has the idea of one of the parts of this machine, that idea will have the same relation to the adequate idea of the whole machine as that part has to the whole machine. And just as everything which happens in the whole machine follows as evidently as possible from the adequate idea of the machine, in the same way, from the idea of each part will follow everything which will happen as a result of the forces of this part alone. But a mind which has only the idea of this part will not perceive those things which obtain due to other parts of the machine acting on this part clearly, but confusedly, that is, it will sense them, since, although they will not in fact follow from its [i.e., the mind's] idea, they will nonetheless have some relation to it. And if things happen in the machine due to its other parts which do not affect this part, just as the part will not be changed at all by those things, so the mind which has the idea of this part will neither perceive nor sense anything of these things.[141]

It is hard not to think about Spinoza's concept of mind as an idea of the body (as given in his *Ethica*, II, proposition 13), in which inadequate ideas (viz. sensations) correspond to those bodily changes due either to an external cause (viz. affections), or to an internal cause not adequately known (propositions 24–25). Leibniz noted this, as in his reply to De Volder of 3 April 1699, he praised De Volder's explanation of his theory of pre-established harmony, noting that "I think that you agree with me that the soul is one thing, the idea of the body another. For the soul remains the same, but the idea of the body is constantly one way then another."[142] In fact, De

---

viribus, suas operationes peragentem. Supponamus mentem, cui insit huius machinae, omniumque eius partium adaequata idea. Haec enim exactissime repraesentet illam machinam, nihil in ea repraesentabitur, quod non sit in machina, et quidquid mutationis inter partes istius machinae continget eius repraesentamen necessario sequetur ex idea, et vice versa, quidquid ex idea sequetur, illud necessario continget ex machina; ita tamen, ut quae in machina fiunt, fiant ex natura machinae, quae vero in idea, ex natura ideae, quae est repraesentamen machinae. Nam, si quae in machina contingerent, quae ex idea machinae non sequerentur, aut vice versa ex idea quaedam sequerentur, quae in machina non contingerent, sequeretur hanc ideam non esse machinae istius ideam, quod est contra Hypothesin. Quod si nunc porro fingamus dari mentem, quae ideam habeat unius ex partibus istius machinae, habebit illa idea eandem relationem ad ideam totius machinae adaequatam, quam illa pars habet ad totam machinam. Verum ut ex illa idea adaequata machinae, omnia sequuntur quam clarissime, quae in tota machina fiunt, sic ex idea partis sequentur omnia, quae solis huius partis viribus fiunt; quae vero contingunt ab aliis partibus machinae in hanc partem agentibus, ea illa mens quae solam hanc partis ideam habet, cum ex sua idea quidem non sequantur, aliquam tamen ad eam relationem habeant non percipiet clare, sed confuse, id est sentiet. Si quae vero fiant in machina a partibus eius aliis, quae hanc partem non afficiunt, ut ex iis illa pars nihil mutabitur, sic mens quae habet ideam huius partis, nihil istorum aut percipiet, aut sentiet," GP II, 166–167. A similar appreciation is restated in De Volder's letter to Leibniz of 12 November 1699: "[q]uae etiam causa est, cur tua huius coniunctionis ratio mihi placuerit tantopere; quae equidem explicat, non qui anima agat in corpus, et corpus in animam, sed qui mutationes animae et corporis fiant simul, licet singulae fiant a suis principiis," GP II, 198.

[141] Lodge 2013, 63.

[142] Lodge 2013, 77. Cf. the full original text: "[a]tque ita ad Commercium venio inter Animam vel quamcunque Entelechiam corporis organici, et Machinam organorum; de quo Hypothesin meam gaudeo Tibi tanti acuminis et iudicii viro non omnino displicere. Et sane praeclare illam illustras,

Volder had distinguished mind from its ideas; as seen above, moreover, he had distinguished, *à la* Leibniz, mind as a faculty, and thoughts as its mode (thus abandoning Descartes's idea that mind constantly thinks). Yet, his interpretation of Leibniz's theory of pre-established harmony has an unmistakeable 'Spinozist' flavour. What probably prevented De Volder from embracing Spinoza's view of mind as an idea of the body was that for De Volder mind is an active principle, whereas matter is completely passive, so that the series of ideas on the mind do not follow from each other in the same way as the series of movements follow in the machine. In fact, the machine is moved by God Himself. In his reply to Leibniz (13 May 1699), indeed, De Volder dropped the discussion on the mind as an idea of the body, reverting instead to the problem of the internal activity of matter, which according to him should correspond to the internal activity of mind.[143]

---

tribuens Animae ideam adaequatam machinae corporeae; idque ipsum volo cum dico animae naturam esse corporis repraesentatricem. Unde quicquid sequitur ex legibus corporis, necesse est animam sibi ordine repraesentare; alia distincte, alia vero (quibus scilicet multitudo corporum involvitur) confuse quorum illud intelligere est, hoc sentire. Interim te mecum convenire puto, aliud esse Animam, aliud corporis ideam: nam anima eadem manet, corporis vero idea perpetuo alia atque alia est prout mutatur ipsum corpus, cuius semper exhibet praesentes modificationes," GP II, 171–172.

[143] "Ex meo enim exemplo, quod gaudeo tibi non plane displicere, si eo attigi mentem tuam, sequitur animam illam non esse causam activam materiae, sed tantum repraesentatricem, adeoque in materia concipiendam aliam vim sive ἐντελέχειαν, quae materiae propria sit. Hanc si ab extensione separes, explicandum erit, qui agere in extensionem possit; si non separes[,] ex natura extensionis deducenda erit," GP II, 180. Not surprisingly, De Volder would be then labelled of Spinozism by Andala also as to this point. First, in his *Heptas* (1711) Andala positively quotes from De Volder's *Oratio* as having asserted that the problem of the union (of which we have immediate awareness) cannot be solved by rational means, as one cannot pass, by the power of reason, from the properties of body to those of soul and vice versa. So that the nature of such union can be grasped by God only (Andala 1711, 158 and 163). Hence, in his *Pentas* (1712) Andala compares such passage from De Volder's *Oratio*, to Leibniz's 1698 *De ipsa natura sive De vi insita actionibusque creaturarum*, in which Leibniz claimed that the 'commerce' between substances does not take place by interaction, but in accordance to Divine preformation. Andala, thus, reduces De Volder's position to Leibniz's (see Andala 1712, 43). However, in his *Examen* (1716), Andala then reduces De Volder's positions to Spinoza's, for the reason that they both maintained the necessary relation between movements and thoughts, whereas such relation is arbitrarily established by God. We do not know whether Andala came to read De Volder's correspondence. What is more clear is that if in his previous works he recognized in De Volder's 1698 *Oratio* a (positive) reference to God's will as the key factor in the 'commerce' between soul and body (which do not act on each other), in his *Apologia* he underlines the necessitarianism underlying De Volder's positions on such a relation, which he fully expounded in his correspondence with Leibniz, maintaining that ideas follow from other ideas, like in a machine a movement follow from a movement. Knowing, moreover, that for De Volder efficient causality means an absolutely necessary relation between causes and effects: "[n]on tantum hasce subtilitates, tricas, imo absurda paradoxa immiscere voluit Clar. de Volder Philosophiae Cartesianae; sed tandem ei, saltem primariis fundamentis primae philosophiae, quando tempora id ferre videbantur, in Oratione dicta A. 1698 [...]. Scilicet si unum ex alio necessario fluat, sequatur, et a nobis deducatur et concludatur, tum demum notio causae et effecti locum habet, secundum novam hanc et subtilem philosophiam; quam supra vidimus. Sed perversa et praepostera est haec via. Non est hic nexus necessarius vel naturalis, sed arbitrarius, et a divino libero placito pendens. Non ergo necessario ex cogitatione motus, nec ex motu cogitatio sequi debet. Ratio ita progredi non debet. Ita Spinosa vellet. Sed orthodoxa philosophia docet Deum

In the same months, moreover, De Volder was discussing Spinoza's ideas also with Philipp van Limborch (himself a critic of Bredenburg's Spinozism).[144] Also in this case, De Volder faced Spinoza's solution to the problem of dualism, even if Spinoza's monism is not embraced by De Volder. De Volder, indeed, was involved by Van Limborch in the dialogue at a distance between Locke and Hudde, concerning the numerical uniqueness of God in 1697–1699. Hudde had already discussed the matter with Spinoza himself in 1666, as testified to by letters 34–36 of Spinoza's correspondence. In a nutshell,[145] Hudde had asked Spinoza for a demonstration of the numerical uniqueness of God (in a letter now lost). In his answer to Hudde of 7 January 1666 (letter 34), Spinoza provided a demonstration that God, as *causa sui*, did not create more Gods, as this is not entailed in His idea.[146] Hudde was not satisfied by this argument, as testified to by a further letter of Spinoza of 10 April (letter 35, mentioning two lost letters by Hudde of 10 February and 30 March), in which Spinoza proposed a further argument, based on the idea of God as perfect being (viz. eternal, simple i.e. part-less, undetermined i.e. infinite and indivisible), whose perfection would be contradicted by the existence of more Gods.[147] Hudde found such an argument once again unsatisfactory, as reported by Spinoza in the last extant letter of their correspondence (letter 36, undated), in which Spinoza attempted to clarify it.[148] As emerges from this letter, Hudde – as a Cartesian – admitted the pos-

---

Creatorem esse causam nexus huius arbitrarii inter cogitationem et motum. Hoc ergo fundamentum ignorantiae praetensae ad omnem veram Philosophiam evertendam nullum est," Andala 1716, 18–19. The same double-reading is present in his *Apologia*, as this time Andala reads De Volder's words as entailing (even if not overtly embracing) a "perverse way" in understanding the relation of soul and body in terms of interaction: their relation, indeed, is completely dependent on God's will and can by no means be addressed by reason ("[...] ea quae Cl. de Volder habet in limine Orationis A. 1698 [...] haec est via perversa, et natura mentis et corporis adversa, pugnans cum reali inter illas duas substantias distinctione, et possibili earum separatione, adeoque unio inter eas debet esse arbitraria, a placito nostri Creatoris dependens," Andala 1719b, 37–38).

[144] See Simonutti 1984; Scribano 1988; Van Bunge 1990, chapter 5; Fix 1991, chapter 9.

[145] For more details, see Montuori 1983; Klever 1989; Di Biase 2014.

[146] "Quoniam vero [...] necessaria existentia ad Dei naturam pertinet, eius vera definitio necessariam quoque existentiam ut includat, necessum est: et propterea ex vera eius definitione necessaria eius existentia concludenda est. At ex vera eius definitione [...] necessaria multorum deorum existentia non potest concludi. Sequitur ergo unici Dei solummodo existentia," Spinoza 1677, 520.

[147] "Ut denique ad rem accedamus, affirmo non nisi unicum posse ens esse, cuius existentia ad suam naturam pertinet, illud nimirum ens tantum, quod omnes in se habet perfectiones, quodque Deum nominabo. Quippe si aliquod ponatur ens, ad cuius naturam existentia pertinet, illud ens nullam in se continere imperfectionem, sed omnem exprimere debet perfectionem [...]. Et propterea natura illius entis ad Deum, (quem [...] etiam existere statuere debemus) pertinere debet, quia omnes perfectiones, nullas vero imperfectiones in se habet. Nec extra Deum existere potest; nam si vel extra Deum existeret, una eademque eatura, quae necessariam involvit existentiam, duplex existeret; quod, iuxta antecedentem demonstrationem, est absurdum. Ergo nihil extra Deum; sed solus Deus est, qui necessariam involvit existentiam. Quod erat demonstrandum," Spinoza 1677, 522–523.

[148] "Ad tuam, (in causa fuit impedimentum aliquod) nono decimo Maii datam, citius respondere non licuit. Quia vero te tuum super demonstratione mea, tibi missa, suspendere iudicium, quantum ad maximam partem, deprehendi (propeter obscuritatem, credo, quam in ea reperis) sensum eius clarius hic explicare conabor," Spinoza 1677, 523.

sibility that extension and thought can exist as perfect beings, and at once as two numerally distinct ones.[149] So that he did not admit Spinoza's conclusion. In turn, Spinoza objected to him that extension and thought do not contain all the perfections (even if they are undetermined or infinite in themselves), and are rather two attributes of God – i.e. the being to which all the perfections belong.[150] Hudde's concern with substance dualism, in fact, resurfaces once he asked, through Van Limborch, for Locke's solution to the problem of the uniqueness of God, after having appreciated his *Reasonableness of Christianity* (1695) – as declared by Van Limborch in his letter to Locke of 28 August/8 September 1697.[151] Hudde approached Locke through Van Limborch, as required by the theological nature of the issue. Indeed, Van Limborch did not immediately unveil the identity of Hudde himself to Locke,[152] who, in turn, posed conditions of secrecy in providing his argument to Van Limborch in a letter of 21 February 1698 – an argument not meant to be sent to Hudde, but for the use of Van Limborch and for that of Jean Le Clerc.[153] Once he read Locke's argument, based on the idea of the omnipotence, omniscience and omnipresence of God, entailed by His perfection,[154] Van Limborch wrote back to Locke pointing out how such an argument would not convince Hudde, as far as Locke's argument relied on the idea of divine omnipresence – a decidedly non-Cartesian idea – an opinion he shared with Le Clerc, with whom he discussed the letter.[155] Thus, in his subsequent reply of 2–4 April 1698, Locke provided a slightly modified version of his argument.[156] As testified to by Van Limborch's letter to Locke of 6/16 May 1698, this was the argument eventually read by Hudde, who in turn communicated to Locke (*via* Van Limborch) that he would rather prefer a dem-

149 "[...] absolute concedis; et tamen dicis, tuam difficultatem, (quare scilicet non plura entia, per se existentia, natura autem differentia possent esse; quemadmodum cogitatio, et extensio diversa sunt, ac forte propria sufficientia subsistere possunt) totam superesse," Spinoza 1677, 525.

150 "[...] quandoquidem Dei natura in certo entis genere non consistit; sed in ente, quod absolute indeterminatum est, eius etiam natura exigit omne, quod τὸ esse perfecte exprimit; eo quod eius natura alias determinata, et deficiens esset. Haec quum ita se habeant, sequitur non nisi unum ens, Deum scilicet, posse esse, quod propria vi existit. Si etenim, verbi causa, ponamus, quod extensio existentiam involvit, aeterna, et indeterminata ut sit, absoluteque nullam imperfectionem, sed perfectionem exprimat, opus est. Ideoque extensio ad Deum pertinebit, aut aliquid erit, quod aliquo modo Dei naturam exprimit: quia Deus est ens, quod non certo duntaxat respectu; sed absolute in essentia indeterminatum, et omnipotens est. Hocque, quod (pro lubitu) de extensione dicitur, de omni eo, quod, ut tale statuere volemus, affirmandum quoque erit," Spinoza 1677, 525–526.

151 This is letter 2318 of Locke's correspondence (Locke 1976–1989).

152 He does so in his letter to Locke of 18/28 November 1697, i.e. letter 2352 of Locke's correspondence.

153 This is letter 2395 of Locke's correspondence. Locke agreed to provide an answer to Hudde's question in letter 2340 (29 October 1697).

154 The argument is given in letter 2395, Locke to Van Limborch, 21 February 1698. For a commentary, see the sources mentioned *supra*, n. 114. See also letter 2400, Van Limborch to Locke, 1/11 March 1698.

155 Van Limborch answered to Locke's argument, expounded in letter 2395, in two letters: 2406 (11/21 March 1698) and 2410 (1 April 1698).

156 This is letter 2413 of Locke's correspondence.

onstration not based on the assumed definition of God as perfect being, but rather a demonstration by which one can provide the very definition of God. Moreover, Hudde presented to Locke his own 'method' to demonstrate the uniqueness of God, encompassed in three propositions:

1. "there is given an eternal being, independent, existing by the necessity of its own nature, and sufficient to itself."
2. "Such a being is only one, and there cannot be more than one being of the same sort."
3. "That being, because it is unique, embraces all perfections in itself; and this being is God."

Hudde claimed to have a demonstration for proposition 2 (which he labelled as just assumed, but not demonstrated, by Descartes), from which proposition 3 should follow, while proposition 1 could be found in the *Essay Concerning Human Understanding* of Locke, from whom Hudde asked for a commentary upon his method.[157] Subsequently, Locke agreed with the order of the propositions set forth by Hudde, and proposed an a priori argument for the uniqueness of God (namely, for Locke there exists an eternal being which is infinite, and thus it cannot be but numerically one, as there is nothing outside it),[158] which was however rejected by Hudde as presupposing proposition 3 itself.[159] Then, the communications between Hudde and Locke were interrupted.[160] Van Limborch had no occasion to meet Hudde in the coming weeks; however, he met De Volder, during the summer holiday of the same year, with whom he discussed Hudde's 'method'. De Volder, as reported by Van Limborch in his letter to Locke of 2/12 September 1698, had already discussed the issue with Hudde, without coming to a solution. Notably,

[157] "Probat argumenta tua, supposita illa quam adhibes Dei definitione: Ens enim undiquaque perfectum, seu, quod eodem redit, omnes in se complectens perfectiones, non nisi unum esse posse, manifestum est. Verum ille quaerit argumentum, non ex definitione Dei desumptum, sed ex ipsa ratione naturali, et per quod in definitionem Dei deducamur. Hac nempe methodo instituit demonstrationem suam. I. Datur ens aeternum, independens, necessitate naturae suae existens, et sibi ipsi sufficiens. II. Ens tale est tantum unum, et plura istiusmodi entia esse nequeunt. III. Illud ens, quia est tantum unicum, omnes in se complectitur perfectiones: atque hoc ens est Deus. Primam propositionem Vir Magnificus ait te in tractatu tuo de intellectu humano egregie adstruxisse, iisdem plane argumentis, quibus ipse in sua demonstratione usus est, adeo ut suas cogitationes in argumentatione tua expressas viderit. Tanto enixius secundam propositionem a te probatam videre desiderat; qua solide probata, tertia nullo negotio ex duabus prioribus deduci potest. Secundam, ait, omnes theologos ac philosophos, quin et ipsum Cartesium, praesupponere, non probare. Non dubito, quin mihi omnem suam argumentationem communicaturus sit: credo autem non id facturum, antequam tua argumenta viderit; ut tuas cogitationes, quas ipse es meditatus, cum suis conferre possit," Locke 1976–1989, volume 6, 386–387. Translation from page 386. This is letter 2432 of Locke's correspondence. Cf. Locke's *Essay Concerning Human Understanding*, book 4, chapter 10, §§ 3–6.

[158] This is letter 2443 of Locke's correspondence, dated 21 May 1698.

[159] This is letter 2460 of Locke's correspondence, dated 21 June/1 July 1698.

[160] The matter is not discussed in the letter by Van Limborch to Locke of 8/18 August 1698 (letter 2482), where it is reported that Van Limborch did not meet Hudde after his last letter.

De Volder disagreed with Hudde's main tenet, namely, the idea of an 'absolute dualism' between extension and thought, so that "neither can have any knowledge of the other." De Volder rejects this idea, since, insofar as thought is infinite, namely, it is not limited by anything else of its kind, it must have knowledge of extension. Moreover, it is on the ground of the idea that a being "existing per se and sufficient to itself" is infinite in its nature or kind, that De Volder justifies, to Van Limborch, the uniqueness of God, thus rejecting Hudde's method, and agreeing with Van Limborch that proposition 2 follows proposition 3. Or, in Van Limborch's own words:

> The Magnifico [he said] had been long at a stand about this question; he had also proposed it to Benedict de Spinoza [...]. Mr. de Volder added that he did not approve of the Magnifico's reasoning [...] if we suppose that thought existing of itself is given, and further, matter or extension, neither can have any knowledge of the other: extension, he said, will indeed have no knowledge of thought; but it cannot be but that thought will have knowledge of extension, because since thought exists of itself and is sufficient to itself it is also infinite; and accordingly [...] necessarily knows that extension exists. But when I [Van Limborch] rejoined that the Magnifico rejected a way by which other attributes are proved to appertain to a being existing of itself and sufficient to itself before it is proved that that being is unique, he [de Volder] replied that it ought necessarily to be affirmed of such a being that it is infinite, infinite, that is, in its own nature [...]. But I [Van Limborch] inferred that it follows thence that other attributes also can be proved; for infinity having been proved other attributes, without which infinity cannot be conceived, can also be proved to appertain to that being. Which he [De Volder] did not deny. He therefore seemed to think [...] that the second proposition ought to be the third.[161]

[161] Locke 1976–1989, volume 6, 464–465. Cf. the original text of the letter: "[p]ost ultimum meum cum Viro Magnifico colloquium nulla ipsum conveniendi occasio fuit: aliquandiu febricula laboravit. Verum etsi aliquoties eum conveniam, nihil ulterius ex ipso elicere spero. Hisce feriis canicularibus apud nos aliquot diebus fuit D. Volderus, in Academia Lugduno-Batava Philosophiae Professor: illi sermones meos cum Viro Magnifico narravi; addidi, me desiderare audire, quibus ille argumentis unitatem entis per se existentis sibique sufficientis adstructurus sit. Respondebat ille, se iam a pluribus annis similes cum ipso habuisse sermones; sed alia illi non esse argumenta praeter illa quae iam mihi dixit: Diu illum circa hanc quaestionem heasisse; illum etiam eandem proposuisse Benedicto de Spinoza, qui ad eam etiam responderit, eaque responsa inter Epistolas eius exstare, Epist. XXXIX. XL. et XLI. Addebat, minime sibi probari Viri Magnifici argumentationem, qua contendit, si supponamus dari cogitationem per se existentem, et praeterea materiam seu extensionem, quod neutra ullam alterius possit habere cognitionem: Extensionem quidem [...] nullam habituram cognitionem Cogitationis: fieri autem non posse, quin Cogitatio cognitionem sit habiturae Extensionis: quia cum cogitatio per se existat, sibique sit sufficiens, etiam est infinita; ac proinde vi infinitae suae cogitationis necessario cognoscit extensionem existentem. Sed cum regererem, Virum Magnificum improbare methodum, qua Enti per se existenti sibique sufficienti probantur inesse alia attributa, antequam probatum sit, illud esse tantum unicum; Respondebat: Necessario de tali ente debere affirmari, illud esse infinitum, verum in sua natura: Cogitationem quidem esse infinitae scientiae, materiam infinitae extensionis, siquidem per se existat: Sed inde sequi colligebam, etiam alia attributa posse probari: probata enim infinitate, etiam probari posse alia illi inesse, sine quibus infinitas concipi nequit. Quod non negavit. Atque ita mecum sentire videbatur, unitatem eiusmodi entis tali methodo frustra quaeri, sed oportere thesin secundam esse tertiam," Locke 1976–1989, volume 6, 464–465. This is letter 2485 of Locke's correspondence.

As late as 4 October 1698, Van Limborch had still not met Hudde. Van Limborch, actually, suspected that Hudde had no arguments by which he could justify his 'method', and was just eager to know Locke's solution.[162] This is confirmed by the following letters, as Hudde did not summon Van Limborch anymore.[163] Moreover, after Locke had asked Van Limborch to approach Hudde and to give him a letter for him, asking for Hudde's argument,[164] Van Limborch would eventually meet Hudde at a friend's house at the beginning of June 1699, together with De Volder and Abraham van den Ende. There, they could not talk openly, as other people were present. However, Van Limborch managed to hand Hudde Locke's letter. Hudde, in turn, postponed its discussion to another time – and Van den Ende and De Volder suggested Van Limborch not press Hudde further. The day after, De Volder and Van den Ende met Hudde again, but he avoided any discussion, leading De Volder to agree with Van Limborch that Hudde simply had no arguments.[165] After two further letters by Van Limborch in which the affair is not mentioned,[166] Locke eventually agreed with Van Limborch on De Volder's judgment on Hudde's behaviour, and the topic did not resurface anymore.[167] Thus far, from De Volder's contact with Van Limborch there emerges a rejection by De Volder of a 'Spinozist' solution to the problem of the interaction of substances, as thought and matter can be completely independent, and at once thought can have knowledge of matter. In turn, in his very correspondence with Van Limborch – of which four letters are now extant at Amsterdam – De Volder briefly discussed with Van Limborch the quarrel between Van Limborch himself, the two sons of Frans Burman (1628–1679), namely Pieter Burman (1668–1741) and Frans Burman Jr. (1671–1719), Le Clerc and Johannes van der Waeyen (professor of theology at Franeker), raised by Van Limborch's observation, in his *Theologia Christiana* (1686), that Frans Burman had used

[162] Cf. the letter by Van Limborch to Locke of 4/14 October 1698; this is letter 2494 of Locke's correspondence.

[163] See the letter by Van Limborch to Locke of 29 November/9 December 1698 (letter 2516), answering Locke's letter of 4–18 October 1698 (letter 2498) in which Locke showed his perplexity at De Volder's idea of the infiniteness of thought: according to him, indeed, thought is nothing but the action of a substance. In his reply, Van Limborch agreed with him.

[164] See the letter by Locke of 4–5 October 1698 (letter 2557).

[165] See the letter by Van Limborch of 13/23 June 1699 (letter 2596): "[q]uaesivi occasionem Viro Magnifico eas ostendendi, quam commodissimam mihi praebuit D. Volderi Professoris Leidensis adventus: itaque ipso et D. van den Ende iurisconsulto comite conveni in aedes, in quas constitutum fuerat Virum Magnificum ad confabulandum venturum, ut ita inter sermones amicos uberior de tua epistola eiusque argumento institueretur collatio. Verum aderant aliqui, quorum sermonibus propositum meum turbabatur. Tandem de valetudine tua quaesivit: Ego hanc arripiens occasionem ipsi epistolam tuam tradidi; quam ubi legit, dixit, alias se de illius argumento locuturum. Cum ab ipso discederemus, modeste a se amoliebatur quas ipsi tribuis laudes, repetiitque se alias de epistola tua locuturum. Sic non dimisit. Duo illi viri iudicabant, ulterius eum non urgendum, me enim frustra fore; et per dilationem illam responsum honeste declinari. Postridie illis adfuit, verum de epistola tua nullum verbum. Credo iam, Volderum rem acu tetigisse, cum mihi dixit, credere se, Virum Magnificum omnes suas rationes iam aperuisse, aliudque nihil habere quod nobis suggerat; verum id dissimulare," Locke 1976–1989, volume 6, 638–639.

[166] These are letters 2605 and 2606, of July and August 1699.

[167] See the letter by Locke of 5–12 September 1699 (letter 2616).

Spinoza's words from part 2, chapter 9 (*De potentia Dei*) of the *Cogitata metaphysica* while treating the issue of divine omnipotence in his *Synopsis theologiae* (1671–1672), (book 1, chapter 25). In particular, Burman used (§ 9) Spinoza's idea (using his very words) that "all things depend absolutely upon the decrees of God, say that He is really omnipotent,"[168] in order to argue that things are impossible not in themselves, but only as God cannot create them. This idea is criticized by Burman, given its implication that something is not impossible in itself, but God cannot create it.[169] Second (§ 10), Burman argued, following Spinoza, that if we could have a clear and distinct perception of anything, we would recognize the necessity of everything.[170] According to Van Limborch, this follows from the idea that all the essences – and not only existences – depend on the will of God, so that no contingency is possible. Thus far, Van Limborch notes that "Burman agrees with Spinoza."[171] Third (§ 15), Burman maintained that miracles are not above nature,

[168] Spinoza 1974, 159. Cf. Spinoza 1663, 125.

[169] "Mirari autem hic subit, Franciscum Burmannum, nuper in Academia Traiectina Theologiae Professorem, in sua Synopsi Theologiae, lib. 1. cap. XXV. de Omnipotentia Dei, § 9. ipsis Benedicti de Spinoza verbis in Cogitatis Metaphysicis cap. IX ex quo pleraque huius capitis ferme ad verbum descripsit, alios reprehendere, quod res quasdam citra decretum Dei possibiles, alias impossibiles, alias necessarias faciant, cum iuxta ipsum omnia a decreto Dei absolute dependeant, non solum rerum existentia, sed et universa earum natura et essentia. Licet enim verissimum sit, nihil esse necessarium quod Deus non decreverit, et unaquaeque res eam habeat essentiam, in sese spectatam, ut a Deo potuerit decerni; aliarum vero esse eam ad existendum repugnantiam, ut a Deo decerni non potuerint; qualia sunt contradictoria. Unde et ipse nugari credendus est, cum immediate subiungit, cur aliquid impossibile sit, causa non est in ipsa re (cum impossibile mera sit negatio) sed in ipsa Dei potentia, quia Deus illud non potest velle aut facere. Quis enim subtilitatem istam capiat, quod causa non sit in ipsa re quod sit impossibilis, sed quia Deus eam non potest facere? Quasi nempe, quod Deus facere non potest, in se non sit impossibile: aut aliquid in se non impossibile, seu quod in natura sua spectatum tale est ut fieri possit, Deus facere non possit. At, inquit, impossibile est mera negatio. Fatemur, ideoque designat repugnantiam ad existendum. Quod itaque tale est, ut repugnantiam ad existendum habeat, simpliciter impossibile est, nec a Deo fieri, seu, quod idem est, id Deus facere non potest," Van Limborch 1686, 99; cf. Burman 1671–1672, volume 1, 145.

[170] "Verum totus illius discursus, quem hic ex Spinoza instituit, eo tendit, ut ostendendo, non tantum rerum existentia, sed et universum earum conceptum, omnemque essentiam earum a Dei decreto et voluntate dependere, simul omnem e rebus contingentiam tollat, omnibusque eventibus eandem respectu decreti divini doceat inesse necessitatem quae est in veritatibus mathematicis. Ita enim ipsis Spinozae verbis § sequente ratiocinationem suam continuat. Si homines clare totum naturae ordinem intelligerent, omnia aeque necessaria reperirent ac illa quae in mathesi tractantur; quia vero huc usque non penetramus, ideo quaedam a nobis possibilia, vel indifferentia, non vero necessaria iudicantur. Quocirca dicendum est, vel quod Deus nihil possit, quia omnia revera necessaria sunt, vel quod Deus omnia possit, et quod necessitas quam in rebus reperimus, a solo Dei decreto provenerit. Quae omnia eo unice tendunt, ut meram in rebus omnibus dominari evincat fatalitatem, omnesque qui sunt, etiam in rebus humanis, eventus omnipotentia divina determinari, ut aliter quam eveniunt evenire nequeant, non magis quam veritates mathematicae, posito hoc quem Deus instituit ordine, mutari in fatalitates queunt. Quod omnem evertere religionem alibi clare ostensum est," Van Limborch 1686, 100. Cf. Spinoza 1663, 125; Burman 1671–1672, volume 1, 145–146.

[171] "Burmannus cum Spinoza sentit," Van Limborch 1686, 100.

"which is the very same impiety of Spinoza."[172] At first, the association of Burman with Spinoza was attacked by Van der Waeyen in his *Dissertatio de Λόγῳ adversus Johannem Clericum* (1698),[173] in which he criticized Le Clerc's *XVIII prima commata capitis primi Evangelii S. Joannis paraphrasi et animadversionibus illustrata, ubi demonstratur, contra alogos, Evangelium hoc esse foetum Joannis Apostoli; et evertitur sententia Fausti Socini, de sensu primorum eius commatum* (1695). Van der Waeyen aimed at showing that Burman was not a Spinozist, as he did not maintain that everything emanates from God – which is, for Van der Waeyen, the central tenet of Spinoza's philosophy – and also criticized the textual similarities between Burman's and Spinoza's texts.[174] In turn, Van Limborch – who wrote to Locke that he never labelled Burman a Spinozist, and who did not wish to provide an answer[175] – eventually replied to him with his *Defensio contra Joannis van der Waeyen Iniquam Criminationem* (1699), in which he remarked on the textual evidence for Burman's use of Spinoza. In turn, Le Clerc defended Van Limborch in his 1699 *Parrhasiana*,[176] while Van der Waeyen replied further with his *Limborgianae Responsionis discussio* (1699)[177] – to which Van Limborch did not reply further. Eventually, in 1700 the two sons of Frans Burman, i.e. Pieter Burman and Frans Burman Jr., published their *Burmannorum pietas, gratissimae beati parentis memoriae communi nomine exhibita*, to which an edition of the correspondence between their father and Van Limborch was appended (and which was left unreplied to by Van Limborch). In this text, Frans Burman Jr. notes how his father had been following Descartes's philosophy as far as the problem of Creation is concerned,[178] and remarks the difference between the Cartesian and the Spinozist schools in philosophy, where the Cartesian one – represented, as put by Burman, by his teacher De Volder – held the tenet that only the existence of God is necessary, while the exis-

[172] "Nec haec dixisse contentus, Spinozam sine ullo iudicio etiam in illis secutus est, quibus Atheismi sui venenum, sed tectius paulo, insinuat; dum non tantum totam rerum naturam non nisi unicum ens esse affirmat, § XI, sed et § XV miracula, quae extraordinaria Dei potentia perfici ait, non nisi secundum naturae leges fieri scribit: Deus, inquit, ordinaria potentia operatur secundum leges naturae nobis notas; extraordinaria vero secundum eas quae nobis notae non sunt. Quod si verum est, miraculum non est opus vires naturae superans, nec supremum Dei in omnia regimen evincens; sed errore nostro, qui naturae leges ignoramus, pro tali solummodo habetur; revera autem fit iuxta leges semel naturae inditas. Quae ipsissima Spinozae impietas est," Van Limborch 1686, 100. Cf. Spinoza 1663, 126; Burman 1671–1672, volume 1, 145–146.

[173] This text was premised on Van der Waeyen's edition of Johann Stephan Rittangel's *Libra veritatis et de Paschate tractatus.*

[174] See Van der Waeyen 1698, 175 and 178.

[175] See the letter to Locke of 2 September 1698 (letter 2485). See also letters 2494, 2516, 2724 and 2742.

[176] See Le Clerc 1699, 408–416.

[177] This text was premised on Van der Waeyen's edition of Rittangel's *Veritas religionis Christianae in articulis de Trinitate et Christo ex Scriptura, Rabbinis et Cabbala.*

[178] "Cap. IX. De Creatione plane enim consideratione dignissimum puto, cum ab gravissimas res, quas ibi tractat, tum ab modum etiam, quo nos omnes armat, qui veri et non fucati atheismi impugnatores volumus, non videri, sed esse. Quaedam ergo, quae maxime in oculos incurrunt, excerpsisse opere pretium puto," Burman 1700, 81.

tence of any other thing is possible and depends on something else (viz. God) thus undermining the essence of Spinozism,[179] with the same strategy pursued in the disputation of Van den Honert, as seen above. In fact, De Volder himself, in his letter to Van Limborch of 6 May 1699, reported to have received the visit of Pieter Burman (at that time professor of history and rhetoric at Utrecht), asking for his help, which De Volder could however not offer, as he clearly stood with Van Limborch on the matter: "you had already freed Burman from the suspicion of Spinozism in your writing. So that it is neither his right to demand that you deny that use of Spinoza's words, nor could you concede it in any way."[180] This somehow confirms De Volder's attitude towards Spinoza: possibilism in his solutions, but no clear backing of his ideas (at least in metaphysics).

3.1.4 The Reception of De Volder's Metaphysics: Two Early-Eighteenth Century Quarrels over Spinozism

De Volder's ideas on metaphysics, especially his idea that extended and immaterial substance have nothing in common and the resulting impasse as to their relation, prompted two distinct quarrels at the beginning of the eighteenth century. With these, his intellectual heritage came to be closely associated with Spinozism.

---

[179] "Quae omnia, quamvis tralatitia et vulgaria sint in schola Cartesii, et a Cl. Voldero, magno huius Philosophie lumine, tibique non ingrato nomine, mihi vero Praceptore, hisce olim auribus instillata, quum ad eum audiendum et mirandum Athenas Batavas essemus profecti, tamen hic repetere placuit: tum ut ostenderem, quam invicta adversus omnem atheismum sint illa arma, quibus illic Dei necessaria, omnium vero aliarum rerum possibilis et dependens existentia, hoc est, quibus illic discrimen inter Creatorem et ejus opera, propugnatur: tum ut simul pateret, quam implacabilis inimicitia et insociabile divortium inter duas istas scholas, Cartesii scilicet et Spinoza," Burman 1700, 105. Moreover, he recalled De Volder's clarity in exposition: "[v]ides enim viros in hac arte eximios, ita Matheseos hac manu formatos et quasi fictos esse, ut, quidquid mente attingunt, statim nescio quam longe prospiciant, quae alii, dum hi quasi oracula fundunt, non admirantur tantum, sed cum enarrant etiam, vix assequuntur. Quam me pernicem et vividissimam mentis aciem in Cl. Voldero meo, quoties ad eum accedo, mirari solere, te, ut credo, non refragante, nunc etiam recordari licet et lubet," Burman 1700, 301.

[180] "Viro admodum Reverendo / Philippo a Limborch. / S.S. Theologiae Professore / S. P. / B. De Volder. / Parum interesse arbitror, sive <c>ea, quae nosti, acta <huc> per nautas, an vero alio modo hic deferantur. Mihi enim persuadeo, illos utplurimum minus peccare, qui impun<{...}>e nequeunt, quidam qui possunt. / Cum Graevio sentio, quoad judicium de Fratribus attinet. Habet enim Professor quid Ventidii, nec fortasse cum illo tam facile quam cum altero amice transiges. Ille tamen ipse, cum ante 3 menses mihi aderat, videbatur mihi animo commotiore quam expecta<b>veram. Nec dubito tamen quin ubi prior ille aestus deferbuerit, modeste acturus sit causam suam, et rationi facile obsecuturus nisi quod fratres intercedant. Sed tua parum refert. Hoc<{a}>unum me male hac in causa habet, quod non videam, quid a te exigere queant. Burmannum enim jam in scripto tuo liberasti a suspicione Spinozismi. Ut illum Spinozae verbis usum ne<q>ges, neque illi jure exigere, nec tu concedere ullo modo possis. Sed ipsi viderint. Illorum enim res agitur. Tua in vado est. Vale. / Dabam, Lugd. Batav. / Pridie Nonas Maii, CIↃIↃCXCIX," Amsterdam University Library, OTM hs. J 83 a. For a commentary, see Strazzoni 2019.

3.1.4.1 The Bernoulli-Hulsius Quarrel

The quarrel in which De Volder's ideas were first called into question involved Johann Bernoulli, Johannes Braun (correspondents of De Volder, and professors of philosophy and theology respectively at Groningen), on the one hand, and on the other the theologian Paulus Hulsius. The story of the beginning of the quarrel is reported both by Hulsius in his *Spinosismi depulsio, nominatim circa corporis peccata et poenas* (1702), and by Bernoulli in a letter to De Volder himself of 28 March 1702.[181] According to Bernoulli, amongst the *opponentes* of his *Dissertatio philosophica de anima brutorum*, which he presided over on 18 January 1702 (*respondens* Phoebus Themmen), there was a student of Hulsius, Petrus van Venhuisen, "devoted to [...] Bacchus and Venus," who vehemently attacked Bernoulli's theory of the relation of soul and body – according to which these operate harmonically like two clocks – as contrary to Holy Writ.[182] In turn, Bernoulli replied that the idea that "body sins," defended by Van Venhuisen, was absurd, and prepared a corollary to be appended to another disputation, his *Exercitatio philosophica de globo et motu Telluris* (presided over by Bernoulli and defended by Johannes Nysingh on 1 February 1702) in order to defend his views. The corollary states that "bodies neither sin nor they are punished, [and] who states otherwise, [he] makes the soul [...] [a] body and the body [...] [a] soul."[183] In turn, Hulsius, labelled by Bernoulli as "Pope of the Jesuits," asked by letter for the support of Braun one day before this disputation, complaining that the corollary would lead to embracing the thesis of Balthasar Bekker, that spirits do not operate on bodies.[184] Braun, after having replied to Hulsius that negating the corollary would amount to

[181] Please note that already in May 1699 Bernoulli had been involved in a quarrel with Hulsius. As reconstructed by Jonathan Israel, in a disputation on nutrition Bernoulli maintained that body is perpetually renewed, so that in three years it is completely changed. This raised questions as to the responsibility of man with regard to his sins. In turn Hulsius attacked Bernoulli's positions, with no reaction from him. In August 1699, in assuming the rectorate at Groningen, Bernoulli attacked Hulsius's objection, without disclosing his name. Eventually, the quarrel erupted in 1702. See Israel 2001, 437. The affair is mentioned in the letter by Bernoulli to De Volder of 31 May 1699. As to the 1702 phase of the quarrel, see also the letter by Bernoulli to De Volder of 28 March 1702. See *Briefwechsel von Johann I Bernoulli. Korrespondenz zwischen Bernoulli, Johann I und Volder, Burchard de* (http://www.ub.unibas.ch/bernoulli/index.php/Kategorie:Volder_Burchard_de-Bernoulli_Johann_I (accessed 13 January 2019)).

[182] This is the text of the disputation as reported by Hulsius (up to "eandem"; the rest is Hulsius's commentary): "[i]nitium fuit animam operari cum corpore. Animam non operari in corpus, quia non possit corpus movere per impulsum. Quippe animam esse spiritum, spiritum vero non posse operari in corpus, nec mediate, nec immediate. Non magis operari animam in corpus, quam unum horologium in alterum, quando utrumque sonat horam eandem. Deinde in thesibus de anima brutorum expressae sunt voces, omnem animae actionem in corpus consistere in quadam harmonia. Nec ideo brachium moveri, quia anima velit. Nec ullum necessarium esse animae cum corpore commercium, neque nexum," Hulsius 1702, 3.

[183] "Corpora nec peccant nec puniuntur, quis secus dicit, ex corpore facit animam et ex anima corporis," Bernoulli 1702a, corollary 2.

[184] On Bekker, see Fix 1999.

supporting Spinozism,[185] received by Hulsius a counter-reply which Bernoulli recommends De Volder to try "not to die laughing" while reading it. Indeed, Bernoulli attached to his letter copies of all the missives involved in the quarrel (as Hulsius sent on the day of the disputation a further letter, this time to Bernoulli, to which he replied by another letter).[186]

---

[185] "Habes causam, Lector, cur pridie disputationis istius habendae ita scripserim ad collegam in facultate; nec est quod me eius poeniteat, cum probe nossem cui heic animi mei sensum committerem, quanquam non credideram in auditorio et certamine Philosophico eam fore promulgandam. Clarissime D. Brauni Collega honorande. Cum audiamus non tantum articulos a nobis Beckero oppositos infringi, sed graviores quoque consequentias non sine scandalo admitti, scire ex te velim, an inter alia ferendum putes facultati nostrae, quod defendendum in crastinum proponitur, corpora nec peccare nec puniri, an vero in commune ea de re consulendum. Subscripsi non tantum nomen, meum, sed qualitatem Professoris Theologiae, non enim ut decanus facultatem convocare volui ante quam perspectus mihi esset Collegae animus. An bene an male egerim iudicent cordati homines, item an vere an secus scripserim. Agitata erant quae dixi, in aperta hominum luce. Primus articulus et quasi fundamentalis Beckero oppositus, est, Spiritus creatos operari posse in spiritus et in corpora tametsi cum corporibus suppositum non constituant. Hoc negari non posse absque doctrinae pernicie, sacrarum scripturarum torsione multiplici, scandalosis interpretationibus et inverecundis subsannationibus. Proinde profitetur examinis candidatus se theses istis thesibus oppositas toto pectore aversari, quas et promittit se ut heterodoxa sese oppugnaturum. [...] Respondit itaque Clar. Ioh. Braunius SS. Th. Dr. et Professor. Clarissime Hulsi, Collega honorande. Beckero oppositos articulos in nostra Academia infringi, ignoro. Graviores consequentias cum scandalo admitti, mihi quoque latet. Quid crastino die defendendum sit, nondum examinavi. Mentem in corpore et per corpus peccare, in eoque puniri, certissimum est; at corpus ipsum cum sua natura nullam habeat actuositatem, nullam voluntatem, nullumque sensum et perceptionem, peccare aut puniri posse, captum meum longe superat. Hoc statuere, magis Spinosismum, quam orthodoxiam redolere mihi videtur. Si tamen hac de re consulendum iudices, paratum me habebis. Hactenus meus Clar. Collega," Hulsius 1702, 5 and 7. These are those letters labelled in Bernoulli's letter to De Volder (quoted in the next note) as letters A and B.

[186] "En eius brevem historiolam quam nactus hanc bonam occasionem Tibi lubens perscribere volui. Erat illo disputationis die inter alios opponentes certus quidam Petrus a Venhuysen, qui se dicit S. S. Min. Cand. homo omnium bipedum pessimus, impudentissimus, ignorantissimus, homo voluptuarius, Bacchi et Veneris (ut fama fert) exercitiis summe deditus, eamque ob causam adulterii suspectus coram consistorio citatus se sistere debuit, et non tam culpam excusavit, quam non satis probatam palliavit, interim turpissima fama laborat. Tali homine usi sunt nostri Persecutores tanquam instrumento propter singularem qua pollet, impudentiam et audaciam aptissimo ad stratagemata sua, et propositum diabolicum effectui dandum. Hic ergo nebulo opposuit dicens nostram sententiam de unione animae cum corpore esse s. scripturae contrariam, et evertere veram naturam peccati actualis externi quod uti dicebat sit attribuendum corpori item[;] nostram sententiam sibi et omnibus christianis praeripere et eripere solatium et spem vitae aeternae et hujusmodi deliramenta alia; quae quidem pro more nostro et ut digna erant risu et facetis retorsionibus explosimus, imprimis vero lepide exagitavimus ridiculam ejus opinionem qua voluit corpus posse peccare, quam putavit usque adeo absurdam esse, ut nemo sanae mentis existere possit, qui velit aut queat eam serio defendere, ideoque iterato alta voce monueram opponentem meum ut abstineret a nugis, et serio ageret, non enim corpus peccare neque puniri, sed animam peccare et puniri in corpore et per corpus; postea multa ex s. Litteris adducere voluit pro corpore peccante, patiente et puniendo praesertim instantiam fecit urgendo, annon Christus sit passus in corpore, annon per consequens corpus Christi sit passum; respondi Christum fuisse passum in corpore, sed corpus Christi non fuisse passum id est dolore affectum; Hinc omnes illae lachrymae, hinc omnes illi tumultus, ferebar postea per urbem quasi negaverim hominem peccare, Christum fuisse passum, nos ab ipso fuisse redemtos etc. Sed in disputatione paulo post sub meo praesidio habita, ut ostenderemus, qualis nobis

Afterwards, Van Venhuisen defended (under Hulsius) an *Exercitatio theologica de unione animae cum corpore* (15 February 1702), attacking Bernoulli, after which replies and counter-replies followed: Bernoulli's *Theses apologeticae* and their *Continuatio* (25 February 1702), and Van Venhuisen's *Apologia Christiana contra Theses apologeticas* (1702), which – according to Bernoulli – was then confiscated by the University and Groningen town authorities, who had already prohibited pastors to intervene in University quarrels. Hulsius, according to Bernoulli, had indeed involved the pastor Johannes van Leeuwen in the affair as opponent to Bernoulli's *De globo et motu Telluris*, since Hulsius viz. the "Pope" could attend the disputation himself.[187] This prohibition did not end the quarrel. Indeed, after Bernoulli's letter

mens staret, inter alia corollaria hoc erat corpora nec peccant nec puniuntur qui secus dicit ex corpore facit animam ex anima corpus putantes neminem futurum qui sese huic corollario serio opponere posset; interim hoc ipsum fuit, quod totam Pharisaeorum cohortem contra me concitavit, et tanta quidem saevitia ut me nonnulli capite plecti et alii mitius censentes ex Provincia saltem cum infamia ut latronem publicum ejici me voluerint, in concionibus ad populum repetitis vicibus exaggerare non erubuerunt. Et quidem hujus Tragoediae aut si mavis comoediae initium fecit H. noster, utpote nostrorum Jesuitarum Papa, qui pridie ante disputationem eum in modum scripsit ad Cl. Collegam meum Braunium ut hic vides, sub Litera A. Cui respondit (Lit. B) sub vesperam ejusdem diei, replicavit Papa, digna autem est replicatio (Lit. C) quam attente legas, sed contine lacertos ne risu moriaris. Ipso vero disputationis die, cum in procinctu essem ut me in Academiam reciperem ad disputandum, ecce etiam ad me advolant litterae papales (Lit. D) quibus extemplo jam indutus pallio, ut domo exirem pauca reposui ut vides sub Lit. E," transcription from *Briefwechsel von Johann I Bernoulli* (Basel University Library, ms. L Ia 675, 173–175). For the texts of Hulsius's reply to Braun (viz. letter C), the letter sent to Bernoulli (letter D), and the reply to Bernoulli (letter E), see Hulsius 1702, 12–15.

187 "Post eam disputationem Adversarii videntes se viis clancularii nobis non posse nocere, inceperunt insanire et tumultuari ut rabiosi canes, et aperte nos haereseos accusarunt, hunc in finem conscripserunt diabolicam aliquam disputationem de Unione animae cum corpore quam supra dictus Venhuysius tanquam auctor sub praesidio Papae contra nos ventilavit refertam absurditatibus et contradictionibus non tantum, sed et atrocissimis mendaciis et calumniis quibus nos oneravit, quas vero respondens meus primae disputationis Auctor Phoebus Themmen Rev. Pastoris filius ingenuus Juvenis et magnae spei in ipso auditorio mascule adeo refutavit, ut calumniator et ejus Praeses pudibundi fere obmutuerint; Ego vero postea sine mora conscripsi Theses Apologeticas (quarum Cel. Braunius exemplar forte mittet) in duas partes divisas, quas bis in uno die ante et post meridiem defendimus, contra calumniatorem ejusque praesidem, ut ita praeriperemus omnem ansam querendi de non sufficienti tempore opponendi, provocavimus expressis verbis calumniatorem ejusque quemvis patronum, sed nec ipse nec eorum alius ausus est comparere in auditorio nedum ad conflictum descendere, et omnes ut stryges et lamiae lucem fugientes pugnae sese subduxerunt, et ita triumphabundi recessimus victoria sine pugna potiti, nemine sc. opponente praeterquam ex nostris studiosis qui exercitii gratia sese exercuerunt: interim rebus ita gestis Amplissimus Urbis hujus Senatus nefandae huic liti sese interposuit sumto salutari hoc decreto 1.° ut posthac pastores rebus Academicis se non immisceant, 2.° ut in disputationibus publicis non opponant, 3.° ut in concionibus nihil de rebus Academicis nihilque amplius de hac lite nec directe nec indirecte ad populum referant. Hoc fulmine improviso Adversarii mire perculsi, hactenus abstinuerunt a persecutione quorundam gravaminum contra me in Consistorio allatorum quae in disputandi ardore contra van Leuwen mihi excidissent quaedam in praejudicium Pastorum in genere, quae gravamina excerpta ex notulis (uti vocabant) van Leuwen mihi insinuata ita se habent ut vides sub (Lit. F), verum est me ea dixisse, sed non pastores in genere intendebam, quippe inter quos multos habeo amicos et fautores, sed eos tantum qui mei jurati sunt adversarii. Non diu post, nequissimus Venhuysius alium contra nos suo nomine protrusit libellum, cui titulus Apologia Christiana, qui

to De Volder it continued with the publication of Hulsius's *Depulsio*, attacking the aforementioned corollary as detrimental to the idea of the sacrifice of Christ (who was punished in His body), and thus to the idea of the redemption of mankind, and labelling Bernoulli's positions 'Bekkerianism'.[188] Moreover, he turned the accusation of Spinozism round on Braun and Bernoulli, by claiming that negating any interaction of soul and body and upholding their correspondence or harmony in changes was exactly what Spinoza did.[189] Eventually, both Braun and Bernoulli rejected Hulsius's accusations with Braun's *Futilis Spinosismi Depulsionis depulsio necessaria* and Bernoulli's *Spinosismi Depulsionis echo* (both of 1702), vindicating the distance between Spinoza and Descartes (in the case of Braun),[190] and claiming that a substantial difference between soul and body does not constrain Christian doctrine (as to Bernoulli).[191] We not know De Volder's reply to Bernoulli's letter, yet, the fact that he was kept informed in detail about the quarrel reveals that he was involved to some extent, and that the accusation of Spinozism Bernoulli had undergone could possibly have been extended to him – as indeed it would be by Andala. In fact, this accusation was consequent upon an interpretation of Descartes's dualism (intended as preventing any kind of interaction) shared in the late seventeenth century, leading to the assumption of a form of pre-established harmony between soul and body as the only means to solve the problem of their relation.

3.1.4.2 The Wittich-Driessen Quarrel – And Beyond

If in the aforementioned quarrel De Volder's name did not come to the fore, his ideas were nevertheless explicitly discussed and attacked in a further quarrel, which spread all over the Netherlands and to Thuringia, and which led to the appearance of several replies and counter-replies. It can be summarized as follows.[192]

nihil aliud est quam farrago novarum calumniarum et ineptiarum locupletissima in qua Cel. Theologi, et ipsa Tota Facultas Theol. Franequerana misere traducitur et proscinditur. Est vero monstrum hoc in ipso partu feliciter extinctum, ejus enim publicationem primo Nobill. Senatus Urbanus suspendit, et postea a Nob. Curatoribus quibus examen ejus committebatur omnino publicari prohibitum et omnia exemplaria sibi exhiberi, et ita (ut vocant) confiscari jusserunt. Et hic erat in nostri gloriam et laudem Comoedia finis," transcription from *Briefwechsel von Johann I Bernoulli* (Basel University Library, ms. L Ia 675, 173–175). Cf. Hulsius 1702, 14.

[188] See Hulsius 1702, 4–6. For an account, see Klever 2000; Israel 2001, chapter 23.

[189] See Hulsius 1702, 23–24 and 27–28.

[190] See, for instance, paragraph 5 of Braun's book, "[u]bi simul ostenditur, magna malitia comparari Spinozam cum dictis Cartesianis."

[191] See Bernoulli 1702b, 6–12.

[192] The whole story is duly reported in *Bibliotheca Bremensis*, *classis* 1, fascicle 4, 1718 (as article 7: *Scripta eristica inter Viros Clarissimos D. A. Driessenium, D. T. H. van den Honert, et Jac. Wittichium, nunc nuper reciprocata*); *classis* 1, fascicle 6, 1718 (under article 9: *Nova quaedam theologico-historico-literaria*, see pages 939–944); *classis* 2, fascicle 2, 1718 (under article 7:

In 1711, De Volder's former student Jacob Wittich,[193] nephew of Christoph Wittich, presided, as a professor of philosophy at Duisburg, over a disputation *De natura Dei, respondens* Gerhard Nicolaus Brouwer.[194] In June 1717, when Wittich was considered for a chair in philosophy at Groningen, the theologian Antonius Driessen denounced to the University Curators the contents of Wittich's disputation as Spinozist. Hence, Taco Hajo van den Honert (viz. the *respondens* and author of the aforementioned 1686 disputation *De mundi in tempore productione*, and now professor of theology at Groningen), sent private letters to Abraham Trommius, who was going to receive an honorary doctorate at Groningen on 1 December 1717. In his letters, Van den Honert defended Wittich's positions as not Spinozist. Hence, Trommius showed the letters to the Curators, who did not take a position on the matter, so that Driessen communicated his *Scrupuli* on Wittich's disputation by a letter addressed to Van den Honert himself. Having had no reaction from Van den Honert, Driessen sent the text of Wittich's *De natura Dei* to unnamed theologians in Franeker, Leiden and Utrecht, expounding his concerns, and obtaining their approval. The theologians of Utrecht and Leiden sent Van den Honert's observations to Wittich himself, who then provided a *Responsio*.[195] Not satisfied by Wittich's answer, and left without any answer by Van den Honert (to whom he had written again), Driessen made the entire affair public by publishing his first letter to Van den Honert, as well as his observations on Wittich's *Responsio*, in his *Epistola ad Honert, qua petitur solutio quarumdam difficultatum motarum adversus V. Cl. J. Wittichii Disputationem de natura Dei, nec non animadversiones in Cl. Wittichii Responsiones ad Scrupulos quosdam illi propositos* (1718). Hence, a long series of replies, counter-replies, translations and corollary texts followed, which I abridge in Table 3.1, *Texts published during the* controversia Driessenio-Wittichiana. The analysis of this quarrel, which also saw the intervention of the town magistrates of Rotterdam in 1718 – unsuccessfully attempting to prevent Wittich giving his inaugural oration *De evidentia et certitudine, in quibus philosophiae partibus illa reperiatur* at Leiden (where he eventually accepted a chair), as well as a judgment of the faculty of theology of Jena (led by Franciscus Buddeus) in 1719 – falls beyond the scope of the present investigation. Instead, it is worth focusing on the traces of De Volder in the quarrel, as his *Contra atheos* and *Exercitationes* constituted the foundation of Wittich's *De natura Dei*.

Nova theologico-historico-literaria, see pages 359–367); *classis* 2, fascicle 6, 1719 (as article 4: *Controversiae Driessenio-Wittichianae continuatio*). On this quarrel, see Israel 2001, chapter 23; Goudriaan 2003, 2006, chapter 2.

[193] He enrolled at Leiden on 21 December 1702, as student of theology: see *Album studiosorum Academiae Lugduno-Batavae MDLXXV–MDCCCLXXV*, column 773.

[194] The disputation had further editions in 1718, 1720, and two Dutch translations in 1719 (one of them anonym and unauthorized by Wittich).

[195] See, besides the aforementioned parts of the *Bibliotheca Bremensis*, also Wittich's *Abstersio calumniarum* (1718), *Praefatio*. The full text of Driessen's *Scrupuli* and Wittich's *Responsio* can be found Driessen's *Responsionum Wittichianarum refutatio* (1718). The letters of Van den Honert to Trommius can be found in Van den Honert's *Repraesentatio altera*.

Table 3.1 Texts published during the *controversia Driessenio-Wittichiana*

1. Antonius Driessen, *Epistola ad Virum plurimum Reverendum et Celeberrimum D. T. H. van den Honert, S. Th. D. et Professorem Leidensem qua petitur solutio quarumdam difficultatum motarum adversus V. Cl. J. Wittichii Disputationem de natura Dei, a Cl. Honertio laudatam, nec non Animadversiones in Cl. Wittichii Responsiones ad Scrupulos quosdam illi propositos*, Groningen, Ex Officina Georgii Spandaw, 1718. Including:
Alloquium ad lectorem
[*Epistola*] *Viro Celeberrimo Honert*
Animadversiones in Cl. Wittichii Responsiones ad Scrupulos
2. Taco Hajo van den Honert, *Repraesentatio cur ad criminationes quibus Antonius Driessen Jacob. Wittichum ob habitam ante aliquot annos de natura Dei Disputationem a cathedra Groningana arcere conatus est nihil responderit; addita est Wittichii de natura Dei Disputatio*, Leiden, Apud Samuelem Luchtmans, 1718. Including:
Repraesentatio
Jacob Wittich's *De natura Dei.*
3. Jacob Wittich, *Abstersio calumniarum, quibus eius Disputatio de natura Dei ante aliquot iam annos habita, non ita pridem inquinata fuit a Cl. Antonio Driessen, S.S. Th. Professore Groningano*, Leiden, Apud Samuelem Luchtmans, 1718. Including:
Praefatio
Epistola ad Driessen
Refutatio Animadversionum Driessens, ad Responsiones ab illo datas ad Difficultates quasdam contra Dissertatione de natura Dei motas
4. Antonius Driessen, *Dissertatio Cl. J. Wittichii Disputationi de natura Dei opposita, qua difficultates adversus Disputationem motae vindicantur et urgentur. Porro adseritur Dei potentia, qua materiam produxit ex nihilo, atque omnimoda falsitas axiomatis Spinosistici: res quae nihil inter se commune habent una alterius causa esse nequit, ostenditur. Praeambula responsionum Wittichianarum refutationi. Praemittitur Repraesentationis Honertianae excussio*, Groningen, Ex Officina Georgii Spandaw, 1718. Including:
Alloquium ad lectorem orthodoxum
Dissertatio opposita
5. Antonius Driessen, *Responsionum Wittichianarum refutatio. Dissertationi nuper emissae adnectenda praemittitur Alloquium ad ipsum Cl. Wittichium*, Groningen, Ex Officina Georgii Spandaw, 1718. Including:
Alloquium ad Wittichium
Responsionum Wittichianarum refutatio
Scrupuli Wittichio propositi
Wittich's *Responsiones ad Scrupulos*
Responsionum Wittichii Refutatio altera
6. Gr. V. (anonymous), *Vindicatio nominis atque orthodoxiae Wittichianae a calumnis atque iniuriis Driessenianus, ex ipsa Cl. Wittichii Disputatione de natura Dei eruta*, Leiden, Apud Samuelem Luchtmans, 1718. Including:
Taco Hajo van den Honert's *Praefatiuncula*
Vindicatio nominis
7. Taco Hajo van den Honert, *Repraesentatio altera, qua Driessenii rei gestae narratio ad examen revocatur, vera molitionum Driesseniarum historia traditur, ac Repraesentatio prior ab eius cavillationibus vindicatur*, Leiden, Apud Samuelem Luchtmans, 1718. Including:
Repraesentatio altera
Frans Burman Jr.'s, Driessen's, Wittich's *Epistolae mutuae*
Hieronymus Simons van Alphen's *Epistolae ad Driessenium*
Wittich's *Alloquium ad lectorem philosophum*

(continued)

Table 3.1 (continued)

8. Jacob Wittich, *Wysgerige twistredenering over de natuur van God, welke onder des Heeren bystand onder de bescherming van Jacobus Wittichius. Hier is tot een Toegift bygevoegt het Oordell van den Heere Ruardus Andala over het Dispuit van den Heere Jak. Wittichius, alles uit Latyn vertaalt*, s.l., s.n., 1718. Including:
Wysgerige twistredenering over de natuur van God
Toegift, including a Dutch translation of Andala's *Apologia*, 1718, part 3, §§ 32–34
9. Antonius Driessen, *Alloquium alterum ad lectorem orthodoxum rei gestae narrationem vindicans adversus Repraesentationem C. Honerti alteram*, Groningen, Ex Officina Georgii Spandaw, 1718. Including:
Alloquium alterum
Antonius Andreae's *In Academiae Groningo-Omlandicae Phaenomenon*
10. Antonius Driessen, *Responsio ad Vindicationem Gr. V.*, Groningen, Ex Officina Joh. a Velsen, 1718.
11. Antonius Driessen, *Responsio ad Jacobi Wittichii Alloquium ad lectorem philosophum*, Ex Officina Georgii Spandaw, 1718.
12. Antonius Driessen, *Alloquium ad Ecclesiam Belgicam in brevi tabella Disputationem J. Wittichii de natura Dei exhibens et paragraphorum singulorum mentem indicans*, Groningen, Ex Officina Georgii Spandaw, 1718.
13. Antonius Driessen, *Aanspraak aan de Kerk van Nederlant, vertoonende het dispuit van J. Wittichius over de natuur van Godt, en de meining van yder stelling aanwyzende*, s.l., s.n., 1718
14. Taco Hajo van den Honert, *Briev aan een voornaam Heer, wegens het Dispuut De natura Dei van* [...] *Jacobus Wittichius: waar by gevoegt is een korte ontleding van het selve dispuut, en een antwoord op de Aanspraak van* [...] *Antonius Driessen aan de Kerk van Nederland*, Leiden, By Samuel Luchtmans, 1718. Including:
Briev
Gr. V.'s *Korte ontleding van het Dispuut de natura Dei*
Gr. V.'s *Antwoord op de Aanspraak van Antonius Driessen aan de Kerk van Nederland*
15. Gr. V. (anonymous), *Altera vindicatio nominis atque orthodoxiae Wittichianae a calumniis atque iniuriis Driessianis*, Leiden, Apud Samuelem Luchtmans, 1718. Including:
Van den Honert's *Praefatiuncula*
Altera vindicatio
16. Jacob Wittich, *Oratio inauguralis de evidentia et certitudine, in quibus philosophiae partibus illa reperiatur*, Leiden, Apud Samuelem Luchtmans, 1718.
17. Jacob Wittich (*praeses*), Johann van den Honert (son of Taco Hajo; defendant and author), *Disputatio philosophica de essentia et existentia Dei*, Leiden, Apud Petrum vander Aa, 1718.
18. Jacob Wittich, *Wijsgerige verhandeling van de nature Gods sijnde deselve nu door den aucteur in het Nederduyts vertaalt, met aanmerkingen, tot oplossing van des Heeren Driessens beschuldigingen*, Leiden, By Samuel Luchtmans, 1719. Including:
Voorreden
Wijsgerige verhandeling van de nature Gods
Byvoegsel

(continued)

Table 3.1 (continued)

19. Antonius Driessen, *Tegen-berigt van Antonius Driessen Doctor en Professor der H. Godgeleerdheid ter verydeling van enige verdigte vertellingen, door welke de HHren van den Honert, Wittichius, en hunne aanhangeren zyn gedrag omtrent het dispuut van Wittichius de natura Dei tragten verdagt te maken*, Groningen, By Jurjen Spandaw, 1719.
20. Antonius Driessen, *Beter berigt van Antonius Driessen Doctor en Professor der H. Theologie, en Akademi-prediker te Groningen gegeven aan den Nederlandzen lezer en dien weledelen gestrengen heer den welken de heer T. H. van den Honert Professor der H. Theologie te Leiden door zynen gedrukten brief misleid heeft*, Groningen, By Jurjen Spandaw, 1719.
21. V. Gr., *De aanspraak van Antonius Driessen aan de Nederlantse Kerk wegens 't Dispuut van J. Wittichius de natura Dei tegen de vernuftelose Sophisticatien van Gr. V. agter den welken schuilt de Heer T. H. van den Honert, verdedigt door V. Gr. met eene Vorrede van Anton. Driessen*, s.l, s.n., 1719 (lost).
22. Gerard van Ewyck, *Difficultas maxima, orta ex concatenationibus idearum, quas Vir Clarissimus Jacobus Wittichius concinnavit in Disputatione sua de natura Dei, cui difficultati ut publice satisfiat, quum et plurium mentes non parum ea torqueri possint, publicitus illam quoque proponere, nec non simul axioma Spinosisticum primarium penitus subvertere voluit*, Leiden, Apud Johannem vander Linden, 1719.
23. Jacob Leydekker, *De blyde Spinosist en de bedroefde Christenleeraar over de wysgeerige verhandelinge van de natuure Gods, welke Jacobus Wittichius in den jare 1711. te Duisburg uitgegeven en verdedigt, nu in Maart 1719. in 't Nederduits vertaalt en met aanmerkingen verrykt in 't ligt gebragt heeft. Hier by komt het eenparig oordeel van de philosophische Faculteit der Academie van Jenen*, Rotterdam, By Reinier van Doesburg, 1719, including:
Opdragt
De blyde Spinosist
Buddeus's *Epistola*
Iudicium theologicae et philosophicae facultatis Ienensis
24. Heinrich Horch, *Invisibilia Dei in rebus ab Ipso factis facta visibilia, contra Spinozam, cum consectario pacifico inter Cll. Viros, Driessenium et Wittichium, et multiplici analogia naturae spiritualis et corporeae, tanquam genuino theologiae hieroglyphicae fundamento*, Marburg, Typis Philippi Casimiri Mülleri, 1719
25. Bartholomaeus Bolk, *Animadversiones philosophicae in decantatam Spinozae propositionem, quae res nihil commune inter se habent, earum una alterius causa esse non potest*, Amsterdam, Excudit Jacobus Borstius, 1719.
26. Jacob Wittich, *Disputatio philosophica de natura Dei. Accessit J. Fr. Buddei Epistola ad theologos quosdam Roterodamenses et iudicium Theologicae et Philosophicae Facultatis Jenensis de isthac Cl. Wittichii disputatione*, The Hague, s.n., 1720. Including:
De natura Dei
Buddeus's *Epistola*
Iudicium theologicae et philosophicae facultatis Ienensis

As far as this text is concerned, in it Wittich confronts the issue of the nature of God by using on five premises:

1. "the knowledge of things depend only on thoughts and ideas."
2. "All things, which fall under human knowledge, are or thinking, or extended"; indeed, it is impossible to figure out the essence of a third kind of essence (our awareness being the most sure witness of this).

3. "There is the same difference between things themselves, as between the ideas of the same [things]."
4. "Some concepts teach the plurality and diversity of things, some [their] identity"; indeed, some concepts are necessarily related to each other – so that they represent the same thing; other ones are separable, and represent different things.
5. "No [...] idea involves the negation of the very thing [which is represented], but [its] mere infinity" in accordance with De Volder's teachings – as Wittich remarks. The foremost case of this principle is the idea of extension, but it also applies to the very idea of knowledge (*scientia*).[196]

---

[196] "§ V. Ad investigandam essentiam et naturam Dei ut ordine demonstrativo procedamus, quaedam necessaria praemittemus fundamenta. Primo. Ex solis cogitationibus et ideis rerum scientia dependet. Scire enim nihil aliud est, quam rerum intelligere essentias, variaque de illis affirmare et negare attributa, quae sane omnia cogitatione fiunt. [...] § VI. Secundo praemittimus. Omnes res, quae sub cogitatione humana cadunt, sunt vel cogitantes, vel extensae. Adeo ut quicquid intelligamus, affirmemus vel negemus, id omne respiciat cogitationem aut extensionem. Excutiat, cui volupe fuerit, loculos ac scrinia mentis suae, intendat vires omnes, abstrahat animum suum ab his duobus entibus, quantum libuerit, experietur, se nihil posse intelligere, quod non referri debeat vel ad naturam cogitantem aut extensam, nihil se cogitare vel imaginari posse, quod aliam ac diversam a cogitatione et extensione habeat essentiam. Sentiet plane caecum esse, quando tertium aliquod genus rerum, quod et cogitationis et extensionis omnis expers, nihilominus determinatam essentiam ac existentiam habeat, repraesentare sibi velit. Efficacioribus hac in re argumentis quam provocatione ad propriam coscientiam uti non possumus, quandoquidem summopere alienum foret, si a quoddam peterem, ut exquisitis rationibus accurate demonstraret, non quid in rerum natura se haberet, se quid ego cogitarem, conciperemve. Poteritne fieri, ut aliquis longa rationum serie certius me doceat, me trianguli ideam habere aut non habere, quam idem illud absque ullis argumentis me propria mea docet conscientia? [...] § VIII. Tertio loco dicimus, eam esse differentiam inter res ipsas, qualis inter ipsarum ideas reperitur. Idea cum suo ideato exactissime convenit, nam obiectum ab idea separari nequit, cum in tantum sit obiectum, in quantum repraesentationem sui in intellectu habet, repraesentatio vero haec de natura idea est. Porro inter ideam et obiectum necessarius et naturalis est nexus, quandoquidem idea nihil est aliud, nisi ipsa res cogitata et repraesentata, quam late patet ipsa rei natura, eo se et idea extendit. Prout res a se ipsa differre, aut se ipsa arctior latiorve esse nequit, ita a rei natura differre, arctius latiusque se extendere non potest rei conceptus. [...] § IX. Quandoquidem non omnis pluralitas conceptuum, et rerum pluralitatem infert, prout de triangulo quam plurimos formamus conceptus, totidem designantes proprietates, qui tamen uni eidemque rei indistincte conveniunt. Quarto ergo loco discrimen ostendimus, unde diiudicari queat, quinam conceptus pluralitatem et diversitatem, quinam identitatem rerum doceant. Binas quando habemus ideas, observari debet, num inter illas nexus necessarius et indissolubilis sit, aut separabilis. Si nexus separabilis, proculdubio repraesentantur res ab invicem separatae, sed si vinculum necessarium et indissolubile advertero, ita ut in idea priori proprietatem cum posteriori idea necessaria et reciproca ratione cohaerentem perspiciam, tunc utraque res altera cum altera cohaerebit necessario. Cum enim prior res priori conceptui respondeat, quidquid ergo ex priori idea deducitur, convenit rei naturae, atqui ex conceptu priori fluit conceptus posterius, respondebit ergo prius obiectum etiam posteriori conceptui. [...] § X. Quinto asserimus, nullius rei ideam involvere negationem istius eiusdem rei, sed meram infinitatem. Quod iuxta ductum Clarissimi Volderi b. m. (cuius acumen iudicii, et soliditatem eruditionis semper admiratus fui) hoc pacto ostendo. Scientiam, potentiam, cogitationem, extensionem et c. in se spectatas concipimus infinitas, nemo enim opinor, qui naturam scientiae potentiaeque concipit, eam intelligit cum ignorantia, aut impotentia iunctam, scientia autem absque ulla ignorantia nonne est infinita? Potentia cui nulla adsit impotentia, nonne etiam infinita? Scientia finita quid est? Ea, quae quaedam novit, reliqua ignorat. Potentia finita quid est? Ea, quae quaedam potest, reliqua non potest. Porro cogitationis

In fact, not only the fifth principle can be traced back to De Volder (the only one overtly ascribed to him), but all the premises of Wittich's arguments were in fact taken from De Volder's texts. We have seen that principle (1) and (3) were maintained by De Volder since his *Contra atheos*. Hence, they were re-stated in his *Exercitationes*, respectively in section *De idea Dei*, thesis 14,[197] and *De Deo*, thesis 5.[198] In turn, principle (2) is expounded by De Volder in the same way as Wittich, in his *Exercitationes*, section *De Deo*, thesis 13[199] as well as principle (4) and (5), taken by Wittich almost *ad literam* from the section *De Deo*, thesis 14, and *De idea Dei*, thesis 17.[200] Thus far, Wittich built upon De Volder's ideas, so that in his 1719

conceptus non involvit negationem cogitationis, illa autem cogitatio, cui nulla negatio cogitationis est iuncta, infinita est: finita enim negationem involvit. Nec dissimili ratione de extensione ratiocinari licet: habeo ideam extensionis, illa certe non involvit ex sua natura negationem extensionis, idea autem extensionis finitae omnem ulteriorem extensionem negat, omnis igitur finitae extensionis idea involvit extensionis aliquam negationem, quam omnem excludit extensio infinita. Qui ergo extensionem in se et sua natura concipit, is eam concipit absque omni extensionis negatione, adeoque infinitam. Cum igitur finitum quid concipimus, non intelligimus rem in se spectatam, sed negatione quadam involutam, quidquid autem in se intelligimus, id eo ipso spectamus infinitum," Wittich 1718a, 7–12.

[197] See *infra*, n. 200.

[198] "Quod si nunc porro ipsas ideas consideremus, facile patebit, inter eas, aut potius inter earum repraesantamina, tantundem esse discriminis, quantum foret inter res ipsas, quae per illas repraesentantur menti, si forte illa existerent. Hoc enim liquido constat ex eo, quod ideae, aut potius idearum in mente repraesentamina sint similia rebus repraesentatis. Idea trianguli mihi repraesentat naturam trianguli, nec minus idea circuli naturam circuli. Quid igitur evidentius, quam has ideas respectu repraesentaminum tantundem differre, ac differunt natura trianguli et circuli? Quae duo proculdubio tantundem differunt, quantum differrent triangulum et circulus, si forte existerent. Ponamus triangulum et circulum extra nos existere, ponamusque eorundem in nobis dari ideam; nonne manifestum est, ideas trianguli, et circuli, aut mihi repraesentaturas naturam trianguli circulique existentium, aut si eas non repraesentant, non fore ideas trianguli et circuli? Quod est contra hypothesin. Si ergo naturas trianguli et circuli revera existentium repraesentant menti, nonne perspicuum est illa repraesentamina ea repraesentatura, si quae sint, quae habent triangulum et circulus communia, ut et ea quae habent diversa: ideoque nonne liquet harum idearum repraesentamina eiusdem fore diversitatis, cuius sunt triangulum et circulus, quos ponimus existere? Inter ideas itaque trianguli et circuli; nec dispar ratio est de omnibus, cum nihil ratiocinium hoc ingrediatur, quod circulo et triangulo eorumque ideis peculiare est, tantundem est discriminis, quantum foret inter triangulum et circulum si existerent," De Volder 1695, *De Deo*, thesis 5.

[199] Cf. the text quoted *supra*, n. 61.

[200] "Habeam nunc binas ideas in mente, videam alteram cum altera necessario connecti, sive alteram ex altera necessario sequi, hoc est, percipiam clare et distincte, alteram hancce ad rem, cuius ideam in me habeo, necessario pertinere; poterone haerere vel minimum quin illud ad illam rem pertineat? Idea huius attributi necessario pertinet ad ideam illius rei, quam cogito, hunc necessarium nexum in mente mea adverto, poterone dubitare an idem sit nexus inter harum idearum obiecta? Non certe proculdubio enim idea respondent obiectis, sive rebus repraesentatis; cumque hic nexus sequatur non ex communi hoc, quod ideae sint, sed ex repraesentamine idearum, sive ex naturis repraesentatis, quid potest fingi evidentius, quam eundem illum, qui est inter ideas, nexum, inter obiecta idearum illis naturis repraesentatis perfectissime similia esse etiam necessario? Quod aut verum est, aut certum certe est de omni cognitione nostra. Nisi enim ex ideis de rebus iudicare possum, nullum mihi aliud indicium adest. Habeo ego ideam trianguli, habeo anguli recti, habeo aequalitatis, sive extra me dentur triangulum, angulus rectus, sive non dentur; ego ex his ideis concludo ideam trianguli necessarium habere nexum cum angulis binis rectis aequalibus. Quid

De blyde Spinosist (aimed at Wittich's *De natura Dei*) Jacob Leydekker – preacher in Middleburg – remarked that

> the best friends of Mr. Wittich write themselves, that there are some unusual expressions, and that singular words are found here and there [in the disputation], yet that those who have enjoyed Mr. De Volder's teaching can explain their meaning very well.[201]

Indeed, in a letter published by Van den Honert in his *Repraesentatio altera* (1718), which he had addressed to Petrus Bodaan (pastor in Amsterdam) on 2 November 1718, Van den Honert overtly refers to De Volder as the source of Wittich's 'unusual words'.[202] Moreover, in the *Vindicatio nominis atque orthodoxiae Wittichianae a calumniis atque iniuriis Driessenianis* (1718 – appeared before Van den Honert's *Repraesentatio altera*, and after the *Repraesentatio*) authored by an anonymous "Gr. V." and prefaced by Van den Honert, attention is drawn to Driessen's critique (given in his *Responsionum Wittichii Refutatio altera*) to Wittich's premise 5, namely, against "the Great De Volder." The anonymous author "Gr. V." criticizes Driessen's lack of understanding that De Volder's "sentence[,] taken from the secrets of Spinozism," (as Driessen put it)[203] does not mean that there are several

itaque? Fierine potest, ut hic nexus in ideis sit, in re ipsa non sit?" De Volder 1695, *De Deo*, thesis 14; "[n]ititur assertio Cartesii, cum ait priorem quodammodo in se esse perceptionem infiniti quam finiti. Med. 3. Nullius enim rei idea involvit negationem istius eiusdem rei. Res, ut opinor, manifesta est. Habeam ideam extensionis, illa certe non involvit ex sua natura negationem extensionis. Idea autem extensionis finitae, decem ex. gr. pedum ponit extensionem ad 10 usque pedes porrectam, et non ultra, omnem enim ulteriorem extensionem negat. Omnis igitur finitae extensionis idea involvit extensionis aliquam negationem, quam omnem excludit extensio infinita. Qui ergo extensionem in se et sua natura concipit, ille eam concipit absque omni negatione extensionis, adeoque eam concipit infinitam. Qui vero finitam percipit ille huic conceptui extensionis addit negationem aliquam. Nec dissimili ratione de cogitatione finita, infinitaque ratiocinari possumus. Cogitationis certe conceptus non involvit negationem cogitationis. Illa autem cogitatio, cui nulla negatio adiuncta est cogitationis, infinita est, finita negationem aliquam involvit. Quare conceptus extensionis, cogitationisve nullam involvens negationem extensionis, cogitationisve, hoc est, conceptus extensionis, cogitationisve infinitae, prior est natura sc. non tempore, de quo non agitur, et simplicior, quam extensionis et cogitationis finitae, cum in hac praeter extensionem, quae in priori sola concipitur, concipiatur etiam extensionis negatio," De Volder 1695, *De idea Dei*, thesis 17.

201 "De beste Vrienden van den Hr. Wittich schryven zelve, dat daar in wel eenige ongewone spreekwyzen, en zeldzame bewoordingen hier en daar worden gevonden, dog dat die gene, die des Heeren de Volders onderwys hebben genoten, dezelve zeer wel in eenen goeden zin konnen verklaren," Leydekker 1719, 1–2 (unnumbered). I owe this citation to Israel 2001, 439.

202 "Die zoo spreekt, en 't gene hy zegt, zoo bondig bewyst verdient zeker! Geenzins van Spinosistery verdagt gemaakt te worden. Hy heeft hier en daar wel eenige ongewone spreekwyzen, maar die des Heeren de Volders onderwys genoten hebben, kunnen die zeer wel in eenen goeden zin verklaren," Honert 1718a, 41. In a letter by Pieter Burman (from 1715 professor of Greek at Leiden) to Jacob Wittich printed in Van den Honert's *Repraesentatio altera*, De Volder-Wittich's principle 5 is criticized as it entails the infinity of what is represented by an idea: see Van den Honert 1718a, *Epistolae mutuae*, 4. In his *Alloquium ad lectorem*, included in the same work, Wittich defends this principle against Driessen's attacks made in his *Animadversiones in Cl. Wittichii Responsiones ad Scrupulos quosdam illi propositos*, included in his 1718 *Epistola* (viz. the first printed text of the quarrel): see Van den Honert 1718a, *Alloquium ad lectorem*, 20–21.

203 "Ego vere demonstravi, scientiam tuam in te sciente necessario esse finitam, quia animam tua finita est, proptereaque, licet scientia tua finita in se spectata, qua scientia, non dicat ignorantiam,

infinite knowledges (*scientiae*), but only that the idea of understanding or knowledge does not entail the idea of its limits. This criticism is supported by the remark that Driessen did not read De Volder's texts,[204] so that this *Vindicatio* comes to clarify exactly Wittich's text in the light of De Volder's ideas.

Further references to De Volder are present in the subsequent *Altera vindicatio* (1718) of the same anonymous defender of Wittich, where attention is drawn to Van den Honert's complaint (given in his *Repraesentatio altera*) that Driessen had spoken so badly of De Volder after his death.[205] To this complaint, Driessen answered (in his *Alloquium alterum ad lectorem orthodoxum rei gestae narrationem vindicans adversus Repraesentationem C. Honerti alteram*) that he had not been the first in "putting De Volder under suspicion," and that Van den Honert, even if was so respectful of the late De Volder, did not write anything against those who brought to

---

tamen propterea non excludere omnem ignorantiam, neque sic tolli scientiam limitatam. Id nisi admittis, palam est, ideam scientiae, in se consideratae, sc. sine modo eius quodam, infinitatem dicere, quod nosti ex mysteriis Spinosae petitum esse," Driessen 1718a, 34. As to Driessen's criticisms of De Volder, see pages 32–36. The first criticism by Driessen of De Volder's principle can be found in his *Scrupuli*, printed in the same text (Driessen 1718a) at page 16; for Wittich's answer, given in his *Responsiones* (to which Driessen reacts with his *Refutatio* and *Refutatio altera*, both included in Driessen 1718a, as well as with his earlier *Animadversiones in Cl. Wittichii Responsiones ad Scrupulos quosdam illi propositos*), see the same book, page 21.

[204] "Magna minatur Driessenius, Refut. pag. 32–36. ut non Wittichium modo, hominem superstitem, qui, modo vellet, suam famam, ut antea, sic etiam nunc vindicare facile posset, verum (quod mihi maximum nefas videtur) ut etiam manes longe Celeberrimi Viri Burcheri De Volder lacescat, atque eiusdem Spinosismi notam optimi Viri memoriae inurat, quia ab eo mutuatus est Cl. Wittichius, quod nullius rei idea involvat negationem istius rei, cuius est idea, sed meram infinitatem. Vix me contineo, ne quid inclementius in hunc iniquissimum iudicem dicam! Qui cum ipse prae ignorantia profundissima de se haud semel Spinosismi suspicionem moverit, non erubescit, tanti criminis reum peragere virum de Philosophia Christiana tam bene meritum, cuius immortale nomen et decus tot sustinent discipuli, qui maximi Viri eruditionis atque Orthodoxiae nulla, nisi quae ex ipso audivere, et quotidie per universam Europam ipsi docent, asserre documenta necesse habent. Haec vero Driessenii audacia eo magis est culpanda, quia clarissime video: eum ne locum quidem, ubi Magnus De Volder hoc in suis scriptis doceat, unquam inspexisse. Quovis enim pignore contendere audeo, si quis ei diceret, Ubi? Cedo locum! Eum responsurum esse nihil. Sed, quia decet, ab omnibus verbis, quae Driessenio invisa futura sunt, abstinere decrevi. Respondeo autem ad omnia eius argumenta huic sententiae Magni Volderi opposita: Driessenium, qui non dubitat affirmare, illam sententiam ex Spinosismi mysteriis desumtam esse (vid. pag. 34) statum controversiae nequaquam percipere. Status controversiae non est, an praeter unam, plures scientiae in infinitae dentur? Hoc negat Magnus De Volder, hoc negat itidem genuinus et Orthodoxus Orthodoxi viri Discipulus, Cl. Iac. Wittichius. Status autem controversiae, si vere quid videas, est, num finitam alicuius rei naturam, vel si mavis, finitatem alicuius rei concipere queas per solam atque simplicem istius rei ideam? Hoc viri Clarissimi negant, dicuntque aliud quid requiri, quo Idea ista limitatur, antequam finitatis aliquem conceptum habere queas. Hoc autem probant exemplis petitis a scientia atque potentia. Finitatem enim scientiae nemo per simplicem scientiae ideam intelliget, sed debet necessario aliud quid accedere, qua scientia illa limita esse appareat," Gr. V. 1718a, 51–52. Driessen was to criticize De Volder-Wittich's principle 5 in his *Alloquium ad Ecclesiam Belgicam* (1718, appeared also as *Aanspraak aan de Kerk van Nederlant*, 1719): Driessen 1718b, 7–8.

[205] See Gr. V. 1718b, 38–39; cf. Van den Honert 1718a, 66.

light De Volder's "worst principles," viz. his definition of cause and effect.[206] To this attack, the anonymous author remarks that Van den Honert did not want to enter into or bother himself with any quarrel.[207] In fact, as seen above, Van den Honert did not reply to Driessen's attacks until Driessen published his *Epistola* against Wittich's *De natura Dei*. Moreover, Andala had already overtly attacked De Volder's idea of causation as leading to absolute necessitarianism. In the third part of his *Apologia pro vera et saniore philosophia*, published in 1718, moreover, Andala explicitly connected De Volder's teachings to Wittich's 1711 *Disputatio*. As Andala puts it, he had warned of the dangers entailed by De Volder's teachings since 1705 (i.e. when he sent De Volder his aforementioned disputation), while he was hoping that "from his school very harmful men would not come into the Churches and Academies." After some years passed, "for the Grace of God," without problems, in 1711 he heard about Wittich's *De natura Dei*, in which he found "multiple dangerous, absurd, abominable positions, similar to those [...] already mentioned," and referring to those paragraphs of Wittich's disputation expounding De Volder's "intricate positions," viz. those paragraphs in which Wittich expounded his aforementioned premises.[208] Andala, thus, had entered indirectly into the quarrel. When in 1719 an

---

[206] "P. 66 in me invehitur, qui in Cl. de Volder post eius fata debachatus fuerim, neque in convitiis in me proiicendis, pro liberalitate sua, parcus est. Nempe primus ego Cl. de Volder in suspicionem adduxi! Quare calamum non stringit Vir. Cl., si tanta illum pietas agit in Volderum mortuum, in illos, qui id egerunt ex instituto, ut Volderi pessima principia de causa et effectu definitionem detegerent?" Driessen 1718c, 32–33.

[207] "Pag. 66. Cl. Honertius suam indicavit indignationem, quia Driessenius mortuo insuergens Leoni, tam male ubique de Magno Burchero de Volder, post eius fata, loqui solet. Ad quod dum respondendum est, dicit Driessenius, se primum non esse, qui Cl. Volderum in suspicionem adduxit. Egregia sane responsio! Primus eum in suspicionem non adduxit, ergo recte fecit! Primus non peccavit, ergo res bene se habet! Sed addit Driessenius, 'quare calamum non stringit Cl. Honertius, si tanta illum pietas agit in Volderum mortuum, in illos, qui id egerunt ex instituto, ut Volderi pessima principia, de causa et effectu definitionem detegerent?' Respondeo, 1. quia Cl. Honertius, ut passim in hac causa Wittichiana docuit, atque affirmavit, pace delectatur omnino. 2. Quia, uti in Repraes. prim. § VIII dixit, Cl. Honertius aliis tantum iuris in bonas suas, quas Academiae atque Ecclesiae debet, horas concedere non vult," Gr. V. 1718b, 38–39. Then, the anonymous author attacks Driessen's systematic misinterpretation, or lack of understanding of "the Great De Volder's" words (in Driessen's *Responsionum Wittichianarum refutatio*): Gr. V. 1718b, 87. In the *Antwoort op de Aanspraak* (of Driessen), of the anonymous author "Gr. V." (included in Van den Honert's *Briev aan een voornaam Heer, wegens het Dispuut De natura Dei van* [...] *Jacobus Wittichius: waar by gevoegt is een korte ontleding van het selve dispuut, en een antwoord op de Aanspraak van* [...] *Antonius Driessen aan de Kerk van Nederland* (1718)), "the Great De Volder" is defended by referring to the aforementioned *Vindicatio* and *Vindicatio altera*: see Van den Honert 1718b, 46.

[208] "Imo ab anno 1705 quo primum detexi et publice scripsi de periculosis Clar. et summi viri subtilitatibus, metuere coepi, ne tandem ex eius schola prodirent viri valde noxii in Ecclesiis vel Academiis. Sequentibus tamen aliquot annis securis nobis per Dei gratiam esse licuit, licet in quibusdam latitare non potuerit malus erga Cartesianam Philosophiam affectus: at tamen nihil periculosum emanasse vidimus. Sed ecce postliminio legenda nobis exhibetur Disputatio iterato impressa, ante septennium circiter A. 1711, scripta de natura Dei et publice ventilata a Clar. D. Iacobo Wittichio, Phil. et Matheseos in Academia Regia Duisburgensi Professore Ord. Qui eam debito adhibito iudicio legit et saepe relegit, deprehendet plures positiones periculosas, absur-

unauthorised Dutch translation of Wittich's *De natura Dei* appeared, viz. the *Wysgerige twistredenering over de natuur van God, welke onder des Heeren bystand onder de bescherming van Jacobus Wittichius. Hier is tot een Toegift bygevoegt het Oordell van den Heere Ruardus Andala over het Dispuit van den Heere Jak. Wittichius, alles uit Latyn vertaalt*, this was accompanied – as the title explains – by the Dutch translation of those paragraphs of Andala's *Apologia* devoted to Wittich's disputation and to its 'Volderian' roots.[209] Thus far, the link between De Volder's teaching and Wittich's positions had come into full light.

As far as the rest of Wittich's disputation is concerned, the aforementioned five premises were used by him, not to demonstrate the existence of God as the absolute being (*ens summe perfectum*) – which in fact had been "asserted with a full mouth" by Spinoza – but rather to ascertain His nature and essence, which was the actual topic of controversy between Spinoza and the Orthodox.[210] On the ground of his five premises, Wittich argues that God can be an infinite extension, an infinite thought, or a being consisting of the two (as maintained by Spinoza). Yet, he argues, since extension and thought have nothing in common, it is impossible, even "by a liberal admission of Spinoza's thesis" – viz. that thought and extension are attributes of the same substance – to admit this very thesis, since one has to prove that these attributes entail each other in order for them to pertain to the same individual, i.e. to God.[211] On the contrary, given the fact that we can only conceive two attributes, and

das, abominabiles, praecedentibus modo citatis etiam similes, in ea contineri, licet admodum hic et illic incrustatas: [...] § 5, 8, 9, 10 aliquot intricatas theses Cl. de Volder (quem § 10 laudat,) proponit," Andala 1718, 139–140.

[209] Namely, a Dutch translation of Andala's *Apologia*, part 3, §§ 32–34 (including the text quoted *supra*, n. 208). Afterwards, an authorized translation appeared without Andala's text, as *Wijsgerige verhandeling van de nature Gods, sijnde deselve nu door den aucteur in het Nederduyts vertaalt, met aanmerkingen, tot oplossing van des Heeren Driessens beschuldigingen* (1719). Please note that in 1719 a new edition of De Volder's *Contra atheos* was published in Utrecht by Schouten printers: the book is now extremely rare, and extant only at the Universitäts- und Stadtbibliothek Köln.

[210] "Ad [...] quaestionem procedimus: quid sit Deus? Responsio, quod sit ens summe perfectum, non satisfacit, quoniam notio rei summe perfectae est nimis generalis, et vaga. Distinctius ergo inquiri debet, quidnam sit, quod naturam et essentiam talis entis constituat. In hoc etiam proprie versatur status controversiae inter Spinosa et Orthodoxos [...]. Adeo ut multi aberrarint, existimantes entis absolute perfecti existentiam a Spinosa negari, cum tamen illam (prout legenti ipsius libros ignotum esse nequit) rotundo ore asserat," Wittich 1718a, 6–7.

[211] "Vidimus hactenus mentem humanam duo tantum summa (iuxta § 6) rerum genera, eaque ut (ex § 10) infinita concipere posse. Undeque sponte sequitur: naturam Dei sive entis summe perfecti esse aut cogitationem infinitam, aut extensionem infinitam, vel ens quoddam constans utraque natura infinita cogitante et extensa. Quod postremum Spinosa amplectimur. Ut in praecedentibus diximus et ex eius Ethica manifestum est. Spinosa igitur arbitratur Deum esse substantiam infinite cogitantem et extensam, non quasi ex his attributis compositus esset: omnem enim compositionem a Deo removendam monet, sed quod cogitatio et extensio constituant unum simplicissimum individuum. Verum enim vero si ipsas ideas cogitationis et extensionis unicum (iuxta § 5) cognitionis fundamentum examinemus, palam fiet, quid de hac sententia statuendum. Intima conscientia inter conceptum extensionis et cogitationis nullam communionem, affinitatem, aut analogiam esse certissime novimus, quare (iuxta § 8) binae hae ideae res etiam plane separatas, ac inter se omnimodo dissimiles repraesentant. [...] Quam alienum igitur est affirmare, cogitationem ac extensionem

that these have nothing in common, we cannot but admit that these are two different substances. This is a conclusion which respects the "limits of human knowledge," the "boundaries of the sciences," and the "rule of judgment."[212] Hence, Wittich assumes Spinoza's tenet – which he labels as posing the "greatest difficulty" in analysing the nature of God – that "[i]f things have nothing in common with one another, one of them cannot be the cause of the other" (given in Spinoza's *Ethica*, part 1, proposition 3).[213] Wittich justifies this principle as evident from the very ideas of cause and effect, i.e. in accordance with the principle that "nothing could give, what it has not." This kind of justification had been criticized by his uncle Christoph Wittich in his *Anti-Spinoza* (1690), where the equivalence of Spinoza's proposition to such a principle is negated, since one has to distinguish between possession "by form" (*formaliter*) and "by eminency" (*eminenter*): i.e. between possessing something as a real attribute (like a king possesses gold), and as having the right of conferring something to someone else (for instance, a king can confer doctorates, even if he is not a doctor *formaliter*). According to Christoph Wittich, Spinoza's proposition is not valid as God is extended *eminenter*, viz. He can create extension from nothing.[214] According to his nephew Jacob (who explicitly rejects

simplex constituere individuum, nulla communione aut vel minima similitudine intercedente, quicquid de uno affirmo, de altero negare cogor. Sed supponamus, quandam communionem et affinitatem inter haec duo locum habere: in quantum haec concipiuntur attributa eiusdem substantiae, quaeque adeo in ea semper fuerunt, et unumquodque realitatem seu esse substantiae exprimit, ac conveniunt in conceptu infinitatis. Verum quis non facile perspicit, ex liberali mea suppositione Spinosae sententiam nullum foenerari lumen, inter attributa enim unius eiusdem substantiae non solum cognatio et similitudo quaedam, sed omnimoda identitas requiritur. Quo affirmat cogitationem ac extensionem constituere unum et simplicissimum individuum, debet non tantum communionem quandam, sed meram et omnimoda identitatem demonstrare, adeo ut omnes extensionis proprietates eaedem, nulla vel minima diversitate interveniente, sint cum proprietatibus cogitationis," Wittich 1718a, 12–14.

[212] "Sed rursus ponamus, ut liberales simus, illa ignota in cogitatione infinita forte eadem esse cum incognitis realitatibus extensionis infinitae. Fatendum tamen est, ab intellectu humano nihil concipi praeter cogitationem toto genere, tota natura ab extensione discrepantem. Hi limites humanae sunt cognitionis. Hi fines scientiarum. Haec norma iudicii. Quapropter conclusio manet legitima. Ideae cogitationis et extensionis nullam similitudinem, communionem aut affinitatem habent, nullus nexus inter eas comperitur, una ex altera nullatenus fluit, ergo (iuxta § 8 et 9) binae hae ideae non unam eandemque, sed duas diversas et distinctas res sive substantias repraesentant," Wittich 1718a, 15.

[213] Spinoza 1996, 3.

[214] "Quae res nihil commune inter se habent, earum una alterius causa esse non potest. Examen. Haec propositio si sumatur eo sensu, ut per verba commune inter se intelligatur commune aliquod attributum reale, uti procul dubio eam vult intellectam Spinoza, falsa est omnino. Huc enim tendit, uti ex sequentibus patet, ut evincatur, Deum, nisi esset extensus, non posse esse causam corporum, quod falsissimus esse [...] probavimus, quando demonstravimus, Deum ex nihilo omnia fecisse. Videri posset hanc propositionem equivalere isti axiomati: nihil potest dare causa effectui, quod ipsa non habet. Sed si illud axioma rite intelligatur, facile percipietur, hanc propositionem non aequivalere isti axiomati. Sensus enim est: nihil posse dare causam effetus nisi id, quod ipsa causa habet vel formaliter vel eminenter. Quid si habere aliquid formaliter per se patet. At quid si habere eminenter non ita manifestum. Rex habet formaliter nummos, fundum, domum et c. quae mihi donat non habet formaliter doctoratum, munus secretarii, munus professoris, quod in me confert,

his uncle's example),[215] on the other hand, one can understand 'by eminency' in two ways: (1) as a cause contains more than the effect, viz. that God contains extension "in a more excelling way," or (2) (as admitted by his uncle) as having the power of creating something. Both these alternatives are rejected by Jacob Wittich, because if one assumes that God is the most perfect mind, (1) He does not posses extension in any "more excelling way" than a substance is extended, (2) one has to demonstrate whether or not the power of creating something completely different pertains to this most perfect mind. However, admitting that this power pertains to this mind does not solve the original problem of justifying how thought can act on matter.[216] Eventually, on the basis of De Volder's premises and of Spinoza's proposition 3, Jacob Wittich (1) rejects Spinoza's monism, and (2) he argues that the problem of interaction between matter and thought, viz. between God and the material world, cannot be solved by rational means.[217] Lastly, Wittich – aiming at not falling into

---

sed dicitur habere illa eminenter h.e. habet ius illud mihi dandi. Ita Deus eminenter habet extensionem, quam tamen ex nihilo producit," Wittich 1690, 42.

[215] "Quidam sequenti exemplo utuntur: rex in subditum confert munus ex. gr. professoris, quod eminenter, neutiquam formaliter habet, hoc est habet ius illud dandi. Applicemus hoc rei praesenti: Deus possidet eminenter extensione, hoc est, secundum explicationem datam, habet ius ad extensionem, quod profecto nihil dicit, praeterquam quod illud Dei ius in res creatas praevia potentia, qua res produxit, nitatur, de qua vi et potentia tota est quaestio, num quidpiam diversum in Deo a cogitatione designet? Si non, qui mens infinita nihil commune habens cum materia, eius nihilominus causa esse possit? Manet igitur difficultas integra," Wittich 1718a, 17.

[216] "Fundamentum huius ratiocinii maximam comprehendere difficultatem omnes philosophi fateri debent. Quoniam evidenter ex notione causae et effectus sequitur, quae nihil commune habent, unum alterius causa per rerum naturam esse nequit. Omnes facile largiuntur, neminem aliquid dare, quod non habet. Quare multi dixerunt Deum eminenter possidere corporum proprietates, hoc est modo quodam praestantiori, adeo ut, quod est in effectu, contineatur in causa veluti minus in maiori. Verum praestantior et eminens modis, quo Deus materiam continet, nonne affirmatur de mente infinita? Nihil ergo distinctum ab ipsa cogitatione dicit! Quapropter verba allata alium sensum habere non possint, quam quod Deus sive substantia infinite cogitans extensionem producere possit, quo autem modo, cum inter rem cogitantem ac extensam nulla affinitas aut communio reperiatur, profunde ignorari: hoc pacto manet difficultas integra nec ulla lux vocabulo illo peregrino conciliata est. Alii quidam illud eminenter continere sequenti modo explicant: Deus eminenter dicitur comprehendere extensionis realitates, quia ad ideam extensionis habet vim ac virtutem materiam producendi. Sed velim libenter scire num illa vis et virtus in Deo differat a cogitatione perfectissima? Si non, relinquitur difficultas insolita, quomodo cogitatio nihil commune habens cum extensa substantia, eius nihilominus causa esse possit?" Wittich 1718a, 16.

[217] "Quomodo omnia ex perfectissima Dei natura oriantur, qua serie, quo ordine singula facta sint? Quis non accusabit mentis imbecillitatem? Quae lux, qualis splendor non exigeretur ad talium rerum contemplationem? Quale coruscans, et splendidum lumen nostros certe vespertilionum oculos perstringeret et obcaecaret. Optanda sed speranda mortalibus nullatenus est ea veritatis lux, qua penetrare in ipsas primas rerum origines possint. Quemadmodum ignoras, quae sit via venti ut ignoras, quae via ossium in utero gravidae: ita ignoras opus ipsius Dei, qua via faciat haec omnia. Sunt verba Salomonis Regis Sapientissimi in suo Ecclesiaste cap. 11:5 [...] Demonstravimus (§ 5) omnem rerum cognitionem, omnemque de eis scientiam ex solis nostris ideis derivandam esse. Ostendimus autem (§ 8) eandem diversitatem inter ipsas res esse, qualis inter ideas, rerum repraesentationes, observantur. Omnes itaque res, quae uno conceptu repraesentantur, ac se solis percipiuntur, quaeque praecise unicum quid menti tantum offerunt, absque omni dubitandi ratione sunt res per se subsistentes. Ubi enim unicum quid repraesentatur, quique conceptus nullam admittit

scepticism, and at overcoming this "greatest difficulty proposed by Spinoza," which Spinoza faced by "confusing those things [which are] to be separated"[218] – (3) demonstrates that God is not extended, but rather a thinking substance. If He could be extended, indeed, movement would be natural to bodies, whereas it is always imparted by external causes.[219]

As far as the rest of the quarrel is concerned, it is worth mentioning that in his *Dissertatio Cl. J. Wittichii Disputationi de natura Dei opposita*, viz. his most sys-

---

separationem, seiunctionemque, sed omnem rem repraesentat, aut nihil eius, est conceptus substantiarum. Illae autem ideae, in quibus plura menti offeruntur, quorum aliqua separatio ac seiunctio fieri potest, necessario erunt modorum et accidentium ideae. Equidem nullum aliud datur signum, ex quo substantia ab accidentibus discerni queat. Porro (iuxta 6) habemus ideas rei cogitantis et extensae, quas substantiarum ideas esse, ex eo constat: quod singuli hi conceptus unicum quid repraesentent, a quo nihil separari potest, quin eo ipso pereat totum illud, quod hae ideae intellectui offerunt. [...] Quapropter colligimus cogitationem et extensionem ab intellectu humano ut duas diversas intelligi substantias, adeo ut, si velimus iuxta rerum ideas ratiocinari, Spinosae opinio de unica substantia duobus attributis, extensione et cogitatione praedita, reiicienda sit," Wittich 1718a, 18–20.

[218] "Propterea tamen victas neutiquam tradimus manus: nos quidem largimur difficultatem a Spinosa propositam esse maximam, sed vicissim Spinosa se nodo plane insolubili constringi a nobis in praecedentibus nexu, fateri debet. Quid hic igitur agendum? Accusare caliginem et imbecillitatem mentis est frustraneum? Rationem ergo unice consulamus, illa procul dubio nos docebit, quaenam eligenda sit sententia. Ausim quemlibet in hac disciplina exercitatum compellare? Numne magis rationi consonum sit nostram prae Spinosae amplecti sententiam? Quoniam ipsa fundamenta omnis cognitionis et scientiae, videlicet rerum ideas sartas tectas conservamus, illas duces sequimur, ex quibus de rebus solummodo iudicare possumus et debemus, nisi omnis cognitionis certitudinisque fundamentum convellere, atque insanabilem introducere velimus scepticismum. Spinosa vero viam hanc optimam deserit, confundit enim quae separanda, cogitationem ac extensionem ad unam eandemque redigit ideam, cum tantum a se distent, ut uno eodemque conceptu concipi nunquam possint. Insistendum propterea viae regiae arbitramur, ac deserendam esse Spinosae sententiam ratio suadet," Wittich 1718a, 18–19.

[219] "Si Deus substantia infinite cogitans et extensa esse nequeat, sequitur necessario, ut sit vel solum cogitans, aut solum extensus. [...] Demonstrabo ergo extensionem Dei attributum nullo modo esse posse, unde (iuxta § 6 et 10) sequetur, sola cogitatione infinita divinam naturam esse praeditam. [...] Si motus ex natura corporum sequeretur, corpora in aeternum moverentur, nec unquam ad quietem reduci possent, atqui videmus corpora hoc momento mota, paulo post quiescere, igitur motus corporibus non est naturalis. Quis adeo hebes! Ut non percipiat, cum ratione responderi posse, motum quidem corporibus particularibus haud esse naturalem, illa enim et moveri et quiescere posse pro data causa, sed ex ipsa materia motum fluere, eamque semper moveri, adeoque indesinenter unum corpus aliis motum suum communicare, illudque per circulum ex uno corpore in alia transire. Propterea si aequa rationis lance rem ponderare velimus, alia via ostendi debet, motum ex natura corporum haudquaquam fluere, sed ad existendum externa egere causa. [...] Si extensionem attente ac otiose contemplemur, fateri cogimur, eam a nobis quiescentem concipi posse: hoc si aliquis negare ausus fuerit, monstranda erit contradictio, equidem mihi affirmanti, materiam clare et distincte sine motu intelligi posse, certissimis rationibus contrarium ostendi debet, siquidem conceptus meus a vero abluderet, quando igitur materiam quiescentem clare nobis repraesentare possumus, indicandum est, quidnam extensioni huic superaddat motus, ut inde fiat corpus, siquidem de natura corporis sit futurus motus, quod quia frustra expectamus, est ergo necesse, ut causa externa arcessatur, quae materiae motum indat, et certis legibus conservet: propterea extensio naturam entis perfectissimi constituere nequit, utpote, quod operationem causae externae in se pati nullatenus potest," Wittich 1718a, 24, 26–27.

tematic analysis of Wittich's text, Driessen addressed four main points in the disputation (which have been analysed by Aza Goudriaan):[220]

1. Spinoza's principle that things which have nothing in common cannot be the cause of each other.
2. Wittich's rejection of the arguments of the Orthodox on the unity (i.e. the being part-less) of God (which Wittich aimed at clarifying by a method leading to a mathematical evidence).[221]
3. Wittich's rejection of the arguments of the Orthodox arguing that the highest perfection can pertain only to a thinking being (as seen above, as to this point Wittich relied on an argument *ex motu*).
4. Wittich's replacing of the orthodox arguments for the production of matter by an infinite Spirit, with arguments which offered no certainty (as seen above, Wittich reverted to Ecclesiastes, rejecting the idea that God could create matter by eminence).[222]

---

[220] See Goudriaan 2003.

[221] "Quicunque igitur crediderit multum referre, sive veritas, sive falsitas admittatur, necessario in hanc descendet sententiam, in veri indagine neminem versus unam magis quam aliam propendere opinionem, aut ullum assumere dogma nisi sola evidentia debere. [...] Hac methodo examinare constitui de Natura Dei quaestionem gravissimam et maximi momenti, nam si ulla res accuratam meretur contemplationem, certe ea est, quae infinitas Dei spectat perfectiones, quibus nihil nobilius aut praestantius, ac sine qua cognitione omnia fluxa et inania. Ut enim Deus unica omnium rerum est causa, sic eius cognitio ea est, ex qua reliquorum omnium scientia tanquam perenni scaturigine fluit. De qua olim admodum variae atque discrepantes fuerunt sententiae, ut videre est in libris Ciceronis De natura deorum [...]. Nec hodie ea sic pertractata est, quin supersint dissentientes adhuc opiniones. Si affectus hac in re consulendus, protinus quid credendum, quidque reiiciendum appareret. Ast quoniam non eam esse veritatem investigandi viam novimus, non igitur quid desiderium, sed quid ratio suadeat, dispiciendum erit. Summopere etiam laborandum, ut non verisimilibus ratiunculis contenti, tantam vim demonstrationis inveniamus, quae nullam rimam elabendi relinquat, aut mathematicae evidentiae vel hilum cedat," Wittich 1718a, 5–6.

[222] Cf. Driessen's summary: "[a]d capita Disputationis argumentum, quidque in illo culpandum et horrendum sit, reducam. A. Affirmat axioma illud Spinosae, Spinosismi basin, res quae nihil inter se commune habent una alterius causam esse non posse, generaliter considerandum verissimum esse, nec ullam continere falsitatem, § XIII, et XIV. B. Argumenta Orthodoxorum pro adstruenda Dei unitate, quae petuntur ex omniscentia et omnipotentia Dei infirma fore existimat contra talem qui duos statueret Deos unum cogitantem alterum extensum, § XIX. C. Argumenta Orthodoxorum, quibus adstruunt summam perfectionem non posse cadere in rem extensam, sed tantum in Spiritum infinitum illi non apparent eius evidentiae et efficaciae quae reluctantem ad assensum cogere queant, nec hactenus alia firmiora illi occurrerunt § XX. Nempe quia regeri posset extensionem non ideo imperfectionis arguendum esse quia non cogitat, natura enim eius tale nihil poscit utpote in sola extensione consistens, quo respectu duntaxat perfecta aut imperfecta dicenda, § XXIII. D. Ut demonstraret materiam a Spirito infinito, non Spiritum infinitum a materia productum esse, argumenta Orthodoxorum, quibus Enti infinito omnipotentiam adferunt, et materiae negant, negligit propter difficultatem ex axiomate Spinosistico motam, et alias rationes duas adhibet, alteram petitam a motu § XXI, alteram inde petit quod facilius sit concipere extensionem a cogitatione infinita productam esse, quam cogitationem infinitam ab extensione, § XXII. Quae utraque ratio, si obtineat eius hypothesis, invalida est, et sic incertum relinquitur, creaveritne Deus materiam, an materia Deum, quod eo magis incertum erit, si quod dicit in responsoriis suis ad scrupulos illi propositos: si hoc pacto ratiocinaretur: extensio utut infinita nulla habet ideam cogitationis infini-

Thus far, Wittich relied on a purely philosophical approach in the investigation on the nature of God. Or, as Johnathan Israel has put it, Wittich "has researched exhaustively [...] to see how best we can advance towards a cogent, philosophically workable stance which is 'scientific' in De Volder's specific sense of being consonant with mathematical rationality."[223] This investigation, however, ultimately led him to revert to Revelation to solve the problem of spirit-matter interaction.

This reminds us of a path already taken on the eve of the insertion of Cartesianism into Holland by Henricus Regius (who deemed the whole metaphysics of Descartes as unreliable),[224] and more recently by Bredenburg, who in his *Noodige verantwoording op de ongegronde beschuldiging van Abraham Lemmerman* (1684) – written after the unauthorized publication of his *Wiskunstige demonstratie* by Lemmerman, included in his *Eenige bewijzen dat Johannes Bredenburg, staande zijn stellingen, geenszins kan gelooven, dat'er zulk een God is, als de H. Schrift leert* (1684) – reverted to a sort of double-truth strategy while aiming to resolve the contradiction between Revealed truth and the necessitarianism he had exhibited in his *Wiskunstige demonstratie*.[225] However, Wittich was to maintain, as De Volder did,[226] that no contradiction between the truth of faith and that of philosophy can ever be conceived. He did so during a further quarrel, this time with Johannes Alexander Roëll (1688–1751, son of the Cartesian theologian Hermann Alexander (1653–1718), and professor of law at Deventer), which is revelatory of De Volder's opinions as well. This position can be found in Wittich's *Zeedig antwoord op het lasterschrift tegen hem gemaekt door Johannes Alexander Roëll* (1723) – written after Johannes Alexander Roëll had attacked him with his bilingual *Larva detracta Jacobo Wittichio*, or *Jacobus Wittichius ontmaskert* (1723).[227] In his *Larva*, Johannes Alexander Roëll had published in a bilingual edition some of letters exchanged between Jacob Wittich and Hermann Alexander Roëll, who had, together with De Volder, recommended Jacob for a chair in philosophy at Duisburg in 1704 (which

---

tae, multo minus voluntatem efficacissimam, adversarius respondere possit, hoc esse ex hypothesi argumentari, nempe quod Deus sit mens infinita, porro se negare ad productionem rei requiri ideam et voluntatem sufficere quod extensio infinitam vim haberet. Quo haec omnia vergant, adesdum Lector, cense, arbitrare," Driessen 1718d, 2–3.

[223] Israel 2001, 439–440.

[224] On Regius's approach to Descartes's metaphysics, see Wilson 2000; Clarke 2010; Bellis 2013; Bos 2013, Strazzoni 2014.

[225] See Van Bunge 1990, chapter 4; Fix 1991, chapter 9; Israel 2001, chapter 19.

[226] See his commentary on I.76, in Hamburg 273, 70, and commentary on I.25, in Hamburg 274, 12–13.

[227] "Dit getuigd self de Heer Roël, seggende, dat syn Vader dieswegen al eenige jaren voor syn dood my ongenegen is geworden, so als wy boven gehoord hebben. Maer segd hy, dat ik stelle, het geen Philosophisch waer is, dat kan Theologisch, niet vals syn. Dit is noch myn gevoelen, welk ook alle God geleerden bekennen. Maer heb ik ooit gesegd, of eenig Godgeleerde, dat men onderscheidentlyk, en volkoomen begrypt de Godlike Persoonen. Hoe sal dan het natuurlik ligt daer van de valsheid konnen ontdekken? Hoe sal de reden konnen oordeelen van saken, die sy niet volkoomen verstaet? Wanneer men uit Gods Woord duidelik een leerstuk ontdekt, konnen wy sekerlik besluiten, dat het selve tegen de reden niet stryd, als hebbende God so wel de openbaring gedaen, als de reden in den mensch geschapen," Wittich 1723, 23. I owe this reference to Israel 2001, 443.

he was to assume in 1707, after having graduated and lectured there since 1704).[228] Johannes Alexander Roëll had published this attack on Jacob Wittich after having been acquainted with the obituary of his father Hermann Alexander, read by Johannes Marckius at the Synod of South Holland (on an unspecified date).[229] According to Johannes Alexander Roëll's version, Marckius had stated that Jacob Wittich claimed that he once heard his uncle Christoph saying that "my father [viz. Hermann Alexander Roëll], at that time still a young man, one day would be a Churchman to be feared." This judgment is labelled by Johannes Alexander Roëll a "childish story" (in fact, Christoph died when Jacob was not even 11 years old), purportedly used by Jacob Wittich and Marckius to tarnish the figure of his father Hermann Alexander.[230] As part of his attempt to vindicate his memory, Johannes

[228] See the letter of Jacob Wittich to Hermann Alexander Roëll of 12 April 1704, in Johannes Alexander Roëll's *Larva*: "Amstelodami praeterita aestate, Vir Magnopere mihi colende, tuo cum frui dabatur colloquio. Mentionem iniecit Clar. V. Professoratu Duisburgensis. Officii mei itaque iudicavi Clar. V. certiorem reddere. Hebdomade hac Celeb. de Volder Duisburgo per litteras professoratum in philosophia ibidem vacare, quod perceperit. In quibus una consilium eius de substituendo idoneo petebatur, qui pro summa humanitate et benevolentia sua me non solum ad id munus dignum iudicavit, ast et omni, qua potuit, ratione consultoribus proposuit. Quid itaque! In lacertos benevolentiae et favoris, quibus me haud duntaxat Franekerae, sed et nuper Amstelodami tenerrime amplexus, tuo ruo. Si Clar. V. quicquam ad hanc meam promotionem se efficere posse existimet: ne me indignum favore suo repellat, obnixissime rogo. Fortean Cl. Hulsius discipulos Roelli contemnet! Limatum iudicium, summaque tua prudentia, quid hacce in re agendum, satis suadebit. Quam optarem! Divina providentia hunc mihi procuraret locum! Admodum conveniens est indoli meae iterato itaque animitus rogo, ne tua me deserat benevolentia. Pluribus occupationes tuas interrompere nolo. Id duntaxat addam, me continuo vota facere, Deus ter Opt. Max. te incolumem ad seros ut servet annos, in emolumentum Ecclesiae, commodum Academiae, bonum familiae, et revera ornamentum generis humani, et nil magis in votis habere, ac ut Jacobum Wittichium tuum olim forte haud ingratum discipulum, amantissimum tui, solidissimae eruditionis tuae cultorem indefessum, ames. / Lugduni Batavorum / prid. Idu. April. 1704," Roëll 1723, *Epistolae*, 10–11. Notably, Wittich is afraid that Hulsius could interfere.

[229] Marckius was author, together with Van den Honert and two other professors of theology, of a *Laudatio* appended to the *Iudicium Ecclesiasticum, quo opiniones quaedam Cl. Herm. Alex. Röell synodice damnatae sunt* (1723).

[230] "Cum perlatus ad me esset sermo, quem in postrema Synodo Australis Hollandiae Joh. Marckius Theol. Professor in parentis mei memoriam Lugduni habuit, multa mihi quidem in eo ingrata atque acerba acciderunt, verumtamen quae animum minus movebant, quod et aures ad quotidiana illa convicia iam paene occalluerant, et quod illa vel nullius, vel levioris ponderis apud bonos futura esse existimabam, cum infestus viri illius in parentem meum animus omnibus satis notus esset. Illa vero orationis pars, qua sibi a Collega suo Jacobo Wittichio aliquoties commemoratum esse narrabat, patruum ipsius, Celeberrimum illum Christophorum Wittichium, iam olim vaticinatum fuisse, patrem meum, tum ad huc iuvenem, metuendum olim Ecclesiae virum futurum, mihi usque eo indigna visa est, ut existimaverim mihi diutius tacendum non esse. Facile enim perspiciebam, nullum alium in finem totam hanc, quae tamen ad rem nihil faciebat, factam esse narrationem, quam ad conflandam infamiam parentis memoriae, et conciliandam dictis suis fidem atque auctoritatem: cum hoc pacto appareret, idem iam olim, pro ea qua erat perspicacia, de parente praesensisse virum optimum, quique apud omnes propter pietatem, eruditionem, atque integritatem, summo in honore merito suo habetur, quod eventu ipso comprobatum esse, omni molimine ostendere Marckius conabatur: de veritate autem affirmationis huius dubitari nullo modo posse, quippe quam ex ipsius patrui ore hausisset Wittichius. Merito itaque exarsi, non tam in illum, qui causae suae serviens puerilem hanc narravit fabulam, quam in ipsum gravissimae huius calumniae auc-

Alexander published and translated into Dutch the texts of the letters between Jacob Wittich and Hermann Alexander. As testified to by his letter of 14 June 1718, Jacob Wittich found the aforementioned *Scrupuli* of Driessen against his *De natura Dei* laughable, and did not believe that Hermann Alexander Roëll (from 1704 theologian at Utrecht, to whose professors of theology they were sent, as seen above) could ever approve them. However, Wittich was eager to know his opinion, as Driessen claimed to have obtained the approval of his *Scrupuli* from the Utrecht faculty of theology.[231] In his answer, Roëll, with some circumlocution, admitted to having approved Driessen's *Scrupuli*, maintaining that his judgments were not "definitive sentences," and that he would have been very happy if "you [Wittich] could exculpate yourself, and [...] demonstrate that you had refuted in good faith the impious theses of Spinoza." Moreover, he admits that the disputation (which had

torem: cuius pectus internecino erga mortuum parentem, quem tot invidiosis, dum vivebat, praeconiis ornaverat, tam magnificis in coelum laudibus extulerat, cuius amicitiam et favorem omnibus modis demereri conatus fuerat, odio flagrare, manifestis indiciis deprehendebam. Videbam itaque, nisi omnino abiecti atque degeneris animi haberi vellem, tam insignem contumeliam mihi non diutius dissimulandam aut concoquendam, sed, quod iam dudum factum oportuit, hominis petulantiam atque vecordiam retundendam esse, ne veteres patienter ferendo novas quotidie illius provocarem iniurias. Quod si maturius factum fuisset, fortassis non eo ipsius processisset audacia. Cum enim iam statim a parentis mei morte publice illum calumniis appetere, illiusque opiniones maxima cum invidia traducere veritus haud fuerit, ut ita inprudentibus fucum faceret, ac si ab iis semper fuisset alienissimus, neque quidquam ab ullo reponi animadverteret, ipse secum cogitavit, defunctus iam sum, nihil est quod dicant mihi, atque inde omnia sibi deinceps impune licere existimavit. Verum cum neque modum neque finem maledicendi faciat, quantumvis animus meus ab omni contentione abhorreat, tamen non sine scelere me paterni nominis, cuius mihi semper veneranda debet esse eritque memoria, defensionem praetermittere posse animadverti," Roëll 1723, 1–2.

[231] "Non soleo virorum eruditorum occupationes levi de causa interrumpere, imo eorum, quos peculiari ratione colo et veneror, religio mihi pene est. In malam igitur partem haud sumes Vir Pl. Reverende, quod has ad te dem litteras, est enim causa gravissima. Cl. Driessen in nupero scripto, Traiectinis Theologis eosdem subortos esse scrupulos, qui sibi, eosque iustos declarasse, gloriatur. Risi in initio, videbar enim certus de te talia dici non posse, utpote, qui eodem anno, quo disputatio ventilata, mihi in aedibus tuis humanissime excepto, narranti, quod Cl. Melchior apud nos iactaverat, se Traiecto per litteras incitatum ad refutationem disputationis, te haud obscure innuens, respondisti, talia falsa esse, nunquam te ea propter litteras ad Melchiorem dedisse, et quid refutare D. Melchior vult, dum satis abunde rem explicatam dedisti. His fretus, securus esse poteram, scrupulos Driessenii te non potuisse iustos declarare. Attamen quia et hic fama pervulgata est, Cl. Driessenium tua imprimis niti approbatione, idcirco quid de tota hac re sentiendum, ut mihi aperire digneris obnixissime rogo. Postquam per biennium fideli et docta tua institutione maximo cum meo emolumento frui licuit, semper dein summa me benevolentia prosecutus fuisti, quod nunquam non grato animo commemorare soleo, cogitavi itaque, si per imprudentiam aliquid erronei a me commissum fuisset, quod pro maximo tuo in me amore, privatim illud correcturus potius, quam adversariis meis ansam me perdendi suppeditaturus fuisses, qui aliter de te iudicarem? Quem velut patrem semper colui summa cum veneratione. Intelligis igitur Vir. Pl. Reverende te rem gratissimam mihi facturum, si quidem paucis aperire volueris, quid hac de re tenendum. Credas velim Vir multis nominibus mihi colende, nunquam in animum meum inducere me potuisse, ut fidem adiungerem, quae durante hac controversia cum Cl. Driessen mihi de te narrata sunt. Etenim tu me ut filium amasti, ego ut patrem constanter te veneratus sum, nec unquam gratum animum exuam. Vale et me quem hactenus amasti, amare perge. / Duisburgi 14 Iunii 1718," Roëll 1723, *Epistolae*, 20–21.

lain for long in his papers), "to say things as they are, started to be suspect" once he read it carefully.[232] No further answer is reported from Wittich, who in his subsequent *Zeedig antwoord* complains that his correspondence had been misused, as had happened – as De Volder himself reminded him, warning him to be careful in dealing with his correspondence – to Bredenburg (whose correspondence with Van Limborch had been used by Van Limborch himself against him).[233] This digression testifies to how Wittich came to revert to an old strategy in solving metaphysical

[232] "Falleris, Vir. Cl., neque Virum illum satis nosti, si, illum mea imprimis vel auctoritate, vel approbatione, niti existimas. Ipse satis acutum cernit, neque in alieno cerebro habitare solet. Non ignoro, quid malevoli homines, qui ex suo aliorum animos aestimant, sparserint. Sed diu est, quod didicerim istiusmodi calumnias contemnere. Nostrum non est, praestare ea, quae non sunt ἐφ' ἡμῖν, atque adeo, ut alii non sint malevoli et maledici. Cl. Driessenium scrupulos ad nos suos misisse, notum est. Quousque nos eos iustos declaraverimus, ex Cl. Burmanni Epistola, quam sponte sua conceperat et conscripserat, quamque, consciis nobis ac consentientibus, suo et Cl. Alphenii meoque nomine ad te misit, instituta parallelia, facile intelligi potest. Scrupuli non sunt decretoriae sententiae. Noluimus enim, etiam post acceptum responsum tuum, sententiam ferre, sed et Epistolam illam, et responsum tuum iis, ad quos ea res singulariter pertinet, examinanda dedimus. Illorum erit arbitraria sumere. Iucundissimum nobis erit, si te expurgare, et liquido ex ipso Disputationis tuae tenore demonstrare poteris, te bona fide impias Spinosae sententias refutasse. Eorum, quos inter nos domi meae habitos dicis, sermonum prorsus, sancte affirmo, immemor sum. Quando primum tua mihi Disputatio oblata fuit, obiter eam perlustravi. Postea diu inter alias chartas latuit. Cum iterum oculis illa se meis ingereret, accuratius a me lecta atque examinata, et, ut quod res est dicam, suspecta esse coepit," Roëll 1723, *Epistolae*, 24–25.

[233] In his *Zeedig antwoord*, Wittich explains that he had discussed two of his letters to Roëll with De Volder – the only person to whom he ever revealed his difficulties – because Roëll had never sent him an answer. These letters, concerning the sensitive issue of the idea of the Trinity, are the first two letters by Wittich to Roëll published in the *Larva*: they are dated 12 November 1701 and 1 April 1702. In the later letter, Wittich noted that Roëll had not answered him. At the time, De Volder warned Wittich that his letters could be dangerous to him, if they fell into the wrong hands: "[d]aer in tegendeel myne twyffelingen nooit tot myn besonder gevoelen zyn overgaen, gelyk niet alleen sonneklaer te sien is uit die twee brieven, maer ook genoegsaem hier uit af te neemen is, dat ik aen geen mensch in de wereld (uitgenoomen den Heer de Volder) myne swaerigheden hebbe vertoond, geen mensch heeft ooit het geschrift gesien, of de reden uit myn mond gehoord. Een genoegsaem blyk, dat het alleen swaerigheden waren, die myn gemoed ontrusteten. Want de dagelikse ondervinding leerd, dat menschen, welke een verschillend gevoelen aengenomen hebben, plegen het selve de eene met meer behoedseamheid, als de andere, aen den man te brengen. Dat ik die swarigheden aen den Heer de Volder heb vertoont, is by gelegendheid gekoomen, dat ik segde, daer over tweemael aen den Heer Professor Roël te hebben geschreven, sonder dat antwoord had gekregen. By welke gelegenheid die wyse Heer myne onvoorsichtigheid aentoonde, seggende, dat die brieven my naderhand noch veel moeyelikheid souden konnen baren, indien sy in trouwloose menschen handen vielen, bybrengende diergelyk geval gebeurd den Heer Bredenburg (die met seer veel verstand en oordeel tegen Spinosa geschreven heeft) radende voorts, dat trachten sou die brieven met dat geschrift weer magtig te worden," Wittich 1723, 25–26. Bredenburg's correspondence with Van Limborch was published by Van Limborch himself as *Schriftelyke onderhandeling, tusschen den Heer Philippus van Limborg ende Johannes Bredenburg. Rakende 't gebruyk der Reden in de Religie* (1686), to which Van Limborch appended two letters by him to Pieter Smout, aimed at refuting Bredenburg's views. In turn, a further, anonymous commentary from Bredenburg's circle (*Eenige consideratien of aanmerkingen over de Schriftelijke onderhandeling*, 1686) came to emphasize Bredenburg's views on necessitarianism. Hence, Bredenburg himself answered Van Limborch's attacks with his *Korte aanmerkingen op de Brieven van P. van Limburg* (1686): see Colie 1957, 100–101; Van Bunge 1990, 219.

problems raised by Cartesiansim (and Spinozism). Moreover, it testifies further to De Volder's sympathies for Bredenburg, notwithstanding the fact that De Volder himself taught that philosophy and faith are always in agreement.[234]

In conclusion, is worth remarking that, probably as a consequence of the fact that Driessen admitted – as underlined by Wittich – not having read his books, De Volder was not deemed the foremost inspirer of Wittich's theses, traced back by Driessen to Spinoza, and to a (rather generic) philosophical meddling with theology. It was Andala, in fact, who fully unveils Wittich's debts to De Volder, and those parts of his *Apologia* devoted to this were swiftly translated into Dutch, and appended to a Dutch, unauthorized version of Wittich's *De natura Dei*. Hence, in Leydekker's *Blyde Spinozist* – published together with the judgment given by the faculty of theology of the University of Jena, led by Buddeus – De Volder was again attacked, albeit without adding much to the criticisms of Andala.[235] In fact, De Volder limited himself to pointing out, in his 1698 *Oratio*, that the problem of the interaction of soul and body is beyond the powers of our reason, without tracing theological consequences or deriving a solution from this. When he dealt with the problem of the nature of God in his *Exercitationes* he just claimed that we clearly and distinctly perceive, in the idea of God, that He is the "most perfect being" and the "omniscient being" or the "infinite thought," thereby devoting his consideration to the aforementioned demonstrations of His existence.[236] Accordingly, Wittich went much further

---

[234] See *supra*, n. 226.

[235] Leydekker criticizes De Volder's and Wittich's principle 5: see Leydekker 1719, 16–17, 32 and 40–41. Moreover, he attacks – like Andala – De Volder's (supposed) absolute necessitarianism (as given in De Volder's *Contra atheos*): see Leydekker 1719, 60–61.

[236] "Quare talis idea clara erit. Erit quoque distincta, cum illud quod sequitur illi naturae rei peculiare sit, adeoque naturam illam ab omni aliarum rerum natura distinguat accuratissime. Quo indicio posito, age, examinemus ad id ideam, quam habeo Dei, ut appareat, num illa clara distinctaque sit, an secus. Habeo ideam Dei, hoc est, enti, summe perfecti. Nunc quaero ex illis, qui hanc ideam claram distinctamque negant, an non clare, percipiant, illi enti summe perfecto, nihil adiici, nihil ab eo demi posse. Neque enim perfectio adiici potest, fuisset enim ante non perfectum. Nec demi, imperfectum enim posset fieri, quod summe perfecto repugnat. Ita nec potest demi imperfectio, quia nullam habet; nec potest adiici, quia redderetur imperfectius, quod repugnat. Hanc ergo sequelam, qui clare percipit, quod nemo non facit, nonne ille clare percipit ens summe perfectum, ex quo haec sequela manat? Clara itaque est haec idea; nec minus distincta est, siquidem manifestum est hanc sequelam ita convenire enti perfecto, ut nulli alii, adeoque per hanc proprietatem illam rem ab omni alia distingui. Deinde, habeo ideam Dei, hoc est, cogitationis infinitae, entis omniscii, quaero nunc, an non evidenter percipiamus, illam cogitationem infinitam, ens illud omniscium, nulli cogitationum mutationi posse esse obnoxium; illud non tantum ea scire, quae nobis apparent necessaria, sed et ea qua evidentur maxime contingentia; ens illud non uti ratiocinio, sed simplici, ut loquuntur, intuitu omnia percipere, intelligere, scire. Haeccine, authorum unicum, quid autem hisce evidentius? Sequi num percipere quis evidenter potest, licet fundamentum huius sequelae, hoc est, ipsum ens omniscium non percipiat. Hasce proprietates ex conceptu demonstrabit suo, de solo ente omniscio, et tamen conceptus ille, ens illud omniscium ab aliis rebus non distinguet? Qua sane ratione, si quis ideam, quam de Deo habet attente examinet, facile ex hisce effectis eam evidenter percipi, nec Deum eum aliis rebus confundere, hoc est, claram distinctamque esse, opinor, percipiet," De Volder 1695, *De idea Dei*, thesis 12. Other disputations on God presided over by De Volder do not offer more insights on the topic. For instance, in the *Disputatio philosophica de unitate Dei* (1673), presided over and authored by De Volder, and

than De Volder in his argumentations, even if building upon the premises set forth in his metaphysics.[237]

As to De Volder, he did not push his ideas as to the matter-spirit relation to their limits, viz. to invoke a theological solution to them, or to revert to Spinozism. In fact, one of the reasons for his resignation was precisely his being discontent with Descartes's metaphysics, and not finding in any other system – with the partial exception of Leibniz's idea of pre-established harmony – a viable alternative. This is, actually, the judgment of Le Clerc on him, which we may endorse as we consider his correspondence with Leibniz, where such difficulties are more overtly discussed than in his printed texts. Partially alternative standpoints as to De Volder's 'discontent' are those of Andala, who claimed to have prompted De Volder's resignation after having shown him his 1705 disputation on the demonstration of the existence of God (criticizing De Volder's a priori proof, and his theory of causality), and of Jacob Wittich himself, who in his 1718 *Oratio inauguralis de evidentia et certitudine*, given at Leiden, remembered how De Volder, man "of the sharpest judgment and clearest mind, the most versed [...] in every part of philosophy, the subtlest in mathematics, and[,] which [is] very rare[,] the most solid in metaphysics," "could [have] provide[d] excellent [things] in metaphysics, [and] not refused the public view, unless [he] preferred to avoid to create annoyances [...] from malevolent men, [...] scarcely versed in this science. Namely, men who "dare to assert that no new truth can be found about the infinite and most perfect mind."[238] Thus far, Wittich

defended by Bernard Nieuwentijt, De Volder defends the numerical unity of God, on the ground of His perfection (see especially thesis 3). See also the *Disputatio philosophica de aeternitate Dei, prima–secunda* (1681, author and *respondens* Johannes Pap Szathmári) (concerning the ideas of duration and eternity), and the *Dissertationum physiologicarum pars tertia, in qua, existentia Dei ex existentia nostri, seu e stupenda mentis corporisque nostri fabrica, functionumque utriusque natura metaphysice adstruitur, breviter, deducitur* (1693). In the latter disputation, the author sets a natural-theological argument, based on apparent spontaneous motions (theses 31–32), and restates Descartes's a posteriori argument based on the innate idea of God (especially thesis 40): however, he does not consider the nature of causation.

[237] For instance, in the hand of another student of De Volder, viz. Theophilus Copius, the problem of the nature of God came to be treated in an overt anti-Spinozist way. In his *Disputatio philosophica de mentis humanae et divinae existentia et reale a corpore distinctione*, which he authored and defended in 1699 (*praeses* and dedicatee De Volder), Copius followed a 'Volderian' path, as he relied on the tenets (of Cartesian derivation) that all our knowledge comes from ideas only (thesis 4), that mind and body are two different natures or attributes, whose ideas do not involve the idea of each other (thesis 6). Moreover, he formalized the principle of causality as follows: 1) as much as the difference between the causes, as much the difference between the effects, so that the effects of completely different causes have nothing in common; 2) a cause cannot produce an effect more perfect than the cause itself (thesis 17) – using such principle to argue (with an a posteriori argument) that the idea of God cannot have been produced but by God Himself (thesis 20), Who is immaterial, as conceiving Him as extended would hinder His perfection (thesis 21). Yet, even though criticizing those "who admit only one substance" ("qui unicam tantum admittunt substantiam") as holding a contradictory position (thesis 10), he claims, that it can anyway possess extension *formaliter*, viz. he did not come to embrace Spinoza's principle that completely different beings cannot act on each other.

[238] "Vos autem Illustriss. Curatores ac eorum collegae Consules Amplissimi favore haud vulgari, me ornare ac pariter in hanc transferre, Academiam voluistis; cum aditus praeclusus videretur,

positions himself in the path of De Volder, who just preferred to avoid quarrels during his life. With this, however, Wittich did not give a judgment different from that of Andala, viz. he saw in De Volder's positions the potential matter of a quarrel. Le Clerc's judgment, on the other hand, just indicated De Volder's (declared) discontent with Descartes's metaphysics in his later days (while he had supported it in previous years).[239] Anyway, Le Clerc himself points out that after the forced resignation of Heidanus, De Volder taught Cartesian philosophy "with a bit more precaution,"[240] so that all the judgments more or less converge towards the 'heterodox' character of De Volder's positions – who, as seen in Sect. 2.3.5, *De Volder's views on politics and religion*, cannot be labelled but a true believer. In the hands of Andala, Johannes Regius, Leydekker, the Roëlls, and also as a consequence of Gronovius's summary of De Volder's metaphysics – this fame (already established around the turn of the century, as testified to by the Stolle-Hallmann report) increased in the early eighteenth century. So that in 1727 Joachim Lange (theologian at Halle), who devoted propositions 8 and 10 of his *De causa Dei* (1723) to demonstrate that Cartesianism was not a cause but rather the occasion of atheism,[241] of which Spinozism properly consists, labelled De Volder as the "little flower plucked from Spinoza's garden" in the second edition of this work (1727), on the basis of Gronovius's statement – and reported in turn by Poiret.[242]

---

spesque omnis decollaverit ob malignorum obtrectationes toti orbi litterario notas, hic vero non commemorandas, non enim lubet renovare dolorem; videtis itaque per mirabiles Dei vias; hominem in hoc collocatum suggestu, e quo haud multis ab hinc annis audivistis saepius doctissime magno cum vestro plausu dicentem Cl. Volderum virum magnum, iudicii acerrimi ac ingenii perspicacissimi, in omni philosophiae parte versatissimum, in mathesi subtilissimum, et quod valde rarum in metaphysicis solidissimum, virum non minus doctum, quam virtutis studiosum, qui me, quod gloriae et honori duco, amavit, quem ut vivum omni observantiae genere colui, ita manes eius in perpetua apud me erunt veneratione. Utinam viri huius nervoso et gravi dicendi genere. [...] Persuasus sum, Acutissimum Volderum egregia in metaphysicis praestare potuisse, nec detrectaturum lucem, nisi molestias, Vir si quis alius quietis et tranquillitatis amantissimus, ab hominibus malevolis, ac in hac scientia parum versatis creandas evitare maluisset. Quis tam audax ac temerarius! Qui asseverare ausit de mente infinita et perfectissima nullas novas veritates detegi posse!" Wittich 1718b, 2–3 and 15.

[239] "[I]l il explica en public non seulement la Physique, mais encore la Métaphysique de Descartes. Il en étoit en ce tems-là beaucoup plus grand admirateur, qu'il ne l'a été sur la fin de sa vie, qu'il en avoit reconnu le foible; comme il l'a avoüé plusieurs fois à ses Amis, & ainsi que je lui ai ouï dire moi-même. [...] Je l'ai ouï se moquer [...] des Méditations de Descartes, quoiqu'il les eût expliquées pendant long-tems," Le Clerc 1709, 374 and 398.

[240] "Mr. de Volder, à qui l'on fit entendre que ce n'étoit pas à lui, à qui l'on en vouloit, continua d'enseigner son Cartesianisme, mais avec un peu plus de précaution," Le Clerc 1709, 373.

[241] Lange 1723a, 47–50 and 54–69.

[242] "Non autem ignotum est prudentioribus, iam ante nostram aetatem pseudophilosophos Spinozianos, certe Spinozizantes, ex hoc ipso malo mali corvi, Aristotelis, ovo, seu principio de essentiarum, tantum intellectu Dei repreaesentatarum, et ab efficientia voluntatis divinae seiunctarum, aeternitate et necessitate fatum suum fatuum derivasse. Et hinc speciatim profluxisse illum Volderi flosculum, in Spinoziano horto decerptum: nullum liberum arbitrium in mente, nullus casus in corpore. Vide apud Petrum Poiretum in Veris et cognitis omnium primis p. 282 seqq.," Lange 1727, 47. I owe this reference and translation to Israeli 2001, 547; see also Lodge 2005. On Lange's use of Poiret's positions on De Volder, see also *supra*, n. 97.

3.2 De Volder on Human Knowledge

We can thus far summarize De Volder's alleged 'Spinozism' in two key points: (1) the impossibility of explaining the interaction of soul and body by rational means, which led his interpreters – notwithstanding at some point he stood with Leibniz's theory of pre-established harmony – to ascribe to him Spinoza's idea that soul and body do not interact, but they are rather the same substance, as well as to criticize his attempt to solve this problem by rational means (in the case of Andala's accusations). (2) The idea of absolutely necessary causation, exemplified by his *Contra atheos*, where efficient causation is reduced to the geometrical relation between attributes and their properties, and extended to God. To these main issues, we need to add what Klever and Lodge have labelled 'epistemological naturalism'. Some insight as to this aspect of De Volder's metaphysics has been provided in dealing with De Volder's demonstrations of the existence of God. Namely, as seen above, De Volder developed his metaphysics by considering the features of ideas only. Now, I will consider the core part of his 'epistemological naturalism', namely, the idea that for De Volder "it is impossible to have a clear understanding of something and not to adhere to it," (as Klever has put it)[243] and that the demonstration of the existence of God does not serve to allow for the reliability of our knowledge. As Lodge summarizes it: "De Volder eschews any reliance on revelation as a source of truth or any reliance on God as a guarantor for our naturally acquired beliefs."[244]

3.2.1 De Volder on Clarity and Distinctness

For De Volder – as for Descartes and Spinoza[245] – clarity and distinctness are the mark, norm or criterion of the truth (*criterium veri*). In a nutshell, a clear and distinct idea represents something that can exist, *formaliter*, outside the idea itself (regardless of the propositional viz. complex, or simple nature of the idea itself). De Volder was so constant to this principle that Huygens, in September 1691, commenting upon De Volder's *Exercitationes* against Huet, wrote in the minute of a letter to De Volder that "I am not at all in favour of the criterion of Descartes. Because in geometry itself, one often imagines understanding very clearly things which are false."[246] In fact, De Volder himself had noted, in his *dictata*, how Joseph

243 Klever 1988, 196. Cited and discussed in Lodge 2005, 125.

244 Lodge 2005, 121.

245 See Curley 2001; Della Rocca 2008, chapter 3; Schneider 2016; Garrett 2018, chapter 1.

246 "[J]e ne suis pas tout a fait pour le Criterium de des Cartes. Parce que dans la geometrie mesme on s'imagine souvent de comprendre tres clairement des choses qui sont fausses. Il y reste donc tousjours a scavoir si l'on a compris clairement et distinctement, ce que est assez douteux dans des longues demonstrations. Et de la naissent les paralogismes. Je serais donc plus pour les divers degrez de vraisemblance, laquelle dans certaines rencontres est si grande que d'estre quelque fois comme 100000000000 et plus contre un, que le vray ou le faux d'une proposition, et qu'en de

Justus Scaliger and Gassendi had deceived themselves in geometry, namely, as they believed, respectively, to have found a method for the quadrature of the circle, and that any converging lines at some point intersect each other. The latter case, in fact, is not valid for the asymptotes. Gassendi used the latter case to criticize Descartes's *criterium veri* in his *Disquisitio metaphysica* (1644).[247] Also Huet, in his *Censura*, used the same example for the same purpose.[248] In his *Exercitatio de Deo*, in turn De Volder objected to Huet that both the proposition that some converging lines intersect, and some others do not intersect are clear and distinct, and thus true. Namely, De Volder interpreted Huet's objection as concerning only limited sets of lines – thus prompting Huygens's reaction.[249] More generally, De Volder – like, for instance,

certaines choses cela va comme à l'infini," Huygens to De Volder (letter 2701a), September 1691, in Huygens 1888–1950, volume 10, 739. In a letter to Gerhard Meier of 16 November 1691, Huygens had however approved De Volder's refutation of Huet: "Huetii Censuram legi cum primum prodijt ab ipso Authore mihi missam, in qua non pauca mihi probari memini, sed et aliqua notavi quibus responderi posset. Quod negotium et vestrates aliquot et Volderus noster sibi sumserunt. At ille parvi facere haec omnia videtur, prout ex nuperis eius ad me literis intelligo. Ego vero nihil nisi Volderi theses legi, quae non ita contemnendae videntur," Huygens to Gerhard Meier (letter 2711), 16 November 1691, in Huygens 1888–1950, volume 10, 195–196.

[247] See De Volder's commentary on Descartes's *Principia*, I.5: "[s]i quis obiceret scepticorum argumenta eius quae certa {…} testimonia profert, e[ss]e satis efficacia, verum alia dari, q[uae] sunt tam manifesta ac duo et tria sunt 5, nec haec ab illo labefactari po[ss]e: producet scepticus exempla eorum, qui in iis, q[uae] evidentissima iis videbantur, errarunt: ut ipse Gassendus, summus ille mathematicus, qui nil profitebatur se distinctius concipisse, q[ua]m 2 lineas in infinitum productas, et ad se accedentes, se invicem tangere debere, cum tamen id falsum e[ss]e, ipse postea cum geometris fateri coactus fuerit: cumque sic ingeniosissimus in hoc, Scaliger cum multis mathematicis in eo erraverint, ut vellent probare quadratura circuli, cur inquiunt, idem tibi non potest contingere?" Hamburg 273, 5. Cf. his commentary on Descartes's *Meditationes*, in Warsaw *dictata*, 3v. Cf. Gassendi 1644, 100–102. See LoLordo 2005, 2018. On Scaliger, who provided a demonstration of the quadrature of the circle in his *Cyclometrica elementa* (1594), see Hogendijk 2010.

[248] See Huet 1689, 123–124.

[249] "Quod autem exemplum attinet, quo huius propositionis falsitatem comprobat Vir Ill. de eo quaerere ex Ill. Vir. lubentissime velim, an assere ausit se clare et distincte percipere ad qualescumque duas lineas in eadem area plana iacentes, et quo longius producuntur, eo propius ad se mutuo accedentes pertinere, ut concurrant aliquando. Si hoc ita generaliter affirmet, iam concedere necesse habebit, se clare et distincte concipere contradictoria. In hyperbola enim et eius asymptoto ita se rem non habere, cum id demonstratum agnoscat ab Apollonio se clare et distincte percipere non inficiabitur opinor. Clare igitur et distincte percipiet omnes lineas in area plana iacentes, et ad se mutuo accedentes, aliquando concursuras, et tamen etiam clare et distincte concipiet hyperbolam cum sua asymptoto, etsi sint in eodem plano, et ad se mutuo accedant concursuras nunquam. Quod si hoc ita generaliter affirmare non ausit, si fateatur id solummodo verum esse de illis lineis, quas nunc concipit, (hoc enim innuere mihi videntur verba, quarum ideam in mente habeo,) non video, quid hoc exemplum faciat ad propositionem illam arguendam falsi. Nihil enim aliud erit, quam clare percipio dari quasdam lineas, rectas ex. gr. quae si eodem plano ad se invicem accedant concurrant necessario, et tamen clare percipio, quasdam esse alias, quae licet sint in eodem plano, licet ad se invicem accedant, non sunt tamen concursurae, utrumque tamen id, quod percipio, de ipsis illis lineis, de quibus hoc percipio, verissimum est, adeoque comprobat hoc exemplum veritatem potius propositionis, quam ut eam falsi arguat," De Volder 1695, *De Deo*, thesis 15.

Arnold Geulincx[250] – does not reject the 'psychological' or 'provisional' character of clarity and distinctness, but finds in mathematics an instrument to exercise the mind in their appraisal. So, even if he maintains that the clarity and distinctness of concepts as a *criterium veri* cannot be taught, as it is *per se nota*,[251] he argues, in his *dictata* and in his 1682 *Oratio de coniungendis philosophicis et mathematicis disciplinis*, that anyway it can and has to be exercised by mathematics, working better than any exercise in logic.[252] Moreover, regardless of the 'psychological' character clarity and distinctness might have, according De Volder there are no other 'marks' for the truth, and, as he insists in many places that we cannot but assent to some knowledge, and consider such knowledge as true. This, as I am going to show, has consequences both for his account of free will, as well as for his idea of certainty, which diverges much from Descartes's.

3.2.1.1 Clarity and Distinctness as a Pre-Condition in Natural Philosophy

According to De Volder, clarity and distinctness are the foremost criteria in establishing the truth of an idea or a proposition. Per se, however, they do not grant that a given idea is true in the sense that it represents something actually existing outside itself. They grant, in fact, that such a thing can exist, viz. that God can create it. As seen above, a way to justify this possibility is the very fact that we not have creative

[250] See Sect. 3.2.2.1.3, *Geulincx's hypothetical physics*.

[251] See De Volder's commentary on *Principia* I.30 and I.45 in Hamburg 273, 25–26 and 33–34, as well as the commentary on I.6 in Hamburg 274, 5–6.

[252] "In quarum numero cum certa, incertave scientia, clara et distincta perceptio, merito censeantur, eadem opera regularum quarundum adminiculo certum ab incerto dignoscere docebit dialecticus harum assertionum imperitum, ac coecum docere poterit colorum diversitatem. Hisce namque in omnibus sola valet experientia, sola evidentis perceptionis conscientia, quae ut absque demonstrationum firmissimarum comprehensione acquiri, sic absque frequenti in iisdem percipiendis exercitio consumari nullo modo potest. Vel unica profecto Euclidea demonstratio recte percepta multo aperiet evidentius quid sit demonstrare, quid clare et distincte percipere, quam si cum dialecticis omnes syllogismorum formulas enumeraverim, ab iis audiverim, principia demonstrationum immediata esse oportere, non heterogenea, neque communia, et quae plura id genus ad fastidium usque ab illis inculcari solent. Sed agamus liberaliter, demus id quod volunt, fateamur vera, imo et usui nostro ac commodata esse ea quae docent, non video tamen eadem illa, usus si desit, multum momenti habitura. Praecepta siquidem vel optima, nisi blando exercitii calore foecundentur, otiosa sunt, nec spei respondebunt nostrae. Quapropter id nihilominus restabit discriminis inter dialecticum et geometram, quod est inter illum, qui musices quidem praecepta callet, neque tamen aut vocem cantui aut manus chordis unquam applicuit, et eum qui ipso et canendi et chordas pulsandi exercitio simul cum ipso habitu regulas addidicerit," De Volder 1682, 11–12 (unnumbered). See Klever 1988, 209. See also his *dictata* on *Principia* I.43, in Hamburg 274, 17. Also in his *Oratio in obitum Siberti Coeman* (1679), De Volder maintains that mathematics serves better than logic in exercising the intellect, viz. in the acknowledgement of clarity and distinctness as *norma veri*: De Volder 1679, 8–9 (unnumbered); De Volder's positions are discussed by Le Clerc, who stresses De Volder's relinquishment of the old logic, praises the Port-Royal logic (not mentioned by De Volder himself), and claims that a new logic would help to correct mathematics itself, being however not applicable to politics and morals: see Le Clerc 1709, 374–376. Such a differentiation between logic and mathematics is mentioned also in De Volder 1698, 6.

power over simple ideas. Therefore, they postulate a different cause than ourselves, and this points out for their being capable of representing something external to them, even if no demonstration is provided for this. A similar attitude by De Volder can be found in his *De rerum naturalium principiis* – a series of 15 disputations authored and presided over by De Volder between January 1674 and June 1676, where, on the one hand, De Volder labels clarity and distinctness as the first criterion to adopt some ideas as explanatory principles or first causes in natural philosophy.[253] On the other, he assesses that clarity and distinctness are necessary, but not sufficient conditions to assess that such ideas or principles are actually at work in the physical domain as the causes of phenomena. Clear and distinct perception, actually, can be compared to the grasping of the meaning of a sentence: it is necessary but not sufficient to assess the truth of such sentence, viz. that it describes an external situation.[254] In the case of natural philosophy, De Volder considers the (Cartesian) ideas of matter and motion[255] as clear and distinct; however, one cannot demonstrate – by clear and distinct arguments – that God chose them as natural causes of phenomena. At the same time, one cannot doubt that He chose them: this being, as I am going to show, the meta-criterion for accepting them in natural philosophy.

3.2.1.2 Clarity, Distinctness and Indubitability of Simple and Complex Ideas

In order to understand how De Volder provided a physics whose truth could not be demonstrated, and by extension to understand what were his views on the relations of physics and metaphysics (as this demonstration would pertain to metaphysics rather than to metaphysics itself)[256] it is necessary to look further at his justification of clarity and distinctness as norms for the truth. Despite its purely mental characterization, clarity and distinctness are the only criteria to judge phenomena; their 'uniqueness' has its ground, as maintained since De Volder's inaugural disputation *De natura* (1664),[257] in its being indubitable. Namely, we clearly and distinctly per-

[253] "Ut enim phaenomenum quodpiam explanem, nonne requiritur, ut eius causas ostendam? Quae aut primae erunt, et a nulla alia corporea causa dependentes, aut erunt aliarum causarum effecta. Priori in casu quid est manifestius, quam me ipsa demonstrasse principia? Sin vero aliarum causarum effecta sint, quis non videt [...] unquam huius phaenomeni claram distinctamque [...] notionem acquiri posse, nisi huius causae iterum cognoscam causas, idque donec ad primas causas, sive ad ipsa rerum principia devenerim," De Volder 1681, thesis 1.

[254] "Hactenus non requiro, ut demonstrentur, non ut certo vera esse ostendantur, sed illud tantum exigo, ut percipiantur, ut quae et qualia sint cognoscatur. Quae sane duo non parum differunt. Aliud quippe est percipere huius illiusve effati sensum, aliud eius veritatem cognoscere, de qua nisi prius illud percipiatur, constare nemini potest," De Volder 1681, thesis 12.

[255] See Sect. 3.2.2.2, *De Volder on the 'truth' of Cartesian principles*.

[256] See Sect. 3.2.2.2, *De Volder on the 'truth' of Cartesian principles*.

[257] In his inaugural disputation in medicine De Volder stressed the importance of a 'quest for principles' (*quaestio de principiis*) as the crucial means for the development of sciences. Clearly referring to Descartes's revolution in philosophy, De Volder emphasizes that as doubt started to be

ceive something whenever we cannot but assent to its being capable of representing something outside ideas themselves. The foremost case being that of mathematical and metaphysical principles and demonstrations, but also of what De Volder would characterize as non-propositional viz. simple ideas, not including any copulation or resolution. As a matter of fact, De Volder does not provide a systematic theory of judgments. For instance, in his *dictata* on Descartes's *Principia*, as he comments upon those articles concerning the relation of will and understanding (especially I.32–34), he does not systematically distinguish between the affirmation, negation, and the very assent we give to the affirmation or negation concerning further ideas, viz. to their copulation or resolution in judgments.[258] Even if he from time to time does concede the possibility of assenting to propositional contents, for example his discussion (commenting on I.39) of our assenting to mathematical judgments such as 3 + 2 = 5 (in which we affirm that 3 and 2 are 5),[259] he is not clear on how an idea can be perceived as complex or as simple. For instance, he claims that an axiom can be perceived as a complex judgment (*effatum*) or by a single act of perception (*uno intuito*) – such as in the case of the *cogito*.[260] Or, in other words, De Volder is not clear on how one can perceive a complex idea (viz. a judgment or proposition) without having his will connected to its simple, non-propositional ideas or terms. As De Volder sticks to Descartes's theory of judgment as entailing a voluntary act in his *dictata*, it seems that for De Volder any kind of copulation or resolution of ideas entails an intervention of the will, even if he sets forth a distinction between understanding and affirming or negating what is understood.[261]

Insofar as this concerns the theory of the truth of judgments, namely of the relation of what he labels simple ideas, we find this in the disputation *De certitudine clarae et distinctae perceptionis* (1689, *praeses* De Volder, author and *respondens* Gaspar vander Tak).[262] In it, the author appeals, on the one hand, to the immediate

systematically applied to philosophy, allowing a re-foundation of thought, as well as the discovery of more reliable physical principles, natural philosophy underwent considerable progress: see De Volder 1664, thesis 1.

[258] See Hamburg 273, 26–29, Hamburg 274, 15. On early modern theories of simplex and complex ideas, as well as proposition and judgment, see Nuchelmans 1973, 1998; Schuurman 2004.

[259] See Hamburg 273, 30; Hamburg 274, 14.

[260] In commenting upon I.50 in his Hamburg 274 *dictata*, De Volder points out that axioms are of two kinds: either being understood *uno intuito*, or as *effata*, i.e. as more complex ideas: see Hamburg 274, 19. For an introduction to the treatment of the *cogito*, see Hintikka 1962; Markie 1992; Ariew 1999, chapter 10; Sarkar 2003; Lindner 2011.

[261] See especially his commentaries on I.32: Hamburg 273, 26; Hamburg 274, 15.

[262] Klever also draws attention to De Volder's 1679 *Oratio funebris in obitum Siberti Coeman* and to his 1682 *Oratio de coniungendis philosophicis et mathematicis disciplinis* as evidence of De Volder's epistemic naturalism. In both texts, De Volders stressed the self-evidence of mathematics as a norm of the truth: see Klever 1988, 202–203 and 207–210. As argued by Lodge, however, "further evidence can be found in writings that Klever does not mention," Lodge 2005, 121. Please note that as early as 1673 De Volder had presided over and authored a *Disputatio philosophica de criterio veritatis prima*, in which the positions of the ancient sceptics are presented (theses 4–10), and mathematics is assumed as the paradigm of indubitable knowledge (thesis 13): however, the second part of the disputation has not been recovered.

awareness we have of the clarity and distinctness of ideas, this awareness being the first criterion to ascertain it.[263] On the other, he defines clarity and distinctness in terms of the compulsion to the assent. This compulsion concerns the perception of the nexus of several ideas, that is, propositions, as in the case of mathematical and metaphysical principles like "the whole is bigger than its part," entailing the absence of any obscurity in perceiving each idea and their connection or copulation.[264] Thus, clarity and distinctness as *norma veritatis* are grounded in our lack of freedom in assenting or not to such propositions. In fact, for De Volder we cannot doubt that an external force is not deceiving us, as we cannot frankly work this hypothesis out.[265]

In conformity with this theory, in his subsequent *Exercitationes*, viz. in the *Exercitatio de clara et distincta perceptione criterio veri* (which, in part, restates the contents of the aforementioned *De clara et distincta perceptione*),[266] De Volder more systematically defines truth first as *veritas rei*, or the very being of things as what they actually are.[267] Second, he characterizes truth in representative terms, namely, the very possibility that ideas which do not involve any affirmation or negation (or, as he puts it in his *dictata*, simple ideas)[268] have of representing something different from themselves. Namely, when their *obiectum* can have a formal existence outside the idea that represents it. Ideas of this kind are always true – since they represent something – but can be different according to their clarity and dis-

[263] "Quamvis ea sit mentis nostrae natura, ut suas operationes prae caeteris rebus clare et intime cognoscat, cum omnis cogitatio conscientiam sui involvat; attamen verba idonea satis ad aliis indicandum, quae et qualesnam sint, vix excogitare potest," De Volder and Vander Tak 1689, thesis 1.

[264] "Ut ipsas voces aliquo modo determinemus, claram et distinctam perceptionem habere dicimur, ubi aut unam eandemque ideam tam evidenter percipimus, ut nullam ignorantiam cum illa commixtam cognoscamus; aut idearum nexum et relationem ad se invicem, absque ulla confusione cum aliis ideis, tanta cum claritate et evidentia, mentis acie intuemur, aut non possimus, quin assentiamur cum plena voluntatis nostrae lubentia; prout ex. gr. (quia non de unius ideae, sed de idearum evidentia, impraesens agemus) intelligimus, totum esse maius sua parte, factum infectum reddi non posse, et eiusmodi sexcenta; unicuique enim attendenti fit manifestum, se a talium veritati, assensu iudicium suum abstinere non posse; ex quo sequitur cum nullo modo errare; quod hocce exercitio serio et modeste paucis defendere aggredimur," De Volder and Vander Tak 1689, thesis 1.

[265] "Hanc itaque veritatem edocemur primo ipsa experientia; ubi enim clara distincta alicuius rei perceptio adest, tam plene de veritate rei perceptae convincimur, ut certi simus, nos errare ne per ullam quidem potentiam posse; quod si fieri posset, iam semper aliquis nobis remaneret scrupulus, an non falleremur; quotquot vero sumus, experimur, nos de talibus veritatibus praesenti illa clara et distincta perceptione, quicquid etiam moliamur, dubitare non posse: ut cum cogito totum esse maius sua parte, et c.," De Volder and Vander Tak 1689, thesis 2.

[266] He remarks that we deal only with ideas, and that we rely on their properties – like clarity and distinctness – to assess their truth: see De Volder 1695, *De clara et distincta perceptione*, theses 2–4, 7–8.

[267] De Volder 1695, *De clara et distincta perceptione criterio veri*, thesis 11. This is the traditional correspondence of truth and being.

[268] In the Hamburg 274 series of his *dictata*, in commenting upon I.30, De Volder stresses that truth does not concern only affirmation or negation, but also simple ideas: Hamburg 274, 14

tinctness.[269] Non-propositional ideas, therefore, are false whenever they cannot represent anything at all apart from themselves, or whenever they can represent just an affection of the mind, whereas they are believed to represent something else, or vice versa. This is the case of obscure and confused ideas.[270]

Thus far, we may summarize De Volder's idea of truth as concerning simple ideas, not explicitly involving any copulation or resolution of concepts (such as the bare idea of a triangle), and relations between ideas (such as the propositional idea that the sum of its internal angles is 180°). Their truth consists of their being capable of representing something different from themselves. In turn, it appears, especially from his *dictata*, that both the assent we give to their truth (viz. the belief that an angle can exist outside mind, and that the sum of its internal angles is 180°), as well as the very copulation and resolution of their terms, viz. their affirmation or negation, seem to entail the intervention of will. Or, at least, De Volder is not clear on when we can perceive a propositional idea (like $2 + 3 = 5$) without voluntarily connecting 2 to 3 and to 5. In fact, there is evidence that, even if nominally distinct, for De Volder any act of understanding involves the intervention of will – *à la* Spinoza. We may appreciate this by looking again at his idea of indubitability. In his *Exercitatio de dubitatione universali*, besides providing a standard Cartesian refoundation of knowledge on radical doubt, De Volder stresses how we cannot refuse to assent to metaphysical propositions, such as the principle already mentioned in the *De certitudine clarae et distinctae perceptionis*,[271] the axiom *cogito ergo sum*,

---

[269] "Conceptus autem quod spectat, in illis omnibus verum est, ita me affici, me hoc cogitare, me hos illosve conceptus inter se coniungere, disiungereve, haec enim omnia aeque vera ac certa sunt, ac certum est me cogitare [...]. Manifestum equidem est [...] cogitationes [...] plures reperiri alias [...] quae praeter illam mentis meae affectionem, aliquis mihi repraesentant, quod ab ipsa mentis meae affectione, ipsa cogitatione mea plane concipitur distinctum. [...] Hoc quod ita mihi repraesentatur, obiectum huius mei conceptus dico, hoc rem reive modum voco, ipsam vero cogitationem huius ipsius, quod repraesentatur sive rei, sive modi, ideam voco [...]. In hisce autem rerum ideis hanc manifeste invenio discrepantiam, quod quaedam harum maximam et perspicuitatem et evidentiam sibi habent coniunctam, quas claras distinctasque vocat Cartesius, quaedam vero [...] confusas vel obscuras dicit. [...] In hisce autem conceptibus, utpote qui nullam affirmationem aut negationem continent, cum sola repraesentatio locum habeat, et semper verum sit mihi hoc illudve repraesentari, huius enim conscius sum, non video, quid in illis falsum dici queat," De Volder 1695, *De clara et distincta perceptione criterio veri*, thesis 12.

[270] "Si tamen has veras, illas falsas, ut in vulgari usu est, dicere velimus, non video quaenam falsae dici queant, nisi vel quae videntur aliquid menti repraesentare, cum nihil offerant, vel etiam quae affectionem mentis repraesentant tanquam quid ab illa affectione diversum, et viceversa. Qua ratione si ideas in veras falsasque distinxero, clarae nihilominus et distinctae perceptiones omnes certissime erunt verae," De Volder 1695, *De clara et distincta perceptione criterio veri*, thesis 13.

[271] "Progrediatur igitur noster hac methodo, ut primo generatim accuratius dispicienda sibi proponat omnia, ipsa etiam axiomata. Haec enim dum fundamenta futura sunt omnis ulterioris ratiocinii, cavendum summopere, ne in illis ulla fallaciae superesse queat suspicio. Quod dum agit, dum ad ipsa axiomata attendit, dum sibi horum aliquod ob oculos ponit, totum ex. gr. maius esse sua parte, experitur statim hoc inter ea esse, de quibus, quantumcunque etiam dubitationi indulgere studeat, dubitare vel minimum non est in ipsius potestate," De Volder 1695, *De dubitatione universali*, thesis 21.

and metaphysical truths.[272] De Volder moves once again within the boundaries of Cartesianism. The idea that clear and distinct connections of ideas cannot be arbitrarily doubted, viz. that we are forced to assent to them is present in Descartes's metaphysics. As Descartes put it in his *Meditatio quinta*, "I am of such a nature that as long as I understand anything very clearly and distinctly, I am naturally impelled [*non possim non credere verum esse*] to believe it to be true."[273] However, De Volder stresses much more than Descartes that our assent to some propositions is forced. According to Descartes, "I am [...] of such a nature that I cannot have my mind constantly fixed on the same object in order to perceive it clearly, [...] it may happen meanwhile that other reasons present themselves to me, which would easily cause me to change my opinion,"[274] viz. one can doubt on past clear and distinct perception, and in this consists human freedom, as "it is not an imperfection in God that He has given me the liberty to give or withhold my assent from certain things as to which He has not placed a clear and distinct knowledge in my understanding."[275] In other words, regardless of the reality of radical doubt as this is applied to any knowledge by Descartes until the formulation of the cogito,[276] it is out of the question that for Descartes one can doubt past clear and distinct perception at any moment.

In turn, De Volder is much more overt than Descartes on our lack of freedom in assenting to *any* notion which we find clear and distinct, either past or present. In the 1689 *De certitudine clarae et distinctae perceptionis* Descartes's ultimate reason to demonstrate the existence of God, namely, to guarantee the reliability of past evident perceptions,[277] is discarded. In the disputation, it is argued that clarity and distinctness as *norma veritatis* of mathematical truths are so compelling that even past perceptions – contrary to Descartes – cannot be put in doubt.[278] Or, as he declares in commenting upon the first article of Descartes's *Principia* in his *dictata*, as well as in his *Exercitatio de dubitatione universali*, one cannot doubt Euclid's

[272] "Atqui tu nescis, an non fallaris in evidentibus: verum est; sed etsi hoc generatim nesciam, an hoc fieri nequeat, scio tamen hoc in casu me non falli, et experior illam generalem rationem me non posse abducere, quin his effatis, cogito, ergo sum; totum est maius sua parte, et c. absque ulla haesitatione assentiar. [...] Sed concedat tamen nobis, obsecro, si talis quis inter homines reperiatur [...] ut ex ea, quam in nobis experimur mentis nostrae constitutione argumentemur, et ea pro certis habeamus quae nobis certa sunt, et de quibus, quidquid sit de fide sensuum, de discrimine inter somnum et vigiliam, de genio deceptore, et si quid porro est, quod ad scientiam nostram labefactandam potest adferri, conscius ego mihi sum, me ita certum esse, ut quidquid agam, quidquid moliar, ut dubitem, dubitare tamen non possim," De Volder 1695, *De dubitatione universali*, thesis 24. De Volder stresses the fact that the *cogito* is an axiom, and not an inference or *consequentia*, in his *dictata* on Descartes's *Principia*, on I.6: see Hamburg 274, 5–6.

[273] Descartes 1993, 84–85; cf. AT VII, 69.

[274] Descartes 1993, 85.

[275] Descartes 1993, 78.

[276] As to this issue, see the entries mentioned *supra*, n. 260.

[277] See Descartes's fourth set of replies, appended his *Meditationes*: AT VII, 246–247.

[278] "Quia quis aliquando, absente illa clara perceptione, dubius haerere posset, [...] ideo hoc argumentum, adducitur [...] illud omne, quod aut unquam evidenter percipimus, aut in posterum sic percepturi, sumus, certum et inconcussum est," De Volder and Vander Tak 1689, thesis 6.

demonstrations, unless one has not read them.[279] Given his stressing that indubitable truths force our assent to them, De Volder also downplays the role of God as a guarantor of the truth of our clear and distinct perceptions, and, more generally, the role of the idea of God in his metaphysics. As he puts it in his *De dubitatione*, the idea of the power of God can by no means be used to justify or doubt propositions in mathematics and metaphysics.[280] In fact, in De Volder's metaphysics the idea of God serves to grant that what we conceive clearly and distinctly can be created as a nature besides as an *obiectum*. Insofar as God exists (as De Volder puts it in his *Exercitatio de clara et distincta perceptione*), He can create whatever we evidently conceive as possible.[281]

To sum up, De Volder assumed clarity and distinctness as the criteria of the truth largely independently of the existence of God, which works as the ontological backbone rather than as an epistemic 'warranty', as it was for Descartes. In turn, the idea that clarity and distinctness force the will, in fact, corroborates Gronovius's statement that for De Volder "there is no free decision in the mind." If we combine De Volder's theory of 'forced assent' with his explanation of the succession of thoughts as following from the idea of a clock, we may, in fact, conclude that the succession of our thoughts, insofar as they concern ideas of a body, is mechanical such as the body is. Therefore, such ideas would be clear and distinct and our assent to them would be unavoidable. Our freedom would consist, at most, in our possibility to doubt unclear ideas. In fact, this idea is not ultimately different from the Spinozist ideal of freedom as consisting in following clear and distinct perceptions only, in which, however, no difference between will and understanding can be drawn.

This is more evident from De Volder's *dictata*: as he comments in Hamburg 273 on *Principia* I.39, viz. on Descartes's tenet that "it is so manifest that there is freedom in our will, and that we have complete power to either assent or not assent to many things, that this must be numbered among the first and most common notions innate in us,"[282] De Volder agrees with Descartes by claiming that in the case of mathematical propositions like 2 + 3 = 5, man is not forced by any external cause in assenting to its truth, and that such freedom does not consist in any indifference in judgment. However, we are forced by the clarity and distinctness of our perception

[279] See Hamburg 274, 3. See also his commentary on the *cogito*, viz. on I.43: Hamburg 274, 17. See, moreover, De Volder 1695, *De dubitatione universali*, thesis 20.

[280] "Cum cogito duo et tria facere quinque conscius mihi sum, huius evidentiae, quae in mente est, hunc esse effectum, ut plane hac de rei veritate certus fiam, ut sciam, quidquid sit de illa potentia Dei, utut illa me forte in evidentissimis aliquando possit decipere, illam tamen hoc in casu me decipere nequam," De Volder 1695, *De dubitatione universali*, thesis 25.

[281] "Erit fortasse triangulum sola mentis meae affectio? Nihil minus. Quod enim mihi repraesentatur nihil habet cum cogitatione commune. Sed forte nullum dabitur extra me triangulum, nec hoc affirmo, non dico dari, dico concipi. Quid ergo? [...] Siquidem ego, qui iam novi Deum esse, et illum omnipotentem et omniscium esse, facile etiam novi omnia illa, quae hoc modo a me percipiuntur a Deo etiam percipi, illum omnia ea posse, quae possum concipere," De Volder 1695, *De clara et distincta perceptione criterio veri*, thesis 13.

[282] Descartes 1982, 18.

in our assenting to it.[283] Accordingly, as De Volder attempts to solve the issue of the consistency of human freedom with divine omniscience (I.40–41), he claims that even if it may appear paradoxical, the more necessary, the more freely we act. Namely, as we act following reason, the less indifferently we act, and the more forced our will is in assenting to certain mental contents: this being the case of using stairs or jumping in order to descend from a tower, and of assenting to the mathematical truth that 2 + 3 = 5.[284] Coupled with his idea of the forced assent entailed by clear and distinct perception, we may see that De Volder's idea of freedom can be labelled, at most, as the possibility of suspending judgment whenever we not have, and – unlike Descartes – we did not ever have, a clear and distinct perception. However, insofar as we have or had a clear and distinct perception, no freedom of the will, viz. possibility of dissenting, can ever be conceived. This makes sense of Gronovius's summary of his metaphysics. Moreover, as I am going to show, for De Volder indubitability can characterize not only metaphysical and mathematical truths, which embody the paradigm of clear and distinct knowledge, but by which one cannot ascertain whether their truths have actual references outside themselves – viz. one cannot ascertain, through metaphysics and with metaphysical certainty, that some principles are at work in physical reality. In fact, this also characterizes physical knowledge as such, namely, the deduction of effects from causes qua proved by the effects themselves. Accordingly, De Volder extends much more than Descartes the scope of indubitable knowledge, and provides a way, other than metaphysics, to ground the truth of physics itself.

---

[283] "Vidimus nos ultimo errare, nec tamen Deum errorum nostrorum ca[us]a e[ss]e, sed in eo consistere, q[uo]d abutamur usu liberi arbitri nostri progreditur hunc author, ut id, q[uo]d hactenus assumpsit, nobis reddat perspicuum: id itaq[ue] agit, ut demonstret, an gaudeamus libertate arbitrii, an minus, nec putat id longa serie demonstrationum, sed ipsa conscientia probandum e[sse], ea, id enim asserit, nonne e[nim] dicimus 2 et 3 e[ss]e 5, quia id tam clarum est et perspicuum, ut non possit negari? Non tamen cogimur externa vi ad id fatendum, sed voluntarie i.e. libere {id} fatemur: voluntas [n]a[m] non consistit in indifferentia, sed assensu voluntario, cum e[nim] nescimus {q[uo]d} sit bonum {a[ut]} malum, non sumus indifferentes, sed tum cum nescimus {q[uo]d} assequendum sit. Itaque non vi externa, sed ductu consilii {aliq[uo]d} agimus, fugimus mala, cupimus bona, {q[ua]m} libertatem tam claram reperimus, ut 2 et 3 sunt 5," Hamburg 273, 30.

[284] "[C]onsiderandum est libertatem arbitrii cum necessitate convenire, imo liberrime nos agere, cum necesse agimus, quod paradoxum licet, verumtamen est, concipiamus nos o[mn]es res, de quibus deliberamus, certo cognoscere potest, certo demonstrare, a qua parte plus mali, a qua plus boni sit; nonne tunc erimus perfectiores quam nunc, cum non scimus, cumq[ue] hoc sit nullo in casu erimus indifferentes. R. cum esto in turri, et volo descendere, an me e turri proiiciam an vero descendam per scalas? Necesse est qui non videt hinc boni, illunc mali quid consecuturum, eritq[ue] quasi qui iudicare potuerit illud bonum et commodum esse? Sane non, consilii descendimus, non tamen possumus non descendere, si enim quis nescit an bonum hoc an illud malum sit, haerebit. Si enim noscet, duceretur ex statu coactionis in libertate, et voluntarie simul, cum e contra ut dicere non voluntarie ageret licet sit indifferens. Eodem modo non haereo in iudicando duo et tria esse quinque sed statim assentior!" Hamburg 273, 30. Cf. Hamburg 274, 16–17.

3.2.2 De Volder on the Relations of Physics and Metaphysics

As seen above, De Volder's idea of causality imposed a severe necessitarianism on the constitution of the world. In turn, in his *De rerum naturalium principiis* he labelled his natural-philosophical theories hypothetical, namely, once granted the clarity and distinctness of the principles of matter and motion, no demonstration of their truth is required or even possible on a metaphysical basis. This, however, does not amount to be a strategy to 'save' the freedom of God or the freedom pertaining to the order of created things. Plainly, it was the bare result of the impossibility scrutinizing the very decrees of God, and was a consequence of De Volder's 'down-play' of the role of God in granting the reliability of our knowledge. In other words, according to De Volder both the clarity and distinctness of some idea, as well as the benevolence of God (Who does not deceive us) do not grant the truth of physics, contrary, as I am going to show, to Descartes. In availing himself of the hypothetical character of physics, moreover, De Volder built upon a tradition which had in Descartes himself its first initiator, as I show next.

3.2.2.1 Physics and Metaphysics in the History of Cartesianism

3.2.2.1.1 Descartes's 'Hypothetical' Physics

In fact, De Volder's hypothetical physics was nothing new within the very history of Cartesianism. In his *Principia*, III.44, Descartes declares, in proposing his cosmology, that "I nevertheless wish those [causes] I am proposing here to be taken only as hypotheses," and that they are provided with a moral certainty, viz. they can be used in practice: "my hypothesis will be as useful to life as if it were true, {because we will be able to use it in the same way to dispose natural causes to produce the effects which we desire}." As he shows in III.45, the reason for the hypothetical character of his cosmology consists of its contradicting the Genesis account of the Creation, so that, he declares, "I shall even assume here some [principle] which it is certain are false."[285] At the end of his *Principia* (IV.205), thus, he concludes that "there are things which are held to be morally certain, that is, [certain] to a degree which suffices for the needs of everyday life; although if compared to the absolute power of God, they are uncertain."[286] Besides the recourse to a theological argument for the uncertainty of his principles, in this case Descartes sets forth also a philosophical argument, viz. that we can explain phenomena by assigning to them certain causes, as though we are deciphering an encrypted alphabet: even if we assign each symbol of this alphabet to a certain letter, we cannot exclude that its deviser has assigned them to different letters. Yet, as far as we fully understand each phenomena viz. the

[285] Descartes 1982, 105.

[286] Descartes 1982, 286–287.

meaning of the sentences, we can be content with such an explanation.[287] Moreover, on the ground of his demonstration of the existence of God, namely on the ground that "God is supremely good and by no means deceitful," Descartes claims that some demonstrations are "more than morally" certain:

> [s]uch are mathematical demonstrations; such is the knowledge that material things exist; and such are all evident demonstrations which are made concerning material things. These reasonings of ours will perhaps be included among the number of these absolutely certain things by those who consider how they have been deduced in a continuous series from the first and simplest principles of human knowledge.[288]

Descartes's physical theories, based on clear and distinct principles and following clear and distinct demonstrations, are therefore provided with the same certainty as mathematical demonstrations – even if Descartes is not entirely clear on the meaning of 'more than morally certain'.[289] Having provided his metaphysical foundation of physics, viz. the demonstration that God is not a deceiver whenever we have clear and distinct perception,[290] the hypothetical character of Descartes's explanations consists only in the understanding of how this or that part of matter has acquired its figure, movement, and so on, since the ways in which God disposed matter are beyond the capacities of our reason. As to these aspects, only experience can be a source of knowledge, both according to his *Principia*,[291] as well as to the concluding part of his *Discours de la méthode*, in which Descartes's summary is the (alleged) way of proceeding in physics, namely, by (1) finding out the first principles of physics by purely rational means, and deducing from them the most general effects of nature (i.e. the constitution of the cosmos) – thus proceeding from causes to effects – and (2) by providing explanations of more particular phenomena by proceeding from the effects to the causes, i.e. by looking, by means of experience, at how certain phenomena are produced, in a way consistent with the first principles:

> [f]irst I tried to discover in general the principles or first causes of everything that exists or can exist in the world. To this end I considered nothing but God alone, who created the world; and I derived these principles only from certain seeds of truth which are naturally in

---

[287] "[...] for example, if someone wishes to read a message written in Latin letters, to which however their true meaning has not been given and if, upon conjecturing that wherever there is an A in the message, a B must be read, and a C wherever there is a B, and that for each letter, the following one must be substituted; he finds that by this means certain Latin words are formed by these letters: he will not doubt that the true meaning of that message is contained in these words, even if he knows this solely by conjecture, and even though it may perhaps be the case that the person who wrote the message did not put the immediately following letters but some others in the place of the true ones, and thus concealed a different meaning in the message. It would however be so difficult for this to happen, {especially if the message contains many words}, that it does not seem credible," Descartes 1982, 287.

[288] Descartes 1982, 287.

[289] See Dobre 2013a.

[290] On Descartes's theory and foundation of clear and distinct knowledge, see Curley 1977; Alanen 2003; Clarke 2005; Della Rocca 2005; Gaukroger 2008; Patterson 2008; Cunning 2014.

[291] "For, seeing that these parts could have been regulated by God in an infinity of diverse ways; experience alone should teach us which of all these ways He chose," Descartes 1982, 106.

our souls. Next I examined the first and most ordinary effects deducible from these causes. In this way, it seems to me, I discovered the heavens, the stars, and an earth [...]. Then, when I sought to descend to more particular things, I encountered such a variety that I did not think the human mind could possibly distinguish the forms or species of bodies that are on the earth from an infinity of others that might be there if it had been God's will to put them there. Consequently I thought the only way of making these bodies useful to us was to progress to the causes by way of the effects and to make use of many special observations. And now, reviewing in my mind all the objects that have ever been present to my senses, I venture to say that I have never noticed anything in them which I could not explain quite easily by the principles I had discovered. But I must also admit that the power of nature is so ample and so vast, and these principles so simple and so general, that I notice hardly any particular effect of which I do not know at once that it can be deduced from the principles in many different ways; and my greatest difficulty is usually to discover in which of these ways it depends on them. I know no other means to discover this than by seeking further observations whose outcomes vary according to which of these ways provides the correct explanation.[292]

As I am going to show in more detail in Sect. 3.2.3.3, *The role of experience in De Volder's natural philosophy*, the conformity with the first principles of physics grants, for Descartes, the scientific character of explanations of phenomena, even if by these one deduces the causes from the effects rather than vice versa.

3.2.2.1.2 Regius's Rejection of Descartes's Metaphysics

Thus far, Descartes aimed at providing his physics with a metaphysical foundation, and in doing so he aimed at providing physics with a metaphysical certainty. Such a 'metaphysicization' of physics was criticized, in the Dutch context, as soon as it came to the fore. Henricus Regius, the first professor who taught Cartesian ideas at a European university (assuming a chair at Utrecht in 1638),[293] in his letter to Descartes of 23 July 1645 wrote that

many people of spirit and honour have often testified to me that they had too much opinion of the excellence of your mind, to believe that you had not in the depths of the soul feelings contrary to those which appear in public under your name. In order to conceal nothing from it, many are persuaded here that you have greatly discredited your philosophy by publishing your metaphysics. You promised nothing but clear, certain and obvious; but, judging from these beginnings, they claim that there is nothing but obscurity and uncertainty, and the disputes you have had with the skilful people [...] serve only to multiply doubts and darkness. It is useless to claim that your reasoning is ultimately as you have promised. For they reply to you that there is no enthusiast, no impious thing, no buffoon who could not say the same thing about his extravagances and his follies.[294]

---

[292] Descartes 1985, volume 1, 143–144.

[293] See Bos 2002; Strazzoni 2018b.

[294] "Vous ne serez pas surpris de ma conduite, lorsque vous saurez que beaucoup de gens d'esprit et d'honneur m'ont souvent témoigné qu'ils avaient trop bonne opinion de l'excellence de votre esprit, pour croire que vous n'eussiez pas, dans le fonds de l'âme, des sentiments contraires à ceux qui paraissent en public sous votre nom. Et pour ne vous en rien dissimuler, plusieurs se persuadent ici que vous avez beaucoup décrédité votre philosophie, en publiant votre métaphysique. Vous ne promettiez rien que de clair, de certain et d'évident; mais, à en juger par ces commencements, ils

This harsh attack amounts to an accusation of enthusiasm, echoing a similar attack levelled by Martin Schoock in his *Admiranda methodus Renati Descartes* (1643).[295] To it, Regius opposed a scientific methodology based on a sort of 'radical empiricism', endorsing the sensory source of all knowledge,[296] attacking Descartes's metaphysics from its ground viz. the *cogito*, criticizing clarity and distinctness as truth criteria (as they ultimately depend on bodily temperaments)[297] and finding a solution for the metaphysical problems of (1) the immortality of the soul and (2) the reliability of our knowledge in Revelation in his *Fundamenta physices* (1646).[298] In this work, Regius proposed a theory of natural-philosophical knowledge opposite do Descartes, vindicating, in the last section of the treatise, the provisional character of physical explanations, on the ground that human temperaments are various and no argument can convince everyone in the same way.[299] In the second, enlarged edition of this work (*Philosophia naturalis*, 1654), after having remarked that God can

prétendent qu'il n'y a rien que d'obscur et d'incertain, et les disputes que vous avez eues avec les habiles gens à l'occasion de ces commencements, ne servent qu'à multiplier les doutes et les ténèbres. Il est inutile de leur alléguer que vos raisonnements se trouvent enfin tels que vous les avez promis. Car ils vous répliquent qu'il n'y a point d'enthousiaste, point d'impie, point de bouffon qui ne pût dire la même chose de ses extravagances et de ses folies," AT IV 255; cf. Bos 2002, 189–191.

295 See Schoock 1643, 255–257; for a discussion, see Verbeek 1992a, chapter 2; Heyd 1995, chapter 4; Strazzoni 2018a, chapter 2. See also Fouke 1997.

296 "[N]ullis videtur menti ad cogitandum opus esse ideis [...] innatis; sed sola innata cogitandi facultas ipsi ad omnes actiones cogitativas peragendas sufficit; quod in doloris, coloris, saporis, aliorumque similium perceptione est manifestum, quae a mente recte percipiuntur, quamvis nullae eorum ideae menti sint innatae. Nec est ulla ratio, cur unae ideae magis a natura sint insitae, quam aliae. Itaque omnes notiones, quae vulgo communes dicuntur, quales sunt, totum est maius sua parte [...] similiaque omnia alia, ex observationibus rerum didicimus," Regius 1646, 251. See also Regius 1654, 355, 1661, 419. As to Regius's theory of knowledge, see Bellis 2013; Bos 2013; Strazzoni 2018a, chapter 2.

297 See *infra*, n. 299.

298 "[C]um itaque sic a natura mens sit comparata, ut a variis motibus variae perceptiones et iudicia ipsi possint excitari, cumque illi motus non tantum a corporibus veris, sed etiam a causis imaginariis, et a potentissimo directore tantum imaginariis productis, animae offerri queant: hinc sequitur per naturam dubium esse, vera an falsa, seu imaginaria, mente percipiamus et diiudicemus. Verum hoc dubium nobis tollit Divina in Sacris revelatio [...]. Unde patet ea quae recte percipimus, esse res veras, et non imaginarias [...]. Atque ita magna illa dubitatio, quae in animis recte philosophantium per naturam necessario utramque paginam etiam in evidentissimis faceret, per Verbum Dei penitus evertitur. Unde recte quilibet verus philosophus, iam cum propheta canit: verbum Dei est lucerna pedibus meis," Regius 1646, 249–250; Psalms, 119:105. See also Regius 1654, 349; Regius 1661, 414.

299 "Atque ita universae Physicae Fundamenta, brevi, quantum potui, systemate comprehensa [...] absolvimus. Sicut autem, nullius consentientis vel dissentientis habita ratione, libere id proposui, quod mihi rationi maxime consentaneum fuit visum ita hic nemini assentiendi vel dissentiendi legem praefigo; [...] imo, ex terentiano proverbio iam olim puer didici tot esse sententias, quot sunt homines. Neque hoc mirum. Cum enim infinita pene temperamentorum sint discrimina, quae iudiciorum producunt diversitatem, innumerae etiam de rebus humanis iudiciorum debent esse dissimilitudines. [...] Dissentiat igitur quilibet," Regius 1646, 305–306. Cf. Regius 1654, 347; Regius 1661, 399. See Strazzoni 2014.

be a deceiver, Regius concludes that we cannot avoid a natural scepticism.[300] Such scepticism concerns mathematical truths and first principles in their application to an actual state of affairs, whereas we cannot doubt such truths when taken as merely conceptual.[301] Notably, he claims that their very truth, in any case, depends on the actual, extra-mental existence of the objects they represent. For instance "the conception of a triangle entails this property, that its three angles are equal to two straight ones, which however is not true, unless a triangle conceived by mind actually exists in nature."[302]

Such a radical de-metaphysicization of Descartes's physics, as well as a radically empirical standpoint on the source of knowledge and the conditions of truth would later be re-composed. At Leiden, Geulincx and De Raey, in different ways, gave to metaphysics the new function Descartes ascribed to it, namely, to justify the reliability of human intellectual faculties, and providing the first principles of natural philosophy. And yet, their physical theories were to some extent de-metaphysicized.

3.2.2.1.3 Geulincx's Hypothetical Physics

As to Arnold Geulincx – reader of logic at Leiden since 1662, having previously been professor of philosophy at Leuven from 1646 to 1658 –,[303] he justified the hypothetical character of physics on an entirely different basis, namely, in the light of his views on the passivity of man with respect to God. For him, contrary to Regius, physics keeps however a metaphysical foundation. This no longer has, however, the role it had for Descartes.

In the introductory oration (1665) to the second edition of his *Quaestiones quodlibeticae* (first appeared in 1653 at Leuven) Geulincx provides an ordered plan of studies beginning with mathematics (provided with a pedagogical role), logic

[300] "Quicunque [...] omnipotentem et liberrimum cognoscit Deum [...] talem omnium rerum verisimilitudinem, vel scepticismum naturalem [...], qualem proposuimus, negari non potest," Regius 1654, 350–351.

[301] "[N]eque etiam obstat, quod indubitatae per naturam veritatis videantur esse haec axiomata: totum est maius sua parte; centrum est in medio circuli; in omni triangulo sunt tres anguli duobus rectis aequales; idem non potest simul esse, et non esse; et c. Cum enim nullum totum, nulla pars, nullus triangulus, nulla essentia vel existentia, cogitatione et rerum apparentia excepta, sit indubitabiliter cognita. Cum horum cognitio a sola sensuum verisimilitudine dependeat, nihil etiam quicquam de iis certo et indubilitabiliter a quoddam enunciari potest," Regius 1654, 350.

[302] Cf. Regius's full text: "attributa alicuius rei essentialia, in conceptu eius a mente tantum comprehensa, non sunt vera, nisi cum res ipsa mente a nobis concepta actu exisitit. Antequam enim ipsa res existat, quamvis eius ideam in mente habeamus, ipsa tamen est merum nihil, cuius nulla attributa existentiae necessaria vel contingentis, nec ulla alia esse possunt. Ita ex. gr. in conceptu trianguli comprehenditur haec proprietas, quod eius tres anguli sint duobus rectis aequales, quod tamen non est verum, nisi triangulus mente conceptus in rerum natura actu existat," Regius 1654, 357–358.

[303] On Geulincx, see Land 1887, 1891, 1895; Nuchelmans 1988; Nadler 1999; Rousset 1999; Van Ruler 1999, 2000; Aalderink 2009; Strazzoni 2018a, chapter 4.

(encompassing mathematical reasoning itself), and metaphysics, concerning the properties of body and mind, but not those of God. In turn, comes physics, being opened – in a Baconian fashion – by a natural history based on the use of "telescopes, [...] anatomical theatres, alembics, furnaces, magnets" and other means which are combined with our demonstrative skills for the sake of the formulation of physical hypotheses.[304] These, at least according to the first edition of the *Oratio*, have a strong provisional status, as far as their validity consists only in explaining phenomena.[305] A similar agenda is proposed in the dedicatory letter of Geulincx's chief work, namely his *Ethica* (posthumously published in 1675; first, partial edition 1665), which sets off with a metaphor in which Geulincx compares the "encyclopaedia" of philosophy to a house. Logic is the foundation, mathematics and metaphysics are the columns, physics is the floor and the decoration of the house, and ethics is the roof which makes the structure complete.[306] Even if, in this plan,

[304] "Erasis iam, excisis et abolitis coecis illis Geniorum nostrorum commentis, paucula supersunt, quae sola Ratio dictat Principia; exposita in propatulo, cunctis nota, purissimo liquidae Veritatis radio conspicua. Dives Inopia! Immensus in parvo scrinio thesaurus! Eliciemus illum et expromemus; sed sensim et sine sensu. Praecipites hic, prurientes, inconsideratos esse non decet. Gemmas istas tener ordo devinxit; si avide, si temere manum iniicis, turbantur, excidunt, et cum spuriis quae humi iacent, confusae, suspectaeque, non satis internoscuntur. Et hinc ordine mutato, quem tradendis scientiis maiores nostri praescripserunt, expediret iis primum Disciplinis excolendam dare iuventutem, quae sine ulla Hypothesi, mere ex Communibus Notionibus demonstrant suas Propositiones; quales sunt Geometria et Arithmetica. Ante tamen Scientiam Consequentiarum, quae est in Logica, perceperint; sine qua Apodixes Mathematicae non satis feliciter procedunt, et cuius ope compendium in illis non exiguum fieri potest. Quae quidem scientia non ita ex tempore et sparsim (ut hodie sit) velut silva quaedam et rhapsodia tradatur, sed ordine ac tenore Geometrico. Hanc autem, quam dico, Consequentiae scientiam ita corrigi, et in apicem illum evidentiae subvehi posse, cum iam ante certum haberem ex ratione, ipso etiam eventu, et successu, quem nuper publici iuris feci, edoctus sum). Tribus his, Logicae, Geometriae, Arithmeticae, Metaphysicam subnectant, sed probe repurgatam, Mentis et Corporis essentiam ac Proprietatis apodictice perhibentem. Dein ad rerum naturalium Historiam applicentur. Hanc voco congeriem ex diversis Naturae Phaenomenis experimentisque certis et indubitatis, pura narratione sine coniectura rationis, ob quam ita se habere videantur, propositis. Haec Historia a docentibus in communia quaedam capita digeretur; hanc tirones non legent in libris dumtaxat, aut ex pulpito audient, sed oculis ac manibus usurpabunt; tantisper suspecta sunt, quae ab his vel illis Physiologis arcana iactantur," Geulincx 1891–1893, volume 1, *Oratio prima*, 41–42. See also Geulincx's commentary on his *Oratio*, added in the Leiden edition: "at hypothesis in disciplinis huc tota refertur, ut declaret et facem praeferat rebus obscuris," Geulincx 1891–1893, volume 1, *Oratio prima*, 64.

[305] "In hanc ne iuret discipulus; teneatur quoad phaenomenis omnibus respondeat; ubi in puncto deficit, reiiciatur, et alia tentetur verum," Geulincx 1891–1893, volume 1, *Oratio prima*, 42. In the second edition, however, the accent is more Cartesian as reason is conceived as the faculty which decide upon the role of authority and experience, and corrects what experience suggests us: "quomodo enim ratione potior et antiquior auctoritas, experientia, aut aliud huius generis quodcunque, si cur ita sit, dicenda est ratio? Certe quod ratione aliqua suaderi probarique debet, totum id sub ratione est. Primum igitur, o homo, ratio est," Geulincx 1891–1893, volume 1, *Paraphrases et commentarii quibus Oratio prima in editione Lugdunobatava illustratur*, 58. See also 50–51 and 54–55.

[306] "Ea re Libellos vobis in lucem edidi Logicos duos; quorum alter palos et caementa, solidando paviendoque fundo, alter intritam et ferrumen conferret, quibus haec inter se durata vincirentur et coalescerent. [...] Iacta sunt Encyclopaediae fundamenta. Interea vero, dum haec fundamenta sibi

metaphysics comes after logic and mathematics, it is in fact the first science which provides the other disciplines with their conceptual foundation: for instance, metaphysics establishes some principles which are used by logic itself, such as those of whole and part. Logic, in turn, is the instrument for inferences.[307] It is both in his *Ethica*, as well as in his *Metaphysica vera* (posthumously published in 1691), that Geulincx set the ground for a metaphysically-grounded hypothetical physics. The reasons for this status of physics are different from those proposed by Regius, and then by De Volder. Besides the plain reason that metaphysics provides physics with its main concepts, viz. the idea of body and its related properties, whose existence can be then ascertained by the senses only,[308] Geulincx maintains that one cannot reach but a hypothetical knowledge in natural philosophy in the light of the axiom "it is impossible, that he makes [something], who does not know how [he] makes [it]."[309] Accordingly, he presents a theory of knowledge according to which the highest knowledge, or what he calls 'wisdom' (*sapientia*), concerns only our internal acts: such as the knowledge of the nature of affirmation and negation.[310] In turn, all the other kinds of knowledge are somehow imperfect: these are: (1) sense experience, (2) the obscure, albeit certain knowledge of what is not created by us – for instance, that God created the world, and (3) what Geulincx labels *scientia*, viz. what we know with clarity and distinctness, but still 'clothed' by our ways of considering it.[311]

---

esse sino, ut, an dehiscant alicubi vel desciscant, an autem perstent et ferendo sint, explorem [...] quaedam, quae inter exstruendum usui futura videbantur, parabam, aptabam, dolabam: columnas, tigna. [...] Contuli me ad opus amoenum magis: futuri aedificii Coronidem fabricare ingressus sum. [...] Coronis ea, De Virtute et primis eius Proprietatibus Commentatio est. [...] Igitur in Sapientiae sano laquearium et tectum est Ethica. Ut a Logica fundamentum sit firmum et bene fistucatum; a Mathematica et Metaphysica columnae robustae, parietes bene materiati; a Physica pavimentum et opera intestina cuncta concinne eleganterque elaborata; – tamen sine Ethica nunquam sartum tectum est hoc templum. Imo sine Ethica non templum est sed impluvium; non ad sacra, non ad polluctum valet," Geulincx 1891–1893, volume 3, *Ethica*, 3–4.

[307] See his *Methodus inveniendi argumenta* (posthumously published in 1675): "principia generalia spectant ad metaphysicam. Metaphysica enim sola praecedit logicam inter scientias, quamvis nec sine logica tradi possit," Geulincx 1891–1893, volume 2, *Methodus inveniendi argumenta*, 6. These notions are the basis of the so-called logical containment theory, treated in Nuchelmans 1988.

[308] Geulincx 1891–1893, volume 2, *Metaphysica vera*, 158. See also Geulincx 1891–1893, volume 2, *Disputationes physicae*, 489–510.

[309] "Impossibile esse, ut is faciat, qui nescit quomodo fiat," Geulincx 1891–1893, volume 2, *Metaphysica vera*, 150. See Verhoeven 1973; Nadler 1999.

[310] "Sapientia est, quam nemo videtur habere nisi qui rem illum effecerit; talis est conscientia nostra [...] affirmationis, negationis, caeterarumque in nobis actionum, eo quod ipsi eas exerceamus et efficiamus," Geulincx 1891–1893, volume 2, *Metaphysica vera*, 192–193.

[311] "[S]apientia [...] est profunda aliqua penetransque cognitio rei coniuncta cum summa animi delectatione. Hanc sapientiam nemo habet in summo gradu circa rem aliquam, nisi qui rem illam effecerit, et in efficiendo intime possiderit. [...] Nota, varias esse perceptiones quae sapientiam non pertineant, ut imprimis est perceptio sensuum, quae minime rem ipsam attingit, sed tantum illa nobis, in quantum homines sumus, quid commodi vel incommodi afferre possit, demonstrat. Secundo cognitio certa etiam, sed rem non penetrans, seu sine evidentia, seu sine claritate (ut

On this ground, Geulincx develops a philosophical system concerning human beings as subject to God's activity in sensorial experience and bodily activity. As we are not aware of the ways by which we receive sensations or make movements, we are not actively involved in them. They all come from God, a "certain knowing and willing [being] different from me."[312] Thus far, Geulincx adopts a form of occasionalism to account for what appear to be body-mind interactions.[313] However, he also develops a theory of knowledge which sees the active intervention of God in the ascertainment of purely intellectual ideas, in the broader context of a theory on the 'human condition' (*conditio humana*) in which man is considered as a passive subject, inspired both by Descartes and Augustine. Such a theory was favoured by Abraham Heidanus – who supported Geulincx's appointment as lecturer at Leiden in 1662 – as the means to develop a philosophical ethics consistent with Calvinism, in which man is guided by reason as the inner instrument of the Grace in directing us to good.[314]

First of all, in accordance with Geulincx's aforementioned axiom even intellectual ideas are not object of a *sapientia*, but are observed in the divine mind itself, as we cannot create them, and we not know them as products of our reason. As he puts it in his *annotata* to his *Metaphysica ad mentem peripateticam*, "all ideas and eternal truths, like for instance two and three is equal to five, etc., are in the divine mind, not in ours; therefore when we consider such ideas, we consider them in God."[315] In this case, Geulincx follows the traditional, Averroistic strategy to guarantee the uni-

cognitio qua videmus Deum nos homines fecisse; etiamsi enim certa sit, cum tamen modum ignoremus et utique ignorare cogamur, obscura est, inevidens, et rem non penetrans). Tertio etiam scientia seu cognitio cum evidentia, sed quae haeret in cortice et rem non penetrat [...] v.g. scientia qua cognoscimus et scimus res, prout substant operationibus intellectui nostri, seu modi illis ac externis denominationibus, quas ab intellectu nostro eiusque operationibus mutuantur [...]. Tandem est scientia illa, quae rem nude et abstractam ab omnibus modi cogitationum nostrarum denominationibusque proponit; haec vero sapientia, quam nemo videntur habere nisi qui rem illum effecerit," Geulincx 1891–1893, volume 2, *Metaphysica vera*, 192–193. The fourfold distinction of knowledge, like the consideration of the last two as true, is analogous to Spinoza's theory of science as developed in the *Tractatus brevis*, and then turned into a threefold distinction in the *Ethica*. See Mignini 1984.

[312] "Quinta scientia: est sciens aliquis et volens diversus a me," Geulincx 1891–1893, volume 2, *Metaphysica vera*, 150.

[313] "Iam autem eas excitare debet aut mediante me, aut se ipso, aut tertio aliquo; non excitat eas autem mediante me ipso, quia cogitationes sunt diversae, et ego sum res simplex, a quo diversae cogitationes emanare non possunt. Non se ipso, quia est aeque simplex ac ego; est enim aeque volens et sciens, id est cogitans, ac proinde simplex; simplex enim sui qui idem cogitavi de variis. Restat ergo tertium, cuius interventu hoc faciat, quodque variarum mutationum capax esse debet, ut per hoc varia cogitationum obiecta exsurgant; illudque est extensum, quod potest variari, seu corpus; tertium enim praeter cogitans et extensum nec novi, nec est," Geulincx 1891–1893, volume 2, *Metaphysica vera*, 150–151.

[314] See Van Ruler 1999, 2000.

[315] "Ideae omnes et veritates aeternae, ut e.g. duo et tria sunt quinque, etc., sunt in mente divina, non in nostra, cum itaque nos consideramus ideas istas, consideramus eas in Deo," Geulincx 1891–1893, volume 2, *Annotata ad Metaphysicam*, 287. Geulincx's *Metaphysica ad mentem peripateticam* was posthumously published in 1691, together with his *Metaphysica vera* and his *annotata*.

versality of knowledge by recognizing its objects in the Divine intellect. As a consequence, he considers single minds as belonging to God, just as particular bodies are modes of extension.[316] Second, the ways we come to produce a succession of thoughts do not depend on ourselves, as it is God which prompts the mind to reason by linking it to a body and to bodily sensations.[317] This point reveals a Platonic influence, because the body is conceived as the instrument of the soul. Geulincx's appeal is indeed to Augustine and Paul, and it contains a critique of the immateriality of angels – which otherwise could not think – affirmed in the fourth Lateran Council (1215).[318] In this way, the appeal to God also guarantees the reliability of intellectual evidence, which is not an autonomous or self-grounded criterion of truth. For Geulincx, indeed, even present clear and distinct perceptions are subjected to deception, as in the case of the addition of the same number to a odd and an even number, which seem evidently to be still single and double after the addition.[319] Thus far, clarity and distinctness acquires a decidedly psychological status, no matter how evident it appears.

On these premises, the hypothetical character of natural philosophy is somehow unavoidable, not only for the reason that even mathematical truths are not the objects of a *sapientia*, but also because both the ways in which we attain them and the con-

[316] "Sumus enim modi mentis, ut corpora particularia sunt modi corporis [...]; si auferas modum, remanet ipse Deus," Geulincx 1891–1893, volume 2, *Annotata ad Metaphysicam*, 273. See also 237–240, 269 and 293.

[317] "Sed Deus potest successionem causare in nostris cogitationibus sine motu corporum; tempus ergo potest esse sine motu. Resp. Merito praesumimus Deum id non posse; ipse enim unus idemque, eodemque modo se habet; necessum ergo est, ut instrumento diversimode affecto utatur, si diversos in nobis cogitandi modos suscitare certum habeat [...]; atqui nullum est aliud instrumentum quod diversimode se habere potest, quam corpus," Geulincx 1891–1893, volume 2, *Metaphysica vera*, 177.

[318] "[N]on puto Deum posse successionem cogitationum in mentibus efficere nisi illas alliget ad corpora. Unde etiam Augustinus, ut salvaret successionem cogitationum in mentibus angelicis, dixit eos habere tenuia corpuscula, forte aëria et similia. Et forte sic est; neque enim contrarium est in Scriptura revelatum; imo saepe de apparitionibus angelorum sub specie corporea in illa legimus. Certum est (quicquid hac de re sit), ecclesiam non posse temere hoc reiicere; Augustinus enim post Paulum optimus doctor Ecclesiae fuit et omnia eius ex intimis verae Philosophiae penetralibus hausta videntur, tam mirabiliter consentiunt nobiscum. Scio tamen Pontificios id reicisse," Geulincx 1891–1893, volume 2, *Annotata ad Metaphysicam*, 282. See Denzinger and Schönmetzer 1991, § 800.

[319] "Et videtur ratio aliqua convincens militare pro ista persuasione. Cum enim certo aliquo modo se habeant duplum et simplum, quamdiu utrique idem accedit, videntur in eodem statu permanere; cum enim ea, quae certo modo se habent, mutantur, et modum istum amittunt, signum est, fortius aliquid uni eorum quam alteri accessisse. Sic duo parietes albi eodem modo albi manebunt, quamdiu aeque multum albedinis accedit ad utrumque; et aequalia semper manebunt aequalia, quamdiu idem aut aeque multa iis addentur. Unde videmus, falsitatem istius principii, duplum et simplum, cum idem et aequalia accedunt utrique semper duplum et simplum manebunt, agnosci tantum a posteriori, applicando mentem ad exempla dupli et simpli. [...] Quamdiu autem a priori principium illud intuemur, fortissimum esse videtur, et nihil occurrit, quod veritatem eius suspectam reddere posse videatur," Geulincx 1891–1893, volume 2, *Metaphysica vera*, 143. This argument is also used in the *Disputationes metaphysicae* as one of the sceptical arguments, see Geulincx 1891–1893, volume 2, 484.

stitution of the world are subjected to divine freedom, which ultimately leads to the contingency of the created world. In quite traditional terms, according to Geulincx God is free in the creation of the world according to these ideas, for He is not constrained by any rule. The principle of goodness He followed in creation, indeed, does not force God as it comes after His will.[320] As a consequence, according to Geulincx physics is contingent from an epistemological and ontological point of view, because it concerns contingent objects and is based on experience. As he specifies in his *Disputationes physicae* (held in 1663–1668), the laws of motion are also ultimately dependent on the arbitrary will of God, and can be discovered only by experience and are not to be deduced a priori.[321] Or, as he puts it in his *Physica vera* (published in 1688), the ontological contingence of the world finds its counterpart in the epistemic contingence of the hypothesis.[322]

3.2.2.1.4 De Raey's Concealment and Defence of Descartes's Metaphysics

In the hands of Regius and Geulincx, thus, Cartesian physics became more and more open to the use of experience as a source of knowledge, and its hypothetical status was strengthened. While maintaining, with Geulincx, its metaphysical basis (apparently lost with Regius) the metaphysical or absolute certainty of its first principles became severely affected, at least with respect to Descartes's views. Actually, the same process of de-metaphysicization of physics was also undertaken in the 1650s by Johannes de Raey, reader at Leiden from 1648, and then professor from 1653, before assuming a position at the Amsterdam Athenaeum Illustre in 1668.[323] In his *Disputationes ad Problemata Aristotelis* (1651–1652), held at Leiden, and later constituting the main text of his *Clavis philosophiae naturalis* (first edition 1654, second edition 1677) De Raey showed from a Cartesian standpoint how the main tenets of Descartes's physics were (allegedly) present in the Aristotelian corpus, and were then neglected and corrupted by the Aristotelians. The 'concordance'

[320] "Sed dices: bonitas Dei necessitatem hic fecit, atque erat necessum ut mundum crearet, et homines conderet, qui tam bonus erat, Resp. Esto haec necessitas (si qua sit); non impedit contingentiam, non officit libertati [...]. Bonitas enim seu inclinatio faciendi hoc quod praestat, idque semper agendi, quod optimum est, non eripit libertatem, quia voluntate posterior est haec bonitas adeoque cum libertate bene compatibilis, et nulli de libertate decedit, si melior fuerit, atque ad id quod expedire judicaverit propensior," Geulincx 1891–1893, volume 2, *Metaphysica vera*, 194.

[321] "[C]um motus sit contingens, et pendeat ab arbitrio moventis, hactenus etiam incertus est, quod ab essentia et a priori, id est ex ratione proprie dicta procedentibus, lateat. Non mirum igitur, si quidam hac sola lucerna gressum dirigentes, in Naturae meridie, in Mundi foro, Mundum, id est Motum, non invenerint. [...] A posteriori vero procedentibus explorata satis et certa est existentia motus; cum et successionem in nostris cogitationibus, et perceptiones nostras, inter caeteras, tales aliquas observemus, quas conscii sumus a nobis solis non pendere," Geulincx 1891–1893, volume 2, *Disputationes physicae*, 511–512.

[322] "Hypothesium prima condicio est ut sint contingentes; si nempe essent necessariae, ex illis, cum metaphysicae theorematis necessariis pariter, nunquam sequerentur apparentiae, quae contingentes sunt," Geulincx 1891–1893, volume 2, *Physica vera*, 422.

[323] On De Raey, see Strazzoni 2011, 2015, 2018a, chapters 5–6; Del Prete 2018.

shown by De Raey, in fact, cannot be labelled part of the more substantial synthesis of Aristotelian and Cartesian principle embodied by the 'novantique' philosophy given in the same years in Johannes Phocylides Holwarda's *Philosophia naturalis, seu Physica vetus-nova* (1651), by Johannes Clauberg in his *Logica vetus et nova* (first edition 1654, second edition 1658), and by Jean-Baptiste Du Hamel in his *De consensu veteris et novae philosophiae* (1663). De Raey, indeed, offered a specimen of the new physics simply concealed as commentary on Aristotle. He provides his alleged concordance on the basis of four *praecognita* i.e. first principles or axioms grounding Descartes's physics: (1) the existence of an extended matter which is the substance of bodies[324]; (2) the principle of the extrinsic origin of motion with respect to bodies, whose essence does not include movement (whose ultimate cause is God)[325]; (3) the three principles of motion and the rules of impact set by Descartes – and proven by means of Aristotelian, textual evidence by De Raey[326]; (4) the existence of a subtle matter, i.e. Descartes's third matter, which allows the explanation of the apparent autonomous movement of bodies and the existence of a void.[327]

In fact, De Raey is careful to avoid the metaphysical claims set forth, on the other hand, by Regius, having been his disputations and *Clavis* – as he himself declared in the *Epistola ad Curatores* opening the *Clavis* – bestowed on him by the Curators for task of teaching the new philosophy by showing its agreement with the 'original' thought of Aristotle, in order to avoid quarrels at the University.[328] Indeed, in 1647 Cartesianism came under attack during the so-called 'Leiden crisis' – which followed the *querelle d'Utrecht* of 1641–1642. Between February and March 1647 Jacobus Revius – at that time Regent of the theological college of the University of Leiden (Statencollege) – presided over a disputation *De cognitione* accusing Descartes of Pelagianism, as the Cartesian account of free will (more extended than the powers of intellect) would lead one to assume that man is capable of saving himself by his own forces.[329] In March of the same year, the theologian Jacobus Trigland accused Descartes of blasphemy, as Descartes used the hypothesis of a deceiving God.[330] Kept informed by Adriaan Heereboord – who later wrote an *Epistola* to the Curators of Leiden University, giving a report of the whole quarrel (1648)[331] – Descartes himself addressed a defensive letter to the Curators on 4 May

---

[324] De Raey 1654, 50–51 and 53–54, quoting from *Metaphysica*, 1029a10–12 and 20–21; *Physica* 192a31–32, 194b24–25 and 216b4–15.

[325] De Raey 1654, 68–71; see *Metaphysica*, 1071b29–30; *De motu animalium*, 700a16.

[326] De Raey 1654, 106–108; see *Physica*, 215a19–22.

[327] De Raey 1654, 127; see *Meteorologica*, 339b25–26.

[328] "Vos estis, qui me ex doctore privato publicum professorem creastis, et ut eam philosophandi rationem, quam pluris a me fieri atque etiam Aristoteli valde adversam ab aliquibus censeri notum erat, cum Aristotele componerem, haud obscure imposuistis," De Raey 1654, 24 (unnumbered). On De Raey's *Clavis*, see especially Strazzoni 2011.

[329] Revius 1647, thesis 13.

[330] "Internum S. Spiritus testimonium de certitudine salutis ad tempus negare sub quocunque praetextu non licet, multo minus ipsum Sp. S. (seu ipsum Deum ut male Carthesius,) pro impostore ac deceptore habere seu fingere, quod plane blasphemum est," Trigland 1647, corollary 7.

[331] Later published in his *Meletemata*: see Heereboord 1654, 1–20.

1647, after which a ban on the discussion of Cartesian ideas followed in the same month.[332] Yet the situation worsened at the end of the year as De Raey, who had recently obtained his master's degree under Heereboord (15 July 1647), intervened in a disputation *De Deo* presided over by the Aristotelian professor of physics and metaphysics Adam Stuart. In the disputation, Stuart had mentioned the Cartesians – without explicitly naming them but as 'neoterici' – as "some new philosophers, who deprive senses of all certain reliability, and solicit philosophers to negate God, and that it is possible to doubt of His existence."[333] Eventually, in February 1648, the University Curators, summoning Heereboord, Revius and De Raey, proscribed for the second time discussion of the opinions of Descartes, and De Raey was formally forbidden to lecture on the new philosophy.[334]

Thus far, De Raey pursued, even if arguably for political reasons, a de-metaphysicization of Descartes's physics, as he justifies the validity of the aforementioned *praecognita* by claiming their being evident to anyone provided with a healthy mind, in accordance with Aristotle's *Analytica posteriora* and *Topica*.[335] This move, in fact, was essentially political. Indeed, De Raey came to be the strongest defender of Descartes's metaphysics in the late 1660s, after the appearance of a more radical philosophy, namely, the materialist theory of Hobbes – as De Raey was to declare in his 1692 *Cogitata de intepretatione*.[336] Indeed, it was precisely in 1668 that Hobbes's *De corpore* and *Leviathan* were published in Amsterdam.[337] Just

332 See Molhuysen 1913–1924, volume 3, 109–112.

333 "[...] neoterici nonnulli philosophi, qui certam omnem fidem sensibus abrogant, et philosophos Deum negare, et de eius existentia dubitare posse contendunt," Heereboord 1654, *Epistola ad Curatores*, 18–19.

334 Molhuysen 1913–1924, volume 3, 11. See also Verbeek 1992a, chapter 3; Van Bunge 2001, chapter 2.

335 De Raey 1654, 37–38; see *Analytica posteriora*, 76b10–14; *Topica*, 141b7–13.

336 See De Raey 1692, *Cogitata*, 215. For a discussion, see Strazzoni 2015. Hobbes's philosophy was not the only polemical target of De Raey. In a letter to Christoph Wittich of 1680, published in De Raey's *Cogitata*, where De Raey outlines the emergence of his views on the misuse of Cartesian philosophy from the 1660s, he mentions Clauberg's development of an Aristotelico-Cartesian logic (improperly philosophical and 'vulgar' worldviews) Franciscus Sylvius's mixing of philosophy and medicine and the Cartesian exegesis of the Bible by Spinoza and Lodewijk Meijer. In the same letter, Tobias Andreae's application of the philosophical standard to the higher arts is criticized, whereas he admired the distinction between philosophy and theology of Johannes Cocceius and Christoph Wittich (see De Raey 1692, 654–661 (*Epistola familiaris prima*)). In a second letter, dated 25 October 1687 and addressed to an anonymous theologian and published in his *Cogitata*, besides blaming the use of philosophy in theology on Ludwig Wolzogen and Hermann Alexander Roëll (the former being a colleague of De Raey at the Athenaeum Illustre, and the latter a professor at the University of Franeker, where Cartesianism became the philosophical background of theological teaching), De Raey uses as his main polemical target Henricus Regius, considered to be the first misuser of Cartesian philosophy and forerunner of Spinoza (see De Raey 1692, *Epistola familiaris altera*, 661–663). For a discussion, see Strazzoni 2015, 2018a, chapter 5.

337 Hobbes's *De corpore* was first published in London in 1655: a second Latin edition appeared only in the *Opera philosophica* published by Joan Blaeu in Amsterdam in 1668, including also the first Latin translation of *Leviathan* (republished in 1670). Leviathan was translated into Dutch by Abraham van Berkel and published in Amsterdam in 1667 and again in 1672. For full bibliographic details, see Schoneveld 1983, 29–46; Van Velthuysen 2013, 13–15.

before his move to the Amsterdam Athenaeum Illustre at the end of the same year, De Raey set forth the basis of his Cartesian metaphysics – which he was going to develop to its full extent by an analysis of language in his *Cogitata* – in two disputations he held at Leiden in October 1668, namely his *De constitutione logicae* and *De constitutione physicae*, later reprinted in the second edition of his *Clavis* (1677 – also including his *Pro vera metaphysica, quae de principiis humanae cognitionis tractat*), repeated in Amsterdam in 1684, and eventually appended to his *Cogitata*. In such texts, Descartes's metaphysics is systematized and unified with a logical theory which De Raey deems as consisting of Descartes's four rules of the method, as well as of a 'logical science' (*scientia logica*, not developed by Descartes's himself) aimed at wiping out philosophical prejudices and at establishing the foundation of physics.[338] For him, this logic or metaphysics is meant to be a science and a way to science at once (*scientia* and *modus sciendi*) and the leading part of philosophy, "commanding and prescribing its object, in the same time master and architectonic."[339] As an architectonic discipline, it serves to distinguish which are the fields in which philosophical knowledge can take place, and those consisting, on the other hand, in the 'vulgar cognition' – as he put it since his 1651 *Oratio inauguralis de gradibus et vitiis notitiae vulgaris circa contemplationem naturae, et officio philosophi circa eandem*, opening his first disputations and then included in his *Clavis* and *Cogitata* – based on the use of experience, opinion and authority, as well as on imagination and witnesses as the source of explanatory principles: namely, all university disciplines besides mathematics, logic, metaphysics and physics.[340] With his metaphysics, De Raey strove to defend what he thought to be Descartes's thought from its radical interpretations – such as those of Meijer and Spinoza, applying it to theology – which had their first root in Regius's ideas, as well as from the materialist thought of Hobbes.[341] In doing so, he mostly stack with Descartes's arguments in metaphysics,[342] promising, *teste* De Volder himself – as reported by Stolle in his

[338] "Denique philosophiam quod attinet, et physicam imprimis quae praecipua philosophiae pars est, in ea hae regulae maiorem difficultatem habent, quam in mathematicis scientiis. Quia circa ea versatur physica ac tota philosophia, quae instar eorum quae mathematici tractant, simplicia et cognitu facilia non sunt, sed composita et difficilia. Quorum notitia idcirco, non videtur posse certitudinem et evidentiam habere, quae in mathematicis demonstrationibus est. [...] Haec difficultas ut superetur, quantum potest superari, praeter regulas logicas, logicam scientiam requirimus de principiis cognitionis humanae, quae prima scientia, in philosophia summe necessaria sit," De Raey 1692, *De constitutione logicae*, 600–601. Later on he adds: "ut autem sit vera scientia, quantum potest, non sufficere regulas logicas, sed logicam diximus scientiam requiri. Atque eam ostendimus Platonis dialecticam esse, quae seposito sensu, sublata suppositione, et omissa fide, ad primas simplicissimasque veritates adscendit. Atque hae demum verae suppositiones in physica sunt, secundum quas facienda ratiocinatio est, ut vera scientia sit," 606.

[339] "[...] imperans ac praescribens obiectum suum, suo modo domina et architectonica," De Raey 1692, *De constitutione logicae*, 601–603 and 605.

[340] See De Raey 1654, 7–8 and 14–16; De Raey 1692, *De constitutione logicae*, 600 and 605–606; De Raey 1692, *De constitutione physicae*, 608–609.

[341] See *supra*, n. 336.

[342] See Strazzoni 2018a, chapter 5.

Reisejournal and then in his *Introductio in historiam litterariam* (1728) – to pay 5 ducatons for each error that could found in Descartes's books.[343]

3.2.2.1.5 De Bruyn's Defence of Descartes's Metaphysics

In conclusion, De Volder came to teach at Leiden after the link between physics and metaphysics established by Descartes with his *Principia* had already been eroded – with the partial exception of De Raey, who was however to expound his metaphysical theories mostly from 1668, viz. just before De Volder's hiring. In fact, De Volder could have been exposed, as a student, mostly to the Cartesian ideas of Alexander de Bie at Amsterdam, then to those of Regius at Utrecht, as well as to those of Johannes de Bruyn (under whom he graduated in 1660), and then of De Raey and Franciscus Sylvius at Leiden (until 1664). Amongst these, and leaving aside Regius, De Raey, and Sylvius (to whom I will return below), De Bie, even if a Cartesian himself, mostly taught on mathematics and mixed mathematics, viz. astronomy and navigation, touching upon physical topics (such as meteors), while De Volder himself acted as *respondens* in his *Disputatio mathematica de profunditate maris* (1658) and *Disputatio de linea, quam globus per aërem describit* (1659). De Bie however did not provide – as far as his printed disputations are concerned – a systematic treatment of the relations between physics and metaphysics.[344]

De Bruyn, in turn, offered a more systematic treatment of Descartes's metaphysics. Notably, in 1665–1668 he presided over a series of 16 disputations *De naturali Dei cognitione*, defending Descartes's arguments for the existence of God,[345] which he was also to support in his 1670 *Defensio doctrinae Cartesianae, de dubitatione et dubitandi modo, ut et de idea Dei in nobis, deque existentiae eius demonstratione ex ea idea*, aimed against Reinier Vogelsang's 1669 *Contra libellum, cui titulus, Philosophia S. Scripturae interpres, exercitatio paradoxa, indignatio iusta*, including a *Specimen conflictus eundem inter auctorum et celeberrimum virum, D. Johannem de Bruin, super dubitatione Cartesiana. Nec non diatribe de idea Dei secundum Cartesium.*[346] In these texts we find a standard defence of Descartes's theory of clarity and distinctness and its foundation upon radical doubt (which De Bruyn deems as actually applicable to all knowledge, not just as a *fictum*), and on the goodness of God.[347] In turn, in his presentation and defence of Descartes's theory of light in his *Epistola ad Isaacum Vossium, ubi iudicium fertur super ipsius libro De natura et proprietate lucis, et simul Cartesii doctrina defenditur* (1663), against Isaac Vossius's *De lucis natura et proprietate* (1662), he recommended following the chains of reasonings of Descartes, insofar as he offered nothing but cer-

[343] See Sect. 2.3.1, *The correspondence with Thévenot.*

[344] See Van Miert 2009, chapter 8.

[345] See Goudriaan 1999a, b, chapter 8.

[346] See Albrecht 1999.

[347] See Borghero 1998.

tain and evident demonstrations, while no considerations as to their hypothetical character are provided.[348]

3.2.2.2 De Volder on the 'Truth' of Cartesian Principles

Thus far, De Volder came to accept a version of Descartes's philosophy in which physics had already been 'severed' from metaphysics, even if metaphysics – together with logic – was still the discipline by which philosophers discussed the principles of physics itself, as well as the nature and reliability of mental faculties.[349] Within the tradition of 'Dutch Cartesianism', however, De Volder came to detach physics and metaphysics for a reason not touched upon by his predecessors: namely, in order to offer, in physics, explanations based only on natural i.e. created causes. As I am going to show in Sect. 4.1.4, *The search for physical unity and activity: De Volder vis-à-vis Leibniz*, De Volder would refuse to rely on an explanation reverting to the action of God as to the justification of the laws of motion and as to the cause of motion (as he puts it especially in his correspondence with Leibniz), as well as to the explanation of cohesion by the idea of rest (which he labelled more metaphysical than physical). Moreover, as seen above, he criticized occasionalism as a possible solution to the body-mind relation problem, namely the recourse to the continuing intervention of God's action in the explanation of the succession of movements and thoughts, preferring the solution of Leibniz. In his detachment of physics and metaphysics, De Volder was therefore closer to Regius than to Geulincx or De Raey. Unlike Regius, however, De Volder still admitted the possibility solving metaphysical problems, such as the demonstration of the existence of God, by metaphysics itself. This demonstration, however, according to De Volder should play no role in explaining phenomena: namely, God should not serve as an explanatory idea in physics. This attitude goes beyond a mere 'experimental' attitude, intended against speculations over ultimate causes or imperceptible features of matter. In fact, De Volder was eager to admit or discuss Descartes's and also Boyle's corpuscular hypotheses – as I am going to show in the next sections. As he put it in his

[348] "Si Cartesi scripta attentius et sine praeiudicio legisses, sane aliter iudicares. Vel si lubet examina eius scripta [...] et ostende, ubi sua deserat principia, vel minus legitime ex iis sequentia deducat: promitto me ostensurum falsa esse, quae dicturus es. Porro videris hic valde crasse philosophari, dum non vis ut ultra visum sapiamus. An non corpora e tenuioribus contexta sunt particulis, quam quae visu possunt discerni? Exiguissima animalcula organa habent ex subtilissimis particulis constructa, quae omnem visum effugiunt; mediantibus spiritibus invisibilibus motus aliasque actiones animales peragunt; et similiter in aliis se res habet. Corpuscula tua solida et simplicia, quae postea introducis, an illa visibilia sunt? [...] Creditur, inquis, vulgo lumen non adesse ubi nihil est quod luceat, et respondes non succedere ratiocinium hoc, non apparet, ergo non est. Ne in puro quidem aëre lux est visibilis, nemo tamen in eo lumen esse negaverit. Haec scilicet sunt principia visibilia. Quantum ad globulos, particulas striatas etc. illa secundum mechanicae leges facta sunt, et etiamnum continuo fiunt, ut demonstravit Cartesius. Sed ubi oculi et sensum deficiunt, illic incipit fingendi libertas. Fateor, ubi non sequimur filum rectae rationis; atqui illud egregie Cartesius secutus est," De Bruyn 1663, 12–13. Cf. Vossius 1662, chapter 13.

[349] I deal with this topic in Strazzoni 2018a. See also Schuurman 2004.

dictata, one must revert to explanations based on imperceptible corpuscles, as phenomena (such as the imperceptible growth of a tree) cannot be explained otherwise.[350] Rather, De Volder's attitude might be intended as the result of a 'long wave' in Dutch Cartesianism, resulting in a progressive detachment of physics and metaphysics which had, in fact, various reasons: for Regius, the reason for the very impossibility of a rational metaphysics; for De Raey, it was a political reason; for Geulincx, it was a metaphysical or rational-theological reason. For De Volder, the reason is more 'scarne': in physics, explanations are to be given in physical terms only, viz. by considering only the idea of matter and motion (and possibly linking them by demonstrating that matter is active, viz. that motion pertains to its essence). Yet, metaphysics intersects physics as some beings, namely men, are composed of a material and an immaterial substance, which is the object of metaphysics (intended as special metaphysics, or the discipline concerning immaterial substances). Also in this case, De Volder prefers not to revert to divine intervention: rather, to provide explanations derived from the ideas of created substances only. This points to an approach, by De Volder, involving the rejection of the use of the idea of God, viz. the rejection of incursion into a theological field – albeit a rational-theological one. De Volder, as seen above, was involved in the defence of Cartesianism against its 'misuses' viz. theological uses around 1674. This might have been a reason, besides his having been 'imbued' with the Dutch interpretations of Descartes, for his refusal to revert to God as an explanatory concept. And yet, his general attitude, namely, his persistent attempts to find only 'natural' explanations, without reverting to the idea of a supernatural creator, points to a more purely philosophical attitude of De Volder, which was somehow original to him.[351]

As he put it in his correspondence with Leibniz, "I have never liked the opinion of Malebranche and certain Cartesians that motion does not arise from the force of the collision of bodies but from the immediate power of God moving this or that body on the occasion of the collision,"[352] as this would entail that "it will be necessary that we [...] summon a *Deus ex machina* with Malebranche."[353] We may suppose, as Leibniz did, that De Volder deemed "the system of occasional causes [...]

[350] Cf. De Volder's commentary on IV.201, in Hamburg 273, 356: "[s]umunt. Ea frequentissima est obiectio, haec fingitis talia, quae cum non sunt sensibus obvia, merito pro figmentis sunt habenda, sed respondeo neminem ad causas primas posse pervenire nisi perveniat ad insensiles et sola ratione patentes, omnia enim debemus explicare phaenomena, quamdiu autem ista sola scimus quae sensibus sint obvia, profecto nil habemus quam phaenomena et nullas eorum causa, ut si omnia phaenomena magnetis, etc. per sensus explicare deberemus in solis haereremus phaenomenis non eorum causis, unde patet id in nobis culpari non debere. Accesserunt. Et haec alia ratio est, quia sc[ilicet] coguntur fateri dari corpuscula insensilia et illis deberi sensibilia effecta: crescit arbor non subito sed successive accrementis, q[uae] non possunt sensibiles animadverti, nisi tum demum ubi magnam effecerint farraginem, unde patet particulas e[ss]e quae non patent sensibus."

[351] As to the relation of physics and metaphysics in Descartes and his reception, see Hatfield 1990; Garber 1992, 2001, 2006; Fichant 1998; Dobre 2017; Antoine-Mahut and Roux 2018; Strazzoni 2018a.

[352] De Volder to Leibniz, 12 November 1699, in Lodge 2013, 143.

[353] De Volder to Leibniz, 30 October 1703, in Lodge 2013, 271.

not worthy of a philosopher."[354] Not just of a natural philosopher, viz. someone aiming to find natural causes for physical phenomena,[355] but of everyone aiming at using clear and distinct explanatory principles. Plainly, the continuous intervention of God could not serve this purpose. As anticipated above, according to his *De rerum naturalium principiis* even those principles respecting the criterion of clear and distinct perception – viz. the Cartesian principles – cannot be demonstrated as being really at work behind natural phenomena. Namely, as metaphysics (like mathematics) concerns essences only, it is not possible, by considering the idea of any essence, to demonstrate that the phenomena are determined by mechanical causes. Neither from the essence of such causes (from which one cannot even demonstrate that movement follows from matter), nor from the idea of God.[356] After having shown that Descartes's ideas on matter and motion (which are the foremost principles, to which he cautiously adds rest and, as derived ideas, those of size, figure and

---

[354] Leibniz to De Volder, 19 November 1703, in Lodge 2013, 279.

[355] For instance, in his *De rerum naturalium principiis* De Volder points out that it is not the duty of the physician, but rather of the metaphysician to demonstrate the truth of physical principles, viz. their being used by God in creation: "[d]einque, in quinta conditione hisce principiis applicanda multus ut sim necesse non est, [...] quia eius demonstratio metaphysici potius quam physici est fori," De Volder 1681, thesis 186.

[356] See also his 1698 *Oratio de rationis viribus*: "[q]uare, [...] si idearum, quas in me reperio, nexus non appareat, sive quod dependeat aliunde, ex ideis scilicet rerum, quas mens non habet. Sive, licet ex iis, quas habet, sequatur, apparere tamen nequeat, nisi per eo usque continuatam argumentorum seriem, ut eam absolvere mentis non sit humanae. De iis definiendum erit nihil, si ut ratio, ita divinum taceat testimonium. Hoc vero si loquatur, absque omni dubitatione fidem illi habendam, nec de modo, quo ideae inter se nectuntur, nec de consequentibus, quae ex iis fluunt, magnopere laborandum esse, quam disertissime praecipit eadem ratio; certa utique, hoc quicquid est ratiocinii humani obiectum esse neutiquam. Necesse quidem est, ut concipiam, de quo aliquid dicatur, et quidnam de eo dicatur. Quippe haec si ignorem, nequidem enunciationi, quae mihi nulla est, cum nihil eius intelligam, fidem habere potero, sed verba proferam, quibus vis subiecta erit nulla; quo sane absurdius nihil fingi potest. Ex ideis quidem, quae inter se iunguntur, si utraque evidentissime percipiatur, quod in hisce oppido quam rarissimum est, seorsim ut ratiocinemur, nihil impedit. Verum ipsam iunctionis rationem, incognitam licet, tanquam ratiocinii materiam adhibere, eius demum est, qui ita desipit, ut, quae non percipiuntur, idoneum veri ratiocinii, quod totum in perceptione est, fundamentum existimet. Quod si vero hae ideae concipiantur quidem, sed, quod longe frequentissimum est, non satis clare, distincteque, quis erit, qui non perspiciat rationi certae in his locum non relinqui? Quod eo magis obtinebit illis in effatis, in quibus, tametsi ratio repugnantiam non inveniat, invenit tamen quidpiam, quod non modo ex ideis suis percepisset nunquam, sed quod postquam aliunde accepit, difficultatem nihilominus concipiendi secum coniunctam habet. In quibus omnibus, attendite, quaeso, prudentissimum certissimumque, quod ex his emanat, rationis decretum, nimium sapere et ultra divinum testimonium ratione nostra vel tantillulum progredi velle, maxima revera insipientia est. Quam utinam rationis suae moderationem adhibuissent illi, qui harum rerum contemplatione ducuntur. Stabilitos haberemus scientiarum, tam quae ratione nostra, quam quae Deo teste fundatur, terminos. Nec haec quidpiam ab illa metueret detrimenti. Nec illa immisceret iis, quae auctoritatis, sed rationis non sunt; nec in modos, quibus haec se haberent, inquireret tanta cogitationum simul et subtilitate, et inanitate. Omnis equidem praescissa foret ansa innumeris discidiis, quae huic vanae hominum curiositati debentur; qua factum est, ut divina oracula, de quibus certa ratio nihil declarat, incertis ratiunculis, coniecturarum, suspicionumque nostrarum tenebris; aut potius meris cerebri humani figmentis illuminare voluerint; quae profecto insaniens admodum sapientia est," De Volder 1698, 14–16.

disposition (situs))[357] obey to four criteria necessary to establish physical principles, namely

1. that such principles are clear and distinct,[358]
2. that they are not the effect of any other natural cause,[359]
3. that they do not involve the idea of mind,[360]
4. that they can explain every phenomenon,[361]

---

[357] I will deal with these notions, especially with the idea of rest, in Sects. 4.1.1, *De Volder's Cartesian standpoint on the ontology of physics*, and 4.1.3, *De Volder's ideas on cohesion and divisibility*.

[358] The principles matching the first criterion are the notions of matter and motion, to which De Volder adds that of rest, even if with a dubitative formula: "[n]eque vero arduum erit ad priora attendenti coniicere haec nobis principia praeter materiam et motum (nisi quis, de quo non laboramus, addere forte velit quietem) nulla alia esse," De Volder 1681, thesis 161. Such concepts fit the first criterion as physical body is one with mathematical body, and the notion of motion can be mathematically described in terms of variation of distance between bodies: "[q]uid enim clarius, quod distinctius cognoscitur ipsa materia? [...] Per ipsam enim nihil aliud intelligimus quam id, de quo agunt mathematici. Quod corpus mathematicum a physico distinguendo, immane quanto noxae, obscuritatis confusionisque scientiis scholastici attulerunt philosophi. [...] Non ab simili ratione obscurum esse nequit, quid sit motus. [...] Quis enim profecto vel stupidissimus mortalium est, qui ignorat, quid sit corpus alteri vicinum ea ex vicinia recedere, et distantiam ab illo corpore continenter immutare?" De Volder 1681, theses 162–163.

[359] Also the second condition is respected, since matter has in God its only cause, and neither motion nor rest are caused by matter itself, but by God only: De Volder 1681, theses 164–165.

[360] The metaphysical demonstration of the substantial difference between mind and matter fulfils the third criterion: De Volder 1681, thesis 166.

[361] Finally, every kind of phenomenon can be deduced from such principles. Indeed, phenomena are *communia* or *propria* to each sense. *Communia* being the size, figure motion and disposition of bodies, these have to be explained through mechanical principles, grasped by intellect alone or by common sense: "[p]haenomena autem quae observantur vel plurium sensuum sunt communia vel singulorum propria. Quae pluribus sensibus conveniunt, vel in motu, vel in magnitudine, vel in figura, situ similibusque consistunt. Quae vero propria sunt manifeste spectant colorem [...] similesque [...] qualitates. Praeter quae nulla in rerum natura aut dantur aut dari queant phenomena. [...] Qua in re id occurrit primum, quaecunque mutationes vel in motu vel in figura vel in magnitudine occurrunt sensibus, eas qualescunque demum sint hisce principiis deberi. [...] Ex quibus itaque sequitur nullam in corpore aut motus aut magnitudinis aut figurae varietatem dari, quin ea ex iis, quae diximus, principiis sequatur," De Volder 1681, theses 169–172. On the other hand, sense data can be proper to each sense, such as colours and sounds. However, the metaphysical demonstration of the distinction between soul and body proves that no qualities such as pain or delight, nor colours and sounds can be found in the body, but only matter, motion, size, figure and disposition (*situs*): "[i]nter obiecta, quae ad singulos senus spectant, et ea quae plurium communia sunt, licet vulgo confundi soleant, permagnam esse differentiam negabit, ut opinor, nemo, modo attenderit, quam clare intelligat, quid sit in rebus extra se positis, motus, figura, magnitudo, situs [...]. Et quam obscurum ipsi sit, quid sit iisdem in rebus, color, odor, sapor, et c. [...] Ex hac autem distinctione sensuum et qualitatum, quae in corporibus sensum excitantibus revera sunt, id licebit animadvertere, qualiscunque ea corporea dispositio sit, eam ab ipso sensu omnino diversam esse. [...] Verum ne hoc generali ratiocinio, quamquam id vel solum puto rem conficere, sollumodo niti videar; accedamus ad speciales sensus, eorumque peculiaria phaenomena. Cui rei non parum conducet annotasse, ex motu locali [...] nihil posse produci, quam varietatem in figura, magnitudine, celeritate, determinationeve ipsius motus. Hinc enim sequetur, si ostendam sensiles qualitates suam originem debere motui, eas vel in figurae, vel magnitudinis, vel motus diversitate consistere,

De Volder shows that such principles do not fulfil to a fifth criterion, namely, their being true. Nor do they need to do so, as the first four conditions are adequate for choosing explanatory principles for physics, as these can serve to "deduce" every kind of phenomenon, and have indubitable consequences. According to De Volder himself, this would surprise those looking for the same degree of certainty in physics as in metaphysics, plainly, the orthodox Cartesians: in fact, whereas metaphysics concerns mere concepts, in physics one has to rely on the senses and provide hypotheses.[362]

Nevertheless, this hypothetical status of physics is not only justified by our relying on experience, or on the apparent existence of the world: even in the case that the world exists, it is still doubtful whether it obeys the clear and distinct principles of mechanism, as reason also does not allow us to go beyond mere phenomena, which can have more than one possible cause. As he puts it both in his *De rerum naturalium principiis* and in his *dictata*, experience provides only the explanandum of mind, i.e. it only offers sense data: Aristotle's five *sensibilia propria*, and the *communia*. In turn, reason does not demonstrate that mechanical causes are at work, since even if two identical phenomena can only lead to the same hypothesis on their cause, they can actually have different causes. For instance (as I am going to show in more detail in Sect. 4.1.3, *De Volder's ideas on cohesion and divisibility*) the principles of the Aristotelians or those of the alchemists. Therefore, the fifth criterion is not required in the formulation of explanatory principles, since reason, like experience, cannot ascertain that some principles are true according to metaphysical certainty: that is, by showing that others are contradictory.[363]

atque adeo nostris principiis deberi," De Volder 1681, theses 173–176. To that extent every kind of phenomenon can be mechanically accounted for: "[e]x quibus ita constitutis non arduum est elicere, si singula phaenomenum genera nostris principiis debeantur," De Volder 1681, thesis 183.

362 "Quinta denique conditio est, ut principia certo demonstrentur esse vera. De qua tamen, an requiratur necessario, admodum dubitem. Me enim quod attinet, facilem concedam unicuique, ut assumat principia, quaecunque visa fuerint sine ulla ratione, ulla demonstratione, modo ea prioribus conditionibus non repugnent. Quod forte mirum videbitur iis, qui putant omnia certo demonstranda esse, nullibi utendum hypothesi. Quod ut in metaphysicis, ubi omnia per ipsam rerum naturam determinata definitaque sunt, verissimum est, ita in physicis, ubi omnia ad sensus referuntur, paris sit evidentia, non immerito forte quid ambigat," De Volder 1681, thesis 28.

363 "Humana [...] industria si inter varios quibus tunc mundum fieri potuisse supponimus modos, discrimen facere et verum eligere modum posset, profecto id benefici vel rationi deberet, vel experientiae. [...] Rationi autem hoc in negotio nullae reliquuntur partes, quippe quae suam explevit potestatem, si doceat tam hanc quam illam causam mundo efficiendo parem esse; neque vero plus potest. Nam quid evidentius si ex iisdem datis, animadvertamus problema variis modis posse dissolvi, rationem eousque ad summum sese extendere, ut varios hosce modos enumeret, verum neutiquam ut demonstret hoc, non illo modo solutum esse mundum. Quod si faceret illud evidenter sequeretur, eo modo productum esse, nec alio posse, quod est contra hypothesin. Neque etiam huic difficultati enodandae auxilio est experientia, utpote quae tam ex hoc quam ex illo modo una sequitur eademque. [...] Nonne evidentissimum est, illum ad summum nihil aliud posse, quam ut ostendat, aut hoc aut illo modo ortam esse glaciem, non vero quidque etiam moliatur, cum idem sit phaenomenum atque adeo eadem ratione, ex quibus ratiocinetur suppeditet data, ut determinet hoc modo hanc, illo vero alteram lagenam concretam esse. [...] Eodem modo si homini huius artificii vel peritissimo duo proponantur horologia diversis ex rotulis confecta, eadem tamen externa facie

Plainly, if such a demonstration were possible, one would be able to demonstrate, at once, that the world is dominated by an absolute necessity, as De Volder puts it in his *dictata*.[364] Namely, one would be capable of demonstrating that phenomena are mechanically ruled by God, and, given De Volder's idea of causality, that such a mechanical world geometrically flows from His essence. In turn – as he claims in his correspondence with Leibniz – demonstrating the activity of the material substance, viz. that movement pertains to the essence of matter, would mean to demonstrate that physical reality follows an internal necessity, viz. that the internal constitution of the world, let's say, is completely mechanical, as one can demonstrate that motion flows from matter.[365] Thus far, the impossibility of demonstrating a priori, viz. by considering essences only, the reality of mechanical causes, as well as the impossibility of demonstrating the activity of substance become the only limits to the admittance of an absolute necessitarianism on the material world by De Volder. However, as I am going to show next, De Volder admitted the possibility of demonstrating the validity of mechanical principles a posteriori, reaching a degree of certainty comparable to absolute certainty. This issue, of course, pertains to the broader topic of the method of natural philosophy (and related disciplines).

3.2.3 De Volder on the Method of Natural Philosophy

Also as to the method of natural philosophy, De Volder moved from Cartesian positions. First and foremost, he assumed the positions of Descartes as to the purely rational source of the first explanatory principles (viz. matter and motion) – so that these can be assumed as clear and distinct notions – obeying the first condition given in his *De rerum naturalium principiis* – without being proved by phenomena themselves. In his *Principia*, III.4, Descartes claimed that

[...] atque adeo eadem exhibeant phaenomena, in vanum profecto ab ipso expectabimus, ut certo concludat hoc modo unum, illo vero alterum horologium confectum esse, nisi forte, quo nihil est absurdius, existimemus ex iisdem datis diversas conclusiones posse elici," De Volder 1681, theses 21–23. See De Volder's commentary on III.43–45, IV.201 and IV.203, in Hamburg 273, 145–151 and 356–359. See also his commentary on III.43–46 and IV.198–204, in Hamburg 274, 63–64 and 121–123.

[364] See his commentary on IV.204, in Hamburg 274, 123: "[p]luribus diversis modis potuerit efficere. Q[uod] si verum sit perspicuum est, nihil aliud ab humano intellectu posse requiri, quam ut modum inveniant, quo omnia potuerint, superat enim nostras vires ut progrediamur ulterius, cum enim plures modi sunt. Ratio hoc unum potest, ut ostendat aeq[ue] hoc quam illo modo, effectum, de quo agitur, potuisse produci, quo autem producitur sit, eius non est determinare, cum sic necessario subsistat. Nec illud separari potest ab experientia. Quia idem effectum eadem utrobique phaenomena exhibet."

[365] See De Volder to Leibniz, 13 May 1699: "si demonstratum habeam, omnem substantiam a natura esse activam, simul, opinor, demonstratum haberem, extensionem omnem moveri necessario, haberemque veram motus, adeoque omnis mutationis causam, in qua enucleanda hactenus haeserunt, quotquot fuerunt physici," GP II, 178.

> the principles which we have already discovered are so vast and fertile that many more things follow from them than we see included in this visible universe, and even many more than we could mentally examine {in our entire lives}. But let us now set forth a brief description of the principal natural phenomena whose causes are to be investigated here, though not in order that we may use them to prove anything. For we wish to deduce the effects from their causes rather than the causes from their effects. Rather, [we do this] only so that we can consider some, rather than others, of the innumerable effects which we judge can be produced by those causes.[366]

Descartes, thus, aimed at deducing "the effects from their causes." In this deduction, the method of exposition he followed – according to De Volder – was synthetic, namely, going from the causes to the effects, i.e. from the principles to the phenomena, and from time to time reduced to a geometrical exposition in De Volder's presentation of Cartesian theories, and in that of his students.[367] In commenting upon the quoted passage in his Hamburg 273 *dictata*, moreover, De Volder compares (in commenting on III.4) the natural philosopher to a seaman who, in the middle of the ocean, cannot orient himself just by looking at winds, at the ship, or at the water, unless he knows already the "position of the place." In the same way, a natural philosopher can orient himself in phenomena only by already knowing universal principles.[368] Thus far, De Volder availed himself of a Cartesian theory of scientific knowledge: namely, a theory in which clear and distinct principles are used to deduce viz. to explain observed phenomena.[369]

At the same time, De Volder deems the observation of phenomena, including the development of natural histories, as necessary both for the discovery or formulation, and for the application of this or that principle to the explanation of phenomena, as he put it both in his Hamburg 274 *dictata* (in commenting on II.4),[370] as well as in

---

[366] Descartes 1982, 85.

[367] See, for instance, Kölesêri 1681, thesis 11, in which it is claimed that Descartes arranged his *Principia* in a synthetic-geometrical order. Cf. De Volder's commentary on II.40, in Hamburg 273, 106.

[368] "Determinemus. Si q[ui]s in immenso haereat oceano, is nulla ra[ti]onem nec respectu venti, nec navis, nec maris se ad {hanc} illasve partes poterit determinare, nece[ss]e igitur, ut prius situm cognoscat loci illius; eodem modo et nos in immenso principiorum universalium oceano haeremus, q[uae] prius cognoscenda e[ssen]t, anteq[ua]m determineremus ad phaenomena mentem nostram," Hamburg 273, 128.

[369] As to Descartes's 'deductive' physics, see Clarke 1982; Garber 1992, chapters 7–8; Schuster 1993; Gaukroger 2002; Sorell 2010.

[370] "Principia. Principia haec ad duo capita revocari possunt, materiam et motum. Hoc enim unum egit autor, tota parte secunda, ut explicaret materiae et motus naturam, et ut ostenderet per motum infinitam diversitatem, in materiam introduci posse; verum ad haec principia curiosius advertens videt author, ea ita esse faecunda, ut ad infinita diversa phaenomena extendi queant, ac proinde necessarium ei fore, ut aliud quid, ipsam nimirum experientiam, q[uae] hoc sola potest in auxilium {vocet}, ut generalia haec principia, ad certa quaedam phaenomena, nobis in posterum usui fore determinet, nam ut mathematicis ex generalibus suis principiis, qualia sunt, totum esse maius sua parte, si erunt, aequalia, est: nunquam particulare aliq[ui]d problema solvet, nisi per quaedam data determinetur: ita physicus ex materia et motu, q[uae] ita generalia sunt, ut omnibus phaenomenis inserviant, nunquam nisi casu aut fortuito, constitutionem huius mundi et phaenomenum, quae in eo observantur, deducet, nisi per historiam naturalem mentem suam determinet, tanquam per data quaedam, ut has potius quam illas materiae motuumque diversitates considerat," Hamburg 274, 57.

his 1698 *Oratio de viribus rationis* – not to mention two now lost discourses, reported by Le Clerc, according to which "in matters of physics, experience serves more than reasoning," while "in metaphysical and speculative things, reasoning went farther than experience."[371] In the *Oratio*, the relevance of experience as a source of knowledge is particularly stressed. As pointed out by Tammy Nyden,

> [i]n the *Oratio de rationis viribus, et usu in scientiis*, De Volder also makes a distinction between two kinds of certainty, "rational" certainty and a certainty that is obtained through experiment. [...] What De Volder terms rational certainty, is basically Descartes' absolute certainty. [...] For De Volder, this is the domain of *scientia*. However, De Volder does not use the phrase "moral certainty" when speaking about sensory knowledge and experiment. He does not have a handy adjective for this second kind of certainty, but importantly, he does express that it is a type of certainty. Further, the concept is quite different from Descartes' moral certainty. Unlike Descartes, De Volder never appeals to the trustworthiness of our sensations in relation to their utility. De Volder points out that God could have created the world in different ways and chose to create this one. Through reason we can come to understand what worlds are possible, that is, what God could have created. However, reason alone can never tell us which world God did actually create, that is, which world exists, nor can it help us with the particulars about that world, which are of key importance for natural philosophy.[372]

Indeed, in his *Exercitationes* De Volder had already maintained that clarity and distinctness as canons of truth must not be applied to Revelation or to disciplines based on authority or experience, such as medicine: it merely concerns metaphysics and mathematics, which are both based on pure reason. However, contrary to Descartes – who restricted the non-applicability of clarity and distinctness as norms of the truth to disciplines aimed at practice – De Volder extends the field in which it cannot be used as the sole criterion of truth to every discipline besides mathematics and metaphysics, implicitly taking into account physics – besides medicine.[373] In turn, in his 1698 *Oratio* De Volder outlines three main fields or

---

[371] "Ses Amis & ses Disciples se souviennent de lui avoir ouï faire alors un discours, où faisoit voir qu'en matieres de Physique, l'Experience sert plus que le Raisonnement. Une autrefois il en fit un, où il montra que, dans les choses Métaphysiques & de pure spéculation, le Raisonnement alloit plus loin que l'Experience. Soit qu'il n'eût pas écrit ces discours, ou qu'il les ait négligé depuis, ils ne se sont pas trouvez parmi ses Papiers," Le Clerc 1709, 364.

[372] Nyden 2013, 243.

[373] "Unum hocce, antequam ab hac inquisitionis via discedam, ut addam, non erit fortasse inutile; rectissime a Cartesio moneri Lectores suos, ne illa in iis, quae ad praxin vitae spectant, abutantur. Nihil enim ad certae scientiae acquisitionem secuturum noxae, ex eo quod pro veris non assumam, nisi quae omnibus numeris sint certissima, secuturum vero plurimum, si eundem hunc canonem et ad praxin vitae, et ad eas disciplinas, quae vel auctoritate sive Divina sive humana, vel experimentis nituntur extendam. Nam primo ipsa huius inquisitionis ratio evidenter satis ostendit adhibendam eam solummodo in iis rebus, quarum non verisimilis admodum, hoc enim non sufficit, sed quarum certa plane, nec metaphysicis hisce per rationem haberi possit cognitio. Haec autem, ut facile patebit attendenti, haberi tantum in iis scientiis poterit, quae ex solis cogitationibus nostris, et ex ideis, quarum intime conscii sumus, suas conficiunt demonstrationes, nequaquam vero de illis, quae externam quandam, sive auctoritatis sive experientiae opem requirunt. Tales autem scientiae praeter mathematicas, metaphysicamque reperiuntur nullae. Simul enim ab hisce discessero, statim cum ratiociniis meis permiscentur ea, quorum intime conscius non sum et quae idcirco ea evidentia, quam haec

degrees in human knowledge. (1) First, the field of purely rational knowledge, concerning ideas only: this being the field of mathematics and metaphysics. (2) Second, the natural-philosophical level, viz. the science of simple bodies. (3) Third, the science of complex bodies, exemplified by physiology. In accordance with his earlier *De rerum naturalium principiis*, in (2) and (3) one does not consider the real existence of the entities to which one refers. Namely, by using the idea of extension, one does not question whether extension exists outside its very idea. In these fields, the foremost means are hypotheses and experience. Moreover, the more complex bodies are, the less one has to make recourse to the ideas of their imperceptible features, which are beyond the scope of our knowledge.[374]

In this, he specifically refers to physiology, a field in which Cartesian ideas had been applied since Regius's first teachings at Utrecht.

exigit methodus, sciri nequeunt, licet ea possint, quae ad usum vitae sufficit. Dein, ut vel in exemplo huius regulae ultra certarum scientiarum limites productae patefaciam incommoda, si ego nullum omnino cibum velim assumere, nisi de quo secundum hunc canonem nulla possit superesse, vel minima dubitandi ratio, quin ille corpori et valetudini meae sit futurus proficuus, profecto miser ego peribo fame, antequam ullum cibum sumsero: scire enim hoc tam certo, ac haec methodus requirit non possum, nisi penitus omnium corporis mei partium, et ipsius cibi constitutionem et ad corpus meum analogiam, habuero perspectissima. Quod sane nunquam potero, aut si quis sorte, licet iniurium foret, de eo dubitet, non certe potero, antequam, si secundum praescriptum regulae nullum interim cibum assumere velim, in huius rei inquisitione perierim. Deliberandum quidem est de iis quae in usu vitae occurrunt, res certissima est, et eo deliberandum sollicitius, quo res, de qua deliberatio instituitur, maiora vel commoda vel incommoda secum trahit, sed ineptissimum foret, in hac deliberatione haerere perpetuo. Haerebit autem perpetuo propter incerta vitae, et rerum humanarum, qui tamdiu deliberare constituit, donec certam, quam si in ullis, quod vix arbitror, in paucissimis certe habere poterit, acquisiverit scientiam, de eo quod agendum esset. Nec absimili ratione turpiter se daret medicus, si aegro nullam medicinam facere vellet, priusquam et aegri statum et pharmaci conditionem accuratissime haberet exploratam, sciretque quam certissime absque ulla omnino haesitatione hoc pharmacum profuturum aegro. Dum enim in hac disquisitione haeret, ex qua iuxta normam hanc exitus nullus est, iam aut mors occupabit aegrum; aut Naturae beneficio suam ille instaurabit valetudinem. Ex quibus exemplis, plura enim adducere quid refert? Perspicuum esse reor, hanc si inquisitionem illis artibus applicuerimus, hoc effecturos, non ut illas reddamus perfectiores, sed ut omnem quae nunc ex illis utut imperfectioribus ad humanum genus redundat, utilitatem illis demanus. Verum haec quoniam ad rem de qua, agimus proprie non pertinent, obiter indicasse suffecerit," De Volder 1695, *De dubitatione universali*, thesis 28.

[374] "In metaphysicis, mathematisque, quae solas ideas, easque clare et distincte perceptas tractant certa demum vigere ratiocinia. In hisce imperium obtinere et dominari rationem. In physicis vero, utut ex hypothesi quam certissime concludamus, incertum tamen relinqui, eane, quae in ratiocinio assumpsimus, corpora revera sint, an minus. Quod eo incertius est, quo ad corpora accedimus minus simplicia. Quare in rebus medicis, quae circa corpus omnium maxime compositum occupantur, non ex insensilium partium constitutione, quae, ut ignoratur hactenus, sic, nisi me animus fallit, aeternum ignorabitur, ratiocinandum esse, sed ex iis, quae experientia et sensibus de illis parent, non secus ac sua ratiocinia iis, quae per sensus comperta habemus, maximo humani generis bono applicuerunt geometrae," De Volder 1698, 33–34.

3.2.3.1 The Case of Explanations in Physiology

The application of mechanical principles to physiology, viz. the so-called iatromechanics, became a standard in the seventeenth century Dutch context. There, Descartes directly inspired physiologists like Regius, Cornelis van Hogelande, Florentius Schuyl, Sylvius, and Theodoor Craanen (who was to publish an annotated version of Regius's *Praxis medica* (viz. a series of clinical cases published as a monograph in 1657), namely his *Lumen rationale medicum, hoc est Praxis medica reformata sive Annotationes in Praxin Henrici Regii*, in 1686). The champion of iatromechanics, however, was active in Italy: Giovanni Alfonso Borelli, who provided a physiology based on a Galilaeian mechanics in his posthumous *De motu animalium* (1680–1681).[375] Plainly, different approaches can be recognized within this broad movement. For instance, Sylvius relied on chemical principles to account for bodily processes,[376] while Regius and Van Hogelande provided a more marked corpuscular reductionism in their treatises.[377] De Volder, in turn, assumed as his main polemical target in his 1698 *Oratio* Craanen, whose theory of *oeconomia animalis* (an idea already used by Van Hogelande)[378] had for De Volder a highly 'speculative' character.[379] In his 1698 *Oratio*, indeed, De Volder criticizes those aiming to deduce the whole "construction of the human body" (*corporis humani fabrica*) from the first principles of physics, distinguishing themselves from the *empirici*. Even if according to De Volder this attempt still deserves some respect, they nevertheless claimed to deduce the complex structure of the body from a few notions, as if one could deduce Archimedes's discoveries from Euclid's principles.[380] This is a position De Volder also defended in the Hamburg 274 series of his *dictata*, claiming that an uninterrupted deduction of medical notions from the first principles of physics is impossible,[381] as well as in his 1689 *Oratio* for Lucas

[375] See Thurston 1999; Des Chene 2005; Pope 2005.

[376] See King 1970; Underwood 1972; Baumann 1949; Beukers 1980, 1982, 1999; Lech 1993; Ragland 2008, 2012, 2016.

[377] See Verbeek 1989; Gariepy 1990; Bitbol-Hespériès 1993; Fuchs 2001, chapter D; Kolesnik-Antoine 2013b; Manning 2013; Strazzoni 2018c.

[378] See Luyendijk-Elshout 1975.

[379] Please note that De Volder does not adopt this term. On the debates over experimental and speculative philosophy in the seventeenth century, see Anstey 2005; Anstey and Vanzo 2016.

[380] "Quapropter ad alterum propero cogitationum genus, quod in rebus est corporeis, in quarum, prout existunt, cognitione rationi soli ascribenda tantum fortasse peccatur, quantum in metaphysicis eidem abnegandis. [...] Quae ut nequaquam inficior, ita vereor non parum, ne qui ita ratiocinantur, nimium magnifice de nostra scientia sentiant [...]. Quoscunque enim physica recentiorum maxime industria hoc tempore fecerit progressus, tam parum ea provecta est hactenus, ut ex illius inventis ad corporis nostri effecta perpetuam argumentationem deducere qui tentant, multis partibus et absurdius et arrogantius facere nec iniuria videantur, ac faceret ille, qui perlectis omnibus Euclidis notionibus hoc solo instrumentum se putaret abunde, ad Archimedes inde perficienda conclusiones," De Volder 1698, 17–18.

[381] "Quicquid de quibuscumque rebus naturali intellectu lumine cognosci potest, philosophia est, quae proinde tam late se extendit, ut omnes artes scientiasve sub se comprehendat praeter theologiam, quae divina, et iurisprudentiam, quae humana authoritate nititur. Ipsa sane medicina pars

Schacht, whose approach to medicine is praised by De Volder as following a *via media* between a purely empirical and a purely rational method, which is the only method to be used in considering a complex body, like the human one. According to such a method, the first principles of medicine are to be based on experience, which are then dealt with by reason, albeit a reason guided by experience in its application to this or that medical case.[382] De Volder, in any case, does not disclose the names of the 'rationalist' physicians attacked in is 1698 *Oratio*, however, his concomitant

quaedam physicae, ac idcirco etiam philosophiae foret, si nostra de rebus corporeis cognitio eousq[ue] esset provecta, ut possumus ex ipsis rerum principiis, deducere per certas et consequentias evidentes, primo naturam et constitutionem humani corporis, et quibus de causis id ipsum a statu naturali deturbatur, iterumq[ue] ad eum reducatur: deinde naturam et constitutionem alimentorum, quidq[ue] mutationis illa eadem corpori possunt adferre. Hisce enim accurate cognitis liceret ex ipsa ratione certo concludere, hisce vel illis tam alimentis quam medicamentis, utendum esse illi, qui vel sanitatem suam conservare, vel morbus, in quem forte incidit a se depellere studet. Rerum ex primis quae de rebus corporeis cognoscimus principiis, necdum possumus ad accuratam et perfectam omniumq[ue] eius partium cognitionem pervenire, nec ex illis iisdem vires tam alimentorum quam medicamentorum evidenter deducere. Hinc factum est, ut medicus, qui in rebus agendis, non contemplandis versatur, ab illis principiis aversus ad ipsam experientiam deferre erit, sicq[ue] non philosophicam scientiam, quae est rerum per causas, sed, cum illa haberi nondum posset, effectorum quorundam per sensus perceptorum cognitionem hominum valetudini aut conservandae, aut in meliorem statum reducendae utilem sibi comparavit: unde et medicina quae ex sua natura pars quaedam physices specialis est, propter hanc cognitionis nostrae imperfectionem, ab ipsa philosophia separata est, et distinctam quandam ab ipsa philosophia artem constituit. Quod idem in artibus mechanicis obtinet," Hamburg 274, 1–2. Notably, this commentary presupposes that in natural philosophy an uninterrupted chain of reasoning from first principles to phenomena is possible: an idea negated elsewhere by De Volder, who maintained that in physics experiments are however necessary. Moreover, in The Hague *dictata* a marginal note is added to this commentary, as it is explanatory of the traditional saying that medicine starts where physics ends: "hinc vulgare illud dictum ubi physicus desinit, ibi medicus incipit," The Hague *dictata*, 1b (i.e. second column). Commented on in Abou-Nemeh 2012, chapter 1.

[382] "In medicina autem docenda eam sectabatur rationem, quae media est, et inter eos, qui Agyrtarum more nullo cum iudicio experimenta crepant, et inter eos, qui contrariam ingressi viam omnia rationi deberi, corrigendam ex naturae principiis medicinam, arrogantius, quam verius iactitant. Verissime enim noster censuit, medicinae et originem et progressus non rationi ex primis principiis morbos, eorumque causas deducenti, sed soli experientiae deberi. Hanc solidissimam esse basin qua omnia medicorum nituntur praecepta. Hac relicta ad ratiocinia, quae in tanta caligine, cui corporis nostri subiacet notitia non possunt non esse incertissima, confugere temeritatis iudicavit plenissimum. Remediorum vim et efficacia in utendi modum, non ratione, quae ad haecce composita nimium corpora oppido quam caeca est, sed usu cognitam. Nos quid quolibet in casu nociturum profuturum quaesit, hoc prudentissime indicio dignoscere, quia alias simili in casu nocuerit, profueritque. Itane ergo empiricam probavit medicinam, rationique in ea nihil reliquit loci? Ne hoc quidem. Et si enim priores absque dubio experientiae deferendas: censuerit, posteriores tamen rationi partes lubens reliquit. Rationem enim omnino consulendam arbitratus est, sed eam quae experimenta confirmaret, non destrueret; quae quanam optima ratione per experientiam cognita huic illive casui applicanda forent rimaretur sedulo, non quae medicinae faciendae modum tot saeculorum experientia probatum convelleret; certissimum ratus veris artificiosissimi huius nostri corporis automati principiis, licet nobis parum cognitis inniti, ea, quae aegris, sive bene sive male eveniunt; et in hac scientiae nostrae instantia multo tutius ex ipsis eventis, quidnam veris rerum naturalium causis conveniat dignosci, quam ex solertibus licet, non tamen satis firmis coniecturis," De Volder 1689, 15–16.

praise of Archibald Pitcairne (also noted in Gronovius's *Laudatio*),[383] points to iatromechanics, which at Leiden had its foremost interpreter in Craanen.

Pitcairne had become professor of medicine at Leiden in 1692 (which he was going to keep for one year only), delivering his inaugural *Oratio, qua ostenditur, medicinam ab omni philosophorum secta esse liberam*. As the title declares, for Pitcairne medicine and philosophy have to be detached, for two main reasons: first, medicine has to be based on well-acquired results, while philosophy proceeds by discussions and disagreement; second, natural philosophy aims at discovering the first causes of phenomena, whereas medicine is based on the direct observation of the properties of medicaments.[384] As Henri Krop has put it, "[t]he physician, therefore, should not deduce the causes of diseases and the forces of medicaments

---

383 "Quod autem in determinata pororum figura, tanquam primaria secretionum causa sibi applaudunt; id quam frivolum sit ostendit Clariss. ante paucos annos hac in Acad. Professor Pitcarnius," De Volder 1698, 23. Cf. Gronovius 1709, 27: "quanta erat coniunctio nostri cum brevi gaudio nostro Archibaldo Pitkarnio?"

384 "In medicina facienda vel docenda non decet pro vero adfirmare, de quo usque adeo sumus incerti, ut nemo de bonis suis tam incertus, esse velit: non enim vitae minor quam pecuniae habendae debet esse cura, hinc sequitur non licere id in medicina tradenda vel facienda pro principio adhibere de quo disputant viri et mathematicae docti et minimis praeiudiciis irretiti: nemo quippe lubens res suas in eam sortem deduci patitur ut ad eas redimendas ancipiti disputatione opus sit; rerum autem nostrarum vita est carissimum. Ex hisce deduco causarum physicarum investigationem qualem instituere solent philosophi medicis neque utilem esse neque necessariam: hae enim sunt de quibus sectarum patroni fere ab orbe condito in haec usque tempora nequicquam litigarunt. Neque hoc mirum videri debet. Cum enim sectarum patroni rerum naturas absolutas causasque intimas neglecta proprietatum investigatione fuerint aggressi, adeoque postulatis multis et datis paucis uterentur, fieri non potuit ut non in diversas sententias discederent. Cuivis profecto in mathematicis versato vel in medicina facienda statim a principio cuivis profecto in mathematicis paulo magis occupato manifestum est nos nihil aliud in rebus cognoscere quam earum ad alias relationem legesque et proprietates virium quibus eas mutare vel ab iis mutari solent. Loquor de rebus corporeis. Vires autem illae viriumque leges actionibus mutuis et utrinque redditis deteguntur. Nam actiones et earum eventa sunt data illa quorum ope leges virium invenimus; causa vero physica, et tantopere a philosophis quaesita rerum natura est illud in rebus ignotum a quo vi res emanare volunt. Illud autem cum sciri non possit nisi prius agnitis viribus harumque legibus inventis, neque quidquam praestet nisi per vires, sequitur viribus ignotis notitiam illius esse nullam, notis vero esse inutilem. Adeoque medicis incumbit solum ut vires medicamentorum et morborum quae per operationes possunt inveniri expendat et ad leges revocet, non autem ut causis physicis eruendis insudet, quae non nisi ex prius inventis virium legibus possunt deduci, iisque inventis medico non sunt profuturae. [...] Equidem non oportet artem illam quae mortalibus una sanitatem promittit disputantium coniecturis et somniis involvi, non enim prudentis est illi vitam credere cuius rationes plerisque falsae, paucissimis probabiles videntur. Sed medicos oportet astronomorum institutum sibi imitandum proponere. Hi non vulgo gratas aut ab oratoribus acceptas opiniones in artem suam adsciscere, neque motibus syderum expediendis advocare quantumvis populo placituram de mundo fabulam, sed observationes longa temporum locorumque intercapedine dissidentes, motuumque caelestium phaenomena methodo sibi familiari et omnibus utili conferendo. Vires illas definire quibus corpora mota ad alia sive mota sive immota contendunt. Hoc eorum saluberrimum consilium amplectamur fide civitate medica hoc est de genere humano bene mereri sit animus. Observationes quae institutae sunt et ubique instituuntur circa morbos morborumque auxilia conferre, et, nulla habita ratione opinionum quae sensu perceptis certitudine cedunt, ex eo quod fieri solet, quid futurum sit ac faciendum colligere oportet," Pitcairne 1692, 9–16. For a full reconstruction, see Krop 2003d. On Pitcairne, also Guerrini 1987.

from their physical causes but from their actions, that is to say from observation. Medicine should take astronomy as its model, for the laws that apply to the smaller bodies also apply to the stars and planets."[385] In his 1698 *Oratio*, praising Pitcairne, De Volder followed a similar path. Namely, he invoked for explanations in physiology a method based on the application of mechanical laws to observable features of the body only, without the consideration of the imperceptible features of particles and pores. These, according to De Volder, can be discovered neither by geometry nor by experience and have no observable effects.[386] Or, as he summarizes it,

> no reasoning, therefore, can have place in the development of medicine? Quite the contrary [...]. But neither that, which is based on the first known elements of things, nor that, which requires [...] the knowledge of imperceptible particles, of which the parts of our body consist. Between these [parts] and the physical principles there is a difference inexplicable by

[385] Krop 2003d, 184.

[386] "Sed ne nimium in his morer, praestabit primaria huius causae capita invadere, quibus eversis reliqua sponte corruent. Ratiocinia siquidem, quorum hactenus usus fuit, duobus praecipuis nituntur causarum fontibus; fermento scilicet, et pororum in aliis ac aliis partibus alia, certa tamen et determinata figura; quorum alterum nihil explanare, alterum falsissimum esse, si evicero, nemo, opinor, erit, qui mechanicam philosophandi rationem eo usque provectam dixerit, ut rei medica satisfaciat. Etenim fermentum generale admodum nomen est, aut nihil causae explanans; aut nihil aliud quam ope externi cuius viam, quod corporibus admiscetur, partes, vel a se invicem secerni, vel laxius commisceri. Quod si omnem hanc corporum mutationem fermentum appellent, sciant, se nobis effectum obtrudere, non causam. Sin vero aliquid particulare per vocem fermenti denotent, hoc quid sit, qua ratione agat, ut definiant opus erit, si cavere velint, ne ad occultam fermenti qualitatem confugere videantur. Quod autem in determinata pororum figura, tanquam primaria secretionum causa sibi applaudunt; id quam frivolum sit ostendit Clariss. ante paucos annos hac in Acad. Professor Pitcarnius. Etenim, si fluidi specie secernantur humores, cuius unquam sibi persuaserit fluidum, quod facile quibuslibet se meatibus accommodat, determinatam pororum exigere figuram? Nemo ita delirat, ut de figura canalis, per quem aqua fluxura est, vel hilum soliditas sit. Sin vero existiment solidi forma corpuscula transire per poros suae figurae et magnitudini adaptatos, iam egregii hi mechanicis, hoc prudentissimae scilicet natura? In secretionibus faciundis artificium ascripserint, quod unico in casu si succedere forte fortuna queat, infinitis in aliis ut fallat omnino necessum est. Qualemcunque enim hanc pororum figuram effinxerit, exigitur profecto, ut diametro transituro corpori respondeant. Nam si minores sint, transitus denegare manifestissime. Si maiores, patebunt et aliis maioribus, nec in minoribus ulla requiretur figura. Sin vero diameter corporis respondeat diametro pori, habenti id, quod volunt, haec sola transitura, excludi vero omnia eiusdem magnitudinis, sed alterius figurae corpuscula. Verum haec ut permeent, non tantum exigitur, ut sint similis figurae, verum insuper, ut similiter ponantur; quod ultimum non nisi unico in situ locum habet ex innumerabilibus diversis corporum positionibus. Quod ut notissimum est vel leviter in geometricis exercitato, ita vel ignarissimo facili experimento patefiet. Orificium faciat triangulare, et triangulum aut triangulare prisma illi exacte respondens. Animadvertet, ut hoc per orificium illud transeat, hunc praecise requiri situm, quo apex apici, basis basi conveniat accuratissime, quemcunque alium situm habuerit, frustra expectari transitum. Quae cum ita se habent, cogitemus sanguinis partes liquorum instar variis inter se motibus agitari, a motu cordis quaquaversum propelli, a vasis maioribus ad minora meando, infinitas in situ mutationes pati, et simul, si possimus, imaginemur nobis, haec nihilominus corpuscula, ubi ad vasa perventum est, in quibus secretio fieri debet, non obstante omni hac motuum, et quae inde oritur situs varietate, nescio, qua naturali inclinatione [...] ita disponi, ut suo situ poris exactissime congruant. Quod ut ab omni mechanica ratione alienissimum est," De Volder 1698, 22–24.

> our reason. Therefore another method is to be used, which[,] having abandoned these unknown things, it infers [conclusions] from those things, which we know.[387]

Before looking at this method, is worth noting that the idea of a medicine based on the notions of particles and pores was, in fact, that one inspiring Craanen's theories. De Volder's distancing from it, however, also indicates the distancing of De Volder from iatrochemistry. Both in *Oeconomia animalis* (1685) and in the more systematic *Tractatus physico-medicus de homine* (1689), indeed, the movement of the parts of the body is explained as a consequence of the movement given by subtle matter, but this explanation also includes some iatrochemical principles. The *Tractatus* opens with the statement of the substantial difference between soul and body – which allows a mechanical explanation of its functioning – and the comparison of the body to a clock, whose primary function is the concoction of food. This works by fermentation, which is generally defined as the separation and modification of particles of food and chyle. Food is first fermented in the stomach, and then sent to the intestines, where it becomes excrement, or in the heart, where it is transformed into blood by a further process of refinement. Fermentation takes place by means both of heat and of the actions of ferments present in the stomach, which are comparable to the action of *aqua regia* and *aqua fortis* on metals. These are nothing but acids and salts, mixed with alkali.[388] If fermentation takes place in the stomach, in the heart its product, namely, chyle, is subjected to effervescence, which is just a faster fermentation, and leads to the generation of blood. The main factor in this generation is Descartes's first element or subtle matter, rather than acids, salts, and alkalis.[389] The particles of blood are then fit to enter the pores of all the parts of the body. The state of health is determined by the fitting of the pores to the particles, whereas diseases are caused by their obstruction. This is the case, for instance, with inflammations, caused by the positioning of the wrong particles in the pores and tubules.[390] As an alternative to this approach, De Volder advocates a method of discovery more attentive to the combination of experience and mechanism in explaining bodily functions, as Borelli and Lorenzo Bellini – whose *De urinis et pulsis* had been extensively read by Pitcairne, and who dedicated to him his medical *Opuscula aliquot ad Archibaldum Pitcarnium* (1695) – practised in Italy. Thus, De Volder embraces a sort of iatromechanical approach, albeit supported by a robust recourse to experience, and a neglect of the insensible features of the human body. According to De Volder, this method is to be intended as a careful procedure of arguing for

[387] "Nullusne ergo in medicina facienda ratiocinio erit locus? Imo vero quam maxime, sed non illi, quod nititur primis rerum cognitis elementis, nec illi, quod ad sui evidentiam requirit insensilium partium, ex quibus membra corporis nostri constant, notitiam. Inter haec enim et physices principia hiatus est ratione nostra inexplebilis. Alia ergo utendum erit methodo, quae sepositis hisce incognitis, ex iis, quae novimus, concludit," De Volder 1698, 24–25.

[388] See chapter 1: "De corpore eiusque distinctione a mente," chapter 2: "Corpus comparatur com horologio," in Craanen 1689, 1–5. See also 26–31 and 37–38. For a full discussion, see Luyendijk-Elshout 1975; Krop 2003d; Booth 2006, chapter 4.

[389] Craanen 1689, 136–141

[390] Craanen 1689, 273–275.

conclusions, and as the explanation of observed bodily functions through mechanical principles.[391] In physiology, explanations are to be provided through detailed observations of the circulation or motion of fluids (as Borelli and William Harvey did), which can be mathematically described, without considering the imperceptible features of the body.[392] In order to develop these mathematical explanations a new anatomy is thus required, based on vivisection and the observation of fluids in motion. In this way, one can collect those data allowing the formulation of explanatory hypotheses on bodily functions, from which phenomena can flow.[393] Such

[391] "Hanc ad rem perspiciamus, quanam via insistant geometrae, quos unicos habemus certi inveniendi magistros, ubi circa res physicas occupantur, inquisituri, num et illa medicis inserviat. Quod eo propono confidentius, quia mathematicorum more rem medicam novo exemplo tractare iam coeperunt illustres omnium suffragiis viri Borellus, Bellinus. Quorum libros a tam paucis etiam inter eos, qui mechanicam philosophiam sectantur, medicis evolvi, et a paucioribus intelligi, dolendum est summopere. Invaluit siquidem apud eorum nonnullus stulta, ne quid gravius dicam, opinio, ac si artes mathematicae, sine quibus nemo horum scripta intellexerit, ad haec studia plus noxae conferret, quam emolumenti. Quasi accurata omnium, quae ad rem faciunt, consideratio; diligens in conclusionibus formandis circumspectio, ipsius denique rationis rectus usus [...] rei medicae officerent potius, quam prodessent. [...] assumam meo iure, omnia rerum corporearum eventa, adeoque, omnes corporis nostri affectiones secundum legum mechanicarum normam peragi. Quod qui negat dignum non arbitror, cui refutando operam insumam. Qui ergo aperiant mechanicam corporis constructionem absque peritia rerum mechanicarum? Sine quibus mechanicae leges intelligi nequeunt," De Volder 1698, 25–26.

[392] "Quae posterior corporis nostri inspectio, quantum priore et dignitate et usu praestet, liquido apparebit, si quis ea, quae de musculis, quae de bile eiusque ductibus et motu passim anatomici, comparaverit cum iis, quae de iisdem geometrico more industrius ostendit Borellus. [...] Quam ieiuna, quam ad corporis fabricam interius cognoscendam infructuosa prae hisce sunt illa! [...] Quasi valetudinis nostrae salus posita esset in hisce minutiis, quae ut plurimum ad oeconomiam animalem vel nihil omnino vel parum faciunt admodum; dum iidem illi, fluida corpora, quae maximam partem bonam vel malam valetudinem praestant, ac si nullius momenti forent, negligunt, nec in mechanica motuum, qui in illis sunt, ratione exploranda elaborant. Sed qui haec commemoro? [...] Huic enim licet per tot secula plurimum operae impenderint viri, in quorum ingenio, industria, diligentia nihil desideres; nunquam tamen omni sua cura, labore, vigilantia, in lucem protrahere potuerunt hunc, quem dixi, sanguinis motum tam facilem, tam promptum, ut ex fluidorum obvio examine numero quam clarissime innotuerit illustri Harvaeo," De Volder 1698, 28–30. Cf. also page 34: "in rebus medicis, quae circa corpus omnium maxime compositum occupantur, non ex insensilium partium constitutione, quae, ut ignoratur hactenus, sic, nisi me animus fallit, aeternum ignorabitur, ratiocinandum esse. Sed ex iis, quae experientia et sensibus de illis patent." In his *Apologia*, Andala criticized De Volder, on the basis of this passage from his *Oratio*, as upholding a form of scepticism as to the possibility of knowing the structure of the body: Andala 1718, 138–139.

[393] "Ea ergo experimenta anatomica et summo quidem cum iudicio facienda sunt, ex quibus patefiant corporis nostri non mortui membra, sed vivi actiones, qui fiant, qua partium operatione, quo fluidorum motu perficiantur. Quibus si ultimo accedat historia corporis affecti, quae morbos, quibus obnoxiis sumus, eorumque singula symptomata, variasque periodos singulatim describat, in numerato habebimus, ut cum geometricis loquar, data, ex quibus de causis porro ratiocinemur. Huic denique aedificio ut fastigium imponatur, non secus ac astronomi hypotheses effinxerunt, quibus iam cognitos astrorum explicarent motus, rudes in principio, quas dein novis ex observationibus sensim emendando tandem perficerent. Ita et nobis necesse erit hypothesin excogitare, quae structurae partium, motui liquorum, efficaciae spirituum sensili conveniens causas in se contineant mechanicas, ex quibus, quae fieri per experientiam novimus, sequantur," De Volder 1698, 30–31.

hypotheses must fulfil some conditions: not to contradict other assumed hypotheses, be open to correction by new experiments and reasoning, be consistent with experience, and allow explanations for newly observed bodily operations.[394] Even if De Volder did not provide teaching in physiology or medicine, there are traces that he taught such a methodology to his students: *teste* Boerhaave, who in his 1703 *De usu ratiocinii mechanici in medicina* praised both De Volder's abilities in mathematics, as well as the ways in which he "shows beforehand [how] to apply [mathematics] to medical things."[395]

3.2.3.2 De Volder's Methodological Liberalism – And Its Continuity over Time

What is of a broader importance is that De Volder comes to praise the use of the method used by Borelli and Bellini – namely, their systematic application of mechanics to physiology – also in natural philosophy, as according to De Volder it has been done by Huygens, Leibniz and Newton. Indeed, he argues, while in the academies the separation of physics and mathematics led to endless disputes over the nature of gravity, or over the nature of celestial bodies, men could exploit the effects of gravity, or predict the path of planets just by "observations and reasonings."[396] An astronomical example pre-dated by Pitcairne, as seen above, and

[394] "Quod cum in fabrica totius corporis nimiam habiturum sit difficultatem, praestabit seorsum in singulis eius operationibus tentare, modo caveamus, ne quid in una hypothesi assumatur, quod alteri repugnet. Nec expectandum erit, eam, quae ita primo nobis in mentem venit, rei satisfacturam. Sed, ut in omnibus fieri solet, ea novis experimentis et rationibus limanda et perpolienda erit. Inquirendum scilicet porro, num, quae ex ea sequuntur, experientiae congruant, et num eadem paucis hinc inde pro re nata additis demtisve omnibus id genus in corpore operationibus adaptari queat. Quod si minus succedat, immutanda erit, donec tandem invenerimus hypothesin, quae in omnibus cum iis, quae fiunt, consentiat," De Volder 1698, 31–32.

[395] "Habetis, qui secreta quaeque mathesios arcana incredibili perspicui sermonis facilitate, revelet, rebusque applicare Medicis praemonstret, Volderum," Boerhaave 1703, 42.

[396] "Assumam meo iure, omnia rerum corporearum eventa, adeoque, omnes corporis nostri affectiones secundum legum mechanicarum normam peragi. Quod qui negat dignum non arbitror, cui refutando operam insumam. Qui ergo aperiant mechanicam corporis constructionem absque peritia rerum mechanicarum? Sine quibus mechanicae leges intelligi nequeunt. Ostenderunt magna huius seculi nostri lumina Hugenius, Newtonus, Leibnitzius, ne simplicium quidem corporum motus, viresque investigandas unquam, non dicam absque notitia matheseos, sed addam absque recondita harum artium scientia. Qua qui instructus non est, in physicis hospes ut sit, necessum est. Tanta igitur cum inter has disciplinas sit affinitas, eo meliori iure inquiremus, num eadem methodo tractari queant. Litigant inter se physici, et si non alia, quam quae hactenus ex perversa physices absque mathematicis tractatione obtinuit, adhibeatur methodus, aeternum litigabunt, de vera gravitatis causa. Quod tamen non obstitit, quo minus hac causa insuper habita, ex cognitis per experientiam gravitatis effectis mechanici plurima quam solertissime adinvenerint, et usibus applicuerint humanis. Quod gravitatis specimen, eo quod simplici et perspicuo effecto innotescat, ad corporis nostri infinitae varietatis structuram ne quis forte minus recte assimilari putet, eiusdem huius methodi efficaciam lustremus in astronomicis. In quibus corpora tractantur, quae et nobis et sensibus nostris remotissima sunt, et magna motuum admodum intricatorum varietate implexa apparent. Quae tamen, neglectis inutilibus physicorum de hisce disputationibus, ex observationibus et

which was to be praised by Le Clerc, who cites De Volder's *Oratio* as purporting a scientific approach in which no speculation over first causes is forthcoming.[397] In fact, starting with Le Clerc, De Volder's 1698 *Oratio* has been considered as his 'farewell' to Cartesianism,[398] and even as testifying an adoption of Newtonianism by De Volder.[399] Reassessing such claims, however, is not easy, for various reasons.

First of all, the methodological indications he gives in his 1698 *Oratio*, allegedly inspired by a bulk of sources which often had opposing views, do not go beyond the level of generality. If in physiology he sets some criterion to formulate hypotheses, and overtly rejects the recourse to corpuscular explanations in a way consistent with his 1689 *Oratio* for Schact (in which a purely rational approach to medicine is criticized), in natural philosophy De Volder limits himself to praising the use of mechanics and mathematics and observation. However, he does not clarify how to use them in a process of discovery. In fact, he seems to trace back the endless debates over the nature of gravity just to the separation of physics and mathematics, without, however, overtly rejecting the use of corpuscular explanations in natural philosophy. Such as those used, for instance, by Descartes and Huygens in their explanations of the cause of gravity (as I show in Chap. 6).

Second, De Volder had already assumed a similar standpoint as to the method of natural philosophy in his *Oratio de coniungendis philosophicis et mathematicis disciplinis* (1682). In it, De Volder had assumed as a methodological model the case of the fruitful combination of astronomy and mathematics (allowing an explanation of planetary motions which is better than that provided by natural philosophers, viz. by the Aristotelians),[400] as well as the application of mathematics to optics, as done by

ratiociniis suis ita illustrarunt astronomi, ut non modo ad multos etiam annos stellarum positum, Solis Lunaeque eclipses praedixerint. Verum et quamplurima et longe utilissima ad geographiam, chronologiam, artem nauticam facientia patefecerint," De Volder 1698, 26–27.

[397] "Après avoir parlé des idées abstraites, il passe à celles du corps & se retranche dans la seule Médecine; pour montrer, par cet exemple, jusqu'où la Raison peut aller. Il se moque d'abord des Médecins Péripateticiens des siecles passez, qui rapportoient tout à des qualitez occultes, à des Sympathies, à des Antipathies, & à d'autres causes dont il n'avoient aucun idée; mais il n'épargne pas non plus ceux, qui sont tout dépendre de la fermentation & de la figure des pores. La premiere n'est qu'un mot, & la figure des pores ne sert de rien, comme il le fait voir. Il veut seulement q'on établisse certains Principes connus par l'experience, & par l'Anatomie, & qu'on raisonne là-dessus, sans vouloir pénetrer les premieres causes, à la connoissance des quelles on ne par viendra jamais: comme les Astronomes, sans savoir quelle est la nature des Astres, ne laissent pas de prédire exactement leurs situations, pendant aussi long-tems qu'il leur plaît," Le Clerc 1709, 388–389

[398] See Ruestow 1973, chapter 6.

[399] See Sassen 1970.

[400] "Siderum profecto, ab illis ut exordiar, doctrinam animo complectetur nemo, qui eorundem motus ignoraverit, quos ut absque astronomia, sic astronomiam scire absque geometriae et arithmetices notitia non licet. Comparemus paululum inter se tum quae de sideribus astronomi, tum quae de iisdem astronomiae rudes tradunt philosophi. Bone Deus! quam frivola, quam ieiuna sunt quae ab his prae iis quae ab illis docentur. Disputant philosophi, sit ne eadem materia caeli ac sublunarium, sint ne stellae partes densiores suorum orbium, sitne calor coelestis idem cum igneo, quinam sint siderum in haec terrestria influxus, et id genus ineptias, quibus omnem de astris cog-

Descartes and Johannes Kepler.[401] Also, he had already praised, in this *Oratio*, Galileo and Huygens as they applied mathematics to the study of motion, namely, in the description of the law of fall and the pendulum vibration.[402] Eventually, he had labelled the laws of motion as the very causes of motion,[403] in a way consistent with his 1698 scorn for the quarrels over the cause of gravity, which seems directed to its treatment by the Aristotelians, rather than to the causal reasoning followed by Huygens.

As far as the method of natural philosophy is concerned, thus, De Volder does not show a substantial change in his 1698 *Oratio* with respect to his earlier 1682 text. In fact, he remarks the distance between the knowledge of the first principles of physics and that of the imperceptible particles of our body. But this seems just to indicate that the more complex the bodies are, the less clear and distinct are our chains of reasoning, and the more we need to rely on experience and hypotheses.

---

nitionem suam conclusam tenent, de motu autem corporum coelestium, quem scire unice inter erat, praeter paucula quaepiam et tralatitia altum silentium, supina ignorantia. Verum fortasse quae ad motum huncce spectant physici non sunt fori, pertinent ad astronomos. Egregie sane. Astronomus ergo nos docebit ea quae usum habent luculentissimum, physicis vero ea relinquentur, quae praeterquam quod ignorantur, sive maxime scirentur, usum haberent nullum. Sed demus hoc ipsis, quanquam iniurium est. Nonne fatebuntur tamen physici esse, celeberrimam illam de motu et quiete Telluris controversiam dirimere? Negari nequit. Nullam enim videas physices compagem in qua non haec decidatur quaestio. Verum alia longe ratione a philosopho mathesios ignaro, alia a Mathematico. Ille enim Telluri nostrae quietem ut asserat, corpori simplici unicum solummodo motum tribuendum assumet, nubes fabulabitur, Terra si ab occasu versus ortum raperetur a nobis versus occidentem relictum iri, nec lapides casuros ad lineam, aut si forte etiam ignotam sibi astronomi personam induere lubeat, non sine supercilio tuebitur, a nobis extra caelorum centrum constitutis dimidium coeli visum non iri, et quae in eam sententiam plura adduci vulgo solent ineptissima, quae ut facile uno difflantur spiritu, ita sine rubore nemo proferat astronomus. Mathematicus contra, nescio enim qui fiat ut hodie a telluris motu stent quotquot sunt astronomi, ab eius quiete omnes motuum coelestium ignari, ex intimis Astronomiae penetralibus, ex multis partibus faciliori et minus intricata phaenomenum expositione, ex maculis solaribus, ex principiis denique mechanicis tam valide Telluris propugnabit motionem, ut quod minima veri specie responderi possit, relinquatur nihil. Quid de cometis? Nihilne efficiet physicus, quam ut non minus vane, quam superstitiose panicum imperite stupenti vulgo terrorem incutiat? De motu autem eorundem, de situ vario, ex quibus vera cometarum detegenda natura, prorsus sileat," De Volder 1682, 18–20 (unnumbered).

[401] De Volder 1682, 27–28 (unnumbered). It is mentioned in Le Clerc 1709, 376–378.

[402] "Nunquam magnus ille Florentinus Galilaeus de Galilaeo admirabilem illam detexisset in motus acceleratione proportionem, nisi in geometricis demonstrationibus fuisset versatissimus. [...] Haec autem ea motus proprietas est, quae in rebus ad usum vitae pertinentibus spectatur plurimum. Nemo enim absque hac cognita motus indole, aquarum ex fontibus [...] erumpentium quantitatem, nemo proiectorum vim [...] definiet accurate. Hinc elegantissima pendulorum doctrina, et ex hisce accuratissima temporum observatio, sine qua in astronomicis [...] nihil exacti fiet unquam. Hinc accuratiora nuper inventa horologia, quae absque vibrationum in pendulis cognita proportione, absque cycloidis lineae contemplatione vere intelligentur neutiquam. Quod inventum ut Illustri Hugenio debet orbis litteratus, ita illi debuisset nunquam, nisi caeteris cum scientiis [...] coniunxisse mathematicarum artium notitiam," De Volder 1682, 16–17.

[403] "Quae tamen illae leges sunt, quae corporum occursibus moderantes, omnium corporearum mutationum, atque adeo omnium physicorum effectuum verae sunt causae," De Volder 1682, 17 (unnumbered).

This is consistent with his 1689 praise of Schacht's methods, as well as with Descartes's own positions,[404] with the plain difference that De Volder overtly criticizes the recourse to corpuscular explanations, at least as far as composite bodies are concerned. In fact, throughout his career De Volder had been decidedly methodologically uncommitted as to the process of discovery in more general domains of physics (i.e. outside physiology, a field in which, however, he did not leave any dedicated text). In particular, De Volder does not clarify (1) how to discover physical principles understood, for instance, as the laws of motion (such as those discovered by Galileo, Huygens, Newton – notably, not by Descartes), (2) whether the use of corpuscular notions in explaining such laws is allowed: the foremost case being the law of fall, explained by Huygens by reverting to his own version of Descartes's vortex theory. (3) Eventually, he does not address the case of cosmology, a field intermediate between that of general physics (viz. the study of first principles) and physiology. Thus far, provided that reasoning follows some criteria – first and foremost, the clarity and distinctness of the explanatory concepts involved, and their conformity to mechanical principles – any process of formulation of explanations i.e. hypothesis seems to be acceptable. Such a methodological openness, as well as his occasional criticisms of Descartes – which I am going to analyse in the next sections – probably cost De Volder his characterization as an eclectic by Stolle and Hallmann.[405] In any case, he firmly rejected eclecticism as a viable philosophical option in his *De rerum naturalium principiis*, as it does not allow the assumption of a coherent, unified set of physical principles, being attacked for this by Johann Christoph Sturm in his *Physica eclectica* (1697),[406] and praised by Andala in his 1708 *Exercitationes*.[407]

As far as De Volder's 'Newtonianism' is concerned, this will be treated in Sect. 6.2.2.3.3, *De Volder's teaching of Newton's ideas at Leiden*, in which I will show how De Volder came to appreciate Newton's explanation of Kepler's laws in a

---

[404] See Sects. 3.2.2.1.1, *Descartes's 'hypothetical' physics*, and 3.2.3.3, *The role of experience in De Volder's natural philosophy*. On Descartes's uses of hypotheses, see Hacking 1980; Clarke 1982, 1989; Roux 1998; Schouls 2000; Ariew 2011b.

[405] See the closing quotation in Sect. 2.4.2, *De Volder's retirement, death and legacy*.

[406] See Sturm 1697, *Physicae hypotethicae sive electivae praeliminarium articulum II*, § 1 (unnumbered pages): "[e]clecticam ac hypotheticam circa res naturales philosophandi methodum a sectariorum quorundam criminationibus modeste vindicans. [...] Philosophus [...] ingeniosus et experimentandi solertia hodienum famigeratissimus, Burcherus de Volder Med. pariter ac Philos. D. in Disput. philos. de rerum naturalium principiis Thes. VI et seqq. Eclecticam Philosophandi methodum ex prosesso impugnat, et, postquam in antecedentibus fiduciam in docentium authoritate ponendam mascule repudiaverat, nec est tamen, addit, quod magnum in scientiis progressum expectemus ab iis, qui, ne aliorum authoritati fidem habere videantur, sese ἐκλεκτικοὺσ fatentur, et ex omnibus philosophorum sectis id quod verum est seligere et sibi sumere, relictis falsis et erroneis. Qui (sic pergit Vir laudatissimus) licet specie quadam probabilitatis sese tueri posse videantur (quam pluribus in eclecticorum gratiam subiecta parenthesi ipse deducit) ignaros tamen sese verae cognitionis, veraeque methodi eam assequendi ostendunt. Quod equidem suum, assertum cum aliquot firmare rationibus deinceps annitatur, eas ordine nunc excutere operae pretium omnino fuerit." On Sturm, see Palaia 1989; Mercer 1999; Ahnert 2002; Gaab et al. 2004; Secundant 2016.

[407] Andala 1708, 22. See Caroti 2014, 238.

disputation of 1696, and how he did not accept Huygens's empirical evidence against Newton's theory of universal gravitation. In addition, it is worth noting – as anticipated in Sect. 2.3.3, *Mathematician and scientific expert* – that Le Clerc reports that De Volder found what he read of Newton's *Principia* as being true, when Huygens came to Leiden to receive from him clarifications on this work.[408] Newton's *Principia*, as well as the opinion of some "Englishmen," moreover, "opened the eyes" of De Volder on Descartes's theories (especially cosmology), once again according to Le Clerc.[409] However, there is no evidence that De Volder came to accept Newtonianism and to use Newton's ideas in his lectures, at least consistently. As seen in Sect. 2.3.3, *Mathematician and scientific expert*, moreover, De Volder found Newton's *Principia*, according to his extant letter to David Gregory, as nothing extraordinary. Moreover, he came to be annoyed by John Keill's (Newtonian) *Introductio ad veram physicam* (1702).

In fact, there is evidence that De Volder came to find Descartes's cosmology and philosophy no longer reliable around 1698 – without however finding a "new system" alternative to it (as Le Clerc put it).[410] Indeed, in July 1698, after his *Oratio* (read in February) De Volder presided over a disputation *De Telluris motu* endorsing Descartes's cosmology, this being, however, the last disputation he presided over in cosmology.[411] This makes sense, of course, of Le Clerc's report on De Volder's intellectual path, according to which De Volder came to be disenchanted with Cartesianism at the end of the seventeenth century, when he shifted to

[408] "Je me souviens même d'avoir ouï dire à Mr. de Volder, que peu de tems après qu'il parut, feu Mr. Huygens, qui étoit un grand Mathematicien, mais à qui les nouvelles méthodes, dont j'ai parlé, étoient inconnues, le vint voir à Leide & le mit sur le sujet du livre de Mr. Newton. Il avoua à Mr. de Volder,qu'il trouvoit ce livre extrêmement obscur, & lut demanda ce qu'il en pensoit. Nôtre Philosophe lui répondit qu'il n'étoit pas en effet facile de pénetrer les principes des démonstrations de l'Auteur, mais qu'il avoit trouvées veritables celles qu'il avoit examinées. On ne peut pas néanmoins douter que Mr. Huygens ne fût un excellent Mathematicien, & que la Physique & les Mathematiques ne lui soient très-redevables," Le Clerc 1709, 379–380.

[409] "Je ne doute pas que le Livre de Mr. Newton ne lui eût ouvert les yeux, sur la doctrine des tourbillons de Descartes; que ce Livre a entierement renversée, à cela près, qu'il établit le Systeme de Copernic, comme lui; en reconnoissant le Soleil, comme le centre commun des mouvemens de la Terre & des autres Planetes. Cette partie de la Physique de Descartes avoit néanmoins paru à bien des gens la mieux imaginée [...] Il n'étoit nullement entêté de la Philosophie, qu'il enseignoit; sur la fin de ses jours, & même quelques années auparavant, il avoit reconnu le foible du Cartesianisme; autant apparemment, par sa propre méditation, que par le secours des habiles Anglois, qui ont établi d'autres principes," Le Clerc 1709, 382 and 198.

[410] "Je l'ai ouï se moquer, plus d'une fois, d'une bonne partie des Méditations de Descartes, quoiqu'il les eût expliquées pendant long-tems. S'il est permis d'ajoûter cette raison à celles, qu'il a rendues de l'envie qu'il avoit de vivre en repos; je croirois qu'il ne vouloit pas composer un nouveau Système, & qu'il étoit ennuyé d'expliquer ces Méditations, aussi bien que la Physique de Rohaut, qu'il avoit expliquée dans les derniers tems; parce qu'il voyoît combien peu il y avoit de choses assurées, dans ces Livres. Il lui échappoit de tems en tems des plaintes, en s'entretenant avec ses amis, du peu de progrès que l'on avoit fait, dans la connoissance de la Verité, & de la petitesse de nos lumieres," Le Clerc 1709, 398–399.

[411] See Sect. 6.2.2.3.4.3, *De Volder's last disenchantment with Cartesianism* (*via* Huygens)?

Jacques Rohault's *Traité de physique* (1671) for his lectures.[412] It is likely, anyway, that De Volder did not completely stop teaching Descartes's ideas in his lectures, which continued until 1705. Certainly, he still stood for a Cartesian idea of material substance after 1698, as adamantly proved by his correspondence with Leibniz, and *pace* Le Clerc, who commented upon De Volder's remarks that the knowledge of the *human* body is difficult, by stating that "one can see that our philosopher has felt the force of the reasons of those, who maintain that the essence of substance is completely unknown to us,"[413] (probably referring to Locke).

Thus far, De Volder's (scattered) sympathies for Newton seem not to be behind his criticisms of the use of 'speculative' ideas in physiology in his 1698 *Oratio*. Rather, such criticisms seem to be fully consistent with his previous positions, which had always been open to the use of experience.[414] Indeed, as I am going to show next, De Volder came to assign to experience a demonstrative role also with respect to the very first principles of physics, in a way exceeding Descartes's approach.

3.2.3.3 The Role of Experience in De Volder's Natural Philosophy

Even if the use of experience is particularly stressed in De Volder's 1698 oration, De Volder had already embraced an overt, empirical attitude as early as in his inaugural disputation *De natura* (1664). In that case, under the influence of Sylvius, which requires a little digression.

Whereas Descartes, Regius, Van Hogelande had acknowledged in fermentation the cause of blood circulation – as blood ferments, rarefies, and comes to an ebullition, moving itself from the heart to the body[415] – in in his *Disputationum medicarum decas*, held in Leiden (1659–1663), Sylvius provides an account of blood circulation based on two different, observable processes: effervescence and fermentation, respectively the composition and the resolution of the parts of matter, which

---

[412] *Teste* Le Clerc (see *supra*, n. 410) and a reference to De Volder's *dictata* on Rohault from the years 1698–1699, in the catalogue of a private library: see Sect. 2.2.2, *The academic* dictata.

[413] "L'année suivante en quittant le Rectorat, il fit une Harangue publique, selon la coûtume, & y traita de la force & de l'usage de la Raison, dans les Sciences. Il la commence par contredire ce Philosophe, qui pour se moquer des définitions de l'Homme, dont les autres se servoient, disoit que l'Homme est ce que tout le monde sait. Il dit au contraire que l'Homme est une chose, que personne ne connoît; car enfin, ajoûte-t-il, quand nous saurions ce que c'est que le corps humain & ce que c'est que l'Ame; connoissances dont nous sommes très-éloignez; il n'y a personne, qui sâche comment ils sont unis ensemble, en quoi consiste la nature de l'homme. C'est un langage que le commun des Cartesiens n'a pas accoûtumé de tenir, & l'on voit bien que nôtre Philosophe avoit senti la force des raisons de ceux, qui soutiennent que l'essènce des substances nous est entierement inconnue," Le Clerc 1709, 386–387.

[414] See Krop 2003d.

[415] See Grene 1992; Anstey 2000; Fuchs 2001; Strazzoni 2018b. Please note that Regius came to reject this view, and ascribed to the movement of the heart the pushing of blood in the body in the third edition of his *Fundamenta physices*, viz. in his 1661 *Philosophia naturalis*: for a discussion, see Petrescu 2013.

are explained as the action of chemical principles (acid and alkali). Effervescence has two kinds, being either intestinal, and consisting of the action of three humours (alkaline bile, pancreatic acid, and saliva) in the duodenum, and vital, i.e. the effervescence of the blood taking place in the heart, caused by the acidity of the blood and the alkaline bile interacting with the innate heat of the heart.[416] Thus, Sylvius developed an explanation of blood circulation by relying on the experimental principle of chemistry rather than on Cartesian reductionism. As shown by Evan Ragland, for Sylvius "the only way to come to know the quantitative mechanisms of the world was through the senses, and especially though the witnesses of sight, touch, and taste in anatomical and chymical experiments. The experience of working with the sensible changes in bodies—animate or inanimate—moved Sylvius to endorse the approach of the chymists."[417] In his *De natura*, De Volder assumes the notion of effervescence as the general principle capable of explaining the effects ascribed to the nature of the body by physicians.[418] For him, effervescence is the cause of the circulation of the blood as this had been mechanically explained by Descartes and Van Hogelande; however, they did not distinguish it from the similar process of fermentation, as Sylvius correctly did,[419] and which is assessed by experience.[420] Later, in his 1698 *Oratio*, De Volder was to criticize the iatrochemical elements used by Craanen, namely, his idea of fermentation, which De Volder labelled an effect to be further explained.[421] If an evolution is to be found in De Volder's methodology, thus, this is to be seen in the replacement of iatrochemistry by an approach more open to iatromechanics – albeit the iatromechanics of Borelli rather than that, more speculative, of Craanen.[422]

---

[416] For an account, see especially Beukers 1999; Ragland 2008, 2016. The disputations were then re-printed several times in seventeenth century: cf., for instance, Sylvius's *Opera medica* (first edition 1679).

[417] Ragland 2016, 192.

[418] General effects require a general explanatory principle: "naturam non in omnibus [...] mutationibus, sed in primaria, et maxime generali sitam esse. Cum enim generalia sint, quae a natura fieri dicuntur, effecta, et primario ad corporis nostri conservationem faciant, ipsam etiam causam talem esse necesse est," De Volder 1664, thesis 6. The criteria in formulating hypotheses on particular phenomena are not addressed by De Volder. This kind of general principle is required to assess what (1) preserves and feeds our body, (2) causes its functions, (3) heals it, (4) excites fevers in order to recover it, (5) makes every medicaments active, (6) accustom itself to medicaments: De Volder 1664, thesis 5.

[419] De Volder 1664, thesis 11.

[420] "[...] constans est omnium anatomicorum, et experientiae consentiens sententiae," De Volder 1664, thesis 11.

[421] See *supra*, n. 386. In fact, in his *De rerum naturalium principiis* De Volder had already rejected the alchemical model of the world, proposed by Jean Baptiste van Helmont: see Sect. 4.1.3, *De Volder's ideas on cohesion and divisibility*.

[422] As noted by Wiesenfeldt, in 1700 Jacob le Mort, professor of chemistry at Leiden, published a *Chymia ab insidiis, erroribus, et calumniis Philosophorum et Galenicorum vindicata*, which can be interpreted as a defence of chemistry also against De Volder's attacks: see Wiesenfeldt 2002, 216.

Besides De Volder's 1664 *De natura* and 1698 *Oratio* – ideally constituting his first and last scientific text – De Volder made reference to the use of experience elsewhere in his works. As seen above, he does not provide methodological rules for such a use. Provided the purely rational character of the first principles of nature – i.e. matter and motion – everything can serve to formulate explanations for phenomena consistent with such principles, and 'saving' the phenomena themselves. In fact, Descartes himself did not provide more specific methodological indications. Leaving aside the general rules of reasoning of the *Discours de la méthode*, we find a specific tool of discovery – amongst his published texts – only in Descartes's *Dioptrique*, as it has been indicated by Stephen Gaukroger. This method traces back to to Pappus's 'problematical analysis', in which one (1) poses a problem, (2) proceeds by unfolding its sub-problems, until (3) one finds a question whose solution is clear, according to pre-established criteria of truth, such as – for instance – what is evident according to experience. In the hands of Descartes, this method amounts to (1) observing a phenomenon, (2) formulating an hypothesis which is consistent with the first principles of physics, and through which one can derive different effects related to the same phenomenon and (3) testing this hypothesis by experience. This is the case, for instance, with Descartes's law of refraction, which is consistent with his principle of the conservation of the quantity of motion.[423] As a problem-solving approach, this method of discovery does not serve in finding first principles in physics, nor it does consist of the deduction of a particular phenomenon from such first principles – namely, as a synthetic-geometrical derivation of conclusions from a theorem[424] – but it provides explanations of phenomena, consistent with such principles.

De Volder does not mention this kind of methodology, which is however compatible with his approach. In fact, for De Volder experience comes to have a demonstrative value for the hypothesis aimed at its explanation, as it is for Descartes's *Dioptrique*. If in the case of the first principles of nature (matter and motion) their universal explanatory scope is justified by De Volder by considering the very idea of sense perception (as *sensibilia propria* or *communia*), in more specific theories he assigns to experience a demonstrative role – linked to indubitability as a truth criterion. In this way, for him (as noted by Nyden) experience entails a kind of certainty and it has a scientific value, namely it confirms the reliability of explanatory principles. This is evident from De Volder's treatment of cosmology (which I will deal with in more detail in Chap. 6). In the *Disputatio philosophica inauguralis de sys-*

[423] See Gaukroger 1989, 73–88 and 110–114. It is worth quoting Gaukroger's words in full: "the approach, as Descartes outlines it, in the case of the discovery of the sine law, the calculation of the angles of the bows of a rainbow, and the solution of Pappus's locus-problem, is the same, and in each case it consists purely in analysis. In each case we take a specific problem bequeathed by antiquity and solve it using procedures compatible with the basic precepts of Cartesian science. We then try to incorporate the solution within a general system which has as its foundations those truths which we cannot doubt because we have a clear and distinct grasp of them (and because God guarantees those truths of which we have such a grasp)," Gaukroger 1989, 114; cf. AT VI, 97–100.

[424] Cf. Descartes's replies to the second objections to his *Meditationes*: AT VII, 155–156. On the problem of synthesis in Descartes's thought, see Garber 2001, 52–63.

temate mundi (1681) of Samuel Köleséri (dedicated to De Volder) we find an overt use of Descartes as to the explanation of the natural-philosophical underpinnings of the Copernican system. Namely, the use of various positions traced back to Descartes's *Principia*. Eventually, in the *Annexa physica* appended to the disputation, one can read that the "system of the world which had been delineated by [...] Descartes, it is not to be taken as a mere hypothesis, by as the truth of thing."[425] Descartes's cosmology, thus, loses the hypothetical character which it had in the *Principia* for the reason that, as put in the main text of the disputation, by using hypotheses and principles such as those of Descartes one can deduce all the empirical evidence concerning astronomy in an indubitable manner.[426] Moreover, the same *annexum* is appended to another disputation, namely the *Disputatio philosophica inauguralis de aestu marino* (1696) of De Volder's student Gysbertus Casembroot, presenting a Cartesian cosmology.[427] In conformity with De Volder's idea of 'experiential' certainty, thus, such a cosmology could not be but true, viz. experience serves to demonstrate the truth of hypotheses. Again, this move is still Cartesian. That is, also Descartes went beyond the categorization of his conclusions as merely morally certain. In his *Discours de la méthode*, referring to the *Dioptrique* and *Météores*, Descartes states that

> [s]hould anyone be shocked at first by some of the statements I make at the beginning of the Optics and the Meteorology because I call them 'suppositions' and do not seem to care about proving them [...]. For I take my reasonings to be so closely interconnected that just

[425] "Systema mundi quod a Nobiliss. Cartesio delineatur, non pro mera hypothesi; sed pro rei veritate habendum est," Köleséri 1681, *Annexa physica, annexum* 14. Discussed in Ruestow 1973, chapter 6.

[426] "Quantopere omnis antiquitas in certis, evidentissimisque hypothesibus et principiis, ex quibus reliquam Veritatum scientiam deduceret ponendis, iisdemque stabiliendis desudaverit, nemini dubium est opinor. Hypotheses enim et principia ea sunt, quibus bene, firmiterque positis, accuratissime omnes sequentur consequentiis; male positis, actum de omni scientia est: nec unquam certior, evidentiorque ad assequendam veritatis cognitionem via esse potest firmis principiis et hypotesibus; utpote quales etiam ipsi scientiae certissima mathematicae sectatores suis praemittunt problematibus et theorematibus, tantumque hypotheses in astronomia, quantum axiomata in Geometria praestant. Ea siquidem quis invenerit, tum dum invenit maximam veritatum obtinuit partem, nam veritates tam indissolubili et mutuo nexu sunt sibi ipsis nexae, ut una posita reliquis sponte sequantur. Oportet interim ea principia et hypotheses ex quibus consequentias deducere, variasque rerum causas reddere contendit philosophus tam clara et evidentia esse, ut mens earum veritatem in dubium vocare nec possit, nec debeat. Tam concatenata, connexave cum suis consequentiis ut tota consequentiarum cognitio, illis suum debeat ortum, sed non vicissim illorum his. Talibus ita positis, adniti philosophum oportet ut notitia rerum, ex principiis hisce, a quibus dependent ita deducat ut nihil in tota deductionum serie inveniatur, quod non sit manifestissimum. [...] Atque his paucis nostram hypothesin asseruimus, et ab obiectiunculis quibus vulgo impetitur vindicavimus, reliquaque adducuntur argumenta, tanquam talia quae nos non concernunt, sponte praetermisimus [...]. Habes itaque hic Lector veritatis avide, ex hac hypothesi systema ordinatissimum, mirandam Mundi fabricam potiori tum ratione, tum experientia nixam. Quia autem hypothesin et principium invenimus, quo firmiter asserto, et vindicato, evidentes phaenomenum caelestium causas obtinemus, quorum demonstrationem ulterius persequi nunc non licet," Köleséri 1681, theses 1 and 12.

[427] "Systema mundi, quod a Nobilissimo Cartesio delineatur non pro mera hypothesi, sed pro ipsa rei veritate est habendum," Casembroot 1696, *Annexa physica, annexum* 6. Cf. *supra*, n. 425.

as the last are proved by the first, which are their causes, so the first are proved by the last, which are their effects. [...] For as experience makes most of these effects quite certain, the causes from which I deduce them serve not so much to prove them as to explain them; indeed, quite to the contrary, it is the causes which are proved by the effects.[428]

Of course, his presuppositions lacked, in the *Essais*, a metaphysical foundation and a derivation from the theory of matter given in *Le monde*. This notwithstanding, all that is needed is their conformity to experience. Or, as he puts it at the end of his *Principia*, some propositions concerning material reality are "more than morally certain," since God does not deceive us, even if Descartes, in this case, does not explicitly relate the certainty of a theory to its explanatory scope.

De Volder, however, seems to make a further move: if according to him the truth of first principles cannot be proven a priori, in his *De rerum naturalium principiis*, discussing the lack of such a demonstration, he argues that no one can truly doubt that extension or motion exist, and that "the thing is the most clear," i.e. that no a priori demonstration is required as, having demonstrated that Descartes's principles fulfil the first four conditions, the fifth one can be taken as granted.[429] Thus far, De Volder was certainly more committed to the proving force of experience than Descartes himself. Moreover, in Descartes's claims that his hypotheses offer 'more than a moral certainty' the benevolence of God is also taken into account as granting their truth. In De Volder, such a rational-theological justification is dropped: only the conformity to phenomena grants, a posteriori, the truth of hypotheses.

[428] Descartes 1985, volume 1, 150. See Clarke 1982, 258–285.

[429] "Deinque, in quinta conditione hisce principiis applicanda multus ut sim necesse non est, cum nemo diffiteatur, haec in rerum natura locum obtinere, qualiacunque etiam principia sequatur. Quis enim est, qui aut extensionem non admittat, aut qui motum neget? Unde nec puto quenquam fore, qui accuratam huius demonstrationem severe exigat, quam conficere hic supersedeo, [...] quia res planissima est," De Volder 1681, thesis 186.

Chapter 4
The Principles of Natural Philosophy

Abstract In this chapter I consider De Volder's treatment of the idea of material substance and of the problem of its source of activity, i.e. of the cause and laws of motion. These issues dominated not only his correspondence with Leibniz, but also his natural-philosophical disputations and handwritten texts, in which he aimed at not reverting to the metaphysical idea of God in accounting for the physical problem of the movement of bodies. This notwithstanding, De Volder never arrived at a solution for the problem of activity, as he did not accept either Leibniz's views on substance or Malebranche's occasionalism, while at the same time he accepted the idea that the quantity of movement is not necessarily conserved in the world. His detachment of physics from metaphysics, moreover, is observed in his treatment of the problem of the cohesion of bodies – traced back by Descartes to the metaphysical idea of rest, and which De Volder solved by experimental means (viz. by considering the pressure of air). In turn, it was by experiments that De Volder came to accept the correct laws of impact for elastic and non-elastic bodies, already formulated by Huygens, Mariotte, Wallis and Wren.

4.1 The Notions of Matter, Motion and Rest

De Volder's views on matter and its modes (viz. motion and rest) were Cartesian, namely, we can trace most of them to Descartes's *Principia philosophiae* and *Meditationes metaphysicae*, these being the texts that De Volder used as a subject of teaching by providing *dictata* on them. Moreover, De Volder's views were constant throughout his career. As highlighted by Paul Lodge and Anne-Lise Rey, De Volder defended a Cartesian idea of matter up to his correspondence with Leibniz, which came to an end as Leibniz could not provide the a priori demonstration of the internal activity of matter repeatedly asked from him by De Volder.[1] Despite the general Cartesian categorization into which De Volder fits, his views were not uncritical with respect to Descartes's. If, on the one hand, he did not find any alternative

[1] Lodge 1998, 2001, 2004, 2005, 2013; Rey 2009a, b, 2016.

A. Strazzoni, *Burchard de Volder and the Age of the Scientific Revolution*, Studies in History and Philosophy of Science 51,
https://doi.org/10.1007/978-3-030-19878-7_4

worldview, or system of philosophy capable of replacing Descartes's ideas, on the other hand De Volder analysed them, used them up to their limits, and compared them with those of his contemporaries. The reconstruction of De Volder's idea of matter and of its modes, therefore, provides us with a lively view of Cartesian natural philosophy in the late seventeenth century.

4.1.1 De Volder's Cartesian Standpoint on the Ontology of Physics

The theory of matter of De Volder is provided in a systematic form in his first printed texts, namely, two short series of disputations *De corpore* (two disputations) and *De motu* (three disputations) which took place in 1671, 1672 and 1675.[2] Also, he provided his ideas on matter in his much longer *dictata* on Descartes's *Principia*, in his series of disputations *De rerum naturalium principiis*, taking place from 1674 to 1676, in his *Experimenta*, which traces back to the same years (1676–1677), and in his *Exercitationes* against Pierre-Daniel Huet's *Censura philosophiae Cartesianae* (1689), as well as in correspondence with Leibniz. All these texts (with the exception of the *Experimenta*, which were a student's notes) were directly authored by him, so there is no doubt that he approved their contents.

The idea that matter consists only in extension is given, first, in the *De corpore*: this idea is acquired both a priori, i.e. just by considering the clear and distinct idea of matter as extension (as Descartes does in his *Principia*, I.53), and a posteriori, namely by abstracting from actual bodies what they have in common, and that cannot be detached from their idea – such as done by Descartes with the example of the wax in the second of his *Meditationes*, as well as in the *Principia* (II.4).[3] A substance which has the essential properties of being impenetrable, divisible, and moveable or figurable (properties which are in fact one and the same thing in the substance we are considering, as seen in Sect. 3.1.3, *De Volder on substance monism*).[4] By extension, the 'differentiator' is movement, from which the various

[2] De Volder 1671a, b, 1672, 1675.

[3] "In huius autem substantiae essentiam duplici methodo possumus inquirere. Una, [...] supponendo nimirum corporis essentiam consistere in quocumque demum libuerit modo id ipsum in suo conceptu nullam involvit obscuritatem, et omnia rerum phoenomena ex hac posita et assumpta corporis natura deduci queant. Altera, inquirendo quidnam illud sit, quod commune sit omnibus iis substantiis quae vel mediate vel immediate sensus nostros afficiunt," De Volder 1671a, thesis 5; as to Descartes's *Meditationes*, cf. AT VII, 31. Please note that in commenting upon Descartes's *Meditationes*, De Volder referred to Descartes's *Principia* as to the explanation of the essence of corporeal and immaterial things: see Warsaw *dictata*, 36v.

[4] De Volder 1671a, thesis 19. Cf. Hamburg 273, on I.53, claiming that from extension as primary attribute (viz., the attribute which is primary according to our way of conceiving it, see Sect. 3.1.3, *De Volder on substance monism*) one can infer its substantiality, because it cannot be reduced to anything else, and – with an anti-Spinozist stance – that no substance can have more than a primary attribute: "[r]eferuntur. Hac in parte quia plurimi solent falli id accurate explicandum: id solet dici

figures arise, as De Volder explains in his series *De motu*.[5] The key notion used by Descartes and De Volder in explaining how movement and figures (two modes of extension) take place is that of *situs*, namely, the reciprocal and measurable distance between parts of matter, to which De Volder reduces the idea of place (*locus*).[6] For instance, the external figure of a body is determined by the different *situs* of its internal parts.[7] 'Situs' or 'positura', for the Scholastics, meant (1) posture, namely

---

quo plura in re sunt attributa, eo rem esse clariorem, verum attributa cum re concipiuntur, aut non: hinc {q[uae]} attributa ex rei natura sequuntur non differunt ab ipsa rei natura, hinc[que] latet {...}, quod voluit proprietates diversas esse a natura rei, verum concipiamus corpus, sed per eius attributa, quod si extensum sit, quod divisibile, quod impenetrabile sit, quidnam attributum quaeritur ex his rei naturam constituat, recte sequis quaerat cur sit corpus? Ratio est quia extensum est, eadem ratio est, si q[uae]ras, cur divisibile, cur impenetrabile sit, si vero quaerat cur moveatur, recte causa externa q[uae]renda est, quia nunc respiciatur cum re, ut diximus, non possunt separare a et sunt eadem cum ipsa: sic idem est sive dicam corpus extensum aut divisibile esse, aut impenetrabile, quia haec omnia reciprocantur inter se, qui tamen inter res, inter quas nullum discrimen est, mens solet {...} distinctionem facere, hinc mens facit differentiam quod unum ex attributis assumat pro primario ex quo reliqua omnia deducit: sic extensionem assumimus pro primario attributo corporis, licet in corpore non differat ab impenetrabilitate aut divisibilitate, sed solum in nostro ratiocinio, similiter sive [triangulum] concipiam tanquam figuram 3 laterum, sive trianguli aequales sunt duobus rectis [...]. Manifestum. Rem hanc explanat mentis et corporis exemplis, quicquid in iis animadverto est extensio aut cogitatio extensionem in corpore sumo pro primario attributo, ex quo alia deduco, dum itaq[ue] ex eo omnia deducimus et per se concipiamus, sequitur esse substantiam, eodem modo cogitatio concipitur se sola nihil in ea involvitur praeter cogitationem hinc eodem modo substantia est, non quod actionem continuam sed potentiam involvat, itaq[ue] cogitatio et res cogitans extensio et res extensa sunt eadem: id itaq[que] asserimus, non aliud quid esse in re cogitante rem non aliud cogitantem, si enim id quis dicat de re nihil intelligi, si enim duo velit concipere. R[espondeo] quid eis id, quod vocat rerum? Si idem intelligit, quod intelligit cum dicit esse rem cogitantem duo diversissima confundit," Hamburg 273, 43–45. Notably, De Volder conceives mind or *res cogitans* not as a perpetual act of thought, but as a faculty: see Sect. 3.1.3.1, *The relation between material and immaterial substances*. In the Hamburg 274 series, De Volder restates his explanation of the supposed primacy of one attribute over the others (which depends only on our ways of considering them): Hamburg 274, 20–21. On the idea of matter and body in early modern philosophy, with particular attention to the Cartesian discussions, see Garber 1992, chapters 3–5; Woolhouse 1994; Slowik 1996, 2002, chapter 4; Des Chene 1996, chapters 4–5; Garber et al. 1998; Gaukroger 2002, chapter 4; Lennon 2007; Normore 2008; Smith 2010; Anstey and Jalobeanu 2011; Pasnau 2011, chapters 4 and 8; Zepeda 2014.

[5] "Unica enim efficiens omnium quae in corpore observantur mutationum causa motus est. Nam is nisi adsit, nullum inter haec vel illa corpora dari discrimen, ullam iis accidere vel levissimam mutationem ne cogitare quidem licet," De Volder 1671b, thesis 2. See also theses 3 and 6–7; cf. Descartes's *Principia*, II.23. On the Cartesian theories of motion, see Gabbey 1971, 1980; Garber 1992, chapters 6–8, 2001, 2002, 2006; Slowik 1996, 2002, chapters 3 and 5; Des Chene 1996, chapters 2 and 8; McLaughlin 2000; Lennon 2007; Anstey and Jalobeanu 2011; Thomas 2015.

[6] "[...] manifestum est, quod motus localis loci, h[oc] e[st] situs, in quo natura loci consistit, inferat mutationem. Nonne enim, cum corpora moveri concipio, concipio distantiam, quae inter haec et alia corpora intercedit continue mutari? Quid autem aliud est distantiam mutari, quam mutari situm? [...] Situs [...] cum relationem sive ordinem quendam inter diversa corpora, sive inter diversas unius corporis partes designet (quis enim situs unico corpori in partes non discreto nisi aliorum corporum respectu tribui potest?) [...]" De Volder 1675, theses 1–2.

[7] De Volder 1671b, thesis 6.

the reciprocal position of the parts within one body, and (2) the stance of a body: as "to sit, to stay, to lie, being prone, supine."[8] It translated Aristotle's idea of κεῖσθαι, which had, in the *Categoriae*, only the second of the aforementioned meanings, i.e. the being-in-a-position, as laying or sitting of a body and does not denote a relation between different substances, but the condition of one substance.[9] In a geometrical world like that of Descartes and De Volder, the old notion of "place" (*locus*), traditionally understood as the surface of the ambient body,[10] is reduced to the idea of *situs* only (*Principia*, II.13 and II.15). Movement, thus, is meant by both Descartes and De Volder as a change of *situs*, as Descartes overtly stated for the first time in his *Meditationes*.[11] From a Scholastic standpoint, they understood single bodies as the different parts of one body.

This idea led Descartes (and De Volder) to deal with the problem of finding a 'frame of reference' granting the possibility of movement, i.e. a change of distances between parts which would not be absolutely relative to each other. Descartes, in his *Principia*, outlined two main concepts of movement: (1) the movement between two bodies only. In this sense, movement is absolutely relative, namely, if one considers two bodies only, one arbitrarily to be considered in movement, or at rest with respect to the other one (II.13 and II.24). (2) "[P]roperly speaking," however, for Descartes movement is "the transference of one part of matter or of one body, from the vicinity of those bodies immediately contiguous to it and considered as at rest, into the vicinity of [some] others,"[12] (II.25). The absolute relativity of motion (usually interpreted as a means, for Descartes, to 'save' the immobility of the Earth with respect to the Sun),[13] was thus tempered by taking into account a common reference for moving bodies: namely, other bodies which make the two bodies no longer arbitrarily moving or resting with respect to each other, but which makes them either both in movement, or both at rest.[14] De Volder does this in his *De motu tertia*, as well

[8] See Burgersdijk 1626, 46: "[s]itus est ordo partium inter se. [...] Etsi autem situs mutari non possit sine motu locali, qui est motus ad ubi, non est tamen situs cum ubi confundendus. [...] Situs alius naturalis est [...] alius voluntarius: ut sedere, stare, iacere, pronum, supinum esse, et c." For the Conimbricenses *situs* relates more strictly to *locus*: "[u]no modo situs dicitur ordo partium in loco, quo pacto, sedere, stare, [...]. Altero modo definitur situs ordo partium in toto, nulla videlicet ratione loci habita," Conimbricenses 1604, 134. On Scholastic and Cartesian ideas of place, see Des Chene 1996, chapter 8; Schmaltz 2002, 54–64; Ariew 2011a, 87–94.

[9] See *Categoriae*, 1b25–2a4, 6b11–14 and 11b10–14. On the idea of *situs* in Aristotle and the Aristotelian tradition, see Almeida 2014.

[10] See Burgersdijk 1626, 26.

[11] See AT VII, 43: "situm, quem diversa figurata inter se obtinent, et motum, sive mutationem istius situs."

[12] Descartes 1982, 51.

[13] See also III.28–29. On the problem of the relativity of motion in Descartes's natural philosophy, see Earman 1989; Garber 1992, chapters 6–8; Des Chene 1996; Slowik 1997, 2002; Gaukroger 2002, 110–114.

[14] "Situs autem cum relationem sive ordinem quendam inter diversa corpora [...] designet, [...] facile poterit contingere, ut unum idemque corpus respectu variorum, ad quae refertur, situm simul immutet et non immutet. Quibus consectarium est, nisi praecise magis hunc determinemus situm,

as in the Hamburg 273 series of his *dictata*.[15] In the other series of *dictata*, after having remarked while commenting on II.24 that, if one considers movement as in II.13 one cannot ascertain if a body is moving or not, and that the nature of movement is respective, and depending on our understanding,[16] he notes (on II.28) that by the definition of movement given in II.25 Descartes attempted to provide a non-relative idea of motion.[17] Yet, the expression "as at rest" that Descartes discussed in II.29, was not used by Descartes as meaning a "philosophical truth," but only to avoid the absurdity that a body can be considered both as in movement and as at

---

unum idemque corpus posse simul moveri et quiescere. [...] Quod ut effugiamus incommodum, quid superest remedii, quam ut ex innumeris locis, quae ratione situum diversorum uni eidemque corpori tribui queunt, unum eligamus, cuius mutationem motum vocemus, quique talis sit, ut non possit simul mutari et non mutari? Quis autem, quaeso, locus, situsque inter externa corpora erit huic rei magis conveniens quam is, quo refertur hoc corpus, de cuius motu agitur, ad ea corpora, quae ipsi proxime sunt contigua?" De Volder 1675, theses 2–3; cf. Descartes's *Principia*, II.29–30 and II.62.

[15] In commenting on II.13, for instance, De Volder underlines that a fixed frame of reference is needed because one cannot determine the *situs* of bodies, if these are in continuous movement with respect to each other: Hamburg 273, 86. In the other series of *dictata*, he claims that if *situs* is changed, also place (*locus*) has to change, against the 'vulgar' way of speaking: Hamburg 274, 36. In commenting on the articles from 24 onwards in the Hamburg 273 series, De Volder criticizes the obscure definition of movement given by the Aristotelians (as *actum entis in potentia, quatenus in potentia*), as well as the 'restriction' of the idea of movement to that of *locus*, which makes both obscure (II.24; see Hamburg 273, 94); he approves Descartes's definition of movement as transference, and remarks that the differentiation between parts and bodies is given by air (II.25: see Hamburg 273, 94); on II.26, see *infra*, n. 22; he remarks that movement and rest are modes (II.27; see Hamburg 273, 97). More noteworthy is that he claims that finding a frame of reference solves the problem of the linguistic ambiguity arising from attributing both movement and rest to the same body. The choice of this frame of reference is justified because, from it, two bodies cannot be both in movement and at rest with respect to each other, such as in the case of the internal movements of the mass of air, composed by moving particles (II.28–30; see Hamburg 273, 97–98). In commenting on II.31 (where Descartes explains how a single body participates in many different movements, like a clock on a boat), he notes that this "rule" had been more extensively explained by Gassendi and Galileo: Hamburg 273, 98.

[16] "[...] determinatam naturam non habeat, siquidem hoc pacto mutationem loci eodem tempore fieri, et non fieri, necesse etiam erit eodem tempore concipere eamdem rem moveri, et non moveri, quod si fit patet motum non haberem determinatam quandam naturam, quippe hoc si foret, non posset eodem tempore eadem de re et negari et affirmari, sed consistere eius naturam in quidam respecti, qui {exque} ac loci dependet a nostra cogitatione," Hamburg 274, 39.

[17] Notably, De Volder remarks that Descartes used the 'ambiguous' idea of *locus*, because the idea of *situs* entails the relativity of motion: "vicinia corporum contiguorum. Author in hac definitione [...] conatur, non relativam sed absolutam motus determinare naturam, utitur termino ambiguo loci, quia varie potest sumi, {mavult} locum circumscribere, eum {q[uod]} determinare, non per situm, ut antea, quia diverse poterat considerari pro diversis corporibus ad q[uos] refertur, sed per viciniam et {uidem} non quorum cum {q} corporum (haec enim modo, in idem quod evitare vult incideret incommodum) sed tantum contiguorum, {q[uae]} corpora contigua, cum {una} tantum sunt, nec possunt varie intelligi. Sequitur unius corporis unicum tantum fieri motum," Hamburg 274, 40.

rest. Thus, De Volder also came to stress more – in this series – the relative nature of movement.[18]

I will come back to De Volder's approach to the laws of motion in Sect. 4.2, *The laws of motion and impact*. Here, it is more urgent to illustrate, on the basis of De Volder's Cartesian idea of movement as change of *situs* or distance between parts, how he deals with other modes and properties of matter.

The other modes of matter are figure – as seen above – and rest. Figure is determined by motion: a body is figurable if it is moveable, namely we can distinguish between its parts if they are at movement or at rest with respect to each other, and with respect to a third body or fixed reference. A figure, thus, can be reduced to *situs*. Figure, let's say, is a derived mode, as it consists of a proportion of movement and rest in matter. Rest, in turn, is important as much as movement in order to account for how what we label parts, and even single bodies can 'emerge' from indeterminate extension. For Descartes, rest is as positive as movement, namely it is a mode of extension. In the pre-Cartesian tradition, rest was meant in two ways: either only as a privation of movement (i.e. the cessation from natural and violent movements), as maintained by Franco Burgersdijk,[19] or as a positive state of a body, namely, the "perfection of movement," (*perfectio motus*) intended only as the *terminus ad quem* of a movement: namely, the "permanence in a place," sc. in a natural place.[20] Descartes, who leaving aside any distinction between natural and

[18] "Quia tanquam quiescentia spectantur. Haec verba author addidisse videtur, non q[uo]d philosophicam involvant veritatem, sed ne nimium a loquendi consuetudine deviaret, si enim motus tantum esset translatio, et ea translatio, ut manifestum est, sit reciproca, patet eandem mutationem fieri in corpore CD cum ex eius vicinia transfertur corpus AB {q uitur} corpore AC ac proinde si tantum ad absolutam motus naturam attendamus, tantum erit motus in uno corpore quam in alio, et tantundem actionis in uno, q[ua]m in alio, quod quidem, ut verissimum est, ita minimum recederet a communi loquendi lege. {Cogeremu…} enim fateri minus corpis, ut demonstratum in sequenti, posse eodem tempore contrariis moveri motubus, quod prima fronte videtur {…}, quamquam si explicemus, non {…}; verumtamen cum rationi sit consentaneum, ut nos quantum veritas patiatur, accomoderemus etiam vulgari loquendi modo, addit author haec verba q[uae] tanquam quiescentia spectantur. Hoc ipso non obscure invicem posse corpora pro arbitrio spectari quiescientia, ita nempe ut si fiat {qui…dam} maior propinquitas, inter A et B, quam antea fuit, mei arbitri sit res, utriumq[ue] corpus dicere moveri, utriumq[ue] quiescere, quicquid enim reale est in A, quod ad motum spectat, esse in eodem modo in B, et quicquid est in B, etiam est in A. It ut pro arbitrio possum B spectare tanquam quiescens, et sic movebitur A, vel A tanquam quiescens spectare, et sic movebitur B, quo pacto tollitur illa in modo loquendi absurditas," Hamburg 274, 40–41.

[19] See Burgersdijk 1626, 133; Burgersdijk 1652, 16–17.

[20] This idea was maintained first by Averroes: see his commentary to *Physica*, book 5: Aristotle and Averroes 1562–1574, commentary 3, 208D–F. Amongst De Volder's sources, listed in the *Bibliotheca Volderina*, see Sennert 1651, 143: "[m]otui opponitur quies, quae est privatio motus eius, quod est aptum natum moveri; et quatenus propter quietem motus fit, potest etiam dici, quod sit finis et perfectio motus," and Plempius 1654, 440: "[r]ogabis: quomodo statui possit, quietem a facultate fieri, cum et illa et omnis alia quies sit privatio motus: privatio autem non habet causam positivam. Respon. quies dupliciter potest sumi: uno modo prout est privatio motus, altero prout est detentio et permanentia, vel etiam praesentia in loco; et sic est quid positivum, habetque positivam causam: quo posteriori modo hic sumitur quies." Johannes Clauberg, in turn, would combine the idea of rest as a positive permanence in a place with Descartes's replacement of *locus* with

unnatural places, came to mean rest as a positive mode in order to make sense of the resistance of resting bodies to being put into motion: for him no more action is required for putting a moving body at rest, than for moving a resting body (*Principia*, II.26). Roughly speaking, the idea of rest served Descartes in 'mechanizing' the world.[21] Also for De Volder the same force (*vis*) or efficacy (*efficacia*) is required both to bring a body to rest and to move it: rest, in fact, can be considered as a privation of movement only conceptually, as a *modum considerandi*, while in reality it is a positive mode of matter.[22] Moreover, rest had another key function for Descartes and his followers: namely, to account for the cohesion, and the sensory quality of the hardness of bodies. Rest is the glue that keeps bodies together: a body whose parts are at absolute rest with respect to each other is therefore absolutely hard, as only rest, for Descartes, is the cause of the resistance to motion, i.e. to the division of bodies (*Principia*, II.54–55).[23] On this aspect, De Volder came to disagree with Descartes. Moreover, whereas Descartes seems to admit the actual existence of numerically distinct bodies, De Volder adopted a monist standpoint on material substance. In what follows, I will focus on the latter point, which has a more fundamental character, and then I will show how he dealt with the idea of rest in order to account for the individuality of what appears to us as single bodies.

---

situs, as he would define rest as permanence in a *situs* in his *Disputationes physicae* (held in 1656–1661): "[q]uies est permansio in eodem situ. [...] Permansio enim positivum quid est, [...] situs denique positivum," Clauberg 1691, *Disputationes physicae*, 146.

[21] On the idea of rest in Descartes, see Schmaltz 2015.

[22] "[N]on arduum erit exterminare insitum illud plurimis philosophorum [...] animis praeiudicium, multo plus actionis in motu esse quam in quiete, plusque requiri virium et efficaciae in causa, quam motum corpori, quam in ea quam eidem quietem conciliare oportet. [...] Hosce autem interrogatos velim quidnam ipsos moveat, ut credant quietem tantum esse motus privationem? Nonne hoc ipsum quod plus actionis supponant in motu, quam in quiete? [...] Sed considerent velim hoc non tam a rei natura, quam a nostro considerandi dependere modo. [...] Demonstrandum itaque est, tantundem exigi efficaciae, in ea causa, quae corpus motum reducet ad quietem, quam in ea, quae corpus quiescens ad eundem gradum deducet motus. Corpus etenim movetur, causa igitur externa est, quae movet, cum ne dissidentibus quidem adversariis, nihil moveantur a se ipso. Quaecunque igitur causa, quae corpus motum ad quietem reducet, tollere necesse habet vim et efficaciam illius causae, quae confert motum," De Volder 1675, theses 6–7 and 9. In the Hamburg 273 series of *dictata*, commenting upon II.26, several arguments are provided to support this theses, such as that the same action is needed for movement and rest, as one is the negation of the other; that the persistence in a state of rest is an *aliquid*, and it requires a cause like movement does; that both movement and rest do not follow from the essence of matter, so that they both need an external cause: see Hamburg 273, 94–95. The latter argument is provided also in the other series of *dictata* (on II.26): see Hamburg 274, 40. More arguments for this theses are provided in a 1684 disputation *De motu*, which I discuss in Sect. 4.2.3.2, *A 'purely relational' standpoint on impact: the 1684 disputation* De motu.

[23] The notions of hardness and solidity, as I am going to show in Sect. 4.2.1.1, *Descartes's laws of motion and idea of force*, are not the same concept in Descartes's theory.

4.1.2 De Volder's Monist View of Material Substance

De Volder only deals *en passant* with the idea of substance: we can find it in his *dictata*, and in his correspondence with Leibniz. In the *dictata*, commenting upon *Principia* I.51–52 De Volder provides arguments on the ways in which substance can be predicated of created things and God, and on the problem of the creation of substances. For Descartes (I.51), a substance is "nothing else than a thing which exists in such a way, that it needs no other thing in order to exist."[24] In this meaning, therefore, the only substance is God, and can be said only 'equivocally' (*aequivoce*) of both God and creatures, while mind and body are substances 'univocally' (*univoce*) only in the following sense (I.52): "that they are things that, to exist, need only the concourse of God," – in the Latin version,[25] while in the French version a different definition is given, according to which a substance "can exist without the help of any created thing."[26] For Burgersdijk, on the other hand, 'substance' or 'to subsist' denotes just the fact that something does not inhere in a further subject, regardless of the cause of its own existence.[27] Descartes's characterization of substance, therefore, stresses its dependence on something other than itself in order to exist, rather than its not having a subject of inherence.

De Volder is much more cautious than Descartes on the use of the idea of substance. In the Hamburg 274 series of his *dictata*, commenting on Descartes's idea of substance given in I.51, as what "does not need any other thing in order to exist,"[28] De Volder stresses the equivocal meaning of 'substance', by claiming that "the concept of substance is formed only by means of our inadequate ways of conceiving," and that "the attributes of God are wholly different from the attributes of finite things," so that "it is not possible to form a common concept of substance to the extent that it pertains to God and [...] to creatures or finite things."[29] In the Hamburg 273 *dictata*, commenting on I.51, where he takes into account the definition of substance as "what exists by means of nothing, what has by its own nature the founda-

[24] Descartes 1982, 23. Cf. the original versions: "[p]er substantiam nihil aliud intelligere possumus, quam rem quae ita existit, ut nulla alia re indigeat ad existendum," AT VIII-A, 24; "[l]orsque nous concevons la substance, nous concevons seulement une chose qui existe en telle façon qu'elle n'a besoin que de soi-même pour exister," AT IX-B, 47.

[25] Cf. the Latin version: "quod sint res, quae solo Dei concursu egent ad existendum," AT VIII-A, 25.

[26] "[I]l faut seulement, pour entendre que ce sont des substances, que nous apercevions qu'elles peuvent exister sans l'aide d'aucune chose créée," AT IX-B, 47.

[27] "Per se subsistere nihil aliud est, quam non esse in aliquo, ut in subiecto. Ut ergo substantia dicatur per se subsistere, non est necesse, ut ab alio non pendeat: sed hoc tantum requiritur, ut ab alio non pendeat, tanquam a subiecto," Burgersdijk 1626, 12.

[28] "[...] si enim per substantiam velimus intelligere {...} nulla alia re indigeat ad existendum," Hamburg 274, 19.

[29] "Per substantiam. Conceptus substantiae formatur tantum per nostro inadaequatos concipiendi modos [...], videmus sane omnia Dei attributa plane diversa esse ab attributis rerum finitarum, nec posse communem formari conceptum substantiae quatenus competat Deo, et quatenus competit creaturis sive rebus finitis," Hamburg 274, 19.

tion of existence," and "is what it is by its own nature,"[30] With an anti-Spinozistic stance, De Volder warns that, since the attributes of God are His essence, any attribute in common between us and God will make us God Himself.[31] The concept of substance, as it applies both to God and creatures, therefore, is nothing but a confused *modum considerandi*. In fact, the idea that a concept can be used to mean entities different *toto genere* – like mind and body, for instance – had already been, first by Bacon, and then by Cartesians such as Johannes de Raey and De Volder himself (in his *De rerum naturalium principiis*) strongly criticised as a source of error in philosophy, especially as to the idea of substantial form.[32]

Yet, the idea of substance still had, for De Volder, too, some application in natural philosophy. In Hamburg 274 series of his *dictata*, as he comments on I.52 – where Descartes proposes the definition of substance as 'univocally' applicable to mind and body ("that they are things that, to exist, need only the concourse of God") – De Volder draws attention to the distinction between substance and modes (not mentioned by Descartes in the article), and to the French text of the article, in which Descartes "more clearly" provided the aforementioned, new definition, for which a substance is what exists without the help of any created thing.[33] In short, De

[30] "Substantia est, {q[uod]} nullius ope existit, {q[uod]} sua natura habet fundamentum existentiae, ergo unica dabitur, {q[uae]} est Deus, [...] qui sua natura est quod est," Hamburg 273, 41.

[31] "[...] non enim maior differentia est intra vitam nostram et aeternitatem, quam inter no[str]am et Dei voluntatem, ita ut certum sit, aut nulla attributa communicari, aut omnia, vero si verum sit omnia Dei attributa esse eius naturam, hinc sequitur si unum nobis communicetur attributum, communicari essentiam i.e. nos esse Deos," Hamburg 273, 42. In commenting upon I.35, De Volder clarifies that Descartes's idea that will is 'infinite' was subjected to many "misrepresentations," and clarifies Descartes's position by claiming that will is more extended as intellect because intellect works only on its own data, whereas will can combine them, by affirmations and negations: Hamburg 273, 29. In the other series of *dictata*, De Volder claims that anyway will cannot go beyond the data of the intellect: Hamburg 273, 15–16.

[32] See De Volder 1681, theses 46 and 48: "formis omnibus licet diversissimae naturae, id tamen competere, quod dent rebus hanc illamve essentiam, atque hac ratione omnem materiam ad unicum genus, omnesque formas ad unicum formarum conceptum reduci. Sed quis non videt id ipsum admodum metaphysicum sive logicum esse, nec ex physicis principiis aut phaenomenum natura, sed ex abstractis et si ad res referantur, phantasticis logicorum notionibus, desumptum esse? [...] Nec dissimili ratione in rebus incorporeis progredi licebit. [...] Habes [...] formam. Qui igitur conceptus cum non sint corporibus peculiares, quis non videt eos non ex physicis sed ex generalibus quas de rebus formamus, hoc est logicis, metaphysicisve notionibus desumtos esse? [...] Profecto non sine causa Verulamius frequentissime conqueritur, Aristotelem, quem hac in parte longe antecedunt Scholastici, *logicis suis notionibus, et abstractis de rebus conceptibus* totam penitus philosophiam pervertisse." Cf. Bacon's *Novum Organum*, part 1, aphorisms 14–15; cf. De Raey's *Specimen logicae interpretationis* (a series of disputations held in 1669–1671), referring to aphorisms 2, 20–21 and 97: De Raey 1692, *Specimen logicae interpretationis*, 536. On De Raey's criticism to Aristotelian logic and metaphysics, see Strazzoni 2012, 2015.

[33] "Sic autem videbimus illis utrisq[ue] commune esse, quod solo Deo ad existendum indigeant, in quo sane {differunt} a modis, qui praeter Dei concursum requirunt substantiam, in qua insunt, ita ut {...} substantiam definiri posse per id quod sit res solo Dei concursu egere ad existendum. Possunt autem. In gallico textu clarius. Verum cum inter res creatas quaedam sunt, {quae} non possunt existere sine aliis [...]," Hamburg 274, 20. De Volder owned the 1651 edition of the French version of Descartes's *Principia*: see *Bibliotheca Volderina*, 6.

Volder, keeping himself within the boundaries of Descartes's ideas, leans on an argument which does not make recourse to the idea of the intervention of God in accounting for the substantiality of matter, which consists only in its real and (as I am going to show in a moment) conceptual independence from any other created thing. Moreover, the idea that a substance does not need another created thing in order to exist, entails the idea that material substance is numerically one, unless one supposes God creates new material substances. This, however, is impossible for De Volder, as is evident from his criticism of substantial forms deployed in his *De rerum naturalium principiis*. By restating the argument against substantial forms used, in the recent past, by anti-Aristotelian thinkers such as Sébastien Basson, and David Gorlaeus, and discussed by Gijsbert Voet in his *Appendix ad Corollaria theologico philosophica nuperae disputationi de iubilaeo romano, de rerum naturis et formis substantialibus* (held on 23–24 December 1641), De Volder claims that substantial forms cannot be generated or educed from matter, because this entails a generation *ex nihilo* of a new form (as maintained by the Conimbricenses). Against this view, De Volder opposes that one for which movement is the only differentiator between individual "forms," which are therefore just accidents.[34]

De Volder's theory of material substance as independent from anything else, and as numerically one can indeed be found in his other *dictata* (Hamburg 273), as he comments on I.52. De Volder still uses Descartes's definition of substance given in I.52 to distinguish material substance and its modes or accidents, but he provides more detail on this move. First, he claims that body, in order to exist, requires only an efficient cause different from itself. However, if we conceive the body "when it has been produced," (*modo producatur*) we can conceive it "by itself only" (*se solo*), even if its existence depends on the operation of God (*ope Dei*).[35] On the contrary, in conceiving a figure, which "can exist or not exist" – i.e. it is an accident – we need to conceive also the idea of matter: so, a figure cannot be conceived

---

[34] "Tentemus tamen, an ex hisce tenebris quidpiam lucis possimus haurire. Aiunt itaque *ante cuiusque rei productionem formas non existere*, quippe quae producuntur generanturque, quod certe, si iam adessent, fieri non posset. *Neque tamen prorsus non existere*, ne natura quae producit formam ex nihil quidpiam faceret, sed esse *potestate in materia et naturalium causarum efficacia de illius potentia et quasi gremio exire, et ad actum perfectionemque deduci*. Conimbr. in Phys. l. I, c. IX [quaestio XII] art. IV. Quis quaeso haec conciliet? Formae *ante rei productionem non sunt*, neque tamen prorsus non sunt. Quae profecto, si ulla quaepiam, manifesta est contradictio, nec tamen difficultatem solvit. Restat enim explicandum, quo pacto naturalium causarum efficacia ex gremio materiae educantur formae. Quas si accidentia esse demus, plana sunt omnia. Quis enim non facile percipiat, motum, unicam illam mutationum omnium in corporibus causam, capacem esse, ut producat varias in materia figuras, varium ordinem, varia deinque accidentia, quae huius illiusve materiae formas constituant?" De Volder 1681, thesis 85. Cf. Conimbricenses 1592, 205–206. Cf. Gorlaeus 1620, 267, and Basson 1621, 159, referred to in Voet 1643, thesis 4. On the Cartesian and pre-Cartesian rejection of substantial forms, see Van Ruler 1995; Lüthy 1997, 2012, 44–49; Hattab 2009, parts 1 and 3; Manning 2012.

[35] "[...] omnis quid est a se solo, aut se solo indiget ad existendum, estq[ue] Deus, aut ind[ig]et quidem alio, sed nulla praeter Deum quales sunt creatae, aut praeter Deum alio {ad...} indiget, quales sunt modi, et accidentia: res plana est in exemplo concipiamus corpus id nil indiget ut existat praeter causam efficientem quandam a se diversam, sed cum corpus se solo concipitur modo producatur, hinc existere potest sola ope Dei," Hamburg 273, 42.

per se (through itself), but only *per aliud* (through something else). As to movement and rest (which, like figures, are 'accidental'), their ideas involve two concepts: that of body, and that of the "efficient cause of the body" – while body requires only the idea of extension.[36] The idea that in order to conceive movement, we need to conceive the very idea of God, as I am going to show, will be the main reason De Volder had for asking Leibniz for a demonstration of the activity of material substance based only on its idea: namely, a way to conceive movement in purely physical terms, and without any *Deus ex machina*.

The idea of body, to sum up, can be conceived *per se* – i.e. not through its efficient cause (which is God). Moreover, according to his *dictata* (Hamburg 273) the concept of body involves only a "[numerically] unique, simple and not divided nature," which needs only God in order to be brought into existence, but which can exist, "once [...] [it has] been produced," without God, for the reason that it can be conceived independently from the idea of its efficient cause.[37] This does not mean that matter, once created, is the cause of its own persistence, i.e. that it has a divine status, but that De Volder attempts to detach physics from rational theology i.e. metaphysics. The actual existence of what is represented by our concepts, indeed, is not required to be demonstrated, as he states in the last thesis of his *De rerum naturalium principiis*, and such a demonstration would belong to metaphysics rather than physics.[38] In other words, a numerically unique extension is required by the natural philosopher, and it is only the idea of material substance that it is conceivable *per se*, without any other concept. As he put it in his *Exercitationes* against Huet, in extension no actual division into parts can be conceivable: parts are only "distinguished" but not really "discrete and divided."[39] So, material substance is one, and mathematically divisible: namely, we can conceive, as mathematicians,

---

[36] "Sed si concipiam figuram corporis, haec concipi potest existere et non existere, hinc sequitur non existere per se, sed per aliud, {restum itaq[ue]} est non existere posse nisis restat corpus, cuius figura sit: eodem modo res se habet circa motum, et quietem, corpus potest existere sine motu, non vice versa motus sine corpore: corpus unicam tantum extensionem requirit, motus vero duas, et corpus et causam efficientem corporis," Hamburg 273, 42.

[37] "Existendum. Ut enim corpus in suo conceptu involvit unicam, simplicem et individuam naturam [...], idq[ue] est quod solo Deo eget, quid enim evidentius, quam id quod {possum} separatim concipere etiam existere posse seorsim, unde si possum concipere solam extensionem, solam cogitationem, cum illi conceptibus nil praeterea involvant, sane se solis possunt existere modo producantur," Hamburg 273, 42.

[38] See Sect. 3.2.2.2, *De Volder on the 'truth' of Cartesian principles*.

[39] De Volder claims that Descartes was not entirely clear on this point: yet, his idea of extension entailed that real division is impossible: "[s]ed de eo non multum laborabit Cartesius, utrum corporis natura in extensione ponatur, an vero in eo, quod habeat partes a partibus distinctas. Existimabit enim, nec opinor iniuria, dicere corpus extensum esse, vel corpus habere partes a partibus distinctas, esse duobus diversis loquendi modis, rem unam dicere eandemque. Dico autem partes a partibus distinctas, non vero [...] partes a partibus secretas et divisas, ne forte hic loquendi modus fucum alicui faciat, qui distinctas partes etiam in continuo et indiviso corpore facile intelligi putet, non ita vero partes a partibus secretas et divisas," De Volder 1695, *De corpore*, thesis 21. See also *infra*, n. 47. On the uniqueness of substance according to Descartes, see Lennon 2007; Smith 2010; Thomas 2015; Slowik 2001.

sub-volumes of it; from a physical point of view, however, it is not divisible into separate substances (i.e. a physical concept), because each part of it is continuous. Both the conceptual independence of a material substance and its simplicity or numerical uniqueness are in fact defended, in the clearest way, in his correspondence with Leibniz, where De Volder claims that a substance is something which can be perceived, first, (1) without reference to another concept (i.e. by itself, or *per se*).[40] This criterion cost De Volder an accusation of monism from Leibniz and from Johannes Regius (as seen in Sect. 3.1.3, *De Volder on substance monism*), i.e. the idea that God is the only substance, which De Volder successfully rejected, as the idea of a created substance cannot entail that of its supernatural cause, i.e. God.[41] Second, for De Volder a substance is (2) a simple entity, namely, one "from which nothing of what is represented can I take away, without the whole would perish,"[42] so that "if extension is conceived *per se*, namely, it is conceived, in a way that it represent one entity by a concept [...] it is more difficult to conceive really distinct parts in extension, rather than the unity. Indeed, if [...] there is no vacuum, a part cannot be conceived [...] without another one."[43] With a dubitative formula, eventu-

[40] See Sect. 4.1.1, *De Volder's Cartesian standpoint on the ontology of physics*.

[41] As seen in Sect. 3.1.3, *De Volder on substance monism*, indeed, Leibniz noted that by this criterion of substantiality only God is a substance: "[v]erum ut ad alterum veniam circa quod monere aliquid volebam, applicationem nempe ad substantiam, nonne etiam ad substantiae conceptum indigemus attributis? quodsi exigas, ut saltem non alia indigeretur substantia, circulum committeremus in definiendo. Deinde praeter primam substantiam nulla est (credo) quae per se possit concipi," Leibniz to De Volder, 31 December 1700, in GP II, 223. In turn, De Volder maintained that we can conceive created substances by themselves, and without conceiving their cause, because this cause is God and falls beyond our comprehension: "[q]uod de causa substantiae dicis, existimem si nullius substantiae essentia concipiatur absque causa eius possibili, nullius essentiam concipi. Modorum siquidem causas possibiles concipere mihi videor, sed substantiarum nullas. Hinc vulgo eas dicimus creari, hoc est, habere causam a qua producantur, sed modo nescio quo," De Volder to Leibniz, 7 October 1701, in GP II, 229. Leibniz did not reply further on this point. For a thorough discussion, see Lodge 2013, *Introduction*.

[42] See Sect. 3.1.3, *De Volder on substance monism*.

[43] See the letter of De Volder to Leibniz of 18 February 1699: "[e]xtensio[...], si quidpiam, per se concipiatur, hoc est, ita concipiatur, ut unum quid conceptui repraesentet [...] maioris laboris est partes realiter distinctas concipere in extensione, quam unitatem. Nam, siquidem, ut concedis, nullum detur inane, non poterit una pars, quam quis sibi fingat, concipi absque altera," GP II, 166. In his letter of 19 November 1703, Leibniz would criticize De Volder's monism on material substance as a form of Spinozism: "aut admittenda Tibi mea sententia est, aut confugiendum ad ἀλλόγλωσσον illud, quod subjicis: forte totum universum unam tantum esse substantiam; quod qui dicit substantiae vocem a sensu aliorum detorquet. Nec video ullum argumentum verisimile tanti paradoxi, nam quae B. de S. eo attulit, ne umbram quidem (si quid iudico) demonstrationis habent," GP II, 257–258. De Volder would then point out that Leibniz mistook his words, in his letter to Leibniz of 5 January 1704: "[u]bi dixi forte totum universum unam tantum esse substantiam[,] expresse locutus sum de universo corporeo, secutus hac in parte communem tantum non omnium opinionem statuentium corpora non esse substantias, quatenus sunt haec vel illa Corpora, sed quatenus sunt Corpora. Quare non capio, quid te moverit, ut opinionem B. D. S. quasi similem meae hic immiscueris. Quod enim illi peculiare est, non consistit in eo quod de universo corporeo dixi, sed quod cogitationem et corpus pro eadem substantia habeat, quod mihi semper visum est absurdissimum," GP II, 260.

ally, he claims that one cannot conceive a part of extension without considering all the others.[44] So, an individual portion of extension cannot be conceived *per se*, i.e. it is not an individual substance.

It is noteworthy that, on this topic Descartes was much more ambiguous, as, in various places of his works, he seems to admit the existence of individual material substances: for instance, in *Principia* I.64 he claims that "one and the same body can be extended in many diverse ways while retaining its same quantity: that is to say, more in length and less in width or depth at one moment, and, on the contrary, more in width and less in length a little later," so "we consider them as modes of those substances in which they are";[45] as if a single 'body' – in conformity with the definition(s) of substance given in I.52 – could exist and be conceived independently from any other parts of matter.[46] On his part, De Volder just straightforwardly moves to a radically monist position on matter. As he comments on I.64, for instance, he simply avoids any reference to different substances: rather, he claims that a visible body, like wax, can assume this or that figure or "form" without any change in matter.[47] In his eyes, the unchanging, definite quantity of one body mentioned by Descartes, which assumes different shapes in this or that direction, is a mode of the one extension as well itself, and cannot lead to the individuation of one substance, discrete from the continuum.

From a broader perspective, it is worth remembering that De Volder's considerations on substance were, at this time, between physics and metaphysics – understood as ontology. Two fields that De Volder aimed at keeping distinct in his *De rerum naturalium principiis*. For him, the natural philosopher uses concepts which serve him only to deduce i.e. explain phenomena, but he does not demonstrate their truth, i.e. he does not demonstrate that phenomena actually derive from them: this being the role of metaphysics (broadly conceived). Yet, metaphysical i.e. ontological considerations, such as those on the difference between substance and mode, were somehow an unavoidable part of his quest for first principles. Moreover, in the

[44] "Fateor, et iam in prioribus fassus sum, in corpore mathematico non inveniri unitates indivisibiles; sed addam simul, me haerere tamen, an non in infinita extensionis massa unitas haec reperiatur, si quidem partes, quas in hac massa concipimus diversas, realiter distingui non videntur, quoniam nulla pars aut poni aut concipi potest, nisi positis, conceptisque omnibus," De Volder to Leibniz, 14 November 1704, in GP II, 272.

[45] Descartes 1982, 29.

[46] On this topic, see especially Slowik 2001.

[47] "Longitudinem. Si concipiamus ceram {quae} potest omnes formas figurasq[ue] induere, eadem manet materia, licet moveatur aut sit sphaerior eodem modo minus licet nunc his nunc illis {vocet} cogitationibus, tamen est una eademq[ue], et nulla est diversitas. Sic cogitatio et extensio haberi possunt pro modis," Hamburg 273, 65; the other series of *dictata* (see Hamburg 274, 24) does not contain references to different material substances either. In his letter to Leibniz of 31 May 1704, De Volder maintained that Descartes and all those following mechanical philosophy "meant" (*voluerunt*) that material substance is one. See the letter of De Volder: "[u]bi dixi, universum corporeum, hoc enim expresse addidi, esse forte unam substantiam, nihil aliud dixi, quam quod Cartesius, et omnes illi, qui mechanicam philosophandi rationem secuti sunt, voluerunt, rerum omnium corporalium unicam esse substantiam," GP II, 266. See also the quotation from De Volder 1695, *De corpore*, thesis 21, given *supra*, n. 39.

natural-philosophical theories of De Volder a further metaphysical i.e. rational-theological intrusion had to be taken into account: namely, the role of God in the production and conservation of movement and rest. These two concepts were problematic issues in De Volder's natural philosophy. In Sect. 4.1.3, *De Volder's ideas on cohesion and divisibility*, I will deal mostly with De Volder's treatment of rest, as it is through this idea that he could allow for the existence of (apparently) individual bodies within the continuum of extension. In Sect. 4.2, *The laws of motion and impact*, hence, I will focus on his theory of motion as such.

4.1.3 De Volder's Ideas on Cohesion and Divisibility

The treatment of the problem of finding individual bodies in extension went along, in De Volder's theories, with his treatment of atomism. These problems are related because searching for individual 'bodies' means to search for parts or volumes within the continuum that are not in reciprocal movement. In this section, therefore, I will first address De Volder's treatment of atomism, as it offers a privileged perspective on his ideas on material individuality. Moreover, it provides an occasion for understanding what were De Volder's main natural-philosophical polemical targets, besides Aristotelianism. Then, I will move on to his treatment of rest as a factor of cohesion i.e. of individuation of parts of matter.

Atomism as a hypothesis on the constitution of natural things is dealt with mostly in De Volder's *De rerum naturalium principiis*. The text is aimed at showing how the ideas of matter, movement and rest (as they had already been presented in his *De corpore* and *De motu*, and dealt with in theses 163–172) can be the first principles from which any kind of phenomenon can be deduced – even if their truth is not a matter of physics, but rather of metaphysics. The justification of their use in physics is preceded by an analysis of various, alternative positions on first principles. By following Aristotle's and Bacon's divisions of philosophical sects according to the number of principles adopted by past philosophers, followed in Aristotle's *Physica*, and in Bacon's *De principiis atque originibus sive Parmenidis et Telesii et praecipue Democriti philosophia* (posthumously published in 1653),[48] De Volder presents various philosophical standpoints: (1) the monist 'sect', represented, amongst his contemporaries, by Jean Baptiste van Helmont, whose idea that water is the nutriment of plants is criticized by De Volder mostly by showing internal contradictions in his texts.[49] (2) The sect of atomists, following the principles of matter and

[48] De Volder 1681, theses 27–28; cf. *Physica*, 184b15–22; Bacon 1653, *De principiis atque originibus secundum fabulas Cupidinis et Coeli sive Parmenidis et Telesii et praecipue Democriti philosophia*, 223–224. On Bacon's approach to the history of philosophy, see Rossi 1968, chapter 2.

[49] De Volder 1681, theses 35–36; cf. Van Helmont's *Elementa*, in Van Helmont 1707, 51; *Astra necessitant; non inclinant, nec significant de vita, corpore vel fortunis nati*, in Van Helmont 1707, 115; *Progymnasma meteori*, in Van Helmont 1707, 64. These texts were originally published in his *Ortus medicinae* (1648). On Van Helmont, see Redgrove and Redgrove 1922; Giglioni 2000; Clericuzio 2000, 54–60, 89–96, 108–114 and 152–163; Debus 2001, 2002; Pagel 2002; Newman

vacuum, which for De Volder was represented, besides by Leucippus, Democritus, Epicurus and Lucretius, by those "dedicated to corpuscular philosophy, as the Englishmen call it," namely Gassendi, Bacon, Descartes and Boyle, who are praised by De Volder as ascribing any natural change to a change in modes, and not to material substance itself, which is one.[50] (3) The Aristotelian-Scholastic sect, relying on the three principles of matter, form and privation, and to whose criticism De Volder devotes a large part of this series of disputations (i.e. 69 on 186 theses); the alchemist sect (represented in modern times by Daniel Sennert and Joseph Duchesne or Quercetanus – whose principles are criticized as themselves needing an explanation, as in the case of fermentation (this probably being a covert attack on Sylvius).[51] (4) The sect of those admitting more than three principles: in ancient times, Anaxagoras's theory of homoeomerias; in present times, Daniel Voet (1630–1660, son of Gijsbert), who in his posthumously published *Physiologia* (1661), quoted by De Volder, proposed, alongside a traditional, hylomorphic view of material substance, a novel theory of elements by claiming that minerals are not mixtures but simple bodies, created by God. De Volder praises such a standpoint as it got rid of the problem of the origin of forms (a main factor of obscurity in Aristotelian philosophy), but he labels it as not really explanatory, as such 'elementary bodies' are just empirically picked out amongst visible, homogeneous bodies, and themselves require an explanation.[52]

In this overview, De Volder does not expressly stand with the corpuscularians, whose approach, even if praised as being 'mechanical', and as based on the ideas of figure, magnitude and motion (which he retains in his explanations), is criticized by

---

and Principe 2002; Hedesan 2016. De Volder's criticisms to the 'monist sect' were rebutted in a disputation taking place on 22 October 1681 in Copenhagen, presided over by Nicolaus Nicolai Seerup, professor of philosophy and medicine: namely the *Disputationum philosophicarum sive Cogitationibus rationalium de Disputationibus philosophicis sive Cogitationibus, uti inscribuntur, rationalibus Viri Clarissimi et Doctissimi Dn. Burcheri de Volder de rerum naturalium principiis prima qua contra eundem solidis, ut spes est, argumentis ac rationibus, aqua solum materiale omnium corporum principium adstruitur, et simul eadem assertio ab animadversionibus clarissimorum aliquot virorum vindicatur.*

[50] "Ad secundam sectam pertinet Democritus, licet infinita agnovisse principia dicatur {Aristoteli}, cum ex eodem constet infinitatem istam non in diversitate consistere substantiarum, sed in innumeris unius eiusdemque substantiae modis. Quae principia dudum reiecta nostro demum saeculo in lucem revocarunt Gassendus, Verulamius, Cartesius, Boylaeus, et quantum est ingeniosorum hominum, qui corpusculari, ut Angli vocant, addicti sunt philosophiae," De Volder 1681, thesis 30. As to the uniqueness of the material substance according to a mechanical perspective, see *supra*, n. 47.

[51] De Volder 1681, theses 31 and 136–148. See, in particular, thesis 139, referring to Sennert 1629, 128, and Duchesne 1648, 31. On Duchesne, Sennert and early modern alchemic tradition, see Michael 1997, 2001; Clericuzio 2000; Hirai 2001; Debus 2001, 2002; Lüthy 2005; Newman 2006, chapter 2.

[52] De Volder 1681, theses 33 and 149–156. See esp. thesis 155, referring to Voet 1661, 71–72. Daniel Voet also admitted blood circulation: see Voet 1661, 81–83. On Daniel Voet, see Dieckhöfer 1970, 1978. Voet's *Physiologia* was present, in the 1661 edition, in De Volder's library: see *Bibliotheca Volderina*, 5.

him for its atomist outcomes.[53] Nor does he ascribe his own principles (matter-motion-rest), to a particular philosopher. De Volder's critique of the atomists or corpuscularians concerns three main points: (1) the idea that gravity, which was correctly understood as movement by the atomists, is due to an internal, occult principle rather than to God, the author of all movements: this being a deviation from "the mechanical way of philosophy."[54] (2) The idea that a void space may exist: this is a criticism of Gassendi, repeatedly quoted by De Volder, to whom he opposes the idea that nothingness has no properties such as length, breadth and height.[55] Gassendi, on the other hand, admitted in his *Animadversiones in decimum Librum Diogenis Laertii, qui est de vita, moribus, placitisque Epicuri* (1649) and *Syntagma philosophicum* (posthumous ed. 1658) that spatial dimensions are different from corporeal dimensions, and that God did not create space.[56] (3) Eventually, De Volder criticizes atomism as such, i.e. the idea that matter has parts which cannot be divided, i.e. undergo any change. For De Volder, the property of divisibility necessarily follows from the idea of extension,[57] namely – as he would put it in his *Exercitationes* – it is just another way to consider it, because being extended means to have parts which are distinguished from each other.[58] In fact, as seen above, De Volder embraced a monist approach to material substance: parts can be conceptu-

[53] "[N]on possum non laudare mechanicam explicandi rationem, qua per figuram, magnitudinem, et motum rerum explicantur phaenomena. Agnoscere enim ex infinitorum corpusculorum figura et magnitudine discrepantium coalitu dissolutioneque omnes mutationes fieri corporeas, est, ut in sequentibus demonstrabimus, ipsam veritatem profiteri. [...] Verum tamen si dicendum est id quod res est, putem tamen in hac opinione quaedam reperiri, quae merito in reprehensionem incurrant. Primo enim id me male habet, quod haecce corpuscula individua faciant, omnisque mutationis expertia [...]," De Volder 1681, theses 114–115.

[54] "[Q]uod in hac opinione mihi non arridet, est, quod gravitate hasce atomos donet Epicurus. [...] Putem enim sufficere, si ostendam hanc gravitatem non percipi, ut eandem ex principiorum proprietatibus eliminem. Huius autem erroris Epicuri haec forte erit causa, quod non putaret id ipsum sufficere, si cum Democrito motum tribueret materiae, nisi eius motus causam assignaret. Qua in re cum ad Deum verum motus auctorem confugere nollet, utpote quem impie ab omni rerum mundanarum gubernatione secludit, maluit gravitatem quandam, licet incognitam, pro causa huius motus assignare. Qua in re et a se ipso et a mechanica philosophandi via descivit non parum," De Volder 1681, theses 119 and 121.

[55] "Considerabo de nihilo nullas affirmari posse proprietates, atque spatium cum habeat aliquas, etiam rem imo substantiam quampiam esse asseverabo," De Volder 1681, thesis 127.

[56] "Duplices quippe [...] Phys. sect l. I, § 1, fingit dimensiones, alias corporeas corporis, [...] alias spatiales [...]. [...] ut enim huic rationi obviam iret, quod pace tanti viri dictum sit, in longe difficiliorem se coniecti labyrinthum. Hoc enim spatium statuit rem quidem esse veram, sed tamen neque substantiam neque accidens, rem esse a Deo improductam, independentem, immutabilem. [...] Imo, inquit, mirum videri non debet *Deum non esse authorem spatii*," De Volder 1681, theses 130 and 133; cf. Gassendi 1649, 614–619, 1658, volume 1, *Syntagma philosophicum*, 181–182. On Gassendi's theory of matter and space, see Joy 1987, chapter 8; Lennon 1993, chapter 2; LoLordo 2007, chapters 5–6.

[57] This passage follows the quotation given *supra*, n. 53: "[p]rimo enim id me male habet, quod {haecce} corpuscula individua faciant, omnisque mutationis expertia. Quippe nulli divisioni natura repugnat corporis, utpote quod cuiuscunque etiam sit parvitas, cum extensum sit, partes habeat, in quas dividi possit," De Volder 1681, thesis 115.

[58] See *supra*, n. 39.

ally distinguished within it, but not actually divided or separated. Their conceptual distinction, in any case, has a real basis. As a principle of the cohesion of parts, he adopts the idea of rest; as a principle of differentiation or change, he adopts that of movement. This allows him to criticize the ideas that (1) smallness (*parvitas*) and (2) solidity (*soliditas*) will allow the physical indivisibility of some part, namely, that some volumes of extension can never be moved with respect to each other.

Against smallness as a factor of indivisibility, De Volder maintains that as one can move i.e. divide a certain part of a body, one can move i.e. divide a smaller portion of this part, regardless of its smallness.[59] The same applies to solidity, or hardness,[60] assumed in two cases, which are two extremes of a spectrum: from the weak solidity of a porous piece of wood, to the strongest solidity of a perfectly continuous body, in which the parts are at absolute rest with respect to each other.[61] In the first case, if a porous piece of wood can be broken, nothing prevents that a subpart of it can also be broken;[62] by extension, therefore, a perfectly continuous body, such as a "full and coherent part" of such a piece of wood, which may constitute the

---

[59] "Naturam divisionem admitteret, non dari tamen causam satis efficacem, quae hanc divisione sive ob parvitatem sive ob soliditatem possit instituere. Sed frustra. Nihil quippe putem manifestius, quam corpusculorum exiguitatem divisionem reddere, non difficilem. Nonne enim evidens est, ad divisionem corporum id unice requiri, ut pars, quaedam moveatur reliqua immota, aut certe ut moveatur separatim a reliqua? Unde quid aliud sequitur, quam, cum facilius corpus dimidii pedis quam totius moveatur, multo etiam facilius caeteris paribus dividi bifariam id corpus, quod pedem, quam quod duos exaequat?" De Volder 1681, thesis 115.

[60] In his 1695 *Exercitationes* De Volder clarifies how, according to Descartes, hardness (*durities*) is properly speaking the sensation that results from the state of rest of parts of the body: De Volder 1695, *De corpore*, thesis 33. On the concept of solidity, which in fact means more than the mere resistance to division, and which has an important role in cosmology and theory of gravity, see Sect. 6.1.3, *Descartes's theory of planetary stability, weight and pressure*. In the present section, I will take solidity and hardness as synonyms of resistance to division.

[61] In a Cartesian cosmos, such are the globules of second matter, which are perfectly continuous, internally. This idea can be found in the disputation presided over by De Volder, but authored by the *respondens* (Paulus von der Lahr), *De absoluta quiete* (1684): the author maintains that simple bodies are those which are not discrete, namely, they are pure, continuous volumes of extension i.e. they are internally motionless, even if with respect to other bodies they change their distances or situs. Absolute rest can be found in the particles or spheres constituting Descartes's second matter, which are the most solid bodies, in which rest is their glue (thesis 12). This disputation may have been an exercise against Boyle's *Discourse about the Absolute Rest in Bodies* (published in the 1669 edition of his *Certain Physiological Essays and Other Tracts*), and then translated into Latin and published in Amsterdam in 1671 as *De absoluta quiete in corporibus*, where Boyle maintains that according to the Cartesian cosmological hypothesis, based on the idea that subtle matter flows in all bodies, no absolute rest can be found in nature.

[62] "[...] consideremus lignum solidissimum firmissimumque, aut aliud quodcunque corpus ligno solidius. Illud certe vel ab hominibus dividetur, vel ab aëre longo temporis tractu tandem corrumpetur dissolveturque. Constituamus porro in hoc ligno dimidiam partem porosam, dimidiam solidam esse: nonne ut aliud nunc corpus dividam, quod totum sit solidum, quodque sit dimidium huius, eadem vis requiritur, quae ad dividendum hocce lignum; ut enim lignum dividatur, dimidia solum superficiei pars, ut separetur necesse est, quam eandem in altero corpore separare solummodo oportet," De Volder 1681, thesis 117.

100th or 1,000th portion of it, can also be broken.[63] So, the division made by removing virtually discrete parts – i.e. those actually divided by pre-existing pores, cavities or meatuses, but keeping some continuity – and the division made by removing a portion from a continuous volume, are the same kind of physical operation for De Volder.

This whole line of physical argument against indivisibility is opened by De Volder with a justification based on a "mathematical" (in fact, mixed mathematical, or mechanical) principle, which justifies how, in concrete cases of division, one can always remove a smaller part of a body. According to this principle, a small force can *move* a big, whole body, as in the case of the lever. So, a greater force may reasonably be enough to *remove* a part of it, from its whole.[64] Yet, he did not provide more details on the mechanical side of this issue, moving to the example of the breaking of a piece of wood as he did not want "to appear to stick to mathematics."[65] The Cartesian rationale of De Volder's argument is that, as cohesion consists of rest, if a movement is required to move the whole, a certain force of movement present in the universe will be enough to move – and remove – a part of it. In his *De rerum naturalium principiis* De Volder does not address this point further.[66] However, it may have been seen as problematic by him, for the reason that this very case, namely the case of a 'terrestrial' body in which the removal of the 100th or 1,000th portion of it is considered, is used by him to argue against Descartes's theory of cohesion as given only by rest.

We find it in the second of his *Experimenta*, which took place in the span of time of disputation 10 of the *De rerum naturalium principiis*, namely in late 1676 or at the beginning of 1677.[67] The experiment, which is a repetition of experiment 1,

[63] "Deinde si putes centesimam aut millesimam solummodo partem plenam et cohaerentem, reliquas porosas et secretas esse, ea certe vis, quae lignum hocce dividit, dividit quoque corpus solidum aequale centesimae millesimaeve parti istius ligni. Unde manifeste et illud sequitur, quo corpus solidum minus est, eo facilius dividi, et ex inde licet atomi solidae sint, cum tamen sint parvulae, dari vim in natura sufficientem, quae ipsas dividat. Imo certe, si nulla solidi corporis poterit fieri divisio, nunquam fiet ulla. Sumamus enim corpus quam maxime porosum, profecto nulla hic divisio fiet pororum ratione, illic enim partes iam separatas obtinent, sed, ut manifestum est, fiet omnis ratione partium suis superficiebus cohaerentium. Ad quas superficies si respiciamus, illa corpora non porosa sed solida sunt, unde si vera est haec opinio, non licuit unamquamque superficieculam a reliqua separare, quia corpus hic solidum est nullis meatibus distinctum, Unde aut agnoscendum est, nullam unquam divisionem fieri, aut fieri solidi corporis," De Volder 1681, thesis 117.

[64] "Mathematicorum effatum est qualibet potentia etiam minima quodlibet corpus etiam maximum moveri posse. Quid itaque? Si minima vis maximum movere possit corpus, nonne vis maior poterit movere minusculi corporis partem, et eo ipso motu id corpus dividere," De Volder 1681, thesis 117.

[65] "[...] ne in mathematicis videar haerere, consideremus lignum [...]," De Volder 1681, thesis 117. Cf. *supra*, n. 62.

[66] In his *De rerum naturalium principiis* De Volder refrains to deal with the nature of rest: "[n]eque vero arduum erit [...] coniicere haec nobis principia praeter materiam et motum (nisi quis, de quo non laboramus, addere forte velit quietem) nulla alia esse," De Volder 1681, thesis 161.

[67] In the *De rerum naturalium principiis*, the thesis in which the idea of rest and continuity are discussed is the 117th one, present in disputation 10 (now lost). Disputation 9 took place in 1675, while disputation 11 took place in 1676 (they are both preserved at Erlangen University Library).

Fig. 4.1 Cohering marbles used by De Volder. (Source: De Volder 1676–1677, 78v. Courtesy British Library, ms. Sloane 1292)

consists of making two flat, polished cylinders of marble of 2 inches and ⅓ in diameter, and a bit thicker in depth, cohere with each other without any intermingling air but only with a stratum of tallow. (See Fig. 4.1, De Volder 1676–1677, 78v, and, for greater clarity, the depictions of a similar pair of cylinders (slightly different from those described by Christopher Love Morley, now lost) given in the *Descriptio antliae pneumaticae et instrumentorum ad eam inprimis pertinentium* (1700) of Johannes Joosten van Musschenbroek, (used also by Wolferd Senguerd: see Figs. 4.2, Van Musschenbroek 1700, plate 3, figure 2, 4.3, Senguerd 1715, plate 3a, and 4.4, Senguerd 1715, plate 3b).[68]

Disputation 10, therefore, was almost certainly coeval with experiments 1–2, taking place in early 1676 (12–13 March).

[68] The apparatus of Johannes Joosten van Musschenbroek was more advanced than the one used by De Volder in the 1670s (probably built for him by Samuel van Musschenbroek, who died in 1681). The use of Johannes Joosten's model is described in Senguerd's *Rationis et experientiae connubium*, chapter 19, in a long quotation from a *pro gradu* disputation of a student of his, Johannes Franciscus de Witte van Schooten, who "exactissime in Collegiis meis multoties observaverit, annotaverit, atque in elaboratissima disputatione inaugurali (quam pro gradu doctoratus consequendo in hac Academia publice defendit) methodice exposuerit, et ad mentem meam tradiderit." The disputation was held in 1712, and published as *Dissertatio philosophica inauguralis de solido, eiusque partium, nec non hemisphaeriorum concavorum, et cylindrorum solidorum cohaerentia*, and appeared also in the *Acta Eruditorum* of February 1714 (as *Wolferdi Senguerdi* [...] *Annotationes circumstantiarum singularium circa cohaerentiam hemispheriorum concavorum et cylindrorum solidorum*). Senguerd used different pairs of cylinders, with different diameters and materials, and disposing them either vertically or horizontally. E.g., cylinders of white marble with a diameter of 2 inches and 7 lines, 2 inches and 1 line, 0 inches and 10 lines, sustaining respectively 1,150, 900 and 200 pounds; cylinders of black marble of 2 inches and 2 lines diameter, sustaining 900 pounds, and two other pairs of ivory and brass. Senguerd noted that cylinders sustained the same weight even if disposed horizontally: see Senguerd 1715, 164, 166–167 and 171–172. The pair described by Morley (2 inches and circa ⅓ diameter), does not really fit with those ones described by Senguerd – who performed experimental lectures also at home, with his own instrument – which correspond to 2 inches and 4 lines, since, according to the English system, 12 lines = 1 inch.

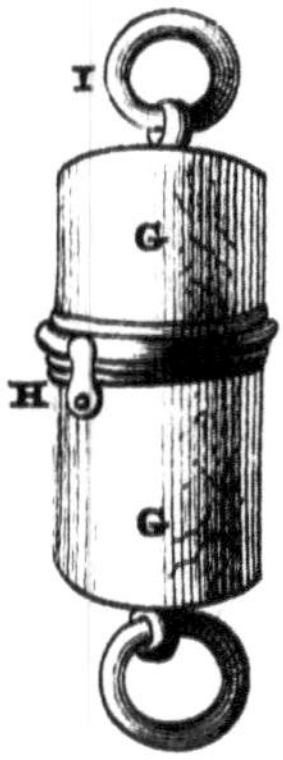

Fig. 4.2 Cohering marbles sold by Johann Joosten van Musschenbroek. (Source: Van Musschenbroek 1700, plate 3, figure 2. Courtesy Hessisches Staatsarchiv Marburg, file UniA 305a Nr 5947)

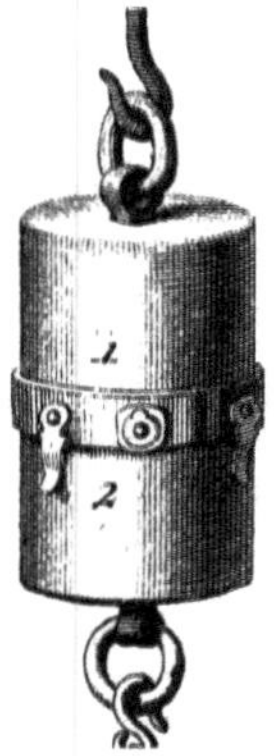

Fig. 4.3 Cohering marbles used by Wolferd Senguerd. (Source: Senguerd 1715, plate 3a. Zentralbibliothek Zürich, signature: NP 1783, https://www.e-rara.ch/zuz/doi/10.3931/e-rara-39787. Public domain mark 1.0)

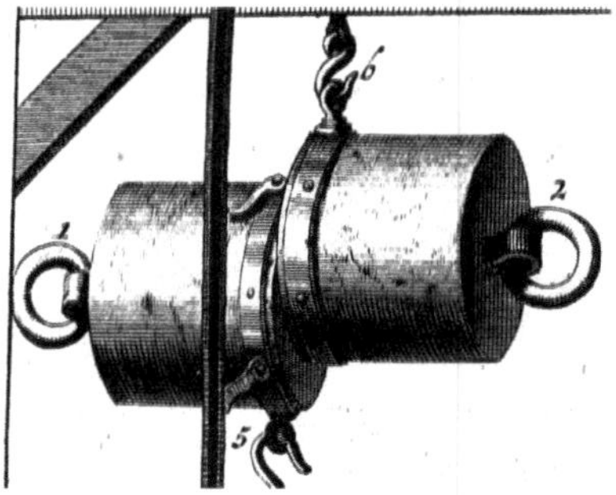

Fig. 4.4 Cohering marbles used by Senguerd, in horizontal position. (Source: Senguerd 1715, plate 3b. Zentralbibliothek Zürich, signature: NP 1783, https://www.e-rara.ch/zuz/doi/10.3931/e-rara-39787. Public domain mark 1.0)

De Volder managed to append to the lower marble weights of up to 285 pounds (in the first experiment, when one of the hooks broke up), and 450 pounds (in the second), before the two marbles separated. In this, he was more successful than

Boyle – who could hang only 80 pounds – as De Volder himself remarked in his lectures.[69]

According to De Volder's explanation, such marbles are kept together by the pressure of the surrounding air.[70] On this basis, De Volder claims that cohesion is to be ascribed also to the pressure of air from all directions on a body, besides the rest of their parts, which alone cannot allow for the cohesion or integrity of bodies. Indeed, according to Morley's report, a round piece of iron can be easily moved "I cannot separate the half, 3rd or 4th part of said things yet I may at least separate the 100 or 1000th part of them, the force of my hands and their bigness being so much bigger than said 100 or 1000th part." "[T]hus" – Morley continues – "it is not so much the quies of said things which maketh them inseparable, for said quies we see we can easily alter, or overcome even in the whole body moving it to and fro."[71] Then, Morley reports an objection of De Volder to Descartes's explanation of this case, given in Descartes's *Principia*, II.63. For Descartes, we cannot break a solid object such as a nail with our hands, because we feel pain, and because only the solid or hard parts of our hands effectively act on the hard or solid structure of the nail. However, if we use a hammer, we can break it by exerting a moderate force, as the hammer has a solider or harder structure than the hands.[72] For De Volder, "this

[69] "Then he hang them up, as before, and into the scales he added more, et more weight, till they disunited and the one fell down with the weight which appeared to be very nigh 450 lb. [...] He said Boyle could only hang on two such cylinders (being even something broader i.e. of 3 fingers breadth) 80 lb weight, which is not the 4th part of this experiment," De Volder 1676–1677, 80v. As Boyle reports in his *Defence* against Francis Line (1662), in open air he could make one marble hang to the other up to a maximum of 1,300 (English) ounces only, which in De Volder's statement are converted into 80 Dutch pounds (with a ratio of 1 pound to 16 ounces, corresponding to the Dutch (i.e. Amsterdam) commercial measurements, rather than to the English apothecary system), as due – according to Boyle – to the fact that their surfaces were not so polished: see Boyle 1662a, 84–86. In experiment 27 (as I am going to show in Sect. 5.1.2, *The weighing of a separate volume of air*) Morley states that De Volder followed the "Amsterdam weight" (De Volder 1676–1677, 138v), which is moreover followed in his calculations on the weight of a determined volume of air in disputation 5 of his *De aëris gravitate*. The same system of measurement was followed by Senguerd in his *Connubium* (see Senguerd 1715, *Ad lectorem*). Notably, De Volder's hanging of weights of 560 and 600 pounds (probably, English ones) to cohering marbles of circa 3 inches diameter are reported also in Dalrymple 1686, 539: "[h]abeo tabulas marmoreas quarum diameter trium pollicum est, quae circiter 500 lib. sustinent. Sunt in Academia Leydensi tabulae marmoreae eiusdem fere diametri, quae uno tentamine a Doctiss. Professore Voldero demonstrato, 560 lib., in alio tentamine ab eodem ultra 600 lib. sustinuerunt."

[70] Notably, one can also find the experiment with marbles in Boyle's *New Experiments* (1660), as experiment 31, which Boyle notoriously performed in what was supposed to be an evacuated container: yet, even in the (supposed) absence of air, the plates of marble did not tear apart, even with an additional weight of 4 ounces appended to the lower one. Boyle justified the failure of the experiment by taking into account a leakage in the pump (Boyle 1660, 230–231). Later, he was attacked by Hobbes in his *Dialogus physicus de natura aëris* (1661): for a full account, see Shapin and Schaffer 1985.

[71] De Volder 1676–1677, 82r and 83r.

[72] "[A]t first glance there seems to be no reason why, for example, an iron nail (or any other body which is not large but extremely solid), cannot be divided into two parts solely by the force of our hands. For each half of this nail may be considered to be an individual body; and since one half is

answer [...] doth not satisfy,"[73] as in fact no parts at all of a piece of iron (as one 1,000th) can be removed from it.

In fact, De Volder does not claim that rest is not a factor in the cohesion of bodies, but rather that 'internal' rest, i.e. the absence of particles in motion in a body, does not allow a body to be disaggregated because the surrounding fluids act on it from outside, and, being not counteracted by internal fluids, keep the solid body together. In turn, we can easily move the whole body, because the pressure is exerted in all directions on it, so that the effect of this pressure on a part of it, can be easily overcame by our hands, because our movement is helped by the pressure of the fluid:

> we cannot of some things separate the 1000th part, although we can move the whole body of them to and fro: for as for the moving the whole body, {they} being in the fluid body of the air, as if it were in water, and the whole body of the air not pressing upon them more than water, nay not so much, hence said pression is not felt by the things, and we can move them: but as for the separation of one part or particle from another, besides the quies of said particles, they are so closely united as these two cylinders, and not the least jot of air being between the pores of said solid things it is extremely hard to pull them asunder: from whence is collected that the more or less pores the body<h> hath, the more or less of air is in them whence the greater or lesser difficulty of separation ariseth.[74]

This solution had the advantage of overcoming the possible objection that we cannot impart motion to the sub-parts of a cohering body, if they are in a condition of rest between each other, because it is always more easy to move the whole body, as it is not in a condition of absolute rest i.e. continuity with respect to the surrounding fluid. Moreover, it had the advantage of making consistent an explanation based on the idea of pressure of air – used at De Volder's time to explain the cases of cohesion of bodies allegedly emptied of air (as the cohering marbles, or Von Guericke's spheres) – with Descartes's general explanation of cohesion as given by rest. Thirdly, let's say, it had the advantage of getting rid of another explanation of cohesion, namely, that one based on the idea of entanglement, which was embraced, amongst others, by Gassendi and by Descartes himself. In order to appreciate De

---

smaller than our hand, it seems that it ought to be possible to move it by the force of our hand and thus separate it from the other half. It must, however, be noted that our hands are extremely yielding, or closer to the nature of fluid bodies than to that of solid ones; for that reason, they are not accustomed to act as a whole against a body which they have to move; only that part of our hands which touches that body brings all its pressure to bear upon it at the same time. [...] It is because this part can more easily be separated from the rest of the hand than the part of the nail from the rest of the nail, and because this separation cannot occur without the sensation of pain, that we cannot break the iron nail by means of our hand alone. If, however, in order to divide the body, we strengthen our hand by applying the force of a hammer, file, pair of cutters, or other tool to a part of the body to be divided which is smaller than the tool being used; it will be easy to overcome its hardness," Descartes 1982, 76. Cf. De Volder 1676–1677, 82r: "[u]nto this Descartes answereth that the softness of our hands the pain we feel and our only touching said things s[ecun]dum extimam superficiem is the cause we cannot separate not so much as the least part of many solid things."

[73] De Volder 1676–1677, 82r.

[74] De Volder 1676–1677, 82v–83r.

Volder's originality with regard to the problem of cohesion, therefore, it is worth comparing his explanation of the case of the cohering marbles with that of some of his nearest predecessors.

4.1.3.1 Four Factors of Cohesion: Vacuum, Rest, Pressure and Entanglement

The phenomenon of the cohering marbles has a central place in the first *Giornata* of Galileo's *Discorsi e dimostrazioni matematiche sopra due nuove scienze* (1638), where it serves to support the idea that the cohesion of solids is due – "at least in part and as a concomitant cause" – to a quantifiable "force of vacuum" (*virtù del vacuo*): a vacuum whose existence is testified to by the fact that, at some point, the marbles can be separated and air flows into the void space between them. This idea served Galileo, for instance, in explaining why water cannot be raised up to no more than 18 *braccia* (circa 10.5 metres) in a syphon: attributing this phenomenon to the fact that a vacuum is not enough strong to raise the water over this height.[75] More generally, he explained the cohesion of bodies by means of the idea of "very small [interstitial] vacuums," which keep together the particles of solid bodies.[76] In commenting

[75] "Dirò prima del vacuo, mostrando con chiare esperienze quale e quanta sia la sua virtù. E prima, il vedersi, quando ne piaccia, due piastre di marmo, di metallo o di vetro, esquisitamente spianate pulite e lustre, che, posata l'una su l'altra, senza veruna fatica se gli muove sopra strisciando (sicuro argumento che nissun glutine le congiugne), ma che volendo separarle, mantenendole equidistanti, tal repugnanza si trova, che la superiore solleva e si tira dietro l'altra e perpetuamente la ritiene sollevata, ancorché assai grossa e grave, evidentemente ci mostra l'orrore della natura nel dover ammettere, se ben per breve momento di tempo, lo spazio voto che tra di quelle rimarrebbe avanti che il concorso delle parti dell'aria circostante l'avesse occupato e ripieno. [...] Tal resistenza, che così sensatamente si scorge tra le due lastre, non si può dubitare che parimente non risegga tra le parti di un solido, e che nel loro attaccamento non entri almanco a parte e come causa concomitante. [...] Dal seguir dunque che fa l'inferior lastra la superiore, si raccoglie [...] che pur tra le medesime piastre resti qualche vacuo, almeno per brevissimo tempo, cioè per tutto quello che passa nel movimento dell'ambiente, mentre concorre a riempiere il vacuo; ché se vacuo non vi restasse, né di concorso né di moto di ambiente vi sarebbe bisogno. Converrà dunque dire che, pur per violenza o contro a natura, il vacuo talor si conceda [...]. Dirò il modo dell'appartar la virtù del vacuo dall'altre, e poi la maniera del misurarla. E per appartarla, piglieremo una materia continua, le cui parti manchino di ogni altra resistenza alla separazione fuor che di quella del vacuo, quale a lungo è stato dimostrato in certo trattato del nostro Accademico esser l'acqua: [...] Ed io mercé di questi discorsi ritrovo la causa di un effetto che lungo tempo m'ha tenuto la mente ingombrata di maraviglia e vota d'intelligenza. Osservai già una citerna, nella quale, per trarne l'acqua, fu fatta fare una tromba, da chi forse credeva, ma vanamente, di poterne cavar con minor fatica l'istessa o maggior quantità che con le secchie ordinarie; ed ha questa tromba il suo stantuffo e animella su alta, sì che l'acqua si fa salire per attrazzione, e non per impulso, come fanno le trombe che hanno l'ordigno da basso. Questa, sin che nella citerna vi è acqua sino ad una determinata altezza, la tira abbondantemente; ma quando l'acqua abbassa oltre a un determinato segno, la tromba non lavora più," Galileo 1890–1909, volume 8, 59–64. On Galileo's discussion of vacuum, see Grant 1981, 60–65; Maffioli 2011; Zepeda 2013.

[76] "E chi sa che altri minutissimi vacui non lavorino per le minutissime particole, sì che per tutto sia dell'istessa moneta quello con che si tengono tutte le parti congiunte?" Galileo 1890–1909, volume

upon Galileo's *Discorsi* in his letter to Marin Mersenne of 11 October 1638 (later published by Claude Clerselier), Descartes claimed that Galileo's explanation of the case of the cohering marbles, and his theory of cohesion as based on it, had to be rejected because it relied on the idea of the fear of a vacuum.[77] For Descartes, as he claims in his letter to Mersenne of 15 November of the same year, the case of cohesion of contiguous bodies can be attributed rather to the "weight of the air." Continuous bodies, on the other hand, are kept together by their unspecified "link and union."[78] This is also discussed in a further letter to Mersenne, of 9 January 1639, where, commenting upon a lost letter of Mersenne, in which he probably used the case of the cohering marbles as a way to measure the weight of the air, Descartes maintains that, when these marbles are divided, for an instant the space between them is filled by subtle matter.[79] For Galileo, this was in fact void: in this, he relied on Lucretius's *De rerum natura*, where the experience of the air progressively flowing in the space between the marbles is used to argue for the existence of a vacuum – which for Lucretius is thus powerless as a factor of cohesion – and where the cohesion of wood and iron is explained by the idea of entanglement, i.e. by "rings and barbs."[80] A similar argument against the fear of a vacuum, actually, is used by De Volder in his *Experimenta* (1–2), as for him also the case of the cohering marble shows that the fear of a vacuum does not prevent the creation of vacuum, and that it is powerless as a factor of cohesion.[81]

It is noteworthy that, the explanation of the properties of iron one can find in Descartes's *Principia* (IV.143) is based on the idea of entanglement of "barbs or hooks" (*hamis vel uncis*), which make iron flexible and soft rather than extremely

8, 66. On Galileo's discussion of cohesion, see Baldini 1976; Le Grand 1978; Redondi 1983, 1985; Shea 1989, 2004; Palmerino 2000, 2001; Bertoloni Meli 2006, 91–95.

[77] "Il donne deux causes de ce que les parties d'un cors continu s'entretienent: l'une est la crainte du vuide, l'autre certaine cole ou liaison qui les tient, ce qu'il explique encore apres par le vuide; et ie les croy toutes deux tres fausses. Ce qu'il attribuë au vuide [...] ne se doit attribuer qu'a la pesanteur de l'air; et il est certain que, si c'estoit la crainte du vuide qui empeschast que deux cors ne se separassent, il n'y auroit aucune force qui fust capable de les separer. [...] Il examine la cole qu'il adiouste avec le vuide pour la liaison des parties des cors, et il l'attribuë a d'autres petits vuids qui ne sont nullement imaginables. Et ce qu'il dit [...] pour prouver ces petits vuids, est un sophisme," AT II, 382. De Volder owned the first edition of Descartes's letters published by Clerselier, in three volumes (1657–1667): see *Bibliotheca Volderina*, 6.

[78] "La seconde partie contient vos remarques touchant Galilée, ou i'avoüe que ce qui empesche la separation des cors terrestres contigus, est la pesanteur du cylindre d'air qui est sur eux iusques à l'athmosphere, lequel cylindre peut bien peser moins de cent livres. Mais ie n'avouë pas que la force de la continuité des cors vienne de là; car elle ne consiste qu'en la liaison ou en l'union de leurs parties," AT II, 439–440.

[79] AT III, 481–482.

[80] *De rerum natura*, book 1, verses 384–398; book 6, verses 1087–1089: "quasi ut anellis hamisque plicata / inter se quaedam possint coplata teneri / quod magis in lapide hoc fieri ferroque videtur."

[81] "Hence also he inferred, as above that it was not the metus vacui which did hinder the separation of the cylinders with this dilemma: aut non datur talis metus vacui, et coë[u]nt illa, talis metu, cum non detur, nequit impedire avulsionem cylindrorum. Aut si datur, et ille sit qui impediat disiunctos, iam profecto disiunximus illos, et coë[u]ntes datur via, et modus, introducendi vacuii in na[tur]am: imo fecimus ipsi mo[do] vacuum, q[uo]d utrumq[ue] negare de[be]nt, qui vacuum tantopere abohrrere na[tur]am aiunt," De Volder 1676–1677, 81v.

hard and fragile, like steel.[82] An explanation which is adopted by De Volder himself in his *dictata*.[83] Generally speaking, recourses to the idea of ramified or branched particles are countless in Descartes's treatises.[84] Yet, Descartes did not use it as the primary factor in the cohesion of bodies, but rather – let's say – as a secondary or derived factor. In the Hamburg 273 series of his *dictata*, commenting upon *Principia* II.55 (where Descartes presented the idea of rest as a factor of cohesion), De Volder addresses the entanglement theory, which is ultimately deficient as it leaves unsolved the issue of the internal cohesion of "barbs and hooks" (*uncis et hamulis*). The theory is ascribed to Gassendi and criticized,[85] who in fact provided a theory of cohesion based on entanglement in his *Syntagma philosophicum*. For Gassendi, rest cannot be adopted as a factor in the cohesion of bodies, for the reason that atoms are provided with an internal, endless impetus: so they cohere by their figure, and keep their fixed position by keeping however a continuous vibration.[86]

Entanglement as a factor of cohesion, such as in the case of the cohering marbles, was moreover discussed by Boyle in his *History of Fluidity and Firmnesse* (published in his *Certain Physiological Essays and Other Tracts*, 1661), in which the causes of "Firmness or Compactness," are hypothesized as being the "Grossness, the quiet Contact, and the Implication of the component parts." The first cause is established by Boyle on an experiential basis: bigger bodies are usually more sol-

---

[82] "[...] these scrapings are no longer so firmly joined together in these droplets, and the droplets no longer adhere to one another by immediate contact, but are fastened together as if by certain hooks or barbs. And therefore the steel becomes soft and flexible rather than extremely hard, rigid, and fragile," Descartes 1982, 248.

[83] "Si rursus ignis admovet globulos, partes, si non avelluntur, {concitiurtur} saltem, et lente si refrigeretur fiunt asperiores, nec cohaerent {superficibus} immediatis, sicque faciunt ferrum flexile {ea it erg} differentia est inter corpora mollia, flexilia, et dura fragilia, quod illa habeant cohaesione. {\.} uncis cohaerent, non a sola quiete, ut dura fragilia quia superficies est admodum parva," Hamburg 273, 520; The same explanation is given also in the other *dictata*: "[u]ncis quibusdam. Ea enim cohaesio q[uae] fiet per uncos, multo minus firma est, quam ea, q[uae] fit immediato contactu, licet plurimis secus videatur," Hamburg 274, 114.

[84] On Descartes's theory of matter, see Gaukroger 2000b, 2002, chapters 5–6; Slowik 2002; Jalobeanu 2002; Lüthy 2006; Anstey and Jalobeanu 2011; Schuster 2013.

[85] "Ab hac authoris nostri opinione dissentit Gassendus, {dicendo} firmiorem connexionem fieri {po...}, si fingamus uncis et hamulis partes sibi invicem implicitas {ea}, sed fallitur profecto, concipiatur {e[nim]} id sic {ea}, ergo omnis cohaesio orietur a firmitate uncinorum et hamulorum, sed tunc {...} inde horum firmitas? Si {requirat} a prioribus, {...} quo unde earum firmitas, donec tandem {...} ad primos, qui nullis aliis connectuntur, cumque sic etiam adest connectio, qua a prioribus ea non posse supponimus, nullum aliud erit glutinum quam quies," Hamburg 273, 114. Cf. the other series of *dictata* on the same article of Descartes's *Principia*: "Ipsarum quies, sunt, qui corporis alicuius cohaerentiam ascribunt hamulis uncisque quibusdam, quibus {coagmentatam}, cohaereantque partes, non vero sola quiete hac {...} existimant, primo quidem ut fieri {...} non puto, sed ab illis petere velim, qua vi istarum hamularum {...} partes coahereant, non hic alias adferent opinor hamulos, quia de iisdem unde redibit quaestio, sed necesse habebunt confiteri, se nihil nimis, quo cohaereant, considerent, quam hanc ipsam iuxta se invicem positarum partium quietem, ni quia proinde sola {...} partium corporis glutinum consistat," Hamburg 274, 47–48.

[86] See Gassendi 1658, volume 1, *Syntagma philosophicum*, 279 and 395. This theory is analyzed in Fisher 2005, 269–270.

id.[87] Their firmness may, however, depend on additional causes, i.e. the rest and the entanglement of parts, which are the causes taken into account, albeit without proof based on "Experiment, or Observation," by Descartes and by Lucretius.[88] The cause that Boyle prefers – as far as the case of the cohering marbles or glass pieces is concerned – is that of atmospheric pressure, on the one hand, and elastic power, on the other, which are two interrelated phenomena[89] (and which I will discuss in Sect. 5.6.3, *De Volder's evolving treatment of the elasticity of the air*). In his more famous *New Experiments Physico-Mechanical* (1660), he provides the same account of this phenomena.[90] De Volder's approach to cohesion is therefore a *via media* between Descartes and Boyle: he overtly avails himself of rest as a factor of cohesion, but he integrates it with a further one, namely, the pressure of air.

De Volder's issues with the idea of rest as a factor of cohesion are in fact confirmed by further propositions discussed during his disputations. Rest is labelled as a primary principle of cohesion of bodies in the corollaries of the *respondens* of the disputation *De criterio veritatis prima* (1673),[91] in the corollaries – apparently of De Volder himself – of the fourth disputation *De aëris gravitate* (1677),[92] and in the section *De corpore* of his *Exercitationes* (thesis 33). Elsewhere, its primacy is doubted. This being the case, in the printed texts, of the propositions of the *respondens* annexed to the third disputation of the sub-series *De corpore* of his *Exercitationes* (viz. disputation 23), where after the anti-Cartesian proposition "[...] whether [God] conserves always the same or equal [quantity of motion], it is uncertain," (to which I will return later), the *respondens* states that "[e]very body whose particles are at rest with respect to each other, is hard. [...] However, such rest of particles is not the only cause of hardness."[93] No further detail is provided in the main text: indeed, the *Exercitationes* were aimed at defending Descartes's (general) ideas rather than those of De Volder himself.[94] If, however, we look again at the handwritten sources, namely at his *dictata* on Descartes's *Principia*, we do find the same position: this being the case of the Hamburg 273 series of his *dictata*. Most likely the *respondens* who appended these propositions to a disputation held on 30 January 1692 had one of the two documents (or their archetypal, or another copy) before his eyes. In commenting upon *Principia* II.63 – i.e. the same article recalled

[87] Boyle 1669a, 208. On Boyle's discussion of cohesion, see Clericuzio 2000, chapter 4; Newman 2006, part 3; Shapin and Schaffer 1985, chapter 2.

[88] Boyle 1669a, 210.

[89] Boyle 1669a, 213.

[90] Boyle 1660, 229–230.

[91] "XI. Nullum [...] datur glutinum efficacius ipsa quiete. XII. A qua quiete omnis pendet vis connexionis, quae fit ab hamulis uncisve, et c." De Volder 1673b, corollaries 11–12.

[92] "VIII. Durities corporum dependet a partium iuxta se invicem positarum quiete," De Volder 1676–1678, disputation 4, corollary 8.

[93] "II. Nullas motus causa assignari potest praeter Deum, qui certam quandam quantitatem produxit, an vero semper eandem sive aequalem conservet, incertum est. III. Omne corpus cuius particulae iuxta se invicem quiescunt durum est. IV. Illa autem quies particularum non est unica causa duritiei," De Volder 1690–1693, disputation 23, *De corpore*, *annexa* 2–4.

[94] See Sect. 2.2.3, *The mid-1670s clash at Leiden and the foundation of the experimental theatre*.

in the *Experimenta* – De Volder notes a "difficulty," namely, that "in order to divide a body, there is nothing [more], than I move a part without moving the other one, but we cannot do this. Therefore, another glue is required besides rest."[95] This is the very position held almost twenty years earlier by De Volder. Rest is still a factor of cohesion, but something more is needed: once again, air, which is excluded from the internal parts of the body, and presses on it: "so, what remains, if cohesion does not consist in rest alone? I answer: I think that [...] air is relevant at most. What therefore happens in an hard body? The parts are at rest with respect to each other, thus air is removed, which therefore presses with a great force that body, [and] which until [it] is not overcome, the cohesion cannot be taken away."[96] Subsequently, by following the mathematical (or rather, mechanical, or mixed-mathematical) approach we saw at work in his *De rerum naturalium principiis*, De Volder devotes an extensive explanation of how, by using the principle of the lever, one can move big weights with a little force: for instance, the force required to raise a body of 100 pounds for 1 foot, is the same force by which one can raise a body of 200 pounds for half a foot. Therefore, in principle, one can divide big bodies with a small force as well, even if no quantification is given, as it was in his *De rerum naturalium principiis*.[97]

[95] "Possint. Haec altera difficultas est, ostendimus in regulis motus corpus minus {moveri} a maiori. {Quod} itaque unde cohaesio corporum, et quid requiritur ut corpus dividam? {Respondemus}, ut moveam, sed {cur cum} totius forcipem moveam, non etiam movere possum partem. Divelli. In {eo secando} difficultatis est, habeo corpus, {ad} ut dividatur, nihil est, ut unam partem moveam non motam altera, sed id non possumus. Ergo requiritur aliud glutinum praeter quietem," Hamburg 273, 121. In the other series of *dictata*, De Volder is more constant to Descartes's text, i.e. he does not challenge his explanation: Hamburg 274, 50.

[96] "Possumus. Primum inquit considerandum est, corpori manibus minori non {possunt totae manus} applicari, sed tantum partem {q[uae]} ei magnitudine est aequalis et contigua. {2. manus} pars illa non tota agit in ferrum, quia non omnes quiescunt iuxta se {invicem}, sed molles sunt, ergo {illae} tantum partes agere possunt, quae {et eis eas} accedunt, non quae recedunt. Sed ut {vera fatemur}, non videntur {haec} sufficere, quid ergo restat, si glutinum non consistat in quiete sola? {R[espondeo]} puto {sic} plurimum conferre {aerem}. Quid ergo fiet in corpore duro? Partes iuxta se invicem {quiescerunt}, ergo excluditur {aer}, qui proinde magna vi premit illud corpus, {q[uo]d} q[uam]mdiu non superatur, non potest {tolli} glutinum," Hamburg 273, 121–122. In the other series of *dictata*, there is no recourse to the action of air as to this point: cf. Hamburg 274, 47–48.

[97] "In ipso {tamen} authoris {R} et hac {...} est, quod vires omnes sunt applicandae {quodque} eadem vis q[uae] potest attollere corpus 100 lb ad {...} unum pedem, possit 200 lb ad pedem dimidium. Et ut {...} regulae veritas pateat, concipio duas vires, quarum una potest corpus 10 lb elevare, ad unum pedem, alia 100 lb ad unum pedem, illa vis posterior est decupla prioris {...}, videamque porro, an non vis quae tollit 1000 lb ad pedis {altitudinem} est decupla {eius q[uod]} 100 lb ad pedem attollit? Unde sequitur illas vires, quarum una, 100 ad {decem} pedes, et altera 1000 ad unum potest tollere {eam aequale}, sed {...} potest tollere 1000 lb, unde {hoc}? Quia omnes vires non applicantur, id {...} est axioma mathematicu{m}, vim minima movere posse corpus maximum modo adhibetur, id itaque {...}, si quis utatur vecte, trochlea etc.: si solummodo vires {applicet} movebit profecto quodlibet corpus. Notissumum est, si quis vecte {velit} uti opus {ea} sustentaculo, {cui} quo corpus movendum propinquius, manus vero remotior est, eo facilius corpus tollitur: quia si tum adhibeantur vires {omnes, sic} ergo {...} frangere non possumus, quia non possumus adhibere vires," Hamburg 273, 122.

4.1.4 The Search for Physical Unity and Activity: De Volder vis-à-vis Leibniz

To sum up, De Volder's explanation of cohesion by the idea of air pressure served him to integrate Descartes's foundation of cohesion on rest only, which De Volder found unsatisfactory on an empirical or experimental basis. Indeed, for De Volder this theory does not really explain why a body can be moved without losing its integrity. Moreover, the idea of pressure served De Volder in not reverting to entanglement as an alternative to Descartes's explanation of cohesion. The idea that rest is the first cause of cohesion, however, did not only leave De Volder suspicious as to its sufficiency in explaining cohesion itself, but also with respect to its metaphysical foundation. This is clear if we look at his Hamburg 273 series of his *dictata*: commenting upon Descartes's *Principia*, II.55 – where Descartes claims that rest is the only glue of bodies – De Volder notes that "if cohesion is given, rest is given, and vice versa, therefore it seems that they [are] both one and the same thing." So, he continues, "the author thus uses here an argument metaphysical rather than physical."[98] De Volder seems to complain that Descartes's explanation of cohesion is a mere tautology, or a *petitio principii*, as no physical cause of rest is given.[99] Descartes, in his *Principia*, II.37, proposing his first law of motion, claimed that a body, as far as it is in itself (*quantum in se est*), "always remains in the same state,"[100] for two reasons: (1) God is immutable and conserves, through His concourse, the same quantity of movement and rest in the world (as stated in II.36),[101] and (2) nothing tends, by itself, to the destruction of itself. Therefore, "each thing [...] always remains in the same state [...] and never changes except by external causes."[102] So, a body keeps itself cohering because it is

---

[98] "Agnoscimus. Quies inquit unicum est glutinum per q[uo]d partes cohaereant, et res manifesta est, q[uo]d {e[nim] q[uae]so}, est cohaerere, nonne quiescere iuxta se invicem? Si {e[nim]} q[uas]dam partes moveantur, q[uas]dam quiescant, {nec} cohaerebunt, sed separabuntur, ergo si posita cohaesione ponatur q[ui]es, et vice versa, patet ambas unum idemq[ue] {...}. Autor {[n]a[m]} hic utitur argumento metaphysico magis quam physico," Hamburg 273, 114.

[99] I discuss the point further in Sect. 6.2.2.3.2, *De Volder's Huygenian theory of weight (and its relation to the idea of cohesion).*

[100] Descartes 1982, 59. Cf. the original versions: "unaquaeque res, quantum in se est, semper in eodem statu perseveret; sicque quod semel movetur, semper moveri pergat," AT VIII-A, 62; "chaque chose demeure en l'etat qu'elle est pendant que rien ne le change," AT IX-B, 89. On Descartes's principle of conservation, see the entries given *supra*, n. 5.

[101] "God Himself [...] in the beginning created matter with both movement and rest; and now maintains in the sum total of matter, by His normal participation, the same quantity of motion and rest as He placed in it at that time," Descartes 1982, 58. On Descartes's 'metaphysical physics', i.e. on the use of rational theology in the foundation of natural philosophy, see Hatfield 1979; Clarke 1982; Garber 1992, chapters 7–8, 2001, 2006; Voss 1993, section 1; Schuster 1993, 2013; Nadler 1993; Gaukroger 2002; Hattab 2007; Dobre 2017.

[102] "[...] from this same immutability of God, we can obtain knowledge of the rules or laws of nature, which are the secondary and particular causes of the diverse movements which we notice in individual bodies. The first of these laws is that each thing, provided that it is simple and undivided, always remains in the same state as far as is in its power, and never changes except by external causes. [...] because experience seems to have proved it to us on many occasions, we are

contrary to its nature to change itself, and because God preserves such state. However – in my reading – for De Volder this is not a physical explanation, unlike that involving the (measurable) pressure of air. In particular, I would add, this explanation does not account for the ways in which rest can be perturbed, and the body separated. As seen above, mere rest does not account for the impossibility of separating, just by the mere pressure of our hands, even one 1,000th of a piece of iron. Yet, De Volder kept rest as a principle of cohesion, because entanglement is an even worse explanatory alternative.

How did De Volder keep rest as a principle of cohesion? First of all, he provided his own justification of Descartes's first law, which explains, even if in metaphysical terms, why a thing which is "simple and undivided" (*simplex et indivisa*) or "particular" (*chaque chose en particulier*) (according to II.37) is preserved in its state.[103] In the Hamburg 273 series of his *dictata*, commenting upon *Principia* II.36, we find no mention of the role the immutability of God – the core of Descartes's argument – in preserving the same quantity of motion in the world. Plainly, God is the cause of movement, because movement does not follow from the nature of matter or extension;[104] however, the same quantity of movement is constant in the universe only because no physical cause can increase or diminish it: neither mind or thoughts, nor matter itself, as movement does not 'flow' from it, and movement cannot diminish movement, because movement is not contrary to movement. Also, he claims that rest cannot diminish movement – without substantiating the claim.[105] In turn, in the Hamburg 274 series of his *dictata*, in commenting on the same article (II.36) De Volder just clarifies how the quantity of movement has to be evaluated by considering the size and speed of a body, and – in a passage not present in the Hamburg manuscript, but in all the others of the same series (viz. in the British Library copy, in column b of The Hague copy, and in the Pretoria copy) he notes that increments

---

still inclined to believe that all movements cease by virtue of their own nature, or that bodies have a tendency toward rest. Yet this is assuredly in complete contradiction with the laws of nature; for rest is the opposite of movement, and nothing moves by virtue of its own nature toward its opposite or its own destruction," Descartes 1982, 59.

[103] See the previous note, and the Latin and French versions of the article: AT VIII-A, 62; AT IX-B, 89.

[104] "[...] quod motus non sit ex natura materiae, et quod ab immediata causa non fluit, necessario a prima causa est sc. Deo, a corpore {a} non pendere ex eo liquet, quia quodlibet corpus possum concipere sine motu, si itaque non fluat ex natura extensionis, et sic non agatur de motu particulari, causa necessario nulla erit praeter Deum," Hamburg 273, 101. The column b of The Hague *dictata*, as to this article, contains the opening phrase of the commentary on the same article of Hamburg 274, 42, and British Library *dictata*, 92v: cf. The Hague *dictata*, 19b.

[105] "Nihil hactenus percepi quam cogitationem, et extensionem. Quaeritur itaque an cogitatio possit augere vel diminuere motus? Resp: nequaquam, quia cogitatio a motu aliena est. Non etiam ex natura materiae fluit, sine iis ne res quidem concipi potest, atqui materia sine maiori minorive motu concipi potest, ergo maior minorve motus ex materiae natura non fluit, non etiam est ex natura motus, quia quemlibet gradum motus sine augmento aut decremento concipere, nec est quod {q[ui]s} dicat, motum tolli a motu {a[ut]} quiete, motus {e[nim]} motui non est contrarius, {q[uae] o[mn]ia} manifeste ostendunt nullis causis naturalibus fieri posse motus in universo augmentum aut decrementum," Hamburg 273, 102.

and decrements in this quantity, in a given body, are not to be attributed to God, who acts in a constant manner,[106] this being his only rational-theological incursion in dealing with this principle, commented on, in the Pretoria copy, in a marginal note referring to Jacques Rohault's *Traité de physique*, in which Rohault provides a simplified version of Descartes's argument for the conservation of the quantity of motion.[107] Notably, a marginal reference to Rohault is provided, by the same unknown hand, also to the commentary on II.37 in the Pretoria copy. In the Hamburg 274 series (i.e. the series to which the Pretoria copy belongs), actually, De Volder does not enter into detail as to the cause of the conservation of motion, as he just explains the meaning of Descartes's phrase "as far as it is simple" (*quantum est simplex*), pointing out how motion is extrinsic to matter.[108] The annotation refers to

[106] "Deum esse causam universalem omnium motuum. Ex eo manifestum est, q[uo]d sit causa materiae, omniumque materiae accidentium. Certam habet quantitatem. Quantitate motuum aeq[ue] possumus mensurare, ac omnia alia quanta, si tamen certam {quidam} mensuram motus ex arbitrio nostro sumamus, ad {quem} tamen omnem aliorum motuum quantitatem expendamus, motus autem quantitas dependet a duobus, vel a magnitudine corporis quod movetur, vel a celeritate motus, sic corpus A quod duplex sit illius B et eadem moveatur velocitate, habebit duplum plus motus quam B etenim si A dividatur in duas partes aequales ipsi B habebit unaquaeq[ue] pars cum {aecq[ue]} velocitate moveatur ac B aequalem cum B motus quantitatem, ergo in duabus partibus simul sumptis, hoc est, in corpore A erit duplex quantitas motus, qui est in corpore B nec minus {manifestam} est fere duplam quantitatem motus, in corpore A, {quam qui est} in corpore B si ambo quidem supponantur aequalia, sed tamen A duplo celerius movetur quam B. [Tantundem motus. Si vel augeretur vel minuetur motus quantitas, fieret illud augmentum vel decrementum, vel ab ipsa materia vel a Deo, non ab hoc, quia Deus semper eodem operat in modo, nec ab illa, quia eius essentia nec motum possit producere nec quietem]," Hamburg 274, 42–43. The part between brackets [ ] can be found in Pretoria *dictata*, 34, British Library *dictata*, 92v–93r, and The Hague *dictata*, 20b. Please note that the original Latin text of Descartes states that "Deum esse primariam motus causam."

[107] The marginal note comes after "accidentium": "[v]id. Rohault Part. 1 cap. 10 § 12–13." In such paragraphs, Rohault relies on a standard Cartesian justification of the conservation of the quantity of motion, and points out – with an anti-metaphysical stance common to De Volder, that the natural philosopher should direct himself to natural causes: "[c]omme il n'y a que les proprietez essentielles d'un sujet qui se puissent deduire de son Essence, quand elle est connuë, ce seroit inutilement que nous tascherions de découvrir, comment le mouvement a pû estre produit la premiere fois dans le corps, puis que ce n'en est pas une Proprieté essentielle; nous ne nous arresterons donc pas à raisonner sur ce sujet; Et comme nous reconnoissons Dieu pour le createur de la matiere, de mesme le reconnoissons-nous pour son Premier Moteur. Mais parce que ce ne seroit pas philosopher que de luy faire faire à tous momens des miracles, & d'avoir perpétuellement recours à sa puissance, nous suposerons seulement qu'en creant la matiere de ce monde, il a imprimé une certaine quantité de mouvement dans ses parties, & qu'ensuitte il ne fait plus que prester son concours ordinaire, pour empescher que les choses ne retournent dans le neant d'où il les a tirées, & conserver ainsi incessamment en la matiere une égale quantité de mouvement; Si bien que ce que nous avons maintenant à faire, est de rechercher les autres circonstances du mouvement, & d'en estudier les causes Secondes, ou Naturelles," Rohault 1671, volume 1, 62–63. See Des Chene 2002; Dobre 2013a; Milani 2015; Spink 2018.

[108] "Quatenus est simplex. Corporis compositi existentia potest constitui in continua mutatione, ut ignis ex. g. essentia consistit in continua agitatione, et continua variatione, sed haec fit, quia ignis non denotat simplex extensione, sed extensionem ita formatam, ita constitutam, ita motam, verum si tamen considerem extensionem, patet ex eius essentia nullam posse oriri mutationem, {q[uae]} sine ea mutatione supponitur existere posse extensio, ergo quaeq[ue] ei accidat mutatio, ea fit a causis externis," Hamburg 274, 43.

Rohault's *Traité*, part 1, chapter 11, §§ 1–3, where Rohault points out the external origin of movement and rest, justifies this tenet by considering how everything tends to its destruction, and by recourse to experience. No rational-theological arguments are deployed, in perfect accordance with De Volder's approach.[109] Also in commenting upon the following article (II.37), in the Hamburg 273 series, De Volder explains Descartes's first law by avoiding any reference to divine immutability, but focusing only on the self-persistence of natures or essences: so, all changes are always due to external causes.[110] It is noteworthy that, in commenting upon *Principia* II.36 in Hamburg 273 De Volder overtly mentioned cases in which this principle is not observed, namely miracles which he excludes from his considerations, so that he claims that, as a consequence, by considering the conservation of the quantity of motion he takes into account only natural causes.[111] Rather than a guarantor of the principle of conservation, God is considered as a factor which can be responsible only for the variations in natural order, as well as a cause to which recourse is necessary whenever a physical cause cannot be adopted. This makes sense, for instance, of the above-mentioned decidedly un-Cartesian proposition appended to his *Exercitationes*, stating that "[…] whether [God] conserves always the same or equal [quantity of motion], it is uncertain."

De Volder, so to speak, attempted to detach his physics from metaphysics as a rational-theology: God has definitely a less important role in his physics than in Descartes's. Yet, De Volder keeps, let's say, an ontological foundation of physics: if the recourse to God is limited, he still keeps the argument for which natures are eternal and do not change themselves. It is on these premises that we can address a core topic of his correspondence with Leibniz, namely, his repeated requests – as highlighted by Lodge – for an a priori demonstration of the activity of a substance. The activity of a substance accounts for the cohesion of parts of matter, i.e. for their

---

[109] "Vid. Rohault part[em] 1. Cap. 11, § 1. 2. 3," Pretoria *dictata*, 34. Cf. Rohault 1671, volume 1, 63–63: "I. Que ce qui est en repos ne peut jamais de soy commencer à mouvoir, et que ce qui a commencé à se muovoir ne peut jamais de soy cesser de se mouvoir. […] II. Que c'est une erreur de croire que les corps qui se meuvent tendent d'eux-mesme au repos, […] III. Que l'opinion d'Aristote n'est pas prouvée par l'experience." The Pretoria copy also contains an additional commentary, in the same hand as the references to Rohault, in a unpaginated sheet between page 34 and 35, in which it is discussed the behaviour of a arrow in air, where it loses its movement by the impact with air itself.

[110] "Prima lex id vult, {q[od]d quaecumque} mutatio non ex natura corporis, sed externis causis fluit, quam rem, ut demonstremus, velim concipi in genere, quicquid mutationis est, esse vel ex natura natura rei, vel a re aliqua ab ea diversa, non potest esse a natura rei, quia […] certum est naturas esse aeternas, ergo si mutetur natura, etiam destruatur, unde (cum nihil tendit in destructionem sui ipsius) sequitur mutationem non esse ex natura rei. […] Motus inquit autor non inducet quietem, quia ipsi est contraria," Hamburg 273, 102. Also in the other *dictata* there is no mention of the role of God, but only to the idea that essences do not change by themselves: Hamburg 274, 43 (the contents of this series, as to this article, are partially given in The Hague *dictata*, 20b).

[111] "Ut haec res clarius evadat, considerandum est a nostra contemplatione aliena esse miracula, i.e. illa phaenomena q[uae] non obtemperant legibus naturae, cum itaque dicimus eandem semper motus quantitatem manere, nihil aliud dicimus, quam nullas esse causas naturales, q[uae] motum augeant vel minuant, easq[ue] non esse sic demonstro," Hamburg 273, 102.

internal rest: rest, as seen above, is indeed a positive state, and accounts for the resistance to being moved, not only for the mere persistence in a state. We find both these ideas – i.e. resistance and persistence – in De Volder's correspondence, as encompassed by the idea of inertia. In his letter to Leibniz of 18 February 1699, De Volder understands inertia as extension itself, which persists in its states (of movement and rest) as nothing changes itself (as seen above).[112] In this meaning, 'inertia' is nothing but simple persistence, by which a body perdures in its state until something external changes it. After the remark of Leibniz, that by this idea one cannot grant a regulated natural system, as matter would be put into motion without exerting any resistance,[113] De Volder agrees that inertia has to be understood as something different from mere extension, but which needs to be ascribed to matter, as it accounts for the resistance to change when an external cause intervenes. Without this resistance to change, De Volder claims, there would be no proportion between causes and effects, as a very little body could move a bigger one, without losing its movement. So, resistance to motion – and so, resistance to separation – is a form of activity rather than being a "mere passion."[114] This exchange between De Volder and

[112] "De inertia materiae [...] non percipio, quid illa notet ab extensione distinctum. Omnis enim res habet ex sua natura vim permanendi in suo statu, quae vis ab ipsa rei natura non differt, et in hoc exemplo extensionis ipsa inertia est," De Volder to Leibniz, 18 February 1699, in GP II, 166. De Volder was commenting upon the letter of Leibniz to him of 27 December 1698, in which Leibniz stated that "[q]uod attinet ea quae attribuenda sunt materiae praeter extensionem, satis manifestum puto, inertiam (de qua supra) inter alia esse aliquid quod extensione nuda non continetur," A II3B, 511. Leibniz was referring to the idea, expressed in the second letter of the correspondence, that mere extension cannot account for its activity; if one considers extension only, as well as its impenetrability, one would conclude that when two bodies collide, they lose their force: see Leibniz to Johann Bernoulli, 12/22 July 1698, in A III7B, 827. I deal with this problem in Sect. 4.2.3.3.1, *The quantification of motive force*: see esp. *infra*, n. 261.

[113] "Fateor unumquodque manere in statu suo, donec ratio sit mutationis, quod est metaphysicae necessitatis principium, sed aliud est statum retinere donec sit quod mutet, quod etiam facit per se indifferens ad utrumque, aliud est multoque plus continet rem non esse indifferentem sed vim habere et velut inclinationem ad statum retinendum atque adeo resistere mutanti. Itaque olim adolescens edito quodam libello, sumens materiam ut indifferentem per se ad motum et quietem, inde colligebam, maximum corpus quiescens moveri debere a minimo quocunque impellente, sine impellentis debilitatione, atque inde abstractas a systemate, motus regulas colligebam. Et fingi posset sane talis mundus, utique possibilis, in quo materia quiescens motori obediret sine ullo renisu; sed is profecto mundus merum chaos foret," Leibniz to De Volder, 24 March/3 April 1699, in A II3B, 546–547. Leibniz refers to his *Theoria motus abstracti* (1670). I will comment upon this passage further in Sect. 4.2.3.4.2, *De Volder on elastic impacts*: see esp. *infra*, n. 319.

[114] "Deinde Inertiam materiae ab extensione differre existimas, quia per extensionem, indifferens est ad motum et quietem. Per inertiam vero etiam resistat mutanti. Me quod attinet, nullam aliam in extensione indifferentiam agnosco, quam ea est, qua suscipit motum et quietem, si causa fit, quae alterum efficiat. Neque putem ex illa sequi, ut quaelibet vis etiam minima quemlibet motum efficiat etiam maximum. Si nulla materiae Inertia ascribatur ab extensione distincta, putasne fore, ut minimum quoddam corpus, quocunque motu praeditum, corpus impelleret maximum sine ulla motus sui debilitatione? Huic rei repugnare videtur ipsa natura causae et effecti, quae certam quandam inter se proportionem servant. Et sane maioris efficaciae est, pari velocitate movere extensum maius, quam minus, vel hoc ipso, quod effectum maius est. Resistentia praeterea ipsa non videtur mera passio," De Volder to Leibniz, 13 May 1699, in GP II, 179.

Leibniz can be interpreted as concerning the meaning of words only. However, as I am going to show in Sect. 4.2, *The laws of motion and impact*, De Volder came to develop his rules of collision without taking into account the idea of resistance to motion, and claiming that elasticity (which presupposes an activity of matter, as it is a force of restitution of movement) is not essential to matter itself. It may be, therefore, that De Volder fully understood the difference between inertia as mere persistence (which is something entailed by his justification of the first law of motion), and as active resistance only after Leibniz's criticism. In any case, his issues with Descartes's theory of cohesion date back well before 1699.

If one supposes that in matter there is an active resistance to motion, as De Volder actually did since his 1670s *De corpore* and *De motu*, one needs to find a source for it. In his correspondence with Leibniz, De Volder asked him for an a priori demonstration that bodies are active, namely, a demonstration based only on the notion of bodily or material substance (*ex ipsa substantiae notione*).[115] This demonstration serves to account for the movement of bodies in a way independent from rational-theology. De Volder did not endorse what he labelled as the "Cartesian" or "Malebranchean" hypothesis on the cause of motion, which is more widely known as a version of occasionalism.[116] According to this hypothesis, at every instant a new cause is required in order to explain movement i.e. the change of *situs*. So, God has to be taken into account as such cause, since movement cannot follow, as seen above, from matter itself.[117] A demonstration of the activity of substance, as De Volder clarifies in his letter of 10 November 1703, would solve all the difficulties concerning not only the movement of bodies, but also their resistance to movement;

115 See, for instance the letter of De Volder to Bernoulli – directed to Leibniz – of 21 November 1698: "[n]am si a priori demonstratum haberemus, omnem substantiam esse activam, facile mihi persuadeo ex hoc foecundissimo veritatum fonte non horum tantum scrupulorum secuturam enodationem, sed et earum difficultatum, quae hactenus omnes quotquot sunt Physicos presserunt. Huius etenim rei ignorantia, compulsi sunt causam motus coniicere in Deum, et nonnulli etiam ad quoslibet corporum occursus Deum e machina arcessere," GP II, 151.

116 On the problem of causality and occasionalism in the seventeenth century, see Nadler 1993, 1998, 1999, 2005; Van Ruler 1995; Clatterbaugh 1999; Perler and Rudolph 2000; Bardout 2002; Schmaltz 2008; Favaretti Camposampiero et al. 2018.

117 "Hoc tantum addam hac occasione, nunquam mihi placuisse illam Malebranchii, et quorundam Cartesianorum opinionem, motum fieri non vi concursus corporum, sed vi immediata Dei occasione concursus, corpus hoc illudve moventis," De Volder to Leibniz, 12 November 1699, in GP II, 199. "Cuius ut exemplum adducam, ex hac assumtione, manifestum erit, illam vim translationem producturam, donec a vi alia, priori contraria inhibeatur, adeoque explicatu erit facile, cur corpora mota suo in motu perseverent, quod mihi, ut lenissime loquar, in Cartesiana hypothesi semper admodum arduum visum est. Nam cum motus non sit ipsis vis quaedam permanens, ex qua translatio sequatur, sed ipsa solummodo translatio, quae consistit in perpetua situs mutatione, debebit ex eadem regula quam ad nodum hunc solvendum adhibent, singulis momentis nova dari huius in situ mutationis causa; non secus ac recte volunt in motu per curvam requiri ut singulis momentis nova sit causa, quae determinationem mutet; qua sublata, ut hoc corpus non amplius determinationem suam, sic illud prius non amplius mutabit suum locum. Ex eadem illa vi, sequitur corporum inter se actio et passio, nullo interveniente ex Machina Deo, et quae plura sunt id genus," De Volder to Leibniz, 25 July 1702, in GP II, 241–242. See also the letter of De Volder to Leibniz, 30 October 1703, in GP II, 254–256.

as to these issues, he found the solution of Leibniz – based on the idea of individual substances provided with an internal activity – "more natural" or "more physical" (φυσικώτερον) than that of Descartes, as it made recourse to only one act of creation by God, and not to His continuous intervention.[118] Yet, Leibniz's claim that individual substances are active by themselves was considered as just assumed a posteriori (i.e. on the basis of the fact that matter shows forms of activity, regardless of its sources – something with which De Volder agreed), and as "gratuitous" by De Volder.[119]

At that time, Leibniz had already made public his theory of substance and dynamics, in its most systematic form, in various essays which appeared in the *Acta Eruditorum*, namely his *De primae philosophiae emendatione, et de notione substantiae* (1694), *Specimen dynamicum pro admirandis naturae legibus circa corporum vires et mutuas actiones detegendis, et ad suas causas revocandis* (1695) – positively mentioned by De Volder in a letter of 1699[120] – and *De ipsa natura sive De vi insita actionibusque creaturarum, pro dynamicis suis confirmandis illustrandisque* (1698) – discussed in a later phase of the correspondence.[121] In his *De emendatione* Leibniz ascribed to bodies an active force, namely the perpetually active entelechy, given by God in the first act of creation, which cannot be rightly understood if one recognizes in mere extension the substance of bodies, and which serves to explain phenomena at a more fundamental level (i.e. as the *ultima*

[118] "Tuam equidem hypothesin Malebranchianae longe praeferrem vel eo quod ille ad singulos actus Deum requirat, tu vero φυσικώτερον ad primum solummodo productionis rerum actum. Hoc tamen utrique videtur commune, quod illam aut primae ἐντελεχειῶν productionis, aut in singulis Divinae actionis rationem relinquat obscuram admodum; deinde et illud, quod per leges Mechanicas, utpote quae utraque assumit, everti neutra possit. [...] Dein ne sic quidem satis percipio, quid id sit, quod substantiam voces. Si quid enim, nihil certe praeter vim activam, et resistentiam percipio, harum vero virium subiectum, quae tibi, ut opinor, ipsa demum substantia est, nequaquam percipio," GP II, 254–255.

[119] See the letter of De Volder to Leibniz of 5 April 1700: "[...] de veritate conclusionis tuae a posteriori stabilitae dubitavi nunquam, verum ex ipsa substantiae notione demonstrationem quaesivi a priori, sed hactenus frustra," GP II, 210; De Volder to Leibniz, 3 April 1702: "[t]ecum equidem sentio per experientiam constare, nequaquam omni vi destitui corpora; sed huius rei per experientiam notae demonstrationem quaero, ex ipsa natura substantiae petitam," GP II, 238; Leibniz to De Volder, 19 November 1703: "[p]orro si mecum agnoscis systema causarum occasionalium, non esse dignum philosopho, si influxum substantiae in substantiam (de veris loquor) inexplicabilem arbitraris; non video quomodo dubitare possis de intrinseca rerum tendentia ad mutationem; cum mutationes adesse in rebus experientia phaenomenorum edoceamur et ab intrinseco mutationes exhibeant vel ipsae operationes mentis. Τὸ ὅτι ergo a posteriori demonstratum puto," GP II, 256; De Volder to Leibniz, 30 November 1703: "[p]raeter quae in tua causa id me insuper male habet, quod videaris gratis assumere omnem substantiam esse activam, cum tuam vocis substantiae definitionem ingrediatur agendi principium. Quare, ne de voce lis sit, mihi {adhucdum} demonstrandum videtur primum illud, quod inter nos quaesitum fuit, nihil posse existere, quod agendi potentiam non habeat," GP II, 254–255.

[120] See his letter to Leibniz of 12 November 1699 (GP II, 196–200).

[121] See the letters of Leibnz to De Volder of 6 July 1701 (GP II, 224–228), 20 June 1703 (GP II, 248–253), 11 October 1705 (GP II, 278–279). On Leibniz's dynamics, see Gabbey 1971, 1980; Iltis 1971, 1974; Costabel 1973; Gale 1973, 1988; Bernstein 1981; Lariviere 1989; Bertoloni Meli 1993; Fichant 1998; Freudenthal 2002; Lodge 2003; Roberts 2003; Jauernig 2008; Rey 2009a, b.

ratio motus in materia) than by mechanical explanations,[122] being therefore praised by De Volder.[123] In his following *Specimen*, Leibniz claims that (1) there is something before and apart from extension; (2) action is the character (i.e. essence) of substance; (3) extension is the "continuation or diffusion" of substance; (4) movement is not real, because, in order to be real, something needs to have its parts as coexisting: in the case of motion this does not happen, because motion takes place successively, in different moments. Therefore, "nothing is real in it [sc. in motion], except that momentary [entity], which has to be found in the force striving to mutation. Therefore, everything which is in the corporeal nature, besides [...] extension, traces back to this [force]."[124] Thus, Leibniz clarifies his ideas on force (*vis*). He distinguishes between an active force which is primitive (or entelechy itself), which inheres in the substance, and a derivative one, which results from the interaction of bodies, and acts as a special cause during their intercourse, whereas the primitive force is only the general cause of motion, and does not suffice to explain phenomena. Analogously, in substance there is also a passive primitive force, which is responsible for the resistance to motion (also called "first matter"), and a passive derivative one (also called "second matter").[125] From such forces, Leibniz

---

[122] "[...] dicam interim, notionem virium seu virtutis, (quam Germani vocant *Krafft*, Galli *la force*) cui ego explicandae peculiarem Dynamices scientiam destinavi, plurimum lucis afferre ad veram notionem substantiae intelligendam. Differt enim vis activa a potentia nuda vulgo scholis cognita, quod potentia activa Scholasticorum, seu facultas, nihil aliud est quam propinqua agendi possibilitas, quae tamen aliena excitatione, et velut stimulo indiget, ut in actum transferatur. Sed vis activa actum quendam sive ἐντελέκειαν continet, atque inter facultatem agendi actionemque ipsam media est, et conatum involvit; atque ita per se ipsam in operationem fertur; nec auxiliis indiget, sed sola sublatione impedimenti. Quod exemplis gravis suspensi funem sustinentem intendentis, aut arcus tensi, illustrari potest. Etsi enim gravitas aut vis elastica mechanice explicari possint debeantque ex aetheris motu; ultima tamen ratio motus in materia, est vis in creatione impressa, quae in unoquoque corpore inest, sed ipso conflictu corporum varie in natura limitatur et coercetur. Et hanc agendi virtutem omni substantiae inesse aio, semperque aliquam ex ea actionem nasci; adeoque nec ipsam substantiam corpoream, (non magis quam spiritualem) ab agendo cessare unquam; quod illi non satis percepisse videntur, qui essentiam eius in sola extensione, vel etiam impenetrabilitate collocaverunt, et corpus omnimode quiescens concipere sibi sunt visi," GP IV, 469–470.

[123] "Quare exultavi gaudio, ubi in Actis Lips. legi te, quem nihil absque summa ratione affirmare, et a quo, si ab ullo tale quid expectari posse, eram persuasissimus, asseverare, *Omni substantiae vim agendi inesse*," GP II, 199.

[124] "In rebus corporeis esse aliquid praeter extensionem, imo extensione prius, alibi admonuimus, nempe ipsam vim naturae ubique ab Autore inditam [...]. Quod si iam Deo per miraculum transcribi non debet, certe oportet, ut vis illa in ipsis corporibus ab ipso producatur, imo ut intimam corporum naturam constituat, quando agere est character substantiarum, extensioque nil aliud, quam iam [...] substantiae continuationem sive diffusionem, dicit; [...] motus (perinde ac tempus) nunquam existit, [...], quia nunquam totus existit, quando partes coexistentes non habet. Nihilque adeo in ipso reale est, quam momentaneum illud, quod in vi ad mutationem nitente constitui debet. Huc igitur redit quicquid est in natura corporea praeter Geometriae obiectum seu extensionem," GM IV, 235.

[125] "Duplex autem est vis activa [...], nempe ut primitiva, quae in omni substantia corporea per se inest [...], aut derivativa, quae primitivae velut limitatione, per corporum inter se conflictus resultans, varie exercetur. Et primitiva quidem (quae nihil aliud est, quam ἐντελέχεια ἡ πρώτη) animae

distinguishes the *conatus* (i.e. the vectorial speed, or velocity, in a certain direction), and the *impetus* (i.e. what the Cartesians labelled the quantity of motion, given by mass and speed).[126] Then, he introduces the ideas of living force (*vis viva*) and dead force (*vis mortua*), both resulting from the active derivative force: dead force is the force exerted, for instance, by a bow when it is drawn, while living force is exerted by it when it is released: so, the living force results from "infinite [and] continuous impressions of dead force."[127]

Some of these ideas are present in Leibniz's "most rounded statement regarding the structure of corporeal substance in relation to his concept of monad,"[128] given in his letter to De Volder of 20 June 1703:

> [...] in the monad or complete simple substance I unite to entelechy nothing but a primitive passive force related to the whole mass of the organic body, of which, indeed, the other subordinate monads are not part, even if they are immediately required [by it], and with the primary monad they concur to the organic corporeal substance, namely the animal or plant. I therefore distinguish (1) the primitive entelechy or soul, (2) the matter namely first [matter] or primitive passive potency, (3) the monad constituted by these two things, (4) the mass or second matter, namely the organic machine, to which innumerable subordinated monads concur, (5) the animal or corporeal substance, which the dominant monad in the machine makes one.[129]

---

vel formae substantiali respondet, sed vel ideo non nisi ad generales causas pertinet, quae phaenomenis explicandis sufficere non possunt. [...] Et quidem vis primitiva patiendi seu resistendi id ipsum constituit, quod materia prima, si recte interpreteris, in Scholis appellatur, qua scilicet fit, ut corpus a corpore non penetretur; sed eidem obstaculum faciat, et simul ignavia quadam, ut sic dicam, id est ad motum repugnatione sit praeditum, neque adeo nisi fracta nonnihil vi agentis impelli se patiatur. Unde postea vis derivativa varie in materia secunda sese ostendit. Sed nostrum est, generalibus illis ac primitivis sepositis suppositisque quibus ob formam corpus omne semper agere et ob materiam corpus omne semper pati ac resistere docemur, nunc quidem pergere ulterius, et in hac doctrina de virtutibus et resistentiis derivativis tractare, quatenus variis nisibus pollent corpora aut rursus varie renituntur; his enim accommodantur leges actionum, quae non ratione tantum intelliguntur, sed et sensu ipso per phaenomena comprobantur," GM IV, 236–237. As seen above, De Volder did not label resistance to motion as a passion: in any case, he agreed with Leibniz that resistance to motion is the result of a force.

[126] "Velocitas sumta cum directione conatus appellatur; impetus autem est factum ex mole corporis in velocitatem, eiusque adeo quantitas est, quod cartesiani appellare solent quantitatem motus," GM IV, 237.

[127] "Hinc vis quoque duplex: alia elementaris, quam et mortuam appello, quia in ea nondum existit motus, sed tantum sollicitatio ad motum, qualis est globi in tubo, aut lapidis in funda, etiam dum adhuc vinculo tenetur; alia vero vis ordinaria est, cum motu actuali coniuncta, quam voco vivam. Et vis mortuae quiddem exemplum est ipsa vis [...] qua elastrum tensum se restituere incipit. Sed in percussione, quae nascitur [...] ab arcu se aliquamdiu restituente, aut a simili causa vis est viva, ex infinitis vis mortuae impressionibus continuatis nata," GM IV, 238.

[128] Antognazza 2009, 423.

[129] "Caeterum in Monada seu substantiam simplicem completam cum Entelechia non coniungo nisi vim passivam primitivam relatam ad totam massam corporis organici, cuius quidem partem non faciunt reliquae monades subordinatae in organis positae, ad eam tamen requiruntur immediate, et cum primaria Monade concurrunt ad substantiam corpoream organicam, seu animal plantamve. Distinguo ergo (1) Entelechiam primitivam seu Animam, (2) Materiam nempe primam seu potentiam passivam primitivam, (3) Monada his duabas completam, (4) Massam seu materiam secundam, sive Machinam organicam, ad quam innumerae concurrunt Monades subordinatae, (5)

In this conceptual framework, extension is by no means material substance: rather, it is a 'derived' idea. We find this idea in Leibniz's *De ipsa natura*, where he claims that conceiving extension as material substance entails that no differentiation between bodies can be conceived. Such differentiation – which results in the visible appearance of bodies[130] – comes only from individual substances or monads. Indeed, since extension is infinitely divisible, it has no "real unities,"[131] as he would repeatedly claim also in his correspondence with De Volder.[132] This idea did not move De Volder at all. The kind of argument that Leibniz offered had indeed been criticized by De Volder as early as 1676, and then reiterated in the *Exercitationes*. In a disputation *De materiae divisibilitate in infinitum*, presided over by De Volder, but authored and defended by a close student of his, namely, Hermann Lufneu, taking place on 27 January 1676, Lufneu argues against Lucretius's claim that if a body can be infinitely divided, two bodies of different, visible size, would be divisible into the same parts, and so they would not be different in size. This claim is rejected on the basis of the idea that not all infinites are equal, so that, for instance, the infinite sub-parts of a body of 100 feet, if divided by following the 1/2 division on a body of 50 feet, would be anyway double the infinite parts of the body of 50 feet. Once recomposed, so, each body would re-acquire its original size.[133] This idea can also be found in the

Animal seu substantiam corpoream, quam Unam facit Monas dominans in Machinam," GP II, 252. On the ideas of substance and monad in Leibniz, see Mugnai 1992; Adams 1994; Rutherford 1995; Fichant 1998; Cover et al. 1999; Look 1999; Phemister 2005; Hartz 2006; Smith and Nachtomy 2011.

[130] "Si nullae essent divisiones materiae in natura, nullae essent diversae res, imo nihil esset nisi mera rerum possibilitas: actualis vero divisio in massis facit res apparentes distinctas, et supponit substantias simplices," Leibniz to De Volder, January 1705 (?), in GP II, 276.

[131] "[...] iudicari potest, debere in corporea substantia reperiri entelechiam primam, tandem πρώτον δεχτιχόν activitatis; vim scilicet motricem primitivam, quae praeter extensionem (seu id quod est mere geometricum) et praeter molem (seu id quod est mere materiale) superaddita, semper quidem agit, sed tamen varie ex corporum concursibus per conatus impetusve modificatur. Atque hoc ipsum substantiale principium est, quod in viventis anima, in aliis forma substantialis appellatur, et quatenus cum materia substantiam vere unam, sed unum per se constituit, id facit quod ego monadem appello; cum sublatis his vere, et realibus unitatibus, non nisi entia per aggregationem, imo quod hinc sequitur nulla vera entia in corporibus sint superfutura. Etsi enim dentur atomi substantiae, nostrae scilicet monades partibus carentes, nullae tamen dantur atomi molis, seu minimae extensionis, vel ultima elementa; cum ex punctis continuum non componatur," GM IV, 511; recalled in Leibniz to De Volder, 6 July 1701, in GP II, 224–228.

[132] "Quod primum attinet, eo ipso quod corpus mathematicum non potest resolvi in prima constitutiva, id utique non esse reale colligitur; sed mentale quiddam, nec aliud designans quam possibilitatem partium non aliquid actuale," Leibniz to De Volder, 30 June 1704, in GP II, 268. "Universum Corporeum compositum esse ex una substantia infinitis diversis modis affecta non dixerim, etsi materiam in se spectatam [...] ubique esse sibi similem dici possit. [...] [I]n extensione Mathematica, qua possibilia intelliguntur, nulla sit divisio actualis nec partes nisi quas cogitando facimus, nec prima Elementa; non magis quam inter numeros fractos minimus datur velut Elementum caeterorum," Leibniz to De Volder, January 1705 (?), in GP II, 276. "[S]patium ut tempus non substantiale est quiddam, sed ideale [...]. Itaque nullae ibi divisiones nisi quas mens facit et pars toto posterior est. Contra in realibus unitates multitudine sunt priores, nec existunt multitudines nisi per unitates," Leibniz to De Volder, 11 October 1705, in GP II, 278–279.

[133] De Volder and Lufneu 1676, thesis 3; cf. Lucretius, *De rerum natura*, book 1, verses 615–622.

propositions of the *respondens* annexed to the third disputation of the section *De corpore* of his *Exercitationes*, discussed above. In the propositions following those already mentioned, the *respondens* states that "[e]very body is divisible in an infinite number of parts. [...] Infinite can be greater than infinite. [...] Therefore, it results that all those reasonings, which are deduced, from that one according to which infinite cannot be greater than infinite, are deficient."[134]

For Leibniz, in conclusion, extension does not constitute the nature of individual substances, being only – as seen above – the "simultaneous diffusion or repetition of a certain nature."[135] Namely, it is the continuity and homogeneity which we conceive – as a mere abstract entity – in more bodies, as he would declare in his letter to De Volder of January 1705 which De Volder just left unanswered.[136] De Volder just could not accept such a theory, as he never abandoned the idea that matter consists of extension only. This is well exemplified in his last letter (5 January 1706, written after an unsolicited letter from Leibniz of October 1705), where he overtly declares to Leibniz that the reason for his silence was his "desperation" in trying to accept Leibniz's ideas of "primitive force" and of the existence of indivisible entities from which extension can be "made" or "appear."[137] In the previous year, moreover, De Volder had resigned from his academic post: according to Jean Le Clerc, this had been a consequence of his incapacity to develop or embrace a new physical theory, alternative to Descartes's. We may suppose, at least as the theory of material substance is concerned, that his frustration with Leibniz's theory may have played a foremost role in all this. Yet, as I am going to show in Sect. 4.2, *The laws of motion*

[134] "V. Divisibile est omne corpus in particulas numero infinitas. VI. Infinitum infinito potest esse maius. VII. Unde liquet omnes illas ratiocinationes, quae ex eo quod infinitum infinito maius esse non possit, deducuntur, mancas esse," De Volder 1690–1693, disputation 23, *annexa* 5–7. On the idea of divisibility in De Volder and Leibniz, see MacDonald Ross 1987.

[135] "Extensio est abstractum Extensi nec magis est substantia quam numerus vel {multitudo substantia} censeri potest exprimitque nihil aliud quam quandam non successivam (ut duratio) sed simultaneam diffusionem vel repetitionem cuiusdam naturae seu quod eodem redit, multitudinem rerum eiusdem naturae, simul cum aliquo inter se ordine existentium," Leibniz to De Volder, 30 November 1704, in GP II, 269. On Leibniz's theory of extension and space, see Fox 1970; Sayre-McCord 1984; Hartz and Cover 1988; Khamara 1988, 1993; Arthur 1994, 2013; Garber et al. 1998; De Risi 2007; Futch 2008.

[136] "Diffusionem quam in extensione concipio et quae Tibi suspicionem nescio cuius paradoxi latentis iniecisse videtur nihil aliud esse volo quam continuationem qua pars est similis toti, ut albedinem concipimus in lacte diffusam et eandem ubique directionem in recta, et in circuli peripheria aequalem curvitatem. Neque vero [...] substantiae simplices diffunduntur [...], aut totum homogeneum constituunt homogeneitas enim materiae non nisi abstractione mentis habetur," Leibniz to De Volder, January 1705 (?), in GP II, 277. This topic is discussed in Lodge 2004.

[137] "Diuturni mei silentii causa non tam est valetudo, quamquam et ea non satis commoda fortasse etiam nonnihil eo contulerit, quam quod ad tuas nihil novi quod reponerem, haberem, verererque ne earundem rerum repetitione tibi nauseam crearem. Nescio enim, qui fiat, ut etiamnum in iisdem, quas in ultimis meis posui, difficultatibus haeream. Neque enim vim illam primitivam mihi videor intelligere, nec capio, qui ex indivisibilibus extensum fieri, aut, si ita malis, apparere queat. Invasit me insuper desperatio quaedam, tuae opinionis rite percipiendae, cum post tot tentamina tam parum promoverim, ut ea requiri ad eius intelligentiam existimem, quae tu instar ultramundanorum ne quaerenda quidem censes," GP II, 279.

and impact, his discontent with Cartesianism, and his dissatisfaction with alternative systems also concerned the laws of motion: namely, such discontent concerned their consistence with the possibility of justifying the conservation of motive force (or quantity of motion) by relying on physical arguments.

4.2 The Laws of Motion and Impact

In Sect. 4.1.4, *The search for physical unity and activity: De Volder vis-à-vis Leibniz* we saw that De Volder (1) retained Descartes's idea of the conservation of the quantity of motion without resorting to the rational-theological argument of the constancy of divine action on the world, and (2) that De Volder had from time to time expressed his doubts on the actual conservation of this quantity. In what follows, these aspects will be investigated further. Namely, I will address the problem of De Volder's approach to Descartes's laws of motion, especially (a) the law of conservation of the quantity of movement, (b) the laws of collisions. First of all, however, it is necessary, for the sake of clarity, to systematically present how such laws were dealt with by Descartes and Huygens, who were the foremost 'interlocutors' of De Volder on this topic.

4.2.1 Descartes on Motion and Impact

4.2.1.1 Descartes's Laws of Motion and Idea of Force

Descartes gave systematic presentations of his laws of nature first in his *Le monde* and then, with some clarification, in the two main editions of his *Principia* (where the French edition of 1647 includes some relevant differences as to the treatment of the rules of impact). In Descartes's *Le monde*, the introduction to the law-like system of natural philosophy is well represented by his discussion of the idea of nature, given in chapter 7 (*Des lois de la nature de ce nouveau monde*), according to which

> [...] by 'Nature' here I do not mean some deity or other sort of imaginary power. Rather, I use the word to signify matter itself, in so far as I am considering it taken together with the totality of qualities I have attributed to it, and on the condition that God continues to preserve it in the same way that He created it. For it necessarily follows from the mere fact that He continues to preserve it thus that there may be many changes in its parts that cannot, it seems to me, properly be attributed to the action of God, because this action never changes, and which I therefore attribute to Nature. The rules by which these changes take place I call the Laws of Nature.[138]

In other words: God, by His continuous preservation of created matter, is responsible for the continuities within the course of material changes, which are due to the

[138] Descartes 2004, 25. On Descartes's idea of nature, see Verbeek 2000.

inherent diversities between the different states of matter (i.e. the different modes of it). If the states of matter would be determined only by the action of God, no change would take place in the world. However, since God himself had imparted different motions to the original, extended continuum of matter, bringing about a sort of 'chaos',[139] the world is conserved by Him in accordance with the various states of matter resulting from the first movement He imparted to it.[140] The continuities i.e. the effects of divine constancy on created matter, in turn, are explained by the laws of nature. In Descartes's *Le monde*, these are presented as follows:

1) "each particular part of matter always continues in the same state unless collision with others forces it to change its state." (Such states are: size, figure, movement and rest).[141]
2) "[W]hen one of these bodies pushes another it cannot give the other any motion except by losing as much of its own motion at the same time; nor can it take away any of the other's motion unless its own is increased by the same amount."[142]
3) "[W]hen a body is moving, even if its motion most often takes place along a curved line [...], nevertheless each of its parts individually tends always to continue moving along a straight line."[143]

Such laws express different effects of divine constancy: (1) the conservation of all the states or modes of matter; (2) the conservation of the quantity of motion; (3) the conservation of the direction of motion (which will be considered in Chap. 6). The first two laws are derived by Descartes from the metaphysical principle of conservation, given at different passages in *Le monde*,[144] and then considered, as seen in Sect. 4.2.1.1, *Descartes's laws of motion and idea of force*, in the *Principia*, where it is part of Descartes's justification of the laws of motion. The constancy of God grants, as Descartes states in *Principia* II.36, that "although motion is only a mode of the

[139] See AT XI, 34–35.

[140] "From this it follows necessarily that from the time they begin to move, they also begin to change and diversify their motions by colliding with one another. Thus, while God subsequently preserves them in the same way He created them, He does not preserve them in the same state. That is to say, if God always acts in the same way and consequently always produces substantially the same effect, many differences in this effect occur, as if by accident. And it is easy to accept that God, who is, as everyone must know, immutable, always acts in the same way," Descartes 2004, 25.

[141] Descartes 2004, 25–26.

[142] Descartes 2004, 27.

[143] Descartes 2004, 29. On Descartes's theory of motion and laws of nature as given in the *Le monde* and *Principia*, see in particular Blackwell 1966; Aiton 1972, chapter 3; Gabbey 1980; Garber 1992; Des Chene 1996, part 3; Slowik 1996, 2002; McLaughlin 2000; Gaukroger 2002, chapter 4.

[144] See *supra*, n. 140, and "these two rules [sc. the first and second one] follow manifestly from the sole fact that God is immutable and that, acting always in the same way, He always produces the same effect. For on the assumption that He placed a certain amount of motion in matter in general at the first instant He created it, we must admit either that He preserves the same amount of motion in it, or not believe that He always acts in the same way," Descartes 2004, 28–29. On Descartes's principle of conservation, see in particular Menn 1990; Garber 1992, chapter 7; Slowik 2002, chapter 5.

matter which is moved, nevertheless there is a fixed and determined quantity of it; which, as we can easily understand, can be always the same in the universe as a whole even though there may at times be more or less motion in certain of its individual parts"[145] – this being the general principle of the conservation of the quantity of motion (to which I will return in a moment). Hence, this principle grants the validity of three laws of motion:

1) "each thing, provided that it is simple and undivided, always remains in the same state as far as is in its power, and never changes except by external causes,"[146] (II.37). This law rephrases law 1 of *Le monde*, and is justified by Descartes also on the basis of the immutability of essences, as seen in Sect. 4.1.4, *The search for physical unity and activity: De Volder vis-à-vis Leibniz*: a part of matter does not change its state, because a state cannot change itself, as everything tends to preserve itself.
2) "[E]ach part of matter, considered individually, tends to continue its movement only along straight lines [...] for He only maintains it precisely as it is at the very moment at which He is maintaining it,"[147] (II.39). This law corresponds to law 3 of *Le monde*.
3) "[A] body, upon coming in contact with a stronger one, loses none of its motion; but that, upon coming in contact with a weaker one, it loses as much as it transfers to that weaker body,"[148] (II.40) corresponding to law 2 of *Le monde*.

The core concept of law 1 and 3 is the conservation of the quantity of motion, an idea that Descartes relates to the more complex ideas of force of motion and resistance to motion (dealt from II.43), and which depends on (1) the quantity of matter of a body (namely, on how much uniform, or continuous matter it is composed of), and (2) on its speed (intended as a scalar quantity). So that quantity of motion can be quantified with the formula 'm·s'). As he states in II.36, "when one part of matter moves twice as fast as another twice as large, there is as much motion in the smaller as in the larger."[149] In turn, the quantity of motion determines what Descartes labels (in II.45) as the "force of motion" of a body, and its "force of resistance to motion."[150] These are concepts dealt with first in II.43, where such forces are related to four factors: (1) the speed, (2) the quantity of matter, (3) the surface which is involved in the interaction between bodies, (4) and the ways in which bodies collide.[151] If the relation of

[145] Descartes 1982, 58.

[146] Descartes 1982, 59.

[147] Descartes 1982, 60.

[148] Descartes 1982, 61.

[149] Descartes 1982, 58.

[150] Cf. the original text: "virium, sive ad movendum, sive ad motui resistendum," AT VIII-A, 67; "force en chacun de ces corps pour mouuoir ou pour resister au mouuement," AT IX-B, 89. On Descartes's theory of force, see in particular Hatfield 1979; Gabbey 1980; Guéroult 1980; Arthur 2007; Hattab 2007; Schmaltz 2008; Manchak 2009.

[151] "In what the force of each body to drive or to resist consists. We must however notice carefully at this time in what the force of each body to act against another or to resist the action of that other consists: namely, in the single fact that each thing strives, as far as is in its power, to remain in the

speed and quantity of matter ground the idea of quantity of motion, the relation between the ideas of quantity of matter and surface ground the idea of solidity, which is crucial in Descartes's dynamics, and is systematically treated only in III.121–123. Solidity, for Descartes, is a complex notion, and it cannot just be equated with the idea of hardness, as opposed to fluidity. As he labels it in III.123, solidity is the "aptitude to retain movement," namely, it is a resistance to change the movement, which has to be understood as a resistance to change the speed, and the direction of a given movement (two ideas which are at the core of Descartes's theory of impact, and which are in turn related to the idea of determination, to which I will reserve a separate treatment).[152] Roughly speaking, for Descartes a body is very solid when it is perfectly continuous (i.e. it has no pores, nor are its parts in motion with respect to each other), and when it has a high ratio of volume and surface (V/S).[153] For instance, a corpuscle such as a globule of second matter can be more solid than a whole planet, because the globule is perfectly continuous in itself – so that it can exert all its force of motion in one direction – and it can have a higher V/S ratio than the planet, because the perfectly spherical shape allows the globule to contain more volume within a smaller surface, while a planet can have a rough surface, and not be perfectly spherical.[154] The V/S ratio is essential to solidity because a spherical body, *caeteris paribus*, is less impacted by other bodies, in comparison with a non-spherical one, so, it retains, in proportion, more force as it undergoes collisions, and has more resistance to the loss of its quantity of motion (III.125).[155]

---

same state, in accordance with the first law stated above. From this it follows that a body which is joined to another has some force to resist being separated from it, while a body which is separate has some force to remain separate. One which is at rest has some force to remain at rest, and consequently to resist everything which can change it; while a moving body has some force to continue its motion, i.e., to continue to move at the same speed and in the same direction. Furthermore, this force must be measured not only by the size of the body in which it is, and by the [...] surface which separates this body from those around it; but also by the speed and nature of its movement, and by the different ways in which bodies come in contact with one another," Descartes 1982, 63.

[152] See Sect. 4.2.1.3, *Descartes's idea of determination, and rules of reflection and refraction.*

[153] This concept is discussed in Schuster 2013, 465–468 and 627–631.

[154] "Thus, it can happen that star N has less solidity, or less ability to continue its movement, than the globules of the second element which surround it; even though it may be very large and covered with fairly many layers of spots. For these globules, in proportion to their size, are as solid as any body can be, because we understand that they contain no pores filled with other [...] matter; and because their figure is spherical; the sphere being the figure which has the least surface area in proportion to its volume, as Geometers know," Descartes 1982, 153–154.

[155] "Finally, it can happen that the same star may be less solid than certain globules of the heaven, and more solid than some other rather smaller ones, both for the reason just stated, {namely, that the forces of several globules are less unified than those of one larger body equal to all of them}, and also because, although the quantity of the matter of the second element in all the globules which occupy a given [amount of] space may be the same whether they are {very} small or {quite} large, the smaller ones have {less force, because they have} more surface area {in proportion to the quantity of their matter}; and therefore they can be drawn off their course and turned aside in other directions more easily than the larger ones, either by the matter of the first element filling the spaces which they leave around themselves, or by any other bodies {which they encounter}," Descartes 1982, 155. The text between brackets thus { } is present in the French version of Descartes's *Principia.*

In any case, the foundation of force of motion, and resistance to motion, is given by the quantity of motion (m·s). The bigger and faster a body is, the more it resists being brought to rest, and the bigger and slower (or resting) it is, the more difficult is to move it. As movement and rest are two opposite states or modes of matter, rest and quantity of matter determine the resistance to acquiring motion, while speed and quantity of matter determine the resistance to losing motion. Finally, there remains to be considered the directions or angles of impact in which bodies collide with each other – a factor which I will discuss below. This consideration, as I am going to show,[156] is used by Descartes in dealing with the problem of reflection and refraction, and involves the idea of determination, which is another key component of Descartes's dynamics.

4.2.1.2 Descartes's Rules of Impact

Besides the three laws of nature, Descartes also presented – in his *Principia* –seven rules of impact, which he derives from law 3. These seven rules are not overtly applied in the other sections of his *Principia*,[157] yet, they are essential in order to understand Descartes's ideas on force of motion and resistance to motion (which are then used in his cosmology). In presenting such rules, Descartes hypothesized, in his *Principia*, the collision between two perfectly hard bodies – namely, perfectly continuous ones – whose movement is not influenced by any external factor. This means that the two bodies are conceived as if they were in vacuum, and as if the surface on which they are standing does not influence their movements. Moreover, their impact is considered only according to straight lines of motion, and without considering their shape, which is, as shown by the engravings in the *Principia*, rectangular prisms (see Fig. 4.5, Descartes 1644a, 60).

The determination of the outcomes of impact is made possible by the calculation of the force of motion and resistance to motion, which makes a body weaker or stronger in producing its effect.[158] Broadly speaking, a body is stronger – in moving

[156] See Sect. 4.2.1.3, *Descartes's idea of determination, and rules of reflection and refraction.*

[157] As argued by Alan Gabbey, the formulation and the addition of the seven rules of impact to the draft of the *Principia* traces back to the last years of the composition of this treatise: see Gabbey 1980. On Descartes's rules of impact, see also Blackwell 1966; Aiton 1972, chapter 3; Garber 1992, chapter 8; McLaughlin 2000.

[158] "In order to determine, from the preceding laws, how individual bodies increase or decrease their movements or turn aside in different directions because of encounters with other bodies; it is only necessary to calculate [in Latin: *calculo subducere*; in French: *calculer*: see AT VIII-A, 67; AT IX-B, 89] how much force to move or to resist movement there is in each body; and to accept as a certainty that the one which is the stronger will always produce its effect. Moreover, this could easily be calculated if only two bodies were to come in contact, and if they were perfectly solid, and separated from all others {both solid and fluid} in such a way that their movements would be neither impeded nor aided by any other surrounding bodies, for then they would observe the following rules," Descartes 1982, 64.

Fig. 4.5 René Descartes's representation of two ideal colliding bodies. (Source: Descartes 1644a, 60. Library of Congress, call number: LC B1860 1644, http://hdl.loc.gov/loc.rbc/Rosenwald.1431.1. Public domain)

or in resisting motion – when after the collision it has changed its state of motion least (in speed or direction), or it has not changed at all (for instance, it is still at rest). As a consequence, the stronger body had produced a greater effect (i.e. a change) in the other body. In other words, the force of motion and resistance to motion is nothing but a force of persistence in a state, or a force of inertia.

In the determination of the force of motion and resistance to motion in the seven rules, Descartes considers four factors: (a) the size of a body; (b) its state of rest or of motion; (c) the swiftness and the slowness of each body (as slowness participates in rest, while swiftness participates in motion; (d) the determination of a motion in a given direction, as one determination can be opposed to another determination.[159] Since in presenting his seven rules the differentiation between direction and determination is not relevant,[160] I will refer here only to direction.

As highlighted by Peter McLaughlin, the guiding theoretical model of Descartes in his formulation of the rules of impact was the theory of the lever.[161] A model in fact overtly used in the treatises of Henricus Regius and Rohault – amongst others – to introduce readers to the treatment of natural laws.[162] In a lever (see Fig. 4.6, *Model of the lever*), indeed, the force required to move a body of weight, mass or

[159] "That movement is not contrary to movement, but to rest; and that determination in one direction is the opposite of determination in another. It must also be noticed that one movement is in no way contrary to another movement of equal speed; but that, strictly speaking, only a twofold opposition is found here. One is between movement and rest, or even between rapidity of movement and slowness of movement (i.e., to the extent that this slowness partakes of the nature of rest): the other is between the determination of a body to move in a given direction and the encounter, in its path, with a body which is either at rest or moving in a contrary manner; and this opposition is greater or smaller according to the direction in which the body which encounters the other is moving," Descartes 1982, 63.

[160] As I am going to show in Sect. 4.2.1.3, *Descartes's idea of determination, and rules of reflection and refraction*, the determination of motion is a concept allowing the decomposition of the quantity of motion in more directions: so, it applies to the case of reflection and refraction, where changes in the angles of impact are considered. In the seven rules of impact, on the other hand, the angle of impact and that of reflection or rebounding are always 90°.

[161] See McLaughlin 2000.

[162] See Regius's *Fundamenta physices* (1646), chapter 1, and Rohault's *Traité de physique*, part 1, chapter 10.

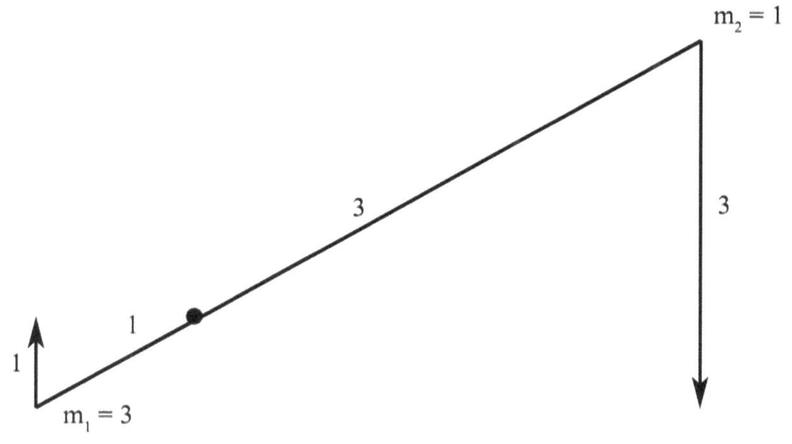

Fig. 4.6 Model of the lever. Ideal rendering of the functioning of the lever. (Figure by the author)

volume (these concepts are overlapping in this case) 3 (in the figure, m_1), for 1 unit of length, is the same force required to move in the same time a body of weight 1 (m_2) for 3 units of length, if the ratio of the length of the two arms of the lever is in an inverse proportion to the ratio of the weights. So, the force of movement, in the case of the lever, is calculated by m·s (where the speed of m_1, or s_1, is 1, and the speed of m_2, or s_2, is 3).[163]

In turn, the "contest model"[164] of impact at work in Descartes's rules can be represented – in Figs. 4.7–4.16, given in Table 4.1, *Visual representation of Descartes's rules of impact (reduced to the model of the lever)* – as follows: (1) the size of each body is represented by the size of its ideal representation in the figure mentioned above, (2) the length of the arm of the lever of each body represents its speed (regardless of its direction) – in those cases in which both bodies move. In the case in which a body is at rest, for the reason that rest is always opposite to motion, the arm representing the speed of the body at rest is equal to that of the body which is not at rest. Indeed, a resting body will always oppose to a moving body, *ceteris paribus*, a resistance equal to the force of motion of the other one, namely, the force calculated by m·s.

Descartes's seven rules can be divided into three main cases: rules 1–3: two bodies A and B move against each other; rules 4–6: body B is at rest; rule 7 (which has three sub-rules): A and B move in the same direction, and B is slower. As follows:

Rule 1 (II.46): two bodies A and B, which are equal in size (m_1 and m_2) and moving at the same speed (s_1 and s_2). In this case, neither body wins over the other, as they have the same value of m·s (= 24). The impact results in a symmetrical

[163] The Cartesian quantification of force or quantity of motion is clearly presented in Rohault's *Traité de physique*, part 1, chapter 10, §§ 7–10, where the model of the lever is explicitly taken into account.

[164] See Gabbey 1980, 137–139.

Table 4.1 Visual representation of Descartes's rules of impact (reduced to the model of the lever)

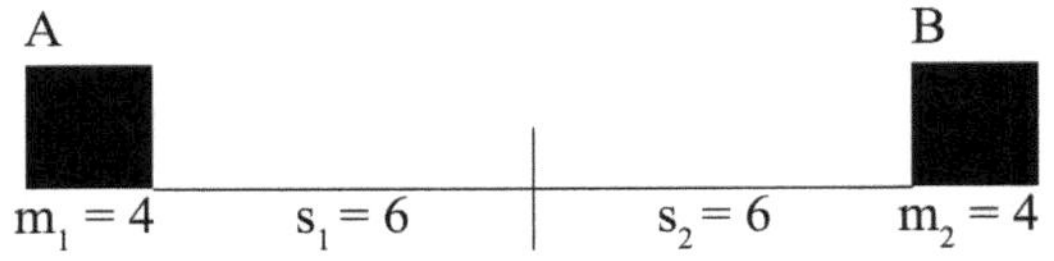

Fig. 4.7 Descartes's rule of impact 1. Reduction of Descartes's rule of impact 1 to the model of the lever. (Figure by the author)

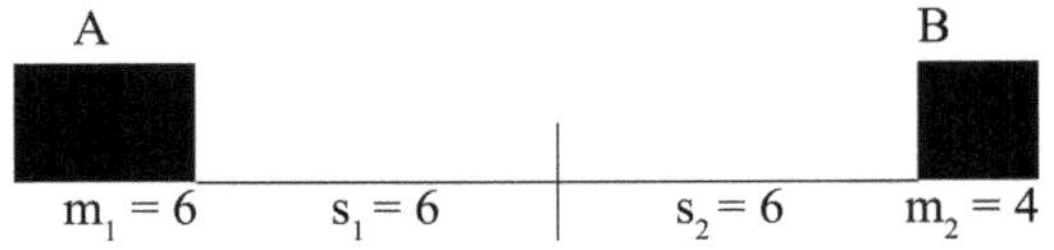

Fig. 4.8 Descartes's rule of impact 2. Reduction of Descartes's rule of impact 2 to the model of the lever. (Figure by the author)

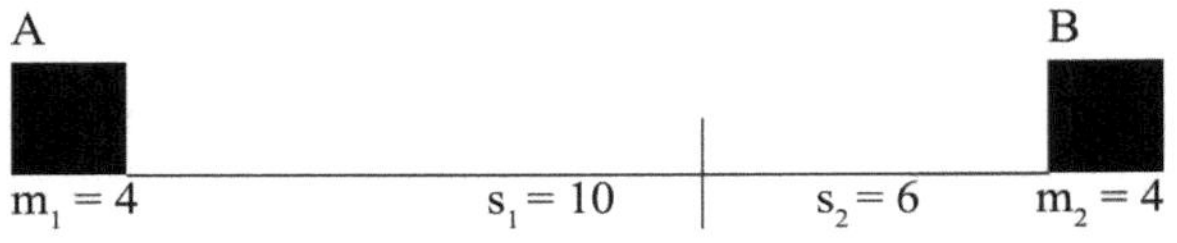

Fig. 4.9 Descartes's rule of impact 3. Reduction of Descartes's rule of impact 3 to the model of the lever. (Figure by the author)

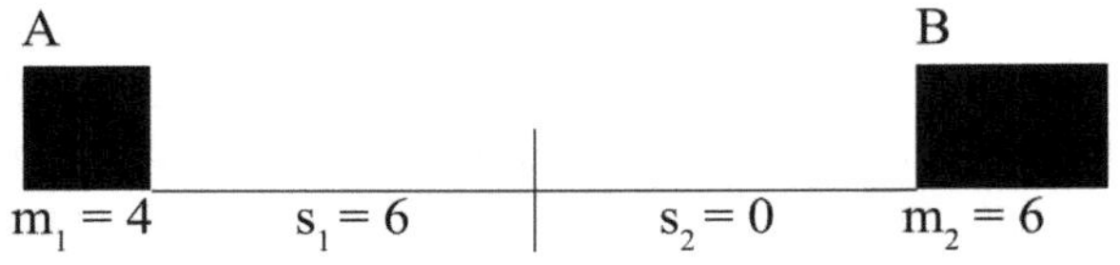

Fig. 4.10 Descartes's rule of impact 4. Reduction of Descartes's rule of impact 4 to the model of the lever. (Figure by the author)

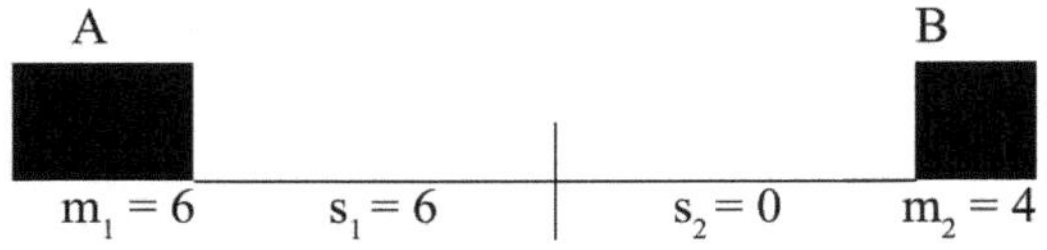

Fig. 4.11 Descartes's rule of impact 5. Reduction of Descartes's rule of impact 5 to the model of the lever. (Figure by the author)

(continued)

Table 4.1 (continued)

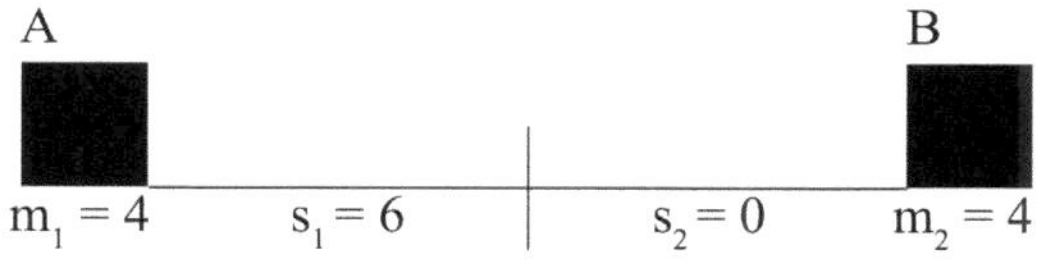

Fig. 4.12 Descartes's rule of impact 6. Reduction of Descartes's rule of impact 6 to the model of the lever. (Figure by the author)

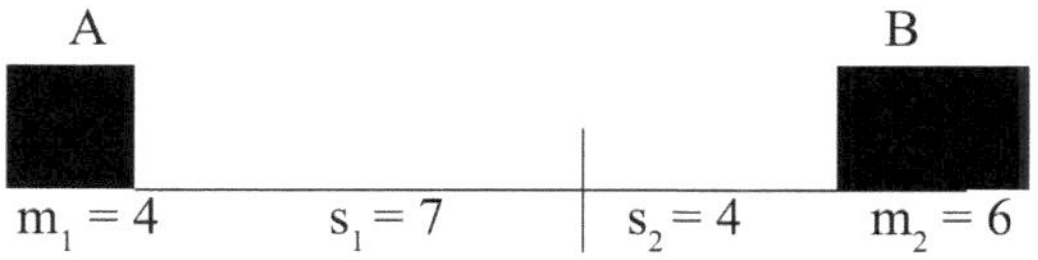

Fig. 4.13 Descartes's rule of impact 7a1. Reduction of Descartes's rule of impact 7a1 to the model of the lever. (Figure by the author)

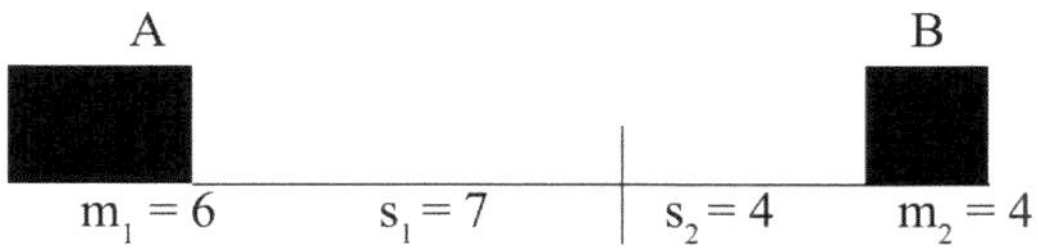

Fig. 4.14 Descartes's rule of impact 7a2. Reduction of Descartes's rule of impact 7a2 to the model of the lever. (Figure by the author)

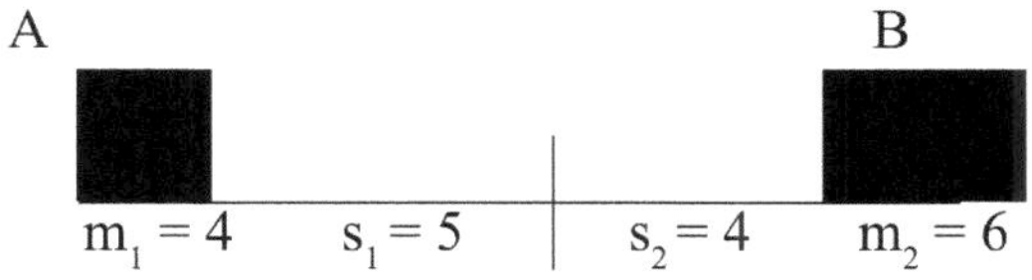

Fig. 4.15 Descartes's rule of impact 7b. Reduction of Descartes's rule of impact 7b to the model of the lever. (Figure by the author)

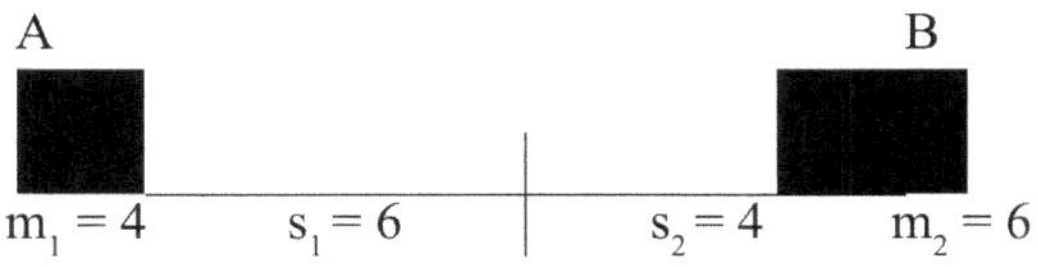

Fig. 4.16 Descartes's rule of impact 7c. Reduction of Descartes's rule of impact 7c to the model of the lever. (Figure by the author)

effect, namely a mutual change of direction, and no change in speed (see Fig. 4.7, *Descartes's rule of impact 1*).[165]

Rule 2 (II.47): in this case, they move at the same speed, but body A is larger, so that on a lever it would raise body B (see Fig. 4.8, *Descartes's rule of impact 2*). In a collision, body A will make body B change its direction, because body A (being stronger) will undergo the least modal change possible. This sort of principle of economy, at work in Descartes's model of impact, will be overtly expressed by Descartes in a letter to Clerselier of 17 February 1645, according to which the changes in the modes of two impacting bodies are always the least required to make their opposite modes – as movement and rest, and directions or determinations of motion – compatible with each other. In the case of motion, Descartes clarifies, there are two modes to be considered (which are thus modes of a mode): speed, and determination in a certain direction.[166] In this case of impact, A will not undergo any change. After the impact, indeed, both bodies will move together (A + B) in the original direction of body A, retaining their original speed. This can be calculated as follows: $(m_1 \cdot s_1 + m_2 \cdot s_2)/(m_1 + m_2)$.[167]

Rule 3 (II.48): in this case the two bodies have the same size, but A is faster (and stronger) (see Fig. 4.9, *Descartes's rule of impact 3*). So, after the impact A will keep its direction of motion, and will move together with B as in the previous case, but it will also lose a part of its speed, for the reason that in order for the total quantity of motion (m·s) to be conserved, the new body (A + B) will move at a speed given by the same formula given in the previous rule. With a little simplification, since $m_1 = m_2$, one can render this as: $(s_1 + s_2)/2$. In this case, of course, body the new body will be slower than body A, and faster than B.[168]

---

[165] "First, if these two bodies, for example Band C, were completely equal in size and were moving at equal speeds, B from right to left, and C toward B in a straight line from left to right; when they collided, they would spring back and subsequently continue to move, B toward the right and C toward the left, without having lost any of their speed. {For, in this case, there is no cause which could take their speed from them, but there is a very obvious one which must force them to spring back; and because it would be equal in each, they would both spring back in the same way}," Descartes 1982, 64–65.

[166] "[C]es regles [...] ne dépendent que d'un seul principe, qui est, Que lors que deux cors se rencontrent qui ont en eux des Modes incompatibles, il se doit veritablement faire quelque changement en ces modes pour les rendre compatibles, mais que ce changement est tousiours le moindre qui puisse estre, c'est à dire, que si certaine quantité de ces modes estant changée ils peuvent devenir compatibles, il ne s'en changera point une plus grande quantité. Et il faut considerer dans le mouvement deux divers modes, l'un est la motion seule ou la vitesse, et l'autre est la determination de cette motion vers certain costé, lesquels deux modes se changent aussi difficilement l'un que l'autre," AT IV, 185. For a thorough analysis of Descartes's own discussion of his rules of impact, see Garber 1992, chapter 8.

[167] "Second, if B were slightly larger than C, and everything else were as previously described, then only C would spring back, and both would move toward the left at the same speed. {For B, having more force than C, could not be obliged by C to spring back}," Descartes 1982, 65.

[168] "Third, if the two bodies were equal in size, but if B were moving slightly more rapidly than C; after their collision not only would {C alone spring back and} both continue their movement toward the left, {that is, in the direction from which C came}, but also one half of B's additional speed would be transferred from it to C, {since B could not move more rapidly than C which would

Rule 4 (II.49): in this case, body B is at rest and bigger than A, which moves towards it. Since the resistance to motion of a body at rest is proportional to the speed of the body impacting it, in this case the length of the arms of the lever are equal (see Fig. 4.10, *Descartes's rule of impact 4*). As body B is bigger, it will always win, regardless of the dimension of A (provided that A is smaller than B). After the impact body B will not undergo any change, while body A will keep its speed, and will change its direction of motion to its opposite.[169]

Rule 5 (II.50): in this case, B is at rest, while A moves towards it and is bigger than B. On a lever, A would win i.e. raise B (see Fig. 4.11, *Descartes's rule of impact 5*). Accordingly, A changes only its speed (i.e. it changes only a mode of a mode), while B is put into motion (i.e. it passes to a mode opposite to rest). The two bodies constitute a new body whose speed, given the conservation of m·s, is calculated as in rule 2: $(m_1 \cdot s_1 + m_2 \cdot s_2)/(m_1 + m_2)$. Namely, since $s_2 = 0$: $(m_1 \cdot s_1)/(m_1 + m_2)$.[170]

---

be ahead of it}. For example, if B had initially been travelling at six degrees of speed, and C at a speed of only four, {B would transfer to C one of its two additional degrees of speed, and} both would subsequently move toward the left at five degrees of speed. {This would occur because it is much easier for B to transfer one of its additional degrees of speed to C than for C to change the course of all the movement which is in B}," Descartes 1982, 65. It is worth to note that also in this case – viz. in the French version of the article – Descartes overtly adopts the principle of least modal change.

[169] "Fourth, if the body C were entirely at rest, {that is, if it not only had no apparent motion but also were not surrounded by air or any other fluid (which makes the hard bodies immersed in such a fluid very easily movable, as I shall show)}, and if C were slightly larger than B; the latter could never {have the force to} move C, no matter how great the speed at which B might approach C. Rather, B would be driven back by C in the opposite direction: because {for B to move C, C would have to be driven as rapidly as B subsequently moves and} a body which is at rest puts up more resistance to high speed than to low speed; and this resistance increases in proportion to the difference in the speeds. Consequently, there would always be more force in C to resist than in B to drive, {because C is larger. For example, if B is one half as large as C and is travelling at three degrees of speed, then because B is only as large as each of C's halves and because it cannot continue in the same direction more rapidly than it pushes C ahead of it; B cannot move C without transferring to it two thirds of its quantity of motion, one third for each of C's halves, keeping for itself only one degree of speed. Similarly, if B has thirty degrees of speed, twenty will have to be communicated to C; if B has three hundred, two hundred will have to be transferred, and so on. But since C is at rest, its resistance to receiving twenty degrees of speed is ten times as great as its resistance to receiving two, and so on. Thus, the greater B's speed, the proportionally greater C's resistance will be. And because each half of C has as much force to remain at rest as B has to drive it, and because both halves resist at the same time, it is obvious that they must succeed in forcing B to spring back. So that, no matter how great the speed at which B approaches C, B can never have the force to move C}," Descartes 1982, 66.

[170] "Fifth, if the body C were at rest and {even very slightly} smaller than B; then, no matter how slowly B might advance toward C, it would move C with it by transferring to C as much of its motion as would permit the two to travel subsequently at the same speed. Thus if B were twice as large as C, it would transfer to C {only} one third of its quantity of motion; because that one third would move the body C at the same speed as the remaining two thirds would move the body B which {we are supposing} is twice as large as C. Therefore, after B had collided with C, its speed would be reduced by one third; that is to say, B would then need as much time to travel a distance of two feet as it previously did to travel a distance of three feet. Similarly, if B were three times as

Rule 6 (II.51): in this case, the two bodies are equal in size, and A moves towards B which is at rest. On a balance, there is a condition of equilibrium, as in rule 1 (see Fig. 4.12, *Descartes's rule of impact 6*). Therefore, there is no a winner: the 'symmetry' in change is calculated by Descartes as follows: (1) body B changes its state of rest and moves in the original direction of A, (2) body A loses some speed, and (3) it changes its direction. The calculation of the speed assumed by both bodies is explained in the French version of the *Principia* (while in the Latin version Descartes just reports the outcome). First, if the two bodies were to move in the same direction after impact (as in rule 2), they will move with half of the speed of A (if m·s is valid); if A does not move B, but will be rebounded by it (as in rule 4), it would retain all its speed. So, in the 'intermediate' case, A gives to B a quarter of its original speed, namely, half of the half of its original speed.[171] As pointed out by Daniel Garber, a sort of "sophisticated model of impact" is at work here. Namely, the outcome of a collision is proportional to the hypothetical effects that it would have in other circumstances.[172]

Rule 7 (II.52): this rule encompasses different cases in which A and B both move in the same direction:

Rule 7a/1: body A is smaller than B, and the excess of speed of A is bigger than the excess of size of B (i.e. $s_1/s_2 > m_2/m_1$). So, A wins, and after the impact they form

large as C, it would transfer to C one quarter of its motion; and so on. {And it is impossible for B to have so little force that it would ever be insufficient to move C; for it is certain that weaker motions must observe the same laws as stronger ones, and must produce, proportionally, the same result. Although we often think we see the opposite on this earth; this is because of the air and other fluids which always surround solid moving bodies and which can greatly increase or decrease their speed, as we shall see later}," Descartes 1982, 67.

[171] "Sixth, if the body C were at rest and exactly equal in size to body B, which was moving toward it; necessarily, C would be to some extent driven forward by B and would to some extent drive B back in the opposite direction. Thus, if B were k) approach C with four degrees of speed, it would {have to} communicate one degree to C, and be driven back in the opposite direction with the remaining three. {Because it must necessarily be the case that either B moves C without springing back, thus transferring two degrees of its speed to C; or that B springs back without moving C, retaining those two degrees of speed as well as the two which cannot be taken from it; or else that B springs back, retaining some portion of those two degrees of speed, and at the same time moves C with the remainder of those two degrees. Since Band C are equal and there is consequently no more reason for B to spring back than to move C; it is obvious that these two effects must be equally shared: that is, B must transfer one of these degrees of speed and spring back while retaining the other}," Descartes 1982, 67–68.

[172] Garber 1992, 242–251. This model is also presented in the letter of Descartes to Clerselier of 17 February 1645, in which Descartes discusses rule 4. In rule 4, B resists being moved by A as much as A strives to move it, regardless of the speed of A: so, the resistance it exerts is proportional to the effect that A would exert on B, if it had managed to move it: see AT IV, 184; cf. Garber's presentation: "[n]ow, the faster B [sc. A] is going, the more motion it would have to transfer to C [sc. B] for the result to obtain. And so, if we think of the force of resisting in C as measured by the quantity of motion it would have if B were to succeed in pushing it after the collision, then the faster B goes, the more the resistance C offers," Garber 1992, 244. Descartes's explanation of rule 4 given in his letter would be then incorporated in the French version of the article: see *supra*, n. 169.

one body, and they follow the same direction with a speed given as in rule 2: $(m_1 \cdot s_1 + m_2 \cdot s_2)/(m_1 + m_2)$ (see Fig. 4.13, *Descartes's rule of impact 7a1*).

Rule 7a/2 (given only in the French edition of the *Principia*): body A is faster and bigger than B (see Fig. 4.14, *Descartes's rule of impact 7a2*). A wins, and after the impact they form one body, moving in the same direction of A, with a speed given as in rule 2: $(m_1 \cdot s_1 + m_2 \cdot s_2)/(m_1 + m_2)$.[173]

Rule 7b: A is smaller than B, and the excess of speed of A is less than the excess of size of B (i.e. $s_1/s_2 < m_2/m_1$) (see Fig. 4.15, *Descartes's rule of impact 7b*). In this case, A loses, and is rebounded after its collision with B. After the impact, B does not undergo any change (as it is stronger in all its modes), and A, accordingly, is rebounded and keeps its speed.[174]

Rule 7c (given only in the French edition of the *Principia*): A is smaller than B, and the excess of speed of A compensates the excess of size of B (i.e. $s_1/s_2 = m_2/m_1$) (see Fig. 4.16, *Descartes's rule of impact 7c*). On a balance, they would be in equilibrium, so, the outcome would be symmetrical as in rule 1. For Descartes, after the impact A changes both its speed (which is decreased), and its direction of motion (which is inverted), while B changes only its speed (which is increased). However, no detail is given by Descartes.[175]

We can summarize all these rules in Table 4.2, *Descartes's rules of impact*.[176]

4.2.1.3 Descartes's Idea of Determination, and Rules of Reflection and Refraction

Together with the ideas of quantity of motion and solidity, the concept of determination is essential to Descartes's dynamics, and to his theory of impact in particular. Determination is, as McLaughlin has put it, "perhaps the most difficult concept of Cartesian physics."[177] It does not consist of the direction of a given body in motion, nor of its speed, nor of the quantity of motion (m·s). Rather, it is a mode of the force of motion exerted in a given direction, or a "directed quantity," which can be resolved into components by the parallelogram rule. According to the parallelogram

---

[173] "Finally, if B and C were travelling in the same direction, C more slowly than B, so that B (which would be following C) would eventually strike it; and if C were larger than B but B's speed exceeded C's by a greater extent than C's size exceeded B's: then B would transfer to C as much of its speed as would be required to permit them both to travel subsequently at the same speed and in the same direction," Descartes 1982, 68.

[174] "However, if, on the contrary, B's speed exceeded C's by a smaller extent than C's size exceeded B's; B would be driven back in the opposite direction, and would retain all its movement," Descartes 1982, 68.

[175] "{And, finally, when the ratio in which C's size exceeds B's is exactly equal to the ratio in which B's speed exceeds that of C, B must transfer some of its motion to C and spring back with the rest}," Descartes 1982, 68.

[176] Analogous tables are given in Aiton 1972, 36; Garber 1992, 255; McLaughlin 2000, 100.

[177] McLaughlin 2000, 87.

Table 4.2 Descartes's rules of impact

	Before impact:	After impact:
Rule 1	A and B move towards each other, and $m_1 = m_2$; $s_1 = s_2$	A and B move in opposite directions, and $s_1 = s_2$
Rule 2	A and B move towards each other, and $m_1 > m_2$; $s_1 = s_2$	A and B move together in the same direction, at speed $(m_1 \cdot s_1 + m_2 \cdot s_2)/(m_1 + m_2)$
Rule 3	A and B move towards each other, and $m_1 = m_2$; $s_1 > s_2$	A and B move together in the same direction, at speed $(m_1 \cdot s_1 + m_2 \cdot s_2)/(m_1 + m_2) = (s_1 + s_2)/2$
Rule 4	A moves towards B, and $m_1 < m_2$; $s_2 = 0$	A is rebounded by B, and moves in the opposite direction without losing speed.
Rule 5	A moves towards B, and $m_1 > m_2$; $s_2 = 0$	A and B move together in the same direction, at speed $(m_1 \cdot s_1 + m_2 \cdot s_2)/(m_1 + m_2) = (m_1 \cdot s_1)/(m_1 + m_2)$
Rule 6	A moves towards B, and $m_1 = m_2$; $s_2 = 0$	A is rebounded by B, and moves in the opposite direction at ¾ of its original speed; B moves in the original direction of A at ¼ of the original speed of A.
Rule 7a/1	A and B move in the same direction, and $m_1 < m_2$; $s_1/s_2 > m_2/m_1$	A and B move together in the same direction, at speed $(m_1 \cdot s_1 + m_2 \cdot s_2)/(m_1 + m_2)$
Rule 7a/2	A and B move in the same direction, and $m_1 > m_2$; $s_1/s_2 > m_2/m_1$	A and B move together in the same direction, at speed $(m_1 \cdot s_1 + m_2 \cdot s_2)/(m_1 + m_2)$
Rule 7b	A and B move in the same direction, and $m_1 < m_2$; $s_1/s_2 < m_2/m_1$	A is rebounded by B, and moves in the opposite direction at its original speed; B keeps its direction and speed.
Rule 7c	A and B move in the same direction, and $m_1 < m_2$; $s_1/s_2 = m_2/m_1$	A is rebounded by B, and moves in the opposite direction at decreased speed; B keeps its direction and acquires some speed.

rule, (see Fig. 4.17, *Model of determination of movement, 1*), the movement, let's say, of a body at A towards D be can be resolved into the movement of line AC towards BD and of line AB towards CD.

So, the moving body A has a direction (represented by the line A to D), a speed, a quantity or force of motion ($m \cdot s$). Its determination cannot really differ from the movement from A to D with a given speed, but it is a quantity which, even if fixed, can be analysed into its components by following the parallelogram rule. For instance, the determination of motion from A to D (reached, let's say, in 1 hour, at 6.3 km/h) can be resolved into a movement from A to B at 2 km/h and a movement from A to C at 4.5 km/h (see Fig. 4.18, *Model of determination of movement, 2*) or into a movement from A to B at 6 km/h and a movement from A to C at 2 km/h (see Fig. 4.19, *Model of determination of movement, 3*).

As McLaughlin has put it,

> Cartesian determination is a hybrid of directionality and the absolute value of a motion. It is a directed quantity with the same absolute value as the motion which it modifies; it is thus only semi-independent. [...] [I]ts quantity cannot vary: it is always given by the quantity of

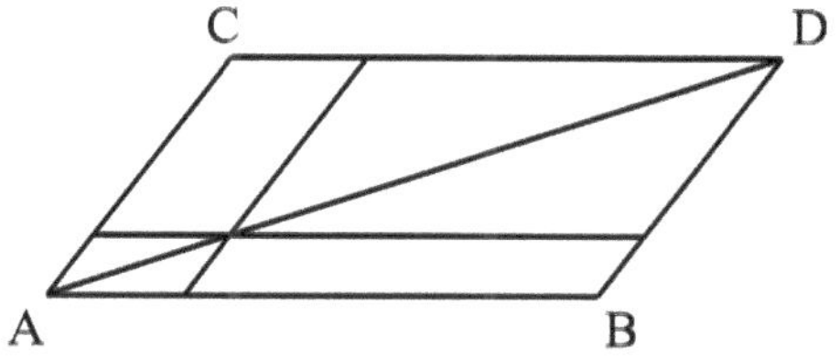

Fig. 4.17 Model of determination of movement, 1. Reduction of the idea of determination of movement to the parallelogram rule. (Figure by the author)

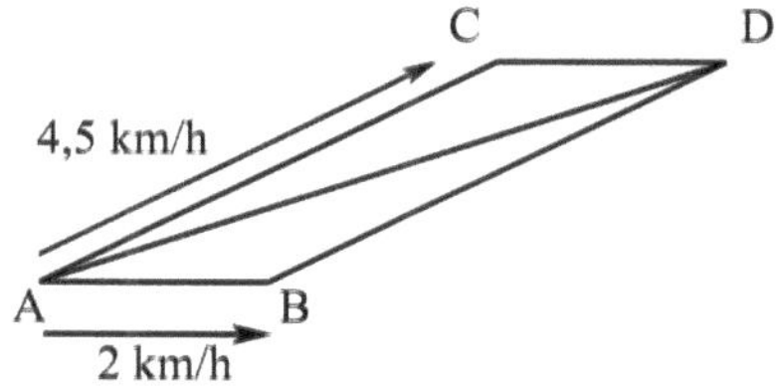

Fig. 4.18 Model of determination of movement, 2. Reduction of the idea of determination of movement to the parallelogram rule. (Figure by the author)

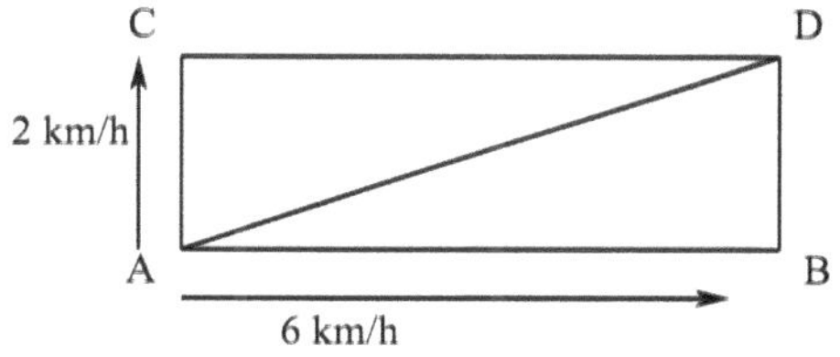

Fig. 4.19 Model of determination of movement, 3. Reduction of the idea of determination of movement to the parallelogram rule. (Figure by the author)

> motion. However, since any determination can be subdivided in various ways according to the parallelogram rule, it is possible for the quantities of the component- determinations to vary a great deal, so long as their vector sum does not.[178]

In the language of the Cartesians, determination was a *modum considerandi* of motion, but a *modum considerandi* well grounded in reality.[179] For instance, in the

---

[178] McLaughlin 2000, 92–93.

[179] As De Volder states in commenting upon Descartes's *Principia* II.32 (where the decomposition of movement is theorized: "[h]ow, properly understood, the single movement peculiar to each body may also be regarded as multiple," Descartes 1982, 55), he notes that this article "is the most worth of attention, because by this multiple phenomena, especially in catoptric and in dioptric, are explained" ("{[h]ic articulum} attentione {dignissimum} est, quia per hunc plurima φαινόμενα praesertim in catoptrica et dioptrica explicantur. Describat. Motus o[mn]is concipi potest instar plurimum etiam {simplicissimus}. Nec est q[uo]d dicunt motum rotae compositum e[ss]e, quia umunq[uo]dq[ue] punctum unam describat lineam sive mixtam sive simplicem, non e[nim] concipi potest linea ut circularis sola, quia recta adest, nec sola recta, quia adest circularis. Ergo motus o[mn]is est compositus, et licet nulla sit linea simplicior recta, composita tamen concipi potest. Circularis. Id vult autor sive concipiamus lineam simp[licem] sive compositam eandem fore conclusionem, o[mn]ia phaenomena eadem, atq[ue] {ad eo non} referre, sive simplicem, sive compositam concipiam vide dioptric. 2. §§ i" Hamburg 273, 98), and underlines that the decomposition of motion "often depends on our understanding, which however as in many things is grounded on the truth of thing [itself], it is indeed certain, that each point describe only one line" ("[i]nstar

case of Descartes's rule 1, body A and B completely lose their determination in the direction of the impact, but they keep their quantity of motion; therefore, they acquire a new determination after the impact, this time in the opposite direction. In other words: all the 'directional' component of their determination is lost, but as soon as their quantity of motion or force of motion is preserved, they acquire a new directional component of their determination. In the case of the laws of reflection and refraction, however, things are different. Descartes, in his *Diotrique*, conceives the movement of light as if it were the movement of a tennis-ball impacting on the surface of a yard (see Fig. 4.20, Descartes 1644b, 81; see also Fig. 4.21, *Descartes's law of reflection*).

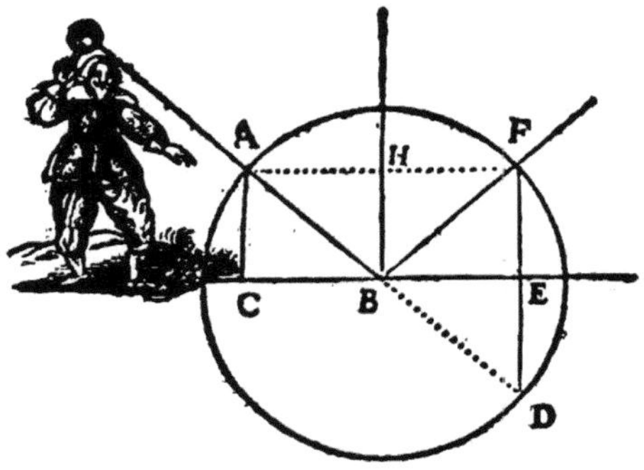

Fig. 4.20 Descartes's law of reflection. (Source: Descartes 1644b, 81. Library of Congress, call number: LC B1860 1644 Copy 2 B1845, http://hdl.loc.gov/loc.rbc/Rosenwald.1432.1. Public domain)

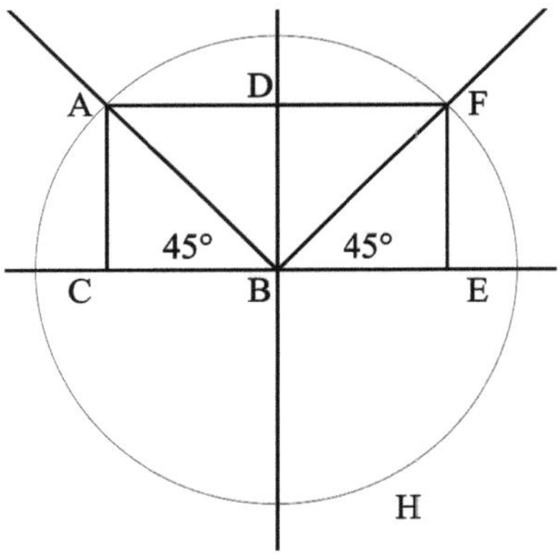

Fig. 4.21 Descartes's law of reflection. (Figure by the author)

plurium. Hoc saepe ab intellectui nostro dependet, qui tamen ut plurimum fundatur in rei veritate, certum enim est, unumquod[ue] punctum tantum describere unam lineam. Et uno moti potest describi, {q[uae]} licet varie sit contorta, tamen una est linea, ita quidem, ut eadem effecta {fiat consectura}, sive concipiamus punctum {rotae}, ut revera sit, moveri per lineam illam contortam, sive concipiamus punctum id describere lineam curvam nimirum {-} circularem, dum ille circulus movetur per lineam rectam, utroq[ue] enim modo eadem describitur curva. Idem sequitur effectus, unde ex hac ipsa distinctionem motus {...} esset ex variis compositus, nullus error {suborri} potest, quod opprime notandum propter usum, quem habet in Dioptricis in {materiae} reflexionum et refractionum. Vide dioptricam cap. 2," Hamburg 274, 41).

In the case of reflection, the ball hit the ground with, let's say, an angle of 45° and is rebounded at the same speed – because the ground does not take away anything of its quantity of motion, nor is the ball broken on impact – in another direction. So, it completely loses a component of its determination, namely, that component of its determination of motion which is perpendicular to the ground. This component is represented by the line AC. However, after the impact the body conserves its overall quantity of motion, and it conserves also the component of the determination which is not affected by the impact, namely the component AD. So, after the impact, the body will continue to move in the 'horizontal' direction AD with the same speed it had before, and it will cover a distance equal to AB (which we can conceive also as the radius of the circle H) with the same speed it had before. In order to find the angle of reflection of the ball, therefore, one needs to (1) prolong the line AD towards right (as the body will still move in this direction after the impact, with unaltered speed) and to find point F on the circle, and (2) connect the point of impact B to the point F – which is the new direction of the body. Or, to put it another way: the length of AB represents the distance covered by the ball, let's say, in a second; we know that after the impact, this speed i.e. length does not change. So, one second after the impact, the ball would have covered the same distance, represented by the radius of the circle H. How to determine the direction of the ball? We need to find a line DF which is equal to the line AD, and to connect the point in which this line intersects the radius, with point B.[180] A determination of motion in a given direction, in conclusion, can be 'picked out' from the quantity of motion which a body exerts in an actual direction or 'path' (as AB or BF), but the determination which has been 'picked out' will still be dependent on the 'path' and quantity of motion of this body.

The same ideas are at work in the case of refraction (see Fig. 4.22, Descartes 1644b, 84; see also Fig. 4.23, *Descartes's law of refraction*).

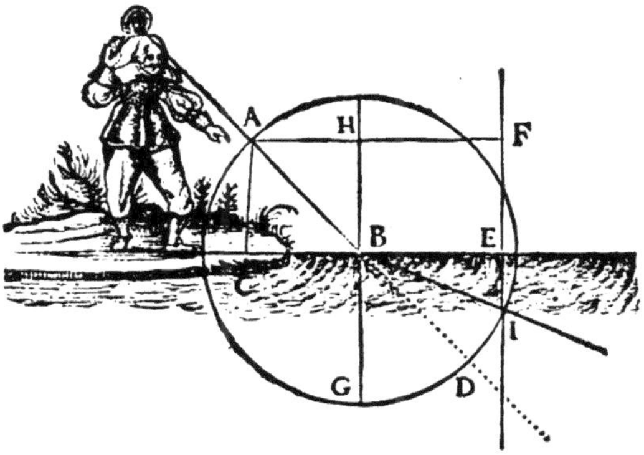

Fig. 4.22 Descartes's law of refraction. (Source: Descartes 1644b, 84. Library of Congress, call number: LC B1860 1644 Copy 2 B1845, http://hdl.loc.gov/loc.rbc/Rosenwald.1432.1. Public domain)

[180] Cf. Descartes's *Dioptrique, Discours second*, AT VI, 93–96.

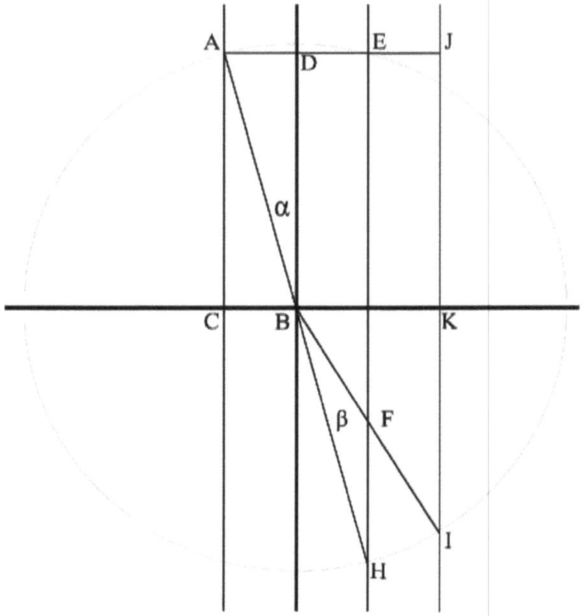

Fig. 4.23 Descartes's law of refraction. (Figure by the author)

Let's suppose that the ball hits the ground at point B, but it is not reflected by the yard: rather, on impact with the ground it loses half of its quantity off motion, namely half of its speed (Descartes assumes that after the impact its movement is not hindered further). So, after the impact the ball will cover a distance equal to AB in double the time it took before. But its horizontal determination (AD) is not affected by the impact, therefore, in the same time that the distance of the radius is covered by the ball (which will be double the time that it took before to cover such a distance), the ball would have covered a horizontal distance which is twice as long as the horizontal distance it covered before the impact, namely, 2AD, or 2DE (where DE = AD): namely, DJ. If we drop a line from J, we can find the point I on the lower semicircle of the circle, which is the point impacted by the ball in a time which is double the time it took to move from A to B. Alternatively, in order to find the new direction of the ball after the impact, we can locate it by using the smaller circle a, whose radius represents the distance reached by the ball in the same time that it took to cover the distance AB before the impact. As the ball lost half its speed, the radius of a is half the radius of the bigger circle. Hence, we need to consider that only the vertical component in the determination of the movement of the ball (i.e. AC) is affected by the impact. So, the horizontal component AD will not vary after the impact. Accordingly, in order to find the new direction taken by the ball after the impact, we need to drop a line from the point E, and to find where it intersects the radius of a below the surface: in our case, at point F. So, the line BF is the new 'path' of the ball after the impact, and β is the angle of refraction.[181]

[181] Cf. Descartes's *Dioptrique, Discours second*, AT VI, 96–100.

As argued by John Schuster, Descartes's natural-philosophical account of refraction is the result of his reinterpretation of the sine law of refraction discovered first by Ibn Sahl (circa 984), Thomas Harriot (circa 1598) and Willebrord Snel (circa 1620), and then rediscovered (with the help of Claude Mydorge) by Descartes himself. This natural-philosophical account was *ad hoc*, namely, it consisted of comparing the non-instantaneous movement of a ball, with the instantaneous 'movement' of a ray of light, which for Descartes is the pressure exerted in an instant by the globules of second matter, as they strive or have an 'instantaneous tendency' (whereas actual movement takes place in multiple instants) to recede from the centre of the vortex which they compose.[182] According to the sine law of refraction, the proportion between the sine (i.e. the segment CB) of the angle α (i.e. the angle between the vertical axis and AB), and the sine (i.e. the segment BK) of angle β (i.e. the angle between the vertical axis and BI) (see Fig. 4.23, *Descartes's law of refraction*), is equal to the proportion of the speeds of the ball after and before the impact: namely, $\sin \alpha / \sin \beta = BF/AB$.[183] As highlighted by Schuster, Descartes most likely came to formulate the law of refraction by using traditional optical means, namely, by (1) the "image placement rule" (in which one finds the point in which a refracted ray falls, (in Fig. 4.22, *Descartes's law of refraction*, F), and the one in which it would fall if not refracted (H), on a line which is perpendicular to the surface of refraction (i.e. the line EFH), and by knowing (2) the actual observational data on the refraction in various media.[184] Schuster suggests that Descartes "took a geometrical picture of a macroscopic phenomenon, garnered by mixed mathematical procedures, and read out of it the underlying dynamical causes," and then he came to formulate laws 1 and 3 as given in his *Le monde*. Indeed, in his theory of refraction Descartes presupposes (1) that the ratio of the quantity of motion (m·s) of a body before and after the impact on a given medium is constant, regardless of its angle of incidence and refraction, and of its size, and (2) the horizontal component of the motion is not affected by the impact, and can be considered independently from the vertical component, which is affected by the impact.[185] In his own words:

> [t]he first law [of Descartes's *Le monde*] establishes the principle of conservation of force of motion, regardless of the direction of the motion or its analysis into components. Hence it appears to transcribe and generalize the assumption that the force of light is related to the nature of the medium and is independent of the path of propagation. The third law specifies the (principal) tangential determination of the force of motion, and, by implication, licenses its analysis into components. It appears to transcribe and generalize the assumption that in the derivation of the refracted ray path, the determination of the light can be manipulated independently of the absolute quantity of the force of the light. The two rules together ground a general dynamical approach to mechanical explanation paralleling that implied in the particular instance of the optical proofs.[186]

---

[182] Schuster 2013, chapter 4. As to Descartes's optics, see also Smith 1987; Wolf-Devine 1993; Ribe 1997; McLaughlin 2000; Dijksterhuis 2004, chapter 4; Osler 2007; Lo 2017; Johnson 2018.

[183] On the sine law, see in particular Schuster 2013, appendix 1.

[184] Schuster 2013, 188–189.

[185] Schuster 2013, 191.

[186] Schuster 2013, 208.

Descartes, in conclusion, based his model of collision on the idea that what is conserved or lost through impact is a value given by m·s, which fitted (1) the explanation of the functioning of the lever, (2) his explanation of the sine law of refraction. Yet, the idea that force of motion can be calculated by m·s, and that this is the value conserved in impacts, led him to formulate laws of collision which were ultimately wrong (with the exception of the first one), namely, they could not predict the actual behaviour of perfectly hard colliding bodies after impact. Moreover, such laws were ultimately inconsistent, because, as clearly shown by rule 4, they contradict the relativity of motion. For instance, according to rule 4 a smaller body cannot move a bigger body which is at rest, and is rebounded by it, but if we conceive the opposite situation, in which a big body collides with a smaller body which is at rest, we find that after the impact they are both in motion (rule 5). This contradicts Descartes's definitions of motion as given in II.13 and II.24, for which motion is a change in distance between two bodies, which can arbitrarily be considered in movement, or at rest with respect to the other one. Both the rules can be consistent, however, if we assume a fixed frame of reference, i.e. if we follow the definition of movement as given in II.25.[187]

However, it will be by building on Cartesian premises – as I am going to show in Sect. 4.2.2, *Huygens's laws of collision* – that Huygens would come to formulate the (correct) laws of impact for hard bodies, and to present them in a communication sent to Henry Oldenburg in January 1669, and published in the April 1669 issue of the *Philosophical Transactions*. This was published after the communications on the same topic by Christopher Wren and John Wallis (1668), and had been published, in French, in the *Journal des Sçavans* (March 1669).[188] Later, the same laws of impact will be presented, with demonstrations, in his *De motu corporum ex percussione*. This treatise, which was completed in 1656, was published without change[189] by Bernard Fullenius and De Volder himself in Huygens's *Opuscula*

[187] See Sect. 4.1.1, *De Volder's Cartesian standpoint on the ontology of physics*.

[188] Wallis communicated his results to the Royal Society by letter in November 1668; Wallis explained them at a session on 17 December 1668, Huygens communicated them by letter in January 1669 (see Huygens 1669a, 925–926). For a full account, see Murray et al. 2011. Please note that, according to the *Bibliotheca Volderina*, De Volder owned the issues of the *Philosophical Transactions* from 1665 to 1705 (13 volumes); the Latin edition published in Amsterdam in 1674 (as 5 volumes are listed, probably De Volder owned the series appearing in 5 volumes between 1671 and 1676, covering the years 1665–1669; in 1681 a *Continuatio*, covering the year 1670, followed); the 1705 abridgement of the *Philosophical Transactions*, by John Lowthorp: see *Bibliotheca Volderina*, 6 and 10. Also, De Volder owned the full series of the *Journal des Sçavans* (1665–1708): see *Bibliotheca Volderina*, 93.

[189] The text used by De Volder and Fullenius coincides with that of a manuscript belonging to Huygens, written by Denis Papin and presenting Huygens's own corrections (ms. HUG 26 A, 91r–108v (Leiden University Library)), and based on a 1656 manuscript of Huygens (in Huygens 1888–1950, volume 16, 137–149; ms. HUG 26 A, 58r–61v): for a commentary, see Huygens 1888–1950, volume 16, 12 and 30; Yoder 2013, 117–119. In their preface to the reader, De Volder and Fullenius do not indicate any intervention by them in the final text: "[...] succedit tractatus de Motu, in quo regulae motus demonstrantur, quas iam ante 30, et quod excurrit, annos in Diario Gallico 18 Martii anni 1669 cum orbe litterario communicavit auctor," Huygens 1703, *Lectoribus*, 11 (unnumbered). The manuscript contains some annotations by the editors: see Yoeder 2013, 118.

postuma (1703), after they had been entrusted by Huygens to publish his posthumous works.[190] It is noteworthy that, De Volder relied only on the latter 1669 communication, until, after the death of Huygens, he was able to read Huygens's unpublished manuscript.[191] Together with Descartes's rules, which De Volder comments upon in his *dictata*, Huygens's rules constitute the basis of De Volder's own laws, which he presented in a *pro gradu* disputation of one of his students in 1688, and then in his correspondence with Leibniz. Therefore, Huygens's rules also deserve separate consideration.

4.2.2 Huygens's Laws of Collision

As testified to by Huygens's first extant manuscript on impact (1652), the starting point of Huygens's formulation of his theory on collision was Descartes's first rule.[192] As to the other rules, Huygens found them contradictory to experience, and as internally inconsistent.[193] Namely, Descartes's rules contradicted the idea of reciprocity of motion, and are valid only in the frame of reference considered by Descartes in providing his rules. This is exemplified by his fourth rule, in which case, if we assume that body A is at rest, and body B (which is bigger) is moving towards it, we find an outcome of the collision which is different if we assume that body B is at rest. In the first case, indeed, we need to apply Descartes's rule 5. Later in his *De motu corporum ex percussione*, Descartes's first rule would become the

---

[190] Besides these two printed expositions of Huygens's rules, there are extant working papers of Huygens on the topic, which have been analysed by Westfall (Westfall 1971, chapter 4), and dating back to 1652, 1654, 1656, 1659 and 1667: see Huygens 1888–1950, volume 16, 92–168. Moreover, Huygens presented to Oldenburg, together with his 1669 communication, an incomplete, larger treatise on collision (published in Murray et al. 2011). See also Andriesse 2005, chapter 6.

[191] See the letter of De Volder to Leibniz of 18/28 February 1699: "[...] post obitum Illustris Viri, libellum ejus de motu videre licuit," A II3B, 529.

[192] "Si duo corpora aequalia perfecte dura aeque celeriter in contrarias partes mota, inter se colliderentur, unumquodque in eam partem unde venit reflecteretur nulla parte [...] celeritatis amissa," Huygens 1888–1950, volume 16, 92.

[193] See his 1656 manuscript on collision, where he considers the cases of rule 2 and 5: "[h]isce Cartesij regulis multi assensum praebuere haud scio verisimilitudine magis an autoritate subtilissimi Viri Philosophi permoti. Sed me primum dubitare coegit nimia illarum cum experimentis discrepantia nam quiescente sphaera ab aequali pulsam hanc ab ictu quiescere motumque omnem in illam transferri saepissime observaveram, atque alia praeterea notaveram percussionis accidentia quae superius recensui longe diversa regulisque istis adversantia. Deinde vero et ipsas inter se dissidere leges ipsius animadverti. quod hic obiter commonstrare expediet. Quinta nimirum docet quod si corpus majus B occurrat minori C quiescenti, aliquid de celeritate sua amittet. At ex lege secunda si occurrat B eidem minori corpori C, venienti ex adverso cum pari celeritate, nihil amittet B de celeritate sua quae quidem inter se pugnant, nisi dicamus corpori moto magis resistere corpus quiescens quam si ipsi ex adverso veniens impingatur. quod absonum est. Hisce itaque regulis minime fidens de alijs cogitare coepi, omnibusque tentatis. Itaque diligentius haec inspicere cepi tandemque veriora, uti existimo," Huygens 1888–1950, volume 16, 139; Westfall 1971, 147.

second hypothesis on which Huygens's theory of collision is based, while Descartes's laws of nature 1 and 2 will become Huygens's first hypothesis.[194] As highlighted by Richard Westfall and Edward Slowik, Huygens's investigation was inspired by Cartesian premises, namely, Huygens constructed his laws of impact on such Cartesian premises, and searched for instances of collision in which the quantity of motion (m·s) is conserved.[195] Even if, on this very basis, he came not only to reject most of Descartes's rules, but also to reject the idea that m·s is conserved during collisions, and came to the conclusion that it is the value m·s² which is conserved in the impacts between hard bodies.

Leaving aside the internal evolution of Huygens's rules – which were presented, as De Volder put it in his correspondence with Leibniz, "without demonstration" in his 1669 communications[196] – we can understand their rationale as follows (with a little simplification). First, the rules of the impact of bodies of equal size were derived by Huygens from Descartes's rule 1.[197] This is made possible by a mental experiment, namely, by imagining all the collisions of such bodies (moving either in the same direction, or in opposite ones), as taking place on a boat, and observed from the shore. By adjusting the speed of the boat, one can foresee the outcome of

---

[194] "I. Corpus quodlibet semel motum, si nihil obstet, pergere moveri eadem perpetuo celeritate, et secundum lineam rectam. II. Cum corpora duo dura, inter se aequalia, aequali celeritate ac directe sibi mutuo occurrunt, resilire utrumque eadem qua advenit celeritate; dicuntur autem directe occurrere, cum in eadem linea recta, utriusque centra gravitatis conjiungente, moventur, punctumque contactus accidit in eadem recta," Huygens 1703, 369. As a third hypothesis, Huygens used a principle of relativity of motion. This principle, stating that the outcome of a collision is not influenced by the movement to which all the involved bodies participate, makes possible to foresee outcomes of collision which are consistent in all frames of reference (*contra* Descartes), for the reason that one can add or subtract to each body the same quantity of speed, without influencing the outcome of the collision: "III. Motum Corporum celeritatesque aequales aut inaequales respective intelligendas esse, facta relatione ad alia corpora, quae tanquam quiescentia considerantur; etsi fortasse et haec et illa alio communi motu involvantur; ac proinde, cùm corpora duo sibi mutuo occurrunt, etiamsi alteri praeterea motui utrumque simul obnoxium fuerit, haud aliter illa se invicem impellere respectu ejus qui eodem quoque motu defertur, ac si omnibus adventitius iste motus abesset," Huygens 1703, 369–370. These three hypotheses served to establish the rules of impact for equal bodies.

[195] See Westfall 1971, 146–158; Slowik 2002, 177–186. Cf. Huygens's 1652 manuscript on collision, stating as an axiom grounding his work on impact between unequal bodies, that "[s]i corpus A majus occurrat B minori, sed velocitas in B sit ad velocitatem in A reciprocè ut magnitudo A ad B, tum utrumque cum eadem qua venit celeritate resiliet. Hoc concesso omnia demonstrari possunt. Cartesius autem concedere cogitur," Huygens 1888–1950, volume 16, 96. This axiom, for Huygens, had to be conceded by Descartes as it conforms to the conservation of m·s, even if it contradicts Descartes's rule 7c. In any case, he added to this axiom the annotation "[s]ed videndum an demonstrari per notiora queat," Huygens 1888–1950, volume 16, 96: see Westfall 1671, 149–150. Indeed, this axiom would become proposition 8 in his *De motu*, demonstrated on the basis of two additional hypotheses he set out in this work: see Huygens 1703, 381. On Huygens's theory if impact, see also Blackwell 1977; Mormino 1990a, b.

[196] "[...] Leges motus Hugenianas, quas absque demonstratione edidit in Diariis," De Volder to Leibniz, 18/28 February 1699, in A II3B, 528.

[197] In fact, in his *De motu corporum ex percussione*, as seen above, Huygens based the deduction of the outcomes of this kind of collision on three hypotheses.

all the collisions of this kind by applying Descartes's first rule. For instance, in the case that, for an observer on the boat, one of the two bodies is at rest and the other one moves at speed 4, if the boat moves in a direction opposite to that of the moving body, at speed 2, for an observer on the shore the two bodies will impact as if they were moving towards each other at speed 2. Or if, for the observer on the boat, they move towards each other at speeds 5 and 4, the observer on the shore will see them as moving with respect to each other at a speed of 4½, if the boat moves in the same direction of the slower one at a speed of ½. If they move (from the perspective of the observer on the boat) in the same direction, at a speed of 5 and 4, it is possible to observe from the shore the slower one as if it is at rest, if the boat moves in a direction opposite to that of its movement at speed 4 (and reduce this case to the first one). With this mental experiment, Huygens could infer that "1. If a hard body at rest is hit by an equal hard body in motion, after impact the impelling body will be at rest, and the velocity that was in the impelling body will have been acquired by the body originally at rest," and "2. But if the first of the two equal bodies is also moved, and is borne in the same straight line, after contact the two of them will be moved with the velocities exchanged," (rules 1 and 2 in his 1669 communication).[198]

Moreover, Huygens devised a way to predict the outcome of collisions involving unequal bodies, by finding a frame of reference or standpoint from which the two bodies maintain, after impact, the same speed they had before impact – even if in opposite directions. This standpoint is that of their common centre of gravity. In doing so, Huygens built upon the work of Evangelista Torricelli, who first conceived, in his *De motu gravium naturaliter descendentium et proiectorum* (published in his *Opera geometrica*, 1644), two different bodies as one body provided with a common centre of gravity, and derived from this idea the so-called 'Torricelli principle' for which two bodies cannot move themselves – i.e. by their weight – unless their common centre of gravity is lowered.[199] Huygens used the idea of the common centre of gravity in his theory of impact, for the reason that a centre of

[198] Murray et al. 2011, 156; cf. the original text: "I. Si corpori quiescenti duro aliud aequale corpus durum occurrat, post contactum hoc quidem quiescet, quiescenti vero acquiretur eadem quae fuit in impellente celeritas. II. At si alterum illud corpus aequale etiam moveatur, feraturque in eadem linea recta, post contactum permutatis invicem celeritatibus ferentur," Huygens 1669a, 927; given also in Huygens 1703, as propositions 1 and 2.

[199] See Torricelli 1644, 99: "[p]raemittimus. Duo gravia simul coniuncta ex se moveri non posse, nisi centrum commune gravitatis ipsorum descendat. Quando enim duo gravia ita inter se coniuncta fuerint, ut ad motum unius motus etiam alterius consequatur, erunt duo illa gravia tamquam grave unum ex duobus compositum, sive id libra fiat, sive trochlea, sive qualibet alia Mechanica ratione, grave autem huiusmodi non movebitur unquam, nisi centrum gravitatis ipsius descendat." From this, it follows the principle for which the motion acquired by two falling bodies cannot raise their common centre of gravity above their point of fall: Huygens overtly used this idea in the demonstration of proposition 8 of his *De motu corporum*: "certissimum enim in mechanicis est axioma, motu corporum qui a gravitate ipsorum proficiscitur, centrum commune gravitatis ipsorum non posse attolli," Huygens 1703, 382. Leibniz would use this idea to criticize the quantification of force of motion by m·s, as I am going to show: see Sect. 4.2.3.3.1, *The quantification of motive force*. For a thorough discussion of Torricelli's principle, see Pisano and Capecchi 2010; Capecchi 2012, chapter 6, 2017, 315–324.

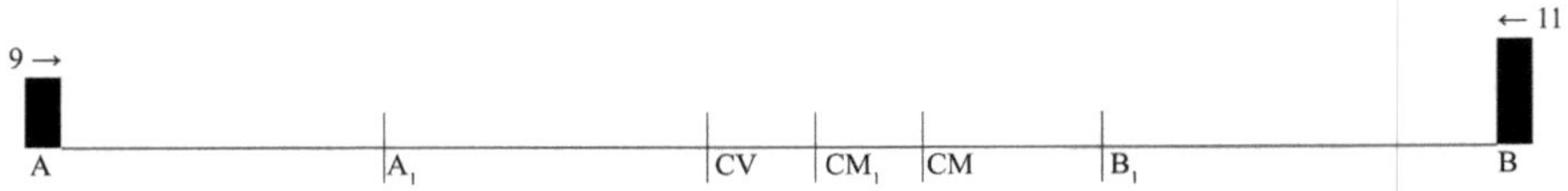

Fig. 4.24 Christiaan Huygens's model of impact. (Figure by the author)

gravity never changes its speed *with respect* to the two bodies, before and after impact. We can clarify this with an example (see Fig. 4.24, *Huygens's model of impact*).

If the size of A is 2, and the size of B is 3, and A moves towards B with speed 9, and B towards A with speed 11, their centre of gravity or mass is located at time 0 at point C (where the distance ACM = 24 and BCM = 16). At time 1, A has covered a distance of 9 (being now at A_1), and B a distance of 11 (being now at B_1). So, CM is now at CM_1, namely, it covered a distance of 3 towards A – in order for it to keep the same ratio of distances between the two bodies. If we suppose that CM is at rest, namely, if we put ourselves at the standpoint CM, A would move towards CM with a speed 12 (which is the relative speed between A and CM), and B would move towards CM with a speed of 8. After the impact, the two bodies, from the perspective of CM, will maintain the same speed. Accordingly, if we move out of standpoint CM, i.e. we cease to move towards the left with a speed of 3, we have to add to the speed of A 3 units of speed, and to subtract from the speed of B 3 units of speed. So that we find that the speed of A is 15, and the speed of B is 5. This method of calculating the outcome of collisions for unequal bodies is given by Huygens in proposition 4 of his *De motu corporum ex percussione*, which states that "[w]henever two bodies collide between themselves, the speed of the separating bodies is the same, with mutual respect, which was in the approaching ones."[200] In our example, the relative speed between A and B is the same before and after impact, namely, 20 units of distance per unit of time (11 + 9 and 15 + 5). This proposition is based on a generalization from Descartes's rule 1, as well as on two additional hypotheses of Huygens, stating that (*contra* Descartes), a body at rest – however big – can be moved by a smaller body (hypothesis 4), and that (*à la* Descartes) if after a collision one of the bodies retains all its original speed, the other body also retains its own speed (hypothesis 5) (see Fig. 4.25, Huygens 1703, *De motu*, plate 2, figure 5).[201]

[200] "Quoties duo corpora inter se colliduntur, eadem est, mutuo respectu, discendentibus celeritas, quae fuit appropinquantibus," Huygens 1703, 375.

[201] "Hypothesis IV. Si corpus maius minori quiescenti occurrat, aliquem ei motum dare, ac proinde de suo aliquid amittere. [...] V. Corporibus duobus duris sibi mutuo occurrentibus, si, post impulsum, contigat alteri eorum omnem quem habebat motum conservari, etiam alterius motui nihil decedere neque adiici," Huygens 1703, 373 and 375. The demonstration of the proposition is as follows (see Fig. 4.25, Huygens 1703, *De motu*, plate 2, figure 5) – (please note that point E is the centre of gravity of the two bodies): "[d]e aequalibus corporibus manifestum est ex propositione II. Sint igitur nunc inaequalia, primumque is casus proponatur, quo, maiori A quiescenti, impingitur corpus minus B, celeritate BA dextram versus pergens. Dico ipsa post contactum eadem celeritate BA separatum iri, adeo ut, si parte una temporis corpus B confecerit spatium BA, post alteram similem temporis partem, rursus spatio, quod ipsi BA aequale sit, separata inveniantur. Constat

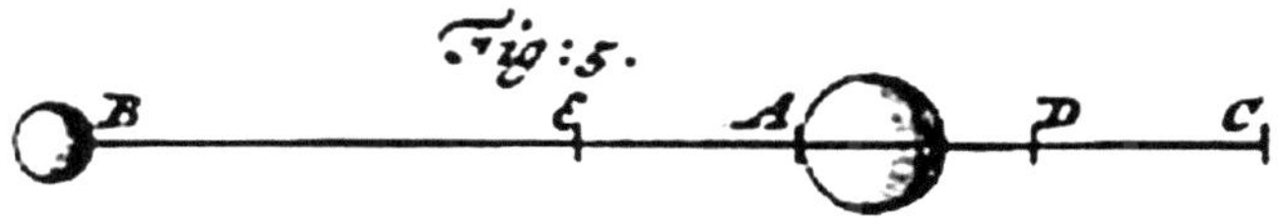

Fig. 4.25 Representation of a case of Huygens's model of impact. (Source: Huygens 1703, *De motu*, plate 2, figure 5. ETH-Bibliothek Zürich, signature: Rar 5219, https://www.e-rara.ch/doi/10.3931/e-rara-4066. Public domain mark 1.0)

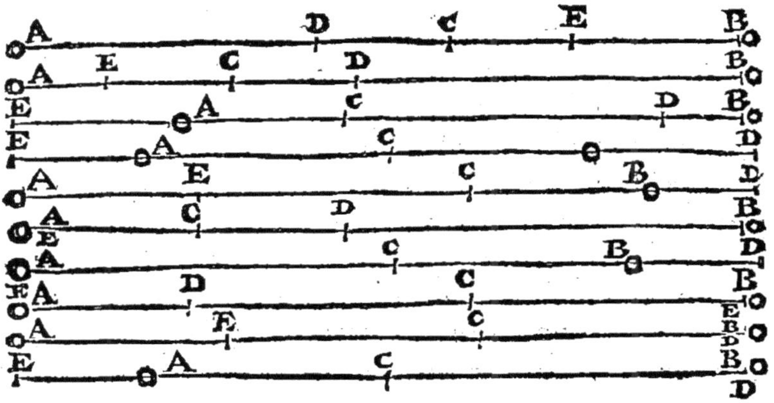

Fig. 4.26 Representation of different cases of Huygens's model of impact. (Source: Huygens 1669a, 928. London, Natural History Museum Library, call number: Q41.L8 R88p, https://www.biodiversitylibrary.org/item/183746#page/3/mode/1up. Public domain)

In his 1669 communication, Huygens avoids these demonstrations and synthetically provides a method to calculate the outcome of any collision (see Fig. 4.26, Huygens 1669a, 928):

> [l]et there be bodies A and B, of which A moves with velocity AD, and B is either moving in the same straight lines with velocity BD, or is at rest in B. Divide the line AB in C, the center of gravity of A and B, and let CE be taken equal to CD. I say that body A after impact

---

enim A celeritatem aliquam accipere impulsu corporis B; sit ea AC, minorem autem esse oportet celeritate BA qua ipsum B movebatur: nam si ipsi B aequale esset A, tum demum celeritatem BA ex impulsu acciperet. Dividatur AC in duo aequalia puncto D, sitque AE aequalis AD. Si igitur in navigio hosce motus contingere existimemus, quod sinistram versus praetervehatur celeritate DA: necesse est ut ante impulsum, corpus A quod in navigio quiescebat, motum fuerit respectu ripae dicta celeritate DA, sinistram versus; post impulsum vero, cum in navigio motum dicatur celeritate AC dextrorsum, ipsum vero navigium celeritate DA in partem contrariam feratur, movebitur A, ripae respectu, celeritate DC seu AD in partem dextram. Itaque, respectu ripae, corpus A ante et post impulsum, eandem servat celeritatem. Quare etiam B, eiusdem respectu, nihil de sua celeritate perdidisse oportet. [...] Hoc casu demonstrato reliqui facile consequuntur, supersunt autem quatuor diversi, nam vel minus corpus quiescit, vel utraque adversis motibus cientur, vel celeriore motu minus insequitur maius, vel contra; quos omnes simul proponere licebit," Huygens 1703, 373–374.

will have velocity EA, and body B the velocity EB, each in the direction indicated by the order of the points EA, EB. Thus if E falls in the point A or B, body A or B will be brought to rest.[202]

According to this model – as Huygens made explicit in the covering letter of his communication of the rules of impact to Oldenburg in 1668 – the movement of the centre of gravity of two bodies does not change after their impact.[203] This is valid, for Huygens, also in the case of the impact of soft bodies (considered by Huygens in a manuscript of 1667), which are not elastic i.e. they do not 'rebound',[204] as is presupposed, on the other hand, for the hard bodies considered in the texts which appeared in 1669 and 1703.

More importantly, according to Huygens's model of impact, the quantity of force as calculated by Descartes by m·s is not conserved in all cases of impact.[205] Rather, it is the total m·s^2 of the two bodies which is conserved. As he put it in his 1669 communication: "[rule] 6. The sum of the products of the mass [*moles*] of each hard body multiplied into the square of its velocity is always the same before and after impact."[206] This position anticipates – as I am going to show – Leibniz's idea of conservation of living force, and may have been derived by Huygens from Galileo's law of free fall, according to which a falling body acquires speeds which are proportional to the square root of the time elapsed (i.e. after 1 second, it would have fallen a distance of 1 unit, with a speed of 1 unit per second; after 2 seconds, it would have fallen a distance proportional to 4 units, acquiring a speed of 2 units per second; after 3 seconds, it would have fallen a distance proportional to 9 units, acquiring a speed of 3 units per second, and so on).[207] According to Westfall, the rejection of the

[202] Murray et al. 2011, 156; cf. Huygens 1669a, 928, 1703, proposition 9.

[203] Cf. Huygens 1669a, 928 (in a paragraph immediately below Huygens's communication): "[...] subiungit, notasse se miram quandam naturae legem, quam demonstrare se posse affirmat in corporibus sphaericis, quaeque generalis ipsi videtur in reliquis omnibus sive duris sive mollibus, sive directe sive oblique sibi occurrentibus, videlicet centrum commune gravitatis duorum, trium, vel quotlibet corporum, aequaliter semper promoveri versus eandem partem in linea recta, ante et post occursum."

[204] See Huygens 1888–1950, volume 16, 160–161. According to Huygens, if two soft i.e. non-elastic bodies with equal m·s impact, they come to rest: "A et B corpora mollia, A movetur celeritate AC B celeritate BC, quae sunt reciproce ut ipsa corpora. Dico post concursum in C, manere utrumque immotum."

[205] See Huygens 1669a, rule 5: "[q]uantitas motus duorum corporum augeri minuive potest per eorum occursum; at semper ibi remanet eadem quantitas versus eandem partem, ablata inde quantitate motus contrarii." Cf. Huygens 1703, proposition 6. This means that, in all collisions, the quantity of motion exerted in each direction (after having subtracted the quantity of motion exerted in the opposite direction) is equal in each body.

[206] Murray et al. 2011, 156; cf. Huygens 1669a, 928, 1703, proposition 11.

[207] Cf. Westfall's own words: "[t]he exit of the Cartesian 'quantity of motion', that is, was only the cue for the entrance of a new quantity (mv^2) onto the stage [...]. In some sense, the quantity had been implicit already in Galileo's kinematics of free fall, in the demonstration that the velocity of a body falling from rest is proportional to the square root of the distance it has fallen [...]. Perhaps Huygens' derivation of the new equation of conservation was connected to his perception of the relation of free fall to the impact of bodies inversely proportional to their speeds. His demonstra-

law of conservation of m·s (which is entailed by Descartes's first rule) led Huygens to avoid references to the idea of force, namely, to switch from dynamics to kinematics – a shift that can be noted across his manuscripts.[208] For instance, in the final version of his treatise on impact (i.e. the version published by De Volder in 1703), he claims (in hypothesis 2, restating Descartes's first rule of collision), that "[w]hatever is the cause of the rebound of hard bodies from mutual contact when they collide with one another, we posit that when two equal bodies with equal speed collide directly with one another from opposite directions each rebounds with the same speed with which it approached."[209] No mention to a force of rebounding is made, nor to any internal elasticity of bodies, i.e. to the features of their matter that could be the cause of such rebounding, or to the resilience of the bodies themselves. This will turn out to be particularly important for De Volder's treatment of collision, which I present in Sect. 4.2.3, *De Volder on motion and impact.*

4.2.3 De Volder on Motion and Impact

De Volder never dealt with the problem of collision in those printed texts bearing his name as author. The only exception is his 1682 *Oratio de coniungendis philosophicis et mathematicis disciplinis*, where Huygens's treatment of collision is praised (in broad terms) as part of De Volder's broader defence of the application of mathematics to natural philosophy, together with Galileo's discovery of the law of free fall.[210] In order to discover more specifically his views on this topic, we need once again to look at the disputations presided over – but not officially authored by him – at his

tion of that case, it will be recalled, employed the kinematics of vertical motion. The source of his insight is purely speculative, however. What is certain is the fact that Huygens first made explicit a quantitative relation destined to play a central role in the future history of mechanics," Westfall 1971, 157.

208 Westfall 1971, 152.

209 Mahoney 1995. Cf. the original text: "[q]uaecunque sit causa corporibus duris a mutuo contactu resiliendi cum in se invicem impinguntur; ponimus, cum corpora duo inter se aequalia, aequali celeritate, ex adverso ac directe sibi mutuo occurrunt, resilire utrumque eadem qua advenit celeritate," Huygens 1703, 369.

210 "Nunquam magnus ille Florentinus Galilaeus de Galilaeo admirabilem illam detexisset in motus acceleratione proportionem, nisi in geometricis demonstrationibus fuisset versatissimus. [...] Quod inventum ut illustri Hugenio debet orbis litteratus, ita illi debuisset nunquam, nisi caeteris cum scientiis, in quibus excellit vir magnus, coniunxisse mathematicarum artium notitiam. Sed non est quod unicam hanc motus proprietatem urgeam, cum ipsae, quas vocant, motus leges, quibus corpora corporibus occurrentia vel subsistunt, vel suis in viis progrediuntur, vel versus contrarias partes repercutiuntur, mathematica agnoscant principia," De Volder 1682, 17 (unnumbered). In his 1698 *Oratio*, De Volder refers to Huygens without overtly mentioning his discovery of the laws of impact: "[o]stenderunt magna huius seculi nostri lumina Hugenius, Newtonus, Leibnitzius, ne simplicium quidem corporum motus, viresque investigandas unquam, non dicam absque notitia matheseos, sed addam absque recondita harum artium scientia. Qua qui instructus non est, in physicis hospes ut sit, necessum est. Tanta igitur cum inter has disciplinas sit affinitas, eo meliori iure inquiremus, num eadem methodo tractari queant," De Volder 1698, 26.

dictata, and at his correspondence with Leibniz. In these texts, in turn, we find a wide range of positions adopted by De Volder. Briefly: (1) through his *dictata* he continued to teach Descartes's rules, intended as describing only idealized cases, and approved the laws of Huygens as describing real situations, namely the behaviour of bodies in fluids; (2) in the disputations he presided over, and in his correspondence with Leibniz, De Volder presented his own laws of collision, which he formulated in algebraic terms after he read Huygens's 1669 communications, and which are consistent with Huygens's rules.

4.2.3.1 Impact in De Volder's *Dictata*

In Sect. 4.1.4, *The search for physical unity and activity: De Volder vis-à-vis Leibniz*, we have seen that in his *dictata* on *Principia* II.36 De Volder provided a demonstration of the conservation of the quantity of motion without referring to the constant action of God, and that he retained only Descartes's metaphysical argument based on the persistence of essences in commenting upon *Principia* II.37. Moreover, De Volder from time to time expressed his doubts on the actual conservation of this quantity. If we look at his comments on the articles of the *Principia* dealing with the problem of the communication of motion between bodies – namely, at Descartes's dealing with the third law of nature and with the seven rules of impact – we find a similar de-metaphysicization of physics; moreover, we also find an attempt to harmonize Descartes's rules with those of Huygens, Wallis, and many others.

The most relevant divergence between Descartes's account and De Volder's commentary on the articles of Descartes's *Principia* relevant to the problem of the communication of motion (II.40–53), can be found in II.42. This article was devoted by Descartes to the demonstration of the third law of nature on the basis of the constancy of God.[211] De Volder, in all his *dictata*, omits any reference to the action of God, and instead remarks (in Hamburg 273) that the third law rules all the natural changes, giving as an example the case of a bigger and more powerful body impacting a smaller and weaker one. In this case, the bigger loses some of its motion, but not its determination, while the smaller (whose direction of motion is not specified) is carried by the bigger body[212] – in accordance with Descartes's rules 2 and 5. As

[211] The article is devoted to the demonstration of the second part of the law (i.e. "that, upon coming in contact with a weaker one, it [i.e. a stronger body] loses as much as it transfers to that weaker body"): "it is obvious that when God first created the world, He not only moved its parts in various ways, but also simultaneously caused some of the parts to push others and to transfer their motion to these others. So in now maintaining the world by the same action and with the same laws with which He created it," Descartes 1982, 61–62.

[212] "Argumentum. Concipio corpus maius et efficacius impingere in minus et imbecillius, certum est c[or]pus q[uo]d efficacius est, progredi debere in suo motu, quia non est q[uo]d determinatio mutetur, nisi a corpore fortiori, sed {…} quod impingit, supponitur imbecillius, sed cum in motu progreditur, debet aliud corpus secum movere, ac per consequens aliq[uo]d sui motu in id transferre. Ex hac {[n]a[m]} lege patet, non po[ss]e dari motum, quis omnia in perpetuo sint constituta motu, ex quo seq[ui]tur muta[ti]onem {q[uam]} ex motu {seq[ue]tur} e[ss]e constantem et con-

to the other relevant articles on movement in general (II.40–41 and II.43–45), De Volder clarifies Descartes's theory of motion without notable divergences, stressing how the total quantity of movement is conserved in collisions (II.40),[213] and that determination can change without changes in the quantity of motion (II.41).[214] The idea of determination (or the decomposition of a quantity of motion exerted in one direction, or in more directions) is then used in commenting on II.43: more specifically, in explanation of the 'modes of impact', in which only the component of the determination of the movement of a body which is perpendicular to the surface of impact is affected (in accordance with the laws of dioptrics).[215] More interesting is

tinuam," Hamburg 273, 106–109. Hamburg 274, 44: "[i]n secunda parte regulae supponitur nullam dari causam efficientem ad mutandam determinationem, ac proinde ea non mutabitur, non potest autem corpus motu eandem servare motus determinationem, nisi id corpus {cui} occurrit, secum propellat, nec potest propellere nisi {quandam} sui {motus} partem ei conferat, quam proinde amittit, ne augetur motus in universo quantitas." The commentary on articles 40–44 present some divergent numbering across the different copies of the Hamburg 274 series of his *dictata*. All the copies of the series lack of a separate commentary on II.40 and II.43, but all their commentaries on II.44 are in fact devoted also to II.43. Moreover, the British Library *dictata* has the commentary on II.41 under the heading of II.39, and the commentary of II.42 under the heading of II.40.

[213] "Perdit. Progreditur sic rursus author methodo geometrica, hoc articulo non demonstrat, sed praeponit, q[uod] in seqq. probat. Duo corpora quiescentia non agunt in se invicem, nec mota in easdem partes, sed unum motum in alterum {quiescentem}, sed diverso modo videamus itaq[ue] quid ex his contingat: si impingens sit imbecillius, resilit, et reflectitur, servatq[que] motum amittendo determinationem. 2. Si {c[or]pus} motum impingi in quiescens imbecillius, tantum amittit de suo motu, q[ua]ntum quiescens recipit. Si vero pila incidat in arenam {a[ut]} lutum, cum luti partes sunt disiunctae potest una moveri immota manente alia, nonnulla igitur luti partes simul cum pila movebuntur, perdetq[ue] tantundem motus pila, q[ua]ntum lutum recipit," Hamburg 273, 106. The commentary on this article is absent in the Hamburg 274 series of *dictata*. The commentary numbered II.40 in the British Library *dictata* in fact comments upon II.42.

[214] "Remanente. Aliud motus, aliud eius determinatio est, si {ex} diversam habeant causam, et motus po[ss]it manere, mutata determinatione, et vice versa, sequitur etiam diversas a se invicem e[ss]e. Motum {[n]a[m] eundem} e[ss]e po[ss]e, mutata determinatione, ex eo sequitur, q[uo]d 2 c[or]pora possint eadem velocitate moveri versus occidentem et orientem. Diversus {[n]a[m]} motus est non mutata determinatione, cum 2 c[or]pora versus easdem plagas inaequaliter moventur. Debere. Pila incidat in parietem, is non opponitur motui pilae, sed determinationi: q[uo]d cum ita sit, quid quaeso manifestius quam parietem sollumodo agere in pilae determinationem. Non {[n]a[m]} opponitur motui quia est corpus fortius, nec a pila moveri potest, sed situ solo eius determinationi contrariatur, servat itaq[ue] pila motum, sed non progreditur versus lineam rectam," Hamburg 273, 106; cf. "[s]aepissime diversas dari causas motus, et determinatio motus patet ex iis, {q[uas]} dicit autor dioptr. 2. Patet ibide enim situm et dispositionem reticuli esse causam determinationis motus, vis vero {q[uae]} reticulum agit in pilam esse causam motus, ita ut eadem vi possit pilae motus conferri, sed in alias partes, modo aliter disponatur reticulum, ut vel maior, vel minor possit imprimi motus pilae, manente eadem determinatione," Hamburg 274, 44.

[215] "Tum a magnitudine. Id e[nim] constat reliquis paribus, quo c[or]pus maius est, eo maiorem superficie: suppono c[or]pus sphaericum et quadratum eiusdem molis, prius ambiatur 5, alterum 6 pedibus, quoniam igitur sphaericum 5 po[ss]it ambiri pedibus, seq[ue]tur {e[ti]am} e[ss]e imbecillius. Modi. Si e[nim] c[or]pus v.g. pila recta feri perpendiculariter incidat in parietem, paries {totus} aget in pilam, si vero oblique non {totus}, sed motu tantum laterali resistet," Hamburg 273, 109; cf. the other series: "[c]um maior requiratur vis ad mutationem inducendam maiori corpori q[ua]m minori, patet etiam vim resistendi maiorem esse omni mutationem, vi corpore maiori qui an vi minori. Sit enim A corpus duplum ipsius B, supponitur enim eandem omnia: in A itaq[ue] erit

De Volder's approach to the rules of impact, in which we find the same clarifying approach,[216] and where he overtly relies on both versions of Descartes's *Principia*, i.e. the Latin version and the French version, which he provides to his students in his own translations into Latin.

First of all (i.e. in commenting on II.45, where Descartes introduces the idea of 'calculation' of force) in the Hamburg 274 series of his *dictata*, De Volder notes that "these rules of motion delivered by the author cannot be either confirmed by experience, or invalidated, because here perfectly hard bodies are assumed, not surrounded by any fluid [...] [which] can be found nowhere." So, Descartes's rules "differ a lot from those of the most noble Huygens, and of other [*sic*] Englishmen, because they reasoned about bodies surrounded by fluid, in which his [sc. Descartes's] rules could [do] nothing; our author [i.e. Descartes] acknowledged, when [he] confesses in § 55 [sc. 56] [that] body B, conceived in fluid, can be moved by a very little force, which then he repeats in § 62 [sc. 61]. Which [...] agrees with the rule of Huygens, however directly contradicted by the fourth rule of the author [i.e. Descartes]."[217] In

vis resistendi cuiuscunq[ue] mutationi duplo maior q[ua]m ea est, q[uae] est in B, non tamen sic spectanda sola magnitudo, sed et externa superficies. Nam ea corpora, q[uae] eiusdem cum aliis sunt molis, plus tamen superficiei habent, non possunt attingi a tot corporibus, ac proinde affici quam illa. Varius etiam occursus {multim} etiam potest variare, quia corpus q[uo]d perpendiculariter in aliud incidit, tota sua mole et omni sua celeritate et determinatione motus in illud agit. Illud vero corpus q[uo]d ad angulos valde obliquos, aliud corpus pertingit, nec tota mole, nec tota motus quantitate in id cui occurrit corpus operetur," Hamburg 274, 44 (in this series the commentary on II.43 is indicated as II.44).

[216] For instance, in commenting on II.44, where Descartes set the various forms of opposition between bodies (i.e. (1) in states of movement and rest, (2) determination or direction of motion, (3) different speeds) De Volder overtly declares that these oppositions guide Descartes's treatment of impact, in three different macro-cases (i.e. rules 1–3, 4–6 and 7): "[m]ovetur. Duplex est contrarietas, si sc[ilicet] 2 corpora pari velocitate, vel in easdem partes moveantur, aut si ambo quiescunt, nulla est contrarietas. Si vero duorum corporum unum moveatur, alterum q[ui]escat, necesse est, ut motus quieti sit contrarius, et in se invicem agunt; aut si 2 corpora in pari celeritate, aut in diversas moveatur plagas, necessaria est contrarietas. In pari velocitate i.e. si id q[uo]d praecedit moveatur tardius, {q[ue]m} id q[uo]d {seq[ue]tur}, si e[nim] praecedens moveatur celerius, nulla erit contrarietas, si vero tardius, sequens necessario impinget in illud, et eadem est contrarietas, ac si unum c[or]pus nempe id q[uo]d {seq[ue]tur}, unicum, praecedens nullum motus gradum haberet. Determinationem. Ut cum 2 c[or]pora moventur in contrarias partes, tunc nulla ratione motus, sed determinationis est contrarietas. Pro ratione partis. Ut si corpus motum incurrat in aliud motum, contrarietas est maior, quia incurrunt directe in superficies mutuas, ut cum concipimus 2 motos circulares, sibi invicem contrarias," Hamburg 273, 109. A commentary on *Principia* II.44 is absent from the Hamburg 274 series of *dictata*.

[217] "Regulas sequentes. Haec regulae motus ab authore allatae nec experientia confirmari, nec infirmari queunt, quoniam sic supponuntur corpora perfecte dura, et nullo fluido {cinta}, quod possit motum multis modis variare, qualia corpora in rerum natura nullibi reperiuntur. Nec mirum est, hasce regulas multum discrepare ab illis nob. Hugenii, aliorum anglorum, quia si ratiocinati sunt de corporibus fluido undiq[ue] cunctis, in quibus suas regulas nihil posse agnoscit author. Cum confitetur §§ 55 corpus B in fluido constitutum, quantumvis exigua vi moveri posse. Q[uo]d denuo repetiti parag. LXII. quod apprime convenit cum regula Hugenii, directe autem contrariatur {regulae 4tae} authoris," Hamburg 274, 44–45. Both the British Library *dictata* and The Hague *dictata* report the correct indication to II.61. The commentary to this article, in the other series of *dictata*, serves only to introduce the next articles: "[v]idimus generalia, quibus autor nunc porro applicat particularia exempla," Hamburg 273, 109.

his 1669 communication, indeed, Huygens stated as rule 3 that a "body however great will be moved if impacted by any other body however small, with any velocity whatever."[218] Moreover, in commenting upon II.53 – where Descartes highlighted the difficulties in applying these rules to phenomena – De Volder provides a Latin translation of a part of this article present only in the French version: "in the French edition the author declares much more clearly that 'in reality often happens that experience [...] seems to contradict [...] these rules [...]'."[219] In reverse order (i.e. in commenting on II.53 only), the same ideas are conveyed in the other series of *dictata*, where De Volder notes that the differences in the rules of Descartes and those of Huygens, Wallis, as well as those of other "Dutchmen, Englishmen and Frenchmen," are due to the fact that Descartes considered bodies in ideal situations, which cannot be found in nature, while Huygens and the others provided rules valid for bodies immersed in fluids.[220] It is worth noting that this kind of accommodation

---

[218] Murray et al. 2011, 156; cf. Huygens 1669a, 927. This rule becomes hypothesis 4 in the *De motu corporum ex percussione.*

[219] "In Gallica editione multo clarius proponit autor. Revera saepe contingit experientiam prima fronte videri repugnare six explicatis regulis. Sed ratio eius rei perspicua est, hae enim regulae supponunt duo corpora B et C plane dura, et ita separata ab omnibus aliis, ut nullum circa se habeant, quod possit eorum motum aut sistere, aut promovere, talia autem corpora in rerum natura nulla observantur, ita ut antequam iudicium ferre possumus, {...} hae regulae in mundo locum habeant an minus, [non] sufficiat {...} quo pacto 2 corpora B et C sibi mutuo occurrentia, in se invicem possunt agere, sed amplius necesse sit advertere, quo pacto corpora, {q[uae]} ambiunt ea, possunt vel augere vel imminuere eorum actionem," Hamburg 274, 46–47; the addition of "non," in brackets, is from British Library *dictata*, 95r. Please note that the part quoted by De Volder includes, in the 1647 French version, some phrases which were translated into French from the 1644 Latin edition; however, De Volder does not use the original Latin, and re-translates into Latin also these sentences: cf. the Latin version: "[s]ed quia nulla in mundo corpora esse possunt a reliquis omnibus ita divisa, et nulla circa nos esse solent plane dura, ideo multo difficilius iniri potest calculus, ad determinandum quantum cuiusque corporis motus ob aliorum occursum mutetur. Simul enim habenda est ratio eorum omnium, quae illud circumquaque contingunt, eaque, quantum ad hoc, valde diversos habent effectus, prout sunt dura vel fluida: quorum ideo diversitas in quo consistat, hic est quaerendum," AT VIII-A, 70, with the French version (the new parts included in the French version are in italics, following to the AT edition): "[*e*]*n effet, il arrive souvent que l'expérience peut sembler d'abord répugner aux règles que je viens d'expliquer, mais la raison en est évidente.* Car elles présupposent que les deux corps B et C sont parfaitement durs, et *tellement* séparés de tous les autres, *qu'il n'y en a aucun autour d'eux qui puisse aider ou empêcher leur mouvement*; et nous n'en voyons point de tels en ce monde. C'est pourquoi, *avant qu'on puisse juger si elles s'y observent ou non, il ne suffit pas de savoir comment deux corps, tels que B et C, peuvent agir l'un contre l'autre, lorsqu'ils se rencontrent; mais* il faut, outre cela, considérer comment tous les autres corps qui les environnent peuvent *augmenter ou diminuer leur action*. Et parce qu'il n'y a rien qui leur fasse avoir en ceci des effets différents, sinon la différence qui 94 est entre eux, en ce que les uns sont liquides ou mous, et les autres durs, il est besoin que nous examinions, en cet endroit, en quoi consistent ces deux qualités d'être dur et d'être liquide," AT IX-B, 93–94. Notably, the Pretoria copy, in this and in other cases in which De Volder dictated his Latin version of the French *Principia*, includes as marginal annotations also the original French text. Moreover, in the second column of The Hague copy, the Latin translations from the French *Principia* (i.e. those given in the Hamburg 274 series) are often provided as an additional commentary.

[220] "Q[uae]rendum. Vidimus has regulas, q[uo]d si consulamus experientiam, o[mn]ia prater has {fient}, super his regulis multi viri docti scripsere, sunt regulae Wallisis, Hugenii, etc. Belgarum,

of Huygens's and Descartes's rules seems to be inconsistent with De Volder's own rules of impact, which agree with those of Huygens, but in which no mention of fluid dynamics is made. Moreover, the rules devised by Huygens (as well as those of Wallis and Wren) concerned bodies in ideal situations, like those of Descartes. As I am going to show, in fact, the main *discrimen* set by De Volder in presenting his rules will be that between elastic and non-elastic collisions (either concerning perfectly hard, or soft bodies). Yet, in presenting his theories on hydrostatics, De Volder justified Huygens's 1669 rule 3 (which contradicts Descartes's rule 4) from a Cartesian perspective.[221] Thus, we may assume that De Volder's attempt to save both Huygens's and Descartes's theories served, if not to aver the validity of the two separate sets of rules, at least to justify the paradoxical nature of Descartes's rule 4.

As far as De Volder's presentation of Descartes's seven rules of impact is concerned, this can be summarized in Table 4.3, *De Volder's treatment of Descartes's rules of impact.*

4.2.3.2 A 'Purely Relational' Standpoint on Impact: The 1684 Disputation *De motu*

In the academic teaching he provided through his *dictata* on Descartes's *Principia* De Volder did not diverge, as far as the rules of impact are concerned, from Descartes's theories, and also took into account those rules alternative to the Cartesian ones. Yet, if we look at the disputations taking place in the 1680s and bearing his name as *praeses* and as dedicatee, we find ideas which diverge from the Cartesian ones. First, this is the case with the disputation *De motu* he presided over on 29 November 1684, which was dedicated to him, whose author and defender was a student from Leiden, Pontianus vander Codde.[222] Second, De Volder certainly had a hand in the writing of the *pro gradu* disputation of Hermann Schuyl – *De vi cor-*

Anglorum, Gallorumq[ue]: q[uae]ritur unde diversitas sententiarum? {R[es]p[ondeo]} considerandum est nonnullos inter quos autor considerare 2 corpora ab o[mn]ibus aliis, motum promoventibus vel impedientibus remota (q[ua]tenus in rerum natura non dantur) si quidem omne c[or]pus per fluidum q[uo]ddam moventur{)} alios contra considerare corpora circumquaq[ue] aliis corporibus cuncta. Ne {[n]a[m]} q[ui]s putet regulas has falsas e[ss]e, quia repugnant experientiae, pergit sic ostendere, quid possint corpora fluida ad movenda alia, {a.} dura ad sistenda, et {c[um]} hoc constet, cur hae regulae falsae appareant; notandumq[ue] sic est c[or]pus in fluido haerere in aequilibrio, adeoq[ue] minimum c[or]pus id movere po[ss]e," Hamburg 273, 113.

[221] See Sect. 5.5.2, *Descartes's corpuscular standpoint on hydrostatics.*

[222] The disputation was followed on 6 December 1684 by the disputation *De absoluta quiete* (which has been mentioned above: see Sect. 4.1.3, *De Volder's ideas on cohesion and divisibility*) covertly criticizing Boyle's *Discourse about the Absolute Rest in Bodies* (1669), as in the 1684 disputation the author defended the idea that, notwithstanding that all parts of matter participate in some motion, some of them do not change their *situs* with respect to each other (this being the case of the sub-parts or sub-volumes of the globules of second matter: De Volder and Von der Lahr 1684, thesis 12). Boyle, on the other hand, supported the idea that the Cartesian theory of matter entails the impossibility of absolute rest, as all parts of matter are moved by subtler particles: Boyle 1669a, *Discourse about the Absolute Rest in Bodies*, 7–9.

Table 4.3 De Volder's treatment of Descartes's rules of impact

	Hamburg 273	Hamburg 274
Rule 1 (II.46)	De Volder stresses the idea of symmetry or identical effect, as a consequence of the identical conditions before the impact. This leans more on the French than on the Latin version of the article.[a]	De Volder remarks on the conditions of symmetry in impact.[b]
Rule 2 (II.47)	De Volder re-states Descartes's second law, with a commentary consistent with both versions (Latin and French).[c]	De Volder illustrates the second rule with a commentary based on the French edition.[d]
Rule 3 (II.48)	De Volder re-states Descartes's second law, with a commentary more compatible with the Latin than with the French version.[e]	De Volder re-states Descartes's third law, with a commentary consistent with both versions (Latin and French).[f]
Rule 4 (II.49)	De Volder notes that this rule is "at first sight very paradoxical, although it is very certain," and then he illustrates it with a commentary based on the French version.[g]	De Volder comments on this rule by simply providing the Latin translation of two additions given by Descartes in the French version of the article; he also re-translates into Latin a sentence, already present in the Latin *Principia*, which Descartes kept in the French version.[h]
Rule 5 (II.50)	De Volder re-states Descartes's fifth law with a commentary based on the Latin version.[i]	De Volder comments on this law by also providing a Latin translation of the French additions to the article.[j]
Rule 6 (II.51)	De Volder does not mention Descartes's quantification of the distribution of speed as ¼ and ¾ in the two bodies, nor its calculation presented in the French version of the *Principia*, and states that "one is reflected, the other is moved, and an average [*medietas*] is given."[k]	De Volder comments on this law by providing only a Latin translation of the French text added to the article.[l]
Rule 7a/1 (II.52)	De Volder comments on all these rules, including 7a/2 (given only in the French edition of the *Principia*); so that he follows both the editions of Descartes's treatise.[m]	De Volder just provides a Latin translation of the French text in which Descartes present sub-rule 7c (absent in the Latin version).[n]
Rule 7a/2 (II.52)		
Rule 7b (II.52)		
Rule 7c (II.52)		

[a]"Considerat author hoc §§ 2 corpora aequalia, et aeq[ue] celeriter mota, sed in contrarias partes, cum itaq[que]sibi invicem occurrunt, quid fiet? a[ut] utrumq[ue] c[or]pus amittet motum, atq[que] ita tolletur motus ex universo, q[uo]d repugnat, a[ut] unum reflectetur, alterum non, q[uo]d e[ti]am repugnat, quia supponantur aequalia et magnitudine et motu, ergo nihil restat, q[ua]m ut ambo aequaliter afficiantur i.e. reflectantur," Hamburg 273, 109. Cf. AT VIII-A, 68: "[p]rimo, si duo illa corpora, puta B et C, essent plane aequalia, et aeque uelociter mouerentur, B quidem a dextra uersus sinistram, et C illi in directum a sinistra uersus dextram, cum sibi mutuo occurrerent, reflecterentur, et postea pergerent moueri, B uersus dextram et C uersus sinistram, nulla parte suae celeritatis amissa." AT XI, 89 (the use of italics indicates the text added in the French edition): "[l]

(continued)

Table 4.3 (continued)

a premiere est que si ces deux corps par exemple B et C estoient exactement égaux et se mouuoient d'egale vitesse en ligne droite l'vn vers l'autre, lors qu'ils viendroient à se rencontrer ils rejalliroient tous deux également et retourneroient chacun vers le costé d'où il seroit venu sans perdre rien de leur vitesse. *Car il n'y a point en cela de cause qui leur puisse oster, mais il y en a vne fort éuidente qui les doit contraindre de rejallir; et pource qu'elle seroit égale en l'vn et en l'autre ils rejalliroient tous deux en mesme façon.*"

[b]"Est aequalis vis agendi in B quam est resistendi in C, ad ideo B non potest secum ferre ipsum C nec ipsum B quiq[ue] hoc si fieret, essent in hoc maiores vires, quam in illo, cum tamen necessario debeat fieri determinationis mutatio fiet illa {...} in B ac in C ac proinde utrumq[ue] corpus reflectitur in contrariam partem," Hamburg 274, 45.

[c]"Habeant inquit 2 corpora parem celeritatem, imparem a[utem] magnitudinem, pars illa superans addit {uni potentiam}, cum itaq[ue] {occurrunt}, minus reflectetur, maius vero in motu procedit," Hamburg 273, 109. Cf. AT VIII-A, 68: "[s]ecundo, si B esset tantillo maius quam C, caeteris positis ut prius, tunc solum C reflecteretur, et utrumque uersus sinistram eadem celeritate moueretur." AT IX-B, 90: "[*l*]*a seconde est, que si B estoit tant soit peu plus grand que C, et qu'ils se rencontrassent auec mesme vitesse* il n'y auroit que C qui rejallit *vers le costé d'où il seroit venu*, et ils continueroient par apres leur mouuement tous deux ensemble vers ce mesme costé. *Car B ayant plus de force que C, il ne pourroit estre contraint par luy à rejaillir.*"

[d]Indeed, De Volder mentions the impossibility of the bigger body being rebounded, given only in the French version: "B cum maius sit, maiori vi aget in C quam resistet, ergo si B aequales resistens ipsi C potest efficere, ut C reflectatur, tanto magis id poterit B, quod iam supponitur maius esse C, ergo necessario reflectitur, B autem non poterit reflecti. Quia propter maiorem magnitudinem plus virium habet ad resistendum actioni corporis C, quam C ad agendum in B," Hamburg 274, 45. Cf. *supra*, n. c.

[e]De Volder does not make recourse to the principle of least modal change, which is evidently adopted by Descartes in the French version of the *Principia*: "[c]onferimus 2 corpora mole aequalia, non vero aeq[ue] celeriter mota, ratione magnitudinis par erit efficacia, sed unum corpus habet {20s} gradus motus plus alio, si profecto eius potentiam augebunt, cum itaq[ue] post occursum aeq[ue] celeriter moveantur, quid inde a posteriori concluditur? Sc. gradus motus aequaliter e[ss]e distributos; i.e. si B ante occursum habuerit 6 gradus moveantur, tunc B unum grad[um] tradidisse C i.e. ambo 5 habere motus gradus," Hamburg 273, 109. Cf. AT VIII-A, 68: "[t]ertio, si mole essent aequalia, sed B tantillo celerius moueretur quam C, non tantum ambo pergerent moueri uersus sinistram, sed etiam transferretur ex B in C media pars celeritatis qua hoc ab illo excederetur: hoc est, si fuissent prius sex gradus celeritatis in B, et quatuor tantum in C, post mutuum occursum unumquodque tenderet uersus sinistram, cum quinque gradibus celeritatis." AT IX-B, 90: "[l]a troisiéme, que si ces deux corps estoient de mesme grandeur mais que B eust tant soit peu plus de vitesse que C, non seulement *apres s'estre rencontrez, C seul rejalliroit* et ils iroiēt tous deux ensemble comme deuant, vers le costé d'où C seroit venu, mais aussi il seroit necessaire que B luy transferast la moitié de ce qu'il auroit de plus de vitesse, *à cause que l'ayant deuant soy il ne pourroit aller plus vite que luy*. De façon que si B auoit eu par exemple six degrez de vitesse *auant leur rencontre*, et que C en eust eu seulement quatre, *il luy transfereroit l'vn de ses deux degrez qu'il auroit eu de plus, et ainsi* ils iroient par apres chacun auec cinq degrez de vitesse: *car il luy est bien aisé de communiquer vn de ses degrez de vitesse à C qu'il n'est à C, de changer le cours de tout le mouuement qui est en B.*"

[f]"In omnium corporum occursu fiet minima mutatio, qua contrarietas omnis possit tolli. Quia nulla causa est, q[uam] maiorem induceret, hoc itaq[ue] in casu aut B movebit ipsum C aut non, si hoc, tollitur tota eius determinatio si illud, conferet tantum in C quandam motus sui partem, q[uae] mutatio cum priori minor sit, necessario obtinebit. Si {A} C moveat, necesse est ut eadem moveantur celeritate, ac proinde ut dimidium sui motus in corpus C conferat," Hamburg 274, 45.

[g]De Volder presents, in accordance with the "sophisticated model of impact," the hypothetical case in which the smaller body moves the bigger one. Following Descartes's French *Principia* (which

(continued)

Table 4.3 (continued)

is however not overtly cited), De Volder shows how a bigger body would exert a resistance proportional to the quantity of movement that the smaller body would transfer to it in order to carry it, and to move with it at the same speed: "[i]mpellendum. C[or]pus C q[ui]escat, {aliquantillum} maius B moveatur, B quavis velocitate, unq[ua]m movebit corpus C. Videtur haec regula primo intuito valde paradoxa, q[ua]mvis sit certissima, incurrat e[nim] B in C, si B moveat C, ambo postea aequali velocitate movebuntur; concipio nunc centuplo plus virium req[ui]ri, ut demus corpori 100 gradus motus, q[ua]m {id unum} demus, si iam ponamus B in C 2bus motus gradibus irruere, si aeq[ua]liter post occursum moveantur, dabit puls q[ua]m dimidium motus, itaq[ue] si ponamus moveri 200 motus gradibus, si C secum moveat, dabit ad min. 100 gradus, si {[n]a[m] det} 100 requiritur 100plo plus virium, q[ua]m ut det unum, prout ergo velocitas unius c[or]poris augebitur, sic etiam resistentia alterius c[or]poris, hocq[ue] est, q[uo]d vult autor. Quod si haec res alicui dubia videatur, q[uae]ro an non po[ss]it concedere, c[or]pus q[uo]d est minus, tam tarde moveri, ut non sufficiat movendo maius? Q[uo]d si concedat (ut concedere debet) sumamus nunc tam tarde moveri ut non po[ss]it movere alterum, illamq[ue] tarditatem voco unicum motus gradum, iam ergo si moveret alterum, debere tradere ad minumum dimidium, itaq[ue] si 100 habere, traderet 50, si itaq[ue] 50 det 100 plus velocius resistet quam si dimidium; et sic porro: unde seq[ui]tur velocitatem sic nihil conferre," Hamburg 273, 110. Cf. AT IX-B, 90–91: "[l]a quatriéme, que, si le corps C estoit tant soit peu plus grand que B et qu'il fust entierement en repos, *c'est à dire, que non seulement il n'eust point de mouuement apparent, mais aussi qu'il ne fust point enuironné d'air ni d'aucuns autres corps liquides, lesquels comme je diray cy-apres, disposent les corps durs qu'ils enuironnent à pouuoir estre meus fort aiséement*, de quelle vitesse que B pust venir vers luy jamais il n'auroit la force de le mouuoir; mais il seroit contraint de rejallir vers le mesme costé d'où il seroit venu. Car *d'autant que B ne sçauroit pousser C sans le faire aller aussi vite qu'il iroit soy-mesme par apres*, il est certain que C doit d'autant plus resister que B vient plus vite vers luy; et que sa resistance doit préualoir à l'action de B, à cause qu'il est plus grand que luy. Ainsi par exemple, si C est double de B et que B ait trois degrez de mouuement, il ne peut pousser C qui est en repos, si ce n'est qu'il luy en transfere deux degrez, à sçauoir vn pour chacune de ses moitiez, et qu'il retienne seulemẽt le troisiéme pour soy *à cause qu'il n'est pas plus grand que chacune des moitiez de C, et qu'il ne peut aller par après plus vite qu'elles. Tout de même si B a trente degrez de vitesse, il faudra qu'il en communique vingt à C, s'il en a trois cent qu'il communique deux cent, et ainsi tous-jours le double de ce qu'il retiendra pour soy. Mais puis que C est en repos il resiste dix fois plus à la reception de vingt degrez qu'à celle de deux, et cent fois plus à la reception de deux cent, en sorte que d'autant que B a plus de vitesse il trouue en C plus de resistence; Et pource que chacune des moitiez de C a autant de force pour demeurer en son repos; que B en a pour la pousser, et qu'elles luy resistent toutes deux en mesme temps, il est éuident qu'elles doiuent preualoir à le contraindre de rejallir. De façon que de quelle vitesse que B aille vers C ainsi en repos et plus grand que luy jamais il ne peut auoir la force de le mouuoir*," and AT VIII-A, 68: "[q]uarto, si corpus C plane quiesceret, essetque paulo maius quam B, quacumque cum celeritate B moueretur uersus C, nunquam ipsum C moueret; sed ab eo repelleretur in contrariam partem: quia corpus quiescens magis resistit magnae celeritati quam paruae, idque pro ratione excessus unius supra alteram; et idcirco semper maior esset uis in C ad resistendum, quam in B ad impellendum."

[h]"Plane quiescerent. Addit in Gallica editione i.e. ut non tantum ad {sensum} non moveatur, sed etiam, ut non cingatur aere, aliisq[ue] corporibus fluidis. {Q[uae]} ut postea dicam, ita disponunt, corpora dura q[uae] ambiunt, ut facili negotio moveri queant. Quia corpus. In Gallica editione addit ampliorem demonstrationem his {ae} verbis. Etenim B quia non potest pellere C, {quum} post occursum eadem moveantur celeritate: certum est C tanto maiorem habere vim resistendi, quanto maiori celeritate B ei occurrit, atq[ue] ad eius resistentiam maiorem fore actione ea qua agit B: ita e.g. si C duplo maius sit corpore B, et si B moveatur tribus {motibus} gradibus, non poterit B movere ipsum C quod quiescit, nisi in id transferat duos motus gradus. Uno videlicet, quo unoquoq[ue] dimidio, et sibi solummodo tertium retineat gradum, quia maius non est dimidio ipsius C, et quia post occursum celerius moveri non potest. Eodem modo si B triginta habeat gra-

(continued)

Table 4.3 (continued)

dus celeritatis, necesse erit, ut cum C communicet 20 gradus, et si habeat 300, ut 200 conferat in C, et sic porro, tu semper transferat duplum istius {motus quem retinet}, sed cum C quiescat, resistet duplo plus viginti motus gradibus, quam 2bus et centuplo plus {200tis}, ita, ut quanto celerius movetur B tanto maiorem in C inveniet resistentiam, {quia autem unumquodq[ue]} dimidium ipsius C tantum habet virium ad perservandum in quiete, q[ua]ntum B habet ad propellendum, et quia ambae partes ipsius C eodem tempore resistunt ipsi B, manifestum est C plus habere virium ut cogat reflectere corpus B, quam B ut secum ferat corpus C, itaq[ue] ut quacumq[ue] celeritate demum B movetur versus C plane quiescens, et ipso B maius, non possit tamen movere C," Hamburg 274, 45–46. Cf. *supra*, n. g.

[i]Indeed, in the French version Descartes adds a sentence remarking that fluids can change the behaviour of bodies with respect to such laws; this idea is not mentioned in De Volder's commentary: "[m]aius. Concipio C minus e[ss]e B, et quiescere, B a[utem] moveri, dico, quacunq[ue] tarditate moveatur debere tamen impellere C nec potest impellere, quin post occursum aequali celeritate moveantur, si itaq[ue] B antea habuerit 3 motus gradus, postea habebit 2, atq[ue] adeo q[ua]ntum temporis antea req[ui]rebatur ut moveretur per spatium 3 ped[um]: nunc requ[ui]ritur ut per 2 pedes," Hamburg 273, 110. Cf. AT VIII-A, 69: "[q]uinto, si corpus quiescens C esset minus quam B, tunc, quantumuis tarde B uersus C moueretur, illud secum moueret, partem scilicet sui motus ei talem transferendo, ut ambo postea aeque celeriter mouerentur: nempe, si B esset duplo maius quam C, transferret ipsi tertiam partem sui motus, quia una illa tertia pars tam celeriter moueret corpus C, quam duae aliae residuae corpus B duplo maius. Et ita, postquam B ipsi C occurrisset, una tertia parte tardius moueretur quam prius, hoc est, tantundem temporis requireret, ut moueretur per spatium duorum pedum, quam prius ut moueretur per spatium trium. Eodem modo, si B esset triplo maius quam C, transferret ipsi quartam partem sui motus; et sic de caeteris." AT IX-B, 91–92: "[l]a cinquiéme est, que si *au contraire* le corps C estoit *tant soit peu* moindre que B, celuy-cy ne sçauroit aller si lentement vers l'autre *lequel je suppose encore parfaitement en repos*, qu'il n'eust la force de le pousser et luy transferer la partie de son mouuement qui seroit requise pour faire qu'ils allassent par apres de mesme vitesse, à sçauoir si B estoit double de C, il ne luy transfereroit que le tiers de son mouuement, à cause que ce tiers feroit mouuoir C aussi vite que les deux autres tiers feroient mouuoir B, puis qu'il est supposé deux fois aussi grand, et ainsi apres que B auroit rencontré C il iroit d'vn tiers plus lentement qu'auparauant, c'est à dire qu'en autant de temps qu'il auroit pû parcourir auparauant trois espaces il n'en pourroit plus parcourir que deux. Tout de mesme si B estoit trois fois plus grand que C, il *ne* luy transfereroit *que* la quatriéme partie de son mouuement, et ainsi des autres; *et B ne sçauroit auoir si peu de force qu'elle ne luy suffise tous-jours pour mouuoir C, car il est certain que les plus foibles mouuemens doiuent suiure les mesme loix, et auoir à proportion les mesme effets que les plus forts, bien que souuent on pense remarquer le contraire sur cette terre à cause de l'air et des autres liqueurs qui enuironnent tous-jours les corps durs qui se meuuent, et qui peuuent beaucoup augmenter ou retarder leur vitesse, ainsi qu'il paroistra cy-apres*."

[j]"Si B non moveret C, reflectentur, ac ideo maiorem haberet vim, quam B, quod est contra {hypothesin}, cum C supponatur minus, ergo B non reflectitur, sed secum feret corpus C, et ideo, tantum ipsi de suo motu dabit, quantum requiritur, ut eadem postea moveantur celeritate, nec obstat motus tarditas, nam addit haec verba in editione Gallica. Certum est motus tardissimos easdem cum celerrimis observare leges, et eosdem secundum motuum proprietatem producere effectis, etsi saepe numero videamus contrarium observare in nostra terra, id[que] propter aerem, aliosq[ue] liquores, qui continenter corpora dura ambiunt, quaeq[ue] se moveant, et q[uae] possunt multum augere vel minuere ipsorum velocitatem, ut apparebit postea," Hamburg 274, 46. Cf. *supra*, n. i.

[k]"Hactenus concepimus 2 c[or]pora inaequalia, nunc vero concipimus aequalia, q[ui]d fiet? R[es]p[ondeo] par in utroq[ue] muta[ti]o, non e[nim] potest unum {mutari} non mutato altero, si e[nim] c[or]pus q[uo]d movetur reflecteretur, et alterum q[ui]esceret, e[ss]et quiescens fortius, si contra imbecillius, q[uo]d utrumq[ue] imbecillius ergo necesse est, ut unum reflectatur, alterum moveatur, et detur medietas," Hamburg 273, 110. Cf. AT VIII-A, 69: "[s]exto, si corpus C quiescens esset

(continued)

Table 4.3 (continued)

accuratissime aequale corpori B uersus illud moto, partim ab ipso impelleretur, et partim ipsum in contrariam partem repelleret: nempe, si B ueniret uersus C cum quatuor gradibus celeritatis, communicaret ipsi C unum gradum, et cum tribus residuis reflecteretur uersus partem aduersam." AT IX-B, 92: "[l]a sixiéme, que si le corps C estoit en repos et parfaitement égal en grandeur au corps B, qui se meut vers luy, *il faudroit necessairement qu'*il fust en partie poussé par B, et *qu'*en partie il le fit rejallir, en sorte que si B estoit venu vers C auec quatre degrez de vitesse, *il faudroit qu'*il luy en transferast vn, et qu'auec les trois autres il retournast vers le costé d'où il seroit venu. *Car estant necessaire ou bien que B pousse C sans rejallir et ainsi qu'il luy transfere deux degrez de son mouuement; ou bien qu'il rejallisse sans le pousser et que par consequent il retienne ces deux degrez de vitesse auec les deux autres qui ne luy peuuent estre ostez; ou bien enfin qu'il rejallisse en retenant vne partie de ces deux degrez et qu'il le pousse en luy en transferant l'autre partie. Il est éuident que puis qu'ils sont égaux et ainsi qu'il n'y a pas plus de raison pourquoy il doiue rejaillir que pousser C, ces deux effets doiuent estre également partagés: c'est à dire, que B doit transferer à C l'vn de ces deux degrez de vitesse, et rejallir auec l'autre*."

[l]"Versus partem diversam. Addit autor in Gallica editione haec verba: cum necesse sit aut B non reflecti, aut pellere ipsum C (quo pacto in C transfer{r}et duos gradus motus sui) aut B pellere C, sed reflecti. Et ut per consequens retineat hos duos gradus motus cum aliis rebus, q[uae] ab eo tolli nequeunt, aut deniq[ue] ut reflectatur B retinendo unam partem duorum istorum graduum, et alteram transferendo in C, manifestum est, cum haec corpora sunt aequalia, ac proinde nulla maior ratio sit, propter quam B potius reflectatur, quam C pellat. Haec 2 effecta aequi aliter necessario dividi, i.e. necesse esse, ut B conferat ipsi C minus ex duobus illis gradibus, et reflectatur cum altero," Hamburg 274, 46.

[m]"Restat nunc unicus casus, quo c[or]pora moventur versus easdem partes. Retinent. Id vult author concipio 2 c[or]pora in easdem partes mota, si moveantur par celeritate, non agent in se invicem, non etiam si id q[uo]d pracedit, celerius moveatur, si a[utem] praecedens minus sit et tardius moveatur, cum sequens maius sit et celerius moveatur, necesse est, ut prius impellatur, ad eoq[ue] aequaliter postea progrediatur, si vero praecedens maius sit, sequens reflectetur, si e[nim] movetur {una} cum altero, deberet superare prioris vires, q[uo]d repugnat, quia supponitur minus. Iudicandum. In hypothesis res certissima est, si excessus celeritatis ab una parte det plus virium, quam excessus magnitudinis det alteri parti, ut posterius vincat prius, debet habere ad minimum duplum excessus, qui est in altero," Hamburg 273, 110–113. Cf. AT VIII-A, 69–70: "[d]enique, si B et C uersus eandem partem mouerentur, C quidem tardius, B autem illud insequens celerius, ita ut ipsum tandem attingeret, essetque C maius quam B, sed excessus celeritatis in B esset maior, quam excessus magnitudinis in C: tunc B transferret tantum de suo motu in C, ut ambo postea aeque celeriter et in easdem partes mouerentur. Si autem e contra excessus celeritatis in B minor esset, quam excessus magnitudinis in C, B in contrariam partem reflecteretur, et motum omnem suum retineret. Atque hi excessus ita computantur: si C esset duplo maius quam B, et B non moueretur duplo celerius quam C, ipsum non pelleret, sed in contrariam partem reflecteretur; si uero magis quam duplo celerius moueretur, ipsum pelleret. Nempe, si C haberet tantum duos gradus celeritatis, et B haberet quinque, demerentur duo gradus ex B, qui translati in C unum tantum gradum efficerent, quia C est duplo maius quam B: quo fieret ut duo corpora B et C cum tribus gradibus celeritatis postea mouerentur; et ita de caeteris est iudicandum. Nec ista egent probatione, quia per se sunt manifesta." AT IX-B, 92–93: "[*l*]*a septiéme et derniere regle* est, que si B et C vont vers vn mesme costé, et que C precede mais aille plus lentement que B en sorte qu'il soit en fin atteint par luy, il peut arriuer que B transferera vne partie de sa vitesse à C, *pour le pousser deuant soy; et il peut arriuer aussi qu'il ne luy en trãsferera rien du tout, mais rejallira, auec tout son mouuemẽt vers le costé d'où il sera venu, à sçauoir non seulement lors que C est plus petit que B, mais aussi* lors qu'il est plus grand, pouruu que ce en quoy la grandeur de C surpasse celle de B, soit moindre que ce en quoy la vitesse de B surpasse celle de C, jamais B ne doit rejallir mais pousser C en luy transferant vne partie de sa vitesse; Et au contraire lors que ce en quoy la grandeur de C surpasse celle de B, est plus grand que ce en quoy la vitesse de B surpasse celle de C il faut

(continued)

Table 4.3 (continued)

que B rejallisse sans rien communiquer à C de son mouuement; *et enfin lors que l'excez de grandeur qui est en C est parfaitement égal à l'excez de vitesse qui est en B, celuy-cy doit transferer vne partie de son mouuement à l'autre et rejallir auec le reste*, ce qui peut estre supputé en cette façon. Si C est justement deux fois aussi grand que B, et que B ne se meuue pas deux fois aussi vite que C, *mais qu'il en manque quelque chose*, B doit rejallir sans augmenter le mouuement de C; et si B se meut plus de deux fois aussi vite que C, il ne doit point rejallir *mais transferer autant de son mouuement à C qu'il est requis pour faire qu'ils se meuuent tous deux par apres de mesme vitesse.* Par exemple, si C n'a que deux degrez de vitesse, et que B en ait cinq *qui est plus que le double*, il luy en doit communiquer deux de ses cinq, lesquels deux estant en C n'en feront qu'un à cause que C est deux fois aussi grand que B, et ainsi ils iront tous deux par apres auec trois degrez de vitesse. Et les demonstrations de tout cecy sont si certaines *qu'encore que l'experience nous sembleroit faire voir le contraire, nous serions neantmoins obligez d'ajouter plus de foy à nostre raison qu'à nos sens*."

[n]"Motum omnem suum retineret. Addit in Gallico deniq[ue] si excessus magnitudinis in C sit accurate aequalis excessui celeritatis in B, tunc debet transferre partem sui motus in alterum, et cum reliqua parte reflecti," Hamburg 274, 46. Cf. Sect. 4.2.1.2, *Descartes's rules of impact*, n. 175.

porum elastica – dedicated to him and taking place on 2 July 1688. In this disputation, indeed, we find the rules of impact that De Volder would later present in his correspondence with Leibniz – which I will treat below.

As far as the 1684 disputation is concerned, it presents a notable divergence from the short series *De motu* De Volder had written and presided over in the 1670s. In particular, the 1684 disputation has decidedly un-Cartesian tones. First, the author stresses the relativity and the reciprocity of motion (as had been done by De Volder in the Hamburg 264 series of his *dictata*).[223] As the change in distance or *situs* between two bodies involves both bodies, they both have to be considered as in reciprocal motion.[224] The consideration of one of them as at rest depends on our

[223] See *supra*, n. 18.

[224] "IV. Ex his sic detectis quaedam deinceps de motu sequuntur proprietates, quae ex legitima eius definitione ordine ratiocinando sumus deducturi: quarum prima sic se habet: motum esse mutuum, hoc est mutationem situs utriusque corpori esse communem, tam illius quod moveri, quam illius quod vulgo quiescere concipitur: quod etsi prima fronte videatur quid esse paradoxum, ut tamen melius innotescat, consideremus ex. gr. duo corpora A et B sibi invicem contigua, B non potest dici ipsum A tangere nisi ipsum A tangat B, atque ut iste contactus sit mutuus sic et omnis translatio illorum corporum erit mutua vel reciproca: ut pateat quid velim, supponamus rursus illa duo corpora A et B sibi mutuo esse vicina, et illam viciniam in posterum mutari; nimirum quando A movetur a B, atque ideo situm suum (respectu habito ad corpus B, cui prius vicinum erat) immutat, nunc autem ab ipso removetur et distat, eamque distantiam recedendo continue mutat: verisimile erit maxime, illam situs variationem, aequali modo in corpore B obtinere quam in A, hoc et, corpus B, aeque situm suum mutat respectu ipsius A, ac A respectu ipsius B: quod idem est ac si dicas B aeque distat ab A ac A a B, ex quo tunc patet evidentissime, eam distantiae variationem aequali modo prorsus fieri inter B et A ac inter A et B. V. Cum igitur manifestissimum est motum esse reciprocum seu relativum, tamen ut illam relationem evidentius percipiamus supponenda sunt haec duo corpora A et B prius adhuc continua, nullis aliis annexa esse, sed quasi poni in vacuo: atqui tum evidentissime patebit si consideremus vim aliunde accedere A non posse divelli a B in medio ac alio transferri, nisi in B vis quaedam esset nitendi et tendendi in contrarium, nam sine vi illa

arbitrium, i.e. it is an abstraction by which bodies maintaining the same *situs* are arbitrarily picked out by the mind.[225] On this basis, he justifies the claim for which the same force is required both to bring a body to rest, and to move it, as in fact they are the very same action. Second, the author proposes his own solution for the case of impact considered in Descartes's rule 6, namely, that in which two bodies of the same size are colliding, and one of them is at rest. For Descartes, as seen above, the moving body is rebounded and maintains ¾ of its speed, while the other is moved with as speed of ¼ of the original speed of the other body (this being a *via media* between a complete rebounding of the moving body – such as in rule 4 – and its carrying the body at rest, as in the other rules). In the disputation, the outcome is the following: after the impact, the two bodies move in opposite directions, but they move at equal speeds (i.e. ½ of the initial speed of the moving body), because

contraria impetu translato in alterum vel in A vel in B, non a se mutuo divellerentur, sed tota moles simul verteretur versus unam aliquam partem, vel versus dextram vel versus sinistram. Quod ut rudi exemplo declarem, consideremus duo navigia sibi obviam ferri; periculum quod utriusque imminet aequale erit, et ut propulsetur indifferens erit in utro horum consistat aliquis homo, qui harpagine motum sui et alterius navigii sistat simul et eadem opera; eodem quoque prorsus modo si aliquis stans in terra fune navigium stipiti alicui attrahat, sive alius stans in navi mota funem fixum stipiti alicui attrahat, eaedem plane hoc in casu requiruntur vires. Sic in illis duobus, ille qui est in terra alter in mota navi, manibus simul eundem funem utrimque trahentes, simul et eadem vi navigium ripae, et ripam navi appropinquare faciunt. VI. Evidentissime igitur patet, motum inter corpora esse relativum, seu eius naturam esse reciprocam, quod ex eius definitione clarissime patet: motus enim essentia nihil considerandum praeter vicinitatem et distantiam nobis proponit, atque hae ambae non aliter quam mutuae possunt concipi, cum enim A concipitur distare a B simul etiam necessario B distat ab A et cum A vicinum sit ipsi B, sic eodem tempore debet esse vicium ipsi A [...]. Primum itaque quod in motu concipimus est quod mutuus sit, quia quicquid ad naturam eius pertinet mutuum est," De Volder and Vander Codde 1684, theses 4–6. It is noteworthy that, in the theses 1–3 the author stresses that movement induces the succession of our thoughts (thesis 1), that movement, understood as a mutation of *situs*, is a matter of absence or presence of one body with respect to another (thesis 2), and that movement, which does not take in place in an instant, is a succession of our thoughts ("motus igitur natura [...], quod consistat nempe in tali relatione nostrarum cogitationum, qua consideramus aliqua prius sibi invicem vicina deinde a se mutuo distantia, evidentissime sequitur motum non posse concipi sine tempore; hoc est [...] hoc ipso quaedam successio inde resultat, quae nobis tempus dicitur," thesis 3). The author, however, does not avail a purely mental characterization of movement, which is a real change in distance.

[225] "VII. Motum itaque in se considerando, seu illud solummodo attendentes, quod revera inter corpora quae dicuntur moveri deprenditur, et quapropter proprie moveri considerantur, quoties duo corpora sibi mutuo vicina, unum in hanc alterum in illam transferuntur partem, atque a se invicem separantur, concipimus esse tantundem motus in uno quam in altero: considerare tamen assueti sumus aliqua corpora quiescere: ast cum illa situs mutatio seu relatio utrumque aeque spectat, penitus a nostro dependet arbitrio, cuinam corpori motus sit assignandus, cuinam vero quies, cum eadem fiat relatio respectu utriusque corporis, eadem distantiae mutatio. VIII. Quod ut evidentius innotescat, cogitandum est, etsi verum sit motu introducto nihil immotum posse reperiri, cunctaque vero moveri: possumus tamen corpus aliquod particulare abstrahere, illudque sic abstractum cogitare, quod comparatum cum uno pluribusve aliis, relationem sui situs non mutet, et eatenus illud corpus (comparatione sic instituta) in illa mea cogitatione quiescet. Cum interea dum, alia quaedam alterius cuiusdam hominis cogitatio illud conferat cum alius, quibuscum (alio respectu habito) moveri concipietur. Sic ex. gr. homo navigio vectus movebitur respectu littorum, si vero rationem habeamus respectu partium navigii, revera censebitur quiescere," De Volder and Vander Codde 1684, theses 7–8.

"[body] B has as much of forces [*sic*] to resist, as [body] A to push."[226] This can be understood as a consequence of the author's 'relativist' approach. First, the outcome of the impact does not change if we consider A at rest, and B moving towards it; after the impact, they would anyway have a speed of ½ and ½ each. In other words, the hypothetical outcome devised in Descartes's rule 4 is not considered. Second, the author seems to reduce this case to Descartes's rule 1, in which in fact a condition of perfect symmetry applies. Namely, the outcome of this collision is the same as in Descartes's rule 1, and, given the perfect reciprocity of movement, nothing changes if we consider the two bodies to be moving as in Descartes's rule 1, before the impact (i.e. both with a speed of ½ and ½ towards each other).

In any case, the rule devised by the author contradicts Huygens's first rule of impact given in his 1669 communications, according to which the two bodies will 'exchange' their states of movement and rest (i.e. A will be at rest, and B will move at the speed that A had before the impact). The result given in the disputation is wrong because, if we consider the two bodies as moving differently, but maintaining the same relative speed (i.e. we can consider B as moving towards A, and A as at rest, or both bodies as moving with a speed of ½), we need to subtract, from the speed they have after the impact, the speed in which they both participate (such as the movement of a boat) by which an observer outside this frame (such as a man on the shore) could 'manipulate' the speeds of the two colliding bodies.[227] It seems, in fact, that the author of the disputation tried to correct Descartes's rule 6 by making it consistent with the reciprocity of motion avowed by Descartes in *Principia* II.13 and II.24, and with the idea that the resistance exerted by body B is equal to that it would exert if it were moving towards A at the same speed of A – since movement and rest can be conceived as the same state. This disputation, as I am going to show,

[226] "IX. His sic detectis pergam ad alte nobis infixum praeiudicium, quo extimamus plus realitatis esse in motu quam in quiete, quod praeiudicium ut eximatur, considerandum velim omnia esse in motu, adeoque necessario causam aequipollentem requiri, quae aliquod corpus iam in motu positum ad statum quietis reducet, maneret utique alias illud corpus in motu, nec unquam quieturum esse conciperetur. Quod cumque igitur fuerit, quo corpus motum ad quietem sit reducendum, necesse habet vim et efficaciam illius causae tollere qua motus introductus est, ad hoc autem praestandum quid quaeso valeret nisi minimum tanta vi polleret, quanta pollet causa movens. Ex. gr. consideremus duo corpora A B, perfecte dura, nullis aliis corporibus annexa. Concipiamus iam ambo esse magnitudine et soliditate aequalia, dein consideremus A moveri, B quiescere. Aio ubi A attigerit ipsum B, moturum quidem B, sed rursus repelletur ab B, adeo ut ambi aeque celeriter pergant moveri in partes contrarias, quia sunt ambo magnitudine et soliditate aequalia. Nam B tantundem virium habet ad resistendum quam A ad impellendum. X. Dein et hoc meretur considerationem, quod putemus plus actionis requiri ad motum quam ad quietem: quod praeiudicium facile exuemus; si in memoria revocemus [...] quietem requirere aeque realem causam, quam motum. Vel hoc evincet, quod extensio non maiorem habet conatum ad motum quam ad quietem. Corpus igitur aeque quiescit a causa externa quam a causa externa movetur. Sed haec est causa cur putemus quietem non requirere tantundem actionis quam motum, quod quies in corporibus mundi sublunaris producatur ob concursum plurium causarum quae saepe nos latent, e contra vis et actio manifestius deprehenditur ad motum inter eadem corpora producendum," De Volder and Vander Codde 1684, theses 9–10.

[227] This condition is stated by Huygens only in his *De motu corporum ex percussione*, as hypothesis 3.

will be important in reconstructing De Volder's own path in the formulation of his own laws (presented in the 1688 disputation and in the correspondence with Leibniz). Indeed, as will be clear from Sect. 4.2.3.4.2, *De Volder on elastic impacts*, in the 1684 *De motu* the author considered, as a dynamic factor underlying the impact of these bodies, only what he will then present as their elastic forces. For the sake of clarity, I anticipate here the quantification of such forces as given in the following texts by De Volder, by considering the case proposed in the 1684 disputation: A (whose speed is, for instance, $s_1 = 4$ and size is $m_1 = 3$) is rebounded by B (whose speed is $s_2 = 0$ and size is $m_2 = 3$) with a speed quantified as $(m_2 \cdot (s_1 + s_2))/(m_1 + m_2)$, i.e. 2, or ½ of its original speed; in turn, B is rebounded by A with a speed $(m_1 \cdot (s_1 + s_2))/(m_1 + m_2)$, i.e. 2, or ½ of the original speed of A.

4.2.3.3 Dynamics in De Volder's Correspondence with Leibniz and Bernoulli

The problem of impact and the related ones of the conservation of the quantity of motion and of living force play a pivotal role in De Volder's correspondence with Leibniz and Johann Bernoulli, who also acted as intermediary between De Volder and Leibniz (i.e. he reports, in his letters, De Volder's opinions). In particular, such problems play a pivotal role in what Lodge has described as the first and second phase of the correspondence, running from the letter of Bernoulli to Leibniz of 5/15 July 1698, to the letter of Bernoulli to Leibniz of 6/17 April 1700.[228] In these phases, both Leibniz and Bernoulli attempted to convince De Volder that the motive force of a body is measured by $m \cdot s^2$, and not by $m \cdot s$. Accordingly, in order to understand De Volder's views on motion and impact in his late years, we need to have an overview on the main topics dealt with in this part of his correspondence with Leibniz and Bernoulli. We can summarize these topics, and the main theoretical 'requests' advanced on both sides, as follows.

On De Volder's side, his main 'expectation' in his correspondence with Leibniz was (as seen in Sect. 4.1.4, *The search for physical unity and activity: De Volder vis-à-vis Leibniz*), that of finding an a priori demonstration of the activity of substance. This would grant, for De Volder, that bodies can move themselves – namely that there is a change in *situs* or distance between bodies – without any direct divine intervention. At the broadest level, therefore, De Volder's problem was that of finding a principle of action in physical terms, i.e. by considering the idea of material substance only. However, this was not his only problem. If we look at the first letter of his correspondence (i.e. the letter of Bernoulli to Leibniz of 5/15 July 1698), we find that De Volder was an admirer of Leibniz, but also that he was a supporter of the idea that the quantification of motive force is provided in the way as Denis Papin conceived it, namely, as the quantity of motion ($m \cdot s$) itself. The correspondence between Leibniz and De Volder, indeed, started after the so-called Leibniz-Papin controversy (1689–1691), in which Papin supported the idea that motive force can

---

[228] Lodge 2013, *Introduction*, xxvii–xlv.

be quantified as quantity of motion. On the other hand, Leibniz maintained that motive or living force is quantified as m·s^2, in accordance with his criticism of the Cartesian theory of force he displayed in his *Brevis demonstratio erroris memorabilis Cartesii* (1686).[229] According to Bernoulli's report, De Volder stood with Papin, as he was "saying that [...] the elasticity of bodies is gratuitously supposed [by Leibniz], indeed, it is only accidental to bodies," since it can happen that perfectly hard bodies collide and – as it was for Huygens's soft bodies – they come to a stop, and lose all their movement.[230] According to Bernoulli, therefore, De Volder stood with Papin – and with Descartes – because not all bodies are elastic, and some come to a complete stop after impact. As seen above, in the elastic collisions described by Huygens (and all the others who stood with him, such as Wren, Wallis, and Edme Mariotte), only m·s^2 is conserved. So, it seems that De Volder opposed Leibniz in the "estimation" (*aestimatio*) of force not because he maintained that m·s is the measure of force, but because m·s^2 is not conserved. Namely, for the reason that if one calculates forces as m·s^2, one would find that this quantity is not always conserved in impacts. Yet, as reported in the letter, in non-elastic collisions m·s is also not conserved, so that it is not very clear why De Volder stood with Papin. This lack of clarity can be resolved by looking at De Volder's own letters. These reveal two main problems: (1) the problem of the quantification or calculation of motive force by the formula m·s or m·s^2 – as De Volder declared that, in fact, he was not certain on which side to stand in the Leibniz-Papin controversy,[231] and (2) the problem of conservation of any quantity (m·s and m·s^2), for the reason that in non-elastic impacts both are lost, and there are no physical causes for the increase of motion (letter of 21 November/1 December 1698).[232] As far as the correspondence is con-

[229] The Papin-Leibniz controversy has been magisterially reconstructed in Freudenthal 2002.

[230] "Cum nuper Leyda transirem, Volderum conveni saepius [...]. Vidi praeterea illum aliquoties apud alios Professores, sensi illum esse summum Tui admiratorem. [...] Intellexi quoque quod a Papini parte stet circa aestimationem virium, dicens a Te gratis supponi corporum elasticitatem, hanc enim corporibus tantum esse accidentalem; si corpora perfecte dura supponantur Tua ratiocinia amplius locum non habere: duritie perfecta supposita sequi ex. gr. quod duo corpora aequalia et aequali celeritate sibi mutuo centraliter occurrentia, non debeant reflecti sed sisti in ipso momento occursus; et quod generaliter corpora duo perfecte dura cujuscunque molis et cujuscunque celeritatis post conflictum separari non debeant, quia separatio proveniat ab elasticitate, quam autem supponendam dicit," Bernoulli to Leibniz, 5/15 July 1699, in A III7B, 205.

[231] "[C]um denuo inspiciens, quae his de rebus in Actis Lipsiensibus habentur perspexerim primariam meam difficultatem a Clar. Papino praeoccupatam. Quibus accedit quod perpendens argumenta hinc inde allata, ea talia inveniam, ut me utrinque omni scrupulo liberare nequeam, adeoque neutram partem pro certa amplecti audeam," De Volder to Bernoulli for Leibniz, 21 November/ December 1698, in A II3B, 477. De Volder was referring to the exchange between Papin and Leibniz in the *Acta Eruditorum*, i.e. to Papin's *De gravitatis causa et proprietatibus observationes* (1689), Leibniz's *De causa gravitatis et defensio sententiae suae de veris naturae legibus contra Cartesianos* (1690), Papin's *Mechanicorum de viribus motricibus sententia* (1691), Leibniz's *De legibus naturae et vera aestimatione virium motricium contra Cartesianos* (1691). Also Leibniz's *Brevis demonstratio erroris memorabilis Cartesii* appeared in the *Acta Eruditorum* (March 1686).

[232] "Me quod attinet, de quantitate virium an eadem an diversa, quod certo asseverem, nihil habeo. Causam, quae vires augeat, nullam reperio; verum cum vires sibi invicem sint contrariae, quid impediat, quominus altera alteram tollat non video. Corpora mollia aequalia, et aequali velocitate

cerned, the first issue was more important for Leibniz and Bernoulli – who wanted to 'convert' De Volder to the idea of living force; the second was more important for De Volder, who was concerned with the problem of the cause of movement, and so with the possibility of its conservation by physical causes. De Volder, indeed, was already convinced that in elastic collisions m·s^2 is conserved:[233] this agreed with Leibniz's positions, and was demonstrated by Huygens's rules,[234] and by De Volder's own rules (as I am going to show in Sect. 4.2.3.4, *Collision in De Volder's late thought*).

Yet, De Volder was also – let's say – not convinced that Leibniz was right in claiming that force can be measured by m·s^2. It was in order to solve both these problems that he (repeatedly) asked from Leibniz a demonstration of the activity of matter, this being, for him, the key to settle both the disputes on the quantification of motive force, and on its conservation, "whatever [the] laws [of motion] we follow."[235] For De Volder, the problem of activity was the "primary" issue, all the

sibi invicem occurrentia cessant a motu suo. Idem de corporibus duris non elasticis, quidquid etiam supponat Hugenius, ut credam admodum proclivis sum. Mihi quippe verisimile apparet, corpus quod aequali vi, sive eam nunc demum accipiat, sive jam ante aliunde acceperit, in contrarias pellitur partes, quieturum. Quod si verum est corpora aequalia sibi invicem aequali motu occurrentia, quiescent ambo. Nec liquet, cur non et quantitas motus, et vires hoc in casu dicantur deperditae," De Volder to Bernoulli for Leibniz, 21 November/December 1698, in A II3B, 480–481.

[233] See the letter of De Volder to Leibniz of 12/22 November 1699: "[c]onstat autem jam inter nos, id quod in corporum elasticorum concursu conservatur, esse ut quadratum celeritatum ductum in molem," A II3B, 611.

[234] In his letter to Leibniz of 21 November/1 December 1698, De Volder points out how Leibniz's theory of the conservation of living force (m·s^2) agrees with Huygens's rules of collisions presented in the *Philosophical Transactions*, which, for De Volder, Huygens based on the hypothesis – which he could read, as a hypothesis, in the manuscript of *De motu corporum ex percussione* – that Descartes's first rule of collision is true: "[q]uod autem monet Vir Ill. se vires vocare, quae semper maneant pari quantitate, cum motus quantitas augeatur minuaturve, id consentit cum Regulis motus Hugenii, quas in Transact. Anglicis videre est, in quibus, quantitas motus licet augeatur, minuaturve, *summam tamen productorum factorum a mole cujuslibet corporis duri, ducta[m] in quadratum suae celeritatis, eandem semper esse ante et post eorum occursum*, affirmat. Deducit has regulas Hugenius ex hypothesi, corpora dura aequalia, aequaliterque in se invicem mota, post occursum reflecti singula, eadem qua prius, velocitate, qua tamen cum Hypothesi, qui consentiat id, quod ait Reg. V. *semper remanere eandem quantitatem motus versus eandem partem, ablata inde quantitate motus contrarii*, perceptu mihi arduum videtur," De Volder to Bernoulli for Leibniz, 21 November/1 December 1698, in A II3B, 480. Cf. Huygens 1669a, 928, 1703, 369. In claiming that Descartes's first rule contradicts Huygens's 1669 rule 5, De Volder (probably) pointed out that in Descartes's first rule the difference between the two quantities of motion (which are equal, and exerted in opposite directions) is 0. In other cases, Huygens's rule 5 is valid, i.e. the quantity of motion is always equal in both directions, once the contrary quantities are subtracted.

[235] "[...] si a priori demonstratum haberemus, omnem substantiam esse activam, facile mihi persuadeo ex hoc foecundissimo veritatum fonte non horum tantum scrupulorum secuturam enodationem, sed et earum difficultatum, quae hactenus omnes quotquot sunt, Physicos presserunt. Hujus etenim rei ignorantia, compulsi sunt causam motus conjicere in Deum, et nonnulli etiam ad quoslibet corporum occursus Deum e machina arcessere," De Volder to Bernoulli for Leibniz, 21 November/December 1698, in A II3B, 481. See also the letter of De Volder to Leibniz of 12/22 November 1699: "[a]n mens mea a falsis notionibus vulgi, et Cartesianorum circa materiam et motum et substantiam corpoream satis libera sit, asserere non ausim. Id tamen me satis intelligere

other questions touched on in the correspondence being only accidental with respect to it.[236] By extension, this demonstration would have also solved the problem of elasticity of bodies. Namely, to demonstrate that movement does not come to a stop during collision, for De Volder intends to demonstrate that bodies rebound when they collide – something that Leibniz 'gratuitously' assumed, according to De Volder, and that cannot not be reduced either to the extension or to the impenetrability of matter.[237] Or, as De Volder put it in his letter to Leibniz of 12/22 November 1699, "if it is demonstrated that every substance is active from its nature, and so that the active force which originates from the nature of substance is necessarily conserved, it will be clear that all bodies that collide with each other moving in opposite directions are elastic."[238]

For De Volder, thus, the so-called '*vis viva* controversy' was subordinated to the problem of finding a physical principle of activity. For Leibniz (and Bernoulli), on the other hand, the solution of the *vis viva* problem was preliminary to that of finding an internal principle of activity. Leibniz refused to discuss the problem of substance until De Volder was convinced that motive force or *vis viva* is quantified by $m \cdot s^2$.[239] For Leibniz, this was a physical problem, upon the solution of which all the

existimo, virium actionumque regulas ex eorum principiis deduci non posse, quascunque etiam leges sequamur, adeoque vel cum iis ad Deum confugiendum esse [...], quod minime probo, vel vires corporum ex substantia corporea deducendas, quod utinam possem," in A II3B, 612.

[236] "Accedo nunc ad id quod primarium est, quod sc. spectat substantiae activitatem; quo discusso, reliqua de quibus per occasionem mentio facta est, facile, ut opinor, enodari poterunt," De Volder to Leibniz, 5 April 1700, in A II3B, 632.

[237] "Te scilicet praeter extensionem et impenetrabilitatem tertium quid requirere ad essentiam corporis quod consistat in vi ingenita ad conservationem virium, unde necessario vim elasticam omnibus corporibus ex natura sua competere. Ille autem regessit Te aliquid statuere quod concipere non possis; an illud tertium sit substantia an modus? si modus, nihil novi esse. Sin substantia an spiritus an corpus? aut forte tertium? hoc autem tertium explicari non posse, nisi cum veteribus ad formam substantialem diu explosam recurrere velis," Bernoulli to Leibniz, 23 July/2 August 1698, in A III7B, 847. See also the letter of De Volder to Leibniz of 18/28 February 1699, commenting upon Huygens's rules of impact: "[v]erum, licet hoc ita eveniat in corporibus perfecte elasticis, hanc tamen virium aestimationem omnibus corporibus qui applicare possis, non video; idque eo minus, quia hae ipsae vires in elasticis sequuntur ex quadam virium amissione in non elasticis, mea quidem in hypothesi. Ubi agitur de viribus corporum, non id spectandum est, quod ex peculiari corporum proprietate sequitur, sed id, quod ex generica omnium corporum fluens natura, omnibus, cujuscunque de caetero sint conditionis, corporibus ex aequo inest. Quare, Vir Amplissime, si velis ut absque omni scrupulo tibi assentiamur, descendendum erit, opinor, ad notionem substantiae, demonstrandumque eam ex natura sua necessario esse activam, aut certe eam esse naturam substantiae corporeae, ut necesse sit ejus vires conservari semper. Hoc enim nisi a priori demonstratum sit, non ita facile, mihi certe, erit concipere, corporis mathematici existentiam plane esse contradictoriam," De Volder to Leibniz, 18/28 February 1699, in A II3B, 530.

[238] Lodge 2013, 143; cf. A II3B, 611: "[...] si demonstratum sit, omnem substantiam esse ex natura sua activam, adeoque eam, quae ex substantiae natura manat vim activam necessario conservari, patebit omnia quae contrariis motibus in se invicem impingunt corpora, esse elastica."

[239] "Ac primum non magis mihi gaudeo, quam Tibi gratulor repertam veritatem, et comprobatas meas aestimandarum virium leges. [...] Desideras postremo, Vir Nobilissime, Substantiae activitatem ex se, non tantum velut a posteriori (ut vocant) et ex eo probari quod ab alia creata substantia nihil accipere potest, et agit tamen, ut experientia constat; sed etiam a priori seu ex ipsa substantiae

subsequent metaphysical discussions over material substance could then be discussed. For him, in fact, physics (or mixed mathematics) was the door to metaphysics,[240] and before moving to more complicated metaphysical matters they needed a common ground in physics.[241] Accordingly, all the letters between De Volder, Leibniz and Bernoulli up to that of Leibniz to De Volder of 6 September 1700, are dominated by the problem of the *vis viva*, and by the attempts of Leibniz and Bernoulli to 'convert' De Volder to Leibniz's positions – which Bernoulli expected that De Volder (regarded by him as a 'defector' from Cartesianism) would teach at Leiden University.[242] As we have already apprehended, however, De Volder was never 'converted' to Leibniz's metaphysics of substance.

The *vis viva* controversy in Leibniz's correspondence with De Volder, Bernoulli and Papin, and the arguments and counter-arguments employed by all the figures involved have been thoroughly analysed in the secondary literature.[243] It is worth here however presenting Leibniz's main argument, given in his 1686 *Brevis demonstratio*, his subsequent presentation of the idea of elasticity in the correspondence with De Volder, and how De Volder was eventually 'converted' to the idea of *vis viva*, i.e. that motive force is not quantity of motion (m·s) but *vis viva* (m·s^2).

4.2.3.3.1 The Quantification of Motive Force

Leibniz's 1686 demonstration is based on the premises that "a body falling from a certain altitude acquires a force [enabling it] to ascend up to that point," (in accordance with Torricelli's principle) and that "as much force is needed to raise a body A of one pound up to the height [...] of four ells, as [much] it is needed to raise a body

---

natura derivari. Verum ita vides constituendam prius esse Substantiae definitionem ac de ea nobis conveniendum, quod etsi non exiguae fortasse operae, non tamen fateor sine operae pretio foret," Leibniz to De Volder, 6 September 1700, in A II3B, 654–656.

[240] "Sed haec ipsa virium aestimatio mentem praeparat, ut facilius sustineat majorem illam lucem, qua perstringuntur nimis velut [...] caligine, qui vulgi et Cartesianorum etiam notionibus circa naturam materiae, motus, et virium sunt assueti. Itaque per hanc portam a Mathesi ad metaphysicam transeundum censeo," Leibniz to Bernoulli, 1/11 September 1699, in A III8, 227.

[241] "D. Volderus postulat demonstrationem activitatis substantiarum; jam vides quam difficile sit demonstrare res Mathematico-physicas quales mechanicae sunt, quid de Metaphysicis futurum putas? Demonstratio est ratiocinatio, cujus vis sit evidens, et a qua indubitatam Tibi adversariorum convictionem promittas; talem etiam cum omnia animo jam comprehensa habemus exprimere verbis difficile est. Quanto magis cum nondum ipsi satis ordinavimus cogitationes nostras?" Leibniz to Bernoulli, 12/22 January 1700, in A III8, 301.

[242] For this reason, in his letter to Leibniz of 6/16 December 1698, Bernoulli encouraged Leibniz to compose a system of philosophy: "non dubito quin systema si quod componeres, felicissimum successum habiturum esset, sunt enim in Batavis nonnulli egregii Viri qui hunc Tuum ingenii foetum fortiter foverent et defenderent, interque illos ipse Dn. Volderus qui cum Cartesiana principia tanquam insufficientia et plurimum falsa ut ipse mihi fassus est jam a longo tempore deseruerit, si Tua semel probe percepisset et imbibisset, dubium non est quin ea gnaviter propagaturus suisque Discipulis quorum semper insignem numerum habet adeo esset inculcaturus," A III7B, 960.

[243] See *supra*, n. 229.

B of four pounds up to a height of [...] one ell" (in accordance with the law of the lever). Then, Leibniz adds to these premises that "all this is accepted by the Cartesians."[244] In this way, Leibniz 'forces' the Cartesians to support the following argument: since according to Galileo's law of acceleration of falling bodies, the distance covered by the fall of the body is proportional to the square of the time it takes for it to fall, the force required to raise a body A (whose size is 4) for 1 metre cannot be quantified by the quantity of motion acquired by body B (whose size is 1) after it has fallen for 4 metres. Indeed, after 4 metres the quantity of motion of B is 2, and the quantity of motion of A is 4 (see Fig. 4.27, *Leibniz's demonstration of living force*). When B acquires a quantity of motion 4 (i.e. after 4 seconds), its actual force would make it able to raise A more than one metre: so, Leibniz claimed that the Cartesian calculation of force as quantity of motion would make perpetual motion possible.[245]

In his letter to Leibniz of 21 November/1 December 1698 (i.e. the first one written by De Volder, in which he started to deal directly with the issue of the quantification of force), De Volder attacks Leibniz's positions as this entails a denial of the principle of the lever (in which the forces which grant the equilibrium of the two bodies have to be calculated as m·s, and not as m·s^2).[246] In his reply (27 December 1698), Leibniz uses two important concepts in order to explain the principle of the lever: the idea of *vis mortua* (dead force), which he had already introduced in the *Specimen dynamicum* (1695) and the law of continuity, also expressed by the saying *natura non facit saltus*. In the case of the lever, Leibniz maintains that the equilibrium of two bodies of different sizes is granted by the fact that both bodies are acting on each other by

[244] "Ego vero, ut ostendam quantum inter haec duo intersit, suppono, primo corpus cadens ex certa altitudine acquirere vim eousque rursus assurgendi, si directio ejus ita ferat, nec quicquam externorum impediat: exempli causa, pendulum ad altitudinem ex qua dimissum est praecise rediturum esse, nisi aeris resistentia similiaque impedimenta exigua alia nonnihil de vi ejus absorberent a quibus nos quidem nunc animum abstrahimus. Suppono item secundo, tanta vi opus esse ad elevandum corpus A unius librae usque ad altitudinem CD quatuor ulnarum, quanta opus est ad elevandum corpus B quatuor librarum, usque ad altitudinem EF unius ulna. Omnia haec a Cartesianis pariter ac caeteris Philosophis et Mathematicis nostri temporis conceduntur," A VI4C, 2028. As he would clarify in his letter to De Volder of 27 December 1698, the second premise opens Descartes's *Traité de la mécanique* (published in 1668): "[e]quidem Cartesius eam admiserat eaque erat usus in Mechanico suo Schediasmate, quod et fecerat Pascalius in tr. de aequilibrio liquorum," Leibniz to De Volder, 27 December 1698, in A II3B, 495. Cf. AT II, 432: "[l]'invention de tous ces engins n'est fondée que sur un seul principe, qui est que la même force qui peut lever un poids, par exemple, de cent livres à la hauteur de deux pieds, en peut aussi lever un de deux cents livres à la hauteur d'un pied, ou un de quatre cents à la hauteur d'un demi-pied, et ainsi des autres, si tant est qu'elle lui soit appliquée."

[245] For a thorough discussion, see Guéroult 1934; Iltis 1971; Freudenthal 2002; Rey 2009a, b; Liu 2014; Tho 2017; Garber and Tho 2018; Adomaitis 2019.

[246] "Dissimulare tamen non possum, in illo exemplo, quo Vir Ill. utitur, in quo vires quae corpus 4 lb. attollere poterant ad pedem, postmodum tollunt ad quatuor, esse quidpiam, ex quo me satis commode extricare nequeo. Verum huic oppono illud, quod ex Ampl. Leibnitzii opinione sequitur, vires corporum admodum inaequales sibi tamen invicem in statera aequipollere. Paradoxum hoc vocat Vir Ill. sed nescio qui fiat, ut hoc paradoxum mihi aeque absurdum videatur, ac illud, quod modo attuli," A II3B, 479–480.

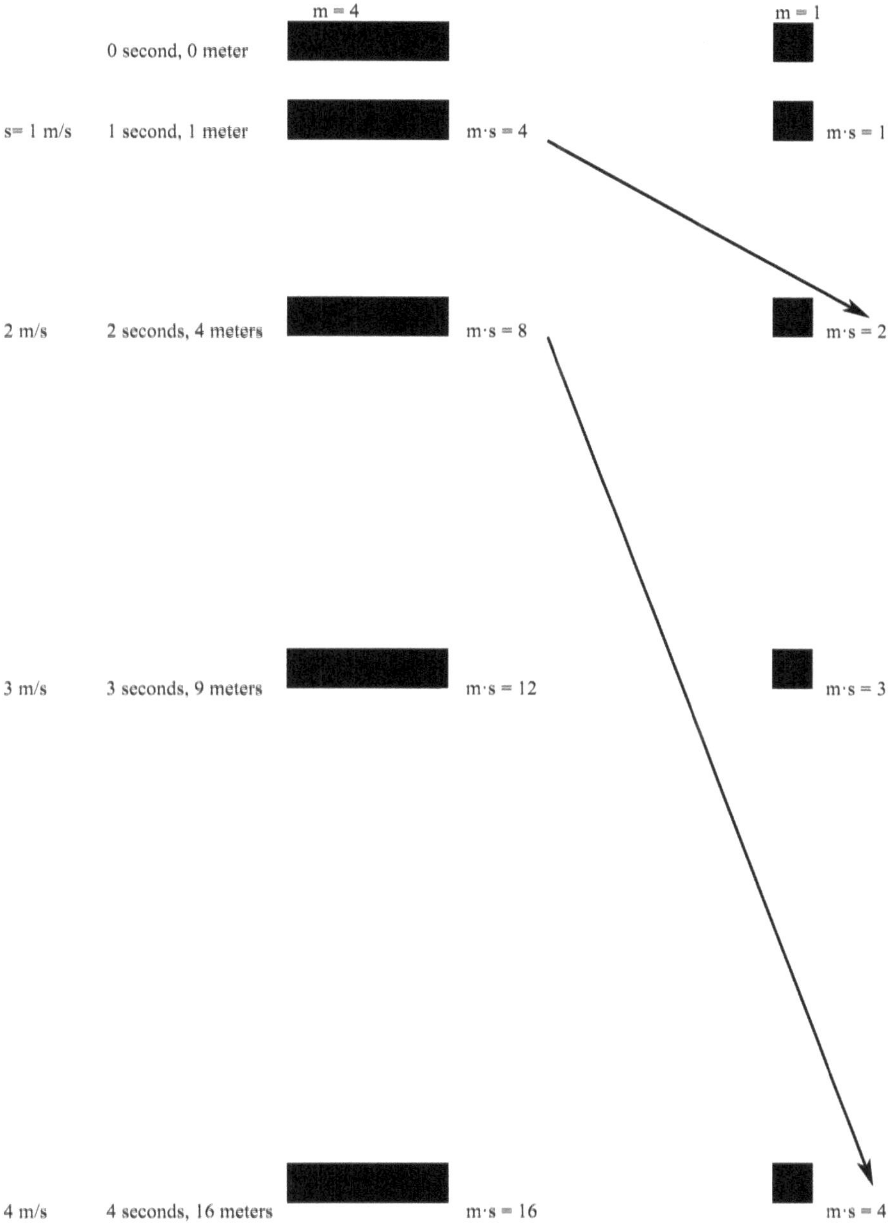

Fig. 4.27 Leibniz's demonstration of living force. (Figure by the author)

infinitely small descents, taking place in instants – namely, there is a continuous movement in both the bodies, in which however no acceleration takes place.[247] Accordingly, in these descents Galileo's law of free fall does not apply, and the descent of each body counteracts the descent of the other body, with a force quantifiable by m·s, and labelled as 'dead force'.[248] The law of continuity, thus makes possible the equilibrium of bodies on a lever (ruled by the 'law of equilibrium'), and entails that the matter of the lever is elastic, namely, that it can communicate the 'dead force' of a body to another one, and vice versa:

> since the descent of the bodies is impeded by their equilibrium, the fulcrum of the scale supports whole weight of both, and it suffers some pressure, and bending or stretching of its fibres. So, whatever force is lost by the bodies is transferred to the elastic elements of those very bodies or to the bond by which they are joined together; or if they are not elastic enough (like soft bodies) it would be not returned to the motion of the parts of the wholes (which would happen if it were elastic) but is completely lost in relation to them. [...] It is a consequence of this that bodies A and B in the present case finally bring each other to a halt [...]; this follows from the law of equilibrium through the continuous subtraction of dead force or of speeds inversely proportional to the bulks.[249]

---

[247] As Tzuchien Tho has put it, "[*v*]*is viva* describes the action of vis in extended motion while *vis mortua* only expresses momentaneous solicitation to move. In other words, the concept of *vis mortua* allows us to apply our understanding of gravitational solicitation as the "starting point" of an extended motion. [...] [J]ust as a point constitutes the extremum of a line segment, *vis mortua* constitutes the "impetus" of a motion that has not yet begun to move. In turn, any motion can be analyzed into smaller line segments at every scale. In turn each of those line segments, smaller than can be assigned, will have (at least one) extremum. In this sense, the distinction between *vis viva* and *mortua* allows for the analysis of the continuity of motion without the reciprocal idea of the composition of motion from tendencies to move (tendencies which are themselves immobile)," Tho 2017, 82. As seen in Sect. 4.1.4, *The search for physical unity and activity: De Volder vis-à-vis Leibniz*, indeed, Leibniz claimed in his *Specimen dynamicum* that movement is not 'real' because it cannot be considered as taking place in a single instant; however, dead force is exerted in instants.

[248] "Transitio enim nunquam sit, nisi per inassignabilia incrementa seu infinite parva, adeoque per vires mortuas: et aequilibrii lex nunquam nisi quoad vires mortuas exercetur, idque sive corpora gravia impetu adhuc careant, ut fit in statera, ubi utrinque tantum descendere conantur, sive impetum jam conceperint, ut in concursu: nunquam enim aliud agunt invicem, quam quod mutuo sibi detrahunt aequalem motus (sed infinite parvi seu mortui) quantitatem. Si enim corpus A sit mole ut 4 celeritate ut 1, et corpus B mole ut 1 celeritate ut 4, tunc in conflictu sive staterae cum tantum conantur, sive ictus vel concursus, cum se impetu suo urgent, continue perit ipsi A gradus celeritatis infinite parvus ut 1, ipsi vero B gradus celeritatis infinite parvus ut 4, reciprocus scilicet corporibus. Nempe ipsi A quantitas motus mortui 4 in 1. seu 4, et ipsi B quantitas motus mortui 1 in 4 seu itidem 4. Idque ex lege aequilibrii Archimedea; neque enim demonstrationes Archimedis et aliorum nisi in conatibus istis primis per se mutuo tollendis locum habent. Quemadmodum fit cum pondera 4 et 1. ita in statera posita sunt ut nequeant descendere nisi illud incipiat descendere vel ascendere celeritate ut 1. hoc ascendere vel descendere celeritate ut 4," Leibniz to De Volder, 27 December 1698, in A II3B, 497–498.

[249] Lodge 2013, 31. Cf. A II3B, 498–499: "[u]t autem cum corporum descensus impeditur aequilibrio, totum utriusque pondus sustinet fulcrum librae et pressionem aliquam flexionemque seu fibrarum tensionem accipit: ita quicquid corporibus perit virium transfertur in Elasma ipsorum corporum vel vinculi quo conjunguntur vel etiam si non satis Elastica sint (ut mollia), in motum partium toti (quod Elastrum faceret) non reddendum, sed plane ipsius respectu pereuntem. [...] Hinc consequens est ut corpora A et B in casu praesenti se tandem mutuo sistant [...]. Nam cum eandem habeant quantitatem (vivi) motus, et aequalia utrinque sint inter confligendum decrementa

Second, the law of continuity grants also the conservation of living forces ($m \cdot s^2$) in impacts, which follows the 'law of equipollence' between causes and effects. Thus the living force of a body does not come to an end, i.e. movement never ceases, but it is always communicated between bodies, without 'leaps'. The law of equipollence also, therefore, entails that all impacts are elastic.[250] Accordingly, for Leibniz the law of continuity allows for the "conciliation" of two other foremost laws of nature,[251] and postulates that in those collisions in which no rebounding takes place, movement is conserved in the particles of the bodies involved in the collision. This straightforwardly leads us to De Volder's experimental lectures at Leiden, because Leibniz, in order to illustrate his points, claims that this is what happens in the case of the cohering marbles that De Volder performed as one of his experiments at Leiden.[252] It is not clear what Leibniz's source was for De Volder's experiment with the cohering marbles. We know that Leibniz knew De Volder's experimental lectures *via* Huygens, as reported by Leibniz in a letter of 1691 to Johann Jacob Spener, where a case of the sucking up of mercury by a syphon is briefly mentioned (but not discussed).[253] As far as the case of cohering marbles is

---

quantitatis motus ex lege aequilibrii per detractionem continuam virium mortuarum seu celeritatum molibus reciprocarum, hinc simul exhauriri utriusque motum necesse est."

[250] "Interim uti reperitur in ipsis differentiis seu crementis semper servari legem aequilibrii, ita reperitur mira naturae arte, in ipsis terminis integris seu viribus vivis, computando tam id quod in corporibus residuum est, quam totum id quod est detractum et in Elasma translatum, semper eandem conservari vim vivam secundum legem aequipollentiae. Idque maxime apparet post restitutionem Elastri, cum Corpora a se invicem denuo recedentia, vim totam recepere. Tunc enim si fingas ea impetum suum sursum convertere ut si progrediendo se pendulo implicarent, reperitur semper eadem ascensus totalis quantitas, quae initio ante concursum obtineri potuisset; posito nihil accidentibus absorberi. In medio autem conflictu si corpora vim residuam sursum convertere ponas, fingendum praeterea erit Elasma corporum vel aliud elasma aequipollens vim suam in concursu acceptam totam exercere in aliquod corpus impositum, sursum ejaculandum. Ita rursus reperietur in summa eadem quae ante concursum ascensus quantitas," Leibniz to De Volder, 27 December 1698, in A II3B, 499.

[251] "Hinc jam porro deprehenditur, naturam elegantissime conciliare legem aequilibrii confligentium, quae est respectiva, cum lege aequipollentiae causarum et effectuum, quae est absoluta. Idque mediante lege transitionis gradualis saltum omnem evitantis," A II3B, 497–498.

[252] "Sed quantumcunque exiguum sit corpus tamen rationem habet ad magnum, et est aliqua vis conflictus, licet fateor exigua et corporis utriusque renitente Elastro eorum fit inflexio, per quam vis conflictus paulatim absumitur et in elastrum transfertur similiter paulatim ab eo restituenda. Caeterum Mollia corpora absorbent vim non destruendo, sed recipiendo in suis particulis exiguis, aut impendendo in abruptiones filorum ex quibus ipsa leviter connectantur; quemadmodum si globus per multa folia chartae trajiciatur. Vis autem filo rumpendo impensa iterum non perit, sed recipitur in materiam adhuc subtiliorem, prorsus ut vis quae [impenderetur] duabus Tabulis politis a se invicem divellendis, de quibus tam praeclara a Te experimenta facta accepi," A II3B, 509.

[253] "Celeberrimus Hugenius, qui Te Hanovera transiisse credebat in Epistola quadam sua inter alia et de Te quaesivit, meminitque nescio cuius experimenti a Te insinuati, quod nondum sibi Volderoque successerit Mercurium per Siphonem attrahere volentibus, credo mente tua non satis percepta," A V3, 16. Cf. the letter of Christiaan Huygens to Leibniz of 18 November 1690, claiming that the experiment, according to Spener and De Volder, could not be successful: "[j]e ne scay s'il vous a debitè une experience avec du mercure attirè par un siphon, que je ne pus croire et que j'ay aussi trouvée fausse et Mr. de Volder de mesme," Huygens 1888–1950, volume 9, 540.

concerned, Leibniz probably knew it from James Dalrymple's 1686 *Physiologia nova experimentalis*,[254] yet, if he knew this experiment, most probably he also knew De Volder's explanation of it, namely, that the two bodies cohere because they are constantly pressed by the ambient air (as reported by Dalrymple, who settled in Leiden in October 1684). Leibniz, therefore, provided an interpretation of this case of cohesion aimed at reducing De Volder's explanation to a case of elastic impact: namely, the pressure undergone by the two marbles is nothing but the force exerted by air, through the continuous movement of its particles. This kind of explanation, however, could not convince De Volder (who does not provide any comment on it), as for him – according to the statements given in his correspondence – when two non-elastic bodies collide and come to rest, they reach a condition of equilibrium because they lose their movement and force; in the case of the cohering marbles (which are, moreover, not supposed to have undergone a collision) they are in equilibrium because they are constantly pressed by other bodies. In one case, the equilibrium is reached by an annihilation of force; in the other one, by a constant exertion of force.

The main point of disagreement with De Volder that would emerge from Leibniz's letter of 27 December 1698, in fact, was that of the use of the law of continuity by Leibniz: by extension, that of the use of a metaphysical principle in physics. In his subsequent letter to Leibniz (18/28 February 1699) De Volder conceded Leibniz's explanation for the equilibrium of the lever by the *vis mortua*, yet, he found that Leibniz's explanation relied on a principle – i.e. the law of continuity – itself requiring justification.[255] In turn, in a subsequent letter to De Volder (26 June/6 July 1699) Leibniz claimed that for him the ideas of good, perfection and order (grounding the law of continuity) are known through reasons as evident as those of number and shape. So, for him the fact that in the world living force is conserved is a necessary consequence of the wisdom of God, which has ruled the choice of this world amongst various possible ones. In turn, this postulates that all bodies are elastic: namely, that once the wisdom of God is granted, elasticity becomes the necessary means (albeit a means which is necessary according to a hypothetical necessity)[256] by which the rules of divine wisdom are respected in nature. So, Leibniz claims in his letter that even if other worlds are possible, the creation of this world (which conforms to rational-theological principles) entails its creation by only a possible

---

[254] De Volder's experimental activities are mentioned in a letter of Bernoulli to Leibniz of 8/18 January 1698, but no detail is provided: "[o]rdines nostri novam mihi imposuerunt docendi provinciam, atque in eum finem certam decreverunt summam ad emenda instrumenta experimentalia, ut exemplo Volderi Lugdunensis Studiosos nostros etiam experimentis Mathematico-physicis exerceam et delectem," A VIIB, 702. As to Dalrymple, see *supra*, n. 69.

[255] "Quod ad difficultatem meam, de staterae, aut etiam, ut recte addis, corporum reciproca ad magnitudinem celeritate in se invicem incurrentium aequilibrio attinet, eam (si quidem constiterit, aequi[li]brii legem nunquam nisi quoad vires mortuas exerceri, aut potius, si quidem demonstratum sit axioma tuum, nullam transitionem fieri per saltum, (ex hoc enim prius illud sequi video,) solutam esse agnosco," De Volder to Leibniz, 18/28 February 1699, in A III8, 527–528.

[256] As to the idea of hypothetical necessity in Leibniz's philosophy, see Mugnai 1990; Adams 1994.

set of 'instruments', such as the property of elasticity of every body.[257] For instance, for Leibniz the existence of atoms is possible (i.e. non contradictory): however, as atoms are perfectly hard bodies deprived of any elasticity, their existence would contradict the law of the conservation of force, granted by "the order of things and metaphysical principles."[258]

For De Volder, all this was much too 'metaphysical': namely, the (physical) properties of bodies are unwarrantedly derived from rational-theological principles, just as the Cartesians – such as Malebranche – did.[259] Indeed, for De Volder – who followed Descartes's explanation of elasticity in his *dictata*, and also in the *pro gradu* disputation of Schuyl (as I am going to show) – elasticity was not an essential property of matter; rather it was the effect of a complex 'mechanism', i.e. the result of the flowing of subtle matter in the pores of visible bodies, which, once bent, are forced to re-acquire their previous shape by such flowing.[260] So, for De Volder no demonstration of the elasticity of all bodies could be provided so far. For Leibniz, too, elasticity was by no means an 'essential' property of matter: in his letter to De Volder of 27 December 1698, indeed, he claimed that elasticity, even if granted by metaphysical principles, consisted in the action of subtle matter on hard bodies, as it was for Descartes.[261] In 'Leibnitian' terms, elasticity, as a means of the divine order

---

[257] "Idem praecise praesens systema per omnia et in omnibus non nisi uno modo produci posse arbitror, nec causas ejus diversas possibiles agnosco. Electio igitur mea sententia locum habet inter diversa systemata possibilia, non inter diversos ejusdem pleni systematis producendi modos. Bonum, perfectionem, ordinem putem non minus liquidis rationibus constare quam numeros et figuras. Finge triangulum produci debere, nullam autem esse rationem aliam qua species trianguli determinetur, haud dubie aequilaterum producetur," A II3B, 570.

[258] "Quod autem corpora inflexibiliter dura attinet, quales Atomos fingimus, Tecum plane sentio, in illis nullam esse posse reflexionem, neque etiam virium conservationem, sed vicissim nulla talia in natura esse arbitror, quod non aegre credo admittes. Elasma ego corporibus essentiale puto ex rerum ordine et metaphysicis principiis," Leibniz to De Volder, 27 December 1698, in A II3B, 509. Cf. also Leibniz to Bernoulli, 20/30 September 1698, in A III7B, 912–913: "[h]aec autem evitatio saltus in mutationibus corporum obtinetur per vim elasticam ipsis inexistentem. Ita enim fit ut corpora in concursu sese comprimentia et mox restituentia paulatim sibi cedant et graduali translatione directiones, viresque et ipsas ut demonstratum vidisti actionum motricium quantitates (longe a vulgo intellecta Quantitate motus diversas) conservent."

[259] See the letter of De Volder to Leibniz of 18/28 February 1698, referring to Leibniz's idea of a 'principle of order' (expressed in the letter of Leibniz to De Volder of 24 March/3 April 1699, as grounding the law of continuity: "[a]xioma quo utor: nullam transitionem fieri per saltum. Id fluere arbitror ex ordinis lege," A II3B, 545): "[h]aec obstant, quo illa argumenta, quae ex ordine Malbranchius imprimis, desumit, minus me afficiant, ut et illa, quae ex beneplacito Dei deducuntur, hoc est, ut mihi videtur, ex principio obscurissimo, et quod saepenumero nullum habet fundamentum praeter ignorantiam nostram," A II3B, 560.

[260] See Sect. 4.2.3.4, *Collision in De Volder's late thought.*

[261] "Elasma ego corporibus essentiale puto ex rerum ordine et metaphysicis principiis: etsi in natura non aliter quam per fluidum intercurrens pergatur. In quo plane assentior Cartesio," Leibniz to De Volder, 27 December 1698, in A II3B, 509. Cf. the quotation given *supra*, n. 258. The same position was also given in the letter of Leibniz to Bernoulli of 12/22 July 1698, commenting upon the first letter of the De Volder-Bernoulli-Leibniz correspondence (5/15 July 1698), where Bernoulli reported De Volder's idea that when two hard bodies collide, they lose all their movement (see *supra*, n. 230). In his letter, Leibniz points out that De Volder was following the same kind of rea-

of nature, was a necessary means, albeit a 'hypothetically necessary' one. Indeed, in his letter to De Volder of 24 March/3 April 1699, Leibniz criticised De Volder for reverting, by asking for an a priori demonstration of the activity of matter (which for De Volder would have been a question pertaining to natural philosophy rather than to metaphysics – understood as rational theology), to a metaphysical demonstration. For Leibniz, this meant imposing an absolute necessity on natural order:

> [a]s for the claim that this measure of force can be applied to all bodies, [...] [a]ccording to my hypothesis, to the extent that bodies are not perfectly elastic, the force is taken up by their internal parts which are elastic themselves, and so it does not perish but just becomes insensible. You will not deny that these things are consistent with what is natural and orderly [...]. Without elasticity one could not obtain the axioms of avoiding leaps and of the conservation of forces, [...] [n]or could one obtain the reconciliation of the laws of dead and living force [...]. And indeed these things cannot be demonstrated except from the supreme law of order, for they are not absolutely necessary with contraries that imply a contradiction. The universe could have been constituted in innumerable ways, but the one which was supported by the greatest reason prevailed. However, the activity of substance is more of a metaphysical necessary, and unless I am mistaken, it would have had a place in every universe.[262]

soning (neglecting elasticity) that he, Leibniz, adopted in his *Theoria motus abstracti* (1670–1671) – in which Leibniz came to the same conclusion as De Volder. This idea of collision in which forces are lost, for Leibniz, is derived from the consideration of the extension and impenetrability of matter alone, without considering elasticity, which he anyway conceives not as an internal quality of matter, but rather as a specific structure of it (*à la* Descartes): "[e]go ipse olim adolescens cum de legibus motus scriberem libellum, in ea eram sententia quae nunc est Domini Volderi, duo corpora aequalia et aequivelocia sua natura post concursum directum non debere reflecti, sed potius se sistere mutuo. Idque etiam sequitur ex vulgari notione materiae cum nempe nihil aliud in ea concipitur quam extensio et ἀντιτυπία seu impenetrabilitas. Sed ex his ipsis et similibus postea agnovi longe aliam esse naturam materiae in systema mundi redactae quam vulgo creditur; et vim Elasticam omni corpori esse essentialem, non ita quasi ea vis sit aliqua qualitas inexplicabilis, sed ex eo quod omne corpus utcunque parvum est machina ex cujus structura resilitionem, ubi opus ea est ad virium conservationem, oriri oportet," Leibniz to Bernoulli, 12/22 July 1698, in A III7B, 827. Cf. Leibniz's *Theoria motus abstracti*: "[t]heoremata. 1. Si corpus impingit in aliud quiescens, vel tardius directe occurrens vel tardius antecedens, secum abripit (id est movet in eandem plagam) differentia celeritatum. [...] 12. Si non detur angulus, qui sit bisectilis (non datur autem angulus omnino in occursu recto, non datur angulus bisectilis in alio impactu, si impingitur linea motus recta et curva, vel curva et curva figurae dissimilis aut inequalis) et impactus aequivelox est, utrumque quiescet," GM VI, 72–73.

[262] Lodge 2013, 73. Cf. the original text: "[c]aeterum quod ista virium aestimatio omnibus corporibus applicari possit, nunc quidem confirmo, satisfaciendo difficultatibus quae contra opponi possunt. Nempe hypothesi mea, quatenus corpora perfecte elastica non sunt, vis intestinis partibus quae et ipsae Elasticae sunt, recipitur, neque adeo perit, sed tantum sensibus subducitur, quae quidem consuetudini naturae et ordini id est experientiae et rationi consentanea esse non negabis. Nec sine Elasmate axiomata aut vitandorum saltuum, aut conservandarum virium tam absolutarum quam respectivarum vel conciliationes legum vis mortuae et vivae, compositionisque motuum cum quantitate virium obtineri possent. Et vero ista non nisi ex lege ordinis supremi demonstrari possunt, neque enim sunt absolutae necessitatis, ut contrarium implicet contradictionem. Innumerabilibus modis poterat constitui systema rerum, sed praevaluit, quod majore ratione nitebatur. Substantiae tamen activitas magis est metaphysicae necessitatis, et locum ni fallor habitura erat in systemate quocunque," A II3B, 545–546. In his letter to Leibniz of 18/28 February 1699, De Volder replied that, if one supposes a law of order, one cannot drawn but absolutely necessary

The kind of demonstration that De Volder was asking for, therefore, would have undermined the broader philosophical project of Leibniz (which De Volder could not know in the late 1690s, when Leibniz had published only communications in dynamics), aimed – as shown in his later *Essais de Théodicée* (1710) and *Monadologie* (1714, posthumously published in 1720) – at a reconciliation of the freedom of God and man, and the order of natural phenomena.[263] For De Volder, as suggested above, this kind of demonstration would not have hindered the freedom of God, because it was by no means a metaphysical demonstration. Namely, a demonstration concerning the real existence of the property which is demonstrated as following from the idea of material substance. In his *De rerum naturalium principiis*, as shown in Chap. 3, De Volder maintained that the principles of physics which he chose – i.e. those of 'mechanical philosophy' are hypothetical, namely, God could have used different principles (such as those of the Aristotelians, or of the alchemists) to create this world. In fact, he was simply not concerned with the problem of Creation and of the principles guiding God in His establishing the real natural causes from which phenomena flow. All these problems, for him, fell beyond the scope of natural philosophy itself.

Not surprisingly, over the entire course of their correspondence Leibniz could not provide the kind of demonstration De Volder was asking for, nor he could 'convert' De Volder to his own views on substance (which to De Volder would seem to be nothing but Aristotelian), as seen in Sect. 4.1.4, *The search for physical unity and activity: De Volder vis-à-vis Leibniz*. The foremost 'result' that Leibniz and Bernoulli achieved, indeed, was only that of convincing De Volder that the measure of motive force is given by $m \cdot s^2$, after which Leibniz agreed to discuss the problem of substance with De Volder. Yet, De Volder was not convinced by Leibniz's arguments, but only by an argument proposed by Bernoulli in his own correspondence with De Volder (and given also in Christian Wolff's *Elementa matheseos universae* (1713)). According to this, a body with speed 2 progressively impacts with four springs, which can absorb from it 1 unit of speed each. As the motion of the body can be decomposed in various determinations, the springs are disposed in such a way that they absorb only the quantity of motion related to a given determination (see Fig. 4.28, Wolff 1741–1756, volume 2, plate 4, figure 43). Such determinations are determined by applying Pythagoras's theorem: before impacting the first spring, for instance, the body will have a speed 2 (namely the length of its hypotenuse is 2), decomposed into a speed of 1 and $\sqrt{3}$ (which are the lengths of the catheti). After the first impact, the will have a speed $\sqrt{3}$ (which is the hypotenuse), divided into determinations (or catheti) of 1 and $\sqrt{2}$ – and so on until the body loses all its motion by impacting with the last spring.[264] Leibniz, who received a report of the

conclusions: "[p]rimum equidem non capio, qua ex lege ordinis determinare possimus ea, quae absolutae necessitatis non sunt," A II3B, 560.

[263] See Adams 1994; Mercer 2001; Antognazza 2008; Garber 2009.

[264] See the letter of Bernoulli to De Volder of 13 February 1700, published also in A III8, 413–414. Cf. Wolff 1713, volume 1, 594–595. See also the letters of Bernoulli to De Volder of 3 January 1699, 8 April 1699, and 31 May 1699. The argument was re-formulated and presented by De Volder

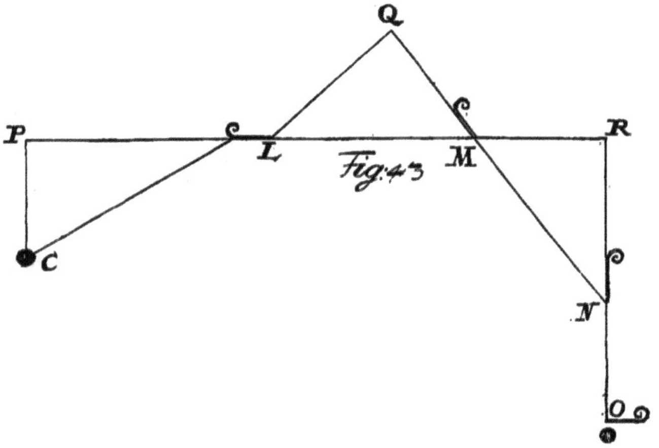

Fig. 4.28 Johann Bernoulli's demonstration of living force. (Source: Wolff 1741–1756, volume 2, plate 4, figure 43. ETH-Bibliothek Zürich, signature: Rar 5507, https://www.e-rara.ch/doi/10.3931/e-rara-10422. Public domain mark 1.0)

demonstration from De Volder himself,[265] found it not completely satisfactorily, as it made recourse to "special assumptions about such things as weight, elasticity, and oblique collision."[266] Yet, through this De Volder would finally agree with Leibniz on the measure of the motive force, and they could move to the discussion of the activity of substance.

4.2.3.4 Collision in De Volder's Late Thought

The correspondence between Leibniz, Bernoulli and De Volder sets the framework for De Volder's presentation to Leibniz, in his letter of 18/28 February 1699, of his rules of collision. The same rules had been presented also in the 1688 *pro gradu* disputation of Hermann Schuyl: accordingly, the two documents have to be analysed together.

to Leibniz in his letter of 5 April 1700 (A II3B, 630–633). Bernoulli was later to discuss Huygens's laws of collision, as formulated in his posthumous treatise, in his letters to De Volder of 10 July 1703, 11 December 1703, 5 January 1704, 7 January 1704. The demonstration is also mentioned in the letters of Bernoulli to Jakob Hermann of 21 July 1718, to Wolff of 27 March 1722, to Willem Jacob 's Gravesande of 31 October 1722 (see Sect. 2.4.2, *De Volder's retirement, death and legacy*) – also printed in the article *'s Gravesande* in the *Dictionnaire historique* of Prosper Marchand (see Marchand 1758–1759, volume 1, 234) –, to Georg Bernhard Bilfinger of 30 September 1722.

[265] See De Volder to Leibniz, Leiden, 5 April 1700, in A II3B, 630–633.

[266] Lodge 2013, 179–180; cf. Leibniz to De Volder, 6 September 1700: "[i]nterim haec quae veritatem optime comprobant, et contraria ad absurdum redigunt, cum 5 tamen specialibus assumendis indigeant velut gravitate, elastro, concursu obliquo, non satis aperiunt fontes causasque, quod sola facere potest mea actionis aestimatio; quam Tibi jam arridere non dubito, cum illud pulcherrimum appareat dum eadem servatur quantitas virium, etiam eandem aequalibus temporibus quantitatem Actionis motricis in universo conservari: quod Cartesius velut per nebulam cum transvidisset expetissetque, non bene est assecutus, celeritate pro viribus substituta," A II3B, 655.

In these texts, De Volder distinguishes between elastic and non-elastic hard bodies, and soft bodies (which are non-elastic). In Huygens's 1669 communication, the bodies involved in the impact are considered as hard (namely, elastic hard bodies, while in his 1667 fragment he considers soft ones, as seen above). In turn, in his 1668 communication Wren did not mention the properties of the bodies involved in elastic impacts,[267] while Wallis (in 1668) considered non-elastic absolutely hard bodies, which are not deformed by impact: elastic bodies, indeed, are those which are deformed by impact, and that then are able to restore themselves (exerting a force of rebounding).[268] Mariotte, in his *Traité de la percussion ou choc des corps, dans lequel les principales règles du mouvement, contraires à celles que Mr. Descartes sont démonstrées par leurs véritables causes* (1673) – whose results conform with those of Huygens, Wren and Wallis – provided an extensive experimental treatment of impact by considering two kinds of bodies: elastic flexible bodies (i.e. bodies which rebound), and non-elastic flexible ones.[269] With the exception of Wren (who simply did not consider cases other than elastic impacts), all of them acknowledged that different properties of bodies – i.e. elasticity, softness and hardness – can give rise to different outcomes of collisions. Descartes, as seen above, considered elastic collisions (i.e. collisions in which bodies rebound, at least in some cases) involving perfectly hard bodies (in part 2 of his *Principia*) while in the fourth part he considered the features of elastic bodies, namely, those bodies provided with a resilience once bent, such as bows and glass. In IV.132, whose titles is "[w]hy it [viz. glass] is rigid like a bow; and, in general, why rigid things, when they have been bent, spontaneously return to their former figure," Descartes explains this property as follows:

> [...] the property of springing back [...] generally exists in all hard bodies whose particles are joined together by immediate contact rather than by the entwining of tiny branches. For, since they have innumerable pores through which some matter is constantly being moved [...], and since the shapes of these pores are suited to offering free passage to this matter [...], such bodies cannot be bent without the shapes of these pores being somewhat altered. As a result, the particles of matter accustomed to passing through these pores find there paths less convenient than usual and push vigorously against the walls of these pores in order to restore them to their former figure.[270]

Such features are not ascribed by Descartes to the bodies involved in collisions. In commenting upon this article in his *dictata*, De Volder does not provide an alternative account for elasticity, yet, in the Hamburg 273 series, before presenting

---

[267] See Wren 1668, 867; Huygens 1669a, 927.

[268] "Si quae sic impingunt corpora, intelligantur non absolute dura (prout hactenus supposuimus) sed ita ictui cedentia, ut elastica tamen vi se valeant restituere, hinc fieri poterit ut a se mutuo resiliant ea corpora [...]," Wallis 1668, 866. I will deal with Wallis's treatment of collision in his *Mechanica* (1670–1671) below.

[269] Mariotte 1673, 1–2. Wren also, according to his 1668 communication, carried out several experiments on collisions: Wren 1668, 857. Huygens and Wallis do not mention any experimental confirmation or development of their theories. De Volder probably met Mariotte during his visit to Paris in 1681; he owned Mariotte's *Traité*: see *Bibliotheca Volderina*, 10. As to De Volder's experiments with collisions, see Sect. 4.2.3.4.2, *De Volder on elastic impacts*.

[270] Descartes 1982, 242.

Descartes's explanation, he states that "Q[uaestion]: from where [does] that force of resilience [come], as it seems [that] it opposes the rule of nature, according to which all bodies strive to remain in the state in which they are? I answer: the cause is to be searched in air, which produces this not only in glass, but in all [things] that it touches immediately."[271] We may interpret these words in two ways: first, De Volder might be referring to the fact that it contradicts the first law of nature that a body, once bent, changes its shape without external causes; accordingly, we need to suppose that air provokes this change. However, it is not clear why De Volder did not just make recourse, like Descartes, to the action of subtle matter. Second, we may hypothesize that De Volder was considering the case of elastic impacts devised by Huygens. Indeed, in the 1688 disputation De Volder (*via* Schuyl) grounds his own laws of elastic impacts (whose conformity with those of Huygens is stressed at the end of the text) on the elastic properties of bodies, explained in a perfect Cartesian manner.[272] Huygens's elastic impacts contradict Descartes's first law of nature, because they contradict Descartes's fourth rule of impact. According to Huygens's third rule (1669), a little body can always move a bigger one, which in the Cartesian model would however always exert a stronger resistance to it. Therefore, in order to 'save' Descartes's first law – which is contradicted by the fact that the bigger body does not keep its state of rest, notwithstanding its greater force of resistance – a recourse to an additional mover like air is required.[273] As seen above, according to De Volder's *dictata* Huygens's laws work only in fluids: so, the elastic impacts

[271] "Ductum est. {Q[uaero]}: unde resiliendi illa vis, videtur {e[nim]} id repugnare regulae naturae, qua o[mn]ia c[or]pora conantur manere in statu in quo sunt? {R[espondeo]} causam q[uae]rendam esse in aere, {q[uae]} haec non solum in vitro, sed o[mn]ibus {q[uae]} se immediate conting[un]t, efficit. Poss[un]t. Id vult, utamur arcu laxato, dum eius partes cohaerent immediate, multi in iis sunt pori materiae {i. expervi}, ubi itaq[ue] inflectitur arcus, ita tamen ut non frangatur, {q[uo]d} fit? Pori mutantur, et una diameter fit altera maior, accedat itaq[ue] mat[eria] subtilis, haec per maiorem diametrum facile ferri potens impingit in minorem, donec cessante externa vi excluduntur latera, et arcus ad priorem statum reducitur. Aptantur. Si vero lignum sit nimis durum, {par[ticu] lae advellentur, q[uae]} transitum impediebant, sicq[ue] continuatur tensio," Hamburg 273, 315. See also the text of the other series of *dictata*: "[e]x iis, q[uae] hoc articulo dicuntur, causae pendent, propter quas omnia corpora, q[uae] flecti nequeunt, resiliant, hoc enim ea fit de causa, quia corpora ea rigida non possunt omnibus ita disponi modis, ut liberum transitum praebeant materiae caelesti per ea fluenti, sed cum eorum {pori a} materia ea formati, eiusq[ue] magnitudini ac motui accomodati sint, si imminuentur, aliquod materiae motui impedimentum possunt praebere, cum autem constent {particulis} rigidis, non facile earum figura mutari potest, hinc in eas agere caelestis materia omnes suas intendit vires, ut eas in pristinum statum reducat, inde vis illa resiliendi dependet, sensim vis illa perit, quia ea materia, quum non possit totas in pristinum statum reducere, tracti temporis hoc agit, ut quantum singulis paulum immutandum tales poros efformet, per quos sine difficultate fluere queat," Hamburg 274, 113–114.

[272] "Idemque est ratiocinium, in quacunque durorum inflexione, aut expansione, nam licet non semper pori angustentur, figura saltem ipsorum mutatur, ita ut v. gr. quae antea erat rotunda, expansione ista, fiat elliptica, quare, cum in generatione pori ipsorum, figurae materiae caelestis sint accommodati, hac mutatione in suo per eosdem transfluxu, impeditur," Schuyl 1688a, chapter 3, 9 (unnumbered). As to the references to Huygens, see *infra*, n. 295.

[273] In fact, Huygens adopted Descartes's first law of nature as part of his first hypothesis in his *De motu corporum ex percussione*: De Volder, however, had no access to this text until the death of Huygens.

devised by Huygens are 'aided' by the action of the surrounding air.[274] On the other hand, Descartes's rules, which describe elastic impacts as well (namely cases in which bodies rebound), but which involve non-elastic bodies (i.e. perfectly hard ones) are valid in 'ideal' cases. In such ideal cases, the rebounding of bodies is not justified by their internal elasticity (i.e. as theorized in IV.132), but in the light of Descartes's first and third laws of nature. The problem, of course, was that in order to accept these laws without recurring to the idea of the elasticity of the colliding bodies, one needs to revert to the action of God in preserving movement, as no physical means can justify such rebounding and such conservation.

In sum, in De Volder's account of impact given in his *dictata* there are three ideas of elasticity at work: (1) the 'physical' elasticity of bodies explained in IV.132 (i.e. the bow-like structure of certain bodies); (2) the 'occasionalist' elasticity entailed by Descartes's rules (i.e. the elastic effect of collisions between bodies deprived of 'physical' elasticity); (3) the elasticity underlying Huygens's rules, which is the 'physical' elasticity itself (as overtly maintained in the 1688 disputation), but whose effects are 'aided' by the action of ambient air. It is not entirely clear whether De Volder was really convinced of the consistency of this model. Indeed, in his correspondence with Leibniz he clearly stated that the quantity of motion, which he still regarded as the measure of the force of motion, is not conserved in impacts (both elastic and non-elastic), while $m \cdot s^2$ is conserved in elastic impacts. In non-elastic impacts, motion and force are just lost. Moreover, in presenting his own rules in the correspondence De Volder never mentions the action of ambient fluids, nor distinguishes between ideal and actual situations. It may be that his attempt to establish a concordance between Descartes and Huygens on impact had mostly a didactic function (i.e. it served to provide students with an introduction to natural-philosophical theories), and that he reserved a more advanced discussion of these topics to his correspondents, and, to some extent, to the students who graduated under him (like Vander Codde and Schuyl). Also, in his *dictata* De Volder does not deal with the problem of divine intervention entailed by Descartes's cases of 'elastic impacts between non-elastic bodies'. As a matter of fact, De Volder's struggle to obtain from Leibniz a demonstration of the activity of matter – i.e. a demonstration of the conservation of motion or force – resulted from the problem of making the conservation of motion or force consistent with the non-elastic nature of all bodies. That is, from his attempt to reduce all collisions to elastic ones, notwithstanding the fact that not all bodies are 'physically' elastic, and so to save Descartes's rules without reverting to divine action. When he could not obtain or find a solution to this problem, he quitted academic teaching.[275]

---

[274] As I am going to show, in his letter to Leibniz of 18/28 February 1699 De Volder stated that Huygens's rules are valid only for elastic bodies: see Sect. 4.2.3.4.1, *De Volder on non-elastic impacts*.

[275] De Volder could not find a solution to this problem, as far as collision is concerned, in the other treatise that he used in his lectures, namely Rohault's *Traité de physique*. Most probably, he taught on this in the late 1690s, as his *dictata* on it (which have so far not been found) are dated 1698–1699. In chapter 11 of the first part of his treatise, Rohault avails himself of Descartes's idea of conservation of $m \cdot s$, and considers only Descartes's rules 5 and 2 or 7a/2 (i.e. when one of the two bodies is bigger and faster than the other, regardless of the direction of movement), which are avowed: see Rohault 1671, volume 1, 65–66; (in Rohault and Le Grand 1691, 47–48, this section

As far as De Volder's rules of impact are concerned, they can be summarized as follows.

4.2.3.4.1 De Volder on Non-elastic Impacts

First of all, as he declares in his letter to Leibniz of 18/28 February 1699, De Volder "deduced" the same rules for elastic bodies presented by Huygens "without demonstration" in 1669.[276] He did so by using two hypotheses which he thought were consistent with those used by Huygens himself. One of these hypotheses assumed that the bodies involved in the collision are elastic.[277] However, once he could look at Huygens's manuscripts after his death (1695), he found that Huygens, in his *De motu corporum ex percussione* – which was still unpublished in 1699 – did not overtly rely on any hypothesis on the elasticity of bodies, which, however, had to be "tacitly" supposed, according to De Volder, in order to make his hypothesis 2 (i.e. a restatement of Descartes's rule 1) acceptable.[278] As seen above, indeed, in his *De motu* Huygens was uncommitted to the cause of the rebounding, as he opened hypothesis 2 with the formula "[w]hatever is the cause of the rebound of hard bodies from mutual contact [...]."[279] It seems, in fact, that De Volder criticized Huygens for the lack of a foundation for the rules of impact on an adequate physical theory (such as the mechanical explanation of elasticity), and, more importantly, the unwarranted extension of the applicability of his rules to the impact between any body, i.e. the unwarranted attribution of elasticity to any body, *à la* Leibniz.[280] On the other hand, De Volder deduced

---

is not commented on by Antoine Le Grand. Samuel Clarke comments on it only in the 1710 edition of this work, proposing, in order to correct it, algebraic rules which perfectly fit with those of Wallis and De Volder: see Rohault and Clarke 1710, 45–48). Please note that the two cases considered by Rohault do not involve any rebounding, rather they are elastic as far as force ($m \cdot s$) is conserved. In chapter 13, as he considers reflection, Rohault maintains that when one body cannot move another, it is necessarily rebounded by it, regardless of its features (i.e. softness or shape): see Rohault 1671, volume 1, 103; in his commentary, Clarke notes that this is not true, as the colliding body will stop: see Rohault and Clarke 1710, 74. Rohault does not reserve any separate treatment to elasticity understood as the capacity of bodies to restore their shape.

276 "Memini me ante multos annos, cum inquirerem in Leges motus Hugenianas, quas absque demonstratione edidit in Diariis, easdem deduxisse in corporibus elasticis ex duplici hac hypothesi [...]," De Volder to Leibniz, 18/28 February 1699, in A II3B, 528.

277 See *infra*, n. 295.

278 "Ex qua conclusione [i.e. De Volder's own finding that $m \cdot s^2$ is conserved], cum sequi animadverterem regulas motus Hugenianas, neutiquam dubitabam, quin meae hae hypotheses cum Hugenianis consentirent, in quo tamen me falsum esse deprehendi, postquam post obitum Illustris Viri, libellum ejus de motu videre licuit. Illic enim nulla elasmatis facta mentione omnia sua deducit ex hac Hypothesi, duo corpora dura aequalia, et aequaliter mota, directe in se invicem impingentia repercuti eadem utrinque velocitatis aequalitate servata. Nec mirum ex ea hypothesi sequi leges elasticorum corporum, cum ea vera non foret, nisi ipsam elastri naturam tacite involveret," De Volder to Leibniz, 18/28 February 1699, in A II3B, 529.

279 See *supra*, n. 209. *Ça va sans dire*, this kind of criticism could also be brought against Descartes, as his rule 1 presupposes, for De Volder, the elasticity of the bodies involved.

280 Please note that Huygens, as seen above (see Sect. 4.2.2, *Huygens's laws of collision*) devoted to non-elastic impacts only a very short handwritten note (1667).

the laws of Huygens by distinguishing elastic and non-elastic bodies, and – as he declares – he never doubted that Huygens had used premises which do not fit his own.[281] This interpretation of De Volder's comments on Huygens's *De motu* can be confirmed by Leibniz's subsequent answer. Having known from De Volder that Huygens tacitly used the idea of elasticity in his rules, Leibniz asked him if Huygens's treatise could published along with his "dynamics" (probably, the second part of Leibniz's *Specimen dynamicum*, which was published only posthumously).[282]

The first hypothesis by which De Volder deduced the rules of impact of Huygens is presented in his letter to Leibniz (18/28 February 1699), as a two-fold hypothesis:

1a) the quantities of motion – i.e. the forces of motion – of two equal bodies moving at equal speeds against each other are reciprocally destroyed during impact;
1b) the quantity of excess motion in one of the two bodies is conserved and is distributed in the two bodies, which then move together as one body.

From this twofold hypothesis, which concerns non-elastic bodies, De Volder deduced the law of collision of soft bodies.[283] These, in fact, are the rules for all non-elastic bodies, which form one body after impact, whose speed is calculated by the formula $(m_1 \cdot s_1 \pm m_2 \cdot s_2)/(m_1 + m_2)$ (according to the direction of the bodies before the impact: if they follow the same direction, an addition is to be made; if they move against each other, one needs to subtract forces, as these destroy each other).[284]

In the 1688 disputation, the same algebraic rule is justified on the basis of a more developed natural-philosophical account, even if in a not entirely consistent way. The disputation is based on a series of definitions, axioms, and postulates, and is divided into three chapters: in the first one the author deals with the problem of gravity; in the second, with the elasticity of air (to which I will return in Sect. 5.6.3.3, *De Volder's assumption of Boyle's law in the works of his students*). In the third chapter, he deals with the elasticity of hard bodies (as seen above),[285] while in the fourth with the rules of collision. As axiom 4, the author states that "a body, which is [...] equally pressed [from each side], or which has, towards opposite

[281] See *supra*, n. 278.

[282] "[...] miratus saepe sum Hugenium nostrum aliunde ista quam ex Elastro deducere sperasse, quod tamen ut recte mones non potuit quin tacite supponeret. Libellum ejus de motu fortasse aliquando, si vobis ita videbitur, quibus scripta sua testamento credidit, possem addere dynamicis meis, et tam praeclaro emblemate exornare opus meum, favente connexione rerum, et similitudine argumenti," Leibniz to De Volder, 24 March/3 April 1699, in A II3B, 545.

[283] "Quantitatem motus binorum corporum, (quam ego vires vocabam) quae esset versus contrarias partes, se invicem destruere (quod ex aequalitate virium contrariarum planum existimabam) reliquam vero quantitatem, ut et quae esset versus easdem partes, conservari, et in utrumque corpus, pro ratione magnitudinis corporum distribui. Ex qua hypothesi Leges motus in corporibus mollibus deducebam," De Volder to Leibniz, 18/28 February 1699, in A II3B, 528.

[284] "Quibus positis ita ratiocinabar. Sit corpus a velocitate c, corpus b velocitate f, quae in se invicem impingant sive in contrarias partes, sive in easdem ferantur. Si corpora non forent elastica progrederentur ambo post occursum velocitate (ac ± bf)/(a + b) ab A versus B, siquidem pono in casu determinationis versus contrarias partes, ac majorem esse quam bf," De Volder to Leibniz, 18/28 February 1699, in A II3B, 529.

[285] See Sect. 4.2.3.4, *Collision in De Volder's late thought.*

directions, an equal inclination [sc. *conatus*] to move, is necessarily at rest, however it can be moved towards any side by a very little force" (*contra* Descartes's rule 4). In definition 2 he clarified the meaning of 'conatus' as "that disposition to movement in bodies, by which, even if they are at rest, they actually start to move, if just the obstacles are removed." In turn, as axiom 1 he states that "nothing is moved by itself."[286] These principles are used to explain the outcome of the collision of non-elastic bodies in chapter 3, in which movement can be (apparently) completely lost, in the case that they move as in Descartes's rule 1.[287] Yet, according to these principles the conservation of movement is granted (at least in principle), for the reason that the cases of collision are considered as taking place in fluids. The *conatus* that resting bodies have, indeed, consists (for axiom 1) of the pressure of ambient fluid. When two non-elastic bodies collide and come to rest, they are just equally pressed by fluid. From this perspective, Leibniz's reading of De Volder's experiment with the cohering marbles is justified. Yet, in the disputation the author also states, as he analyses more systematically this case in chapter 4, that one of the bodies "destroys the forces" in the other.[288] In short, in the disputation we find a sort of tension between the teachings of De Volder (i.e. the 'concordist' strategy one finds in the *dictata*), and his much more puzzled approach to the problem of conservation of motion which he showed in his correspondence with Leibniz.[289]

---

[286] "Definitiones. [...] II. Per conatum ad motum talem in corporibus intelligo dispositionem, qua, licet quiescant, actu tamen modo removeatur obstacula, moveri incipiunt. [...] Axiomata. I. Nihil movetur a se ipso. [...] IV. Corpus, quod ex Diametro aequaliter premitur vel aequalem, versus oppositas ex Diametro partes, ad motum conatum habet, necessario quiescit, minima tamen vi versus utramvis partem moveri potest," Schuyl 1688a, 1–2 (unnumbered).

[287] "Ponamus enim duo corpora, A et B, perfecte dura, hoc est, nullis poris, ac proinde, nec vi elastica praedita, aequali vi ab opposita parte in se invicem impingere, dico, quod in puncto concursus ambo debeant quiescere. Utrumque enim hoc in puncto aequalem versus oppositam ex diametro partem, progrediendi conatum acquirit, in quantum enim A propria vi motus versus orientem v. gr. in tantum praecise a B determinatur versus occidentem. Idemque verum est de B, ergo [...] ambo in puncto concursus quiescere debent. Q.E.D.," Schuyl 1688a, chapter 3, 9 (unnumbered).

[288] In chapter 4, indeed, the author considers all three possible cases of impact (1. when they both move against each other; 2. when they move in the same direction; 3. when one is at rest). As to the first case, he states: "[s]it autem ac [namely, $m_1 \cdot s_1$] = 3 | 2 = bd [namely, $m_2 \cdot s_2$]: certe, cum B destruat in A vires, viribus suis aequales, [...] non poterit A post concursum cum corpore B maiori vi progredi, et B secum abripere, quam quam praecise vim ipsius B superat, quapropter ambo simul spectata pergent moveri versus B, vi ac – bd [namely, $m_1 \cdot s_1 - m_2 \cdot s_2$]." As to the second: "[s]i vero ambo versus eundem terminum progrediantur, ut obtinet in caus secundo, quia hic utriusque vires combinantur, pergent moveri ambo simul spectata, vi ac + bd," Schuyl 1688a, chapter 4, 10–11 (unnumbered). In the third case, of course, one just needs to consider $m_2 \cdot s_2 = 0$ (see page 11). "|" is a separation mark. I have partially modernized the mathematical notation.

[289] This can be confirmed by considering, on one hand, that the author claims that in the world perfect solidity cannot be found, so that all bodies are provided with an elastic force: "[c]um vero nulla dentur corpora perfecte dura, sed omnia vi elastica sint praedita, hinc fit, ut non diu illa in quiete haereant [...]," Schuyl 1688a, chapter 3, 9 (unnumbered). This statement leaves unsolved the problem of the conservation of motion: as is evident from corollary 8: "[q]uaestioni, cur corpus in motu constitutum, aliquamdiu pergat moveri, a nemine hactenus est satisfactum," Schuyl 1688a, *Corollaria*.

4.2.3.4.2 De Volder on Elastic Impacts

The second hypothesis by which De Volder deduced his rules of impact is presented in his letter to Leibniz (18/28 February 1699), and involves the idea of elasticity:

2) the force of the strike (*vis ictus*) between the two bodies is independent from the ratio of their speeds, but it depends on their relative speed of approach (which is $s_1 + s_2$ if they move against each other; $s_1 - s_2$ if they move in the same direction). So, the elastic force exerted in the impact is to be estimated by considering this relative speed, and as "given back" reciprocally, in proportion to the size of the bodies.[290]

In the disputation, this hypothesis is not declared. Instead, the author relies, as Huygens did in his manuscript, on the case considered in Descartes's rule 1, as according to the author "experience testifies to this very thing." In this case the elastic force, or the force of restitution is equal to the force exerted in the collision (calculated as $m_1 \cdot s_1 + m_2 \cdot s_2$), and is re-distributed in the two bodies after the impact, in a way proportional to their size. Namely, as is declared in both texts, A is rebounded by B with a speed quantified as $(m_2 \cdot (s_1 + s_2))/(m_1 + m_2)$; B is rebounded by A with a speed $(m_1 \cdot (s_1 + s_2))/(m_1 + m_2)$. If they move in the same direction, the relative speed has to be calculated by a subtraction i.e. as $s_1 - s_2$.[291]

This result, however, contradicts Huygens's rules of impact. If the two bodies have the same size (i.e. $m_1 = m_2$), their speed will always be identical after impact, such as in the case presented in the 1684 disputation *De motu*. In that case, as seen above, the speeds of the bodies after impact could indeed have been calculated as

[290] "Altera erat, cum vis ictus inter eadem corpora eadem sit, quaecunque inter ea sit velocitatis ratio, modo intermedia distantia eodem superetur tempore, elastri vim ex hac celeritate aestimandam, eamque idcirco restitui reciproce pro corporum magnitudine," De Volder to Leibniz, 18/28 February 1699, in A II3B, 528–529.

[291] "Quod si sint elastica, erit vis elastri ut c ± f, quae idcirco celeritas distribuenda erit in utrumque corpus reciproce pro ratione magnitudinis. Quare si fiat ut a + b ad c ± f, ita b ad celeritatem, quam elastrum reddit corpori a, et ita a ad celeritatem, quam elastrum reddit corpori b; habebit A ex occursu velocitatem ab A versus B, ut (ac ± bf)/(a + b) et ex elastro acquiret velocitatem a B versus A, ut (bc ± bf)/(a + b)," De Volder to Leibniz, 18/28 February 1699, in A II3B, 529. Cf. Schuyl 1688a, chapter 4, 11–12 (unnumbered): "[c]onsideremus iam corpora, pro ut vi elastica sunt praedita, ut, quanta in singulis ex mutuo ipsorum impulsu, vel actione, secundum tres illos iam nominatos casus, oriatur vis elastica, determinemus. Primum ergo ab opposita parte in se invicem impingant: quo in casu, cum ambo totis suis viribus in se invicem agant, erit ut (a + b)/(c + d) = 2 | 2 = a/((ac + ad)/(a + b)) vis elastica ipsius A, qua propellit B. b/((bc + bd)/(a + b)) vis elast. ipsius B, qua propellit A. Si vero ambo versus eandem partem moveantur, quia hoc in casu non agunt in se invicem, nisi quatenus velocitas ipsius A superet velocitatem ipsius B, hinc erit, ut (a + b)/(c – d) = 2 | 2 = a/((ac – ad)/(a + b)) vis elast. ipsius A, qua propellit B. b/((bc – bd)/(a + b)) vis elast. ipsius B, qua propellit A." A description of an experimental proof of Descartes's first rule is given in Mariotte 1673, 8–22. Mariotte certainly used an elastic and inelastic collision apparatus, of which however he provides no depiction in his *Traité de la percussion ou choc des corps*. I discuss below the fortunes and adoption of such apparatus by De Volder. Wren, Robert Hooke and Lawrence Rooke performed experiments on collision at the Royal Society: see Bertoloni Meli 2006, chapter 8. As to De Volder's collision apparatus, see below.

$(m_1 \cdot (s_1 + s_2))/(m_1 + m_2)$ and $(m_2 \cdot (s_1 + s_2))/(m_1 + m_2)$. One can hypothesize, therefore, that De Volder came to progressively realize that in impact both forces, i.e. the elastic force, and the force of impact (both calculated as m·s) are to be taken into account, in order to fit with Huygens's rules. In presenting the formulas to calculate the final speeds of elastic bodies, indeed, he claims that the speed of body A (which moves towards the right i.e. towards B), and B (which moves towards the left, right, or it is at rest) are to be calculated as follows. First, one needs to consider that if body A is moving towards B with a force $m_1 \cdot s_1$, and B is moving with a force $m_2 \cdot s_2$, after a non-elastic collision A will carry body B with a speed $(m_1 \cdot s_1 \pm m_2 \cdot s_2)/(m_1 + m_2)$ – (if we consider two bodies as moving towards each other, we need to make a subtraction; if we consider two bodies as moving in the same direction, we need to make an addition). Second, the speed by which A is rebounded by the elastic force of B is $(m_2 \cdot (s_1 \pm s_2))/(m_1 + m_2)$ – (in this case, we need to make the addition if we consider the bodies as moving in opposite directions, and a subtraction if they move in the same direction). Hence, if we subtract the speed of A resulting from the elastic force of B only, from the speed by which A would carry B without such force, we obtain: $((m_1 \cdot s_1 \pm m_2 \cdot s_2)/(m_1 + m_2)) - ((m_2 \cdot (s_1 \pm s_2))/(m_1 + m_2))$, and we get this formula: $(m_1 \cdot s_1 - m_2 \cdot s_1 \pm 2m_2 \cdot s_2)/(m_1 + m_2)$.

This "denotes the velocity by which A, in the presence of the elastic force, after impact with body B, is moved towards e [i.e. the right]."[292] In simpler terms: A impacts B, and its resulting speed is given by its own force (or quantity of motion), minus the elastic force of B. As to the speed of B after impact, this is given by the combination of the force of motion by which A would carry B after impact if no elastic force is considered – i.e. $(m_1 \cdot s_1 \pm m_2 \cdot s_2)/(m_1 + m_2)$ – and the speed assumed by B as a consequence of the elastic force of A, i.e. $(m_1 \cdot (s_1 \pm s_2))/(m_1 + m_2)$ – (such forces are to be combined because they are exerted in the same direction). Accordingly, we get: $(2m_1 \cdot s_1 \pm m_1 \cdot s_2 \pm m_2 \cdot s_2)/(m_1 + m_2)$. Of course, if B is at rest, it is enough to pose $s_2 = 0$ in these formulas.[293]

[292] "Cum autem A propria vi motus, posita ac = 3 | 2 = bd secum abripiat ipsum B, velocitate (ac ± bd)/(a + b) in utroque sc. casu, a vi vero elastica ipsius B repercutiatur versus contrariam partem, velocitate (bc ± bd)/(a + b). Hinc detracta hac ab illa remanebit (ac – bc ± 2bd)/(a + b) quae proinde quantitas denotat velocitatem qua A, posita vi elastica, post concursum cum corpore B, moveri debeat versus e," Schuyl 1688a, chapter 4, 12 (unnumbered).

[293] "B vero cum a vi elastica ipsius A, versus eandem impellatur partem, versus quam ab eodem A motu suo progressivo impellitur, hinc, combinatis illis quantitatibus, erit velocitas, qua B post concursum cum corpore A, moveri debet versus e. (2ac ± ad ± bd)/(a + b). Quod si B ponatur quiescens, deleatur duntaxat d, eritque (ac – bc)/(a + b) velocitas ipsius A. Velocitas vero ipsius B, erit 2ac/(a + b)," Schuyl 1688a, chapter 4, 12 (unnumbered). In the letter to Leibniz De Volder provides these quantifications in a more synthetic way (following the quotation given *supra*, n. 291): "[q] uarum ergo differentia [sc. between the 'non elastic' speed of A, and its 'elastic' speed after impact] erit velocitas ipsius A = (ac ± 2bf – bc)/(a + b) idque ab A versus B, si haec quantitas sit affirmata, a B vero versus A si sit negata. Itidem B ex occursu habebit velocitatem ab A versus B ut (ac ± bf)/(a + b), ex elastro vero acquiret velocitatem etiam ab A versus B = (ac ± ad)/(a + b) quarum ergo summa erit velocitas ipsius B = (2ac ± bf ± af)/(a + b), qua pergit ab A versus B," De Volder to Leibniz, 18/28 February 1699, in A II3B, 529.

To sum up, De Volder, as testified to by the 1688 disputation, generalizes the case of the elastic collision between two elastic bodies established on the ground of experience, extending it to bodies of different sizes. In the correspondence, De Volder is more specific, namely, it is expressly stated that the force of restitution or elastic force is equal to the total force (quantified in terms of $m \cdot s$) exerted in the impact (i.e. it is a *vis ictus*), which is estimated by considering the relative speed between the two bodies, and is proportional to the size of the two bodies. This means that body B is moved by the force that body A would have if it were moving at the relative speed between two bodies (i.e. body B is imagined as if it were at rest): namely, it is moved by the force $m_1 \cdot (s_1 \pm s_2)$. On the other hand, A is moved by the force that B would have in the same situation, i.e. $m_2 \cdot (s_1 \pm s_2)$. The force of restitution (equal to the total force exerted by the body before impact, i.e. it is $m_1 \cdot (s_1 \pm s_2) + m_2 \cdot (s_1 \pm s_2)$, or $(m_1 + m_2) \cdot (s_1 \pm s_2)$) gives rise to a speed, in the two bodies, which is quantified by $((m_1 + m_2) \cdot (s_1 \pm s_2))/(m_1 + m_2)$, namely $(s_1 \pm s_2)$. As the force of restitution is exerted in both bodies, the elastic forces of A and B have to be considered as distributed in, or divided by the quantity $(m_1 + m_2)$. However, it is not clear how De Volder came to formulate his hypotheses. If we take into account the 1684 disputation (in which only elastic forces are considered), as well as his own statement, given in the correspondence, that De Volder deduced Huygens's rules while inquiring into his 1669 communication, we may suppose, as anticipated above, that De Volder 'combined' elastic and non-elastic forces in order for them to fit with Huygens's findings, published "without demonstration" in 1669. After having presented his rules (both in the 1688 disputation, and in the letter to Leibniz), indeed, De Volder claims that these fit with Huygens's rules, i.e. with the conclusion that $m \cdot s^2$ is conserved after impact,[294] and with Huygens's 1669 rules.[295]

[294] "Ex his autem ipsis, quod ad rem nostram multum facit, sequitur in corporum perfecte elasticorum concursu, remanere semper idem productum corporum in celeritatum quadrata. Nam si quadr. $(ac \pm 2bf - bc)/(a + b)$ ducatur in a, et quadr. $(2ac \pm bf \pm ab)/(a + b)$ ducatur in B, horum productorum summa aequabitur $acc + bff$. Juxta hanc autem mensuram vires corporum elasticorum aestimandas confirmatur ex eo, quod si illa corpora a et b celeritatibus post primum occursum acquisitis in se invicem iterum impingant, corpus A, post alterum hunc occursum, habiturum sit celeritatem ut c, B celeritatem ut f, adeoque hic causa effectui aequipolleat exactissime, cum alterum ex altero praecise sequatur, et vice versa," De Volder to Leibniz, 18/28 February 1699, in A II3B, 529.

[295] In the disputation, the cases involved in Huygens's first and second rule are explicitly considered, and the rules themselves are referred to. However, the cases in which the two bodies are not equal are considered: "[e]x quibus [i.e. from the algebraic formulas given in the disputation] innumera de motu corporum ex mutuo ipsorum impulsu orto determinari possunt, et variae construi regulae cum experientia convenientes. Patet enim evidenter, quod, si corpori quiescenti duro B v. gr. aliud aequale corpus durum A occurrat, post contactum hoc quidem quiescat, quiescens vero eandem, quae fuit in impellente, acquirat celeritatem a enim eum, hoc posito, sit aequalis b, erit et $ac = bc$, ac proinde A quiescere debet, cumque $a + b$ sit $= 2a$, movebitur B celeritate c, quae est celeritas ipsius A. Quod si a ponatur maior quam b, progredietur quidem A versus eundem terminum cum B, sed B semper ipso celerius movebitur; et si a minor ponatur, quam b, reflectetur A versus contrariam partem. Patet etiam, quod, si alterum illud corpus aequale etiam moveatur, sive ab opposita parte in se invicem impingant [...] sive versus eandem partem ambo progrediantur [...] quod, inquam, post contactum permutatis invicem celeritatibus sequerentur. Ita ut A moveatur

Moreover, De Volder's rules are also labelled (in the 1688 disputation) as conforming to experience: this might have been a covert reference to Mariotte's *Traité de la percussion*, where rules of impact conforming to Huygens's rules are demonstrated on an experimental basis. More probably, however, it was the result of De Volder's own experimental activities. In fact, there is no textual evidence referring to the details of De Volder's experimentations on collisions: however, De Volder devised a collision apparatus, which is mentioned in De Volder's inventory of 1705, as "a wooden instrument standing on its pedestal, with below two pieces of copper, circles divided and marked in grades to demonstrate the rules of motion," together with "16 ivory and 3 wooden spheres," and a "little wooden gutter for the same purpose"[296] – probably an inclined plane or an U-shaped gutter in which the pendular motion of spheres could be observed (or, at least, the acceleration of falling bodies). The first instrument is described by Johannes Joosten van Musschenbroek in various letters sent to Marburg, and was extant at Leiden as late as in 1711, when Zacharias Conrad von Uffenbach reported the presence of "one [...] small wooden sextant, to make the movement of pendulums with it."[297] This instrument consisted of a wooden board, with the shape of a sextant (viz. the sixth part of a circle), to which two balls of different materials could be appended to a metallic arc, as if they were pendulums, and their movement could be observed on another metallic arc: see Fig. 4.29, Hessisches Staatsarchiv Marburg, file UniA 305a Nr 5947, 14. According to a catalogue of Van Musschenbroek of 1694, the instrument was constructed upon De Volder's design, and cost as much as 60 guilders.[298] The final cost, however, was 30 guilders (as it did not include ivory balls).[299]

Van Musschenbroek repeatedly refers to Mariotte's model as the first source of the apparatus and of its use, however, he also mentions the *Journal des voyages* of Balthasar de Monconys (first edition 1665–1666), in which a letter by a certain

celeritate d; B vero celeritate c, sic tamen ut A in priori casu ea celeritate reflectatur, in posteriori vero secundum lineam rectam progrediatur. Vid. has duas regulas Societati Regiae Angl. inter alias communicatas ab Illustrissimo Viro Christiano Hugenio in Philos. transaction. pag. 927. Quia autem vel ei, qui prima duntaxat didict calculi algebraici elementa, reliqua omnia hinc deducere sit facillimum, ulteriori ipsorum deductioni supersedebo," Schuyl 1688a, chapter 4, 12–13 (unnumbered). Cf. *supra*, n. 272.

[296] "Een hout instrument staande op syn kruck, onder met twee stucken van kopere afgedeelde circuls ad regulas motus. Hier toe 16 ivore en 3 houte bollen. Nogh een hout gootje tot het selfde gebruik," Molhuysen 1913–1924, volume 4, 105∗.

[297] "Ein kleiner Sextans von Holz, den motum pendulorum damit zu machen," Uffenbach 1753–1754, volume 3, 426.

[298] In an unpaginated sheet entitled *Catalogus van alle d'instrumenten die by mij gemaakt werde*, inserted into a letter of Van Musschenbroek dated 4 March 1694, Van Musschenbroek noted that "[m]ijn heer de Volder heeft een instrument om de wetten van de beweegingh te toonen dat heel fray is kost ontrent 60 guld[en]," Hessisches Staatsarchiv Marburg, file UniA 305a Nr 5947, 1–3 (transcribed also in De Clercq 1991, 119–120). According to a letter of 20 September 1694, the apparatus was not yet ready: file UniA 305a Nr 5947, 23. I thank Dr. Katharina Schaal, librarian at the Universitätsarchiv – Philipps-Universität Marburg, for her help in locating these manuscripts.

[299] File UniA 305a Nr 5947, 17. This price is included in a letter dated 11 May 1695. As to the ivory balls, see *infra*, n. 300.

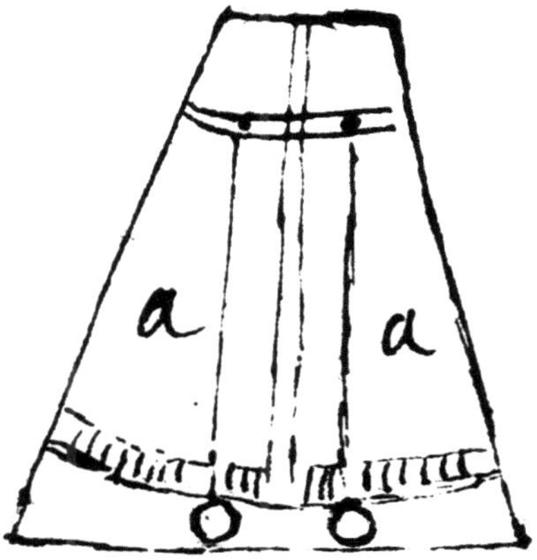

Fig. 4.29 De Volder's collision apparatus, sold by Johannes Joosten van Musschenbroek. (Source: courtesy Hessisches Staatsarchiv Marburg, file UniA 305a Nr 5947, 14)

Monsieur Regnault, concerning experiments with pendulum collisions are described (1655),[300] as well as the *Cours entier de philosophie, ou Systeme general selon les principes de M. Descartes, contenant la logique, la metaphysique, la physique, et la morale* (1691) of Pierre-Sylvain Régis, in which one can find a depiction of Mariotte's apparatus (see Fig. 4.30, Régis 1691, volume 1, 366).[301] A similar

[300] In the sheets extant at Marburg, ending with a letter dated 11 May 1695, one can find the description of the collision apparatus: "[l]'instrument pour voir les le loix de mouvement et de telle figure aa est un planche dans le quelle sont fait deux pieces d'un {mesh} circle de cuivre marque avec degrez en haut le mesme, en haut pendent <{...}> \en/ un petit {corde} deux boules d'yvoire le quelle on laisse egalement tombe{r} et se recontrent on void a quelle degrez se rejettent, on laisse un tomber est l'autre en repos on void combien de mouvement qu'il donne a l'autre et combie qu'il peut, {on} prend aussi des boules de diverses grandeurs et aussi de bois de metail, de terre, de tout sortes," file UniA 305a Nr 5947, 14; see Fig. 4.29, Hessisches Staatsarchiv Marburg, file UniA 305a Nr 5947, 14. In the letter itself, it is implied that this instrument was based on Mariotte's apparatus, to whose *Traité*, as well as to the *Journal des voyages* of De Monconys (viz. to the 1695 edition), Van Musschenbroek re-directs the reader: "[l]'instrument pour le choq du corps l'usage trouvera vous dans le livre Mariotte Le choq du corps. J'ay seulement pendu 2 boules du bois dur. L'yvoire est mieux parce qu'ils sont plus dur, mais ils sont fort cher, ils voudrent bien 20 florin pour les deux. Et vous pouvez tujours laissez faire. Et aussi de toutes sortes de bois, de cuivre \de fer/ pour voir la difference de reflexions. Pu pend cet instrument contre une muraille <{...}> en un clou a: le piece du plomb pour le pendre au niveau. Si Monsr desire les boules d'yvoire le lasseray bien faire. [...] En le Voyage de Monconijs est l'instrument de choq du corps ecrit dans tom 5 fol 232," file UniA 305a Nr 5947, 15 and 18. The place of De Monconys's *Journal* mentioned by Van Musschnebroek is in fact a letter of a certain Monsieur Regnault (not to be mistaken with Valère Regnault (1545–1623) and with Noël Regnault (1683–1762)), dated 21 December 1655, in which Regnault describes his experiments on collision performed with pendulums (De Monconys 1695, volume 4, 231–244). Regnault was also the author of a *De humidorum aequilibrio in siphonibus* (De Monconys 1695, volume 4, 176–191), based on the principle of Torricelli. On De Monconys's *Journal* (which included several observations on scientific instruments), see De Clercq 2016.

[301] In a letter dated 28 July 1696, Van Musschenbroek refers again to Mariotte, noting that in his treatise there are no illustrations of the apparatus. This illustration can however be found, as Van Musschenbroek notes, in the *Cours entier de philosophie, ou Systeme general selon les principes de M. Descartes, contenant la logique, la metaphysique, la physique, et la morale* (1691) of

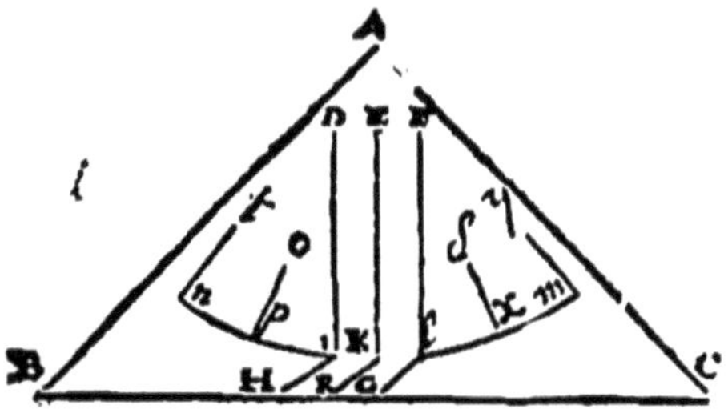

Fig. 4.30 Edme Mariotte's collision apparatus. (Source: Régis 1691, volume 1, 366. Naples, Biblioteca Nazionale, signature: B. Branc. 045B 38-44, https://archive.org/details/bub_gb_goRde-u4pJdgC/page/n6. Public domain mark 1.0)

instrument, built by Jan van Musschenbroek, was later used for the same purpose by Willem Jacob 's Gravesande, who then came to modify it as a pendulum apparatus (still extant at Leiden).[302] It is however unclear whether De Volder – who generally used his apparatuses for didactic rather than for discovery purposes – came to formulate his laws after his experiences with collision. In fact, De Volder could have found in Wallis's rules of impact for elastic bodies – given in the third volume of his *Mechanica sive De motu* (published in 1671) – a decisive 'help' in the development of his own rules. According to the *Bibliotheca Volderina*, De Volder owned at least those parts of Wallis's *Mechanica* printed in 1670 (i.e. the first and second volumes): indeed, this treatise is mentioned twice in the *Bibliotheca*, as two separate items both dated 1670. Moreover, De Volder owned the complete edition of the

---

Pierre-Sylvain Régis, in which it is depicted exactly like Mariotte's apparatus: "touchant une explication pour l'instrument le choq du corps je vous a envoye le livre de Monsr Mariotte il a \2/ figures qui manquent en ce livre nous avons achette ces deux livres \Du choq et mouvement des eaux/ en un auction car on ne puis en Hollande avoir ces livres, mais la description de l'instrument est en \Le/ Physique de Regis tom 1 chap 21 pag 366: j'espere que Monsr par ces deux livres aurez contentement, car ces choses sont trop long en une lettre {sil} tost que je puis trouver une livre de Monsr Mariotte je vous envoyer par le poste en une lettre les deux figures," file UniA 305a Nr 5947, 20–21. See Régis 1691, volume 1, 366–367, containing Fig. 4.30, Régis 1691, volume 1, 366. On Régis's physics and impact rules, see Clarke 1989; Des Chene 2002, 2005; Milani 2010, 2013, 2015. In a letter of 12 December 1696, Van Musschenbroek comments again upon the use of the instrument: "j'ai le mesure pour les boules de l'instrument de mouvement. Et Monsr le puis prendre aussi selon son desin, par exemple prenex une boule de 3 ponces de diametre et prennez un de 2 pounce et faite l'experience combien ils se reflechent, on prend deux egales boules aussi bien inegaux. Mais quand on veut faire exactement l'experience le boules doivent estre si grand que le <deux> distance des {elon} qui sont en {hout} de l'instrument car ils faut que le boules ne sont plus grand autrement on ne puis exactement mesurer, car ils douvent seulement se toucher quand ils sont en repos. Mais on le puis prendre bien petit et Monsr le faut faire deverses grandeurs de cette grandeur comme j'ay envoye jusque a une pouce de diametre, et aussi de diverses matiere, de terre, du bois, yvoire est le mieux pour le peu de ressort car ils reflechent le mieux j'ay fait ma delegence pour les figures de Mariotte du choq des corps, et je n'ay pas trouvex, la guerre mons empeche pour avoir ces livres mais si tost que je puis trouver je ne manqueray pas pour Monsr le faire avoir," file UniA 305a Nr 5947, unnumbered page.

[302] See De Clercq 1989, 30–31; De Clercq 1997b, 43.

Mechanica included in the first volume of Wallis's *Opera omnia mathematica* (published in 1695).[303] The catalogue offers no evidence, therefore, that De Volder owed the third part of Wallis's *Mechanica* before 1695. This is interesting because the rules that De Volder presented can be found, in the same algebraic form, in the third part of Wallis's treatise, namely chapter 11 (*De percussione*, on non-elastic impacts) and 13 (*De elatere, et resilitione seu reflexione*, on elastic ones). In chapter 11 (proposition 1, scholium), Wallis distinguishes between what is "perfectly hard [...], which never yields to a hit (*ictus*), and so [it is] neither soft, nor elastic," and what is "elastic [...], which, even if for some time [it] yields to a hit, however returns to its original form by itself [*suopte marte*]."[304] Wallis does not consider the bodies involved in collisions as immersed in fluids (as in the 1688 disputation). Yet, the way in which he formulates the rules of non-elastic impacts are perfectly consistent with De Volder's hypotheses. Wallis maintains that forces, i.e. momenta (calculated by the formula weight·speed) exerted in opposite directions destroy each other (as he declares in chapter 1, propositions 8–9 (in the first volume)),[305] and from this he deduces the laws of non-elastic collisions. For instance, in the case involving two bodies A and B, where "mP" is the weight of A, "rC" is its speed, "nP" is the weight of B, and "sC" is its speed, and sC = 0 (proposition 2), after the impact they form one body whose speed is (according to Wallis's notation) mrPC/(mP + nP), namely (m/(m + n))rC.[306] The other formulas for non-elastic bodies are given accordingly (i.e. by adding or subtracting momenta): (mrPC + nsPC)/(mP + nP) and (mrPC − nsPC)/(mP + nP).[307]

In the case of elastic impacts, dealt with in chapter 13, Wallis (unlike De Volder) proceeds as follows. First, when a body collides with a wall with a momentum mrPC, mrPc is its "impelling or compressing force" to which the wall

---

[303] "Mathematica. Libri in folio. [...] 26. Joh. Wallis *Opera omnia Mathematica*, Oxon. 1699. 3 voll. [...] Mathematici in quarto. [...] 99. [Joh. Wallisii] [...] *Mechanica sive de Motu*, Lond. 1670. [...] 101. [...] *de motu*, Lond. 1670," *Bibliotheca Volderina*, 13, 14–15. 1699 is the date of the last volume of Wallis's *Opera*, published from 1693 (when also the second volume was published), to 1699 (third volume).

[304] "Perfecte durum, appello, quod ictui nequaquam cedit, adeoque nec molle, nec elasticum. Molle, appello, quod ictui ita cedit ut pristinam figuram amittat [...]. Elasticum appello, quod, utut ictui aliquantisper cedat, se tamen in pristinam formam suopte marte restituit," Wallis 1670–1671, 661. The numbering of the three volumes is continuous.

[305] Wallis 1670–1671, 16–17.

[306] "Prop. II. Si grave motum, gravi quiescenti directe impingat, sed ita constituto, ut aliunde ne moveatur non impediatur: utrumque iunctim movebitur, ea celeritate quam calculus (ponderum rationis et pristina celeritate rite computabis) indicabit. Nempe, si momentum (ex moti gravis pondere et celeritate compositum) per utriusque simul pondus dividatur, habebitur futura celeritas. [...] Cum itaque vi mrPC, movendum sit pondus mP + nP, id fiet celeritate m/(m + n)rC (divisio scilicet momento mrPC, per pondus mP + nP)," Wallis 1670–1671, 662–663. In proposition 1 Wallis considers the case in which a body is thrown onto an unmoveable object, like a wall. In this case, he uses the idea of resistance, which is equal to the momentum of the body in movement: see Wallis 1670–1671, 660.

[307] Cf. propositions 3–4: Wallis 1670–1671, 664–665.

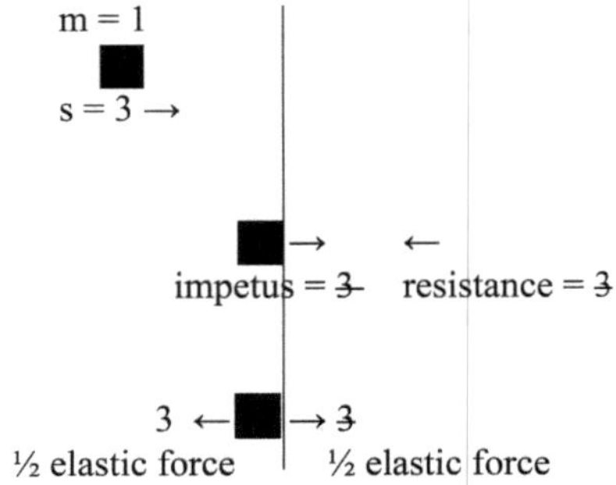

Fig. 4.31 Model of impact on a wall. (Figure by the author)

(supposed to be unmoveable) exerts an equal resistance.[308] (See Fig. 4.31, *Model of impact on a wall*).

Therefore, the total "compressing force," exerted by the body on the wall, and by the wall on the body, is 2mrPC. If only these forces are considered, the moving body will come to rest, and the momentum is not "restituted."[309] However, if the bodies are elastic, once the "impelling or compressing motion" ceases, the elastic force – which Wallis assumes as equal to the whole compressing force, i.e. double the momentum of the colliding body (2mrPC) – is freed, and is exerted in both directions: namely, half of it (or: mrPC) is exerted in the direction of the wall, which absorbs it ("however with an invalid *conatus*"), and the other half of it in the direction of the moving body, which is rebounded with a momentum mrPC, i.e. with its initial speed rC.[310] Wallis's conclusion is valid if both bodies are elastic, and also if

[308] This foreruns Newton's third law of nature. Newton, indeed, cites the rules of impact of Huygens, Wren, Wallis and Mariotte to support his account: see Home 1968; Cook 1986.

[309] "Est autem ea vis comprimens, dupla momenti gravis impingentis. Nam vis ab impingente gravi illata, est ipsius momento aequalis (utpote cuius motus totus sistitur,) puta mrPC. Sed et, propter aequalem obicis resistentiam, (quae tantundem ad compressionem confert ac si aequalis vis contraria occurreret,) tantundem inde advenit, puta alterum mrPC, (sed cuius hic minor habenda ratio, utpote quod aliunde compensatur). Adeoque vis tota compressiva, 2mrPC. Dico porro, vim impellentem seu comprimentem, ubi ad aequipollentiam res redigitur cessare (utpote quae tota flectendo elatere impensa est). Vis enim extrinsecus impressa, qualem supponimus impellentem (secus quam vis insita, qualis est gravitas, et vis elastica compressa,) ubi semel ad quietem redigitur, perit, neque valet seipsam restituere," Wallis 1670–1671, 688.

[310] "Dico itaque tertio, cessante [...] motu impellente, seu comprimente, vim elasticam, iam liberam, et [...] vi pridem comprimenti aequalem, aequaliter utrinque se explicare nitentem, vim dimidiam seu alterum mrPC in obicem impendere (utut irrito conatu) alterum vero mrPC in advectum grave. Quod itaque eadem vi, adeoque et eadem celeritate, corpus A repellit, qua advectum fuerat. (Nempe, vi mrPC, adeoque, propter pondus mP, celeritate rC)," Wallis 1670–1671, 688. Wallis does not provide a natural-philosophical explanation of how the elastic force is exerted: rather, in definitions 1–2 of chapter 13, he defines elasticity as the force by which a body re-aquires its previous shape, and provides a short etymological analysis of the term 'elater': Wallis 1670–1671, 686–687. In any case, an elastic force is required, for Wallis, to explain rebounding, as he clarifies in the scholium of the first proposition of chapter 13: "[n]on ignoro, plerosque mechanicorum vim nescio quam in motu incepto ponere, qua [...] mutata directione (etiam absque nova causa) resiliat. Cum vero hoc gratis dictum videatur (ut novus motus absque nova causa incipiat,) et postulatum, atque multis porro incommodis urgeri possit," Wallis 1670–1671, 690. See also

just one of them is elastic: in all cases, the elastic force exerted by one of the bodies, or both, leads to a re-doubling of the total momentum of the bodies.[311] From this, Wallis derives the formulas to calculate the outcomes of the collisions involving two bodies moving against each other or in the same direction (proposition 9–10). Given the weight of A = mP; speed of A = rC; weight of B = nP; speed of B = sC, the speed of A after its impact with a body B moving in the same direction (proposition 9) is (mPrC − nPrC + 2nPsC)/(mP + nP) and the speed of B after impact is ((2mPrC – mPsC + nPsC))/(mP + nP). In the case that the bodies move against each other (proposition 10): speed of A = (mPrC − nPrC − 2nPsC)/(mP + nP); speed of B = ((2mPrC + mPsC − nPsC)/(mP + nP).[312]

These formulas correspond to those presented by De Volder, who shared with Wallis the ideas that forces exerted against each other annihilate each other (according at least to his correspondence), and overtly required that bodies are elastic, in order to undergo elastic collisions. However, what is lacking in De Volder's theory of impact is a reference to resistance, which in Wallis's account ideally serves as a factor of redoubling the 'force of pushing' exerted by the bodies, and thus of redoubling, *via* the elastic force which is freed after impact, the initial, total momentum of the bodies. De Volder, as seen in Sect. 4.1.3, *De Volder's ideas on cohesion and divisibility*, found Descartes's idea of rest as problematic in granting the cohesion of bodies; moreover, in his correspondence he acknowledged in inertia the persistence of bodies in their state.[313] In other words, he found problematic the idea that bodies can have a sort of internal force of resistance to movement. In the case of Wallis's example, the way in which the resistance of the wall gives rise to a doubling of the momentum of body A is not explained, i.e. the source of this resisting force, which is always equal to the force of A – exactly as in Descartes's rule 4 – is not explained. De Volder could not be keen to accept this premise. It is noteworthy that, the idea of the re-doubling of the initial impetus is not always used in Wallis's demonstrations: as he declares in the scholium to proposition 5 (where he considers the case of A = B, and B as at rest), he offered more demonstrations for the same proposition (as he does), because different readers have different 'tastes'. For instance, he proposed, for the first proposition of chapter 13 (which has been reconstructed here), a dem-

---

supra, n. 304. Wallis might have appropriated the idea that elastic force doubles the initial momentum of the moving body from Giovanni Alfonso Borelli, who held similar positions in his *De vi percussionis* (1667). Borelli maintained that the force of collision of a body (calculated by considering size and speed) is first destroyed, and then 'restituted' as doubled in the case where one of the bodies is elastic. He provided a corpuscular explanation of collision not involving the idea of subtle matter: according to him, the particles of elastic bodies are flexible *machinulae*, whose interactions are however not explained by him: see Borelli 1667, 144 and 148–149. De Volder owned only the second edition of this treatise, published in Leiden in 1686: *Bibliotheca Volderina*, 3.

[311] "Quodque iam demonstratum est, posita in obice vi elastica: perinde obtinet si vis illa sit in corpore impingente, vel etiam in utroque. Quippe illud utcunque obtinet, non prius sisti vim impellentem, quam aequipolleat elateris (sive singularis, sive gemini,) adversa resistentia, sed neque ultra perseverare. Unde reliqua sequuntur," Wallis 1670–1671, 688.

[312] See Wallis 1670–1671, 705–707.

[313] See Sect. 4.1.4, *The search for physical unity and activity: De Volder vis-à-vis Leibniz.*

onstration explaining "more the physical cause of the thing, which however in the subsequent ones I follow more sparely, as these depend on those [sc. in the former demonstrations]."[314] The conclusions of De Volder coincide with those of Wallis, but De Volder adopted different physical premises, namely, he avoided any reference to the idea of resistance (both in the 1688 disputation and in the correspondence), entailing an augmenting of forces (which he found absurd).[315] His main issue with Huygens's rules of impact, and with non-elastic impacts, indeed, was that of the loss of forces (as calculated by m·s). Wallis's supposition of a doubling of momentum would – in De Volder's eyes – just worsen the problem of conservation. Accordingly, he adopted a 'sparer' version of Wallis's rules. The way in which he adopted these rules is obscure: on the one hand, he could have owned the third volume of Wallis's *Mechanica* (not listed in his *Bibliotheca*), and have – let's say – plagiarized it. Moreover, he could have learned Wallis's laws of collision of elastic bodies from Wallis's own mouth during his sojourn in England in 1674. Yet, as De Volder skipped any reference to the idea of resistance, it could be that De Volder used Wallis's ideas, but only those he found in his 1668 communication in the *Philosophical Transactions*, which concerns non-elastic impacts,[316] but in which Wallis outlined a method to provide rules of collision for elastic bodies, which would be the method then followed by De Volder:

> [i]f the bodies that impinge are understood to be not absolutely hard (as we have previously supposed) but ceding to the blow so that by their elastic force they are able to restore themselves, it can happen that these bodies mutually rebound, which would otherwise proceed together, and indeed more or less, according as this restituting force is greater or less, namely if the impetus from the restituting force is greater than the progressive.[317]

This could have provided De Volder with the idea of considering both elastic and non-elastic forces (let's say) in his rules, which, moreover, fitted with the Cartesian idea of the "sophisticated model of impact" – stressed in the French edition of the *Principia*, and used by De Volder in his *dictata*. In his rules, indeed, De Volder takes into account the hypothetical, non-elastic outcome of a collision.

This reading of De Volder's elaboration of his rules of impact can be confirmed by considering Leibniz's comments on his rules of non-elastic collisions: as he

---

[314] "Plures hic demonstrationes congessi [...] non quod diffidam singulis [...], inest enim singulis sua vis. Sed, cum pro vario lectorum gustu modo hic modo ille demonstrandi modus placeat magis, ut quam magis velit lector eligat. Prima quidem, aut etiam secunda [...] physicam rei causam magis explicant, quam tamen in sequentibus parcius prosequor, ut quae ab his dependent," Wallis 1670–1671, 699.

[315] See Sect. 4.2.3.4.2, *De Volder on elastic impacts*.

[316] This makes sense of De Volder's remark, given in the *dictata*, that rules of impact alternative to Descartes's were provided by Huygens and Wallis: Huygens, indeed, provided rules for elastic bodies, Wallis provided rules for non-elastic ones. No other author is specifically mentioned by De Volder.

[317] Murray et al. 2011, 179. Cf. the original text: "[s]i quae sic impingunt Corpora, intelligantur non absolute dura (prout hactenus supposuimus) sed ita ictui cedentia, ut Elastica tamen vi se valeant restituere, hinc fieri poterit ut a se mutuo resiliant ea corpora, quae secus essent simul processura," Wallis 1668, 866.

stated as early as the second letter of the correspondence (Leibniz to Bernoulli, 12/22 July 1698), De Volder's idea that forces are annihilated in collision arises from a wrong conception of matter, according to which it does not possess a form of activity. As he declares in his letter to De Volder of 24 March/3 April 1699 – commenting upon De Volder's identification of inertia with extension itself,[318] – according to this idea matter offers no resistance to any kind of movement, leading to chaos.[319] Accordingly, in commenting on De Volder's rules of impact in the same letter, he praises his recourse to the idea of elasticity, as through this the forces which are destroyed in the conflict, are restituted. Anyway, for Leibniz De Volder's rules had to be integrated with a "principle of more fruitful comprehension" (*uberioris intelligentiae causa*), i.e. with the principle of continuity, according to which forces are never destroyed.[320] Leibniz did not make the same criticism of Wallis, whose positions, however, Leibniz found to rely on the 'prejudice' of the equilibrium of the lever, according to which motive forces are estimated by the quantities of motion, which are lost in the impacts.[321]

---

[318] These passages have been considered in Sect. 4.1.4, *The search for physical unity and activity: De Volder vis-à-vis Leibniz.*

[319] See Leibniz to De Volder, 24 March/3 April 1699, quoted *supra*, n. 113.

[320] "Modum quo regulas motus binorum corporum concurrentium, ex destructione motus per conflictum, et restitutione per Elastrum deducis, puto esse rectissimum; si modo uberioris intelligentiae causa accedat, quemadmodum jam inter nos convenit, utrumque fieri successive ex lege aequilibrii interventu virium mortuarum," Leibniz to De Volder, 24 March/3 April 1699, in A II3B, 545.

[321] "Quae Joachimus Jungius, Marcus Marci, Joh. Alph. Borellus, Hugenius, Wrennus, Wallisius, Mariotus de his recte (pro parte) meditati sunt, inaedificata ab ipsis partim experimentis, partim hypothesibus particularibus, ea non tantum ex veris fontibus explico, sed et longius profero ad casus quos vel nullo modo vel non bene attigere. [...] Omnibus praejudicio fuit, lex aequilibrii ex qua sumsere, ut vires simpliciter aestimarent in ratione composita corporis et celeritatis neque adeo (explosa Cartesiana conservatione quantitatis motus) virium conservationem aliquam animadverterent; quae tamen superest mirifica et multimoda. Unde Wallisius in libro de motu, etiam agnoscens quantitatem motus non servari, nihilo minus tamen per eam vim aestimavit," Leibniz to De Volder, 27 December 1698, in A II3B, 503–504. Moreover, according to Leibniz Huygens accepted the idea that motive force ($m \cdot s^2$) is conserved only after Leibniz himself had pointed this out to him, but meant by this only the "ascensional forces." These are dealt with, but not overtly labelled as 'vires', in the second part of Huygens's *Horologium oscillatorium* (1673), concerning the law of free fall. Cf. the same letter: "[i]pse Dn. Hugenius de virium conservatione fassus est se non cogitasse, postea a me admonitus vires quae conservantur appellavit Ascensionales, non male quidem, sed tamen non satis plene. Idem enim in omni viva vi locum habet, sive ad gravitatem, sive ad Elastrum, sive simpliciter ad motum, aliumve effectum qualemcunque applicetur," A II3B, 503. The idea that motive force ($m \cdot s^2$) is ascensional force itself is given by Huygens in his marginalia to Leibniz's *Demonstratio brevis*, as he comments upon Leibniz's words "[h]inc sequitur, corpus A delapsum ex altitudine CD, praecise tantum acquisivisse virium, quantum corpus B lapsum ex altitudine EF," with "virium ascensionalium, nam si acquisitis celeritatibus sibi mutuo occurrant, non aequaliter resistunt," Huygens 1888–1950, volume 22, 790, and Leibniz's article *Schediasma de resistentia medii, et motu proiectorum gravium in medio resistente* (*Acta Eruditorum*, January 1689), as he comments on Leibniz's words "sunt enim vires ut quadrata velocitatum" with "vires ascensionales. Sed his hîc nihil opus," Huygens 1888–1950, volume 22, 793. Leibniz could know the contents of these marginalia, as Bernoulli communicated them to him in a letter of 12/22 September 1696: see A III7A, 130–131. However, no direct communication between Leibniz and Huygens on ascensional forces is extant in the *Akademie Ausgabe* edition of Leibniz's correspon-

De Volder, to sum up, strived throughout his life to find an accommodation between Descartes's principles and the new theories of collision, which he acknowledged as valid under certain conditions – i.e. in fluids, or in non-elastic frameworks. In fact, he came to accept these theories from a natural-philosophical standpoint: yet, these new theories could not solve the problem of the conservation and 'restitution' of forces. From a metaphysical standpoint, this problem would never be solved by De Volder (as seen in Sect. 4.1.4, *The search for physical unity and activity: De Volder vis-à-vis Leibniz*). From a physical standpoint, this problem was partially solved by taking into account (as he does in the *dictata* and in the 1688 disputation) the action of fluids, to which movement is communicated and preserved in the case of non-elastic impacts. Accordingly, Chap. 5 is devoted to a more detailed analysis of De Volder's hydrostatics, and hydrodynamics, namely to the ways in which he explained the behaviour of bodies in fluids – either air or water.

dence. See also the letter of Leibniz to Guillaume de L'Hôpital, 4/14 December 1696, in A III7A, 214: "[q]uant aux Dynamiques je croy que M. Hugens estoit de mon sentiment dans le fonds; et qu'il reconnoissoit qu'il se conserve tousjours la meme force, comme j'avois avancé. Apres avoir examiné mon sentiment, il trouva à propos d'appeller cette force Ascensionale, parce qu'il se conserve tousjours autant qu'il faut precisement pour faire montrer le même poids à la même hauteur. Mais comme cette même force a lieu, soit qu'on employe des corps pesans ou des ressorts ou autre chose, parce qu'il se conserve tousjours autant qu'il faut précisement pour faire monter le même poids à la même hauteur." *Pace* Leibniz, Huygens understood the scope of the idea of the conservation of $m \cdot s^2$, well beyond the case of ascensional forces: see the letter of Huygens to Henri Basnage de Beauval of July 1690 (letter 2606), in Huygens 1888–1950, volume 9, 463: "[…] ma raison est, qu'il se perd souvent du mouvement, sans qu'on puisse dire qu'il s'est consumé à rien, comme dans plusieurs cas du choc de deux corps durs, suivant ce que j'ay remarqué en publiant les loix de ces sortes de mouvements dans le Journal des Sçavans en 1669. au mois de Fevrier): de sorte que ce n'est pas une necessité que la quantité de mouvement se conserve toûjours, si elle ne se consume à quelque chose; mais c'est une loy constante, que les corps doivent garder leur force ascensionelle, et que pour cela la somme des quarrez de leurs vitesses doit demeurer la même. Ce qui n'a pas seulement lieu dans les poids des pendules, et dans le choc des corps durs, comme je l'ay remarqué au même endroit, mais aussi en beaucoup d'autres recherches de Mechanique."

Chapter 5
The Theory of Hydrostatics and Pneumatics

Abstract In this chapter I focus on De Volder's treatment of hydrostatics, i.e. the core of his teaching activities at Leiden experimental theatre. The chapter is devoted to discussion of De Volder's attempt to combine two main models in hydrostatics. First, the Cartesian – which Descartes expounded through his theory of gravity – according to which bodies (either solid or fluid) provided with the same specific weight constantly nullify their reciprocal pressure, so that, for instance, there is no increasing pressure in water. Second, the Archimedean model (assumed by Boyle) according to which the conditions of the floatation of bodies are determined by the different pressures exerted within a fluid. I show that these models are ultimately inconsistent with each other, so that in refraining from publishing his disputations, De Volder might have been partially justified by this problem. Moreover, I consider De Volder's application of such models to the explanation of the effects of the pressure of air, and its partial overcoming by his assuming, as a key factor determining pneumatic phenomena, the idea of the elasticity of air. This was a notion that De Volder also applied to physiology, and which testifies to his assumption of the Boyle-Mariotte law.

In Chap. 4 it has been ascertained that De Volder appropriated Huygens's laws of impact, and interpreted them – in his *dictata* – as explaining the behaviour of bodies in fluids (such as water or air). The action of fluids, indeed, turned out to be crucial in making acceptable Huygens's violation of Descartes's fourth rule of impact, namely Huygens's idea that a small body can always move a bigger one. In turn, in his correspondence with Leibniz De Volder used only the idea of elasticity to ground Huygens's rules, while in the 1688 disputation *De vi corporum elastica* the action of a surrounding fluid is used to explain the equilibrium of forces resulting from non-elastic collisions, and their apparent annihilation. In the case of the impact between non-elastic bodies of equal size and equal speed, indeed, their post-collision state of rest is explained as due to the equal pressure exerted by fluids on them, to which their quantity of movement is completely transferred. In this chapter, I will deepen the analysis of De Volder's statics and hydrostatics (as according to him all solid bodies are immersed in fluids such as water, air or subtle matter), by

A. Strazzoni, *Burchard de Volder and the Age of the Scientific Revolution*, Studies in History and Philosophy of Science 51,
https://doi.org/10.1007/978-3-030-19878-7_5

considering chiefly his experimental activities, which concerned mostly hydrostatic phenomena.

In order to shed light De Volder's treatment of hydrostatics, we need first of all a map of his texts on the topic. First, we need to consider his actual experiments, i.e. those described in his *Experimenta* (1676–1677), *Disputationes de aëris gravitate* (1676–1678), in Hermann Lufneu's 1685 and 1687 reports (referring to the years 1677–1678), and in Jakob Bernoulli's 1683 *Dissertatio de gravitate aetheris* (referring to an experiment by De Volder in 1682). Moreover, we need to consider the 1688 *pro gradu* disputations of Hermann Schuyl (*Disputatio philosophica de vi corporum elastica, Disputatio physico-medica de respiratione*), the 1690 *pro gradu* disputation of Joseph Serrurier (*Disputatio philosophica de gravitate aëris* – dedicated to De Volder). Also, we need to consider a bulk of disputations on cosmology he presided over, and which are listed below. Along with all these texts, we need to consider De Volder's commentaries on those articles of Descartes's *Principia* devoted to hydrostatics: chiefly, II.56–62 (concerning the behaviour of a solid in a condition of equilibrium with the surrounding fluid), III.39, III.57–60, IV.20–27 (presenting a 'hydrostatical' theory of gravity), IV.45–47 (on the properties of air) and IV.48–49 (on the properties of water). By taking into account different interrelated topics in De Volder's treatment of hydrostatics, one can divide De Volder's texts as follows:

1. On the idea that air has weight like any other 'element': experiments 24 and 26; *Disputationes de aëris gravitate* (disputation 1).
2. On the measurement of the weight of a limited quantity of air: experiment 27; *Disputationes de aëris gravitate* (disputation 5).
3. On the ascertainment of the pressure of the atmosphere: experiments 1–4, 13–14 and 26–27; *Disputationes de aëris gravitate* (disputations 3 and 5).
4. On the *metus vacui*: experiments 1–2, 4, 11, 13–14 and 26; *Disputationes de aëris gravitate* (disputation 4).
5. On the conditions of equilibrium of fluids: experiments 15 and 17–24; *Disputationes de aëris gravitate* (disputations 2–3); De Volder's *dictata* on Descartes's *Principia*; Serrurier's *Disputatio de gravitate aëris* (1690).
6. On the idea that elements gravitate on themselves: experiments 23–24; De Volder's *dictata* on Descartes's *Principia.*
7. On pressure exerted along vertical lines, and on the relation of the height of a fluid and its pressure: experiments 3 and 25; De Volder's *dictata* on Descartes's *Principia*; on pressure exerted in all directions: experiments 1–2; *Disputationes de aëris gravitate* (disputation 2); Schuyl's *De vi corporum elastica* (1688).
8. On of the lack of feeling of pressure in water: experiments 1–2; *Disputationes de aëris gravitate* (disputation 3); De Volder's *dictata* on Descartes's *Principia.*
9. On the hydrostatic paradox: experiments 25 and 28; Lufneu's reports.
10. On the extension of the hydrostatic model to air, and the elasticity of the air: experiments 1–3, 5–9 and 21; *Disputationes de aëris gravitate* (disputations 2–3); Bernoulli's report; Schuyl's 1688 disputations; Serrurier's 1690 *De gravitate aëris*; De Volder's *dictata* on Descartes's *Principia.*

11. On respiration: experiments 6–9; Schuyl's 1688 disputations; Stephanus Huszthi's 1693 *De hominis automato seu corpore animali* (presided over by De Volder).
12. Other properties of air (fire and transmission of sound): experiments 10, 12 and 16.

In what follows, I will analyse De Volder's theories on all these issues by placing them in the relevant historical and theoretical context.

5.1 The Gravity of the Air: A Complex Idea

The idea that air is not absolutely light, namely, that it is not characterized by a tendency to move upwards in all circumstances, but that it has gravity, or weight, or a tendency to move downwards and to exert pressure on the lower bodies (in some circumstances), traces back to Aristotle himself. As De Volder recalls in the first of his *Disputationes de aëris gravitate* (14 October 1676), before being ascertained by Galileo and Evangelista Torricelli,[1] the gravity or weight of air was admitted in *De coelo*, book 4, chapter 4, where Aristotle claims that air is light, with respect to earth and water, and is heavy with respect to fire. Water, in turn, is heavy when in air and fire, and light when in earth[2] Thus, fire is the only element which is absolutely light, while earth is absolutely heavy:

[1] "Quam suam firmiter persuasum fuerit omnibus ab omni antiquitate Philosophis, aërem nostrum, quem haurimus, gravem esse, vel ex eo coniici potest, quod hoc ipsum nequidem ab illis in controversiam vocatum fuerit. Imo tam certe id creditum fuit Aristoteli eiusque sequacibus, ut non dubitarint, huic fundamento, omnem de corporum sub Luna degentium cognitione doctrinam superstruere. Neque vero haec res dubia habita fuit, nisi postquam quam experimentis Galilaei, Torricellii, Robervallii, Pascalii, Guericke, Boylaei, Hugenii, aliorumque excellentium hac in parte virorum, gravitas ipsius aëris adeo manifestae demonstrata fuit, ejusque effectus adeo notabiles animadversi, ut qui eam nihilominus negare velit, non tantae demonstrationum Mathematicarum claritati tenebras offundat, quam quidem suam in certissimis ratiociniis percipiendis prodat ignorantiam, tarditatemque," De Volder 1676–1678, disputation 1, thesis 1.

[2] "Qua in re, ut omnem verborum effugiamus controversiam, praemonendum censeo, non id agi ut quaeratur de gravitate levitateve aëris ad alia corpora comparati, an nimirum aër levis possit dici si ad aquam et terram aliave corpora graviora conferatur, licet et ille gravis forte dici quaeat si referatur ad ignem ipso Aëre leviorem; quae res tantae est evidentiae ut non minus ridiculum mihi videatur eam, (quasi tam ineptus quispiam foret qui de ea, dubium moveret,) experimentis aut rationibus operose, ostendere velle, quam aëris, de qua haec agemus, gravitatem certis ab experientia petitis ratiociniis demonstratam convellere. Quanquam enim hoc effugium prima fronte speciem quandam veri mentitur, qua ab errore Aristoteles eiusque sequaces salvari possint, cum ille multis in locis expressis verbis asserat, nec aërem nec aquam simpliciter graves esse, ἁπλῶς μὲν γὰρ οὐδ ἕτερον τούτων κοῦφον ἢ βαρύ γῆς μὲν γὰρ ἄμφω κουφότερα, πυρὸς δὲ βαρύτερα. *Simpliciter enim neuter eorum aut levis, aut gravis, ambo enim terra leviores, igne vero graviores* l. IV. de coelo c. IV. et passim alibi," De Volder 1676–1678, disputation 1, thesis 2; cf. *De coelo*, 311a25–26.

> I use the term 'absolutely', in view of the generic character of 'light' and 'heavy', in order to confine the application to bodies which do not combine lightness and heaviness. It is apparent, I mean, that fire, in whatever quantity, so long as there is no external obstacle moves upward, and earth downward; and, if the quantity is increased, the movement is the same, though swifter. But the heaviness and lightness of bodies which combine these qualities is different from this, since while they rise to the surface of some bodies they sink to the bottom of others. Such are air and water. Neither of them is absolutely either light or heavy. Both are lighter than earth-for any portion of either rises to the surface of it-but heavier than fire, since a portion of either, whatever its quantity, sinks to the bottom of fire; compared together, however, the one has absolute weight, the other absolute lightness, since air in any quantity rises to the surface of water, while water in any quantity sinks to the bottom of air.[3]

This theory served Aristotle in criticising the Platonic idea, given in the *Timaeus*, that all bodies are absolutely heavy, and that lightness is to be intended only as a form of upwards extrusion of a heavier body on a lighter one (which will be later assumed by Epicurus). According to Aristotle's *De coelo*, book 4, chapter 2, this theory does not apply to those bodies which always move upwards or downwards, namely fire and earth.[4] Moving from Aristotle's 'relativist' theory of elements given in *De coelo*, book 4, chapter 4, De Volder develops an argument against the characterization of elements in terms of 'gravity' and 'levity', by claiming that since there are mixed bodies – like gold – which are heavier than any 'pure' element, the determination of the nature of a mixed body on the basis of the consideration of its gravity or levity has to be relinquished, as all elements are lighter than gold.[5] Hence, De

---

[3]Aristotle 1922a, 311a15–30 (this edition follows the Bekker numbering). For an introduction to Aristotle's theory of elements, see Crowley 2008a, b; Gill 2009.

[4]"Those of our predecessors who have entered upon this inquiry have for the most part spoken of light and heavy things only in the sense in which one of two things both endowed with weight is said to be the lighter. And this treatment they consider a sufficient analysis also of the notions of absolute heaviness, to which their account does not apply. This, however, will become clearer as we advance. One use of the terms 'lighter' and 'heavier' is that which is set forth in writing in the Timaeus, that the body which is composed of the greater number of identical parts is relatively heavy, while that which is composed of a smaller number is relatively light. As a larger quantity of lead or of bronze is heavier than a smaller-and this holds good of all homogeneous masses, the superior weight always depending upon a numerical superiority of equal parts-in precisely the same way, they assert, lead is heavier than wood. For all bodies, in spite of the general opinion to the contrary, are composed of identical parts and of a single material. But this analysis says nothing of the absolutely heavy and light. The facts are that fire is always light and moves upward, while earth and all earthy things move downwards or towards the centre," Aristotle 1922a, 308a30–308b15.

[5]"Facile tamen patebit attendenti quam frivolum id sit, et quam, si hoc admittatur, principia rerum sublunarium instabili nitantur fundamento. Eam enim ob causam si liceat Aëri levitatem assignare, quia levior terra, certe et licebit aquam levem dicere, quam tamen omnes gravem clamitant. Nec aqua solummodo levis erit, sed et terra, sed et metalla, et mineralia, imo omnia corpora excepto solo auro, utpote quod omnium hactenus cognitorum est gravissimum. Quo ipso si forte praeterea aliquod corpus reperiatur gravius, actum erit prorsus de auri gravitate, idque quod iam sua natura grave audit, mox post talis, si forte detur, corporis inventionem, licet natura sua neutiquam immutatum sit, iuxta hunc Philosophicum sc. loquendi morem de hoc gravitatis solio deturbabitur, et dicendum erit leve. Quo ipso non difficulter animadvertitur, gravitatem corporum hoc sensu sumtam nihil de eorum natura determinare, quippe qua manente eadem corpus possit ex gravi mutari in leve, et ex levi in grave: unde et illud clarum est, quam solidis et constantibus fundamentis

Volder presents the most recent evidence against the idea of the 'levity' of air, namely, Torricelli's experiment with a tube filled with mercury,[6] which for De Volder "shakes the foundations of Scholastic philosophy not a little,"[7] as it is Scholastic philosophy, rather than Aristotle, that is his polemical target. However, De Volder's mention of Aristotle's theory of elements omits what would then be a key point in the early modern discussions of the gravity and lightness of air, and which he will himself discuss. Namely, the question whether a 'relatively' light or heavy body like air (and by extension water, too) has gravity in its natural place. Namely, whether a volume of air or water has a tendency to move downwards and exert a pressure on a lower volume of the same element. According to Aristotle's *De coelo*, book 4, air has a weight or pressure on itself which is revealed by the difference in weight between a bladder containing compressed air, and a bladder in which air is not compressed: "in its own place each of these bodies has weight except fire, even air. Of this we have evidence in the fact that a bladder when inflated weighs

superstruatur Aristotelis et Peripateticorum de elementis rerum sublunarium doctrina, utpote quae si non in totum, magna certe ex parte hac gravitate levitateve nititur, quae tamen, ut ostendimus, ipsorum corporum naturam non attingit," De Volder 1676–1678, disputation 1, thesis 2. The use of the case of gold – whose properties cannot be reduced to any of the four elements – in order to discard the Aristotelian theory of elements can also be found in an anti-Aristotelian thinker such as David Gorlaeus, and later in Henricus Reneri: see Lüthy 2012, chapter 4; Buning 2013, chapter 5.

[6] In Torricelli's *esperienza*, is worth recalling, a tube filled with mercury, and closed on one side, is put into a container full of mercury: at that point, the mercury present in the tube does not descend completely in the container, as it is sustained by the pressing air, and leaving a 'void' space in the upper part of the tube itself. Torricelli provided an account of his experiment in two letters to Michelangelo Ricci of 1644, in Galluzzi and Torrini 1975, volume 1, 122–123 and 130–132.

[7] "Hanc autem posteriorem partem multis experimentis quae vulgo commentitio cuipiam vacui metui ascribi solent confirmarunt recentiores. Inter quos primum Torricellius non in hunc quidem finem, sed ut spatium inane ostenderet, tubo usus est vitreo, trium pluriumve pedum longitudine, quem ab altera solum parte apertum et mercurio repletum, aperto orificio post repletionem digito obturato, inversoque tubo, in aliud vasculum eodem mercurio refertum immisit ea lege, ut orificium tubi, antequam aperietur, depressum foret intra superficiem mercurii in vase stagnantis. Quibus ita constitutis, remotoque digito, quid fit? Mercurius nonne quod expectandum videretur, sua gravitate ex Tubo aperto effluet omnis? Praesertim cum in vulgata opinione nulla vis sit in aëre, huic descensui resistendo, imo e contrario videatur aër potius ei adiumento fore quam impedimento; nam quo delabitur plus mercurii, eo plus ascendit Aëris, qui motus ipsi fingitur naturalis. An vero providae Naturae iussibus obsecundans, suìque ipsius boni negligens, summo hoc sapientiae gradu, quo plerique mortalium carent, praeditus Mercurius in ipsa haerebit summitate? Ne forte suo lapsu in summitate Tubi, ad quam Aëri aditus non patet, monstrum illud horrendum producatur, quod Inane vocamus? Alterutrum certe horum fieri oporteret, si vulgaris Philosophiae Natura inhaereret principiis. Verum illa harum legum prorsus oblita, nec sese ad commentitias Philosophorum regulas astringi permittens, mercurium deprimit, non quidem ad fundum usque, sed ad altitudinem 28. aut 29. digitorum circiter, ad quam eundem subsistere cogit. Quod profecto experimentum Scholasticae Philosophiae fundamenta non parum conquassat. Corpus enim tantae gravitatis quanta est mercurius aperto orificio, per quod possit effluere, ad eam altitudinem sustineri, sine ullo, quod sustineat, aequipondio, vel novae cuiusdam intelligentiae opus est, utpote quam, ut ex coelorum motibus addiscimus, eo in casu, quo causas rerum ignoramus, in subsidium vocandam suo nos exemplo docuit Aristoteles; aut forte nova quaepiam in hoc rerum statu forma mercurio adiumento est, ne lapsu graviore ruat," De Volder 1676–1678, disputation 1, thesis 4.

more than when empty."[8] De Volder does not mention this point, at least not in the printed version of the text of the first disputation. Yet, he related the thesis of the gravity or weight of all elements to the idea that each of them also gravitates in its own 'place' in experiment 24 (28 January 1677), where Christopher Love Morley refers to what he assumed to be the printed contents of a disputation that De Volder had presided over: "this experiment did relate, unto this thesis [...] printed by him: aquam, et aërem esse corpora gravia, cum r[ati]o, tum exper[ienti]aa[9] docet, ac proinde falsum est, elementa in suis locis non esse gravia," ("water, and air are heavy bodies, as both reason and experience teach, and therefore it is false that elements are not heavy in their places").[10] If the first part of the thesis was sustained by De Volder in disputation 1,[11] the second part (i.e. the thesis that elements gravitate in their natural places) is never mentioned in the printed text of the disputation. We may suppose that De Volder's main reason for doing so was in order to distance himself as much as possible from the 'Aristotelians', and to number himself among innovators such as Galileo and Torricelli. Indeed, if De Volder had openly defended the thesis that elements gravitate on themselves, it would not have been easy for him to use the authority of Galileo, nor that of Descartes (whose theory of weight, as I am going to show and discuss in Sects. 6.2.2.3.1, *De Volder's Cartesian cosmology and theory of weight* and 6.2.2.3.2, *De Volder's Huygenian theory of weight (and its relation to the idea of cohesion)*, was accepted by De Volder). Indeed, both Galileo and Descartes came to endorse, in different ways, the tenet that elements do not exert weight or pressure on themselves.

As a matter of fact, De Volder could not have read the treatise in which Galileo denied that elements gravitate on themselves, namely his unfinished *De motu antiquiora* (written between 1589 and 1592, but published only in 1854 in Eugenio Alberi's edition of Galileo's works).[12] Yet, Galileo's treatment of the weight of the air in his *Discorsi e dimostrazioni matematiche intorno a due nuove scienze* (1638) – a text which De Volder knew well[13] – builds upon his earlier, undisclosed theory of the pressure of air on itself, and bodies specifically heavier than air. So that Galileo's natural-philosophical premises, as to the treatment of the weight of the air, were at odds with De Volder's. Moreover, the consideration of Galileo's treatment and practical ascertainment of the weight of the air sheds light on De Volder's immediate background regarding these topics: thus, a consideration of Galileo's positions is required to understand De Volder's own intellectual path. In Sect. 5.1.1, *Galileo on the weight of the air and his Aristotelian predecessors*, I will consider these posi-

[8] Aristotle 1922a, 311b5–15.

[9] *Sic*.

[10] De Volder 1676–1677, 130r–v.

[11] Probably, it was dictated in the quoted form during the experiment itself, as it is reported in Latin, and as it cannot be found in this form in the text of the disputation.

[12] See Fredette 2001.

[13] As I am going to show, it is cited in De Volder's *dictata*: see Sect. 6.2.1.2, *Huygens's quantification of centrifugal force*.

tions. Later (see Sect. 5.1.3, *De Volder's 'didactic' ascertainment of the pressure of the atmosphere – and related experiments*) I will consider De Volder's positions.

5.1.1 Galileo on the Weight of the Air and His Aristotelian Predecessors

Galileo was certainly a founding father in seventeenth-century experimental researches on the weight of air, first of all, by inspiring Torricelli. Yet, he also criticized the idea that elements gravitate on elements, which had been a widely discussed topic since Antiquity, and on which the interpreters of Aristotle had not come to an agreement even as late as the seventeenth century.[14] Moreover, the method used by De Volder to measure the weight of the air – i.e. by putting a sealed container on a balance, before and after having emptied it with an air-pump – would have seemed unfeasible to Galileo, who did not admit that air has weight i.e. that it can exert pressure on a heavier body, such as the arms of a balance, unless it is compressed.

In the Conimbricenses commentary on book 4 of *De coelo* (1592),[15] this thesis is supported by the authorities of Archimedes, Ptolemy, Simplicius, and Themistius, for whom (according to the commentary) elements do not gravitate on elements because (1) if a body lighter than a given fluid gravitates on this fluid – i.e. it exerts a downwards pressure – it would sink. Moreover, (2) one cannot feel the weight of water when one is underwater. Yet, it is true that elements gravitates on lighter elements, as shown by experience: once a heavy body is immersed in a lighter fluid, indeed, it moves downwards, and the fluid moves upwards.[16] Other interpreters of

[14] For an introduction to these issues, see Gliozzi 1930; Knowles Middleton 1963; Taton 1963; Zouckermann 1981; Nonnoi 1988; Conti 1997; Palmieri 2005; Shank 2012; West 2013.

[15] Owned by De Volder: see *Bibliotheca Volderina*, 5. The positions of the Conimbricenses on the heaviness and lightness of elements are reported in the *Idea philosophiae naturalis* (1652) of Franco Burgersdijk, without mentioning, however, the problem of the gravity of air on air. Burgersdijk reports the idea that levity and gravity are derived properties of elements (like being dense and rarefied), and that such properties ground the natural movement of elements (upwards or downwards), which, once they reach their natural place, naturally come to rest: Burgersdijk 1652, 41–44. As to the positions of the Conimbricenses, see the next footnote.

[16] "Caeterum non solum Archimedes et Ptolemaeus, sed etiam Simplicius, Themistus, Syrianus, Alexander, ac Peripatetici fere omnes contrariam sententiam amplexi sunt, quae hisce duobus assertionibus continetur. Prima. Nullum elementum in suo loco naturali constitutum gravitat aut levitat, si modo gravius subsideat, levius emineat. Probatur: quia ideo elementa tanto naturae impetu praescriptam sibi a natura sedem petunt, ut inibì tranquillitate adepta consistant, nec iam ulterius cieantur. Deinde, quia si aër et aqua in patria regione deorsum niterentur, degravaret nostra capita aëris pondus, et ii qui sub aquis natant, onus magnum sentirent, quod tamen experientiae repugnat. Cur autem adhibeamus moderationem illam, si modo gravius, et c. ex sequenti assertione constabit. Secunda assertio: corpora gravia supra minus gravia, aut supra levia, ubicunque sint, gravitant: similiterque levia infra minus levia, aut infra gravia, levitant. Hanc assertionem probat experientia. Videmus enim ubicunque gravia aut minus gravia ponantur, confestim superiorem

Aristotle (whose books are extant in De Volder's *Bibliotheca*) were more conciliatory towards *De coelo*, book 4. For instance, in his *De rebus naturalibus* (1590) Jacopo Zabarella differentiated between the nature and the operation of elements, namely between their gravity (*gravitas*), intended as the natural tendency that elements have to move towards their natural place, and gravitation (*gravitatio*) which is the pressure (*pressio*) that they exert downwards, due to the "excess of motive force upon the resistance of the medium." In the case of air, according to Zabarella, Aristotle rightfully maintained that it 'gravitates' when it is in its natural place because, even if its nature is that of lightness, it can be more easily moved downwards than upwards. In other words, according to its action it is more heavy than light; according to its nature, however, it is more light than heavy.[17] This distinction was widely adopted in late-Scholastic and Aristotelian treatises, such as in the *De motu* (1591) of Francesco Buonamici (lecturer at the Pisa Studium in the years Galileo studied there), whose theory of floatation would be discarded by Galileo in his *Discorso intorno alle cose che stanno in su l'acqua* (1612) (as I shall show in Sect. 5.3.1, *Galileo on floatation (1612)*).[18] Girolamo Borri – who lectured at Pisa

locum levibus aut minus gravibus, si haec infra sint, deferre: quod certe non fit, nisi mutuo gravium descendentium, leviumque ascendentium conatu atque impulsu, quem gravitandi et levitandi actum diximus," Conimbricenses 1603, *In de Coelo*, *quaestio* 2, article 1, columns 529–530.

[17] "Ego igitur respondendum puto gravitatem sumi posse duobus modis. Cum enim sit qualitas quaedam et propensio ad inferum locum, quam consequitur hic effectus, haec operatio, ad eum locum moveri, ideo accipi potest tum secundum se pro actu primo, tum pro operatione et actu secundo [...]. Sed altero modo accepta augeri ac minui potest, et dicitur gravitatio potius quam gravitas. Cum enim nil aliud sit quam excessus virtutis motricis supra resistentiam medii, unde fit descensus et pressio substantis corporis resistentis, hic maior fit dum resistentia imminuitur. [...] Ipsa vero gravitatis ac levitatis distinctio, quam posuimus, tum per se manifesta est, tum ab Aristotele posita, nam gravitatem pro gravitatione sumimus quando dicimus: grave in suo loco non est grave, id est, non gravitat, nec premit deorsum. Primo enim modo falsum diceremus, quoniam elementum grave habet ubique naturam gravis, etiam in loco suo, ob id Aristoteles in contextu 16 et 27 quarti libri De caelo sumit priore modo gravitatem et levitatem, dum dicit levissimum esse id quod omnibus supereminet, et gravissimum id quod omnibus substat. Sic enim etiam in locis suis dicit esse gravissimum atque levissimum. In contextu autem 29 et 30 eiusdem libri sumit gravitatem et levitatem secundo modo, dum dicit lignum unius talenti in aëre maiorem habere gravitatem quam plumbum unius librae, contra vero in aqua rem sese habere. Sumit enim gravitatem pro operatione, quae notat respectum ad externa corpora per quae fit motus. Eo namque respectu variato etiam gravitas variatur, non quidem natura ipsa, sed gravitatio. Dicit etiam in contextu illo 30 omnia elementa in suis locis aliquam habere gravitatem praeter ignem, nam aër quoque in suo loco existens facilius deorsum pellitur, quam sursum, quia magis gravitat, quam levitat; attamen secundum eius naturam magis est levis, quam gravis," Zabarella 1617, 341–342; cf. *De coelo*, 309a26–309b17, 311a21–29 and 311b2–12. See also Zabarella 2016, 441–442.

[18] "Illud sane constat q[uod], non est idem penitus utrobiq[ue] gravitatis officium: namq[ue] extra proprium locum efficit motum; in proprio loco magis studet quieti, ut videatur naturae consilium potius esse, ut ibi maneat corpus, neque inde facile detrudi pariatur, quanq[ue] centri satagat, q[uod] cum in proprii loci latitudine versetur; bona etiam ex parte finem suum sit consecutum. Nunc pondus ex ea propensione cognoscitur quae cum motu coniuncta est, aut affinis est motus, velut oppressio, non ex ea quae cum quiete consistit, neq[ue] n[am] pondus rei quae in trutina appensa est, apparet, si quis ea sustineat, sed si liberae naturae suae dimittatur. Itaq[ue] et o[mn]e impedimentum {evariat} pondus. Hoc igitur, cum extra proprium locum eveniat; fit, ut gravitas extra proprium locum, non in ipso sentiat. Poterat igitur Alexander occurrere dubitationi, negans

during Galileo's studies – was also constant to Aristotle's idea that air gravitates on air. In his *De motu gravium et levium* (1575), he considers the evidence brought to support this idea in *De coelo*, book 4, chapter 4: first, the above-mentioned case of an inflated bladder; second, sample concerning mixtures. According to the *De coelo*, "[i]n air, for instance, a talent's weight of wood is heavier than a mina [i.e. 1/60 of a talent] of lead, but in water the wood is the lighter." Aristotle used this sample to argue for the 'relativity' of weight, namely, that "a body which is regarded as light in one place is regarded as heavy in another, and vice versa."[19] Borri used this sample to argue for the thesis that an element always weighs on the same element, when he considers the proportions of air in wood and in lead: as a piece of wood contains more air, in proportion, than a piece of lead, once it is put in water the quantity of air in it makes it lighter.[20] Once in air, however, the weight of the wood is also determined by the quantity of air it contains. A piece of wood, indeed, falls more rapidly in air than a piece of lead of the same weight, even if it contains more air (in proportion) than lead does. In other words, the quantity of air it contains does not make it weightless in air, but contributes to its weight.[21]

---

elementa in proprio loco gravitatem, levitatemve habere, et id ipsi dari poterat, si eam facultatem accepisset quae pondere deprehenditur; nos illam vim accipere voluimus, qua rerum natura definitur, et ei perpetuo manet affixa," Buonamici 1591, 480. See Conti 1997.

[19] Cf. the full text: "[i]t is due to the properties of the elementary bodies that a body which is regarded as light in one place is regarded as heavy in another, and vice versa. In air, for instance, a talent's weight of wood is heavier than a mina of lead, but in water the wood is the lighter. The reason is that all the elements except fire have weight and all but earth lightness. Earth, then, and bodies in which earth preponderates, must needs have weight everywhere, while water is heavy anywhere but in earth, and air is heavy when not in water or earth. In its own place each of these bodies has weight except fire, even air. Of this we have evidence in the fact that a bladder when inflated weighs more than when empty. A body, then, in which air preponderates over earth and water, may well be lighter than something in water and yet heavier than it in air, since such a body does not rise in air but rises to the surface in water," Aristotle 1922a, 311b1–15.

[20] "Lignum ergo talenti unius, quod aëris in se, plus habeat, quam aquae, ac terrae, gravius est in aëre, quam plumbum eminae tantum unius in aquam, quod longe plus terrae quam aquae, et longe plus aquae quam aëris habeat plumbum ocissime mergetur: aquae extimam planiciem lignum vix attingit, et eamdem postquam attingerit, supernatans illico quiescet: ita ut eam ob rem maiorem levitatem praeseferre videatur. Id Aristoteles quarto libro de Caelo [...] ideo verum esse est ratus, quod aër in suo proprio naturalique loco aliqua gravitate sit praeditus," Borri 1576, 213.

[21] "[...] ad experientiam omnium rerum magistra, perinde ac ad sacram ancoram confugimus, et duobus adinventis, tum ligni, tum plumbi frustulus aequalis, ut ex aspectu coniicere licebat, ponderis, neque enim nos ad lancem illa expendere necessarium esse duximus, sed periculo quod facturi fueramus, sat esse rati fuimus, si ad oculum expenderentur, eadem ergo duo adinventa aequalis ponderis frustula, ex altiore nostrarum aedium fenestra par impulsu, eodemque tempore proiiciemus: plumbum segnius descenderet, super lignum enim, quod prius in terram ceciderat, omnes quotquot ibi, rei exitum expectabamus, illud praeceps ruere vidimus: idque non semel, sed saepenumero eodem successu tentavimus. Cuius rei experimento ducti omnes in eamdem nobiscum pedibus {iuere} sententiam. Ergo tum rationi, tum experimento, tum auctoritati consentaneum est, aëris in suo proprio, naturalique loco, nonnullam esse gravitatem quo fit, ut lignum in quo plus est aëris, quam in plumbo, aequalis ponderis per aerem medium velocius descendat, et super aquam natet, quod in aqua aër sit levius, perinde ac levis est in terra, levitate licet existente minore," Borri 1576, 215.

In short, as has been highlighted by Lino Conti, "when [...] the tricky question of the weight of air in its natural place had to be faced, [...] an endless disagreement reigned among the commentators of Aristotelian physics."[22] Once Galileo came to deal with the problem of the weight that elements have on themselves in his *De motu antiquiora*, he developed a theory of hydrostatic equilibrium according to which elements do not gravitate or press on elements. In his *De motu*, in the unnumbered chapter devoted to criticism of Aristotle's idea that air gravitates on air, he claims that this idea would entail a perpetual motion of masses of air, and of all fluids. A pressure exerted by a mass of air, or water, on the lower layers of the same element would not find any resistance by the medium, and perpetual motion would follow. Also, according to Galileo, Aristotle's experience with a bladder does not prove that air weighs on air, for the reason that the bladder contains compressed air, the weight of which is higher than that of uncompressed air.[23] As indicated by Paolo Palmieri, Galileo's positions contradicted Archimedes's model of floatation, which in the 1580s was known through Niccolò Tartaglia's and Federico Commandino's editions of Archimedes's *De insidentibus aquae* (according to Tartaglia's title, 1543), or *De iis, quae in aqua vehuntur* (in Commandino's version, 1565), which were based on a Latin translation by William of Moerbeke.[24] Archimedes's model of floatation (referred to by the Conimbricenses to argue for the weightless nature of water in water) presupposes that the pressure exerted by a solid on the water in which it floats, is equal to that exerted by a portion of water of the same weight. In Fig. 5.1, Archimedes and Commandino 1565, 4r, illustrating Commandino's edition, proposition 5, the solid HTEF weighs on the water below it as much as the portion of water RSQY weighs on the water below it. Otherwise, the semicircle XO would undergo a pressure different from that undergone by the semicircle OP.[25]

[22] Conti 1997, 13.

[23] "Et, primo quidem, omnino inexcogitabile videtur, quomodo aër et aqua in proprio loco gravitent. Nanque aliqua pars aquae in loco aëris, hoc est in aëre ipso, gravitat, et deorsum quidem fertur quia gravitat; sed quis unquam mente concipiet, aliquam partem aquae in aqua descendere? Si enim descendet, quando erit in fundo, necesse est ut locus, in quem intrat, iam evacuetur ab alia aqua, quae coacta erit ascendere unde alia recessit; et sic iam illa pars aquae erit levis in proprio loco. [...] Ad exemplum autem Aristotelis de utre, respondeo quod, si foramen utris seu follis inflati sit apertum, ita ut aër, non vi compressus, in folle detineatur, non erit iam uter gravior quam non inflatus: sed si vi multum aëris in eo comprimatur, cui dubium erit quod gravitabit? Aër enim tunc, vi constrictus, gravior est aëre libero et vaganti: sicut si uter lana repleatur, deinde vero alterum tantum lanae superaddatur, vi comprimendo, quis anceps erit an gravior fiet uter necne?" Galileo 1890–1909, volume 1, 286.

[24] Commandino's version was based on Tartaglia's. For a thorough account of the editorial history of Archimedes's treatise, and of the rediscovery of the original Greek version in 1906 by Johan Ludvig Heiberg, see Archimedes and Fleck 2016.

[25] "Proposition 5. Of solids one which is lighter than the fluid, when thrown into the fluid, will sink down until a volume of the fluid equal to the volume of the immersed portion has the same weight as the whole solid. [...] [L]et [HTEF] now be the given solid which is specifically lighter than the fluid, [RSQY] a volume of the fluid equal and similar to the immersed portion [BCHT]. If the fluid is at rest, the weight of [HTEF] must be equal to that of [RSQY]," Dijksterhuis 1987, 375–376. Cf. Archimedes and Commandino 1565, 4. As to David Rivault's version, see Fig. 5.2, Archimedes and Rivault 1615, 496.

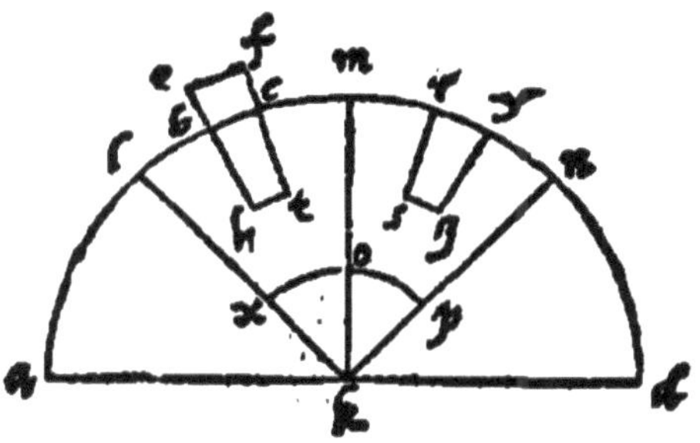

Fig. 5.1 Representation of the condition of floating of bodies, in support of Archimedes's propositions 3 and 5. (Source: Archimedes and Commandino 1565, 4r. Rome, Biblioteca Nazionale Centrale, signature: †[4]A-K[4]L[6], https://archive.org/details/bub_gb_RR5685AlyVAC. Public domain mark 1.0)

Fig. 5.2 Representation of the condition of floating of bodies, in support of Archimedes's propositions 3–5. (Source: Archimedes and Rivault 1615, 496. ETH-Bibliothek Zürich, signature: Rar 3843, https://www.e-rara.ch/doi/10.3931/e-rara-18792. Public domain mark 1.0)

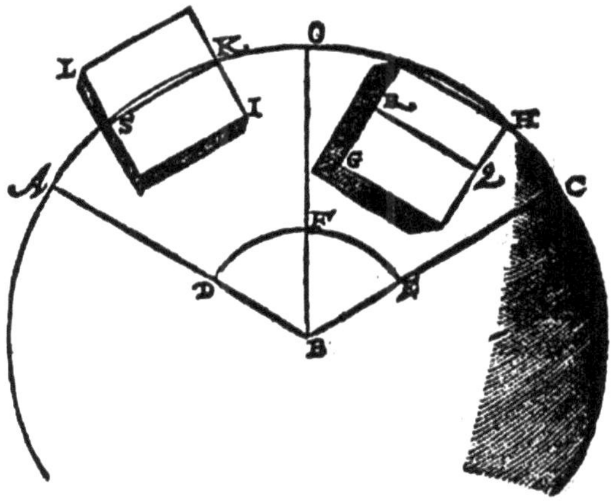

If they undergo the same pressure, however, they will be in equilibrium, in accordance with Archimedes's first postulate:

> [l]et it be granted that the fluid is of such a nature that of the parts of it which are at the same level and adjacent to one another that which is pressed the less is pushed away by that which is pressed the more, and that each of its parts is pressed by the fluid which is vertically above it, if the fluid is not shut up in anything and is not compressed by anything else.[26]

[26] Dijksterhuis 1987, 373. Cf. Archimedes and Commandino 1565, 1: "[p]onatur humidi eam esse naturam, ut partibus ipsius aequaliter iacentibus, et continuatis inter sese, minus pressa a magis pressa expellatur. Unaquaeque autem pars eius premitur humido supra ipsam existente ad perpendiculum, si humidum sit descendens in aliquo, aut ab alio aliquo pressum."

The postulate – which would be widely used by De Volder – presupposes that there is a pressure of water on water. Yet, in the way that the Conimbricenses read it, this is not the case. For them – as for Galileo – if layers of water add weight to the layers below them, there would no longer be any equilibrium in the fluid, but rather a perpetual movement. As argued by Palmieri, in an attempt to overcome this problem, Galileo came to conceive the balancing of volumes of water with each other by considering – as it appears from the different draft versions of the figures representing floating bodies in *De motu* – the water as contained in artificial containers, i.e. outside its natural place, whereas Archimedes considered a water-world, i.e. a planet composed only by water, which is thus contained only by itself.[27]

To sum up, according the Conimbricenses and to Galileo's *De motu*, the weighing, pressing or gravitating of water on itself would result in a perpetual movement or displacement of water. The way in which the Conimbricenses read – according their brief reference – Archimedes's theory of flotation presupposes that once in equilibrium, the weight of water is no longer exerted as pressure. Galileo, too, came to embrace a 'phenomenic' understanding of weight and gravitation, namely, that when there are no apparent effects of pressure – such as the displacement of water – pressure is no longer exerted. As he declares in his *De motu*, heavy bodies (*gravia*) are to be intended as bodies which *actually* move downwards, and light bodies (*levia*) are those bodies which *actually* move upwards. Water, in water, is weightless, i.e. it does not weigh or press, for the reason that it does not move downwards.[28] For Galileo, gravity and levity are to be considered only comparatively: bodies are absolutely heavy only in a vacuum, where the weight or gravity of a body is not affected by the surrounding medium. In a vacuum, all bodies would move downwards, whereas in a medium they gravitate or have weight only to the extent that their 'specific weight' exceeds the absolute weight of the volume of the medium they displace. In the case of levity, this is determined by the upwards extrusion of a comparatively lighter body by a heavier medium, which presses upon it and moves it upwards.[29] Galileo does not deny that one body can press or gravitate on another

[27] According to Paolo Palmieri, Galileo was then ultimately led to develop his own theory of buoyancy – which he could not justify in an Archimedean way, i.e. by considering the weight of water on water – and presented it in his *Discorso intorno alle cose che stanno in su l'acqua o che in quella si muovono* (1612): see Palmieri 2005.

[28] "Haec, meo iudicio, quicquid dicant alii, est vera problematis explicatio. Cum igitur nec aër nec aqua deorsum in suis regionibus ferantur neque sursum, ne dicantur esse aut gravia aut levia; cum gravia definiantur ea esse quae deorsum feruntur, levia vero quae sursum. Et cum de motu loquimur, semper non solum gravitatis aut levitatis mobilis, sed gravitatis et levitatis medii etiam, ratio est habenda: non grave deorsum movebitur, nisi medio per quod ferri debet gravius erit; nec leve ascendet, nisi levius fuerit medio per quod movetur. Quod cum ita sit, aqua non descendet in aqua, cum aqua gravior non sit quam aqua; et cum non descendat, non erit aqua in aqua gravis," Galileo 1890–1909, volume 1, 289.

[29] "Grave et leve non nisi in comparatione ad minus gravia vel levia considerarunt qui ante Aristotelem; et hoc quidem, meo iudicio, iure optimo. [...] Cum enim in omni medio gravium gravitates tantum imminuantur, quantum illius medii pars aequalis moli solidi ponderaret, patet quod in illo solum medio integrae et non imminutae solidorum habebuntur gravitates, cuius nulla fuerit gravitas: tale autem solum est vacuum," Galileo 1890–1909, volume 1, 289 and 295.

without displacement of one of the two bodies: for him, we are "weighed" (*gravari*) when we feel a pressure to which we oppose a resistance.[30] In the case of water on water, or air on air, however, such gravitation or pressure does not take place. Or, in other words: we can be pressed by a body without being displaced by it because we can resist it; water, on the other hand, cannot resist such pressure. This explains why, for Galileo – as for the Conimbricenses – we do not feel actual pressure of air in air, and of water in water. Upper layers of water do not 'transmit' their weight to lower layers, just as a man immersed in water does not feel the pressure of the incumbent mass of water, but he feels only the difference, in weight, between his body and an equal volume of water.[31] As argued by Conti, a consequence of this theory is that since pressure is exerted only by heavier bodies on lighter ones – which are extruded upwards – a mass of water like that of the sea does not press on that part of the Earth which is the seabed: nor does air actually press on earth, being lighter than it.[32]

In short, Galileo does not deny that air has a weight: rather, he affirms that it is weightless when surrounded by air, or when in contact with a heavier body (either solid or fluid). In fact, he devised a method to measure the proportion of the weight of air and that of water, which he described in a letter to Giovanni Battista Baliani of 12 March 1614, and then in his *Discorsi*. According to this method, (1) one measures the difference in weight between a bottle containing compressed air, and one containing air at its usual pressure; (2) then, one measures the volume that the air forcefully inserted in the bottle occupies when it is not compressed, by joining the bottle with compressed air to another bottle, containing water. The volume is found by observing how much water is expelled by the second bottle (through a hole) by the air in the first bottle. Through this method, Galileo found that the proportion of the weight of air and water is 1 to 400.[33] Yet, in the *Discorsi* Galileo did not take into

---

[30] "[D]icimur gravari, quando super nos incumbit aliquod pondus quod sua gravitate deorsum tendit, nobis autem opus est nostra vi resistere ne amplius descendat; illud autem resistere est quod gravari appellamus," Galileo 1890–1909, volume 1, 288.

[31] "Quomodo ergo solvetur problema, nisi dicamus, aquam et aërem non gravare in suis regionibus? [...] Si autem in aqua existentibus aliquod corpus aeque grave ac aqua nobis immineat, neque ab illo gravabimur neque attollemur, quia neque sursus neque deorsum tale corpus feretur. At non invenitur corpus aliquod, quod magis aquae in gravitate aequetur quam ipsamet aqua: non ergo mirum est si aqua in aqua non descendat et gravet; diximus enim, gravari esse resistere nostra vi corpori deorsum petenti. Et eadem prorsus ratio de aëre habenda est. Haec, meo iudicio, quicquid dicant alii, est vera problematis explicatio," Galileo 1890–1909, volume 1, 288–289. Galileo was responding to Borri's solution to the problem of the absence of the feeling of pressure underwater; for Borri the mass of water behaves as a solid body: cf. Borri 1576, 232.

[32] Conti 1997, 20–21.

[33] "Ho preso un fiasco di vetro assai capace e col collo strozzato, al quale ho applicato un ditale di cuoio, legato bene stretto nella strozzatura del fiasco, avendo in capo al detto ditale inserta e saldamente fermata un'animella da pallone, per la quale con uno schizzatoio ho per forza fatto passar nel fiasco molta quantità d'aria; della quale, perché patisce d'esser assaissimo condensata, se ne può cacciare due e tre altri fiaschi oltre a quella che naturalmente vi capisce. In una esattissima bilancia ho poi pesato molto precisamente tal fiasco con l'aria dentrovi compressa, aggiustando il peso con minuta arena. Aperta poi l'animella e dato l'esito all'aria, violentemente nel vaso conte-

account the pressure of air in order to explain the raising of water in syphons. As shown in Sect. 4.1.3.1, *Four factors of cohesion: vacuum, rest, pressure and entanglement*, Galileo used instead the idea of the force of the vacuum in order to account for the maximum height reached by water in syphons (18 *braccia*, circa 10.5 metres). Galileo probably did so – without expressly confronting the issue of the pressure of air on water, neither in the *Discorsi*, nor in the *Discorso intorno alle cose che stanno in su l'acqua* – in accordance with his theory of weight, the consequence of which is that air, being lighter *in specie* than water and earth, does not exert a pressure or gravitate on them in a way leading to ascertainable effects. For him, the only way to measure the weight of air was to compress it and to weigh it separately from the surrounding body of air. Later – as I show in Sect. 5.1.2, *The weighing of a separate volume of air* – the techniques of weighing air were perfected.

5.1.2 The Weighing of a Separate Volume of Air

Over the following years, Descartes proposed the same method to ascertain the ratio of the weights of air and water, finding a proportion of 1 to 145 (in a letter to Marin Mersenne of 19 January 1642).[34] In turn, in his *Cogitata*

nuta, e rimessolo in bilancia, trovandolo notabilmente alleggerito, sono andato detraendo dal contrappeso tant'arena, salvandola da parte, che la bilancia resti in equilibrio col residuo contrappeso, cioè col fiasco: e qui non è dubbio che 'l peso della rena salvata è quello dell'aria che forzatamente fu messa nel fiasco e che ultimamente n'è uscita. Ma tale esperienza sin qui non mi assicura d'altro, se non che l'aria contenuta violentemente nel vaso pesò quanto la salvata arena; ma quanto resolutamente e determinatamente pesi l'aria rispetto all'acqua o ad altra materia grave, non per ancora so io, né posso sapere, se io non misuro la quantità di quell'aria compressa: ed a questa investigazione bisogna trovar regola, nella quale ho trovato di potere in due maniere procedere. L'una delle quali è di pigliar un altro simil fiasco, pur, come 'l primo, strozzato, alla strozzatura del quale sia strettamente legato un altro ditale, che dall'altra sua testa abbracci l'animella dell'altro, e intorno a quella con saldissimo nodo sia legato. Questo secondo fiasco convien che nel fondo sia forato, in modo che per tal foro si possa mettere uno stile di ferro, con il quale si possa, quando vorremo, aprir la detta animella per dar l'esito alla soverchia aria dell'altro vaso, pesata ch'ella sia: ma deve questo secondo fiasco esser pieno d'acqua. Apparecchiato il tutto nella maniera detta ed aprendo con lo stile l'animella, l'aria, uscendo con impeto e passando nel vaso dell'acqua, la caccerà fuora per il foro del fondo; ed è manifesto, la quantità dell'acqua che in tal guisa verrà cacciata, essere eguale alla mole e quantità d'aria che dall'altro vaso sarà uscita. Salvata dunque tale acqua, e tornato a pesare il vaso alleggerito dell'aria compressa (il quale suppongo che fusse pesato anche prima, con detta aria sforzata), e detratto, al modo già dichiarato, l'arena superflua, è manifesto, questa essere il giusto peso di tanta aria in mole, quanta è la mole dell'acqua scacciata e salvata; la quale peseremo, e vedremo quante volte il peso suo conterrà il peso della serbata arena, e senza errore potremo affermar, tante volte esser più grave l'acqua dell'aria: la quale non sarà dieci volte altrimenti, come par che stimasse Aristotele, ma ben circa quattrocento, come tale esperienza ne mostra," Galileo 1890–1909, volume 8, 123–124. For the letter to Baliani, see Galileo 1890–1909, volume 12, 33–37.

[34] See AT III, 484. Discussed in Cottingham 1997.

physico-mathematica (1644–1647) Mersenne proposed the measurement of the weight of air by means of an aeolipile, in which air is expelled by means of fire, and which is weighed afterwards, finding a ratio of 1 to 1356.[35] In his *Almagestum novum* (1651), Giovanni Battista Riccioli proposed a measure for the proportion of the weight of air and water by weighing a bladder full of water, and comparing it with the weight of a bladder containing air: he found a ratio of 1 to 10,000.[36] In 1660, through a similar procedure, Giovanni Alfonso Borelli found a value of 1 to 1,175, in an experiment carried out before the Accademia del Cimento.[37] Boyle also used an aeolipile to measure the ratio of the weight of air and water in experiment 36 of his *New Experiments*, providing a ratio of 1 to 938.[38] De Volder, however, was the first who used an air-pump to measure the weight of air separately from the pressure exerted by the atmosphere. As later noted by Christian Wolff, the description of the method used by De Volder can be found only in Otto von Guericke's *Experimenta nova* (1672). However, Von Guericke did not perform it as

> some very learned scholars have tried to establish the comparative weights of air, water, and mercury. But since, in fact, the lower the air is, the more compressed and consequently the

---

[35] Mersenne 1644, 151–152, 1647, 101–105. Discussed in Sturm 1676, 65–66; Schott 1657a, 169–170.

[36] Riccioli 1651, volume 2, 652.

[37] Borelli 1670, 251. Please note that in Magalotti 1666, 254–255, is reported the value of 1 to 1,179.

[38] "We took then an Aeolipile made of Copper, weighing six ounces, five drachms, and eight and forty grains: this being made as hot as we durst make it, (for fear of melting the mettle, or at least the Sodar) was removed from the fire and immediately stopped with hard Wax that no Air at all might get in at the little hole, wont to be left in Aeolipiles for the fumes to issue out at: Then the Aeolipile being suffer'd leasurely to cool, was again weighed together with the Wax that stopt it, and was found to weigh (by reason of the additional weight of the Wax) six ounces, six drachms, and 39 grains. Lastly, the Wax being perforated without taking any of it out of the Scale, the external Air was suffered to rush in (which it did with some noise) and then the Aeolipile and Wax, being again weighed amounted to six ounces, six drachms, and 50 grains. So that the Aeolipile freed as far as our fire could free it, from its Air, weighed less than it self when replenished with Air, full eleven grains. That is, the Air containable within the cavity of the Aeolipile amounted to eleven grains and somewhat more; I say somewhat more, because of the particles of Air, that were not driven by the fire out of the Aeolipile. And by the Way (if there be no mistake in the observations of the diligent Mersennus) it may seem strange that it should so much differ from 2 or 3 of ours; in none of which we could rarefie the Air in our Aeolipile (though made red hot almost all over, and so immediately plung'd into cold Water) to half that degree which he mentions, namely to 70 times its natural extent, unless it were that the Aeolipile he imploy'd was able to sustain a more vehement heat than ours (which yet we kept in so great an one, that once the Soder melting, it fell asunder into the two Hemispheres it consists of.) The fore-mentioned way of weighing the Air by the help of an Aeolipile, seems somewhat more exact than that which Mersennus used, In that in ours the Aeolipile was not weighed, till it was cold; whereas in his, being weighed red hot, it is subject to lose of its substance in the cooling, for (as we have elsewhere noted on another occasion) Copper heated red hot," Boyle 1660, 286–288. All these values are far from the proportion of 1 to 3,000 and 1 to 4,000 mentioned as erroneous in experiment 27; probably, De Volder was referring to the proportion of 1 to 400 given by Galileo.

heavier it is, and contrarily the higher it is, the more expanded and hence the lighter, all endeavors in this direction are disappointing and generally unsuccessful.[39]

De Volder performed his measurement[40] by weighing a sphere (composed of two hemispheres) with a diameter of 8 inches and a volume of 1/6 cubic foot, and by using a balance capable of ascertaining a difference in weight of 1 grain between weights of 25 or 30 pounds.[41] In Table 5.1, *De Volder's measurements in hydrostatics*, I summarized the values given by De Volder and by all his reporters, in Amsterdam trade measures.[42] According to this system of measurement,

---

[39] Von Guericke 1993, 156. Cf. the full explanation: "[f]or example, hang a glass of this kind from a balance and weigh it, (with the stopcocks open, however, so that you may be certain that the glass is filled with air). Then evacuate all air from it and you will discover that after this has been done, it weighs one or two ounces less than its capacity. Thus after my receiver has been exhausted and is surrounded by the pressure of the atmosphere, it will be found, when examined on the balance, to be two ounces lighter than before, being equal in weight to two imperial thalers. Then prepare to let air enter once again (only a little at a time, however, so that it does not rush in so forcefully that it breaks the glass) and not only will you hear air enter with a hissing sound but you will even see the glass gradually taking on its former weight. This is a very clear demonstration of the weight of air," Von Guericke 1993, 156; cf. Von Guericke 1672, 100–101. In his *Elementa matheseos universae* (first edition 1713–1715), Wolff reported the De Volder's measurement given in the 1681 editions of the *Disputationes de aëris gravitate*, stating that De Volder followed the method described first by Von Guericke: see Wolff 1733, volume 2, 369. On Von Guericke, see Slaby 1907; Van Helden 1991; Hackmann 1979; Bazerman 1993; Harsch 2007; Conlon 2011; Schneider 2013.

[40] It is described as experiment 27 (22 March 1677) and in disputation 5 of his *De aëris gravitate* (29 June 1678), where he refers to an experiment performed a little more than a year before.

[41] "Hoc autem experimentum, quod ante annum et quod excedit publice ostendi, eam ob causam propono lubentius, quia non tantum in genere gravem aërem demonstrat, sed et data cuiuscunque massa aërea certam gravitatem et definitum pondus determinat. Qua in re usus sum bilance accurata, quae etiam si vel 25., aut 30. et c. lb. utrique imponerentur lanci, grano uno alterove addito demtove, in hanc illamve partem manifeste propenderet. Nisi enim ex aëris utamur bilancibus frustra certe erimus in examine tantilli ponderis, quod exhibet nobis aër. Hisce itaque exploravi pondus binorum hemisphaeriorum cera occlusorum, idque aëre nondum educto, ut nimirum fi inter hocce sphaerae pondus aëre repletae, et eiusdem aëre vel in totum, vel maxima ex parte vacuae discrimen quoddam animadverteretur, certo constaret id aëri educto deberi. Sphaerae itaque aëre refertae pondus fuit lb. vii. ℥i. ʒ ii. gr. xlviii. Hinc nulla omnino re sive addita sive demta, ope Antliae Pneumaticae aërem eduxi, occlusamque, ut ante, sphaeram imposui uni lanci, dum alteri adderetur idem pondus, quod sphaerae aëre refertae aequiponderare exploraveram. Profecto si nullum aër obtineat pondus necesse foret ut rebus ita constitutis in aequilibrio haereret bilanx. Nihil enim sphaerae decesserat praeter aërem quem ponderis expertem volunt. Sed quid factum? Lanx illa, quae pondus continebat,. deprimi, illa vero, quae sphaeram, attolli coepit, manifesto indicio, cum altera lanx non facta esset gravior, alteram demto aëre factam esse leviorem. Aëremque idcirco gravitatis esse participem Neque vero ad aequilibrium reduci res potuit, nisi demta parte ponderis, quod in altera lance continebatur, et demere quidem oportuit ʒ i. gr. xvii. Huius quippe sphaerae aëre tantum non prorsus vacuae pondus exploratum, fuit in eadem bilance lb. vii. ℥ i. ʒ i. gr. xxxi. Differentia itaque est, ut dixi, ʒ i. gr. xvii. quod certe, pondus, fatis notabile est, si pauculam illam Aëris quantitatem spectes, quae in sphaera, cuius diameter vix viii. digitorum est, contineri potest. Tota enim haec sphaerae cavitas, si eam perfecte rotundam. Et penitus aëre exinanitam, vix accedit ad sextam pedis cubici partem, ut notum est geometris," De Volder 1676–1678, disputation 5, thesis 2.

[42] Cf. experiment 27; see also and the conversion of the weights in grains reported in De Volder 1676–1678, disputation 5, thesis 6.

Table 5.1 De Volder's measurements in hydrostatics

	Weight of the emptied sphere	Weight of the sphere full of air	Weight of the air contained in the sphere	Weight of the sphere full of water	Weight of the water contained in the sphere	Weight of a cubic foot of air	Weight of a cubic foot of water (most probably by using the brass cube with the ½ foot side, mentioned in the 1705 inventory)[a]	Weight of a cube of water with a side of ½ foot, i.e. 6 inches (using the brass cube with the ½ foot side, mentioned in the 1705 inventory)	Ratio of the weight of air and water
Experiment 15								7 pounds, 14 and ½ ounces.	
Experiment 27	Between 7 and 8 pounds; or 7 pounds, 1 ounce, 4 drams, 1 scruple, 5 grains.		1 dram and ½ = 85 grains (*sic*; it should be 90 grains).		9 pounds, 13 ounces, 2 drams, 2 scruples = 7,035 grains (*sic*; it should be 75,520 grains.				Less than 1 to 1000.

(continued)

Table 5.1 (continued)

	Weight of the emptied sphere	Weight of the sphere full of air	Weight of the air contained in the sphere	Weight of the sphere full of water	Weight of the water contained in the sphere	Weight of a cubic foot of air	Weight of a cubic foot of water (most probably by using the brass cube with the ½ foot side, mentioned in the 1705 inventory)[a]	Weight of a cube of water with a side of ½ foot, i.e. 6 inches (using the brass cube with the ½ foot side, mentioned in the 1705 inventory)	Ratio of the weight of air and water
De Volder's *Disputationes de aëris gravitate* (disputation 5)	7 pounds 1 ounce, 1 dram, 31 grains.	7 pounds, 1 ounce, 2 drams, 48 grains.	1 dram, 17 grains = 77 grains.	16 pounds, 12 ounces, 7 drams, 14 grains.	9 pounds, 11 ounces, 5 drams, 43 grains = 74,743 grains.	1 and 1/18 ounces.	circa 64 pounds.		1 to 970 + 53/77, or 1 to 970.
James Dalrymple's *Physiologia nova experimentalis* (1686)						10 grains, in *mensura Hollandica* (page 255);[b] 6 drams in apothecary system (at page 604).[c]	63 pounds, 4 ounces, 5 drams, in *mensura Hollandica* (at page 255), as well as in the apothecary system (*sic*: see page 604), by which it is rendered as 6,000 drams. This measurement is also reported in De Frankenau 1696.		1 to 1,000.

Pieter van Musschenbroek's *Institutiones physicae* (1748), page 383							63 pounds, 4 ounces, 7 drams, 36 grains, of *aquae putealis Leydensis*, measured on 22 September 1686 (Van Musschenbroek's source is unknown: it might be Johannes Joosten van Musschenbroek, who sold the cube with the side of ½ foot: see Sect. 5.2, *The conditions of hydrostatic equilibrium in De Volder's experimental lectures*). Johannes Lulofs, in his *Grond-beginselen der wynroey- en peil-kunde, ten dienste der landgenooten* (1764), refers to Van Musschenbroek's report: see Lulofs 1764, 123–124. Weighings of different kinds of water are also mentioned in the letter of Henri Justel to Locke of 24 July/3 August 1681 (letter 651).		
Dirk Rembrandtsz. van Nierop's *Tweede deel op de Wiskonstige rekening* (1680), page 156								7 pounds, 14 and ½ ounces.[d]	

(continued)

Table 5.1 (continued)

[a] In De Volder's *Experimenta* there is described the actual measurement of a cube of water with a side of ½ foot (in experiment 15), and the determination of the weight of a cubic foot of water is referred to (in experiment 27) as having been accomplished in a previous experiment, even if the way in which this measurement had been carried out is not specified: "[…] remembering by his precedent experiments the weight of a foot cubic of water," De Volder 1676–1677, 139r. In any case, the 1705 inventory lists a "hollow copper cube of half a foot, intended to determine the weight of a foot of water." The cube with a side of ½ foot is now lost.

[b] Dalrymple mentions that this is the weight of the air contained in a container of a volume of a cubic foot. As this is largely inconsistent with the value given by De Volder himself, it might be that Dalrymple mistook the actual measurement of the weight of air De Volder carried out with the usual sphere (with a diameter of 8 inches), and his subsequent deduction of the weight of a cubic foot of air. Cf. Dalrymple's own words: "D. Volderus publice demonstravit pedem aquae continere 63 libras 4 uncias 5 drachmas, mensurae Hollandicae, vas etiam cuius cavitas fuit pedalis exhaustum, fuit levius per 10 grana, unde pes aëris paulo plus 10 granis continet, quia aër nequit totus exantlari, unde proportio gravitatis aquae et praedictorum corporum facile infertur, pes cubicus aut alia quaevis quantitas auri continet gravitatem similis molis aquae novemdecies, mercurii tredecies et ultra duas tertias, plumbi undecies fere, argenti fere decies, aëris novies, ferri octies," Dalrymple 1686, 255.

[c] This value seems to be deduced from the air-to-water ratio established by De Volder (1 to 1,000) and from the value of 6,000 drams for a cubic foot of water given by Dalrymple. Cf. Dalrymple's own words: "[v]ideamus ergo quodnam sit pondus pedis solidi aquae, et pedis solidi aëris, ut proportio gravitatis utriusque appareat, et quia mensurae medicorum sunt communissimae, per eas procedemus, drachma ergo continet 60 grana, uncia continet 8 drachmas, libra continet 12 uncias, pes solidus aquae ut (a Doctiss. D. Voldero professore Leidensi publice saepius demonstratum est) continet 63 lib. 4 uncias et 5 drachmas quae comprehendunt circa 6000 drachmas, pes autem solidus aëris continet sex drachmas praeter propter, ergo proportio gravitatis unius pedis solidi aquae continebit gravitatem 6000 pedum solidorum aëris, ergo proportio gravitatis aëris ad aquam erit ut 1 ad 1000," Dalrymple 1686, 604.

[d] Cf. Sect. 5.2, *The conditions of hydrostatic equilibrium in De Volder's experimental lectures*, n. 118.

1 pound = 16 ounces = 128 drams = 384 scruples = 7680 grains; i.e.: 1 pound = 16 ounces; 1 ounce = 8 drams; 1 dram = 3 scruples; 1 scruple = 20 grains.

This system was the main one in the late seventeenth century, together with the apothecary system (followed, in one case, in Dalrymple's report), for which

1 pound = 12 ounces = 96 drams = 288 scruples = 5760 grains; i.e.: 1 pound = 12 ounces; 1 ounce = 8 drams; 1 dram = 3 scruples, 1 scruple = 20 grains.

By proposing his measurement of the weight of a mass of air as if it were a mass of earth or water, De Volder demonstrated that air has a weight, or a tendency to move downwards, like any other body. By using the air pump, moreover, he could weigh a container 'emptied' of air more easily than with an aeolipile. Eventually, with this method he avoided the problem of weighing a mass of air more than usually compressed, namely, he could weigh a volume of air which has just the same degree of condensation as the ambient air. For De Volder, the differentiation between the weight of the air (which one can ascertain, roughly speaking, by putting a container filled with air on a balance) and the pressure it exerts as a result of its condensation was very clear. De Volder highlights this point, the difference between these two factors, in the second of his *Disputationes*. Namely, between 1) the weight measurable in the way he did it, and 2) the pressure that a volume of air can exert in *all directions*, and which is preserved, for instance, in two hemispheres sealed by wax. If we suppose (as De Volder does in his *Disputationes*) the case of such a sphere, ideally suspended in a column of air, it is the condensation of the air present in the sphere that allows it to exert a pressure equal to that of the whole column of air above it, and which is also exerted, *via* a lateral transmission of pressure, in the lower hemisphere. This explains that only the wax is responsible for the 'suspension' of the lower hemisphere itself.[43] By the idea of the condensation of

[43] "VII. Quae ut scopo nostro applicentur, concipiamus binos aëris cylindros, quarum latitudines et inter se, et cum ea quae est nostrae Spherae accuratissime aequales ponantur, ab ipsa Telluris superficie ad ultimos usque athmosphaerae limites extentos. Facile autem est percipere unamquanque horum aëriorum cylindrorum superficiem modo horizonti parallela sit, et in aequali altitudine, premi necessario aequaliter, cum aequalem aëris quantitatem et pressionem supra se undiquaque sustineant. Ponamus porro in horum cylindrorum altero suspendi duo haec sibi invicem cera coniuncta hemisphaeria, aëre adhuc dum referta. Manifestum est, utramque tam internam quam externam huius sphaerae superficiem aequaliter ab aëre premi, eo quod internus ille aër aeque ac externus condensatus sit. VIII. Neque obstat parvula illa aëris quantitas, quae in sphaera est, quae vix videtur tanta vi premere posse internam quanta immensa aëris externi moles externam premit sphaerae superficiem. Etenim si rem rite consideremus, nulla pars aëris agit in sphaeram, nisi ea quae eam tangit proxime, quae tamen eo agit efficacius, quo a mole superioris aëris validius comprimitur. Unde si artificio quopiam fieri posset, ut aër externe sphaeram ambiens aequaliter ac nunc comprimatur, etsi nullus aër superior eum comprimeret, nonne manifestum erit hunc aërem licet in minori copia, quia tamen aeque compressus supponitur ac si omne pondus totius athmosphaerae sustineret, aequali vi acturum in ipsam sphaeram. Ex quo sane conficitur non tam aëris quantitatem quam eiusdem compressionem magis minusque validam spectandam esse. Haec autem cum aequalis sit aëri in sphaera contento, cum eo, qui extra sphaeram est, quid est evidentius quam utriusque vim et actionem in sphaeram aequalem esse? Cui consequens est hoc, in rerum statu ad separationem efficiendam nihil aliud requiri, quam ut parvula ea, quae per ceram est con-

the air in the sphere, plainly, De Volder meant the elasticity of the air, namely, the rebounding or pressure that air exert when compressed, and which can counterbalance the pressure of the external air. As far as De Volder's teachings in the 1670s are concerned, however, elasticity is dealt with only in the *Experimenta* (if we suppose that both series of the *dictata* were used after the *Disputationes*). I will come back to the problem of the elasticity of air in Sect. 5.6.2, *Air as an elastic fluid.* Here, it is just worth pointing out that (1) the pressure in all directions exerted by a limited amount air as a consequence of its condensation (or elasticity) is ultimately due, according to De Volder's second disputation (thesis 8), to the compression that this air undergoes by the weight, or downwards pressure, exerted by the whole column of air above it. So that one has to distinguish between (i) the pressure as the effect of the condensation or elasticity, and (ii) the weight of the air itself, by which it compresses its lower volumes. In doing so, De Volder contributed to the historical evolution of the idea of pressure, as described by Alan Chalmers, shaped by the works of Simon Stevin, Pascal, Boyle and Newton. Namely, De Volder also came to develop a 'technical' idea of pressure as exerted by a fluid in all directions, from the earlier, vague idea of pressure as gravitation or downwards thrust.[44] (2) The differentiation between pressure and weight is not related only to the elasticity of the air, as one can also observe it – as I am going to show in Sect. 5.4.2, *The hydrostatic paradox* – in an incompressible fluid like water. As a matter of fact, De Volder seems to suggest that water can experience

nexionis vis et efficacia superetur vincaturque. Ea enim aëris superficies, quae infra sphaeram est, aequaliter premitur cum superficie aëris laterali, nisi quod sphaerae pondus addat hic cylindro aliquod praepondium," De Volder 1676–1678, disputation 2, theses 7–8; "[u]t autem nunc porro ad alterum obiectionis caput accedamus, quo si vel maxime aër premat, eam tamen pressionem gravitatem male dici: quia scilicet omnis pressio non est à gravitate. Quis negat: Sed tamen hanc pressionem, gravitatem dicendam esse existimavi semper, quae oritur ex descensu corporum sublunarium versus centrum Telluris. Sic merito eam aquae vim qua descendendo deorsum aliam aquam minus pressam cogit aut ascendere aut Tubum saepius nobis memoratum superiora versus premere, gravitatem dici et eam pressionem qua aër lateralis deorsum ruens, aërem sphaerae orificio subiectum in sphaeram cogit, gravitatem merito vocari censui. Quod si tamen hanc vocem huic pressioni applicare quis nolit, cum eo litem non movebo, modo fateatur hanc qualemcunque pressionem motui aëris deorsum deberi," disputation 3, thesis 9. See also experiment 27: "[l]astly to show how great the pression of the air was, he said that that globe whose hemispheres could not otherwise [139v] be separated with less than 700 lb weight, he could separate it, when full of air <when> with only his breath, the which he easily performed by blowing it asunder," De Volder 1676–1677, 139r–v.

[44] See Chalmers 2017, 6: "[a] key difference between pressure in the common sense and pressure in the technical sense is that the former relates to forces on bounding surfaces between media whereas the latter refers to forces within the body of media. Another is that, from the technical point of view, pressure is a scalar not a vector. Directed forces, such as those that occur at the boundary of a liquid are determined by variations of pressure, the gradient of pressure in technical terms, rather than by pressure itself. The technical concept of pressure in fluids breaks from the directedness implicit in the verb 'to press' from which 'pressure' originally derived and which is presupposed in the concept of pressure in its common sense." On the idea of pressure, see also Calero 2008.

degrees of condensation and rarefaction. Yet, he does not use the idea of elasticity to explain what is the clearest evidence of the differentiation between weight and pressure, namely, the so-called 'paradox of Stevin', which he explains by Archimedean means. (3) For the sake of clarity, it is also worth noting that in weighing a sealed container on a balance, De Volder measured its weight, not its internal pressure. Indeed, the internal pressure is equally discharged on all the internal surfaces of the container, so that it has no effect on the balance, as the pressure exerted in one direction compensates for the pressure in the opposite direction. On the balance, thus, only the pressure exerted by the quantity of air as its weight has a discernible effect. Of course, the weight depends on the condensation of air, insofar as a volume of condensed air has a higher specific weight. Or, from a Cartesian perspective, it has a greater quantity of the matter of air which acts downwards as if it were one body. From this, it follows that air at sea level, for instance, is heavier than air on high mountains, and that it exerts a greater pressure in all directions, in both cases because it is more compressed. This is not the case with water, which is very much less compressible than air (something which De Volder, in any case, could not have satisfactorily grasped). De Volder, thus, focused mostly on the variations in density or compression of air, in a number of issues, such as the explanation of respiration, maintenance of fire, and propagation of sounds. The first and main one, however, is the measurement of the weight of a whole column of atmosphere, which I consider in Sect. 5.1.3, *De Volder's 'didactic' ascertainment of the pressure of the atmosphere – and related experiments.*

5.1.3 De Volder's 'Didactic' Ascertainment of the Pressure of the Atmosphere – And Related Experiments

Besides ascertaining the weight of a separate volume of air with the aforementioned procedure, De Volder aimed also at showing his students the effects of the pressure of a whole column of open air, namely, of atmospheric pressure. By the idea of pressure, moreover, he could disprove the idea of the fear of a vacuum. In order to ascertain such pressure of air, De Volder repeatedly used four main kinds of instruments – besides the air pump:

1. 'Torricellian tubes', namely (a) tubes of glass similar to those ones used by Torricelli (used in experiment 24); of these, one was provided with a piston moveable by a screw (experiment 26) – now lost; (b) a syringe (used in experiment 4) – probably like the one still preserved at the Museum Boerhaave,[45] whose "embulus" (namely, the piston of the syringe) could be moved by turning the head of the bottle in which it was inserted.

[45] See De Clercq 1997b, 76; Molhuysen 1913–1924, volume 4, 105*.

2. The aforementioned cylinders of marble (used in experiments 1–2).
3. Two pairs of Magdeburg hemispheres (used in experiments 3 and 27, and mentioned in all the *Disputationes*).
4. Various cupping glasses which can be installed on the top of the glass container or receiver of the pump, and which can be moved downwards and upwards by a screw (used in experiments 13–14).

First of all, De Volder's performing Torricelli's *esperienza* or *experimentum Torricellianum* – including some of its variants – does not pose particular interpretative problems, as far as its background is concerned. In 1676, indeed, Torricelli was unanimously considered as the founding father of the study of vacuums and atmospheric pressure.[46] So that when Morley noticed – in experiment 26 – that "this day, we were detained so long by Professor De Maëts, in the public laboratory, about some experiments of chemistry, that we all came too late, to hear his proposal at the beginning [of] the experiment."[47] Consentinus, the author mentioned as the chief source for this kind of experiment by Morley, was the Italian natural philosopher Tommaso Cornelio (1614–1684), who decisively contributed to disseminating the account of Torricelli's *esperienza* in late 1640s. He intervened, with his *Timaei Locrensis Epistola, qua motuum illorum, qui vulgo ob fugam vacui fieri dicuntur, vera causa per circumpulsionem ad mentem Platonis explicatur* (1648, then included in his *Progymnasmata physica* (1663)), in the debate over the paternity of the *esperienza*, which would have been then definitely settled by the *Lettera a' Filaleti della vera storia della famosissima esperienza dell'argento vivo* (1663) of Carlo Dati, which included the texts of the letters of Torricelli to Michelangelo Ricci (1644), which are the only extant sources for Torricelli on his experiences with mercury.[48] De Volder performed a variant of

[46] See Knowles Middleton 1964. In the *Experimenta*, for instance, the air pump itself is labelled as 'instrumentum Torricellianum'. Usually, however, the phrase 'instrumentum Torricellianum' was used to mean barometer. In the eighteenth century, a distinction between a 'Boylean' and a 'Torricellian' vacuum had been well established: see, for instance, Pieter van Musschenbroek's *Elementa physicae conscripta in usus academicos* (1734): "[i]nventor Antliae Pneumaticae, circa elapsi seculi medium, in Germania fuit Otto Guirikius, qui pulcherrima cum ea fecit pericula. Haec impulerunt Nob. Boyleum, adiutum opera [...] Hookii, et Papini, ad similem antliam in Britannia construendam, quacum plurima instituendo experimenta philosophiam naturalem summopere promovit, hinc machina, vel antlia Boyleana, saepe vocatur. Eadem tempestate Leydae a Cl. Voldero alia inventa fuit, quacum an. 1675 laboratorio physico experimenta instituta Leydae fuerunt. Hanc machinam, nostra tempestate Cl. s' Gravesandius ad magnam perfectione et simplicitatem reduxit, ita ut levi opera, et brevissimo tempore, ex recipientibus maxima aëris copia exantliari possit," Van Musschenbroek 1734, 381.

[47] De Volder 1676–1677, 134r. Morley refers to Carel de Maets, professor of chemistry at Leiden.

[48] The history of the invention of barometer has been duly reconstructed in Thirion 1907–1909; Knowles Middleton 1963, 1964; Taton 1963; Prager 1981; Zouckermann 1981; Jones 2001; Shank 2012; Grosslight 2013; West 2013. However, it is worth providing a brief summary of the early dissemination of the story of Torricelli's *esperienza*. After Torricelli described his *esperienza* in his letters to Ricci of 11 June and 28 June 1644, the news of the *esperienza* was disseminated, and it was repeated, first, in Rome, where Ricci resided (February 1645). Then, it spread in France, *via* a

Torricelli's *esperienza*, namely, he used a pipe provided with a piston moveable by a screw, by which he could 'suck' out the air in the pipe itself (where it could also be let in again), and make the mercury ascend to the level allowed by the pressure of the atmosphere.[49] The experiment had a particular didactic function. In experiment 26, indeed, Morley reports a version of what he thought to be Torricelli's *esperienza*, narrated by De Volder, which cannot be found in any other printed source in the seventeenth century:

> [De Volder] added, that Torricellius or the author of this experiment to these precedent proofs did adjoin, that having left the mercury casually in the pipe, at the highest degree it would ascend unto, for some days, he observed that some days the mercury of its own accord would ascend higher than the mark, and some days it would be lower than the mark which observation he said could proceed from no other principle, but from the different ponderosity, or pression of the air, more in one day, than in another, which accordingly did cause the mercury to ascend sometimes higher, and other times lower, \as/ if it were, of its own accord.[50]

These details cannot be found either in Torricelli's letters to Ricci, or in any other subsequent account. Nor can they be found anywhere else. Indeed, the very aim of Torricelli – as declared in his letters – was not to ascertain that a vacuum can be created (against the idea of the *metus*), nor that air has a weight, but rather to ascertain the variations in pressure of the atmosphere. So he could not have discovered them 'by chance'. For Torricelli, it was already well assessed that "we live in a sea of air" (*pelago d'aria*), which – following Galileo – he maintained weighs, at sea level, 1/400 of the weight of water, while at higher altitudes it is more rarefied, and

report of some of the contents of the letters written by a friend of Ricci, François Du Verdus, who sent it to Mersenne in 1644 (see Mersenne 1977, volume 13, 177–183). Then, in 1644 Mersenne visited Torricelli in Florence and witnessed a repetition of the experience by Torricelli himself (Mersenne 1647, 216). Eventually, Mersenne himself unsuccessfully attempted to repeat the experience in Paris in July 1645, together with Pierre Chanut (see the letter of Pascal to Antoine de Ribeyre of 25 June 1651, in Pascal 1923, volume 2, 482). Hence, in November 1646 Pierre Petit – a correspondent of Mersenne – successfully repeated the experience (see Petit 1647, 1–24), and in the same year Pascal undertook various researches on the topic (see Thirion 1907–1909). Also, the experience was performed in Poland in the same year (see Petit 1647, 27–68). Up to this point, it seems that the main source for Torricelli's own experience had been Du Verdus's report, thus raising a debate over its actual, first author (see Shank 2012, 167). In March 1648, moreover, Mersenne received from Raffaello Magiotti (from whom he probably asked for information on Torricelli's experiment) – a friend of Ricci – the report of a previous experiment, carried out by Gustavo Berti in Rome (around 1640–1643), in the presence of Athanasius Kircher, Emmanuel Maignan, and Niccolò Zucchi (who later included accounts of it in their books: see Zucchi 1649, 101–115; Kircher 1650, 11–13; Maignan 1653, volume 4, 1885–1940). This experiment was, in principle, the same one conducted by Torricelli, with water in place of mercury, and using a tube of circa 22 *braccia*. Eventually, in June 1648, Cornelio – correspondent of Torricelli – published the whole story of the *esperienza*, decisively contributing to its ascription to Torricelli. Later, Dati gave the final word on the issue. On Cornelio, see Torrini 1977.

[49] In fact, his direct source seems to be Pascal's *Traité de la pesanteur de la masse de l'air*, chapter 3, where a similar experiment, carried out with a syringe, is used to criticize the fear of a vacuum.

[50] De Volder 1676–1677, 136r–v.

lighter.[51] In the first letter, however, Torricelli claimed that his *esperienza* failed, as the variations in the level of mercury that he observed were, for him, due to variations in the temperature of the air.[52] Therefore, we may conclude De Volder gave his own narration of Torricelli's *esperienza* as a didactic device: in the same way, one finds resurfacing, in contemporary popular histories of science, the story of the 'casual' discovery of variations in the pressure of the air.[53]

De Volder's didactic aims, however, did not prevent him from considering the variations in the pressure of air more systematically. These are mentioned in the first disputation *De aëris gravitate*, thesis 6, and in the fifth disputation, thesis 8. De Volder mentions John Wallis's measurement of maximum variations of atmospheric pressure around sea level, and quantified as 2 inches i.e. a variation between 28 and 30 inches (given by Wallis in a 1666 communication in the *Philosophical*

---

[51] "Le accennai già che si stava facendo non so che sperienza filosofica intorno al vacuo, non per far semplicemente il vacuo, ma per far uno strumento che mostrasse le mutuazioni dell'aria, hora più grave e grossa, et hor più leggiera e sottile. Molti hanno detto che il vacuo non si dia, altri che si dia, ma con repugnanza della natura e con fatica; non so già che alcuno habbia detto che si dia senza fatica e senza resistenza della natura. Io discorreva così: se trovassi una causa manifestissima, dalla quale derivi quella resistenza che si sente nel voler fare il vacuo, indarno mi pare si cercherebbe di attribuire al vacuo quella operazione, che deriva apertamente da altra cagione, anzi che, facendo certi calcoli facilissimi, io trovo che la causa da me addotta (cioè il peso dell'aria) doverebbe per sé sola far maggior contrasto che ella non fa nel tentarsi il vacuo. [...] Noi viviamo sommersi nel fondo d'un pelago d'aria elementare, la quale per esperienze indubitate si sa che pesa, e tanto che questa grossissima vicino alla superficie terrena, pesa circa la quattrocentesima parte del peso dell'acqua. Gli Autori poi de' crepuscoli hanno osservato che l'aria vaporosa e visibile si alza sopra di noi intorno a cinquanta, overo cinquanta quattro miglia, ma io non credo tanto, perché mostrerei, che il vacuo doverebbe far molto maggior resistenza che non fa, se bene vi è per loro il ripiego che quel peso scritto dal Galileo s'intenda dell'aria bassissima che ve praticano per l'homini e gli animali, ma che sopra le cime degl' alti monti l'aria cominci ad esser purissima e di molto minor peso che la 1/400 parte del peso dell'acqua," Galluzzi and Torrini 1975, volume 1, 122–123.

[52] "La mia intenzione principale poi non è potuta riuscire, cioè di conoscer quando l'aria fusse più grossa e grave e quando più sottile e leggiera collo strumento EC, perché il livello AB si muta per un'altra causa (che io non credevo mai) cioè per il caldo e freddo e molto sensibilmente, apunto come se il vaso AE fusse pieno d'aria," Galluzzi and Torrini 1975, volume 1, 123. It may be that De Volder was referring to Boyle, who in experiment 33 of the *New Experiments* recalled that the variations in pressure of atmosphere can be related not only to different altitudes, but to different circumstances, such as seasons, and latitudes, "seeming capable to alter either the height or consistence of the incumbent Atmosphere," which, "might most of them be more exactly try'd by the Torricellian Experiments, if we could get Tubes so accurately blown and drawn," Boyle 1660, 123. Besides, no narration of a 'casual' discovery in variations of atmospheric pressure can be found among De Volder's sources, nor it could have been, as Torricelli purportedly used his *strumento* as a barometer. On the various attempts to collect data on variations of pressure by means of 'Torricellian tubes', see, for instance, the communications of Wallis appeared in the *Philosophical Transactions* (see *infra*, n. 54).

[53] The idea that Torricelli casually discovered variations in the level of mercury can be found in Haven 2006, 18; Long 2012; Allaby 2014, 80. And yet, De Volder's 'popularized' version of the *esperienza* of Torricelli corroborates the view of J. B. Shank that Torricelli did not actually 'invent' the barometer, as claimed in Knowles Middleton 1964, 29: see Shank 2012.

Transactions).[54] Moreover, he discusses the different degrees of condensation that air can assume in caves and on mountains, where its pressure is inferior (as shown, for the first time, by the experiment directed by Pascal on the Puy de Dôme in September 1648). Such variations are in fact explained by De Volder by considering the different degrees of compression that air has at different altitudes, which give rise to different pressures exerted on the surface of the mercury. Later (as I am going to show in Sect. 5.6.3, *De Volder's evolving treatment of the elasticity of the air*) De Volder will more clearly distinguish between the weight and the pressure of air due to its elasticity. Already in the *Disputationes*, however, he ascribes to compression, rather than to weight, the effect of the action of air on mercury. He also addresses the relation between the compression and weight of a separate volume of air (namely, that which he weighed with a balance), claiming that the different degrees of compression of such a volume are hardly discernable, since Wallis had shown that the greatest observed variation in the pressure of the atmosphere results in a variation of 1/15 of the weight of the column of air which is being considered.[55] The

[54] Wallis's measurements are provided in his *Relation concerning the late earthquake neer Oxford; together with some observations of the sealed weatherglass, and the barometer both upon that phænomenon, and in general*, communicated to the Royal Society on 29 January 1666, and published in the issue 10 of the first year of the *Philosophical Transactions* (following the English style, March 1665): "[i]n my Baroscope, I have never found the Quicksilver higher than 30. inches, nor lower than 28. (at least, scarce discernably, not 1/16 of an inch higher than that, or lower than this;) which I mention, not only to shew the limits, within which I have observed mine to keep, vid. full 2 inches, but likewise as an Estimate of the Clearness of the Quicksilver from Air. For, though my Quicksilver were with good care cleansed from the Air," Wallis 1665, 169–170. The same data are reported in Wallis 1670–1671, volume 3 (which probably was not owned by De Volder), 728.

[55] Cf. De Volder 1676–1678, disputation 2, thesis 8: "[n]eque obstat parvula illa aëris quantitas, quae in sphaera est, quae vix videtur tanta vi premere posse internam quanta immensa aëris externi moles externam premit sphaerae superficiem. Etenim si rem rite consideremus, nulla pars aëris agit in sphaeram, nisi ea quae eam tangit proxime, quae tamen eo agit efficacius, quo a mole superioris aëris validius comprimitur. Unde si artificio quopiam fieri posset, ut aër externe sphaeram ambiens aequaliter ac nunc comprimatur, etsi nullus aër superior eum comprimeret, nonne manifestum erit hunc aërem licet in minori copia, quia tamen aeque compressus supponitur ac si omne pondus totius athmosphaerae sustineret, aequali vi acturum in ipsam sphaeram. Ex quo sane conficitur non tam aëris quantitatem quam eiusdem compressionem magis minusque validam spectandam esse. Haec autem cum aequalis sit aëri in sphaera contento, cum eo, qui extra sphaeram est, quid est evidentius quam utriusque vim et actionem in sphaeram aequalem esse? Cui consequens est hoc, in rerum statu ad separationem efficiendam nihil aliud requiri, quam ut parvula ea, quae per ceram est connexionis vis et efficacia superetur vincaturque. Ea enim aëris superficies, quae infra sphaeram est, aequaliter premitur cum superficie aëris laterali, nisi quod sphaera pondus addat hic cylindro aliquod praepondium." Cf. De Volder 1676–1678, disputation 5, thesis 8: "[h]aec itaque fuit aëris gravitas circa Tellurem eo quo experimentum feci tempore. Quae addo eam ob causam, quia athmosphaerae gravitas atque ideo etiam aëris compressio vel maior vel minor est, prout vel in Terrae cavitatibus profundius descendimus, vel in montium cacuminibus altius assurgimus. Deinde, quia eadem illa athmosphaera, atque ideo et aër eius gravitatem sequens pro diversis tempestatibus, nunc gravior nunc levior est, ut ex barometris, quae Mercurio suspenso Aëris gravitati aequiponderant, liquere potest. Nonnunquam enim in ilis descendit mercurius si pondus athmosphaerae fiat minus, nonnunquam ascendit, si pondus aëris fiat maius. Quae tamen differentia gravitatis licet in tota athmosphaerae massa satis sit notabilis, in ea tamen quantitate pedis cubici, de cuius gravitate determinata hic loquimur, considerari vix meretur. Id etenim observatum est maxi-

reference to the idea of compression was most probably included by De Volder in order to 'save' those measurements conducted in sealed environments, or in those cases in which they are not conducted in the open air (as in houses). In such cases, the communication of pressure between the external and internal air – exerted in all directions – results in an equal compression of internal and external air. This justifies those instances in which, apparently, one cannot measure the weight of a whole column of atmosphere which is perpendicular to the surface of the mercury – as this column is interrupted by the roof. As I am going to show,[56] this is also the case with the pressure of water, where, however, there is no elasticity. In that case, De Volder would solve the problem of increasing pressure without reverting to the idea of the compression of water (which nevertheless he did not deny), but by purely Archimedean means.

Besides being repeated in this form in experiment 26, the *experimentum Torricellianum* was also attempted by De Volder (experiment 4) in a container from which air had been pumped out, where – of course – it did not succeed (a depiction, albeit idealized, of this experiment was provided by Wolferd Senguerd in Fig. 5.3, Senguerd 1681, 65).

The fact that in a vacuum the *esperienza* does not succeed offers a proof *a contrario*, i.e. it shows that when there is no pressing air, the mercury does not ascend, so that in the vacuum a vacuum is created. The production of a vacuum in a vacuum had been first described by Gilles Personne de Roberval in his *De vacuo narratio prima* (September 1647, published by Valeriano Magni in his *Admiranda de vacuo*, 1647),[57] *De vacuo narratio secunda* (May–June 1648, unpublished until the Brunschwig-Boutroux edition of Pascal's works),[58] and reported – as performed by Roberval – in the *Gravitas comparata, seu Comparatio gravitatis aëris cum hydargyri gravitate* (1648) of Étienne Noël. Roberval used a curved tube, then improved and illustrated in Pascal's *Traité de la pesanteur de la masse de l'air* (1663), chapter 6, (see Fig. 5.4, Pascal 1663, 105) – in which the mercury present in the curved part of the tube ascends if the finger is removed from the hole – (a different instrument, i.e. a tube provided with a large chamber (functioning in the same way), was then used at the Accademia del Cimento and illustrated in the *Saggi di naturali esper-*

---

mam altitudinis, ut Wallisius refert, differentiam fuisse duorum digitorum atque adeo, vix confecisse circiter 1/15 totius gravitatis. Ex quo sequeretur si ad illas minutias attendere velimus, maximam differentiam gravitatis, quae intercedit inter aërem levissimum, et gravissimum in pede cubico fore circite; drachmae dimidiae. Haec enim unciae quin tam decimam circiter conficit partem".

[56] See esp. Sects. 5.2, *The conditions of hydrostatic equilibrium in De Volder's experimental lectures*, 5.5.3, *De Volder's Archimedeo-Cartesian hydrostatics: an impossible synthesis?*, and 5.6, *De Volder's treatment of the properties of air*.

[57] Cf. the full title: *Admiranda de vacuo, scilicet Valeriani Magni demonstratio ocularis de possibilitate vacui. Eiusdem altera pars demonstrationis ocularis. D. De Roberval Narratio de vacuo. Valeriani responsio ad D. de Roberval. Responsio eiusdem ad Peripateticum Cracoviensem*. On it, see Nejeschleba 2017.

[58] See Pascal 1923, volume 2, 3–35 and 283–340. For a commentary on the different versions of the experiment, see Webster 1965; Koyanagi 1989; Shea 2003, 165–170.

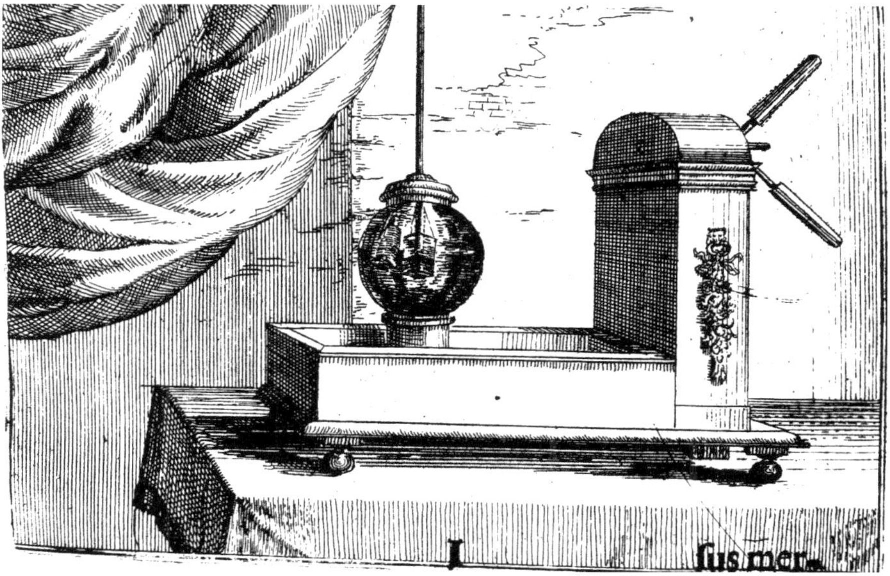

Fig. 5.3 The model of the air-pump built by Samuel van Musschenbroek for De Volder in 1675. (Source: Senguerd 1681, 65. Smithsonian Libraries, Special Collections (Dibner), call number: QC17.S47. Public domain)

ienze fatte nell'Accademia del Cimento (1666),[59] written by Lorenzo Magalotti (see Fig. 5.5, Magalotti 1666, 27).

Adrien Auzout, who produced a vacuum in a vacuum in 1648 (reported by Jean Pecquet in his 1651 *Dissertatio anatomica de circulatione sanguinis et chyli motu*), used a Torricellian tube (see Fig. 5.6, Pecquet 1654, 58) in which he put another Torricellian tube, and showed that, once air enters into the larger one, the mercury in the smaller one ascends.[60] This version of experiment, actually, had already been performed by Pascal prior to Roberval's one, before 15 November 1647.[61] Eventually, Jacques Rohault provided his own version of the experiment, by using a complex instrument now called 'la chambre de Rohault' (see Fig. 5.7, Rohault 1671, volume 1, 93), and described in the *pro gradu* disputation *De gravitate aëris* (1690) of Joseph Serrurier.[62] De Volder, by using an air pump, just used a syringe: in doing so, he repeated the same procedure followed by Boyle in experiment 34 of his *Continuation of New Experiments Physico-Mechanical* (1669b), showing that as "without the Pressure of the Air the Quick-silver would not be elevated, we thought it seasonable to shew, that by the Pressure of the Air it would."[63]

[59] See Magalotti 1666, *Esperienza* 31. For a commentary, see Favino 2008.

[60] Pecquet 1651, 56–58. For a commentary, see Webster 1965.

[61] See Fouke 2003.

[62] "[…] est experimentum Rohaultii quod maxime facit ad rem nostram," Serrurier 1690, thesis 7; cf. Rohault 1671, part 1, chapter 12, §§ 49–51. See Shea 2003, 170–171; Dobre 2013a.

[63] Boyle 1669b, 114.

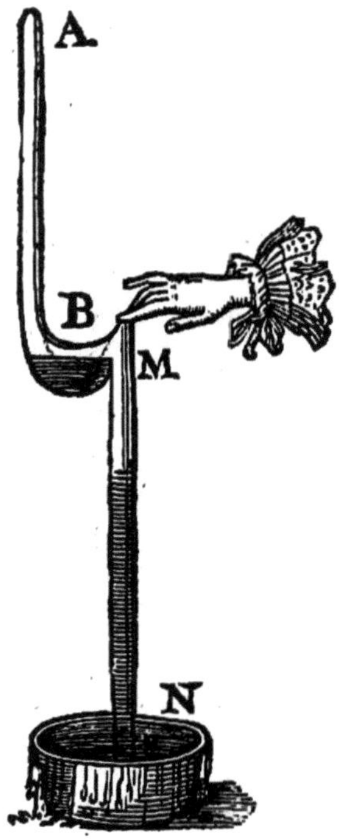

Fig. 5.4 Instrument to perform the experiment of the vacuum in a vacuum devised by Gilles Personne de Roberval and improved by Blaise Pascal. (Source: Pascal 1663, 105. ETH-Bibliothek Zürich, signature: Rar 4668, https://www.e-rara.ch/doi/10.3931/e-rara-20296. Public domain mark 1.0)

In experiment 24, in turn, De Volder used water instead of air, repeating an experiment described by Pascal in his *Traités de l'équilibre des liqueurs* (1663), chapter 3, which is nothing but the performance of Torricelli's *esperienza* by inserting the tube containing mercury in a basin full of water.[64] Pascal devised, for this experiment, a tube of 14 feet immersed in a river. Boyle, in his *Hydrostatical Paradoxes* (1666) (paradox 2, scholium), considered this and other experiments described by Pascal as generally unfeasible[65] – both views being reported by the writer of the *Experimenta*.[66]

[64] Pascal 1663, 18–21.

[65] "It tempts me much to suspect, that Monsieur Paschall never actually made the Experiment, at least with a Tube as big as his Scheam would make one guess, but yet thought he might safely set it down, it being very consequent to those Principles, of whose Truth he was fully perswaded. And indeed, were it not for the impetus, the Quicksilver would acquire in falling from such a height, the Ratiocanation were no way unworthy of him. But Experiments that are but speculatively true, should be propos'd as such, and may oftentimes fail in practise; because there may intervene divers other things capable of making there miscarry, which are overlook'd by the peculator that is wont to compute only the consequences of that particullar thing which he principally considers," Boyle 1666, 64–65. The writer of the *Experimenta* mentioned a tube of 20 feet, used by Pascal with regard to another experiment, described in chapter 4 (concerning the equilibrium of solids and fluids).

[66] "Torricellius was the first that found the ascension of mercury by the pression of the air: but Pascalius a very ingenious [. . .] Frenchman by reason rather than expressa [*sic*] pretended to prove that in a tube of 20 foot long the mercury ought to ascend only thus, et thus, by the pression of the

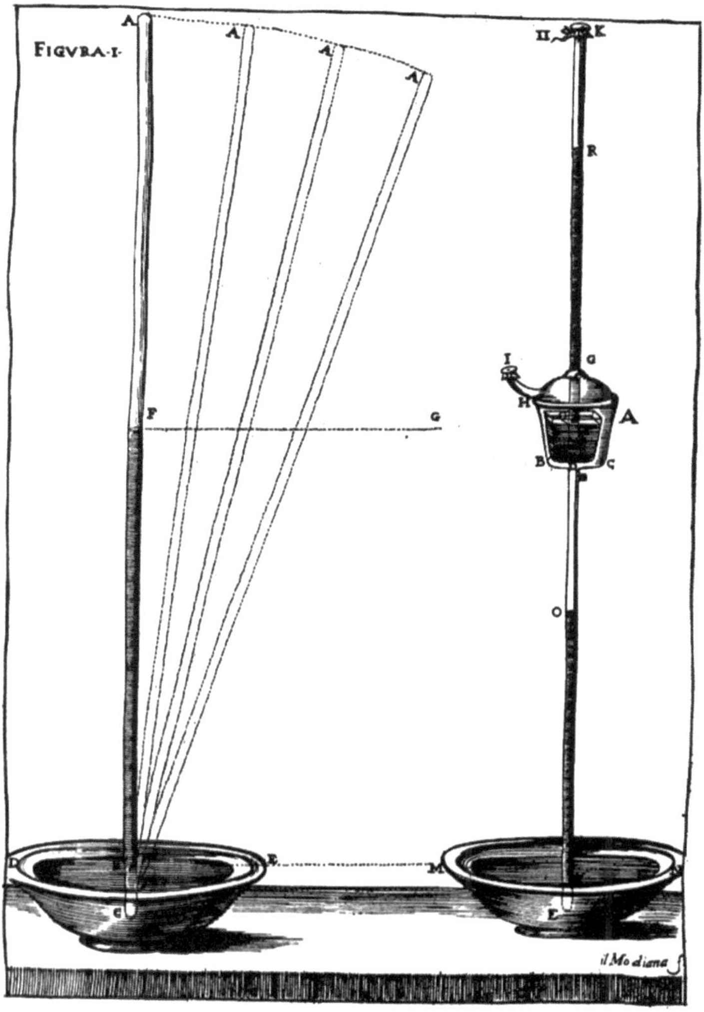

Fig. 5.5 Instrument to perform the experiment of the vacuum in a vacuum used at the Accademia del Cimento. (Source: Magalotti 1666, 27. Naples, Biblioteca Nazionale, signature: B. Branc. 085F 2, https://archive.org/details/bub_gb_BcCpGiXkRKsC/page/n5. Public domain mark 1.0)

Eventually, De Volder performed experiments with a 'Wittenberg syphon' (also listed in the theatre catalogue) in 1686, which are given by Jacob Leupold in his *Theatrum machinarum hydraulicarum* (1724). This kind of syphon was presented in 1684 and 1685 by Salomon Reisel (physician to the Duke of Wittenberg) and

water. Boyle first found that he had never tried those experiment, as never being at the bottom of the water thereby to see how far it did ascend or not. Hence he refuteth Pascalius: v[id]e ipsum Boilium," De Volder 1676–1677, 132r–v.

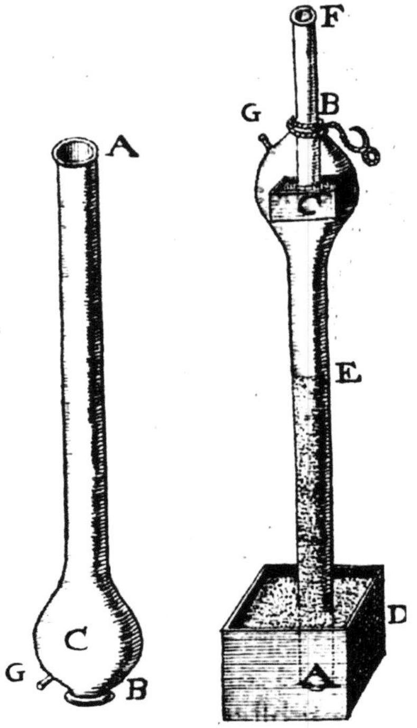

Fig. 5.6 Instrument to perform the experiment of the vacuum in a vacuum devised by Adrien Auzout. (Source: Pecquet 1654, 58. Rome, Biblioteca Nazionale Centrale, signature: ā⁴ē⁴A-2I⁴, https://archive.org/details/bub_gb_XzZOLZQE1qIC. Public domain mark 1.0)

Denis Papin in a book and in some articles in the *Philosophical Transactions.*[67] An illustration is provided in Leupold's treatise (see Fig. 5.8, Leupold 1724, volume 1, plate 4, figure 4). When this syphon is filled with water, and one of the two recipients (for instance, the one on the left) is filled with water as well, the higher pressure exerted on the leg on the left causes a movement of water from this leg to the other one (whose water is pressed only by air). According to Reisel and Papin, with this kind of syphon water can be raised more than 30 feet. According to De Volder, on the other hand, this was impossible (however, he could not test this, as his syphon had legs 3 feet long).[68] Leupold's source for De Volder's experiment is unknown. His rejection of the idea that water could be raised more than 30 feet, in any case, is consistent with his sticking to his Archimedean explanation of the equilibrium of fluids.

[67] See Papin 1685a, b, c; Reisel 1685, 1689.

[68] "Nachdem auch diese Beschreibung der Königl. Societät in Engelland zu Gesichte kommen, haben sie gleichfals die Würckung untersuchen wollen, und es dahero Herrn D. Dionysio Papino aufgetragen, welcher auch durch Verfertigung der Machine (Fig. IV) befunden, daß das Wasser über 30 Fuß gestiegen. Die Würckung von des Papini Heber, hat Herr Volderus, Prof. Phys. zu Leyden, Anno 1686 öffentlich gezeiget und erklähret, an Cruribus von 3 Schuh lang, hat aber nicht zugeben wollen, daß das Wasser über 30 Schuh hoch steigen könne; weil solches die Kräffte der Atmosphaer nicht zuliessen," Leupold 1724, 11.

Fig. 5.7 Instrument to perform the experiment of the vacuum in a vacuum devised by Jacques Rohault. (Source: Rohault 1671, volume 1, 93. ETH-Bibliothek Zürich, signature: Rar 5463, https://www.e-rara.ch/doi/10.3931/e-rara-4795. Public domain mark 1.0)

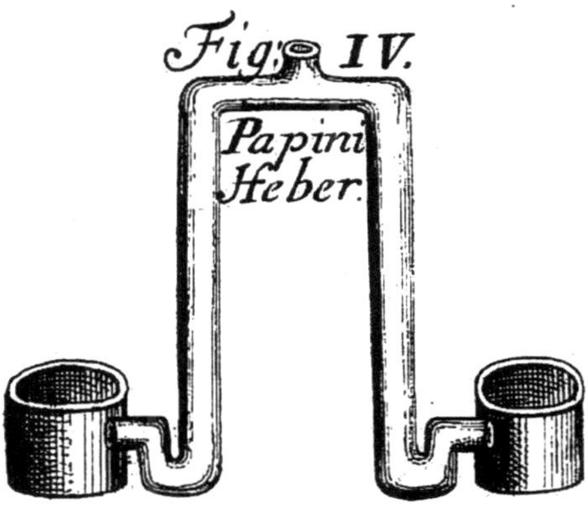

Fig. 5.8 The syphon devised by Salomon Reisel and Denis Papin, allegedly capable of raising water above 30 feet. (Source: Leupold 1724, volume 1, plate 4, figure 4. ETH-Bibliothek Zürich, signature: Rar 5463, https://www.e-rara.ch/zut/doi/10.3931/e-rara-51374. Public domain mark 1.0)

To sum up, through his experimental didactics De Volder disseminated recent empirical discoveries that students could otherwise find only in books, and show some of their variants. Moreover, through his activities he aimed at criticizing Scholastic philosophy, whose fundamentals, as seen above, were 'shaken' by Torricelli's *esperienza*. De Volder assigned to experiments a function that Descartes and some of his followers such as Johannes Clauberg and Johannes de Raey assigned to radical doubt, namely, to eradicate the prejudices of Scholastic philosophy.[69] In the case of De Volder, these were the lightness of air and the idea of the *metus vacui*. The treatment of these ideas between the sixteenth and seventeenth centuries, however, was not as linear as De Volder presented it in his *Disputationes*. As seen above, indeed, some commentators on Aristotle, and Aristotle himself, maintained that air has a weight or gravitates even in its proper place. Moreover, an anti-Aristotelian source such as Galileo adopted the idea of the *metus vacui* to the raising of water in syphons. As he came to criticize the *metus vacui*, moreover, De Volder also attacked (in his *Experimenta*) the Cartesians, as I show in Sect. 5.1.4, *De Volder's criticism of the* metus vacui.

5.1.4 De Volder's Criticism of the Metus Vacui

The arguments against the idea of the *metus* that De Volder assumed from the first principles of physics were based on standard, Cartesian positions. In the fourth of his *Disputationes*, De Volder criticized this idea as it entails an action at a distance,[70]

[69] See Savini 2011; Strazzoni 2018a, chapters 3 and 5.

[70] Cf. De Volder 1676–1678, disputation 4, theses 5–6: "V. Dum itaque, nullum corpus movetur a semet ipso, patet causam motus, qui in uno est corpore, necessario in alio corpore quaerendam esse. An enim, et quo pacto res incorporales motum dare possint inquirere huius loci non est, cum de motibus solummodo agamus corporeis. Ex quo sequitur, id quod movet, necessario quoque moveri. Nemo enim comprehendet, ut opinor, corpus quiescens aliud impellere, movereque ulla posse ratione. Etenim aut distabit corpus quiescens ab eo, quod in motum concitat, aut continget proxime. Prius certe non fiet, quia nulla actio in distans, ut vulgatum et certum habet axioma, dari potest. Neque posterius. Et si enim contingat proximne, cum tamen sua quiete in id non agat, non certe movebit. Nam quid certius, quam eodem ritu ac motus causa est omnis in corporibus mutationis, ita quietem causam esse, cur corpora in eodem statu perseverent. Ex quo illud conficitur, priores species tum attractionis, tum pulsionis, in quibus corpus quiescens aliud vel ad se allicit vel a se propellit, manifestae repugnare rationi. VI. [...] Posteriorem vero quod spectat attractionis speciem ea facillima intellectu est, et revera a pulsione non differt, si duo corpora, quorum alterum alterum trahere et tractione ad motum concitare dicitur, vinculo quopiam inter se connexa sint. Nam quid clarius, quam si catenula duo corpora inter se vincta consideres, non posse alterum propelli, quin ea ipsa propulsione moveatur alterum? Quid praeterea manifestius quam hanc attractionem a pulsu non differre? Ut enim hic fiat motus necesse, ut alterum impellatur idque, ea efficacia, qua utrumque impelli et moveri possit. Hanc speciem attractionis si defensam velint, facile patior, sed alteram illam, quae proprie suctio appellatur, qua corpus quod movetur sequitur corpus movens, quocum nullo vinculo, nullaque communione coniunctum est, rerum naturam nullo modo pati firmiter assevero."

and as it is incompatible with the idea that movement does not pertain to the essence of matter – being just communicated by impact through parts of matter.[71] Thus, no body can move itself to 'fill' an empty space, nor it can increase the overall quantity of motion, which is constant in the world.[72] Moreover, the attribution of a *metus* to matter would entail its animation.[73] Above all, however, since space and matter are the very same thing, neither a vacuum, nor the 'danger' of a vacuum is possible.[74] For De Volder, as well as for Descartes and Cartesians like De Raey at Leiden, or Rohault, what appears as an empty space is filled with a subtle matter. In particular, De Raey used the case of Torricelli's tube in his *Disputationes physicae ad Problemata Aristotelis* (1651–1652) at Leiden (later collected in his *Clavis philosophiae naturalis* (first edition 1654, second edition 1677)), to 'illustrate' (building

[71] Cf. De Volder 1676–1678, disputation 4, thesis 4: "[q]uibus de rebus quid censendum sit, ut constet, considerari primum velim nullum corpus a semet ipso moveri, Nam cum in corporis natura motus non involvatur, cumque quodlibet corpus cuiuscunque figurae aut magnitudinis absque motu concipere valeam, quid certius quam ex nullius corporis natura motum fuere? Si enim ullum corpus sit quod causa sit sui motus, nonne ut effectum causam, sic motus huiusce corporis naturam sequeretur necessario? Cui rei consequens est illud corpus in perpetuum motum iri, nec sine motu concipi posse. Ut enim nulla res sine sua natura, aut esse aut omnino cogitari potest; sic certe nec corpus, si eius naturam sequatur, eamque ideo necessario comitetur motus, sine illo motu aut a se, aut percipi posset. Quae cum de corpore absurdissima sint, patet sane de natura corporis motum non esse, sed omnem, qui inest corpori ab alio, hoc est, a causa externa esse."

[72] Cf. De Volder 1676–1678, disputation 4, thesis 7: "[...] [n]eque difficilis est demonstratio, si consideremus nullum corpus aliud movere posse, nisi cum eo partem quandam fui, quem habet, motus communicet, neque posse cum altero motus sui communicare partem, quin tantundem ei decedat. Si enim corpus quod habeat 10. gradus motus alterum moveat, cum nulla causa sit, quae motum augeat, manifestum est eosdem 10. gradus permanere, et cum antea uni deberentur corpori, nunc duobus illisi m partiri, atque adeo, quantum alteri accedit motus, tantundem priori decedere."

[73] Cf. De Volder 1676–1678, disputation 4, thesis 8.

[74] Cf. De Volder 1676–1678, disputation 4, thesis 9. In commenting on Descartes's criticisms of the possibility of vacuum in *Principia* II.16, in the series Hamburg 273 of his *dictata* De Volder mentions the raising of water in pumps up to 33 feet as evidence against the supposed efficacy of the *metus vacui*: "si concedamus vacuum posse dari, cum illum fingunt eius metum in natura? Illud fieret causis naturalibus, nisi ubiq[ue] velint assignare miracula, profecto stolidissima e[ss]et natura q[uae] produceret id q[uo]d metuit. Si vero dicant id fieri miraculo, possum ego dicere, eodem miraculo muros po[ss]e collabi, etsi fiunt causae naturales, cur illas natura metuit, si non, cur metuit ea q[uae] non sunt. Deinde experientia contrarium evincit, si e[nim] aqua assurgat in antliis propter metum vacui, ergo tandiu assurget, quamdiu natura metuit, sed notum est {embijlus} (belgicae [*sic*] dictus de {stijger}) non po[ss]e attolli ultra 32 a 33 pedes, neq[ue] ulterius ascendit aqua, unde ra[ti]o cur in fodinis, non una sed pluribus antliis indigeant quarum non indigerent, si vacuum natura metueret, nam si aqua tamdiu ascenderet, quamdiu {embijlus} attolleretur, certum e[ss]et aquam nonnunquam attolli po[ss]e ultra 100 pedes, q[uo]d repugnat ante dictis," Hamburg 273, 89. None of this prevents Morley, however, from using the term 'vacuum' countless times in his notes, nor using the phrase "the vacuum within the cupping-glass, was not sufficient, or strong enough, to draw, retain or bear any longer the weight of the bladder, and its ring," in experiment 13 (De Volder 1676–1677, 111r).

upon the identification of space and body) that the tube is in fact filled by an aether (see Fig. 5.9, De Raey 1654, 194).[75]

This matter, however, according to De Volder, does not have any effect, if taken alone, on the equilibrium of fluids, nor does De Volder ascribe to it a discernible pressure or gravitation on the surface of the fluid. In his *Disputationes*, indeed, he declares his suspension of judgment on the actual weight of such matter.[76] In the Cartesian model of gravity, which, as I am going to show in Sect. 6.2.2.3.1, *De Volder's Cartesian cosmology and theory of weight*, De Volder would also adopt in the corollaries to his *Disputationes*, subtle matter is what makes 'gross' bodies heavy, i.e. it pushes them downwards and, consequently, it makes them press on other bodies (for example, the heavy mass of air presses upon the heavy mass of water). However, such matter has no effect besides causing the property of the weight or gravity of such bodies. In a Torricellian tube, for instance, subtle matter (which also permeates the volume of mercury) is what makes mercury heavy, but it does not exerts a discernible pressure on its surface. Accordingly, such matter is weightless, even if, as I am going to show in Sect. 6.1, *Descartes's theory of planetary stability and weight*, particles of subtle matter can be pushed downwards by other particles of the same kind, and to be heavier than them.

Moreover, in his *Experimenta* and *Disputationes* De Volder offers experimental as well as theoretical evidence. Besides the experiments with Torricelli's tubes (as

---

[75] Descartes discussed the matter with Pascal in a meeting on 23–24 September 1647, as reported in a letter by Jacqueline Pascal to Françoise Gilberte Périer, née Pascal (both sisters of Blaise) of 25 September 1647: see Pascal 1923, volume 2, 42–48. Cf. De Raey 1654, 193–197: Rohault 1671, volume 1, 73–74. Rohault devoted an experiment conducted with a syringe to demonstrate his thesis: he supposed that the hole of the syringe is closed, and that the syringe has no pores of any kind. In this case, the external air that would be displaced if the piston were raised, will not find any space to move to, as everything is filled. Therefore, no force in nature can raise such piston. Therefore, if the piston can be raised, we must admit that the syringe has some pores, allowing the passage of a matter subtler than air (discussed in Dobre 2013a). Boyle, in turn, devoted experiment 38 of his *Continuation of New Experiments Physico-Mechanical* (1669b) to attempt to ascertain the existence of subtle matter, without success: see Boyle 1669b, 127–132. Boyle's experiment consisted in trying to ascertain whether a sheet, placed around a piston moving in a vacuum, is moved by the subtle matter displaced by the piston. As to this issue, see Chalmers 2009, chapter 6.

[76] See De Volder 1676–1678, disputation 3, thesis 4: "[e]t certe quis non miretur, huic demonstrationi, quae tam clare, si credere fas sit, ostendit nec sphaeram ab aëre nec tubum ab aqua premi, experientiam repugnare. Qua stolidus ego (qui, ut ingenue fatear quod res est, ratiocinii huius subtilitatem non assequor, aut si assequar, futilitatem simul deprehendo) perfusus opinabar, non secus ac ille tubus nullam ab aqua pressionem sentit, quando interne aeque aquam ac externe habet, ita quoque sphaeram ab aëre non premi, ubi tam ab interna quam ab externa parte aeque compressum habet aërem. Eademque stoliditate inductus aeque ac demta ex tubo aqua, succedente aëre ipsa aqua multo minus gravi, omni gravitate aquae premitur versus superiora tubus, credidi, demto aëre, succedenteque materia subtiliori ipso aëre multo minus gravi, (si modo ullius gravitatis sit particeps, quod hic non determino) hemisphaerium inferius omni gravitate aëris versus superius propelli. Quod superius cum ab eadem gravitate deorsum pellatur, putavi demens huic hemisphaeriorum ad se in vicem pressioni cohaesionem hanc deberi."

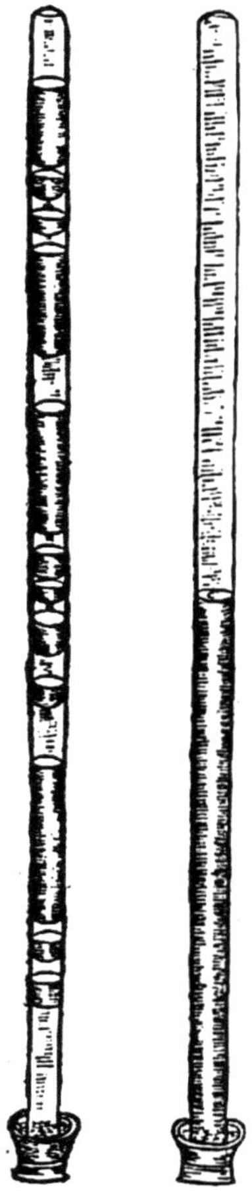

Fig. 5.9 Torricelli's tubes. (Source: De Raey 1654, 194. The University of Oklahoma Libraries, History of Science Collection, signature: 18459-1001. Public domain)

seen in Sect. 5.1.3, *De Volder's 'didactic' ascertainment of the pressure of the atmosphere – and related experiments*), De Volder offers the case of the detachment of the pieces of the marbles (experiments 1–2), which was used to show that the *metus* is not sufficient to keep the two pieces together, and that a vacuum would be created, if one admits its possibility. The case of the separation of the Magdeburg hemispheres, discussed in the *Disputationes*, is treated in the same way. De Volder adopts in both cases the same argument or *dilemma* against *metus vacui*, namely (experiment 2) that "either such fear of vacuum does not exist, and they [i.e. the marbles] cohere by such fear, [but] as it does not exist, it cannot impede the separation of the cylinders. Or, if it exists, and it is what impedes the disjunction, yet we [...] disjoined them, and [...] a way is given to introduce a vacuum in nature,"[77] and (disputation 4), that if one supposes that the space in the hemispheres is void, their separation should be easy, in order for it to be filled by the external air. If it is filled with some matter, the 'danger' of a vacuum would take place only in the moment of their separation (as in the case of the separation of the marbles). So that (1) if their separation can take place without a vacuum, it is senseless to suppose any 'fear' of it; or (2) if it happens with a vacuum, there is a natural power – like that of the weights hung on the sphere – that is more powerful than the vacuum itself – which takes place as they are separated.[78] Apart from Lucretius's *De rerum natura*,[79] this kind of criticism of the *metus vacui* can be found in the *Experimenta nova* of Von Guericke, who addresses the use of the sample of the marbles in order to argue for the *metus vacui* by the Italian Jesuit Niccolò Zucchi,[80] and claims that, by knowing the weight of a cylinder of air on a given circular superficies (which he quantified as equal to the weight of a column of water of about 19 Magdeburg cubits), one can know the force required to separate the tables of marbles – which according to Zucchi could be separated only by moving them horizontally, as in this case no vacuum is created.[81] In fact, the experiment with marbles could serve as a way to measure the pressure of a column of air without an air-pump, as it is the pressure of air – according to seventeenth-century interpretations – that hinders their separation.[82] In any case, no systematic measurements of atmospheric pressure by De Volder are extant, neither with cohering bodies, nor with barometers.[83] Eventually,

[77] "Hence also he inferred, as above that it was not the metus vacui which did hinder the separation of the cylinders with this dilemma: aut non datur talis metus vacui, et coë[u]nt illa, talis metu [*sic*], cum non detur, nequit impedire avulsionem cylindrorum. Aut si datur, et ille sit qui impediat disiunctos, iam profecto disiunximus illos, et coë[u]ntes datur via, et modus, introducendi vacuii in na[tur]am: imo fecimus ipsi mo[do] vacuum, q[uo]d utrumq[ue] negare de[be]nt, qui vacuum tantopere abohrrere na[tur]am aiunt," De Volder 1676–1677, 81v.

[78] Cf. De Volder 1676–1678, disputation 4, theses 10–11.

[79] See Sect. 4.1.3.1, *Four factors of cohesion: vacuum, rest, pressure and entanglement.*

[80] Reported in Schott 1657, 25–26; see Gorman 1994.

[81] Von Guericke 1672, 101–103 and 119.

[82] In fact, the cohering glasses or marbles are kept together, not only by the pressure of the air, but also by the Van der Waals's force, i.e. forces of molecular cohesion. Senguerd performed measurements of the weights sustained by cohering cylinders disposed both vertically and horizontally, claiming that they could sustain the same weight in both positions (see Sect. 4.1.3, *De Volder's ideas on cohesion and divisibility*). Locke would use this kind of evidence to criticize the recourse to atmospheric pressure as a factor of cohesion of bodies, which he found in the *De gravitate aetheris* (1683) of Jakob Bernoulli: see Sect. 6.2.2.3.2, *De Volder's Huygenian theory of weight (and its relation to the idea of cohesion).*

[83] As I am going to show, his measurements were rather aimed at showing the ways in which fluids press: see Sect. 5.4.1, *"[...] les liqueurs pesent suivant leur hauteur"*. Notwithstanding this, his performance of experiments with the Magdeburg hemispheres were frequent. According to his *Experimenta*, once he came to measure the weight sustained by two void hemispheres, or by one hemisphere joined to a plate, he measured it as 700 pounds with hemispheres of a diameter of 8 inches, while by using hemispheres of 6 inches' diameter he could hang a weight of 425 pounds (in experiments 3 and 27, and disputation 5). De Volder's performance of the experiments with the hemispheres was also reported by Dalrymple, mentioning the cohesion of hemispheres capable of

in experiments 13–14 (the same experiment performed twice) De Volder used a cupping-glass within the air-pump container. With a screw, he could move the cupping-glass downwards over a bladder. The bladder sticks to the glass in the case where the container (but not the glass) is emptied of air, as the effect of the air pressure, and which does not stick if both the container and the glass are empty. Even if in the latter case "there was a greater vacuum, or metus vacui."[84] This experiment was a repetition of experiments 35–36 of Boyle's *Continuation of New Experiments Physico-Mechanical*, in which the same cause is advocated to explain the phenomenon, and in which the same criticism of the *metus vacui* is brought forth.[85]

In the experiments with marbles and in the *Disputationes*, De Volder generically refers to the "Aristotelians" and the "Scholastics" as his polemical targets. However, in experiment 11 – aimed at measuring the volume of air drawn from the pump at every movement of the piston – he declares that the Cartesians themselves used the idea of *metus vacui*. In this experiment we find the only discussion of the structure of the pump in a text which can be traced back to De Volder (whereas we can find a short description of it in Senguerd's *Rationis atque experientiae connubium* (1715), noting that it required two men for its operation).[86] According to Morley's account, the volume of air expelled by the pump at each 'suction' is measured by De Volder by filling the basin of the pump with water:

> [h]e set a glass pretty big, and empty, on the pump: the basin of the pump was all full of water: and through the glass there did go up the {wuall} brass pipe, through which the air is drawn always out. [...] He told us that at every exuction we should see the water to ascend through that pipe, and fill the glass, and how much water did ascend, just so much air was drawn out, and that that was the rule, and measure of the air drawn out.[87]

sustaining a weights of 655 pounds (Dalrymple 1686, 542), and by Van Musschenbroek in his *Dissertationes physicae*, referring to De Volder's *Disputationes* (Van Musschenbroek 1729, 453).

[84] De Volder 1676–1677, 112r. De Volder appended to the bladder weights of 7 or 8, 10 and 17.5 pounds.

[85] Boyle 1669b, 118–121 and 122–124.

[86] "[...] Nobilissimus Robertus Boyle [...] in meliorem, experimentorumque tentaminibus peragendis commodiorem formam redegit antliam, multisque accessionibus locupletavit. Hinc et aliis nata occasio inventis nova adiiciendi; novae fabricae, felicioris, atque expediti magis usus, antlias construendi. Donec tandem circa annum 1675 consummatae magis structurae, experimentis minori cum molimine, et periculo peragendis, antlia, ut et apparatus ad illa requisiti, in lucem prodiret. Antliae tubus perpendiculariter erectus erat, et immobilis; inferiori eius parti ad angulum rectum affixus erat minor tubulus, cui recipientia, campanulae, tubi, quaeque porro evacuanda veniebant, adaptabantur. Huius formae antliam illo tempore, ad experimenta molienda, adhibui. Verum, cum et illius: structurae, situs, ac constitutionis antliam multis obnoxiam difficultatibus, atque incommodis; exquisitissimae aëris evacuationi minus convenire experirer, (duplicis enim, quo gaudebat, verticilli usus, plus involvit molestiae, et vitio inficitur facilius, quam unius: binos ad experimenta tentanda requirebat operarios, alterum ad antliam agitandam, et aërem subducendum; qui versurae epistomii vacaret alterum) antliam, quae minorem exigeret apparatum, exquisitiori aëris exhaustui, validiori eius constipationi, pluribus experimentis perficiendis, quibus prior impar erat, inservire aptam, minusque vitiis obnoxiam, excogitavi, atque ut mihi construeretur, anno 1679 cum artifice conveni," Senguerd 1715, 3–4.

[87] De Volder 1676–1677, 106r.

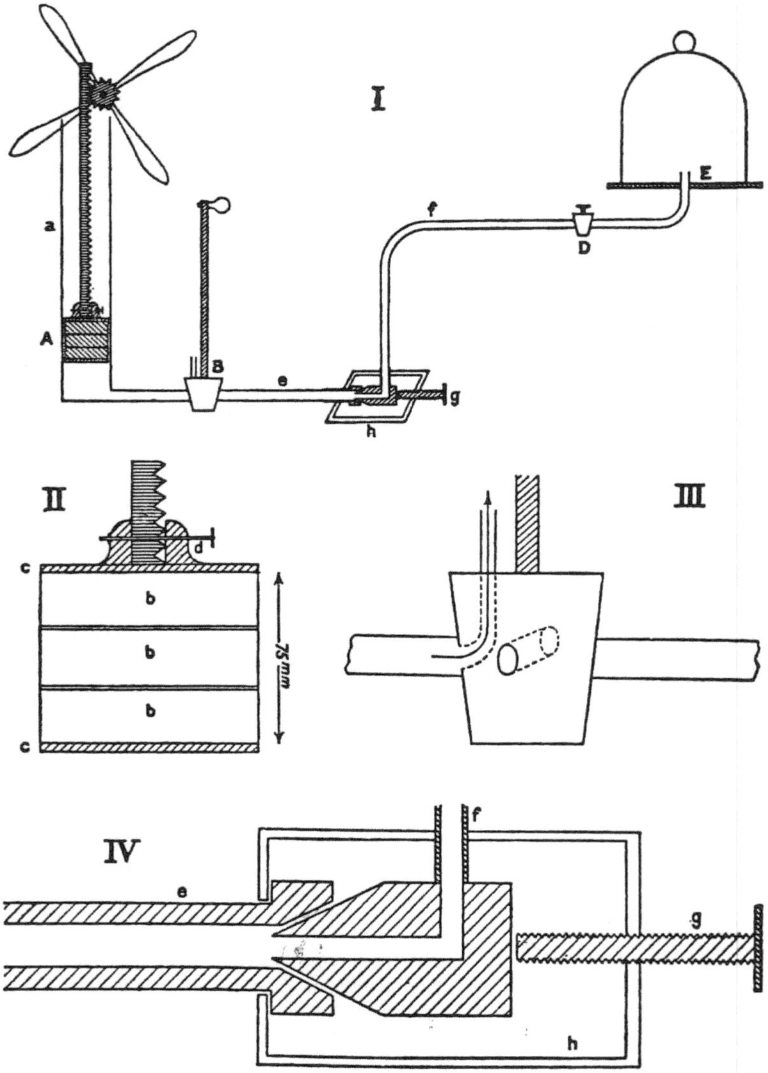

Fig. 5.10 The internal structure of De Volder's air-pump. (Source: Daumas 1972, 86. © Pavilion Books Group Limited)

The air pump worked as follows (see Fig. 5.10, Daumas 1972, 86): when the piston ascends in the cylinder, the stopcock B on the tube which connects the cylinder allows the flow of air between the container and the barrel of the pump. In this way, the air contained in the receiver is rarefied, as it enters into the barrel. As the piston is completely raised, the stopcock is put in a position closing the container, and opening the barrel to external air. Hence, the piston is moved downwards, and the operation is repeated.[88]

Usually, in Robert Hooke's pump – on which De Volder's model was based – the upper section of the cylinder was filled with water, in order to prevent leakage (see Fig. 5.11, Boyle 1669b, volume 3, 180, plate 1). In this experiment, the whole of the basin of the pump is filled with water, which then fills the container by ascending through the tube by which air is 'sucked'. The reported scheme, actually, depicts the pump in its present state, namely, after Willem Jacob 's Gravesande, in 1720s, removed a bigger container or trough which enclosed the tubes, in order to prevent leakages of air.[89] Such a container can be found in the depiction of the pump provided by Senguerd in his *Philosophia naturalis* (see Fig. 5.3, Senguerd 1681, 65). So that this container (now lost) was the 'basin' mentioned by Morley; water, in turn, probably could enter into the container *via* a valve opened in the tubes.

According to the report, the pressure of air on the water in the basin is the cause of the ascension of water, which he also labels as 'circumpulsion'. On the other hand, according to the report, Cartesians still made recourse to the *metus vacui*: "[t] he ascending of the water, he referred unto the circumpulsion, or pressure of the air, and not ad replendum vacuum, as the Cartesians do usually do."[90] It may be that the writer just mistook 'Cartesians' for 'Aristotelians': indeed, the text of the *Experimenta* contains plenty of errors. This interpretation is corroborated by the fact that "circumpulsion" or circular thrust was a sort of trademark of Cartesian explanations, as it was the result of the idea of the plenum. As far as the interaction of solids and fluids is concerned, circumpulsion was used, by the immediate followers of Descartes (as Henricus Regius, Cornelis van Hogelande, and Cornelio) to explain respiration, the raising of water in syphons, and Torricellis's *esperienza*. Originally, the idea of circular thrust (περίωσις) was found in Plato's *Timaeus*, where it is used to explain respiration. The argument is as follows: the chest is dilated, so that it pushes the external air, which is then pushed into the lungs because all the external space is already filled:

> [s]eeing that there is no such thing as a vacuum into which any of those things which are moved can enter, and the breath is carried from us into the external air, the next point is, as will be clear to every one, that it does not go into a vacant space, but pushes its neighbour

---

[88] Cf. Daumas's explanation: "[t]he air-pump of Samuel van Musschenbroek. Figure I is a schematic representation of the air-pump. The single cylinder (a) is of brass. The piston A (Fig. II), of almost 9 cm diameter, is of cork, covered with three strips of leather, and held between two brass plates (c). The piston rod consists of an iron rack attached by means of a pin to a small cylinder fixed to the upper of the two brass plates of the piston, there are no valves. After each movement of the piston the stopcock B is opened or closed (Fig. III). In the position illustrated the pump is in communication with the air; in the reversed position, with the receiver. The tube (f) between the pump and the plate E is of copper. Near the receiver is a tap D. The junction of the tube (f) with the tube (e) of the pump is shown in Fig. IV. The interior of the end of the tube (e) is ground away in conical form to receive the conical end of the tube (f). The latter is held in place by means of a screw passing through the collar (h)," Daumas 1972, 86.

[89] See Sect. 2.2.3, *The mid-1670s clash at Leiden and the foundation of the experimental theatre.*

[90] De Volder 1676–1677, 107r.

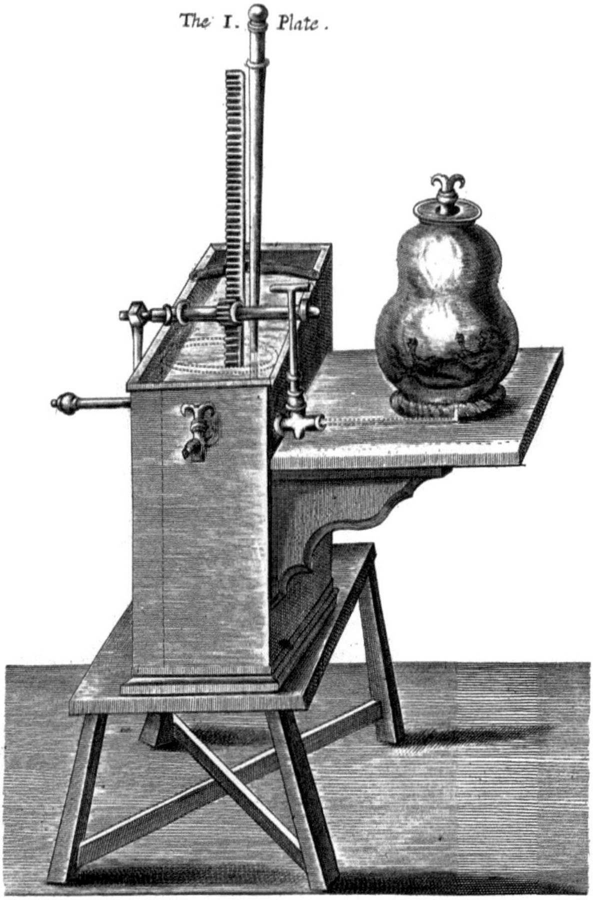

Fig. 5.11 The second model of the air-pump built by Hooke. (Source: Boyle 1669b, volume 3, 180, plate 1. London, Wellcome Collection, shelfmark: 14991/B, https://wellcomecollection.org/works/hnjqcvh6. Attribution 4.0 International (CC BY 4.0))

> out of its place, and that which is thrust out in turn drives out its neighbour; and in this way everything of necessity at last comes round to that place from whence the breath came forth, and enters in there, and following the breath, fills up the vacant space; and this goes on like the rotation of entering a wheel, because there can be no such thing as a vacuum.[91]

The same argument is at the basis of Descartes's justification of the circular nature of every movement in his *Principia* (II.33). This is justified on the ground that all space is already 'filled', so that the displacement of a part of matter would not be possible, if a whole circle of parts were not made:

[91] Plato 1892, volume 3, 501–502. Cf. *Timaeus*, 79a–e.

> [...] all places are full of bodies and that the size of each part of matter is always exactly equal to that of its place [...]. From this it follows that no body can move except in a {complete} circle {of matter or ring of bodies which all move at the same time}; in such a way that it drives another body out of the place which it enters, and that other takes the place of still another, and so on until the last, which enters the place left by the first one at the moment at which the first one leaves it.[92]

Descartes did not maintain that circular movement is necessary in order not to allow a vacuum, but rather that since there is no vacuum, all movement must be circular. This argument was widely used by the followers of Descartes. Regius adopted it (probably following Plato's account, which was, however, perfectly compatible with Descartes's texts that appeared before the *Principia*) as early as his 1640 disputation *Pro circulatione sanguinis* – and then in all his major works. In the hands of Van Hogelande (viz. in his *Cogitationes quibus brevis historia oeconomiae corporis animalis, proponitur, atque mechanice explicatur*, 1646), and Jan Swammerdam (in his *Tractatus physico-anatomico-medicus de respiratione usuque pulmonum*, 1667), this kind of explanation became famous as the 'Cartesian circle'.[93] This argument is also used by Descartes, in his letter to Mersenne 16 October 1639 (published by Claude Clerselier, and discussed further below (see Sect. 5.5, *De Volder's Cartesian hydrostatics: a mixed approach*)), to explain the raising of water in pumps. For Descartes water is raised in pumps because the cylinder of the pump displaces, during its rising, a volume of air which comes to press on the water and to make it ascend in the pump itself.[94] Notably, once Cornelio came to explain the apparent vacuum at the top of the *tubum Torricellianum*, as filled with subtle matter, in his

[92] Descartes 1982, 55–56.

[93] "Aër itaque in inspiratione pectus ingreditur, non propter fugam vacui attractus; cum attractio ob illius fugam nulla detur, nec dari possit: sed quia thoracis dilatatione vicinus aër [...] de loco deturbatur, [...] et cum omnia corporibus plena sint [...] necessario aër a pectore et alio aëre sic pulsus, in thoracem dilatatum per asperam arteriam adigitur," Regius 1640, thesis 9. Also in Regius 1654, 276–277. "Sciendum itaque, in rerum natura vel dari vacuum, et haec praetensa explicatio per fugam vel metum vacui, cessabit; vel non dari, id est, omnia corporibus plena esse. Si plena sint (uti revera sunt) non potest in universo dari locus aliquis novus, nisi qui vel aëris, vel alterius (locum complentis) materiae, eam recipiat quantitatem, quae capacitati ipsius sit commensurata. Itaque, thorax in maiorem molem distendi nequit, nisi tanta aëris portio, quanta ampliori isti (quem per distentionem suam, maiorem quam antea, in universo occupat) loco exacte respondet, in pulmones impellatur," Van Hogelande 1646, 264–265. Regius's and Van Hogelande's texts were owned by De Volder (see *Bibliotheca Volderina*, 3, 10). Swammerdam, in his *Tractatus physico-anatomico-medicus de respiratione usuque pulmonum* (1667) (dedicated to Melchisédech Thévenot), provided a similar explanation: "[i]n inspiratione igitur pulmones explicantur, quia aër tum a pectore sursum moto, tum etiam ab extrorsum propulso abdomine, de loco propulsus, in ipsos impellitur. Minime vero movetur in pulmones, seu eos ingreditur aër, quia ipsi primo explicantur seu extenduntur; cum in tantum explicentur ac extendantur pulmones, in quantum a propulso abdomine, atque a dilatato simul pectore, aër de loco movetur ac in ipsos impellitur," Swammerdam 1667, 17. Please note that in his *Traité de l'homme* (posthumous editions 1662 and 1664), chapter 16, Descartes does not focus on the movement of air in and out the lungs, but only on the causes of the movement of the thorax. On the 'Cartesian circle', see Ragland 2016; Schmaltz 2016, chapter 5.

[94] "L'eau des pompes monte avec le piston qu'on tire en haut, à cause que n'y ayant point de vide en la nature, il ne s'y peut faire aucun mouve ment qu'il n'y ait tout un cercle de corps qui se meuve en même temps," AT II, 588–589. See Nonnoi 1994.

1648 *Epistola*, he used the idea of Platonic circumpulsion, justified on the basis of the identification of matter and space.[95] For him, circumpulsion was a Platonic as much as a Cartesian idea, thus Morley could not ignore that circumpulsion was, in the seventeenth century, another way to refer to Descartes's circular motion, based on a metaphysically-grounded idea of the plenum. It seems, thus, that Morley mistook 'Aristotelians' for 'Cartesians'. Yet, since De Volder's *Experimenta* were elsewhere influenced by Boyle, we may venture to argue that De Volder was actually referring to the Cartesians.[96] Let's note, first of all, that De Volder was a Cartesian himself, namely, he was a plenist throughout his career, as he always favoured a Cartesian theory of material substance. So that he himself favoured circumpulsion.[97] Yet, he might be criticizing the Cartesian explanation of the raising of water in the pump as metaphysically-grounded, namely, as leaning on the idea of impossibility of a vacuum instead of on the consideration of the differences in pressure between different kinds of fluids. This latter kind of account does not make recourse to the idea of the plenum, but takes into consideration only the conditions of equilibrium of fluids. The criticism reported by Morley (if we take it *ad literam*), in fact, echoes that which Boyle moved against the 'plenists' in the 33th of his *New Experiments* (which I touched upon above).[98] The 'plenists' are criticized by Boyle as explaining the ascension of water in tubes – such as in drinking straws – not as the effect of the diminishing in pressure of air in the lungs, but as the result of the rushing in of water in order prevent a vacuum:

> [...] if by a Vacuum we will understand a place perfectly devoid of all corporeal Substance, it may indeed then, as we formerly noted, be plausibly enough maintained that there is no such thing in the world; but that the generality of the Plenists, (especially till of late years some of them grew more wary) did not take a Vacuum in so strict a sense, may appear by the Experiments formerly, and ev'n to this day imploy'd by the Deniers of a Vacuum, to prove it impossible that there can be any made. For when they alledge (for Instance) that when a man sucks Water through a long Pipe, that heavy Liquor, contrary to its Nature, ascends into the Sucker's mouth, only, to fill up that room made by the Dilatation of his Breast and Lungs, which otherwise will in part be empty. [...] When (I say) they alledge such Experiments, the tendency of them seems plainly to import, that they mean, by a Vacuum, any space here below that is not filled with a visible body, or at least with Air though it be not quite devoy'd of all Body whatsoever. For why should Nature, out of her detestation of a Vacuum, make Bodies act contrary to their own tendency, that a place may be fill'd with Air, if its being so were not necessary to the avoiding of a Vacuum.[99]

[95] Cornelio 1663, 122–123.

[96] Or, as Wiesenfeldt has put it, commenting upon this experiment: "[t]atsächlich erschienen Descartes und ‚die Cartesianer' in de Volders Experimentalvorlesungen nur sehr selten und wurden dann beständig dafür kritisiert, daß sie Phänomene, die de Volder vom Luftdruck verursacht sah, anders zu erklären versuchten," Wiesenfeldt 2002, 115.

[97] In commenting upon *Principia* II.33 in his *dictata*, for instance, De Volder does not deal with the problem of *metus vacui*: rather, in both the series he criticizes the atomist theory (ascribed to Epicurus and Lucretius) according to which movement is not possible in a plenum; against this theory, De Volder opposes the idea that all movement is ultimately circular: Hamburg 273, 98–101; cf. Hamburg 274, 41–42.

[98] See *supra*, n. 52.

[99] Boyle 1660, 246–247.

Boyle does not disclose the identity of these plenists: he was probably referring to Hobbes, or Descartes and the Cartesians – as he refers to the idea of subtle matter. In fact, in his *Dialogus physicus de natura aëris* (1661) Hobbes adopted the idea of a circular thrust to explain the apparent vacuum in the container of the pump: namely, the piston, once withdrawn from the barrel, pushes the external air; since all the space is 'full', however, a sort of aether or the purest part of air is pushed into the container through its pores.[100] Of course, Hobbes did not make recourse to the idea of *metus vacui*, but his explanation – like that provided by Descartes in his letter to Mersenne (which De Volder knew well)[101] is based only on the presupposition of the plenum. Moreover, the explanation of the raising of fluids by pumps entails the idea that water is not raised by atmospheric pressure, but rather by the displacement of a separate volume of air which comes to press on the surface of the water itself. So that Boyle – and De Volder – might have been labelling this kind of argument as (still) relying on the fear of a vacuum, rather than on an accurate consideration of the pressure of air. Boyle, in fact, did not commit himself, neither to the thesis that all the space is filled up, nor to the existence of a vacuum.[102] Eventually, Theodoor Craanen, professor of medicine at Leiden from 1673 to 1686, and a former student of Regius, in his posthumous *Tractatus physico-medicus de homine* (1689) listed, amongst possible explanations for respiration a theory according to which inspiration is due to a circular movement of air, which enters the lungs "since it opposes that vacuum is given," and then fills the lungs themselves, and makes the thorax move.[103] Thus far, Morley, who – amongst others – brought to

---

[100] "[...] dum suctor retrahitur, quanto relictus locus major fit, tanto minus loci relinquitur aëri externo, qui retrusus a suctore moto versus externa, proximum sibi aërem similiter movet, et hic alium, et sic continue: ita ut necesse sit aërem tandem compelli in locum desertum a suctore, et intrare inter superficiem suctoris convexam et cylindri concavam. Supposito enim aëris partes esse infinite subtiles, impossibile est ut via illa, qua retrahitur suctor illae non se insinuent. [...] Esto locus ille relictus plenus [...] aëre puro, id est [...] corpore aethereo," Hobbes 1668, *Dialogus physicus, sive de natura aëris*, 12.

[101] Indeed, as I show below, it is discussed, through Boyle's mentioning of it, in his *dictata*: see *infra*, n. 296.

[102] Cf. Boyle's *Examen of Mr. T. Hobbes his Dialogus Physicus de Natura Aeris* (1662b): "Mr. Hobbs proves the space deserted by the Quicksilver or the Air to have no Vacuity, because according to his Supposition the World is full; and not by any sensible Phaenomena that prove the Space in Question to be perfectly full [...]. For no less Fulness is requisite to the truth of his Hypothesis [...]. From all which it seems probably deducible, That 'tis a very hard thing, by. Mr. Hobbs's way of managing the Controversie, to prove that there can be no Vacuum, But as for the Cartesian's more subtile and plausible way of asserting a Plenum, it concerns me not here to Dispute against it, or Declare for it," Boyle 1662b, 8–9. As to the Boyle-Hobbes controversy, see Shapin and Schaffer 1985.

[103] "Respirationis plane aliud est negotium, sed cum hic de ea ex professo agere non possimus, quia nervos nondum vidimus, attendemus ad illa solummodo, quae experientia nos docet; videmus enim hic, ventrem intumescere, quod dicimus fieri ab aëre: sed quomodo ille hoc facit? Quia aër primo a ventre et thorace inflatis propellitur, et cum vacuum dari repugnet, debet ille aër ingredi pulmones, eosque inflat et deprimit sua gravitate, hi diaphragma, hoc viscera abdominis, illa ventrem, et sic porro. Alias causas respirationis videbimus, cum ad nervos respirationi inservientes pervenerimus: tunc videbimus, quod pulmones sint tantum passiva respirationis instrumenta, qui

England some *dictata* on Craanen's still unpublished *Tractatus physico-medicus de homine*, as well as the full manuscript of Craanen's *Lumen naturale medicum, hoc est Praxis medica reformata sive Annotationes in praxin Henrici Regii* (1686)[104] could just have read De Volder's explanation of the experiment as directed against Craanen – who was to be attacked, for his speculative approach, in De Volder's 1698 *Oratio de rationis viribus, et usu in scientiis*. As for De Volder, his explanation of the ascension of fluids relies solely on the consideration of their differences in pressure, as it was for Boyle. Sects. 5.2, *The conditions of hydrostatic equilibrium in De Volder's experimental lectures* and 5.6.4, *The application of Boyle's pneumatics to physiology* are devoted to its discussion.

5.2 The Conditions of Hydrostatic Equilibrium in De Volder's Experimental Lectures

The conditions of the equilibrium of fluids are established by De Volder by overtly following Archimedes's first postulate, which has been briefly discussed above, and which can be rendered as follows:

> it is supposed that the nature of fluid is such as, that among the parts of it which are at the same level and continuous to one another, the less pressed part is expelled by that which is more pressed.

In his *Disputationes*, De Volder quotes the postulate from David Rivault's edition of Archimedes's works (1615), and overtly uses it to explain two main cases: the 'forced' floatation of a body lighter than water, forcibly inserted in a basin full of water, and the cohesion of the Magdeburg hemispheres. Moreover, De Volder would use this to explain the hydrostatic paradox, as I show in Sect. 5.4.2.4, *De Volder's Archimedean approach to the paradox*. De Volder's explanations are based on two laws he derives from the postulate:

> the first is, in each fluid at rest, each surface of the fluid parallel to the horizon is pressed equally. The second one, if each surface of a fluid parallel to the horizon is unequally pressed, the part [which is] more pressed expels that one, which is less pressed.[105]

---

aërem propulsum solummodo excipiunt, eique repulso exitum concedunt. Activa vero instrumenta respirationis, dicemus, esse musculos tam abdominis, quam diaphragmatis, qua de re nemo medicorum unquam antehac cogitavit," Craanen 1689, 264–265.

[104] Cf. mss. Sloane 1274–1276.

[105] "Prima sit, in fluido quovis stagnante unamquamque fluidi superficiem horizonti parallelam qualiter premi. Altera vero, si superficies quaevis fluidi horizonti parallela inaequaliter prematur, partem magis pressam expellere eam, quae premitur minus. Quae sane leges eius sunt evidentiae, ut eas Archimedes supponere sine ulla demonstratione veritus non fuerit. Sic enim ait in libro περὶ τῶν ὀχουμένων. ὑποτιθέσθω τοῦ ὑγροῦ τοι αὑταν φύσιν εἶμεν ὥστε τῶν αὐτοῦ μερέων ἐξ ἴσου κειμένων, καὶ συνεχομένων ἐπ' ἀλλήλων, ἔλαττον πεπιεσμένον, ὑπὸ τοῦ μᾶλλον πεπιεσμένου ἐκβάλλεσθαι, *ponatur fluidi eam esse naturam, ut partibus eius aequaliter iacentibus, et continuatis inter sese, pars minus pressa a magis pressa expellatur*. Hisque iisdem fundamentis usi sunt Galilaeus, Stevinus, Boylaeus in hydrostaticis, uno verbo omnes rerum Hydraulicarum scriptores,"

The underlying model is that of the balance, in which the two pressures or weights are at rest until one of them overcomes the other, and a displacement occurs.[106] Moreover, De Volder claims that the postulate is self-evident, and that it is just an instance of the law of causality, so that Archimedes did not need to justify it.[107] De Volder's demonstrations, thus, do not diverge from those of Archimedes, which thus need to be considered in order to have a grasp of De Volder's model.

As seen above, in Archimedes's demonstrations the existence of a water-world is assumed, in which an ideal surface below the level of the water undergoes the same pressure from the bodies above it: either volumes of water, or bodies floating on it (see Fig. 5.1, Archimedes and Commandino 1565, 4r). In order to maintain the equilibrium or the state of rest which is supposed in the first postulate, one has to suppose that such bodies have the same weight: otherwise one part will displace the other. In the case of Archimedes's proposition 5, for instance, the pressure exerted on surface XO will expel the surface OP. All the cases of sinking or floating are deduced accordingly by Archimedes. This theory, of course, presupposes that water exerts pressure on water like any other body – as noted by Rivault in his edition of Archimedes's treatise, where he adds to the original text a further postulate, stating that "every humid [body] has weight," with his comment that air also weighs on air

---

De Volder 1676–1678, disputation 2, thesis 2; cf. Archimedes and Rivault 1615, 491: "Υποθεσισ A. Υποτιθέσθω τοῦ ὑγροῦ τοιούταν φύσιν εἶμεν, ὥστε τῶν αὐτοῦ μερέων ἐξ ἴσου κειμένων, καὶ συνεχέων ἐπ' ἀλλήλων, ἔλαττον πεπιεσμένον, ὑπὸ τοῦ μᾶλλον πεπιεσμένου ἐκβάλλεσθαι. [...] Positio I. Ponatur humidi eam esse naturam, ut partibus ipsius aequaliter iacentibus, et continuatis inter sese, minus pressa a magis pressa expellatur [...]." De Volder also quotes the Greek version of the postulate, which Rivault had provided in his 1615 edition (which includes variations with respect to those of Tartaglia and Commandino), by re-translating the postulate from Latin into Greek. Archimedes's treatise on floating bodies (*Περί τών οχουμένων*), indeed, was until the discovery of the Archimedes's palimpsest by Johan Ludwig Heiberg (1906) known only in the Latin translation provided by William of Moerbeke, on which Tartaglia and his followers based their edition. Another important seventeenth century edition was the Latin edition of Isaac Barrow (1675). As I am going to show (see Sect. 5.3.2, *A theory of matter for floating bodies*), De Volder also relied on Rivault's commentary on the postulate.

[106] Cf. disputation 2, thesis 4: "[s]uperficie imaginamur partibus pressis inaequaliter sese res habet ac cum duabus lancibus ex eadem libra suspensis. Ut enim earum altera non descendit nisi ascendente altera, ita in hoc vase stagnantis aquae, altera eius pars non labetur deorsum, nisi altera sursum moveatur. Ut autem ea lanx necessario descendit, alteramque assurgere efficit, quae maiori premitur pondere, sic sane et ea pars aquae, quae maiori efficacia deorsum pellitur, descendit, et per consequens ad ascensum cogit alteram aquae partem, quae utpote imbecillior huic alterius partis descensui non est resistendo. Ut denique tum demum statera in aequilibrio haeret, quando in utraque lance eadem est pressio, sic et aquae Superficies tunc demum in eadem ubique consistet altitudine, quando omnes eius partes aequaliter pressae sunt, nec potet idcirco altera alteram ex loco, in quo est, expellere."

[107] "Et sane huius effati veritas ex fluidorum natura fluit evidentissime. Nonne enim ea horum proprietas est, ut vel minimae cuivis cedant pressioni? Immittat quis sine ulla vi aut impetu manum in aquam; nonne confestim aqua cedit manui, et contra suam, ut vulgo creditur, naturam adcendit?" De Volder 1676–1678, disputation 2, thesis 3.

(following Aristotle's *De caelo*).[108] Moreover, Archimedes's theory of floatation presupposes that pressure is exerted both vertically and laterally, i.e. that surface OP, for instance, can be raised by the pressure on XO, *via* a sort of upwards turn of the pressure exerted on XO, which is communicated from below to OP (albeit in a way not clarified by Archimedes).[109] Eventually, Archimedes's model presupposes that the vertical pressures are continuously exerted both downwards and upwards, both in the case of equilibrium, and displacement. This is evident from the case of a forced floatation of a solid lighter than water, forcibly inserted into a basin full of water. In this case Archimedes's proposition 6 would apply, according to which "[s] olids which are lighter than the fluid, when forcibly immersed in the fluid, are thrust upwards by a force equal to the weight by which a volume of the fluid equal to the solid exceeds that solid."[110] Archimedes demonstrates this by reducing it to the case of propositions 4–5, concerning bodies freely floating on water.[111] Namely, he shows that in the case of a freely floating body the same forces involved in the forced immersion and resulting expulsion of a solid are at work. He considers the case of a body half immersed in water, and whose weight is equal to the weight of the volume of water it displaces. According to Rivault's version – where the speed and resilience of the fluid on the solid body is much more explicitly taken into account than in Commandino's – the water below the body exerts a resistance to the downwards pressure of this body, i.e. it exerts a resilience equal to the weight of the volume of displaced water above it (in accordance with the former propositions), which results in a condition of equilibrium. If the pressure or weight exerted by the solid body is diminished, the solid body will be moved upwards with a speed proportional to the weight of the displaced volume of fluid.[112]

---

[108] "Positio III. Humidum omne pondus habere. Σκολιον. Praeter Archimedis positiones, et hanc addendam putavimus, nequis iis quae dicentur deinceps inficias ierit. Quod autem pondus habeat humidum omne, et grave sit, ipsimet sensus docent, ut non sit alio recurrendum in hac veritate determinanda. Nam ipsemet aër gravis est: quod utribus inflatis deprehenditur, qui si aliquando Aeolios Ithacis inclusimus utribus Euros, ut habet Poeta, gravius pendent quam si vacui aëre sint, etiam teste Aristotele [...] cap. 4. l. 4 De caelo," Archimedes and Rivault 1615, 492.

[109] For a discussion, see Chalmers 2017, 2.

[110] Dijksterhuis 1987, 376. Cf. Archimedes and Commandino 1565, 4; Archimedes and Rivault 1615, 497.

[111] "Proposition 4. Of solids one which is lighter than the fluid, when thrown into the fluid, will not sink down altogether, but a portion of it will project above the surface of the fluid. [...] Proposition 5. Of solids one which is lighter than the fluid, when thrown into the fluid, will sink down until a volume of the fluid equal to the volume of the immersed portion has the same weight as the whole solid," Dijksterhuis 1987, 375–376. Cf. Archimedes and Commandino 1565, 2–4; Archimedes and Rivault 1615, 496. See Fig. 5.2, Archimedes and Rivault 1615, 496.

[112] "Sit humidum ABCD et aliqua magnitudo FE eadem mole levior quam humidum: et ipsius quidem FE pars BE ponderet ut HI. Humidum vero molis quidem ipsius BE ponderet ut GI. Tandem tota FE gravitet eodem pondere GI. [...] Et quoniam cuilibet ponderi in premendo resistentiae est: quanta erit GH gravitas obnitetur, prementi vi, et demergenti in humidum magnitudinem FE ita ut si vis illa cessaverit, resiliat ex humido FE magnitudo eadem pressa gravitate GH quae quo maior fuerit eo celerior erit repulsio, donec CF pars, qua magnitudo levior est, ex humido sese extulerit, quod erat probandum," Archimedes and Rivault 1615, 497. See Fig. 5.12, Archimedes and Rivault 1615, 497. Cf. Commandino's version: "[s]it enim magnitudo a levior humido: et sit magnitudinis

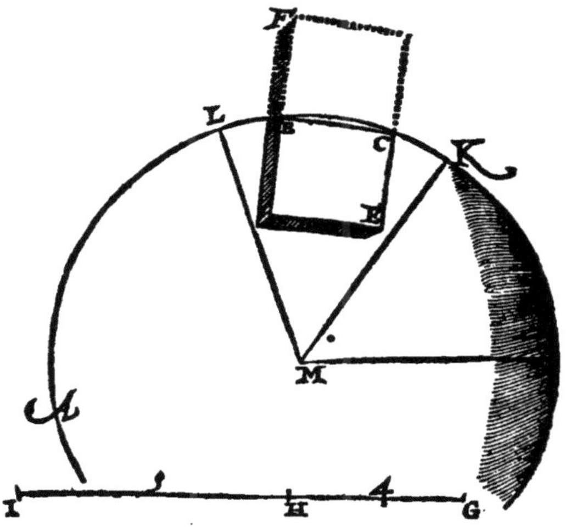

Fig. 5.12 Representation of the condition of floating of bodies, in support of Archimedes's proposition 6. (Source: Archimedes and Rivault 1615, 497. ETH-Bibliothek Zürich, signature: Rar 3843, https://www.e-rara.ch/doi/10.3931/e-rara-18792. Public domain mark 1.0)

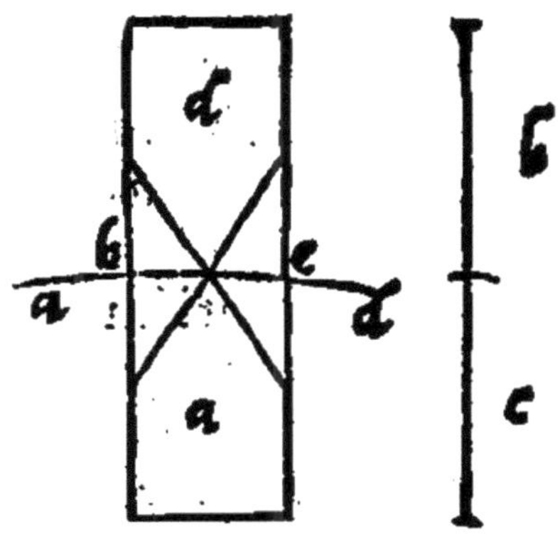

Fig. 5.13 Representation of the condition of floating of bodies, in support of Archimedes's proposition 6. (Source: Archimedes and Commandino 1565, 4v. Rome, Biblioteca Nazionale Centrale, signature: ✝[4]A-K[4]L[6], https://archive.org/details/bub_gb_RR5685AlyVAC. Public domain mark 1.0)

---

quidem a gravitas b: humidi vero molem habentis aequalem ipsi a, gravitas sit bc, demonstrandum est magnitudinem a in humidum impulsam tanta vi sursum ferri, quanta est gravitas c. Accipiatur enim quaedam magnitudo, in qua d habens gravitatem ipsi c aequalem. Itaque magnitudo ex utrisque magnitudinibus constans, in quibus ad, levior est humido: nam magnitudinis quidem quae ex utrisque constat gravitas est bc; humidi vero habentis molem ipsis aequalem gravitas maior est, quam bc: quoniam bc gravitas est humidi molem habentis aequalem ipsi a. Si ergo dimittatur in humidum magnitudo ex utrisque ad constans; usque eo demergeretur, ut tanta moles humidi, quanta est pars magnitudinis demersa eandem, quam tota magnitudo gravitatem habeat. Hoc enim iam demonstrandum est. Si autem superficies humidi alicuius abcd circunferentia. Quoniam igitur tanta moles humidi, quanta est magnitudo a gravitatem habet eandem, quam magnitudines ad: perspicuum est partem ipsius demersam esse magnitudinem a; reliquam vero d totam ex humidi superficie extare. Quare constat magnitudinem a, tanta vi sursum ferri, quanta deorsum premitur ab eo, quod est supra; videlicet a d, cum netura ab altera expellatur, sed d fertur deorsum tanta gravitate, quanta est c: ponebatur enim gravitas eius, in quo d ipsi c aequalis," Archimedes and Commandino 1565, 4–5. See Fig. 5.13, Archimedes and Commandino 1565, 4v. Cf. Dijksterhuis 1987, 376. This theory can be exemplified by what we usually know as Archimedes's principle, viz. "any object, wholly or partially immersed in a fluid, is buoyed up by a force equal to the weight

By leaning on Archimedes's model, in the second disputation De Volder explains the case of the forced floatation of a body lighter than water immersed in water in the following way: the water which is around it – i.e. the lateral portion of water – exerts a downward pressure which, after an upwards turn, presses from below the lower part of the body, as can be experienced by keeping it immersed.[113] De Volder follows Archimedes's model of floatation by expressly taking into account the pressure of lateral water on a given surface, which is exerted from above and below the surface itself, and which results from a turn of the downwards pressure. The ascending movement of lighter bodies in water and in any fluid, thus, is the result of a downward pressure exerted by the weight of a volume of the fluid itself.[114] As seen above, the underlying model is that of the balance: a body is in equilibrium with the volume of water it displaces, when they have the same weight, as is repeatedly reported in the *Experimenta*, where several hydrostatic samples are given. After having stated or dictated[115] Archimedes's postulate in experiment 15, in experiment 17 De Volder stated the thesis that a "container [immersed in water] expelled exactly as much water from its place, as much as the same container weighed." The postulate and the thesis serve him in justifying his observations on the conditions of floating of bodies: namely, on the weighing of a ½ cubic foot of water in experiment 15, which he performed by filling a brass container (whose use was described also by Johannes Joosten van Musschenbroek)[116] floating in water,

of the fluid displaced by the object." A formulation which was never provided by Archimedes himself, but which is the most widely present in contemporary textbooks of physics. I have taken it from Chacón Rebollo and Lewandowski 2014, 30.

113 "Sumamus cylindrum concavum aëre plenum, altero orificio apert, altero obturato in aquam mitti, orificio obturato versus inferiora posito, nonne manifestum est, si hic tubus cylindricus, ad 5. pedes infra aquam deprimatur, superiori orificio aperto extra aquam eminente, magna vi, imo omni gravitate istorum 5. pedum ab aqua subiacente versus superiora premi oportere? Quod ipsum certe fieri non posset, nisi aqua proxime inferius: tangens orificium et fundum Tubi, omni vi et gravitate lateralis aquae versus superiora premeretur? Si enim nulla gravitate, prensioneve deorsum rueret aqua, quae à lateribus est, nec illa aqua, quae fundo subiacet, ulla vi superiora versus deduceretur," De Volder 1676–1678, disputation 2, thesis 5.

114 "Ex quo id notari velim, in fluidis corporibus motum gravitatis quasdam deprimendo fluidi partes alias attollere eadem vi, qua reliquas deprimit, unde conficitur non omném superiora versus motum gravitatem tanquam causam excludere," De Volder 1676–1678, disputation 2, thesis 5.

115 As Morley reports them in Latin, it is probable that he wrote them down as De Volder uttered them.

116 In the sheets extant at Marburg, ending with a letter dated 11 May 1695, and describing also the collision apparatus, one can find the description of the use of the cube: "[l]'usage de cube. Mette le cube dans un vaisseau d'eau avec le tuyaux a en haut, et mette autant de grelle de plomb {…} petites pieces de fil de cuivre, par le tuyaux a dans le <{…}> cube jusque a quand que la superficie de l'eau est egale avec la poincte de cube aupres b. C'est evident que cette cube quand il est avec la pointe egale avec l'eau qu'il fait rejetter autant d'eau, qu'il est grande, otez le cube et pese le, autant qu'il pese, autant pese un demy pied d'eau. Cela est le plus exact maniere pour peser l'eau. En un coste de cette cube est un vis pour ouvrir le cube quand il est trop pesant pour oter le boules," Hessisches Staatsarchiv Marburg, file UniA 305a Nr 5947, 13–14. See Fig. 5.14, Hessisches Staatsarchiv Marburg, file UniA 305a Nr 5947, 13.

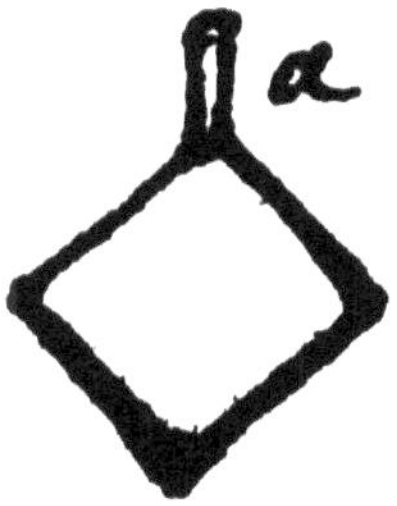

Fig. 5.14 Cube aimed at weighing ½ cubic foot of water, devised by De Volder and sold by Johannes Joosten van Musschenbroek. (Source: courtesy Hessisches Staatsarchiv Marburg, file UniA 305a Nr 5947, 13)

and observing when it is in a condition of equilibrium with the surrounding fluid.[117]

A kind of experiment also mentioned by Elias Camerarius,[118] even if it is unclear if this was the only method De Volder had followed in measuring the weight of water. Moreover, the thesis served De Volder in explaining the various floating situations of glasses filled with sand (experiment 17).[119] In turn, in experiment 24 this thesis is related to the aforementioned idea that elements gravitate on elements, which is the condition of possibility of De Volder's Archimedean model of

[117] The cube of water weighed 7 pounds and 14½ ounces, as seen above. Archimedes's postulate is used to justify the condition of equilibrium between the cube of brass and a certain volume of water. Cf. the text of experiment 15: "[t]his he explicated with the comparison of an exact balance, or pair of scales. Presupposing that the element of water, was a thing, so easily loco mobilis, so easily fluid one way, as other, and so easily giving way to the least poise, or weight, as the exactest pair of scales, could be: the which he said was grounded for a principle in hydrostatic matters among the ancients: nim[irum] aërem circumambientem premere superficiem aquae s[ecun]d[um] m o[mn]es p[rop]ter aequale: nisi ii ita prement illa loco cederet, et fieret commotio, et perturbatio, et inaequalitas in superf[ici]e acquae, cum {I} nimirum {violet$_9$} cedat imaginabili," De Volder 1676–1677, 115r. In his *Tweede deel op de Wiskonstige rekening* (1680), Dirk Rembrandtsz van Nierop reported that De Volder weighed ½ cubic foot of water, namely 7 pounds and 14½ ounces: see Van Nierop 1680, 156.

[118] See his *Dissertationes Taurinenses epistolicae physico-medicae* (1712), where is provided a summary of the experimental lectures of Pierre Polinière, among which is mentioned De Volder's ascertainment of the weight of 1 cubic inch of water: "Tent. 21. [...] [n]os simul incidenter viam ostendimus, qua Clar. Volderus in specificam aquae et dehinc aliorum corporum gravitatem inquisivit, per aequilibrationem scil. cubi pollicaris aenei exacte aequilibrati per plumbeos globulos immissos cum aqua et c. quae plenius tum deducta sunt," Camerarius 1712, 362. The experiment cannot be found among Polinière's *Expériences de physique* (1709). Most probably, Polinière referred to the usual cube of ½ foot side.

[119] Cf. experiment 17: "he inferred the same which in the first hydrostatic experiment: viz. q[uo]d tantumdem aquae exacte o[mn]i[n]o illud vas loco suo pepulerat, quantum ipsum vas ponderabat: et coë[u]ntes in o[mn]ibus granib[u]s quae in liquida immittunt, tantumdem sc[ilicet] liquidi illa loco suo pellere, qua[nt]um ipsa ponderent. Which by the by he said was the only way, and fundament for the gaging of ships or measuring how much weight they carry et cetera," De Volder 1676–1677, 118r. This experiment is comparable to experiment 15: in experiment 17, however, De Volder used glass containers instead of the cube of brass, and filled them with sand.

floatation,[120] and which is demonstrated in experiment 23 (which is the performing of a demonstration repeatedly given by Boyle in his *Hydrostatical Paradoxes* (1666)).[121] In experiment 23, De Volder partially immerses a tube opened on both sides in a basin containing water. As he adds water to the basin (not, however through the tube itself), the level of water in the tube also rises, so, water "doth ponder, and press, and gravitate" on water. Second, he repeats this operation by adding different kinds of liquids in the basin. If coloured water is added, this rises both in the basin and in the tube. If oil is added to the water in the basin (with which it does not mix), it makes the water rise in the tube, as it presses it, albeit the level reached by the water in the pipe is lower than the level reached by the oil outside it (which is lighter than water).[122]

De Volder also devoted further experiments (18–20 and 22) to show – albeit without providing natural-philosophical insights – the behaviour of solids immersed in water, weighing them with a balance in and outside water, and in different kinds of mixed liquors. So, in experiment 18 he showed how solids lose weight in different liquids, annotating such losses "upon a counting board."[123] In experiment 19, he furthered this kind of investigation, showing "the demonstration of Archimedes his crown. Of this point he had touched something the last experiment: but reserved the true proof, or demonstration, thereof unto this."[124] In this experiment, indeed, De Volder puts into practice the method, ascribed to Archimedes, of ascertaining the composition of alloys (or mixtures of metals) by noting how much weight they lose in water. Historically, two methods were ascribed to Archimedes, and are both described by De Volder. According to the first – reported first by Vitruvius and Plutarch – one ascertains the different volumes of the three crowns (or three solid bodies) of the same weight, namely a gold crown, a silver one, and the supposedly mixed one, by measuring the volume of water they displace. Since the volume of the silver one is the greatest, and that of the gold one is the least, one can know whether a crown is made of silver, gold, or is a mixture. According to the second method, described for the first time in the *Carmen de ponderibus et mensuris* (circa fourth–sixth centuries), and in early modern times in Giovanni Battista Della Porta's *Magia naturalis* (ed. 1589) and in *La bilancetta* of Galileo (written in 1586, published in the Bologna edition of his *Opere* (1655–1656)), one uses a (hydrostatic) balance in order to ascertain how much weight different bodies lose in water.[125] This method is

[120] "This experiment was made with mercury: relating to the former ones, viz. the equiponderosity of material things with water, or rather as Volder printed these days in some theses which he presided, omnia corpora, quae innatant liquori, t[antu]m h[ab]ent ponderis, quantum liquor a corpore suo ex loco expulsus ponderat. As also this experiment did relate, unto this thesis likewise then printed by him: aquam, et aërem esse corpora gravia, cum r[ati]o, tum exper[ienti]aa [*sic*] docet, ac proinde falsum est, elementa in suis locis non esse gravia," De Volder 1676–1677, 130r–v.

[121] See Sect. 5.3.4, *Stevin's 'geometrization' of hydrostatics, and Boyle's experimental approach.*

[122] De Volder 1676–1677, 127v–129v.

[123] De Volder 1676–1677, 120r. A "big [writing] slate" is indeed mentioned in the 1705 inventory of the theatre.

[124] De Volder 1676–1677, 120v.

[125] See Crombie 1996, chapter 10. A set of balls aimed at performing this experiment was sold by Johannes Joosten van Musschenbroek, who duly describes this method to Johannes Dorstenius in

actually followed by De Volder, by using three balls, respectively of lead, brass and their league. Yet, no natural-philosophical explanation is provided in Morley's report.[126] Eventually, in experiment 22 De Volder again shows the different conditions of equilibrium, by adding weights to a floating body (i.e. a globe-like bottle filled with water).

The explanation offered by De Volder for these cases – as well as for the so-called hydrostatic paradox, as I am going to show in Sect. 5.4.2.4, *De Volder's Archimedean approach to the paradox* – was Archimedean. By adopting this kind of explanation, however, he adopted an account of hydrostatical equilibrium which was at odds with Descartes's explanation of gravity. An explanation which De Volder favoured at least until 1688, when he reverted to Huygens's account (which was, in turn, developed from Descartes's theories). Roughly speaking, one can divide the explanation of floatation in the late sixteenth and seventeenth century into three main categories:

1. the Archimedean model, according to which a body floats as it is sustained by the pressure exerted by other volumes of water, by a pressure exerted through the whole volume of the water. According to this model, heavy bodies extrude lighter ones upwards.
2. The Cartesian model, according to which only the difference in weight between a body and the water it displaces matters, and no pressure exerted across the whole mass of the fluid is taken into consideration. According to this model, light bodies extrude heavier ones downwards.
3. The Aristotelian model, according to which floatation depends on the shape of the floating body, and on the cohesion of the fluid.

De Volder adopted the first model of floatation. Moreover, he extended this model to the explanation of the pressure of air, as well as to the hydrostatic paradox, providing a model for the explanation of the conditions of the equilibrium of solids and fluids more comprehensive than those of his main source on these topics, namely Boyle. Yet he assumed a Cartesian model in his explanation of gravity, according to which solid bodies immersed in a fluid medium (such as subtle matter, air or water) are extruded downwards by those volumes of the medium rising in their places – rather than being pressed on by its upper parts. De Volder, as I am going to show, attempted to find a synthesis between these two models. In order to shed light on such an effort, and, by extension, on De Volder's role in seventeenth century debates

a letter of 13 March 1702: Hessisches Staatsarchiv Marburg, file UniA 305a Nr 5947, unnumbered page. This letter is extant in two copies (apparently by two different hands).

[126] It is worth noting that the reporter provides wrong data on the weight that the different bodies have in water. Indeed, according to the manuscript, in water the ball of brass weighs 5 ounces, 4 drams, 5 grains; that of lead weighs 5 ounces, 6 drams, 5 grains; their league weighs 5 ounces, 4 drams, 3 grains. However, in water the ball of brass, which has the lowest specific weight, would lose more weight than the league, which is specifically heavier than the brass, and lighter than the lead.

over fluid stability, pressure, and the theory of weight, we need thus to look at the theories of floatation he dealt with.

5.3 Theories of Floatation Before De Volder

In the second disputation (thesis 2) of his *De aëris gravitate* De Volder mentions Galileo, Stevin and Boyle as the chief figures who adopted Archimedes's postulate as the basis of their theories in hydrostatics. A postulate which, according to De Volder, was moreover adopted by "all the writers on hydrostatics."[127] This was decidedly an oversimplification. Yet, even though Galileo, Stevin and Boyle provided different kinds of demonstrations for the various cases of hydrostatic equilibrium, and only Boyle overtly relied on Archimedes's postulate, their demonstrations did all make use of it. In a nutshell: Galileo, Stevin and Boyle considered specific weight as the ultimate factor in hydrostatic equilibrium, and explained the conditions of floatation in terms of the different pressures exerted by the mass of water and that of the solid body immersed in it.[128] In what follows, I will provide an analysis of their theories, as well as those of other figures, which is necessary to shed light on how De Volder inserted himself into seventeenth century discussions on hydrostatics and on its natural-philosophical implications. From the 're-discovery' of Archimedes's treatise on floating bodies, up to Boyle's and De Volder's dealing with hydrostatics through experiments, indeed, one can ascertain a process of re-thinking the physical foundations of this discipline (in itself a branch of mixed mathematics). This process is noticeable in (1) the development of a theory of matter capable of supporting Archimedean accounts of floatation (considered chiefly in Sects. 5.3.1, *Galileo on floatation (1612)*, 5.3.2, *A theory of matter for floating bodies* and 5.3.3, *Pascal's theory of floatation and the hydrostatic lever*, where I consider some of the theories of Galileo, Buonamici, Rivault, Descartes and Pascal), and (2) the 'empiricization' of hydrostatics, treated by considering Stevin's and Boyle's

[127] Cf. the text quoted *supra*, n. 105.

[128] Galileo, Pascal and Boyle repeatedly used the idea of specific gravity or weight (for Galileo, *gravità in specie*), a notion which is also theorized in Rivault's commentary on Archimendes – namely, in his scholium to the postulate that Rivault added to the original text (see *supra*, n. 108): "[c]aeterum hoc humidi pondus, non tam metimur quantitate molis, quam genere naturae: nam, ut ait Vitruvius [*De architectura*, book 7, chapter 8], non amplitudine ponderis, sed genere singularum rerum gravitatem esse, non est negandum. Et hoc deinceps erit animadvertendum cum dicemus aliquod grave se habere ad humidum in gravitate in aliqua ratione. Sensum enim erit, de densitate et rei expositae naturali compactione, secundum quam aestimamur ipsius humidi gravitas, et cuius causa res ipsa et humidum paris molis, vel differunt vel aequantur in ponderositate," Archimedes and Rivault 1615, 492. Stevin used this concept in the third definition of his *De hydrostatices elementis* (1605, first Dutch edition as *De beghinselen des waterwichts,* 1586): "[m]ateria ponderosius corpus, quod magnitudine aequalibus praeponderat," Stevin 1605, 111.

approaches in Sect. 5.3.4, *Stevin's 'geometrization' of hydrostatics, and Boyle's experimental approach.*[129]

5.3.1 Galileo on Floatation (1612)

We have already considered the case of Galileo's unfinished *De motu*. According to the model proposed in this treatise, the only factor determining the floating or sinking of a body is the difference in specific weight between it and its medium. For Galileo, the lightness of a solid body is a form of upwards extrusion exerted by the heavier medium pressing upon a lighter body, and displacing it upwards. This theory, as seen above, entails the idea that water does not gravitate on itself (as declared by Galileo), as otherwise a perpetual displacement of liquid would arise. Galileo, then, dropped the project of the publication of his *De motu*, and came back to the problem of floatation in his *Discorso intorno alle cose che stanno in su l'acqua* (1612). With this work, Galileo faced the problem of floatation as in 1611 a group of Aristotelians, led by Ludovico Delle Colombe, had criticized Galileo's idea that ice floats as it is more rarefied than water – so that it has a lower specific weight (*gravità in specie*, according to the terminology of the *Discorso*) – and defended the idea that it is the shape of bodies, rather than their specific weight, that determines the conditions of floatation.[130]

This idea can be found in the *De motu* (1591) of Francesco Buonamici, the main polemical target of Galileo's *Discorso*. In the second chapter of his treatise, Buonamici supports a theory according to which the phenomena of floatation are explainable in terms of penetrability of different bodies in water, due to their figure, rather than to their differences in specific weight. Leaning on the Aristotelian definition of humid as what is easily contained by other bodies, but not by itself, Buonamici claims that humid bodies such as water and air are more penetrable than solid bodies: for this reason, they are lighter.[131] In turn, Buonamici criticizes the Archimedean

[129] In what follows, I assume a systematic rather than a chronological perspective on these topics. For a more comprehensive account of the historical evolution of the theories of floatation between the sixteenth and seventeenth centuries, see Chalmers 2017.

[130] For a full account, see Palmieri 2005.

[131] "Sunt et caussae gravitatis, aut levitatis per se: huiusmodi [...] sunt naturae elementorum, quae quattuor modis variatae efficiunt et quattuor simplicia corpora, et quattuor item affectiones, sive habilitates ad motum quae per gravitatem, et levitatem menti nostrae subiiciuntur: nam gravia levia quae sunt corpora simplicia, et singula ex his, vel absolute, vel aliqua ex parte, sed et multae sunt aliae caussae comites propter quas haec enatant, illa merguntur. Principem vero locum inter ipsas corporis figura sibi vendicavi: illa n[am] quae latiora sunt, per aërem difficilius cadunt, aegriusq[ue] merguntur in aquis, et si terrae plurimum habent, veluti ferrum. [...] Ac primi quidem caussas explicat Arist. hae vero resolvuntur in duo principia patiens et efficiens; patiens autem est corpus continuum, quod dividitur ab eo quod per ipsum fertur; illud aute[m] humidum esse oportet; na[m] sic facile ab alio terminos accipit: prout igitur magis minusve dividuum fuerit: sic et magis minusve aditum praebet corporib[us] quae feruntur: ita per aërem citius, q[uam] per aqua aliquid fertur; propterea q[uia] aqua minus humida est, et terminatur difficilius, q[uam] aër, quo plus fuerit, aut

theory of floatation – based on the idea that floating or sinking depends on the specific weight of bodies – as reinstating the Platonic idea that all bodies are heavy, and that lightness is, accordingly, a form of upwards extrusion.[132] In turn, in his *Discorso* Galileo does not come back to the problem of the internal pressure of water on water, nor to the problem of the lack of feeling of pressure by swimmers. Rather, he re-instates the idea that the cause of floatation or sinking consists of the specific gravity of a body, namely in the difference in weight between a body and the surrounding medium, "which had been subtly demonstrated by Archimedes" – whose demonstrations are however not discussed by Galileo, who prefers to "conclude the same thing by reducing the causes of such effects to more intrinsic and immediate principles."[133] Moreover, he criticizes various (alleged) counter-examples to Archimede's theory of floatation, such as the case, moved by Buonamici, that a container filled with earth (heavier *in specie* than water) is afloating. To this case, Galileo objects that the container is not entirely filled by earth, but it also contains air (lighter than water).[134] Moreover, Galileo criticizes the objection (whose author is not specified) that a plate of ebony or gold (specifically heavier than water) does not sink, while a ball of the same wood sinks – because a ball can penetrate the continuity of the water. This is due to the fact, for Galileo, that in-between this a plate and the water there is a small quantity of air which makes the plate a mixture

minus, ita quoq[ue] facilius, aut aegrius se dividi patitur," Buonamici 1591, 491. On Aristotle's definition, see *De generatione et corruptione*, 329b30–31.

[132] "Huc accedit q[uod] perspicue veteres confutavit Arist. [cf. *De caelo*, book 1] quia corporum levium motus sursum ad suum locum ex graviorum pulsione fieri contenderent, et profecto videret illud necessario consequi, ut omnia corpora naturalia forent gravia secundum naturam, nulla omnino levia, {q[ui]n} prorsus evenire idem, si aër et ignis in ima aquae sede collocarent. Et quamvis ab Arist. pulsio concedat in elementis, unde terrae partes contrahant in orbem, non est, mea quidem sententia, talis, ut de suo loco naturali gravia pellat, sed ut magis centrum versus deducat," Buonamici 1591, 494.

[133] "[L]a cagione per la quale alcuni corpi solidi discendono al fondo nell'acqua, esser l'eccesso della gravità loro sopra la gravità dell'acqua, e, all'incontro, l'eccesso della gravità dell'acqua sopra la gravità di quelli esser cagione che altri non discendano, anzi che dal fondo si elevino e sormontino alla superficie. Ciò fu sottilmente dimostrato da Archimede, ne' libri *Delle cose che stanno sopra l'acqua*. [...] Io con metodo differente e con altri mezzi procurerò di concludere lo stesso, riducendo le cagioni di tali effetti a' principii più intrinsechi e immediati," Galileo 1890–1909, volume 4, 67.

[134] "[...] intendendo d'un vaso di terra, inferisce contro Archimede così: Tu di' che i solidi che galleggiano, sono men gravi dell'acqua; questo vaso di terra galleggia; adunque tal vaso è men grave dell'acqua, e però la terra è men grave dell'acqua. Se tale è la illazione, io facilmente rispondo, concedendo che tal vaso sia men grave dell'acqua, e negando l'altra conseguenza, cioè che la terra sia men grave dell'acqua. Il vaso che soprannuota, occupa nell'acqua non solamente un luogo eguale alla mole della terra della quale egli è formato, ma eguale alla terra e all'aria insieme nella sua concavità contenuta; e se una tal mole, composta di terra e d'aria, sarà men grave d'altrettanta acqua, soprannoterà, e sarà conforme alla dottrina d'Archimede: ma se poi, rimovendo l'aria, si riempierà il vaso d'acqua, sì che il solido posto nell'acqua non sia altro che terra, né occupi altro luogo che quello che dalla sola terra viene ingombrato, allora egli andrà al fondo, per esser la terra più grave dell'acqua; e ciò concorda benissimo con la mente d'Archimede," Galileo 1890–1909, volume 4, 81–82; cf. Buonamici 1591, 494–497.

lighter than water.[135] Galileo is thus constant to an Archimedean model of floatation, namely a model taking into account the specific weight of bodies rather than the penetrability of the medium. For Galileo, moreover, this model does not invalidate Aristotle's criticism of Plato, who maintained that all bodies are heavy, and lightness is a form of upwards extrusion. Indeed, the model of Archimedes does not concern fire, the only absolutely light body according to Aristotle.[136]

Yet, the way in which Galileo justifies this model of floatation does not make recourse, *pace* De Volder, to Archimedes's first postulate. Rather, Galileo makes recourse only to the "more immediate" principle of the lever, and a theory of momentum (*momento*), according to which the force of a body can be measured by its weight multiplied by its speed, so that A) two equiponderant bodies have the same momentum if they have the same speed, and B) two bodies with different weight have the same momentum if the ratio of their speeds is equal to the ratio of their weight – such as in the Cartesian model seen in Sect. 4.2.1.2, *Descartes's rules of impact*.[137] Galileo

[135] "[...] quello che in questo caso discende e vien locato nell'acqua, non è la sola lamina o tavoletta d'ebano, o di ferro, ma un composto d'ebano e d'aria, dal quale ne risulta un solido non più in gravità superiore all'acqua, come era il semplice ebano o 'l semplice oro. E se attentamente si considererà, quale e quanto sia il solido che in questa esperienza entra nell'acqua e contrasta con la di lei gravità, scorgerassi esser tutto quello che si ritrova sotto alla superficie dell'acqua; il che è un aggregato e composto d'una tavoletta d'ebano e di quasi altrettanta aria, una mole composta d'una lamina di piombo e dieci o dodici tanti d'aria," Galileo 1890–1909, volume 4, 98–99.

[136] "[...] parermi che 'l Sig. Buonamico imponga ad Archimede e deduca dal suo detto più di quello ch'egli ha proposto e che dalle sue proposizioni si può dedurre: avvegnaché Archimede né neghi né ammetta la leggerezza positiva, né pur ne tratti, onde molto meno si debbe inferire ch'egli abbia negato che ella possa esser cagione e principio del moto all'insù del fuoco o d'altri corpi leggieri; ma solamente, avendo dimostrato come i corpi solidi più gravi dell'acqua discendano in essa secondo l'eccesso della gravità loro sopra la gravità di quella, dimostra parimente come i men gravi ascendano nella medesima acqua secondo l'eccesso della gravità di essa sopra la gravità loro; onde il più che si possa raccorre dalle dimostrazion d'Archimede è che, sì come l'eccesso della gravità del mobile sopra la gravità dell'acqua è cagion del suo discendere in essa, così l'eccesso della gravità dell'acqua sopra quella del mobile è bastante a fare che egli non discenda, anzi venga a galla, non ricercando se del muoversi all'in su sia o non sia altra cagion contraria alla gravità. [...] Sieno dunque indirizzate l'armi del Sig. Buonamico contra Platone e altri antichi, li quali, negando totalmente la levità e ponendo tutti li corpi esser gravi, dicevano il movimento all'insù esser fatto non da principio intrinseco del mobile, ma solamente dallo scacciamento del mezo; e resti Archimede con la sua dottrina illeso, poi che egli non dà cagion d'essere impugnato," Galileo 1890–1909, volume 4, 84–85.

[137] "[...] io piglio dalla scienza meccanica due principii. Il primo è, che pesi assolutamente eguali, mossi con eguali velocità, sono di forze e di momenti eguali nel loro operare. Momento, appresso i meccanici, significa quella virtù, quella forza, quella efficacia, con la quale il motor muove e 'l mobile resiste; la qual virtù depende non solo dalla semplice gravità, ma dalla velocità del moto, dalle diverse inclinazioni degli spazii sopra i quali si fa il moto, perché più fa impeto un grave descendente in uno spazio molto declive che in un meno. [...] Come, per esemplo, due pesi d'assoluta gravità eguali, posti in bilancia di braccia eguali, restano in equilibrio, né s'inclina l'uno alzando l'altro; perché l'egualità delle distanze di ambedue dal centro, sopra il quale la bilancia vien sostenuta e circa il quale ella si muove, fa che tali pesi, movendosi essa bilancia, passerebbono nello stesso tempo spazii eguali, cioè si moverieno con eguali velocità, onde non è ragione alcuna, per la quale questo peso più di quello, o quello più di questo, si debba abbassare; e per ciò si fa l'equilibrio, e restano i momenti loro di virtù simili ed eguali. Il secondo principio è, che il

Fig. 5.15 Representation of the condition of floating of bodies according to Galileo Galilei. (Source: Galileo 1655, 14. The University of Oklahoma Libraries, History of Science Collection, signature: †[2], †[2], A-G[4], H[6], https://repository.ou.edu/uuid/4226aac2-2516-5262-815f-845d50788252#page/1/mode/2up. Public domain)

uses the first principle to demonstrate that the volume of water corresponding to the volume of a body below the surface of the water has the same weight as the whole floating body.[138] This is a pure application of the principle of the lever, without considering the ways of transmission of forces through the fluid. Hence, the second principle is used by Galileo to account for two instances of the so-called hydrostatic paradox (which I discuss further in Sect. 5.4.2, *The hydrostatic paradox*). Namely, (1) the case in which a solid body floats in a volume of water which is smaller than its own volume (see Fig. 5.15, Galileo 1655, 14), and (2) the case in which in an U-shaped container of water, one of the horns is bigger than the other one. So that the water contained in the smaller horn, balances the one contained in the bigger one (see Fig. 5.16, Galileo 1655, 15) – even if their weights are different. Galileo accounts for these cases by using the principle of the lever with unequal arms (principle B). For instance, when a solid is released in a basin containing less water than that equal to the volume of the solid itself, the (small body of) water rises around the solid with a speed which makes its momentum capable of balancing that of the solid which moves downwards. This is valid also in the case of the U-shaped tube.[139]

---

momento e la forza della gravità venga accresciuto dalla velocità del moto; sì che pesi assolutamente eguali, ma congiunti con velocità diseguali, sieno di forza, momento e virtù diseguale, e più potente il più veloce, secondo la proporzione della velocità sua alla velocità dell'altro. Di questo abbiamo accomodatissimo esemplo nella libra o stadera di braccia diseguali," Galileo 1890–1909, volume 4, 68. On Galileo's theory of *momento*, see Galluzzi 1979.

[138] "[...] la mole dell'acqua che s'alza nell'immersion del solido, o che s'abbassa nell'estrarlo, non è eguale a tutta la mole del solido che si trova demersa o estratta, ma a quella parte solamente, che nell'immersione resta sotto il primo livello dell'acqua, e nell'estrazione riman sopra simil primo livello," Galileo 1890–1909, volume 4, 72.

[139] "Séguita in oltre che, posto un solido men grave dell'acqua in un vaso di qual si voglia grandezza, e circunfusagli attorno acqua sino a tale altezza, che tanta acqua in mole, quanta sia la parte del solido sommersa, pesi assolutamente quanto tutto il solido, egli da tale acqua sarà giustamente sostenuto, e sia l'acqua circunfusa in quantità immensa o pochissima. [...] Il che a molti potrebbe, nel primo aspetto, aver sembianza di grandissimo paradosso, e destar concetto che la dimostrazione di tale effetto fosse sofistica e fallace; ma per quelli che per tale la reputassero, c'è la sperienza di mezo, che potrà rendergli certi: ma chi sarà capace di quanto importi la velocità del moto, e come ella a capello ricompensa il difetto e 'l mancamento di gravità, cesserà di maravigliarsi, nel considerare come all'alzamento del solido M pochissimo s'abbassa la gran mole dell'acqua ABCD, ma assaissimo ed in uno stante decresce la piccolissima mole dell'acqua ENSF come prima il solido M si eleva, benché per brevissimo spazio; [...] E per amplissima confermazione e più chiara esplicazione di questo medesimo, considerisi la presente figura [...] nella quale al vaso

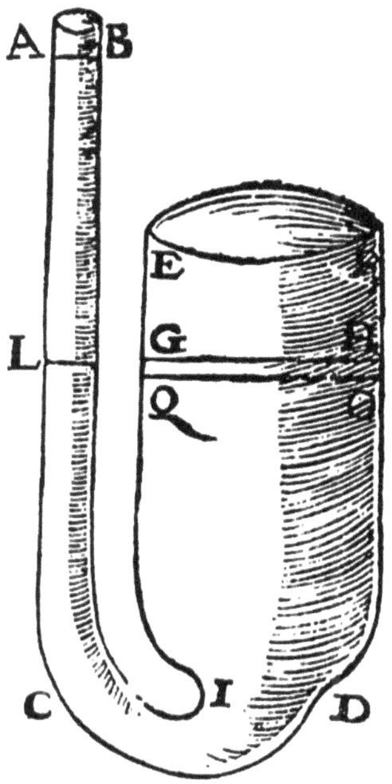

Fig. 5.16 Inverted asymmetrical syphon. (Source: Galileo 1655, 15. The University of Oklahoma Libraries, History of Science Collection, signature: †[2], †[2], A-G[4], H[6], https://repository.ou.edu/uuid/4226aac2-2516-5262-815f-845d50788252#page/1/mode/2up. Public domain)

Galileo, thus, avoided any consideration of the ways in which a body of water and that of a solid counterbalance themselves: namely, he does not explain how the weight of water and that of the solid are in balance through any medium. It is worth recalling how Leibniz would explain the balancing of bodies on a lever, *via* their *vires mortuae*: namely, by a constant communication of such forces *via* the arms of the balance, which are elastic. De Volder, in turn, explained the balancing of bodies in water by a communication of forces *via* the lateral columns of water, and by considering the whole volume of the fluid. Galileo refrains from entering into these issues, and reduces the phenomena of floatation to instances of static equilibrium, or, as he puts it, by "more intrinsic and immediate principles." He does not provide,

larghissimo EIDF, vien continuata l'angustissima canna ICAB, ed intendasi in essi infusa l'acqua sino al livello LGH; la quale in questo stato si quieterà, non senza meraviglia di alcuno [...]. Ma tal meraviglia cesserà, se noi cominceremo a fingere l'acqua GD essersi abbassata solamente sino a QO, e considereremo poi ciò che averà fatto l'acqua CL. la quale, per dar luogo all'altra che si è scemata dal livello GH sino al livello QO, doverà per necessità essersi nell'istesso tempo alzata dal livello L sino in AB, ed esser la salita LB tanto maggiore della scesa GQ, quant'è l'ampiezza del vaso GD maggiore della larghezza della canna LC, che in somma è quanto l'acqua GD è più della LC. Ma essendo che il momento della velocità del moto in un mobile compensa quello della gravità di un altro, qual meraviglia sarà se la velocissima salita della poca acqua CL resisterà alla tardissima scesa della molta GD?" Galileo 1890–1909, volume 4, 76–77.

moreover, a theory of fluidity, i.e. a theory of matter which can justify how weights are in balance between each other, which I consider in Sect. 5.3.2, *A theory of matter for floating bodies*.

5.3.2 A Theory of Matter for Floating Bodies

In fact, an issue in the acceptance of Archimedes's model of floatation was that of the introduction of a theory of matter capable of supporting it, i.e. to justify how different volumes (of fluid and solid bodies) could interact. This was by no means a minor topic, insofar an Aristotelian like Buonamici challenged Archimedes's theory of floatation precisely on the ground of the theory of matter. As just seen, Buonamici used the Aristotelian definition of humid as that which is not easily contained by its own boundaries, but is contained by other bodies. Water, in this perspective, is a continuous body which can be divided by another body, if the latter is provided with an appropriate shape. However, according to Buonamici water is by no means a substance in which different pressures can be transmitted and balance each other as in Archimedes's model (which is duly reported by Buonamici in chapter 29 of his *De motu*).[140] Against this model Buonamici considers the case, in chapter 30, of a wooden container floating on water, which sinks once it is filled with water. For him, this case disproves Archimedes's theory of floatation, as a body lighter than water (such as the container and the water in it), once it breaks the continuity of water, can sink. More specifically, it can break the continuity of water if the shape of the container allows the pressure of the water present in the container to be exerted in a way suited to penetrating the rest of the body of water.[141] For Galileo, on the other hand, the sinking of such a container is not due to its shape, but only to the fact that it contains iron and other metals (such as a ship's armament), which

[140] See, for instance, his treatment of Archimedes's proposition 5, which I summarized above: "[h]is stantibus apparet XO et OP humidi partes ex aequo positas esse et continuatas, non premi vero similiter, quod XO premitur a magnitudine EHTF et a residuo pyramidis. OP vero premitur a solido RSQV, quod est aequale ipsi BHTC parte ipsius EHTD, et ab altero residuo superiori residuo aequali. Magis ergo premitur OP. Itaque altera pars OP ab ea pelletur. Non igitur consistet humidum, quod tamen consistere ponebatur. Proinde ex humidi superficie nihil extat ipsius EHTF magnitudinis. Verum neque demensa descendet inferius: necesse est enim, ut in descensu pariter expellantur: at ibi solum expulsio succedit, ubi una pars premitur magis, quam proxima; sed hic cum solida magnitudo et humidum sint aeque gravia, nulla pressio maior, propterea neque pulsio, neque descensus igitur," Buonamici 1591, 493. See Fig. 5.1, Archimedes and Commandino 1565, 4r.

[141] "[...] ab experientia significamus nempe vas, quod caeteroquin eminebat in aquis, ubi plenum aqua fuerit, nihilominus ad imum proficisci, quod sane difficiles extricatus habet; q[ui]n neque id vasis naturae po[ssi]t adscribi, quod {et} si dissolvatur ex aqua extabit; nec illae aquae, quae recipitur, cum ipsa propter aequalem potestatem, quam habet cum reliqua, in eadem superficie manere debeat. [...] Arbitror [...] quam aqua per illud vas, a quo continetur, a toto segregata est: ideoque totius beneficio non {polles}, tanquam aliud quid, ab aqua diversum, per proprium pondus habet, et reliquam aquam, tanquam medium per quod feratur, opprimit, simul dividit, neque vas, ut lignum simpliciter descendit, sed tanquam unum quid cum eo factum, quod continet, in aqua pondus habet figurae beneficio quae multum alieni comprehendit," Buonamici 1591, 494–495.

makes it heavier than water (and that can float only if it contains air as well).[142] For Galileo, a fluid medium may slow the descent or ascension of bodies: but the reason for the descent or ascension of a body through it cannot be reduced to it, as otherwise every body which can fall in air, can also descend in water, even if at a different speed.[143] So that according to Galileo "water has not any renitence to division, and it is not possible to find any solid body, of any figure one wants, to which, once put on water, its movement is prohibited by the crassitude of that [water]." The only cause of floating or sinking is the difference in specific weight.[144]

The acceptance of a theory of floatation based on the idea of specific weight, in short, needed a new theory of matter capable of supporting it. Namely, a theory of fluidity which renders useless the recourse to the idea of penetrability or impenetrability in accounting for floatation. Moreover, Archimedes's first postulate (albeit not overtly employed by Galileo) needed a theory of matter capable of supporting the idea that pressure is exerted across the entire volume of a fluid. The first evidence of the development of such a theory of matter can be found in the commentary that Rivault provides to his edition of Archimedes's treatise on floating bodies – namely, the edition used by De Volder. In order to support a theory of floatation which encompasses Archimedes's first postulate, both Rivault and De Volder adopted a corpuscular theory of matter, according to which different sections of water can be easily moved, and their movement can be communicated through all the parts of the water. So, Rivault comments on Archimedes's postulate by quoting the pseudo-Aristotle's *Problemata*, in which it is explicitly stated that water, unlike air, is a continuous body ("σῶμα συνεχές"), so that a movement imparted to a part of it is transmitted through all its parts; for the pseudo-Aristotle this does not happen in air, which is discontinuous and movements are not always easily transmitted.[145] Rivault uses the notion of the discontinuity of air to characterize water itself, in order to

[142] "Io non dico già che non si possano, di legno che per sua natura galleggi, far barche, le quali poi, piene d'acqua, si sommergano; ma ciò non avverrà per gravezza che gli sia accresciuta dall'acqua, ma sì bene da' chiodi e altri ferramenti, sì che non più s'avrà un corpo men grave dell'acqua, ma un composto di ferro e di legno, più ponderoso d'altrettanta mole d'acqua. Cessi per tanto il Sig. Buonamico di voler render ragioni d'un effetto che non è," Galileo 1890–1909, volume 4, 83.

[143] "Può ben l'ampiezza della figura ritardar la velocità, tanto della scesa, quanto della salita, e più e più secondo che tal figura si ridurrà a maggior larghezza e sottigliezza: ma ch'ella possa ridursi a tale, ch'ella totalmente vieti il più muoversi quella stessa materia nella medesima acqua, ciò stimo essere impossibile. [...] Che se l'incremento e 'l decremento della tardità o velocità non avessero altro rispetto che alla grossezza o sottilità de' mezzi, ogni mobile, che scendesse per l'aria, scenderebbe anche per l'acqua: perché qualunque differenza si ritrovi tra la crassizie dell'acqua e quella dell'aria, può benissimo ritrovarsi tra la velocità dello stesso mobile nell'aria e qualche altra velocità; e questa dovrebbe esser sua propria nell'acqua: il che tuttavia è falsissimo," Galileo 1890–1909, volume 4, 88 and 130.

[144] "Ricevasi, dunque, per vera e indubitata conclusione, che l'acqua non ha renitenza alcuna alla semplice divisione, e che non è possibile il ritrovar corpo solido alcuno, di qualunque figura esser si voglia, al quale, messo nell'acqua, resti dalla crassizie di quella proibito e tolto il muoversi in su o in giù, secondoché egli supererà o sarà superato dall'acqua in gravità, ancorché l'eccesso e differenza sia insensibile," Galileo 1890–1909, volume 4, 108.

[145] *Problemata*, 931a38–931b8.

grant its mobility, which however does not hinder the connection (series) between its parts.[146] In other words, for Rivault water is mobile like air, but it is continuous as well. *Ça va sans dire*, the Archimedean postulate implies that movement is transmitted across the volumes of water, namely, that the mobility of water still allows a continuous transmission of movement, such as on a balance. This is clearly expressed in De Volder's *Experimenta*, as he introduces Archimedes's postulate: "[p]resupposing that the element of water, was a thing, so easily loco mobilis, so easily fluid one way, as other, and so easily giving way to the least poise, or weight, as the exactest pair of scales, could be: the which he said was grounded for a principle in hydrostatic matters among the ancients."[147] In commenting upon Archimedes's postulate in his *Disputationes*, moreover, De Volder mirrors parts of Rivault's commentary, besides mentioning more overtly the corpuscular nature of fluids.[148] The main change adopted by De Volder with respect to Rivault's version of the postulate is the replacement of 'humid' (*humidum*) with 'fluid' (*fluidum*), translating 'υγρό': a conceptual change which can be noted, for instance, as early as the *Exercitationes philosophicae* (1620) of the Dutch atomist David Gorlaeus, and then – probably through the mediation of Henricus Reneri, who taught some of Gorlaeus's ideas – in the theories of Dutch Cartesians such as Regius, Johannes de Bruyn and De Raey. According to them, the traditional definition of humidity as what is more easily contained by another body, than by itself, fits better the notion of fluidity, and applies also to bodies which do not contain water, i.e. dry bodies. Among these, Gorlaeus, Reneri and Regius counted also air, traditionally labelled as a humid substance.[149] On this ground, it is possible to advance the hypothesis that the re-discovery – let's say – of Archimedes's theory of floatation brought about a new conceptualization of solidity and fluidity in corpuscular terms (replacing the old differentiation between humid and dry bodies), which will have in Descartes's *Principia* (II.54) a more articulated formulation. According to Descartes,

---

[146] "Ratio positionis seu assumpti huius a natura, qualitatibusque humidi dependent. Ea enim ipsius natura est, ut facile mobile sit: tum ut facile cedat, quia non partibus constat arcte compactis, quin potius leviter divisibilibus, similis fere aëri, cuius flexibilitatis ergo, dixit Aristoteles, ἐπ' οὔτε ἐστὶν ἓν σῶμα συνεχές. Propterea si prematur quocumque modo etiam levissime, facile [...] disiungitur [...]. Tum si motu [...] extima illa superficies facta [...] inaequalis, nativo pondere restituetur, donec ad aequilibrium ierit. Imo id acciderit [...] si totius corporis non disiuncta fuerit series," Archimedes and Rivault 1615, 491. Cf. *Problemata*, 931b, 4–5.

[147] De Volder 1676–1677, 115r.

[148] "Et sane huius effati veritas ex fluidorum natura fluit evidentissime. Nonne enim ea horum proprietas est, ut vel minimae cuivis cedant pressioni? [...] Cum enim agatur solummodo de separatorum [...] corpusculorum motu et agitatione, quis non videt hunc illumve motum facile quibusdam fluidi partibus imprimi posse, qui non imprimatur aliis, a prioribus omnino seiunctis avulsisque?" De Volder 1676–1678, disputation 2, thesis 2.

[149] Gorlaeus 1620, 332–333; Reneri 1635, disputation 4, thesis 29; Regius 1646, 155 and 158; De Bruyn and Posahazi 1654, thesis 5; De Raey 1666, thesis 3. Traditionally, air was labelled as humid and hot, while water was humid and cold; by considering air as dry, Gorlaeus was dismantling the traditional, four-fold theory of the elements, following Girolamo Cardano and Henricus de Veno, who claimed that air cannot truly mix with other bodies: see Lüthy 2001.

> the parts of fluid bodies easily move out of their places, and consequently do not resist the movement of our hands into those places; while, on the contrary, the parts of solid bodies adhere to one another in such a way that, without sufficient [...] force to overcome their cohesion, they cannot be separated. [...] From this we may conclude that those bodies which are divided into very small parts which are agitated by a diversity of {independent} movements, are fluid; while those bodies whose particles are all contiguous and at rest, are solid.[150]

For Descartes and his followers, the particles of fluid bodies – regardless of their shape – are always in movement with respect to each other, and in all directions: that is why it is easy to separate them (i.e. it is enough to change their direction of motion). The adoption of corpuscular conceptualizations of fluidity, in sum, favoured the acceptance of Archimedean theories of floatation, and vice versa. These were theories in which a fluid medium never impedes the sinking, or the ascension, of a solid body which is heavier or lighter than it. In other words, such theories granted mobility and separability within the continuity of a body – a feature not accounted for in the (old) theories of humidity.

5.3.3 Pascal's Theory of Floatation and the Hydrostatic Lever

The theories of fluidity that appeared in the seventeenth century did not only favour the acceptance of theories of floatation based on the idea of specific weight. In fact, they also served as a means to explain hydrostatical paradoxes, namely, those cases of floatation in which (as in the cases addressed by Galileo) a small volume of water counterbalances a greater volume of the same substance, or a heavier solid. This is the case with the theory of floatation of Pascal, who is mentioned in De Volder's *Experimenta*,[151] who deployed an idea of fluidity comparable to that of Rivault (and of De Volder himself) in his *Traité de l'équilibre des liqueurs*. In this treatise, written in 1654 and published with his *Traité de la pesanteur de la masse de l'air* in 1663, Pascal does not mention Archimedes's first postulate. Yet, in his theories one can find the idea of equilibrium between pressures exerted on ideal and real surfaces, which is the core of the postulate. In chapter 2 of his treatise, Pascal considers a variant of the hydrostatic paradox considered by Galileo, namely, what we know as the hydraulic press or lever. Let us take the second case considered by Galileo (see Fig. 5.15, Galileo 1655, 14), in the version used by Pascal (see Fig. 5.17, Pascal 1663, figures 6–7).

---

[150] Descartes 1982, 69–70. De Volder does not provide a relevant variant of Descartes's account in his *dictata* on II.54, as he remarks only that fluid bodies can be easily separated as their particles move in all directions: cf. Hamburg 273, 113–144; Hamburg 274, 47–48. Moreover, for Descartes – and for his followers – there is no dichotomy between fluids and solids: rather, they are 'different states' of matter, and there are degrees of solidity and fluidity, as in a spectrum. This idea is expressly given in *Le monde*, chapter 3. I discuss Descartes's theory of floatation – which was part of his theory of gravity – in Sect. 5.5, *De Volder's Cartesian hydrostatics: a mixed approach*.

[151] See *supra*, n. 66.

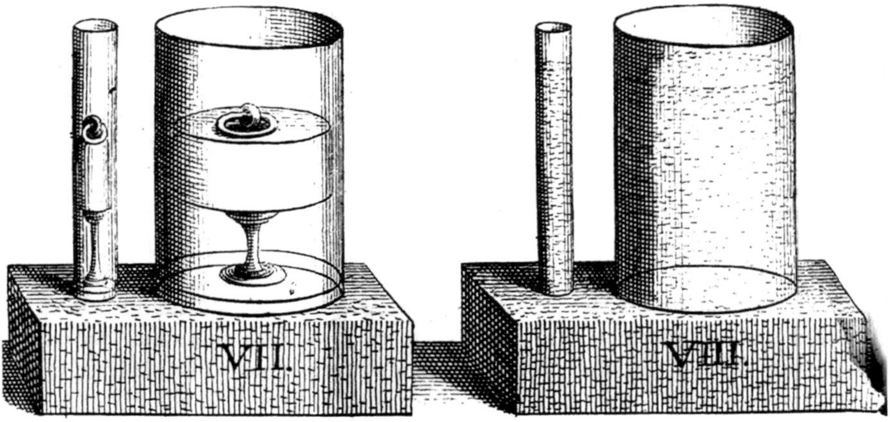

Fig. 5.17 Inverted asymmetrical syphon reduced to the model of the lever. (Source: Pascal 1663, figures 6–7. ETH-Bibliothek Zürich, signature: Rar 4668, https://www.e-rara.ch/doi/10.3931/e-rara-20296. Public domain mark 1.0)

As Pascal illustrates it, a pressure of 1 pound applied on the water in the smaller tube, can raise a weight of 5 pounds, posed upon the water in the larger tube, if it is 5 times larger than the smaller tube. Of course, this also works if the pressure of 1 pound applied on the water in the smaller tube, is applied by putting 1 pound of water on it. This appears to be a paradox, because it seems that a multiplication of forces is taking place. However, as for Galileo, this is nothing but an instance of the principle of the lever, namely, we should conceive the two sections of the fluid contained in the tubes as the two plates of a lever, in which the centre of gravity is placed at a distance proportional to the ratio of the two weights. What allows the transmission of pressure, or weight between the two plates, or surfaces of the fluid is for Pascal the "continuity and fluidity" of water, which makes "all these things equal and indifferent," i.e. it allows for a sort of multiplication of forces operated in accordance with the principle of the lever.[152]

[152] "[...] si un Vaisseau plein d'eau n'a qu'une seule ouverture, large d'un poulce, par exemple, où l'on mette un Piston chargé d'un poids d'une livre, ce poids fait effort contre toutes les parties du Vaisseau generalement, à cause de la continuité et de la fluidité de l'eau: mais pour determiner combien chaque partie souffre, en voicy la regle: Chaque partie large d'un poulce, comme l'ouverture, souffre autant que si elle estoit poussée par le poids d'une livre (sans compter le poids de l'eau dont je ne parle pas icy, car je ne parle que du poids du Piston), parce que le poids d'une livre presse le Piston qui est à l'ouverture, et chaque portion du Vaisseau plus ou moins grande, souffre precisément plus ou moins à proportion de sa grandeur, soit que cette portion soit vis à vis de l'ouverture ou à costé, loin ou prez; car la continuité et la fluidité de l'eau rend toutes ces choses là égales et indifferentes [...]. Voicy encore une preuve qui ne pourra estre entenduë que par les seuls Geometres, et peut estre passée par les autres. Je prends pour principe, que jamais un corps ne se meut par son poids, sans que son centre de gravité descende. D'où je prouve que les deux pistons figurez en la Figure VII. sont en Equilibre, en cette sorte; car leur centre de gravité

Moreover, for Pascal the application of the principle of the lever to fluids works for each surface of the fluid, enabling the construction of hydraulic presses. Namely, a pressure of 1 pound exerted on the surface of the smaller tube (with an area, let's say, 1) is exerted on each area of size 1 of the whole surface of the mass of water. This pressure is thus exerted through the whole mass of the fluid. This idea, actually, conforms to Archimedes's first postulate, which we can find implicitly applied in chapter 5, *Des corps qui sont tout enfoncez dans l'eau*, where Pascal explains the case of the forced floatation of a body specifically lighter than water. According to Pascal, a body immersed in water is pressed from all the directions. Moreover, the lower part of the body is pressed with more force on its lower surface, than on the upper surface. This is so because the lower surface is pressed on by more water than the upper one, namely, it is pressed on by the surrounding columns of water, which, after a sort of 180° upwards turn, exert their weight from below. So that if the body immersed in the container is specifically lighter than water, i.e. it exerts less downwards pressure than the water below itself, such a body will rise.[153]

In sum, Pascal applied Archimedes's postulate both to the conditions of normal floatation, as well as to instances of hydrostatic paradox, coupling it with the principle of the lever – already used by Galileo. Moreover, he adopted a theory of matter according to which water is both continuous and fluid: namely, it is separable without resistance, and at the same time it can transmit force through all its parts. All these elements will be found in De Volder's hydrostatics, especially in his explanation of the hydrostatic paradox – which he addresses in the cases discussed by Stevin, and which were, in turn, experimentally attempted by Boyle. These were the two other direct sources for De Volder in hydrostatics.

---

commun est au point qui divise la ligne, qui joint leurs centres de gravité particuliers, en la proportion de leurs poids; qu'ils se meuvent maintenant, s'il est possible: donc leurs chemins seront entre eux comme leurs poids réciproquement, comme nous avons fait voir: or, si on prend leur centre de gravité commun en cette seconde situation, on le trouvera precisément au mesme endroit que la premiere fois," Pascal 1663, 8–11. Pascal refers to Torricelli's principle, dealt with in Sect. 4.2.2, *Huygen's laws of collision*. This explanation, notably, will be assumed also by Rohault, who in chapter 10 the first part of his *Traité* de *physique* (1671) addresses the equilibrium of fluids in an asymmetrical, inverted syphon (in § 11) exactly after having treated the principle of the lever (§ 10).

[153] "Nous voyons par là que l'eau pousse en haut les corps qu'elle touche par dessus; qu'elle pousse en bas ceux qu'elle touche par dessous; et qu'elle pousse de costé ceux qu'elle touche par le costé opposé: d'où il est aisé de conclure que, quand un corps est tout dans l'eau, comme l'eau le touche par dessus, par dessous et par tous les costez, elle fait effort pour le pousser en haut, en bas et vers tous les costez: mais comme sa hauteur est la mesure de la force qu'elle a dans toutes ces impressions, on verra bien aisément lequel de tous ces efforts doit prevaloir. Car il paroist d'abord que comme elle a une pareille hauteur sur toutes les faces des costez, elle les poussera également; et partant ce corps ne recevra aucune impression vers aucun costé, non plus qu'une giroüette entre deux vents égaux. Mais comme l'eau a plus de hauteur sur la face d'en bas que sur celle d'en haut, il est visible qu'elle le poussera plus en haut qu'en bas, et comme la difference de ces hauteurs de l'eau est la hauteur du corps mesme, il est aisé d'entendre que l'eau le pousse plus en haut qu'en bas, avec une force égale au poids d'un volume d'eau pareil à ce corps. Un corps dans l'eau est contrepesé par un volume d'eau pareil, de là vient que quelques corps y tombent. [...] D'autres y montent. S'il est de bois, ou d'une autre matière plus legere que l'eau en pareil volume, il monte avec toute la force dont le poids de l'eau le surpasse," Pascal 1663, 25–26.

5.3.4 Stevin's 'Geometrization' of Hydrostatics, and Boyle's Experimental Approach

Stevin is usually known as having systematically dealt with the so-called hydrostatic paradox, i.e. with the differentiation between the pressure that a fluid can exert, independently of its weight, and with relation only to its height (which I consider in detail in Sect. 5.4.1, *"[...] les liqueurs pesent suivant leur hauteur"*). Stevin's treatment of this issue was part of his broader theory of hydrostatics, to which he devoted his *De beghinselen des waterwichts* (1586), later translated into Latin and published by Willebrord Snel as part of Stevin's *Hypomnemata mathematica* (1605) where it occupies book 4 (*De hydrostatices elementis*) and book 5 (*De initiis praxis hydrostatices*) of the section on statics (*Statica*).[154] In his treatise, Stevin offered the first systematic treatment of hydrostatics after Archimedes, also dealing with two topics which were not touched upon by the latter, namely the paradox (in its variants), as well as the exertion of water pressure in lateral directions. This turns out to be of particular importance in seventeenth century theories of hydrostatics, because it was entailed by Archimedes's theory of floatation, but it had never been the subject of a demonstration.

The exertion of pressure in lateral directions – namely not only in directions perpendicular to the surface of the water – was an empirical fact which Stevin most probably dealt with as director of water management of the Dutch Republic, to which he was appointed by Maurice of Nassau in 1592. As one of his tasks, he worked on the underwater doors of dikes.[155] Indeed, as noted by Chalmers, as Stevin proposed the demonstration of proposition 11 of his *De hydrostatices elementis*, he simply presupposed that water exerts a lateral pressure.[156] According to proposition 11, the pressure on the surface ACED (see Fig. 5.18, Stevin 1605, 121) is equal to the weight of the volume of water ACEDH, where CD = HD. So, for instance, the point V is pressed on by the column of water Vγ, while point D is pressed on by the column DH. This means that the lateral side of a container of water undergoes a pressure proportional to the height of the water, namely, that the lateral pressure undergone by D is equal to the weight of the columns of water CD and HD. In other words, water exerts the same pressure laterally, downwards and upwards, and this pressure depends only on the height of the fluid.[157]

---

[154] Stevin's *De beghinselen des waterwichts* was published as part of his *De beghinselen der weeghconst,* together with a treatise *De weeghdaet* and an appendix. These were all later included in the *Hypomnemata* as the section on statics (as tome 4). In 1634, a French edition of Stevin's *Hypomnemata* was published as *Oeuvres mathématiques.*

[155] On Stevin, see Dijksterhuis 1970; Grabow 1985; Chalmers 2017, chapter 3.

[156] "It is true that the force on a lock gate follows from Proposition XI, but it does so only because the relevant knowledge has been assumed in its proof. As we have noted, Steven assumes that the force on a vertical surface of a given area at some level is equal to the force on a horizontal surface of the same area at the same level," Chalmers 2017, 40.

[157] This topic is considered in Sect. 5.4.1, *"[...] les liqueurs pesent suivant leur hauteur"*.

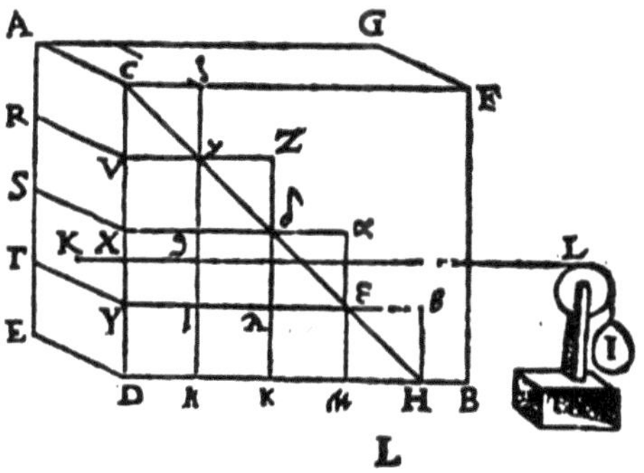

Fig. 5.18 Ideal representation of the horizontal pressure exerted by water. (Source: Stevin 1605, 121. ETH-Bibliothek Zürich, signature: Rar 9069, https://www.e-rara.ch/doi/10.3931/e-rara-9068. Public domain mark 1.0)

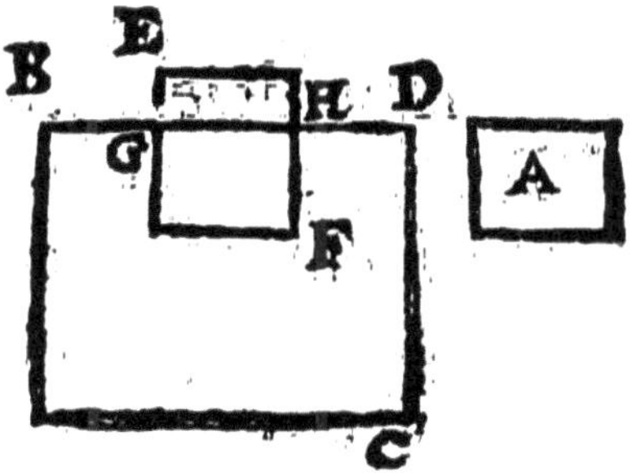

Fig. 5.19 Representation of the condition of floating of bodies according to Simon Stevin. (Source: Stevin 1605, 114. ETH-Bibliothek Zürich, signature: Rar 9069, https://www.e-rara.ch/doi/10.3931/e-rara-9068. Public domain mark 1.0)

Yet – as Boyle will remark in his 1666 *Hydrostatical Paradoxes* – Stevin lacked a ‘physical account’ of floatation. In his geometrical demonstrations on the conditions of floatation, Stevin relies on the idea of specific weight,[158] and on the definition of the conditions of stability for this idea. For instance, he defines a weight by which a surface is more depressed, as heavier than a weight by which a surface is less depressed.[159] In proposition 2 (see Fig. 5.19, Stevin 1605, 114), he applies these concepts, demonstrating that if a body A, occupying the volume GF is lighter than water, it will emerge because A depresses less of the surface of the water.[160] As a

[158] Cf. *supra*, n. 128.

[159] See definition 7: "[v]as superficiarum est superficies corporis cogitatione ab eo separabilis," and postulate 3: "[p]ondus a quo vas minus alte deprimitur, levius; quo altius, gravius; quo aeque alte, aequipondium esse," Stevin 1605, 111–112.

[160] See proposition 2: "[s]olidum corpus materia leviore quam sit aqua non omnino mergitur, sed eminet aliqua sui parte. [...] Demonstratio. Cum igitur A ex hypothesi materiam habeat leviorem aqua GF, et haec dato corpori A sit aequalis, GF necessario ipsi A praeponderabit, iam in vase superficiario loco aquae substituitur corpus A ipsi congruum, nam ex fabrica parti GF aequale est et simile, cumque A corpus levius sit aqua effusa propterea vas superficiarum EF [...] non tam alte mergetur ab A atque ab FG: atqui quanto minus alte corpus superficiarum EF subsidit, tantum corporis A supra aquam extare necesse est. Conclusio. Quamobrem corpus solidum materiae levioris quam aqua non totum mergitur, sed aliqua sui parte supereminet. Quod demonstrasse oportuit," Stevin 1605, 114.

matter of fact, we find an undeclared use of Archimedes's postulate, but no explanation of the ways in which pressures are balanced across the fluid is provided. We do find, as Boyle remarked, a geometrical rather than a (natural-)philosophical treatment of hydrostatics:

> [t]hose Mathematicians, that, (like Marinus Ghetaldus, Stevinus, and Galileo) have added anything considerable to the Hydrostaticks have been [...] very few, and those too, have been wont to handle them, rather as Geometricians, then as Philosophers, and without referring them to the explication of the Phaenomena of Nature.[161]

Instead, in his *Hydrostatical Paradoxes* Boyle provided "Physical Enquiries," namely he provided explanations for observed hydrostatical phenomena and proved them by experiment. In this, he came to criticize some of the conclusions established by more abstract, or geometrical treatments of hydrostatical problems (such as, as I am going to show, the hydrostatic paradox).[162] Please note that Boyle does not rely on principles entirely different from those of Stevin or Galileo: rather, he applies such principles (i.e. mechanical ones) to the explanation of actually observed phenomena, and on this experimental basis he confirms or rejects their validity, as far as their application to phenomena is concerned. This is what he labels a 'physical enquiry'. For Boyle, moreover, it is hardly surprising that demonstrations following a geometrical order are then refuted by experience, as hydrostatics is a branch of mixed mathematics, so that its results are to be proved or disproved by experience, and have not the "exactness" of pure mathematics:

> [...] [b]ut this is not all for which I am to excuse my selfe to Mathematicall Readers. For some of them, I fear, will not like that I should offer for Proofs such Physical Experiments, as do not alwayes demonstrate the things, they would evince, with a Mathematical certainty and accuratenesse; and much less will they approve, that I should annex such Experiments to confirm the Explications, as if Suppositions and Schemes, well reason'd on, were not sufficient to convince any rational man about matters Hydrostaticall. In Answer to this I must represent, that in Physical Enquiries it is often sufficient that our determinations come very near the matter, though they fall short of a Mathematical Exactness. [...] For though, in pure Mathematicks, he that can demonstrate well, may be sure of the Truth of a Conclusion, without consulting Experience about it: Yet because demonstrations are wont to be built upon Suppositions or Postulates; and some things, though not in Arithmetick or Geometry, yet in Physical matters, are wont to be taken for granted, about which men are lyable to slip into mistakes; even when we doubt not of the Ratiocination, we may doubt of the conclusion, because we may, of the Truth of some of the things it supposes. And this Consideration, if there were no other, will, I hope, excuse me to Mathematicians, for ventring to confute some reasonings that are given out for Mathematical demonstrations. For I suppose it will be consider'd, that those whose presum'd Demonstrations I examine, though they were some of them Professours of Mathematicks, yet did not Write meerly as Mathematicians, but partly as Naturalists: so that to question their Tenets, ought not to disparage those, as well certain, as excellent and most useful Sciences, pure Mathematicks.[163]

[161] Boyle 1666, *Preface*, 6 (unnumbered). Please note that in his *Paradoxes* Boyle deals with the positions of Stevin only; Galileo and Ghetaldi are mentioned only in this passage.

[162] See Sect. 5.4.2.2, *Boyle's experimental approach to the paradox.*

[163] Boyle 1666, *Preface*, 7–15 (unnumbered).

Even in the case of Pascal, as briefly seen above, Boyle found his samples not to be actual experiments, but rather idealizations.[164] Thus, Boyle aimed at experimentally testing, for the first time, what he considered the chief matters in hydrostatics, reduced to 11 'paradoxes'. Like De Volder would do, Boyle does so by relying on a chief "postulate" or "lemma," which is nothing but a re-statement of Archimedes's first postulate. He divides it into three parts: (1) a volume of water can be divided into infinite imaginary surfaces parallel to the horizon; (2) each area of each surface is equally pressed on by the column of water perpendicular to it; (3) if an area of a surface is pressed on more than the others, this area moves down, and the other areas are moved upwards, and the fluid is displaced.[165] Boyle offers immediate evidence of this postulate, by using the same kind of experimental proof performed by De Volder as experiment 23 of his *Experimenta*, and demonstrating, thus, that water gravitates or presses on water. He does so both in the preface, and in paradox 1 (stating that "in Water, and other Fluids, the lower parts are press'd by the upper,") by using different oils, suspended in pipes, and kept in equilibrium by the water in the rest of the container (see Fig. 5.20, Boyle 1670, figures 1–2). Such water increases in pressure once one moves the pipe downwards, namely, the level of the oil

[164] "[...] [the] difficulties making the Experiments propos'd by Monsieur Paschall more ingenious then practicable, I was induc'd on this occasion to bethink my self of a far more Expeditious Way, to make out, not only most of the Conclusions wherein we agree, but others that he mentions not; and this with so much more ease and clearnesse," Boyle 1666, 6. See also the *scholium* to paradox 3.

[165] Or, in Boyle's own words: "[t]his postulatum or Lemma, consists of three parts; the first of them more, and the two last, less principal. Suppose we then, (First) That if a Pipe open at both Ends, and held perpendicular to the Horizon, have the lower of them under Water, there passes an Imaginary plain or Surface, which touching that Orifice is parallel to the Horizon; and consequently parallel as to sense to the upper Surface of the water, and this being but a help to the Imagination will readily be granted. Secondly, To this it will be consonant, that each part of this designable surface, will be as much, and no more press'd, as any other equal part of it, by the water that is perpendicularly incumbent on it. For the water or other Fluid being supposed to be of an homogeneous substance, as to gravity, and being of an equal height upon all the parts of the imaginary Surface; there is no reason why one part should be more press'd by a perpendicular pillar of that incumbent fluid, then any other equal part of the same Surface by another perpendicularly incumbent pillar of the same or equal Basis and height, as well as of the same Liquor. But Thirdly, Though whilst our imaginary Surface is equally press'd upon in all parts of it, the Liquor must retain its former position; yet if any one part comes to have a greater weight incumbent on it, then there is upon the rest, that part must be displac'd, or depress'd, as it happens, when a stone or other Body heavier then water sincks in water. For wherever such a Body happens to be underneath the water, that part of the imaginary plain that is contiguous to the lower part of the stone, having on it a greater weight then other parts of the same Surface, must needs give way, and this will be done successively till the stone arrive at the Bottom; and if, on the other side, any part of the Imaginary Surface be less press'd upon then all the rest; it will by the greater pressure on the other parts of the Surface be impell'd upwards, till it have attain'd a height, at which the pressure (of the rais'd water, and the lighter or floating Body (if any there be) that leans upon it, and gravitates together with it, upon the subjacent part of the Imaginary Surface) will be equal to that which bears upon the other parts of the same Surface," Boyle 1666, 8–11.

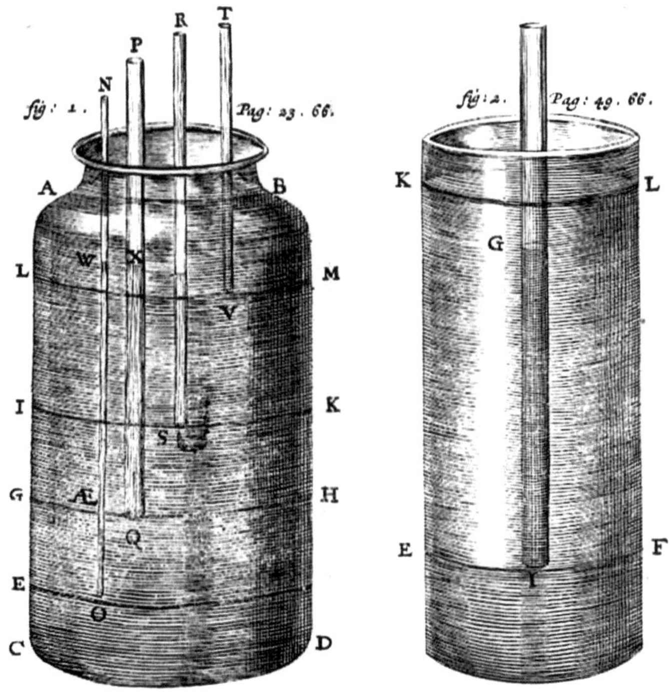

Fig. 5.20 Representation of the varying pressure of water on liquids contained in pipes. (Source: Boyle 1670, figures 1–2. Zentralbibliothek Zürich, signature: NP 2815, https://www.e-rara.ch/zuz/doi/10.3931/e-rara-47228. Public domain mark 1.0)

contained in the pipe increases or decreases, insofar as the water which touches it from the lower orifice is more or less deep.[166]

[166] "From what has been hitherto shewen, we may safely infer the Proposition, upon whose occasion all this has been delivered. For since the oyle in a Pipe, open at both Ends, may be kept suspended in any part under water, as at Q, because it is there in an Aequilibrium with the External water; and since being lifted up in the water, as from Q to S, the oyle can no longer be kept suspended, but by its own gravity will runne out. And since, in a word, the deeper the water is, the greater weight and pressure is requir'd in the Cylinder of oyle, to be able to countervail the pressure of the water, and keep it self from being lifted up thereby; there seems no cause to doubt but that the parts of the water incumbent on the Superficies G H, do more press that Superficies, than the parts of the water contiguous to the Superficies J K do press that; and consequently, that the parts of the water that are under the uppermost Surface of it, are press'd by those of the same Fluid that are directly over them: As we saw also that the upper parts of the oyle, whil'st the pipe was in raising from Q to S, depress'd the lower so much, as to force them quite out of the Pipe; there being in these cases no reason why the lowermost parts of a Liquor should press more, or have a stronger Endeavour against any other Liquor (or any other Body) the higher the Liquor incumbent reaches, if these inferiour parts deriv'd their pressure only from their own particular Gravity, (which is no greater then that of the other Homogeneous parts of the Liquor) and therefore they must derive the great force wherewith they press from the weight of the Incumbent parts, which consequently must be allow'd to press upon them," Boyle 1666, 33–35.

The conditions of equilibrium established on the basis of the postulate, and declared in paradox 3, indeed, prescribes that "the lower part of the Body will be press'd upward by the water that touches it beneath," as shown by the very experiment used to justify the postulate or lemma, where the oil contained in the tube is kept floating, or suspended, by a pressure from below the tube itself.[167] Boyle assumes all this as being a "physicall account," whereas Stevin, according to him, provided a "not physical" demonstration of the conditions of equilibrium of solids and fluids:

> [a]nd so we may prove, [...] That a Body acquiponderant in specie to water, will rest in any assignable place of the water where 'tis put. [...] Which is a Physicall Account of that grand Theorem of the Hydrostaticks, which I do not remember that I have seen made out in any Printed Book, both solidly and clearly; The Learned Stevinus himself, to whom the later Writers are wont to refer, having but an obscure (and not Physical) demonstration of it.[168]

The 'physical' explanation provided by Boyle for the case of a floating solid is the same one later followed by De Volder: namely, in a body immersed in a fluid, the lower part of the body itself is pressed upwards by the pressure exerted by the columns of fluid pressing on the other areas of the same ideal surface that this lower part describes.[169] Thus far, he accepts the idea of the lateral exertion of pressure: as he shows further in paradox 7, "a Body immers'd in a Fluid, sustains a lateral pressure from the Fluid; and that increas'd, as the depth of the immers'd Body, beneath the Surface of the Fluid, increaseth," which he demonstrates by using a tube represented in Fig. 5.21, Boyle 1670, figure 13, in which the pressure of the water on the horizontal arm of the tube (containing oil) increases or decreases if the whole pipe is lowered or moved upwards in the basin filled with water.[170]

---

[167] Boyle 1666, 67. See also 8–23

[168] Boyle 1666, *Preface*, 16 (unnumbered).

[169] "This may be prov'd by what has been already delivered in the Explication of the first Experiment: For where ever we conceive the lowest part of the Body, which is either totally, or in part, immers'd in water, to be there the imaginary Superficies being beneath the true Superficies, every part of that imaginary Superficies must be press'd upwards, by vertue of the weight of the water incumbent on all the other parts of the same Superficies, and so that part of it, on which the immers'd Body chances to leane, must for the same Reason have an endeavour upwards. And if that Endeavour be stronger then that wherewith the weight of the Body tends downwards, then [...] the Body will be buoy'd or lifted up. And though the Body be heavier then so much water, and consequently will subside, yet that Endeavour upwards of the water, that touches its lower part, is onely rendred ineffectual to the raising or supporting the body, but not destroyed; the force of the heavy Body being from time to time resisted, and retarded by the water, as much as it would be if that Body were put into One Scale, and the weight of as much water, as is equal to it in bulk, were put into the other," Boyle 1666, 67–68.

[170] "[...] this pressure of the perpendicular Cylinder doth really urge the oyle in the shorter leg to flow out; you may learn by slowly lifting the Syphon (without changing its, former posture) towards the Surface of the water. For as the lower leg comes nearer and nearer to that Surface, (to which, as I newly intimated, it is still to be kept parallel) the oyle in the Horizontal leg will be driven out in drops, by the pressure of the other oyle in the perpendicular leg. That likewise before you begin to raise the Syphon, the lateral pressure of the water against the lower Orifice of it is, at least in such Experiments, near about the same with what would be the perpendicular pressure of a Cylinder of water, reaching from the same Orifice G (or some part of it) to the top of the water, may be gather'd from hence, That the Surface of the oyle in the longer leg will be a litle higher then that of the external water, as (by reason of the often mention'd comparative levity of the oyl) it

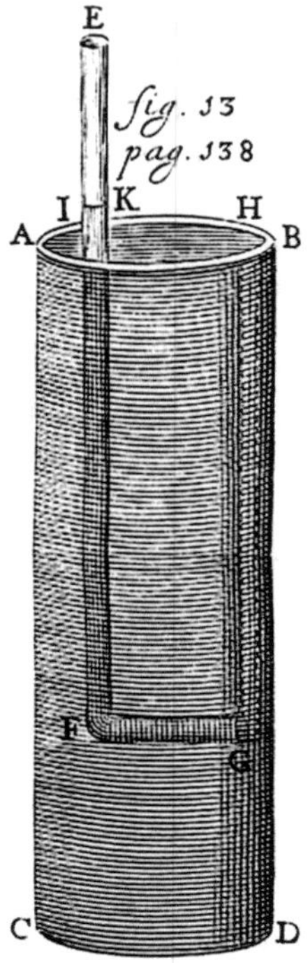

Fig. 5.21 Representation of the varying pressure of water on a liquid contained in an L-shaped pipe. (Source: Boyle 1670, figure 13. Zentralbibliothek Zürich, signature: NP 2815, https://www.e-rara.ch/zuz/doi/10.3931/e-rara-47228. Public domain mark 1.0)

To sum up, Boyle overcame the abstract treatment of hydrostatic equilibrium by Stevin by applying mechanical principles to actual situations. In doing so, he provided a theory considering – as Pascal did – the communication of pressure across the whole fluid involved in the experiment. De Volder, through his experimental lectures – which we now see as clearly based on Boyle's *Hydrostatical Paradoxes* – would then follow the same path. This path, eventually, would lead De Volder to provide an explanation in purely Archimedean means of the chief case of hydrostatic paradox, i.e. the so-called paradox of Stevin, which was not accepted by Boyle on an experimental basis. Before addressing De Volder's solution of the paradox, and his experimental proof of it, we need to consider a further hydrostatical tenet discussed between the sixteenth and seventeenth centuries. Namely, the aforementioned idea that the pressure of fluids depends only on their height.

would be, if we suppose, That a pipe of Glass of the same bore, and reaching to the top of the water, being fitted to the Orifice of the Horizontal Leg (as in the annex'd figure the Cylinder, G H) were fill'd with water," Boyle 1666, 144–145.

5.4 The Explanation of the Hydrostatic Paradox

5.4.1 *"[...] les liqueurs pesent suivant leur hauteur"*

The idea that the pressure exerted by fluids depends on their height, rather than on their quantity, can be intended in two ways. On the one hand, it can be intended as meaning that pressure is exerted only according to lines perpendicular to the surface of the fluid, and that no lateral pressure is exerted anyhow. On the other hand, it can be intended as meaning that pressure is exerted also in lateral or oblique directions, but that the degree of such pressure depends only on the height of the fluid. As a matter of fact, the second interpretation is the correct one: at least, as far as its acceptance in the sixteenth and seventeenth centuries is concerned. Yet, one can find interpretations of the tenet following the second meaning, including Stevin, Rivault and De Volder, who, in some places, seem to accept this kind of reading.

The tenet, actually, was formulated for the first time by Stevin himself. We have seen above that he theorized that the pressure exerted on the side of a container depends on the height of the water above the point of pressure – and not on the quantity of water in the container. In a more systematic form, the tenet is demonstrated in proposition 10 of his *De hydrostatices elementis*. It states that

> [o]n any bottom of the water being parallel to the horizon there rests a weight equal to the gravity of the water the volume of which is equal to that of the prism whose base is that bottom and whose height is the vertical from the plane through the water's upper surface to the base.[171]

The 'paradoxical' character of this proposition consists in its being applicable also to those cases in which, on the bottom, there does not lean a quantity of water whose weight can exert such a pressure. I will return to this issue (overtly treated by Stevin in the corollaries of the proposition), namely on the 'paradox of Stevin', in Sect. 5.4, *The explanation of the hydrostatic paradox*. Here, it is worth considering how Stevin demonstrated such a proposition. He does so by conceiving a basin ADCB filled with water (see Fig. 5.22, Stevin 1605, 119a), and showing that if, for instance, on the bottom EF there is a pressure stronger than that exerted by the portion of the water EFGH, this pressure would be exerted on it by the lateral columns ADEG and HBFC. However, at this point, on the whole bottom DEFC there would be a pressure greater than that exerted by the water ABCD. So, a surface is pressed on only be the water perpendicularly above it.[172]

---

[171] Dijksterhuis 1970, 66. Cf. Stevin 1605, 119.

[172] "Si fundo EF plus ponderis insideat aquae GHFE, id erit ab aqua finitima, atque ideo si fieri possit est ab AGED et HBCF, quibus positis fundo DE quoque, propter aquam finitimam GHFE (cum utrobique sit par ratio) plus ponderis incumbet quam sit aquae AGED, perinde quoque basi FC plus insidet ponderis quam aquae HBCH; quare toti fundo DC maius quoddam pondus insidet quam aquae totius ABCD, quod tamen, cum ABCD corpus rectangulum sit, absurdum fuerit. Eadem ratione evinces fundo EF non minus pondus sustentari quam sit aquae GHFE; quare tantundem duntaxat ponderis necessario ipsi incumbet," Stevin 1605, 119. See Schuster 2013, chapter 3.

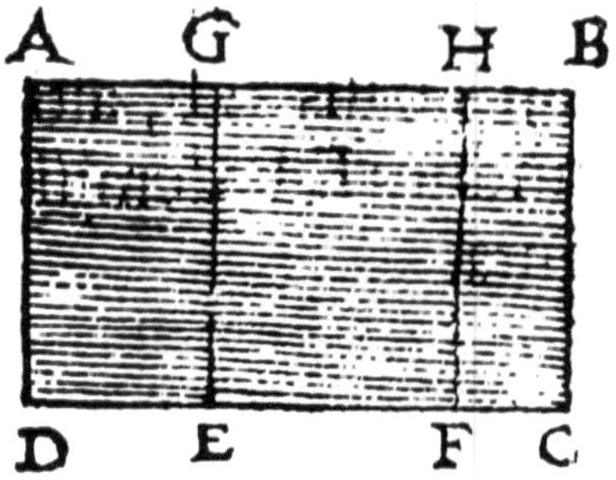

Fig. 5.22 Container of water aimed at demonstrating that the pressure of water depends only on its height. (Source: Stevin 1605, 119a. ETH-Bibliothek Zürich, signature: Rar 9069, https://www.e-rara.ch/doi/10.3931/e-rara-9068). Public domain mark 1.0

In this case, the first interpretation of the tenet can be applied. Of course, in his *Statica* Stevin did not maintain that for each shape of a container filled with water, the bottom of the container is pressed on only by the water perpendicularly above it. However, his demonstration of proposition 10 leans on this idea. Or, rather, that a lateral pressure is exerted, but this does not lead to an increase of pressure on the bottom of the container.

Later in the seventeenth century, namely in the hands of Pascal and Boyle, the idea that pressure depends on the height of a fluid, and that it is transmitted in every direction became a main tenet in hydrostatics, and was interpreted according to the first meaning presented above. The first chapter of Pascal's *Traité de l'équilibre des liqueurs* bears the title *Que les liqueurs pesent suivant leur hauteur*: in it, Pascal shows that the pressure exerted on the lower part of the first container represented in Fig. 5.23, Pascal 1663, figures 1 and 3, is the same as that exerted on the lower part of container 3, which includes far more water, so that, as seen above, if one connects the two containers with an inverted syphon, the pressure of water in container 1 will balance that in container 2. In the second chapter, *Pourquoy les liqueurs pesent suivant leur hauteur*, as seen above, Pascal explains that fluids press according to their height, using the aforementioned idea of their centre of gravity, and applying to them the principle of the lever – made possible by the idea of the "continuity and fluidity" of water.[173] In turn, Boyle repeatedly uses this idea in his *Hydrostatical Paradoxes*, such as in the aforementioned paradox 7 (where the pressure exerted on the hole G is equal to the weight of the ideal column of water GH (see Fig. 5.21, Boyle 1670, figure 13)), and more overtly in paradox 6, where he re-states Stevin's proposition 10. According to it,

> [i]f a Body be plac'd under water, with its uppermost Surface parallel to the Horizon; how much water soever there may be on this or that side above the Body, the direct pressure sustain'd by the Body [...] is no more then that of a Columne of water having the Horizontal superficies of the Body for its Basis, and the perpendicular depth of the water for its height.[174]

[173] See *supra*, n. 152.

[174] Boyle 1666, 117.

Fig. 5.23 Containers used to demonstrate that the pressure of water depends only on its height. (Source: Pascal 1663, figures 1 and 3. ETH-Bibliothek Zürich, signature: Rar 4668, https://www.e-rara.ch/doi/10.3931/e-rara-20296. Public domain mark 1.0)

He demonstrates this by using, first, an inverted syphon (see Fig. 5.24, Boyle 1670, figure 8), filled with oil, and put into a basin full of water: in this case, the weight of the oil in the syphon is kept in equilibrium with the water, so that one can calculate that the volume of water counterbalancing the oil is the ideal column of water perpendicular to the orifice.[175]

Boyle's paradox 6 is valid also for tubes provided with various shapes (see Fig. 5.25, Boyle 1670, figures 9–11).

In particular, in the case of the asymmetrical, inverted syphon (i.e. the first tube on the left), already considered by Galileo and Pascal, Boyle put oil of turpentine (specifically lighter than water) in the uniform leg; in the other leg, he put water. By this, he discovered that "all this water was [not] able to keep up the oyle in the uniform leg, at a greater height then if the leg that contain'd the water had been uniform too; as much of the water in the Globe, as was not directly over the lower Orifice of

[175] "Take a slender Glass pipe, of an even Bore, turn'd up at one end like the annexed Syphon. Into this Syphon suck oyl of Turpentine till the Liquor have fill'd the shorter leg, and be rais'd 2 or 3 Inches in the longer. Then nimbly stopping the upper Orifice with your finger, thrust the lower part of the Syphon so farre into a deep Glass full of water, That the Surface of the oyle in the longer leg of the pipe, may be but a little higher then that of the External water; and, upon the removal of your finger, you will find the Surface of the oyle to vary but little, or not at all, its former Station. And as, if you then thrust the pipe a little deeper, you will soe the oyle in the shorter leg to begin to be depress'd; so, if afterwards you gently raise the pipe toward the top of the water, you shall see the oyle not only regain its former station, but flow out by degrees in drops that will emerge to the Top of the water. Now, since the water was able, at first, to keep the oyl, in the longer leg of the pipe, suspended no higher, then it would have been kept by a Cylinder of water equal to the Orifice of the shorter leg of the pipe, and reaching directly thence to the Top of the water; (as may be easily cried, by making a Syphon, where the shorter leg may be long enough to contain such a Cylinder of water to conterpoize the oyl in the longer;) & since, when once, by the raising of the pipe, the height of the incumbent water was lessen'd, the oyle did more then Counter-ballance it; (as appears by its flowing out of the Syphon,) we may well conclude; That, though thence were in the Vessel a great deal of water, higher then the immers'd Orifice of the Syphon, (and it would be all one, though the Syphon were placid at the same depth in a pond or lake;) yet, of all that water, no more did gravitate upon the Orifice, then that which was plac'd directly over its, which was such a pillar of water, as the Paradox describes," Boyle 1666, 121–123.

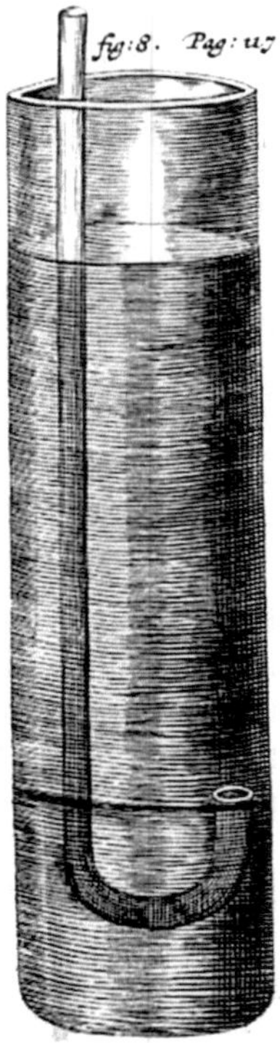

Fig. 5.24 An inverted syphon used to demonstrate that the pressure of water depends only on its height. (Source: Boyle 1670, figure 8. Zentralbibliothek Zürich, signature: NP 2815, https://www.e-rara.ch/zuz/doi/10.3931/e-rara-47228. Public domain mark 1.0)

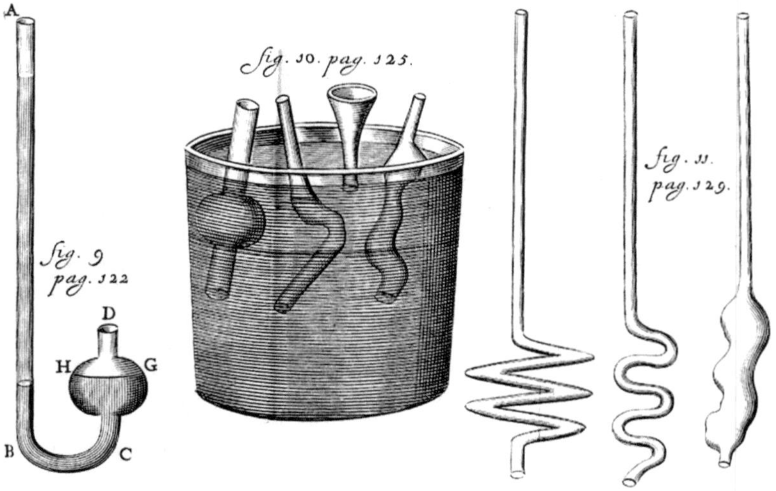

Fig. 5.25 Different kinds of pipes used to demonstrate that the pressure of water depends only on its height. (Source: Boyle 1670, figures 9–11. Zentralbibliothek Zürich, signature: NP 2815, https://www.e-rara.ch/zuz/doi/10.3931/e-rara-47228. Public domain mark 1.0)

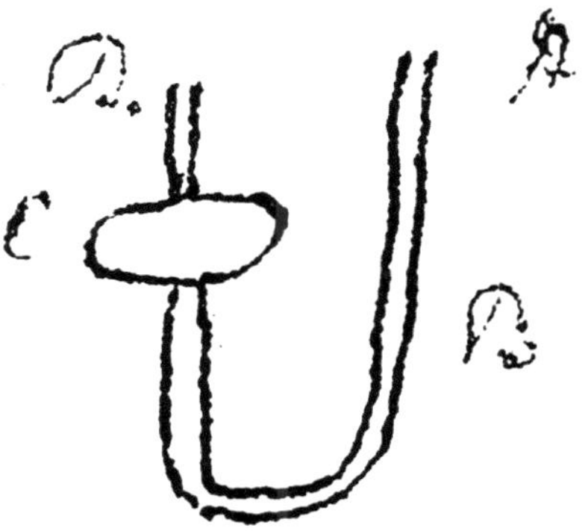

Fig. 5.26 An inverted asymmetrical syphon used to demonstrate that the pressure of water depends only on its height. (Source: De Volder 1676–1677, 133r. Courtesy British Library, ms. Sloane 1292)

it, being supported by the lateral parts [...] of the same Globe."[176] However, no mechanical explanation of this phenomenon (such as those of Pascal and Galileo) is provided by Boyle, who proves his assertion by empirical means only, without any further reduction to other principles, such as Archimedes's postulate.

It is worth here drawing attention to De Volder's treatment of the same case, as well as to his more general consideration of the tenet. Prima facie, De Volder seems to interpret it in the first meaning, namely, by considering only the volume of water exactly above the surfaces considered. De Volder considers the case of an inverted, asymmetrical syphon in experiment 25, which is gradually filled with water, reaching the same level in both horns of the syphon (see Fig. 5.26, De Volder 1676–1677, 133r).[177]

From this, De Volder infers that this shows "a received tenet by all those, that treat of hydrostatics [...] viz. aquam, dum premit, non premere nisi per lineas rectas, nullo {vò}[178] per laterales, transversas, aut obliquas,"[179] ("water, while it presses, does not press but through straight lines, in no way through lateral, transverse, or oblique [lines]"). Thus far, one can consider – instead of an equilibrium maintained by the principle of the lever, as in Galileo's and Pascal's treatments (later followed by Rohault)[180] – ideal columns of water which sustain themselves, and on which pressure is exerted only by those parts of those water exactly above the other parts, as no pressure is exerted through or by means of lateral or oblique directions. This, of course, entails that the water in the lowest part of the syphon is not pressed by that present in the horns, namely in each ideal column, as the water

[176] Boyle 1666, 125.

[177] Please note that the 1705 inventory includes a more complex U-shaped instrument to show the pressure exerted by the height of water: i.e. a hydrostatic lever: "a big glass with a copper lid, with a pipe [attached] at the side, to sustain the weight of the water in this side-pipe standing [rising] above the weight." See Molhuysen 1913–1924, volume 4, 106*.

[178] Probably: "m[od]o."

[179] De Volder 1676–1677, 133r.

[180] See *supra*, n. 152.

behaves as if it is separated by the other columns, as in Stevin's demonstration of proposition 10. The same explanation, in fact, can be found in Rivault's commentary on Archimedes's first postulate, where the case of the equilibrium of fluids in an inverted syphon – regardless of the shape of its 'horns' – is explained as a consequence of the fact that "water is not supposed to have a force of pressing other parts of its body, unless by that quantity, which stands perpendicularly above as a suspended weight" – but no further detail on the lateral communication of pressure is provided.[181] In short, De Volder (who seems to be, once again, following Rivault) appears to interpret the tenet that water presses according to its height in its strictest sense, as meaning that water presses on the water which is exactly below it. This interpretation of the description of the experiment is confirmed by the Latin used by Morley: most probably, he followed De Volder's own words at the lecture, where the use of "per" clearly points to the idea that pressure is not exerted through or by means of lateral or oblique directions. That is, Morley's report suggests that pressure is never communicated but in directions perpendicular with respect to the highest surface of the fluid.

This interpretation can corroborated by De Volder's *dictata* on Descartes's *Principia*, and by experiment 3. In the Hamburg 273 series of *dictata*, commenting upon IV.26 – where Descartes provided his treatment of the internal pressure of water, to which I will devote separate discussions in Sects. 5.5.1, *An introduction to Descartes's hydrostatics* and 5.5.3, *De Volder's Archimedeo-Cartesian hydrostatics: an impossible synthesis?* – De Volder defends Descartes's idea that "in a vessel [...] nor are individual parts of the bottom of the vessel pressed upon, except by as many drops as rest upon them perpendicularly."[182] He does so by the authority of Stevin, as "this meant according to Stevin, that water exerts all the pressure according to perpendicular lines, and the individual parts of the bottom sustain the water [which is] leaning perpendicularly [on them]"[183] – in accordance with the demonstration by Stevin of proposition 10 (as seen above). Hence, De Volder comments upon Descartes's claim that only the particles of water perpendicularly above a certain section of the bottom actually press upon it, because, if the bottom descends,

[181] "Impleatur vas ABC aqua per A in elatiori loco: possitque per C effluere aqua si abundaverit, totumque vas censeatur plenum: certissimum est vase existente pleno, aquam fore elatiorem in A quam in C et propterea, ut novimus ex quotidiana experientia, effluxuram aquam per C quousque superficies aquae quae in A redierit ad D quod est in eodem plano cum C ut sic aqua tota sua elatiori superficie stet ad libramentum. Et hoc quidem contigerit, licet sit maior, licet minor sit, moles aquae in parte vasis AB quam in BC. Unde aqua non censetur habere vim premendi alias corporis sui partes, nisi ea quantitate, qua ad perpendiculum libramento supereminet. Eo etenim constituto, totum corpus quiescit et subsidet. Quod de omni humido intelligendum est," Archimedes and Rivault 1615, 492.

[182] Descartes 1982, 193.

[183] "Incumbunt. Id vult iuxta Stevinum, aquam omnem pressionem exercere secundum lineas perpendiculares, et singulas partes fundi sustinere aquam perpendiculariter incumbentem," Hamburg 273, 235. In the other series of *dictata* Descartes does not consider this issue.

only such particles can follow the section in the moment in which it is released.[184] As De Volder applies this idea to analyse the conditions of equilibrium of water in a funnel (see Fig. 5.27, Hamburg 273, 235), he claims that the speed of descent of the water, or any other fluid, both in a funnel ABCD, and in a tube EFBD is equal, if the bottom is removed.

This proves that the pressure exerted on the bottom BD is (in static conditions), identical in both the tubes, i.e. that in a fluid "it is to be considered always the height, not [...] the breadth."[185] Thus far, De Volder does not make recourse to the principle of the lever – as Galileo did – but to Stevin's proposition 10, and to Descartes's 'dynamical' rendering of a static situation – which, to some extent, becomes inconsistent with the principle of the lever, where different speeds of descent compensate different weights. In fact, this is not the case of the water in the funnel and in the tube, as in both cases two different quantities of fluid pass through the bottom at the same speed,[186] something which De Volder himself could have measured in front of his students, as in the 1705 catalogue of the theatre there is listed a "copper tube marked in feet to weigh and measure the perpendicular pressure of the water as well as its speed."[187] By avoiding applying this principle to the explanation of the paradox, somehow, De Volder revealed he had grasped an essential difference between solids and fluids: namely, that in solids pressure or 'gravitation' is equal to their weight; in fluids, pressure depends on their height. However, his insistence on the idea that in fluids pressure is exerted only by parts leaning perpendicularly on each other, and his avoidance of dealing with the lateral communication or exertion of pressure, reveals that he still did not realize, in full,

[184] "[...] no more drops than {are in the cylinder} 1, 2, 3, 4, or others equivalent, press upon this same part B of the vessel; because at the same moment of time at which this part B can descend, no other drops can follow it," Descartes 1982, 193; see Fig. 5.57, Descartes 1644a, 203. This is a dynamical rendering of a static situation, which is used by Descartes to explain the hydrostatic paradox, which I discuss in Sect. 5.5.1, *An introduction to Descartes's hydrostatics*.

[185] "Possunt. Id vult, si q[ui]s utatur infundibulo (Belg. e[ss]e trechter) sive tubo aequali, cuius orificium aequale sit angustissimae parti infundibuli, pari celeritate aquam aliumve liquorem per utrumq[ue] descensurum, concipiamus ex. gr. infundibulum ABCD et tubum EBFD dico aquam aeq[ue] cito descensuram per tubum EBFD ac per infundibulum ABCD. Nec e[nim] q[uo]d q[ui]s dicat aquam lateralem G a H in infundibulo premere[,] quia id falsum e[sse], sola e[nim] perpendicularis premit, itaq[ue] s[e]mp[er] respicienda altitudo, [n]o[n] vero latitudo," Hamburg 273, 235.

[186] Please note that both Descartes and De Volder accepted the principle – first treated by Benedetto Castelli – according to which a fluid passing from a large to a narrow passage increases its speed (I discuss it in Sects. 6.1.2, *Descartes's cosmogony and cosmology* and 6.2.2.1, *Newton's critique of Descartes's vortex theory*). This might not be at odds with the arguments provided here, because the principle of Castelli applies to the continuous movement of fluid, which takes place in several instants, while in a Cartesian framework a static condition, such as that of the fluid in the funnel, is rendered as a dynamic condition by considering a movement taking place in the first instant. In Descartes's perspective, this is as a tendency to motion rather than an actual movement (taking place in more instants): I deal with this in Sect. 5.5.1, *An introduction to Descartes's hydrostatics*.

[187] "Een copere buys verdeelt in voeten om de perpendiculare pressie van 't water als ook de snelheyt van die te wegen en te meten," Molhuysen 1913–1924, volume 4, 105*.

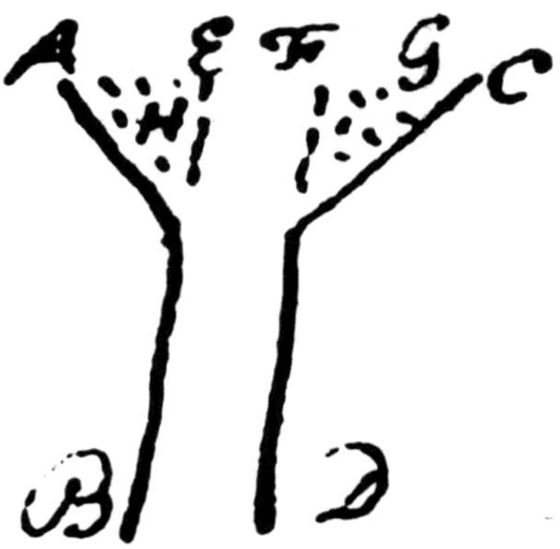

Fig. 5.27 A funnel used to demonstrate that the pressure of water depends only on its height. (Source: Hamburg 273, 235. Courtesy Staats- und Universitätsbibliothek Hamburg Carl von Ossietzky, Cod. philos. 273)

what Chalmers has labelled as the "isotropy of hydrostatic forces that had been identified by Stevin."[188]

Eventually, De Volder's 'strict' interpretation of the tenet that fluids press only according to their height, i.e. his lack of full understanding, as Boyle put it, of the 'differing pressure of heavy solids and fluids',[189] is confirmed by look at the third of his *Experimenta philosophica naturalia.* In this, De Volder repeats Von Guericke's experiment with hemispheres, to which weights are appended as in Fig. 5.28, Von Guericke 1672, 106.

De Volder uses pairs of hemispheres of different diameters (6 and 8 inches), and then replaces the lower hemispheres with a plate (cf. Senguerd's depiction of an analogous instrument: Fig. 5.29, Senguerd 1715, plate 3c).[190]

Namely, De Volder reduces the evacuated volume of ½ with respect to the original volume. However, this does not affect the amount of weight which can be sustained, namely 425 and 700 pounds, respectively for each diameter. This amount thus depends only on the diameter of the hemisphere, as "otherwise the whole globe would have sustained twice or much weight as the hemisphere did viz. 850 lb and not only 425 et a little more." Accordingly, De Volder "inferred [...] [t]hat the air did press only perpendiculariter, else it would have pressed more on the \whole/globe than on the half and consequently the whole globe would have sustained more weight than the half, which by this experiment we did see it was not so. [...] [H]ere was a great uniformity in this experiment with the other, made \which/ the other

[188] Chalmers 2017, 59; cf. also the text quoted *supra*, n. 44. On the understanding of pressure, see also Bertoloni Meli 2006, chapter 6.

[189] See Sect. 5.5.3, *De Volder's Archimedeo-Cartesian hydrostatics: an impossible synthesis?.*

[190] As in the case of the cohering marbles, discussed in Sect. 4.1.3, *De Volder's ideas on cohesion and divisibility*, the hemispheres are also discussed in the *pro gradu* disputation of Johannes Franciscus de Witte van Schooten. Senguerd used hemispheres with a diameter of 4 inches and 4 lines – viz. those represented in the figure –, to which he managed to append a weight of 275 pounds, and of 8 inches and 6 lines, undepicted (maybe the same ones used by De Volder), to which he appended weights of 850 pounds: see Senguerd 1715, 166.

Fig. 5.28 Magdeburg hemispheres sustaining weights. (Source: Von Guericke 1672, 106. ETH-Bibliothek Zürich, signature: Rar 5021, https://www.e-rara.ch/doi/10.3931/e-rara-9099. Public domain mark 1.0)

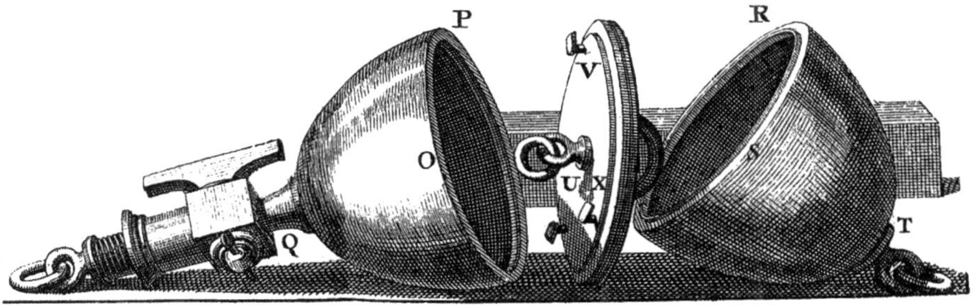

Fig. 5.29 Magdeburg hemispheres and the plate closing them. (Source: Senguerd 1715, plate 3c. Zentralbibliothek Zürich, signature: NP 1783, https://www.e-rara.ch/zuz/doi/10.3931/e-rara-39787. Public domain mark 1.0)

globe the proportion of 425 to 6 inches corresponding very well to that of 700 to 8 inches broad."[191] In order to clarify De Volder's conclusion, we need to note that the hemispheres are actually kept together by the pressure of air exerted perpendicularly to their line of conjunction, regardless of the direction of the line of conjunction, viz. whether the sphere is made to roll, or if it is kept hanging. Air, indeed, presses in all directions. The pressure exerted in lines other than this does not affect the force of conjunction, because they nullify each other. De Volder might have been considering this idea, by which he then could also justify the cohesion of hemispheres when they are rotated, such as in the case of Von Guericke's famous exhibition with horses. Yet, De Volder does not overtly consider this case, as he just appends weights to hemispheres cohering as in Fig. 5.28, Von Guericke 1672, 106, in a "static machine."[192] Thus, his conclusion seems rather to be that, by considering a line of cohesion which is parallel to the horizon, the force exerted on this line depends only on the vertical component of the column of air in which presses on it – exerted upwards and downwards. This idea can be also found, *in nuce*, in experiments 1–2, where he claims that the cylinders of marble are kept united because air "weigheth and compresseth both above us, and beneath us,"[193] i.e. considering only air from above and below, and not from lateral directions – by which it has however to be communicated in order to allow an upwards pressure. In turn, according to this possible interpretation of De Volder's reasoning, the pressure exerted in directions other than in the vertical ones does not increase the overall force of cohesion keeping the hemispheres united. In the light of this reasoning, however, if he had rotated the spheres, he would have needed to explain the exertion of pressure in lateral and oblique directions – which probably fell beyond the scope of his lectures in 1676–1677. As a matter of fact, De Volder performed at Leiden experiments overtly aimed at showing and measuring the oblique pressure of water, as testified by the presence, in the 1705 inventory, of "copper boxes of one and a half foot, with lids, for use in the wooden container, to measure the horizontal pressure of water, and at any angle," and a "glass with an opening in the middle to demonstrate the lateral pressure of the air."[194] Furthermore, De Volder's experiments on the pressure of water on oblique surfaces (following Stevin's tenets) are mentioned – but not detailed – by Dirk Rembrandtsz. van Nierop in his *Tweede deel op de Wiskonstige rekening* (1680), and reported, in turn, by Pieter Hellingwerf.[195] Thus far, we might interpret De Volder's

---

[191] De Volder 1676–1677, 86r. See *supra*, n. 83. In fact, the ratio 425/6 is 70.8: lower than the ratio 700/8, i.e. 87.5. De Volder does not relate this proportion to the weights of 450 pounds he appended, in experiment 2, to the cylinders of 2 inches and ⅓ diameter: see Sect. 4.1.3, *De Volder's ideas on cohesion and divisibility*.

[192] De Volder 1676–1677, 85r. This kind of apparatus, probably a tripod, is not mentioned in the 1705 inventory.

[193] De Volder 1676–1677, 79r.

[194] See Molhuysen 1913–1924, volume 4, 105*–106*.

[195] "[…] Waer uit genoegh te besluiten is dat om den bodem reghthoekigh af te trekken so komt dit geheel met Symon Stevin over een: en nogh boven dit so is dit ook tuighwerkelijk ondersoght by den B. de Volder Professor in de Philosophia en Mathematise Konsten tot Leyden en is hier mee ook heel effen over een gekomen," Van Nierop 1680, 153. Mentioned, in turn, in the *Wiskonstige oeffening, behelsende eene verhandeling over veele voorname zaken van de mathesis* (1718) of

ambiguous statements, in the *Experimenta*, as resulting either from Morley's own interpretation of his words (which is unlikely as he seems to have written down De Volder's explanation *verbatim*), or as resulting from the 'early stage' of his lecturing (viz. he had not yet introduced his students to the more complex issue of lateral pressure), or – as I am going to develop – from the inherent difficulties of interpreting, from a Cartesian standpoint, the phenomenon of the lateral pressure of water. In any case, in order to make De Volder's claim that fluids presses only in perpendicular directions consistent with his explanation of the cohesion of the hemispheres as given in his *Disputationes* – where lateral pressure is taken into account – and to justify the upwards thrust of air in a way not invoking its elasticity (such as in the case of water) we need to infer De Volder's arguments as entailing the idea that lateral pressure is nevertheless communicated, but only its vertical component affects the force of pressure itself. This idea, as I show in Sect. 5.4.2.4, *De Volder's Archimedean approach to the paradox*, is at the core of his explanation of the hydrostatic paradox, which he solves by purely Archimedean means.

5.4.2 The Hydrostatic Paradox

What is commonly known as the hydrostatic paradox, is a variant of Stevin's proposition 10: namely, the case in which the pressure exerted on the bottom of a container is greater than that exerted by the bare weight of the fluid. This can be exemplified by considering an inverted T-shaped container, such as that represented in Fig. 5.30, De Volder 1676–1677, 140v (taken from De Volder's experiment 28, where the paradox is presented), or in Fig. 5.31, *Hydrostatic paradox*, in which the pressure exerted on the bottom CD is equal to the weight of a volume of water CDGH, even if only the volume ABCDEF is filled with water.

Variants of Stevin's proposition 10 have already been considered: it is the case of the inverted, asymmetrical syphon and the case of the floatation of a solid in a small volume of water. In these cases, different weights are balancing each other, and the underlying 'mechanism' is the same: namely, pressure depends on height. Accordingly, the 'paradox of Stevin' represented in Fig. 5.31, *Hydrostatic paradox*, is no more 'paradoxical' than the two aforementioned cases. For instance, Galileo labelled the second case he dealt with as "grandissimo paradosso." Nor is it different, in principle, from the hydraulic lever described by Pascal. And yet, if, for instance, Boyle accepted the fact that in an inverted, asymmetrical container the water is balanced in the two horns, he did not accept the fact that in an inverted T-shaped container the pressure on the bottom is greater than that exerted by the bare weight of water. Or, as Morley put it in experiment 28,

---

Pieter Hellingwerf: "[e]n de hoegrootheyt van de perpendiculare perssing tegens een scheve bodem, vint men by Dirk Rembrantsz. van Nierop aangetekent in zyn tweede deel van de Wiskonstige Reekening, pag. 153 dat de Heer Professor de Volder tot Leyden, door ondervinding heeft gemeeten en de uytkomst van die effen met de 10 Propositie, 3 Afdeeling, 1 Boek, bevonden over een te komen," Hellingwerf 1718, 348.

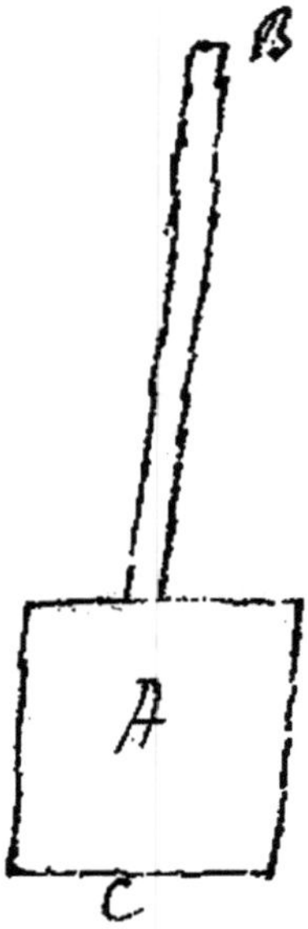

Fig. 5.30 An ideal, inverted T-shaped container aimed at showing the hydrostatic paradox. (Source: De Volder 1676–1677, 140v. Courtesy British Library, ms. Sloane 1292)

Fig. 5.31 An ideal, inverted T-shaped container aimed at showing the hydrostatic paradox. (Figure by the author)

[De Volder] commended, first, Stevinus very much, for his exact inquisition into hydrostatics, affirming, he had given the t<two>\rue/ grounds, and rules, of this art, having examined all the chief things, which herein did occur. [...] Next he said, that notwithstanding this commendable exactitude of Stevin he had one proposition, which had seemed very strange, and paradox, to the virtuosi of this art viz. that water did press more when elevated in a small tube upon a cylindrical vessel, than the weight of the water contained in the tube did require. [...] This problem of Stevinus had rendered itself incredible to almost all authors that had treated since of hydrostatics yea perhaps some had laughed at it, as bearing the

> appearance of a ridiculous thing. Boyle could not believe it, yea, said he conceived Stevinus had never tried it by any true experiment.[196]

Boyle, actually, devotes to the discussion of the paradox the scholium appended to paradox 6 of his *Hydrostatical Paradoxes*. There, he discusses previous attempts to prove experimentally the paradox, and describes his own unsuccessful attempt, which led him not to accept its validity – even if he accepted Stevin's proposition 10. In turn, De Volder devised his own experiments demonstrating the validity of the paradox, whose performance is duly described and explained in the 1685 and 1687 communications of Hermann Lufneu to the *Nouvelles de la République des Lettres*, referring to experimental lectures performed by De Volder "when I had the honour of studying under him seven or eight years ago," namely, in 1677–1678, when Morley was also a student at Leiden.[197] Such reports, thus, are a sort of completion of De Volder's lectures, which were reported by Morley only up to experiment 28 (25 March 1677), where the paradox is just presented, but not proved. Before addressing De Volder's treatment of the paradox, however, it is necessary to shed light on its background, namely, on how it was treated before De Volder (apart from the cases discussed above). In fact, as testified by Morley's report, as late as in the 1670s its very possibility was still debated.

5.4.2.1 Early Discussions of the Paradox: Stevin, Galileo and Schott

The paradox was systematically presented for the first time by Stevin, as part of his treatment of proposition 10 of his *De hydrostatices elementis*. From this proposition, Stevin draws 5 corollaries (*consectaria*). First, he considers (corollary 1) the case of a body floating in a little water – which is explained by considering that, in principle, both the solid and the water will press on the bottom in the same way, i.e. that the solid IKML presses as the volume of water NMLO does (see Fig. 5.32, Stevin 1605, 119b). Namely, Stevin reduces this case to a simple case of floatation.[198]

---

[196] De Volder 1676–1677, 140r–141r.

[197] "Or puis que cette expérience a été révoquée en doute par tant d'habiles gens, je serai bien aise de dire ici en peu de mots ce que j'ai vû faire sur cela à M. de Volder, Prof. célebre en Philosophie & en Mathématique à Leyde, lorsque j'avois l'honneur d'étudier sous lui il ya sept ou huit ans," Lufneu 1685, 384. Most probably, the experiments took place after 19 February 1675 (when Lufneu enrolled at the University), and before 20 June 1679 (when he graduated in medicine). Lufneu's 1685 report was reprinted – without the last two paragraphs (quoted *infra*, n. 204, and *infra*, n. 253) – in volume 28 of the *Nouveau choix de pieces tirées des anciens Mercures, et des autres journaux*, published by Jean-François Marmontel (1759). The report was also included in the various reprints of the *Nouvelles*.

[198] "1 consectarium. Immitto in aquam ABCD huius propositionis solidum IKLM, materiae levioris quam aqua, quodque ideo ipsi innatet parte NOLM immersa, reliqua NOKI supereminente, ut in subiecta figura apparet. Iam solidum IKLM [...] est aequale tantae aqueae moli, quanta est pars sui demersa NOLM. Quare solidum IKLM cum reliqua ipsum ambiente aqua pondere aequat corpus aqueum magnitudinis ABCD. Itaque etiamnum asserimus secundum propositionis sententiam, fundo EF inniti pondus aequale corpori aqueo magnitudinis columnae, cuius basis sit EF, altitudo perpendicularis GE a summa superficie aquae AB ad imum fundum EF demissa. Unde efficitur a materia qualibet aquae innatante fundum nec magis nec minus affici, quam ab aqua in eadem altitudine constituta," Stevin 1605, 119.

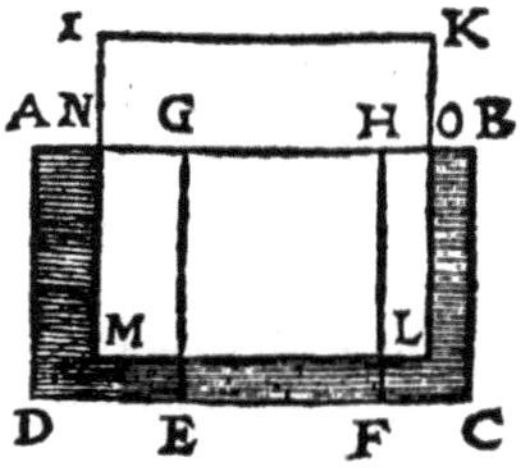

Fig. 5.32 Representation of the condition of floating of bodies in a small quantity of water according to Stevin. (Source: Stevin 1605, 119b. ETH-Bibliothek Zürich, signature: Rar 9069, https://www.e-rara.ch/doi/10.3931/e-rara-9068. Public domain mark 1.0)

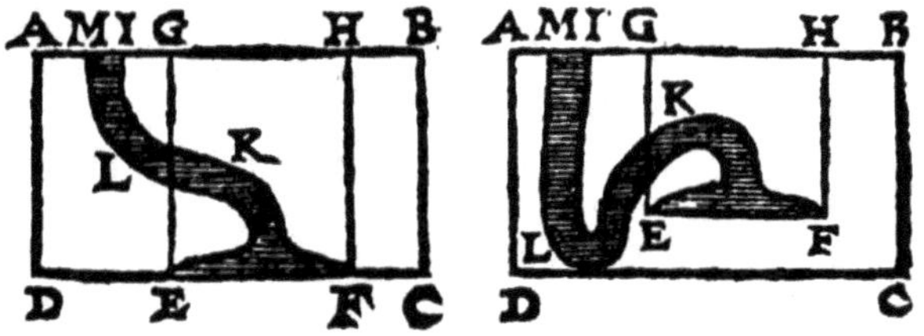

Fig. 5.33 Instances of the hydrostatic paradox according to Stevin. (Source: Stevin 1605, 120a. ETH-Bibliothek Zürich, signature: Rar 9069, https://www.e-rara.ch/doi/10.3931/e-rara-9068. Public domain mark 1.0)

Hence, he comes to the 'paradoxical' conclusion (corollary 2) that the pressure exerted on EF (see Fig. 5.33, Stevin 1605, 120a) by the water MILKEF is equal to the weight of the water included in GHEF. He justifies this idea with the following mental procedure. First, by conceiving the volume ABCD as filled with water; then, by considering the volume MILKEF as a solid equiponderant, *in specie*, with water; eventually, by 'removing' all the remaining water. After this process of abstraction, Stevin claims that the bottom EF is pressed no less than before. The whole process, basically, is grounded on the idea that one can abstract this or that volume of water, and to conceive it as a solid, without any change in the conditions of equilibrium of the whole volume of water.[199] On this basis, he also justifies (corollary 3) the idea that a portion of water MIKLEF presses upwards, in order for it to be in equilibrium with the portion EFGH (in Fig. 5.34, Stevin 1605, 120b), with the same pressure

[199] "2 consectarium. Secundo in aquam ABCD immititor corpus solidum, solidave quotcunque materia aquae aequipondia, inter quae, reliqua omnia aqua expulsa, tantum comprehendatur IKFELM; quae cum ita sint, haec corpora fundum EF nec aggravant necque relevant ab eo pondere quo aqua prius ipsum afficiebat. Quare etiamnum ex sententia propositionis dicimus, fundo EF insidere pondus aequale aqueo ponderi, magnitudine columnam aequante, cuius basis EF, altitudo perpendicularis GE ab aquae summo AB seu MI ad imum EF demissa," Stevin 1605, 120.

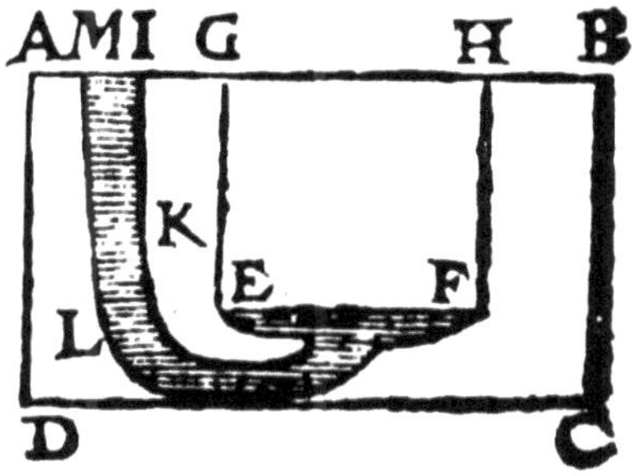

Fig. 5.34 An instance of the hydrostatic paradox according to Stevin. (Source: Stevin 1605, 120b. ETH-Bibliothek Zürich, signature: Rar 9069, https://www.e-rara.ch/doi/10.3931/e-rara-9068. Public domain mark 1.0)

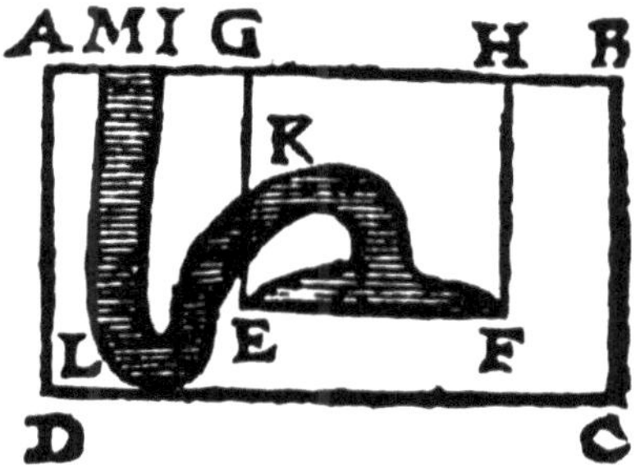

Fig. 5.35 An instance of the hydrostatic paradox according to Stevin. (Source: Stevin 1605, 120c. ETH-Bibliothek Zürich, signature: Rar 9069, https://www.e-rara.ch/doi/10.3931/e-rara-9068. Public domain mark 1.0)

exerted by the weight of the volume of water EFGH.[200] So that (corollary 4) the surface EF undergoes the same pressure both in the case that the whole container ABCD is filled with water, and if only the space IKFELM – considered in both Figs. 5.34, Stevin 1605, 120b, and 5.35, Stevin 1605, 120c – is filled.[201]

Eventually (corollary 5) Stevin re-states proposition 10, by re-considering the cases of corollary 2 and 3 (see Fig. 5.36, Stevin 1605, 120d).[202]

---

[200] "3 consectarium. Si tertium ABCD mera aqua, et in ipsa EF fundum horizonti parallelum. Quibus positis, aqua sub fundo EF tam potenter ipsum sursum premit, quam aqua supra insidens deorsum; secus enim per 1 propos. infirmius validiori concederet, quod hic non fit quia utrumque loco suo permanet. Iam corpus solidum isti aquae pondere homogeneum ita collocator ut aqua IKEFLM ab inferiori parte presset fundum EF, ut hic. Quibus positis, aqua subter EF nunc tam valide premit fundum EF, sive ipsum solidum, quam prius ipsam aquam oppositam: sed impressio tanta tunc erat quanta superioris aquae ad EF depressio, ut supra patuit, superioris autem aquae depressio aequalis erat ponderi columnae aqueae cuius basis EF, altitudo perpendicularis GE, a superficie AB seu MI ad fundum EF demissa. Itaque aquae subter EF constitutae impressio erit quoque tanta," Stevin 1605, 120.

[201] "4 consectarium. Corpora solida secundi tertiique consectarii istic ita firmentur, effusaque aqua spatium IKFELM vacuum nullo amplius pondere efficiet fundum EF; unde apparet aqua in vacuum locum rursum infusa fundum EF tam valide premi, ac si intregrum vas ABCD, sublato isto corpore solido, aqua repletum foret," Stevin 1605, 120.

[202] "5 consectarium. At vero quia immissa solida 2 et 3 consectarii sunt suo loco defixa, ipsorum materia extrema nec gravitate nec levitate ulla afficiet fundum EF, quamobrem sublata omni aquam ambiente materia, relinquentur internae istae aqueae figurae MIKFEL, quales hic vides. Atque hic

Fig. 5.36 Instances of the hydrostatic paradox according to Stevin. (Source: Stevin 1605, 120d. ETH-Bibliothek Zürich, signature: Rar 9069, https://www.e-rara.ch/doi/10.3931/e-rara-9068. Public domain mark 1.0)

For Boyle, as seen above, Stevin's demonstrations were not physical. In fact, he relies on a procedure of abstraction and manipulation i.e. 'solidification' of portions of water. Yet, Stevin annexed to proposition 10, in his *De initiis praxis hydrostatices* (proposition 2), five "pragmatical samples" i.e. experimental samples which could "prove the truth of proposition 10."[203] It is worth drawing attention to those samples later dealt with in the seventeenth century: namely, sample 1 and 4 (which I will present in a moment) which according to Lufneu's communication of April 1685 were performed by De Volder at Leiden. Notably, Lufneu reports that the main experiment performed by De Volder, namely Stevin's sample 4, had already been performed and described by Boyle in paradox 6, where he also maintained (for Lufneu) that this was the only sample of Stevin which had been put into practice, albeit unsuccessfully. Yet, as Lufneu proudly reports, De Volder also performed Stevin's sample 1 with success.[204] If we look at Boyle's *Hydrostatical Paradoxes*, indeed, we do find the description of the performing of Stevin's sample 4, as well as the mention of another experiment, unsuccessfully attempted by people whose identity is not disclosed by Boyle. Or, in his own words:

> [...] though the Learned Stevinus's aims were to be commended; who finding this Proposition doubted, seems to have had a been mind to give an Experimental Demonstration of it, and therefore proposes no less then five pragmatical Examples [...] to make out the truth of what he asserts; yet in this he hath been somewhat unhappy, that that Experiment, which alone (for ought I can find) has been try'd of all the five, is rejected as incompetent, by those that profess to have purposely made tryal of it. And indeed, by reason of the difficulty of bringing them to a practical examen, I have somewhat doubted whether or no this useful writer did ever make all those Tryals himself; rather then set down the events, he

etiam ex sententia propositionis dicimus base EF subnixum esse pondus, aequale ponderi aqueae columnae cuius basis EF, altitudo perpendicularis ab MI aquae summo in fundum EF demissa. Atque ita in caeteris omnibus figuris quarum fundum sit in plano horizonti parallelo. Conclusio. Itaque in fundo horizonti parallelo, et c.," Stevin 1605, 120–121.

[203] Stevin 1605, 145.

[204] "Avant que de finir, il faut que je dise encore que M. Boyle, dans l'endroit que j'ai cité de son paradoxe 6. remarque, que l'expérience que je viens de décrire est la seule des cinq de Stevin, qu'il sache avoir été examinée, et mise à l'essai. Mais je suis témoin qu'outre celle-ci, qui est la troisieme en rang dans l'Hydrostatique Pratique de Stevin, M. de Volder a executé la premiere avec un pareil succès," Lufneu 1685, 389.

Fig. 5.37 Instrument aimed at demonstrating the hydrostatic paradox (*libra hydrostatica mirabilis*). (Source: Stevin 1605, 146a. ETH-Bibliothek Zürich, signature: Rar 9069, https://www.e-rara.ch/doi/10.3931/e-rara-9068. Public domain mark 1.0)

> suppos'd they must needs have; [...]. Wherefore though another of the Experiments, he proposes, be not free from difficulty, yet having, by the help of an Expedient, made it practicable, we are induc'd by its plainness and clearness to prefer it to what else he proposes to the same purpose.[205]

A detailed report on the performance of the unspecified sample mentioned by Boyle can in fact be found in Gaspar Schott's *Mechanica hydraulico-pneumatica* (1657), a text which Boyle knew well.[206]

This sample is nothing but Stevin's sample 1. In presenting it, Stevin describes the use of an instrument like the one represented in Figs. 5.37, Stevin 1605, 146a, and 5.38, Stevin 1605, 146b, namely an empty vessel M capable of containing 10 pounds of water. It is appended to a balance, and it is placed under the unmoveable solid P, in such a way that the solid is in the basin itself, and that a lateral space is left in-between the solid and the basin. On the other side of the balance, a weight of 10 pounds Q is placed, so that the bottom of vessel M presses upwards the solid P

[205] Boyle 1666, 136–137.

[206] In the third volume of his *Magia*, Schott proposed five arguments against the idea that elements do not gravitate on elements: see Schott 1657–1659, volume 3, book 5, *syntagma* 3, *erotemata* 3–4. All these arguments – whose author is not disclosed by Boyle, are systematically criticized by Boyle in appendix 1 to his *Hydrostatical Paradoxes*. The arguments, in Boyle's versions, are as follows: "(1) Because else the inferiour parts of the water would be more dense then the Superior," (2) "Because Divers feel not, under water, the weight of the water that lyes upon them," (3) "That ev'n the slightest Herbs growing at the bottom of the water, and shooting up in it to a good height, are not oppress'd or lay'd by the incumbent water," (4) "That a heavy Body ty'd to a string, and let down under water, is supported, and drawn out with as much ease, as it would be if it had no water incumbent on it; nay, with greater ease, because heavy bodyes weigh less in water then out of it," (5) "Because a Bucket full of water, is lighter in the water, then out of it; nor does weigh more when full within the water, then when empty out of it," Boyle 1666, 202–206.

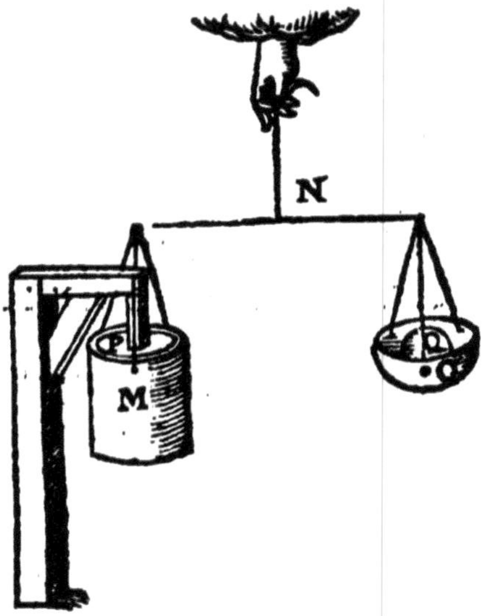

Fig. 5.38 Instrument aimed at demonstrating the hydrostatic paradox (*libra hydrostatica mirabilis*). (Source: Stevin 1605, 146b. ETH-Bibliothek Zürich, signature: Rar 9069, https://www.e-rara.ch/doi/10.3931/e-rara-9068. Public domain mark 1.0)

with a force of 10 pounds. Afterwards, this lateral space is completely filled with water, so that (1) the basin is slightly depressed, (2) the space in-between the solid and the basin is filled with 1 pound of water, and (3) the basin is in perfect balance with the weight Q. The pressure exerted on the bottom of the basin (measured by the balance), at that point, would not be equal to the weight of the water, but it will be equal to the weight of a volume of water equal to that of (a) the actual water (1 pound), and (b) the volume of water equal to the volume of the solid submerged by the water.[207]

[207] "MNO libra esto, cuius lances M, O; atque M quidem figurae cylindraceae aequalis exposito GCD ideoque 10 librarum aquae capax; tum P solidum simile lanci M et minus, scapo affigatur ut hic vides. Inseratur igitur solidum P in lancem M, ut in secunda figura, lancique O imponatur pondus Q 10 lb; iam fundum M tam valide impinget in corpus P quam a 10 lb impelletur. sit autem corpus P decima parte minus quam M, ut vacuus inter utrumque locus 1 lb aquae expleatur, hoc est aquea mole aequante corpus EAB. Itaque 1 lb aquae in lancem infusa hanc deprimet, reliquamque attollet, id ipsum testante experientia, et 10 propositionis demonstratione approbante. Quare 1 lb aquae in lance M istic tantae erit potentiae, quam 10 lb plumbi ferrive aut alterius materiae solidae eidem lanci M affixae. Atque eadem ratione 1 lb aquae, huiusmodi partium dispositione maioris erit efficaciae, quam millae librae materiae alterius. Quae cum ita sint, aqua quae inter utriusque fundum, corporis P lancisque M intercessit, fundum M nunc tam valide pressat ac prius fundum corporis P, hoc est, ac 10 lb; cum pondus Q 10 lb in reliqua lance O immissum sit. Itemque contra aqua tanta vehementia premit fundum lancis M, quanta est efficientia 10 lb Q. Ponamus autem aquam in fundo M aequari ipsi KLBA, reliquam autem ipsi P circumfusam, reliquae IE. Quare aqua EAB tam potenter pressat fundum A B, quam haec aqua fundum M, ideoque EAB premit suum fundum AB aequivalenter 10 lb, sed tantus est item pressius aquae GCD contra fundum CD. Quamobrem, quod pragmaticè confirmare statueramus, aqua EAB pondere 1 lb suum fundum A B aequè validè premit, atque GCD 10 lb fundum CD. Pari ratione evinces, vel 1 lb pressare potentius mille libris," Stevin 1605, 146.

According to Lufneu, as seen above, De Volder successfully performed this sample, even if no further detail is provided. Before Schott (whose report I analyse below), we find a description of this sample in Galileo's *Lettera a Tolomeo Nozzolini*, namely a short tract addressed in January 1613 to the Professor at Pisa Studio Tolomeo Nozzolini (who had written a report on Galileo's *Discorso* for the Archibishop of Florence Alessandro Marzi Medici).[208] Galileo did not provide a purported repetition of Stevin's sample 1, rather, he proposed the analogous case, in which a basin filled with water is gradually filled by a solid attached to a wall (as in Stevin's case), and the water is progressively spilled out of the basin itself. Yet, notwithstanding there is progressively less water in the vessel, and that the solid is sustained by an arm attached to the wall, the weight of the basin, or the effort of the holder to maintain it, does not change.[209]

Galileo explained this case by considering the force of buoyancy of water. The submerged solid, indeed, will press on the water with the same force, or weight, of a volume of water equal to the volume of the submerged part of the solid.[210] In this

[208] The letter was printed for the first time in the Florence edition of Galileo's works of 1718.

[209] "Ma che dirà V. S. se io mostrerò che un vaso che galleggi sendo anco ripieno di acqua, non farà mutazione alcuna se con l'imposizione di un solido nel modo detto si scaccierà quasi tutta l'acqua che in esso vien contenuta? Ma per ben dichiarare il tutto, ed insieme accrescer la maraviglia, intendasi un cilindro solido AB, fermato e sostenuto in A, di poi intendasi il vaso CDE, capace della mole AB e di poco di più, il qual vaso, sendo separato e allontanato da esso cilindro AB, sia ripieno di acqua, della quale ne capisca, per esempio, 100 libre: di poi postolo sotto 'l solido fisso AB, lentamente s'innalzi verso esso solido, in guisa che, entrandovi egli dentro, faccia a poco a poco traboccar fuori l'acqua, secondo che esso vaso CDE si anderà elevando. Or io dico, che quella persona che anderà alzando detto vaso contro il solido AB, sempre sentirà il medesimo peso, ben che di mano in mano vadia uscendo fuori l'acqua; né meno si sentirà aggravare dopo che nel vaso non sarà rimasto più di due o tre libre d'acqua, di quello che egli sentisse gravarsi quando era del tutto pieno, ancor che il solido AB non tocchi il vaso, ma sia, come si è supposto, fissamente ed immobilmente sostenuto in A," Galileo 1890–1909, volume 4, 306–307. See Fig. 5.39, Galileo 1744, volume 1, 263. For a thorough discussion, see Palmieri 2005.

[210] "Ciò potrà per esperienza esser fatto manifesto ad ognuno, ma oltre all'esperienza non ci manca la ragione. Imperocchè considerisi come la potenza sostenente il solido in A, mentre esso era fuori di acqua, sentiva maggior peso, che dopo che il solido B è immerso nell'acqua, perchè non è dubbio alcuno, che se io reggerò in aria una pietra legata ad una corda, sentirò maggior peso, che se alcuno mi vi sottoponesse un vaso pieno d'acqua, nel quale detta pietra restasse sommersa; scemandosi dunque la fatica nella virtù che sostiene il solido AB, mentre e' si va immergendo nell'acqua del vaso CDE, che lo va ad incontrare, nè potendo il peso di questo andare in niente, è forza che s'appoggi nell'acqua, ed in conseguenza nel vaso CDE, ed in quella virtù che lo sostiene; e perchè noi sappiamo, che ogni solido più grave in specie dell'acqua, e che in essa si de merge, va di mano in mano perdendo di peso, tanto quant'è il peso d'una mole d'acqua uguale alla mole del solido demersa, facilmente intenderemo tanto andare scemando la fatica della virtù sostenente il solido AB in A, quanto l'acqua va scemando la gravità di esso solido: adunque il solido AB va gravando sopra alla forza sostenente il vaso CDE tanto quanto è il peso d'una mole d'acqua uguale alla mole del solido demersa: ma alla mole del solido demersa è di mano in mano uguale la mole dell'acqua che si spande fuori del vaso, adunque per tal'effusione d'acqua non si scema punto il peso che grava sopra la virtù che sostiene il vaso. Ed è manifesto che il solido AB, se ne scaccia l'acqua del vaso, nientedimeno con l'occuparvi il luogo dell'acqua scacciata vi conserva tanto di gravità, quanta appunto è quella del l'acqua che si versa," Galileo 1890–1909, volume 4, 307–308.

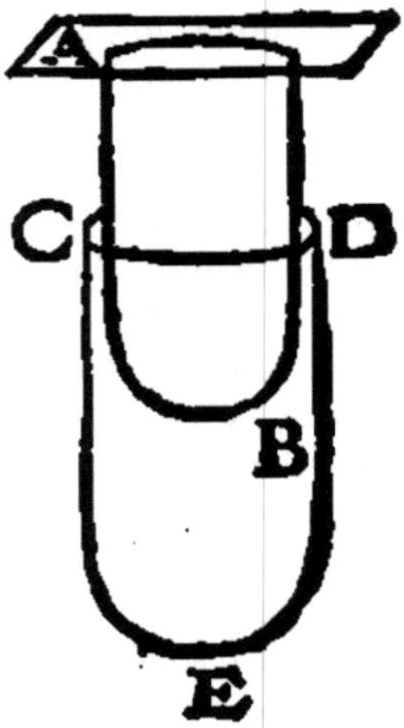

Fig. 5.39 Instrument aimed at demonstrating the hydrostatic paradox. (Source: Galileo 1744, volume 1, 263. Zentralbibliothek Zürich, signature: NE 687|G – NE 690|G, https://www.e-rara.ch/zuz/doi/10.3931/e-rara-49850. Public domain mark 1.0)

case, Galileo does not use the reduction of this 'paradox' to the case of the lever, but to an instance of forced floatation. *À la* Stevin, Galileo ideally 'liquifies' the submerged volume of the solid, and vice versa, i.e. he 'solidifies' the volume of water equal to the submerged part of the solid. One year later, Stevin's sample 1 is discussed in the *Eorum quae vehuntur in aquis experimenta ad Archimedis trutinam examinata* (1614) of Giovanni Bardi – who overtly mentions Stevin – where it is labelled as 'incredible', and is explained by considering the solid immersed in water as exerting the same pressure of an ideal volume of water equal to its submerged part.[211]

Thereafter, the case is described by Schott in his 1657 *Mechanica hydraulico-pneumatica*, presenting the instrument described by Stevin, and labelled by Schott as 'libra hydrostatica mirabilis' (see Fig. 5.40, Schott 1657, *iconismum* 27, figure 8)[212] and in the third part of his *Magia universalis* (1657–1659) (book 5, *erotema* 14), where he discusses Stevin's proposition 10 of his *De hydrostatices elementis*, and proposition 2 of his *De initiis praxis hydrostatices*. In the first text, Schott describes the *libra*, which is the same instrument described by Stevin, and refers – in his commentary on the description – to the *Thaumaturgus mathematicus* (1636) of Gaspar Ens, in which the same experience of Stevin is labelled as incredible, "if I [...] had not performed [it]" (*periculum fecissem*).[213] Schott, moreover, reports – in

[211] Bardi 1614, 16.

[212] See Schott's *Mechanica hydraulico-pneumatica*, part 2, *classis* 1, *machina* 11.

[213] Schott 1657, 318–319. Cf. Ens 1636, 215 (i.e. 217)–218: "[m]irabile est aquam vase quolibiet conclusam, et violenter divisam tantum acquirere ponderis, ac si locus a corpore dividente occupatus uniformiter aqua refertus esset. Varia liceret ad propositionis huius probationem experimenta congerere, at egregia duo tantummodo in medium proferam; quae equidem vera esse nunquam crederem, nisi ipsemet eorum periculum fecissem. Primum est lapidem ingentem, locum 10, 100 aut 1000 librarum aquae occupantem, vel corda, vel catena suspensum, in vas aquae plenum dimit-

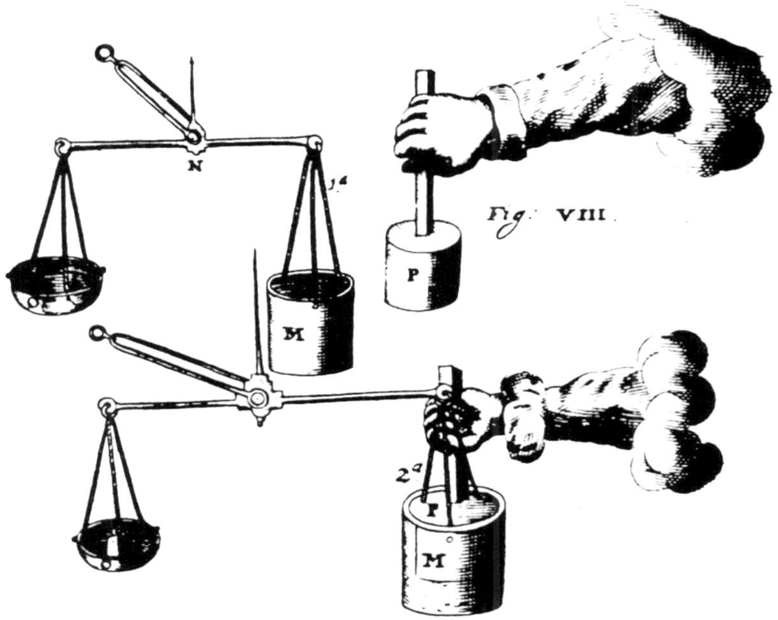

Fig. 5.40 Instrument aimed at demonstrating the hydrostatic paradox (*libra hydrostatica mirabilis*). (Source: Schott 1657, *iconismum* 27, figure 8. ETH-Bibliothek Zürich, signature: Rar 8034, https://www.e-rara.ch/zut/doi/10.3931/e-rara-31789. Public domain mark 1.0)

both texts – that this experience was referred to during a public disputation at the Jesuit Collegio of Palermo (where Schott taught until 1652) concerning the immobility of the Earth. According to Schott's *Mechanica*, in this disputation (whose date is not mentioned), the *viceré* of Sicily and skilled mathematician Juan Carlos de la Faille cited this experience as evidence that Earth is moving, against a student of Schott who defended the geocentric theory. According to Schott, La Faille mentioned that it had been successfully performed in the presence of Albert VII

---

tito; aut ferro alicui muro adhaerenti saxum immobiliter affigito, illique vas aqua refertum supponito: porro vas ita collocandum est, ut et saxum immissum summa ex parte contingat, et inter utrumque locus unius librae aquae intercedat; quae iniecta (si lapis locum 100 librarum aquae occupet) 100 libris gravior evadet; adeo, ut prae aquae ponderositate, vas suppositum aegre sustineri possit. Secundum longe mirabilius est. Bilancem communibus aliis simillimum parato, hoc tamen uno excepto, quod huius cratere licet aequalis ponderis, inaequalis tamen capacitatis sint, in uno quippe pondera ponenda, in alterum aqua infundenda erit. Affige igitur muro (ut antea dictum est) corpus aliquod, sive solidum, sive intus vacuum, capax locum 9 librarum aquae occupare: deinde imposito in craterem debitum 10 librarum pondere, et in oppositum infusa libra aquae; craterem libram aquae continentem dicto corpori submittito, ut corpus cratere, nulla ex parte contracto, comprehenduntur: quo peracto, libram aeque infusam, 10 libri plumbi aequipollere disces."

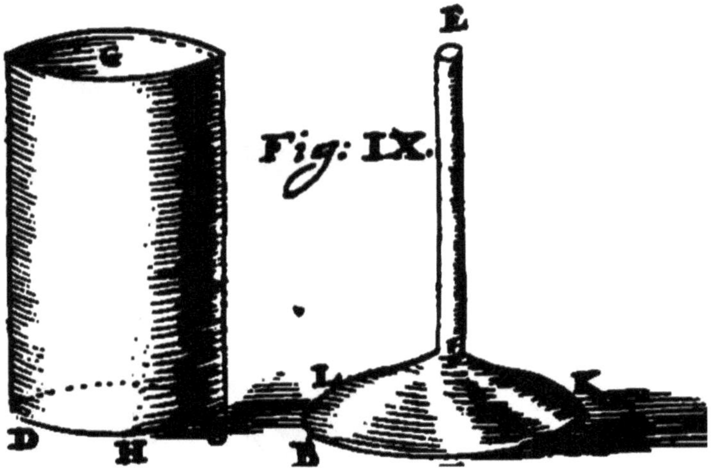

Fig. 5.41 Containers aimed at demonstrating the hydrostatic paradox. (Source: Schott 1657–1659, volume 3, *iconismum* 15, figure 9. Berlin, Max-Planck-Institut für Wissenschaftsgeschichte, call number: Rara S375m -3, http://echo.mpiwg-berlin.mpg.de/MPIWG:G6HSU7XR. Attribution-ShareAlike 4.0 International (CC BY-SA 4.0))

(Governor of the Spanish Netherlands).[214] In his *Magia*, Schott provides more details on the disputation, describing that the thesis of his student was that the Earth cannot be moved except by a force equal to that exerted by a weight equal to that of the Earth itself. In criticizing this thesis, La Faille – according to Schott's *Magia* – did not use the case of the libra *hydrostatica mirabilis* (which he nevertheless cited), but rather the case of the pressure exerted on the bottom of an inverted T-shaped container (see Fig. 5.41, Schott 1657–1659, volume 3, *iconismum* 15, figure 9), such as that one used by Schott to present Stevin's proposition 10 in this *erotema*. According to La Faille, as reported by Schott, using this kind of container would enable obtaining a pressure sufficient to move the Earth.[215]

[214] "Res haec est omnino mirabilis, sed verissima, et experientia saepius comprobata quam quilibet facile poterit facere. Affirmabat doctissimus mathematicus P. Ioannes Carolus la Faille, cum Panormi in disputatione publica predictam experientiam contra quendam discipulum meum, Terrae immobilitatem, non obstante continua centri gravitatis mutatione, mathematice demonstrantem attulisset; illam olim Serenissimo Alberto Belgii Gubernatori fuisse exhibitam; eumque, cum versatissimum esset in rebus mathematicis, et in sumendis experimentis curiosissimus, tantopere tamen obstupuisse, ut asseruerit, nunquam se rem mirabiliorem spectasse," Schott 1657, 319.

[215] "Memini, cum Panormi in Sicilia in disputatione publica meus quidam in mathematicis discipulus demonstraret, Terram non ad quamlibet novi ponderis ex parte una additionem, et inde secutam centri gravitatis mutationem, titubare, nec locum mutare, nisis fortassis accederet pondus haemispherii terraquei pondere non minus [...], memini inquam, virum doctissimum P. Joannem Carolum La Faille, Serenissimi Joannis ab Austria nunc Belgii Gubernotoris tunc vero Sicilia Proregis Mathematicum, opposuisse hoc Stevini Exemplum [...] asseruisseque, summa omnium admiratione, vel unam libram aquae vasi praedicta ratione efformati conclusam sufficere, ad globum tarraqueum loco suo dimovendum," Schott 1657–1659, volume 3, 455. On La Faille's mentioning of Stevin's sample 1, see *infra*, n. 218.

Then, Schott reports that at the disputation there was present Carlo Ventimiglia, a Palermo nobleman, who just after the disputation filled with water two containers like those represented in Fig. 5.41, Schott 1657–1659, volume 3, *iconismum* 15, figure 9, and weighed them on a balance in the presence of Schott. In this case (of course) the experiment failed.[216] In fact, according to Schott, La Faille maintained that in this way the pressure exerted on the bottom cannot be revealed, but that it is nevertheless exerted: however, he could not convince Schott, as Ventimiglia's experiment disproved such a claim.[217] For Schott, in this case or sample there is no increased pressure on the bottom of the container: so that for him Stevin's proposition 10 is true, but it can be demonstrated only by the experiment with the *libra mirabilis*. In this case, as he explains both in his *Mechanica* and *Magia*, the increased pressure is the effect not of the weight of the water, but of the force or impetus by which the water pushes upwards the solid body P – provided that it is specifically lighter than water itself – and which is then reflected on the bottom of the container.[218] Schott cites Archimedes's treatise on floating bodies in order to support his claim. He does not enter into details; however, as seen above, Archimedes (like De

[216]As a matter of fact, indeed, in this way one can just measure the weight of the water in the containers, but not their internal pressure. This can on the other hand be revealed by the use of the *libra mirabilis* – in which the pressure of water in-between the bottoms of the solid and the container, keeps them separated, and can be measured with a balance. In the case used by Ventimiglia, indeed, the pressure exerted at the level of the bottom of the inverted T-shaped container is not revealed just by putting the container on a balance, because this pressure is also exerted upwards, so that its effects are nullified.

[217]"Aderat disputationi illustrissimum Dominus Carolus Vintimiglia, Eques Panormitanus nobilissimus, et eximius mathematicus, et statim absoluta disputatione fieri curavit duo vasa praedicto modo efformata, et me praesente experimentum fecit, sed sine ullo effectu; manifeste enim deprehendimus, vas ABE cum aqua sua non gravare tantum manum, aut bilancem sustentantem, quantum eam gravabat vas GCD, unde conclusimus, nec fundum utriusque aequaliter premi ab aquis inclusis. Fatebatur ingenue dictus P. La Faille [...] nos in bilancibus vel manibus non totum illud vasis pondus sentire fundum tamen illo premi; sed non persuadebat, repugnante manifesta experientia," Schott 1657–1659, volume 3, 455–456.

[218]"Aio itaque, propositionem, illam X Stevini, prout ab ipso ibidem proponitur et explicatur, veram esse, et bene ab ipso probari, et confirmari etiam experientia; at falsum esse allatum exemplum, et ex propositione illa nequaquam deduci, et insuper experientiae manifesto repugnare. Legatur proposito apud ipsum, et cum ea conferatur exemplum. Confirmare nititur Stevinus loco cit. libri 5 exemplum praedictum alia experientia, quam ibidem adducit, et nos ex ipso adduximus in Mechanica hydraulico pneumatica parte 2 classe 1 cap. 6 machina 2 et appellavimus libram hydrostaticam mirabilem. Eandem experientiam in citata paulo ante disputatione afferebat P. La Faille, experientia est verissima, et ab aliis saepius probata, sed non est ad rem: nec enim fundum ancis librae premit tam valide modica aqua, sed impetus ab aqua cylindro ligneo impressus, et a cylindro in aquam et fundum lancis reflexus, ut ibidem nos explicavimus," Schott 1657–1659, volume 3, 456. Cf. Schott 1657, 319: "[q]uaenam igitur huius experimenti est ratio? An quia corpus P aquae lancis M immersum pellitur ab aqua sursum, utpote aqua levius, ut supponitur: et cum palo seu scapo affixum sit, et cedere non possit, impetus ab aqua ipsi impressus reflectitur in fundum vasis M, illudque deorsum premit tanta vi, quantae deprimeret aquae moles corporis P aequalis, nempe in casu posito moles aquae librarum novem? Scimus enim experientia, corpora aquae intrusa, si leviora sint quam aqua eiusdem molis, sursum pelli, et quidem tanta vi, quanto aqua molem habens corpori intruso aequalem gravius est ipso corpore, ut Archimedes demonstrat lib. I de iis quae vehuntur in aquis."

Volder would do) used the case of the forced floatation of a body lighter than water in order to show that it reveals a 'force of buoyancy', exerted, according to him, by the weight of an equal volume of water in his 'waterworld'. In other words, for Archimedes water does not press on a body just because it is lighter than it, but because this body is counterbalanced by an equal volume of water. Indeed, *pace* Schott, water also presses with the same force on a solid heavier than water, once immersed in it. Schott's explanation, actually, falls short in explaining the case of the floatation in a small quantity of water, which is solved by Galileo and Pascal, as seen above, by the principle of the lever. Thus far, Schott instrumentally used what he labelled as an idea of Archimedes, bypassing the difficulty that a more careful consideration of Archimedes's argument would have revealed. Moreover, Schott's views were at odds with the very Archimedean model of floatation, as he admitted elsewhere in his *Magia* that water does not press on water.[219] Thus far, he had no other means to save Archimedes's account but to refer to the idea that pressure is exerted only between bodies with a different specific weight – regardless of the height of the fluid. It is noteworthy that, in his *Magia*, Schott even claims that if Stevin had known the very explanation given by Schott, he would not have used his sample 1 to confirm proposition 10, which for Schott cannot be confirmed by a sample such as this; for him, it is rather the effect of the impetus of the water on a lighter container, rather than the effect of the height of water.[220] In short, Schott accepts Stevin's proposition 10, but he does not label it as valid in the case of sample 1, or in the case of the pressure exerted on the bottom of an inverted T-shaped container. It seems, actually, that he aimed at 'mitigating' its more paradoxical consequences, rejecting the idea that on an inverted T-shaped container the bottom is pressed as if the container were cylindrical, and explaining Stevin's first sample without reducing it to proposition 10.

5.4.2.2 Boyle's Experimental Approach to the Paradox

Schott's reading of the experimental evidence provided by the *libra mirabilis*, thus, fits with Boyle's claim that the "Experiment, which alone (for ought I can find) has been try'd of all the five, is rejected as incompetent, by those that profess to have purposely made tryal of it." Boyle, in turn, performed Stevin's sample 3, namely the one then repeated by De Volder (with some variations, as I am going to show). According to Boyle's *Hydrostatical Paradoxes* (paradox 6, scholium),

> we could not but look upon the success, as that, which though it did not answer what the reading of Stevinus might make a man expect; yet may deserve to be further prosecuted, that whether or no the Paradox of Stevinus (which not only some others, but the Learned Dr. Wallis himself question) wil hold.[221]

[219] See *supra*, n. 206.

[220] "Quod si scivisset Stevinus, nunquam experientiam in exempli sui confirmationem adduxisset," Schott 1657–1659, volume 3, 456; see also the quotation from the same text, given *supra*, n. 218.

[221] Boyle 1666, 140.

I could not find any place in Wallis's published treatises (such as the section on hydrostatics of his *Mechanica*), in his communications to the Royal Society, and his correspondence, where he deals with the paradox.[222] So that it might just be that he discussed the topic privately with Boyle. In turn, Boyle conceded the 'success' of the experiment (namely a refined version of Stevin's sample 3. Yet, he did not find it adhering fully to Stevin's theory, for the reasons that (1) the very theoretical suppositions of the experiment were not entirely clear,[223] and (2) the experiment could not provide an exact confirmation of the theory, as Stevin just proposed a mental exemplification rather than a description of a practical procedure. In Boyle's own words:

> [...] the Inquiry he has started, may be so persued, as to occasion some improvement of this part of Hydrostaticks: where, to define things with certainty, will perhaps be found a difficulter Task then at first glance one would think; both because divers speculative things must be taken into consideration, whose Theory has not perhaps yet been clear'd, and because of the difficulty that will be found in practice by them that shall go about to make Stevinus's Experiments, or others of that sort with all requisite Accurateness: As indeed, it is far easier to propose Experiments, which would in likelyhood prove what we intend, in case they could be made, then to propose practicable Expedients how they may be made.[224]

As far as the experiment itself is concerned, Stevin proposed to use an inverted T-shaped container IKL (see Fig. 5.42, Stevin 1605, 147a), whose bottom OP, which is lighter than water, is moveable and is connected to the weight T with a rope passing through the vertical tube. The weight T, as the container is filled with water, counterbalances the pressure exerted by the water on OP, namely a pressure which

---

[222] See the third part of his *Mechanica*, chapter 14, *De hydrostaticis*. Wallis proposes a theory of floatation based on Archimedes's postulate 1 (which is not overtly mentioned): see proposition 1 ("[s]i grave fluidum [...] superne vel non prematur, vel aequaliter prematur ubique: retinebit illud, supernam superficiem quod spectat, situm horizontalem, seu potius sphaericum, Telluri concentricum [...]. Atque inde deturbatum, suapte se gravitate eo restituet. Sin superne prematur inaequaliter, [...] ea parte quae maiori pressu urgetur, descensus fiet; partibus ita pressis, alias, quae vel non pressae sunt, vel minus pressae, loco suo detrudentibus, quae itaque, pro illo descensu, assurgent," Wallis 1670–1671, volume 3, 708. According to this theory, a body lighter than water, once put in water, is bounced upwards as the effect of the different pressure exerted on the (ideal) surfaces of water: see proposition 3: "[s]i illud D [i.e. a solid body] plus gravitet quam tantundem circumpositi aëris [...], minus autem quam tantundem fluidi subiecti: innatabit D; sed depressis eousque partibus C subiectis, donec eum situm obtinuerit D, ut aequigravitet aggregato aëris et subiecti fluidi quorum locum occupat," Wallis 1670–1671, volume 3, 711. Probably, Wallis found it difficult to justify, on such an Archimedean basis, the fact that the pressure of a fluid can be independent from the quantity of water which is above it. In fact, he does not even consider the case of floatation on a little water, and dedicates to the conditions of floatation only 7 of the 15 propositions of the chapter, the others being dedicated to the explanation of the suspension of mercury in Torricellian tubes.

[223] As to this point, one can also find it in Lufneu's 1685 report. In the opening, Lufneu claims that physics had made little progress because of the lack of experiments, and, even if in recent times progress had been made by Stevin, Bacon, Galileo, William Gilbert, Descartes and Boyle, no agreement had ever been reached as to the explanation of the experiments they provided, nor on the conditions of the possibility of such experiments, namely, on the principles allowing for their very possibility. A foremost case, as to this, was the paradox of Stevin: see Lufneu 1685, 381–382.

[224] Boyle 1666, 140–141. Cf. the Latin version: Boyle 1669c, 117.

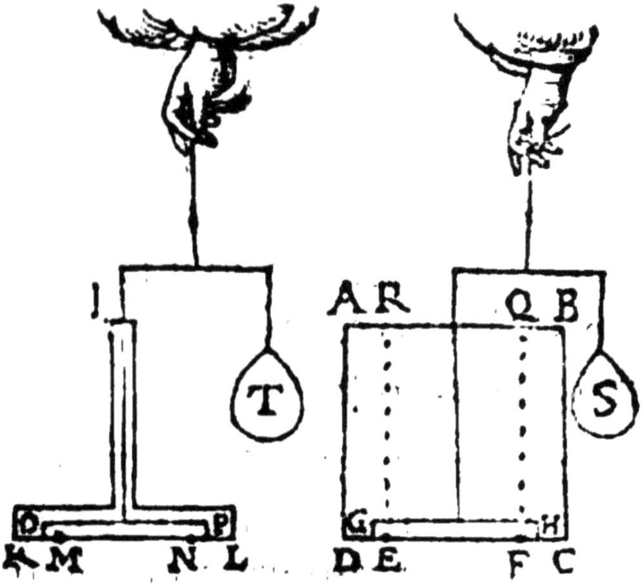

Fig. 5.42 Instrument aimed at demonstrating the hydrostatic paradox. (Source: Stevin 1605, 147a. ETH-Bibliothek Zürich, signature: Rar 9069, https://www.e-rara.ch/doi/10.3931/e-rara-9068. Public domain mark 1.0)

is equal to that exerted by the column of water EFRQ in the container ABCD. In both the container the pressure exerted on the moveable bottom (OP or GH) is measured by connecting the bottom itself to a balance holding the weight S (equal to the weight of the volume of water EFRQ).[225]

Stevin uses moveable bottoms made of a material lighter than water, so that, in water, this bottom loses all its weight and only the pressure exerted on it by the

[225] "Vas ABCD aquae plenum, cuius fundum DC horizonti parallelum rotundo foramine pertundatur, quod ligneus tegat orbis GH materie quam aqua levior. Exponatur deinde vas alterum IKL superiori quidem aeque altum, sed minus et aquae item plenum, cuius fundum ad MN perforatum aequaliter antecedenti EF, et orbe quoque OP ipsi GH aequali obtegatur. Quibus positis, orbis GH contra communem ligni naturam vimque ingenitam ex aqua non emerget, sed foramini EF incumbens tam valenter premet quam columna aquea EFQR multatum differentia ponderum lignei orbis GH et aquae ipsi aequalis. Et qui experimento hoc cognoscas, orbi GH libram affigito, cuius pondus S ponderi dicto aequale sit, eritque orbis GH ipsi aequilibris. Similiter firmato ad orbem OP libram, cuius pondus T superiori S aequeponderet, orbisque hic OP ponderi T manebit aequilibris. auctis autem ponderibus S, T, orbes GH, OP attollentur, atque adeo hac via deprehendes orbes istos in funda subiecta aequalem impressionem facere; unde propositi veritas perspicitur videlicet minorem aquae molem IKL tam valide quam maiorem ABCD premere fundum sibi subiectum," Stevin 1605, 147. He does not specify if the container is placed somewhere, or if its provided of some kind of support (as it was to be for De Volder's instrument).

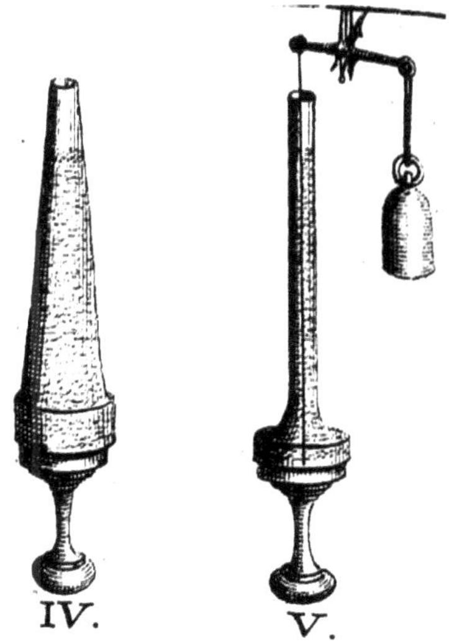

Fig. 5.43 Instruments aimed at demonstrating the hydrostatic paradox. (Source: Pascal 1663, figures 4–5. ETH-Bibliothek Zürich, signature: Rar 4668, https://www.e-rara.ch/doi/10.3931/e-rara-20296. Public domain mark 1.0)

water above it is measured.[226] Later, in his *Traité de l'équilibre des liqueurs*, chapter 1, Pascal used the same sample (see Fig. 5.43, Pascal 1663, figures 4–5) to demonstrate that fluids press according to their height, however without adding much detail with respect to Stevin's presentation.[227] On his own, Boyle – who doubted that

[226] In the *Notato* to the sample, Stevin specifies that if the moveable bottom GH is made, for instance, of lead, the weight S will counterbalance a weight equal to that of the column EFRQ plus the weight that the bottom GR has in water: "si discus GH esset e plumbo, ferrove, aut alia materia graviore quam aqua formatus, eius in subiectum foramen impressionem fore tantam, quanta sit columnae aqueae EFQR auctae differentia ponderis, quae inter H orbem dictum et aquae molem sibi aequalem intercedit," Stevin 1605, 147.

[227] "[...] l'autre estroit, comme en la quatriéme; l'autre qui ne soit qu'un petit tuyau qui aboutisse à un Vaisseau large par en bas, mais qui n'ait presque point de hauteur, comme en la cinquiéme Figure; et qu'on les remplisse tous d'eau jusques à une mesme hauteur, et qu'on fasse à tous des ouvertures pareilles par en bas, lesquelles on bouche pour retenir l'eau; l'experience fait voir qu'il faut une pareille force pour empescher tous ces tampons de sortir, quoy que l'eau soit en une quantité toute differente en tous ces differents Vaisseaux, parce qu'elle est à une pareille hauteur en tous: et la mesure de cette force est le poids de l'eau contenuë dans le premier Vaisseau, qui est uniforme en tout son corps; car si cette eau pese cent livres, il faudra une force de cent livres pour soûtenir chacun des tampons, et mesme celuy du Vaisseau cinquiéme, quand l'eau qui y est ne peseroit pas une once. Pour l'éprouver exactement, il faut boucher l'ouverture du cinquiéme Vaisseau avec une piece de bois ronde, enveloppée d'étoupe comme le piston d'une Pompe, qui entre et coule dans cette ouverture avec tant de justesse, qu'il n'y tienne pas, et qu'il empesche neanmoins l'eau d'en sortir, et attacher un fil au milieu de ce Piston, que l'on passe dans ce petit tuyau, pour l'attacher à un bras de balance et pendre à l'autre bras un poids de cent livres: on verra un parfait Equilibre de ce poids de cent livres avec l'eau du petit tuyau qui pese une once; et si peu qu'on diminue de ces cent livres, le poids de l'eau fera baisser le Piston; et par consequent baisser le bras de la balance où il est attaché, et hausser celui où pend le poids d'un peu moins de cent livres," Pascal 1663, 1–3. See Fig. 5.43, Pascal 1663, figures 4–5.

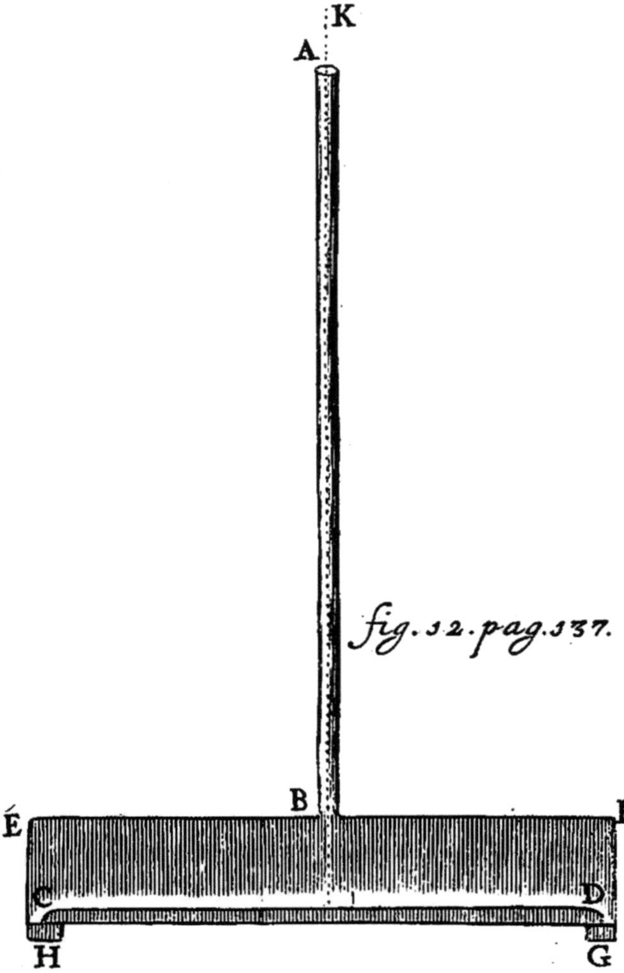

Fig. 5.44 Instrument aimed at demonstrating the hydrostatic paradox (*libra hydrostatica*). (Source: Boyle 1670, figure 12. Zentralbibliothek Zürich, signature: NP 2815, https://www.e-rara.ch/zuz/doi/10.3931/e-rara-47228. Public domain mark 1.0)

Pascal himself ever performed any of the samples he proposes[228] – describes his performing of an actual experiment (see Fig. 5.44, Boyle 1670, figure 12).

In it, the container is progressively filled with water, and once filled the bottom – covered with a bladder to prevent leakage – is connected to a balance holding weights. Boyle finds that the weight needed to raise the bottom is much heavier than

[228] See *supra*, n. 65.

the volume of water contained in the container itself.[229] So that he "could not but look upon the success," but, as seen above, he did not accept the validity of Stevin's paradox, and just claimed that further investigations were needed.

5.4.2.3 ...and De Volder's Refinement of It

In his 1685 communication, Lufneu built upon Boyle's account (as he refers to Wallis's doubts concerning the paradox, not reported elsewhere), and – probably – upon his student's notes on the experiment. In his communication Lufneu remarks that Boyle's experiment was not successful, for the reason that the pressure exerted on the bottom was greater than that exerted by the bare weight of the water filling the container, but it did not correspond to the "proportions of Stevin."[230] It is unclear whether this was the opinion of De Volder himself, yet the instrument he used was far more developed than the one used by Boyle, and enabled a more accurate measure of the pressure exerted on the bottom. The instrument used by De Volder (see Fig. 5.45, Lufneu 1685, 385) was similar to those of Stevin and Boyle. Some nota-

---

[229] "We provided then a vessel of Laton, of the figure express'd in the Scheme, and furnished it with a loose Bottom C D, made of a flat piece of wood cover'd with a soft Bladder and greas'd on the lower side neer the edges, that leaning on the rim of wood G H, contiguous every where to the inside of the Laton it might be easily lifted from off this Rim; and yet lye so close, upon it, that the water should not be able to get out between them: And to the midst of this loose bottom was fastned a long string, of a good strength, for the use hereafter to be declared. The Instrument thus fitted, the water was poured in apace at the Top A B, which, by its weight pressing the false Bottom C D against the subjacent Rim, G H, contributed to make the Vessel the more tight, and to hinder its own passing. The Vessel being fill'd with water we took the forementioned string, one of whose ends was fastned to I, the middle part of the loose Bottom; and, tying the other end K to the extremity of the Beam of a good pair of Scales, we put weights one after another into the opposite scale, till at length those weights lifted up the false Botom C D from the Rim G H; and, consequently, lifted up the Incumbent water; which presently after ran down between them. And having formerly, before we poured in any water, try'd what water would suffice to raise the Bottom C D, when there was nothing but its own proper weight that was to be surmounted; we found, by deducting that weight from the weight in the scale, and comparing the Residue with the weight of as much water, as the cavity of the broad, but very shallow Cylinder B E C H G D F would have alone (if there had been no water in the pipe A I) amounted to; we found, I say, by comparing these particulars, that the pressure upon C D was by so very great odds more, then could have been attributed to the weight of so little water, as the Instrument pipe and all contain'd, in case the water had been in an uniform Cylinder, and consequently a very shallow one, of a Basis as large as that of our Instrument," Boyle 1666, 137–140.

[230] "L'illustre M. Boyle, tout consommé qu'il est dans l'art de faire des expériences, a douté de celles de Stevin; car ayant prouvé la 10. proposition de Stevin dans son paradoxe 6. d'Hydrostatique, il y joint une remarque, par laquelle il rend douteux les corollaires de Stevin, & principalement le cinquieme, & il ajoûte que M. Wallis & quelques autres Savans en doutoient aussi. Il tâche pourtant d'éprouver s'il y avoit quelque chose de vrai dans l'expérience; & pour cet effet, il fit faire un vase semblable à celui de Stevin, mais le succès ne fut pas tel qu'il l'eût voulu. Il trouva bien à la verité que la pression étoit plus grande que celle qu'on pouvoit attribuer au poids de l'eau, mais non pas selon les proportions de Stevin, ainsi il laissa la chose pour être plus soigneusement examinée," Lufneu 1685, 383–384.

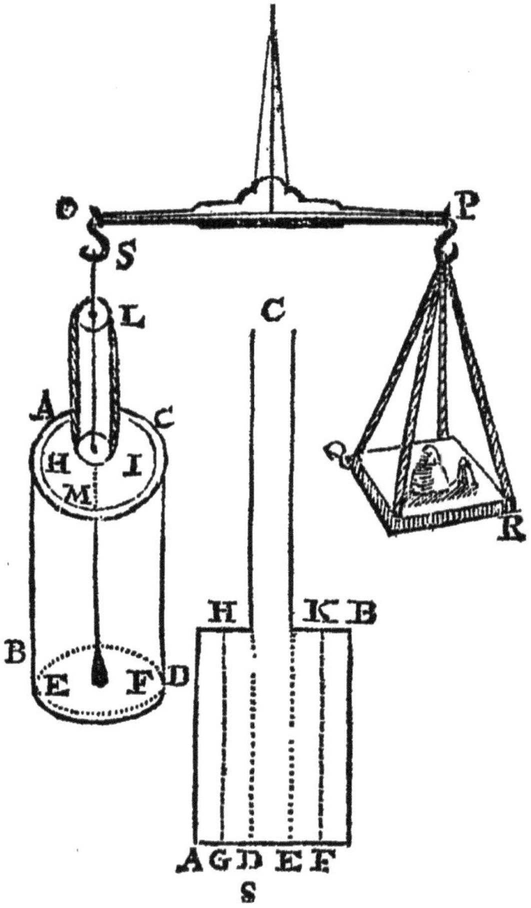

Fig. 5.45 Instrument aimed at demonstrating the hydrostatic paradox (De Volder's cylinder). (Source: Lufneu 1685, 385. Yale University Library, LSF- Beinecke, call number: WA 10332. Public domain)

ble differences are however to be considered, namely, (1) the bottom of the container (made of copper)[231] has the same diameter of the container itself, namely all the water contained in the container presses on the bottom; (2) not only is the bottom moveable, but so is the upper part of the lower cylinder, namely the cover HMI, which is fixed to the container by means of four screws.[232]

[231] See *infra*, n. 242. The material was chosen in order to avoid leakages of water.

[232] "Il prit le tuyau ou cylindre ABCD de la hauteur d'un pied representé dans la 1. Figure, au bas duquel BD il y avoit en dedans une bordure d'environ six lignes, qui servoit à empêcher que le fond EF ne tombât hors du cylindre, au dedans duquel il pouvoit monter & descendre. Ce fond avoit un crochet au milieu pour les usages que nous dirons ci-dessous, & son diametre étoit presque aussi grand que celui du cylindre, aufin qu'il ne restât point d'intervale par où l'eau pût s'échaper; tous ceux qui resterent furent exactement bouchez avec de la cire. On appliqua à l'ouverture AC le couvercle HIM avec quatre vis bien fortes, pour empêcher que l'effort de l'eau enfermée dans le cylindre, ne le soûlevât. Au milieu du couvercle étoit le tuyau ML haut de cinq pieds, & large

The first feature was probably aimed by De Volder at measuring exactly the pressure exerted by all the water contained in the container, rather than of a partial volume of it (as in Boyle's instrument). The second feature, as I am going to show, served De Volder in measuring the upwards pressure exerted by the water in the lower section of the container. The instrument seen by Lufneu was most probably the same one listed in the 1705 inventory of the theatre, mentioning a "copper cylinder with a plunger of ½ foot diameter inserted, with several copper pipes and a chain going from the mobile plunger through the pipe up to its end."[233] This instrument, as reconstructed by De Clercq, came to be widely known as the *cylindrum Volderi* and was depicted and its use described by Johannes Joosten van Musschenbroek in various letters sent to Marburg (see Figs. 5.46, Hessisches Staatsarchiv Marburg, file UniA 305a Nr 5947, 9, and 5.47, Hessisches Staatsarchiv Marburg, file UniA 305a Nr 5947, unnumbered page).[234]

---

d'environ 6. lignes, autant qu'il m'en peut souvenir. On passa une corde par ce tuyau, laquelle étoit attachée d'un côté au crochet du fond EF, & de l'autre à l'une des branches de la balance OP. J'ai voulu décrire un peu amplement cette machine, parce que ni Stevin, ni M. Boyle ne l'ont point fait," Lufneu 1685, 384–386.

[233] "Een copere cylinder met een ingesette boom van ½ voet diameter, met verscheyde kopere pypen en een kettingh gaande van de losse boom door de pyp naar boven tot de selfde eynde," Molhuysen 1913–1924, volume 4, 105*.

[234] The first figure is included in a catalogue probably sent as a letter dated 11 May 1695, where the cylinder aimed at demonstrating the hydrostatic paradox is extensively described (please note that a marginal note, in a different hand, refer to Valentini 1714, 43, commented in De Clercq 1991, 101–102): "[l]e cilindre de Monsr Le Profr de Volder. A. Le cilindre haut 1 pied. B. Le couvercle. CC. Quatre vis pour fermer le couvercle B. D. Un tuyaux de 6 <de> pied de longue. E. Cinq vis a chaque pied un. F. Un vis pour faire sortir l'air et remplir le cilindre. GGG. Trois pied. H. Un vis qui comprime l'instrument I pour tenir ferme le couvercle en bas. K. Le couvercle en bas en le quel est fonde un anneau, en le quel le crochet est attache en ce crochet et une chaine qui passe par le tuyaux D. L'usage. Applique le couvercle K qui est au fond avec un peu de cire dans le cilindre, attache le crochet a le couvercle, et mette l'anneau L a un bras de balance, et pese combien de poids qu'on faut pur attirer le couvercle quand il est soulement attaché avec cire. Mette derechef le couvercle avec de cire, et applique dans le cilindre, et attache l'instrument I qui on comprime avec le vis H, {est} sert quand on tourne le cilindre que le couvercle ne tombe car on faut bien fermer que l'eau ne puis sortir, <{...}> car on faut dehors aussi {graisser} le couvercle. Rempli le cilindre <{...} le tuyau> d'eau, <est> et applique le couvercle B avec le cire sur le cilindre et {ferre} bien avec le quatre vis CC, est <{...}> pese combien de poids qu'on faut pour attirer le couvercle K qui \est/ 6 pouces de diameter <est> et le cilindre est un pied de hauteur. Mais on faut de compter combien que le cire <{...}> a tenu la premiere fois. Mette pur le troisieme fois le couvercle K: et applique le tuyaux \D/ sur le couvercle B et rempli le tuyaux et le cilindre bien exact avec l'eau, vous verrez qu'on faut avoir autant de poids pour attirer le couvercle comme si le cilindre estoit egal{e}ment si haut que le tuyaux, le vis E de tuyaux {sont}, quand on veut remplir le tuyaux a trois pied on ouvre le vis a trois pied l'eau qui est dessus coule dessous, et on peut peser a chaque pied. Le chaine passe per le tuyaux quand on pese. C'est une chose incroyable si on ne voyait pas l'experience que c'est petit peu d'eau a autant de force, car il {n'y a} point de proportion entre le largeur de cilindre et de tuijau et le chaine passe encor. N.B. on ne fait pas oublier pour detacher l'instrument I quand on pese. Secondement on fait encor des autres experiences avec cette cilindre. Le couvercle en bas et en haut bien ferme et le tuyau applique mais le chaine extrait, rempli le cilind[re] et tuyaux plein d'eau et mette de poid sur le couvercle B et detache le vis CC que vous verrez que c'est petit peu d'eau qui est dans le tuyau a un incroyable force pour elever le couvercle

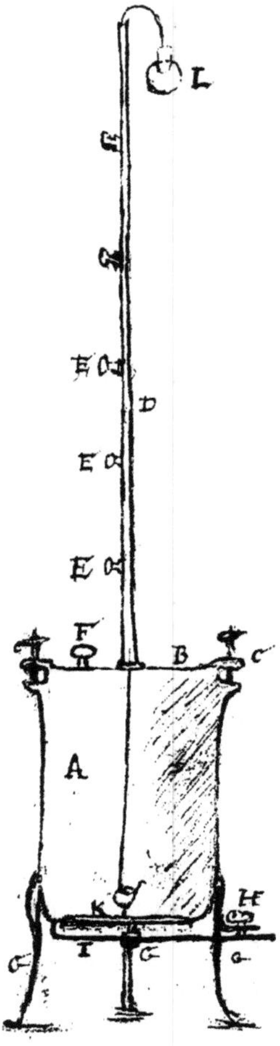

Fig. 5.46 Instrument aimed at demonstrating the hydrostatic paradox (De Volder's cylinder). (Source: courtesy Hessisches Staatsarchiv Marburg, file UniA 305a Nr 5947, 9)

Moreover, it was described in the *Armamentarium naturae* (1709) of Michael Bernhard Valentini (1657–1729) (see Fig. 5.48, Valentini 1709, plate 5, figure 2), to whose German edition of 1714 a sales catalogue of the Van Musschenbroek workshop, reporting the cylinder, was appended (it costed 50 guilders).[235]

avec tout le poid, mais si tost que seulement une gutte ou deux d'eau puis couler entre le couvercle et le bord de cilindre, la force est fait avec un pied de hauteur de tuyau il puisse elever le couvercle avec 40 <{. . .}> \livres/ si on prend donc 6 pied il fera un incroyable force," Hessisches Staatsarchiv Marburg, file UniA 305a Nr 5947, 9–12. The description is repeated in the following, unnumbered sheets, which all end with a letter dated 11 May 1695 – so that one of the two descriptions is a copy (apparently, they are in two different hands). The file UniA 305a Nr 5947 contains two more unpaginated sheets (put at the end of the file), in which the apparatus is described two further times, in German: the second figure is provided in the last sheet. For a commentary, see De Clercq 1991, 104 (see also 90 and 101–102), and De Clercq 1989, 32–33.

[235] Valentini 1709, 59–60; cf. Valentini 1714, 43, 104. The instrument sold by Van Musschenbroek and described by Valentini had slightly different dimensions from the one described by Lufneu (see

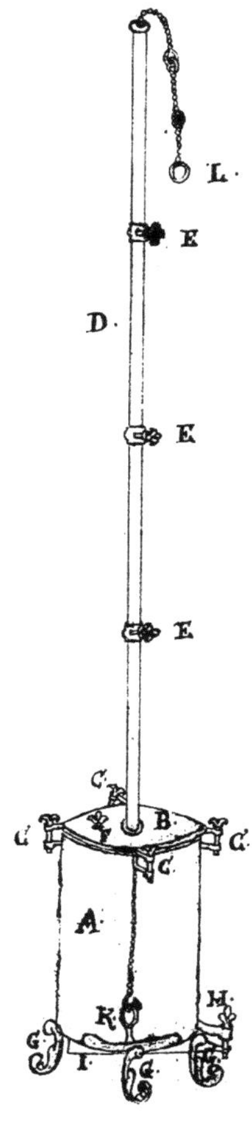

Fig. 5.47 Instrument aimed at demonstrating the hydrostatic paradox (De Volder's cylinder). (Source: courtesy Hessisches Staatsarchiv Marburg, file UniA 305a Nr 5947, unnumbered page)

According to Lufneu, De Volder repeated the experiment as carried out by Boyle, i.e. he progressively filled the container until the level AC (which was 1 foot high, and whose diameter is unspecified),[236] with 10 pounds of water, then the smaller tube above the container until L (5 feet high, and with a diameter of 6 lines),[237] with 2 or 3 pounds of water, and then he put on the other side of the balance a weight of up to 60 pounds: after that, the bottom EF started to be raised by this counterpoise. It is hard to say whether 'Stevin's proportions' were accurately measured in the

Table 2.2, *De Volder's natural-philosophical instruments*). Valentini's description was probably based on Van Musschenbroek's manuscripts, as argued in De Clercq 1991.

[236] Lufneu only claims that it had, at the bottom, a border of around 6 lines to prevent the bottom to fall downwards.

[237] Where 12 lines = 1 inch, and 12 inches = 1 foot. I consider the Parisian *ligne*; the same duodecimal proportions were used in England, while in the Netherlands there were various kinds of conversion, based on proportions of 8, 10 or 12 lines (*lijnen*) per inch (*duim*).

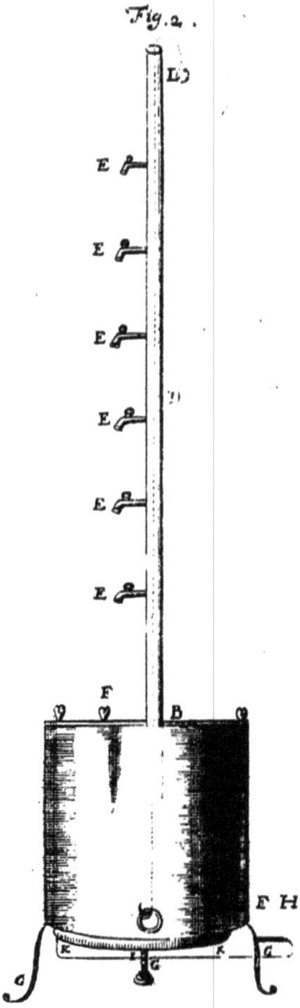

Fig. 5.48 Instrument aimed at demonstrating the hydrostatic paradox (De Volder's cylinder). (Source: Valentini 1709, plate 5, figure 2. Zentralbibliothek Zürich, signature: NP 1035,2, https://www.e-rara.ch/zuz/doi/10.3931/e-rara-38518. Public domain mark 1.0)

experiment, as we do not know the diameter of the bigger cylinder, and the measures given by Lufneu are quite vague. In any case, he specifies that De Volder, after having observed that the bottom is raised by a weight of no less than 60 pounds – thus completing the experiment of Stevin and Boyle, and having confirmed, according to Lufneu, the experiment of Stevin – continued the experiment by placing a weight of 20 or 30 pounds on the cover HMI. After having released the screws, the water bounced this weight up.[238] So that at the surface HMI, the water had an

[238] "En premier lieu M. de Volder remplit d'eau le cylindre jusques à l'ouverture AC. Il falut dix livres d'eau pour cela. Ensuite il remplit d'eau le tuyau LM jusques au bout L. Puis il chargea le plat de la balance QR de divers poids. Il falut y mettre 60 livres avant que le fond EF montât, ce qui montre évidemment que ce fond, qui ne soûtenoit que le poids d'environ 12 ou 13 livres d'eau (car le tuyau ML n'en contenoit qu'environ deux ou trois livres) étoit néanmoins aussi pressé que s'il eût soûtenu 60 livres, d'où paroît que l'expérience de Stevin est très-certaine. Pour mieux faire voir que l'eau contenuë dans le tuyau M L augmentoit extrêmement la pression, M. de Volder

upwards pressure of more than 20 or 30 pounds, this being due to the pressure exerted by a column of water of 5 feet. In fact, in order to provide the finding of an exact confirmation of Stevin's paradox, Lufneu would have need to specify that the ideal cylinder or volume of water with base EF and height LF is of (about) 60 pounds. He claims, in any case, that the experiment conformed to Stevin's theory.[239] Valentini also, actually, provided a description of the functioning of the machine, without however providing a natural-philosophical account of it.[240]

Before addressing the explanation of the experiment reported by Lufneu, it worth mentioning that De Volder did not perform only Stevin's samples 1 and 3. According to Lufneu's 1687 report – aimed against some criticism by Moïse Pujolas, who published an attack on Lufneu's first report in the *Nouvelles* of January 1687 – De Volder performed also what is nowadays known as the experiment of Pascal's barrel, or *crève tonneau*, in which a barrel is made to explode by the pressure that the water filling a tube above it exerts on its lower surfaces. This experiment is never mentioned in Pascal's works: rather, one finds it for the first time in nineteenth century popular textbooks on physics, such as *Les phénomènes de la physique* (1868) of

relâcha les quatre vis qui servoient à tenir ferme le couvercle, ayant mis premiérement au dessus de ce couvercle un poids de 20 ou de 30 livres. Je remarquai que les vis étant relâchées le couvercle s'éleva, & qu'il s'écoula de l'eau, preuve évidente qu'une assez petite quantité d'eau peut produire une fort grande pression," Lufneu 1685, 386–387.

[239] "Il est donc constant que tout le fond BD de la 1 Figure est autant pressé par l'eau, qui est dans le cylindre ABCD & dans le tuyau ML, que si le cylindre s'élevoit selon toute sa largeur jusques au sommet du tuyau, & qu'on y versât 60 livres d'eau, d'où on peut recueillir ce principe universel, que l'eau y toutes les autres liqueurs pressent le fond qui les soutient, non pas selon leur masse ou leur pesanteur specifique, mais selon la hauteur qu'elles ont par-dessus ce fond, ce qu'il faloit demontrer," Lufneu 1685, 388.

[240] "[...] iuxta [...] Mariotti experimenta minima quoque fluidorum portio tubo gracilior CD dolio vastissimo AB aut cylindro perpendiculariter erectis annexo immissa vel emissa, non tantum omnem in dolio AB contentum humorem [...] commovere, sed et magnam eius partem, cum pondere E in fig. 3 tab. V imposito, elevare, aut et operculum annexum vehementer attrahere valent, ut in Volderi cylindro tab. V fig. 2 delineato videre licet, cuius sequentes sunt enchireses: applica operculum KK (quod in fundo cylindri est) ipsi machinae mediante cera. Insere hamum catenae, per medium tubi D decurrentis, modo dicto operculo: annulum vero L in summitate catenulae applica bilanci et observa, quantum ponderis requiratur ad operculum attrahendum, quando cera tantum fundo agglutinatum est. Abhinc operculum de novo mediante cera applica cylindro, combinando simul fulcrum I verticillo HF comprimendum: hoc enim impedit, quo minus operculum, si machina invertatur, excidat. Optime enim obturandum est, ne aqua immittenda effluat, eumque in finem externe operculum quoque axungia quadam oblinendum erit. His ita praeparatis cylindrum aqua imple illique operculum superius B mediante cera applica omniaque bene claude firmaque quatuor verticillis CC. Pondera de novo, quantum ponderis requiratur ad operculum K (6 digitos in diametro habens) per cylindrum unius pedis attrahendum. Ubi tamen observetur, ut tantum de pondere detrahatur, quantum sola antea cera tenuerat. Applica porro tertia vice operculum K et desuper impone tubum aeneum D operculo B. Imple tubum aeque ac cylindrum aqua et videbis, quod tantum ponderis requiratur ad operculum attrahendum, ac si cylinder altitudine aequaret tubum D quod proportione quadam habita minuitur, si reseratis verticillis EEE plus minus aqua detrahatur. Admiratione sane dignum est, tantillum aquae adeo valide reniti, cum nulla sit proportio inter latitudinem cylindri et inter latitudinem tubi superimpositi, a catena transeunte iamdum ex parte occupatam," Valentini 1709, 59–60. I discuss De Volder's debt to Edme Mariotte below.

Amédée Guillemin.[241] According to Lufneu, De Volder performed it with a cubic container made of four wooden boards, into which he put a tube. The whole instrument was as high as 5 feet, and it was completely filled with water, and the boards started breaking, spilling the water. Lufneu refers to it as to the "first machine" used by De Volder,[242] so that its performance might have just followed the presentation of the paradox, given in experiment 28, where De Volder proposes the case of a "cylindrical vessel"[243] (see Fig. 5.30, De Volder 1676–1677, 140v) such as that described by Stevin in sample 1,[244] and then by Boyle.[245]

With this "first machine" it was demonstrated, according to Lufneu's report, that the spilling of water was due to the pressure exerted on the sides of the container. Using this case, Lufneu criticized Pujolas's objection that, before the bottom of the container used by Boyle (and De Volder) is raised, there is no more than the pressure equal to the weight of water being exerted on it.[246]

---

241 See Guillemin 1868, 79.

242 "Vidi primam machinam, quam conficiendam curaverat D. de Volder e quatuor asseribus sibi mutuo quam fieri potuit accuratissime coagmentatis, ne aqua efflueret. Fundo, ut in praecedenti experimento, rite disposito replebatur vas istud quadratum aqua. Sed nulla hactenus per commissuras effluebat aqua: operculo autem ei coaptato et tubulo aqua ad 5 pedum altitudinem repleto, tantam vim pressionis exercebat aqua in fundum et in vasis latera, ut undique per commissuras asserum efflueret. Quas dein pice accuratissime obturari, sed frustra, aqua nihilominus undiquaque perfluente, curavit; unde coactus fuit eam quam supra descripsimus e cupro fieri machinam. E quibus iterum evidenter constat aquae pressionem in fundum et latera (secundum leges hydrostatices) non ab elevatione fundi, sed ab aquae incumbentis pressione dependere," Lufneu 1687, 242–243. In his criticism, Pujolas maintained that the upwards pressure of water on the upper part of the container was due to its being raised: so that it was not exerted but in this case: see Pujolas 1687, 26–27. Cf. Lufneu 1687, 242: "[n]otavi quippe postquam cylindrus ut et tubulus ML repleti erant aqua Cl. Virum D. de Volder operculo imposuisse 20 aut 30 libras, dein relaxasse quatuor cochleas, quibus operculum cylindro annectebatur, atque eo ipso aquam elevato operculo effluxisse. Quam circumstantiam si vel solam respexisset, qui potuisset D. Pujolasius negare veritatem experimenti, et hypothesin Stevini tanquam falsam reiicere? Et quonam pacto illius causa in sola fundi elevatione quaesivisset?"

243 De Volder 1676–1677, 140v.

244 "Fundum AB fundo CD similis esto et aequalis, itemque altitudo EF altitudini GH; sed pars IE insistens suiectae aquae KLBA minor sit, quam pars ipsius GCD sibi respondens, pendatque aqua EAB 1 lb, GCD 10 lb, sitque GCD cylindrus, is igitur ipsius EAB erit decuplus, huius tamen in fundum AB impressum esse tantum asserimus, quantus sit totius GCD in fundum CD," Stevin 1605, 145. See Fig. 5.49, Stevin 1605, 145.

245 "The Learned Stevinus, having demonstrated the Proposition we lately mention'd out of him, subjoyns divers consectaries of which the truth hath been thought more questionable, then that of the Theorem it self. And therefore he thought fit to add a kind of Appendix to make good a Paradox, which seems to amount to this. That If, in the Cover of a large Cylindrical Box, exactly closed, there be perpendicularly erected a Cylindrical Pipe open at both ends, and reaching to the Cavity of the Box; this Instrument being fill'd with water, the circular Basis of it will susteine a pressure, equal to that of the breadth of the Basis and height of the Pipe," Boyle 1666, 135.

246 "[...] ut luce meridiana clarius constet fundum, dum est in CD seu ante sui elevationem, actu premi ea gravitate ac vi, quam conatus sum probare, alteram, quam primo non annotavi, circumstantiam paucis apponam. Vidi primam machinam [...]," Lufneu 1687, 242–243. See *supra*, n. 242.

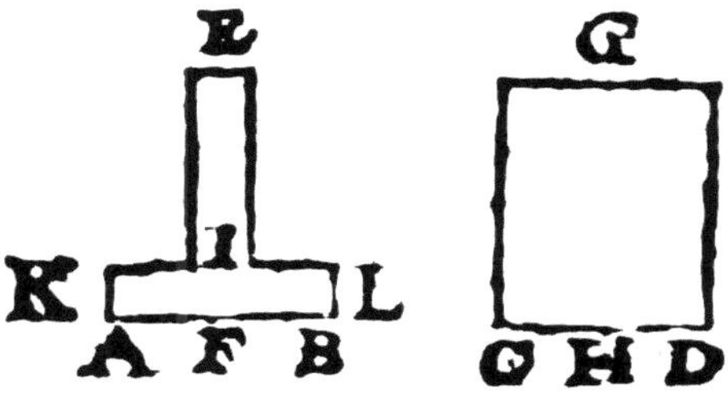

Fig. 5.49 Ideal representation of the hydrostatic paradox. (Source: Stevin 1605, 145. ETH-Bibliothek Zürich, signature: Rar 9069, https://www.e-rara.ch/doi/10.3931/e-rara-9068. Public domain mark 1.0)

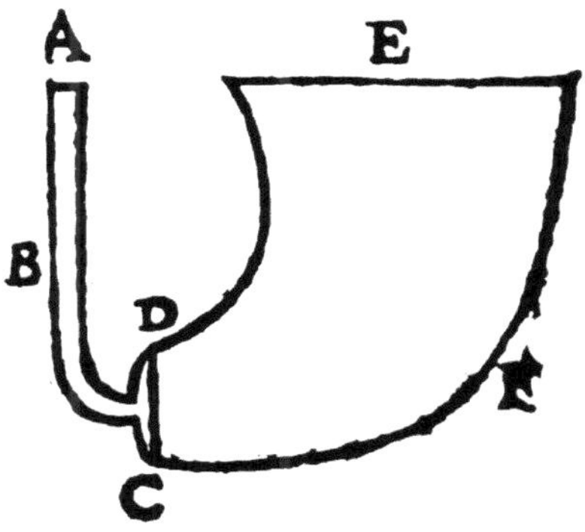

Fig. 5.50 Representation of the horizontal pressure of water. (Source: Stevin 1605, 146c. ETH-Bibliothek Zürich, signature: Rar 9069, https://www.e-rara.ch/doi/10.3931/e-rara-9068. Public domain mark 1.0)

It is not clear whether the reference to the lateral pressure exerted on the sides of the container was part of De Volder's explanation of the phenomenon, or it was just a comment by Lufneu. But it might be that De Volder started to introduce students to the idea of lateral pressure only when he was dealing with the paradox: in fact, among Stevin's five samples one can find, such as sample 2, the measurement of pressure on a lateral surface, depending only on the height of the water pressing on the surface itself (see Fig. 5.50, Stevin 1605, 146c).

De Volder might have been following Stevin's samples. To these, and as a preamble, he could have used the experiment we know as 'Pascal's barrel' which had been widely known, at a literary level, at least since Plinius's description of the *ruina montium* mining technique. With this technique, cavities in mountains were filled with water, until their lower chambers literally exploded, as the effect of the pressure of water (independent of its weight). Moreover, the fact that De Volder actually showed the pressure exerted by water in a lateral or oblique direction is indirectly confirmed by a variety of evidence – as seen above. Namely the presence, in his inventory of the Leiden experimental theatre, of two instruments aimed at showing and measuring the lateral and oblique pressure of water and air, as well as Van Nierop's report on De Volder's experiments with lateral pressure.[247] Thus far, we might interpret Morley's reports on De Volder's lectures on perpendicular pressure as concerning only the principle that fluids press with a force proportional to

[247] See *supra*, n. 195.

their height. This is a way to make the evidence for De Volder's teaching consistent. However, as seen above, it might be that De Volder's dealing with pressure was complicated by his reliance on Cartesian ideas, which were, as put by Chalmers, "ill-equipped to deal with the isotropy of hydrostatic forces that had been identified by Stevin."[248]

5.4.2.4 De Volder's Archimedean Approach to the Paradox

Leaving aside the practical performance of the experiments, what is more interesting – at least according to Lufneu – is the explanation of this phenomenon. It is unclear whether Lufneu reports the very explanation given by De Volder, yet the fact that the explanation is decidedly Archimedean strongly suggests this hypothesis. The explanation is based on the principle that "no liquid spreads itself on a plane bottom without pressing it equally everywhere." So that, by considering the abstract representation of a vessel given in Fig. 5.45, Lufneu 1685, 385, the central column of water CDE presses the bottom DE with a pressure equal to its weight, and, given the aforementioned principle, also the other parts of the bottom AGDEF will be necessarily subjected to the same pressure. Otherwise, as Lufneu put it, there will be a difference in pressure and the central column will displace the water of the other columns, which is against the fact that the liquid is at rest in the container.[249] That De Volder himself could have been providing this explanation, and that he moreover aimed at experimentally confirming it, is evident from the second part of the experiment. In it, he showed the upwards pressure exerted by water at level HKB, just by releasing the screws fixing it to the cylindrical vessel. This second part conforms to the case of the hydraulic press widely considered by Pascal, and by Stevin in sample 4 (see Fig. 5.51, Stevin 1605, 147b) aimed at confirming the case simplified by Fig. 5.34, Stevin 1605, 120b, discussed above, as corollary 3 of proposition 10.[250]

By the second part of the experiment, De Volder not only aimed at showing the pressure on the bottom, but also the upwards pressure which, in the Archimedean model, is exerted by the other columns of the fluid.

[248] Chalmers 2017, 59.

[249] "Je suppose pour fondement ce principe d'Hydrostatique, qu'aucune liqueur ne se répand sur un fond plat, sans le presser également par tout. Prenons garde dans la 2. Figure à la colomne CDE égale en hauteur au cylindre ABCD & au tuyau LM de la premiere. L'eau de cette colomne presse le fond DE par toute sa pesanteur, mais afin qu' elle puisse demeurer en équilibre, il est nécessaire que la colomne d'à côté HD presse le fond GD de la même force que CDE presse DE, autrement l'eau qui est en CDE descendroit, & celle qui est en HD monteroit, ce qui est contre l'expérience, donc, &c. Par la même raison on prouve que l'eau de la colomne KEF presse le fond EF, de la même force que CDE presse DE, & la même chose faut-il dire de toutes les colomnes laterales," Lufneu 1685, 387–388. At Leiden, Senguerd did not deal with the paradox, even if he maintained that the pressure of a fluid depends on its height, mostly by considering the case of Torricellian tubes disposed obliquely (in which the vertical component of the mercury reaches the same level): see his *Philosophia naturalis* (ed. 1685), chapter 8, and *Rationis atque experientiae connubium*, chapter 16.

[250] See Sect. 5.4.2.1, *Early discussions of the paradox: Stevin, Galileo and Schott.*

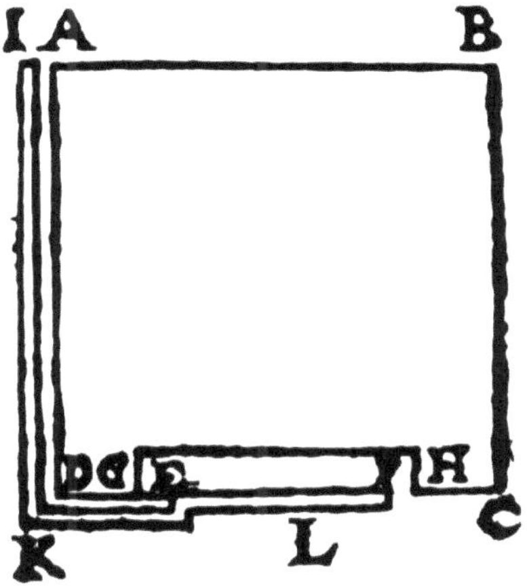

Fig. 5.51 Hydraulic lever. (Source: Stevin 1605, 147b. ETH-Bibliothek Zürich, signature: Rar 9069, https://www.e-rara.ch/doi/10.3931/e-rara-9068. Public domain mark 1.0)

In the case of the paradox, actually, this is the very idea which is put into question, as the bare weight of the upper column of fluid is not sufficient to exert such a pressure. Yet, De Volder seems to have overcome this issue, just by reducing the experience he presented to that of the forced floatation of a solid in little water (namely, as in Stevin's sample 1, which he performed too), in which the same conditions of 'free' floatation of a body in open water apply. Namely, one can consider the plate HMI, either as fixed to the cylinder by the screws, or as 'freely floating' on the water with weights put on it, as exerting the same pressure of the volume of water HILM. This, of course, goes along with a process of abstraction from the real components of the apparatus: wood, water, copper, and so on. In this, he leaned more on Stevin than on Boyle. In fact, Boyle did not come to accept Stevin's paradox because he also labelled his theoretical premises as unclear, and instead, against Stevin's geometrical approach, he proposed a "Physicall account" in which, in principle, one should be able to reduce effects – such as the exertion of pressure – to causes one can grasp by experience, such as the weight of a column of water. By putting weights on the plate HMI, De Volder partially conformed to this approach, namely he showed that this plate behaves as if it were a floating body. It behaves in the same way, however, when it is also 'forcibly' put in water, namely when it is fixed to the rest of the container by screws. In the latter case, it is enough to consider, as Lufneu reports, the uniformity of pressure on each ideal surface of the fluid, regardless of the cause of such pressure.

Actually, Boyle himself provided a sort of 'geometrization' of fluids and floating solids. He did so in answering a criticism moved by Henry More to Boyle's Archimedean explanation of flotation, in his *Enchiridion metaphysicum* (1671). In chapter 13, More considers the case of a wooden plate, of a breadth very close to that of the container (see Fig. 5.52, More 1671, 155), in which it is forcedly submerged to the bottom of the container, and then released. More objects that in this case, the 'lateral columns' of water in-between the plate and the container, being very thin, cannot exert an upwards thrust on the plate, which has to overcome the pressure of the huge amount of water above it. This being, for More, one of the

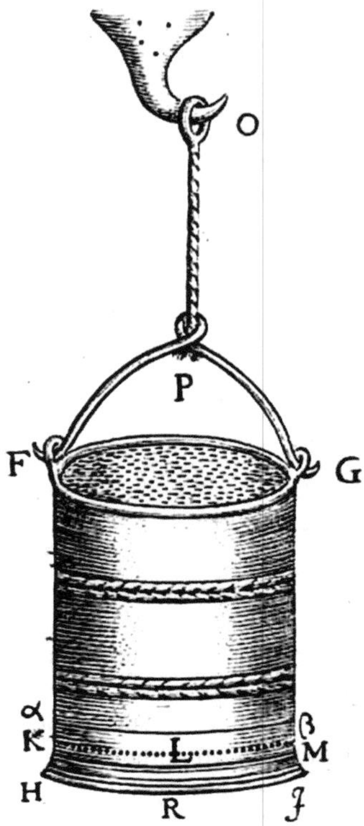

Fig. 5.52 Container used to criticize Archimedes's model of floatation. (Source: More 1671, 155. Yale University Library, LSF- Beinecke, call number: K8 M80 Hv671. Public domain)

numerous cases in which the application of mechanical principles to natural philosophy was not sufficient to explain phenomena, for which he reverted, instead, to an 'hylarchic' or 'plastic' principle.[251]

In his subsequent *Hydrostatical Discourse* (opening his 1672 *Tracts Containing New Experiments*), Boyle replied by re-instating Archimedes's first postulate, which for him works regardless of the composition of the (ideal) volumes of what is contained in the vessel: either fluids or solids, provided that the pressure exerted on

---

[251] "Nam cum diameter laminae ligneae KM partes 61 habeat quales diameter vasis HI habet 62, manifestum est quod superficies fundi vasis ad superficiem laminae se habet ut 3844 ad 3721; quorum differentia est 123. Itaque rotundum intervallum inter latera vasis et marginem laminae ligneae habet se ad aream laminae ut 123 ad 3721, hoc est, area laminae ligneae excedit aream dicti intervalli plus quam triginta vicibus. Ac proinde aqua incumbentem dicto intervallo inter margine laminae et later vasis plus quam triginta vicibus, pondusque, sive pressione huius alterius pondus pressionemque vincit plus quam triginta vicibus. Adeo ut impossibile sit ut aqua incumbens praedicto intervallo ita premat aquam ipsi subiectam, ut huius vi sublevetur lamina quam vis tricies maior deprimit. Quod aeque absonum atque absurdum phaenomenon esset, atque si pondus lanci iniectum alteri ponderi ipso tricies maiori in oppositam lancem imposito praeponderare," More 1671, 155–156. More challenged Boyle also to the case of in which the quantity of water sustaining a body is more or less than the water displaced by the floating body (see More 1671, 169–170): in this case, however, Boyle did not answer. On More's hylarchic principle, see in particular Hall 1990; Hutton 1990.

every imaginary surface is equal, both from above and below (such as, for instance, in the case of the floatation of a body in a little water). What matters, according to Boyle, are therefore not the 'absolute' weights of the fluid and the solid counterbalancing each other, but their specific weight, that grants the uniformity in pressure.[252] Thus far, this approach was akin to the 'geometrical' one of Stevin – followed by De Volder as well. Notably, Lufneu closes his 1685 communication by claiming that the experience he reported can be used against the objection that More moved, in chapter 13 of his *Enchiridion*, to Boyle's mechanical explanation of hydrostatic phenomena, as well as against all More's criticisms of Descartes and Boyle.[253] Thus far, for Lufneu De Volder pursued and furthered Boyle's approach to hydrostatics.

More light on De Volder's dealing with the paradox, eventually, can come from what appears to be a further source for him, namely, Edme Mariotte, who performed an experiment similar to that one described by Lufneu. The experiment of Mariotte took place at the Collège de Bourgogne, and was then reported (without date) in the third volume of the *Philosophia vetus et nova* of Jean-Baptiste Du Hamel (1678), in the review of it which appeared in the issue of 23 May 1678 of the *Journal des Sçavans* (probably De Volder's earliest source for Mariotte's experiment),[254] and eventually in Mariotte's posthumous *Traité du mouvement des eaux et des autres*

---

[252] "I thus argue: 'Tis manifested by Hydrostaticians after Archimedes, that in water, those parts that are most press'd, will thrust out of place those that are less press'd: which both agrees with the common apprehensions of men, and might, if it were needful, be confirm'd by Experiments. 'Tis also evident, that that part of the above-mention'd imaginary Plane, that is cover'd by the woodden Plate, must be pressed by a less weight than the other part of the same Plane; because the wood being bulk for bulk lighter than water, the aggregate of the wood and water incumbent on the cover'd part of the same Plane must be lighter in specie, than the water alone that is incumbent on the uncover'd part of the same Plane; and consequently this uncover'd part being more press'd than the other part of the Plane, the heavier must displace the lighter, which it cannot do but by thrusting up the board, as it does, when the external force that kept it down is removed. And, to add this upon the by, this greater pressure against the bottom than against the top of bodies immers'd in water specifically heavier than they, is a true reason of their emersion, as I have elsewhere shewn. So that there happens no more in this case than what usually happens in the ascension of bodies in liquors specifically heavier than themselves, on the account of the newly mention'd difference of Pressure. [...] 'Tis true, that according to the Doctors supputation, if the solid Cylinder, consisting of the woodden Plate, and all the water directly incumbent on it, were put into an ordinary ballance, it would there many times out-weigh the hollow Cylinder of water alone that leans upon the uncover'd part of the imaginary Plane. And that is it that seems to have deceiv'd the Learned Doctor. But there are divers Hydrostatical Cases, wherein the Phænomenon depends not so much upon the absolute weight of the compared Bodies, as upon their respective and their specifick Gravity," Boyle 1999–2000, volume 7, 111.

[253] "On pourroit déduire plusieurs choses de cette expérience, si la briéveté que je me suis proposée me le permettoit. J'ajoûterai seulement que l'on ruïne par là de fond en comble, l'objection qu'un célebre Théologien, nommé Henri Morus, a proposée contre les loix Méchaniques de l'Hydrostatique. Je ne m'arrêterai pas à le montrer, car tous ceux qui examineront la chose attentivement, pourront d'eux mêmes satisfaire à cette objection, & à toutes les autres qui sont répandues contre M. Descartes & M. Boyle, dans le Livre de ce Théologien," Lufneu 1685, 388–389.

[254] De Volder owned the 1681 edition of Du Hamel's treatise, so that his earliest source for Mariotte's experiment seems to be the *Journal* (which he owned in a complete series): see *Bibliotheca Volderina*, 10 and 93.

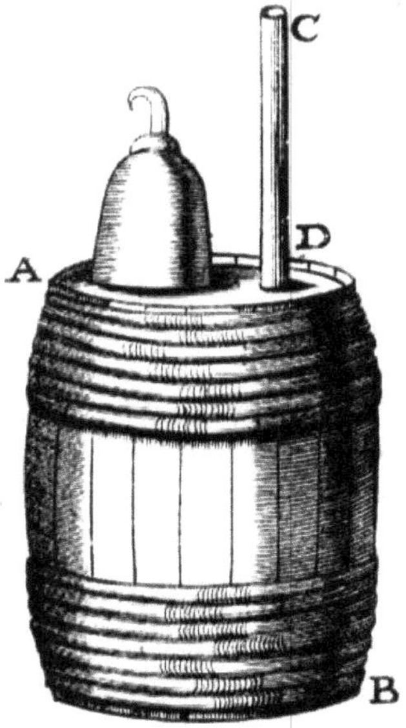

Fig. 5.53 Barrel aimed at demonstrating the hydrostatic paradox. (Source: Du Hamel 1682, volume 2, 208. Zentralbibliothek Zürich, signature: NP 882: 1–2, https://www.e-rara.ch/doi/10.3931/e-rara-25724. Public domain mark 1.0)

corps fluides (1686). Mariotte used a barrel on which he put a tube (see Figs. 5.53, Du Hamel 1682, volume 2, 208, and 5.54, Mariotte 1686, 107), and whose cover was moveable (being the very same instrument also represented in Valentini's *Armamentarium naturae*, alongside De Volder's cylinder).[255] On it, as De Volder did at Leiden, he put weights counterbalancing the upwards pressure of water.[256] Most

[255] See Valentini 1709, plate 5, figure 3.

[256] "On voit un effet surprenant de l'équilibre dans l'expérience suivante: Aïez un tonneau de bois large de deux ou trois pieds ABCD, plein d'eau, enfoncé par les deux bouts: faites une ouverture au fond d'en-haut comme en E, pour y mettre un tuyau d'un pouce de largeur, si bien joint avec de la poix & de la filasse ou avec quelqu'autre matière, que l'air n'y puisse entrer, & que ce tuyau étroit, sçavoir EF, ait 12 ou 15 pieds de hauteur: emplissez d'eau le tonneau par quelques trous qu'on fera au fond supérieur, & posez sur le fond sept ou huitcent livres de poids, qui le feront courber en concavité, comme AMD. Si l'on met une marque blanche au dehors du tuyau, comme au point H, & à côté un peu plus haut une régle IL, plantée dans le mur voisin, & affermie de manière qu'elle demeure immobile; en versant de l'eau en suite peu à peu dans le tuyau étroit EF, vous verrez que quand il sera plein, le fond AMD sera élevé avec les poids de 800 livres dont il est chargé, non seulement à son premier état AED, mais même qu'il aura pris une courbure convexe, & que son élévation dans le milieu sera autant élevée par-dessus le point E, que le point M étoit au-dessous auparavant; ce que l'on connoîtra parce qu'on verra élever la marque blanche H, & passer peu à peu plus haut que la régle IL, dont on pourra mesurer la différence. Que si le tuyau est encore plus haut, l'élévation des poids sera encore plus grande: d'où l'on juge que le peu d'eau qui est dans le tuyau, a autant de force pour élever ce grand poids & courber le fond du tonneau en convexité, que si ce tuyau étoit de même largeur que le tonneau," Mariotte 1686, 106–108. The report given by the anonymous reviewer is very synthetic, and does not include any explanation: "[e]n traitant de l'Equilibre des liqueurs d'où il prend occasion d'expliquer les Principes d'Hydrostatique, il rapporte une autre Experience de M. Mariotte, quia esté faite au College de Bourgogne en présence de grand nombre de personnes d'un tonneau AB dont le fonds chargé d'un poids de six cens livres fut élevé visiblement à la hauteur d'un pouce après qu'on eut fait couler dans ce tonneau 4 ou 5 livres d' eau seulement par le moyen d'un tuyau CD assez étroit & d'une hauteur de 10. ou 12. pieds, comme on peut voir dans la 2. figure," Anonymous 1678, 106.

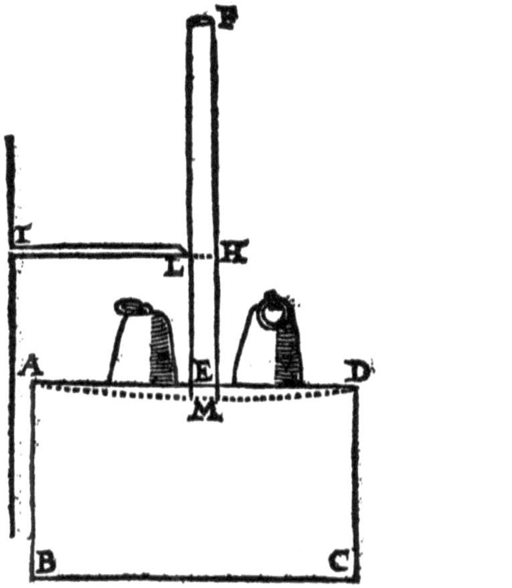

Fig. 5.54 Idealized inverted T-shaped container aimed at demonstrating the hydrostatic paradox. (Source: Mariotte 1686, 107. ETH-Bibliothek Zürich, signature: Rar 1405, https://www.e-rara.ch/doi/10.3931/e-rara-13780. Public domain mark 1.0)

probably, the idea to put weights on a moveable cover was inspired in both Mariotte and De Volder by their Archimedean solution to the problem of floatation, according to which there is always an upwards pressure exerted on all the ideal surfaces of a fluid.

It is by considering the differences in pressure on a given surface, indeed, that Mariotte, like De Volder, explained the floatation of bodies (see Fig. 5.55, Mariotte 1686, 116): for Mariotte, for instance, the body AFGH, specifically lighter than water, floats if the column AFIL (i.e. the column composed by the water GHIL and the solid AFGH) weights as much as the column BKEI.[257]

As reported both in Du Hamel's and Mariotte's treatises, their common strategy to solve the problem of the equilibrium between different amounts of water (such as the case of the asymmetric inverted syphon), is that followed by Galileo and Pascal, namely, by considering the differences in the ascent and descent of unequal volumes of water. The case of the barrel is an instance of such equilibrium.[258] De Volder, actu-

[257] "Soit dans la figure suivante BCDE de l'eau dont la surface superieure soit BC, contenuë dans quelque vaisseau, & soit AFGH un corps cubique plus leger spécifiquement que l'eau, & plus pesant que l'air; je dis qu'il ne demeurera pas sur la superficie de l'eau: car la colomne quarrée d'eau KRLI seroit plus pressée qu'une colomne égale BEIK, puisque le poids du corps AH y seroit de plus. Donc le poids descendra, & entrera dans l'eau, mais il ne s'y cachera pas entièrement, parce qu'alors la colomne KRIL composée de ce corps & d'eau, seroit plus legére qu'une égale colomne d'eau BEIK. Soit donc son enfoncement jusques en KR, & que l'eau qui l'environne se soit élevée jusques en BC, qui sera plus haute qu'elle n'étoit auparavant à cause que la portion KGHR du corps occupe la place d'une partie qui est obligée de s'élever: je dis que l'eau contenue en KGHR, dont le corps occupe la place, sera d'un poids égal au poids de tout le corps, c'est-à-dire, que si une quantité d'eau égale en volume à KGHR pése autant dans l'air que le corps entier AFGH, il demeurera dans cette situation la portion KRGH de ce corps sera au total, comme la pesanteur spécifique de tout ce corps sera à celle de l'eau," Mariotte 1686, 116–117.

[258] See Du Hamel 1678, volume 3, 413–416; Mariotte 1686, 102–103.

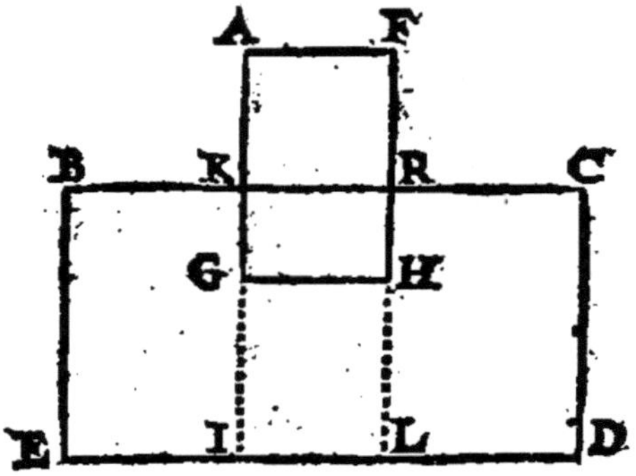

Fig. 5.55 Representation of the condition of floating of bodies according to Mariotte. (Source: Mariotte 1686, 116. ETH-Bibliothek Zürich, signature: Rar 1405, https://www.e-rara.ch/doi/10.3931/e-rara-13780. Public domain mark 1.0)

ally, never used the sample of the balance with unequal arms: thus far, his approach was more akin to that of Stevin and Boyle – who only considered the uniformity of pressure on given surfaces. Yet, his solution was perfectly consistent with the explanation of Galileo, Du Hamel and Mariotte, as he relied throughout on the comparison with balance in his hydrostatical explanations. In fact, as the experiment reported by Lufneu took place in 1677–1678, and it was probably performed in the weeks following the last experiment reported by Morley (25 March 1677), De Volder most probably devised and performed his experiment before hearing about Mariotte's.

Eventually, both the experiments were presented by Valentini in his *Armamentarium naturae*, who provided depictions of both De Volder's cylinder and of the barrel,[259] and they were discussed by one of the most famous students of De Volder (who enrolled at Leiden University in 1675, after having already acted as *respondens* in a disputation of De Volder in 1673),[260] and later his correspondent, namely Bernard Nieuwentijt, who in his *Het regt gebruik der werelt beschouwingen, ter overtuiginge van ongodisten en ongelovigen aangetoont* (1715) (or, in its 1718 English edition, *The Religious Philosopher, or the Right Use of Contemplating the Works of the Creator*) compares the two cases (see Fig. 5.56, Nieuwentijt 1747, plate 20), mentioning Mariotte but without disclosing his source for the first one, who most probably was his former teacher De Volder. His main aim being that of remarking its 'wonderful' character, noting that as late as in the works of Mariotte, Pierre Varignon and William Whiston no satisfactory explanation could be found for it.[261] Moreover, in the reprint of Lufneu's 1685 report in volume 28 of the *Nouveau*

[259] Cf. Valentini's commentary, *supra*, n. 240.

[260] See *Album studiosorum Academiae Lugduno-Batavae MDLXXV–MDCCCLXXV*, column 595. He acted as *respondens* in De Volder's *Disputatio philosophica de unitate Dei*, 25 March 1673. On him, see Vermij 1991. See also Jorink 1999; Jorink and Maas 2012.

[261] Cf. the French translation, *L'existence de Dieu démontrée par les merveilles de la nature* (1725, the treatise had many translations and editions): "M. Varignon, que tout le monde reconnoît pour un si grand Méchaniste, l'appelle un fameux Paradoxe; tous ne différent que quant à la maniere d'expliquer comment cela se fait dans les fluides. M. Mariotte l'appelle un effet surprenant de l'équilibre. M. Whiston, Prelect. Phys. pag. 247, dit, en parlant de cette Loi, qu'elle est parfaitement connuë dans l' Hydrostatique; mais jusqu'à présent à peine en a-t-on trouvé aucune preuve naturelle ou Mathématique," Nieuwentijt 1725, 594; see also 591–595. Cf. Varignon 1692; Whiston 1710, 247;

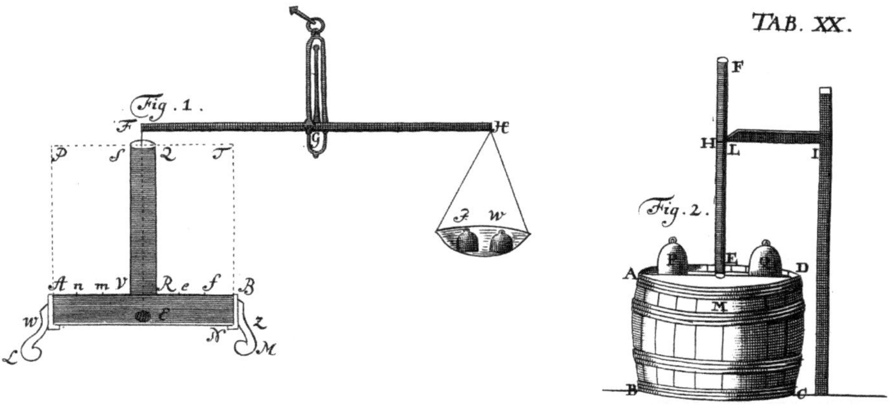

Fig. 5.56 Instruments aimed at demonstrating the hydrostatic paradox. (Source: Nieuwentijt 1747, plate 20. The University of Oklahoma Libraries, History of Science Collection, signature: 19732-1001. Public domain)

choix de pieces tirées des anciens Mercures, et des autres journaux, published by Jean-François Marmontel (1759), a short note accompanies the text: according to it, De Volder's experiment could be easily explained by reducing it to the case of the equilibrium of fluids in an inverted, asymmetrical syphon, and that was performed by De Volder in the same period in which Mariotte performed his experiment in Paris.[262] Thus far, De Volder certainly had a crucial role, with his experimental demonstrations, in disseminating the idea of the validity of the paradox in the late seventeenth century, and in providing a feasible instrument to prove it. What, however, is lacking in his extant lectures is a treatment of the lateral pressure of water, which is mentioned *en passant* only in Lufneu's second report, and which can be proved by the presence, at Leiden, of two apparatuses to ascertain the oblique pressure of water and air.[263] As I am going to show in Sect. 5.5, *De Volder's Cartesian hydro-*

[262] "Quoi qu'il en soit, cette proposition est vraie, & elle est aujourd'hui admise si unanimement par tous les Méchaniciens, qu'il fa lieu d'etre surpris que quelques grands hommes en aient douté. On peut dire qu'en cela ils consultoient davantage leur imagination que les principes de l'Hydrostatique; car il est facile de démontrer la vérité dont on parle, à priori, ou du moins en s'appuyant seulement de la plus simple des expériences hydrostatiques; sçavoir l'équilibre qui se fait dans les branches d'un siphon d'inégal diametre, lorsque le fluide est dans l'une & l'autre à la même hauteur. S'il y avoit cependant encore quelques lecteurs peu convaincus de cette vérité, nous lui alléguerions une curieuse expérience que M. Mariotte faisoit à Paris vers le même-tems que M. de Volder faisoit la sienne en Hollande. Le Physicien François prie un tonneau de deux à trois pieds de diametre qu'il dressa sur un de ses fonds. Il perça celui de dessus d'un trou auquel il ajusta solidement un tuyau d'un pouce de diametre, & de plusieurs pieds de hauteur. Il chargea aussi ce fond d'environ sept à huit cens livres, ce qui le fit courber considérablement vers le dedans. Alors il mit de l'eau dans le tonneau, & quand il fut rempli, il en versa dans le tuyau; à mesure qu'il le faisoit, on la vit non-seulement redresser & soulever le fond avec les poids dont il étoit chargé, mais il fallut encore y en ajouter de nouveaux pour l'empecher de le courber en sens contraire," Lufneu 1759, 198–199.

[263] I.e. a "glass with an opening in the middle to demonstrate the lateral pressure of the air," and "several copper boxes of one and a half foot, with lids, for use in the wooden container, to measure

statics: a mixed approach, the lack of a treatment of oblique pressure in De Volder's extant texts might have been the result of his (partial) acceptance of Descartes's model of floatation and theory of gravity. As briefly seen above, indeed, Descartes's model entailed the idea that pressure is exerted only by perpendicular volumes of fluids, and that it is not internally exerted within water itself.

5.5 De Volder's Cartesian Hydrostatics: A Mixed Approach

5.5.1 An Introduction to Descartes's Hydrostatics

As seen above, in the Hamburg 273 series of his *dictata*, De Volder provided an explanation of the tenet that fluids press only according to their height, based on Descartes's account of the pressure of fluids. By considering the case of the funnel, De Volder claimed that only the volume perpendicularly above its bottom presses on the bottom itself. Indeed, if the bottom is released, in the first instant of descent, only this volume would follow the bottom itself. Accordingly, for De Volder the 'instantaneous' speed of the descent of the water is equal in both cases. This explanation is Cartesian, because it leans upon Descartes's idea that the pressure exerted by fluids can be rendered in terms of an instantaneous tendency to motion. According to Descartes, a movement does not take place in an instant, while a tendency to motion is exerted in an instant. Yet, it is by the idea of instantaneous tendencies that he judges the pressure of a fluid on its surroundings. The chief application of this idea can be found in Descartes's theory of light, in which the instantaneous centrifugal tendency of rotating particles of luminous bodies is transmitted instantaneously from the stars to our eyes – and exerted as an actual pressure. In more imaginative terms, we can work out Descartes's explanation of light by the example of the stick he used in his *Dioptrique*: light behaves like a stick, in which the extremities move at the same time.[264]

---

the horizontal pressure of water, and at any angle." See Table 2.2, *De Volder's natural-philosophical instruments*.

[264] "[...] mais considérez-la en ceux qui, étant nés aveugles, s'en sont servis toute leur vie, et vous l'y trouverez si parfaite et si exacte qu'on pourrait quasi dire qu'ils voient des mains, ou que leur bâton est l'organe de quelque sixième sens qui leur a été donné au défaut de la vue. Et, pour tirer une comparaison de ceci, je désire que vous pensiez que la lumière n'est autre chose, dans les corps qu'on nomme lumineux, qu'un certain mouvement ou une action fort prompte et fort vive qui passe vers nos yeux par l'entremise de l'air et des autres corps transparents en même façon que le mouvement ou la résistance des corps que rencontre cet aveugle passe vers sa main par l'entremise de son bâton. Ce qui vous empêchera d'abord de trouver étrange que cette lumière puisse étendre ses rayons en un instant depuis le soleil jusqu'à nous; car vous savez que l'action dont on meut l'un des bouts d'un bâton doit ainsi passer en un instant jusqu'à l'autre, et qu'elle y devrait passer en même sorte, encore qu'il y aurait plus de distance qu'il n'y en a depuis la terre jusqu'aux cieux," AT VI, 84. On Descartes's treatment of the idea of tendency to motion, see Garber 1992, chapter 7; Schuster 2013, chapter 3.

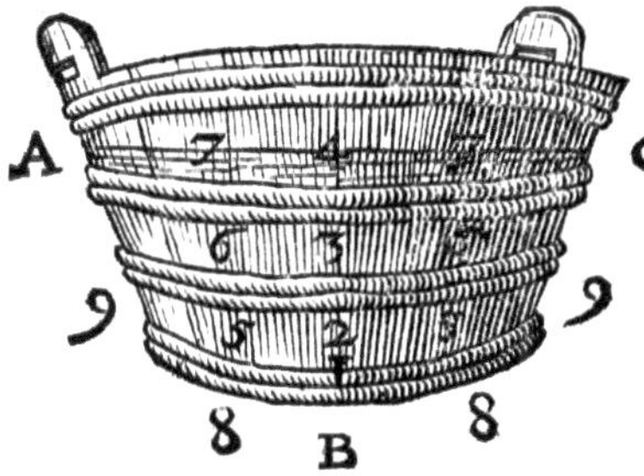

Fig. 5.57 Container aimed at demonstrating the conditions of hydrostatic equilibrium according to Descartes. (Source: Descartes 1644a, 203. Library of Congress, call number: LC B1860 1644, http://hdl.loc.gov/loc.rbc/Rosenwald.1431.1. Public domain)

In the case of the pressure exerted on the bottom of a container, the underlying dynamics is the same. As Descartes put it in his *Principia*, IV.26 – where he provides his account of the traditional tenet that "bodies do not gravitate [...] in their natural places,"[265] – (see Fig. 5.57, Descartes 1644a, 203), the column of fluid composed by drops 1, 2, 3, 4 exerts pressure on the bottom because they unite their forces and, if the bottom is released, they will follow it. In other words, for Descartes this volume of fluid behaves as if it were a solid i.e. continuous body, such as in the case of light. This is part of his broader treatment of hydrostatic equilibrium (which, in turn, informs his theory of gravity and his cosmology, as I show in Sect. 6.1, *Descartes's theory of planetary stability and weight*), according to which the downwards pressure exerted by a fluid on a fluid always leads to a displacement of the pressed part – such as in Galileo's *De motu antiquiora*, and as remarked by De Volder in the Hamburg 273 series of his *dictata*, where he notes, on IV.26, "[what] is pressed i.e. is expelled."[266] Or, in Descartes's own words,

> in vessel ABC, drop of water 1 is not pressed upon by the others 2, 3, 4, situated above it, because if these were carried downward, other drops 5, 6, 7, or similar, would have to ascend into their place; and since these drops are equally heavy, they {hold the former in equilibrium and} prevent their descent. However, these drops 1, 2, 3, 4, having united their forces, press upon part B of the bottom; because, if they cause it to descend, they also will descend, and parts of air 8, 9, which are lighter than they are, will ascend.[267]

Thus far, what keeps the drops in equilibrium is their equal weight (on whose nature I will be specific in Sect. 6.1, *Descartes's theory of planetary stability and weight*), which results in a lack of internal pressure once the equilibrium is re-established, after the displacement of drop 1.

The rationale of Descartes's explanation of pressure can be found, actually, in his famous 1619 manuscript on hydrostatics, in which he proposed to Isaac Beeckman an explanation of the Stevin's hydrostatic paradox (a topic on which he did not

[265] Descartes 1982, 193.

[266] "Premitur. I.e. expellitur," Hamburg 273, 235.

[267] Descartes 1982, 193.

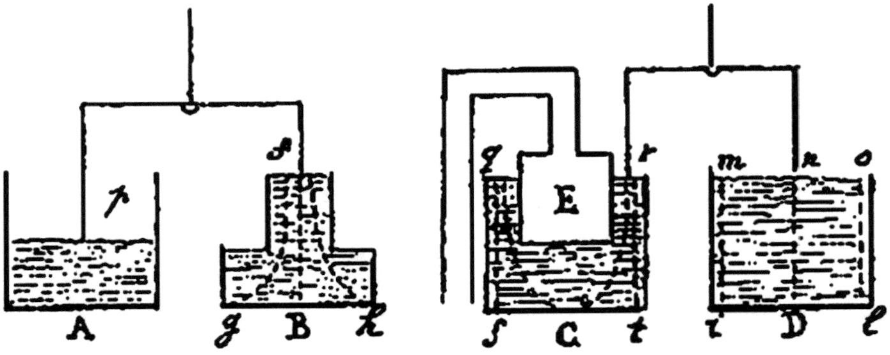

Fig. 5.58 Containers aimed at demonstrating the hydrostatic paradox. (Source: AT X, 69. University of Connecticut Libraries, Babbidge Library, signature: B1833 1897, https://archive.org/details/uvresdedescartes09desc/page/n10 and https://fr.wikisource.org/wiki/Livre:Descartes_-_Œuvres,_éd._Adam_et_Tannery,_IX.djvu. Attribution-ShareAlike 3.0 Unported (CC BY-SA 3.0))

return in his following texts and correspondence). Published only after the rediscovery of Beeckman's *Journal* in 1905, in the manuscript Descartes confronts Stevin's sample 1 (see Fig. 5.58, AT X, 69). As he considers container B, Descartes claims that, in a condition of equilibrium, the particle of water f presses all the points on the bottom, such as g and h, with the same force with which it presses the point of the bottom perpendicularly below itself. So that the pressure undergone by the bottom of B is equal to those exerted on C and D. Descartes's explanation leans, like that later given in the *Principia*, on the hypothetical behaviour of the fluid once the bottom is released from the container.

In this case, the particle f can fall by following any of the ideal lines connecting it to the bottom, and, in the first instant of motion, this particle descends three times faster than the particles m, n and o in container D, for the reason that it would simultaneously occupy the three points of the bottom: g, h, and the one exactly below itself. Moreover, for Descartes such pressure increases only according to the vertical component of such lines, so that the bottom undergoes a uniform pressure, viz. the same exerted by the drop f on the point of the bottom perpendicularly below it.[268]

[268] "Verbi gratia, in fundo unius determinentur puncta g, B, h, in alterius i, D, l; dico omnia illa puncta aequali vi premi, quia scilicet premuntur lineis aquae imaginabilibus eiusdem longitudinis, nempe a suprema parte vasis ad imam. Neque enim fg linea hic longior censenda est quam fB vel aliae: non premit enim punctum g iis partibus quibus curva est et longior, sed iis tantum quibus deorsum tendit, quibus aequalis est aliis omnibus. Probandum autem est solum punctum f aequali vi premere tria puncta g, B, h atque tria distincta m, n, o premunt alia tria i, D, l. Quod fit hoc syllogismo: Res graves aequali vi premunt omnia circumquaque corpora, quibus expulsis aeque facile inferiorem locum occuparent. Atqui solum punctum f aeque facile occuparet inferiorem locum, si posset expellere tria puncta g, B, h atque tria puncta, m, n, o, si expellerent alia tria puncta i, D, l. Ergo solum punctum f aequali vi premit tria simul puncta g, B, h atque tria puncta distincta, m, n, o premunt alia tria i, D, l. Maior videtur esse tam clara et evidens ut possit esse principium scientificum. Minor ulterius probatur: Imaginentur omnia inferiora puncta g, B, h et i, D, l eodem momento aperiri vi gravitationis corporum suprapositorum, certe eodem instanti concipiendum

To sum up, Descartes's explanation is based on (1) the idea that there is a communication of pressure from the surface to the bottom in all directions, (2) that this pressure depends on the height, and (3) on the dynamical rendering of a static situation. His solution, actually, is not very different from that of Galileo, in which the speed of the (hypothetical) fall of a volume of water compensates for the weight of a greater volume of water (and which is at odds with De Volder's explanation of the pressure of fluids in a funnel, which relied on Descartes's *Principia* only). As noted by John Schuster and Stephen Gaukroger, Descartes's explanation of the paradox was a first instance of that corpuscular reductionism which will then become the standard in Descartes's mature writings. In particular, in his manuscript Descartes used the idea of instantaneous tendency to motion giving rise to a pressure – an idea later used in his theory of light.[269]

In his later discussions of hydrostatics, Descartes drops the idea of the exertion of pressure along oblique lines – such as those represented in container B – and considers only the pressure exerted in vertical lines, perpendicular to the pressed surface. That is the case in *Principia* IV.26, as well as of his explanation of the vexed issue of the lack of feeling the weight of water by swimmers and divers (*urinatores*), which he provides in a letter to Mersenne of 16 October 1639. This letter anticipates the explanation then given in the *Principia*: a man at B (see Fig. 5.59, Descartes and Clerselier 1657–1667, volume 2, 184) does not feel the pressure of an incumbent column of water, because if he descends in the water a volume of water will rise in his place, which will be in equilibrium with the water of such columns.

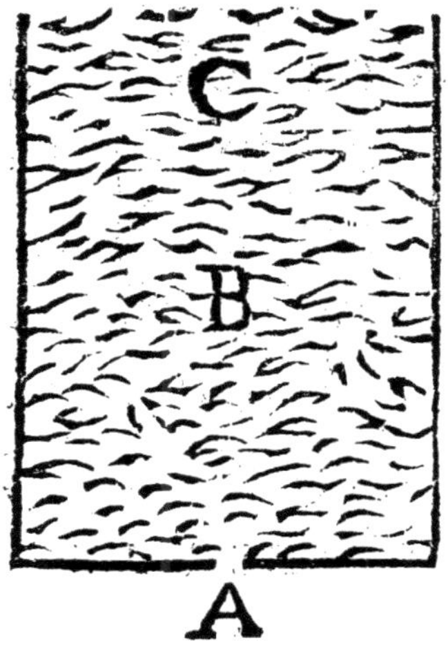

Fig. 5.59 Container aimed at demonstrating the conditions of hydrostatic equilibrium according to Descartes. (Source: Descartes and Clerselier 1657–1667, volume 2, 184. The University of Oklahoma Libraries, History of Science Collection, signature: 17612-1001. Public domain)

erit solum punctum f triplo celerius moveri quam unumquodque ex punctis m, n, o. Illi enim tria eodem momento loca erunt explenda, quo momento unum tantum cuilibet ex punctis m, n, o erit occupandum. Ergo vis qua solum punctum f premit inferiora, aequalis est vi trium simul punctorum m, n, o," AT X, 70–71.

[269] See Gaukroger and Schuster 2002.

In other words, volumes of water counterbalance each other and, as Descartes was to put it in his *Principia*, once they are in equilibrium they do not press on each other. The combined pressure of a column of water, however, is exerted on the bottom of the container, and if a man is placed above a hole in the bottom, he would feel all the pressure of the water above him. The only pressure that a man feels in descending is that of the volume of water rising in his place, by which – as will be clearer in the light of Descartes's theory of weight – he is displaced downwards, since for Descartes weight can be inferred to be a form of downwards displacement by the lighter.[270] In a more concise form, Descartes's theory of pressure is presented in his *Le monde*: a man does not feel pressure since he is not displaced by a whole column of water above him, but only by that volume of water which strives to occupy the place of the human body itself, if not 'balanced' with it.[271]

To sum up, in Descartes's hydrostatics (1) the equilibrium between volumes of fluids entails the lack of exertion of internal pressure: so that for Descartes 'elements do not gravitate on elements', and divers do not feel pressure in water; (2) the pressure a swimmer feels is that exerted during his displacement by a lighter fluid; (3) the cause of his descent is the downwards extrusion exerted by this lighter fluid, and determined by the difference in specific weight between his body and an equal volume of fluid. By the way, this entails that, *stricto sensu*, no pressure is exerted on the bottom of a container specifically lighter than water, as weight is a form of downwards displacement exerted by the lighter on the heavier body. All this,

[270] "Ie ne me souuiens pas de la raison de Steuin, pourquoy on ne sent point la pesanteur de l'eau quand on est dessous; mais la vraye est qu'il ne peut y auoir qu'autant d'eau qui pese sur le cors qui est dedans, ou dessous, qu'il y en auroit qui Pourroit descendre, en cas que ce cors sortist de sa place. Ainsy, par exemple, s'il y avoit un homme dans le tonneau B, qui bouchast tellement de son cors le trou marqué A, qu'il empeschast que l'eau n'en pust sortir, il sentiroit sur soy la Pesanteur de tout le Cylindre d'Eau ABC, dont ie suppose la EF base de mesme grandeur que le trou A, d'autant que, s'il descendoit en bas par ce trou, tout ce cilindre d'eau descendroit aussi; mais s'il est un peu plus haut comme vers B, en sorte qu'il n'empêche plus l'eau de sortir par le trou A et que ce trou soit bouché, il ne doit sentir aucune pesanteur de celle qui est sur luy entre B et C, d'autant que, s'il descendoit vers A, cette eau ne descendroit pas avec lui; mais au contraire une partie de l'eau qui est sous lui vers A, de même grosseur qu'est son corps, monterait en sa place, de façon qu'au lieu de sentir que l'eau le presse de haut en bas, il doit sentir qu'elle le soulève de bas en haut: et cela s'accorde à expérience," AT II, 587–588.

[271] Descartes provides this theory in explaining his theory of gravity, which in his hands becomes a sort of branch of hydrostatics: "[t]hat is to say, just as one side of the balance can only be raised or lowered if the other side, at exactly the same instant, moves in the opposite direction, and just as the heavier always makes the other rise, so too the stone [...], for example, is so opposed to the quantity of air above it, which is exactly the same size as it and whose place it would have to occupy if it were to move further away from the centre T [i.e. the centre of the Earth] that this air would of necessity have to descend to the extent that the stone rose. [...] In this way you can see that every part of terrestrial bodies is pressed towards T, not indifferently by all the matter surrounding it but only by an amount of the matter exactly equal to the size of that part and which, being underneath the part, can take its place if that part moves down. This is the reason why, among the parts of any single body that we call 'homogeneous', such as those of air or water, the pressure on the lowest is not notably more than that on the highest, and why a man at the bottom of very deep water does not feel it pressing on his back any more than if he were swimming right on top," Descartes 2004, 48–49. De Volder owned Descartes's *Le monde* as well: see *Bibliotheca Volderina*, 9.

moreover, contradicts Archimedean explanations taking into account the pressure exerted across entire volumes of fluids, and the idea that on each ideal surface of a fluid there is a uniformity in pressure from above and below: indeed, there is no pressure of water on water.

5.5.2 Descartes's Corpuscular Standpoint on Hydrostatics

Beyond the idea that, at a phenomenic level, there is an equilibrium in which volumes of water counterbalance themselves and no internal pressure is exerted, Descartes also dealt with the underlying, corpuscular interactions of particles of fluids, and of the solid bodies floating in them. His explanations, as to the underlying corpuscular dynamics, however, were far less monolithic than his theory of hydrostatics.

The corpuscular features of water and air – viz. the foremost 'terrestrial' fluids – is extensively dealt with by Descartes in his *Météores*, for the obvious reason that this treatise is devoted to all meteorological phenomena, including the formation of vapours, winds, rain, sweet and salty waters. In his *Météores*, Descartes describes water as composed by oblong and slippery particles, which can be compared to eels and, even when they connect to each other, can be easily separated. Also the particles of air have a similar figure, but they are less connected to each other, as they are only juxtaposed.[272] This account is kept in the *Principia*, where he refers to his *Météores*, and according to which air is composed by particles which are "thin and separated from one another," and where their shape is labelled as not really relevant, provided that they are oblong and able to describe, once put in a rotating movement by subtle matter, small circles (this being the foundation of his explanation of the elasticity of the air in IV.47, as I am going to show in Sect. 5.6.2.1, *The idea of*

[272] "[...] je suppose que les petites parties, dont l'eau est composée, sont longues, unies et glissantes, ainsi que de petites anguilles, qui, quoiqu'elles se joignent et s'entrelacent, [...] puissent aisément être séparées; et [...] que presque toutes celles tant de la terre que même de l'air et de la plupart des autres corps ont des figures fort irrégulières et inégales, en sorte qu'elles [...] se lient les unes aux autres, ainsi que font les diverses branches des arbrisseaux [...]. Et lorsqu'elles se lient en cette sorte, elles composent des corps durs, comme de la terre, du bois, ou autres semblables; au lieu que, si elles sont simplement posées l'une sur l'autre, sans être que fort peu ou point du tout entrelacées, [...], elles doivent [...] composer des corps liquides fort rares et fort légers, comme des huiles ou de l'air," AT VI, 233–234 and 236. De Volder owned the 1637 edition of Descartes's *Essais*,: *Bibliotheca Volderina*, 4. In his *Bibliotheca*, no Latin edition of Descartes's *Essais*, namely the *Specimina*, is listed. He owned the 1670 Latin edition of the *Meditationes* and *Principia*, namely the first two tomes of the fifth edition of Descartes's *Opera philosophica* (1670–1672), printed in Amsterdam, in which the *Meteora* are printed in the third tome. Yet, De Volder referred to it several time in his *dictata* (for instance, as to the explanation of the elasticity of air: see *infra*, n. 333); as he refers to chapters (*capita*) of this treatise, rather than to discourses (in which the French editions are divided), he probably had a Latin version at his disposal. Latin editions of the *Meteora* appeared in 1644, 1650, 1656, 1664, 1672, 1685 and 1692.

elasticity of air before De Volder).[273] Water, in turn, is composed of particles which, according to the French edition of the *Principia*, are "long and united."[274] What they have in common, beside the oblong shape, is the fluids they are composed of, i.e. bodies which are easily separable as they are "agitated by a diversity of {independent} movements," in accordance with the theory of fluidity given by Descartes in *Principia* II.54, as seen above. On this ground, in II.56–57 Descartes shows "[t]hat the particles of fluids {tend to} move with equal force in all directions; and that a solid body, immersed in a fluid, can be set in motion with very little force."[275] The key idea beneath this claim is that each particle of a fluid continuously moves in all directions, so that for each particle of fluid one can draw infinite circles along which such movement is exerted (see Fig. 5.60, Descartes 1644a, 63). Accordingly, a solid body immersed in water, and with which it is in a condition of perfect hydrostatic equilibrium, is continuously pressed from all directions with the same force – and a small push from one side makes it move very easily, since this is enough to change or re-direct the determination of motion of the particles surrounding it, in one direction.[276] This, of course, overcomes the difficulty entailed by Descartes's fourth rule

---

[273] "45. What the nature of air is. [...] [A]ir must be nothing other than an accumulation of particles of the third element, so thin and separated from one another that they obey all the movements of the heavenly globules; and that therefore air is a very rare, fluid, and transparent body and can be formed of particles of any figure at all. [...] Since they are disunited, each one is moved independently of its neighboring ones; and thus occupies that whole small spherical space which it requires for its circular movement around its own center, and drives all neighboring particles out of that space. This is why it does not matter {for this effect} what figures the particles of air may have," Descartes 1982, 203–204.

[274] "[...] longues et unies," AT IX-B, 227; in Descartes 1982 it is translated as "long and smooth," Descartes 1982, 205. This phrase is not given in the Latin edition.

[275] Descartes 1982, 70.

[276] "[...] as far as fluid bodies are concerned, even though our senses may not inform us that their particles move, since they are too small, this is nonetheless easily deduced from effects; especially in the cases of air and water, because many other bodies are destroyed by them: for no physical action of that kind can occur without {the fluid particles} moving [...] Yet there is a difficulty here, because these particles of fluids cannot all move at the same time in every direction; [...] In order to resolve this difficulty, we must recollect that it is not movement but rest which is contrary to movement; and that the determination of a movement in one direction is contrary to its determination in the opposite direction, as was stated earlier. Furthermore, [we must remember that] all moving bodies always tend to continue their movement in a straight line. Now from these things, it is obvious [...] that as far as determination is concerned, the fact is that there are as many particles of D moving from C to B as in the opposite direction: for the same ones which come from C strike the surface of the body B and are driven back toward C. Although some of these particles, considered individually, strike B and drive it toward F (and thus more greatly impede its movement toward C than if they were at rest); an equal quantity of particles also move from F toward B, and drive B toward C. The result is that B is no more driven in one direction than in another, and therefore remains at rest, unless something else intervenes. No matter what we suppose the shape of B to be, it is always driven by exactly the same number of particles of the fluid coming from one direction as from the other. [...] Thus far, however, we have been considering B to be at rest; if we now suppose it to be driven toward C by some force coming from elsewhere, this force, however slight, would suffice, not indeed to move, B by itself, but to unite with the [force of the] particles of the fluid body FD and to enable them to drive B toward C and to transfer to B some of their

Fig. 5.60 Idealized representation of the transmission of pressure in water according to Descartes. (Source: Descartes 1644a, 63. Library of Congress, call number: LC B1860 1644, http://hdl.loc.gov/loc.rbc/Rosenwald.1431.1. Public domain)

of motion. In a fluid, a minimal force is sufficient to move a big body at rest, since this force makes the determination of the whole fluid changing and pushing the body.[277] In the hands of De Volder, Descartes's discussion of the situation of a solid immersed in a fluid serves to justify Huygens's rules, especially rule 3 (contradicting Descartes's rule 4) as he overtly put it in his *dictata*.[278]

Thus far, Descartes's idea of fluidity – as anticipated above – entails the idea of transmission of pressure in all directions which characterizes the theories of hydrostatics of Stevin, Pascal, Boyle, and is perfectly compatible with accounts of floatation assigning to the difference in specific weight the role of cause in sinking and floating. With the remarkable exception that for Descartes – as for Galileo – water does not exert pressure on water, so that the conditions of equilibrium are not accounted for in terms of a progressive increase in pressure, as it was for the models

motion," Descartes 1982, 70–71. In II.57, Descartes provides another demonstration of the same matter by considering more overtly the circles *aeiouy* described by the particles of the fluid. Please note that the particles do not 'sustain' the body in the fluid: its equilibrium is solely determined by its being equiponderant to water, i.e. of being exactly light as an equal volume of water, which thus does not displace it downwards.

[277] "From the preceding, it is clearly perceived that a solid body, immersed in a fluid and at rest in it, is held there as if in equilibrium. Further, no matter how large it may be, it can always be driven in one direction or another by the least force; whether this force comes from elsewhere [...]. When this occurs, it is absolutely necessary for a solid body situated in such a fluid to be carried along with it: nor is this contradicted by the fourth rule; according to which, as I stated before, a body which is at rest cannot be set in motion by any smaller," Descartes 1982, 75.

[278] On II.56: "[e]xigua. Haec regula est nobilissimi Hugenii," Hamburg 273, 117; cf. Hamburg 274, 48: "[h]aec vis quantumvis exigua. [...] [E]st idem cum regula nobiliss. Hugenii, qui affirmat nullum corpus quantumvis magnum in sua quiete permansurum, si aliud corpus, utut exiguum, occurrat."

of Boyle and De Volder. Moreover, no criteria for the quantification of the pressure exerted laterally, on the side of a solid are provided. In principle, one can describe infinite circles of matter pressing on a solid. Accordingly, this theory fell short when Descartes attempted to explain – in his aforementioned letter to Mersenne of 16 October 1639 – the raising of water in pumps by the idea of circumpulsion, rather than the idea of atmospheric pressure as such. Namely, for Descartes what raises water in pumps is not the pressure of a whole column of air on its surface, but the displacement of a certain volume of air by the cylinder of the pump – which then presses on the water. This, of course, is problematic as to the explanation of the maximum level reachable by a fluid in a pump – determined by atmospheric pressure. As he came to comment on Galileo's idea that water cannot be raised but up to 18 *braccia* in syphons, in his letter to Mersenne of 11 October 1638, he claimed – without any substantiation – that this "has not to be related to vacuum, but or to the matter of the pumps or to that of water itself, which flows between the pump and the pipe, instead of rising itself higher."[279] Thus far, no pressure of air is taken into account, even if the idea of the weight of the air is used, in the same letter, to explain the cohesion of the cylinders of marble.[280]

Elsewhere, however, Descartes adopted the idea of the pressure exerted by a whole 'column' of atmosphere to explain the phenomena of raising of fluids – albeit combining it with the theory of circular thrust – and also provided an alternative explanation for the lack of feeling of pressure in fluids. In his letter to Henricus Reneri of 2 June 1631, Descartes discusses the problem of the weight of the air in a way consistent with that of Boyle and De Volder. By comparing air to a mass of wool (with some contradiction of the theory of matter of his *Météores*), Descartes maintains that there is an increasing pressure within the mass of air, as the lower parts are pressed by those above it. Such pressure, however, is not felt by us because, for example, if we raise air at E (see Fig. 5.61, Clerserlier 1657–1667, volume 3, 603), immediately this air displaces other air F (above it) and a circle of air GHI towards E is formed. The idea of circular thrust, in this case, is used to explain the lack of feeling of pressure in a way different from that then used in the *Principia*, as in the *Principia* the idea of circular displacement of water (i.e. the idea of the replacement of the space occupied by our body with other water) is used to argue for the continuous balancing of equal volumes of water. In this letter, the idea of circular thrust is used to argue for the fact that the air we displace comes back to its former place from below, and sustains itself.[281] This is a sort of *via media* between

[279] "[...] l'obseruation que les pompes ne tirent point l'eau a plus de 18 brasses de hauteur ne se doit point rapporter au vuide, mais ou a la matière des pompes ou a celle de l'eau mesme, qui s'escoule entre la pompe & le tuyau, plutost que de s'eslever plus haut," AT II, 382. For a discussion, see Nonnoi 1994.

[280] See Sect. 4.1.3.1, *Four factors of cohesion: vacuum, rest, pressure and entanglement.*

[281] "[...] imaginez l'air comme de la laine, et l'æther qui est dans ses pores comme des tourbillons de vent, qui se meuvent çà et là dans cette laine, et pensez que ce vent qui se joüe de tous costez entre les petits fils de cette laine, empesche qu'ils ne se pressent si fort l'un contre l'autre, comme ils pourroient faire sans cela; Car ils sont tous pesans, et se pressent les uns les autres autant que l'agitation de ce vent leur peut permettre; Si bien que la laine qui est contre la terre est pressée de

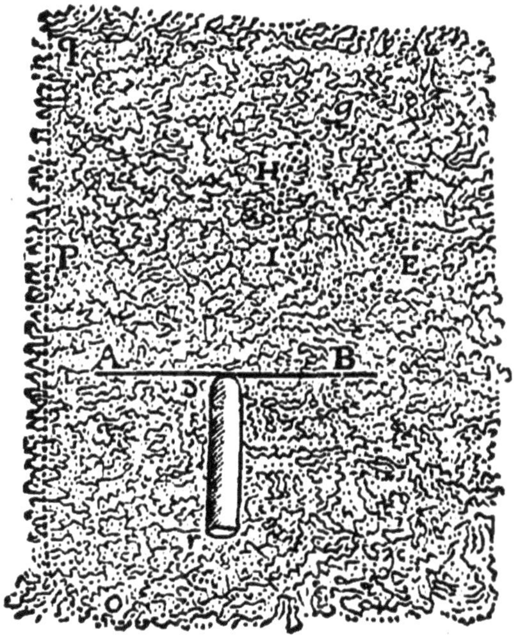

Fig. 5.61 Idealized representation of the transmission of pressures in water according to Descartes. (Source: Clerserlier 1657–1667, volume 3, 603. The University of Oklahoma Libraries, History of Science Collection, signature: 17612-1001. Public domain)

the idea of hydrostatic equilibrium involving entire columns of air (or water), and that one considering only separate volumes – as in Descartes's *Principia*.

The idea of a 'whole column' of atmosphere is however immediately used by Descartes as he proposes to Reneri his explanation of the suspension of mercury in a pipe, closed at the upper extremity. In this case, no circle can be formed and the whole atmosphere comes to press with all its force on the lower surface of the mercury suspended in the tube.[282] This explanation conforms to those of Torricelli and Boyle, who, however, did not consider the circular displacement of air but only the exertion of its pressure. Eventually, as Descartes came to comment upon Torricelli's and Pascal's attempts to ascertain atmospheric pressure, or the weight of the air at

toute celle qui est au dessus iusques au delà des nuës; Ce qui fait une grande pesanteur; En sorte que s'il falloit élever la partie de cette laine, qui est, par exemple, à l'endroit marqué O, avec toute celle qui est au dessus en la ligne OPq, il faudroit une force tres-considerable. Or cette pesanteur ne se sent pas communément dans l'air, lors qu'on le pousse vers le haut; pour ce que si nous en élevons une partie, par exemple celle qui est au point E, vers F, celle qui est en F va circulairement vers GHI, et retourne en E, et ainsi sa pesanteur ne se sent point; non plus que seroit celle d'une roüe, si on la faisoit tourner, et qu'elle fût parfaitement en balance sur son aissieu," AT I, 205–206.

[282] "Mais dans l'exemple que vous apportez du tuyau dr, fermé par le bout d, par où il est attaché au plancher AB, le Vif-argent que vous supposez estre dedans, ne peut commencer à descendre tout à la fois, que la laine qui est vers r, n'aille vers O, et celle qui est vers O n'aille vers P et vers q, et ainsi qu'il n'enleve toute cette laine qui est en la ligne OPq, laquelle prise toute ensemble est fort pesante; Car le tuyau estant fermé par le haut, il n'y peut entrer de laine, ie veux dire d'air en la place du Vif-argent, lors qu'il descend," AT I, 206.

different altitudes, he claimed that their experience conformed to his ideas – which of course was easy to admit as Descartes had a theory of the pressure exerted by a fluid on the bottom of its container.[283]

Thus far, Descartes assumed a standpoint, on the issue of the pressure of fluids, according to which the only ways to account for their upwards pressing, are (1) hypothesizing that particles of fluids moves in all directions, composing endless circles (this being consistent with the idea that the pressure of fluids is exerted also upwards and laterally); (2) considering the differences in specific weight between a solid suspended in mid-air, or mid-water, heavier than the air or the water itself. In this second case, the fluid, being lighter, would exert a constant pressure on the solid, derived from its being lighter than it. However, such a solid would not be pressed by an entire column of air or water, but only by the volume of fluid 'displaced' by itself. Descartes's second model, thus, could not account for the suspension of Magdeburg hemispheres in open air. Accordingly, De Volder could not accept Descartes's hydrostatic model given in the *Principia*, in order to account for phenomena such as the cohesion of the Magdeburg hemispheres, or the cylinders of marble. Rather, he provided his own synthesis between Cartesian and Archimedean principles. This synthesis, as I show in Sect. 5.5.3, *De Volder's Archimedeo-Cartesian hydrostatics: an impossible synthesis?*, was not easy, and probably came to determine De Volder's concerns to make his theories circulate outside Leiden University.

5.5.3 De Volder's Archimedeo-Cartesian Hydrostatics: An Impossible Synthesis?

De Volder's treatment of Descartes's ideas on hydrostatics can be found, of course, in his *dictata*. Coupled with the consideration of his treatment of the issue of the lack of feeling of pressure in water as given in his *Disputationes de aëris gravitate* and *Experimenta*, the analysis of De Volder's views shows that he was struggling with two alternative models in hydrostatics. On the one hand, the Archimedeo-Boylean model, according to which water presses on itself, there is an increasing internal pressure in water, and the lack of feeling of such pressure is due to its being exerted in all directions. On the other hand, the Cartesian model, in which the pressure of water on water is nullified between each volume of the fluid, so that we do not feel the pressure of an incumbent column of water, as seen above.

As far as the corpuscular features of air and water are concerned, De Volder shows ambivalent attitudes. In the series Hamburg 274 of his *dictata*, for instance, on the one hand he follows the theory of water and air given by Descartes (according to which the particles of air are juxtaposed, and those of air are entangled): in commenting upon IV.45, he notes that the figure of the particles of air is not deter-

[283] See the letters of Descartes to Mersenne of 13 December 1647 and to Carcavi of 11 June 1649, where Descartes claims that he himself had suggested to Pascal the performing of his experiment with mercury: AT V, 99 and 366.

mined, provided that such particles are tenuous, separable and disjoined.[284] On the other hand, in commenting upon the article III.148, where Descartes dealt with the generation of solar atmosphere from the dissolution of sunspots, forming a sort of "air or aether," whose "particles [...] having very irregular branching figures, become attached to one another in such a way that they cannot be moved individually like the globules of heavenly matter,"[285] De Volder claims that the particles of our air are to be considered in the same way, and as cohering to each other so that by this one can explain the raising of water and mercury in syphons and tubes.[286] This model is much more emphasized in the other series of *dictata* (Hamburg 273). As he comments on IV.45 – where De Volder criticizes the traditional theory of elements – he claims that the particles of air are linked, so that (1) they press upon each other (an idea not mentioned in the other series), and (2) they press having united their forces, on the surface of the Earth.[287] Plainly, this is the same model of the

[284] "Cuiuslibet figurae. Nonnulla certa et determinatur [*sic*] figura, {...} aquaeq[ue] particulae separatim moveatur, ab aliis, cuius cumq[ue] enim figurae sint, si modo nimia non impediat magnitudo, hoc pacto a globulis moveri potest, ita ut forma aeris non consistat in certa figura, sed tantum in tenuitate particularum, {earumdemq[ue]} a se invicem secretione," Hamburg 274, 102.

[285] Descartes 1982, 172.

[286] "Sibi invicem sic adhaerent. Hoc sane in aere nostro locum habere ostendunt plurima effecta, q[uae] ex aeris gravitate dependent. Q[ua]lia sunt ascensus aquarum in antliis, ad pedes circiter 33, mercurii in tubis residentia ad pedes circiter 2½. Quae sane effecta tam notabilia non forent, nisi particulae ramosae in aere haerentes tota sui mole in ea corpora agerent, ac proinde non omnino disiunctim moveantur," Hamburg 274, 82.

[287] "Componi. Si vulgatos philosophos qui in 4 elementis omnem operam pereunti consulamus, dabunt aere pro corpore simplici, humido, et calido, nec quid praeterea, ad haec omnia referunt q[uo]d ipsum praeterquam quod nil sit ab ere alienum aliis potius competit corporibus, in aqua enim est humiditas, in eadem calida, calor, in eadem simplicitas, unde sequeretur aerem idem esse quod aqua, quod falsum est. E contra, cum manifestum sit, si experientiam consulamus, solem omnia effluvia attollere ex corporibus q[uae] in aere haerent, quid certius et evidentius est, quam aerem non simplicem, sed omnino compositum e[ss]e, nec est q[uo]d dicant, se non dicere id de aere nos cingente, sed alio quem purum concipiunt, id e[nim] nil ad nos, is e[nim] si detur, imaginarius plane est, nullumq[ue] nobis usum prahebet. Quo itaq[ue] errore ut nos liberemus, videamus quidnam hic noster aer sit: si generationem aetheris antea descriptam videamus et respiciamus, constituunt aerem e[ss]e mobilia, et ex iis, q[uae] detruduntur, versus aquam et terram, calore attolli, frigore [n]a[m] rursus decidere, quodcunq[ue] enim sit attollitur, quamdiu est in illo motu, est aer, unde patet e[ss]e congeriem diversarum particularum et diversi generis, cum autem obsequantur motui globulorum caelestium constituere debent corpus valde rarum, cum globuli per quaelibet spatia moveantur, cumq[ue] omnia corpuscula q[uae] impedimento sunt, globulis expulsa sint, globuli ubiq[ue] sibi vias fecere, sicq[ue] aer factus est pellucidus, fluidus autem est ex eo, q[uo]d particulis crassioribus detrusis reliquae disiunctae et a {se} invicem remotae, cuilibet motui obsequantur. Figurae. Dicit particulas disiunctas e[ss]e, non quod una nos contingat alteram, et peculiarem habeat sphaeram, id e[nim] nec ex natura rei nec aeris fluit, separatae ergo sunt; et diversimode moventur, sed ita tamen ut se invicem tangant et premant: nonne enim antea ostendimus aetherem congregari circa sidus, quia globuli maiore habent recedendi vim eodem modo globulis premunt par[ticu]las ad motum in {eptas} deorsum, non itaq[ue] cohaerent, sed tamen sibi mutuo incumbunt, quemadmodum particulae in aqua quidem sunt separate, et tamen, se mutuo tangunt. Id itaque non concipiendum est quasi aer habeat particulas divulsas, sed ita, ut non cohaerentes tamen sibi mutuo incumbant, q[uo]d sequitur a priori ex natura generationis, et a posteriori, ex ipsa earum gravitate, si enim avulsae essent, non premerent se invicem, patet itaq[ue] quidem separatas esse, sed tamen sibi mutuo incumbere, et premere inferiora corpora," Hamburg 273, 248–251.

letter of Descartes to Reneri, where Descartes used the analogy of wool, instead of that of eels, to explain the action of air in its entirety. In this way, however, both De Volder and Descartes introduced a differentiation between fluids: namely, cohering fluids – such as air and water – and non-cohering fluids, such as subtle matter, whose particles are not entangled with each other. Subtle matter makes terrestrial bodies heavy and pressing on themselves and on other bodies, while the subtle matter is, in itself, always light and not pressing on itself. According to this model, the downwards pressure exerted by subtle matter takes place as a form of downwards extrusion by the lighter, and the downwards pressure of air and water takes the form of a pressure exerted by the weight of a body.

As to the macroscopical level, the series of *dictata* Hamburg 274, in commenting upon Descartes's explanation of the lack of feeling of pressure (IV.26), De Volder just uses Descartes's explanation given in the letter to Mersenne (not overtly mentioned by De Volder). Namely, that a man, sinking in water, does not feel the weight of the whole column above him as this will not descend with him, but his place is progressively occupied by a portion of water equal to his volume, by which the incumbent water reaches a condition of equilibrium.[288] In the other series of *dictata*, the account is once again Cartesian, but far more developed. As seen above, in commenting on IV.26 De Volder points out that pressure is exerted according to perpendicular lines, as demonstrated by Stevin and as shown by the case of the funnel. Hence, as he addresses the issue of the lack of feeling, he discusses the positions of Stevin (given in his *De initiis praxis hydrostatices*) and Boyle, given in appendix 2 of his *Hydrostatical Paradoxes* – where the very solutions of Stevin and Descartes are analysed, and rejected. The solutions of Stevin and Boyle can be summarized as follows. For Stevin, as he argues in proposition 3 of his *De initiis praxis hydrostatices*, we do not feel the pressure of the water as it presses uniformly on all parts of the body: so that no part is more dislocated (*luxatur*) by water, than any other. Only in the case where a man is put above a hole (see Fig. 5.62, Stevin 1605, 149) – like in Descartes's sample – such man would feel the pressure of the water.[289]

---

[288] "[...] deduci potest istius vulgaris experimenti in quo enodando plurimi frustra insudarunt, cur nimirum fiat, q[uo]d urinatores aliisq[ue] homines sub aqua degentes nullum sentiant aq[uam] supra incumbentis pondus: huius enim ratio ex iis, q[uae] hoc §§ adferuntur, manifesta est. Etenim si homo constitutus in puncto ex. gr. 3, premeretur aqua 3 4 deorsum ut occuparet locum 2, necessario aqua, q[uae] in 2 est, eadem mole 1 est corpus hominis ascenderet versus 3, quae cum aeq[ua] li impetu huic ascensui resistat, efficit etiam, ut homo in 2 nullo pacto deorsum premi queat, unde nec mirum est, eum nullum suora se sentire pondus, quippe cuius tota vis in ea consistit pressione," Hamburg 274, 100. See Fig. 5.57, Descartes 1644a, 203.

[289] "3. Propositio. Causam reddere cur homo alte infra aquam natans maximo eius pondere non opprimatur. Humani corporis planum occupet pedes 10, is infra aquam 20 pedes demersus, aquae pede cubico 65 lb aestimato, sustinebit per 10 et 11 prop. de Elem. Hydrostat. 13,000 lb. Quamobrem qui potest, ut tanto pondere pressus non opprimatur; Causa autem haec est: Omni pressu quo corpus dolore afficitur, pars aliqua corporis luxatur. sed isto pressu nulla corporis pars luxatur. Isto igitur pressu corpus dolore nullo afficitur. Assumptio syllogismi reipsa manifesta est, nam si pars aliqua ut caro, sanguis, humor aut quodlibet denique membrum luxaretur, in alium locum concedat necesse esset; atqui locus ille non est extra corpus, cum aqua undique aequali pressu circumfusa sit (quod vero pars ima per 11 propos. Hydrostat. paulo validius prematur supe-

Fig. 5.62 Idealized representation of the condition of feeling of pressure in water by the human body according to Stevin. (Source: Stevin 1605, 149. ETH-Bibliothek Zürich, signature: Rar 9069, https://www.e-rara.ch/doi/10.3931/e-rara-9068. Public domain mark 1.0)

This solution is criticized by Boyle in his *Hydrostatical Paradoxes*, as it does not account for the obvious case in which "the bare compression of two small parts, one against another, is sufficient to produce a sense of pain."[290] Moreover, Boyle criticizes the solution of Descartes, for two reasons. First – as noted above – because in Descartes's model the pressure exerted by water on another body depends on the difference in specific weight between the water, and this body. However, in accordance with Boyle's Archimedean model of floatation, water just presses on every kind of body, regardless of its specific weight.[291] Second, Descartes's solution works only in the case in which the volume of a body is 'replaced' by water rising in its place, namely, it works only as a consequence of the "fulness of the Vessel."[292] In

riore, id hoc casu nullius momenti est, quia tantula differentia partem nullam sede sua dimovere potest) neque item intra ipsum corpus concedit, cum istic corpore omnia oppleta sint, unde singulae partes singulis partibus aequaliter resistunt, namque aqua undiquaque eadem ratione corpus totum circumstat. Quare cum locus is nec intra neque extra corpus sit, absurdum imo impossibile fuerit partem ullam suo loco emoveri, ideoq[ue] nec corpus hinc afficietur ullo dolore. Sed in exemplo clarius ita intelliges, esto A B C D aqua cuius fundum D C, in quo foramen E habeat epistomium sibi insertum, cui dorso incumbat homo F. Quae cum ita sint, ab aquae pondere ipsi insidente nulla pars corporis luxari poterit, cum aqua ut dictum est, undique urgeat aequaliter. si vero eius veritatem explorare libeat, eximito epistomium E, tumque tergum nulla re fultum sustinebitur, ut in locis caeteris; ideoque istic tanto pressu afficietur, quantus 3 exemplo 2 propositionis huius demonstratus est, videlicet quantam efficit columna aquea cuius basis sit foramen E altitudo autem eadem quae aquae ipsi insidentis. Quo exemplo propositi veritas manifesto declaratur," Stevin 1605, 148–149.

[290] Boyle 1666, 242.

[291] "[...] this subtil Philosopher: for whose Ratiocinations though I am wont to have much respect, yet I must take the liberty to confess my self unsatisfy'd with this. For haveing already sufficiently prov'd, That the upper parts of water press the lower, and the bodies plac'd beneath them, whether such bodies be lighter in specie then water or heavier; we have subverted the Foundation, upon which Monsieur Des Cartes's ingenious, though unsatisfactory, Explication is built," Boyle 1666, 231–232.

[292] "And yet I shall add ex abundanti, That supposing what he sayes, That in case the solid B should descend towards A, the incumbent water would not descend with it, but a part of the subjacent water, equal in bulk to the solid, would ascend, and succeed in its room; yet that is but accidental, by reason of the steinchness and fulness of the Vessel," Boyle 1666, 232.

proposing his own solution, Boyle reverts to Stevin, and proposes as the cause of the lack of feeling of pressure (1) the pressure of water in all directions, and (2) the "texture" of the bodies of animals, which makes them apt not to feel pressure, as confirmed, for Boyle by an experiment in which he presupposes that water can be compressed. Namely, by forcibly compressing the air in a vessel containing both air and water, and observing that no visible effect was exerted on the tadpoles swimming in the water, notwithstanding that for Boyle the water was also compressed – as "the Tadpole seem'd to be compress'd into a little less Bulk then it was of before."[293] This explanation, actually, is then also given in Boyle's *New Experiments*

[293] "VVe took then a somewhat slender Cylindrical pipe of Glass, seal'd at one end, and open at the other; and to this we fitted a Rammer, which (by the help of some thongs of soft leather, that were carefully wound about it) did so exactly fill the pipe that it could not easily be mov'd to and fro; and would suffer neither water, nor aire, to get by betwixt it, and the internal surface of the Glass. VVe also provided some small Tad-poles (or Gyrini) about an Inch long or less; which sort of Animals we made choice of before any other, partly because they could, by reason of their smalness, swim freely to & fro in so little water as our pipe contain'd; & partly because those Creatures, being as yet but in their Infancy, were more tender, and, consequently, far more expos'd to be injur'd by compression, then other Animals of the same Bulk, but come to their full age and growth, would be, (as indeed such young Tad-poles are so soft and tender, that they seem, in comparison to the bigger sort of flies, to be but organiz'd Gelly.) One of these Tadpoles being put into the water, and some Inches of aire being left in the pipe, for the use anon to be mention'd; the water and aire, and consequently the Tadpole, were by the intrusion of the plug or rammer, with as great a force as a man was able to imploy, violently compress'd; and yet, though the Tadpole seem'd to be compress'd into a little less Bulk then it was of before, it swom freely up and down the water, without forbearing sometimes to ascend to the very top, though the Instrument were held perpendicular to the Horizon. [...] And having repeated this Experiment several times, & with Tadpoles of differing ages; we may, I presume, safely conclude, That the Texture of Animals is so strong, that, though water be allowed to weigh upon water, yet a Diver ought not to be opprest by It: Since, whether or no water weighs in water, 'tis manifest that in our Experiment, the water, and consequently the Tadpole, was very forcibly by an External Agent compress'd betwixt the violently condens'd aire, and the rammer," Boyle 1666, 243–246. The solution of Stevin and Boyle was followed also by Von Guericke, who provides a short statement on this issue in his *Experimenta nova*, book 3, chapter 1: "[b]ecause of its own weight, air presses not only upon itself but upon all things beneath it with an essentially constant pressure. We do not feel this because we live in this air which surrounds us with equal pressure on all sides and at the same time penetrates us. For as fish have no perception of pressure in the water, so animals perceive it even less in the air," Von Guericke 1993, 113. The reference to the fact that air penetrates us, seems to rely particularly on Boyle's idea that air makes us resisting to its pressure as it is provided with an elastic power, given in his *New Experiments about the Differing Pressure of Heavy Solids and Fluids*: "I have elsewhere proved by Experiments, that a cubick inch of Air, for instance, has as strong a spring as suffices to enable it to resist the weight of the whole Atmosphere, as far as it is exposed thereunto; for else it would be more compress'd than de facto it is. And 3. I have also shewn, that a very little portion of Air, though it will much sooner loose its spring by expansion than a greater, yet 't will resist further compression as much as a greater: And 4. I have also shewn, that in the pores of the parts of Animals, whether fluid or consistent, as in their Blood, Galls, Urines, Hearts, Livers &c. there are included a multitude of Aereal corpuscles, as may appear by the numerous bubles afforded by such Liquors, and the swelling or expansion of the consistent parts in our exhausted Receiver. 5. To this we may add, that, besides the Bones, whose solidity is not questioned, a much greater part of the humane body than is wont to be imagined, does really consist of Membranes and Fibers, and the coalitions and contextures of these; and that these substances are by the Providence of the

about the Differing Pressure of Heavy Solids and Fluids (1672), where he reports evidence that water actually presses on the human body: namely, reports from deep-sea divers, experiencing, besides the direct pressure of water on the thorax, collateral effects such as nausea, due to the pressure of water on the nerves.[294] For Boyle, this confirms that the constitution of our body is a key factor in preventing us from feeling the effects of pressure.[295]

In the Hamburg 273 series of *dictata*, in commenting on IV.26, De Volder reports Stevin's solution – which he claims was endorsed by Boyle – and rejects it, since it does take into account that the human body contains many spaces filled only by air, which therefore cannot sustain the pressure of water. Against Stevin's solution, De Volder reverts to a solution criticized by Boyle, namely Descartes's, which (1) is restated almost *ad literam*, namely, by considering that the pressure exerted by the incumbent water is counterbalanced by the water rising in the place of the human body which is displaced downwards. Moreover, this solution (2) is explained by reducing it to an abstract, static model: De Volder claims that we do not feel pressure in water, in the same way as when we 'sustain' a weight on a balance on which equal weights are appended, we do not feel any pressure from it. Indeed, in both cases equal weights sustain themselves.[296] Incidentally, this second solution is akin

most wise Author of Things made of a much closer and stronger Texture, than those, that have not tried, will be apt to think; as I could make probable by the great force that Bladders will endure, and the very great weight that Tendons of no great thickness will lift up or sustain, and by other things that I shall not now insist on. Lastly, There is a far greater difference, than men are wont to suspect, between the effects of the Pressures made upon Bodies by incumbent or otherwise applied solid weights, and those that they suffer from heavy but every way ambient fluids; as will appear by the Experiments to be mention'd by and by," Boyle 1999–2000, volume 7, 217.

[294] "[...] having met with a Learned Physician, that living by the Sea-side in a hot climate, delighted himself much in diving; and inquiring of him whether he felt no compression, when he passed out of the Air into the Water, he answered me, that when he div'd nimbly as others use to do, he took not notice of it, but when he let himself sink leisurely into the water, he was sensible of an unusual pressure against his thorax, which he several times observed. A man that gets his living by fetching up goods out of wrack'd Ships, complain'd to me, that if with his diving Bell he went very deep into the Sea, and made some stay there, he found himself much incommodated; which though he imputed to the coldness of the water, yet by the symptoms he related I was inclin'd to suspect, that the pressure of it upon the Genus Nervosum might have an interest in the troublesome effect. And I have been assured by an eminent Virtuoso of my acquaintance, that he was lately informed by a person, whose profession it is to fetch up things from the bottom of the Sea by the help of a diving Bell, that several times when he descended to a great depth under the surface of the water, he was so compress'd by it, that the blood was squeez'd out at his Nose and Eyes; which Relation seems to favour our conjecture, and would much more confirm it, if I were sure, that the effect was no way caus'd by some fermentation or other commotion in the blood it self, occasioned by the great density or other alterations of the Air he breath'd in and out, or by some other operation of the ambient Medium distinguishable from the compression of the water, though perhaps conjoyn'd with it," Boyle 1999–2000, volume 7, 220–221.

[295] See the text quoted *supra*, n. 293.

[296] "Explicare. Id hac in re situm, po[ss]e hic rationem dari rei, de qua tam acriter disputatur, id q[uod]: cur urinatores non sentiant infra aquam eius gravitatem. Stevinus fingit vas quodpiam, et in eius fundo hominem horizonti parallelum, vasiq[ue] aquam infundit, q[uo]d fit? Non {sentis} o[mnin]o aquam, si vero in fundo sit orificium, per q[uo]d aqua descendere potest, idq[ue] si cor-

to the one proposed by Descartes to Reneri in his letter of 2 June 1631 (present in Clerselier's edition of Descartes's correspondence), where Descartes claims that air, composing a circle, comes to sustain our body from below, so that we are ease in 'holding' the air above us. Paradoxically, it is also akin to the one of Stevin himself, as the simple equilibrium of pressures is regarded as a key factor in our lack of feeling them. By extension, De Volder's argument does not challenge Boyle's objection to Stevin that this model does not explain what happens when an individual part of the body is pressed equally by large weights. In fact, De Volder reduces the explanation to "mechanical principles" only, as he put it in his *dictata*. The same approach is to be found as he comments upon Descartes's explanation of the behaviour of a body in equilibrium with a fluid (II.56–57), in which it is surrounded by particles pressing in all directions, and forming infinite circles around it. As seen above, Descartes explained the conditions of rest, as well as the fact that a solid in a fluid can be moved with very little force, by considering the circular motion of any part of the fluid – and without addressing the issue of the degree of pressure exerted by any given circle. According to De Volder, probably aware of this, all the "force of the demonstration" of Descartes consists only in the idea of the equilibrium between the parts. As he puts it in the Hamburg 273 series, one can avoid the difficulty of hypothesizing infinite circles of matter in motion just by considering that matter, in a fluid at rest, is in a condition of equilibrium among its parts, as in the case of a balance. Or, as he puts it in the Hamburg 274 series, Descartes's explanations of the conditions of stability of a solid in a fluid, indeed, are valid not because they are based on subtleties (*hamulis*) such as the existence of infinite circles, but first of all because they encompass the idea of mechanical equilibrium between equal forces.[297]

pore tegat, o[mnin]o tunc demum sentiet aquam, is, eumq[ue] secutus Boijle, putat aquam premere quidem, sed quia undiq[ue] premit aequaliter, non sentire, q[uo]d si verosimile sit, {ea} et inverosimile erit, q[uo]d sc[ilicet] o[mnin]o a torculari pressus non sentiet pressionem, verum quidem {e[nim]} si c[or]pus solidum e[ss]et, non sensurum pressionem, {[n]a[m]} si o[mn]ia in homine e[ss]ent aqua repleta, {c[um]} vero cavum sit corpus. {X} non aqua sed aere repletum, o[mnin]o sentiet. Alia ergo causa e[ss]e debet, quam licet Boijle voluerit impugnare. Verissima tamen {e[st]}, et quamvis Boijle testatur sibi ab autore nostro {[n]o[n]} e[ss]e satisfacturum, nil tamen, q[uo]d aliquomodo urget, profert. Concipiamus itaq[ue] hominem in aquis mediis constitutum, non sentiet (si premeretur) aquam, nisi deorsum orientem, si vero ostendamus hanc pressionis vim, non ab o[mn]e sed alia aqua sustineri, causa certa {e[st]}, concipiamus ergo cylindrum, {aq[uae]} ipsi incumbentem, si {hic} descenderet, alius ipsi aequalis ascendere deberet, c[um] vero sint aequilibrio, non potest unus descendere, nec {prom} premere, si vero concipiamus ap[er]turam q[ua]m tegit o[mnin]o, sentiet pressionem quia tunc una aqua descendere potest non ascendente alia, sed aere, ut si 2 habeamus bilances, et utrisq[ue] imponamus 100 ll, si q[ui]s manui sustineat alterutram, non sentient pressionem, quia ab altera parte est aequalis, si vero, 100 ll addat ʒ 1 tunc demum sentiet, quia utraeq[ue] non amplius sunt in aequilibrio, sed una praevalet, et magis premet, quia proinde causa sequitur ex principiis mechanicis," Hamburg 273, 235–236.

[297]As to II.57: "[m]utare. B ergo movebitur sed hoc motu non tolletur motus ab iis versus a, quia non opponitur huic motui, sed determinationi, nec est q[uo]d opponat quispiam, tales annulos authorem hic pro modulo fingisse, atq[ue] adeo hijpothesis usum e[ss]e. Moveantur. Ut haec difficultas tollatur, dico demonstrationem nostram non uti talibus circulis, sed fluidi aequilibrio, quod si non facere, pars q[uae] praevaleret, tolleret motum particularum, in contrarias partes motarum, unde haec aucta viribus reliquas secum aget, versus easdem partes, atq[ue] adeo non stagnans

With this reduction, however, he does not consider whether volumes of water sustain themselves by the arms of a balance (as in his samples), through our bodies (as if we were in-between a press, as for Stevin and Boyle), or by opposing each other outside our body, as in Descartes's model, where the volumes of water upon us sustain each other, as if they were an umbrella preventing us from being pressed by the weight of the fluid.

The very same 'combination' of Cartesian and Stevinian models, actually, is also present in the *Experimenta*, as he explains the lack of feeling of pressure underwater, in experiment 1,

> which is not felt when we are in a river, or sea, be cause all the water compresseth equally on all sides: but were there in the bottom of the sea a tubulus, hole, or cavity, which had not one drop of water in it, we should feel all the weight of the water then to press, and ponder, upon us, and with huge violence to force us into that hole, whereas were the hole full of water we should feel no such violence forcing us into it: even so the air.[298]

So far, this is Stevin's model, integrated, in experiment 2, with a reduction to the case of the lever:

> [...] the compression of the air which presseth and compresseth one part against the other which it findeth empty of all kind of air, even as water presseth against a hole that is empty of water, et not against a hole that is full of water, this being the nature of fluid bodies so to press, and yet to leave the bodies which are totally totally surrounded by them so easily moveable by virtue of the nature of the same fluid body, which maketh a contrary pressure ad aequilibrium. Even just of a scale with 100 lb in it presseth heavily on my hand but if you put into the other scale another 100 lb weight, then my hand which is under the scale will feel no pressure at all as it did before.[299]

amplius, sed motum erit fluidum, sic quoscunq[ue] fingamus circulos, demonstratio sequitur. Si itaq[ue] ponamus 2 circulos, sive plures ambientes B s[e]mp[er] eadem erit demonstratio, si enim tot sunt par[ticu]lae, q[uae] impellunt versus orientem ac versus occidentem, necessario manebit eodem in loco. Si vero plures ab oriente, movebitur versus occ[identem]: q[uo]d in stagnante e[ss]e nequit. Res haec commodius explicari nequit quam exemplo bilancis," Hamburg 273, 117–118. As to II.56–57, in the other series of *dictata*: "[...] ratio autem huius rei manifesta est, ut enim bilanx, in cuius utraq[ue] lance aequale sit positum pondus, manebit in aequilibrio; si vero alterutri addatur vel minimum pondus, statim ex ea parte praeponderabit. Sic etiam corpus B in fluido stagnante, sive cuius exiguae particulae non in hanc magis quam in illam feruntur partem, constitutum, pellitur a particulis q[uae] a {C} B versus F feruntur versus B, et idem corpus pellitur a totidem particulis aequali vi versus C, unde necessarium est, ut in eo loco, in aequilibrio quiescat, si vero alteri parti huius aequilibrii vis quaedam, licet minima addatur, {illae} vis efficiet ut nunc maiori vi propellatur B versus f quam versus C, cum autem maiores vires semper sortiatur effectus, necesse etiam est ut B versus F moveatur. §§ LVII. Hoc paragrapho autor aliq[ua]nto clarius explicat id quod priori proposuerat, et hunc in finem eligit in fluidi particulis certum quoddam motus genus, in quo evidentius id, quod prius universum demonstraverat, demonstrat, non tam existimandum est, autoris demonstrationem semper talibus hamulis fundatam esse, cum si accurate attendamus, tota vis {nitator illis}, q[uo]d partes fluidi in diversas aequali impetu ferantur partes, deinde quod omnis impetus vel sit circularis, vel circularibus aequipollere, tandem q[uo]d ex primo sequitur corpus B aequalibus viribus propelli versus C, quam versus F, ex quibus tribus, q[uae] in omni fluido locum habent, totam suam conficit demonstrationem," Hamburg 274, 48–49. See Fig. 5.60, Descartes 1644a, 63.

[298] De Volder 1676–1677, 81r.

[299] De Volder 1676–1677, 83r.

Please note that in this case not only is the lack of feeling of the pressure reduced to a pure static model, but also that Morley reports that water does not press on a hole full of water, exactly as in Descartes's model.

Eventually, this model is defended in De Volder's *Disputationes de aëris gravitate*, disputation 3, theses 7–8, in which De Volder claims that the water ascending in the place occupied by a man who sinks in water counterbalances – as on two plates of a balance – the pressure of the column of water above him, constituting a sort of "wall" (*obex*) protecting the man. In turn, a human body is not carried downwards by the pressure of water, but by its own weight, which is the cause of the ascension of the lighter medium, as happens in every fluid.[300] A Cartesian 'wall' which was also used as an explanatory model in a further disputation on the gravity of air De Volder presided over, namely in the *Disputatio inauguralis de gravitate aëris* (1690) of one of his foremost students, Joseph Serrurier, in whose *Annexa hydrostatica* it is clearly stated that "the superior parts of water gravitate on the lower ones," (this being in perfect accordance with the Archimedean model), that water presses "downwards, laterally, and upwards," and "only according to [its] height[,] whence it follows, that swimmers feel no weight under [...] water," (in accordance, in turn, with the Cartesian explanation of the lack of feeling of pressure as given by the model of the funnel, in the Hamburg 273 series of his *dictata*), on which water "a body lighter than water however gravitates."[301]

Thus far, for De Volder the ascension of 'light' bodies is just due to the downwards pressure exerted by heavier ones, which displaces them upwards, as he declares in disputation 2 of his 1676–1678 *De aëris gravitate*:

---

[300] "Verum ut pateat simul et quam imperite de hisce rebus loquantur, qui aëris hancce pressionem et gravitatem impugnatam volunt, et quando pressio illa debeat animadverti, age excutiamus huius effecti manifestam causam. Consideremus hunc in finem bilancem ex aequilibrio suspensam, ita ut utrique lanci mille imponantur pondo. Supponat quis uni lanci manum. Proculdubio nullam pressionem sentiet, non quod illa lanx non omni suo pondere deorsum ruat, sed quod huic deorsum pressioni obsequi nequeat, quin altera lanx paribus ad descensum viribus pollens assurgat. Qua ratione dum hae vires sibi invicem impedimento sunt, neutra lancium vel ascendit vel descendit, neutra suppositam manum vel minimum premit. Eadem ratione aquae aut cuiuscunque alterius liquoris quae in vase stagnat, una pars nequit descendere, quin ex altera parte tantundem aquae ascendat, quae cum parium cum priori si virium, huic descensui obicem ponit, efficitque pressionis aequalitate, ut duae quaelibet aquae partes eiusdem altitudinis eiusdemque quantitatis suas invicem premendi vires tollant et in aequilibrio sint. Ponatur nunc porro infra hasce aquas corpus quodcumque, id sane descendere nequit ea pressione, qua aqua superior imminet, quia dum descendit tantundem inferioris aquae, et parium idcirco cum superiore virium ascenderet. Hinc si vel descendat vel ascendat, id non fiet a pressione ipsius aquae, sed ab excessu gravitatis levitatisve quo aquam illud corpus superat. Cui consequens est, hanc aquae pressionem cum in aequilibrio sit a manu aut corpore nostro aquis supposito nequaquam animadverti posse. Quae demonstratio non tantum in aquis locum habet, sed in fluidis quibuscunque corporibus aqua sive levioribus sive gravioribus, sive minus sive magis compressis. Non enim nititur ea peculiari aquae natura, sed universali lege gravitatis et natura fluidorum corporum," De Volder 1676–1678, disputation 3, thesis 7.

[301] "I. Aquae partes superiores in inferiores gravitant. II. Premit deorsum, ad latera et sursum. III. Premit tantum secundum altitudinem unde sequitur, urinatores sub aquis nullum incumbentis pondus sentire," Serrurier 1690, *annexa* 1–3.

in fluid bodies the movement due to gravity, depressing some parts of the fluid, raises other parts, with the same force with which it depresses the other ones; whence it is demonstrated that not all the upwards movements exclude gravity as cause.[302]

This is consistent with the Archimedean model, according to which the parts of fluid or the solid bodies which are less pressed are brought upwards by the more pressed parts, by a lateral communication of pressure. However, it is not consistent with the Cartesian model, where each volume of matter strives to move upwards, and gravity is the result of being 'less light'. It worth noting here, that in his explanation of the lack of feeling of pressure in water (IV.26), Descartes himself considers weight as the cause of the descent of a solid in water ("no body is borne down by its own weight unless another lighter body, equal to it in size, is borne upward at the same moment of time"),[303] whereas in his *Le monde* he was far more consistent with his overall theory of gravity, as he claimed that

every part of terrestrial bodies is pressed towards [the centre of the Earth], not indifferently by all the matter surrounding it but only by an amount of the matter exactly equal to the size of that part and which, being underneath the part, can take its place if that part moves down. This is the reason why, among the parts of any single body that we call 'homogeneous', such as those of air or water, the pressure on the lowest is not notably more than that on the highest, and why a man at the bottom of very deep water does not feel it pressing on his back any more than if he were swimming right on top.[304]

The presentation given in the *Principia*, thus far, is more ambiguous than that offered in *Le monde*. This ambiguity is just accentuated in De Volder's disputations, where, (1) in accordance with his Archimedean approach, he considers downwards pressure, exerted by weight, as the cause of the upwards displacement of bodies, and (2) he does not enter into the cause of gravity as such, namely subtle matter, as he remarks "[...] the subtler matter is much less heavy than air (just in case it participates of any gravity, which I do not determine here)."[305] In other words, De Volder takes into account two separate models: (a) the interaction of 'terrestrial' i.e. 'heavy' bodies, whose interactions are determined by weight, and the heavy displaces upwards the lighter; (b) the interaction of terrestrial and subtle matter, which in Descartes's model of gravity determines the heaviness of terrestrial bodies. The two models are not unified in the *Disputationes*, nor, actually, were they unified in Descartes's *Principia* (IV.25), where the Frenchman does not really distinguish between the role of subtle matter and terrestrial fluids in determining the weight of a body, as he considers air as acting exactly like subtle matter:

in order that nothing may be omitted, it must also be noticed that, by 'heavenly matter', I mean here not only the globules of the second element, but also the matter of the first mingled with them. And those terrestrial particles which are following the course of the heav-

[302] "[...] id notari velim, in fluidis corporibus motum gravitatis quasdam deprimendo fluidi partes alias attollere eadem vi, qua reliquas deprimit, unde conficitur non omnem superiora versus motum gravitatem tanquam causam excludere," De Volder 1676–1678, disputation 2, thesis 5.

[303] Descartes 1982, 193.

[304] Descartes 2004, 49–50.

[305] See *supra*, n. 76.

enly matter (and thus are more rapidly moved than the rest) are also to be placed in this category; such are those which form the air.[306]

Eventually, a last difficulty is entailed by De Volder's mixed approach to the lack of pressure: namely, surfaces act as "walls" protecting the lower volumes of water – and of the human bodies in it – from being pressed by the upper volumes. Thus far, no increase in pressure can take place across the body of water, as water does not gravitate on itself. This may explain why we do not find, in the disputations, the negation of this traditional thesis, which is reported, as seen above, in Morley's account of experiment 24.[307] More importantly, this undermines the very Archimedean model used by De Volder to explain floatation, according to which a solid body, lighter than water, is immersed by as much as the pressure it undergoes from below counterbalances its overall weight. So that an increasing pressure in the water has to be admitted for the model to hold. In fact, in the Cartesian model entailed by De Volder's explanation it is not the depth of the submerged part which matters in explaining flotation (as pressure does not increase across the fluid), but only the volume of fluid displaced by the solid, which strives to occupy the 'place' of the solid, and keeps it floating. These internal inconsistencies in De Volder's 'mixed' model (not to mention his explanation of the pressure exerted in a funnel, which is inconsistent with the principle of the lever), in fact, might have been a cause of his lack of willingness to circulate his ideas outside his academic lectures and disputations.

5.6 De Volder's Treatment of the Properties of Air

5.6.1 The Extension of the Hydrostatic Model to Air

The adoption of a theory of floatation based on the idea that pressure is exerted by and across entire columns of fluids had a primary function in De Volder's theories: namely, to explain the cohesion of heavy bodies like the cylinders of marble, and the Magdeburg hemispheres. As he put it in his *Disputationes de aëris gravitate*, the laws of hydrostatics are valid for any fluid (at least terrestrial ones, since he does not advance claims on the heaviness, and the kind of pressure exerted by subtle matter, as seen above).[308] His extension of the hydrostatic model to air is evident – besides his direct analysis of the cohesion of the hemispheres and the marbles in the *Experimenta* (experiments 1–2 and 3) and *Disputationes*, – in experiment 21, where

[306] Descartes 1982, 192. Please note that De Volder does not address the role of air, in commenting upon this article: cf. Hamburg 273, 235; Hamburg 274, 99–100.

[307] See Sect. 5.1, *The gravity of the air: a complex idea.*

[308] "Quae demonstratio non tantum in aquis locum habet, sed in fluidis quibuscunque corporibus aqua sive levioribus sive gravioribus, sive minus sive magis compressis. Non enim nititur ea peculiari aquae natura, sed universali lege gravitatis et natura fluidorum corporum," De Volder 1676–1678, disputation 3, thesis 7.

he shows how a sphere of glass loses some weight when weighed in air, with respect to when it is weighed in a vacuum. In other words, the behaviour of a body in water or air is exactly the same. However, De Volder's criticisms of some of the objections brought against the idea of the weight of the air in disputation 3 show, at full extent, his positions on the nature of air.

In this disputation, De Volder takes into account four mains objections, whose source is not disclosed, and – alas – has not been identified to date. In fact, they could have been devised by De Volder to forestall criticisms of his model. These can be summarized as follows:

1. since subtle matter flows into evacuated hemispheres, the external air, occupying the space left vacant by the subtle matter, does not press on the container itself (thesis 3).
2. Air does not press, as we do not feel this pressure (thesis 5).
3. The pressure of air is not due to its weight (thesis 9).
4. It is the volume of air expelled from the hemispheres that sustains the weights appended to them, since air presses equally from above and below (thesis 10).

To objection 2, De Volder replied by proposing his argument about the lack of feeling of pressure in water, as seen above. To objection 3, he deploys a theory of the compression of air – which I analyze below. Against the other two objections, De Volder proposes his Archimedean theory of flotatation: which works well against them as, in fact, objection 1 is rather directed against the theory of circumpulsion, according to which it is only the weight of the expelled air that raises water in pumps, rather than to his Archimedean model. Objection 4, in turn, works both against (1) the idea of circumpulsion – since it entails that, pressure being equal on the hemispheres, the weights appended to them are sustained only by the weight of the small amount of air expelled from the hemispheres, which increases this pressure, allowing the suspension of further weight – and (2) against the idea that hemispheres 'float' in air as if they were a body lighter than air itself. So that the appended weight is sustained by a force equal to that exerted by a volume of water equal to the hemisphere. An idea, in fact, that was discussed in 1670s, after the appearance of Francesco Lana de Terzi's *Prodromo, ovvero Saggio di alcune inventioni nuove, premesso all'Arte maestra* (1670) (not owned by De Volder),[309] in which the Italian Jesuit proposed the model of an aerostat, sustained in mid-air by evacuated copper spheres (see Fig. 5.63, Lana de Terzi 1670, plate 4), and considered the difference in specific weight between the air and the spheres.[310] This idea was then criticized in the first issue of Robert Hooke's journal *Philosophical Collections* (1679) – where

[309] De Volder owned the first two volumes of Lana de Terzi's *Magisterium naturae et artis* (1684–1686; the third volume was published in 1692): see *Bibliotheca Volderina*, 1.

[310] Lana de Terzi established the ratio of the weights of air and water *via* the method proposed by Galileo in his *Discorsi*, quantifying it as 1 to 640, and then claimed that the diameter of the copper spheres capable of floating in air should be of 14 feet: see Lana de Terzi 1670, 53, 56–57.

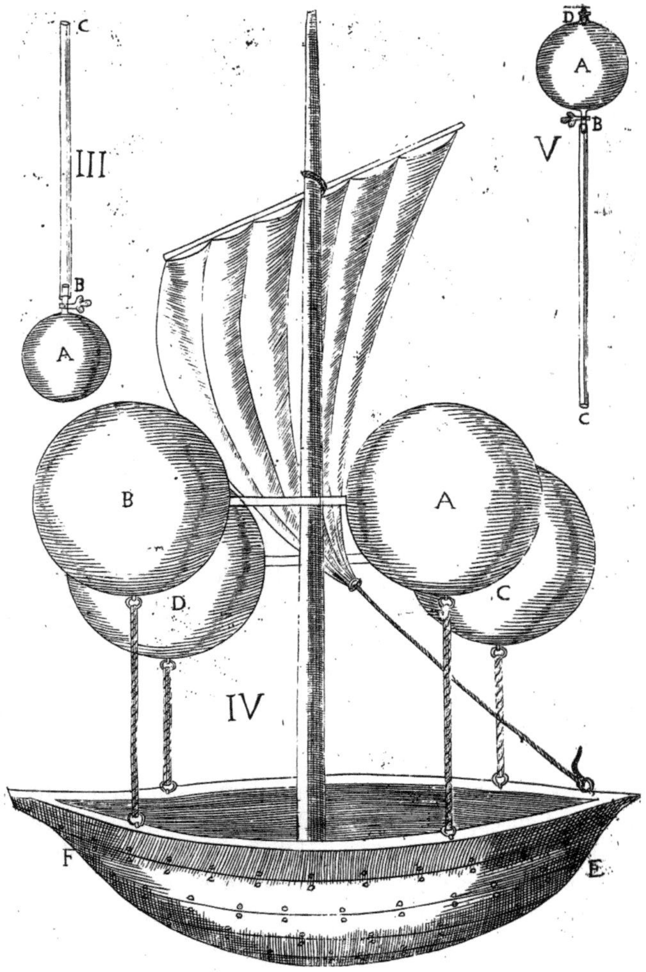

Fig. 5.63 Model of a flying chariot. (Source: Lana de Terzi 1670, plate 4. ETH-Bibliothek Zürich, signature: Rar 9005, https://www.e-rara.ch/zut/doi/10.3931/e-rara-30705. Public domain mark 1.0)

Hooke deemed it unfeasible, since such spheres cannot be built, as they would be too thick (and heavy) to resist to air pressure.[311]

In addressing these criticisms, thus far, De Volder re-instates a model of floatation involving entire columns of fluid, and not only the volume of fluid displaced by the solids immersed in it. Such criticisms, too, work well against the idea of

[311] See Hooke 1679, 21–24. Lana de Terzi's idea was also discussed in Leibniz's *Hypothesis physica nova* (1671). In 1671 a review by Henry Oldenburg of the *Prodromus* appeared in the *Philosophical Transactions*, but no discussion of the idea of the aerostate is made. Johann Christoph Sturm labelled Lana's flying chariot as realizable: Sturm 1676, 56–66.

circumpulsion, defended by Hobbes and by Descartes. So that we may infer that their discussion by De Volder is an indirect criticism of this theory. In fact, the only place in which Descartes supported a theory of 'suspension' of a heavy body (such as mercury) by air is his aforementioned letter to Reneri, where the pressure of a whole column of air is taken into account. In his overall model of floatation, on the other hand, only 'separate' volumes of fluid are considered.

5.6.2 *Air as an Elastic Fluid*

And yet, this extension of the hydrostatic model had to deal with features that the air has, and that water does not. Namely, compressibility and elasticity. Generally speaking, in the seventeenth century no experiments on the compression of liquid water could be successfully carried out, as water has a very limited degree of compressibility. Boyle, as seen above, claimed that water can be compressed – but he does not devote specific experiments to this topic. In fact, direct observations on the compression or rarefaction of water were carried out only as to ice, such as in the *Saggi di naturali esperienze*, where several observations on ice are reported. De Volder does not deal with this topic, even if the rationale of his theory of hydrostatics is that all fluids (at least, terrestrial ones) obey the same laws, so that water, too, should be compressible.[312] The case of elasticity was clearer, as the elastic power of air (which water does not posses) was a foremost topic in the seventeenth century, and could be easily demonstrated, especially with the air pump. Water, on the other hand, was generally labelled as not elastic.[313] De Volder inserted himself into a debate over the nature of elasticity and its effects, which had started when Roberval, in 1648, observed what happens to a bladder, filled with air, once it is placed in an evacuated space, such as the top of a Torricellian tube.

5.6.2.1 The Idea of Elasticity of Air Before De Volder

The idea that compressed air exerts a force to dilate itself had been well known at least since the *Pneumatica* of Hero of Alexandria, in which he described what we know as 'Hero's fountain', working by air compressed by water, and which was fully translated into Latin and published for the first time by Federico Commandino in 1575.[314] In the theoretical preamble to his treatise, Hero labelled the effort of air

[312] See, in particular, disputation 3, thesis 5: "[e]stne igitur aqua non gravis? An vero ea ipsa aëre minus coarctata compressave est?" Also, De Volder – as seen in the comparative table on the weights of air and water (Table 5.1, *De Volder's measurements in hydrostatics*) – performed measurements on the weight of the water, ascertaining its changes.

[313] See Bertoloni Meli 2006, chapter 6.

[314] Further editions of Commandino's translation, both owned by De Volder (see *Bibliotheca Volderina*, 5) appeared in 1583 and 1680. Another popular device, working by air compression, was the arquebuz, or wind-gun, allegedly invented by Hero's master, Ctesibius. Before

to dilate as an effect of the "preternatural compression" of the particles of which it is composed – compared to Hero to sand grains – and made possible by its interstitial vacuums. After their compression, such vacuums try to expand and restore their previous condition. According to Hero, this is what happens in sponges as well.[315] In fact, Hero's treatise – as shown by Matteo Valleriani[316] – definitely had a key role in dismantling Aristotelian theories of the rarefaction of air. According to Aristotle's *Physica*, rarefaction and condensation are not quantitative alterations of substances. Rather, they are different qualities that a given quantity of matter of a substance assumes, once the substance itself undergoes a substantial mutation. For instance, when a certain quantity of water is heated by the Sun, it rarefies and becomes air, but the quantity of matter does not change:

> [t]he same matter also serves for both a large and a small body. This is evident; for when air is produced from water, the same matter has become something different, not by acquiring an addition to it, but has become actually what it was potentially, and, again, water is produced from air in the same way, the change being sometimes from smallness to greatness, and sometimes from greatness to smallness. Similarly, therefore, if air which is large in extent comes to have a smaller volume, or becomes greater from being smaller, it is the matter which is potentially both that comes to be each of the two.[317]

In this process, for Aristotle, no interstitial vacua are produced, but the same quantity of matter occupies a greater volume. With this change in quality, moreover, the same quantity of matter becomes more or less heavy, as what is rarefied (like air) is light, and what is condensed (like water) is heavy, as Aristotle continues by remarking:

> [t]he dense is heavy, and the rare is light. [...] There are two types in each case, both in the dense and in the rare; for both the heavy and the hard are thought to be dense, and contrariwise both the light and the soft are rare; and weight and hardness fail to coincide in the case of lead and iron.[318]

With the appearance of the full text of Hero's *Pneumatica*, this view (which was supported by Zabarella and the Conimbricenses)[319] came to be progressively aban-

---

Commandino's edition, Hero's ideas were spread by their discussion given in Girolamo Cardano's *De subtilitate rerum* (1550).

[315] "[A]ëris corpora inter sese quidem coaherent, non tamen ex omni parte, sed interiecta habent intervalla quaedam vacua, sicut arena, quae est in littoribus. Itaque minimo concipiendum est arenae particulas corporibus aëris similes esse, aërem vero, qui inter particulas arenae intericientur, similem vacuis intra aërem contentis. Quemombrem vi quadam accedente aërem densari contingit, et in vacuorum loca residere, corporibus praeter naturam inter sese compressis: remissione vero facta rursus in eundem ordinem restituitur, ob naturalem corporum contentionem, quemadmodum et in cornuum ramentis, et in spongiis siccis: quae si compressa remittantur, rursus in eundem locum redeunt, eandemque accipiunt molem. Similiter si aliqua vi aëris particulae a se invicem distractae fuerint, et maior praeter naturam locus vacuus fiat, rursus ad sese recurrunt," Hero and Commandino 1583, 8–9. Descartes used the model of the sponge to explain rarefaction in *Principia* II.6–7. His idea of elasticity, however, does not rely on the comparison with the sponge.

[316] See Valleriani 2010, 173–176.

[317] Aristotle 2014, volume 1, 369; cf. *Physica*, 217a26–33.

[318] Aristotle 2014, volume 1, 369; cf. *Physica*, 217b11–19.

[319] See Zabarella 1590, column 711; Conimbricenses 1592, 500–504.

doned. As argued by Edward Grant, indeed, Aristotle's theory of condensation was silent on the relation between the quantity of air, its heaviness, and its condensation. Thus, for Aristotle there can be a change in density, in heaviness, but not in the quantity of matter filling a given volume.[320] Hero's *Pneumatica*, on the other hand, offered a theory of matter accounting for rarefaction and condensation in terms of quantity of matter, and moreover explained what would then be labelled as the 'elasticity' of air, or the effects of its decompression. As early as 1589, for instance, Giovanni Battista Aleotti explained the phenomena of decompression of air in his *Aggiunta* to *Gli artifitiosi et curiosi moti spiritali di Herrone* by reverting to the idea of a vacuum, which he labelled as possible, even if with a violent motion. According to Aleotti, what we now label as the elasticity of air was the effect of its filling, after it had been compressed, the violently created vacua. With his explanation, Aleotti no longer relied on the idea of rarefaction of air as due to a substantial change, but on the "packing" of air (*aria constipata*).[321]

Later, Hero's machines would be presented in very popular books such as Della Porta's *Magia naturalis* (book 19), and Ens's *Thaumaturgus mathematicus* (*problema* 93), where it is claimed that air "dilates itself with all the force," but no detail is given as to the underlying processes.[322] Such dissemination of Hero's pneumatics prompted the further elaboration of corpuscular and atomic theories, such as that of Bacon, who overtly used Hero's idea of a disseminated vacuum in order to explain rarefaction and dilatation in his *Cogitationes de natura rerum* (*De sectione corporum, continuo, et vacuo cogitatio prima*)[323] and Beeckman, proposing an atomist and vacuist theory in his *Journal*.[324] Eventually, Hero's ideas on matter can be found the *Prefatio* of Mersenne's *Cogitata physico-mathematica* (1644), where Hero's preface to the *Pneumatica* is used to defend the idea of disseminated vacuums, as allows an explanation of the condensation and rarefaction of the air – whose force of rebounding is compared to that of a bow.[325]

The real turning point in the study of the 'elastic' properties of air, however, was the experiment of Roberval, showing that air has a power of dilatation which is exerted in any condition, not only when it is forcibly pressed. In other words, Roberval's experiment concerned air in what was supposed to be its 'natural' state. From a theoretical standpoint, his experiment was grounded on the underlying idea that air behaves like a mass of wool which, in its lower strata, is always compressed

[320] See Grant 1981, 72–73.

[321] Hero and Aleotti 1589, 8. For a thorough discussion, see Valleriani 2014.

[322] "Quibus ita peractis aqua fundum occupante, aër valide pressus, et tota se virtute dilatare nisus, aquam ita coarctat, ut eam semel aperto epistomio violenter exprimens, in altum prosilire cogat," Ens 1636, 204. On the fortune of Hero's pneumatics, see Boas 1949.

[323] Posthumously published in his *Scripta in naturali et universali philosophia* (1653). Discussed in Manzo 2003. See also Gemelli 1996.

[324] For a thorough discussion, see Gemelli 2002; Van Berkel 2013, chapter 4.

[325] See Mersenne 1644, *Praefatio*, 8 (unnumbered). In turn, in his *Mechanica hydraulico-pneumatica* Schott (who criticizes to the idea of interstitial or disseminated vacuum) traces back the resilience of air after compression to the resistence that bodies exert to their penetration by other bodies: see Schott 1657, 45–46.

by the incumbent air. Moreover, it was made possible by the use of the *instrumentum Torricellianum*, namely a tube at whose top a vacuum is created, where Roberval put a sealed bladder containing air. According to his *De vacuo narratio secunda*, addressed to Pierre des Noyers in May–June 1648 (and published only in 1923), this bladder, once freed by the compression of the surrounding air, is dilated by its own "appetite" or "force" to dilate or rarefiy, which for Roberval is equal to the force by which it has been compressed, such as happens for solid bodies which are elastic.[326] In accordance with this idea, Roberval also showed that the pressure exerted by a separate volume of air can be equal to the pressure of the atmosphere itself: for instance, a certain quantity of air left at the top of a Torricellian tube depresses the mercury more than an equal volume of water would do (a theory which he shared with Étienne Noël, who discussed this case in his *Gravitas comparata, seu Comparatio gravitatis aëris cum hydargyri gravitate* (1648)).[327]

Roberval's experiment with the bladder – for which he does not provide a corpuscular explanation – was then reported by Gassendi in his *De facto novo circa inane experimento* (October 1648).[328] For Gassendi – who echoes Hero's explanation – the elastic power of air depends on the compression of the compound of atoms, which makes them fill their interstitial voids and, once released, tend to reassume their previous disposition.[329] Hence, Roberval's experiment was widely reported by Jean Pecquet, who provided a depiction of Roberval's instrument in his

[326] "Debet autem ad hoc intelligi aër hic noster quem respiramus, tantum possidere appetitum, seu tanta vi pollere ad sese dilatandum atque rarefaciendum, quanta est vis naturae elementaris ipsum comprementis seu condensantis. Quod etiam ipsius naturae legibus apprime quadrat, et constantissime servatur in omnibus illis corporibus quibus resiliendi facultas ab ipsa natura concessa est, ut in arcubus, et aliis innumeris: quae quidem omnia corpora, donec tenduntur aut vi comprimuntur, modo non ultra metas virium suarum, nunquam cessant reniti; ac tanta vi ad resiliendum feruntur innato appetitu, quanta ab aliis corporibus trahuntur, aut impelluntur: ita tamen ut initio sui resultus, vis illa fit maxima, inde vero sensim languescat fiatque minor ac minor, ac tandem nulla; postquam scilicet corpus resiliens ad debitum sibi statum redierit. Eodem modo, idem aër tubo inclusus, quandiu tubus ille clausus remanet ex utraque parte, tantum premitur atque condensatur quantus est nisus naturae elementaris ad premendum hunc nostrum aërem; ideoque ille idem vi rursus pari, tali compressioni renititur, dilatari appetens, ut locum ampliorem sibi naturaliter debitum acquirat. Neque enim ille tubo inclusus minus premitur aut condensatur quam extra; quia cum iam pressus atque condensatus assumptus sit, nullam exinde libertatem nactus est adhuc ut dilatari ac rarefieri posset," Roberval 1923, 315. For an extensive account, see Webster 1965.

[327] "At vero, si praeter hydrargyrum aut aquam, admittatur in tubo portio aliqua hujusce nostri aëris pressi atque condensi, ut supra diximus, is aër libertatem nactus, ac suis omnibus partibus resiliens ad sese rarefaciendum, impellet ipsum hydrargyrum aut aquam, quae ideo deprimentur infra praedictam altitudinem, vel magis vel minus, prout aër ipse majores aut minores ad rarefactionem vires obtinebit. Eodem prorsus modo quo aqua in aequilibrio constituta: dum scilicet ejus superior superficies in mari vel stagno aliquo, vel etiam in vase, ad horizontem librata est," Roberval 1923, 317–318. Cf. Noël 1648, 55–56. I owe most of the references to Webster 1965, who provides a full-fledged account of the first experiences on the elasticity of air.

[328] This text is included in Gassendi's *Animadversiones in decimum librum Diogenis Laertii*: see Gassendi 1649, *Appendix*, iii–x. Gassendi had also reported and commented upon Roberval's experiment with vacuum in a vacuum: see Mersenne's *De nupero circa inane coacervatum* (April 1647): in Gassendi 1649, 424–444.

[329] See Gassendi 1649, *Appendix*, v–vi.

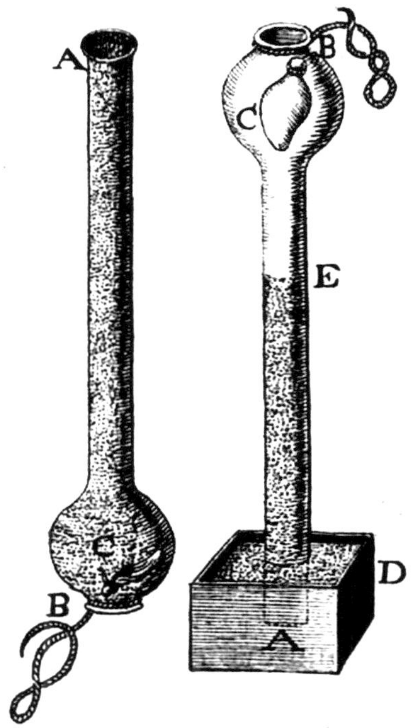

Fig. 5.64 Bladder in a vacuum aimed at demonstrating the elasticity of air. (Source: Pecquet 1654, 52. Rome, Biblioteca Nazionale Centrale, signature: ā⁴ē⁴A-2I⁴, https://archive.org/details/bub_gb_XzZOLZQE1qIC. Public domain mark 1.0)

1651 *Dissertatio anatomica de circulatione sanguinis et chyli motu* (see Fig. 5.64, Pecquet 1654, 52), and supposed that air has an elastic power (*elater*) besides a weight (*pondus*).[330] For Pecquet, as for Torricelli, air can be compared to a mass of wool, or to a sponge whose lower parts are compressed by the upper ones, and once freed shows their resilience.[331] Moreover, he remarks that Descartes's theory of elasticity does not fit with Roberval's experiment. Descartes, actually, had provided a theory of elasticity of air in his *Principia* (IV.46–47), where he claims that air is composed of flexible, non-rigid particles (so that they are not elastic in themselves), which are comparable to feathers. Their elastic force consists in their being agitated by subtle matter, which, once flowing in-between them, it puts them into a rotational

[330] See chapter 8: "[e]sse non pondus tantum, sed et rarefactorium aëri elaterem experimentis demonstratur," Pecquet 1651, 48.

[331] "Hunc fingito tibi velut spongiosi vel lanei potius cumuli terraqueum orbem ambientis molem; cuius proinde partes superiores ab inferioribus, compressione gradatim incedente, sustineantur, sicut, quo Terris accedunt vicinius, eo etiam incumbentium nisu et pondere compactius opprimantur: et ob id spontaneo dilatatu (quem elaterem nuncupo) utcunque graviter aggestarum onus, subiectae nitantur, si libertas adsit, rarescere. Hinc infero eiuscemodi partium infimam, ut toti subiectam oneri, sic omnium esse maxime condensatam; et per hoc ab eadem non solo duntaxat Pondere, sed et elasteris, cuius tum validissimus ad rarefaciendum nixus est, virtute terraqueae sphaerae premi superficiem," Pecquet 1651, 48–49. The idea of the wool can be found also in the letter of Torricelli to Ricci of 28 June 1648, where, however, no elasticity of the air is considered, but only its compression: see Galluzzi and Torrini 1975, volume 1, 130–132.

movement so that they slam into each other. For Descartes, this explains both the decompression of air in machines, and its rarefaction due to heat.[332] This theory would be then adopted by De Volder, who in both series of his *dictata* comments upon IV.46 by redirecting the reader to the *Météores*, where Descartes had used the same kind of explanation to account for the dilatation of vapours, whose particles are like cords, and whose circular movement, due to the flowing of subtle matter, makes them rigid as a consequence of the centrifugal force exerted by their extremities, which lead them to slam into each other.[333] Pecquet denies that a theory based

[332] "46. Why it is easily rarefied and made dense. Air, however, is easily made dense by cold and rarefied by heat: for, since almost all of its particles are flexible, like soft feathers or thin cords; the more rapidly they are driven, the more widely they extend themselves, and therefore require a larger sphere of space for their movement. [...] 47. Concerning its forcible compression in certain machines. Finally, air which is forcibly compressed in some vessel has the {same} force to spring back {as was employed to compress it} and to extend itself immediately into a wider space. On the basis of this, machines are created [...]. The cause of this is that, when air has thus been compressed, each of its parts does not have that small spherical space which it needs for its movement to itself, but other neighboring particles enter into that space. Since the same heat or the same agitation of these particles is meanwhile maintained by the heavenly globules constantly flowing around the particles of air, the latter strike one another with their extremities, and attempt to drive one another out of their place, and thus together produce the force to occupy a greater space," Descartes 1982, 204–205.

[333] In the Hamburg 273 series, notably, De Volder makes recourse to the case of closed thermoscopes: "aeris phaenomenum est, quod frigore condensatur, et calore rarefit, ut patet in thermoscopiis, in quorum superiore parte est aer, in inferiore aqua fortis, colore q[uo]dam tincta, ubi enim aer calore maiore movet in latius se extendens premit aquam adeo ut effluere debeat, in frigore contra, contrarium fit in thermoscopiis clausis, non tam a rarefactione aeris, quam, aquae pendentibus. Duo enim eorum sunt genera, quorum prius solummodo huic nostrae rei confert, aer sc[ilicet] rarefit, et plus occupat spatii, quod ex natura eius evidentissime sequitur. Part[ticu]lae enim eius sunt flexiles et molles, quae flaccidae quidem haerent extenduntur et maius requirunt spatium, instar funiculi cuius exempli author utitur metor. c. 2. § 3. Sicq[ue] maior motus impellit corpuscula, aut per lineam rectam aut circularem q[uae] regula generalis est per circulare in orbem feruntur, si transferantur in maiori motu q[ua]m ut ei per lineam possunt obsequi, sic radii solares ita impingunt et movent corpora, cumq[ue] non possunt omnia obsequi motus per lineam rectam ferri debent per circularem, quod cum fit vim habent alia corpora excludendi, sicq[ue] maius spatium occupandi, ubi autem motus c[i]eat pori rursus sunt minores et aer condensatur," Hamburg 273, 251–252. Cf. the other series: "[q]uo pacto molles plumulae et tenues folliculi celeri agitatione possunt rigiditatem acquirere, accurate explicatum vide in meteoris p. 3 cap. 2," Hamburg 274, 102. Cf. Descartes's *Météores*, *Discours second*: "[r]emarqués aussy que les vapeurs occupent tousiours beaucoup plus d'espace que l'eau, bienqu'elles ne soient faites que des mesmes petites parties. Dont la raison est que lorsque ces parties composent le cors de l'eau, elles ne se meuuent qu'assés fort pour se plier, et s'entrelacer, en se glissant les vnes contre les autres, ainsi que vous les voyés representées vers A. Au lieu que lors qu'elles ont la forme d'vne vapeur, leur agitation est si grande, qu'elles tournent en rond fort promptement de tous costés, et s'estendent par mesme moyen de toute leur longeur, en telle sorte que chascune à la force de chasser d'autour de soy toutes celles de ses semblables, qui tendent à entrer en la petite sphere qu'elle descrit. Ainsi que vous les voyés representées vers B. Et c'est en mesme façon que si vous faites tourner assés viste le piuot LM, au trauers duquel est passée la chorde NP, vous verrés que cette chorde se tiendra en l'air toute droite et estendue, occupant par ce moyen tout l'espace compris dans le cercle NOPQ, en telle sorte qu'on n'y pourra mettre aucun autre cors, qu'elle ne le frappe incontinent auec force, pour l'en chasser: au lieu que si vous la faites mouuoir plus lentement, elle s'entortillera de soy mesme autour de ce piuot, et ainsi n'occupera plus tant d'espace," AT VI, 241–243.

on subtle matter could fit with Roberval's experiment, as the subtle matter, entering into the tube through the pores of the glass, would rather compress the bladder – namely in Pecquet's perspective subtle matter would just act as air does.[334]

To put it briefly, after Roberval's experiment, three main 'schools' emerged as to the explanation of the elasticity of air. First, those explaining the elasticity of air by some sort of subtle matter flowing in-between its particles, and rarefying it: this explanation was that of Descartes and his followers, such as Regius, Rohault, and, in England, of Henry Power, who discussed it in his *Experimental Philosophy* (1664).[335] Second, those explaining elasticity by considering the features of the particles of air itself, i.e. by conceiving them as elastic in themselves, like Pequet (who adopted the model of the wool), Gassendi (who ascribed elasticity to compounds of atoms), Pascal and Mariotte (who both conceived of air as a mass of wool, or as a sponge), and later Senguerd.[336] Third, the non-committal approach of some experimentalists, like those of the Accademia del Cimento, as well as Von Guericke and Johann Christoph Sturm, who did not provide insights into the causes of elasticity, but just demonstrated it by rehearsing Roberval's experiment.[337]

The chief figure in the treatment of the elasticity of the air, however, was Boyle, who repeated the experiment with the bladder in a container evacuated by the air pump, and discussed it in his *New Experiments* (1660), viz. in experiments 4–6. For Boyle, the experiment with the bladder shows that air has an elastic power, which may be due, on the one hand, to the agitation of its particles, as evident from experiments on the heating of air carried out by Mersenne (who labelled the problem of rarefaction as the "most difficult," and rejected the idea of subtle matter as explaining it) with the aeolipile.[338] On the other hand, Boyle maintains that air is elastic because its particles are like little springs, so that "in case they be bent or compress'd by the weight of the incumbent part of the Atmosphere, or by any other Body, they do endeavor [...] to free themselves from that pressure, by bearing against the contiguous Bodies that keep them bent."[339] Such springs are like "flexible Hairs; each of which, may indeed, like a little Spring, be easily bent or rouled up; but will also, like a Spring, be still endeavouring to stretch it self out again," and are provided with a "Principle of self-Dilatation."[340] Descartes's explanation of the phenomenon

[334] "Nam, quam afferunt, ratio futilis est. Penetrat, inquiunt, vitream aetherea subtilitas densitatem, et affusa quaquaversum complicatae vesiculae, per illius tandem poros intus irrepit, ac totam in pristinum distendit tumorem. Sed meminerint, quod autumant, pervium esse circumquaque vitrum ingressuro aetheris et proinde versus vesicam [...], aetherem aequali undique radiorum irruentium impetu ferri; sic ut eam potius interceptam comprimat, opprimatque, quam eidem in seipsam reacturus, inutilem dilatationem curet imponere," Pecquet 1651, 52.

[335] See Regius 1654, 63–64; Power 1664, 101–103; Rohault 1671, volume 2, 150.

[336] Pascal 1663, 18–21; Senguerd 1681, 178; Mariotte 1686, 139–140. See also Charleton 1654, 55–57, discussed in Webster 1965.

[337] See Magalotti 1666, 31–34; Von Guericke 1993, 179; Sturm 1676, 22.

[338] Boyle 1660, 28–29, 32–33. Cf. Mersenne 1644, 149.

[339] Boyle 1660, 22.

[340] Boyle 1660, 23.

of the dilatation of air after its compression in some machines is also reported by Boyle, who however declares that his own solution is "more easie."[341]

Hence, in his subsequent *Defence of the Doctrine Touching the Spring and Weight of the Air* (1662) – aimed against the objections of Francis Line to his *New Experiments*[342] – Boyle provided, *in nuce*, what we know today as the 'law of Boyle-Mariotte', according to which the volume and pressure exerted by air are inversely proportional. In fact, Boyle did not express the law in a formula: however, he provided a table reporting the values of the pressure of air according to its degree of condensation, which he measured by using an U-shaped tube, in which one of the horns was shorter than the other, and which was kept closed. By filling the longer tube with mercury, Boyle could measure the different volumes, the degrees of compression of air in the shorter one, and the elastic force it exerts – capable of counterbalancing 29 inches of mercury. In this way, Boyle – after Roberval – ascertained that the pressure of air, due to its "spring," can 'weigh' like a whole atmospheric column, and that its spring or elastic force restitutes the pressure exerted by such a weight.[343] Notably, the corpuscular theory supporting Boyle's theory, as given in his 1662 *Defence*, was a sort of *via media* between the theory he had provided in the *New Experiments*, and that of Descartes, as in Boyle's *Defence* it is maintained that particles of air are "very long, slender, thin and flexible laminae, coyled or wound up together as a Cable, piece of Riband, Spring of a Watch, Hoop, or the like, are," which extend themselves as a consequence of a rotational movement which is innate to them.[344] As reconstructed by Antonio Clericuzio, this theory was not devised by Boyle, but by Hooke: in fact, it postulated the Epicurean idea that movement is

---

[341] Boyle 1660, 25–27.

[342] I.e. to Line's *Tractatus de corporum inseparabilitate* (1661). See Shapin and Schaffer 1985, chapter 5.

[343] "[W]e began to pour Quicksilver into the longer leg of the Siphon, which by its weight pressing up that in the shorter leg, did by degrees streighten the included Air: and continuing this pouring in of Quicksilver till the Air in the shorter leg was by condensation reduced to take up but half the space it possess'd (I say, possess'd, not fill'd) before; we cast our eyes upon the longer leg of the Glass, on which was likewise pasted a list of Paper carefully divided into Inches and parts, and we observed, not without delight and satisfaction, that the Quicksilver in that longer part of the Tube was 29. Inches higher then the other. Now that this Observation does both very well agree with and confirm our Hypothesis, will be easily discerned by him that takes notice that we teach, and Monsieur Paschall and our English friends Experiments prove, that the greater the weight is that leans upon the Air, the more forcible is its endeavour of Dilatation, and consequently its power of resistance, (as other Springs are stronger when bent by greater weights.) For this being considered, it wil appear to agree rarely-well with the Hypothesis, that as according to it the Air in that degree of density and correspondent measure of resistance to which the weight of the incumbent Atmosphere had brought it, was able to counter-balance and resist the pressure of a Mercurial Cylinder of about 29. Inches, as we are taught by the Torricellian Experiment; so here the same Air being brought to a degree of density about twice as great as that it had before, obtains a Spring twice as strong as formerly," Boyle 1662a, 58–59.

[344] "We will further suppose each of these so coyled up to have such an innate circular motion [...]. By this Circular motion the parts of the laminae endeavouring to recede from the Centre or Axis of their motion, acquire a Springiness outward like that of a Watch-Spring, and would naturally flye abroad untill they were stretch'd out at length," Boyle 1662a, 94–95.

innate to corpuscles: an idea elsewhere criticized by Boyle, who does not adopt it in his other works.[345]

Eventually, Boyle's results were to later find confirmation in Mariotte's 1679 *Discours de la nature de l'air*[346] and *Traité du mouvement des eaux* (1686), where the same procedure followed by Boyle is described (see Fig. 5.65, Mariotte 1686, 142), and the same conclusion is inferred: the more air is compressed, the more it exerts pressure, or *ressort*.[347]

---

[345] For instance, in his *The Origins of Forms and Qualities* (1666). As detailedly reconstructed in Clericuzio 1998, Huygens criticized Boyle's explanation in a letter read at the Royal Society on 30 July 1662, in which he objected to the idea that the particles of air are provided with an innate movement. To it, Huygens opposed the Cartesian theory of elasticity. In his answer, read in the same day, Boyle pointed out that the theory was Hooke's, and that the printers omitted his name.

[346] The *Discours* or *Essay de la nature de l'air* dates back to 1676, and it was printed as the second of his *Essays de physique* (1679–1681). Cf. "[...] il faut avoir un tuyau recourbé, dont les deux branches soient parallèles, et dont l'une soit d'environ huit pieds de hauteur, et l'autre de douze pouces; la grande doit être ouverte en haut, et l'autre scellée exactement. On commencera à verser un peu de mercure pour remplir le fond où est la communication des deux branches, et on fera en sorte que le mercure ne soit pas plus haut dans l'une que dans l'autre, afin d'être assuré que l'air enfermé n'est pas plus condensé ou dilaté que l'air libre. On versera ensuite peu à peu du mercure dans le tuyau, prenant garde que le choc ne fasse entrer de nouvel air avec celui qui est enfermé; et on verra, comme je l'ai vu plusieurs fois, que, lorsque le mercure sera élevé à quatre pouces dans la petite branche, le mercure sera dans l'autre quatorze pouces plus haut, c'est-à-dire, dix-huit pouces au-dessus du tuyau de communication; ce qui doit arriver, si l'air se condense à proportion des poids dont il est chargé, puisque l'air enfermé est alors chargé du poids de l'atmosphère qui est égal au poids de vingt-huit pouces de mercure, et encore de celui de quatorze pouces, dont la somme 42 pouces est à 28 pouces premier poids qui tenait l'air à douze pouces dans la petite branche, réciproquement comme cette étendue de douze pouces est à l'étendue restante de huit pouces. Si l'on verse du nouveau mercure jusqu'à ce qu'il soit monté à 6 pouces dans la petite branche, et qu'il n'y reste que 6 pouces d'air, le mercure sera dans l'autre branche plus haut de 28 pouces que le haut de ces six pouces; ce qui doit arriver suivant la même hypothese: car alors l'air enfermé sera chargé de 28 pouces de mercure, & de la pesanteur de l'atmosphére qui en vaut aussi 28, dont la somme 56 est double de 28, comme la première étenduë de 12 pouces d'air est double des 6 pouces qui restent; & lorsqu'en continuant de verser du mercure dans la grande branche, il sera dans la petite a 8 pouces de hauteur, il y aura 56 pouces de mercure au-dessus, dans la grande branche; ce qui fait encore la même proportion," Mariotte 1679, 25–30. De Volder did not own, according to his *Bibliotheca*, these works. He owned Mariotte's *De la nature des couleurs* (1681), namely the fourth of his *Essays*, the *Traité de la percussion ou choc des corps* (1673), and the *Traité du mouvement des eaux et des autres corps fluides* (1686), wrongly attributed to Philippe de La Hire, who edited it: see *Bibliotheca Volderina*, 10 and 12.

[347] See part 2, discourse 2, *De l'équilibre des corps fluides par le ressort*: "[o]n verra donc manifestement dans cette expérience, que l'air EC aura suivi en sa condensation la proportion des poids. On trouvera la même proportion dans les autres expériences en faisant le calcul en cette sorte. Il faut prendre pour premier terme la somme du poids de l'Atmosphére & du mercure qui sera monté plus haut que le bas de l'air dans la branche EC; pour second terme, del'Atmosphére, c'est-à-dire 28 pouces de mercure; pour troisiéme, la distance EC; & le quatriéme proportionne sera l'espace ou hauteur où se réduira l'air enfermé dans le tuyau EC: comme si l'air étoit seulement reduit à l'espace IC de 8 pouces, on trouveroit que le mercure seroit en l'autre tuyau seulement 14 pouces plus haut que la ligne horizontale IL. Or ces 14 pouces avec les 28 del'Atmosphére sont 42: il faut donc dire suivant cette régle, comme 42 pouces est à 28 pouces, ainsi l'étenduë de l'air EC est à l'étenduë IC. Que. si on vouloit reduire ce même air en l'espace MC de 3 pouces, qui est le quart de EC; il faudroit mettre 84 pouces de mercure dans la branche DA au dessus de la ligne horizontale MN, & on trouveroit cette proportion par le calcul suivant: Comme MC 3 pouces est à ME 9

Fig. 5.65 An idealized inverted, asymmetrical syphon used to demonstrate that the pressure of water depends only on its height. (Source: Mariotte 1686, 142. ETH-Bibliothek Zürich, signature: Rar 1405, https://www.e-rara.ch/doi/10.3931/e-rara-13780. Public domain mark 1.0)

This, of course, represented the clearest conceptual detachment of the idea of the pressure exerted by a fluid like air, from the idea of its weight. An idea that did not pass unnoticed by De Volder, who can be labelled as a witness of the passage from a unified approach to the laws of fluids, to the assumption of two separate models: an Archimedean model for water, and a model based on the idea of elasticity and Boyle-Mariotte's law for air.

pouces, ainsi 28 pouces poids de l'Atmosphére, est à 84: car en changeant, 84 sera à 28 comme 9 à 3; & en composant, 84 plus 28, c'est-à-dire112, sera à 28 comme 9 plus 3, c'est-à-dire EC 12, à 3. Et si l'on veut savoir quelle hauteur de tuyau il faudroit pour reduire cet air en l'espace OC d'un pouce, il faut dire, comme OC un pouce, est à OE 11 pouces, ainsi 28 pouces de mercure poids de l'Atmosphére, à 308; & 308 sera la hauteur verticale qu'il faut donner au mercure au dessus du point O ou P: par où l'on connoîtra que pour faire cette expérience il faut que la branche DA soit plus haute que 308 pouces, c'est-à-dire qu'il faut qu'elle soit d'environ 320 pouces afin qu'il reste un espace au dessus du mercure pour empêcher qu'il ne verse," Mariotte 1686, 142–143.

5.6.3 De Volder's Evolving Treatment of the Elasticity of the Air

5.6.3.1 De Volder's Approach to Elasticity in the 1670s

De Volder's treatment of the elasticity of the air can be found in, besides his *dictata* (where he follows Descartes's *Météores*), experiment 5 of his *Experimenta*. This is a repetition of Roberval's experiment with the bladder, in a variant devised by Boyle in his *New Experiments*, namely, by putting an animal bladder in a container from which air is progressively evacuated. As a result, the bladder inflates and can raise the weights hung on it, proving the elastic power of the air contained in the bladder itself. For De Volder, this effect is not due to the air itself (which has no "intrinsecall principle whereby it could move, and expand, and dilate its own self,"[348]) but rather to the celestial matter flowing in it. The process is explained in Morley's report as follows. First, the evacuation of the container does not hinder the expansion of the air contained in the bladder. Then, the subtle matter or aether filling the container from which the air has been evacuated, penetrates the bladder and dilates the air contained in it, by breaking it into its "minute parts,"[349] which we may infer, if we do not assume that De Volder used Aristotle's idea (given in the *Physica*) that air is a continuous body,[350] as concerning entangled corpuscles, such as those of wool. Thus far, the key idea used by De Volder in his explanations is that of subtle matter, which for him is crucial in justifying how a small amount of air increases its power once decompressed, i.e. it works, as for Descartes, as an external mover. What is more interesting in De Volder's treatment of elasticity, however, is his gradual assumption of a Boylean model. In the third of his *Disputationes de aëris gravitate*, as he replies to objection 3, according to which the cause of the cohesion of Magdeburg hemispheres is not the weight but only the pressure of air, he concedes the objection and claims that such pressure is in fact exerted only by the air sur-

[348] De Volder 1676–1677, 89v.

[349] "[T]he aether or m[ate]r[i]a caelestis, which at the exuction of the common air, penetrating, et piercing through both the glass, and the bladder, did in the bladde[r] comminuere in minutas partas [*sic*] aërem coe[unte]m, insinuate itself into the pores of said common air and so dilate it and make it occupy more room than it did before [...]. [H]e inferred [...] that the air which was within the glass beforehand had hindered by its pression the bladder from expanding and the ingress of the aether, which was the efficient cause of said expansion," De Volder 1676–1677, 88v–89v.

[350] See book 5: "[t]hose who try to show that the void does not exist do not disprove what people really mean by it, but only their erroneous way of speaking; this is true of Anaxagoras and of those who refute the existence of the void in this way. They merely give an ingenious demonstration that air is something by straining wine-skins and showing the resistance of the air, and by cutting it off in clepsydras. But people really mean that there is an empty interval in which there is no sensible body. They hold that everything which is in body is body and say that what has nothing in it at all is void (so what is full of air is void). It is not then the existence of air that needs to be proved, but the non-existence of an interval, different from the bodies, either separable or actual – an interval which divides the whole body so as to break its continuity, as Democritus and Leucippus hold, and many other physicists-or even perhaps as something which is outside the whole body, which remains continuous," Aristotle 2014, volume 1, 362–363; cf. *Physica*, 213a22–213b3.

rounding the hemispheres, but it is ultimately caused by the weight of a whole column of atmosphere.[351] This causes a compression of this surrounding air, as it is evident from the fact that if the sphere contains air, this can counterbalance atmospheric pressure:

> [...] nor does it contradict [our theory] that a small quantity of air, which is in the sphere, which seems scarcely capable of pressing with so much force the internal surface of the sphere, as much as the immense bulk of the external air [presses] the external [surface]. And indeed if we consider this thing carefully, no part of air acts on the sphere but that one which touches it closely, which [...] the more forcibly it is compressed by a bulk of air above it, the more efficaciously it acts. Hence, if by some artifice it can be done, that the air externally touching the sphere is compressed as it is now, even if no air above it compresses it, would it not be manifest that this air, even if in less quantity, because nevertheless it is equally compressed, [...] as if it were sustaining all the weight of all the atmosphere, would act with the same force on the sphere itself[?] From this it is safely deduced that it is not to be considered as a more or less powerful quantity of air, but its compression.[352]

If the hemispheres are evacuated, thus, the lower one is pressed by air only from below, so that air pressure can keep it 'floating', as it does not undergo pressure from above. If the hemispheres contain air, its condensation counterbalances the pressure from below, and they fall apart.[353]

---

[351] "Ut autem nunc porro ad alterum obiectionis caput accedamus, quo si vel maxime aër premat, eam tamen pressionem gravitatem male dici: quia scilicet omnis pressio non est a gravitate. Quis negat? Sed tamen hanc pressionem, gravitatem dicendam esse existimavi semper, quae oritur ex descensu corporum sublunarium versus centrum Telluris," De Volder 1676–1678, disputation 3, thesis 9. Cf. thesis 7: "[q]uae demonstratio non tantum in aquis locum habet, sed in fluidis quibuscunque corporibus aqua sive levioribus sive gravioribus, sive minus sive magis compressis. Non enim nititur ea peculiari aquae natura, sed universali lege gravitatis et natura fluidorum corporum."

[352] "Neque obstat parvula illa aëris quantitas, quae in sphaera est, quae vix videtur tanta vi premere posse internam quanta immensa aëris externi moles externam premit sphaerae superficiem. Etenim si rem rite consideremus, nulla pars aëris agit in sphaeram, nisi ea quae eam tangit proxime, quae tamen eo agit efficacius, quo a mole superioris aëris validius comprimitur. Unde si artificio quopiam fieri posset ut aër externe sphaeram ambiens aequaliter ac nunc comprimatur, etsi nullus aër superior eum comprimeret, nonne manifestum erit hunc aërem licet in minori copia, quia tamen aeque compressus supponitur ac si omne pondus totius athmosphaerae sustineret, aequali vi acturum in ipsam sphaeram. Ex quo sane conficitur non tam aëris quantitatem quam eiusdem compressionem magis minusque validam spectandam esse," De Volder 1676–1678, disputation 2, thesis 8.

[353] "Ponamus [...] suspendi duo haec sibi invicem cera coniuncta hemisphaeria, aëre adhuc dum referta. Manifestum est, utramque tam internam quam externam huius Sphaerae superficiem aequaliter ab aëre premi, eo quod internus ille aër aeque ac externus condensatu sit. [...] Verum educto iam aëre, quid quaeso fiet? Externa sphaerae superficies eandem habebit, quam antea, ab aëre pressionem, pauxillum enim illud educti aëris hic in censum suam ob exiguitatem non venit. Verum interna superficies educto aëre nihil habet quo resistat. Unde quid sequitur evidentius, quam cum aër immediate subiacens inferiori hemisphaerio a gravitate aëris lateralis superiora versus feratur, nec quidquam in Sphaera sit quod huic pressioni resistere valeat, oportere sane hoc hemisphaerium ad alterum, quod eadem aëris superioris vi et gravitate deorsum premitur, comprimi tota vi et pressione subiacentis aëris, quae cum tota dependeat ab ea qua aër lateralis deorsum ruit gravitate, quid est certius quam omni huius gravitatis vi et impetu, haec hemisphaeria ad se invicem compelli," De Volder 1676–1678, disputation 2, theses 7 and 9.

5.6.3.2 The Background of De Volder's Theories

All this, of course, complicates De Volder's Archimedean theory of floatation, which can be conveniently used in the case of bodies floating in water, but which in the case of air serves only to account for the degree of pressure of air. Implicitly, thus, De Volder accepted Boyle's idea that air exerts a pressure proportional to its degree of compression, which in turn 'releases' the force exerted by the weight exerted on it. His positions, therefore, encompass an idea of pressure more advanced than that of the experimentalists and natural philosophers of his time. Von Guericke, for instance, extensively dealt with the elasticity of air and its effects in his *Experimenta nova* (book 3, chapter 33), and admitted that air presses on itself, being thus more or less rarified.[354] Yet he did not take the further move to use elasticity to account for the cohesion of the hemispheres (in chapters 23–24) – which he traced back only to the weight (*pondus* or *gravitas*) of air, exerted from all directions on the bodies it surrounds.[355] Pascal, who deals with the cohesion of marbles in his *Traité de la pesanteur de la masse de l'air*, also does not connect the pressure exerted by air to its elastic force – which he himself admitted at the beginning of this treatise.[356] According to Pascal, the inferior piece of marble is sustained by the pressure or weight of the air from below, as happens in water, which keeps it cohering with the upper part as it does not undergo direct pressure from the air above this part.[357] Rohault, who noted that the compression of air and its elasticity have effects

[354] "Air, moreover, has the characteristic that it can be more and more condensed through violent compression as well as expanded by being given greater space, as will be seen in our experiments. [...] The air surrounding the earth presses upon itself because it is corporeal and has a certain weight. Indeed, the upper air presses down increasingly heavily on the lower. Thus it follows that the lower air here, which surrounds us, is much denser than the upper air. And whatever is denser, has more mass, and whatever has more mass, is heavier. Therefore we have more and heavier air here on the earth's surface than that found in towers and mountains; to be sure, the higher the air is, the lighter and more rarefied it is," Von Guericke 1993, 112–113.

[355] "Because of its own weight, air presses not only upon itself but upon all things beneath it with an essentially constant pressure. We do not feel this because we live in this air which surrounds us with equal pressure on all sides and at the same time penetrates us. For as fish have no perception of pressure in the water, so animals perceive it even less in the air. The weight of air on the earth's surface is as great as the weight of water about 20 Magdeburg cubits deep. In other words if water should rise 20 cubits above the earth's surface, the pressure it would exert on all things beneath is the same as the pressure of air," Von Guericke 1993, 113.

[356] See chapter 1: "si on prenoit un balon à demy plein d'air seulement, et non pas tout enflé, comme ils le sont d'ordinaire, et qu'on le portât sur une montagne, il devroit arriver qu'il seroit plus enflé au haut de la montagne, et qu'il devroit s'élargir à proportion de ce qu'il seroit moins chargé," Pascal 1663, 50.

[357] "La mesme chose doit s'entendre de toute autre liqueur; et par consequent si deux corps sont polis et appliquez l'un contre l'autre, en tenant celuy d'en haut avec la main, et en abandonnant celuy qui est appliqué, il doit arriver que celuy d'en bas demeure suspendu, parce que l'Air le touche par dessous, et non pas par dessus; car il n'a point d'accés entre deux: et partant il ne peut point arriver à la face par où ils se touchent; d'où il s'ensuit par un effet necessaire du poids de toutes les liqueurs en general, que le poids de l'Air doit pousser ce corps en haut, et le presser contre l'autre; en sorte que si on essaye de les separer, on y sente une extréme resistance: ce qui est tout conforme à l'effet du poids de l'eau," Pascal 1663, 67.

comparable to those exerte by its weight,[358] did not use the idea of elasticity to account for cohesion of bodies – but only for well-known effects like those of air-guns.[359] Even Mariotte, in his *Essays de physique* (1679–1681) and *Traité du mouvement des eaux* does not link the topic of elasticity of air to the cohesion of solid bodies.[360] Senguerd, eventually, even though devoting extensive considerations to the elasticity of the air in his works, ascribed to circumpulsion the effect of the cohesion of marbles in his *Philosophia naturalis*, as he observed that different kinds of marbles can sustain different weights, since different kinds of marbles are more or less porous, and allow the passage of different quantities of air.[361]

In this *panorama*, only a few figures explicitly combined the ideas of weight and spring of the air. In his *History of Fluidity and Firmnesse* (1661), for instance, Boyle explains the cohesion of marbles by atmospheric pressure, on the one hand, and elastic power, on the other, which are two interrelated phenomena:

> the pressure of the Atmosphere, proceeding partly from the weight of the ambient Air, [...] and partly from a kind of Spring, by vertue of which the Air continually presses upon the bodies contiguous to it.[362]

The explanation is the same as that he provided in his *New Experiments*, as he discussed (experiment 2) why the stopper of an evacuated container cannot be easily removed: namely, because the rarefied air it contains has less "spring" to counterbalance the spring of the external air.[363] However, it is in his *History* that Boyle discusses this phenomenon in more detail, supposing that air is made of "little springs," i.e. that the corpuscles of air can bear each other, and that they undergo a

[358] See Rohault 1671, volume 1, 86–87.

[359] See Rohault 1671, volume 2, 150–153.

[360] Mariotte devotes considerations only to the resistance of bodies to being broken, rather than to the cohesion of different bodies: see Mariotte 1686, 370–388.

[361] "Circulo, enim et successioni hunc effectum attribuendum esse, non vero gravitati, aut pressioni aëris, sequentia evincunt experimenta. 1. Quod eo facilior est hemisphaeriorum separatio, quo facilius circulus dari, aut successio perfici potest; hinc minima admissa apertura, per quam successio perfici potest, facili negotio separantur. 2. Quia quo difficilior est successio, eo difficilior eorundem est separatio; et eo arctius cohaerent hemisphaeria, quo materia eorum minus porosa extiterit; hinc hemisphaeria, eandem habentia latitudinem, e marmore si confecta fuerint, triplo plus ponderis sustinent, quam si ex ebore; et si eburnea fuerint, duplo plus ponderis, ut a se invicem avellantur, requirunt, quam si ex aëre facta sint. 3. Quia hemisphaeria saepe multo plus ponderis sustinent, quam gravitas, vel pressio aëris, illis incumbentis, est; cum tamen ab aëre arctior non possit resultare connectio, quem quae aequivalet gravitati cylindri aërei, eandem latitudinem habentis cum hemisphaeriis connexis. Confirmat dicta cohaesio hemisphaeriorum e solido marmore confectorum, diametrum duorum digitorum cum dimidio habentium, quae tanta est, ut sibi imposita, interiecta pinguedine, ne quid inter illa se insinuet, septigentae et ultra requirantur librae ad eorum separationem efficiendam; cum aëris cylindrus, eandem cum hemisphaeriis latitudinem habens, a terra ad extremos athmosphaerae terminos protensus, ducentarum librarum gravitatem non superet," Senguerd 1681, 45–46. See also Senguerd 1715, 173. On the elasticity of air, see Senguerd 1681, part 3, chapter 2, and Senguerd 1715, chapters 4–8, explaining the elasticity of air by the features (as their being branched and flexible) of its particles.

[362] Boyle 1669a, 213.

[363] See experiment 2 of Boyle 1660. For an extensive discussion, see Chalmers 2017, chapter 8.

"recoyl" from the surfaces of the Earth, so that they can sustain the inferior part of the two cylinders of marble.[364] Later, the combination of the ideas of weight and elasticity, as to the explanation of the cohesion of bodies, would be adopted by Sturm – who in turn relied on the dialogue *Ars nova et magna gravitatis et levitatis* (1669) of George Sinclair.[365] De Volder, thus, was among the few figures who explicitly connected the issues of weight, elasticity and used these ideas to explain the cohesion of bodies.

5.6.3.3 De Volder's Assumption of Boyle's Law in the Works of His Students

In fact, De Volder did not just combine Boyle's idea that the pressure exerted by air is proportional to its compression with his Archimedean explanations. At some point, indeed, he came to get rid of Archimedean explanations, and to use only the idea of elasticity of air in addressing phenomena such as the cohesion of marbles or the raising of water in syphons. This evolution of his thought, actually, can be found only in the works of his students, even if there is compelling evidence that he was the source of their contents. We do find evidence of this change first in the *Dissertatio de gravitate aetheris* (1683) of Jakob Bernoulli (1654–1705), elder brother of De Volder's correspondent Johann Bernoulli, and then in three inaugural disputations of De Volder's students. Namely, the already discussed *De vi corporum elastica* (1688) of Hermann Schuyl, his subsequent *De respiratione* (held on the same day as the previous disputation), and the *De gravitate aëris* (1690) of Joseph Serrurier, who was to be the author of a *Physicae experimentis innixae compendiosa tractatio, in usum scholarum domesticarum* (1700), – in which he does not refer to De Volder – and then (1705) professor of philosophy and mathematics at Utrecht.[366]

Jakob Bernoulli enrolled at Leiden University in May 1682 as a "candidate in theology," without taking any degree and during a long tour which led him, after having studied at the University of Basel – where he had obtained a licentiate in theology in 1676 – to France (where he was acquainted with Descartes's works and Malebranche's *Recherche de la vérité* (1674–1675)), Germany, Holland, and England (where he met, among the others, Boyle and Adriaan van Beverland). In Holland, Bernoulli lived first in Amsterdam (where he had contact with Alexander de Bie) and then in Leiden, where he spent some time at the University, where "[having] discarded the shadows and prejudices, in which he was immersed so far, [and being] delighted by the sweetness of a more healthy philosophy and of the mathematical demonstrations, which he saw being publicly exhibited by the most excellent masters of this science, he himself [...] started, after their sample, to choose the way of an higher degree of doctrine." Accordingly, it seems that the lec-

[364] Boyle 1669a, 212–213. Cf. Boyle 1660, 229–230.

[365] See Sturm 1676, 46–48; cf. Sinclair 1669, 336–337. De Volder owned Sinclair's book: see *Bibliotheca Volderina*, 5.

[366] See Sect. 2.4.2, *De Volder's retirement, death and legacy*.

tures of De Volder – among the others – had a decisive role in shaping Bernoulli's intellectual path. After his visit to England in the same year, when he came back to Basel, he started to give the first, public "experimental physico-mechanical" course at the University[367] (while his brother Johann did the same at Groningen, as seen in Sect. 2.4.2, *De Volder's retirement, death and legacy*). In the same year, he published his *Conamen novi systematis cometarum* (dedicated to Johannes Hudde and Bernard Fullenius) and then his *Dissertatio de gravitate aetheris*, whose "occasion of writing" (*occasio scripti*) had been his reading of De Volder's *Disputationes de aëris gravitate* (in the edition of 1681).[368] In 1686, moreover, he published his

[367] "Ex illo ergo Gallico itinere domum redux, suasu amicorum Cel. Malebranchii Scrutinium veritatis et Cartesii scripta tum primum coepit evolvere, cuius scriptoris methodum potius quam principia approbabat. Haec ei lectio ad id profuit, ut iam in philosophia ultra consueta compendia sapere inciperet. Dum in his est, cometae, qui per id tempus in caelo formidandae magnitudinis effulsit, occasione, quendam ingenii lusum de futura eius nova apparitione in publicum edidit: pauloque post A. CIƆIƆCIXXXI m. Aprili secundo Rheno alteram in Belgium et Angliam profectionem instituit, certus, id quod priori itinere a se peccatum fuerat, in hoc emendare. Et in Belgio quidem Amstelodami aliquanti temporis moram traxit, ibique Alexandrum de Bie, Matheseos Professorem, res mathematicas in gratiam nautarum vernacula lingua explicantem aliquoties audiit: ac per otium in vicinas urbes et provincias excursionem fecit. Inprimis autem Lugduno-Batava Universitati penitius lustrandae aliquod tempus dedit, in qua Celeberrimis Viris, Wittichio, le Moine, Theologis: Bockelmanno JC. et Woldero philosopho innotuit. Et huic certe Belgicae commorationi illud se debere sepe praedicabat, quod excussis quibus hactenus immersus erat, tenebris atque praeiudiciis, sanioris Philosophia: et demonstrationum mathematicarum, quas a praestantissimis eius scientiae magistris publice videbat exhiberi, dulcedine inescatus, ipse quoque ad illorum exemplum ad altiorem aliquem doctrinae gradum viam affectare coepit. [...] Postquam autem in literario illo Batavicarum Academiarum mercatu animum mathematica eruditione egregie instruxerat, et gravissima scorbuti aegritudine fuerat defunctus, discendi aviditate provectus per praecipuas Brabantiae, Zeelandiae, Flandriaeque civitates, Caletum usque continuato itinere in Angliam traiecit, in qua insula Ill. Boylium, Isaacum Vossium, Robertum Hookium, Justellum, Stillingfleetum, Baxterum, Galium, aliosque Celeberrimos Viros, salutare non intermisit. Inter alios compellavit Adrianum quoque Beverlandium, Virum ab impiis, quas etiam scriptis publice editis Orbi manifestas esse voluit, sententiis quam eruditione sua celebriorem: qui tunc ex Belgio relegatus in Isaaci Vossii familia degebat. Non quod malae frugis hominem vel tanti aestimaret, sed ut ex perspecta bonorum malorumque, et eorum qui vere, quique ad speciem tantum eruditi edent, indole, qua nativa esset eruditionis facies, pollet internoscere: secutus in ea re exemplum prudentium familiae patrum, qui, dicente Plinio, pluribus saepe veris denariis adulterinum emunt, ut verus agnoscatur. Ex Anglia Hamburgum est transvectus, unde brevi per Germaniam transitu in patriam A. CIƆIƆCIXXXII rediit. Quanquam ne tum quidem prius sibi certandum existimavit, quam bimestri itinere Helvetia pagos omnes in duorum amicorum, et inter eos dilectissimi Fratris mei, comitatu esset emensus. Ex eo tempore stabilem in patria pedem posuit, et mathematica studia, cum principia tam pulchre ipsi se dedissent, maiori etiam latiore urgenda sumsit: ad quorum amorem et diligentem tractationem ut popularium animos, hactenus in ea re segniores, excitaret, Collegium, quod vocant, experimentale Physico-Mechanicum publice aperuit, primusque rerum harum pulcherrimarum in urbe nostra vel autor vel evulgator extitit," Bernoulli 1744, volume 1, *Vita*, 14–17. Cf. *Album studiosorum Academiae Lugduno-Batavae MDLXXV–MDCCCLXXV*, column 352. On Jakob Bernoulli, see Fleckenstein 1977; Bell 1986, chapter 8; Dunham 1987; Sierksma 1992.

[368] "Occasio scripti. Quia vero maximi momenti esse et ad maiorem intelligentiam quamplurimum conducere iudico, si qua primum occasione quave via in cogitationes tuas incideris, enarres, quod alias per modum praefationis fieri solet; non abs re erit, si tribus id verbis hic innuam. Incidi nuper

Methodus ratiocinandi, sive Usus logicae in praeclaro aliquo phaenomeno physico enodando, namely the text of a special lecture in which he reconsiders some of the ideas already expounded in his *Dissertatio de gravitate aetheris*, and gives evidence of his dramatic encounter with the orthodox Cartesianism of Johannes de Raey.

In his *Dissertatio*, Bernoulli reported the contents of one of De Volder's experimental lectures which he witnessed in 1682, as he deals with the issue of the "passive resistance of air," which for Bernoulli is not "so much some quality hidden in air itself, and far from our knowledge, but rather a defect in the force [*virtus*] in the liquid pressing [on] the air, which has to be regarded as not having enough force to displace or to condense the air." This is a feature different from the elasticity of the air, (i.e. the *conatus* of air to dilating itself, on whose causes Bernoulli does not advance any hypothesis),[369] and which is used by Bernoulli in order not to rely on any 'active' power of air, such as elasticity itself.[370] In fact, the way in which he

in Scriptum aliquod de Gravitate aëris, Auctore Cl. Voldero, Professore in Academia Lugduno-Batava Celeberrimo. Id cum secunda vice evolverem, coepi attentius ruminare, quae sect. 37. seqq. de receptis duobus motus generibus [i.e., thrust and attraction], docte et solide scripsit. Quae quia sequenti tractatui ortum dedere, non posum, quin totidem pene Auctoris verbis, quantum ad propositum meum faciunt, huc transferam; quod citra plagii notam interpretabitur benignus lector," Bernoulli 1683, 7.

369 "Quid sit elaterium aëris? Haec dicta sunt de motu gravitatis. Postquam autem Naturae Consulti vidissent, hunc solum motum non sufficere explicandis omnibus circa suspensionem liquorum phaenomenis; quippe qui non explicat, quare suspensus haereat in tubo liquor, obturato vasculo, ubi totius tamen atmosphaerae gravitatio intercepta: hinc alium adhuc acri peculiarem, atque a motu graviatis independentem, ascripsere motum, quo aëris particulae conatum quendam (non communem aquae, fluidisque aliis crassioribus) habeant sese expandendi, dilatandi, remotoque obstaculo maius occupandi spatium; quique conatus, in aëre incluso, par fit sustentando tanto ponderi liquoris alicuius, quantum sustinere valeat gravitate sua tota atmosphaerae moles; non nunquam minori, aliquando etiam maiori, pro re rata. Illud mirari subit, quod cum omnes hydrostaticorum Scriptores hanc aeris vim elasticam unanimi fere fateantur ore, plerique eam ostendisse sint contenti; pauci vero in naturam et causam illius penitius inquirere sustinuerint, aut solliciti fuerint, ut certas illi regulas praescriberent, atque omnes evolvendo casus aëris liberi, inclusi, condensati, rarefacti, exponerent, quantum in singulis horum casuum effectum sortiri aër debeat. Eius causa Obscura. Enimvero unde istud elaterium sive conatus sese dilatandi in aëris particulis proficiscatur; an ex eo, quod singulae illarum circa proprios axiculos rotentur, vel plures aliquot in unum motum circularem conspirantes, infinitos parvos vortices constituant; dumque ab horum centris recedere conantur, ambientes particulas loco pellendi ac se dilatandi vim acquirant: an vero procedat ex peculiari harum particularum figura vel textura; quod forsan sint graciles, flexiles, intortae ac conglomeratae spirae, instar taeniae, funis, aut elaterii horologii portatilis: an quod, ab agitatione materiae primi et secundi Elementi inter corpuscula aëria rapidissime discurrentis, illis hic elaterii motus communicetur, quo in continua quasi conserventur bullitione, ut qua licet sese diffundant: an denique quod ista sese dilatandi virtus, absque adminiculo causae externae, immediate a Primo Motore in creatione illis indita olima fuerit: hoc, inquam, negotium est tam arduum, coniecturis ubique aequali difficultatum numero laborantibus; ut inter abstrusissima naturae mysteria iure merito referatur. Quocirca, hac de re quicquam determinare non sustineo; praesertim cum unusquisque, salvis forte phaenomenis, hic suo sensu abundare possit," Bernoulli 1683, 81–82.

370 "Quid sit aëris resistentia passiva? Quod vero effectum spectat huius Elaterii; illi paulo distinctius excutiendo inhaerebimus; et quia multis id videtur comprehensu valde difficile, qua ratione pauxillum aëris etiam non compressi, ingentis atmosphaerae pressionem, aequivalente pressione et

deals with this idea is revelatory of an intermingling of the sources behind his *Dissertatio*, on the one hand De Volder, on the other, Senguerd. In fact, the idea of the passive resistance of air can be found in no other sources than in Senguerd's *Rationis atque experientiae connubium* (which was to appear in 1715), as well as in a disputation presided over by Senguerd, whose contents were used in the *Connubium*, namely the *Exercitium experimentale septimum decimum, quod est de aëreae elasticitatis et resistentiae passivae effectis* (1698). In fact, this kind of resistance is nothing but the mere inertia of matter – as Senguerd would put it – and consisting in its extension.[371] An idea which, to some extent, had also been embraced by De Volder as he admitted, as seen above, that inertia consists in extension itself. Indeed, Bernoulli illustrates this idea with a typical Volderian sample: namely, by considering what happens in a non-elastic collision between two bodies A and B. If they collide with the same force, they come to rest. In this case, Bernoulli argues, one cannot ascertain whether they are constantly pressing each other, or if they have lost all their force, unless one moves one of the two in a direction opposite to that from which it came. If the other one follows it, it means that they did not lose all their force. Hence, Bernoulli considers the case in which body A is thrown against B, which is fixed to a wall. In this case, regardless of the force of the two bodies, they will come to rest. Third, he considers the case in which A is pushed by a force 1,000 times greater than that it had before: also in this case, B will not be moved, and both will be at rest after impact. In all these cases, A is not resisted by the elastic force of B (which is considered as made by a non-elastic material, like ivory), but only by its "interposition," which makes ineffective the force of A: this is the 'passive resistance' of a body.[372] Notably, as Senguerd came to formulate his own laws

actione efficaci (talem enim activam efficaciam significatio vocis in illis rebus, quibus tribui solet, requirit) repellere irritamque reddere valeat: hinc ad captum illorum nos accommodaturi, atque elaterium hocce mitigaturi, aliud quiddam praeterea in aëre considerabimus, quod resistentiam vocabimus passivam, atque ita effectum soli hactenus elaterio tributum bipertiemur, partem relinquendo actioni elaterii, partem vero asserendo resistentiae illi passivae, monstrabimusque, quo pacto idem sequi debeat effectus, omniaque allata experimenta non minus, sed forte intelligibilius, solvi possint, etiamsi aër longe minori, quam vulgo creditur, elatere foret praeditus, caetera vero mere passive se haberet, resistentia supplente elateris vicem. Notandum vero ante omnia, per hanc aëris resistentiam passivam me non tam intelligere qualitatem aliquam in ipso aëre latitantem, et a nostra cognitione remotam, quam vero defectum virtutis in liquore aërem premente, ut non satis habet virium ad aërem loco movendum vel condensandum," Bernoulli 1683, 82–83.

371 "[...] aëri competit resistentia quae passiva dicitur. Haec non consistit in virtute, impetu, efficacia, aut conatu aëri impresso, quo eiusdem particulae in se mutuo, vel in vicina irruendo corpora, suamve in illa exercendo vim, actuosus foret, eorumve operationes promoveret, aut contraria virtute easdem irritas redderet: sed passive se habendo, sola sui substantia, ac mole corporea, indeque impenetrabili, excipiendo vim, et actuositatem corporum in se incidentium, efficaciae ac motui eorum, proportionaliter molis suae magnitudini, ac convenienter naturae legibus, quantum in se est, resistit, illorum introitum in spatium a se occupatum avertit, et tamdiu impedit, donec devita aëris resistentia, iis, utpote potentioribus, cedere cogatur; quod tamen fieri nunquam poterit, nisi alia ipsi cedant corpora, eorumque loca occupare aëri concedatur," Senguerd 1715, 90–91.

372 "Ut vero distinctius cognoscamus, quae possint esse partes huius Resistentiae passivae, consideremus duo corpora se invicem prementia, (puta duos luctatores, vel duas pilas) sitque primo utrumque libero aëri, (id est, loco ubi nihil vel adiuvat, vel impedit illorum motum) expositum,

of impact in his *Philosophia naturalis*, he maintained, following Descartes, that when two bodies are perfectly hard, they will collide with the same force (m·s), i.e. they do not come to rest but their quantity of motion is conserved, so that they are rebounded in opposite directions.[373] De Volder denied this conclusion, unless one admits that the bodies are elastic.

---

nullique innixum sustentaculo occurratque corpus A corpori B; quo si fortius est, illud propellet; si debilius, pelletur ab ipso in contrariam partem; si aequali denique vi premat et renitatur utrumque, sublatis ex aequo viribus eodem loco tanquam quiescentia spectabuntur ambo. Hoc unico proinde in casu, non constabit ex sola loci consideratione, utrum alterum in alterum aequali pressionis conatu agat, an vero ambo inertia et otiosa iuxta se quiescant; quod postmodum demum cognoscere datur, cum alterum loco moveris; si pone enim sequatur alterum, concludes sese pressisse antea: si immotum maneat, indicio est, antea quievisse utrumque; quamvis interim utrobique praestet illa considerare, ut omnibus viribus destituta, cum si quas habent, tantundem iis efficiant, ac si non haberent. Sit vero etiam porro corpus B (luctator vel pila) innixum solido alicui fulcro, puta luctator parieti cuidam, vel pila lateri mensae tudiculariae; faciatque corpus A impetum in corpus B sussultum, quid fiet? hoc quidem illud in contrariam adhuc partem repellet, ubi plus illo impendit virium: sed sive vires utriusque sint aequales, sive vires corporis B. sint debiliores, sive plane nullae; in omnibus his tribus casibus neutrum corpus loco suo expellet alterum, sed iuxta se quiescent, adeo ut hactenus nulla pateat ratio, quae nos cogat ad credendum, corpus B. ad impetum corporis A infringendum et sufflaminandum, aequalem potius conatum adhibere, quam vel debilius vel plane non reniti. Sed ubi porro consideraverimus, etiamsi mille praeterea homines aut pilae in directum positae essent, quae omnes vires suas iungerent cum luctatore vel pila A, ad pellendum corpus B, illas tamen omnes non plus effecturas, quam antea fecerat solum corpus A iustam habebimus suspicandi ansam, obstaculum, quo impediebatur paulo ante corpus A, ne propellere posset corpus B non provenisse a renitentia et repulsione aequivalente facta a corpore B id est ab aliquo eius elaterio (quale praecipue in pila eburnea concipere absurdum foret), cum non sit verosimile, eandem hanc vim corporis B, postea parem esse potuisse repellendo impetui millies maiori: sed a mera interpositione corporis B. quae sola sufficiens esse possit sistendo impetui corporis A totiusque seriei corporum istud iuvantium. Pergat enim, si possit, corpus A moveri in directum post contactum corporis B; aut penetret necesse est dimensiones huius, quod omnino impossibile; aut faciat, ut hoc permeet solidum fulcimentum, cui innixum esse supponimus: sed sic vel integrum corpus B deberet traiicere, quod idem involvit absurdum, vel deberet prius in minutissimas partes conteri, eaeque dein per poros muri adigis quod cum non fiat, concludendum, corpus B esse talis texturae, cui dissolvenda impar sit conatus quantumvis maximus corporis A. Omnium quae reliquorum vires suas, huic adiungentium, atque in hoc illud ipsum consistere puto, quod vocare soleo resistentiam passivam," Bernoulli 1683, 83–87.

[373] "[S]i corpora aequalia in motu constituta in se mutuo impingant, utrumque versus oppositum terminum reflectetur," Senguerd 1681, 35. Senguerd's theory of motion requires a little digression. He conceived motion in quasi-Cartesian terms, namely, he still intended quantity of motion as something which is conserved through impact, but he 1) criticized the idea that movement is respective, or reciprocal, for the reason that it is a real mode of bodies, so that one cannot arbitrarily attribute it to this or that body. On this basis, he criticizes Descartes's characterization of motion as given in *Principia* II.25 because it is contradictory to his idea of motion (given in II.24) as relative (Senguerd 1681, 25, §§ 4–5). Being an absolute mode of bodies, a body can be said to be in motion even without considering all the other bodies (Senguerd 1681, 28, § 11), and intends it as a 'translation from space to space', rather than a 'translation from place to place', because place is defined as the space occupied by a body already present in it (Senguerd 1681, 29, § 13). Accordingly, he distinguishes between space and body. For him, the quantity of motion is only probably conserved (because God is free in the conservation of such quantity) – (Senguerd 1681, 31, § 19), and motion is distinguished from the *vis* or force to motion (which pertains to God only), from the impetus impressed by divine force in bodies: so that by 'motion' only the translation of

On the ground of this idea, Bernoulli devised an explanation for an experiment carried out by De Volder himself, which is duly described in the *Dissertatio*, which is alternative to De Volder's own explanation, based on the idea of elasticity of air. The experiment was the last one performed by De Volder in 1682,[374] and involved the use of a container (see Fig. 5.66, Bernoulli 1683, figure 18) in which another container and a syphon were placed.

De Volder first completely filled the bigger container with water, then he made the water flow from hole e, until it reached the level il in the bigger section of the container, and o in the smaller container a. At this point, the rest of the apparatus (including the syphon c) is void, so that no pressure of air is exerted on the surfaces of water. Hence, De Volder opened the hole g, from which air entered the apparatus and, pressing on the water, made it rise in the syphon both from b and d. As he closed g, he observed that water continued to flow in the syphon from b to d, "because," as reported by Bernoulli, "by the elasticity of the column of air qr, [qr] is resisted more by the weight in the longer leg, than by the elasticity of the column op, [op] is resisted by the minor weight of the shorter leg, [so that] it happens that the stronger liquid driven in the shorter leg, repels the other one pressed more weakly, and so it endures in a continuous flowing." In fewer words: the columns of water in the legs of the syphon (cb and cd), and the columns of air in the container (qr and po) exerts both a downwards pressure (which I indicate as "F"), so that Fqr < Fcd, Fpo > Fcb. Accordingly, the water cb is pressed upwards and flows into

bodies has to be intended, which is the effect of the impetus (Senguerd 1681, 32–33, § 21). The quantity of motion, in turn, can be quantified as m·s (i.e. it depends on the size and speed of a body) – (Senguerd 1681, 35, § 26), and is conserved through collisions. These are the rules of impact of Senguerd, which he conceived as between perfectly hard bodies (Senguerd 1681, 39, § 32): 1) the same situation as in Descartes's rule 1 (two equal bodies moving against each other at equal speed), which is valid for Senguerd as well. 2) The situation of Descartes's rule 6 – namely the same case considered in De Volder's 1684 *De motu* (two equal bodies, one of which is at rest) – which Senguerd resolved in another way, namely, by reducing it to Descartes's rule 4. For Senguerd, the body at motion cannot move the resting body, and is rebounded by it, because there is a condition of equilibrium between the force of motion of the moving body, and the force to resist the motion of the resting one (Senguerd 1681, 35, § 27). 3) The case of Descartes's rule 5 (i.e. a bigger body impacting on a smaller body, which is at rest). In this case, the outcome is as it was for Descartes (i.e. the two bodies form a new body, in which m·s is conserved) – (Senguerd 1681, 37, § 30). 4) The case of Descartes's rule 3 (i.e. two equal bodies move against each other, but one of them has more speed): also in this case the outcome is 'Cartesian' (i.e. the move in the same direction, with m·s conserved) – (Senguerd 1681, 38, § 31). 5) The case, not expressly considered by Descartes's rule 7, in which two equal bodies move in the same direction at different speeds: in this case, the faster one 'carries' the preceding, slower one, with m·s conserved (Senguerd 1681, 38–39, § 31). Such rules presuppose that motion is not reciprocal: for instance, if we consider Senguerd's third case by assuming that the smaller body is moving, and the bigger one is at rest, the outcome does not follow the rule he sets for this case, as the smaller body would be rebounded in the opposite direction.

374 "Fluxum liquoris per siphonem in loco clauso explicatur per resistentiam aëris passivam. [...] Hanc in rem autem opportune incidit experimentum, a Cl. Voldero pro ultimo superioris anni specimine in Theatro Physico Academiae huius publice ostensum, quo elateristae admodum gloriantur," Bernoulli 1683, 107–108.

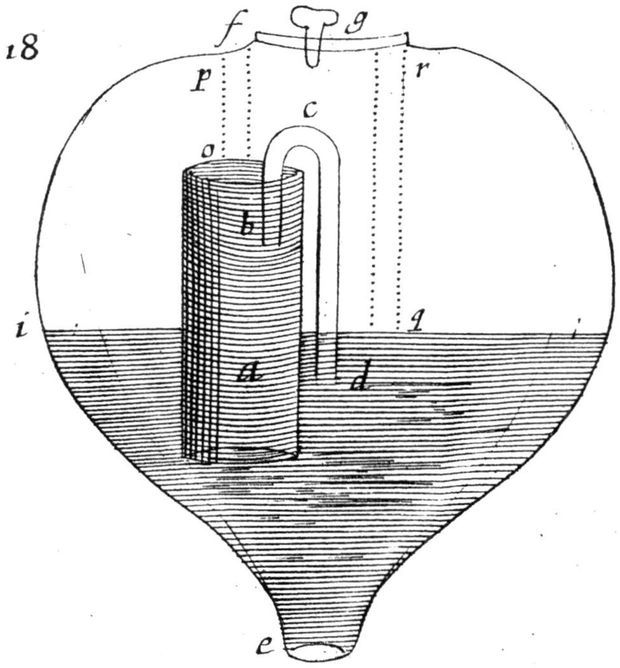

Fig. 5.66 Instrument aimed at demonstrating the elasticity of the air. (Source: Bernoulli 1683, figure 18. ETH-Bibliothek Zürich, signature: Rar 5600, https://www.e-rara.ch/doi/10.3931/e-rara-9000. Public domain mark 1.0)

the other leg of the syphon. Of course, as the container is closed, only the elasticity of air accounts for its downwards pressure.[375]

[375] "Sumsit vitrum cylindricum a. aqua subrubido colore tincta impletum, eique immisit crus brevius siphonis bcd atque ore admoto longiori adsuxit per siphonem aquam, qua fluente protinus vitrum cum siphone demisit in recipiens efg quod pariter mox aqua subrubida ad summam usque oram adimplevit, ne quid in illo remaneret aëris, atque tandem operculo admoto clausit, et cera undiquaque probe munivit. Quo facto coeptum est, agitatione emboli evacuari recipiens, extracto per eius collum e liquore, usque ad superficiem circiter il quo subsidente sensim, subsidit pariter liquor in cavitate siphonis contentus, mansitque subinde in eodem plano cum superficie liquoris extra siphonem, descendens in breviori quidem crure ad summam usque oram vitri cylindrici; in longiori vero, quousque subsederat reliquus in recipienti liquor; propterea quod, praeter materiam subtilem, nihil aderat quod ponderare super liquore in recipienti, eumque in siphonem impellere, vel in eo suspensum tenere potuisset. Evacuato sic maximam partem recipienti, intromisit per apertum obstructorium aërem, qui irruens in liquorem vitri cylindrici, eum pondere suo impellebat in crus siphonis brevius, et exinde porro in longius; nec cessabat liquoris per siphonem fluxus, quantumvis postea obstructorium loco suo iterum intrusum fuisset. Cuius rei quidam ratio, supposito aëris elaterio, reddi potest facile; cum enim per obstructorium intromissus aër, virtute sua elastica, premat cum super liquore in vitro cylindrico contento, tum super reliquo extra cylindrum; fit ut liquor in utroque crure sursum impellatur, usque ad mutuum occursum in flexura siphonis c ubi quia in contrarias tendit partes, species quaedam luctae oritur inter liquores utriusque cruris, adeo ut neuter alteri praevaleret, sed immoti haererent, si aequalibus ambo viribus fuissent impulsi:

As a matter of fact, the explanation reported by Bernoulli is not entirely correct, or at least not complete. By considering the very ideas used by De Volder in late 1670s, the pressure exerted on the surface il by the water in the longer leg is greater, because such leg is longer (as fluids press according to their height). So that surface il undergoes the pressure, from below, exerted by an ideal volume of water encompassed by this surface and the ideal surface at the same height of o, while on the real surface o, in the smaller container, only air pressure is exerted. Since, however, from above the surface il undergoes only the pressure of air, this surface is raised by the water in the leg of the syphon, which flows into the container as it finds less pressure than the pressure of air which it undergoes at o. This notwithstanding, what is clear is that De Volder, as early as in 1682, already dismissed the Archimedean model in his explanation of the pressure of air, and instead had reverted to the idea of elasticity.

Bernoulli's next move, actually, is to demonstrate that we can explain the passage of water from the bigger to the larger container if we also suppose that air is not elastic, namely by considering only its "passive resistance." What he does, actually, is nothing but an application of the idea of circumpulsion. According to him, the pressure exerted by the weight of water cd is communicated, *via* the water and then *via* the air, to the surface o. Since Fcb > Fcd, cd is pressed upwards by cb, *via* a communication of force through air.[376] By using the same reasoning, Bernoulli

Verum, quia elaterio columnas aëris qr a pondere liquoris in longiori crure magis resistitur, quam elaterio columnae op resistitur a minori pondere cruris brevioris; fit ut liquor fortius adactus in crus brevius, alterum debilius impulsum repellat, et ita in continuo fluxu perduret, ex vitro cylindrico ascendendo in crus brevius, ex breviori descendendo in longius, et ex longiori in recipiens. Atque sic quidem elaterio res conficitur: idem vero absque elaterio demonstrari quoque posse, forsan videbitur nonnullis prima fronte impossibile; cum facile quidem intelligi possit, qua ratione aëris inclusi resistentia passiva suspensum teneat in cruribus siphonis liquorem, eiusque descensum impediat; non vero identidem, quo pacto fluxu continuo novus subinde liquor in crus brevius assurgat, nisi supponatur aliquid supra liquorem vitri cylindrici, quod eum efficaci pressione in crus illud intrudat, quae pressio aliunde procedere posse non videtur, quam ab aëris elaterio," Bernoulli 1683, 108–111.

[376] "Atque sic quidem Elaterio res conficitur: idem vero absque elaterio demonstrari quoque posse, forsan videbitur nonnullis prima fronte impossibile: cum facile quidem intelligi possit, qua ratione aëris inclusi Resistentia passiva suspensum teneat in cruribus siphonis liquorem, eiusque descensum impediat; non vero identidem [...] Consideremus vero nunc recipiens iterum obstructum, et liquorem in utroque siphonis crure suspensum; quid fiet? cessabit atmosphaerae gravitatio, nec aëris inclusi pondus ullius erit momenti; et siquidem elaterio nullum quoque locum hic tribuimus, cessabit omnis premendi in illo conatus: sed non cessat pariter gravitatio liquoris in cruribus suspensi, qui naturali suo pondere subinde descensum moliens, liquorem in recipiente et vitro cylindrico sublevare, et cum liquore aërem imminentem versus supremam recipientis cavitatem attollere conabitur, longioris quidem cruris liquor columnam qr; brevioris columnam op; ille nisu maiori, quia ponderosior, hic minori, quia ut brevior, ita minus gravis. Cedet ergo fortiori pressioni columna po unaque deprimet liquorem sibi subiectum, eumque in crus brevius ascendere faciet, non vi propriae elasticitatis, sed vi adventitia, communicata sibi a pressione praevalente liquoris in crure longiori: adeo ut, quemadmodum prius ex Elateristarum mente considerabamus pugnam duorum liquorum ab elaterio aëris impulsorum in flexura siphonis c factam, fortioremque ex parte posita vice versa eandem nunc contemplari conveniat, tanquam inter duas columnas aërias a pondere liquorum impulsas, in superficie concava operculi g gestam, fortioremque ex altera parte qr," Bernoulli 1683, 111–113. Not surprisingly, as Senguerd came to explain respiration in his

applies Boyle's law, avoiding any reference to the elasticity of air, but considering only its passive resistance. We need, at this point, to recall his use of the case of non-elastic collision with a wall: in that case, the resistance exerted by the wall is proportional to the force of the impacting body. In that case, no elasticity is considered, but the rebounding force exerted by the wall is equal to the force of motion of the impacting body. For Bernoulli, air behaves in the same way, by its passive resistance. The rebounding force of air is nothing but the effect of the pressure which it sustains, and by which it is condensed. On the basis of this idea, he formulates two main rules: "the densities of two portions of air are to each other, [...] as the weights they sustain," and their "rarities are, to each other, in a relation inversely proportional to the sustained weights." He uses these laws to explain why, in a Torricellian tube in which some air is left at the top, the mercury neither ascends to circa 29 inches, nor descends completely. As to this problem, he had a dramatic encounter with De Raey, to whom he proposed it after having read De Raey's *Clavis philosophiae naturalis*. In the *Clavis* De Raey used the case of Torricelli's instrument (see Fig. 5.9, De Raey 1654, 194) to show that some space can be deprived of all terrestrial matter, and that the air left in-between the mercury behaves as "the most hard body."[377] As Bernoulli reports in his 1686 *Methodus ratiocinandi*, according to De Raey – who did not acknowledge Boyle's law – the mercury reaches the same height even if some air is left at the top of the tube. As soon as Bernoulli objected to this idea, De Raey inveighed against him. Or, in Bernoulli's own words:

> when four years ago I was free to write the above mentioned Dissertation, in the Netherlands, [...] I was stuck about the issue of this same phenomenon, which now we have at hand. I went to Amsterdam, in order to consult there, on this topic, that famous oracle [viz. De Raey]. He, [having] understood the reason of my coming, first kept silence [for a moment], but in order not to show that he did not know that [solution], he asserted, [posing himself] as a master, that the mercury would not fall in the tube, but it would remain at the same altitude as before. I, [as I] knew already that it would certainly descend, and [as I] required at least to know, whether a more or a less quantity of air left in the tube would displace downwards that [mercury], modestly reported [this] to the philosopher. At this, he immediately looked at me with a harsh face, then he asked who I was, [and] after [this] he was indignant, irritated, inveighed against experimental philosophy, and labelled it as histrionic, and he almost [...] threw me out of his house by force. This was, actually, the solution of that famous Cartesian. We, on the other hand, in order to address this topic less militarily, and more philosophically, establish before anything else some principles [...].[378]

Connubium by the idea of the passive resistance of air, he relied on the idea of circumpulsion as well: see Senguerd 1715, 155–161. Bernoulli's solution, besides to Senguerd's, is particularly similar to Schott's explanation of the functioning of Hero's fountain, which he attributed to a rebounding of air caused by the resistance to penetration by water: see Schott 1657, 46–48.

377 "Si tubus satis angustus fuerit, saepe videbis descensum gravissimi illius liquoris impediri ab incluso aëre; utpote qui nonnihil quidem comprimi potest, sed mox tamen ob vim interfluentis materiae subtilis, quae singulas eius particulas seorsim movendo, eas omnino coire non sinit, durissimi corporis instar renititur ac resistit, et vel in tubi fundo universam mercurii quantitatem sustinet, vel in locis intermediis eundem in duas pluresve portiones divulsum ostendit," De Raey 1654, 193–194.

378 "Si quis enim me iuberet divinare numerum, quem quis mente concepit, neque adiiceret conditiones, quibus vestitus esse debeat, quibusque ceu characteribus ipse mihi se prodat, is αδύνατον

Thus far, Bernoulli witnessed a clash, internal to Dutch Cartesianism, between the 'orthodox' approach of De Raey, who would devote much effort to defend the idea that sense experience cannot provide philosophical principles, and the experimental attitude of De Volder, with whom, in any case, De Raey had friendly relationship.[379]

In fact, in explaining this phenomenon in his *Dissertatio*, Bernoulli still did not revert to the idea of elasticity of air. Quite the contrary, he stood with the idea that extension, in itself, resists to the pressure of bodies. In other words, he considered – like De Raey – air to be a solid.

As he considers the first horn of the problem of the determination of the level of mercury in a tube containing air, namely, the case in which mercury is supposed to completely fill the tube, he does not make recourse to the elasticity of air, as for him the air at the top of the tube constitutes a "wall" (*obex*) impeding its ascension. As to the second horn, i.e. the descent of the mercury, he claims that the air present in the tube is not pressed by a whole column of atmospheric air, but only by the pressure, calculated by subtracting (1) the weight of the column of mercury, from (2) the weight of a column of air.[380] On this basis, Bernoulli calculates the density assumed

profecto mihi perciperet. Pariter quoque, antequam quis perspectam habeat naturam aëris, reliquaque Principia hydrostatica ad praesentem quaestionem necessaria, is illius solutioni frustra insudabit; et si paulo sit morosior, usu fere illi veniet, quod antehac celebri cuidam Professori Amstelodamensi, primo Cartesii Discipulo, iuxtaque Defensori acerrimo Clave Philosophica claro, cuius tanto libentius, quanto opportunius mentionem hic iniicio. Cum ante quadriennium scribendae modo dictae dissertationi in Belgio vacarem, atque inter peregrinandum, experiundi destitutus ipse copia, haererem circa eventum huius ipsius, quod nunc prae manibus habemus, Phaenomeni; Amstelodamum concessi, consulturus ibidem hac de re celebre illud Oraculum. Ille, intellecta mei adventus causa, subticuit primo, sed ne quid nescire videretur, argentum in tubo non descensurum, sed in eadem, qua prius, haesurum altitudine, magistraliter asseveravit, Ego, qui descensurum certo praenoveram, et scire saltem efflagitabam, utrum maior minorve aëris copia in tubo relicta illud humilius detrusura esset, modeste Philosopho regessi; ad quae ille torvo me statim intueri vultu, dehinc percontari quis essem, postea indignari, stomachari, in Philosophiam Experimentalem invehi, eamque histrionicam nuncupare, et me tantum non vi ex aedibus suis expellere. Haec erat tum solutio celebris istius Cartesiani. Nos vero ut minus militariter, ac magis philosophice rem aggrediamur, stabiliemus ante omnia principia quaedam," Bernoulli 1686a, 257–258.

379 See Sect. 2.3.1, *The correspondence with Thévenot.*

380 "Unius adhuc superest phaenomeni ut evolvamus causam (quod intellectis nostris de elaterio et resistentia passiva regulis non erit arduum) quare viz. si in tubi mercurio impleti summitate relictum fuerit pauxillum aëris, argentum vivum nec omne effluere, nec omne in tubo suspensum haerere, sed notabiliter tamen descendere debeat, etiamsi argentum ad longe minorem altitudinem 29 digitis infusum fuerit. Notandum autem, duo hic distincte quaeri posse, semel cur argentum, quamvis ad minorem altitudinem infusum, non ascendat; columna enim atmosphaerica vasculo imminens illud altius impulsura esset in tubum, sine interventu aëris in summitate relicti: dein, cur praeterea etiam notabiliter descendat. Prioris causam reiiciemus non in elaterium, sed passivam tantum resistentiam inclusi aëris, qui cum naturalem habeat consistentiam, et a summa base tubi suffultus sit, iuxta reg. 2 toti atmosphaerae ponderi, argentum altius subinde impellere conanti, obicem ponere potis est. Quod vero argentum non tantum non ascendat, sed et descendat, exinde est, quoniam aër inclusus non premitur a tota cylindri atmosphaerici mole, sed a tanta duntaxat illius portione, quae correspondet excessui, quo totum eius pondus superat pondus cylindri mercurialis inclusi: ex. gr. si altitudo mercurii infusi fuerit 20 digitorum; aër, inter tubi summitatem et mercurium interceptus, sentiet tantum pondus 9½ digitorum mercurialium; quanta videlicet est

by the air included in the tube (which he assumed to be, at first, as dense as external air), and thus the volume it occupies, and the level of the mercury. According to him, one can in this way infer the same values given by Boyle in his table, without any consideration of the elasticity of air.[381]

As a matter of fact, even if Bernoulli does not consider, as to this case, the elasticity of air, a force of rebounding is nevertheless supposed by him, even if he considers the resistance to pressure, rather than an active force of rebounding, as the cause of it. Indeed, as he will come back to discuss the problem of the suspension of mercury in the tube in his *Methodus ratiocinandi*, he would completely abandon the idea of the passive resistance of air, and explain this phenomenon[382] by considering some principles, by which, as seen in the quotation, he could address this problem more philosophically than De Raey. Namely, (1) Archimedes's first postulate, (2) the weight of the air, (3) the elasticity of the air, (4) Boyle's law, for which "the pressures of air are directly proportional to its densities, or inversely proportional to its rarities."[383] On these premises, Bernoulli analyzes three alternative situations concerning the case in which a Torricellian tube, containing air in its top, is placed in a

---

differentia inter pondus atmosphaericum, aequivalens 29½ digitis mercurialibus, et pondus mercurii inclusi: quoniam enim atmosphaera ab una parte, tota sua mole sursum impellere conatur argentum, ab alter vero argentum naturali sua gravitate, contra nititur; fit ut aequalibus, illinc pellendi sursum, hinc descendendi, viribus sublatis, aër inclusus ea tantum pressione afficiatur, qua pondus cylindri mercurialis inclusi superatur a simili cylindro atmosphaerico," Bernoulli 1683, 115–117.

[381] "Et quoniam per reg. 4 [...] volumen aëris inclusi debet esse ad volumen aëris rarefacti, ut vicissim Pondus sustentandum ab hoc, ad Pondus totum atmosphaericum sustentabile ab illo. [...] Certum autem est, a parte nostri nullam posse esse hallucinationem, eo quod suppositio haec, Densitates aëris esse ad invicem, ut pondera sustentata, e qua calculus noster immediate fluxit, non alia est quam ipsissima Clar. Boylii hypothesis, egregiis insuper experimentis, gemina tabula in Tractatu eius contra Linum exhibitis, stabilita: quodque in toto hoc negotio nullum aliud inter nos discrimen est, quam quod ille Elaterio tribuit, quae sola nonnunquam aëris resistentia passiva explicari posse autumo; caetera vero iidem omnino utrinque effectus expectandi," Bernoulli 1683, 118, 121–122.

[382] "Phaenomenum [...] nunc [...] ut methodus ususque utriusque logicae eo clarius patesceret, prolixius enodandum mihi proposui, istud est. Si fistula cylindrica 29 pollicum longitudinem non excedens, una extremitate clausa, altera patula, repleta sit ex parte mercurio seu argento vivo, reliquo spatio aëri concesso, eaque postmodum obstructo digiti pulpa orificio invertatur, atque erecta perpendiculariter immergatur cum obstruente digito in stagnantem alicubi mercurium; explorandum est, an et quousque remoto digito mercurius in fistula descensurus sit?" Bernoulli 1686a, 256.

[383] "Principia autem sunt sequentia: I. omnes partes cuiuscunque liquoris aequaliter a centro Terrae remota, a pondere perpendiculariter sibi incumbente premi debent aequaliter: et si premuntur aequaliter, eo situ quiescunt, sin minus, non prius componuntur ad quietem, quam res ad aequipondium reducta fuerit, assurgentibus hinc partibus quibusdam, subsidentibus inde aliis. Celebratissimum hoc principium hydrostaticum fluit ex ipsa natura liquidi. [...] II. Atmosphaericus noster, quem spiramus, aër, non minus atque argentum vivum, pondere seu gravitate aliqua instructus est. [...] III. Aër, secus quam alia fluida, praeter gravitatem insigni quoque praeditus est elaterio, seu virtute sese expandendi et contrahendi. [...] IV. Neque vero [...] natura incerto hic agit motu, sed certam observat Legem et Proportionem in aëris magis minusve densati pressionibus. Deprehendit enim Illustris Boylius eleganti Experimento, nobis etiam feliciter tentato, quod Pressiones aëris sint in ratione directa densitatum, vel reciproca raritatum illius," Bernoulli 1686a, 258–261.

vessel of mercury. In the first case, its level does not change. This alternative is to be excluded because the pressure exerted by this air, given its compression, is equal to that exerted by the external atmosphere. Second, the mercury ascends: in this case, such air is compressed even more, so that its elastic force is increased, and it must descend. Third, the mercury descends: it does not descend completely since, as far as such air is rarefied, it loses its elastic force and the pressure it exerts. At some point, it stops, namely when the pressure exerted by its elastic force, plus the force exerted by the weight of the mercury are in balance with the atmospheric pressure – which can be calculated by following Boyle's law.[384]

Leaving aside Bernoulli's 'interpretation' of Boyle's law, what is interesting for our survey is De Volder's use of the elasticity of air only, which points to his probable assumption of Boyle's law. In fact, if Bernoulli himself reverted to this idea only in his 1686 *Methodus*, his earlier re-interpretation of De Volder's experiment testifies that De Volder was consistently using the idea of elasticity in his experimental courses at Leiden. As a matter of fact, De Volder did not teach Boyle's law in his *Disputationes*, *dictata* and *Experimenta*. However, there is extant evidence that he carried out observations with thermometers, namely with two kind of instruments: (a) the open thermometer, which is basically a barometer as, in it, the level of fluid ascends or descends as a consequence of (1) the heat of air – viz. its dilatation – and (2) its pressure; (b) the closed thermometer, also called the Florence thermometer (and referred to in his *Bibliotheca* as "Grand thermomètre de Florence")[385] – which was described in detail for the first time in the *Saggi di naturali esperienze* (1666) (see Fig. 5.67, Magalotti 1666, 3).

The use of the two kinds of thermometers is reported in De Volder's Hamburg 273 series of *dictata*, as he comments upon the compression of air (IV.46), and claims that in open thermoscopes the air at the top is dilated by heat, so that the *aqua fortis* is depressed; in closed ones, it is the *aqua fortis* which is rarefied by heat, so that its volume grows in proportion to its temperature.[386] Bernoulli also, in his *Dissertatio*, notes that De Volder reported to him his observations with the two kinds of instruments.[387] Furthermore, two disputations, now lost, presided over by De Volder in 1693 concerned thermometers,[388] as well as the *Disputatio philosoph-*

[384] Bernoulli follows a syllogistic reasoning, by which he excludes all these alternative, and infers his conclusion: see Bernoulli 1686a, 261–267.

[385] See *Bibliotheca Volderina*, 95.

[386] Cf. the text quoted *supra*, n. 333.

[387] "Ita quoque mihi retulit Cl. Volderus, sibi observatum aliquando fuisse circa duo thermoscopia, quorum unum utrinque sigillatum erat, alterum infima sua extremitate cum aëre externo correspondebat, quod videlicet eodem die aestivo liquor in utroque thermoscopio ascenderet; notum autem est, solius thermoscopii utrinque sigillati genium esse, ut liquor in eo aestate ascendat, reliqui vero naturam esse, ut liquor in eo aestate deprimatur. Causam ergo anomaliae istius adiecit hanc fuisse, quod gravitas aëris externi solito fuerit maior, adeoque aëri superius incluso impedimento fuerit, ne per calorem sese dilatando, liquorem deprimeret," Bernoulli 1683, 214.

[388] Both defended or possibly authored by Antonius van Houten, who also acted as *respondens* of the 28th of De Volder's *Exercitationes* against Pierre-Daniel Huet's *Censura philosophiae Cartesianae* (1689): namely, Burchard de Volder (*praeses*), *Disputatio philosophica de causis*

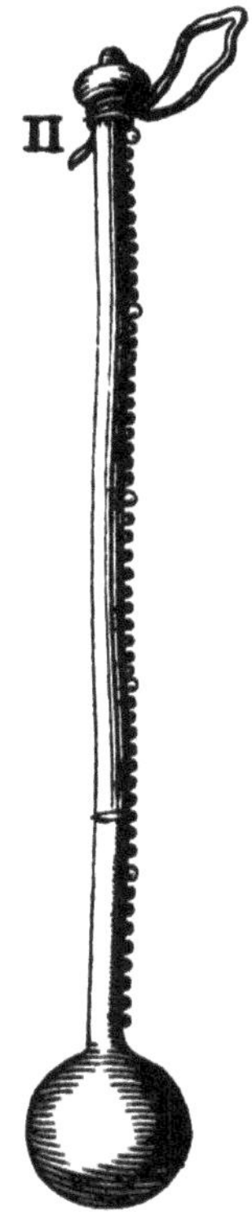

Fig. 5.67 Closed thermoscopes (or *thermomètres de Florence*). (Source: Magalotti 1666, 3. Naples, Biblioteca Nazionale, signature: B. Branc. 085F 2, https://archive.org/details/bub_gb_BcCpGiXkRKsC/page/n5. Public domain mark 1.0)

ica inauguralis de vitri thermometrici of Jacobus Rattrey (1679), where he acted as promotor but where he is never mentioned.[389] De Volder, therefore, probably dealt with the link between the pressure and the condensation of air, i.e. with Boyle's law, in his lessons, even if no direct trace of the teaching of the law is extant.

More substantial, indirect evidence of this teaching can eventually be found in the aforementioned disputations of Schuyl (*De vi corporum elastica*, 1688) and Serrurier (*De gravitate aëris*, 1690), whose texts reveal the same, common source, as they both provide a corpuscular theory of elasticity which was unheard in the

variationum thermometrorum, Leiden, Elsevier, 1693; *Disputatio philosophica de thermometris*, Leiden, Elsevier. Mentioned in Bierens de Haan 1960, 131.

[389] In the disputation the two kinds of thermoscopes are presented: the thermoscope whose lower part is opened and inserted in a vessel of water or *aqua fortis* (theses 3–4), and whose level increases when it is cold, and the thermometer of Florence, where the liquid dilates when it is hot (thesis 15). However, since Rattrey enrolled at Leiden just few days before his graduation (see *Album studiosorum Academiae Lugduno-Batavae MDLXXV–MDCCCLXXV*, column 628), it seems that behind this disputation there was, rather, the hand of Petrus Hoffwenius, professor at Uppsala, who is positively mentioned in the disputation and who had studied under De Raey at Leiden. De Raey's *Clavis*, in turn, is cited by Rattrey in his account of rarefaction (theses 9 and 22).

seventeenth century, and they ascribe to elasticity only the foremost phenomena involving air. In fact, in these disputations it is claimed that phenomena like the cohesion of hemispheres and the ascension of mercury are due to the elasticity of air rather than to its gravity, and that the pressure of air depends on its compression.[390] This amounts to an implicit acceptance of Boyle's law. Thus far, while in his *Disputationes de aëris gravitate* De Volder taught that the cohesion of hemispheres had in the weight of the air its principal cause, and unified the theory of hydrostatics with his pneumatics – providing a model of cohesion valid both for elastic and non-elastic fluids – in these later texts he overtly ascribed to elasticity only – and to the pressure it exerts – those effects beforehand treated by 'Archimedean' means. Of course, the accounts are compatible with each other: already in 1676–1678, as seen above, De Volder ascribed the cohesion of hemispheres to the pressure actually exerted by a limited portion of atmosphere, in turn compressed by the weight of the whole atmosphere. However, in his later disputations the idea of elasticity comes to be at the centre of De Volder's explanations, while the weight or gravity of air is a sort of remote cause of the pressure exerted by air itself.

As to the corpuscular theory of elasticity, in both texts it is claimed that the corpuscles of air are not elastic in themselves, and have a conical form, so that, once

[390] "Iam enim ostendi, quod, quo aër magis sit compressus et condensatus, eo vis eius elastica sit maior, rarefactione vero minuatur, ita ut, modo eiusdem maneat condensationis, eandem etiam vim elasticam retinere debeat. Cum ergo infimus aër a vi totius athmosphaerae sit compressus, eiusdemque maneat condensationis, sive in recipiente, sive extra ipsum constitutus, patet evidenter, quod aër, sive in recipiente contineatur undique clauso, sive totam habeat incumbentem athmosphaeram, easdem vires, eandemque pressionem necessario habere debeat. Unde simul patet, quod causa omnium illorum effectuum, qui a pressione aëris dependent, ut sunt: duorum haemisphaeriorum firmissima cohaesio, adscensus mercurii alteriusve liquoris in baroscopii, et antliis, et c. non gravitati aëris, sed vi eius elasticae sit adscribenda," Schuyl 1688a, chapter 2, 7–8 (unnumbered); cf. Serrurier 1690, theses 9–10: "IX. Diximus antea aërem altitudine sua premere cui refragari videtur experientia, in recipiente scilicet undique aëri impervio mercurius aeque suspenditur ac in aëre aperto. Ex quo videtur sequi aëris gravitatem hic nihil facere, tantillulum enim quod est inclusi aëris pondus ne millesimam quandoque mercurii suspensi partem aequat. Huic autem difficultati ut obviam eamus, cum doctissimis viris aëri vim elasticam tribuimus, talem videlicet qua postquam a vi externa compressus est sese elateris instar dilatat. [...] X. Considerari velim aërem qui circa nos est premi ab incumbente atmosphaera, et sic impediri in motu suo vorticoso, aetherem autem qui interfluit aëris interstitia sollicitare continuo ad liberiorem motum quod ipsum in posterum dicetur vis elastica. Haec autem dupliciter considerari debet, vel quatenus remoto pondere atmosphaerae aërem dilatat, vel quatenus compressioni aëris resistit. Notetur insuper aërem qui nobis est contiguus tanta vi elastica praeditum quanta aequivalet comprimenti atmosphaerae; ideoque quamdiu manet idem pondus tandiu idem vigebit elaterium. Quid autem evidentius quam quod liceret atmosphaerae intercipiatur a corpore satis firmo aër mansurus sit eiusdem compressionis ac si cum incluso aëre externus communicaret; illud enim ponitur ita firmum ut dilatationi aëris queat resistere, manebit itaque eiusdem condensationis et simul elaterii quod [...] habebat; sed illud valebat atmosphaerae pondus, huius autem pondus aequivalet mercurio 29 digitos alto, ergo et aër inclusus, quem per consequens sustinebit in eadem altitudine qua potest gravitas totius atmosphaerae. Respondemus itaque ad propositam difficultatem mercurium ideo suspendi a pauco illo aëre quoniam si descenderet deberet aërem [...] comprimere in angustius spatium sed coarctationi huic resistit aequali vi, qua mercurius eam molitur, conatus itaque descendendi et lapsum impediendi in mercurio et aëre aequalis est, et idcirco primus effectum non sortietur i.e. mercurius manebit suspensus." Cf. also thesis 13.

subtle matter flows in-between them, they start to rotate as their centre of gravity is located in their thicker side, so that they always describe circles.[391] This idea was

[391] "Cuius causa ut detegatur, considerari velim, quod corpora sphaerica, sive quiescant, sive moveantur circa suum axem, quicunque demum ille sit, aequale semper sua circumferentia describant spatium. E contrario vero, quod corpora angulosa, et praecipue oblonga in orbem moveri nequeant, quin multo plus spatii motu suo describant, quam, cum quiescunt, occupant. Adeoque cum eadem pars aëris modo maius, modo minus spatium occupare queat, patet evidenter, aëris particulas non sphaericam, sed oblongam habere figuram. Figura autem oblonga ad duas hasce primarias species reduci potest, quod nimirum sit, vel cylindrica, vel conica. Quantum ad priorem, non videtur illa ad hanc rem accomodata, non potest enim talis figura in orbem moveri, nisi una eius extremitas, altera vel immota, vel saltem fortius impellatur, si enim utraque eadem vi versus eandem partem impellatur, secundum lineam rectam progredietur, cum sint eiusdem [...] ubique crassitiei. Contra vero posterior eo ipso, quod utraque eius extremitas aequali vi impellatur, necessario in orbem moveri debet; alter enim eius extremitas [...] altera est tenuior, quae propterea [...] altera celerius moveri debet, quod fieri non potest, cum partes maius sint continuae, nisi moveatur in orbem. Quare, cum nulla hic appareat ratio, cur una particularum extremitas, altera fortius impelleretur, sed contra, partim propria gravitate, partim actione materiae caelestis, per ipsarum intervalla dispersae, aequaliter premantur, [...] aëris particulas non cylindricam, sed conicam figuram haere concludo. Quo positis omnia, quae circa vim hanc aëris elasticam observantur phaenomena, facillimo negotio explicari possunt. Cum enim figura haec conica apice suo circa basim, tanquam centrum sui motus diversae amplitudinis gyros describere possit, patet evidenter, cur aër varios per gradus condensari, et rarefieri queat. Docet enim experientia, quod machinarum ope multo plus aëris intra idem spatium, quam naturaliter in ipso continetur, cogi possit, vel contra multum aëris ex ipso educi queat, ita ut multo minor quantitas idem spatium, vel eadem multo maius occupet. Hac autem aëris condensatione, vis eius elastica intenditur, rarefactione vero minuitur, ut innumera probant experimenta, quae, quia hodierno tempore sunt vulgatissima, ipsorum relatione chartas hasce meas implere, supervacaneum existimo. Ratio autem huius rei non potest iam esse obscura. Hac enim condensatione, intervalla inter aëris particulas interiecta, redduntur minora, quo fit, ut non tantum singulae minores apicibus suis gyros describere debeant, sed et materia caelestis multum in suo per eadem tranfluxu impediatur, quo mutua ipsarum pressio non parum augetur, et ut, simul ac detur exitus, cum praevaleat aëri externo, ex recipiente expellatur, tamdiu, donec cum eodem ad aequilibrium sit redactus," Schuyl 1688a, chapter 2, 5–6 (unnumbered). Notably, in Serrurier's disputation this idea is defended after having presented Boyle's corpuscular theory of the elasticity of air, including Boyle's own judgment on Descartes's theory (described above): see Serrurier 1690, thesis 9: "Illustris Boyle experimentis suis Physico-mechanicis vim hanc aëris elasticam ita manifestavit, ut nullus doctior ulterius dubitet. Elaterium hocce nullo alio in fluido observarunt curiosissimi quique, quocirca hoc aërearum particularum speciali figurae adscribendum censuerunt physici. Nobiliss. Boyle duplicem huius elaterii proponit caussam. Primam constituit in figura aëri ramosa et flexili instar velleris seu spongiae, haec enim manu compressa, simulac desinis premere in priorem statum sese expandit; pili velleris similiter compressi ubi dimittuntur sponte sua, vel per caussam sensus latentem sese restituunt. Eodem ritu concipit flexiles inferiori aëris particulas comprimi ab incumbente atmosphaera; quod pondus atmosphaerae si artificio tollatur aër antea compressus nunc autem pondere aëris liberatus instar spongiae sese explicat. Tandem Nobilis Author cum Cartesio concipit particulas aëris vi aetheris circumrotatas gyris suis parvulam quandam circa se sphaeram constituentes, hae autem particulae circa Tellurem ab incumbentis in motu suo impediuntur, quod impedimentum continuo amovere nituntur, donec eo sublato, aether subtilior qui aëris permeat poros ita circumrotet partes aëris ut maiorem describentes circulum ex sphaera suae activitatis expellant obvias quasvis particulas. Posterior caussa maxime mihi arridet, interim illum conatum maiorem describendi circulum pono in figura conica quam aëris particulis adscribo. Si enim cylindraceam ponamus quoniam in his centrum magnitudinis cum centro gravitatis coincidit, cum moventur circa axem uniformi gyrant motu, neque plus spatii quam quod sua mole dimetiuntur requirunt pro circulo suo motus absolvendo." It is worth mention-

probably inspired in De Volder by the *De motionibus naturalibus a gravitate pendentibus* (1670)[392] of Giovanni Alfonso Borelli, whose mathematical approach was intended as a careful procedure for arguing for conclusions, and as the explanation of phenomena through mechanical principles in physiology, is praised by De Volder in his 1698 *Oratio de rationis viribus, et usu in scientiis*, together with Lorenzo Bellini.[393] Borelli was a vacuist: for him, elasticity does not depend on any subtle matter provided with a filling function, and in rarefaction there is an actual vacuum between parts.[394] Elasticity – dealt with in propositions 123–125 of his *De motionibus* (where however Boyle's law is not taken into account), depends on the features of single particles of air, which, by following an hypothesis capable of 'saving' phenomena, and not based on "fantastic and fanciful hypotheses," are rigid, concave cones, which during the condensation of air are inserted into each other (see Fig. 5.68, Borelli 1670, 259).[395]

De Volder, thus far, refined his older, Cartesian model of elasticity by using a mechanical explanation of the rotation of particles, based on the idea of a displaced centre of gravity. Since this model is given, in Serrurier's disputation, after a careful discussion of Boyle's and Descartes's model,[396] we may suppose that Boyle's ideas were still at the centre of De Volder's interests and teachings in late the 1680s.

---

ing that De Volder is mentioned in two other disputations concerning the elasticity of air: namely the *Dissertatio de vi elastica corporum solidorum et fluidorum* (May 1704), presided over by Johann Georg von Bergen and authored by Daniel Fremaut (who was to enrol at Leiden in September 1704, but who apparently had already attended lectures there), defended at the Viadrina University (Frankfurt (Oder)) in which De Volder appears as a dedicatee together with Frederik Ruysch, both labelled as Fremaut's 'preceptors'. In the disputation, the positions of a large bulk of sources, including Boyle and Huygens, are discussed. Moreover, De Volder's discussion of the elasticity of the air is mentioned (but not detailed) in the *Disputatio medica inauguralis de functione pulmonis* of Georgius Remus, held at Leiden in September 1711.

[392] De Volder probably read this book in its second edition (Leiden, 1686), which he owed: see *Bibliotheca Volderina*, 3.

[393] "Hanc ad rem perspiciamus, quanam via insistant geometrae, quos unicos habemus certi inveniendi magistros, ubi circa res physicas occupantur, inquisituri, num et illa medicis inserviat. Quod eo propono confidentius, quia mathematicorum more rem medicam novo exemplo tractare iam coeperunt illustres omnium suffragiis viri Borellus, Bellinus. Quorum libros a tam paucis etiam inter eos, qui mechanicam philosophiam sectantur, medicis evolvi, et a paucioribus intelligi, dolendum est summopere," De Volder 1698, 25. De Volder probably referred to Borelli's *De motu animalium* (1680–1681), and Bellini's *De urinis et pulsibus* (1685), *Opuscula aliquot ad Archibaldum Pitcarnium* (1695) (which he owned: see *Bibliotheca Volderina*, 3, 58).

[394] See proposition 265 of his *De motionibus naturalibus a gravitate pendentibus* (1670): Borelli 1670, 535–537.

[395] Borelli 2015, 125–128. See especially proposition 125: "[t]he properties of air are better explained if its minute particles are hard, flexible and resilient like springs and if they have the shape of tubes or hollow cylinders made of smooth or multiple sheets or threads obliquely coiled," Borelli 2015, 127. Cf. Borelli 1670, 257–261. Alternative hypotheses Borelli conceives are more Boylean: namely, air can be composed by sub-particles, entangled as flexible tubes, which can assume, if compressed, an elliptic shapes. According to the third, they have a spring-like shape, i.e. a spiral shape, and made by a flexible, but elastic matter: Borelli 1670, 261–262.

[396] See *supra*, n. 391.

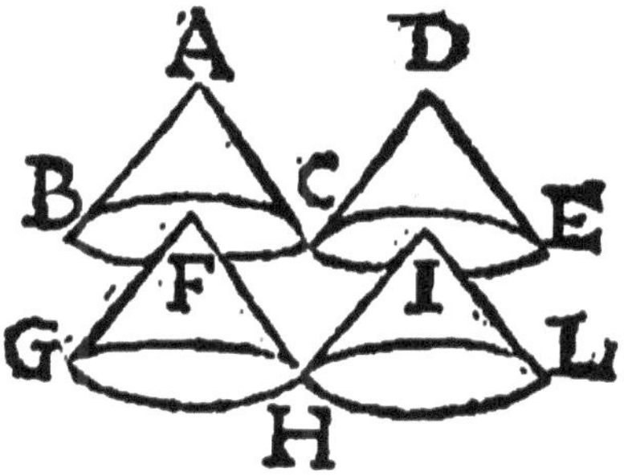

Fig. 5.68 Hypothetical corpuscles of air, aimed at explaining its elasticity according to Giovanni Alfonso Borelli. (Source: Borelli 1670, 259. ETH-Bibliothek Zürich, signature: Rar 3338, https://www.e-rara.ch/zut/doi/10.3931/e-rara-37434. Public domain mark 1.0)

5.6.4 The Application of Boyle's Pneumatics to Physiology

The ascertainment of De Volder's appropriation of Boyle-Mariotte's law can, eventually, shed light on an entire body of his experiments, namely, his studies on respiration, of which the ideas of *metus vacui*, circumpulsion and air pressure are an essential component. As testified by the *series lectionum* of the University, De Volder did not teach on medicine or medical topics at Leiden. Physiology, intended as the theoretical part of medicine, traditionally pertained to the medical curriculum.[397] However, from 1688 he started to preside over, being the dedicatee of, disputations with relevant entailments in physiology, viz. the study of the functioning of animal bodies. This is the case of Schuyl's doctoral disputation *De respiratione* (1688), Petrus Voogd's *Exercitium medico-physico aphoristicum: de morborum causis* (1691), Stephan Huszthi's *Dissertatio physiologica prima quae est de hominis automato seu corpore animali*, 1693 (part of a broader series of disputations, covering topics going from physiology to rational theology),[398] Elias Petrus de Beaumont's *Dissertatio medico-physica de circulatione sanguinis* and *Dissertatio physica de circulatione sanguinis in foetu* (1698). Also, De Volder appears as dedicatee of the inaugural disputations of Johannes Zoutmann, namely his *Disputatio medico-philosophica inauguralis de imaginationis maternae viribus in foetum* and *Disputatio medica inauguralis physico-anatomica de spirituum animalium vera existentia et operatione* (1702).[399] For the sake of the present discussion – concerning natural philosophy and its branches – it is worth looking at how De Volder dealt

[397] See Horstmanshoff et al. 2012.

[398] See the bibliography for more details.

[399] De Beaumont later graduated in medicine at Utrecht, with a *Disputatio medico-physica inauguralis de natura sanguinis*, 1699. No traces of his magisterial graduation are extant. Other disputations dedicated to or presided over by De Volder, touching upon the functioning of human body, such as Bernard de Mandeville's *Disputatio philosophica de brutorum operationibus* (1689), Hermann Oosterdijk Schacht's *Disputatio philosophica inauguralis de sensibus internis memoria et imaginatione* (1693), Johannes Robberghtsten's *Disputatio philosophica de sensu brutorum* (1694) and Elias Petrus de Beaumont's *Dissertatio physica de carentia sensuum et cognitionis in brutis* (1698) were in fact concerning the theory of knowledge rather than physiology.

with the topic of respiration in these disputations, as in addressing this topic he applied and discussed his ideas on pneumatics.

In fact, De Volder devoted vast attention to the problem of respiration and the nature of air as early as 1676–1677, as testified to by his *Experimenta*. In experiments 6–9, he demonstrated the necessity of air for respiration by putting various kinds of animals in the container of the air pump, and asphyxiating them by pumping out the air. He was not the first who attempted this. While Von Guericke devoted limited attention to this topic, as he putted in a container a bird and fishes, but without entering into detailed explanations, and Sturm, in turn, reserved a few scattered comments to the topic,[400] the members of the Accademia del Cimento and Boyle devoted more substantial attention to these kind of experiments. For instance, we find in the *Saggi di naturali esperienze* a detailed account of what happens to the bodies of different kinds of animals, such as birds, fishes, reptiles, a crab, and frogs, and reporting the different reactions they had to the lack of air (in the case of the frog, for instance, its swelling is duly described).[401] It was Boyle, however, who ventured to discuss natural-philosophical issues concerning respiration, in his *Digression Containing Some Doubts Touching Respiration*, appended to experiments 40–41 of his *New Experiments*. While in such experiments he put in the receiver insects, as well as a lark, a sparrow, and a mouse, demonstrating that they die "from the want of air," and without providing other conclusions as to their physiology,[402] in the *Digression* he discussed others's positions on breathing. Even if he did not want to "meddle with those Controversies so much agitated among the Moderns, namely, Whether the motion of the Lungs in Respiration be their own, or but consequent to the motion of the Thorax, Diaphragme,"[403] he defends the idea that lungs are passively dilated by air (as they are not attached to any muscle).[404]

[400] In part 3, chapter 16 of his *Experimenta*, Von Guericke claims that "[o]ne could very clearly see from this that all bodily movement ceased in the animal with the loss of air and that its life-force, residing in the heart, had been extinguished (like a flame of wine spirits)," Von Guericke 1993, 143. See Sturm 1676, 106–107.

[401] See the *Racconto degli accidenti vari di diversi animali messi nel voto*, in Magalotti 1666, 113–125.

[402] For instance, as to the cause of the convulsions observed in such animals, or the colour of their dissected lungs (which De Volder, on the other hand, was to address): "I forgot to mention, that having caus'd these three Creatures to be open'd, I could, in such small Bodies, discover little of what we sought for, and what we might possibly have found in larger Animals; for though the Lungs of the Birds appear'd very red, and as it were inflam'd, yet that colour being usual enough in the Lungs of such winged Creatures, deserves not so much our notice, as it does, That in almost all the destructive Experiments made in our Engine, the Animals appear'd to die with violently Convulsive Motions: From which, whether Physicians can gather any thing towards the Discovery of the Nature of Convulsive Distempers, I leave to them to consider," Boyle 1660, 332.

[403] Boyle 1660, 336–337.

[404] "But that it may appear what kinde of service it is that may be expected from it on this occasion, we must premise a few Words to shew wherein the strength of the Objection we are to answer, lies: In favor then of those that would have the Lungs rather passive then active in the business of Respiration, it may against the common opinion be alledg'd, That as the Lungs being destitute of Muscles and of Fibres, are unfit to dilate themselves, so it appears, that without the motion of the Thorax they would not be fill'd with Air," Boyle 1660, 337.

Moreover, Boyle furthers his discussion of the cause of death in animals in a vacuum, and on the function of respiration. For instance, he hypothesizes that animals are killed by the dilatation of the air contained in them, or by the phenomena of fermentation observed in fluids in a vacuum,[405] however omitting to declare his positions as "I must not now dwell upon matters of this nature, because I think it high time to proceed to the consideration of the principal subject of our Engine, namely, The use of Respiration; or rather, The use of the Air in Respiration." In dealing with this topic, Boyle criticizes the idea that it serves to cool the blood (as cold-blooded animals also breathe), and supports the idea that it serves to carry excrement out of the body, ventilate the blood in the lungs, on the grounds of the evidence that a far too "thick" air, such as that found in mines, does not allow respiration.[406] In turn, a too much rarefied air leads to death, as shown by the experiment with the animals in the container, and with the case of travellers on mountains, such as José de Acosta, who claimed in his *De natura novi orbis libri duo et de promulgatione Evangelii apud Barbaros sive De procuranda Indorum salute libri sex* (1589, also known in the translation as *Historia natural y moral de las Indias*, 1590) that on the high mountains of Peru (much higher than "ordinary houses" like the Alps or the mountains of Spain) felt severe nausea.[407] Boyle, however, does not concede that there are difficulties in breathing at lower altitudes, such as on the Alps, where he himself did not notice any change in the quality of air on the Alps, or on Mount Teides, or Tenerife Peak, of which he read the report of an ascension. With this, he criticizes the idea, attributed to Aristotle, that on Olympus – far lower than Mount Teides – there can be difficulties in breathing.[408] In fact, if we look at the *Relation of*

[405] Boyle 1660, 342–345.

[406] Boyle 1660, 348, 350–353.

[407] "That (on the other side) an Air too much dilated is not serviceable for the ends of Respiration, the hasty death of the Animal we kill'd in our exhausted Receiver, seems sufficiently to manifest. And it may not irrationally be doubted, whether or no, if a Man were rais'd to the very top of the Atmosphere, he would be able to live many minutes, and would not quickly dye for want of such Air as we are wont to breath here below. And that this Conjecture may not appear extravagant, I shall on this occasion subjoyn a memorable Relation that I have met with in the Learned Josephus Acosta, who tells us, That when he himself past the high Mountains of Peru, (which they call Pariacaca) to which, he says, That the Alps themselves seem'd to them but as ordinary Houses, in regard of high Towers, he and his Companions were surprised with such extreme Pangs of Straining and Vomiting, (not without casting up Blood too) and with so violent a Distemper, that he concludes he should undoubtedly have died, but that this lasted not above three or four hours, before they came into a more convenient and natural temperature of Air," Boyle 1660, 355–357; cf. Acosta 1589, book 3, chapter 9.

[408] "That, which some of those that treat of the height of Mountains, relate out of Aristotle, namely, That those that ascend to the top of the Mountain Olympus, could not keep themselves alive, without carrying with them wet Spunges, by whose assistance they could respire in that Air, otherwise too thin for Respiration: (That Relation (I say) concerning this Mountain) would much confirm what hath been newly recited out of Acosta, if we had sufficient reason to believe it: But I confess, I am very diffident of the truth of it; partly because cause when I pass'd the Alps, I took notice of no notable change betwixt the consistence of the Air at the top and the bottom of the Mountain; partly because in a punctual Relation made by an English Gentleman, of his ascension to the top of the Pike of Tenariff (which is by great odds higher than Olympus) I find no mention of any such difficulty of breathing," Boyle 1660, 357–358. One cannot find such a statement in Aristotles's

the Pico Teneriffe (circa 1646), read by Boyle before being printed in Thomas Sprat's *History of the Royal Society* (1667), we find that "[i]n this miles ascent some of our company grew very faint and sick, disorder'd by fluxes, vomitings, and Aguish distemper," even if, contrary to Acosta's experience, such symptoms then passed, and the ascension continued.[409]

In his *Experimenta*, De Volder went further than Boyle, and used the air pump to provide insights on physiological topics omitted by Boyle, who, among others, did not enter into the causes of the convulsions observed in these animals, or the colour of their dissected lungs, which De Volder was to address.[410] First of all, De Volder used a larger animal to perform his experiments. In experiment 6, or "the trial of quadrupeds viz. the little dog,"[411] the animal, after having been asphyxiated by evacuating the container, was observed as swelling and deswelling according to the presence of air in the container itself, as it was for the bladder in experiment 5 – referred to in this experiment – and as often observed in several other experiments mentioned (but not reported) by Morley.[412] As he dissected the body of the dog, he observed different colours in its lungs, namely a red colour, and a paler one – characterizing animals which are healthy – and provided confirmation of the idea that the lungs of newborns which are born already dead, without breathing, collapse – a topic widely discussed in the seventeenth century, not only in medicine (where it was treated by Theodoor Craanen) but also used as a method to settle disputes on legal inheritances.[413] Moreover, in experiment 7 – where he used birds – De Volder also explained the phenomena of convulsions in animals in a vacuum, as due to the extraction of air and animal spirits from the container, leading to the entrance of subtle matter agitating the rest of animal spirits present in the muscles.[414]

works: however, it was attributed to Aristotle, among others, by Kepler: for a discussion, see West 1998, 3–4.

[409] Sprat 1667, 201.

[410] See the text quoted *supra*, n. 402.

[411] De Volder 1676–1677, 95r.

[412] "Even just as we had seen in the last experiment of the bladder, and the weight, wherein the air being taken out of the great bottle glass (which did compress bladder, et constipate the air within the bladder) that which was within did presently dilate itself, and so much even to the lifting up of such a weight: and the said air being let in again did condense and constipate the air in the bladder whereby it de-swelled, et crumpled and let down the weight to its place where it was before: and this was also the reason of several other experiments before-seen, and rehearsed in this bask," De Volder 1676–1677, 93v.

[413] "The dog being cut up he took one the lungs and showed unto us: the which were found to be of two colours: the one was of a darkish red colour which he called color fuscus. The other part were of a whitish colour such as we see in the lungs of other animals that are sound. [...] [H]ence might easily be examined, if an infant was born dead or alive: for if born dead the lungs will certainly sink: if it was born alive, and breathed never so little while, they will never sink do what you can," De Volder 1676–1677, 91v–94r. See Craanen 1689, 254–255.

[414] "Of these convulsive fits he said it was very remarkable that all animals that [...] ever he had put into the vacuum did die all alike by convulsions whereof he said the cause was not easy to be rendered, convulsions being caused by contraction of the muscles which is performed by the affluxion unto them of the animal spirits: {now} why the extraction of the air should cause the animal spirits

However, the most relevant natural philosophical-physiological topic dealt with in De Volder's *Experimenta* is that of respiration. On the one hand, he must have noted that elasticity plays some role in it, as he explicitly connects experiment 5 to experiment 6. Moreover, in experiment 9, where he put insects and a frog in the container, he noted the swelling of the body of the frog (such as that observed at the Accademia del Cimento).[415] Probably, however, he still did not adopt the idea of elasticity in explaining respiration for the reason that, in the 1670s, he was puzzled by the source of the powers of air, as on the one hand he explained the elastic power of air by the idea of subtle matter, which penetrates the bladder and makes it capable of lifting weights. On the other, he explained other phenomena, namely its being capable of sustaining life, by its density i.e. its being "thick" rather than by including a certain amount of subtle matter. For instance, in experiment 6 he recalls the difficulties in respiration at high altitudes, such as on Tenerife Peak, or Mount Teides (which he interprets in a way opposite to Boyle),[416] and, in experiments 12 and 16, by showing that rarefied air cannot sustain a flame, and in experiment 10 that rarefied air does not transmit sounds. Namely, he still did not relate the compression of air to its elasticity, insofar as it is a surplus of subtle matter, rather than the compression and decompression of air, that causes its forceful expansion and lifting of weights appended to the bladder.

Thus far, in experiment 7 – where he put birds in the container, and asphyxiated them – he provided a theory of respiration still based on the idea of circumpulsion. After having dissected the sparrows, he made the students note that both in birds and in men the cause of respiration is the movement of the thorax, while lungs are passively filled by the air entering into them by circumpulsion, like bladders:

> [h]e premitted, we should see this necessity of air in those volatills notwithstanding that volatills have the organs of respiration aliter disposita q[ua]m in hoc. For birds et volatills,

---

to conflux unto the muscles after such a {nattra} and there to convell the muscles he said again it was not easy to explicate. Let he thought in some manner it might be explicated thus: viz. that these spirits are of the nature of liquid things as being of engendered of the subtlest, and purest part of our blood even as spirit of wine is the purest, and subtlest part of the wine, although the liquidity thereof seemeth to be only the vehicle of said spirit of wine, and not so properly the spirit itself," De Volder 1676–1677, 96v–97r.

[415] "Of the frog he told us afterwards that he had left him in that posture for a whole hour, and at 12 o' clock he found the glass, as it should be without air but the frog still alive: at 2 he found him not with any motion at all, and the glass without air: yet he dared not let in the air fearing he would come to himself again: at five he come again and found the frog again without the least motion, and then he let in the air, and presently his sides, which were much swelled, did fall down flat, yea more flat then they were before he put the frog into the glass: yea in his side appeared some small rupture either with too much swelling, or else with the force of the air <g>rarified which in his body, or of the air when it was let in," De Volder 1676–1677, 103r–v.

[416] "[W]hen men ascend unto some very high mountains as el pico de Tenerife they experiment squeamishness of stomach and vomits and other symptoms whereby they are feign to make use of several preparatives the air, there, by reason of its thinness, subtility, and too much rarefaction, being scarce sufficient for life, and some ascending too high have died for want of breathing for although there \is/ much and open air enough, yet it is not of such a consistency as is requisite for human breathing, and life," De Volder 1676–1677, 93r.

> saith he, have not their lungs, as men, hanging loose in the thorax but costis affixos, atque coalitos, et coadunatos. Which proves saith he that the lungs are not the cause of respiration, as they are thought to be i.e. that they are like bellows, that drawn in the air by their dilatation, but rather that they are like bladders into which the air is thrust, and inflated by circumpulsion, and that their dilatation dependenth on the dilatation of the thorax et non e c[ontrari]o being we see in dissection that many men have great parts of their lungs united and coagulated to the ribs of the breast, and yet these men no otherwise than these birds nullo cum damno respira[ti]o[n]is respirant, q[ua]mvis h[ab]eant pulmones ita costis affixos.[417]

De Volder criticizes the idea that respiration is due to the movement of the lungs, which was given by Aristotle in his *De respiratione*, and was followed – among De Volder's more recent sources – in Daniel Sennert's *Institutiones medicinae* (1611). For Sennert, human lungs dilate by means of an inner faculty, and attract air in order to avoid a vacuum.[418] Apparently, according to De Volder this theory would work only in the case of humans, whose lungs are detached from the thorax, so that they can dilate freely. As far as birds are concerned, indeed, the idea that their lungs cannot dilate, as they are attached to their ribs had been used, for instance, by William Harvey in his *Exercitationes anatomicae de generatione animalium* (1651).[419] In turn, in his *De respiratione* Swammerdam maintained that birds breathe without any dilatation of the lungs, which adhere too strictly to ribs: in fact, for him their respiration takes place by means of "hidden bladders," while their thorax is moved by muscles, and displaces the air generating a circular thrust.[420] Thus, by comparing

---

[417] De Volder 1676–1677, 95r–v.

[418] "[T]horax plus vel minus dilatatur, prout plus vel minus aëris attrahendum est. Sed non fit eius dilatatio, quia thorax extenditur; verum sua virtute pulmo dilatatur, et impletur tantum, quantum par est, ut vacuum evitetur. [...] [L]icet per se solus pulmo hunc motum non absolvat, sed etiam thoracis motus necessarius sit: tamen quia totus motus hic in eum finem dirigitur, ut aër attrahatur, atque a pulmone fiat haec eius attractio: merito pulmo praecipuum respirationis organum statuitur: cui etiam peculiaris vis, eodem modo ut cordi, a facultate animali diversa, indita est. [...] Pulmo non dilatatur, quia impletur. Sed impletur, quia dilatatur. [...] si dilataretur ob repletionem, sicut vesica, quae inflatur: aliunde aërem in pulmonem immitti, et quasi inflari necessum esset: sed repletur, quia dilatatur suapte natura, ad modum follis scilicet," Sennert 1667, 102–103. According to the *Bibliotheca Volderina*, 55, Sennert's *Opera omnia* (in the edition of 1656) was extensively annotated by De Volder. Sennert's medical theories were taught by De Raey: a copy of his *dictata* on Sennert's *Epitome institutionum medicinae* (1644) is extant at the British Library, ms. Sloane 1733.

[419] See Harvey 1651, 4–5.

[420] "Hactenus a musculorum respirationi ancillantium inaequalitate, atque a diaphragmate antagonista carente, (ad quae porro et alia illa, quae iam enumeravimus, accedunt) thoracis motum perpetuum, alternatum, atque infinitis quoque modis mutatum, deduximus. [...] Sed quoniam in pennatis, serpentibus, et c. diaphragma desideratur, iure quaerat aliquis, quo modo in iisdem respiratio peragatur? Respondemus, quod in pennatis aër vesiculas aliquas, in abdomine absconditas, intret, idque mediantibus pulmonibus, quos tamen ob arctam connexionem haud expandi videmus. Quae forsan ratio est, cur in aquatilibus, imo in aliis etiam avibus, meatus in venis arteriisque, etiam foetui humano peculiares et c. haud claudantur, attamen in ipsis totum pecus, secus ac in aliis animalibus, a validioribus musculis, eos, qui exspirationi ancillantur longe superantibus, valde movetur, et sonum atque stridorem in costulis, cum pectore per peculiarem articulationem unitis, (nam cartilagines non habent), in vehementi inspiratione excitari notamus," Swammerdam 1667, 83–84. Cf. *supra*, n. 93. Also Van Helmont maintained that the lungs of bird do not dilate in respiration: see Van Helmont 1652, 355.

men to birds De Volder shows that the lungs are not the key or active organ in respiration – which works by circumpulsion – as some men can breath even if their lungs are attached to the ribs.

This account, however, would change in the 1688 inaugural disputation in medicine of Schuyl, *De respiratione*, dedicated to De Volder. The disputation bears witness to De Volder's own experiments, as the case of the long-lasting life of a frog in a vacuum is reported, which is also noted in experiment 9.[421] In the disputation, Schuyl claims that automatic inspiration is not due to the movement of muscles, but rather only to the passive dilatation of the lungs, which are dilated by the elastic force of the air, which enters the lungs at the moment of birth (when the lungs do not contain air) and, by its increased pressure within them, it dilates the lungs and pushes the thorax outwards (as happens to a bladder in rarefied air).[422] Expiration, in turn, is due to a two-stage process. First, the air in the lungs is heated by the body: as the heated air increases its elastic force, it flows out from the lungs. This causes a rarefaction of air itself, namely – in accordance with Boyle's law – a decrease in its pressure. This takes place both in the air present in the cells (*cellulae*) of the lungs, which start to contract by their own elastic force, and in the air contained by the lungs, which loses pressure.[423] The process never stops, namely an equilibrium

[421] "[...] ranae recipienti cuidam impositae, licet ex ipsa educatur aër, non tamen, nisi admodum difficulter extinguuntur, cuius contrarium obtinet in aliis animalculis, quae in solo degunt aëre," Schuyl 1688b, chapter 4, 7–8 (unnumbered); cf. "[o]f the frog he told us afterwards that he had left him in that posture for a whole hour, and at 12 o' clock he found the glass, as it should be without air but the frog still alive," De Volder 1676–1677, 103r.

[422] "[A]nte respiratione, compressis adhuc pulmonibus a vi incumbentis aëris hae partes introrsum deprimuntur, quocirca, si aër in pulmones irruens, eosque ita distendens, ut quaquaversum costas, et diaphragma premant, vel aequalem duntaxat cum externo haberet vim, ad eam certe altitudinem elevari deberent, in qua naturaliter, priusquam foetus aëri exponeretur, constitutae erant. Quare, cum aër hac sua in pulmones invasione maiores acquirat vires [...], patet evidenter, quod thorax et abdomen in inspiratione elevari, et possint, et debeant. Adeo ut haec thoracis dilatatio mihi non causa, sed effectus sit inspirationis, et pulmones in tantum active se habere dici possint, in quantum thoracem, diaphragma, hinc et abdomen distendunt, passive vero, quatenus ipsi ab irruente aëre expanduntur. Eodem ferme modo se res habet, ac si ex vesica recipienti cuidam inclusa per tubulum ipsi insertum omnis quantum pote educatur aër, concidet enim hoc in casu et comprimetur vesica a pressione aëris in recipiente contenti, et deinde, clauso istius tubuli orificio, ne quid aëris vesicam ingredi queat, omnis ex recipiente aër educatur. Si enim hoc in casu orificium tubuli vesicae inserti aperiatur, hoc solo aër externus omni sua vi in vesicam irruet, ipsamque quaquaversum distendet. Cum autem totum hoc sit, quod in inspiratione contigit, nihilque assumpserim, nisi quod necessario ex vi aëris elastica sequi debet, nullum potest remanere dubium, quin revera hoc pacto in respiratione automatica peragatur inspiratio," Schuyl 1688b, chapter 2, 2–3 (unnumbered).

[423] "Restat igitur altera duntaxat respirationis pars, quae consistit in eiusdem aëris exspiratione. Certum est, quod, quamdiu aër in pulmonibus iam contentus eadem vi partes illas, quas dilatavit, et expandit, premat, non possint nisi a fortiori quadam vi comprimantur, in pristinum reduci statum, eodem modo, ac arcus vi quadam incurvatus, resilire non potest, quamdiu eadem illa causa perseverat, a qua incurvatus est [...]. Videamus ergo, num ipsas ab aëris illa pressione liberare queamus. Hunc in finem in memoriam revocanda sunt ea, quae in praeced. disp. [...] dicta sunt [i.e. in Schuyl's *De vi corporum elastica*], nimirum quod aëris vis elastica calore augeatur, quo fit ut ex corporibus, qua data via, expellatur. Magnum autem in thorace adesse calorem nemo dubitat, nemo est, qui negat. Debet igitur aëris in pulmonibus contenti vis elastica augeri, quo cum praevaleat aëri

between internal and external air is never reached, because, as in the moment of birth, once the air stops to flow out, its elasticity makes it expanding again, the external air – being less compressed – does not counterbalance it, and the whole process restarts.[424] This, of course, presupposes that at some point the air starts to enter the lungs again, i.e. the impossibility of an equilibrium is presupposed by the argument. Schuyl does not deal much with the case of voluntary respiration: in that case, he maintains that muscles move the thorax, so that it is expelled as the thorax is contracted.[425] What is clear, in any case, is that circumpulsion no longer plays a role in the theory of respiration, since only air is active, through its compression and elastic force, in the whole process.

By this account, eventually, De Volder (and Schuyl) would oppose the theory of respiration of Craanen. In 1688 his posthumous *Tractatus physico-medicus de homine* (1689) was still to be published. Yet, as mentioned above, its contents must have been well known at Leiden, where he taught medicine until 1686. In his texts, Craanen criticized the idea that the lungs serve respiration, since in older men they barely move (viz. in his *Oeconomia animalis*, 1685),[426] and, in his *De homine*, claimed that it had two main functions: (1) cooling blood, and (2) allowing the fermentation of blood, since, as shown by De Volder himself at his theatre, fermentation takes place only with the presence of air. Craanen's source for De Volder's experiment is Hermann Lufneu, who in his *Disputatio chemico-medica inauguralis de fermentatione* (1679) reported that De Volder mixed *aqua fortis* with *oleum tar-*

---

externo, ex iisdem per os, et nares, cum hic, nec ullibi alias pateat exitus, expelli debet. Quo fit, ut exigui illi ramuli, qui in pulmonum cellulas terminantur, quique ratione trunci in quem omnes coeunt, admodum sunt exiles, et quo trunco sunt propiores, eo ampliores sunt, omni destituantur aëre, vel saltem ita rarefiat aër, qui in ipsis remanet, ut omni sua privetur vi elastica, ita ut hac expulsione aëris pulmones ab eiusdem pressione non parum liberentur. Debent ergo pulmones vi propria elastica, interim dum hoc pacto aër ex iisdem expellitur, denuo subsidere, eandemque ob causam et thorax, et diaphragma, et hinc abdomen in pristinum statum reduci debent, in quo tamen non parum a pressione aëris externi adiuvari debent, ita ut in exspiratione hae partes ultra elaterem introrsum nonnihil comprimantur," Schuyl 1688b, chapter 3, 3–4 (unnumbered). As to the relation between temperature and pressure, see Schuyl 1688a, chapter 2.

[424] "Omnia ergo in eundem statum reduci debent, in quo fuere ante primam inspirationem, hac sola cum differentia, quod non omnis, ut iam vidimus, ex pulmonibus expellatur aër, postquam enim ad aequilibrium perventum est, nihil amplius ex iisdem expelli potest. Hoc autem aequilibrium ne per momentum quidem subsistere potest. Simul ac enim eo perventum est, ut nihil amplius ex pulmonibus expellatur aëris, statim vis eius elastica, quae, quamdiu expellebatur, maxima ex parte eo versus erat determinata, cessante iam illa expulsione, quaquaversum aequaliter vires suas exercere incipit. Quare cum pulmones eodem sint in statu, in quo erant ante primam inspirationem, et vis aëris elastica sit eadem (haeret enim cum externo in aequilibrio, quod certe nunquam contingere posset, nisi eandem cum ipso haberet vim elasticam) hinc necessario pulmones, partesque reliquae circumiacentes, eodem modo, ac contigit in prima inspiratione, dilateri, et expandi debent," Schuyl 1688b, chapter 3, 4–5 (unnumbered).

[425] "[...] etsi in respiratione automatica omnis, quantum pote, aër sit expulsus, ita ut iam iam in pulmones denuo sit illapsurus, voluntaria tamen pectoris constrictione multum insuper aëris ex iisdem expellere possum," Schuyl 1688b, chapter 3, 4 (unnumbered).

[426] See *Quaestio* 82: "Pulmones an ad respirationem sint necessari?" Craanen 1685, 206–207.

tari (namely nitric acid and potassium carbonate), as well as *oleum tartari* and *spiritus vitrioli* (sulphuric acide) both in a vacuum and in the presence of air. In the first case, the fermentation, or effervescence was minor.[427] This idea, actually, finds an indirect confirmation in Schuyl's disputation, where it is stated that by respiration "the circulation of blood is promoted."[428] In Cartesian physiologies such as those of Regius and Craanen himself, indeed, blood circulation was due to the fermentation and expansion of blood.[429] According to Craanen, in turn, respiration may be due to various causes: namely, (1) the circular movement of air, by which the lungs are dilated: these, in turn, propel air itself; (2) the weight of the air, which fills the lungs and dilates the thorax; (3) the very movement of the muscles of the thorax – which he avers – make the lungs a passive instrument in respiration.[430] All these alternatives disappear from Schuyl's account, which is also distant from Senguerd's, who

---

[427] "Aëris absentiam occasionem suppeditare pororum isti dispositioni, quam [...] affirmavimus. Cuius asserti veritatem experimentis ab ingeniosissimo viro [...] B. De Volder in theatro physico publice coram frequenti studiosorum corona factis, confirmare est animo. Coepit igitur vir Cl. phialam aqua forti dimidia fere parte plenam; dein et alteram oleum tartari continentem: quas vitro ad id adaptato imposuit. Omnibus rimis bene clausis (ne aër externus irrepere posset) extrahendum per antliam pneumaticam (cuius iconem si quis desiderat, adeat Boylaei Tractatum de vi aëris elastica, ubi parum diversam reperturus est figuram) curavit aërem. Quo extracto quantum fieri potest, oleum tartari aquae forti commiscuit, quid fit? Continuo ebulliunt liquores, usque adeo ut vix intra vitri limites contineri possent. Eosdem liquores in aëre libero commiscuit, sed diverso cum successu. Etenim effervescentiae excitata multo minor, nec tam efferata, liquores qui ante ad supremam, nunc vix ad mediam fecit ascendere partem. Idem tentavit cum oleo tartari et spiritus vitrioli, eadem φαινόμενα, quae in praecedenti comparuere," Lufneu 1679, thesis 12. Cf. Craanen 1689, 263.

[428] "Quantum ad usum, quem respiratio praestat oeconomiae animali, nemo certe est, quin fateri, ipsum esse permagnum, necesse habeat, cum sola hac cohibita, moriatur homo, et quodvis aliud animal, quod respirat; sed in quonam praecise usus ille consistat, et quidnam ad conservationem vitae conferat, in eo dissentiunt ab invicem authores. Non autem varias illorum opiniones hic recensere, multo minus refutare est animus, sed meam solummodo hac de re sententiam proferam. Dico ergo, quod per respirationem promoveatur sanguinis circulatio, ita ut, cohibita respiratione, sensim retardetur et tandem cesset," Schuyl 1688b, chapter 4, 6 (unnumbered).

[429] See Regius 1640 (whose text was later included in Regius's major treatises); Craanen 1689, 134. See Anstey 2000; Fuchs 2001; Petrescu 2013; Ragland 2016.

[430] "Respirationis plane aliud est negotium, sed cum hic de ea ex professo agere non possumus, quia nervos nondum vidimus, attendemus ad illa solummodo, quae experientia nos docet; videmus enim hic, ventrem intumescere, quod dicimus fieri ab aëre: sed quomodo ille hoc facit? Quia aër, primo a ventre et thorace inflatis, propellitur, et cum vacuum dari repugnet, debet ille aër ingredi pulmones, eosque inflare, qui inflati propellunt diaphragma deorsum, hoc viscera id abdomine contenta, quae, cum cedere non possint multum, ventrem extendunt, ac proinde prima causa intumescentiae ventris et inspirationis est aëris motus circularis. Altera causa potest esse ipsa gravitas aëris, quam multis experimentis in physica probatam vidimus: sic enim aër propter pulsionem sui ruit cum vi in pulmones, eosque inflat et deprimit sua gravitate, hi diaphragma, hoc viscera abdominis, illa ventrem, et sic porro. Alias causas respirationis videbimus, cum ad nervos respirationi inservientes pervenerimus: tunc videbimus, quod pulmones sint tantum passiva respirationis instrumenta, qui aërem propulsum solummodo excipiunt, eique repulso exitum concedunt. Activa vero instrumenta respirationis, dicemus esse musculos, tam abdominis, quam diaphragmatis," Craanen 1689, 264–265.

still reverts to the idea of circumpulsion.[431] None of this, in fact, means that De Volder had started to systematically apply Boyle's ideas to physiology and to teach them as part of a new system of philosophy. As Jean Le Clerc pointed out, De Volder never did develop such a system. As late as 1693, for instance, another student disputing under De Volder, Stephan Huszthi, in his *Dissertatio physiologica prima quae est de hominis automato seu corpore animali*, presented a perfectly standard Cartesian account of respiration, arguing in thesis 8 that the thorax is the active part in the process of respiration. This notwithstanding, we may claim that De Volder's scattered teaching on the physiology of respiration embodied the state of the art on the topic at Leiden in the 1680s: at least, for some of his students.

[431] Senguerd's theory of respiration is given in his *Philosophia naturalis* and then in his *Inquisitiones experimentales* (first edition 1690, second edition 1699), reporting experiments from 1687 onwards. According to Senguerd's *Philosophia naturalis*, respiration – which has a cooling function – takes place as follows: (1) air enters in the lungs of the newborn, and dilates them for the first time – (as shown by the fact that the lungs of a fetus who was born dead do not float); (2) once in lungs, air is expelled by the contraction of thorax, ribs, and lungs themselves (not explained by Senguerd) and by its own rarefaction, which is caused by the subtle matter in the body. (3) The air left in the lungs cannot resist to the pressure of external air, so that lungs are filled once again: see Senguerd 1681, 289–290. In turn, in the third experiment described in his *Inquisitiones*, and taking place in December 1687, Senguerd argued that the air is propelled into the lungs by the expansion of thorax, thus still relying on the idea of circumpulsion: see Senguerd 1690, 27 and 31.

Chapter 6
Cosmology and Theory of Weight

Abstract In this chapter I consider De Volder's approach to cosmology, which probably led him to relinquish Cartesianism around 1700. I argue that the appearance of Newton's *Principia* – containing a striking criticism of Descartes's vortex theory of planetary motion – and the failure of Huygens to provide an experimental confirmation of his own alternative model, which he used to correct Descartes's model of gravity and oppose Newton's idea of universal gravitation, eventually led De Volder to renounce the teaching of natural philosophy, as he could accept neither Newton's, nor the Cartesian models. In particular, I analyse De Volder's role in evaluating the results obtained by Huygens in two distinct trials to calculate longitude at sea by means of pendulum clocks, the shortening of which according to latitude was calculated by Huygens using his model of gravity. The outcome of these trials led De Volder to reject Huygens's theory, and ultimately to dismiss Descartes's model. Moreover, I consider De Volder's attempt to combine Newton's and Descartes's ideas in the explanation of the movement of the Moon, a case in which Descartes's vortex theory was immune to Newton's criticism, but which could not be extended to the explanation of planetary movements.

In Chap. 5 it has been shown how De Volder and Descartes dealt with hydrostatics. Also, it has been anticipated that Descartes's hydrostatical model relies on the idea that every body is light, and that heaviness is a form of downwards displacement of the 'less light' by the 'more light'. In order to shed more light on De Volder's theories, it is necessary to look further at these issues, namely, at how he dealt with (1) the problem of the cause of weight, and, (2) by extent, at his cosmology – since, from a Cartesian standpoint, the weight of a body is determined by the constitution and structure of the planetary vortex – and (3) at how his ideas on weight and cosmology relate to hydrodynamics – besides hydrostatics. Indeed, from a Cartesian standpoint, it is by considering bodies in movement or at rest in a fluid that one can explain the conditions of equilibrium in planetary orbits. In other words, for Descartes the theory of weight served as the basis for the explanation of the conditions of hydrostatic equilibrium, and a hydrostatic model is an instance of a broader theory of planetary stability, used by Descartes to explain such hydrostatic stability.

A. Strazzoni, *Burchard de Volder and the Age of the Scientific Revolution*, Studies in History and Philosophy of Science 51,
https://doi.org/10.1007/978-3-030-19878-7_6

Descartes's theories on these topics may appear circular, as a theory of weight explains planetary stability and vice versa, and the theory of hydrostatic equilibrium serves to explain and illustrate the former theories and vice versa. Or, as John Schuster has put it,

> [t]he former [viz. the model of floatation] is a 'volume' phenomenon, grasped through classical hydrostatics; the latter [viz. the theory of planetary stability] a 'surface to volume ratio' phenomenon, grasped through the new Cartesian vortex mechanics. And despite Descartes' clear indication that the latter approach will subsume and explain the former, there is no simple or necessary reduction of the one to the other. [...] Descartes' difficulties here are indicated by slippages and occlusions involved in several passages where he attempts to smooth the articulation from classical statics/hydrostatics to his own corpuscular vortex mechanics.[1]

In fact, there is a way to exit from this circle and to find an idea common to these three issues (weight, cosmology, hydrostatics), which unifies them and explains them all. This is Descartes's idea of the tangential tendency of motion, from which I will start my elucidation. All these topics, of course, have had an immense secondary literature devoted to them.[2] It is thus enough, here, to consider its salient points, which will then be connected to their treatment by Descartes's commentators and critics, and by De Volder.

6.1 Descartes's Theory of Planetary Stability and Weight

6.1.1 Descartes's Theory of Tangential and Centrifugal Tendencies to Motion

The foundation point of Descartes's theory of weight and cosmology is his second law of motion, provided in his *Principia*, II.39, and corresponding to *Le monde*'s third law of nature. According to this (in the version of the *Principia*),

> all movement is, of itself, along straight lines; and consequently, bodies which are moving in a circle always tend to move away from the center of the circle which they are describing.[3]

This law – which rules the behaviour of bodies in a fully constrained system, whereas the two other laws concern a single body, or two bodies in collision – is represented by Descartes with the image of the sling (see Fig. 6.1, Descartes 1644a, 56). In a constrained system i.e. as being hindered in its rectilinear movement by the sling, the ball moves according to a circular line ABF. Moreover, being impeded by

[1] Schuster 2013, 489.

[2] These three topics have been comprehensively dealt with in Schuster 2013. See also, as to Descartes's cosmology and theory of weight, Aiton 1957, 1958a, b, 1959, 1972; Gaukroger 2002, chapters 5–6. As to Descartes's hydrostatics, see Gaukroger 2000a; Gaukroger and Schuster 2002; Chalmers 2017, chapter 5.

[3] Descartes 1982, 60.

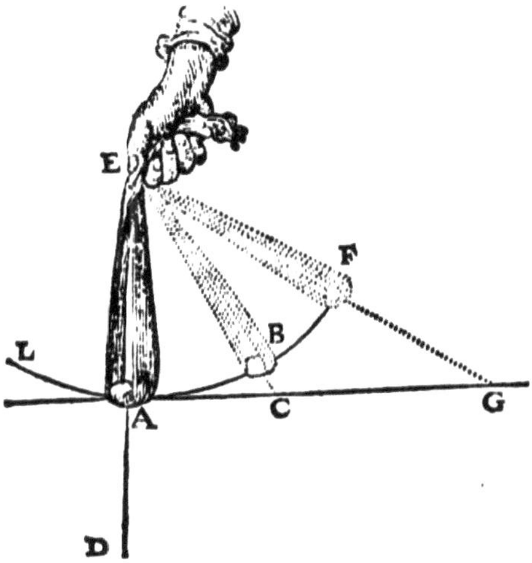

Fig. 6.1 Representation, by the example of the sling, of tangential and centrifugal tendencies of motion, according to Descartes. (Source: Descartes 1644a, 56. Library of Congress, call number: LC B1860 1644, http://hdl.loc.gov/loc.rbc/Rosenwald.1431.1. Public domain)

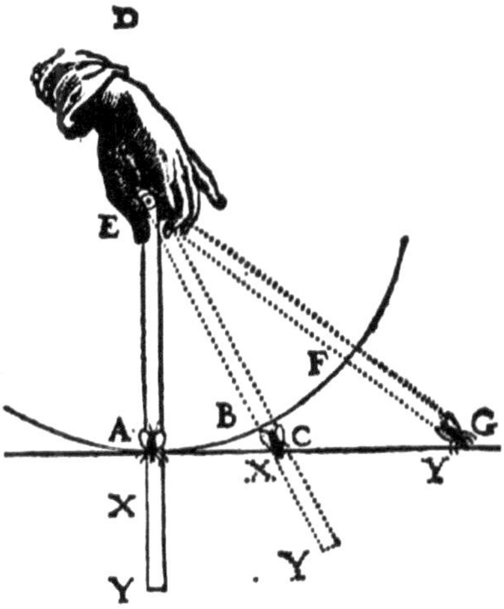

Fig. 6.2 Representation, by the example of the bee, of tangential and centrifugal tendencies of motion, according to Descartes. (Source: Descartes 1644a, 100. Library of Congress, call number: LC B1860 1644, http://hdl.loc.gov/loc.rbc/Rosenwald.1431.1. Public domain)

the sling in moving in the direction of the radius of the circle itself– in accordance with Descartes's break down of movement in various determinations – the rotating ball exerts a tendency to recede from the centre in a radial direction. So that when it is at point B, it exerts a centrifugal (a term not used by Descartes himself) force towards C, for the reason that if, when it was at point A, it had been released from the sling, it would have pursued its movement in a tangential path along ACG, so that when it is (actually) at point B, it would have been, if released at A, at point C; when at F, it would have been at G, and so forth. Descartes renders this reasoning (in III.58–59) with the example of the movement of an ant (see Fig. 6.2, Descartes 1644a, 100), moving along the radius of the 'sling' – which one can represent also by a rod EY on which the ant moves – in a way that when the sling or rod is at point B, the ant would have covered the distance AC, or AX.[4]

From this idea, Christiaan Huygens and De Volder himself were to develop their quantification of centrifugal force (as I am going to show). Descartes, on the other hand, did not venture to deal with quantification of such force, albeit he based his theory of light on this idea, which is nothing but the instantaneous tendency of rotating balls of first matter transmitted to our eyes.[5] Moreover, it is from the third rule of motion that he develops his cosmology and theory of weight, for the reason that, according to his idea of matter as extension, all motion is ultimately circular, so that in the circles *par excellence*, namely planetary vortices, the conditions of orbital stability are provided by the force that each part of matter possesses to recede from the circle according to tangential lines. In a nutshell: a body is lighter i.e. it moves towards the periphery of the vortex, when it has more force to recede from the centre than the matter than surrounds it. This force is determined by a key factor, namely its solidity (which, as seen above, depends on the quantity of a given matter and its internal cohesion, and its volume/surface ratio) and by its speed. These factors determine the force of motion that a body has in the vortex, and that makes it balanced with the matter that surrounds it at a certain 'layer' of the vortex, namely with the force of motion according to the tangent exerted by a volume of matter equal to the body we took into consideration. Translating this into hydrostatic terms (as Descartes did) a body is moved upwards, in water, when it has a greater force of 'upwards' motion (i.e. tangential or moving according to the tangent of the planetary vortex) than a volume of water equal to it, which is displaced as soon as the body moves upwards. Or, vice versa, a 'heavier' body, in water, viz. a body with a specific weight greater than water, is displaced downwards by a volume of water equal to its own. On a cosmological scale, a planet is kept in its orbit when it has no more and no less force to move according to the tangent, than an equal volume of subtle matter (constituting the heavens), which it would displace if it had more force, or by which it would have been displaced if it had less force. Accordingly, they are in equilibrium. Thus far, one can once again notice that this model is far from De Volder's Archimedean explanation of hydrostatic equilibrium, in which forces are exerted throughout the fluid, and columns of fluid balance each other. To put it more succinctly, in De Volder's Archimedean model all bodies are heavy and light bodies are extruded upwards if there is not an equal pressure across the ideal surfaces of the fluid. According to Descartes's model, on the other hand, all bodies are light, and heaviness is due to the downwards extrusion of light bodies moving upwards.

[4] Descartes 1982, 113–114. On Descartes's theory of centrifugal force, see Bertoloni Meli 1990; Garber 1992, chapter 7.

[5] On Descartes's theory of light, see Sabra 1981; Smith 1987; Schuster 2013, chapters 4 and 10.

6.1.2 Descartes's Cosmogony and Cosmology

In order to understand Descartes's vortex theory of planetary motion, it is worth recalling that Descartes grounds his cosmology on a cosmogony i.e. an account of the formation of the cosmos up to its present state. This account presents some differences between the version given in his *Le monde* and *Principia*. It can be summarized as it follows.[6] The starting point of the cosmogonic process is the creation of a continuum of matter or extension.[7] As he created it, God put this continuum in motion, i.e. He put different motions in different volumes of matter, dividing them from each other, so that a sort of chaos arose, albeit ruled by the laws of nature.[8] As a consequence of the plenitude of all space (let's say), such parts started to move in circles – so that the 'chaotic' extension started to be divided into vortices – and, as a consequence of the impacts between different parts matter, this started to compose two main types of body. Namely, a rounded shape type of particle, namely the balls or globes constituting the second element (the fluid in which planets and other solid bodies would then be carried and float), which assume such a shape as the effect of the continuous scratching of their more irregular parts, and the debris of such scratching, namely the first element, which constitutes the matter of stars. A third type of element, namely that parts of matter constituting visible or 'gross' bodies' (like planets), is supposed to be that part of matter not involved in this scratching (in *Le monde*), or in the *Principia*, as resulting from a further stage of the cosmogonic process, as I show in a moment.[9]

In his monograph *Descartes-Agonistes*, John Schuster has reconstructed with remarkable clarity how, according to Descartes, the solar vortex assumed its present form, in accordance to the law of motion ruling circular movement. As Schuster put it, one needs to distinguish between two phases in the early life of the vortex. A) First, a pre-star condition, in which, basically, there is no clear differentiation in the parts of matter rotating around a centre, and which start to distribute themselves by following the second rule of motion. B) Second, a post-star condition, namely the condition of the vortex when a star, composed by first matter, has been formed at its centre.[10] In the pre-star condition, the matter disposes itself in circles, so that the particles with more force of motion (m·s), namely those which have more force to pursue their movement according to a straight line, manage to reach more external layers in the vortex. Since the larger the bodies are, the more difficult it is for them to be moved, the smaller particles acquired more speed than the bigger ones, and moved more easily to the peripheral or external layers of the vortex. Thus one can

---

[6] The reader can find thorough discussions of Descartes's cosmogonic processes in Aiton 1972; Lynes 1982; Jalobeanu 2002; Schuster 2013, chapters 9–12.

[7] On the problem of an actual differentiation of individual bodies in the Cartesian continuum, see Taliaferro 1986; Grosholz 1994; Normore 2008; Thiel 2011, chapter 1; Ariew 2012; Reid 2014. See also Sect. 4.1.2, *De Volder's monist view of material substance*.

[8] See Descartes's *Le monde*, chapter 6, and *Principia*, III.46–47.

[9] See Descartes's *Le monde*, chapter 8, and *Principia*, III.48–52.

[10] See Descartes's *Le monde*, chapter 8, and *Principia*, III.53–54.

observe, from the centre to the periphery of the vortex, an increase in the force of motion, an increase in the speed of the particles, and a decrease in their size. As soon as 'elements' started to be differentiated, however, the smaller parts or the debris of the *boules*, being indefinitely little and thus provided, in proportion, with less force of motion than the *boules*, started to be extruded towards the centre of the vortex, forming a star. For Descartes a star is composed by the first element, it has a spherical shape (for the reason that the boules of the second matter strive to recede from the centre of the vortex in all directions),[11] and the first matter moving in it changes the distribution of particles in the rest of the vortex, according to their size and shape. Indeed, the subtle matter forming the Sun provides additional speed to the 'layers' of *boules* closer to the Sun itself: so that, given the fact that the force of motion (m·s) has to increase across the layers of the vortex itself, in a post-star situation, in which elements 1 and 2 are formed, there is (1) a decrease of speed from the Sun outwards (since the Sun accelerates those particles closer to itself), and an increase in the size of boules, this being consequent upon the increase in the value m·s across the vortex. (2) As far as the influence of the Sun, i.e. the 'additional' agitation or speed communicated by the Sun to the boules vanishes – this being, in the solar vortex, figured out by Descartes in his *Principia*, the line HNQR (see Fig. 6.3, Descartes 1644a, 110). At that point, the speed of the *boules* stops decreasing, and starts to increase, while their size decreases or is equal across the layers of the vortex, provided that the force of motion increases across the layers.[12]

At this point, the vortex is in a condition of stability. Three main types of body, however, are still 'missing' from the picture: namely, planets, moons and comets, which are constituted by the so-called third matter. As far as their formation in the cosmogonic process is concerned, Descartes provides two different accounts in *Le monde* and *Principia*. If in chapter 9 of *Le monde* he assumed that the third matter had existed since the birth of the cosmos, namely that it is a portion of the initial continuum which has been reduced neither to the second nor to the first matter, in *Principia* (III.94–126) he provided a much more detailed account of the formation of the planets, moons and comets, encompassing a history of the Earth in the various phases of its formation (IV.1–14) and grounding his extensive account of magnetism (IV.139–184). Briefly,[13] in *Principia* Descartes assumes that a flow of particles of the first matter always crosses the vortices, entering from the poles, reaching the star, and there forming solar spots. If the force of the star's matter is strong enough, such spots are periodically washed out by it, constituting an 'aether' i.e. the atmosphere of the Sun (III.98–100); in some cases, they can occlude the entire star before being wiped out (III.102–104 and III.111–114) – giving rise to the phenomena of variable stars and *novae*. In other cases, however, the matter of the star does not wipe out the spots: so that the equilibrium of the vortex (determined by the

[11] See Descartes's *Principia*, III.54 and III.61.

[12] See Descartes's *Le monde*, chapter 10, and *Principia*, II.82–85.

[13] Again, Schuster has provided a thorough treatment of how Descartes, in providing his cosmogony, developed an all-encompassing account of these phenomena: see Schuster 2013, chapter 12. On Descartes's theory of magnetism, see Georgescu 2017, chapter 3.

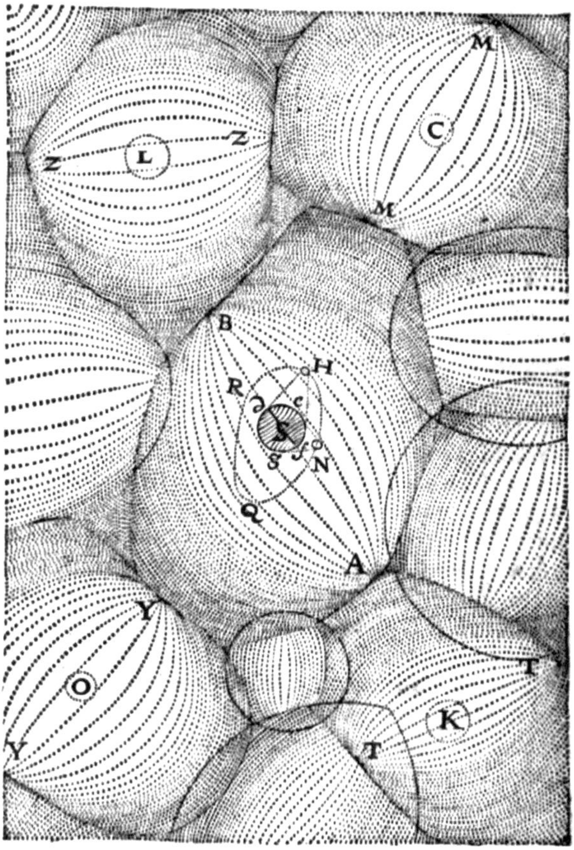

Fig. 6.3 Representation of Descartes's vortex theory of planetary motions. (Source: Descartes 1644a, 110. Library of Congress, call number: LC B1860 1644, http://hdl.loc.gov/loc.rbc/Rosenwald.1431.1. Public domain)

action of the star itself) collapses, and the matter of the vortex is absorbed by a neighbour vortex. The 'dead star', at this point could be a planet, a moon or a comet (III.118–126 and III.146–153).

In *Principia*, thus, Descartes provides an account of an ever-evolving universe, in which stars become planets, comets and moons as a result of the constant flow of the first matter through the vortices. As far as the shape of the vortex is concerned, Descartes is not entirely clear. The shape of the central layer of the solar vortex – i.e. the vortex of our Sun – is *grosso modo* elliptical (accommodating the actual orbits of the planets). In fact, however, the various vortices have a shape determined by the shapes of the other vortices, so that we can deduce that the vortices are bubbles filling a space (see Fig. 6.4, *Image of soap bubbles*).

Descartes's vortices are therefore not exactly spherical or ovoidal – this being later remarked by Huygens – even if they behave as if they have a rounded or globular shape. In fact, inter-vortical relations are determined by their being disposed in such a way that their movements do not hinder those of each other (for Descartes, this means that the parts where they touch each other both follow the same direction:

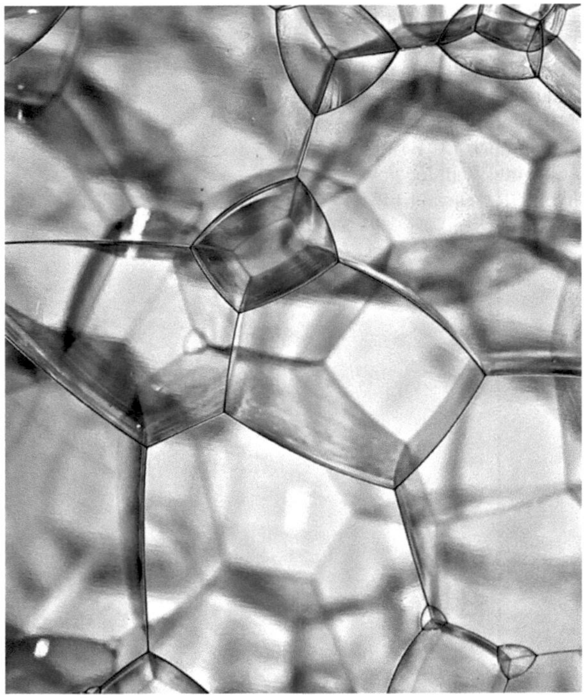

Fig. 6.4 Image of soap bubbles, aimed at illustrating Descartes's vortex theory. (Source: WikimediaCommons, https://commons.wikimedia.org/wiki/File:More_Bubbles_2.jpg. Attribution 2.0 Generic (CC BY 2.0). The picture has been rendered in black and white with respect to the original)

III.65, and that two vortices cannot touch each other at their poles or ecliptics, otherwise they will merge into one vortex: III.67), while the second matter rotating at the poles of the vortex does not hinder the entrance of the first matter from the ecliptic of another vortex (where they touch each other), the reason being that it has less force of motion than the matter rotating at the ecliptic of the vortex itself (III.69). Namely, we can deduce that the vortex is composed by various 'discs' i.e. strata of particles, which in turn can be represented, from above, as in Fig. 6.5 (Descartes 1644a, 142), ideally representing the central 'disc' of the vortex, in which the star I is located. The overall three-dimensional figure of the vortex, however, has to fit the other vortices, so that, as seen from above, it looks like a mass of bubbles.

Eventually, as a consequence of the elliptical shape of the central disc or stratum of particles of our vortex, Descartes claims (III.83) that the more external particles moving, for instance, in the area close to points M and Y (i.e. at the ecliptic of the vortex, and where the vortex itself is narrower) they move faster than at point HNQR – where, as seen from above, their speed ceases decreasing after the speeding effect of the star vanishes. Descartes explains this as a consequence of the fact that "they must change their order so as to pass from a wider path to a narrower one." Descartes exemplifies this by considering two balls moving together in a narrowing path: for him, it is granted that the preceding ball becomes faster than the other (see

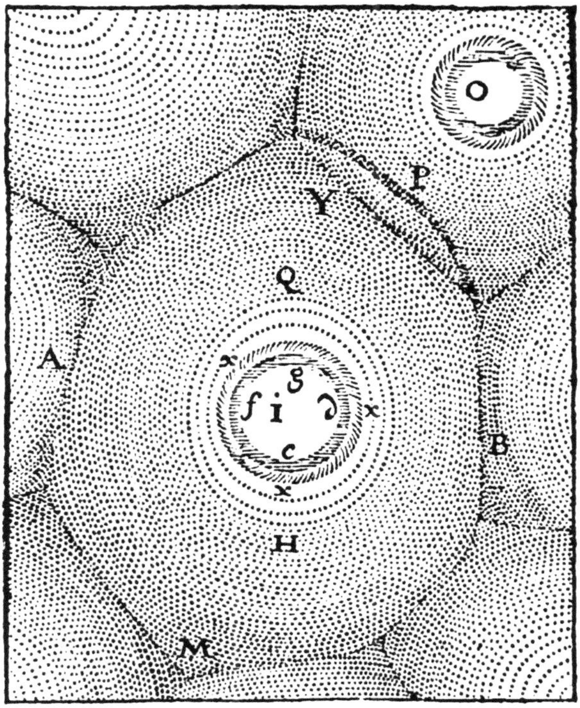

Fig. 6.5 Representation of Descartes's vortex theory of planetary motions. (Source: Descartes 1644a, 142. Library of Congress, call number: LC B1860 1644, http://hdl.loc.gov/loc.rbc/Rosenwald.1431.1. Public domain)

Fig. 6.6, Descartes 1644a, 128): as they cannot pass together, one of them must move faster.[14] More generally, he relied on a hydrodynamic principle which – as shown by Domenico Bertoloni Meli – was widely used by Benedetto Castelli in his *Della misura dell'acque correnti* (1628), according to which water, when forced into a narrower tube, moves faster because it cannot be compressed. Notably, this principle is illustrated by De Volder, in the series Hamburg 273 of his *dictata*, as he comments upon II.33 (where Descartes discusses it),[15] by re-directing the reader to Spinoza's *Principia philosophiae Cartesianae* (1663), where such a principle is given as axiom 14, before being used to explain the movement of water in a narrowing passage in

[14] "[...] some of these parts must move more quickly than others when they must change their order so as to pass from a wider path to a narrower one [...]: the two balls which are between points A and B cannot pass between the other two points C and D {which I am supposing to be closer together}, unless one moves ahead of the other; and it is obvious that [...], the one which precedes must move more rapidly than the other," Descartes 1982, 129.

[15] "[...] we can suppose that speed of movement compensates for narrowness of space. Thus, in any given length of time, the same quantity of matter will pass through one section of this circle as through another," Descartes 1982, 56; see also III.98: "the rapidity of rivers is always greater in shallow and narrow places than in deep and wide ones," Descartes 1982, 138. Cf. Castelli 1628, 9–12. For a thorough discussion, see Bertoloni Meli 2006, 84–85 and 157.

Fig. 6.6 Representation of the movement of particles in a narrowing passage, used to demonstrate that a fluid increases its speed as it passes from a larger to a narrower tube. (Source: Descartes 1644a, 128. Library of Congress, call number: LC B1860 1644, http://hdl.loc.gov/loc.rbc/Rosenwald.1431.1. Public domain)

Fig. 6.7 Two tubes used to demonstrate that a fluid increases its speed as it passes from a larger to a narrower tube. (Source: Spinoza 1663, 44. The University of Oklahoma Libraries, History of Science Collection, signature: 14293-1001. Public domain)

Fig. 6.8 Two tubes used to demonstrate that a fluid increases its speed as it passes from a larger to a narrower tube. (Source: The Hague *dictata*, 38a. Courtesy Koninklijke Bibliotheek, The Hague, ms. 72 A 7)

proposition 9.[16] Moreover, De Volder illustrates this with a figure of his own invention (based on a figure given by Spinoza to illustrate the axiom): see Figs. 6.7, Spinoza 1663, 44, and 6.8, The Hague *dictata*, 38a.[17] This idea is also used by Descartes in order to explain the tides – which are due to the rapid passage of subtle matter between the narrower channel between the Moon and the Earth (IV.49).

[16] "Si canalis A sit eiusdem longitudinis, ac C, at C duplo latior, quam A, et aliqua materia fluida duplo celerius transeat per canalem A, quam quae transit per canalem C, tantundem materiae eodem temporis spatio per canalem A transibit, quantum per C; et si per canalem A tantundem transeat, atque per C, illa duplo celerius movebitur," Spinoza 1663, 44–45. See Fig. 6.7, Spinoza 1663, 44.

[17] "Compensari. Vide Spinozam in principior. Carthes. partem secundam axiomate XIV," Hamburg 273, 101; cf. The Hague *dictata*, 38a, providing the illustration.

6.1.3 *Descartes's Theory of Planetary Stability, Weight and Pressure*

Having illustrated how Descartes's cosmos is formed, we can now address the problem of orbital stability, and by extension his 'hydrostatical' explanation of gravitation. As for the equilibrium or stability of the planets and moons in their orbits, Descartes explains this by leaning (1) on the idea of the above-mentioned distribution of particles across the layers of the vortex, and (2) on the idea of solidity as dependent on the a) size, b) surface, and c) the internal homogeneity of a body. As seen above, a solid body can exert a certain force of motion with all its component or sub-volumes (which are homogeneous and can exert their force all together), and, if it has a high volume/surface ratio, it loses less force as, in proportion, it impacts with fewer bodies than a body with a lower volume/surface ratio. As anticipated above, a planet finds its own orbit when it encounters a layer of particles, in which a volume of such particles equal to the volume of the planet has the same force for proceeding in a straight movement as it has. In this way, an equilibrium of forces is reached, as no parts of the matter have more force in moving according to straight lines than the others. So, for instance, a dead star i.e. a newborn planet, coming closer to a star under layer HNQR (see Fig. 6.3, Descartes 1644a, 110), finds boules which are faster and faster, and less and less big. Such boules have less and less force of motion, and, being faster and faster, they communicate more and more speed or quantity of motion to the planet. Thus the planet will acquire the force of motion and will rise in the vortex, until it finds a layer of boules which, being bigger as well as slower, hinder its movement towards the periphery of the vortex i.e. they make it progressively lose its force of motion. Schuster effectively points out that the increasing size of the particles plays a central role, for the reason that the bigger the boules are, the more a planet is 'enveloped' by particles whose matter requires more force to be moved by the planet, which therefore loses more force of motion.[18] In more imaginative terms, the matter from the star to HNQR is progressively denser and slower, and it 'absorbs' more of the planet's force. If the planet stops 'rising' before level HNQR, it becomes effectively a planet i.e. it is 'locked' (as Schuster puts it) in a layer where the boules have the same force of motion i.e. force to proceed in a straight line, or to rise in the vortex (III.140).[19] If it passes beyond HNQR,

[18] Schuster 2013, 465–466.

[19] "Let us suppose that star N is less solid, or capable of less agitation {or force to continue its movement in a straight line} than the globules of the second element near the circumference of our heaven, but that it has somewhat more than those close to {the center in which is} the Sun. Given these conditions, we shall understand that as soon as N has been carried away by the vortex of the Sun, it must continuously descend toward the center, until it reaches {the point at which are [found]} those heavenly globules which are equal to it in solidity, or in ability to continue their movement along straight lines. When it is finally at that point, it will neither move closer to nor farther from the Sun (unless driven slightly this way or that by some other causes), but will constantly revolve around the Sun among those heavenly globules {which are equal to it in force}, and will be a Planet. For if it descended closer to the Sun, it would there find itself surrounded by slightly smaller heavenly globules which it would exceed in force to recede from the center around which it revolves. These parts would also be more rapidly moved, which thus would increase its

it will find particles which are faster and faster, and with a decreasing or an equal size. Such particles behave like those closer to the Sun, i.e. they provide the 'planet' with more speed, and can no longer hinder its rise. So that the planet will be a comet, passing from one vortex to another (III.126).

Please note that, in any case, what matters is not the force of a single particle in impeding the rise of a planet. That is, a planet is not locked into an orbit when it finds *boules* which, taken singly, have the same force of motion of the whole planet so that no one of them prevails in the rise (according to a tangential path). Rather, when it finds a layer of boules which constitute volumes equal to that of the planet, which have the same force of motion of the planet itself, it cannot displace a volume of boules – i.e. of fluid – equal to itself, and therefore rise in the vortex. As he puts it in III.123:

> Although there is a very great disparity between their tininess and the magnitude of a star, this is in part compensated for by the fact that it is the force of many of these globules taken together, rather than individually, which opposes the force of this star. So that, while they are rotating with some star around the center S, and they, as well as the star, are all striving to recede from that center; if it happens that this force in the star is greater than the united forces of all the little globules which are required to fill its place, the star must move away from the center and cause these globules to descend into its former place. On the other hand, if the star has less force, it will be driven by the globules toward S.[20]

In hydrostatical terms, a planet rises when it is 'specifically lighter' than the fluid in which it is immersed – i.e. when it has more force of motion than the celestial matter in which it is placed – so that it can displace a volume of fluid equal to its own, until it reaches a layer where the forces are in equilibrium. This means that there are no pressures acting across the vertical sections of fluid i.e. sector of the vortex (as in the Archimedean model), and that the 'floatation' of planets is determined only by the volume of fluid with which it is in an equilibrium of the forces of motion (this being closer to the Galilean model of floatation discussed above). This makes sense of Descartes's idea that 'elements do not gravitate on elements', namely, that in a fluid in equilibrium, such as water, there is no displacement of parts, and when this happens it concerns only volumes of fluids and solids equal to each other.

As is well known, Descartes grounds his mechanical explanation of weight or gravity on his 'vortex theory', presented in *Le monde*, chapter 11, and *Principia* IV.20–27. In *Le monde*, the gravity of any kind of body towards the Earth is explained in a way no different from that in which Descartes explains the orbital stability of the planets themselves. For Descartes, the Earth is surrounded by a sort of planetary vortex within the solar vortex, which is caused by the fact that the Earth is slower than the matter of its layer, so that (see Fig. 6.9, Descartes 1664, 151):

---

own agitation along with its force, causing it to ascend. If, on the other hand, it receded further from the Sun, it would encounter there heavenly globules which were somewhat less rapidly moved and would thus decrease its agitation, and which were slightly larger and would thus have the force to drive it back toward the Sun," Descartes 1682, 168–169.

[20] Descartes 1682, 154.

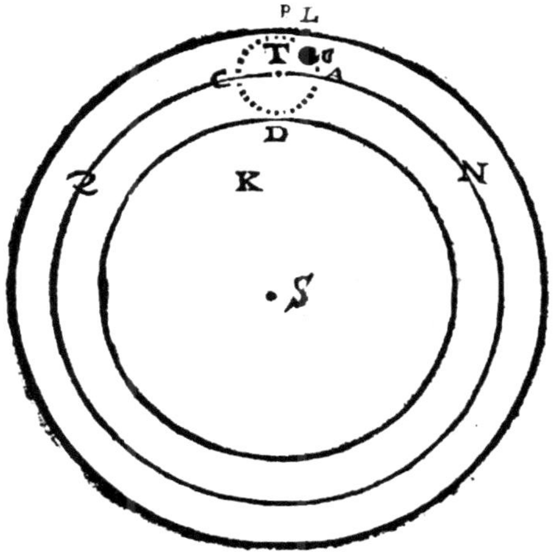

Fig. 6.9 Representation of the solar system, according to Descartes. (Source: Descartes 1664, 151. ETH-Bibliothek Zürich, signature: Rar 3862, https://www.e-rara.ch/doi/10.3931/e-rara-18973. Public domain mark 1.0)

> the parts of the heaven that are, say, at A move faster than the planet marked T, which they push towards Z, it is evident that they must be diverted by it and constrained to take their course towards B. I say toward B rather than toward D. For having an inclination to continue their motion in a straight line, they must go toward the outside of the circle ACZN that they describe, rather than toward the centre S. Now passing in this way from A to B, they force the planet T to turn with them about its centre, and conversely this planet, in so turning, gives them occasion to take their course from B to C, then to D and to A, and thus to form about the planet a particular heaven.[21]

The Moon is located in the outermost layer of this vortex or heaven. According to the model of the *Principia*, on the other hand, the planetary vortex of matter follows the rotational movement of Earth on its axis, which for Descartes is due to a sort of inertial, rotational motion which the Earth has had since its formation ((III.33 and III.144).[22] This of course contradicts Descartes's first rule of motion, as 'inertial' motions are only rectilinear.[23] With the presence of planetary vortices, Descartes could grant, in *Le monde* (chapter 11), that bodies are pushed downwards (in fact,

[21] Descartes 2004, 44–45.

[22] "For we see that a spinning top acquires enough force, merely from the fact that a boy twirls it once, to continue subsequently to spin on its own for several minutes, and to rotate during that time several thousand times {around its axis}, even though it is very small and even though both the air which surrounds it and the earth on which it presses oppose its movement. Similarly, one can easily believe that if a Planet had been set in motion from the moment of its creation, that alone would be sufficient to allow it to continue its rotations from the beginning of the universe up to the present time without any significant decrease in speed, because {the greater a body is, the longer it can retain the agitation which has been communicated to it in this way; and because} the five or six thousand years for which the universe has existed are a much shorter time compared to the size of a Planet than a minute is compared to the tiny bulk of a spinning top," Descartes 1982, 170.

[23] For a discussion, see Garber 1992, chapter 7; Slowik 2002, chapter 3. This is a sort of 'anomaly', or one of the "deviations" (see Descartes 1982, 169) in what would be the path of a planet which Descartes admits in order to explain the actual phenomena observed in astronomy: see III.141–145.

towards the axis of the planetary vortex, as his critics were to note) as a consequence of their lack of a tangential tendency. For Descartes, all bodies are in principle light, namely, they strive to recede from the centre of their rotation, according to tangential directions. So that a 'heavy' body such as a stone is pushed downwards by a volume of matter equal to its own, provided with more force of motion – due in turn to the presence of subtle matter which provides it with more force. The overall model, of course, is that of the balance:

> just as one side of the balance can only be raised or lowered if the other side, at exactly the same instant, moves in the opposite direction, and just as the heavier always makes the other rise, so too the stone R, for example, is so opposed to the quantity of air above it, which is exactly the same size as it and whose place it would have to occupy if it were to move further away from the centre T, that this air would of necessity have to descend to the extent that the stone rose. [...] Now it is evident that, since much more terrestrial matter is contained within this stone than is contained in an amount of air of equal extent, and that to counterbalance this it contains much less celestial matter, and since its terrestrial parts are far less agitated by celestial matter than those of that air, the stone should not have the force to rise above it; but on the contrary this amount of air should rather have the force to make the stone fall downwards.[24]

So that, as seen in Sects. 5.5.1, *An introduction to Descartes's hydrostatics*, and 5.5.3, *De Volder's Archimedeo-Cartesian hydrostatics: an impossible synthesis?*, "[t]his is the reason why, among the parts of any single body that we call 'homogeneous', such as those of air or water, the pressure on the lowest is not notably more than that on the highest, and why a man at the bottom of very deep water does not feel it pressing on his back any more than if he were swimming right on top."[25] Or, has he puts it in *Principia*, only the quantity of subtle matter exceeding that present in the other body, effectively acts on the portion of terrestrial matter of the other body exceeding that of the first body: according to IV.24, what "remains of the matter of the heaven in this mass of air acts against what remains of the terrestrial parts in body B; and B's weight consists in this alone."[26] This, actually, leads to the consequence that, in water, we would feel as acting on us only the difference in weight between us and a volume of water equal to us, for the reason that such water acts on us only to the extent that it is lighter than us: a consequence not made explicit by Descartes.

The model of gravity, in fact, is elaborated further in the *Principia* (IV.20), where an additional factor is considered: namely, if in *Le monde* only tangential tendencies of motion are considered, in the *Principia* Descartes conceives of a subtle matter – which is impeded in its movement by the body of the Earth – as striving to recede from the body of the Earth in all directions, so that 'heavy' bodies are pushed towards its centre (rather than to the axis of rotation, as implied by *Le monde*):

[24] Descartes 2004, 49.

[25] Descartes 2004, 50.

[26] Descartes 1982, 192.

> when they [sc. the particles of subtle matter] have been prevented, by encounter with the whole bulk of the Earth, from being transported along straight lines; these globules drive all its parts toward the middle: and in this consists the weight of terrestrial bodies. [...] In order that the nature of weight may be perfectly understood, it must be noted that if all the spaces around the Earth which are not occupied by the matter of the Earth itself were empty, [...] all those parts of the Earth which were not very firmly attached to one another would fly off in all directions toward the heaven. [...] Finally, it must be noted that, although the particles of heavenly matter are agitated at the same time by many diverse movements, yet all of their actions harmonize and, as it were, counterbalance one another in such a way that, due solely to their encounter with the bulk of the Earth which resists their movements, they strive to move away equally in all directions from its vicinity, as if from its center.[27]

According to Descartes, terrestrial gravity is due to the same cause as the roundness of drops of water (IV.19), namely (1) the impact of celestial matter on the surface of the drop, (2) the striving of the celestial matter present in the drop to move out of the drop itself:

> [t]he third effect of the heavenly globules is that, as I have already explained in the [...] *Meteorology*, they make drops of water or other fluids, which are suspended either in the air or in another fluid different from them [...]. [S]ince these heavenly globules have very different paths in a drop of water than in the surrounding air, and always, as far as is in their power, are moved along lines as straight as possible; it is evident that, upon encountering a drop of water, those globules which are in the air are less impeded from continuing their movements along lines as straight as possible if that drop is perfectly spherical than if it has any other figure whatever. For if there is some part on the surface of this drop which protrudes beyond the spherical figure, the heavenly globules travelling through the air will strike against that part with greater force than against the others, and therefore will thrust it toward the center of the drop. And if some part of the surface of the drop is nearer to the center than the rest, the heavenly globules contained in the drop itself will drive that part away from the center with greater force; and thus all the heavenly globules contribute to making the drop spherical.[28]

The same explanation can indeed be found in Descartes's *Météores*, where a further factor (later omitted in the *Principia*) is considered, namely, the pressure of air, which being thoroughly equal makes the drop spherical.[29]

As an intermediate conclusion, we can now highlight two different kinds of pressure exerted on 'terrestrial' bodies, and an ambivalence in Descartes's theory of weight. As to the ambivalence in the theory of weight, (1) on the one hand, a body is heavy because a certain amount of subtle matter extrudes it downwards: such as a stone falling through the air, or in water, or in space beyond the atmosphere. In this case, the stone is displaced or carried downwards simply by its weight i.e. by its being 'less light' than an equal volume of subtle matter, rising in its place. (2) On the other hand, the stone is also carried down by an equal volume of terrestrial

[27] Descartes 1982, 193.

[28] Descartes 1982, 189.

[29] "[...] il n'y a point de raison qu'aucune des parties de leur circonference s'esloigne ny s'approche de leurs centres plus que les autres en ce sens là, vuqu'elles ny sont ne plus ne moins pressées d'vn costé que d'autre par l'air qui les enuironne," AT VI, 281.

fluid – such as air or water – displacing it downwards, like the body of a man sinking. In both these cases, we do find the same kind of pressure, which, being exerted either by air or subtle matter, works by displacing a body downwards in virtue of its being lighter. Thus far, weight is ambivalent because, in Descartes's theory, the medium – like air – can act as if it were subtle matter. Or, as he puts it in the *Principia* (IV.25),

> in order that nothing may be omitted, it must also be noticed that, by 'heavenly matter', I mean here not only the globules of the second element, but also the matter of the first mingled with them. And those terrestrial particles which are following the course of the heavenly matter (and thus are more rapidly moved than the rest) are also to be placed in this category; such are those which form the air.[30]

Descartes, actually, does not clearly distinguish between the 'absolute weight' of a body, quantifiable by the pressure exerted by an amount of subtle matter equal to its volume, and the downwards pressure exerted by terrestrial light matter, i.e. the relative weight of a body with respect to its medium.

Second, we need to distinguish between two kinds of pressure. (1) The above-mentioned pressure exerted by subtle matter and other mediums concerns only volumes of fluids equal to the body being pressed, which is pressed a) externally – in the case of terrestrial fluids which do not penetrate the body, and b) internally, i.e. by subtle matter, which penetrates every other body and expels it downwards. (2) The pressure exerted by the 'bare weight' of a mass of fluid, which acts on the bottom of a given container (as the seabed, or as land) – as seen in Sect. 5.5.3, *De Volder's Archimedeo-Cartesian hydrostatics: an impossible synthesis?* – only if a certain amount of a lighter body can rise in its place, once the bottom is released, such as the air below the hole of a vessel. This idea was criticized by Boyle as it assumed that pressure is exerted, as a bare weight, only by a heavier on a lighter body, by which the heavy body is potentially displaced downwards. In fact, according to this perspective, subtle matter is always lighter than any other terrestrial body (an issue left purportedly unanswered by De Volder in his *Disputationes de aëris gravitate*),[31] so that the only pressure it can exert is of the first kind, even if, in Descartes's cosmology, certain particles of subtle matter can be heavier than others – in accordance with their distribution in the vortex. To this picture, we may add a third meaning of pressure, namely that one at work in Descartes's explanation of the roundness of drops: namely, the impact of particles of fluids – either subtle or terrestrial – on other bodies, such as the particles of subtle matter pressing, as if they were air, i.e. not penetrating it, the body of the drop. As I will show, De Volder would also adopt this idea of pressure, and Jakob Bernoulli with him.

---

[30] Descartes 1982, 192.

[31] See *infra*, n. 132.

6.2 The Criticisms of Descartes's Cosmology and the Corrections to It

Descartes's vortex theory and the related explanations of gravity did not go uncontested. On the one hand, the publication of Descartes's *Principia* and *Le monde* (published in 1664 and again, in Latin, in 1677) prompted the assumption of Descartes's cosmology (and cosmogony) in a wide number of natural-philosophical treatises, as well as in works dedicated to cosmogony and cosmology as such – some of which aimed at showing agreement of Descartes's theories with Genesis.[32] On the other hand, we find several works in which the Cartesian cosmology and theories of gravity were criticized for their internal inconsistencies, and as to their agreement with accepted laws of nature and astronomical evidence. I summarize here the main arguments deployed against them: (1) according to Descartes's theory of gravity (viz. the one given in *Le monde*), falling bodies would not be displaced perpendicularly towards the centre of the Earth, but in a spiralling movement towards the axis of rotation of the terrestrial vortex; (2) Descartes's theory of gravity does not explain the increasing acceleration of bodies (in accordance with Galileo's law of falling bodies); (3) Descartes's theory of tides – according to which the Moon's perigees and syzygies (i.e. the alignment of Moon, Earth and Sun) take place at the same time – contradicts astronomical evidence; (4) Descartes's vortex theory is inconsistent with Kepler's laws of planetary motion.

In what follows, I will explore such criticisms by focusing mostly on the main philosophical interlocutor of De Volder, Christiaan Huygens, who leaned upon Descartes's ideas in proposing his own theory of weight and cosmology.

[32] Namely Henricus Regius's *Fundamenta physices* and *Philosophia naturalis* (1646, 1654, 1661, 1687), Henry More's *Coniectura Cabbalistica* (1653), Daniel Lipstorp's *Specimina philosophiae Cartesianae* (1653), Christoph Wittich's *Dissertationes duae, quarum Prior de S. Scripturae in rebus philosophicis abusus examinat; Altera dispositionem et ordinem totius universi tradit* (1653) and *Consensus veritatis in Scriptura divina et infallibili revelatae cum veritate philosophica a Renato des Cartes detecta* (1659), Louis de Beaufort's *Cosmopoea divina: seu Fabrica mundi explicata* (1656), Johannes Amerpoel's *Cartesius mosaïzans, seu Evidens et facilis conciliatio philosophiae Cartesii cum historia creationis primo capite Geneseos per Mosem tradita* (1669), Géraud de Cordemoy's *Copie d'une lettre ecrite à un sçavant religieux de la Compagnie de Jesus, pour montrer, I. Que le systeme de M. Descartes, son opinion touchant les bestes, n'ont rien de dangereux. II. Et que tout ce qu'il en a écrit semble estre tiré du premier chapitre de la Genese* (1668), Jacques Rohault's *Traité de physique* (1671), Claude Gadroys's *Le système du monde selon les trois hypothèses, où conformement aux loix de la mechanique l'on explique dans la supposition du mouvement de la terre* (1675), Claude Mallement de Messange's *L'ouvrage de la Création: Traité physique du monde* (1679), Thomas Burnet's *Telluris theoria sacra* and *Sacred Theory of the Earth* (1681, 1684, 1689), Théodore Barin's *Le monde naissant ou La création du monde, démontrée par des principes tres simples et tres conformes à l'histoire de Moyse* (1686), Fontenelle's *Entretiens sur la pluralité des mondes* (1686), Pierre-Sylvain Régis's *Système de philosophie* (1690) and *Cours entier de philosophie ou Système général selon les principles de Descartes* (1691). See Blair 2000; Harrison 2000; Anstey 2011, chapter 5, 2018.

6.2.1 Huygens Confronting Descartes's Theory of Weight

6.2.1.1 Huygens on Descartes's Vortex Theory (1669–1690) – And His Background

In fact, such criticisms of Descartes's cosmology emerged from the late 1660s, after the publication of Descartes's *Le monde*. In particular, from 1669, when at various meetings of the Académie Royale des Sciences in Paris the topic of gravity came under discussion in a series of sessions from7 August to 20 November. Amongst others, this topic was discussed by Gilles Personne de Roberval, Edme Mariotte, Christiaan Huygens and Claude Perrault.[33] Whereas Roberval and Mariotte defended the ideas of occult qualities and natural inclinations to motion as explanatory concepts,[34] Huygens and Perrault proposed corrections to Descartes's explanation, without rejecting the idea of a vortex as such. In his communication of 28 August (which he would later include in his *Discours de la cause de la pesanteur*, published as an appendix to his *Traité de la lumière* (1690)), Huygens moves various objections to Descartes's theory of weight. First of all, as Descartes explained gravitation "by the movement of a certain matter that rotates around the Earth," according to his model (1) we would feel the pressure of such matter, (2) bodies would be displaced horizontally with respect to the surface of the Earth, as the subtle matter surrounding Earth is faster than Earth itself, (3) falling bodies would fall with a spiralling movement, not to the centre of the Earth but to the axis of the planetary vortex.[35]

Criticism (2), in fact, was to be found in the *Enchiridion metaphysicum* (1671) of Henry More[36] and it was later developed – and integrated criticism (3) – by Jakob Bernoulli in his correspondence with Johann Christoph Sturm. Bernoulli's argument is reported in the second part of Sturm's *Collegium experimentale* (1685) (viz.

---

[33] See Huygens 1888–1950, volume 19, 628–645.

[34] For an account, see Aiton 1972, chapter 4; Gabbey 1986; Snelders 1989; Dijksterhuis 2004, chapter 6; Boantza 2011.

[35] "L'on sçait que Mr. Descartes a aussi taché dans la Physique d'expliquer la pesanteur par le mouuement de certaine matiere qui tourne autour de la terre. [...] [S]i l'on supposoit que la matiere celeste tournast du mesme coté que la terre mais auec beaucoup plus de vitesse il s'ensuiuroit que ce mouuement rapide d'une matiere qui se meut toute vers un mesme coté se feroit sentir et qu'elle emporteroit avec elle les corps qui sont sur la terre de mesme que l'eau emporte la poudre de bois dans notre experience, ce qui pourtant ne se fait nullement, mais outre cela ce mouuement circulaire a l'entour de l'axe de la terre, ne pourroit en tout cas chasser les corps qui ne suiuent pas le mesme mouuement que vers ce mesme axe, de sorte que nous ne verrions pas tomber les corps pesants perpendiculairement vers l'horizon mais par des lignes perpendiculaires a l'axe du monde, ce qui est encor contre l'experience," Huygens 1669c, 631–632; cf. Huygens 1690, 134–135.

[36] "Hoc tantum adnotabo, non esse verisimile, si gravitas corporum terrestrium ex actione materiae colestis oritur, quin descensus eorum fieret ad Terram non perpendicularis, sed in angulos satis obliquos ab oriente occidentem versus, propter impactiones materiae coelestis in superficiem Terrae tardius ipsa materia coelesti in orientem circumrotatae. Sed abunde satis iam demonstratum est totam doctrinam Cartesianam de gravitate et levitate puram putam est imaginationem, ex eo animi morbo quem ego mechanicam credulitatem appellare sole genitam et prognatam," More 1671, 114.

in Sturm's *Ad virum celeberrimum Henricum Morum Cantabrigensem epistola qua de ipsius principio hylarchico seu spiritu naturae et familiari modernis hydrostaticis aëris gravitatione et elatere libere disseritur* against More's *Enchiridion metaphysicum*, appended to the *Collegium*), and then in Bernoulli's own *Dubium circa causam gravitatis a rotatione vorticis terreni petitam* (appeared in the *Acta Eruditorum* of February 1686). According to Bernoulli, the rotation and falling of objects takes place according to lines parallel to the equator of the vortex, so that the vortex itself would have a cylindrical rather than a spherical shape.[37] An objection which was moved, amongst others, also by Robert Hooke.[38] Criticisms (2) and (3), notably, more affect Descartes's theory of weight as proposed in *Le monde*, where, with a more unifying approach, the orbital stability of planets, their descent to the centre of the vortex, or their rising outwards, are treated in the same way as the falling of bodies on Earth. On the other hand, as seen above, in his *Principia* Descartes would change this model, assuming that the weight of bodies on Earth is due to subtle matter receding from the surface of the Earth in all directions, thus pushing heavier matter towards the centre of the Earth rather than towards the axis of rotation of the planetary vortex. Huygens, however, also attacks the model given in Descartes's *Principia*, as he notes that, criticism 4, if subtle matter were to be reflected by the body of the Earth – as claimed by Descartes – such matter would not permeate the Earth itself and bodies under the Earth's surface would be lighter than those at its surface.[39]

[37] "Evidenter enim demonstrasti [...] id ipsum quod nuper admodum amicus aliquis his verbis Basilea ad me perscriptis in Cartesiana quoque desideravit: posita hac hypothesi certum esse puto, corpora gravia secundum illa lineam detrusum iri, secundum quam materia subtilis a Terra recederet; recedere conatur autem quaelibet particula [...] secundum lineam talem quae in eodem iacet plano cum circulo per rotationem particulae descripto, i.e. secundum lineam parallelam aequatori. [...] Quod si ergo vorticum aut orbium coelestium fluidorum talem gyrationem supponamus, qualis in orbibus aut globis solidis contingit, ut partes singulae singulos etiam circulos circa singula centra describant, totusque adeo vortex non tam circa centrum quam circa axem alique convolvatur, non solum hoc, de quo nunc sermo nobis, incommodi sequetur, sed necessum etiam foret, hoc motu vorticolos stellas et corpora mundana non sphaerica, sed cylindrica potius, facta, sub polis gravitatem nullam esse et c.," Sturm 1685, *Epistola*, 22–23. The same text is reported in Bernoulli 1686b, 92–93.

[38] In the posthumously published *Account of Dr. Isaac Vossius's Hypothesis of Gravitation, with Some Animadversions thereupon*, Hooke maintained that heavy bodies would be displaced towards the axis of the vortex, and that both the vortex and the Earth would have a cylindrical shape, according to the Cartesian model of gravity: see Hooke 1705, 202.

[39] "Dans l'explication que Mr. Descartes a donnée de la pesanteur je ne trouue pas qu'il ayt assez faict de reflexion sur cette action de la matiere fluide sur les parties du dedans des corps terrestres, et il semble mesme qu'il n'a point admis la liberté de son mouuement a trauers du composé de ces parties puis qu'il veult que par la rencontre de la terre elle soit empeschée de continuer ses mouuements en ligne droitte, et que pour cela elle s'en esloigne autant qu'elle peut. En quoy il ne semble pas auoir pensé aux proprietez de la pesanteur que je viens de remarquer. Car si le mouuement de cette matiere est empesché par la terre, elle ne penetrera non plus librement les Corps des metaulx, ny du verre, donc il s'ensuiuroit que du plomb enfermé dans une phiole, perdroit son poids ou que du moins il seroit diminué comme aussi que les parties du dedans d'un corps solide ne contribueroient point a sa pesanteur, comme ne ressentant pas l'action de la matiere que les doibt pousser vers en bas; de plus en portant un corps pesant au fond d'un puit ou de quelque mine profonde, il y

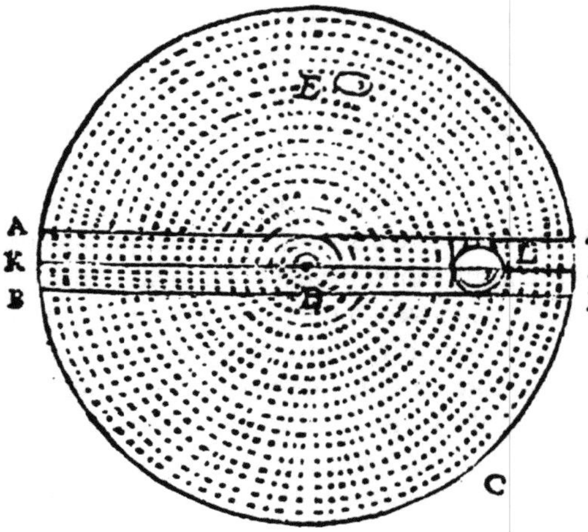

Fig. 6.10 Illustration of Huygens's experiment aimed at demonstrating the centrifugal motion of a rotating body. Source: Huygens 1690, 460. ETH-Bibliothek Zürich, signature: Rar 5085, https://www.e-rara.ch/doi/10.3931/e-rara-3766. Public domain mark 1.0

In order to resolve these issues, Huygens developed his own theory of weight, based on a reworking of the vortex theory. First of all, Huygens provided an experimental confirmation of the vortical model of gravity, by using a vessel containing water (see Fig. 6.10, Huygens 1690, 460), in which he placed a solid body equiponderant to water. This solid, made of wax, is put between two ropes A and B, which are fixed to the container. Then, Huygens started to rotate the vessel itself. As the water also rotates, a vortex is formed. Hence, when the vessel is stopped (together with the ropes and the solid), the water continues to move in a circle, while the solid body moves perpendicularly towards the centre in the 'track' of AB.[40]

deuroit perdre sa pesanteur ce qui ne se trouue point par experience," Huygens 1669c, 637; cf. Huygens 1690, 140–141.

[40] "L'on peut voir cet effect par une experience fort aisée mais qui est digne de remarque parce qu'elle nous faict voir a l'oeil quelque image de la pesanteur, car en faisant tourner de l'eau dans quelque vaisseau qui ait le fond plat apres y avoir mis dedans des petites parcelles de quelque matiere un peu plus pesante que l'eau, asin qu'elle puisse aller au fond, l'on verra qu'au commencement ces petits corps flottant dans l'eau a cause de son agitation et suivant son mouvement circulaire ne s'approcheront nullement vers le centre du vaisseau, mais aussy tost qu'ils commenceront a toucher au fond et que leur mouuement circulaire sera par la interrompu ou diminué ils s'amasseront tous alentour dudict centre y tendants par des lignes spirales parce que ils suivent encore en partie le mouuement de l'eau. Que. si l'on y met un corps ajusté en sorte qu'il ne puisse point suiure du tout le mouuement de l'eau mais seulement s'approcher vers le centre il y sera poussé tout droit, comme si L. est une petite boule qui puisse rouler librement entre les filets A, B, C, K, et un troisiesme un peu plus eslevé F. H, tendus par le milieu du vaisseau pres du fonds, lesquels filets soient arrestez immobiles pendant que l'eau tourne (ce qui se peut faire en arretant subitement le vaisseau apres l'auoir fait tourner car l'eau continuera encore quelque temps le mouuement circulaire qu'elle a conceu) l'on verra qu'aussi tost cette boule s'en ira vers le centre D. et s'y tiendra arrestée. Et il faut noter que dans cette experience l'on peut rendre le corps L. de la mesme pesanteur que l'eau, et que mesme l'experience en succedera mieux, de sorte que sans aucune difference de pesanteur des corps qui sont dans le vaisseau le seul mouuement en produit icy l'effect," Huygens 1669c, 633; cf. Huygens 1690, 132–133.

With this experiment – which was duly described and disseminated by Jacques Rohault in his 1671 *Traité de physique*[41] – Huygens improved an example given by Descartes in his letter to Mersenne of 16 October 1639, in which he illustrated his vortical explanation of weight, by considering that in a vessel filled with small balls of lead (representing subtle matter) and containing some pieces of wood (representing 'gross' matter), the wood is displaced towards the centre when the lead is rotating.[42] Or, in a simpler form (described in a letter to Marin Mersenne of 30 July, 1640) by using only rotating water and wood – which moves to the centre of the vessel.[43] Such examples are used to prove that the pieces of wood (or straw) have less force of motion than an equal volume of water or lead, and therefore they are moved to the centre of the vortex.

Huygens re-worked the cases descried by Descartes by making two important changes: (1) he created, with the ropes, a track in which the solid body could move – in order to simulate the perpendicular falling of a body; (2) he chose a body he thought to be as specifically heavy as water, namely, wax. He did so for the reason that, according to him, in Descartes's example with wood and lead the wood might have been displaced towards the centre of the vortex because of the different specific weight between the two materials, which is, in fact, what the experiment itself was meant to simulate on a dynamic basis. In other words, using bodies of different specific weight would hinder comprehension of the cause of weight itself, as the phenomenon of weight is exactly what the experiment is meant to reproduce. By using bodies with the same specific weight, namely what would be, in Descartes's perspective, homogeneous bodies (even if one is a fluid, and the other is a solid), and by observing their interaction in a dynamic situation, Huygens could ascertain whether the factors considered by Descartes (solidity and speed) make one of two bodies heavier than the other one. For Huygens, the use of wood in water cannot

[41] Rohault accepted Descartes's theory of weight as presented in *Le monde*, namely as based on the idea of a tangential tendency to movement only. To support it, he described Huygens's experiment: see Rohault 1671, volume 2, part 2, chapter 28. Rohault does not deal with the problem of the displacement of bodies towards the axis of the vortex.

[42] "Pour entendre comment la matière subtile qui tourne autour de la terre chasse les corps pesants vers le centre, remplissez quelque vaisseau rond de menues dragées de plomb, ayant mêlé parmi ce plomb quelques pieces de bois, ou autre matiere plus legere que ce plomb, qui soient plus grosses que ces dragées; puis, faisant tourner ce vaisseau fort promptement, vous esprouuerez que ces petites dragées chasseront toutes ces pieces de bois, ou autre telle matiere, vers le centre du vaisseau, ainsy que la matiere subtile chasse les cors terrestres, & c.," AT II, 593–594.

[43] "Ie voy bien que ie ne me suis pas assez expliqué en vous disant ce que ie prens pour la Pesanteur, que ie dis venir de ce que la Matiere subtile tournant fort viste autour de la Terre, chasse les Cors Terrestres vers le Centre de son Mouvement, ainsi que vous pourrez experimenter en faisant tourner de l'eau en rond en quelque grand vaisseau, et iettant dedans quelques petits morceaux de bois; vous verrez qu'ils iront vers le milieu de l'eau, et s'y soustiendront comme fait la Terre au milieu de la Matiere subtile," AT III, 134–135. Later in his *Principia* Descartes would overtly compare planets to straws and leaves in water, moving faster as they are closer to the centre of whirlpools (III.30), thus confirming his vortical model, according to which planets closer to the Sun move faster than those far away. The comparison of planets to straws is used also in Regius's *Fundamenta physices* (see Regius 1646, 55) and in Antoine Le Grand's *Institutio philosophiae secundum principia Renati Descartes* (see Le Grand 1675, 266; first edition 1672).

really show why wood is moved to the centre, since (1) if the wood floats, i.e. it is specifically lighter than water (as most of kinds of wood are) the pieces do not gather at the centre of the vortex.[44] (2) If the wood sinks (as ebony does), it is displaced towards the centre because it is progressively slowed down as it touches the bottom of the vessel. This was another factor not considered by Descartes, and that ultimately hinders the comprehension of the phenomenon.[45]

In fact, Descartes and Huygens had not been the first to compare planets to bodies in water. For instance,[46] Kepler argued in the first book of his *Epitome astronomiae Copernicanae* (1618–1621) that the displacement of straw towards the centre of a vortex of water, in a vessel, proves that vortical motion cannot be the cause of the displacement of heavy bodies towards the centre, since straw is lighter than water and has less "impression" of motion than water, which is heavier.[47] Descartes would adopt this argument, that is, he used the evidence first invoked by Kepler (viz. that lighter bodies are displaced inwards, or towards the centre, by a heavier medium) explaining this phenomenon by their lack of force of motion, with respect to the medium. Later, Descartes was accused of plagiarism by Leibniz (in his *Tentamen de motuum coelestium causis*, 1689, where Descartes's cosmology is also described as inconsistent with Kepler's laws)[48] and by Pierre-Daniel Huet in his

[44] Probably Huygens had in mind the case in which not only the water is rotating, but also the vessel: in this case, floating bodies are displaced outwards: this point is discussed in Palmerino 2007.

[45] "Ce qui n'est pas ainsy dans l'experience que Mr. Descartes propose dans une de ses lettres imprimées, car il remplit le vaisseau A. B. C. de menue dragée de plomb et y mesle parmy quelques pieces de bois ou d'autre matiere plus legere que le plomb, et faisant tout tourner ensemble, il dit que les pieces de bois seront chassées vers le milieu du vase, ce que je puis bien croire, mais c'est un effet de la differente pesanteur du bois et du plomb au lieu qu'il faut expliquer la pesanteur sans en supposer aucune et en considerant tous les corps comme faicts d'une mesme matiere. Il propose encore dans une autre lettre de jeter dans de l'eau tournante des petits morceaux de bois et dit qu'ils s'en iront vers le milieu de l'eau, auquel endroit s'il entend du bois qui nage sur l'eau, comme il y a apparence, il ne se fera point de concentration, mais s'il veut qu'il aille au fond, ce sera veritablement la mesme experience que j'ay proposée un peu auparauant, et le bois s'amassera au centre, mais ce sera a cause qu'en tombant au fond du vase son mouuement circulaire en sera retardé, de laquelle raison Mr. Descartes n'a point parlé," Huygens 1669c, 633–634; cf. Huygens 1690, 133–134.

[46] See Palmerino 2007.

[47] "Proba, neque violentia motus mundani excuti gravia in medium. In motu circulari violento si qua petunt medium totius rei mobilis, illa oportet esse leviora reipsa mota, ut in vorticibus ligna et paleae sunt leviora, quam est aqua ipsa rotata in gyrum; ibi namque maior a rotatione fit impressio in corpus aquae, quod gravius est, ut impetu ruat et rectitudinem affectans extima circuli petat, centrumque veluti exhauriat, quo facto leviora, innatantia, cum propter minorem impressionem motus in ipsa tardioremque motum destituuntur et ab aquis velocioribus introrsum repelluntur, tum etiam propter declivitatem centri in medium naturaliter influunt," Kepler 1618, 59

[48] "Nam ipsi primum indicium debetur verae causae gravitatis, et huius naturae legis, a qua gravitas pendet, quod corpora rotata conantur a centra recedere per tangentem, et ideo si in aqua festucae vel paleae innatent, rotato vase aqua in verticem acta, festucis densior, atque ideo fortius quam ipsae, excussa a medio, festucas versus centrum compellit; quemadmodum ipse diserte duobus, et amplius locis, in Epitome Astronomiae exposuit; quanquam adhuc subdubitabundus, et suas ipse opes ignorans, nec satis conscius quanta inde sequerentur, tum in Physica, tum speciatim in Astronomia. Sed his deinde egregie usus est Descartes, etsi more suo autorem dissimulavit.

Censura philosophiae Cartesianae (1689),[49] where the objection that heavy bodies move to the axis of the vortex – as shown by the very example of the vessel filled with water – is restated.[50] Descartes, in fact, assumed this case to show that water is stronger i.e. more solid than straw, so that in his example it represents subtle matter, which is in fact – according to his overall model – lighter than 'terrestrial' bodies. Such bodies, in turn, are represented by the straws, which however are lighter than water. In other words, Descartes did not consider the possible objection that, in accordance with his cosmology, a body lighter than its medium (for instance, a straw in comparison with water) would instead move towards the periphery of the vortex. An objection which was in fact moved at Leiden by Wolferd Senguerd in his *Philosophia naturalis* (1681).[51] Moreover, a further objection concerning the repre-

---

Miratus autem saepe sum, quod Descartes legum coelestium a Keplero inventarum rationes reddere ne aggressus est quidem, quantum constat; sive quod non satis conciliare posset cum suis placitis, sive quod felicitatem inventi ignoraret, nec putaret tam studiose a natura observari," Leibniz 1689, 83.

[49] "Nonnulla quoque Cartesianis haud absimilia prodidit Keplerus, insignis aetatis huius mathematicus, a quo multa se didicisse Cartesius profitetur. Is licet fixis singulis suos non attribuit vortices, in eoque et in mundi infinitate dissentire se a Bruno et Gilberto saepius affirmet, suum tamen dat Soli verticem, Cartesianis et Brunianis vorticibus parem in quo Terra et planetae volvantur," Huet 1689, 216.

[50] "Nihilo felicior fuit Cartesius in assignanda caussa gravitatis corporum terrestrium, quam confert in motum globulorum secundi elementi; qui cum a vortice in quo sunt, circumagantur in orbem, motu hoc tendere illos necesse esse, ait, ad extimum vorticis ambitum secundum lineas rectas; atque idcirco si quae terrestria corpora occurrant, cum tanta in iis non insit ad fugiendum centrum et extima petenda propensio, de loco suo decedunt, urgentibus globulis, et ab his ad alta contendentibus retruduntur ad inferiora; atque eo magis, quo plus habent materiae terrestris, minusque coelestis. Id ut commodius explicet, utitur exemplo ingentis alicuius vasis aqua pleni; quae aqua, si in orbem vertatur, et minutae alique iniiciantur ligni particulae, confluent illae statim ad centrum, et suspensae in medio aquae vortice natabunt, uti medium vorticis sui innat Terra. [...] Haec quantumvis belle et ad speciem veritatis composita, fidem tamen non obtinebunt a cordatis philosophis, si observaverint aquam hanc in vase circumactam, ferri circa axem motus huius, et iniectas in eam particulas ad axem illum tendere, et quidem unamquamque particulam ad id axis punctum, in quo secatur axis ab area plana vertiginis particulae huius; non vero tendere ad centrum vasis. Nam si aqua impleatur vas aliquod in cylindri formam excavatum, et aqua in orbem agatur, tum in eam iniiciantur particulae aliquae ligneae et lapideae; conspicientur illico particulae lignae ad supremam axis partim confluere, lapideae ad infimam; neutrae vero ferri ad centrum cylindri. Nempe singulae petent centrum areae planae, in qua moventur; ac proinde cum in suprema parte aquae ligna, in inferiore lapides circumagantur, centra utriusque vertiginis in variis erunt axis partibus, in iis nempe punctis, in quibus areae vertiginis utriusque axem secabunt. Hinc sequitur, corpora terrestria a globulis materiae caelestis depressum iri, non versus Terrae centrum, sed versus centra vertiginum suarum, nempe unumquodque versus centrum vertiginis globulorum, a quibus deprimitur; hoc est, versus punctum illud axis Terrae, in quo secatur axis ab area vertiginis huius: unde fiet, ut ea sola corpora terrestria, quae sunt sub aequinoctiali, cadant in Terram ad perpendiculum [...]; relique vero ferantur quidem ad perpendiculum versus axem Terrae," Huet 1689, 160–162.

[51] "Alii vero gravitatis rationem petunt a materia subtili, inter Terram et Lunam interiecta, quae in gyrum circum Terram acta, eam, ut et Lunam, omniaque corpora interiecta, secum abriperet; et cum illius materiae motus, utpote celerior quam Terrae et c. nonnihil a Terra impediatur, quantum potest ab eadem recederet, ac recedendo versus inferiora propelleret particulas terrestres, inter materiae caelestis particulas haerentes, utpote quibus tanta vis recedendi a centro non competeret;

sentation of heavy bodies and planets by straw is highlighted – and criticized – by De Volder himself in the Hamburg 273 series of his *dictata*, in commenting upon III.140, where it is attributed to a certain "Voetius." According to the "argument of Voetius" reported by De Volder, if straws are displaced to the centre of a vortex of water, planets would also be displaced to the centre of their vortex of subtle matter, as both the bodies follow the same mechanical laws. According to De Volder, this just "indicates the ignorance of the thought of our author [viz. Descartes]," as for De Volder only the straws are carried to the centre, because they have less force to recede from the centre than water – whereas planets can find their proper orbit once they meet the subtle matter with the same force of motion that they have.[52] In fact, the only person of the Voet family[53] who dealt with Descartes's theory of weight and cosmology was Daniel Voet, in the main text of his *Physiologia* (1661) – attacked in De Volder's *De rerum naturalium principiis*[54] – and in the disputations appended to it (originally held in 1654–1660). However, Daniel Voet did not consider the case of straws in a vessel, neither in the *Physiologia* nor in the appended *Disputatio de causis descensus gravium* (held in 1654), where he instead criticizes Descartes's theory of weight as presented in the *Principia*, namely a non-vortical theory, which he criticizes as, according to it, subtle matter would carry other bodies upwards in its effort to distance itself from the centre of the planet.[55] In fact, Voet did not used

---

simili ratione, qua in vase rotundo, minutioribus globulis plumbeis, quibus alterius materiae plumbo levioris fragmenta permista sunt, replet, quod celeriter circa centrum suum convertitur, a plumbo, per quem materia subtilis repraesentatur, versus centrum propelluntur aliorum corporum leviorum fragmenta: aut ut frusta ligni coniecta in aquam in gyrum actam, versus eius medium feruntur, ibidemque haerent. Verum nec horum explicatio naturae gravitatis satisfacit. Quantum enim ad citata exempla, illa arguunt gravia a Terra sursum potius propelli debere, quam versus eam, si gravium motus a materiae subtilis propulsione deducendus esset; cum in iis a centro propellantur quae graviora sunt, leviora vero versus centrum ferantur," Senguerd 1681, 57–58.

[52] "Planeta. Si concipiamus argumenta eorum, qui impugnant authoris philosophiam nulla inveniemus mechanica, nisi circa hanc materiam, id inquiunt sc[ilicet]: si omnes vortices et materia caelestis in orbem vertantur, eas habebunt leges quas materia caelestis, siq[ue] in orbem vertentur vortex et si innatent planeta omnes accedent ad centrum, ut festuca innatans aq[uae] vortici ad eius centum confluit sed argumentum illud Voetii indicat ignorantiam authoris nostri sententiae, fatemur quidem id fieri in festuca leviori, negamus vero id de corporibus aqua gravioribus, quod non parum confirmat ea, q[uae] antea {dixi}, cur enim levia accedunt ad centrum? Quam quia sunt minus solida, sic in aquam si iniiciamus festucam accedet ad centrum, quia aqua habet maiores vi recedendi a centro: obiici quidem potest planetas esse minus solidos sed R[espondeo] aquam ubiq[ue] aeq[ue] solidam e[ss]e et ex par[ticu]lis constare homogeneis, q[uo]d si in vortice contingeret planetae ad centrum accederent, sed in vortice globuli in extremitate sunt maiores, {...} {sensim} fiunt minores minores[que], verum itaq[ue] q[ui]dem est planetam e[ss]e minus solidum respectu superiorum globulorum alias enim e[ss]et cometa, progressus autem planeta invenit minus solidos, donec omnes qui eum cingunt sunt aequales," Hamburg 273, 208.

[53] I intend here Gijsbert Voet (1589–1676) and his sons Daniel Voet (natural philosopher, 1629–1660) and Paul Voet (writer of law and theology, 1619–1677), and Johannes Voet (1647–1714), son of Paul and writer of law. Some disputations on physical topics of Paul Voet were appended to the first edition of the *Physiologia*, yet, they do not deal with the argument discussed by De Volder.

[54] See Sect. 4.1.3, *De Volder's ideas on cohesion and divisibility*.

[55] See Voet 1661, 49 and 138.

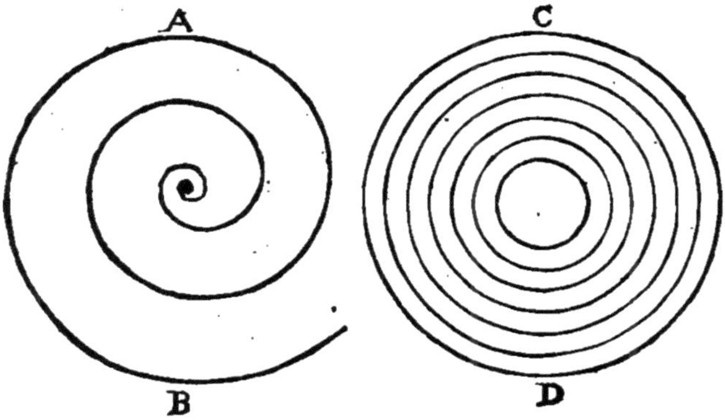

Fig. 6.11 Two kinds of vortices. (Source: Voet and De Vries 1678, 42–43. Yale University Library, Medical Historical Library, signature: *8 A-R8 S4. Public domain)

the argument cited by De Volder, but he maintained (in book 1, chapter 6, *De motu*) that there are two kinds of vortices, namely spiral ones – where all bodies are carried to the centre, so that in this case the argument reported by De Volder might apply – and those composed only by circles (where all bodies move out from the centre) (see Fig. 6.11, Voet and De Vries 1678, 42–43).

The difference in the vortex consists in the fact that the spiralling vortices are those resulting from a violent percussion or deflection of the rotating matter (which strives to recede from the centre, in accordance with the model of the sling, overtly used by Voet) by its surroundings. In circular vortices this matter is reflected more "softly" (*leniter*). Hence, in book 3, chapter 2, *De corporibus celestibus*, he claims against Descartes that it is absurd to suppose celestial bodies to be carried by celestial matter moving in vortices, since the carried bodies would move away from the centre of the vortex, and such vortices would just mix with each other[56] – a criticism

[56] "Substantia haec coelestis non in orbem movetur, licet in orbem moveantur astra; cum omnia corpora mota, quantum in ipsis est, conentur recedere a centro sui motus, consequenter et haec a suo recederet centro, quod prorsus absurdum: neque enim quippiam ad incrustandum hoc absurdum facit, si dixeris partes huius materiae exteriore interioribus crassiores et maiores esse; cum tale quid duntaxat fingatur, non vero ulla probabili ratione probetur. [...] Ut autem absurdum est hic admittere orbes realiter distinctos, ita non minus absurdum admittere [...] machinam coelestem ex diversis vorticibus fluidis conflatam esse: tum quod mox illi vortices in unum (nisi per miraculum impediatur) coalescere deberent, cum fluidum in fluidum impingens, ob non simultaneam cuiusque partis illius fluidi actionem, eidem permisceri debeat, et permisceri experientia doceat; tum quod vorticum extremus (nisi mundum infinitum posueris) ob rationem ante datam, videl. Conatum cuiusque corporis recedendi a centro, dissilire, sicque consequenter et reliqui debeant," Voet 1661, 56. In commenting upon these passages in the second edition of the *Physiologia* (1678), Gerard de Vries notes that according to Descartes's vortex theory stars would have a cylindrical shape: "[a]ccedit, quod facile sit demonstrare, secundum hanc hypothesin vorticosam, stellas esse debere figurae, non rotundae, sed oblongae, ac protensae ab uno vorticis polo ad alium," Voet and De Vries 1678, 133.

which was later to be found in Huygens's *Cosmotheoros* (1698).[57] In fact, Voet's direct argument against the vortex theory goes in a direction opposite to that criticized by De Volder, who seems rather to refer to Voet's description of spiral vortices.[58]

Leaving aside this digression, having shown that a vortical model can support a theory of gravitation, Huygens attempted to correct it in a way that overcame some of the criticisms he himself had proposed against it. In fact, his idea was to justify that heavy bodies do not fall to the axis of the vortex with a spiralling movement (viz. his objection (2) and (3)) by finding a 'track' – which he simulated with the ropes in his experiment – perpendicular to the centre of the vortex. He ideally 'replaced' the ropes which in his experiment functioned as a track, by devising a movement, in subtle matter, capable of making it push a heavy body to the centre of the Earth. Huygens assumes the existence of a subtle matter surrounding the Earth and all the bodies around it, and enclosing terrestrial bodies as if they were in a sphere of such matter. This matter is moved at a great speed, and, as it cannot go out from this spherical space because it is hindered by other celestial bodies around it, it moves circularly, without, however, forming a vortex moving in one direction, rather, spherical layers of subtle matter moving circularly in all directions. Accordingly, heavy bodies are extruded downwards, but not towards the axis of the planetary vortex, since the ever-changing directions of the layers of subtle matter compensate each other.[59]

This solution was similar to the solution to the problem of the axial falling of heavy bodies proposed, in the same series of communications at the Académie, by Perrault, who in his *Mémoire* of 20 November 1669 assumed a system of 'twin vortices' in which the downward spiralling extrusion of a body, caused by one vortex (moving from west to east), is corrected by a second vortex moving from south to north, and carrying the Earth itself and all celestial bodies. According to Perrault,

---

[57] See *infra*, n. 202.

[58] De Volder could have found Voet's argument in Daniel Voet's *Meletemata philosophica, in quibus antiquae et receptae philosophiae fulcra magis stabiliuntur, novatorum philosophandi methodus evertitur*. This book – not listed in the *Bibliotheca Volderina* – was however extremely rare and is today mentioned only in the catalogues of the Universitätsbibliothek Greifswald – where it is labelled as lost – and the Bibliothèque publique et universitaire of Neuchâtel, where it is reported as printed in Utrecht in 1661. Unfortunately, the book could not be accessed during my research.

[59] "Pour parvenir donc a une cause possible de la pesanteur, je supposeray que dans l'espace spherique qui comprend la terre et les corps qui sont autour d'elle jusqu'a une grande estendue il y a une matiere fluide qui consiste en des parties tres petites et qui est diversement agitée en tous sens avec beaucoup de rapidité, laquelle matiere ne pouuant sortir de cet espace qui est entouré d'autres corps, je dis que son mouuement doibt deuenir en partie circulaire a l'entour du centre, non pas tellement pourtant qu'elle vienne a tourner toute d'un mesme sens, mais en sorte qua la plus part de ses mouuements differents se fassent dans des surfaces sphériques a l'entour du centre dudict espace, qui pour cela deuient aussi le centre de la terre. [...] [D]ans la matiere celeste que j'ay supposée [...]il n'y a pas de raison pourquoy le mouuement d'une partie de la matiere l'emporteroit sur celuy des autres pour faire que toute la masse tournast vers un mesme centre [*sic*; in Huygens 1690 this is corrected into "mesme sense"]," Huygens 1669c, 634–635; cf. Huygens 1690, 135–136.

the combined movements of the vortices is enough to grant the perpendicular falling of bodies.[60] However, Huygens's solution was far more developed, as he completed it with the calculation of the angular speed of the subtle matter surrounding the Earth at the equator, by taking into account his own quantification of centrifugal force (which I consider in Sect. 6.2.1.2, *Huygens's quantification of centrifugal force*).

6.2.1.2 Huygens's Quantification of Centrifugal Force

Huygens came to provide 13 theorems on centrifugal force (*vis centrifuga* – a denomination not used by Descartes, who moreover did not quantify it) in the short appendix *De vi centrifuga ex motu circulari theoremata* to the fifth part of his *Horologium oscillatorium* (1673).[61] This is the only publication of Huygens, on this topic, which appeared during his lifetime. After his death, De Volder and Bernard Fullenius edited and published a manuscript of his entitled *De vi centrifuga* (dating back to 1659), which they included in their 1703 edition of Huygens's *Opuscula posthuma* – even if this treatise was not mentioned by Huygens in his testament as being amongst those to be published (unlike his *Dioptrica*, *Tractatus de vitris figurandis poliendisque*, *Dissertatio de parheliis*, and *Tractatus de motu corporum ex percussione*).[62] As De Volder and Fullenius found that Huygens's sheets were in no order, they adjusted the text to the order of the theorems given in Huygens's 1673 *Horologium oscillatorium.* Moreover, they added demonstrations "from the foundations posed by […] Huygens" to some of the propositions (7 and 9–11) given in the manuscript they found.[63] Huygens's 1673 theorems, as well as

[60] "Je suppose donc la terre au milieu du monde et au milieu des estoiles fixes et des Planetes. je fais que la terre est emportée comme il a este dict par un tourbillon du Corps aetheré d'occident en orient, et je luy en donne encore un autre pareil en force et vistesse qui va du midy au septentrion, ou du septentrion au midy, causé par un autre tourbillon qui emporte generalement tout le monde de ce sens là, cest a dire la Terre les planetes et les Estoiles fixes. Ce mouvement qui ne change rien et qui ne sçauroit repugner a aucun Phenomene parce qu'il est commun a tous les corps qui peuvent faire des Phenomenes, ne laisse pas de faire son effect sur les Corps grossiers er solides qui composent le Globe Elementaire; et le flus rapide du corps aetheré qui emporte tout le monde du midy au Septentrion qui croise celuy qui emporte la terre du couchant au levant, produit l'égalité du mouvement des Corps pesants et fait que des poles de la Terre a son centre il y a un mouvement egal a celuy qui a desia esté monttré se devoir faire de la ligne aequinoctiale au mesme centre," *Académie des Sciences*, *Procès-verbaux*, volume 6, 220v–221r. I owe the citation to *Christiaan Huygens in het Nederlands*, https://adcs.home.xs4all.nl/Huygens/ (accessed 28 January 2019).

[61] On this topic, see Yoeder 1988, chapter 3, 1998; Barbour 1989, chapter 9; Bertoloni Meli 1990; Erlichson 1994; Chareix 1996, 2004; Slowik 1997; Ducheyne 2008.

[62] See Sect. 2.4.1, *De Volder as editor of Huygens's posthumous works*; cf. Molhuysen 1913–1924, volume 4, 58∗–59∗.

[63] They also corrected some errors they found in the demonstrations provided in 1701 by Antoine Parent, in a communication to the *Journal des Sçavans*: "[…] sequitur libellus de Vi centrifuga, quem in scedis auctoris, sed nequaquam convenienti ordine dispositum invenimus. Quare factu optimum et auctoris menti convenientissimum extimavimus, si, quantum pote, sequeremur ordinem theorematum de Vi centrifuga, quae ad finem Horologii oscillatorii divulgavit iam ante multos

the opening section of the posthumous *De vi centrifuga* (which was the last part of the original text of 1659)[64] are the texts to be considered in order to understand the quantification of centrifugal force given in his *Mémoire* of 1669 and in the *Discours de la cause de la pesanteur* of 1690.

In the texts of 1669 and 1690, Huygens intends centrifugal force as – let's say – specular to gravity: namely, a piece of lead would be in equilibrium with subtle matter rotating around the Earth, and thus it would appear as suspended in air, if it has the same angular speed as such subtle matter. Namely, if "its effort to distance itself from the centre is exactly equal to the effort of its bare weight"[65] – determined by the centrifugal force of subtle matter. Provided, as Huygens supposed, that the lead, the air and the subtle matter do not differ in anything, whereas Descartes introduced the complex idea of solidity to explain gravity and planetary stability. In accordance with this idea, in the theorems on centrifugal force given in the *Horologium oscillatorium*, and in the introduction to the 1703 *De vi centrifuga*, the centrifugal force of a body moving on a circumference is quantified as being proportional to its size, to the square of its (angular) speed, and inversely proportional to the radius of the circle on which it is moving.[66] That is, it can be calculated with the formula (not explicitly provided by Huygens) $m \cdot \omega^2/r$, where 'ω' is the angular

annos; licet ad ipsum necessarium fuerit, quaedam ex iis theoremata, ut puta VII, IX, X, XI, prout hic numerantur, ex fundamentis ab Ill. Hugenio positis deducere, sed ne sic quidem praecise licuit eundem numerum propositionum obtinere, ne quidpiam, quod de his conscripserat auctor, publico invideremus. Quae dum scribimus incidi forte fortuna in manus nostras Diarium Gallicum anni 1702, quod dum obiter pervolvimus, animadvertimus 23 Maii 1701 adduci demonstrationes horum 13 theorematum; et simul in iis erroris accusari auctorem nostrum [...]," Huygens 1703, *Praefatio*, 11–12 (unnumbered). In the light of De Volder's and Fullenius's editing, Joella Yoeder has asked, in her analysis of Huygens's posthumous works, whether "[i]s the edited *De Vi Centrifuga* a treatise by Huygens? He did not reorder the manuscript, insert additional proofs, and have it copied [...]. Nor did he acknowledge its existence in his will. In fact, Fullenius and de Volder felt the need to justify what they had done, particularly defending their added proofs as in the style of Huygens. The editors of *Oeuvres completes* point out that the proofs are not that Huygenian," Yoeder 1998, 103.

64 See Yoeder 1998, 103.

65 "Reprenant donc la figure dont je me suis servy cy dessus puisque la pesanteur du corps E est justement egale a l'effort qu'a une portion aussi grande de la matiere fluide a s'esloigner du centre D, ou que c'est plustost la mesme chose; il fault dire qu'une liure de plomb par exemple icy sur terre pese autant vers le centre, qu'une masse de la matiere fluide de la grandeur de ce plomb pese vers en haut pour s'en esloigner par la vertu de son mouuement circulaire. Or puisque la matiere de plomb et la matiere fluide ne different en rien selon notre hypothese l'on peut dire que la liure de plomb, pese autant vers en bas qu'elle peseroit vers en hault si elle tournoit autour du centre de la terre et a la distance qu'elle en est auec autant de vitesse que faict la matiere fluide, mais je trouve par ma Theorie du mouuement circulaire qui s'accorde parfaictement auec l'experience qu'un corps tournant en cercle, si l'on veut que son effort a s'esloigner du centre egalle justement l'effort de sa simple pesanteur [...]," Huygens 1669c, 638–639; cf. Huygens 1690, 142–143.

66 See theorems 2 and 3 in the 1673 version: "II. Si duo mobilia aequalia, aequali celeritate ferantur, in circumferentiis inaequalibus; erunt eorum vires centrifugae in ratione contraria diametrorum. III. Si duo mobilia aequalia in circumferentiis aequalibus ferantur, celeritate inaequali, sed utraque motu aequabili, qualem in his omnibus intelligi volumus; erit vis centrifuga velocioris, ad vim tardioris, in ratione duplicata celeritatum," Huygens 1673, 160.

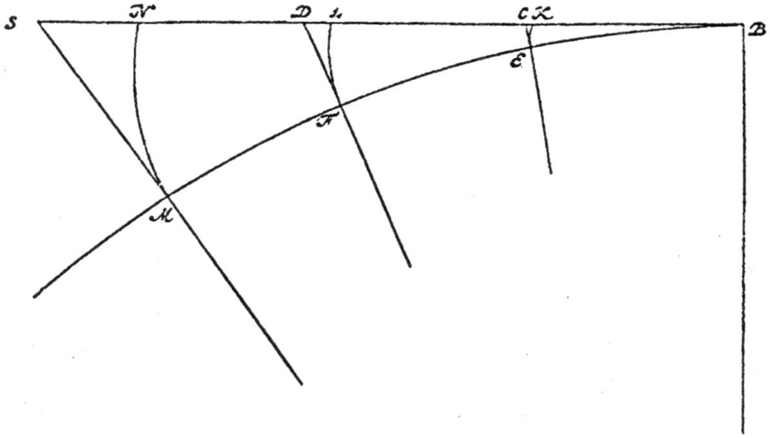

Fig. 6.12 Image used to demonstrate the quantification of centrifugal force by Huygens. (Source: Huygens 1703, *De vi centrifuga*, plate 2, figure 6. ETH-Bibliothek Zürich, signature: Rar 5219, https://www.e-rara.ch/doi/10.3931/e-rara-4066. Public domain mark 1.0)

speed. Huygens's path to the quantification of centrifugal motion has been reconstructed by Joella Yoeder.[67] Here, it is worth drawing attention to the justification of this quantification by considering that, as Huygens put it in what would appear as his posthumous *De vi centrifuga*, a body in circular motion along the curve BEFM (see Fig. 6.12, Huygens 1703, *De vi centrifuga*, plate 2, figure 6) tends to move along the tangent BS. So that if this body is attached by a string to the centre of the circle (supposed) A, this string will tend as though the body were moving along the tangent BS. That is, it will tend as though, after being released, the body will progressively distance itself from the curve BEFM along the tangent BS. From an observer moving on the circle itself, such a body will progressively cover the lines EK, FL, and so on. Such lines are conceived by Huygens as equal to BC, CD, as he considers infinitely small arcs BE, BF. In other words, the body moves along the curve BEFM 'as if' it were moving along the tangent BS, where in the same time that it covers the arc BE, it covers the distance BK (equal to BE) on the tangent – which can, however, be equated to BC as one is considering very small arcs.

Accordingly, the body moving on the tangent BS distances itself by lines whose lengths follow the proportion of the odd numbers 1, 3, 5, and so on, that is according to a uniform acceleration, by which the distances covered follow the square of the time: that is, at instant 1 a distance 1 (EK) is covered, at instant 2 a distance 4 (EK + FC), and so on. As highlighted by Huygens himself at the beginning of his demonstration, this is the same acceleration of falling bodies, moved only by their gravity, and first described by Galileo.[68] Centrifugal force is thus 'specular' to grav-

[67] Yoeder 1988, chapter 3.

[68] Cf. Huygen's account: "[c]aeterum ad illorum demonstrationem, quae nos hic tractabimus, sufficit in minimis quamlibet spatiis a quietis puncto accelerationem crescere secundum impares

ity, and can be calculated in the same way as what would be, in Leibniz's hands, the living force, namely $m \cdot s^2$. Moreover, given the fact that the distance between the circumference and the tangent is proportionally greater if the radius of the circle is less, one obtains the above-mentioned formula $m \cdot \omega^2/r$.

Huygens's result is important not only for De Volder's involvement in its publication in 1703, but also for his own appropriation of it. As I will show in Sect. 6.2.2.3.2, *De Volder's Huygenian theory of weight (and its relation to the idea of cohesion)*, we can find it underlying Hermann Schuyl's 1688 disputation *De vi corporum elastica* (which we can characterize as a Huygenian disputation, *via* the intermediation of De Volder), and, likely before that, in the Hamburg 274 series of De Volder's *dictata* – even if not in the Hamburg 273 series, where a different quantification is proposed. This allows us to establish a chronological priority of the Hamburg 273 series over the Hamburg 274, as in commenting upon those articles of the *Principia* Descartes devoted to the centrifugal tendency of bodies, chiefly II.57–60, De Volder states that the distances between the positions of a body rotating on its circle, and those he would have reached in the same time on a tangential line (in Descartes's own example (see Fig. 6.2, Descartes 1644a, 100), the distances BC and FG), increase – according to Hamburg 273 series (on III.59) – by following the progression 1, 2, 3 and so on. In the other series (Hamburg 274), commenting upon the same article and overtly referring to Galileo's theory of falling bodies (which is referred to as expounded in his *Libellus de motu*, i.e. the third *Giornata* of his *Dialoghi*, translated into Latin in the 1656 edition of Galileo's works as *Giornata terza de motu locali*), De Volder claims that the ratio of such distancing follows the series of odd numbers 1, 3, 5, and so on – exactly as in Huygens's theory.[69] De

numeros 1, 3, 5, 7, ut Galileus statuit [...] Cum itaque globus cum rota circumactus curvam describere conetur respectu radii in quo situs est, ac talem quidem quae radium contingat, apparet conatu hoc filum cui alligatus est, non secus tendi debere quam si secundum ipsum radium productum globus ire conetur. Sunt autem et spatia quae in dicta curva globus peracturus esset temporibus aequaliter crescentibus, sicut series quadratorum ab unitate 1, 4, 9, 16, etc. si videlicet principium motus spatiaque minima attendamus; quod apposita figura ostendit, ubi arcus aequales in circumferentia rotae accepti sunt BE, EF, FM; et in tangente BS rectae dictis arcubus aequales BK, KL, LN; lineae vero ex centro sunt EC, FD, MS. Hic itaque si globus a circumeunte rota divelleretur in B; tum ubi punctum B pervenisset in E globus esset in K, particulamque curvae supra descriptae percurrisset EK. post secundum vero tempus exactum cum B venisset in F, globus in L reperiretur, iamque curvae partem FL percurrisset. Similiterque cum B venisset in M, globus peregisset curvae portionem MN. Hae vero curvae lineae partes tanquam eaedem cum rectis EC, FD, MS, quas contingunt, in principio separationis globi ab rota, considerandae sunt. Quoniam tam parvi arcus a puncto B accipi possunt ut differentiola quae est inter rectas curvasque hasce, minorem rationem habeat ad ipsarum longitudinem, quavis ratione imaginabili. Proinde igitur et spatia EK, FL, MN tanquam crescentia secundum seriem quadratorum ab unitate 1, 4, 9, 16, spectanda sunt. Atque ita conatus globi in circumeunte rota retenti haud alius erit, ac si secundum rectam quae ex centro per ipsum ducitur progredi contenderet, idque motu accelerato quo aequalibus temporibus crescentia percurreret spatia secundum numeros 1, 3, 5, 7, etc.," Huygens 1703, 402–407.

[69] "Momentis. Concipit sic autor canalem rotundum, in quo globus contentus non potest ab eo recedere, sed in eo tubulo manere debet, manifestum enim, est globulum si sibi relinquatur, per rectam lineam permoturum, si sc[ilicet]: non moveatur circulariter, sed per rectam lineam ad tubi extremitatem, canalis enim non esto impedimento, quominus pergat recta, ideoq[ue] movebit[ur]

Volder might either have been using Huygens's 1673 *Horologium oscillatorium*, or he might have understood on his own that centrifugal force is opposite to the force of gravity, or, furthermore, he could just have discussed these topics with Huygens himself – even if no traces of this are extant in their surviving correspondence. A further hypothesis is that De Volder came to provide an exact measurement of centrifugal force by using the "wooden circle rotating on an ax aimed at demonstrating centrifugal force"[70] with which his experimental theatre was provided. In any case, what is clear is that the Hamburg 274 series of *dictata* was prepared after the Hamburg 273 series, and that De Volder had a clear grasp of the idea that the centrifugal force is 'specular' to gravity.

On the basis of this theory of centrifugal force, Huygens calculated in his 1669 *Mèmoire* (and *Discours de la pesanteur*) the speed of subtle matter rotating at the equator of the Earth, required for it to counter-balance its weight. He did so by considering that the radius of the Earth (as calculated by Willebrord Snel in his *Eratosthenes Batavus*, 1617) is 19,595,154 Parisian feet, and that subtle matter, in order for it to be in a condition of equilibrium, completes a turn around the Earth in the same time that a pendulum, whose length is equal to the radius of the Earth, completes two oscillations. The time in which this pendulum completes two oscillations can be calculated by knowing that a pendulum of 3 feet and 8½ lines completes a full oscillation in 1 second (at the latitude of Paris). A pendulum as long as the radius of the Earth, thus, completes two oscillations in 1 hour and 25 minutes, so that subtle matter rotating at the equator turns around the Earth 17 times faster as the rotating movement of the Earth itself (since the Earth completes a full circle in 24 hours).[71]

---

per lineam AY, ex quibus certum est, si ostendamus in vorticibus tales e[ss]e tubulos aut latera, omnes e[ti]am in iis globulos recedere a centro: qua in parte notandum est tangentem in A circulo e[ss]e proximam, adeo ut inter illam et circulum nulla linea dici possit, adeoq[ue] ut parva sit recessio, sed cum ideam motus perseveret, accelerabitur, et si consideremus divisibilibus, gradatim augeri debet, primo itaq[ue] momento si habeamus unicum gradum motus, secundo habebimus duos, tertio tres, et sic porro, adeo ut tandem evadat celerrimus," Hamburg 273, 156. Cf. the other series of *dictata*: "[d]um sortitur effectum, augetur. Manente enim eadem causa, necesse est, ut continuo novus accedat conatus: sic primo ex. gr. est momento vis q[uae]dam recedendi a centro. Secundo momento, quia durat adhuc motus circularis accedit nova q[uae]dam vis priori quia iam aderat, tertio adhuc alia, et sic porro: unde motus is, qui huic virium augmento respondet, continuo maior maiorq[ue] fiet. Et ut loquuntur uniformiter accelerabitur. Demonstravit autem Galileus in suo libello De motu, ea spacia, q[uae] tali moti absolvuntur esse inter sese, ut quadrata temporum, hoc est, si uno momento absolvatur spacium unius E[xempli] G[ratia] pedis, duobus {absolvitum} iri spacium 4 pedum. Tribus, 9 pedum, quatuor deniq[ue] momentis 16 pedum, et sic porro, ita ut velocitas acceleretur secundum ordinem numerorum imparium, 1, 3, 5, 7, 9, 11, 13 [...]," Hamburg 274, 66.

[70] "Een houte circul drajende op een ax ad demonstrandam vim centrifugam," Molhuysen 1913–1924, volume 4, 105*.

[71] Huygens provides his demonstration by considering the oscillations of a pendulum whose length is equal to the radius of the Earth: "si l'on veut que son effort a s'esloigner du centre egalle justement l'effort de sa simple pesanteur, il faut qu'il fasse chaque tour en autant de temps qu'un pendule de la longueur du demy diametre de ce cercle employe a faire deux vibrations. Il faut donc voir en combien de temps un pendule de la longueur du demy Diametre de la Terre feroit ces deux vibrations; ce qui est aisé par la proprieté connüe des pendules et par la longueur de celuy qui bat

6.2.1.3 The Accomodation of Descartes's Theory of Weight to Galileo's Law of Falling Bodies

This calculation of the speed of subtle matter was not the only one provided by Huygens, as it emerges from a 1689 communication of Denis Papin to the *Acta Eruditorum*, namely Papin's *De gravitatis causa et proprietatibus observationes*. In this, it is reported that Huygens quantified the speed of subtle matter rotating at Earth's equator as 24,000 times that of a body placed and rotating at the same latitude, on the basis of an experiment he carried out in order to measure the weight sustained by a circularly moving body. Huygens used a circular table, provided with a hole in the middle, under which a weight was suspended by of a rope attached to a body placed on the table. As the second body was put in circular movement around the table, Huygens ascertained the speed necessary for this second body to sustain, through the tension of the rope, the appended weight.[72] Hence, by knowing that centrifugal force is inversely proportional to the radius of the circle of rotation, he calculated the speed necessary for a body to have the same centrifugal force required to keep suspended a body equal to it, as it turns with the Earth at the equator. According to Papin, he found it to be 24,000 times the rotation of Earth at the equator – which is also the speed of the subtle matter capable of counterbalancing it.[73]

---

les secondes qui est de trois pieds 8½ lignes. Et je trouue qu'il faudroit pour ces deux vibrations une heure 25 la mesure de Snellius le demy Diametre de la terre de 19,595,154 pieds. La vitesse donc de la matiere fluide a l'endroit de la surface de la Terre doibt estre egale a celle d'un corps qui feroit le tour de la Terre dans ledict temps d'une heure 25. minutes, laquelle vistesse est a peu pres 17. fois plus grande que celle d'un poinct de la Terre situé soubs l'Equateur qui faict le mesme tour en 24. heures, comme il paroist par la proportion entre 24. heures et une heure 25. minutes," Huygens 1669c, 639; cf. Huygens 1690, 143–144.

[72] "[P]aravit mensam, eo artificio constructam, ut in horizontali situ circa suum centrum circumrotari posset: centrum tamen ita erat perforatum ut transitum praeberet filo, cuius extremis alligati erant duo globi aequales quorum alter infra mensam pendulus in linea axis sustinebatur; alter vero in mensa iacebat, ita aut ab alterius globi pondere statim versus centrum attraheretur, nisi ipsum vis aliqua retineret: tum circumrotabatur mensa circa axem velocissime, dum autem ad certum velocitatis gradum pervenerat globus in mensa iacens, observabatur quod tanta vi tendebat a centro sui motus recedere, ut posset globi alterius infra mensam dependentis gravitari resistere: tunc enim non descendebat pendulus ille globus, quanquam a nulla alia potentia retineretur, nisi a vi centrifuga globi alterius cum mensa circumacti: tunc igitur, mensurando circumferentiam, a globo illo descriptam dicebar per experientiam ostendere, quanta velocitas, in data circumferentia centrifugam habeat vi gravitatis aequalem," Papin 1689, 184–185.

[73] "Demonstravit praeterea Cl. Hugenius, quod in magnis circumferentiis maior quam in parvis celeritas requiratur, ut aequalis utrobique centrifuga vis habeatur: in circumferentiis igitur maiorum in superficie Telluris circulorum, quae circumferentias in mensa descriptas tam immani superant intervallo, necesse est ut materia gravitatem efficiens celerrime moveatur, quo possit tantam vim centrifugam habere: illa enim materia non deprimit gravia versus centrum Telluris nisi eadem vi qua ipsa ab eodem centro conatur recedere: ac revera, calculo rite subductum venit idem Cl. Hugenium quod materia gravitatem efficiens tanto debeat impetu moveri, ut singulis horis totum Telluris ambitum millies ferme percurrere possit: illa autem velocitas, respectu celeritatum a gravibus cadentibus acquisitarum infinita meritissime censeri potest," Papin 1689, 185.

Both these calculations served both Huygens and Papin in justifying Galileo's law of free fall on a mechanical basis. A law that, as underlined by Papin, nobody had explained so far, and that Huygens addressed for the first time on a Cartesian basis. The justification of Galileo's law, proposed by Huygens in his texts of 1669 and 1690, and then by Papin, is as follows: a falling body is uniformly accelerated by the impact with subtle matter, because in equal time it will receive equal, additional increases of speed, as the difference in relative speed between the particles and the body is negligible, i.e. the body can be considered as at rest.[74]

6.2.2 Newton's Critique of Descartes's Vortex Theory and its Reception by Huygens and De Volder

To sum up, Huygens developed a theory of planetary vortices consistent with the law of accelerated fall towards the centre of the Earth, and the quantification of centrifugal motion. In doing so, he corrected Descartes's theory of gravity, and made it consistent with Galileo's theory of falling bodies. Eventually, after the appearance of Newton's *Principia* in 1687 – where further criticisms of Descartes's vortex theory are put forth – Huygens would develop a more complete vortex theory, concerning the whole solar vortex, and critical of both Newton and Descartes. This will appear in his *Cosmotheoros*, of which De Volder supervised the publication (and which I consider in Sect. 6.2.2.3.4.3, *De Volder's last disenchantment with Cartesianism* (via *Huygens)?*). In what follows, I will reconstruct the further debates over Descartes's vortex theory by focusing on these authors, the role played by De Volder in the debate, and its consequences for his own ideas.

[74] "[L]a mesme vistesse de cette matiere jointe a la maniere dont nous auons dict qu'elle agit sur les corps qu'elle rend pesants faict voir la raison du principe que Galilée a pris pour demonstrer la proportion de l'acceleration des corps qui tombent qui est que leur vitesse s'augmente également en des temps egaux. Car ces Corps estant poussez successiuement par les parties de la matiere voisine, qui tachent de monter en leur place et dont le mouuement est tousiours infiniment plus viste que celuy qu'ils peuuent auoir acquis par des cheutes qui tombent soubs notre experience, cela fait que l'action de la matiere qui les presse peut estre considerée tousiours aussi forte que lors quelle les trouue au repos, d'ou l'on conclud ensuitte assez facilement l'accroissement des vitesses proportionné a celuy des temps," Huygens 1669c, 640; cf. Huygens 1690, 144–145. Cf. Papin 1689, 183–184: "Galilaeus in Dialogis suis mechanicis supponit, quod gravi cadenti aequales velocitatis gradus aequalibus temporibus accrescant: et [...] eandem illam veritatem experimentis multis conatur a posteriori evincere, quasi nulla dari possit demonstratio a priori. Verum illustrissimus Hugenius praestantiori rem methodo peregit: illam enim propositionem, quam alii pro primo principio assumunt, demonstravit ipse a priori per aliam veritate, quae sic se habet: potentia quae gravitatis causa est, celeritatem habet infinitam prae velocitatibus gravium cadentium, quas nobis observare licet." For a thorough account, see Freudenthal 2002.

6.2.2.1 Newton's Critique of Descartes's Vortex Theory

The appearance of Newton's *Principia* caused Huygens to become discontent with Descartes's cosmology. As Huygens wrote in his private notes on 14 December 1688, "vortices destroyed by Newton. Vortices of spherical movement in place. Correct the idea of vortices. Vortices [are] necessary," and "Newton removed all the difficulties [viz. on variations of the speed of planetary motions], together with Cartesian vortices; and taught that planets are retained in their orbits by the gravitation towards the Sun. And the eccentricities necessarily become of an elliptic figure."[75] Probably, De Volder himself – whom Huygens asked for clarifications of Newton's book – reinforced this idea in Huygens. As is well known,[76] in his manuscript *De motu corporum in gyrum* (sent to Edmond Halley in 1684) and then in his *Principia* Newton provided a solution to what has been labelled the 'Kepler problem', i.e. a justification of the laws that Kepler had set out on a purely empirical basis in his *Astronomia nova* (1609) and *Harmonices mundi* (1619), and stating (in modern form) that

1. the orbit of a planet is elliptical, and one of the two foci of the ellipsis is occupied by the Sun.
2. A line connecting the Sun to a planet covers equal areas in equal times.
3. The square of the revolution period of a planet around the Sun is proportional to the cube of its average distance from the Sun.

In particular, the second law, or the law of areas, constitutes the object of proposition 1 of book 1, section 2 of Newton's *Principia*, stating that "[t]he areas, which revolving bodies describe by radii drawn to an immovable centre of force [...] are proportional to the times in which they are described,"[77] and demonstrated on the basis of the idea of centripetal force – such as gravity is – labelled (in definition 5) as that force "by which bodies are drawn or impelled, or any way tend, towards a point as to a centre."[78] Hence, in section 3, proposition 11 – "[i]f a body revolves in an ellipsis; it is required to find the law of the centripetal force tending to the focus of the ellipsis" – the law of the inverse of the square is expressly given as the law of centripetal force, stating that "the centripetal force is [...] reciprocally in the duplicate ratio of the distance,"[79] (i.e. proportional to $1/d^2$, where d = distance) between

[75] Cf. the original texts, in *Varia astronomica*, in Huygens 1888–1950, volume 21, 437–439: "[t]ourbillons détruits par Newton. Tourbillons de mouvement sphérique à la place. Rectifier l'idée des tourbillons. Tourbillons nécessaires, [...]" and in some marginalia written in the same day on the manuscript *Projet de 1680–1681, partiellement exécuté à Paris, d'un planétaire tenant compte de la variation des vitesses des planètes*: "[...] omnes difficultates abstulit Clar. vir. Neutonus, simul cum vorticibus Cartesianis; docuitque planetas retineri in orbitis suis gravitatione versus solem. Et excentricos necessario fieri figurae Ellipticae," Huygens 1888–1950, volume 21, 143.

[76] The reader can find a thorough account in Westfall 1981, chapter 10; Pourciau 1991; Weinstock 1992; Wilson 1994; Gal 2005; Guicciardini 2005; Nauenberg 2005; Ducheyne 2011, chapter 3.

[77] I relied on the translation of Andrew Motte: Newton 1803, volume 1, 40.

[78] Newton 1803, volume 1, 2.

[79] Newton 1803, volume 1, 54.

an orbiting body and the focus of the ellipses around which it orbits. Newton's demonstrations of these propositions are geometrical, in the sense that they take into account geometrical points and resolve their relations by means of proportions between quantities. In the terse words of Michael Mahoney, discussing Newton's proposition 1,

> Newton concentrated on the orbit or trajectory of motion. Keeping before his and the reader's eyes the path of the body's motion, he located within the geometric configuration structural elements proportionally representative of the dynamical, non-geometrical parameters determining that motion.[80]

For the sake of our discussion (in which this idea will recur), it is enough to mention here Newton's idea of gravitation, which is the centripetal force *par excellence*, and which he deals with in book 3 of his *Principia* (in its more systematic form, in propositions 7–8). As Niccolò Guicciardini has put it,

> the theory of universal gravitation: every two masses attract each other with a force proportional to the product of the masses, and inversely proportional to the square of the distance between the masses. This attractive force, responsible for long-range interaction between Sun, planets and stars, is the very force which attracts bodies towards the centre of the Earth; it is the force which we commonly experience as weight. The argument in favour of universal gravitation contained in Book 3 is extremely complex: it is a masterpiece of mathematical insight, physical intuition and experimental accuracy. Newton has recourse to the Moon test, to astronomical observations on the motion of planets and comets, to experiments with pendula aimed at proving the equivalence between gravitational and inertial mass, and to data concerning the shape of planets and the flow and ebb of the sea.[81]

Actually, Kepler's laws – and by extent Newton's theory of universal gravitation – are at odds with Descartes's vortex theory. In particular, they are at odds with a key idea used by Descartes to explain both the recurrence of and anomalies in tides, as well as the higher speed of particles of subtle matter at the periphery of the vortex. Namely, the above-mentioned principle of Castelli. Newton had deployed, against Descartes's theory of planetary motion, a devastating criticism in the last scholium at the end of book 2 of his *Principia*, showing that Descartes's vortex theory was in contrast with the idea that a planet rotating around its star sweeps equal areas in equal times. So that if the orbit is an ellipsis, when the planet is at the aphelion, i.e. it is most distant from the Sun, it will slow down (otherwise it will describe unequal areas in equal times), while at the perihelion (when it is closer to the Sun) it will speed up. Yet, as seen above, for Descartes a planet is faster when it is in the aphelion – in accordance with a principle of hydrodynamics fully described by Castelli – since in the aphelion the subtle matter pushing a planet passes through 'narrower spaces' (*angustiae*), and its increased speed compensates for the narrowness of the space it passes through. In particular, in illustrating this idea in *Principia* II.33, Descartes uses the case of a circle (represented in Fig. 6.13, Descartes 1644a, 51), in which

[80] Mahoney 1993, 187. I owe this reference to Guicciardini 1999, 51.

[81] Guicciardini 1999, 90.

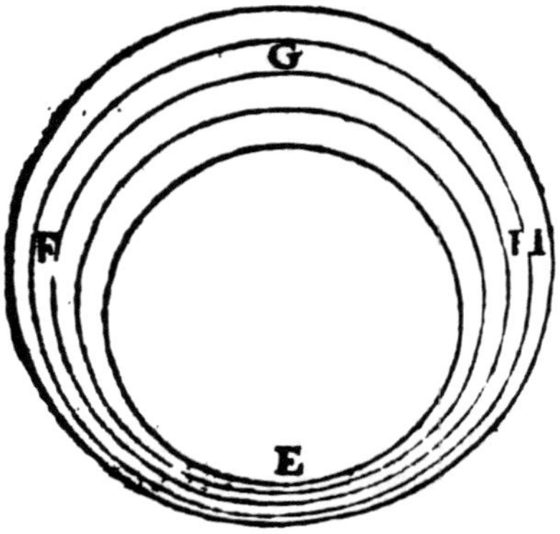

Fig. 6.13 Image used to demonstrate that a fluid increases its speed as it passes from a larger to a narrower passage, according to Descartes. (Source: Descartes 1644a, 51. Library of Congress, call number: LC B1860 1644, http://hdl.loc.gov/loc.rbc/Rosenwald.1431.1. Public domain)

> [...] without there being any condensation or vacuum; all the matter contained in the space EFGH can move in a circle. The part of it which is near E can move toward G and that which is near G can simultaneously move toward E, provided only that (since we are supposing the space at G to be four times as wide as at E, and twice as wide as at F and H) we also suppose the movement to be four times as rapid at E as at G, and twice as rapid as at F and H. Similarly, in all remaining places, we can suppose that speed of movement compensates for narrowness of space. Thus, in any given length of time, the same quantity of matter will pass through one section of this circle as through another.[82]

As far as the course of a planet in it elliptical orbit is concerned (discussed in III.141), the same reasoning applies:

> [t]he first of these other causes which drive a Planet (thus suspended around the Sun) slightly in some direction, is that the space in which it is revolved along with all the matter of the heaven is not perfectly spherical; for the matter of the heaven must necessarily flow more slowly where this space is wider than where it is narrower, {thus permitting the Planet to move further away from the Sun}.[83]

Accordingly, Descartes's vortex theory was at odds with astronomical observations, Kepler's second law, as well as with Newton's demonstration of Kepler's law of areas, according to which a planet is faster in the perihelion. Or, according to Newton's final scholium of book 2 of his *Principia* (in which is used a figure similar to Fig. 6.13, Descartes 1644a, 51, i.e. Fig. 6.14, Newton 1687, 400):

> [h]ence it is manifest that the planets are not carried round in corporeal vortices; for, according to the Copernican hypothesis, the planets going round the sun revolve in ellipses, having the sun in their common focus; and by radii drawn to the sun describe areas proportional to the times. But now the parts of a vortex can never revolve with such a motion. Let AD, BE, CF, represent three orbits described about the sun S, of which let the utmost circle CF be concentric to the sun; and let the aphelia of the two innermost be A, B; and their perihelia D, E. Therefore a body revolving in the orb CF, describing, by a radius drawn to the sun, areas proportional to the times, will move with an uniform motion. And, according to the

[82] Descartes 1982, 56. For De Volder's commentary, see Sect. 6.1.2, *Descartes's cosmogony and cosmology*.

[83] Descartes 1982, 169. De Volder comments upon this article only in the Hamburg 273 series of his *dictata*, without adding anything substantial to it: see Hamburg 273, 208.

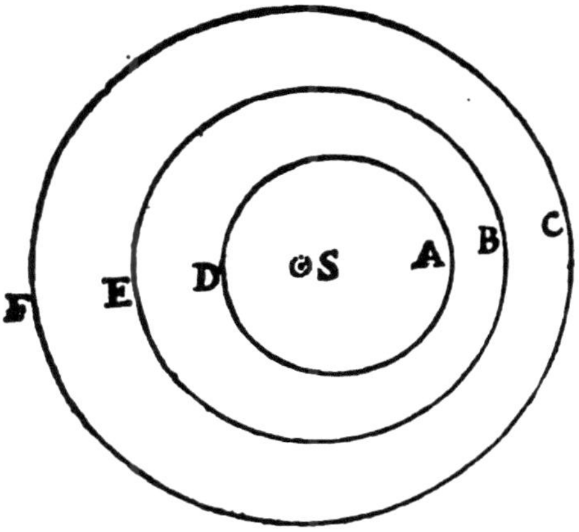

Fig. 6.14 Image used to demonstrate that a fluid increases its speed as it passes from a larger to a narrower passage, according to Isaac Newton. (Source: Newton 1687, 400. ETH-Bibliothek Zürich, signature: Rar 4011, https://www.e-rara.ch/doi/10.3931/e-rara-440. Public domain mark 1.0)

> laws of astronomy, the body revolving in the orb BE will move slower in its aphelion B, and swifter in its perihelion E; whereas, according to the laws of mechanics, the matter of the vortex ought to move more swiftly in the narrow space between A and C than in the wide space between D and F; that is, more swiftly in the aphelion than in the perihelion. Now these two conclusions contradict each other.[84]

In fact, this was not the only criticism levelled, on the basis of astronomical evidence, against Descartes's vortex theory. Ten years before – as noted by Eric John Aiton – in his *Lettre à M. l'abbé Bourdelot, dans laquelle il démontre que les raisons que M. Descartes a données du flux et du reflux de la mer sont fausses* (1677), Alexandre Tinelis de Castelet criticized Descartes's idea, expounded in his *Principia*, IV.51 (and before that in chapter 12 of *Le monde*), that "the ebb and flow of the ocean must be greater when the Moon is full or new than during the intervening times,"[85] namely that when the Moon is new or full, i.e. it is at its syzygies (at positions B and D in Fig. 6.15, Descartes 1644a, 222) or in alignment with the Sun, according to Descartes it is also at its perigee, or the shortest distance from the Earth.

In this position, the 'channel' between the Moon and the Earth is narrower, and tidal ebbs are greater as subtle matter passes faster and more forcefully in the chan-

[84] Newton 1803, volume 1, 378–379. Cf. the original text: "[h]inc liquet planetas a vorticibus corporeis non deferri. Nam planetae secundum hypothesin Copernicaeam circa solem delati revolvuntur in ellipsibus umbilicum habentibus in sole, et radiis ad solem ductis areas describunt temporibus proportionales. At partes vorticis tali motu revolvi nequeunt. Designent AD, BE, CF, orbes tres circa solem S descriptos, quorum extimus CF circulus sit Soli concentricus, et interiorum duorum aphelia sint A, B et perihelia D, E. Ergo corpus quod revolvitur in orbe CF, radio ad solem ducto areas temporibus proportionales describendo, movebitur uniformi cum motu. Corpus autem, quod revolvitur in orbe BE, tardius movebitur in aphelio B et velocius in perihelio E, secundum leges astronomicas; cum tamen secundum leges mechanicas materia vorticis in spatio angustiore inter A et C velocius moveri debeat quam in spatio latiore inter D et F; id est, in aphelia velocius quam in perihelio. Quae duo repugnant inter se," Newton 1687, 399–400.

[85] Descartes 1982, 207.

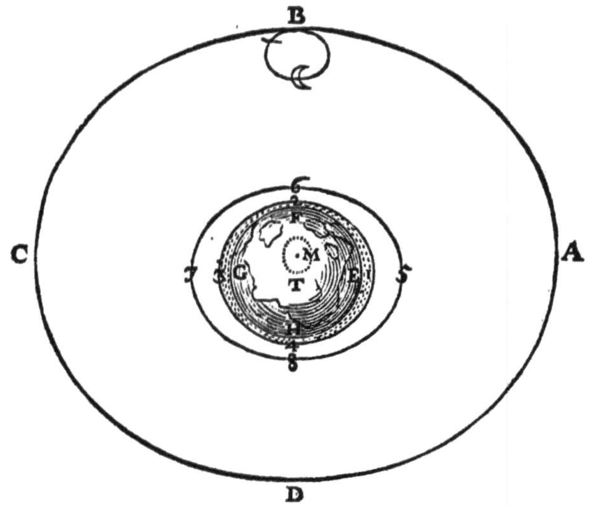

Fig. 6.15 Representation of the Earth-Moon system, according to Descartes. (Source: Descartes 1644a, 222. Library of Congress, call number: LC B1860 1644, http://hdl.loc.gov/loc.rbc/Rosenwald.1431.1. Public domain)

nel itself, depressing the surfaces of seas and oceans.[86] This effect, in reality, is because when the Moon is new or full, it is in line with the Sun (syzygies) and the gravitational forces of the Moon and the Sun are combined. However – as Tinelis de Castelet noted – the syzygies can also occur when the Moon is not at its perigee. Thus was Descartes's theory of ebbs proved to be false.[87] We do not find traces of this objection in Newton's *Principia* – where tidal ebbs and their variations are explained as the effect of gravitational forces and of the alignments of celestial bodies (in book 3, proposition 24, theorem 20)[88] – nor in the subsequent review, by Locke, of Newton's *Principia*, which appeared in Jean Le Clerc's *Bibliothèque universelle et historique* in the March 1688 issue, where Newton's criticism of Descartes's vortex theory (and Descartes's use of Castelli's principle) was paraphrased – with the inclusion of the same figure already used by Descartes and Newton – and widely disseminated.[89] It might be that this was related to the fact

[86] Cf. his *Le monde*: "[t]o this end, consider the Moon to be at B for example [...], where you can assume that it is stationary in comparison to the speed at which the celestial matter below it moves. Consider also that this celestial matter, having less space to traverse [...] and consequently having to move a little more quickly there, cannot but have the force to push the whole Earth [...]. In this way, since the surface [...] of the Earth remains round, because it is hard, that of the water [...] and the air [...], which are fluid, must take the shape of an oval," Descartes 2004, 52.

[87] "[L]'on peut voir malgré Descartes que le tourbillon de la Lune n'est pas disposé comme il s'imagine, & que le Diametre apparent de cette Planete est quelque fois plus grand aux Quadratures qu'aux Syzygies," Tinelis de Castelet 1677, 12. See Aiton 1955a, 1972, chapter 4.

[88] See Aiton 1955b.

[89] "[L]'auteur preuve que les planetes ne sont pas emportées par des turbillons corporels. La raison en est que, selon l'hypothese de Copernic, elles font leurs révolutions par des ellipses dont le Soleil est le foyer, & décrivent des aire proportionnelles aux temps, par des rayons tirez jusqu'au Soleil. Or les parties d'un Tourbillon ne peuvent pas se mouvoir de cette maniere; come on le voit dans la figure suivante. Qu'AD, BE, CF, soient trois Orbes decrits autour du Soleil S. que l'exterieur CF

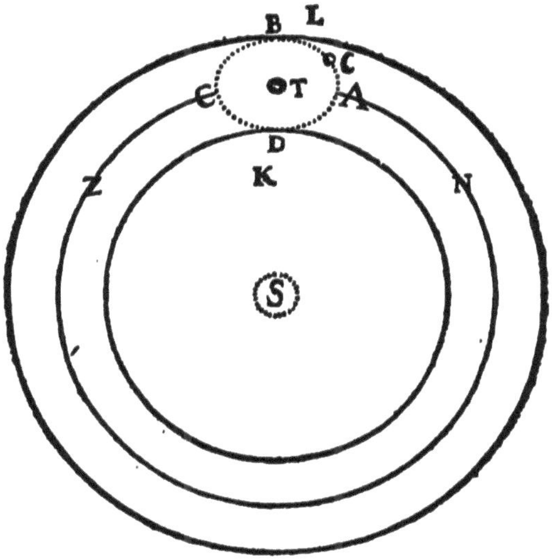

Fig. 6.16 Representation of the solar system, according to Descartes. (Source: Descartes 1644a, 187. Library of Congress, call number: LC B1860 1644, http://hdl.loc.gov/loc.rbc/Rosenwald.1431.1. Public domain)

that, given how Descartes explained the elliptical shape of the planetary vortex of the Earth (in *Principia* III.153, referred to in IV.51), he claimed that the Moon is faster in the syzygies (in accordance with the observation of lunar variations by Tycho Brahe), namely at its perigees (in his perspective), because at the perigees it meets particles of subtle matter which are different in shape and speed from the particles composing the planetary vortex, which are of equal size and speed, and in which the Moon is placed at the outermost layer.

The consequence, for Descartes, is that "they spread more freely toward A and C than toward B and D," (see Fig. 6.16, Descartes 1644a, 187) namely that when the particles of the planetary vortex are at B and D (or at the Moon's syzygies), they start to move towards a space where the particles of the solar vortex are equal to them and they speed up. Accordingly, the Moon (pushed by such particles) also moves faster.[90] Or, more briefly, according to Descartes's mode Earth is faster at its

soit consentrique au Soleil, & que les Aphelies des deux interieurs soient AB, & leurs perihelies DE. Le Corps, qui fait sa revolution dans l'Orbe CF, décrivant des aires proportionnelles aux temps par un rayon tiré jusqu'au Soleil, aura un mouvement uniforme. Mais le corps qui fait son tour dans l'Orbe BE, se mouvra plus lentement dans l'Alphelie B & plus prontement dans le perihelie D, conformément aux Loix Astronomiques: au lieu que selon les Loix des Méchaniques, la matiere du tourbillon, qui est en AC l'espace le plus étroit, doit se mouvoir plus vite que celle qui est en DF, où elle se peut étendre davantage: de sorte qu'il faudroit qu'une Planete fut emportée par son tourbillon avec plus de rapidité, lorsqu'elle est le plus éloignée du Soleil, & qu'elle allât plus lentement lorsqu'elle en est le plus proche," Locke 1688a, 440–442. See, for more details, Anstey 2011, chapter 5.

[90] "153. Why the Moon travels more rapidly, and deviates less from its mean motion [when] in conjunction than in quadrature, and why its heaven is not spherical. Neither shall we marvel at the fact that the Moon is seen to be somewhat more rapidly moved, and to deviate less from its course in any direction when it is full or new (that is, when it is near parts B or D of the Heaven); than when only half of it is visible, that is, when it is near A or C. Since the heavenly globules in the

aphelion (in accordance with Castelli's principle) and slower at the perihelion – contrary to Kepler's laws – while the Moon is faster at its perigee and slower at the apogee (in a way consistent with Kepler's laws and with Newton's theory). Thus far, Descartes's vortical theory of the movement of the Moon accorded with Kepler's laws, while his theory of solar vortices did not. In fact, Newton would ultimately refute Descartes's vortex theory only in the *Scholium generale* in the second edition of his *Principia* (1713), where he pointed out that Descartes's vortex theory is at odds with the regular movements of comets (in fact ruled by the law of gravitation) – which in a system of vortices would just wander from one vortex to another.[91]

6.2.2.2 Huygens's Criticisms of Universal Gravitation – And the Voyage of the *Alcmaer*

As Huygens read Newton's *Principia,* he published his correction to Descartes's vortex theory, which he had first proposed in 1669. Moreover, he criticized Newton's idea of universal gravitation, i.e. the idea that any body will attract any other body, regardless of the presence of subtle matter pushing one body towards another. Indeed, as he published his *Traité de la lumiere* in 1690, Huygens appended to it his *Discours de la cause de la pesanteur*, which is the same text of 1669 (with minor corrections), to which is added (1) a brief account of the variations of the periods of oscillation of seconds pendulums registered at different latitudes (*De l'observation du racourcissement du pendule à secondes pres de la ligne equinoctiale*), and (2) an *Addition* where he discusses and criticizes some of Newton's results. First of all, Huygens accepts from Newton the idea that planets have a sort of centripetal force (*vis centripeta*) towards the Sun which however is nothing but a matter of fact, i.e. it is heaviness itself, exerted also by bodies towards the Earth:

> [...] I do not have anything against the centripetal force [*vis centripeta*], as Mr. Newton calls it, by which he makes the planets gravitate towards the Sun, and the Moon towards the Earth, but I remain in agreement without difficulty: because not only one knowns by experience that there is such a way of attraction or impulse in nature, but that it is also explained by the laws of motion, as we have seen in what I have written above about gravity. Because nothing hinders the cause of this centripetal force towards the Sun, to be similar to that which pushes the bodies, which are called heavy, to descend towards the Earth.[92]

space ABCD differ in size and motion from those which are below D near K and from those above B near L, but are similar to those near N and Z; they spread more freely toward A and C than toward B and D. From this it follows that the orbit ABCD is not a perfect circle, but closer to the figure of an ellipse; and that the matter of the heaven is transported more slowly when in the regions near C and A than when near B and D. Therefore, the Moon, which is carried along by this matter of the heaven, {must also move more slowly and deviate more from its course near C and A and} must move closer to the Earth if it is waxing and further away if it is waning; that is, further away when it happens to be near A or C, than when it is near B or D," Descartes 1982, 175.

[91] See Snow 1924; Rosenfeld 1969; Anstey 2005; Kochiras 2017.

[92] "Je n'ay donc rien contre la Vis Centripeta, comme Mr. Newton l'appelle, par la quelle il fait peser les Planetes vers le Soleil, & la Lune vers la Terre, mais j'en demeure d'accord sans diffi-

In this way, Huygens accepts the idea that planets can be kept in their orbits by such a force, which in principle can be explained in the same way as terrestrial gravity. This, however, would imply a reformulation of the model of vortices, as far as the model Huygens himself had devised in his 1669 communication and in his *Discours* concerns only terrestrial gravity i.e. a planetary vortex. Or, as he puts it:

> it had been a long time since I had imagined that the spherical figure of the Sun could be produced like the one which, according to me, produces the sphericity of the Earth. but I had not extended the action of gravity to such great distances, as from the Sun to the planets, or from the Earth to the Moon; because the vortices of Mr. Descartes, which had once seemed very likely to me, and which I still had in my mind, were against this.[93]

As seen above, in Descartes's own model – especially that of *Le monde* – solar and terrestrial gravity are unified, namely, Earth gravitates towards the Sun as the Moon gravitates to Earth, and like any 'heavy' body either gravitates or falls on one of these celestial bodies (provided that they are surrounded by a vortex). In the *Principia*, on the other hand, the system of vortices allows for orbital stability – so that it determines a form of weight – while the heaviness of terrestrial bodies with respect to the centre of the Earth is explained by the 'drop' model. Namely, by considering the movement of particles of subtle matter in straight lines outwards with respect to the Earth. In turn, in the model of Huygens, only the action of a vortex turning around the Earth in many different directions is considered. As to planetary orbits around the Sun, Huygens was silent, i.e. he did not provide a cosmological theory – which, as I will show below, he was going to provide in his *Cosmotheoros*.

So far, Huygens's acknowledgement of Newton's idea of centripetal force had little impact on his own model of gravity, which is kept unaltered in the *Discours* of 1690. Of more impact was his acceptance of Newton's inversed square law, namely, the idea that the gravity or weight of celestial bodies is inversely proportional to the square of their distance – as seen above. Huygens accepts that this idea could be the foundation of a cosmological theory that can resolve all the difficulties – such as those related to the elliptical orbits of planets, their speeds, and the variations of orbits with respect to the ecliptic – that one can find in Descartes's vortex theory. Or, as he puts it:

> I had not thought either about this regulated diminution of gravity, namely that it was in a proportion reciprocal to the square of the distances from the centre: which is a new and very remarkable property of gravity, of which it is well worth looking for the reason. But seeing now by the demonstrations of Mr. Newton, that supposing such a gravity towards the Sun,

culté: parce que non seulement on sçait par experience qu'il y a une telle maniere d'attraction ou d'impulsion dans la nature, mais qu'aussi elle s'explique par les loix du mouvement, comme on a vû dans ce que j'ay écrit cy dessus de la pesanteur. Car rien n'empêche que la cause, de cette Vis Centripeta vers le Soleil, ne soit semblable à celle qui pousse les corps, qu'on appelle pesants, à descendre vers la Terre," Huygens 1690, 160.

[93] "Il y avoit long temps que je m'estois imaginé, que la figure spherique du Soleil pouvoit estre produite de mesme que celle qui, selon moy, produit la sphericité de la Terre); mais je n'avois point etendu l'action de la pesanteur à de si grandes distances, comme du Soleil aux Planetes, ni de la Terre à la Lune; parce que les Tourbillons de Mr. Des Cartes, qui m'avoient autrefois paru fort vraisemblables, & que j'avois encore dans l'esprit, venoient à la traverse," Huygens 1690, 160.

> and which diminishes according to the said proportion, it counterbalances so well the centrifugal forces of the planets, and precisely produces the effect of the elliptic motion, which Kepler had guessed, and was verified by the observations, I can doubt neither that these hypotheses concerning gravity are true, nor that the system of Mr. Newton, so far as it is based on such [premises], is the same. Which must seem so far more probable, since there is the solution of several difficulties, which are a problem in the vortices supposed by Descartes. We now see how the eccentricities of the planets can remain constantly the same: why the planes of their orbits do not unite, but keep their different inclinations towards the plane of the ecliptic, and why the planes of all these Orbs necessarily pass through the Sun. How the motions of the planets can accelerate and slow down by the degrees which one observes there; [all] which could hardly be so, if they swam in a vortex around the Sun.[94]

Huygens, thus, accepted Newton's law of the inverse square as it applies to planetary orbits (and for which he aimed at looking for a cause), but, as testified to by the *Addition*, he did not accept Newton's idea of universal gravity, namely the idea that bodies attract each other.

As duly and fully reconstructed by Eric Schliesser and George E. Smith, Huygens did not just reject Newton's idea of universal attraction or gravitation because it was not explained by Newton in mechanical terms. Rather, he offered empirical evidence for his claim, as he (1) refuted Newton's theory of gravitation on an empirical basis, namely, by taking into account the data on the periods of oscillation of seconds pendulums at different latitudes at sea, and (2) he justified their differing periods of oscillation by his own theory of weight. He could do so as he has instructed a chief mariner of the Dutch East India Company (Vereenigde Oostindische Compagnie, or VOC), namely Thomas Helder, how to take care of and draw data from two pendulum clocks on a ship at sea. The ship was the *Alcmaer*, and the data were annotated during her voyage to Texel from the Cape of Good Hope (10 April–15 August 1687), where Helder assumed her command (the clocks had been sent ahead to the Cape on board another ship). As Helder died during the voyage, the data were collected by his assistant, Johannes de Graaf, from 10 May.[95] The overall

[94] "Je n'avois pas pensé non plus à cette diminution reglée de la pesanteur, sçavoir qu'elle estoit en raison reciproque des quarrez des distances du centre: qui est une nouvelle & fort remarquable proprieté de la pesanteur, dont il vaut bien la peine de chercher la raison. Mais voyant maintenant par les demonstrations de Mr. Newton, qu'en supposant une telle pesanteur vers le Soleil, & qui diminue suivant la dite proportion, elle contrebalance si bien les forces centrifuges des Planetes, & produit justement l'effet du mouvement Elliptique, que Kepler avoit deviné, & verifié par les observations, je ne puis guere douter que ces Hypotheses touchant la pesanteur ne soient vrayes, ni que le Systeme de Mr. Newton, autant qu'il est fondé la dessus, ne le soit de mesme. Qui doit paroitre d'autant plus probable, qu'on y trouve la solution de plusieurs difficultez, qui faisoient de la peine dans les Tourbillons supposez de Des Cartes. On voit maintenant comment les excentricitez des Planetes peuvent demeurer constamment les mesmes: pourquoy les plans de leurs Orbes ne s'unissent point, mais gardent leurs differentes inclinaisons à l'égard du plan de l'Ecliptique, & pourquoy les plans de tous ces Orbes passent necessairement par le Soleil. Comment les mouvemens des Planetes peuvent s'accelerer & se ralentir par les degrez qu'on y observe; qui malaisement pouvoient estre tels, si elles nageoient dans un Tourbillon autour du Soleil," Huygens 1690, 160–161.

[95] For a thorough account, see Schliesser and Smith 2000. See also Mahoney 1980; Schliesser and Smith 1996; Howard 2008; De Grijs 2015. Please note that Huygens had already organized a sort of pre-trial in the Zuiderzee in 1685.

end of the entire operation was actually that of finding a secure method to calculate longitude at sea, by observing the time counted by the clocks.

The main problem with this sort of method was that of the variation of the periods of oscillation of pendulums at different latitudes, given the centrifugal force they have at different latitudes, which is due to the rotation of the Earth. This effect was well known since in 1672–1673 Jean Richer led an expedition to La Cayenne, organized by the Académie Royale des Science, through which he found that a seconds pendulum set in Paris, in order to count the seconds at La Cayenne (where it is slower), had to be shortened by 1 and ¼ lines.[96] Huygens, by considering both the effects of weight as well as those of the centrifugal forces on the clock, devised corrections to the lengths of the pendulums to compensate for the variations in the centrifugal force according to the latitude of the ship. As he received De Graaf's data, namely the uncorrected measurements of longitudes at sea, he corrected them and discovered two main results: (1) that without corrections, the measurements of the longitudes obtained by the clocks would have put the route of the ship directly through Ireland (as the ship circumnavigated Ireland, Scotland and England before arriving in Texel), and (2) with the corrections, the 'virtual' final point reached by the ship was just 5 or 6 miles from the actual longitude of Texel. Or, in Schliesser and Smiths's words, "[a]t the end of the 117 day voyage, with the correction to the clock accumulating the entire way, the calculated location of Texel was only 17 arc-minutes to the east of the location Huygens had inferred from Guy Tachard's observations. This is a mere 19 km."[97] Therefore, in his report to the directors of the VOC (i.e. his *Rapport aengaende de lengdevinding door mijne horologien op de reys van de Caep de B. Esperance tot Texel A^o. 1687*, sent on 24 April 1688, and later reviewed by De Volder, as I will show), Huygens acclaimed the operation as

[96] Similar observations were carried out in the same years by Jean Picard (1671), Edmond Halley (1677), Varin, Jean Deshayes and Guillaume de Glos (1682): see Olmsted 1942; Débarbat 1986; Matthews 2000; Howarth 2007; Dew 2013.

[97] Schliesser and Smith 2000, 16. In fact, the original difference between the 'virtual' final point of the voyage and the actual location of Texel (i.e. what Huygens thought it to be on the basis of Tachard's estimations, to which I shall return below) calculated by Huygens was circa 0° 25′, resulting from the calculated last point of the *Alcmaer* (14° 1′ west of the Cape, in the first calculation), and the location of Texel (14° 25′ west of the Cape). The version reproduced in the *Oeuvres*, i.e. Huygens's own copy, in fact, was modified by Huygens after De Volder provided some corrections on Huygens's table in his report. In the report (which I analyse below), De Volder provided corrections to the calculated positions of the *Alcmaer*, locating its final point 14° 8′ west to the Cape: "[...] [w]aer bij komt dat de horologien soodanigh verschil van lenghte tussen de Caep en Texel geven, [...] sijnde dit verschil volgens de Horologien, niet 14 gr. 1 minuut, als de Hr. Huijgens seyt, tgeen uijt het voorseijde abuijs in 't adderen komt te spruijten maer 14 gr. 8 m.," De Volder to the Directors of the VOC (letter 2547), 22 July 1689, in Huygens 1888–1950, volume 9, 343. Huygens wrote a marginal note in his copy of his own report about this: "[m]ijn misrekening in de 7e colomme en 't geen in de verdere colommen daer uijt volgden, is hier van daen gecorrigeert volgens de rekening van de Hr. de Volder," Huygens to the Directors of the VOC (letter 2519), 24 April 1688, in Huygens 1888–1950, volume 9, 280; see Schliesser and Smith 2000, 2 (n. 3), 16 (n. 41) and 50 (n. 108). Of course, De Volder's correction made Huygens's result even more compelling.

a success.[98] Moreover, in his letter to Hudde – himself a director of the VOC – which accompanied his report, as well as in the 1690 *Addition* to his *Discours*, Huygens claimed that the 'virtual' positions calculated on the basis of his own theory of gravity and centrifugal force were more correct than those calculated on the basis of Newton's theory of universal gravitation.[99] As he put it in his letter to Hudde (who, in turn, also received De Volder's report):[100]

> concerning the said effect of the rotation of the Earth[,] perhaps you may have seen what recently has been written about this by Professor Newton in his book with the title *Philosophiae naturalis principia mathematica*, which proposes various hypotheses that I cannot approve, from which he therefore draws a different conclusion than that resulting from my calculation. The foundations on which I have built are mentioned in my report, which, as far as they relate [to] the weight of bodies, differ little from those of Descartes and Rohault.[101]

In order to support his claim (in fact, in his report to the VOC Huygens did not provide insights into his explanation of weight, but he referred to his still unpub-

[98] "Doch eyndelijck, nae langh wachten, kan ick seggen seer goede tijdinghe te brengen aengaende dese Inventie, als bevonden hebbende dat door middel der voors, horologien de Lengdens tusschen de Caep de Bonne Esper.[ce] en Texel doorgaens seer wel sijn afgemeten, en de geheele Lengde tusschen dese twee plaetsen soo perfect, dat het maer 5 à 6 mijlen en verscheelt, 't welck ick bekenne met sonderlingh vergenoegen gesien te hebben; als sijnde een seeckere preuve van de moghelijckheydt deser soo lang nae gewenschte saeck," Huygens to the Directors of the VOC (letter 2519), 24 April 1688, in Huygens 1888–1950, volume 9, 272.

[99] Cf. the beginning of the *Addition*: "[q]uelque temps aprés que j'eus achevé d'escrire ce qui precede, ayant receu & examiné le journal du voiage, qui, par ordre de Messieurs les Directeurs de la Compagnie des Indes Orientales, a esté fait, avec nos Horloges à pendule, jusqu'au Cap de Bonne Esperance; & du dépuis ayant encore lû le tres sçavant ouvrage de Mr. Newton, dont le titre est Philosophiae Naturalis principia Mathematica; l'un & l'autre me fournit de la matiere pour étendre d'avantage ce Discours. Et premierement, quant aux differentes longueurs des Pendules dans divers Climats, dont il a aussi traité, je crois avoir, par le moien de ces Horloges, non seulement une confirmation évidente de cet effect du mouvement de la Terre, mais aussi de la mesure de ces longueurs, qui s'accorde tres bien avec le calcul que je viens d'en donner. Car ayant corrigé & rectifié, suivant ce calcul, les Longitudes qu'on avoit mesurées par les Horloges, au retour du Cap de B. Espe. jusqu'au Texel en Hollande, (car en allant elles n'avoient point servi) j'ay trouvé que la route du vaisseau en estoit beaucoup mieux marquée sur la Carte, qu'elle n'estoit sans cette correction; & si bien, qu'en arrivant à ce Port, il n'y avoit pas 5 ou 6 lieues d'erreur dans la Longitude ainsi rectifiée," Huygens 1690, 152–153.

[100] See Schliesser and Smith 2000, 4 (n. 8): "[a] copy of de Volder's review can be found in Hudde's personal archives at the Algemeen Rijksarchief (signature: 1.10.48 no. 44) in The Hague."

[101] "Aengaende het gemelte effect van het draeijen der aerde sal VEdt. misschien gesien hebben 't geen onlanghs daervan geschreven is door den Professor Newton in sijn boeck genaemt Philosophiae Naturalis principia Mathematica, stellende verscheyde hypotheses die ick niet en kan approberen, waeruyt dan oock ander besluijt treckt als mijn rekeningh uytgeeft. De fondamenten waerop ick gebouwt hebbe sijn in mijn Raport vermelt, welcke voor soo veel der Lichaemen swaerheydt aengaet, weynigh verschelen van die van des Cartes en Rohault," Huygens to Hudde, 24 April 1688 (letter 2517), in Huygens 1888–1950, volume 9, 267. Hudde did not write a review on Huygens's report.

lished *Discours*),[102] Huygens provides in his *Addition* calculations of the shortening of the lengths of the pendulum on the basis of (1) his own views, and (2) Newton's theory of universal gravitation.

As to (1), Huygens provides a calculation on the premise that celestial matter rotates, at the equator, at 17 times the speed of the Earth – as seen above. In other words, a body at the equator would be weightless if the Earth rotates 17 times faster than it does, carrying such a body. Since centrifugal force is proportional to the square of the speed of a rotating body – in accordance with Huygens's quantification – the weight removed by the centrifugal force of a body rotating at the equator is $(1/17)^2$ i.e. 1/289 of the weight of a body at the poles, where such force is not exerted (as at the poles a body can be considered at rest). So that a pendulum set at the poles has to be shortened of 1/289 in order to count time at the equator.[103] As calculated by Huygens, a pendulum moving on a ship from the pole to the equator would have, therefore, a total delay of 2 and ½ minutes, with respect to a clock at the pole. This is the coefficient of delay used by Huygens in his corrections of De Graaf's data.[104]

Hence, (2) Huygens calculates the coefficient of delay on the basis of Newton's theory of universal gravitation – and the law of the inverse of the square. Namely, the idea that the force of gravitation is inversely proportional to the square of the distance between the centres of two bodies, so that a body placed at the equator – provided that the Earth is not a perfect sphere – is not only lighter, with respect to a body at the pole, as a consequence of Earth's rotation i.e. of its centrifugal force, but also as a consequence of the greater distance of this body from the centre of the Earth. In turn, according to Huygens, the Earth is not perfectly spherical but is a spheroid, flattened at the poles. This is a consequence of the centrifugal force. In

[102] "En wat aengaet het gansche fondament deser Calculatie, en wat ick daer ontrent in acht genomen hebbe, 't selve heb ick verhandelt in een apart Tractaet van de Oorsaeck der Swaerte), 't welck ick geerne wil, en oock voor genomen hebbe te onderwerpen aen alle Ervarene Mathematici haer examen, om te doen sien dat ick in mijn stellingh nochte uytrekeningh niet in 't minste toegegeven hebbe om de gevonden Lengdens goedt te maecken," Huygens to the Directors of the VOC (letter 2519), 24 April 1688, in Huygens 1888–1950, volume 9, 276.

[103] Cf. Huygens's own words: "ayant trouvé, comme on a vû, que, si la Terre tournoit 17 fois plus viste qu'elle ne fait, la force Centrifuge sous l'Equateur seroit égale à toute la pesanteur d'un corps; il faut que le mouvement de la Terre, tel qu'il est maintenant, oste une partie de la pesanteur, qui soit à la pesanteur entiere comme 1 au quarré de 17, c'est-à- dire 1/289; parce que les forces des corps, à s'éloigner du centre autour du quel ils tournent, sont entre elles comme les quarrez de leurs vitesses, suivant mon Theoreme 3e. de Vi Centrifuga). Chaque corps, sous l'Equateur, estant donc moins pesant de 1/289 de ce qu'il seroit si la Terre ne tournoit point sur son axe; il s'ensuit, par les loix de la Mechanique, que la longueur d'un Pendule, en cet endroit, doit aussi estre diminuée de 1/289, pour faire ses allées dans le mesme temps qu'il les feroit sur la Terre immobile," Huygens 1690, 146–147.

[104] "[...] l'on trouvera que ces retardemens, entre eux, suivent assez precisément la mesme proportion que les diminutions de la longueur du pendule: & que le plus grand retardement, tel que seroit celuy d'une Horloge sous l'Equateur, lors qu'elle auroit esté reglée sous le Pole, seroit par jour fort prés de 2½ min. En ayant donc calculé des Tables, on pourroit corriger, par leur moien, le mouvement des Horloges, & s'en servir avec la mesme sureté que si ce mouvement estoit par tout égal," Huygens 1690, 150.

order to illustrate his point, he conceives the Earth as traversed by two canals: one cutting it across the poles, and another across the equator. Since at the equator matter is lighter by 1/289 than it is at the poles, as a consequence of the centrifugal force, there Earth is 1/578 larger than at the poles, i.e. the ratio between the polar and the equatorial canal is 577/578. Indeed, the value 1/289 has to be divided by 2, for the reason that matter is lighter by 1/289 only at the equator, while at the centre of the Earth it has the same weight which it has at the poles – so that an average needs to be made.[105] Accordingly, in order to calculate the 'lightness' of a body at the equator, one needs to consider both its centrifugal force – which makes it 1/289 lighter – as well as its increased distance from the centre of the Earth, namely another 1/289. This provides a slowing of the pendulum at the equator of circa 5 minutes of delay, i.e. double the delay it derives from the centrifugal force alone. As Huygens points out very clearly, however, this kind of delay contradicts the empirical data he obtained from the *Alcmaer* voyage.[106]

[105] "Le canal qu'il suppose est representé dans nostre figure par ECP, faisant un angle droit au centre de la Terre. Il faut le concevoir comme ayant quelque peu de creux, & rempli d'eau. Ce qui estant, il est certain que les deux jambes, EC, CP [respectively, the canal connecting the equator and a pole to the center of the Earth], se doivent tenir en équilibre, si l'on suppose que la Terre, estant toute composée d'eau, prend une figure, dont les diametres soient EA & PQ [respectively, the 'equatorial' canal and the 'polar' canal]: parce qu'autrement, cette eau du canal, ne demeureroit pas non plus dans son assiette en la concevant sans canal, contre ce qu'on suppose. d'où il est aisé de trouver la raison de EA à PQ. Car en posant EC = *a*; CP = *b*, & representant la pesanteur absoluë par une ligne *p*; & la force centrifuge en E par la ligne *n*; le poids du canal PC est *pb*, sçavoir ce qui se fait en multipliant toutes les parties de ce canal egalement par la ligne *p*. Mais le poids du canal EC, qui seroit *pa*, est diminué par la force centrifuge de toutes ses parties, des quelles la plus élevée, qui est en E, a la force *n*; & toutes les autres parties l'ont proportionée à celle cy, suivant leur distances du centre D. ce qui fait 1/2 *na* pour toute la force centrifuge de l'eau du canal EC, qui estant ostée de son poids *pa*, reste *pa* – 1/2 *na*; qui doit estre égal à *pb* poids du canal PC. d'où il paroit que *a* est à *b* comme *p* à *p* – 1/2 *n*. C'est-à-dire que le diametre EA de la Terre, est à son axe PQ, comme 289 à 288 1/2, ou comme 578 à 577; car la raison de *p* à *n* estoit comme 289 à 1," Huygens 1690, 155. On the other hand, Newton discovered a difference, between the equator and the poles, of 1/231: he used the example of the canals later used by Huygens, applying to them the effect of centrifugal force and of gravitation: see his *Principia*, book 3, propositions 18–20, where Newton proposes various adjustments to the lengths of pendulum clocks, for different latitudes.

[106] "[...] cet exemple de la Lune prouve si bien la diminution du poids, suivant la raison reciproque des quarrez des distances du centre de la Terre [Huygens had just (successfully) applied Newton's law of inverse square to the gravitation of the Moon]; on pourroit douter s'il n'y auroit pas aux Pendules une autre inégalité, outre celle qui estoit causée par le mouvement journalier. Car si la Terre n'est pas spherique, mais assez pres spheroïde, & qu'un point sous l'Equateur est plus eloigné du centre, que n'est un point sous le Pole, dans la raison de 578 à 577, comme il a esté dit cy-devant; les pesanteurs estant en ces endroits en raison contraire des quarrez de ces distances, il faudroit aussi que le pendule sous l'Equateur fust plus court, que celuy dessous le Pole, dans cette mesme raison contraire). C'est à dire que ces pendules seroient comme 288 à 289; ou que le pendule sous l'Equateur seroit plus court de 1/289 de ce qu'il seroit sous le Pole. Qui est justement la mesme difference, qui provenoit cy dessus du mouvement journalier, ou de la force centrifuge. De sorte qu'une Horloge, avec la mesme longueur de pendule, iroit plus lentement sous l'Equateur que sous le Pole, du double de ce qu'elle retardoit par le mouvement de la Terre; & ainsi cette difference journaliere sous l'Equateur seroit de pres de 5 minutes. Et sous les autres paralleles, on la

6.2.2.3 De Volder vis-à-vis Descartes, Huygens and Newton

De Volder played a crucial role in the pendulum trials organized by Huygens, both in the one conducted on the *Alcmaer*, and a second, conducted in the years 1690–1692 (as I will show). De Volder was indeed a referee for the VOC of Huygens's 1688 report, on which he wrote a review dated 22 July 1689, and he also commented upon the results of the second trial in 1693.[107]

In 1689, actually, De Volder had already been involved – together with Huygens himself – in the evaluation of a proposal for a method of calculating longitude at sea, namely in the famous case of Lieuwe Willemsz. Graaf, a Frisian maritime navigator, and then Mennonite preacher in Harlingen and Amsterdam (where he also opened a school for navigators),[108] proposed to the States of Holland a patent concerning a method to establish longitude at sea. Graaf had various meetings before the States in the Hague. There, he defended his method before four examiners, namely, Huygens and De Volder first, and then Jan Stampioen and Abraham de Graaf. He had previously discussed his method with Balthasar Bekker (who had occasioned Lieuwe Willemsz. Graaf's devising of his own method with his *Ondersoek van de betekeninge der kometen*, 1683) and Fullenius, with both of whom he was to engage in a quarrel in 1688–1690.[109] Indeed, in his meeting with Fullenius in autumn 1688 (to whom Graaf had been re-directed by Bekker) he did not reveal his method – fearing that it would be divulged. Thereafter, he met the Deputy States of Friesland, and then the States of Holland in March and April 1689, to which he presented his method. The method, as it has been reconstructed by Arjen Dijkstra on the basis of two of Graaf's booklets, namely his *Eenvoudig en onvervalscht verhaal van 't voorgevallene in 's Gravenhage ter vergaaderinge van de H. M. Heeren, de Staaten Generaal der Vereenigde Nederlanden over 't aanwijsen van 't vinden der lengte van Oost en West, en de waare tijdreekeninghe* (1689),

---

trouveroit par tout plus que double de ce qu'elle y estoit auparavant. Mais je doute fort que l'experience confirme cette grande variation), puisque j'ay vû que, dans le voiage dont j'ay fait mention, la seule premiere équation suffit, & que la plus que double mettroit, vers le milieu du chemin, trop de difference entre la route du vaisseau, calculée sur le Pendule, & celle qu'il tenoit par l'Estime des Pilotes," Huygens 1690, 166–167. Please note that if the theoretical delay calculated by Huygens on the basis of the idea of universal gravitation amounts of 5 minutes, according to Newton's *Principia* it amounts in fact to a delay of 3 minutes and 7 seconds, as Newton takes into account local variations of gravity, due to the spheroidal shape of the Earth, and to its internal density: see Newton 1687, 425–426, discussed in Schliesser and Smith 2000, 21. Also by considering this variation, however, Huygens's calculation of the longitude at the ship's arrival in Texel would have been more precise than Newton's: see Schliesser and Smith 2000, 18.

[107] See De Volder to the Directors of the VOC, 22 July 1689 (letter 2547), and De Volder to Huygens, 6 April 1693 (letter 2800), in Huygens 1888–1950, volume 9, 276, 339–343, volume 10, 435–436. For Huygens's reaction to De Volder's comments, see the letters of Huygens to De Volder of 14 March 1693 (letters 2798–2799), 19 April 1688 (letters 2802–2803), and the letter of Huygens to the Directors of the VOC of 6 March 1693 (letter 2796). No official report for the VOC on the second trial could be retrieved.

[108] On him, see Dijkstra 2012, chapter 10.

[109] On it, see Dijkstra 2007.

in which Graaf gives his version of the meetings at The Hague, and his *D'eerste opening en gewis onderwys van de waere vinding der lengte van Oost en West* (1691), which is rather obscure. As Dijkstra has put it,

> [i]n order to calculate longitude an observer would measure the distance between the moon and a designated constellation: this would give figure 'x'. With x the position on the globe could be calculated from Lieuwe Willemsz's tables. It was as simple as that. The theory behind this stunningly straightforward plan is much less clear. Distance x in the sky correlated with a distance y in Lieuwe Willemsz' tables. That distance y would be the difference on the globe between the position of the observer and the biblical Garden of Eden. The backbone to his method was that by calculating x into y, the local time in that Garden of Eden was found. Strangely, however, Lieuwe Willemsz found that the observer did not need to calculate the time of the place where the observation was made. If he had included this in his method, and if his tables had been true and reliable(!), his method would have worked. However, without this he seems to have been playing with the evidence. This, at least, was the conclusion of his contemporaries. Because x was the only variable that was taken from an observation, all others came from his tables.[110]

In fact, Graaf never showed anyone his tables. In the first meeting at The Hague on 29 March 1689, as Graaf reports in his *Eenvoudig en onvervalscht verhaal* (1689), De Volder asked him some questions, which Graaf did not satisfactorily answer, and provoked De Volder's annoyance.[111] Indeed, the second meeting, as reported by Graaf, took place only 14 days after because of the "the slow and reluctant return of Mr. Professor De Volder,"[112] who not surprisingly could not be convinced by Graaf, neither in this nor in the next session, on 13 April, the last one

---

[110] Dijkstra 2012, 322.

[111] De Volder asked him how he determined the longitude, and Graaf answered him that he could do so by the movement of the moon in the firmament. As De Volder asked whether he could observe such movement, Graaf asked the members of the States of Holland whether they would not misinterpret what he was going to answer, and stated that De Volder had posed a misleading question, because longitude cannot be found by observation. Eventually, De Volder looked at him severely, without words: "[w]aar op d'Heer Professor Volder hem liet hooren vragende; Waar door ik de langte zoude vinden? Ik antwoorde hem, door het verachteren van de Maan aan het Sterredak. Daar op hy heel for vraagde, wat Starredak? Ik zeide, dat het even veel was, of ik zeyde firmament of uitspansel. Daar op hy weder heel spijtig sprak, en vraagde; of ik dat wel eens konde observeren? Waar op ik verzochte, dat Haar Ho: Mog: my niet qualijk geliefden af te neemen, 't geen ik daar op nu dacht te zeggen: die my verlofgaven, om vry uit te spreken. Dies ik zeyde, dat de Heer Professor my vraagde 't geene niet ter zaake dien de. Alzoo men door het observeren die lengte niet konde bekomen [...]. Ik dan sprak tegen Volder: Laat ons een goed fundament leggen, op dat het Hemelsch gebouw niet om verre raakt, wanneer wy ten halven gekoomen zijn: Ook moeten wy betoonen, dat wy mannen zijn in 't verstand; en geen woorden spreken, die in den wind of lucht vervliegen: Maar die tot de zaak en des zelfs waarheid dienen. Dewijl hy my daar op sterk aanzag, zeide ik vorder tot hem; Laat ons nu eerst het geen men 't lichtste meint te wesen bezien, of het met de waarheid over een komt, te weeten, de beweginge van de Son. Ik vrage dan, hoe groot de cyclus of cirkel van de Son is? Hy my noch al sterk aanziende, en sprak niet. Ik zeide, vraag my: Hy sweeg noch al stil," Graaf 1689, 2–3. See also the commentary of the editors of Huygens's *Oeuvres*: Huygens 1888–1950, volume 9, 315 and 317.

[112] "In de volgende, of tweede Sessie, ontrent 14 dagen daar na vermidts de traage en weerwillige wederkomst van d'Heer Professor Volder, is niets besonders voorgevallen, [...]" Graaf 1689, 9.

attended by De Volder.[113] Notably, Jakob Gronovius reports that an *Epistola* of De Volder on the issue was read before the States of Holland – a document, however, no longer retrievable, nor mentioned by Graaf or Fullenius during their quarrel.[114] Huygens – who was going to attend further meetings with Graaf (who also presented his 'method' before the Amsterdam City council) – was also to come to severely reject his theories.[115] Despite their negative opinions, the States granted Graaf a remuneration of 2,000 guilders for his patent (which was then revealed as useless).

The method proposed by Huygens, and its philosophical entailments, on the other hand, had a more relevant role in shaping De Volder's intellectual path. As I show in Sect. 6.2.2.3.4, *De Volder's criticisms of Huygens's empirical evidence*, De Volder's own remarks on Huygens's reports – including Huygens's commentary on the results of the second expedition – finalized his 'disenchantment' with Descartes's vortex theory, which according to Le Clerc's *Éloge* was determined by De Volder's reading of Newton's *Principia*, whose contents were found as true by him during a meeting he had with Huygens in order to discuss and clarify its contents.[116] Yet, as I

---

[113] See Graaf 1689, 12–16.

[114] "Quanto in silentio audivimus lectam fuisse in consessu Potentissimorum Hollandiae Ordinum Epistolam Volderi, quum de oriente et occidente excitatum esset negotium?" Gronovius 1709, 28.

[115] See, for instance, his letter to Lodewijk Huygens of 14 May 1689 (letter 2538). See Dijkstra 2012, 321–331.

[116] "Pendant que Mr. de Volder s'appliquoit à étudier & à enseigner les Mathematiques, de trés-habiles gens découvrirent ces nouvelles Méthodes aux quelles on a donné les noms de Calcul differentiel & de Calcul integral, & faisoient de tems en tems paroître quelque chose en public, qu'ils avoient découvert, par ces Méthodes. Nôtre Philosophe s'applica aussi à ces nouvelles manieres, & dès que le livre de l'Illustre Mr. Newton, des Principes Mathematiques de la Philosophie Naturelle, eut paru en MDCLXXXVII. Il s'attacha fortement à cette lecture, pour découvrir les principes sur les quels l'Auteur s'étoit fondé; & l'on a encore trouvé parmi ses papiers les calculs de ses démonstrations. On sait que ce livre suppose une connoissance des Mathematiques, qui n'est pas commune. Je me souviens même d'avoir ouï dire à Mr. de Volder, que peu de tems après qu'il parut, feu Mr. Huygens, qui étoit un grand Mathematicien, mais à qui les nouvelles méthodes, dont j'ai parlé, étoient inconnuës, le vint voir à Leide & le mit sur le sujet du livre de Mr. Newton. Il avoüa à Mr. de Volder, qu'il trouvoit ce livre extrémement obscur, & lui demanda ce qu'il en pensoit. Nôtre Philosophe lui répondit qu'il n'étoit pas en effet facile de pénetrer les principes des démonstrations de l'Auteur, mais qu'il avoit trouvées veritables celles qu'il avoit examinées. On ne peut pas néanmoins douter que Mr. Huygens ne fût un excellent Mathematicien, & que la Physique & les Mathematiques ne lui soient très-redevables. [...] Je ne doute pas que le Livre de Mr. Newton ne lui eût ouvert les yeux, sur la doctrine des tourbillons de Descartes; que ce Livre a entierement renversée, à cela près, qu'il établit le Systeme de Copernic, comme lui; en reconnoissant le Soleil, comme le centre commun des mouvemens de la Terre & des autres Planetes. Cette partie de la Physique de Descartes avoit né an moins paru à bien des gens la mieux imaginée. On voit en cette rencontre, comme en plusieurs autres, l'exemple d'un grand Mathematicien, qui ne raisonnoit point mathematiquement, ni même conséquemment," Le Clerc 1709, 379–380 and 382. Le Clerc also more generally refers to some "Englishmen" as having brought De Volder to reject Cartesianism (may be referring also to Locke: see Sect. 2.2.3, *The mid-1670s clash at Leiden and the foundation of the experimental theatre*): "[i]l n'étoit nullement entêté de la Philosophie, qu'il enseignoit; sur la fin de ses jours, & même quelques années auparavant, il avoit reconnu le foible du Cartesianisme; autant apparemment, par sa propre méditation, que par le secours des habiles Anglois, qui ont établi d'autres principes," Le Clerc 1709, 398.

am going to show, De Volder had also embraced, at some point, Huygens's own version of the vortex theory and explanation of weight, which was then itself relinquished as De Volder's expectations of Cartesian philosophy (in all its emendations) collapsed. Before analysing such a late evolution of De Volder's ideas, however, we must look at his general ideas on cosmology.

6.2.2.3.1 De Volder's Cartesian Cosmology and Theory of Weight

De Volder devoted to cosmology (and the related idea of weight) a large number of disputations which he presided over. Besides his *dictata*, as well as the corollaries of his *Disputationes de aëris gravitate*, he devoted to such topics the two last disputations of his *Exercitationes* against Huet (viz. his *De mundo* and *De tribus Cartesii elementis* (all of 1693)), and a considerable number of disputations presided over or dedicated to him. Namely, Samuel Köleséri's *Disputatio mathematico-physica de lumine* and *Disputatio inauguralis de systemate mundi* (1681); Pál Derecskei's *Exercitatio coelorum, siderumque lucidorum originem, et phaenomena, methodo synthetico-mathematica demonstrans* (1682); Hieronymus Simons van Alphen's *Exercitatio utrum Sol, an vero Tellus, in planetarum numerum referenda sit, inquirens* (1684); Henricus van Bronchorst's *Disputatio de vera gravitatis causa* (1685); Hermann Schuyl's *De vi corporum elastica* (1688); Gysbertus Casembroot's *Disputatio de mundi systemate* (1694) and *Disputatio philosophica inauguralis de aestu marino* (1696); Joseph Serrurier's *Disputatio inauguralis de gravitate aëris* (1690); Fredericus Gulielmus Westhovius's *Dissertatio mathematico-philosophica de fluminibus* (1698); Henricus Jacobus van Bashuysen's *Disputatio philosophica de Telluris motu* (1698).

With few, remarkable exceptions (above all Schuyl's *De vi corporum elastica* and Casembroot's *De aestu marino*) all the disputations mentioned and the *dictata* stick to the Cartesian cosmological theory, and theory of gravity. If we look at a two-part disputation authored by the Hungarian student Samuel Köleséri[117] (who also acted as *respondens* in two of De Volder's disputations *Contra atheos*), and presided over by De Volder in March 1681, namely, Köleséri's *Disputatio mathematico-physica de lumine, pars prima–secunda*, as well as at Köleséri's *Disputatio philosophica inauguralis de systemate mundi* (June 1681), we do find an overt use of Descartes as to the explanation of the natural-philosophical underpinnings of the Copernican system. Namely, the use of various positions traced back to Descartes's *Principia*,[118] culminating, in the *De systemate mundi*, with the appeal to the first three books of Descartes's *Principia* as the foremost defence of

[117] On him, see Font 2014; Balázs et al. 2016. See also Réka 2009.

[118] See, for instance, the defence of Descartes's matter theory in the *De lumine, Pars prima*, thesis 3, which goes along with an overall presentation of Descartes's theory of light developed in the whole text, and the use of Descartes's definition of relative motion as given in *Principia* II. 25 (*De systemate mundi*, thesis 8).

Copernicanism (together with Copernicus's *De revolutionibus orbium coelestium* (1543), Galileo's *Dialogo sopra i due massimi sistemi del mondo* (1632), Daniel Lipstorp's *Copernicus redivivus* (1653), and others),[119] and the claim, given in the *Annexa physica* appended to the disputation, that the "system of the world which had been delineated by [...] Descartes, it is not to be taken as a mere hypothesis, but as the truth of thing."[120] Descartes's cosmology, thus, loses the hypothetical character which it had in the *Principia* (overtly stated in III.44–45) – according to which, such a cosmology just provided an explanation alternative to the revealed truth established in the book of Genesis[121] – for the reason that, as it was put in the main text of the disputation, by using hypotheses and principles such as those of Descartes one can deduce all the empirical evidence concerning astronomy in an indubitable manner.[122] In conformity with De Volder's idea of certainty,[123] thus, such a cosmology could not but be true.

---

[119] Köleséri 1681, thesis 11.

[120] "Systema Mundi quod a Nobiliss. Cartesio delineatur, non pro mera hypothesi; sed pro rei veritate habendum est," Köleséri 1681, *Annexa physica, annexum* 14. See Ruestow 1973, 80–81.

[121] "44. That I nevertheless wish those [causes] I am proposing here to be taken only as hypotheses. [...] That I shall even assume here some which it is certain are false. Indeed, in order to better explain natural things. I may even retrace their causes here to a stage earlier than any I think they ever passed through. For example, I do not doubt that the world was created in the beginning with all the perfection which it now possesses; so that the Sun, the Earth, the Moon. and the Stars existed in it, and so that the Earth did not only contain the seeds of plants but was covered by actual plants; and that Adam and Eve were not born as children but created as adults. The Christian faith teaches us this, and natural reason convinces us that this is true; because, taking into account the omnipotence of God, we must believe that everything He created was perfect in every way. But nevertheless, just as for an understanding of the nature of plants or men it is better by far to consider how they can gradually grow from seeds than how they were created [entire] by God in the very beginning of the world," Descartes 1982, 105. On the topic, see Harrison 2000; Ariew 2011b; Torero-Ibad 2011.

[122] "I. [...] Hypotheses enim et principia ea sunt, quibus bene, firmiterque positis, accuratissime omnes sequentur consequentiae; [...] nec unquam certior, evidentiorque ad assequendam veritatis cognitionem via esse potest firmis principiis et hypothesibus; utpote quales etiam ipsi scientiae certissimae mathematicae sectatores suis praemittunt problematibus et theorematibus, tantumque hypotheses in astronomia, quantum axiomata in geometria praestant, Ea siquidem quis invenerit, tum dum invenit maximam veritatum obtinuit partem, nam veritates tam indissolubili et mutuo nexu sunt sibi ipsis nexae, ut una posita reliquae sponte sequantur. Oportet interim ea principia et hypotheses ex quibus consequentias deducere, variasque rerum causas reddere contendit philosophus tam clara et evidentia esse, ut mens earum veritatem in dubium vocare nec possit, nec debeat; tam concatenata, connexave cum suis consequentiis ut tota consequentiarum cognitio, illis suum debeat ortum, sed non vicissim illorum his: talibus ita positis, adniti philosophum oportet ut notitia rerum, ex principiis hisce, a quibus dependent ita deducat, ut nihil in tota deductionum serie inveniatur, quod non sit manifestissimum. [...] XII. [...] Habes itaque hic lector veritatis avide, ex hac hypothesi systema ordinatissumum, mirandam mundi fabricam potiori tum ratione, tum experientia nixam. Quia autem hypothesin et principium invenimus, quo firmiter asserto, et vindicato, evidentes phaenomenωn caelestium causas obtinemus, quorum demonstrationem ulterius persequi nunc non licet," Köleséri 1681, theses 1 and 12.

[123] See Sect. 3.2.3.3, *The role of experience in De Volder's natural philosophy*.

Hence, in the lengthy, four-parts *Exercitatio philosophica, coelorum, siderumque lucidorum originem, et phaenomena, methodo synthetico-mathematica demonstrans* (March and May 1682), authored and defended by another Hungarian student, Pál Derecskei,[124] and presided over by and dedicated to De Volder (in its fourth part along with Christoph Wittich), the vortex theory of Descartes is presented – in accordance with the title – by a series of lemmata, definitions, axioms, and 29 propositions. In this *Exercitatio*, however, only stars and skies – including the differences in the dimensions and speeds of globules across the vortex (especially propositions 26–27) – are considered, while the consideration of the planets and their equilibrium is overtly omitted (probably as it would entail the consideration of the evolution of the vortex i.e. the formation of a third element after the first and second).[125] The very hand of De Volder is difficult to ascertain: yet, it is worth noting that in this disputation we do find – as in both series of De Volder's *dictata*, where De Volder comments on Descartes's second law of motion (II.39) – an overt appeal to Archimedes in the illustration of how particles moving in a circle have a tendency to move according to the tangents (namely, that a circle can be rendered in terms of an infinite series of tangents: see Fig. 6.17, Hamburg 273, 105).[126]

Further, explicitly Cartesian disputations on cosmology presided over and authored by De Volder were those pertaining to the long series of disputations against Huet, namely, the 1693 disputation 26, *De mundo*, and disputations 27–28, *De tribus Cartesii elementis*. If the *De mundo* is a defence of Descartes's idea that the world, i.e. extension, is indefinite in its boundaries, the other two disputations

---

[124] On him, and on the presence of Hungarian students at Leiden, see Réka 2009.

[125] Cf. the closing scholium: "[...] sideribus lucidis (de quibus solis nobis sermo est)," De Volder and Derecskei 1682, *Scholium*. In the *Annexa*, it is reported a standard Cartesian proposition on gravity: "Omnia copora terrestria sunt levia, et ea tantum dicuntur gravia, quae sunt minus levia," De Volder and Derecskei 1682, *annexum* 9.

[126] Cf. De Volder and Derecskei 1682, lemma 10: "[o]mne corpus circulariter motum continuo determinatur, ut secundum tangentem pergat moveri, idque obtinet, nisi a causis externis impediatur. Circulus, (iuxta Archimedem) est polygonum infinitorum laterum seu est figura constans ex infinitis lineis rectis. Adeo ut quamdiu numerus laterum certus est, eousque semper est differentia, inter circulum, et polygonum. Si vero numerus laterum sit infinitus, seu omni numero maior, iam illud polygonum erit circulus. Unde sequitur. Corpus circulariter motum, in quolibet puncto sit; semper haerere in aliqua linea recta, nec ab ea deflectet, nisi a causa externa cogatur [...]. At illa linea recta, si producatur est tangens circuli." See Hamburg 273, 105–106: "[t]angens circuli. Tangere linea dicit[ur] circulum {q[uod]} circulum tangit ut AB, non vero secat ut CB. Circulus nihil aliud est quam infinita congeries tangentium ex definitione Archimedis. Cum itaq[ue] corpus in orbem movetur, ex una tangente in alias progreditur, ex quo sequitur, si in illa tangente in qua est, sibi relinquatur, moveri oportere per eandem." See Fig. 6.17, Hamburg 273, 105. See also Hamburg 274, 43: "[t]angens circuli. Hoc equidem patet, quia circulus potest considerari, ut fit ab Archimede aliisq[ue] mathematicis, tanquam poligonum infinitorum laterum, unde lapis in A haerebit in uno latere istius poligoni, et si a funda dimittatur per illud {latus} perget moveri, {latus} illud autem productum ut manifestum est, est tangens circuli, q[uae] demonstratio non {...} omnibus cuiusq[ue] generis curvis potest applicari, cum ea eodem modo possum {...} constare ex infinitis tangentibus pro curvarum varietate varie dispositis." In the copy of the *dictata* extant in Pretoria, consistent annotations are added in an unnumbered page between pages 34 and 35.

Fig. 6.17 Representation of tangential lines. (Source: Hamburg 273, 105. Courtesy Staats-und Universitätsbibliothek Hamburg Carl von Ossietzky, Cod. Philos. 273)

are aimed at criticizing Huet's idea that Descartes did not account properly for the proportion in the quantities of the three elements, especially the particles of the first element, resulting from the scratching and polishing of those of the second element. Against this criticism, De Volder points out, following Descartes, that the ways in which the parts of matter are formed can be grasped only by experience.[127] Notably, the first of *De tribus Cartesii elementis* was defended by Gysbertus Casembroot, who would then author and defend a *Disputatio philosophica de mundi systemate* under De Volder in 1694, and graduate with the *Disputatio de aestu marino* in 1696. Casembroot's *De mundi systemate* is aimed at defending the Copernican model of the cosmos, mostly on the basis of the principle of economy, namely by showing, in thesis 11, that it saves all the astronomical observation.[128] Descartes is overtly used to support the idea that attributing rest to Earth is absurd – following *Principia*

[127] "Si hoc incrementum qualibet alia proportione procedat, potest autem qualibet, neque enim ulla ratio est, quae hanc potius stabiliat, quam illam, iam quaelibet potest esse Solis magnitudo, et concidit omnis vis mathematicae huius, si dis placet, demonstrationis, Certe cum nihil assumat Cartesius nec assumere debeat, quam ex angularibus corpusculis, aequalibus et, si ita velis, inter se similibus exsculpi globulos, longe plurimis in corpusculis maximos, in pauculis vero reliquorum si spectemus multitudinem, aliquantulum minores, hoc enim tantum requiritur, ut plura fiant ramenta, quam ut globulorum intervallulis replendis sufficiant, nec quicquam cogatur determinare de huius incrementi quantitate, imo nec ratio sinat determinare, quamtum hoc sit incrementum, cum per data possit esse quodlibet; multo rectius ratiocinatus foret Vir. Ill. si ex hisce conclusisset, quamlibet posse sequi ex principiis a Cartesio positis Solis magnitudinem; nec posse certam determinari ulla alia ratione quam per experientiam," De Volder 1695, *De tribus Cartesii elementis*, thesis 13.

[128] The author lists 9 astronomical *difficultates* raised by the attribution of rest to Earth, and resolves them by considering Earth as a planet. In this, the disputation is kindred to a shorter disputation presided over by De Volder in 1684, authored by Hieronymus Simons van Alphen, *Exercitatio philosophica utrum Sol, an vero Tellus, in planetarum numerum referenda sit, inquirens*, in which, as in the 1694 disputation, the Tychonic system is criticized as presupposing an enormous speed in the stars (thesis 9), and it is supposed that planets are carried in a vortex (thesis 12).

III.38, where Descartes criticizes Tycho's model –,[129] and the idea that planets are carried by fluids.[130]

So far, in these disputations, as well as in both series of his *dictata*, De Volder does not show appreciable divergences with respect to Descartes's own cosmology. Nor does he do so as to the theory of weight – at least until 1688. In commenting upon *Principia* IV.22, in both series of the *dictata*, De Volder sticks to Descartes's explanations: subtle matter recedes from the Earth as Earth reflects it; this effect, in the Hamburg 273 series, is labelled as combined with that of the centrifugal motion of subtle matter, so that De Volder combines the models of *Principia* and *Le monde*.[131] Also, in the *corollaria* to his disputations *De aëris gravitate*, a model of gravity based on the centrifugal force of celestial matter is overtly made use

[129] "Atque ita iuxta ipsorum hypothesin ratio, cur motus sit Terrae tribuendus, est certa et evidens, ratio vero, cur illum coelo tribuant et Terrae quietem, est incerta, et a sola illorum imaginatione efficta: Phil. pr. parte 3 § 38," De Volder and Casembroot 1694, thesis 11. Cf. Descartes's text: "[t]hus, according to their hypothesis, the reason for which we ought to attribute motion to the heaven and immobility to the Earth is uncertain and depends entirely on their imagination; while, on the other hand, the reason for which they should say that the Earth moves is obvious and certain," Descartes 1982, 102.

[130] "Obi. Tellurem a materia coeli ambiente deferri non posse, cum secundum ipsam Cartesii hypothesin tardius ipso fluido moveatur, adeoque fore, ut tractu temporis ei obtemperet, ac eadem cum celeritate feratur. R. Utinam adduceretur locus, in quo ita ratiocinatur Cartesius, sic enim melius pateret, an vere an falso haec illi adscribantur: memini quidem me alicubi legisse, illud fluidum secundum partes celerius ipsa Tellure, aliove planeta moveri, id est, nunc has nunc illas fluidi particulas a Terra recedere, et ad ipsam rursus accedere; sed tamen illius fluidi motum, quo simul cum eo ferri, accurate aequare," De Volder and Casembroot 1694, thesis 13. See Descartes's *Principia*, III.28: "[t]hat the Earth, properly speaking, is not moved, nor are any of the Planets; although they are carried along by the heaven. And it is important to remember here what was said earlier concerning the nature of movement; i.e., that (if we are speaking properly and in accordance with the truth of the matter) it is only the transference of a body from the vicinity of those bodies which are immediately contiguous to it, and considered to be at rest, into the vicinity of others. However, in common usage, all action by which any body travels from one place to another is often also called movement; and in this sense of the term it can be said that the same thing is simultaneously moved and not moved, according to the way we diversely determine its location. From this it follows that no movement, in the strict sense, is found in the Earth or even in the other Planets; because they are not transported from the vicinity of the parts of the heaven immediately contiguous to them, inasmuch as we consider these parts of the heaven to be at rest," Descartes 1982, 94.

[131] "Terra non fertur suo motu, defert[ur] vero a materia coelesti eius poros pervadente, si itaq[ue] co[rpor]is {motus} telluri et materiae coelesti nil efficiet in tellure, et ratione {suiis} nulla in eo fiet mutatio, a materia coelesti, manet haec ratione eius in eodem loco, praeter hunc {[n]a[m] motum alium} adest materiae subtili per quam a tellure recedit, in tellurem ergo eius motum impedientem incidens, ab ea recedit, et erit levis. Sicq[ue] apparet q[uo]d ante in o[mni]bus vorticibus levior habere diximus, o[mn]ia sc. corpora recedere conari a centro, circa quod moventur, c[um]q[que] non possint o[mn]ia, subtiliora saltem, et celerius mota, sicq[ue] sola materia coelestis a tellure recedit, non a[utem] alia corpora, quia sic daretur vacuum, aliud[que] in eius locum succedere debet, q[uo]d vocatur grave," Hamburg 273, 232. Cf. the other series: "[q]uantum possunt recedunt. Nam cum in terram incidant, et ab ea in determinatione sua mutentur, necesse est, ut reflectantur, aeq[ue] ac pila, q[uae] incidit in parietem, non potest autem reflecti, {quin} a terra recedat, ut evidens est," Hamburg 274, 99.

of.[132] Eventually, in the 1685 disputation presided over by De Volder, but authored by the *respondens*, Henricus van Bronchorst, *De vera gravitatis causa* (dedicated to Hudde), a Cartesian model based on the idea of an equilibrium between the centrifugal tendencies of boules of rotating matter is duly expounded, and where – as Descartes did – a comparison with the condition of equilibrium in water is given. This is an equilibrium which is due to the internal balancing of different volumes of water which do not weigh on themselves, namely in which there is not an increasing pressure within water, as "identical things cannot act [on each other]."[133] This was, *ça va sans dire*, the same model at work behind De Volder's hydrostatics – at least as given in his *dictata*.

6.2.2.3.2 De Volder's Huygenian Theory of Weight (And Its Relation to the Idea of Cohesion)

From 1688, however, different ideas were taken into account by De Volder in regard to his cosmology and theory of weight. The first and foremost example of this change is the already extensively discussed disputation *De vi corporum elastica*. As far as cosmology and the theory of weight are concerned, in this disputation two Huygenian ideas are at work: (1) the idea that gravity is determined by matter circularly moving in every direction (and not just in one direction); (2) the idea that orbital stability, and by extension the phenomena of gravity itself is determined by

[132] See the *corollaria respondentis* (i.e. of Jacob Copper, who would translate Descartes's *De homine* into Dutch (as *De verhandeling van den mensch en de makinge van de vrugt in 's moeders lichaam*, 1682)) appended to the third disputation, and the *corollaria* (whose author is not specified) of the fourth one: "[...] quia in omni motu est circulus, alia corpora minoris motus capacia, deprimuntur versus centrum materia caelesti recedente," De Volder 1676–1678, disputation 3, corollary 11; "gravitas corporum dependet a minori vi qua pollent a centro recedendi", De Volder 1676–1678, disputation 4, corollary 7. The latter corollary is compatible both with *Le monde*'s and *Principia*'s accounts. In the main text of disputation 3, De Volder is uncommitted on the gravity of subtle matter: "[...] materia subtiliori ipso aëre multo minus gravi, (si modo ullius gravitatis sit particeps, quod hic non determino)," De Volder 1676–1678, disputation 3, thesis 4.

[133] "X. Quam rem si paulo inspiciamus accuratius, reperturi sumus corpora, quae sua natura nec gravia nec levia sunt, suam vel gravitatem vel levitatem debere causae externae; imo magis proprie et simplicius dici corpora terrestria omnia esse levia et ea quae vocantur gravia respectu aliorum esse tantum minus levia et ea quae vocantur levia esse tantum minus gravia: ex rotatione enim Telluris circa axem concludere licet, omnia a centro quantum possent recessura, quod autem non recedant, sed contra versus centrum deprimantur, hoc oriri ex alia causa, quia sc. dantur corpora ad motum aptiora, maioremque propterea vim habentia ad recedendum. XII. Sed ubi omnis materia aequalem a centro recedendi vim habet nullus potest esse gravitatis effectus; aequalia enim in aequalia agere non possunt: quod clarissime exempli vasis aqua referti illustrabitur: in vase enim aqua pleno omnis illa aqua tam quae est in superficie, quam quae est prope fundum aequali vi premit deorsum; si enim illa quae est in superficie maiori vi premeret quam quae est prope fundum, eo ipso descendere et alia in eius locum adscendere deberet. Hinc sequitur, si ponamus omnia corpora aequalem habere vim recedendi, quod nulla eorum nec adscendere, nec descendere ulterius possint, nulla enim est ratio, cur unum magis descendat alterum," De Volder and Van Bronchorst 1685, theses 10–11.

an equilibrium between centripetal and centrifugal forces, rather than between (as in Descartes's model) different centrifugal tendencies only.

In the disputation, the vortex theory of planetary motion is given as a postulate, and as demonstrable only a posteriori, namely by observation.[134] Hence, on the basis of the idea that all movements are circular, De Volder demonstrates that, as from a circle infinite tangents can be drawn, celestial matter does not strive to recede from the centre according only to radiuses, namely in directions drawn from the centre to the circumference, but according to lines which can be drawn from the diameter of the vortex in all directions. From this, De Volder derives the idea that a body is in equilibrium when there is a perfect balance between its effort or force of moving out from the centre, and moving towards the centre. Of course, gravity or weight is still justified in Cartesio-Huygenian terms: centripetal forces are by no means the result of an internal property such as that postulated by Newton. And yet, whereas Descartes explained planetary stability simply as an equilibrium between the tendencies of moving in a straight line, Huygens – and De Volder with him – more overtly take into account the idea of an equilibrium between the forces of moving out and towards the centre of the vortex. Moreover, De Volder claims that since in a circle bodies strive to recede from the centre according to tangential paths, and since infinite tangents can be drawn from a circle, a body in a vortex is not only moved according to the radiuses of the vortex itself, i.e. lines connecting the centre and the circumference, but according to lines proceeding from the diameter of the vortex, and drawn in all the possible directions. So that a body is equally pressed from all directions by subtle matter.[135] This makes sense, however, only if we suppose that the matter, in a vortex, does not rotate only in one direction: in that case, indeed, each section of the circumference would be pressed only by the matter coming from one of the sides of the vortex. Therefore, we need to suppose that the vortical model assumed in the disputation is such as that of Huygens, in which subtle matter rotates in all directions, and displaces bodies towards the centre of the vortex. Indeed, for

[134] "Circa Solem, tanquam centrum, ingens materiae moles, in orbem movetur; similiter et circa Tellurem nostra, quae in vortice illo Solis continetur, alius quidam minor vortex datur, quicum et circa proprium axem in orbem movetur. Quod postulatum, a posteriori saltem, si non a priori, ex motu Lunae circa Tellurem, ut et reliquorum planetarum circa Solem, tanquam ipsorum centra in orbem, demonstrari apodictice potest," Schuyl 1688a, *Postulatum*.

[135] "Posito (per postulat. Praeced.) quod omnia in orbem moveantur, sequitur [...] quod omnia corpora conatum habeant recedendi a centro, secundum tangentem circuli, quare [...] actu, nisi quid impediat, a centro recedent. Et licet quidem impediatur, quod omnis materia simul a centro recedeat, ea tamen corpora, quae ad recedendum plus virium habent longius [...] a centro recedere debent, quam quae habent minus virium. Unde sequitur, quod, licet in principio merum fuerit chaos, tandem nihilominus omnia ad aequilibrium reduci debuerint, ita ut unaquaeque pars eum sibi vindicaverit locum, in quo aequalem tum a centro recedendi, tum ad centrum accedendi conatum habet, quare et in eo [...] haerere debet, ex eoque nec a centro, nec versus centrum excedere potest, prement tamen se invicem [...] cum mutuis obstent conatibus, sed ex diametro aequaliter, quae pressio, cum non tantum fiat secundum lineam rectam e centro eductam, sed versus quascunque alias imaginabiles plagas (circulus enim ex infinitis tangentibus aliter, et aliter continuo determinatis est compositus). Hinc unaquaeque pars istius materiae, quae vortices istos constituit, quaquaversum ex diametro premetur aequaliter," Schuyl 1688a, chapter 1, 2–3 (unnumbered).

the author this theory can explain the phenomenon of the weight of bodies, and their falling – viz. exactly the problem raised by Galileo's law of falling bodies, and solved by Huygens in his 1669 communication and 1690 *Discours*.[136]

As to the means by which De Volder could have heard about Huygens's theory of weight, we may advance the hypothesis that Huygens himself expounded his ideas to him (personally, when Huygens came to Leiden and discussed the contents of Newton's *Principia*), or that he could have read a version of Huygens's 1669 communication before 1688. In any case, De Volder showed he knew his peer's theory well in his referral to the VOC, dated 22 July 1689, as he claims that Huygens "supposes that the cause of the gravity of bodies consists in a very subtle matter of one and the same nature with all other bodies, which moved very rapidly around the centre of the Earth in all directions."[137] Indeed, De Volder received a copy of Huygens's *Discours* on 16 May 1689, in order for him to better examine Huygens's report, as testified to by Huygens's correspondence with Hudde, and by Huygens's notes (*Adversaria*).[138]

To sum up, this disputation, as abundantly seen in the previous sections, represents an extreme point in De Volder's elaboration of advanced ideas in natural philosophy. A further confirmation of this is that, as he had provided this Huygenian theory of movement of subtle matter, De Volder uses it to account for the cohesion of hard bodies. Besides accounting for the fluidity of parts of matter (as they are continuously pressed by subtle matter from any direction, as in Descartes's model in *Principia* II.56) – and thus for all "hydrostatical paradoxes," the movement of subtle matter ideally replaces what according to his *dictata* and *Experimenta* was performed by particles of air, namely constantly pressing on hard bodies – and hard particles – making them cohere.[139] Therefore, for De Volder subtle matter exerts a

136 "Ex eodem etiam ratiocinio sponte sua se prodit causa, tum gravitatis, et descensus corporum gravium," Schuyl 1688a, chapter 1, 3 (unnumbered).

137 "De Hr. Huijgens onderstelt dat de oorsaeck van de swaarte der lichamen bestaet in een seer subtile materie van een en selfde natuyr met alle andere lichamen, dewelcke rondom het centrum van de aarde naer alle kanten seer snel bewogen wort," De Volder to the Directors of the VOC (letter 2547), 22 July 1689, in Huygens 1888–1950, volume 9, 339–340.

138 According to Huygens's *Adversaria*, a copy was sent on 17 February: "Mr. Huet Evesque de Soissons, Mr. Cuper St. Gen. M. Papin prof.r a Marpurg. Dierkens. de Volder. Fullenius correctum. 17 feb.," Huygens 1888–1950, volume 9, 379. However, it seems that De Volder got his copy only on 16 May: see the letter of Huygens to Hudde of 25 May 1689 (letter 2539): "[a]en de Heer Hudde genotificeert mijn reys naer Engelandt. en dat het tractaet de la Cause de la Pesanteur aen de Volder behandight hebben," Huygens 1888–1950, volume 9, 319; cf. Huygens's *Adversaria*: "16 Maj. aan Burchardus de Volder Prof. matheseos tot Leyden mijn Discours de la cause de la Pesanteur gebracht om te examineeren 't geen daer uyt geallegueert werdt in mijn Bericht aen de Bewindthebbers van de O. Indische Compagnie, aengaende de Vindingh der Lengden door mijn Horologien, die aen de Caep de B. Esperance geweest zijn. Welck bericht bij Haer Ed. aen gemelte de Volder in handen gestelt is om het zelve te examineeren," Huygens 1888–1950, volume 9, 319.

139 "Unde non tantum patet, cur fluidorum partes sine ulla sensibili resistentia quaquaversum moveri, et ab invicem separari possint, sed et, cur, simul ac ab una parte fortius prematur, quam ab altera, eo statim cedant, quaquaversum illud sit, sive sursum, sive deorsum sive versus latera; imo ex unico hoc fundamento omnes fluidorum proprietates, omniaque hydrostaticorum paradoxa facillimo negotio explicari possunt. Ex eodem etiam ratiocinio sponte sua se prodit causa [...]

pressure on 'gross' matter in two ways: (1) by permeating it and, striving to recede from the centre of the vortex, extruding it downwards and making it 'heavy'. In this meaning, subtle matter is always lighter than the matter that it presses. (2) By pressing it like any other heavy fluid, like water and air. In this sense, the difference in specific weight between the subtle matter and the 'gross' bodies is irrelevant – provided that it presses on them from every direction. Subtle matter, thus far, presses on a gross body like a column of air presses, for instance, on land: i.e. subtle matter, too, is heavy. In principle, it should be made heavy, and pressed by further, lighter subtle matter, in an infinite regress: a regress which is indeed overtly considered in the disputation as to the cause of motion, which is the action of subtle matter itself, whose particles are moved by subtler and subtler ones.[140]

It is unclear when De Volder began to assume that subtle matter, too, exerts pressure as air does, making other bodies hard and cohering. The same solution, actually, had been adopted by Malebranche as early as in 1675, namely in book 6 of his *Recherche de la vérité* (published with books 4 and 5, while books 1–3 were published in 1674). Malebranche assumed this kind of explanation as he – like De Volder – labelled Descartes's theory of cohesion as determined by rest, as contradicting a number of experiences.[141] This explanation of cohesion would then be assumed by a foremost student of De Volder, namely Jakob Bernoulli, who in his *De gravitate aetheris* explained the cohesion of all bodies by means of the action both

duritiei, et cohaesionis, quae profecto nulla alia est, quam sublatum hocce aequilibrium, ita ut corpus, vel partes ipsius ex diametro premantur inaequaliter. Premuntur enim omnia, ut ex circulari omnium corporum motu iam demonstravi. Illa autem pressio vel est ex diametro aequalis, vel inaequalis. Si aequalis, quiescet quidem corpus illud [...] sed minima tamen vi moveri potest. Sin vero inaequalis, movebitur [...] nisi a parte, qua minus premitur maior sit resistentia, quam est ab altera vis pressionis, quo in casu quiescet non tantum, sed nec moveri poterit [...] nisi a causa, quae pressionem, vel resistentiam illam superare potest," Schuyl 1688a, chapter 1, 2–3 (unnumbered).

[140] "Quin imo cum haec demonstratio sit universalis, eaque proprietas, quod sc. unaquaeque ipsorum pars minima vi quaquaversum moveri queat, omnibus fluidis sit essentialis, sequitur omnium fluidorum partes, quaquaversum ex diametro premi aequaliter. Quod cum verum sit de quacunque fluidorum parte, partesque ipsorum singulae seorsim omnem effugiant sensum, in minutissimas partes fluida ut sint divisa oportet, hoc est, ut singulae separatim ab invicem, et diversis inter se agitentur motibus (nihil enim aliud per divisionem in materia intelligo) cumque nihilo moveatur a se ipso [...] debent hae partes ab alia quadam materia fluida moveri per ipsarum intervalla fluenti, quae rursus ab alia, et alia, in infinitum usque," Schuyl 1688a, chapter 1, 3 (unnumbered).

[141] "Or comme il y a toujours beaucoup de parties de cette matière invisible qui passe par les pores des corps durs, elles ne les rendent pas seulement durs comme nous venons d'expliquer, mais de plus elles sont causes qu'il y en a quelques-uns qui font ressort et se redressent, d'autres qui demeurent ployés, d'autres qui sont fluides et liquides; et enfin, elles sont cause non-seulement de la force que les parties des corps durs ont pour demeurer les unes auprès des autres. Mais aussi de celle que les parties des corps fluides ont de s'en séparer, c'est-à-dire que c'est elle qui rend quelques corps durs, et quelques autres fluides," Malebranche 1842, 598. De Volder owned Malebranche's *Recherche* in the Paris edition in 12° (1678–1679), whose third volume (in which book 6 is contained) appeared in 1679: see *Bibliotheca Volderina*, 51. On this ground, Malebranche rejected the validity of Descartes's rules of collision involving bodies at rest: for a discussion, see Nadler 2000; Schmaltz 2015.

of air and of subtle matter,[142] once again refuting Descartes's explanation as based on the idea of rest.[143] Notably, Bernoulli mentions that he was able to read an explanation of cohesion in terms of air pressure in Boyle's *History of Fluidity and Firmnesse* (1661) only after having finished his *Dissertatio* (whose dedication is dated "Leiden, 11 July 1682" – viz. before his journey to England).[144] Second, in the

[142] "Postquam itaque satis, ut opinor, constiterit, cohaesionem durorum corporum nulli alii causae, quam compressioni alicuius corporis externi deberi, non multum porro illi determinando insudabimus, cum praeter aërem nullum detur, quod corpus durum immediate tangat et ambiat," Bernoulli 1683, 51. As to air, Bernoulli draws his conclusion from the example of the cohering marbles, explained, according to him, as an effect of the pressure of the lateral columns of atmosphere – i.e. in an Archimedean manner, as in De Volder's and in Boyle's explanations: "[c]ui assertioni stabiliendae opportune incidit celebre illud experimentum de duobus marmoribus politis et laevigatis, quae sibi iuncta, ut nullus aër intermediare possit, quovis caemento tenacius cohaerescunt, adeo ut nisi maxima adhibita vi avelli a se mutuo nequeant. Quod phaenomenum, ubi per pressionem vel gravitatem atmosphaerae explicant scriptores hydrostaticorum, hoc volunt: cum duo marmora ita sibi iuncta in altum elevantur, vel ex alto suspenduntur, per hanc elevationem vel suspensionem fit, ut marmor inferius nullum amplius supra se pondus habens, quo deprimatur, a pondere lateralis aëris sursum impelli debeat contra superficiem inferiorem superioris marmoris, atque ita suspensum teneri, quamdiu una cum pondere annexo, si quod annexum fuerit, non praeponderat simili cylindro aërio [...] a marmoribus ad ultimos atmosphaerae limites protenso," Bernoulli 1683, 54–55. Against the idea that the mere pressure of air can explain cohesion, however, Bernoulli notes that very heavy bodies can be appended to chains or nails for years, even if they are much heavier than the column of air which would supposedly sustain them. Accordingly, he supposes that this is due to the action of subtle matter or aether, which being much higher than the columns of atmosphere, can exert enough pressure to keep bodies cohering: "[f]ateor, nos primo ne cogitando quidem assecuturos, quae unquam possit esse causa, quale caementum, qualeve gluten, quod has ingentis ponderis catenas a lapsu sustentet; cum absurdum sit, effectum hunc proficisci posse a tantillo pondere cylindri atmosphaerici toties minori. Sed quoniam ex superioribus clare quoque percepisse nobis persuademus, pertinacem hanc cohaesionem partium duri corporis nulli alii deberi causae, ne quieti quidem ipsi, praeterquam soli pressioni corporis alicuius externi, concludere non dubitamus, omnino necessum esse, ut suspensio et cohaesio partium baculi proficiscatur quidem a pondere corporis alicuius externi, sed a pondere longe maiori, quam est pondus solius atmosphaerae: unde in suspicionem hanc incidemus, non solum aërem crassiorem, sed aetherem ipsum, omnemque materiam subtiliorem, longe supra atmosphaerae limites diffusam, aliqua quoque gravitate praedita esse, quae iuncta cum gravitate atmosphaere effectum producat, quem haec sola producere nequibat," Bernoulli 1683, 129–130.

[143] To this idea, Bernoulli opposes empirical arguments: for instance, he considers the case of a piece of wood, such as a rod, which can be broken by pushing it according to a line perpendicular to its centre, but it cannot be broken by pulling it from its two extremities: "[n]on parum etiam gestio audire rationem, ob quam lignum satis crassum applicato genu vel etiam sola manu [...] facile dirumpi queat, conatu frangendi facto iuxta lineam [...] perpendicularem longitudini baculi [...] cum tamen tenuis bacillus, aut corpus adhuc fragilius nullis humanis viribus diffringi possit, ubi conatus adhibetur in directum [...] tametsi enim obtendi posset, id fieri ob naturae a vacuo abhorrentiam, eo quod facta hoc ultimo casu separatione aër non posset eodem momento a lateribus bacilli ad eius medium irruere, non tamen ita respondebunt vacuistae, nec inter plenistas illi qui quietem cohaesionis causam statuunt," Bernoulli 1683, 41.

[144] See the footnote in Bernoulli 1683, 57: "[c]um hanc dissertationem ad umbilicum fere perduxissem, incidi in Exc. Dn. Boylii Tractatum de historia firmitatis corporum, e quo perspexi, Nob. Authori iam olim suboluisse vim aëris in connectendus duobus corporibus sensibili mole constantibus [...]. Tametsi vero non parvam inde lucem istis afferre potuissem, et nonnulla alio disponere ordine; consultius tamen iudicavi, nihil mediationibus meis addere, easque hic recensere, quo primum naturalissimo ordine sese menti meae obtulerunt."

Monitum ad lectorem opening his *Dissertatio*, Bernoulli points out that he developed his ideas on aether or subtle matter as a factor in the cohesion of bodies independently from Malebranche, as he "did not observe, or at least remember" the argument provided by Malebranche, whose *Recherche* he read only "with a fugitive eye" (notwithstanding the opinion of his biographer, Jakob Battier).[145] In the *Monitum*, Bernoulli specifies that a "friend" with whom he was having a conversation on the topics of the *Dissertatio*, brought his attention to Malebranche's theory.[146] It is unlikely that the person whom Bernoulli labelled as a "friend" was De Volder – in that case, he would probably have been more formal. Yet, the environment in which Bernoulli came to develop his theory of cohesion was undoubtedly shaped by the teachings of De Volder, who had already provided a similar solution to the problem of cohesion, and whose explanations on a number of topics are included (as seen above) in Bernoulli's *Dissertatio* and in his *Methodus ratiocinandi*. In turn, De Volder might have finally accepted the idea that aether is a factor in the cohesion of hard bodies after having heard about Huygens's model of weight, according to which particles of subtle matter turn around the Earth not only in one direction, but surround bodies from every direction. Weight (i.e. the downwards extrusion exerted by the centrifugal force of aether) and cohesion (i.e. the pressure exerted by such matter from every direction) are thus unified by De Volder on a Huygenian basis. It might be the reason, at any rate, of why he made such a theory public (even if concealed under Schuyl's name) precisely in July 1688 (when the disputation took place). In January of the same year, indeed, in Le Clerc's *Bibliotheque universelle* appeared the *Extrait d'un Manuscrit Anglois, intitulé Essai Philosophique, concernant l'entendement*. This is better known as the *Abrégé* of Locke's *Essay Concerning Human Understanding* (at that time unpublished), by means of which its contents circulated widely in Europe. In his *Abrégé*, Locke criticized Bernoulli's theory of cohesion by pointing out, first, that the use of the example of the marbles does not take into account the fact that it is easier to separate the marbles if they are disposed horizontally. For Locke, this disproves that they are kept joined together by the pressure of the air, which acts in the same way in all directions. Second, that Bernoulli's solution implies an infinite regress, as in order to account for the cohesion of particles of aether one needs to postulate the exis-

[145] See Sect. 5.6.3.3, *De Volder's assumption of Boyle's law in the works of his students*.

[146] "Cum bona pars huius dissertationis sub prelo prodiisset, ferebat occasio, ut inter confabulandum cum amico, varii praecipue de praesentibus, ut fit, studiis miscerentur sermones, interque alia mentio incideret causae firmitudinis corporum durorum, ubi monuit amicus, eam a Malebrancio compressioni aetheris ambientis ascribi. Ego illum per iocum haec dixisse, postquam in adversaria mea, vel impressa dissertationis folia forte casu quodam incidisset, suspicatus, sesquiannus est, regessi, ex quo auctoris istius Scrutinium veritatis (Recherche de la verité) fugitivo quidem, fateor, oculo perlustravi; sed non observavi, saltem non memini, illum circa cohaesionem partium duri corporis ultra Cartesii quietem penetrasse. Amicus vero asseverare dicta sua, et ne dubitem, commodare librum, ac monstrare locum, qui fidem dictis faceret. Quo sane perlecto mirabar, non tam quod dictus auctor in cohaesione partium duri corporis iam ante me pressionem aetheris repererit, sed praecipue quod in cognitionem huius veritatis eodem filo Ariadnaeo deductus, eique comprobandae iisdem rationibus, iisdemque adeo exemplis mecum usus sit; uti e Scrutinii veritatis et Dissertationis nostrae collatione patere poterit," Bernoulli 1683, *Monitum*, 1–2 (unnumbered).

tence of further particles.[147] In Schuyl's disputation the idea that the explanation of cohesion, as well as that of the source of any movement is subtler and subtler matter, in an infinite regress, finds a place. Indeed, leaning on the idea that nothing can be a self-mover, the author also explicitly commits himself to the idea that each part of matter is moved by a subtler one, *in infinitum*. De Volder might have read Locke's *Abrégé*,[148] his 1688 disputation being thus the occasion of a discussion, if not of a defence of Bernoulli's positions on the topic.

6.2.2.3.3 De Volder's Teaching of Newton's Ideas at Leiden

If the 1688 disputation represents the most advanced product of De Volder's teachings at Leiden, the subsequent disputations presided over by De Volder touching upon cosmology and theory of weight do not include the discussion of as much

[147] "M. Bernouli qui a tâché de rendre raison de la cohésion des particules de tous les corps, par la pression de l'Ether, a omis deux choses de grande importance. 1. Il n'a point consideré que quelque grande que puisse être la pression d'un ambiant fluide, s'il n'ya autre chose pour tenir jointes les particules des corps, encore qu'on ne les puisse pas éloigner l'une de l'autre perpendiculairement, néanmoins on peut démontrer qu'on peut pousser l'une de dessus l'autre, aussi aisément que s'il n'y avoit point de semblable pression. L'experience de deux marbres polis posez l'un sur l'autre, que la pression de l'Atmosphere tient en cet état, fait voir à l'eouil ce que je veux dire, puis qu'on les peut séparer fort aisément, en les poussant de côté, au lieu qu'ils ne le peuvent être perpendiculairement. 2. Il n'a aucun égard aux particules de l'Ether, qui étant aussi bien des corps formez d'autres particules, doivent avoir quelque chose qui les tienne uniës, ce qui ne peut pas venir d'elles mêmes; car il est aussi difficile de concevoir comment les parties du moindre Atome de matiere demeurent uniës les unes aux autres, que celles des plus grosses masses," Locke 1688b, 76–77. Notably, experiments with marbles disposed vertically and horizontally were performed and described by Senguerd in chapter 19 his *Rationis atque experientiae connubium* (1715), where he noted that they can sustain the same weight in both cases (see Sect. 4.1.3, *De Volder's ideas on cohesion and divisibility*) and criticized the idea of an infinite regress entailed by the recourse to the pressure of subtle matter as a factor of cohesion. Rather, he claimed that circular thrust of corpuscles can account for the cohesion of discrete bodies such as the cylinders of marbles: "[n]eque iuvat pressione aëris ac materiae caelestis et aetherea combinare: quia hoc posito pressionis progressus in infinitum admittendus foret. Etiam, si immensa aetheris, cohaesionem procurantis, pressio supponeretur, nullius corporis soliditas exaequaret potentiam huius pressionis. Porro, cum subtilissimi corporum pori caelestis materiae transfluxum sistere nequeant, quantum particulae solidum constituentes earum pressione introrsus et versus invicem adigerentur, indeque consolidantur, aequali impetu a particulis transfluentibus dissociarentur, ac soliditatis fieret eversio. Ne etiam quaeram unde extremorum corpusculorum, ut et ipsarum caelestium particularum soliditatis ratio petenda foret, quave sit premendi vis, qua et ipsae ex subtilioribus particulis coaluere. Rectius iudico, ut quorumvis corpusculorum, et particularum [...] non cohaerentium concretio, mutua agglutinatio, et in unum corpus solidum, aut durum conversio, fluidorumque in summe dura transmutatio, unice exoritur ex earundem approximatione, superficierum adsimilatione, congruentia, et immediata contiguitate. Cuius causa aliorum corpusculorum, in locum illorum quae separanda veniunt, successio, ac circularis propulsio, in quavis corporum translatione requisita, difficilior redditur, vel etiam impeditur," Senguerd 1715, 174–175. In his *Philosophia naturalis*, Senguerd explained the cohesion of cylinders by the circumpulsion of air, and explained the different degrees of cohesion of cylinders of different materials as a consequence of their porosity: see Senguerd 1681, 45–46; discussed in Sect. 5.6.3.2, *The background of De Volder's theories*.

[148] Le Clerc's *Bibliotheque* is present in the *Bibliotheca Volderina*, 93.

advanced natural-philosophical positions. Amongst Serrurier's *Disputatio inauguralis de gravitate aëris* (1690), Casembroot's *Disputatio philosophica de mundi systemate* (1694) and *Disputatio inauguralis de aestu marino* (1696), Van Bashuysen's *De Telluris motu* (1698), and Westhovius's *De fluminibus*, only Casembroot's *De aestu marino* includes a short reference to Newton, which is however worthy of a separate analysis. Indeed, it constitutes the first mention of Newton in a public disputation at Leiden, and even within the context of a Cartesian explanation of tides, it still reveals some connection with Huygens's positions on Newton's theory of gravity.

Together with Hermann Lufneu, Bernoulli and Schuyl – not to mention more famous students of De Volder, such as Bernard de Mandeville, Hermann Boerhaave and Bernard Nieuwentijt – Casembroot can be counted amongst De Volder's more 'advanced' students. In fact, he was somehow a specialist on cosmological topics, as he acted as *respondens* for De Volder's first disputation *De tribus Cartesii elementi* (in the series of disputations against Huet), and as author and *respondens* of the *De mundi systemate* (1694), before his inaugural disputation *De aestu marino*.[149] This disputation presents a theory of tidal ebbs overtly based on the cosmology of Descartes,[150] so that – repeating the words of Köleséri's *Disputatio inauguralis de systemate mundi* (1681) – in the *Annexa* appended to the disputation it is stated that Descartes's system of the world is not an hypothesis but a truth.[151] Moreover, the disputation overtly mentions and presents Descartes's explanation of sea tides,[152] including what we have labelled the principle of Castelli.[153] And yet, we do also find

[149] On 29 February 1696, moreover, Casembroot was the defender (but not the author), of the *Exercitium experimentale decimum de aëris fluiditate*, part of 17 disputations presided by Senguerd between 1687 and 1698. As reconstructed by Wiesenfeldt, some of these *Exercitia* would be the basis of Senguerd's *Inquisitiones experimentales* (second edition, 1699): see Wiesenfeldt 2002, 162. Please note that Senguerd rejected, in his *Philosophia naturalis*, the Copernican system (preferring to it the Tychonic one) so that Casembroot cannot be labelled as defending Senguerd's ideas in his 1694 and 1696 disputations. Moreover, both the disputations are dedicated to De Volder.

[150] "Ut autem haec phaenomena exponamus, supponimus mundi systema ex innumeris constare vorticibus, atque in his unum esse, cuius Sol centrum est, circa quod omnes planetae, ipsaque Terra, et cum Tellure Luna circumvolvitur, uti Cartesius demonstrat in suis Philosophiae principiis, et ego quoque pro virili [...] etiam clare in nostra De mundi systemate dissertatione ostendi," Casembroot 1696, thesis 9.

[151] "Systema mundi, quod a Nobilissimo Cartesio delineatur non pro mera hypothesi, sed pro ipsa rei veritate est habendum," Casembroot 1696, *Annexa physica, annexum* 6. Cf. Sect. 3.2.3.3, *The role of experience in De Volder's natural philosophy.*

[152] "Nobilissimus non modo natalium splendore, sed et eruditionis fama Renatus Descartes Lunae in aquas subiacentes pressionem aestus et recessus marini causam iudicavit, sed ita tamen, ut eas difficultates, quas modo attulimus, plane sustulerit. Incredibili enim mentis sagacitate veras causas indagavit, ex quibus omnia fluxus refluxusque phaenomena explicari possunt. Certe non potui non, caeteris reiectis opinionibus huic uni assensum praebere: quapropter incomparabilis philosophi vestigiis insistens, hanc aestus marini causam explanare et quantum potero, ab omni difficultate liberare conabor," Casembroot 1696, thesis 7.

[153] "Hisce ita positis, notandum est 1. Lunam in vortice Terrae circa ipsam singulis mensibus deferri, atque ipsum vorticem, qui Terram Lunamque continet, annuo spatio a vortice solari circum

a use of Newton in regard to the explanation of the anomalies of tidal ebbs, namely, in De Volder's justification of the elliptical figure of the planetary vortex in which the Moon is carried, and which results in a narrowing of the channel of fluid matter between the Earth and the Moon. De Volder, indeed, claims that as "Kepler supposed, but Newton elegantly demonstrates that this vortex has an elliptical figure."[154] We may assume that Newton's elegant demonstration is nothing other than Newton's demonstration of Kepler's law of areas, which he could have read in 1687, when he received a copy of Newton's *Principia.* In fact, De Volder (or Casembroot) then reverts to a Cartesian explanation of the elliptical figure of the vortex. As seen above, for Descartes (*Principia*, III.153), since the planetary vortex is composed by subtle matter of the same size, whereas the matter in the solar vortex changes its size and speed (and force of motion), the particles of the planetary vortex (see Fig. 6.16, Descartes 1644a, 187) "spread more freely toward A and C than toward B and D. From this it follows that the orbit ABCD is not a perfect circle, but closer to the figure of an ellipse."[155] This also explains the increased speed of the Moon at B and

---

Solem abripi. 2. Tellurem in medio sui vorticis haerere, ubi aequalis ab omni parte existit pressio. Agedum, concipe ita libere in vorticis sui centro haerere Terram, et per eiusdem vorticis extremitates vagari Lunam. Quid fiet? Evidentissimum est, spatium vorticis, quod inter Lunam et Terram est, coarctari, et materiam caelestem velocius per angustiam inter illa duo corpora interceptam fluere, motusque celeritate spatii angustiam compensare; quum id in omni naturali motu constat, ut via coarctata, fluida corpora celerius fluant, et impetu sic satis valido in latera impingant. Ex quo rursus clare intelligitur, Terram, quae tantum in aethere ob aequalem materiae ambientis pressionem subsistit, hoc impulsu cessuram versus oppositum vorticis latus donec via ab utraque parte aeque angusta evadat, et aequalis reddatur pressio: quum igitur in duabus et quidem oppositis Telluris partibus tanta pressio sit, necessum est, ut aër et aqua, quae corpora fluida sunt, et pressioni faciliae [*sic*] cedunt, huic etiam pressioni cedant, et subsidant in ea, cui Luna imminet, et huic opposita Terrae parte, atque e contrario versus latera extollantur," Casembroot 1696, thesis 10.

[154] "Hunc quidem vorticem ellipticam habere figuram Keplerus supposuit, sed Newton eleganter demonstrat," Casembroot 1696, thesis 12. Wiesenfeldt labels this as the earliest evidence of the teaching of Newton's ideas at Leiden: "[...] [d]amit stellt diese Disputation den frühesten Nachweis einer inhaltlichen Auseinandersetzung mit den Arbeiten Newtons im Rahmen philosophischer Lehrveranstaltungen in Leiden dar," Wiesenfeldt 2002, 182–183.

[155] "[...] nec adeo obscura huius rei ratio est, quippe quo propriores sunt Soli coelestes globuli, eo minores; et quo remotiores eo maiores existunt: nam secundum leges circularis motus magis a centro recedet ea materia quae plus habet virium, et ad centrum confluet ea, quae minus habet. Quum autem omnes lunaris vorticis globuli in orbem vertantur, laterales, ubi ad superiores pervenerunt, versabuntur inter globulos paulo maiores, et habentes plus virium ad recedendum a centro: ubi vero tetigerunt inferiores, versabuntur inter globulos minores, et consequenter minori vi pollentes: unde necessarium est ut lunaris vortex non circuli perfecti, sed ellipseos figuram induat. Quae cum ita sint, manifestum est, quamvis Luna sui mole spatium aequale tam in maiori, quam in minori diametro occupet, spatium tamen in minori diametro, ubi Luna plenilunii tempore versatur, magis coarctari, (ubi enim a minori quantitate idem decrementum tollitur, quod a maiori, differentia: ratio in minori maior, quam in maiori erit) ac per consequens materiam caelestem per illas angustias transeuntem maiori celeritate maiores aestus efficere," Casembroot 1696, thesis 12. Cf. Descartes's *Principia*, III.153: "[s]ince the heavenly globules in the space ABCD differ in size and motion from those which are below D near K and from those above B near L, but are similar to those near N and Z; they spread more freely toward A and C than toward B and D. From this it follows that the orbit ABCD is not a perfect circle, but closer to the figure of an ellipse," Descartes 1982, 175. On the composition of the planetary vortex, see *Principia* III.151: "both [i.e. Earth and

D, namely at the perigees.[156] In fact, since at the perigees the Moon does not slow down, but increases its speed, such a theory is consistent with Kepler's law of areas, whose application to the movement of the Moon can thus be justified on a Cartesian basis. Or, in other words, De Volder's use of Newton is consistent with his use of the principle of Castelli, whose application – in turn – to the explanation of movements of celestial bodies in the solar vortex contradicts Kepler's laws.

All this has a sort of Huygenian tone, since, as seen above, in his *Addition* Huygens appreciated Newton's idea of a centripetal force as well as the law of the inverse of the square – albeit rejecting the idea of universal gravitation.[157] This kind of appreciation and refusal of Newton's ideas is in fact repeated by Huygens in his *Cosmotheoros*, of which De Volder took care of the publication and wrote a short preface to it in 1698. At the same time, we do not find the use of Huygens's theory of weight which was given in the disputation of 1688. In what follows, I will explain this apparent contradiction by taking into account De Volder's review of Huygens's report to the VOC (1688), concerning the first trial at sea of his clocks (1686–1687), and his examination of the results of the subsequent trial organized by Huygens (1690–1692).

Moon] are transported by the same heavenly matter, which probably moves at least as rapidly in the vicinity of the Earth as in that of the Moon," Descartes 1982, 174.

[156] See Sect. 6.2.2.1, *Newton's critique of Descartes's vortex theory*. We have seen above that Tinelis de Castelet criticized Descartes's idea that anomalies in tides occur when the Moon is closer to the Earth, namely when it is in the perigee, for the reason that such tides are independent from the distance between the Earth and the Moon, as the perigee (i.e. the closest distance between the Moon and the Earth can occur also when the Moon is at the quadratures – i.e. when it is not on the same line with the Earth and the Sun, and thus the tides are weaker). This kind of objection is taken into account in Casembroot's *De aestu marino*: the answer to the objection is that Moon can be equally far from the Earth both at the syzygies and at the quadratures: yet, from this it does not follow that when it is at the syzygies, it is not in the part of the diameter of the vortex closer to Earth: "[o]bi. Supponitur Lunae apogaeum in quadraturis, perigaeum in coniunctione et oppositione; sequitur ergo, Lunam, quando in coniunctione et oppositione est, versari in minori vorticis diametro, et in maiori, ubi Luna in quadraturis haeret: sed constat ex diligentissimis astronomorum observationibus, Lunam in oppositione et coniunctione saepe tantum a Terra distare, quantum ab ea nonnunquam distat in quadraturis: unde aperte colligitur, in oppositione et coniunctione eandem fore Lunae pressionem, quae est in quadratis Lunae aspectibus. R. Me lubenter dare, Lunam quandoque in coniunctione et oppositione aeque remotam, quam in quadraturis, imo interdum remotiorem esse. Sed nego et pernego, hinc sequi, Lunam eo tempore non esse in minori vorticis diametro: nam quamvis sit Luna in apogaeo tamen in minori sui vorticis diametro quandoque haerebit, ut etiam potest esse in maiori diametro, licet sit in perigaeo: quo in casu eadem quoque pressio in aquis erit, quia, ut hanc Luna efficiat, non opus est, ut ipsa in sui vorticis extremitate versetur, sed tantum in minori diametro, in qua ubi existit, habita proportione, plus de ea sua mole tollit, quam si in maiori versaretur, et consequenter spatium illud, per quod materia subtilis celerius fluit, perangustum reddit," Casembroot 1696, thesis 16.

[157] In the *Annexa*, also, we do find a reference to Huygens's *Systema Saturnium* (1659): "Solum Systema Hugenianum mirandis Saturni phaenomenis satisfacit," Casembroot 1696, *Annexa astronomica, annexum* 2. A reference to Huygens's *Systema Saturnium* is given also in De Volder's *dictata*, commenting upon *Principia* III.146: cf. Hamburg 273, 212.

6.2.2.3.4 De Volder's Criticisms of Huygens's Empirical Evidence

6.2.2.3.4.1 De Volder's Review of Huygens's Report on the First Trial at Sea

Direct evidence of De Volder's confronting Huygens's theory of gravity – besides his 1688 disputation – is the review or referral he wrote for the VOC on Huygens's 1688 report on the trial of the clocks on board the *Alcmaer*. In order to appreciate De Volder's judgment on Huygens's report, it is worth here recalling that by correcting the data of his clocks, Huygens – as reconstructed by Schliesser and Smith – located the final point of the voyage of the *Alcmaer* only 5–6 miles from the actual longitude of Texel, which Huygens calculated as 14° 25′ west with respect to the longitude of the Cape of Good Hope. In his calculations, indeed, the final point of the *Alcmaer* was at 14° 1′ west to the Cape, and after some corrections by De Volder (provided in his report, and added by Huygens to his own copy), even closer to Texel, namely 14° 8′ west of the Cape.[158] According to his report, Huygens had followed in his calculation the longitudes of Paris, the Cape, and Texel known at this time. First of all, Huygens assumed a distance of 18° between the Cape and Paris, discovered by the Jesuit Guy Tachard in 1685 by observing an eclipse of the moons of Jupiter.[159] Indeed, the ascertainment of the different times at which the same phenomenon could be observed at different longitudes was at that time the best means to ascertain the distance between two places. Huygens himself, in his report to the VOC, suggested (in vain) to its directors the financing of a campaign of observations in order to find out the actual distances between places by observing the eclipses of the moons of Jupiter.[160] Second, Huygens calculated the difference in longitudes between Paris and Texel as 3° 35′, by considering the longitudes of European cities given by Giovanni Battista Riccioli in his *Geographia et hydrographia reformata* (1661). In his report, Huygens cites page 378 of this treatise.[161] At this page, however, only the longitudes of Amsterdam and Paris are provided (respectively, 27° 55′ and 24° 30′ east of the island of La Palma at the Canaries,

---

[158] See *supra*, n. 97.

[159] On Tachard, see Vongsuravatana 1992. See *supra*, n. 159.

[160] "En alhoewel dese fauten wel haest door het gebruijck der horologien sullen konnen verbetert werden, soo waer het nochtans seer dienstigh dat men van eenighe voornaeme plaetsen de rechte Lengde ten respect van de Meridiaen van Texel of Amsterdam ondersocht, door observatie aen de omloopers van Jupiter, waer van hier te voren mentie gemaeckt is. Sijnde dese manier van Lengdevindingh van vaste plaetsen onfeilbaer, besonder als men een Eclipsis, of wederverschijningh van de binnenste omlooper, op een selfde tijdt, komt te observeren op beyde de plaetsen daer van de Lengde tusschen beijde gesocht werdt; en daer nae de uren deser observatien met malkander compareert," Huygens to the Directors of the VOC (letter 2519), 24 April 1688, in Huygens 1888–1950, volume 9, 290.

[161] "[...] de Lengde van de Caep ten respect van Parijs was effen van 18 graden ten Oosten dewijl nu Texel 3 graden 35 minuten oostelijcker leght als Parijs, gelijck bij Riccioli in sijn Geographie pag. 378 door neerstigh ondersoeck der observatien van Eclipsen werdt bethoont soo komt de Caep oostelijcker als Texel 14 graden 25 minuten," Huygens to the Directors of the VOC (letter 2519), 24 April 1688, in Huygens 1888–1950, volume 9, 274.

assumed by Riccioli as meridian 0).[162] In fact, it seems that Huygens provided the value of 14° 25′ by considering the more detailed tables following, in Riccioli's *Geographia*, the page he mentioned, and providing the longitude of Texel as 28° 5′ east of La Palma.[163] From this, the distance between Texel and the Cape as 14° 25′ follows, as only the distances between Paris and the Cape (18°), and Texel and Paris (28° 5′ − 24° 30′ = 3° 35′) are considered: namely, 18° − 3° 35′ = 14° 25′. This value is crucial because it serves Huygens to show that, by considering the effect of the centrifugal tendency on the length of the clocks, he calculated the point of arrival of the *Alcmaer* more precisely than by considering Newton's theory of universal gravitation. Moreover, he could claim that his clocks were reliable in the calculation of longitudes at sea – as he did in his report to the VOC.

De Volder, however – besides providing a correction to the data of Huygens's clocks (and placing the final point of the *Alcmaer* even closer to what Huygens supposed to be the real location of Texel, i.e. 14° 25′ west of the Cape) – came to contest the very value of 14° 25′ as the distance between the Cape and Texel. De Volder takes for granted that the distance between Texel and Amsterdam is 0° 17′, namely, Texel is 0° 17′ more to the west than Amsterdam, as reported "in some maps." (According to Riccioli's longitudes, on the other hand, Texel is 0° 10′ east to Amsterdam, as he locates Amsterdam and Texel, respectively, at 27° 55′ and 28° 5′). Moreover, by following, as he declares, Riccioli's tables to set the distance between Paris and Amsterdam, De Volder discovers that the distance between Paris and Amsterdam is 3° 52′ (a value not provided by Huygens). From these two values (in which, please note, Texel lies in between Paris and Amsterdam), De Volder infers the Paris–Texel distance of 3° 35′, as resulting from 3° 52′ – 0° 17′ = 3° 35′, which is the value given by Huygens as the distance Paris–Texel. On this basis, De Volder infers the distance Cape–Texel which Huygens gave on the basis of Riccioli's Tables (14° 25′), by considering the distance between Amsterdam and Texel of 0° 17, and the distance between Amsterdam and the Cape (14° 8′ – obtained by subtracting what he considered the distance Amsterdam–Paris (3° 52′) from Paris-Cape (18° – 3° 52′)): namely, 14° 8′ + 0° 17′ = 14° 25′. In De Volder's words:

> [...] [it] is necessary that one knows the true distance of the longitude between the Cape and Texel; for which it is very convenient that, from the observations of the eclipses of the satellites of Jupiter, the distance of the longitude between the Cape and Paris has been found to be 18 degrees[;] so it remains only to determine the distance between [the longitudes of] Paris and Texel. From the eclipses of the Moon, which Riccioli cites in the tables referred to by Mr. Huygens, about the distance between Paris and Amsterdam one finds that it is 3 degrees [and] 52 minutes, from which follows that the difference between the Cape and Amsterdam, is [of] 14 degrees [and] 8 minutes[.] [I]n which case, one should take [the] 17 minutes, more or less[,] which according to some maps Texel lies more to the west than

[162] Riccioli 1661, 378.

[163] See Riccioli 1661, 423. As signalled by Schliesser and Smith, in his workbook Huygens overtly followed such longitudes in his calculation: see Schliesser and Smith 2000, 34 (n. 77), referring to ms. HUG 1, f. 163 (Leiden University Library).

> Amsterdam, [and] one should find the difference [in longitude] between Texel and the Cape to be 14 degrees [and] 25 minutes[,] such as stated by Mr Huygens.[164]

And yet, if we look at Riccioli's tables referred to by De Volder (see also the next quotation), we see that the difference in longitudes of Paris and Amsterdam is not 3° 52′, but 3° 25′: for the reason that he repeatedly gives a longitude of 24° 30′ east for Paris, and 27° 55′ for Amsterdam.[165] Namely: 27° 55′ – 24° 30′ = 3° 25′. It is surprising that De Volder came to include such a mistake in his report. Hence, he refers to "some maps" as reporting a difference of 0° 17′ between Amsterdam and Texel (placed to the west of Amsterdam). If we look at the map used by Huygens to prepare his report to the VOC, namely a map by Dirck Rembrantsz. van Nierop, whose pupil and nephew Albert Jansz. van Dam – a cartographer himself – was a correspondent of De Volder,[166] we do clearly find that Texel lies slightly to the east of Amsterdam.[167] Other maps to be considered are those of Nicolaes Visscher II,

[164] "[...] is nootsaaekelijck dat men weete het waare verschil der lenghte tusschen de Caap en texel, waertoe seer wel te pas komt, dat door de observatien van de Eclipsen der omloopers van jupiter gevonden is het verschil der lenghte tusschen de Caep en Parijs van 18 graden soodat maer rest te weten het verschil tusschen Parijs en Texel. Uijt de Eclipsen van de maan, die Riccioli op de plaets bij den Hr. Huijgens geciteert, aenhaelt vint men omtrent het verschil tusschen Parijs en Amsterdam van 3 gr. 52 minuten, en dien volgens het verschil tussen de Caep en Amsterdam, 14 graden 8 min. Waerbij soo men nu doet 17 min. Die texel na sommige caarten omtrent westelijcker leyt als Amsterdam, sal men het onderscheyt tusschen Texel en de Caep vinden op 14 gr. 25 minuten gelijck het de Hr. Huijgens stelt," De Volder to the Directors of the VOC (letter 2547), 22 July 1689, in Huygens 1888–1950, volume 9, 341.

[165] See, for instance, Riccioli 1667, 378, 380, 403 and 419.

[166] Only one letter (from Albert Jansz. van Dam to De Volder) has survived, dated 8 November 1677. In his letter, Van Dam asks De Volder a solution to the criticisms raised by Henry More to Descartes's cosmology: "[...] ik omtrent de beginselen der wysbegeerte heb doet dit zoo daanig verminderen dat ik my eyndeling verstoute om met alle beleeftheyd UE dezelve zwarigheden voor te stellen dewelke Hendrik Morus eertyds den zeer doorlugtigen Heer en voortreffelykste wysbegeerige Renatus des Cartes voorgesteld heeft, te weeten; hoe 't komt dat alle de Dwaalsterren niet in een zelfde plat omgevoert worden, te weten, in 't plat van 't Taanront, gelyk ook de Zonne vlakken, of ten minsten in platten die gelykwydig met het Taanront zyn, ja de Maan selfs of in de Evenaar of in een plat dat met de Evenaar gelykwydig is, dewyl zy van geen inwendige kragt bestiert, maar alleenlyk van een inwendige drift gedreven worden. Dit zelve heb ik voor dezen andere wel meer voorgestelt, maar heb nooyt geen oplossinge konne bekomen, en alzo ik tot dezelve zeer begeerig ben, en geen ander reden hebbe konnen bedenken, om daar aan te geraaken, ten zy ik myn toevlugt t' u waarts wende, vriendelyk aan UE verzoekende, my hier op een toeverlatig antwoord te wille laaten toekomen," Rijks 2012, 249. No answer by De Volder is extant.

[167] Huygens included a map of his own in his report, which was based, for the European part, on Van Nierop's *Wassende graade paskaart, vertonende alle de zeekusten van Europa, de geheele Middelandsche Zee, als oock ten Noordwesten, en Noordoosten* (circa 1660), corrected, in its African part in order for it to fit with the distance between Texel and Cape calculated by De Volder: it is discussed in Schliesser and Smith 2000, appendix 1; see also Schliesser 1997 and Schliesser 2000. The map can be found, at high resolution, on the website of the University of Amsterdam: hdl.handle.net/11245/3.1326 (accessed 3 December 2018). De Volder owned Van Nierop's *Nederduytsche astronomia* (1653), *Des Aertrycks beweging en de Sonne stilstant* (1661), *Eenige oefeningen in Godlijcke, wiskonstige en natuerlijcke dingen* (1669): see *Bibliotheca Volderina*, 24–25. According to the *Bibliotheca Volderina*, De Volder owned the apt instruments to measure the longitudes on maps: namely, a "[d]oekmeter om een platte Globus of Landkaart te meten," *Bibliotheca Volderina*, 95.

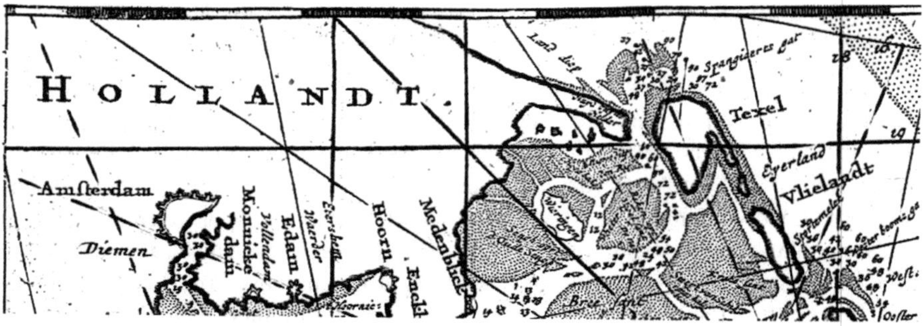

Fig. 6.18 Map of the Netherlands (extract). (Source: De Wit 1688. Universitätsbibliothek Basel, signature: Kartenslg Mappe 247:35, https://www.e-rara.ch/bau_1/misc/content/titleinfo/4026881. Public domain mark 1.0)

mentioned by De Volder in a further letter to Huygens, and listed in the *Bibliotheca Volderina*.[168] Visscher's *Germaniae Inferioris XVII Provinciarum accuratissima tabula* (circa 1680) – more recent than that of Van Nierop – however, still places Texel slightly to the east with respect to Amsterdam.[169] These maps are thus to be excluded. A better candidate is a map by Frederik de Wit, author of a monumental *Atlas maior* (a series of maps published between circa 1662 and 1671) owned by De Volder,[170] and whose map entitled *Daniae, Frisiae, Groningae et Orientalis Frisiae littora* (circa 1688) clearly places Texel to the West with respect to Amsterdam – as a meridian is depicted as crossing Texel (see Fig. 6.18, extract from De Wit 1688).[171]

I suspect, in any case, that De Volder might have used the value of 0° 17′ to compensate for the wrong value of 3° 52′ he had drawn from Riccioli's tables, or that, having found the value Texel–Amsterdam 0° 17′, elsewhere he corrected the actual value of Riccioli (3° 25′) in order for it to fit with Huygens's final result i.e. the distance Cape–Texel 14° 25′.

However, regardless of the actual path De Volder followed in 're-working' Huygens's distance Cape–Texel as 14° 25′ – i.e. as 'composed' by the distance Paris–Cape as 18°, Paris–Amsterdam as 3° 52′, and Texel-Amsterdam as 0° 17′ (or: 18° – 3° 52′ + 0° 17′ = 14° 25′) – what is crucial is to note that De Volder contested the distance Paris–Amsterdam as 3° 52′, which he deemed necessary to obtain

---

[168] Mentioned in a letter to De Volder of 6 April 1693 concerning the second expedition with Huygens's clocks: see *infra*, n. 187. De Volder owned certain "Groote Geographise Caarten van Holland en Zeelend door Nic. Visscher," *Bibliotheca Volderina*, 65

[169] The map can be found, at high resolution, on the website of the Vrije Universiteit van Amsterdam: http://imagebase.ubvu.vu.nl/getobj.php?ppn=330047426 (accessed 3 December 2018).

[170] *Bibliotheca Volderina*, 65; De Volder also owned De Wit's *Nieuw Kaertboeck van de XVII Nederlandse Provinciën* (circa 1667): see *Bibliotheca Volderina*, 65.

[171] The map can be found, at high resolution, on the website of the national Spanish Ministerio de Educación, Cultura y Deporte: http://bvpb.mcu.es/es/consulta/registro.cmd?id=477833 (accessed 3 December 2018). In any case, De Volder was right in placing Texel west to Amsterdam, as the actual longitude of Den Burg (the main town of Texel) is 4° 48′ 31.205″ east, Oudeschild (the VOC's port, in the eastern part of the island) is 4° 50′ 59″, and Amsterdam is 4° 53′ 42.605″.

Huygens's distance Cape–Texel. He does so, in his referral, by considering the distance in longitudes between Paris and Amsterdam reported by Philippe de La Hire (correspondent of Huygens) as 2° 32′ or 33′:

> [...] which[,] however[,] in my opinion, is not so certain, that it could not well differ some minutes or even more, considering that not all the observers have practised the same [degree of] precision to determine the exact time of the eclipse of the Moon in relation to the difference in longitude, as appears from the various observations of the eclipses. Riccioli postulates the difference between Paris and Amsterdam at 4 degrees[,] others such as La Hire, at 2 degrees and 32–33 minutes[;] so that, in order to be sure, it would have been desirable that one had observed this difference of longitude based on the observations of the satellites of Jupiter precisely.[172]

La Hire did not provide this value, but he provided the difference in time between Paris and Amsterdam, as 10 minutes and 10 seconds in his *Tabulae astronomicae* (1687),[173] which, converted from time to longitudes (i.e. as 1 minute of time = 15 minutes of arc in longitude), is 150′ 150″ i.e. 2° 32′ 30″. This measure, actually, displaces Texel – as shown by Schliesser and Smith – from 14° 25′ to 15° 44′ west of the Cape, i.e. of 1° 25′ (on a contemporary scale, circa 47 nautical miles, or circa 87 kilometres). We saw that Huygens's calculations – after De Volder's correction – provided a difference between the 'real' Texel, and the final point of the voyage of the *Alcmaer* as distant 0° 17′, or circa 19 kilometres on a contemporary scale.[174] This value changes if Texel is 'displaced' towards west of 87 kilometres. As a philosophical consequence, Huygens's calculations, based on his theory of gravity and centrifugal force, become as unreliable as their natural-philosophical premises. As underlined by Schliesser and Smith, "La Hire's value would have placed Texel 15 degrees 44 minutes west of the Cape according to de Volder's numbers, and 15 degrees 17 minutes west of the Cape according to the numbers Huygens had taken from Riccioli – either way thoroughly undermining Huygens's argument."[175]

Leaving aside this consideration by De Volder, it is worth noting that even if he did not reject Huygens's corrections to the data reported by the clocks – rather, he improved such a result by placing the final point of the *Alcmaer* even closer to what Huygens supposed to be the real location of Texel – De Volder pointed out his doubts about the very natural-philosophical foundations of Huygens's calculations.

---

[172] "T'welck evenwel mijns aghtens soo seecker niet gaet of soude wel eenige minuten en mischien wel meerder konnen verschillen soo ten respecten dat alle de observateurs niet een ende selfde precies heijt hebben gebruijckt om de nette tijt van de Eclipsis van de maan vast te stellen als ten respecten van het onderscheyt der lenghte, tgeen men uyt diverse observatien der Eclipsen bevint. Riccioli stelt het onderscheijt tussen Parijs en Amsterdam op 4 gr. andere als de la Hire, op 2 gr. en 32 a 33 min. Soo dat het om seecker te gaen wel te wenschen was, dat men dit onderscheyt van lenghte door de observatien van de omlopers van jupiter nauwkeurigh geobserveert hadt," De Volder to the Directors of the VOC (letter 2547), 22 July 1689, in Huygens 1888–1950, volume 9, 341. This follows the previous quotation (see *supra*, n. 164).

[173] La Hire 1687, 4.

[174] See Sect. 6.2.2.2, *Huygens's criticisms of universal gravitation – and the voyage of the* Alcmaer; or, in the first version of the report, 0° 25′ (see *supra*, n. 97). Schliesser and Smith have pointed out how Huygens did not challenge De Volder's result, and that in his report to the VOC he complained about the reliability of the maps at his disposal (see Huygens 1888–1950, volume 9, 287).

[175] Schliesser and Smith 2000, 35.

In particular, he claims that besides the rotation of the Earth other causes could be at work in determining the length of the pendulums, such as the effect of heat on metals. Moreover, in order to test such natural-philosophical foundations (i.e. whether only a circular movement is the cause of weight), De Volder himself recommended the performing of a second trial:

> all this, though it gives very good hope of success in finding the true longitudes by means of clocks corrected in this way, I nevertheless would doubt that, from the result of this one single voyage, one might conclude absolutely, that no other causes could be found in nature that could either impede, or modify the effect of the rotation of the Earth on the weight of bodies, upon which this correction of the clocks is based[;] and also whether the slowing down in the region of the Equator, in respect to more Northern or Southern regions[,] might not depend on some other causes; and whether, amongst other things, the change in temperature – through which even many solid bodies expand, and become longer – might not contribute something to this [effect]. Which, when this happens to the pendulum in the region of the Equator, because of the heat over there, would necessarily make the clocks go slower, though irregularly slower. But which of these, or of other [possible] causes, might at present still be unknown to us, and whether these might have any impact in this matter, cannot be determined except by experience, in which matter a second trial, such as you are intending to undertake with the ships that are due to set sail for India in September, could provide greater elucidation and certainty.[176]

In fact, De Volder did not question Huygens's theory of weight based on the idea of subtle matter: rather, he questioned Huygens's idea that only centrifugal force is to be taken into account in determining the length of the pendulums. In any case, De Volder could not deem Huygens's empirical findings as decisive evidence against Newton's theories.

6.2.2.3.4.2 De Volder's Review of Huygens's Commentary on the Second Trial (1693)

If De Volder showed some doubts as to the actual distance between the final point of the *Alcmaer* and the actual location of Texel in his first report, and expressed also his natural-philosophical doubts, his opinions (expressed in a letter to Huygens) on

---

[176] "Twelck alles, hoewel seer goede hoop geeft van succes om door middel van Horologien soodanigh gecorrigeert de waere lenghtens te bekomen; soo soude ick evenwel twijffelen of men uijt het succes van dese eenige reijs soude mogen absolut concluderen, dat er geen andere oorsaeken in de natuur gevonden worden, die het effect van 't drayen der aarde in de swaarte der lichamen op welck dese correctie der Horologies steunt soude konnen of beletten of veranderen, als oock of niet het langhsamer gaen onder de linie, als onder noorderlijcker of zuijderlijcker plaetsen van eenige andere oorsaecken soude mogen dependeren, en of misschien onder andere hier toe niet wel iets soude konnen contribueren de veranderingh der hitte door dewelcke veele oock harde lichamen uijt geset, en langer gemaeckt worden. Twelk omtrent het pendulum onder de linie gebeurende door de hitte aldaer nootsaekelijck een langsamer dogh irregulier langsamer gangh der Horologien soude maeken. Maer wat van dese ofte andere oorsaeken ons misschien nu nogh onbekent soude konnen sijn, en of die eenige ingressie in dese saek soude konnen hebben is niet als door de ervarentheijt te determineren, sullende een tweede proef, die uwelEd. met de scheepen die tegen September naer jndia staen te gaen van meeningh sijn te nemen, hier van meerder elucidatie en seeckerheyt konnen geven," De Volder to the Directors of the VOC (letter 2547), 22 July 1689, in Huygens 1888–1950, volume 9, 343.

the conclusions reached by Huygens with respect to the second trial he organized in the years 1690–1692 were far harsher.

After the voyage of the *Alcmaer* the VOC agreed to the performing of a second trial,[177] with two pendulums carried by the ship *Brandenburg* en route between Texel–Santiago on the Cape Verde–Cape of Good Hope route, and then by the ship *Spierdijck* from the Cape of Good Hope to Texel (where the ship arrived in October 1692).[178] Yet, this time the trial was unsuccessful, as on the Texel–Santiago route the ship met severe bad weather, and Johannes de Graaf (Huygens's deputy, in charge of maintaining the clocks and registering the data obtained) did not prepare the clocks properly on the *Spierdijck* for its route Cape–Texel, and systematically miscalculated the delays on the Santiago–Cape route (the only part of the voyage on which some data were obtained, from one clock only). Yet, according to his letter of 6 March 1693 to the Directors of the VOC, Huygens found the data collected by De Graaf in his journal (now lost, and on which Huygens prepared a separate commentary)[179] – once corrected – as fitting the "newest maps and globes" (i.e. Visscher's *Africae accurata tabula* (circa 1689) and Willem Blaeu's globes)[180] placing, according him, the Cape 48° east of Santiago of Cape Verde:

> what the Lord of Blocquerij has had the pleasure to communicate to me in his [...] letter of the 16th November of last year, concerning the little success of my clocks in the last test [on the voyage to] the Cape of Good Hope, does not permit me to doubt that Your Excellencies will be well persuaded of the imperfection of this [method of] to determine longitude; and not without reason, because the person who has supervised the trial is himself of this opinion. [...] However, seeing that it was necessary, both for Your Excellencies' satisfaction, as for my own, having heard the oral report of Mr. de Graef, I have examined with attention his Journal, sent to me on 19 November; both of which have given me completely different thoughts concerning the success of this trial. Having thus found that, the performance of the

[177] Huygens briefly discusses De Volder's first report – which he received on 9 September 1689 (see the letter of Huygens to the Directors of the VOC of 9 September 1689 (letter 2546), in Huygens 1888–1950, volume 9, 338.) – in a letter to the Directors of 10 May 1690, in which the performing of a second trial is treated: "[s]edert U WelEd. mij in Sept. des voorleden jaers de Horologien tot de Lengdevindingh gedestineert, beneffens het oordeel van de Hr. Prof. de Volder daer ontrent hebben gelieven te laeten toekomen en met eenen te kennen gegeven VWelEd. intentie van een naeder Preuve deser Inventie te nemen, soo hebbe het gheene noodigh was aen deselve doen repareren, oock met eenen iets tot verbetering daer aen doen veranderen, ende voorts door gedurighe observatie haer gangh geexamineert, om te sien hoe nae deselve over een konde brengen, waer in niet sonder effect gearbeijt hebbende soo twijffele oock niet of men sal sich op de reys noch beter daerop konnen vertrouwen als voor desen," Huygens to the Directors of the VOC, 10 May 1690 (letter 2588), in Huygens 1888–1950, volume 9, 418. In a letter to his brother Constantijn of 9 September 1689, Christiaan Huygens's reported De Volder's favourable opinion: "[i]l semble que sur l'avis du Professeur de Volder, a qui ils ont donnè a examiner le rapport que je leur avois fait du premier essay, ils ont conceu bonne opinion de cette affaire," Christiaan to Constantijn Huygens, 9 September 1689 (letter 2545), in Huygens 1888–1950, volume 9, 337.

[178] On the whole trial, see again Schliesser and Smith 2000.

[179] Namely, Huygens's *Verklaeringh en aenmerckingen op het Journael van Jo. de Graef en 't geen ontrent de Horologien is voorgevallen in de laetste proeve der Lengdevindingh Ao 1690, 1691 en 1692*, in Huygens 1888–1950, volume 18, 643–651.

[180] Huygens owned Visscher's map, and attached it to his letter to De Volder of 24 March 1693 (letter 2798): see Huygens 1888–1950, volume 10, 433–434. See also Schliesser 2000.

> clocks not having improved, because of unavoidable accidents or misunderstanding, or has been wrongly understood because of miscalculation, they have completed the measurement of longitude very well and precisely: namely, on the outbound voyage from the Island of Santiago down to the Cape of Good Hope, when alone the clocks have been able to function: [the result] according perfectly with the newest maps and globes, which set the [difference in] longitude between these two places at about 48 degrees.[181]

Huygens, actually, provided a distance of 48° 14′ in his commentary on De Graaf's journal.[182] In the correction of De Graaf's data, Huygens was helped by De Volder himself, who is thanked in a letter from Huygens of 24 March 1693 (accompanying the journal of De Graaf and the map of Visscher).[183] Notwithstanding this, in his answer to Huygens (6 April 1693) De Volder declares he does not trust the delays calculated by Huygens, for various reasons. First, (1) the daily delay of the clock at Santiago was unknown (probably because it was uncertain if the clock was well calibrated at the initial point of the trial). Second, (2) some data from De Graaf's journal provided a distance between Santiago and the Cape 2 and ½ degrees more than the one which he deemed as being in accordance with the newest maps. As Huygens himself admitted, this might be due either to a defect of the clock, or in the map. So that new maps, based on the calculations based on the observation of eclipses should be provided.[184] For De Volder, in turn, the actual distance between

[181] "Het gheen de Heer van de Blocquerij mij heeft gelieven te doen weten bij sijn [...] schrijvens van den 16 Novembr. des voorleden jaers, aengaende het weynigh succes van mijne Horologien in de laetste proeve naer de Caep de B. Esp.[ce] laet mij niet toe te twijffelen of UEd. G. achb. Sullen seer gepersuadeert sijn van de onvolmaecktheydt deser Lenghdevindingh en niet sonder reden, dewijl de Persoon selfs die het bewint daer van gehadt heeft van die opinie is. [...] Doch siende evenwel dat het noodigh was soo tot UEd. Gr. Achtb.[re] als mijn eyghen satisfactie, soo hebbe nae het mondelingh raport gehoort te hebben van Mr. de Graef, voorts sijn Journael, op den 19 Nov. mij toegesonden, met aendacht geexamineert. 'T welck beyde mij geheel andere gedachten ontrent het succes deser proeve gegeven heeft. Alsoo bevonden hebbe dat daer het effect der horologien bij onvermijdelijcke toevallen os misverstandt niet en is beter geworden, of door misrekeningh verkeert verstaen, sij seer wel en precijs de Lengdemetingh hebben volbracht: Te weten op de uyt reyse van het Eylandt S.[t] Jago af tot aen de Caep de B. Esp.[ce] daer d horologien alleen hebben konnen dienen: accorderende perfect met de nieuwste Caerten en Globen die de Lengde tusschen dese twee plaetsen stellen van ontrent 48 graden," Huygens to the Directors of the VOC, 6 Marc 1693 (letter 2795), in Huygens 1888–1950, volume 10, 423–424. Huygens refers to letters 2773 and 2774.

[182] "Alle deze noodsaeckelijcke verbeteringen in de Tafel van 't Journael gemaeckt en bewesen hebbende, soo siet men dat de Lenghde tusschen 't Eyland S. Jago en de Caep de B. Espe. door de horologien is gevonden van 48 gr. 14 min. 3," Huygens 1888–1950, volume 18, 646.

[183] "Ick send UE hier neffens weder het Journael van M. de Graef met mijne aenmerkingen ende gepointeert de Caert van Africa nae dat ick volgens UE goede advisen verbetert hebbe 't geen ik bij abuis anders gestelt hadde als mijn eygen raisonnement mede bracht," Huygens to De Volder, 24 March 1693 (letter 2798), in Huygens 1888–1950, volume 10, 433. In the letter, Huygens explains to De Volder his own corrections.

[184] "[...] ick voor vast houde, dat het schip, ontrent dese daghen van den 23 en 23[en] Apr. door de vloedt uijt den Oosten seer verre vervoert is geweest. Ick weet wel dat volgens dese gestipte Cours de Lengde tusschen St. Jago en de Caep ontrent 2 ½ gr. meerder soude komen als de 48 gr. die ick te vooren, met de caert accorderende, bevonden hebbe, en ick beken dat het horologie voor soo veel kan gemanqueert hebben, maer het kan oock de faut van de Caert sijn, daer men niet vast op

Santiago and the Cape could be even shorter than the one provided in the maps. Indeed, (3) according to De Volder the very map of Visscher could be wrong. De Volder points out that Huygens, by using the map of Visscher, admitted a distance between the Cape and Tenerife (2° to the east with respect to El Hierro, assumed by De Volder as meridian 0)[185] different from that provided by Tachard, which was probably the most accurate. Tachard located the Cape 40° 30′ east to El Hierro, i.e. 38° 30′ east of Tenerife. On the other hand, by following Visscher's map (placing Santiago 7° west to Tenerife) Huygens deduced that the Cape was 41° 14′ east of Tenerife, as by his clocks he discovered that the distance between Santiago and Cape is 48° 14′.[186] Or, in De Volder's own words:

> I did [indeed] receive your pleasant [letter] of the 24th of March, together with the annexes, but because of an indisposition I could not examine [them] until today, and yesterday. Looking into the matter now, I hardly know what conclusion to formulate. For, being uncertain of the diurnal retardation of the clock at St. Iago, and even uncertain to such a degree, that [in this manner] this might imply a difference of 2½ or even more degrees, how can it be possible to conclude with any certainty about the rest? It is rather probable, as you point out, that a mistake has been made in the observation of the 1st of April, and this may be one such as you propose, but these seem to me merely guesses [...]. It is also true, that the longitude of the Cape[,] as indicated on the map[,] is not so certain, that one could attribute this difference of 2½ degrees with certainty to a defect of the clocks; but, on the other hand, it is also true, that the map can err just as much in positioning the Cape too much to the East than is correct, as too much to the West. If this should be true, and the Cape indeed were positioned less easterly from St. Iago than indicated in the map by Visscher, the mistake of the clocks would be even greater. Moreover, the observation of the French Jesuits travelling to Siam, which is perhaps the most reliable we have as to the longitude of the Cape, appears to favour the clocks not at all. For they determine the longitude of the Cape in relation to the meridian passing through the Isle de Fer at 40½ degrees[;] that is, in relation to the meridian passing through Teneriffa in our maps, a little more than 38 degrees, with which the maps of the Compagnie appear to agree. So that, according to these observations, the clock would have indicated a difference in longitude between St. Iago and the Cape which differs from the true [difference in longitude] of more than 5 degrees; at least, always assuming that St. Iago is correctly positioned at 7 degrees to the west of the meridian of Teneriffa. From all of which I see no way to conclude anything else, but that this trial at best leaves the question more or less in the same state as before, because the observer, in not having observed the diurnal retardation of the clock accurately enough, has made it impossible for us to draw conclusions from this trial with any certainty as to the bad or good functioning of the clocks.[187]

---

gaen kan, soo langh men door onfeilbare observatien, gelijck die aan de omloopers van Jupiter de voorsz. Lengde niet perfect heeft gedetermineert," Huygens to De Volder, 24 March 1693 (letter 2798), in Huygens 1888–1950, volume 10, 434.

[185] Huygens, following Riccioli, assumed La Palma as meridian 0; De Volder assumes El Hierro. Riccioli located El Hierro 44 minutes east to La Palma: Riccioli 1661, 322.

[186] "En om dat volgens de Caert van N. Visscher hier nevensgaende St. Jago 7 graden westelijcker leght als de Pico de Teneriffe, daer het begin der Lengden gestelt werdt, soo komt de lengde van de Caep 41 gr. 14 min. t welck seer net met de voorsz. Caert over een komt," Huygens 1888–1950, volume 18, 646. Cf. *supra*, n. 182.

[187] "Ick hebbe UEdts aangename van den 24ste maart nevens de bijlagen wel ontfangen, maar door indispositie niet konnen examineren als heden, en gisteren. Nu de saack insiende, weet ick bijna niet wat conclusie te formeren. Want onseecker sijnde hoe groot de dagelyxse vertragingh van 't

This is the last extant letter of De Volder on the matter. In his last reply to De Volder on this topic (19 April 1693), Huygens claimed again that the distance between Santiago and the Cape, as shown by his clocks, is circa 48°, in conformity with the new maps – even if he conceded his previous mistake as to the westernmost location of the ship. Once again, he eventually claimed that the solution to these problems could come only from the establishment, by the observation of the eclipses of the satellites of Jupiter, of the actual longitude of different locations:

> I thank you again for your corrections to my calculation, and your further comments, which in this present examination of the undertaken trial of the measurement of longitude are very considerable. Nevertheless, it remains certain that according to the observations of Mr. de Graaf (his miscalculations being corrected, and my instruction simply followed), the longitude between Santiago and the Cape of Good Hope has been measured by the clock as very close [to] 48 degrees, and that this agrees very well with the maps of Visscher and Blaeu. My mistake in the locating of some of the longitudinal positions of the ship does not preclude, as you know, that this conclusion is true, even though – this mistake being corrected by me in the accompanying map – its course now results having come much closer to the coast of Brazil than I had thought. [...] That from this one now could decide that the perfection of the measurement of this longitude has been demonstrated sufficiently, or [even] more precisely than by the previous trial of the year 1687, I do not want to pretend, for not only your remark concerning the observed changeable diurnal retardation of the clock from Santiago, does not allow this, but also the uncertainty of the positioning of the longitudes in the maps would make such decision doubtful, even had the clock functioned as well as possible; and this will always be the case, unless one would find occasion to measure the longitude of two places on the outbound and inbound voyage [in exactly the same spot], or that the difference in longitude is determined by secure observation of the satellites of Jupiter.[188]

Horologie is geweest op St. Jago, en wel sodanigh onseecker, dat het een verschil van 2 ½ of oock wel meer graden op dese wijs soude komen te importeren, hoe kan men met eenige seeckerheijd van de rest concluderen? T'is wel waarschijnelyck, gelijck UEdt it aanmerckt, dat in de observatie van den 1ste April een misslagh is begaan, en misschien is 't oock wel soodanigh een, als UEdt. bybrengt, maar dit schijnen mij ten minsten altijdt het laatste, alleen gissingen. Tis oock waar, dat de lengde van de Caap, die de caart aanwijst soo vast niet is, dat men dese 2 ½ graden verschil, seeckerlyk tot een fout aan de Horologien soude konnen toeschrijven; maar aan de andere kant is 't oock waar, dat de caart soo wel kan missen met de Caap oostelijcker te leggen, als Westelijcker alst behoort. Twelck soo waar mocht sijn, en dat de Caap inderdaat minder ten oosten van St. Jago verscheelde, als de caart van Visser medebrengt sou dit de fout der Horologien noch grooter maacken. Waarbij komt, dat de observatie van de Franse Jesuiten gaande naar Siam, die misschien de seeckerste is, die wij omtrent des Caaps lengde hebben, den Horologien gansch niet schijnt te favoriseren. Want die determineren de lengde van de Caap naar de meridiaan gaande door l'isle de fer op 40 ½ gr. dat is, naar de meridiaan van onse caarten gaande door Tenariffe, wat meer als 38 gr. waarmede de caarten van de Compagnie schijnen overeen te komen. So dat na dese observatien het Horologie een verschil van lengde tusschen St. Jago en de Caap soude hebben aangewesen, 'tgeen van het waare over de 5 gr. soude verschillen, ten minsten altydt incas dat St. Jago, te recht 7 gr. westelijcker gestelt wert als de meridiaan van Tenariffe. Uyt welck alles ick dan niet anders sie te concluderen, als dat dese proeve ten besten genomen de saacke laat genoechsaam in deselfde staat als voorheen, als hebbende de observateur door de daaghelyxse vertragingh vant Horologie niet accuraat genoech geobserveert te hebben, ons buijten postuur gestelt, om met eenige seeckerheijt vant qualyck of wel uijtvallen der Horologien, uijt dese proeve te konnen oordelen," De Volder to Huygens, 6 April 1693 (letter 2800), in Huygens 1888–1950, volume 10, 436–437.

188 "Ick bedanck UE. nochmaels van UE gedaene Correctie in mijn Rekening, en UE verderene aenmerckingen die in dat tegenwoordigh Examen der genomene Proeve van Lengdemetingh seer

In fact, no further communications between the two on this topic is extant. As late as in 1697, De Volder would once again be a referee for the VOC, evaluating another method to calculate longitudes at sea, proposed by a certain Antonius de Saint Julien-Pottier.[189] According to his letter of 9 November 1697,[190] the method was based on the observation of the position of the Moon with respect to the pole. In fact, this method calculating the longitudes would become feasible only in the late eighteenth century, when precise mapping of the movement of the Moon in the sky became available.[191] De Volder, indeed, noted in his referral that the perfect knowledge of the path of the Moon – required by this method – was not available, and that from a ship the precise observation of the position of the Moon is not possible:

> in this consists, in my judgment, the argument of M. Pottier, as to which one should consider that, in all the [various] ways one has ever attempted to find the longitude by observing the Moon, two problems have presented themselves, which have made them fall into abeyance[:] the first is, that the course of the Moon is not known so perfectly, as is necessary for determining longitude. The second, that even if this orbit were known with the greatest accuracy, on board ship, however, one cannot observe so clearly and certainly, or one would have sailed a few minutes, and therefore make great mistakes in establishing the longitude.[192]

---

considerabel sijn. Evenwel soo blijft dit seecker dat volgens de Observatien van M. de Graat (sijn misrekeningen verbetert sijnde, en mijne Instructie simpelijck naegekomen) de Lengde tusschen S. Jago en de Caep de B. Esp.[e] van seer nae 48 gr. door het horologie is afgemeten, en dat dit met de Caerten van Visscher en Blaeuw seer wel overeenkomt Mijn abuijs in 't pointeren van eenighe Lenghdens van 't schip en beletten niet, gelijck UE weet, dat die conclusie waer zij, alhoewel dit abuijs bij mij geredresseert sijnde in de nevens gaende Caert, de cours nu veel naeder aen de Cust van Brasilien komt te vallen dan ick gemeent hadde. [...] Dat men nu hier uijt soude konnen besluyten de perfectie deser Lengde metingh genoegsaem gedemonstreert te sijn, of naerder als door de voorgaende proef van A°. 1687 wil ick niet pretenderen want niet alleen UE remarque ontrent de veranderlycke daghelycksche verachtering van thorologie van S. Jago waergenomen, en laet sulx niet toe, maer oock de onsekerheijdt der stellinge van Lengden in de Caerten soude sulk besluyt twijffelachtigh maecken al hadde het horologie noch soo wel gegaen. en sal altydt soo doen behalven als men gelegentheijdt heeft om de Lengde van 2 selfde plaetsen, op de heen en weerreys te meten, of dat het Lengde-verschil door seeckere observatie aen de omloopers van Jupiter geobserveert zij," Huygens to De Volder, 19 April 1693 (letter 2803), in Huygens 1888–1950, volume 10, 443–444.

[189] He was the author of a *Circuli quadraturae nodum per helicem enodatum* (Utrecht, 1698).

[190] Preserved, like a copy of his referral on Huygens's first report, among the papers of Hudde (The Hague. Algemeen Rijksarchief, signature: 1.10.48 no. 44). It is given as appendix in Mac Lean 1971.

[191] See Howse 1989.

[192] "Hier in bestaat, mijns oordeels, het voorgeven van Monsr Pottier, omtrent het welke te considereren valt, dat in al de manieren, die men ooyt of ooyt heeft bijgebracht om door waarneminge van de maan de lenghtens te vinden, sigh twee swarigheden hebben opgedaan, die die manieren 't onbruyck hebben gemaakt; de eerste is, dat de maansloop so volmaakt niet bekent is, als tot de lenghtevindingh van noden is. De tweede, dat als dese loop selfs al ten accuraatste bekent was, dat men evenwel te scheep so net en seeker niet kan observeren, of men sal eenige minuten komen te seylen en daardoor bij gevolgh groote fijl in de lenghtevindingh begaan," Mac Lean 1971, 159.

After his previous exchange with Huygens, in which he could not completely convince his correspondent, De Volder notes in the letter that he had delayed in providing his report because

> [...] I would have wished, to have heard that Gentleman speak first, before I gave my judgment on his case, so if not to take away all his complaints (which, because of the love inventors usually have for their ideas, seems almost impossible if one does not approve the [their] invention); at least always that [complaint] of not having been heard; and thus also to prevent all further personal contact that he might wish to undertake, if he afterwards should find out that his invention was not approved also by me.[193]

This might be interpreted as a reference to his previous evaluation of Huygens's method of calculating longitudes at sea, indirectly remarking again a distance from Huygens's own positions.

6.2.2.3.4.3 De Volder's Last Disenchantment with Cartesianism (via Huygens)?

To sum up, De Volder had no reason to accept Huygens's criticisms of Newton on the ground of the empirical data provided by the two expeditions at sea. If we look at the disputations on cosmology presided over by De Volder after 1689, we find no further 'Huygenian' trace. Leaving aside the 1696 disputation *De aestu marino*, the other disputations presided over or dedicated to De Volder, i.e. Serrurier's *De gravitate aëris* (1690), Casembroot's *De mundi systemate* (1694), Westhovius's *De fluminibus* (1698), and Van Bashuysen's *De Telluris motu* (1698), show either an unconcern about the cause of gravity (as in the case of Serrurier or Westhovius's disputations) which is a concept anyway touched upon in such disputations,[194] or an explanation based on Descartes's *Principia*, as in the case of Casembroot's disputa-

[193] "[...] ik wel gewenscht had, dien Hr eerst te hebben hooren spreeken, eer ik mijn oordeel over sijn saak gaf, soo om hem te beneemen ist niet alle klachten ('t welck door de liefde die de Inventeurs ordinary tot haar concepten hebben genoegsaam onmogelijk schijnt, als men de inventie niet komt te approberen;) ten minsten altijt die, van niet gehoort te sijn, als ook om hier door voor te komen alle verdere aanloop, die hij misschien sal willen doen, als hij daar naar sal komen te vernemen, dat sijn Inventie door mij mede niet goedgekeurt is," Mac Lean 1971, 157.

[194] In Serrurier's disputation the problem of the cause of weight is just not mentioned. In turn, Westhovius's *De fluminibus* concerns the various conditions of depth, breadth, and direction of water in a river. Amongst the natural-philosophical premises of the demonstrations of 13 propositions on the topic, it is stated that bodies heavier than water sink, and lighter ones float; the cause of their weight, however, is assumed as "already demonstrated." In fact, as noted by the author, this topic exceeds the scope of the disputation: "[n]otiones communes. I. Notum cuilibet est, aquam habere gravitatem, qua ad Terrae centrum prolabi conatur. II. Eandem porro graviorem his, leviorem aliis corporibus esse, item graviora in aqua subsidere, leviora vero ei innatare. Supervacaneum prorsus foret, et a scopo nostro alienum, si causam gravitatis corporum hic demonstrare susciperemus: nobis illam diu demonstratam supponere sufficiat," De Volder and Westhovius 1698, 5. Westhovius was also the *respondens* of one of Senguerd's *Exercitia*, namely the *Exercitium experimentale quintum decimum, quod est de aëris gravitate, secundum* (1698). He would later become professor of eloquence at the Latin School of Gorinchem. No evidences on his graduation at Leiden could be found.

tion.[195] Eventually, in the last disputation on cosmology presided over by De Volder and dedicated to him, i.e. *De Telluris motu* (July 1698), we do find a standard, Cartesian cosmology.[196]

We may thus advance a concluding hypothesis. Namely, that De Volder did not fully accept Huygens's ideas, as he found Huygens's empirical evidence against Newton's theory of universal gravitation unreliable. At the same time, De Volder did not come to accept Newton's idea of universal gravitation, even if he accepted – like Huygens – the law of the inverse of the square. That is, he implicitly accepted Newton's demonstration of the elliptical figure of the vortex. Such an elliptical figure, as well as the idea that a celestial body is faster when it is closer to the body around which it orbits, is anyway justifiable on a Cartesian basis, as far as the case of the planetary vortex is concerned. In this precise case, indeed, Descartes's theory does not contradict Kepler's laws. However, De Volder rather preferred not to touch cosmological topics after 1698, nor did he come to accept Huygens's 'extension' of the model of gravity to the whole solar vortex, which he could read in Huygens's *Cosmotheoros*, which deserves a little analysis.

Huygens's *Cosmotheoros* was published in 1698 with a short preface by De Volder himself, who took care of the last phase of its publication, after the brother of Christiaan, namely Constantijn Huygens Jr. passed away in 1697 and could not complete the publication.[197] The greater part of the *Cosmotheoros* presents *coniecturae* on the plurality of inhabited worlds, based on astronomical evidences concerning the orbits, movements and distances of planets.[198] Yet, in the closing section of the second book, entitled *Every Sun has a vortex round it, very different from*

[195] "Si omnia circa Terram spatia vacua essent, id est quae aliorum corporum motum ne iuvarent, nec impedirent, illud ipsum fore; iam vero omnes Telluris partes a coelesti materia undique comprimi, impedirique, ne dissolvantur: cum enim globuli coelestes per lineas rectas, vel parum a rectis devias a Terra recedant, ac proinde efficiunt, ut aequali vi omnes eius partes ad centrum protrudantur, impediunt, ne versus coelos desiliant," De Volder and Casembroot 1694, thesis 12.

[196] See De Volder and Bashuysen 1698, thesis 2 (providing a Cartesian definition of motion), thesis 9 (defending Descartes's theory of tides), thesis 15 (endorsing Descartes's theory solar spots). Besides Descartes, the author reports the astronomical observations of Galileo and Huygens (viz. on the phases of Venus and on the rings of Saturn: see theses 9 and 15–16).

[197] The very short preface reports no name, however, in his *Éloge* Le Clerc reveals that De Volder took care of the last phase of the publication, and wrote the preface itself (Le Clerc 1709, 389). The editors of Huygens's *Oeuvres*, however, do not mention De Volder in their reconstruction of the genesis of this work: see Huygens 1888–1950, volume 21, 655–657.

[198] On the *Cosmotheoros*, see Ait-Touati 2011, chapter 4; Van der Schoot 2014. Notably, in the preface De Volder notes that the work does not contain mathematical demonstrations, being however based on conjectures grounded on reason: "[d]emonstrationes equidem mathematicas non invenient ubique, neque enim res patitur, sed, quo in his rebus nihil ultra desiderari iure posse videtur, verisimiles et ingeniosas coniecturas. Quae ex coelorum notitia depromi potuerunt, ea hic videbunt ratione demonstrata; quae ex iis non patent, ex coelestium corporum cum Tellure nostra affinitate solerter coniecta," Huygens 1698a, *Lectori*, 2 (unnumbered). On Huygens's use of conjectures, see Elzinga 1971. De Volder, as seen above, was already warned by Huygens that clarity and distinctness could not be deemed as absolute signs of truth: see Sect. 3.2.1, *De Volder on clarity and distinctness*.

those of Cartes,[199] Huygens furthers his previous remarks on a cosmology based on the idea of the vortex, extending his vortical model of gravitation to stars, criticizing further Descartes's theories, and implicitly criticizing Newton. In a few paragraphs, Huygens re-states some of the criticisms on Descartes's vortex theory we have seen above, for instance, (1) the shape of the vortices, hindering the circular movements of each other, or (2) the fact that Descartes's model does not account for the sphericity of the Sun, as his vortices cannot be conceived but as cylindrical, and therefore their central stars.[200]

More importantly, Huygens criticizes Descartes's idea of orbital stability, as he took into account only the centrifugal tendencies of planets, rather than the equilibrium or balancing between such forces and their gravity, as done by Plutarch, Giovanni Alfonso Borelli and Newton – who, once again, is praised as, on the basis of this idea, he had demonstrated the elliptical shape of orbits.[201] Yet, Huygens does not accept Newton's idea of universal gravitation. Rather, by following his previous *Discours de la pésanteur*, he still sticks to a vortical-mechanical model, in which (1) the model of gravitation concerning planets and 'terrestrial' bodies is extended to

---

[199] Huygens 1698b, 157.

[200] For Descartes (*Principia*, III.61), the Sun is spherical because the matter of the vortex strives to recede from it: yet, in his argument – and in his illustration of it (see Fig. 6.19, Descartes 1644a, 102) – Descartes assumes that the particles of the vortex describe circular paths: "from the fact that all the globules which rotate around S in the vortex AEI attempt to move away from the center S, as has already been shown, we can conclude that those situated on the straight line SA all push one another toward A, and that those situated on the straight line SE push one another toward E. and so on; so that if there are not enough of them to occupy all the space between S and the circumference AEI, all the unoccupied space will be in the vicinity of S. And inasmuch as those which rest upon one another (for example, those which are situated on the straight line SE) do not rotate in a body, like a rod, but complete their revolutions in varying lengths of time (as I shall explain later); the space which they leave around S must be spherical," Descartes 1982, 115.

[201] "I am of opinion then that every Sun is surrounded with a Whirl-pool or Vortex of Matter in a very swift Motion; tho not in the least like Cartes's [...] manner of Motion. For Cartes makes his [...] as every one of them to touch all the others round them, in a flat Surface, just as you have seen the Bladders that Boys blow up in Soap-suds do: and would have the whole Vortex to move round the same way. But the Angles of every Vortex will be no small hindrance to such a Motion. Then the whole matter moving round at once, upon the Axis as it were of a Cylinder, did not a little puzzle him in giving Reasons for the Roundness of the Sun: which however they may satisfy some People that do not consider them, really prove nothing of the matter. In this æthereal matter the Planets float, and are carry'd round by its motion: and the thing that keeps them in their own Orbs is, that they themselves, and the matter in which they swim. Equally strive to fly out from the Center of this Motion. Against all which [...] I touch'd upon in my Essay of the Causes of Gravity. Where I gave another account of the Planets not deserting their own Orbs; which is their Gravitation towards the Sun. [...]. Plutarch in his Book of the Moon above mentioned says, that some of the Antients were of opinion, that the reason of the Moon's keeping her Orbit was, that the force of her Circular Motion was exactly equal to her Gravity, the one of which pull'd her to, as much as the Other forc'd her off from the Centre. And in our Age Alphonsus Borellus, who was of this same opinion in the other Planets as well as the Moon [...]: Which Mr. Isaac Newton has more fully explained, with a great deal of pains and subtilty; and how from that cause proceeds the Ellipticity of the Orbs of the Planets, found out by Kepler," Huygens 1698b, 157–158. As to Plutarch, see Furley 1989, chapter 16; Russo 2013, chapter 11. As to Borelli, see Koyré 1973; Hall 1983, chapter 11; Capecchi 2017, chapter 5.

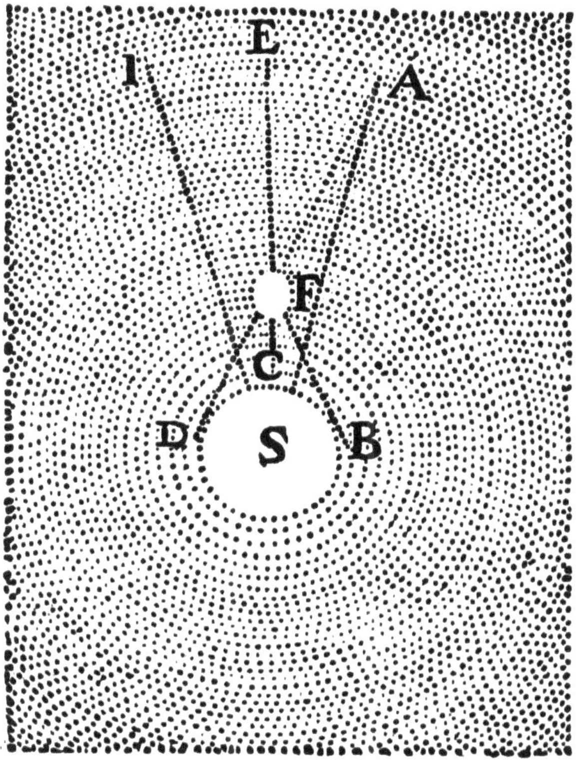

Fig. 6.19 Representation of Descartes's vortex theory of planetary motions, aimed at demonstrating the sphericity of the Sun. (Source: Descartes 1644a, 102. Library of Congress, call number: LC B1860 1644, http://hdl.loc.gov/loc.rbc/Rosenwald.1431.1. Public domain)

the relation between suns and planets, i.e. the whole solar system is conceived as a vortex in which subtle matter rotates in all directions; (2) he supposes that the vortices do not hinder the movement of each other, by distancing them from each other, and supposing the existence of an aether separating them; (3) he does not conceive of comets as 'dead stars' coming from other, very distant vortices, as this plainly contradicts the fact that we can easily see them in their path. Accordingly, he attacks the whole cosmogenesis of Descartes, i.e. the 'dynamical' theory of generation and death of stars and planets expounded in the *Principia*, as badly grounded.[202] This

[202] "According to my Notion of the Gravitation of the Planets to the Sun, the matter of his Vortex must not all move the same way, but after such a manner as to have its parts carry'd different ways on all sides. And yet there is no fear of its being destroy'd by such an irregular motion, because the æther round it, which is at rest, keeps the parts of it from flying out. With the help of such a Vortex as this I have pretended in that Essay to explain the Gravity of Bodies on this Earth, and all the effects of it. And I suppose there may be the same cause as well of the Gravitation of the Planers, and of our Earth among the rest, towards the Sun, as of their Roundness: a thing so very hard to give an account of in Cartes's System. I must differ from him too in the bigness of the Vortices, for I cannot allow them to be so large as he would make them. I would have them dispers'd all about the immense space, like so many little Whirl-pools of Water, that one makes by the stirring of a stick in any large Pond or River, a great way distant from one another. And as their motions do not

model, however, was nothing but an 'extension' of his previous ideas on gravity: notably, he still accepts the law of the inverse of the square, and does not accept Newton's idea of universal gravitation.

Thus far, De Volder could not have been particularly 'moved' by the *Cosmotheoros*, after having already found Huygens's empirical evidence to be unsatisfying – and, as appears from his review of 1696 – his insistence as bothersome. As to the disputations between 1688 and 1698, he sticks to the Cartesian model – and just briefly mentioned Newton's demonstration of the elliptical shape of the vortex. As later noted by Le Clerc in his *Éloge*, De Volder became disenchanted by Descartes's vortex theory through some 'Englishmen': but he rather preferred to resign from his academic duties in 1705, than to develop or accept a new system – such as that of Huygens.

all intermix or communicate with one another; so in my opinion must the Vortices of Stars be plac'd as not to hinder one anothers free Circumrotations. So that we may be secure, and never fear that they will swallow up or destroy one another; for that was a mere fancy of Cartes's; when he was showing how a fix'd Star or Sun might be turn'd into a Planet. And 'tis plain, that when he writ it, he had no thoughts of the immense distance of the Stars from one another; particularly, by this one thing, that he would have a Comet as soon as ever it comes into our Vortex, to be seen by us. Which is as absurd as can be. For how could a Star, which gives us such a vast Light only from the Reflection of the Beams of the Sun, as he himself owns they do; how I say could that be so plainly seen at a distance ten thousand times larger than the Diameter of the Earth's Orbit? He could not but know that all round the Sun there is a vast Extensum; so vast, that in Copernicus's System the magnus Orbis is counted but a point in comparison with it. But indeed all the whole story of Comets and Planets, and the Production of the World, is founded upon such poor and trifling grounds, that I have often wonder'd how an ingenious man could spend all that pains in making such fancies hang together. For my part, I shall be very well contented, and shall count I have done a great matter, if I can but come to any knowlege of the nature of things, as they now are, never troubling my head about their beginning, or how they were made, knowing that to be out of the reach of human Knowlege, or even Conjecture," Huygens 1698b, 158–160.

Appendix: Burchard de Volder's *Experimenta philosophica naturalia* (1676–1677)

[78r] Experimenta philosophica naturalia, auctore M[a]gis[tro] De Valdo Lugd[uni] ann[o] 1676.[1]
1um 12 Martii

His subject was the ponderosity, or rather the compression of the air over us, which he had proved by some other experiments before now. Yet to assert it more authentically, and to give the auditory to understand that it was not the metus vacuii[2] which did make nature to work so violently, and potent in many operations, but rather the compression of the air itself, he used the following demonstration. First he had made two little cylinders of white stone perfectly round, and perfectly plane on the superficies, where they were to touch one [78v] another, and as perfectly equal to each other as he could get them made. Their diameter was <about> \just/ two inches and ⅓d each, and the depth, <mi> or length, might be a little more Next he joined these two cylinders together with a little tallow, and no other artifice, and so close that not the least atom of air could get in betwixt them: and this he said he did least any might calumniat[3] that the superficies were not perfectly plain and consequently some air must be got in betwixt them, for at least, quoth he, those very little

[1] Conventions adopted in the transcription: (1) the text deleted in the manuscript has been put between brackets thus < >; whenever possible, I have provided the deleted text; otherwise, I have used dots instead of the illegible letters; (2) the text in the margin or between the lines is put between the symbols \ /; (3) doubtful text is put between brackets { }; whenever possible, I have provided the text, otherwise, I have used dots instead of the illegible letters; (4) my additions are put between brackets [ ]; (5) the text has been modernized, i.e. forms like "y.e" have been rendered as "the", plurals given in the form "cylinder's" have been rendered as "cylinders", the ligatures have been dissolved, obsolete and archaic spellings and forms of words have been rendered into modern terms, punctuation has been fixed. Latin has been translated in footnotes, whenever necessary, and Latin abbreviations have been completed. Doubtful text has been kept in its original form.

[2] "Fear of vacuum."

[3] "Calumniate."

A. Strazzoni, *Burchard de Volder and the Age of the Scientific Revolution*, Studies in History and Philosophy of Science 51,
https://doi.org/10.1007/978-3-030-19878-7

Fig. 1 Cohering marbles used by De Volder. (Source: De Volder 1676–1677, 78v. Courtesy British Library, ms. Sloane 1292)

holes which might be left by the unevenness of the superficies, are closely staying with the tallow.

3^{ly}. He hung them up (for which purpose they had brass hooks at both ends thus [see Fig. 1, De Volder 1676–1677, 78v])

[79r] Then he put weights to the lowermost end, which was first a huge scale weighing 35 lb with all its {tanr'}, and then into the scale he went adding by degrees more and more weight: to the intent he might experiment how much weight would tore those two cylinders asunder, which were only joined with a little tallow. Has he added to the weight of 285 lb, and then the lower brass hook of the cylinders broke, which had it not happened it would in all probability have sustained much more weight.

This experiment he said did prove the compression of the air, and its ponderosity of whose nature it is as in fluid bodies that it weigheth and compresseth both above us, and beneath us: where by the oppression of the air upon the uppermost cylinder meeting with the oppression of the air, or heaving under the lowermost \{uider fª}/ did cause that reluctancy to a disunion of the cylinders: here being no air drawn out, nor left in, not the least, between [79v] the two cylinders, which might be said to be the cause of their own separation propter metum vacuii[4] which is the common solution of Aristotelians for such reluctances in such cases.[5]

[80r] Experim[en]tum 2um 13 Martii.

His subject was the same viz. the pressure or ponderosity of the air.

1. For this purpose he had refitted again his cylinders with better hooks (the just breadth of said cylinders being 2 digitos cum 3^{a} parte,[6] as he declared again more expressly this day), and joined them together with tallow two days before very close by means of the heat of a lamp, or candle, which joining, after the divulsion of the

[4] "Because of the fear of vacuum."

[5] See Schott 1657, 25–26.

[6] "Two inches and ⅓."

cylinders, appeared to be so thin, or superficial as that are scarce could perceive any tallow upon the superficies of one of the cylinders, which I had in my hand.

2. Then he hang them up, as before, and into the scales he added more, et more weight, till they disunited and the one fell down with the weight which appeared to be very nigh 450 lb.

[80v] 3. He said Boyle could only hang on two such cylinders (being even something broader i.e. of 3 fingers breadth) 80 lb weight, which is not the 4th part of this experiment. Let this he attributed to the unevenness of the cylinders of Boyle, and partly to this gluten which was only a certain oil.[7]

4. His conclusion was the above-said viz. that the compression of the air, both from above, et below, as being a fluid body like unto water, did hinder their separation, et there not being the least jot of air betwixt them, whereby the compression of the air being equal on both sides must needs equally force, and compress, on cylinder against the other, just as when two weights of equal ponderosity do compress two boards they cannot be separa- [81r] ted till the force, which is to separate them, be stronger than the weights: but if any thing were betwixt the two boards which might press on one side, or other, or both together, then certainly the separating force need be much lesser.

5. Hence he explicated the compression of the air to be like the compression of the water, which is not felt when we are in a river, or sea, be cause all the water compresseth equally on all sides: but were there in the bottom of the sea a tubulus,[8] hole, or cavity, which had not one drop of water in it, we should feel all the weight of the water then to press, and ponder, upon us, and with huge violence to force us into that hole, whereas were the hole full of water we should feel no such violence forcing us into it: even so the air.[9]

6. Hence [81v] also he inferred, as above that it was not the metus vacui which did hinder the separation of the cylinders with this dilemma: aut non datur talis metus vacui, et coë[u]nt illa, talis metu,[10] cum non detur, nequit impedire avulsionem cylindrorum. Aut si datur, et ille sit qui impediat disiunctos, iam profecto disiunximus illos, et coë[u]ntes datur via, et modus, introducendi vacuii in na[tur]am: imo fecimus ipsi mo[do] vacuum, q[uo]d utrumq[ue] negare de[be]nt, qui vacuum tantopere abohrrere na[tur]am aiunt.[11]

---

[7] See Boyle 1662a, 84–86.

[8] A small tube.

[9] See Stevin 1605, 148–149; AT II, 587–588; Boyle 1666, 231–232 and 242, 1999–2000, volume 7, 220–221.

[10] *Sic.*

[11] "Either this fear of vacuum does not exist, and those [cylinders] cohere, [but] since that fear does not exist, it cannot impede the separation of the cylinders. Or, if it exists, and that is what hinders that they [are] separated, [and] we actually have already separated them, since they [were] cohering, it is given a way and manner to introduce a vacuum in nature: what is more, we created a vacuum in this very way[.] Those who say that nature abhors a vacuum to such an extent, have to negate both [these alternatives]."

7. By the way he explicated the opinion of Descartes, which holdeth that all the cohesion or union, of bodies, proceededh only from their quies.[12] But unto this he said he could not assent. For certainly quoth [82r] he if this quies be so strong that I have not force to separate two things, how comes it that I can so easily move the whole v.g. a whole iron, or board, which with my hands I cannot tore in pieces, yet I can rowle it to and fro? Or if I cannot separate the half, 3rd or 4th part of said things yet I may at least separate the 100 or 1000th part of them, the force of my hands and their bigness being so much bigger than said 100 or 1000th part.

8. Unto this Descartes answereth that the softness of our hands the pain we feel and our only touching said things s[ecun]dum extimam superficiem[13] is the cause we cannot separate not so much as the least part of many solid things.[14]

9. But this answer, said he, doth not satisfy, for the said argument may be again and again instantiated against it. Wherefore besides the things [82v] specified by Descartes he thinketh that the experiment now mentioned is the chief cause of our impotency of separating solid things from each other, yea and that we cannot of some things separate the 1000th part, although we can move the whole body of them to and fro: for as for the moving the whole body, {they} being in the fluid body of the air, as if it were in water, and the whole body of the air not pressing upon them more than water, nay not so much, hence said pression is not felt by the things, and we can move them: but as for the separation of one part or particle from another, besides the quies of said particles, they are so closely united as these two cylinders, and not the least jot of air being between the pores of said solid things it is extremely hard to pull them asunder: from whence is collected that the more or less pores the body<h> hath, the more or less of air is in them whence the greater or lesser [83r] difficulty of separation ariseth. And thus it is not so much the quies of said things which maketh them inseparable, for said quies we see we can easily alter, or overcome even in the whole body moving it to and fro as the compression of the air which presseth and compresseth one part against the other which it findeth empty of all kind of air, even as water presseth against a hole that is empty of water, et not against a hole that is full of water, this being the nature of fluid bodies so to press, and yet to leave the bodies which are totally surrounded by them so easily moveable by virtue of the nature of the same fluid body, which maketh a contrary pressure ad aequilibrium.[15] Even just of a scale with 100 lb in it presseth heavily on my hand but if you put into the other scale another 100 lb weight, then my hand which is under the scale will feel no pressure at all as it did before.

[83v–84r] Experim[en]tu[m] 3um 23° Martii.

1. His subject the pressure of the air viz. to prove that it did only press per lineas perpendiculares:[16] and that the greater capacity of a vessel exhausted did not make

---

[12] "Rest."

[13] "By the outermost surface."

[14] See Descartes's *Principia philosophiae*, II.54–55 and II.63.

[15] "To an equilibrium."

[16] "Through perpendicular lines."

the air to press more or less, thereby to hinder the separation of the uppermost part of the vessel from the lower.

2. It is to be presupposed, that a week or two before this experiment he had made use of a brass globe divided in two equal hemispheres which had just 8 digitos diameter, and these hemispheres being conjoined with tallow, they had sustained without separating 700 lb weight.

3. He made use this time of a lesser brass globe divided likewise into two hemispheres whose diameter was 6 digitos: so that the quadrate of [84v] this globe mathematical or cubic was 36 and of the former 64 and this was the proportion of one to the other, or as of 9 to 16.

4. First he tried what weight the one hemisphere of this 2nd globe would sustain by applying unto it a brass plate, plane, and even, and very thick (least the pression of the air in the exhaustion might bend it, or else the weight of the scales in the divulsion) and with tallow {...} united it to the hemisphere, and so the hemisphere was close shut up that no air could get in or go out, and it remained a perfect hemisphere only.

5. Then he exhausted by degrees every jot of air contained within said hemisphere viz. till his pump did suck little or no air at all: and here we saw ad oculum,[17] as it were, the pression of the air for after the 2nd or 3rd draughts of the pump the plate did squeeze the tallow [85r] out of the sides of the hemisphere betwixt the plate, and the half globe, and so made it be very strongly united to the half globe, whereas before it was but slightly set on.

6. After exhausting all the air he <p>hung it in the static machine, and laid weights in the scales and it sustained 425 lb weight before the plate separated from the half globe.

7. Then he joined the other half globe to the first closed them both with tallow, exhausted the air of the whole globe to the same degree, as before, very little more, or less.

8. Hence he hung the whole globe in the machine, and by degrees added weight in the scale: it sustained clearly the 425 lb above-said, but having added 23 lb more it presently separated, I mean, one half globe from the other: so that the difference was very little.

[85v] 9. Hence he inferred 1° that the difference was so little as it might be attributed to the more or less perfect exhaustion of the air as also something to the more or less closure of the tallow: whereby he said that from both these causes, or any other slight accident, he sometimes had experimented some such small difference, yea sometimes greater, and yea sometimes not the least at all, which was only when all things hit well together.

10. 2ly. That it was not the greater capacity of the vessel exhausted that did make the air to press more or less: otherwise the whole globe would have sustained twice or much weight as the hemisphere did viz. 850 lb and not only 425 et a little more.

---

[17] "By eye."

11. 3^{ly}. That the air did press only perpendiculariter,[18] else it would have pressed more on [86r] the \whole/ globe than on the half and consequently the whole globe would have sustained more weight than the half, which by this experiment we did see it was not so.

12. That only the greater diameter of the vas was the cause of greater pression of the air for so the other of 8 inches broad could sustain 700 lb and this must needs be so, for this pression being perpendicular the air must needs lie more heavy on a thing of 8 than 6 inches broader.

13. That here was a great uniformity in this experiment with the other, made \ which/ the other globe the proportion of 425 to 6 inches corresponding very well to that of 700 to 8 inches broad.

[86v–87r] Experim[en]t[um] 4um 4° Iunii.

1. His subject was to prove that metus vacuii was but inanis et chimaericus.[19]

2. He had placed on the anthlia or instrum[en]tum torricellianum,[20] a great round bottle, and in it was a smaller glass bottle with mercury. Over this small bottle was placed a syringe whose embulus[21] he could draw up by turning round the head of the great glass bottle.

3. This experiment had not succeeded formerly, some day that I was not present. Now he with the pump sucked out all the air that was in the great glass bottle et then turning <round> round the head of said great glass bottle, he drew up the embulus of the syringe, as high as it would go, thereby to see if p[ro]p[te]r metum vacui the mercury would ascend: but it did not budge, nor stir at all.

[87v] 4. This done he let in the air again into the great glass bottle, and presently the mercury ascended, as high, as it could go.

5. Hence he inferred that it was not the metus vacuii that did make the mercury to ascend, seeing when it was all drawn out, I mean the air, and there was the greatest cause to fear a vacuum yet even then the mercury did not stir: but rather it was the pressure of the air newly let in which did cause the mercury to ascend, seeing immediately upon its being let in, when as one should think then there should be not so great fear of a vacuum the mercury did ascend.

[88r] Experim[en]tum 5um 8° Iunii.

1. His subject was to prove the elastic power of the air.

2. He had placed on the pump the great bottle glass of which we spoke in the last experiment. In it he had a pretty big bladder the which by the uppermost end was tied by a string to the upper part of said glass. And at the lower end was tied to the bladder a weight of 5 lb which was surrounded by some cancells of brass, on which it stood least falling on a sudden it might break the great glass.

---

[18] "Perpendicularly."

[19] "Foolish and chimerical."

[20] Of course, the actual *instrumentum Torricellianum* was not the air pump, but the barometer or tube.

[21] This is the plunger of the syringe.

3. This weight he said might be bigger, but that the inferior orifice of the glass was not capable of letting in a bigger weight and besides the cancells of brass were not capable of surrounding a bigger weight, than that.

[88v] 4. Then by degrees i.e. several suctions he drew out all the air that was in the great glass, and at every suction the bladder, (which before was all flaccid, and wrinkled, and had very little air in it), began to expand and dilate itself that air which was within though but a little expanding and dilating itself, by the force (as he said) of the aether or m[ate]r[i]a caelestis,[22] which at the exuction of the common air, penetrating, et piercing through both the glass, and the bladder, did in the bladde[r] comminuere in minutas partas [*sic*] aërem coe[unte]m,[23] insinuate itself into the pores of said common air and so dilate it and make it occupy more room than it did before.

5. This was not all. But as the bladder did extend itself, so also it did by degrees lift up the 5 lb weight at every suction: so that at last the weight was at the top of the cancells [89r] of brass: which elevation of this weight, was only caused by the force of the expansion of the air within the bladder, though but a very little, as hath been said.

6. This done, he let in by degrees, and softly, the air into the pump, and presently the bladder (which had not been extended to the full, but might have been expanded much more, and consequently would have lifted up the weight much higher, or a greater weight than that but that he was unwilling the weight should much surpass the limits of the cancells) became flaccid again, and crumpled, as it was before, and the weight descended to the bottom, or resting plan of the cancells made to sustain it.

6. Hence he inferred, 1° that only the air i.e. its expansion within the bladder could be the cause of [89v] of the elevation of the weight. 2° that it was very remarkable that such a small quantity of air, as was within the bladder, should have such great strength, as to elevate such a weight. 3° that the air which was within the glass beforehand had hindered by its pression the bladder from expanding and the ingress of the aether, which was the efficient cause of said expansion. 4° that only the aether could cause said dilatation seeing the bladder did not dilate its own self: nor the air within the bladder had no intrinsecall principle whereby it could move, and expand, and dilate its own self to such a degree, and much less with such a force as to cause the elevation of such a weight.

6. 6°. That the air ought not to be considered in its whole body, and taken altogether, [90r] as being the greatest body, extant in the world, and that hence it may have indeed a great strength, but rather that a very small quantity thereof, when expanded, and dilated, by the aether, hath in itself a very main strength, as appeareth by this experiment.

[90v] Experim[en]tum 6um 11° Iunii.

1. His subject was the necessity of air for respiration et cetera.

2. He put a little dog into a big glass bottle, which was placed upon the pump. This dog had been in once, or twice, before, and so in laughter he said that he was used to those kind of diseases.

[22] "Celestial matter."

[23] "[...] to break in little parts the cohering air."

3. The phiala being closed, and stopped, with a kind of bitumen black something like to pitch, though it was not pitch: he began to draw out by several suctions the air.

4. Upon the very first suction the animal began to pant, and be sickly et faintish, and to breath [91r] faster and harder than before: and likewise his body began to swell, and appear considerably bigger, than when it was well, and in its natural state.

5. Presently after, at the next exscutions, he began to gasp, and be troubled with convulsive fits: he cried a little with a very weak, and faintish voice which could scarce be heard, made so shrill doubtless by the rarefaction of the air; and then he lay as it were a dying after having tumbled up and down in the glass sometimes with the pangs of death.

6. To the end he might not suddenly depart, the air was let in again, and the glass opened at the top, whereby presently he began to come to himself again: was taken out, and went up et down [91v] the room, though reeling and as it were lightheaded, et seasick.

7. Then he was put in again: the air was drawn out, and his body swelling, as before he was troubled with the same fits till after some few exuctions more then, at first, he died, and lay without the least motion.

8. The air being let in again he came not to himself again, but {yett} notwithstanding his body which had swelled, as before, immediately after the letting in of the air de-swelled, and returned to his natural state.

9. The dog being cut up he took one the lungs and showed unto us: the which were found to be of two colours: the one was of a darkish [92r] red colour which he called color fuscus.[24] The other part were of a whitish colour such as we see in the lungs of other animals that are sound.

10. The several lungs, or lobes, he divided asunder, and threw them into a pale of water: and three of the lobes which were of the whitish colour, did swim and would not sink: the other lobe which was of the darkish red colour sank presently, and did not swim at all.

11. From the lobes which were of the <darkish{…}> whitish colour, and did swim, he cut of a small piece or two, which had the same darkish red colour, with that lobe that did sink: he threw this particle or two, into the water, and they did sink likewise, whereas all the rest of the lobe which was whitish, being thrown in again did still swim, as before.

12. Hence he concluded, and inferred 1o that res- [92v] piration was necessary to life, and the inspiration continually of fresh air: seeing that this animal for want of it {had}, the first time, been so soon brought so nigh his end, and then so soon had again recovered, when it was let into him anew, et afterwards it being taken away from him again he quit, et suddenly, died.

13. 2°. That not any sort of air was sufficient to keep life in men and beasts: seeing that all the air was never drawn out of the glass, nay peradventure is never, or can never be drawn out by the pump but that some is still left within. But that the air necessary for life ought to be of a sufficient consistence, thickness, or crassitude, for

[24] "Dark colour."

want of which this animal had perished, the air within in the glass, by the subduction of the rest, being too much rarefied, and subtilized, whereby it was rendered unfit for animals to live upon.

[93r] 14. 3°. That this was the true, and natural reason, of another experiment or phenomenon of nature: viz. when men ascend unto some very high mountains as el pico de Tenerife they experiment squeamishness of stomach and vomits and other symptoms whereby they are feign to make use of several preparatives the air,[25] there, by reason of its thinness, subtility, and too much rarefaction, being scarce sufficient for life, and some ascending too high have died for want of breathing for although there \is/ much and open air enough, yet it is not of such a consistency as is requisite for human breathing, and life.

15. 4°. That the swelling of the dog's body, both times, when the air was drawn out, and de-swelling of it again when it was let in again, proceeded from this principle that the compression of the air within the glass being taken away, the air remaining within the body of the dog as in the muscles, veins, parenchymas, and fibres, in all which place he said were some real quantities of air; this, I say, did expand and dilate itself and [93v] so cause the body to swell, but the air being let in again did compress, and condense, and constipate the air within the body of the dog, and this cause it to de-swell. Even just as we had seen in the last experiment of the bladder, and the weight, wherein the air being taken out of the great bottle glass (which did compress bladder, et constipate the air within the bladder) that which was within did presently dilate itself, and so much even to the lifting up of such a weight: and the said air being let in again did condense and constipate the air in the bladder whereby it de-swelled, et crumpled and let down the weight to its place where it was before: and this was also the reason of several other experiments before-seen, and rehearsed in this bask.

16. 5°. That the two different colours seen in the lungs did proceed from hence that the 3 lobes which were whitish had some air in them which caused [94r] them to swim, but the other lobe by some accident had none at all which caused it to sink: and the same ought to be said of the particle of the reddish dark colour cut of from one of the lobes which swam: said particle having no air in it which caused it to sink: even as bladders do when they have no air in them.

17. 7°. That hence might easily be examined, if an infant was born dead or alive: for if born dead the lungs will certainly sink: if it was born alive, and breathed never so little while, they will never sink do what you can.

[94v] Experim[en]t[um] 7um 15° Iunii.

1. His subject was the same with the precedent experiment viz. the necessity of respiration and air for life in all sorts of animals.

2. To this effect he put into the phiala after having placed it on the pump a couple of birds, which were sparrows.

[25] See Sprat 1667, 201.

3. He shut the phiala very close, with his ordinary bitumen made, as Margrave judged, of wax, and tallow.[26]

[95r] 4. He premitted to the experiment by way of preamble, that we should see in those volatills the same necessity of air and respiration for respiration towards the preservation of life, as we had seen the last experiment in the trial of quadrupeds viz. the little dog. Also that the air necessary for life was not any kind of air, (seeing in the pump some air was always left though very much attenuated, rarefied, and subtilized) but only air of such a crassities,[27] as he called it, or consistency, as was requisite for life, without which no animals could subsist.

5. 2ly. He premitted, we should see this necessity of air in those volatills notwithstanding that volatills have the organs of respiration aliter disposita q[ua]m in hoc.[28] For birds et volatills, saith he, have not their lungs, as men, hanging loose in the thorax but costis affixos, atque coalitos, et coadunatos.[29] Which proves saith he that the lungs are not the cause of respiration, as they are thought [95v] to be i.e. that they are like bellows, that drawn in the air by their dilatation, but rather that they are like bladders into which the air is thrust, and inflated by circumpulsion, and that their dilatation dependenth on the dilatation of the thorax et non e c[ontrari]o being we see in dissection that many men have great parts of their lungs united and coagulated to the ribs of the breast, and yet these men no otherwise than these birds nullo cum damno respira[ti]o[n]is respirant, q[ua]mvis h[ab]eant pulmones ita costis affixos.[30]

6. 3o. He premitted that volatills had in pulmonibus aperturas quasdam satis manifestas per quas transmittebat inspiratus aër usq[ue] in abdomen ipsum[31]: which did seem to prove that the same aperturae ought to be men, although as yet not found, yet that that reason did seem to convince it, seeing that many [96r] that have matter, and pus, in their breast, and not their lungs at all vitiated, yet per asperam {asperians}[32] they do often bring up, and spit out said pus, which could not be well understood which way it could come up that way except in the <breast> lungs some such apertures were considered to be into which said pus could enter and come up by the aspera arteria.

7. Hence he proceeded to his experiment: he drew out the air and after some exuctions the birds began to have convulsion fits, they {had} vomited, as if they were sick, panted much, and tumbled up and down gasping et holding up the heads for air, and thus they did lie a dying when he let in the air again, et suddenly they

[26] Christopher Love Morley refers to Christiaan Marggraf, who was a private teacher of chemistry and medicine in the city of Leiden. His judgment on De Volder's bitumen could not be retrieved.

[27] "Thickness."

[28] "Disposed differently than in this."

[29] "[...] attached to the ribs, and adhering, and united".

[30] "[They] breath without any damage to respiration, even if they have lungs attached to the ribs in this way".

[31] "[They had] in the lungs certain holes sufficiently visible through which the inspired air [is] conveyed up to the abdomen itself."

[32] Likely: "arteriam."

were well again, et revived, though not quite so lively as before: their bodies also swelled, as that of the dog had done, et de-swelled when the air was let in again.

[96v] 8. Afterwards he drew out all the air again, as before, and the birds returned to their convulsive fits, tumbling, vomits, till at last they died.

9. Of these convulsive fits he said it was very remarkable that all animals that \X/ ever he had put into the vacuum did die all alike by convulsions whereof he said the cause was not easy to be rendered, convulsions being caused by contraction of the muscles which is performed by the affluxion unto them of the animal spirits: {now} why the extraction of the air should cause the animal spirits to conflux unto the muscles after such a {nattra} and there to convell the muscles he said again it was not easy to explicate. Let he thought in some manner it might be explicated thus: viz. that these spirits are of the nature of liquid [97r] things as being of engendered of the subtlest, and purest part of our blood even as spirit of wine is the purest, and subtlest part of the wine, although the liquidity thereof seemeth to be only the vehicle of said spirit of wine, and not so properly the spirit itself.

10. Now in all liquid things is found, by the trial of the pump, to be contained some air: for oil, water, wine, and other wine being set on the vacuum and the air extracted, at the first exsuctions do ascend some bubbles, afterwards more and more, till at last is made a huge commotion of the whole liquor, which doubtless, quoth he, is caused by the extraction of what air was contained in the liquor and rarefaction of what remaineth by the ingress into it of the aether, or ma[teri]a caelestis.

[97v] 11. Hence, saith he, [a]dd[end]um {vs},[33] that a[nim]al spirits being, as I have said, of the nature of liquid things, the air when extracted out of the pump is also extracted out of the a[nim]al spirits contained in the body, and the aether entering incontinently doth rarefy extenuate expand and dilate the rest remaining which causeth a commotion and revolution amongst the animal spirits even such as is experimented in liquid things, and they thus {wruell} the muscles disorderly every time an an[ima]l so dyeth for want of respiration.

12. At last he inferred the huge necessity of air for humane life, and of thick and gross air not of any sort, et so ended.

[98r] Experim[en]tum 8um 22° Iuni

1. His subject was the former concerning the necessity of air, and respiration for life.

2. He set an indifferent big glass upon the anthlia, and filled it with water up to the very top of the brass pipe, through which the wind is drawn out. Then he put in several fishes of different sorts, and sizes and amongst the rest an indifferent big eel.

3. The glass being well closed, he drew out the air, et loe at the very first or 2nd suction the fish felt the want of the air which was in the water, and from the very bottom of the glass where most of them lay, they all mounted suddenly to the top of the water, and there they lay gasping for breath.

[98v] 4. At these very first exsuctions an infinite number of small bubbles did begin to ascend from the bottom of the glass to the superficies of the water, where

[33] Probably: "est." Read: "it is to be added."

they did stick, and an huge number of these did stick to the sides of the fishes, which made them appear as if they had been silvered over, especially the eel was a pretty spectacle the bubbles sticking so thick, and close all along the two rows of her fins. These bubbles continued ascending at every suction, and at last they were so small, as that they were almost imperceptible, and did seem to obscure a little, and trouble the water.

5. The fish continued gasping, and the eel was she that last of all ascended, as seeming to endure the absence of the air more than the rest: now and then with pangs, and as it were convulsive motions they did flutter, and tumble [99r] up and down to the bottom of the glass, but presently, as it were drawn by some force, they did ascend again unto the superficies of the water. These bodies did not swell at all to the sight, as the bodies of other animals had done before, which he said must be because they doubtless had not so much air in their muscles, and bowels as other creatures had.

6. Every one of the fish did gasp much for air, and so did the eel likewise, and the bubbles did seem to stick most above their fins which he said might given suspicion, that at their fins, they did breath, and there cause those bubbles by emission of some small air out of their bodies. Let that those bubbles were not all caused by the fish (if any were) being in that water he said was evident, for that clear water being put into a glass, and the air exhausted, did send up the same bubbles and to the appearance in the same quantity.

[99v] 7. When the fish had to the appearance lost the greatest part of their motion, and life, and did lie almost for dead with their bellies all upwards: then he let in the air again et suddenly all the fish tumbled down to the the bottom of the glass again, as fast as they had come up before. From whence he inferred the pressure of the air which pressing upon the water did make them so to descend.

8. With the air let in they did come to themselves again as appeared manifestly when the air was drawn out again although at present they lay as it were \X/ being stand, and made not any very great motion. All the bubbles likewise of the water disappeared presently and were seen no more neither in the water nor sticking to the fish.

[100r] 9. After he had let them revive and breath for a while, he drew out the air again: and presently at the very first suction of the air the fish flew all up again, as before, to the superficies of the water and began to gasp, and flutter up and down as they had done before, and, at last, after some more suctions to be convulsive, and turn their bellies upwards, till at last the same tragedy being acted as before they lay for dead.

10. Notwithstanding he believed, they were not quite dead, and so he said, he would leave them for a while, about half an hour, before he would let in the air again, and then see if they made any further motion. The eel did still gasp when we came away. And seemed to let out air below the mouth a fingers breadth where she had a kind of a bag that did extend and compress, as the lungs in men do, of which he bad us to take notice.

[100v] 11. Note that at this second time, when he drew out the air not so many bubbles by many did ascend as at the first time he drew it out, and of this he bad us take notice saying it was a constant experiment of all liquors.

12. After half an hour he found the <air> eel still gasping which he attributed to some air that was got into the glass but the rest gave no tokens of life. He let in the air upon them, and presently they were precipitated as before, though not so fast, their tails descending first, and then their heads: but none come to life again except the eel.

[101r] Experim[en]tum 9um 23° Iunii.

1. His subject was the precedent, viz. the necessity of air, and respiration for all sorts of animals, even the most imperfect.

2. He set a small glass upon the pump: and into it he put 3 erucas called caterpillars one butterfly, and another sort of butterfly which he seemed to call<'d> noctua and a pretty big fly, of these that haunt up and down in summer time.

3. He closed the glass, as well, as he could, and then he began to exhaust the air: and it was remarkable that at the very first exuctions, the butterflies lost their flight so that they could fly upon, and down the glass no more, making use of their wings, yet they made a shift to crawl upon and down, and so did the fly (which also lost its wings), and the caterpillars.

[101r] 4. After 10 or 12 suctions they \all/ continued still crawling up and down, which gave him suspicion that some air did still get into the glass, whereupon he examined the case, and closed better than before the said glass. And then after few exuctions they were all brought to no motion at all: so that they did not budge in the least. Let he told us we must not think they were dead, for as much, as those sort of creatures would lie so for dead a long time, and then, when air was let in again they would revive: especially the caterpillars, which as last of all, they would cease from motion, so longer would they lie, and yet certainly come to themselves again.

5. He said more that to be certainly informed how long those things could live in that dead kind of posture, we ought to leave the [102r] glass still standing so upon the anthlia: and moreover now and then every half quarter of an hour give it an exsuction, for it was certain that some air of that which was exhausted, by degrees did get in again, and a little was sufficient to keep those creatures alive. Note that their bodies did not seem to swell as those of other creatures.

6. Hence being to make farther use of the pump he took of the glass wherein these creatures did so lie for dead (shutting it before at the bottom with a screw, so that no more air could get in than when it was upon the pump). He set it by, saying he would try how long they did live.

7. Next, he set a bigger glass on the pump, and into it he put a pretty large frog. He [102v] closed the glass, and drew out the air, and at the very first suctions the frog gave some token of his missing the air by his {trem}bling and crawling. His body after some more exuctions did visibly extend, and above the belly he did seem to breath, or pant, with difficulty.

8. He continued crawling upon the glass, till after many exuctions the air was quite drown out so that the pump would yield no more, and still he was in the same posture, and thus he {told} before he would be. So that to kill him, or to know how long he would live, it was necessary to leave him just so in the glass without any air, and moreover now and then he said he would exhaust the glass anew to the end none might come in to him.

9. Hence he returned to his other animals, [103r] which he had set aside. But he found most or all of them with new motion whereas they had none before: and hence he concluded that some air was got in to them through the glue or bitumen of the glass, which had brought them to life again. Hence he let in the air fully unto them: and presently the fly and butterflies having now air thick enough found their wings again and began to fly up and down and the other animals to crawl, and they were well enough to the appearance in a moment.

10. Of the frog he told us afterwards that he had left him in that posture for a whole hour, and at 12 o' clock he found the glass, as it should be without air but the frog still alive: at 2 he found him not with any motion at all, and the glass without air: yet he dared not let in the air fearing he would come to himself again: at five he come again and found the frog again without the least motion, and then he let in the air, and presently his sides, which were much swelled, did fall down flat, yea more flat then they were before he put the frog into [103v] the glass: yea in his side appeared some small rupture either with too much swelling, or else with the force of the air <g>rarified which in his body, or of the air when it was let in. He gave no signs of motion: he took him out, and laid him in the san, but he came not to himself again but was perfectly dead. Thus long he said he had prolonged this experiment because aliunde[34] he knew how long frogs would lie as it were for dead in the empty glass, and notwithstanding come to themselves afterwards again.

[104r] Experim[en]tum 10um 2° Iulii.

1. His subject was the necessity of air (and indeed of natural breathing air such as men do need for life) for the propagation of sounds.

2. To this end he had a bell with a clapper which before the putting it into the glass did a very clear and distinct sound. This he put into the glass separated from the clapper so that there was no connexion at all betwixt them and there, the glass being closed, we could perceive that through the body of the glass the sound was but half so loud as it was before: yet could clear, and distinctly be perceived as also the {tinning} it left behind every stroke, which did sound for a good space after the clap was given.

3. Then he exhausted the air, and this done, no sound nor {tinening} of the bell, was very scarce perceivable though the sound of the clapper was almost as [104v] before. But the reason of hearing the clapper, he said, was because of the air contained in the clapper and in the epistomium to which it was joined wherewith he did, from the outside of the glass, make it strike, which continuation of things though it were brass, by means of the air within said brass did propagate the sound unto no

[34] "From elsewhere."

out of the glass, where the handle of the epistomium was: and for this cause he said he had separated the bell from the epistomium, and clapper, for otherwise we should not perceive any very great alteration in the sound of the bell.

4. Then by degrees he let in the air again, and loe, as the air came hissing in, so by degrees did increase at every clap the sound of the bell: which proved evidently that sounds depend, for propagation on the air, seeing without it they cannot be heard, and are totally lost.

5. Next to show us how that the continuity of [105r] things doth propagate the sounds by the air that is within them, and make them be heard, at such a distance, as otherwise they would not, (yea and, here, even when the ambient air was drawn out of the glass) he joined the bell to the epistomium (by which also the clapper did hang) he drew out all the air, as before, and loe then sounding the bell we could hear it distinctly, with a perfect tinning after it, though to speak truth the sound was much diminished from what it was before the exhausting of the air: hence he inferred that nothing else could be the cause of this hearing the sound of the bell now, and not before, but the {continuity} of that brass epistomium whose air in the pores did transfer the sound out of the glass unto the handle thereof which was out, and thus we came to hear it: so a long trabes, or beam, doth transfer the <knocke> \sound/ of a little knock unto the other end which otherwise cannot be heard half the way, or the length of the trabes.

6. Lastly he condemned the experiment of Mersennius, who saith that a bell so shut up in a tubulus with quicksilver did sound, and could be heard as much, as before,[35] [105v] which experiment he said could very hardly be made but that in the tubulus, and quicksilver, much air would be remaining, which might be the cause of Mersennius's hearing the bell: whereas this experiment did evidently convince the contrary: and that much air was always in mercury he said was easily to be proved seeing the air doth so often make it run into little and small pieces et round bubbles.

[106r] Experim[en]tum 11um 6° Iulii.

1. His subject was the measure of the air viz. how much was drawn out at every exuction of the pump which he said could easily be measured by water.

2. He set a glass pretty big, and empty, on the pump: the basin of the pump was all full of water: and through the glass there did go up the {wuall} brass pipe, through which the air is drawn always out.

3. He told us that at every exuction we should see the water to ascend through that pipe, and fill the glass, and how much water did ascend, just so much air was drawn out, and that that was the rule, and measure of the air drawn out.

4. Then he drew out one exuction of air: and [106v] presently he unscrewed the glass from of the pump, holding the lower part thereof in the water, to the end the <water> \air/ might not get in, and suddenly the water (as it were ad replendum vacuum)[36] began to ascend through the pipe, and did ascend till it had filled about

[35] See Mersenne 1636, 161.

[36] "To fill the vacuum."

one half quarter part of the glass, and then it ceased running any more, whereby he inferred that just so much air was drawn out, for were any more air drawn out it would continue running still.

5. Then he screwed the glass on again, and gave it another exuction: unscrewed it, and holding it in the water, as before, loe the water ascended, as much again, as it had done before, and then ceased whereby he <{...r}> inferred that just so much more air had been let out at this 2nd exuction.

6. This he repeated 5 or 6 several times [107r] at least, till the glass was full, I mean, as full, as that the water did reach almost, quite, to the top of the glass: but he seemed to say it could be quite filled with water, by reason the air could never be quite drawn out. The ascending of the water, he referred unto the circumpulsion, or pressure of the air, and not ad replendum vacuum, as the Cartesians do usually do.

[107v] Experim[en]tum 12um 29° 7^{bris}

1. His subject was the necessity of <light> \air/ for the preservation of light.

2. He lighted a short candle, and put it into a great glass which did hold 28 stopes of water (every stope maketh at least 3 of our english pints) and then closed the glass: and all the while he was closing the glass the light remained unalterable to the sight, and all outward appearance.

3. This glass he made use of said, there fore ought to be of such a bigness, because having tried this experiment in a small glass, the candle had been stifled before he could close the glass with pitch, to shut out all the air.

[108r] This done he immediately went to work, and began to drew out the air. At the first exsuctions the candle appeared unalterable: but at 4th or fifth the flame began to diminish, growing downwards less, and less, and so contracting itself.

5. At the next exuctions that part of the candle which was all about the elichnium or wicke, changed and appeared of a blueish colour, whereas the flame which was above the wicke conserved still its own proper colour: but yet at every exuction contracting, itself, and diminishing.

6. At twelve exuctions the candle went quite out, there appearing when it went out, a long steam of smoke which ascended in the glass.

7. Then he lighted the candle again, and put it into [108v] the same glass, and closed it as before, with a design to try how long the light would live within the glass by itself it being closed without drawing out the air: he meaning to prove from hence that the extraction of the air was the cause of sooner going out of the said light.

8. The light continued unalterable for a pretty while but the hour striking he was feign to hasten his work: hence he drew out the <light> \air/ again out of the pump and at every exuction the light was like to go out but afterwards when he did let <{...}> \in/ the air, it did blaze and flame again very brightly thus at five exuctions the <air> light went out sending up a steam of smoke as before.

9. Hence he inferred first. That drawing out of the air did doubtless hasten the extinction of [109r] the light. The which he confirmed by the experiment of shining pieces of rotten wood: the which he said he had tried in the pump. And when the air was drawn out, their lustre, and shining did disappear but when the air was let in

again, it did return to its former lustre, and brightness: so this candle the second time did almost go out every time and when it was let in again, it did blaze anew.

10. 2[ly]. That this candle would have lived longer in a thinner air, than in a ticker, for by the blazing of the candle in a close {shuft} up place the light is stifled, which is not so in a thinner air.

[109v] Experim[en]tum 13[um] eod[em] die.

1. His subject was the compression of the air viz. that in cupping glasses, it was not the metus vacui that made skin to ascend within the cupping-glass, but rather the compression, or gravity of the circumambient.

2. He laid a bladder having within it a round brass circle, within the pretty big glass and over the bladder did hang perpendicular a cupping glass. The great glass being set on the pump, and closed, according to custom, he at several exuctions drew out all the air out of the great glass: and then with a screw let down the cupping glass perpendicularly to the very superficies of the bladder, laying it as even, and equally as he could upon the skin of the bladder, to see if the glass did draw up the skin within its belly. But it did not draw at all.

[110r] 3. Hence he inferred, that if in cupping-glasses, it were the metus vacui that which did make the skin to ascend within the cupping-glass, doubtless then likewise that bladder would ascend, and the cupping-glass would draw it upwards, which seeing it was not so, it could not be the metus vacui that which was the true cause of that effect of nature, these being then within the great glass, and within the cupping-glass also a great vacuum sufficient to cause the ordinary effect.

4. Next to try this effect a contrario,[37] he suddenly, or rather by degrees, and slowly, let in the air again into the the great glass (but not into the cupping glass, for as much as that lying with its brim even upon the bladder, the air could not get into that, betwixt the superficies of the bladder, and the brim of the cupping-glass) and presently, as the air came in to the great glass, there being still the same metus vacui in the cupping-glass, so did the bladder ascend, and was drawn upwards, within the cupping-glass, even [110v] just as it happens in all other cupping-glasses, we see.

5. Now to prove that this was the true cause of this effect, and that this bladder was just so drawn within the glass, as in all other cucurbitulae[38] it happens viz. that the bladder did stick with the same force to this cupping-glass, as in all others, it happens to do.

6. He unclosed the great glass, which was on the pump, and with his hand took out the cupping-glass, which drew out the bladder also, and the brass ring within the bladder: and thus they did hang together viz. the bladder, its ring, and the cupping-glass to the sight of all presents, though within a little while they fell asunder.

[111r] 7. This falling asunder, he said was caused, by that the cupping-glass, when the air was let in, did not lie equal, and flat, and close enough, upon the bladder, whereby some air was got in betwixt them, and consequently the vacuum within

[37] "From the contrary."

[38] "Small cupping-glass."

the cupping-glass, was not sufficient, or strong enough, to draw, retain or bear any longer the weight of the bladder, and its ring.

8. But if that no air had been gotten into the cupping glass doubtless that cohesion, or attraction would, have been of sufficient strength to bear some pounds weight, before they could be disjoined. And hence he said, a great care must be taken in this experiment, that no air do get into the cupping-glass, when it is let into the great glass, else it will not sortiri eff[ect]um,[39] or at least, not so manifest, strong, and evident.

[111v] 9. Thus he told us, that morning he had tried the same experiment, in the presence of some few students, and the bladder with its ring could not be separated from the cupping-glass with less, than 10 lb weight.

10. Also that two days before he had experimented the same, and the bladder could not be separated with less, than 7 or 8 lb weight, for which he had also witnesses. And yet he promised to iterate this experiment so often till he had made us all eyewitnesses thereof.

11. This being certainly so, he inferred that only the compression of the circumambient air within the great glass was the cause of the ascending of the bladder within the cupping-glass, and also of [112r] its sticking so fast unto the cupping-glass, as to bear so much weight, and not the ancient reason of metus vacui, for as much, as there was a greater vacuum, or metus vacui before the letting in of the air into the great glass, than afterwards, and yet no such effort was seen: {go} it must needs be the ponderosity of the air, and its compression, which did cause \it/ so to adhere unto the cupping-glass, for no other reason could be excogitated.

[112v] Experim[en]tum 14um 28° 7^{bris}.

1. His subject was the very same, as in the former precedent experiment.

2. His experiment was also the very same, viz. the cupping-glass, and bladders, to prove that in cupping-glasses it was not the periculum, nor metus vacui that made the flesh to ascend, but rather the compression, and ponderosity of the ambient air, when the air within the cupping-glass, by the fire was rarified, attenuated, and in great part expelled.

3. He had already hanging in the room two several bladders with cupping-glasses sticking to them: the one he had took out of the pump two days before, and had continued so quite 30 [113r] whole hours: the 2nd he had took out of the pump but 2 hours before.

3. These he weighed, to see how fast, the letting in of the air into the pump had made the glass stick unto the bladder: and the first of them full presently of, through, I knew not what default so that it bore little, or no weight at all.

4. The 2nd held out as far as 12 lb weight, and then the bladder broke at the bottom with the weight, which had it not happened it would doubtless have held more weight. He mended it, and put it on again, and it bore to 14 lb weight, and then it broke off again, and then he gave this bladder over.

---

[39] "To have effect."

5. He put a new and strong bladder into the pump with [113v] its cupping-glass, and ring, as before and drew out all the air: this done, the cupping-glass being let down upon the bladder it drew it not up at all: but the air being presently let in, the glass drew up the bladder into its belly: he presently took it out, and hung it up, and it bore 17½ lb weight, and then the bladder broke at the end, where the weight did hang, which if it had not happened, doubtless it would have bore more weight, before the glass had been separated from the bladder.

6. He inferred the very same things, as in the precedent experiment.

[114r] Experim[en]tum 15um 15° 8^{bris}.

1. His subject was: how to know exactly the ponderosity of the water, or weight thereof.

2. He had a perfect \cube/ of brass very thin, the which had every way the exact measure of half a foot: so that the was every {...} of way <u>6</u> <ped> digitos s[cilicet] dimidium pedis cubici latus.[40]

3. At one of the corners thereof it had a small brass pipe, through which any thing might be easily with a funnel be conveyed into the body of said cube.

4. He put the said cube into a pretty big cube [114v] of water having already some shot in the belly thereof, which was not so much, but that the said end of the cube, which had the small pipe, did swim about the superficies of the water 2 or 3 fingers breadth.

5. Then one by one he went on putting at the same pipe small pistol shot. And thus he continued, till the cube was perfectly depressed under the superficies of the water: there remaining nothing above the water, the just, and only, the small brass pipe: but all the body of the cube was equally, exactly, and perfectly depressed, as it ought to be.

6. Then he took out the cube and weighed it with all the shot it had with in it, and found it weighed exactly 7 lb 14½ ℥.

[115r] 7. This he inferred was the exact weight of every cube of 6 fingers breadth, and length, of said water, and neither more, nor less.

8. This he explicated with the comparison of an exact balance, or pair of scales. Presupposing that the element of water, was a thing, so easily loco mobilis,[41] so easily fluid one way, as other, and so easily giving way to the least poise, or weight, as the exactest pair of scales, could be: the which he said was grounded for a principle in hydrostatic matters among the ancients: nim[irum] aërem circumambientem premere superficiem aquae s[ecun]d[um]m o[mn]es p[rop]ter aequale: nisi ii ita prement illa loco cederet, et fieret commotio, et perturbatio, et inaequalitas in superf[ici]e acquae, cum {I} nimirum {violet$_9$} cedat imaginabili.[42]

---

[40] "6 inches sc. half a cubic foot broad."

[41] "Locally moveable."

[42] "Evidently the surrounding air presses the surface of the water on all [parts] in the same way: if those [parts of air] do not press in this way, that [surface] would move from its place, and a movement will take place, and a perturbation, and an inequality in the surface of the water, as evidently [the water] would violently move [according to the] imaginary [surface]."

9. His argument was thus. Presupposing water to [115v] be as this balance. Even just as this balance is elevated, or depressed, with the least weight which is added to the other balance. Even just if the water did weigh more than the cube it would not give way to the cube nor permit it to sink, but rather keep it up: or if the cube did weigh more than the water, it would sink and kept up the water: {go} they must needs be of equal weight.

[116r] Experim[en]tum 16um d[ie] 18 8^{bris}

1. His subject was the necessity of air, for fire and light.
2. He put some pieces of shining wood, which were sent him from Amsterdam into the pump. And having drawn out the air, loe at every suction the wood did shine less, and less: till at last the wood grew so obscure, and its light so diminished, that it could not, very scarcely, be seen: yet it was not totally diminished for this he said he had tried, that it would not be done, when the wood was fresh i.e. newly digged out of the earth: but after some days, it could easily be done;
3. He let in the air again into the pump: and loe as fast, as it came in even so fast did [116v] the wood come to its former brightness: so that it did shine as much, as before.
4. His inference was the same with his subject. Viz. the absolute necessity of air for the preservation of light.

[117r] Experim[en]tum 17um d[ie] 26 8$^{br[is]}$

1. His subject was the exact measuring of the ponderosity of water.
2. He took a pretty long glass, much like {...} glasses in England and filled it three quarters full of sand. This glass he put into a bigger glass of water, till it did swim, and not touch the bottom. Then he let it stand till the water was perfectly settled, so that the least floating, or motion, was not to be perceived to be in the water, which was in the great glass and wherein the glass with sand did swim.
3. Next he took out of the water the glass with the sand, and weighed exactly in a pair of scales how much the glass with sand and all did way: setting before exactly upon the glass with water a mark, or scale, justly to know how high [117v] the water had ascended, when the glass, with the sand was in it.
4. This done he weighed also exactly the great glass with the water: and letting it stand in the scales he added <y> the weight thereof, the weight also of the glass with sand.
5. Then he told us, that the water being now sunk, as necessarily it must be, since the glass with the sand was taken out: in case he did add unto that water, just as much as would fill the glass up again unto the scale, which he had set on it: we should \see,/ that the water added would weigh just as much as the weight of the glass with the sand, which weight, as we had seen, he had added to the scale.
[118r] 6. This he did, and added as much water to the former, as did exactly reach the mark: and loe then the scales were, as even, as a dye.
7. Hence he inferred the same which in the first hydrostatic experiment: viz. q[uo]d tantumdem aquae exacte o[mn]i[n]o illud vas loco suo pepulerat, quantum ipsum vas ponderabat: et coë[u]ntes in o[mn]ibus granib[u]s quae in liquida immit-

tunt, tantumdem sc[ilicet] liquidi illa loco suo pellere, qua[nt]um ipsa ponderent.[43] Which by the by he said was the only way, and fundament for the gaging of ships or measuring how much weight they carry et cetera.

[118v] Experim[en]tum 18 d[ie] 29 $8^{br[is]}$

1. His subject was still, how exactly to know the weight of waters, difference in weight betwixt them, et thence how to \X/ infer sundry <{…us}> consequences.

2. He had before him three tubs of several sorts of water: the first with common sweet water: the 2nd with salt water, from the sea: and the 3rd with brine, or water, which he had made very salt, by casting, and dissolving into it great quantity of salt.

3. Then he took a pair of scales, and into the balance thereof he put 12 lb weight: at the other end of the scale he hung [1] lb [119r] weight more of iron, and this by a cord of hair of a horse's tail, the which he said, though dipt in water, could not possibly alter the weight as being not capable of admitting, or imbibing any water, or humidity, into it.

4. The scale, and weight on both sides being exactly even, he bid us observe, et remark, what a great difference we should perceive in {these} scales so even, as soon as he did but dip the 12 lb weight of iron into the common water, for that doubtless those 12 lb of iron would loose much of their weight: and more, or less, in every one of those waters, viz. according to the diversity in weight, and strength, of them i.e. as they were more, or less, or not strengthened at all with salt towards the bearing up of the iron.

5. Then he dipt the iron fully into the [119v] first tub of common water, and loe the 12 lb iron lost of their weight I know not how many ounces i.e. a very considerable number, and if I well remember about 2 lb.

6. After this he dipt the 12 lb iron again into the second tub of water, viz. the brine, which he had made: and loe the iron lost of its weight a considerable number of ounces more, than it had done before.

7. Lastly he dipt it into the tub of seawater, and loe the iron lost much of its 12 lb weight, which though it was not so much as it had lost in the second water, yet was something more, than what it had lost in the first water, which was but common water.

8. All these several weights, which at several [120r] dippings the iron was found to have, he set down exactly upon a counting board, and reducing each loss into ounces: he inferred that as the first weight was to the second, and the second loss to the 3rd even just so was the difference of these waters.

[120v] Experim[en]tum 19 9° 9^{bris}.

1. His subject was by hydrostatics to show the demonstration of Archimedes his crown. Of this point he had touched something the last experiment: but reserved the true proof, or demonstration, thereof unto this.

---

[43] "Because that container pushed out of its place utterly exactly as much water, as how much the container itself weighed: and those [things] which are put in liquids, cohering in all the grains, [they] push from its place as much liquid, as the same [things] weigh."

2. The history was thus. K[ing] Hieron of Syracusa had vowed a golden crown unto \X/ the gods. Having {đđ} out the gold by weight unto the artifices, he put in silver into the crown, and gave it to the full weight unto the K[ing] this being suspected, or crown, afterwards the question was to know how much silver, and how much gold was in the crown without undoing the crown. Archimedes at last found it out by hydrostatics, much [121r] after this manner.

3. He made two crowns more, both of the same weight with that of the artifices the one of gold and the other of silver. Next having silvered, that several balls of divers metals, being put into a glass of water, though all of one, and the same weight, yet proportionably to the different bulk, or body, that each metal has, even so much water more, or less, did they cast out of the glass, and that hence might be inferred the difference exact betwixt one, and the other metal.

4. Hence he put each crown into one, and the same vessel of water, and observing exactly how much water each metal did void out of the vessel: hence he inferred that as the gold by by reason of its lesser bulk did void but 10 lb for example, of water, and [121v] the silver by reason of its greater bulk did void fifteen, and the silver and gold crown did void but twelve for example: or rather, going another way to work,[44] (which is that the Professor used in this experiment,) observing exactly that the first gold {of} crown lost, being put into water, 10 lb of is weight (for example) the silver one lost 13 and the gold and silver crown but 12. Hence, I say, he inferred that even just as the silver and gold crown did loose less weight than the silver one, and more weight than the gold one, even just so much more silver had it than the gold one (which had none at all), and so much less silver than the silver crown, which was all of pure silver.

5. But now let us see, how the Professor proceeded in his demonstration. He had therefore ready made three balls [122r] the first of brass, the second of lead, et the third of brass, and lead: these three balls were all exactly of one, and the same exact weight. The first of brass being put into, a glass with water did way <{bee}> exactly ℥ v ʒ iiii gr. v

The lead in water did way ℥ v ʒ vi gr. v
The third ball weighed ℥ v ʒ iiii gr. iii
The exact weigh of all three balls \before/ put into water was ℥ vi ʒ iii gr. {14}

So that the first of brass lost ʒ vii gr. {35} et so the rest proportionably to the bulk, or body of each metal <the {…}> from whence he inferred that as the two of pure brass, and lead se habebant unto[45] the third, which was <only> of brass, and lead together, even so this third ball had more or less lead mingled with the brass.[46]

[44] For an account of the two methods, the first described by Vitruvius and Plutarch, and the second in the *Carmen de ponderibus et mensuris* (circa fourth–sixth century), as well as by Giovanni Battista Della Porta and Galileo, see Crombie 1996, chapter 10.

[45] "They were with respect to […]."

[46] To sum up, the proportion of metals in the mixed body can be calculated as it follows: the weight of the volume of gold (or lead) in the mixed crown is obtained by subtracting the amount of weight lost by the mixed crown, from that lost by the silver (or brass) one. The weight of the volume of

[122v] Experim[en]tum 20 eod[em] die.

1. His subject was the same with the former.

2. To prove that things, I mean balls of the same weight do loose of their when put into water. That is to say, the one will weigh down the other. He put two brass balls upon a pair of scales I mean two balls of equal weight the one of brass, and the other of lead. Then he held two glasses of water under those two balls dipping the balls in the water so far, as till they were just covered with water: and loe the one did clearly weigh down the other: by reason of its bulk according to the principle already given.

[123r] Experim[en]tum 21 d[ie] 16 9^{bris}

1. His subject was a trial whether as bodies of equal weight did become of an unequal weight being put into water, as hath been proved by precedent experiment whither I say this same would hold in the air, as being likewise a liquid body much like unto water: viz. in water two things have been proved: 1° that two bodies of the same weight, and bigness, the one being put into water they do become unequal, and that which is left out of the water doth weigh down that which is in: 2° that two bodies of the same weigh but of different metals, or bigness, being put into water, that which is biggest, is overcome, and weighed down by the least, by reason of the greater pression of the water upon the bigger body.

[123v] 2. This therefore he desired to try by air, especially in the pump. To this effect he hung in the pump a very small pair of scales for gold weight.

3. In the one of these scales he put a crystal globe, of the bigness of a small apple. In the other scale he the equal, and just weight of said globe. Now whereas these two

---

silver (or brass) in the mixed crown is obtained by subtracting the amount of weight lost by the gold (or lead) crown, from that lost by the mixed one. In the case of De Volder's experiment, their weight outside the water is 6 ounces, 3 drachms, 19 grains (n.b.: number is scarcely readable), in my conversion, 3,079 grains (based on the proportion of 1 ounce = 8 drachms = 480 grains; or: 1 drachm = 60 grains, according to the Amsterdam measures). Their weights in the water are as follows: brass: 5 ounces, 4 drachms, 5 grains (which can be converted into 2,645 grains); lead: 5 ounces, 6 drachms, 5 grains (2,765 grains); mixture: 5 ounces, 4 drachms, 3 grains (2,643 grains). Morley writes that the brass body lost in water 3 drachms and (n.b.: number is scarcely readable) 35 grains. In my conversion, this corresponds to 455 grains, diverging by 21 grains from the difference (in grain) between the initial weight and the weight in the water, which is 434 grains. This error can be attributed to the scarcely readable values reported by Morley. Yet, his values for the loss of weight in the water may be incorrect, as the mixed body loses more weight than the other ones. Lead, of course, has the higher specific weight and loses the least weight in the water, as it is more dense: in this experiment, it corresponds to gold. Brass has a lower specific weight, and it ought to lose more weight than the mixture, which however loses more weight than the two other bodies. Before reporting these actual measurements, Morley notes that gold should lose, for example, 10 pounds, silver loses 13 pounds, and their mixture 12 pounds: this example contradicts the actual values which are reported. Eventually, he claims that each body loses more or less weight according to its "bulk" (which can be either the quantity of matter or the volume, which are not distinguished) and that the proportion between the lead and the brass body with the mixed one (in fact, Morley omits to clarify that what matters is the proportion of the differences between the weights which are lost by these two bodies), corresponds to the proportion of lead and brass in the mixed body (namely, of their weights, and quantity of matter or volume).

being put into water, according to the rule given n.° 1. of two bodies equal weight of a different bigness the least weigheth down the biggest by reason of the pression of the water, as hath been said: so should it proceed e contrario viz. that if that pression be taken away and the water[47] more, and more rarefied, the greatest should weigh down the least.

[124r] 4. Thus having rarified the air, or drawn it out of the pump: loe the scales {...} and the globe weighed down the counterpoise, or weight: then letting in the air again loe the counterpoise weighed down the globe: which was his intent and proof.

[124v] Experim[en]tum 22. d[ie] 19° 9bris.

1. His subject was the same with the former experiments, viz. the ponderosity of water. Let in this he chiefly intended a confirmation of some former propositions, which he had grounded in some of his precedent operations, as {thesei} corpora quae sunt aqua ponderosiora, et graviora, ad fundum aquae subsident. Item, corpora quae sunt leviora, qui[48] ipsa aqua, supernatare, et super fluitare in ipsa aqua. 3° corpora, quae sunt eisudem cum aqua ponderis, neque ad fundum aquae subsidere, neque superfluitare, sed in medio aquae haerere, et si ad fundum aquae detrudant illic haerere, si in medio collocent, illic itidem haerere, sin vi alicuius elevent prope ipsam super- [125r] perficiem aquae, illic itidem, ac eodem m[od]o haerere.[49]

2. To this effect he had standing before him an indifferent big glass full of clear water, which might contain two, or 3 gallons.

3. Then he took a round glass, or ball, of crystal, which seemed to be hermetically sealed. The crystal glass, which was of the bigness of an ordinary apple, or somewhat less, had a neck, unto which any thing might be tied, or hanged.

4. This glass (or any other empty bottle stopped will do) being flung into the great glass of water, would swim, and never sink of its own accord: and this experiment [125v] prooveth one of those 3 propositions viz. that bodies which are lighter,[50] than the water they are cast into, do swim, and not sinke in said water. R[ati]o e[rat], q[u]a cum illa magnitudo aquae, quae corr[espond]et magnitudini ipsius aquae,[51] non possit loco suo pelli ab ipso corpore, q[u]a gravior ipso e[st], hinc in eodem loco suo manet, corpusque in ipsam proiectum superfluitat.[52]

---

[47] Read: "air."

[48] *Sic.*

[49] "[The] bodies which are heavier than water, and weightier, subside to the bottom of the water[.] Also, the bodies which are lighter, [than] water itself, swim and float upon the water itself[.] Third, the bodies, which are of the same weight as water, neither subside to the bottom of the water, nor do they swim upon it, but they remain in the middle, and if they [are] dislodged to the bottom of the water they remain there, if they [are] placed in the middle, they likewise remain there, if on the contrary they emerge by the force of something close to the very surface of the water, they remain there, in the same way."

[50] *Sic.*

[51] Read: "corporis."

[52] "The reason was, because since that quantity of water, which corresponds to the quantity of the water [read: "body"] itself, cannot be pushed out from its place by the body itself, because [the water] is heavier than the same [body], from this [it follows that the water] remains in its place, and the body thrown into it floats upon it."

5. Then he hang by the neck of the crystal glass certain weights of ounces, drams, grains et cetera and loe with these it presently sunk to the bottom, which proved also evidently another of these propositions: et r[ati]o erat, q[u]a cum iam corpus illud, mediantibus illis additis ponderibus redditum, e[ss]et gravius ipsa aqua, ita ut magnitudo aquae, quae corr[ispond]ebat [126r] magnitudinis corporis in ipsam proiecti, iam levior e[ss]et ipso corpore in illam proiecto, hinc aqua loco suo pellebat[ur] ab illo copore, atq[ue] adeo corpus ad fundum subsidebat, et submergebat. Eod[em] m[od]o, quo in bilance contingit illa lanx {ei} quae gravior deprimitur, illa v[er]o, quae levior, perpetuo attol[litu]r.[53]

6. Last of all he added, and diminished, several times by drams, grains, and half grains, of the weight which did hang at the neck of the crystal glass, until that he made it to be of an even weight with the water (which was somewhat tedious to perform) and loe then the glass did neither sink, nor swim but kept still in that place where {y.^{no}} should putt: it: if at the top of the water, there it would hover prope superficiem aquae,[54] (leaving [126v] not leas jot of its own body extant upon, or out of the superficies of the water) if thrust down to the bottom, there it would remain, and not ascend, <and {...}> if brought up again prope superficiem aquae there it would stay, and not descend: deniq[ue] if it were left in the middle of the water, there it would hover, and hang, and neither ascend, nor descend.

7. This did evidently prove unto us the other of those three propositions, viz. q[uo]d corpora eisud[em] gravitatis cum aqua neq[ue] ascendunt, neq[ue] descendunt in aqua.[55] For the reason of this experiment was: qu[i]dem magnitudo aquae corr[espond]ens magnitudini illius corporis in illam proiecti eiusdem o[mn] i[n]o esset cum illo gravitatis, hinc neque illa [127r] pellebat corpus ad superficiem, neq[ue] item ab illo corpore eadem aqua poterat pelli loco suo, unde corpus ad fundum subsideret, ac demergeret in postremum aquae locum. Eodem plane m[od]o, quo cum bilances aequalis o[mn]i[n]o ponderis sunt, tunc nempe altera alteri n[on] cedit, neque una deprimit, ac altera attollit, sed aequales prorsus permanent in aëre.[56]

---

[53] "[...] and the reason was, because that body had been made heavier than the water itself by means of those added weights, so that the quantity of the water, which corresponded to the quantity of the body thrown in the water itself, [was made] lighter than the same body thrown in the water, hence the water [was] pushed [out of] its place by that body, and so the body subsided to the bottom, and sank. In the same way, indeed[,] in which in a balance it happens that the arm which is heavier is depressed, while that one, which is lighter, is perpetually raised."

[54] "Near the surface of water."

[55] "That the bodies of the same weight with the water neither ascend, nor descend in water."

[56] "[I]n fact[,] being the quantity of water correspondent to the quantity of that body thrown in that [water] entirely of the same weight with that [body], hence neither that [water] pushed the body to the surface, neither indeed the same water could be pushed [out of] its place by that body, whence the body would subside to the bottom, and sink in the last spot of water. Plainly[,] in the same way, in which when balances are wholly of the same weight, then obviously the one does not cede to the other, and one does not depress, and the other does not rise, but straightly equal they stand in air."

[127v] Experim[en]tum 23um d[ie] 23° 9^{bris}

1. His subject was against that tenet of the ancient Peripatetics, viz. corpora naturalia non gravitare, nec ponderare in elementis suis,[57] as water, the which if a bucket be let down into a well, or pond, so long as the bucket is within the water, though full of water, yet it seemeth, to weigh at all, until it be drawn out of the water into the air. The contrary of this tenet he intended to prove viz. that corp[or]a na[tur]alia gravitabant in elementis suis, et q[uo]d aqua ponderabat in aqua, et super aquam et cetera.[58]

2. To this effect he had before him 3 or 4 several liquors: in distinct glasses, as 1° aqua pura: 2° aqua pura sangui- [128r] neo colore, ab admixto aliquo, tincta: 3° aqua pura {virididi}[59] colore, ab admixto item a<q>liquo, tincta: 4° ol[eum] terebinthinae lucidum ac aethereum: 5° oleum item terabinthinae sanguineo colore tinctum.[60]

3. First therefore he toke a glass, less than half full of crystal clear water. Into this he put a small crystal tube, (such as are used for weather glasses), and holding it perpendicular in the glass, I mean letting it down to the bottom of the glass, loe, as is commonly known, and seen, the water ascended within the tube to the same height, the water was within the glass. Now, saith he, if that tenet were true, that water doth ponder, nor press, in water, as being in its own element, this water would ascend not higher though {I should} fill the glass up to the brim: [128v] but now so it is, that the more water I put in, the higher it ascends: {go} this is an infallible sign that the water, which is newly put in doth ponder, and press, and gravitate, upon that which was in the glass before, for what else could make the water in the tube ascend higher, except it were the water newly poured in?

4. This same tube he put into the water which was dyed redd, as hath been said: and loe the water ascended within the tube as before. Now, saith he, that any one may not, that the water which is newly put in ascendeth and not that which was not the bottom of the glass: behold upon this red water, I will pour lucid oil of terebinth: and this not mingling at all with the water, but swimming at the top, shall make [129r] by its pression the red water to ascend higher and higher within the tube: thus he did, and so it was, where by the difference of colour, we could perceive all the oil to swim at the top, and not a drop to sink downwards, or ascend in the tube: {go} quoth he this is an evident proof of the pression of this oil of terebinth, upon the water, otherwise what could be imagined to make the red water with in the tube to ascend, but only said pression. Let the red water did not ascend to the total height of the oil of terebinth, I mean to its uppermost superficies, of which he said the cause was that the oil was lighter than the water, and so could not press it totally so higher as the oil itself was.

---

[57] "Natural bodies neither gravitate, nor weigh in their elements."

[58] "Natural bodies gravitated on their [own] elements, and that water weighed on water, and upon water, and so on."

[59] Likely: "viridi" (green).

[60] "First[,] pure water. Second[,] pure water tinged with a bloody colour, from some mixture. Third[,] pure water tinged with a green colour, likewise from some mixture. Fourth[,] ethereal and lucid oil of terebinth. Fifth[,] likewise the oil of terebinth tinged with a bloody colour."

5. This same experiment he tried seve- [129v] rall other ways with those liquors, backwards, and forwards, making of every way a new proof of the same subject: but all came to the same purpose and effect, and all did prove the very same thing, whereby this being the clearest and most intelligible proof of all, I have here omitted the rest.

[130r] Experim[en]tum 24um d[ie] Ian[uarii] 28 an[no] 1677. @[61]

1. This experiment was made with mercury: relating to the former ones, viz. the equiponderosity of material things with water, or rather as Volder printed these days in some theses which he presided, omnia corpora, quae innatant liquori, t[antu]m h[ab]ent ponderis, quantum liquor a corpore suo ex loco expulsus ponderat.[62] As also this experiment did relate, unto this thesis likewise then printed by him: aquam, et aërem esse corpora gravia, cum r[ati]o, tum exper[ienti]aa[63]docet, ac proinde falsum est, elementa [130v] in suis locis non esse gravia.[64]

2. He had standing by him a glass full of mercury or quicksilver, as also several small tubes of glass, such as do commonly serve for weather glasses.

3. His subject was to prove that water could press upon mercury no less than air, and would make it ascend within one of those tubes by its pression, no less than air, if not ad replendum vacuum as hath been proved by sundry experiments at least [131r] by circumpuls[i]o[n]em.

4. Hence he dipped one of the tubes into mercury. This done he stopped the other end of the tube with his fingers, and loe the mercury did hang, as is usual, in the tube to the thickness or depth of an inch or more, (sometimes more, sometimes less). Then putting the tube with the mercury in it, into the water (viz. a cup of water that was standing by) loe the water to that proportion of mercury which was in the tube, being heavier than the mercury, did make the mercury ascend within the tube, he having before taken his finger off from the other end.

[131v] 5. This he repeated several times, and so often the mercury would ascend. Others more, or less, according to the proportion of the mercury to the ponderosity of the water. Others it would stand as it were in aequilibrio. Others when the tube was too much filled with mercury, it would fall flat to the bottom when too heavy for the water.

6. Concerning this experiment he observed that the tube must be neither too small nor too big. If too small the mercury would stick in the [132r] glass, and neither ascend nor descend: if too big the water would get in betwixt the mercury and the glass, and so get above it, and press it downwards.

7. Torricellius was the first that found the ascension of mercury by the pression of the air: but Pascalius a very ingenious \X/ Frenchman by reason rather than

[61] *Sic.*

[62] "All bodies, which float in a liquid, have so much weight, as much liquid expelled by their body from its place[,] weighs."

[63] *Sic.*

[64] "Water, and air are heavy bodies, as both reason, and experience teach, and therefore it is false, that elements are not heavy in their places."

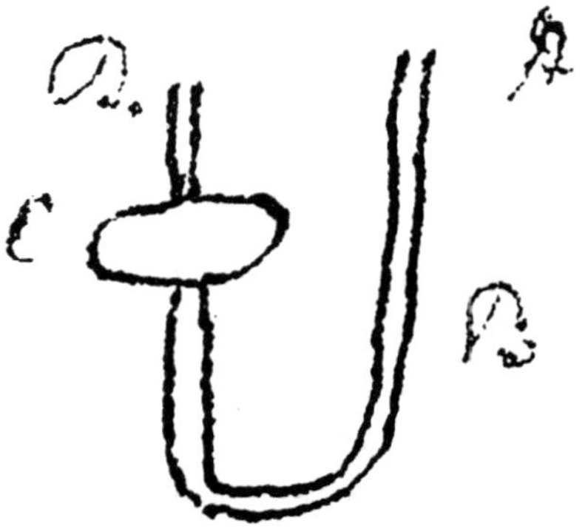

Fig. 2 An inverted, asymmetrical syphon used to demonstrate that the pressure of water depends only on its height. (Source: De Volder 1676–1677, 133r. Courtesy British Library, ms. Sloane 1292)

expressa[65] pretended to prove that in a tube of 20 foot long the mercury ought to ascend only thus, et thus, by the pression of the water.[66] Boyle first found that he had never tried those experiment, as never being at the bottom of the [132v] water thereby to see how far it did ascend or not. Hence \X/ he refuteth Pascalius: v[id]e ipsum Boilium.[67]

[133r] Experim[en]tum 25um d[ie] Martii

1. His subject was hydrostatics or the pression of water.
2. He said this was a received tenet by all those, that treat of hydrostatics \X/ viz. aquam, dum premit, non premere nisi per lineas rectas, nullo {vò}[68] per laterales, transversas, aut obliquas.[69]
3. This he proved by an instrument of glass made much after this nature: [see Fig. 2, De Volder 1676–1677, 133r][.] [133v] Into this instrument he first infused water at A. And as he proceeded in the infusion, the water did equally ascend towards B and towards C, so that on both sides it was equal: but when it came into the round ball at C, then extending itself collaterally in the ball it would ascend no higher towards A. This experiment he tried several other ways beginning the infusion at D and with other liquors but all tended to prove the same thing, and had the same effect.

[134r] Experim[en]tum 26um d[ie] 18 Martii.

1. His subject seemed to be the pression of the air, viz. that it is the pression of the air, and not metus vacui that maketh water, and other liquors to ascend, when the air is drawn out of a pipe: and this seemed to me to be the experim[en]tum

[65] *Sic*. Read: "experientia."

[66] See Pascal 1663, chapter 3.

[67] See Boyle 1666, 64–65.

[68] Likely: "m[od]o."

[69] "Water, while it presses, does not press but through straight lines, in no way through lateral, transverse, or oblique [lines]."

torricellianum of which Consentinus,[70] and others do make so much mention, this being performed by mercury: I say it seemed so to me, for, otherwise, this day, we were detained so long by Professor De Maëts,[71] in the public laboratory, about some experiments of chemistry, that we all came too late, to hear his proposal at the beginning [of] the experiment.

[134v] 2. He had in his hand a long pipe of glass, of such a bigness as your weather glasses though the base seemed bigger, and the pipe was as long again, as your ordinary weather glasses. At the end this pipe had a kind of screw, whereby he easily could suck, or draw out all the air, and make any liquor to ascend through it.

3. This pipe he had placed in a pot of mercury, and was, when we came in actually drawing out the air and making the mercury to ascend, some drops {ther if} <{...}> having already [135r] ascended to about the middle of the pipe, where he had placed a mark with a red silk: and by letting in again the air the mercury was gone down to the bottom.

4. He said, therefore, that according to his former principles of the pression of the the air in opposition to the ancient tenet of the metus vacui, if now really the metus vacui were the cause of the ascension of the mercury, consequently the mercury ought to ascend ad replendum omne vacuum i.e. as far as the vacuum did go, viz. to the very end of the pipe it being all equally void or empty of air, by the exsuctions made [135v] sed sic erat,[72] that no exsuctions, though never so many, could make the mercury to ascend about the mark of that red silk: {go} he concluded that was an evident token, it was not the metus vacui, but the compression of the air, which did make the mercury to ascend, seeing that when the mercury came to be in aequilibrio with the air, then it would ascend no higher.

5. This he confirmed further, by that at the first exsuctions of the air the mercury did ascend v.g. 8 fingers breadth in height at the second and third less, and so less, and less, at every exuction till that coming to the mark it would go no [136r] higher: which proved, said he, that {an ordiny} as the pression of the air was greater and greater, or lesser, and lesser, so much the more, or less, did the mercury at every exuction, according to the equilibrium it was to be reduced unto, for as it came nigher the equilibrium of the air, so did it ascend less, and less, at every exsuction.

6. Then he added, that Torricellius or the author of this experiment to these precedent proofs did adjoin, that having left the mercury casually in the pipe, at the highest degree it would ascend unto, for some days, he observed that some days the mercury of its own accord would ascend higher than the mark, and some days it would be lower than the mark [136v] which observation he said could proceed from no other principle, but from the different ponderosity, or pression of the air, more in one day, than in another, which accordingly did cause the mercury to ascend sometimes higher, and other times lower, \as/ if it were, of its own accord.

[70] See Cornelio 1663.

[71] Morley refers to Carel de Maets, professor of chemistry at Leiden.

[72] "But it was so."

7. All this he proved, lastly, by the experiment itself, we seeing the mercury to ascend by degrees, more at the beginning, and less towards the end, till it would ascend no higher, than the mark of the red silk.

[137r] Experim[en]tum 27 d[ie] 22 Martii.

1. His subject was not only to prove the real <pression> \gravity/ of the air, but also how to demonstrate the just weight of the said air.

2. He set upon the pump a round bowl of brass, divided into two hemispheres, hollow in the middle, the which were joined together only with tallow: and this way was the same bowl which in precedent experiments had demonstrated the <{...}> pression of the air, insomuch as that being so slightly joined, the at the air being drawn out of this bowl, it had sustained 700 lb [137v] weight, before the two hemispheres could be disjoined, or torn asunder.

3. Out of this bowl he, first, drew all the air: then took it off from the pump, and put into a pair of scales: in the <putt> contrary scale he put the just, and exact weight of said bowl which was betwixt seven et eight lb: the bowl being in aequilibrio with the weights, by turning a screw, he let the air again (which he \had/ drawn out) into the same globe, and loe then the globe considerably weighed down the scales, which was an evident proof that the globe did weigh more when [138v] full of air, than when it was void.

4. Next, he placed the globe, the second time, on the pump, and drew out all the air: then weighed \it/ in the scales exactly, and it was found to weigh 7 lb i ℥ 4 ʒ i ℈ 5 gr. Then he let in the air again into said globe, and weighed it exactly, and it was found to weigh exactly one ʒ and an ½ more than before: which was item an evident proof the air had ever eased this weight one whole dram and an half, and that this was the just weight the air contained within the globe of brass.

[138v] 5. After this he filled the globe with common water, and weighed it exactly in the scales, and, full of water, it was found to have 9 lb 13 ℥ ii ʒ and ii ℈. All this was Amsterdam weight, which is the usual in these countries.[73]

6. Hence he made an exact supputation of the proportion betwixt the weight of the air, and the water which was as 85 grains this being the weight of the air unto 7035 grains which was the weight of the water contained in the globe.

[139r] 7. Hence, remembering by his precedent experiments the weight of a foot cubic of water he made the account, and comparison of the proportion betwixt a foot cubic of water, and a foot cubic of air and found by this {an.°} that the disproportion was not half so great as many had conceived, and that air was much heavier and had a greater pression, than was commonly thought the difference not being as one to three or four thousand, but rather not quite so much as one to a 1000.

8. Lastly to show how great the pression of the air was, he said that that globe whose hemispheres could not otherwise [139v] be separated with less than 700 lb weight, he could separate it, when full of air <when> with only his breath, the which he easily performed by blowing it asunder.

[73] See *supra*, n. 46.

[140r] Experim[en]tum 28 d[ie] 25 Martii

1. His subject was hydrostatics.

2. He commended, first, Stevinus very much, for his exact inquisition into hydrostatics, affirming, he had given the t<two>\rue/ grounds, and rules, of this art, having examined all the chief things, which herein did occur.

3. Next he said, that notwithstanding this commendable exactitude of Stevin he had one proposition,[74] which had seemed very strange, and paradox, to the virtuosi of this art viz. that water did press more when elevated in a small tube upon [140v] a cylindrical vessel, than the weight of the water contained in the tube did require. V.g. in this instrument. [See Fig. 3, De Volder 1676–1677, 140v]

Let the vessel A contain 6 lb of water, it will press upon the bottom C even just as 6 pound weight would do: let the tube B contained one lb of water being filled up to the very top, where B is: hence it should follow that all the water both in the vessel, and [141r] the tube should press upon the bottom C no more than as much as 7 lb weight. Now, quoth Stevinus, this is not so, but by reason of that elevation in the tube of the water, it will not only press as 7 lb weight, but even perhaps as much as seven times seven pounds weight; at least it well press incredibly above reason, or the expectation of men's wit.

4. This problem of Stevinus had rendered itself incredible to almost all authors that had treated since of hydrostatics yea perhaps some had laughed at it, as bearing the appearance of a ridiculous thing. Boyle could not believe it, yea, said he conceived Stevinus had never tried it by any true experiment.[75]

Fig. 3 An ideal, inverted T-shaped container aimed at showing the hydrostatic paradox. (Source: De Volder 1676–1677, 140v. Courtesy British Library, ms. Sloane 1292)

[74] See Stevin 1605, 119–121.

[75] See Boyle 1666, 135–140. See also Schott 1657–1659, volume 3, 456.

Bibliography

Primary Sources

Texts in Which Burchard de Volder Figured as Author, Praeses, Respondens, Candidatus *or* Promotor *(in Chronological Order)*

De Bie 1658 = Bie, Alexander de (*praeses*), and Volder, Burchard de (*respondens*). 12 October 1658. *Disputatio mathematica de profunditate maris*. Amsterdam: Apud Ioannem Banningium.

De Bie 1659 = Bie, Alexander de (*praeses*), and Volder, Burchard de (*respondens*). 30 January 1659. *Disputatio de linea, quam globus per aërem describit missus e tormento*. Amsterdam: Apud Joannem Ravesteinium.

De Volder 1660 = Volder, Burchard de (*candidatus*), and Bruyn, Johannes de (*promotor*). 17 October 1660. *Disputatio philosophica inauguralis de simplicitate Dei, echo et republica*. Utrecht: Ex Officina Jacobi a Doeyenborgh.

De Volder 1664 = Volder, Burchard de (*candidatus*), and Sylvius, Franciscus (*promotor*). 3 July 1664. *Disputatio medica inauguralis de natura*. Leiden: Apud Severinum Matthiae.

De Volder 1671a = Volder, Burchard de (*praeses*), and Carceus, Martinus (*respondens*). 28 January 1671. *Disputatio physica de corpore prima et secunda*. Leiden: Apud Viduam et Haeredes Johannis Elsevirii.

De Volder 1671b = Volder, Burchard de (*praeses*), and Baumgartus, Valentinus (*respondens*). 18 April 1671. *Disputatio physica de motu prima*. Leiden: Apud Viduam et Haeredes Johannis Elsevirii.

De Volder and Enyedi 1671 = Volder, Burchard de (*praeses*), and Enyedi, Johannes (*auctor et respondens*). 30 May and 6 June 1671. *Disputatio philosophica de anima humana prima* [*–secunda*]. Leiden: Apud Viduam et Haeredes Johannis Elsevirii.

De Volder 1671c = Volder, Burchard de (*praeses*), and Eperjesi, Michael (*respondens*). 31 October 1671. *Disputatio philosophica de Deo*. Leiden: Apud Viduam et Haeredes Johannis Elsevirii.

De Volder and Van Alphen 1672 = Volder, Burchard de (*praeses*), and Alphen, Johannes van (*auctor et respondens*). 16 March 1672. *Disputatio philosophica de ideis*. Leiden: Apud Viduam et Haeredes Johannis Elsevirii.

De Volder and De la Bassecour 1672 = Volder, Burchard de (*praeses*), and Bassecour, Carolus de la (*auctor et respondens*). 19 March 1672. *Disputatio philosophica, in qua quid de principiis

A. Strazzoni, *Burchard de Volder and the Age of the Scientific Revolution*, Studies in History and Philosophy of Science 51,
https://doi.org/10.1007/978-3-030-19878-7

philosophiae vulgaris sit censendum disquiritur. Leiden: Apud Viduam et Haeredes Johannis Elsevirii.

De la Bassecour 1672 = Bassecour, Carolus de la (*candidatus*), and Volder, Burchard de (*promotor*). 19 March 1672. *Disputatio philosophica, inauguralis, continens positiones ex universa philosophia desumtas*. Leiden: Apud Viduam et Haeredes Johannis Elsevirii.

De Volder 1672 = Volder, Burchard de (*praeses*), and Hoet, Simon (*respondens*). 14 December 1672. *Disputatio physica de motu secunda*. Leiden: Apud Viduam et Haeredes Johannis Elsevirii.

De Volder 1673a = Volder, Burchard de (*praeses*), and Nieuwentijt, Bernard (*respondens*). 25 March 1673. *Disputatio philosophica de unitate Dei*. Leiden: Apud Viduam et Haeredes Johannis Elsevirii.

De Volder 1673b = Volder, Burchard de (*praeses*), and Hannot, Michael (*respondens*). 28 June 1673. *Disputatio philosophica de criterio veritatis prima*. Leiden: Apud Viduam et Haeredes Johannis Elsevirii.

Volder, Burchard de (*praeses*). 1674–1676. *Disputatio philosophica de rerum naturalium principiis, prima*[*–quintadecima*]. Leiden: Apud Viduam et Haeredes Johannis Elsevirii.

De Volder 1675 = Volder, Burchard de (*praeses*), and Bruno, Johannes (*respondens*). 13 November 1675. *Disputatio physica de motu tertia*. Leiden: Apud Viduam et Haeredes Johannis Elsevirii.

De Volder and Lufneu 1676 = Volder, Burchard de (*praeses*), and Lufneu, Hermann (*auctor et respondens*). 27 January 1676. *Disputatio philosophica de materiae divisibilitate in infinitum*. Leiden: Apud Viduam et Haeredes Johannis Elsevirii.

De Volder 1676–1678 = Volder, Burchard de (*praeses*). 1676–1678. *Disputatio philosophica de aëris gravitate prima*[*–quinta*]. Leiden: Apud Viduam et Haeredes Joannis Elsevirii.

Heidanus et al. 1676 = Heidanus, Abraham, Volder, Burchard de, and Wittich, Christoph. 1676. *Consideratien, over eenige saecken onlangs voorgevallen in de Universiteyt binnen Leyden*. Leiden: By Aernout Doude.

Heidanus, Abraham, Volder, Burchard de, and Wittich, Christoph. 1676. *Consideratien, over eenige saecken onlangs voorgevallen in de Universiteyt binnen Leyden*. Amsterdam: By Johannis van Someren, 1676.

De Volder and Helvetius 1677 = Volder, Burchard de (*praeses*), and Helvetius, Johannes Balthasar (*author et respondens*). 10 July 1677. *Disputatio philosophica de magnete*. Leiden: Apud Viduam et Haeredes Johannis Elsevirii.

Heidanus, Abraham, Volder, Burchard de, and Wittich, Christoph. 1678. *Considerationes ad res quasdam nuper gestas in Academia Lugduno-Batava. Cum triplici appendice*. Hamburg: Apud Petrum Grooten.

Rattrey 1679 = Rattrey, Jacobus (*candidatus*), and Volder, Burchard de (*promotor*). 30 May 1679. *Disputatio philosophica inauguralis exhibens vitri thermometrici historicam simul et acroamaticam delineationem*. Leiden: Apud Viduam et Haeredes Johannis Elsevirii.

De Volder 1679 = Volder, Burchard de 1679. Oratio funebris in obitum [...] Siberti Coeman. In Volder, Burchard de, and Coeman, Sibert. 1679. *Orationes duae, quarum altera inauguralis Sib. Coeman* [...] *altera Baruch de Volder funebris*. Leiden: Apud Jacobum Voorn.

Volder, Burchard de (*praeses*). 1680–1681. *Disputatio philosophica contra atheos prima*[*–quarta*]. Leiden: Leiden: Apud Viduam et Haeredes Johannis Elsevirii.

De Volder and Szathmári 1681 = Volder, Burchard de (*praeses*), and Szathmári, Johannes Pap (*auctor et respondens*). 5 March 1681. *Disputatio philosophica de aeternitate Dei prima* [*–secunda*]. Leiden: Apud Viduam et Haeredes Johannis Elsevirii.

De Volder and Köleséri 1681 = Volder, Burchard de (*praeses*), and Köleséri, Samuel (*author et defendens*). 12 and 15 March 1681. *Disputatio mathematico-physica de lumine, pars prima* [*–secunda*]. Leiden: Apud Viduam et Haeredes Johannis Elsevirii.

De Volder 1681 = Volder, Burchard de. 1681. *Disputationes philosophicae sive Cogitationes rationales de rerum naturalium principiis*. Middelburg: Typis Remigii Schreverii.

Volder, Burchard de. 1681. *Quaestiones academicae de aëris gravitate*. Middelburg: Typis viduae Remigii Schreverii.

Volder, Burchard de. 1681. *Disputationes philosophicae sive Cogitationes rationales de rerum naturalium principiis, ut et de aëris gravitate*. Leiden: Apud Jacobum Moukee.

Köleséri 1681 = Köleséri, Samuel (*candidatus*), and Volder, Burchard de (*promotor*). 30 June 1681. *Disputatio philosophica inauguralis de systemate mundi*. Leiden: Apud Viduam et Haeredes Johannis Elsevirii.

Volder, Burchard de (*praeses*), and Vereem, Johannes (*respondens*). 1682. *Disputatio philosophica de dependentia creaturae*. Leiden: Elsevier. (Saint Petersburg, Academy of Science of Russia, not retrieved).

Volder, Burchard de (*praeses*), and Langenhert, Casparus (*respondens*). 1682. *Disputatio philosophica de systemate mundi, prima*. Leiden: Elsevier. (Saint Petersburg, Academy of Science of Russia, not retrieved).

De Volder and Nemethi 1682 = Volder, Burchard de (*praeses*), and Nemethi, Samuel (*auctor et defendens*). March 1682. *Exercitii philosophico-metaphysici de perenni duratione mentis pars prima*[*–tertia*]. Leiden: Apud Abrahamum Elzevier.

De Volder and Derecskei 1682 = Volder, Burchard de (*praeses*), and Derecskei, Pál (*author et defendens*). 16 March–2 May 1682. *Exercitatio philosophica, coelorum, siderumque lucidorum originem, et phaenomena, methodo synthetico-mathematica demonstrans prima*[*–quarta*]. Leiden: Apud Abrahamum Elzevier.

De Volder 1682 = Volder, Burchard de. 17 July 1682. *Oratio de coniungendis philosophicis et mathematicis disciplinis, cum philosophicae professioni adiunctam mathematicam rite auspicaretur*. Leiden: Apud Jacobum Voorn.

Volder, Burchard de (*praeses*), and Wariensis, Michael (*respondens*?). 21 October 1684. *Theorema philosophicum de natura et modo praesentiae rerum intellectualium ducta viae analyticae investigatum*. Leiden: Elsevier. (Budapest, National Library of Hungary, signature: RMK III. 3296. Not retrieved).

De Volder and Van Alphen 1684 = Volder, Burchard de (*praeses*), and Alphen, Hieronymus Simons van (*author*). 19 April 1684. *Exercitatio philosophica utrum Sol, an vero Tellus, in planetarum numerum referenda sit, inquirens*. Leiden: Apud Abrahamum Elzevier.

De Volder and Vander Codde 1684 = Volder, Burchard de (*praeses*), and Codde, Pontianus vander (*auctor et defendens*). 29 November 1684. *Disputatio philosophica de motu*. Leiden: Apud Abrahamum Elzevier.

De Volder and Von der Lahr 1684 = Volder, Burchard de (*praeses*), and Lahr, Paulus von der (*auctor et respondens*). 6 December 1684. *Disputatio philosophica de absoluta quiete*. Leiden: Apud Abrahamum Elzevier.

De Sadeler 1684 = Sadeler, Matthaeus Antonius de (*candidatus*), and Volder, Burchard de (*promotor*). 3 February 1684. *Disputatio philosophica inauguralis de meteoris*. Leiden: Apud Abrahamum Elzevier.

De Volder and Huneken 1685 = Volder, Burchard de (*praeses*), and Huneken, Jacobus (*autor*). 7 July 1685. *Disputatio philosophica de variis quibus ignis generatur modis*. Leiden: Apud Abrahamum Elzevier.

De Volder and Van Bronchorst 1685 = Volder, Burchard de (*praeses*), and Bronchorst, Henricus van (*auctor*). 11 July 1685. *Disputatio philosophica de vera gravitatis causa*. Leiden: Apud Abrahamum Elzevier.

De Volder 1685 = Volder, Burchard de. 1685. *Disputationes philosophicae omnes contra atheos*. Middelburg: Apud Joannem Lateranum.

Wariensis 1685 = Wariensis, Michael (*candidatus*), and Volder, Burchard de (*promotor*). 24 September 1685. *Disputatio philosophica inauguralis de metallo regio*. Leiden: Apud Abrahamum Elzevier

De Volder and Van den Honert 1686 = Volder, Burchard de (*praeses*), and Honert, Taco Hajo van den (*auctor*). 3 July 1686. *Disputatio philosophica prior–posterior de mundi in tempore productione*. Leiden: Apud Abrahamum Elzevier.

De Volder and Serrurier 1688 = Volder, Burchard de (*praeses*), and Serrurier, Joseph (*auctor et respondens*). 19 May 1688. *Disputatio philosophica de mente humana secunda*. Leiden: Apud Abrahamum Elzevier.

Schuyl 1688a = Schuyl, Hermann (*candidatus*), and Volder, Burchard de (*promotor*). 2 July 1688. *Disputatio philosophica inauguralis de vi corporum elastica*. Leiden: Apud Abrahamum Elzevier.

Schuyl 1688b = Schuyl, Hermann (*candidatus*), and Volder, Burchard de (*promotor*). 2 July 1688. *Disputatio physico-medica inauguralis de respiratione*. Leiden: Apud Abrahamum Elzevier.

De Volder and Mandeville 1689 = Volder, Burchard de (*praeses*), and Mandeville, Bernard de (*author et respondens*). 13 March 1689. *Disputatio philosophica de brutorum operationibus*. Leiden: Apud Abrahamum Elzevier.

De Volder and Vander Tak 1689 = Volder, Burchard de (*praeses*), and Tak, Gaspar vander (*auctor et respondens*). 13 July 1689. *Disputatio philosophica de certitudine clarae et distinctae perceptionis*. Leiden: Apud Abrahamum Elzevier.

De Volder 1689 = Volder, Burchard de. 15 March 1689. *Oratio funebris in obitum Lucae Schacht*. Leiden: Apud J. Voorn.

De Volder 1690–1693 = Volder, Burchard de (*praeses*). 1690–1693. *Exercitationum philosophicarum adversum Censuram prima*[*–vicesima octava*]. Leiden: Apud Abrahamum Elzevier.

Serrurier 1690 = Serrurier, Joseph (*candidatus*), and Volder, Burchard de (*promotor*). 18 December 1690. *Disputatio philosophica inauguralis de gravitate aëris*. Leiden: Apud Abrahamum Elzevier.

De Volder and Voogd 1691 = Volder, Burchard de (*praeses*), and Voogd, Petrus (*auctor et respondens*). 9 May 1691. *Exercitium medico-physico aphoristicum: de morborum causis*. Leiden: Apud Abrahamum Elzevier.

De Volder and Huszthi 1693 = Volder, Burchard de (*praeses*), and Huszthi, Stephan (*author et defendens*). May–September 1693. *Dissertatio physiologica* [...] *prima*[*–quarta*]. Leiden: Apud Abrahamum Elzevier.

Oosterdijk Schacht 1693 = Oosterdijk Schacht, Hermann (*candidatus*), and Volder, Burchard de (*promotor*). 6 July 1693. *Disputatio philosophica inauguralis de sensibus internis memoria et imaginatione*. Leiden: Apud Abrahamum Elzevier.

Volder, Burchard de (*praeses*), and Houten, Antonius van (respondens). 1693. *Disputatio philosophica de causis variationum thermometrorum*. Leiden: Elsevier. (Reported in Bierens de Haan 1960, 131. Now lost).

Volder, Burchard de (*praeses*), and Houten, Antonius van (*respondens*). 1693. *Disputatio philosophica de thermometris*. Leiden: Elsevier. (Reported in Bierens de Haan 1960, 131. Now lost).

Volder, Burchard de (*praeses*), and Leeuwen, Henricus van (*respondens*). 1694. *Disputatio philosophica de corporis essentia*. Leiden: Elsevier. (Reported in Wiesenfeldt 2002, 429; from *Apparaat-Van der Woude. Bibliografie van Nederlandse dissertaties en oraties verschenen vóór 1800, op fiches. Samengesteld door prof. dr. S. van der Woude*. Special catalogue, Amsterdam University Library. Not retrieved).

De Volder and Casembroot 1694 = Volder, Burchard de (*praeses*), and Casembroot, Gysbertus (*auctor et respondens*). March 1694. *Disputatio philosophica quae est de mundi systemate*. Leiden: Apud Abrahamum Elzevier.

De Volder and Robberghtsten 1694 = Volder, Burchard de (*praeses*), and Robberghtsten, Johannes (*auctor et respondens*). 14 July 1694. *Disputatio philosophica de sensu brutorum*. Leiden: Apud Abrahamum Elzevier.

De Volder 1695 = Volder, Burchard de 1695. *Exercitationes academicae quibus Ren. Cartesii philosophia defenditur adversus Petri Danielis Huetii Episcopi Suessionensis Censuram philosophiae Cartesianae*. Amsterdam: Apud Arnoldum van Ravestein.

De Volder and Rouxel 1695 = Volder, Burchard de (*praeses*), and Rouxel, Esaias (*auctor*). 20 April 1695. *Disputatio physica quae est de materia variisque eius proprietatibus*. Leiden: Apud Abrahamum Elzevier.

Casembroot 1696 = Casembroot, Gysbertus (*candidatus*), and Volder, Burchard de (*promotor*). 9 July 1696. *Disputatio philosophica inauguralis de aestu marino*. Leiden: Apud Abrahamum Elzevier.

De Volder 1698 = Volder, Burchard de. 8 February 1698. *Oratio de rationis viribus, et usu in scientiis*. Leiden: Apud Fredericum Haringium.
De Volder and Westhovius 1698 = Volder, Burchard de (*praeses*), and Westhovius, Fredericus Gulielmus (*author*). 7 May 1698. *Dissertatio mathematico-philosophica de fluminibus*. Leiden: Apud Abrahamum Elzevier.
De Volder and Van Bashuysen 1698 = Volder, Burchard de (*praeses*), and Bashuysen, Henricus Jacobus van (*auctor*). 9 July 1698. *Disputatio philosophica de Telluris motu*. Leiden: Apud Abrahamum Elzevier.
De Volder and De Beaumont 1698a = Volder, Burchard de (*praeses*), and Beaumont, Elias Petrus de (*auctor et defendens*). 12 July 1698. *Dissertatio medico-physica de circulatione sanguinis*. Leiden: Apud Abrahamum Elzevier.
De Volder and De Beaumont 1698b = Volder, Burchard de (*praeses*), and Beaumont, Elias Petrus de (*auctor et respondens*). 20 December 1698. *Dissertatio physica de circulatione sanguinis in foetu*. Leiden: Apud Abrahamum Elzevier.
De Volder and De Beaumont 1698c = Volder, Burchard de (*praeses*), and Beaumont, Elias Petrus de (*auctor et defendens*). 20 December 1698. *Dissertatio physica de carentia sensuum et cognitionis in brutis*. Leiden: Apud Abrahamum Elzevier.
Gale 1699 = Gale, John (*candidatus*), and Volder, Burchard de (*promotor*). 2 July 1699. *Inquisitio philosophica inauguralis de lapide Solis*. Leiden: Apud Abrahamum Elzevier.
De Volder 1709 = Volder, Burchard de. 1709 (held on 3 July 1699). *Oratio de novis et antiquis*. Leiden: Apud Cornelium Boutestein.
De Volder and Copius 1699 = Volder, Burchard de (*praeses*), and Copius, Theophilus (*auctor*). 8 July 1699. *Disputatio philosophica de mentis humanae et divinae existentia et reale a corpore distinctione*. Leiden: Apud Abrahamum Elzevier.
Zoutmann 1702 = Zoutmann, Johannes (*candidatus*), and Volder, Burchard de (*promotor*). 1702. *Disputatio medico-philosophica inauguralis de imaginationis maternae viribus in foetum*. Leiden: Apud Abrahamum Elzevier.
Vander Ghiessen 1703 = Ghiessen, Jacob vander (*candidatus*), and Volder, Burchard de (*promotor*). 20 September 1703. *Disputatio philosophica inauguralis de mundo*. Leiden: Apud Abrahamum Elzevier.
De Volder 1705 = Volder, Burchard de. 1705. *Oratio qua* [...] *sese laboribus academicis abdicavit*. Leiden: Apud Cornelium Boutestein.
Bibliotheca Volderina = Linden, Johannes vander. 1709. *Bibliotheca Volderina, seu Catalogus selectissimorum librorum Clarissimi, Expertissimique Viri Defuncti D. Burcheri De Volder*. Leiden: Apud Johann. Vander Linden et Johannem Voorn.
Volder, Burchard de. 1719. *Disputationes philosophicae omnes contra atheos*. Utrecht: Schouten.

Handwritten Sources Authored by De Volder (in Chronological Order)

Dictata and Lecture Notes

Pretoria *dictata* = Volder, Burchard de. 1687–1688. *Principia Renati des Cartes dictata a doctissimo celeberrimoque domino Burchero de Volder*. Pretoria, National Library of South Africa, MSD27.
The Hague *dictata* = Volder, Burchard de, and Hinojosa, Adriaan Pieter de (copist). 1690. *Dictata Cl. Viri Dn. B. de Volder Professor. Publici in Celeberrim. Acad. Lugdun. Batava ad Principi. philosoph. Cartesi*. The Hague, Koninklijke Bibliotheek, ms. 72 A 7.
British Library *dictata* = Volder, Burchard de. Undated. *Dictata Viri Clarissimi Burcheri de Volder in Principia Cartesii*. British Library, ms. Sloane 1216, ff. 75–128.

De Volder 1676–1677 = Volder, Burchard de (lecturer), and Morley, Christopher Love (copist/writer). 1676–1677. *Experimenta philosophica naturalia, auctore M[a]gis[tro] De Valdo Lugd[uni] ann[o] 1676[–1677]*. British Library, ms. Sloane 1292, ff. 78–141.

Hamburg 273 = Volder, Burchard de. Undated. *Notulae quaedam in nobilissimi doctissimique viri Dni Renati Descartes Principiorum philosophiae partem primam–quartam scriptae ex ore clar. doctissimique viri Burcheri de Volder medicinae ac philosophiae doctoris, et in Academia Lugduno Batava professoris*. Hamburg: Staats- und Universitätsbibliothek Hamburg Carl von Ossietzky, Cod. philos. 273. https://resolver.sub.uni-hamburg.de/kitodo/PPN1014827108. Accessed 5 February 2019.

Hamburg 274 = Volder, Burchard de. Undated. *D. D. Burcheri De Volder dictata in Carthesii Principia philosophica*. Hamburg, Staats- und Universitätsbibliothek Hamburg Carl von Ossietzky, Cod. philos. 274. http://resolver.sub.uni-hamburg.de/goobi/PPN1014826934. Accessed 5 February 2019.

Leiden *dictata* = Volder, Burchard de. Undated. *B. de Volder Med. et Phil. Doct. eiusque professoris Dictata in Principiorum philosophiae partem quartam de terra*. Leiden University Library, ms. BPL 2841, ff. 1r–32r.

Warsaw *dictata* = Volder, Burchard de. Undated. *B. de Volder Annotationes in Meditationes Renati Descartes. Accedunt quoque notae quaedam Burcheri de Volder in primum librum Principiorum Cartesii De cognitione humana*. Warsaw, Biblioteka Narodowa, ms. BN Rps 3365 II.

Other Sources

Volder, Burchard de. 1682. *Verklaring van Johannes Voet, rechtsgeleerde aan de universiteit van Leiden en D. de Volder, secretaris van deze instelling, dat Cornelis de Witte gedurende twee jaar colleges Romeins recht heeft gevolgd. 1682 juli 7 en 9*. The Hague, Algemeen Rijksarchief, signature: 1.10.72, no. 408.

Volder, Burchard de. Circa 1682–1683. 14 letters to Melchisédech Thévenot. Staats- und Universitätsbibliothek Göttingen, Cod. Ms. Hist. nat. 102, fasc. 25, ff. 44–69.

Volder, Burchard de. 1683. *Testimonium voor Herman Boerhaave (1668–1738), August 1683*. The Hague, Koninklijke Bibliotheek, ms. 121 E 4.

Volder, Burchard de. 1684–1687. *Letter 1684-11-24, Leiden; Letter 1686-08-14* [*sic*], *Amsterdam* [2 letters to Isaac Newton]. Uppsala University Library, Waller mss. benl-00770–00771.

Volder, Burchard de. 1685. *Letter to J[ohannes] Braun*. British Library, Ms. Add. 24712, 76r–77r.

Volder, Burchard de. 1686–1691. *Brie[ven] van Burchardus de Volder aan Abraham van den Ende* [extracts from 2 letters]. Amsterdam University Library, OTM hs. Gs 24.

Volder, Burchard de. 1687–1699. *Burchardus de Volder aan Philippus van Limborch Brieven* [4 letters]. Amsterdam University Library, OTM hs. J 83: a–d.

Volder, Burchard de. 1691. *Methodus tangentium Slusii illustrata a Burchero de Volder*. Edinburgh University Library, ms. Coll-33/Folio C [81].

Volder, Burchard de. Circa 1691. *Methodus Tangentium demonstrata*... Edinburgh University Library, ms. Coll-33/Quarto A [39].

Volder, Burchard de. 1695. *Brief van B. de Volder aan Reinier Leers (1654–1714)*. Leiden University Library, ms. PAP 15.

Volder, Burchard de. 1695. *31 Dec 1696* [*sic*]*: Volder, Burchard de (Dr), 1643–1709 (Leiden, South Holland, Netherlands) to Gregory, David, 1659–1708* [1 letter]. Oxford, Bodleian Library, ms. Tanner 24, f. 189.

Volder, Burchard de, and Almeloveen, Theodorus Janssonius van. 1698. *Brief van Burchardus de Volder (1643–1709) aan Theodorus Janssonius van Almeloveen (1657–1712); Brieven* [2 letters] *van Theodorus Janssonius van Almeloveen (1657–1712) aan Burchardus de Volder (1643–1709)*. Utrecht University Library, Hs. 995, 163–164, and Hs. 996, 458–460.

Volder, Burchard de. 1704–1705. *Brieven van B. de Volder aan Petrus Burman (1668–1741)* [2 letters]. Leiden University Library, ms. BUR Q 27.

Other Primary Sources

Van der Aa 1712 = Aa, Pieter van der. 1712. *Les Délices de Leide: une des célébres villes de l'Europe; le tout enrichi de tailles douces*. Leiden: Chez Pierre vander Aa.

Académie Royale des Sciences. 1693. *Mémoires de mathématique et de physique tirés des registres de l'Académie Royale des Science*. Paris: De l'Imprimerie Royale.

Acosta 1589 = Acosta, José de. 1589. *De natura novi orbis libri duo et de promulgatione Evangelii apud Barbaros sive De procuranda Indorum salute libri sex*. Salamanca: Apud Guillelmum Foquel.

Acosta, José de. 1590. *Historia natural y moral de las Indias. En que se tratan de las cosas notables del cielo, elementos, metales, plantas y animales dellas y de los ritos y ceremonias y leyes y gobierno de los indios*. Sevilla: En casa de Juan de León.

Amerpoel, Johannes. 1669. *Cartesius mosaïzans, seu Evidens et facilis conciliatio philosophiae Cartesii cum historia creationis primo capite Geneseos per Mosem tradita*. Leeuwarden: Pro Haeredibus Thomae Luyrtsma.

Andala, Ruardus (*praeses*), and Vries, Theodorus de (*respondens*). 1705. *Dissertatio philosophica prima, in qua ostenditur existentiam Dei non tantum a posteriore, sive per effectus, sed et a priore, sive per ipsam eius naturam, validissime demonstrari*. Franeker: Apud Franciscum Halmam.

Andala 1708 = Andala, Ruardus. 1708. *Exercitationes academicae in philosophiam primam et naturalem; in quibus philosophia Renati Descartes clare et perspicue explicatur, valide confirmatur, nec non solide vindicatur*. Franeker: Ex Officina Wibii Bleck.

Andala, Ruardus (*praeses*), and Vitringa, Campegius Jr. (*respondens*). 1711. *Dissertatio philosophica prima, in qua vera substantiae notio vindicatur, simulque confutatur vis activa, a G. G. L. cuilibet substantiae, tam corporeae, quam spirituali, in Specimine dynamico attributa*. Franeker: Apud Franciscum Halmam.

Andala 1711 = Andala, Ruardus. 1711. *Dissertationum philosophicarum heptas. Quarum in prima de argumenta pro exsistentia Dei a priore; secunda de essentia mentis humanae; tertia de distinctione mentis a corpore; quarta de unione mentis cum corpore; quinta de voluntatis libertate; sexta de immortalitate mentis humanae; septima de corporum fluiditate et duritie agitur*. Franeker: Ex Officina Wibii Bleck.

Andala 1712 = Andala, Ruardus. 1712. *Dissertationum philosophicarum pentas, cui adiecta est continuatio ephemeridum aëris atmosphaerici variationum: a mense Julio A. 1710 ad mensem Junium A. 1712*. Franeker: Apud Franciscum Halmam.

Andala 1716 = Andala, Ruardus. 1716. *Examen Ethicae Clar. Geulingii sive Dissertationum philosophicarum, in quibus praemissa introductione sententiae quaedam paradoxae ex Ethica Clar. Geulingii examinantur, pentas*. Franeker: Apud Wibium Bleck.

Andala 1718 = Andala, Ruardus. 1718. *Apologia pro vera et saniore philosophia. Pars tertia in qua utilitas et praestantia philosophiae Cartesianae breviter notatur, fusiusque in causas, ob quas eadem sensim negligentius tractari, minusque amari et excoli coeperit, inquiritur, iisque occurritur*. Franeker: Henricus Halma.

Andala 1719a = Andala, Ruardus. 1719. *Cartesius verus Spinozismi eversor et physicae experimentalis architectus*. Franeker: Ex Officina Wibii Bleck.

Andala 1719b = Andala, Ruardus. 1719. *Apologia pro vera et saniore philosophia quatur partibus comprehensa*. Franeker: Ex Officina Wibii Bleck.

Anonymous 1678 = Anonymous. 1678. Philosophia vetus et nova, ad usum scholae accommodata, in regia Burgundia novissimo hoc biennio pertractata Tom. 3 qui Physicam generalem complectitur. In 12. à Paris chez Estienne Michallet. 1678. *Journal des Sçavans* 18 (23 May 1678): 101–107.

Anonymous. 1682. D. Burcheri de Volder Medicinae et Philosophiae Doctoris, huiusque in illustri Academia Lugduno Batava Professoris Ordinarii celeberrimi Disputationes philosophicae rerum naturalium principiis, ut et de aëris gravitate. Lugdun. Batav. 1681 in 8. *Acta Eruditorum*, May 1682: 147–148.

Anonymous. 1685. Clar. V. Burcheri de Volder, Disputationes philosophicae omnes contra atheos. 8. Medioburgi. *Journal des Sçavans* 31 (19 November 1685): 378–379.
Anonymous. 1686. *Eenige consideratien of aanmerkingen over de Schriftelijke onderhandeling*. S.l.: s.n.
Anonymous. 1718. IX. Nova quaedam theologico-historico-literaria. *Bibliotheca Bremensis, classis* 1, fascicle 6: 933–944.
Anonymous. 1718. VII. Nova theologico-historico-literaria. *Bibliotheca Bremensis, classis* 2, fascicle 2: 347–382.
Anonymous. 1718. VII. Scripta eristica inter Viros Clarissimos D. A. Driessenium, D. T. H. van den Honert, et Jac. Wittichium, nunc nuper reciprocata. *Bibliotheca Bremensis, classis* 1, fascicle 4: 550–559.
Anonymous. 1719. IV. Controversiae Driessenio-Wittichianae continuatio. *Bibliotheca Bremensis, classis* 2, fascicle 6: 1090–1103.
Archimedes and Tartaglia, Niccolò. 1543. Liber Archimedis de insidentibus aquae. In Archimedes and Tartaglia, Niccolò. 1543. *Opera Archimedis Syracusani philosophi et mathematici ingeniosissimi per Nicolaum Tartaleam Brixianum multis erroribus emendata, expurgata, ac in luce posita*, 31v–35v. Venice: s.n.
Archimedes and Commandino 1565 = Archimedes and Commandino, Federico. 1565. *Archimedis De iis quae vehuntur in aqua libri duo a Frederico Commandino restituti et commentariis illustrati*. Bologna: Ex officina Alexandri Benacii.
Archimedes and Rivault 1615 = Archimedes and Rivault, David. 1615. *Ἀρχιμηδους παντα σωζομενα. Archimedis opera quae extant. Novis demonstrationibus commentariisque illustrata*. Paris: Apud Claudium Morellum.
Archimedes and Barrow, Isaac. 1675. *Archimedis Opera, methodo nova illustrata et succincte demonstrata, per Isaacum Barrow*. London: Excudebat Guil. Godbid.
Archimedes and Fleck 2016 = Archimedes and Fleck, Heinrich F. 2016. *Archimede – Περὶ τῶν ὀχουμένων - α′ (Sui corpi galleggianti – libro I) – Traduzione commentata*. Quaderni di scienze umane e filosofia naturale, 3–4. http://www.heinrichfleck.net/Quaderni/corpi%20galleggianti.pdf. Accessed 30 January 2019.
Aristotle 1922a = Aristotle. 1922. *De caelo*, trans. John Leofric Stocks. Oxford: Clarendon Press.
Aristotle 2014 = Aristotle. 2014. *Complete Works of Aristotle. The Revised Oxford Translation*, ed. Jonathan Barnes. Princeton: Princeton University Press. First edition 1984.
Aristotle and Averroes 1562–1574 = Aristotle and Averroes. 1562–1574. *Omnia quae exstant opera* [...] *Averrois Cordubensis Commentarii*. Venice: Apud Iunctas.
Augustine 1986 = Augustine. 1986. *Soliloquiorum libri duo*, ed. Wolfgang Hörmann. Vienna: Hoelder-PichlerTempsky. (Corpus Scriptorum Ecclesiasticorum Latinorum 89).
Bacon 1653 = Bacon, Francis. 1653. *Scripta in naturali et universali philosophia*. Amsterdam: Apud Ludovicum Elzevirum.
Bardi 1614 = Bardi, Giovanni. 1614. *Eorum quae vehuntur in aquis experimenta* [...] *ad Archimedis trutinam examinata*. Rome: Ex Typographia Bartholomaei Zanetti.
Barin, Théodore. 1686. *Le monde naissant ou La création du monde: démontrée par des principes tres simples & tres conformes à l'histoire de Moyse Genes. chap. I. & II.* Utrecht: Pour la Compagnie des libraires.
Barrow, Isaac. 1669. *Lectiones XVIII [...] in quibus opticorum phaenomenon genuinae rationes investigantur, ac exponuntur: annexae sunt Lectiones aliquot geometricae*. London: Typis Gulielmi Godbid.
Basnage de Beauval 1695 = Basnage de Beauval, Henri (Ed.). 1695. Extraits des diverses lettres. *Histoire des Ouvrages de Sçavans*, May 1695: 421–430.
Basson 1621 = Basson, Sébastien. 1621. *Philosophiae naturalis adversus Aristotelem libri XII, in quibus abstrusa veterum physiologia restauratur, et Aristotelis arrores solidis rationibus refelluntur*. Geneva: Apud Ludovicum Elzevirium.
Bayle 1685 = Bayle, Pierre. 1685. Catalogue de livres nouveaux accompagné de quelques remarques. *Nouvelles de la République des Lettres*, February 1686: 225–227.

Bayle 1697 = Bayle, Pierre. 1697. Spinoza. In *Dictionnaire historique et critique*, ed. Pierre Bayle, volume 2, part 2, 1083–1100. Amsterdam: Chez Reinier Leers.

Bayle 1983 = Bayle, Pierre. 1683. *Écrits sur Spinoza*, ed. Françoise Charles-Daubert and Pierre-François Moreau. Paris: Berg International.

Beaufort, Louis de. 1656. *Cosmopoea divina: seu Fabrica mundi explicata*. Leiden: Ex officina D. Lopez de Haro.

Beeckman, Isaac. 1939–1953. *Journal tenu par Isaac Beeckman de 1604 à 1634*, ed. Cornélis de Waard. The Hague: Martinus Nijhoff.

Bekker, Balthasar. 1683. *Ondersoek van de betekeninge der kometen, by gelegendheid van de gene die in de jaren 1680, 1681, en 1682, geschenen hebben*. Leeuwarden: Gedrukt bij Hero Nauta.

Bellini, Lorenzo. 1685. *De urinis et pulsibus, de missione sanguinis, de febribus, de morbis capitis, et pectoris*. Frankfurt-Leipzig: Sumptibus Johannis Grossii.

Bellini, Lorenzo. 1695. *Opuscula aliquot ad Archibaldum Pitcarnium* [...] *in quibus praecipue agitur de motu cordis in et extra uterum, ovo, ovi aëre et respiratione. De motu bilis et liquidorum omnium per corpora animalium. De fermentis et glandulis, et c*. Pistoia: Ex nova officina Stephani Gatti.

Bergen, Johann Georg von (*praeses*), and Fremaut, Daniel (defendant and author). 1704. *Dissertatio medico-physica, de vi elastica corporum solidorum et fluidorum*. Frankfurt (Oder): Literis Christophori Zeitleri.

Bernard, Jacques. 1685. Burcheri de Volder Philosophiae et Matheseos Professoris Laudatio ab Jacobo Gronovio peracta ex Decreto Rectoris Magnifici et Senatus Academici. Ad Diem XVII. Calendas Maias. C'est-à-dire, Oraison Funebre de Mr. de Volder Professeur en Philosophie & en Mathématiques, recitée par Mr. Gronovius, conformément au Decret du Recteur Magnifique & du Senat Académique. A Leide, chez Corneille Boutestein. 1709. in 4; pagg. 38 gros caractére. *Nouvelles de la République des Lettres*, May 1709: 558–566.

Bernoulli, Jakob. 1682. *Conamen novi systematis cometarum, pro motu eorum sub calculum revocando et apparitionibus praedicendis*. Amsterdam: Apud Henricum Wetstenium.

Bernoulli 1683 = Bernoulli, Jakob. 1683. *Dissertatio de gravitate aetheris*. Amsterdam: Apud Henr. Wetstenium.

Bernoulli 1686a = Bernoulli, Jakob. 1686. *Methodus ratiocinandi, sive Usus logicae in praeclaro aliquo phaenomeno physico enodando*. Basel: Typis Johan. Conradi a Mechel.

Bernoulli 1686b = Bernoulli, Jakob. 1686. Dubium circa causam gravitatis a rotatione vorticis terreni petitam. *Acta Eruditorum*, February 1686: 91–95.

Bernoulli 1744 = Bernoulli, Jakob. 1744. *Opera*. Geneva: Sumptibus haeredum Cramer et fratrum Philibert.

Bernoulli 1702a = Bernoulli, Johann (*praeses*), and Nysingh, Johannes (*respondens*). 1702. *Exercitatio philosophica de globo et motu Telluris*. Groningen: Typis Catharinae Zandt.

Bernoulli, Johann (*praeses*), and Themmen, Phoebus (*respondens*). 1702. *Theses apologeticae quas* [...] *ad depellendas atrocissimas calumnias cordate et in trepide defendent contra nuperi libelli famosi auctorem illiusve quemvis patronum*. Groningen: s.n.

Bernoulli, Johann (*praeses*), and Themmen, Phoebus (*respondens*). 1702. *Continuatio thesium apologeticarum: quas iuvante Deo contra nuperi libelli famosi auctorem eiusve quemvis patronum intrepide defendent*. Groningen: s.n.

Bernoulli, Johann (*praeses*), and Themmen, Phoebus (*respondens*). 1702. *Dissertatio philosophica de anima brutorum*. Groningen: s.n.

Bernoulli 1702b = Bernoulli, Johann. 1702. *Spinosismi Depulsionis echo id est Depulsoris imputationes depulsae*. Groningen: Typis Johannis Lens.

Bernoulli, Johann. *Briefwechsel von Johann I Bernoulli. Korrespondenz zwischen Bernoulli, Johann I und Volder, Burchard de*. Universitätsbibliothek Basel. http://www.ub.unibas.ch/bernoulli/index.php/Kategorie:Volder_Burchard_de-Bernoulli_Johann_I. Accessed 6 February 2019.

Bibliotheca Barkeyana = Daale, Nicolaus van, and Plaat, Christianum. 1787. *Bibliotheca Barkeyana sive Catalogus libr. Nic. Barkey*. The Hague: Apud Biblipolas Nicolaum van Daalen et Christianum Plaat.

Bibliotheca Boerhaaviana = Luchtmann, Samuel. 1739. *Bibliotheca Boerhaaviana sive Catalogus librorum* [...] *D. Hermanni Boerhaave*. Leiden: Apud Samuelem Luchtmans.

Bibliotheca Heinsiana = Vivie, Johann de. 1682. *Bibliotheca Heinsiana sive Catalogus librorum, quos, magno studio, et sumtu, dum viveret, collegit Vir Illustris Nicolaus Heinsius*. Leiden: Apud Johannem De Vivie.

Bibliotheca Martiniana = Scheurleer, Henry. 1752. *Bibliotheca Martiniana, ou Catalogue de la Bibliothèque de feu Monsr. David Martini*. The Hague: Chez Henry Scheurleer.

Bibliotheca Musschenbroekiana = Luchtmans, Samuel. 1826. *Bibliotheca Musschenbroekiana, sive Catalogus librorum cum impressorum, tum maxime manuscriptorum, mapparum geographicarum* [...] *aliarumque rerum ad antiquitatem, patriam imprimis, facientium, quibus usus est* [...] *Petrus van Musschenbroek*. Leiden: Per S. et J. Luchtmans.

Bibliotheca Nicolaiana = Waesbergen, Johannes van. 1698. *Bibliotheca Nicolaiana, in duas partes divisa; quarum prima libros continet, altera numismatum ac operis prisci thesaurum* [...] *Cornelius Nicolai*. Amsterdam: Apud Janssonio-Waasbergio.

Bibliotheca Schalbruchiana = Wetsten, Rudolf, and Wetsten, Gerard. 1722. *Bibliotheca Schalbruchiana, sive Catalogus exquisitissimorum rarissimorumque librorum* [...] *Schalbruch*. Amsterdam: Apud R. et G. Wetstenios.

Bibliotheca Triglandiana = Luchtmans, Jordan. 1706. *Bibliotheca Triglandiana sive Catalogus* [...] *librorum* [...] *quos* [...] *collegit* [...] *Jacobus Triglandius*. Leiden: Apud Jordanum Luchtmans.

Bibliotheca Vriemoetiana = Brouwer, P. R. 1761. *Bibliotheca Vriemoetiana, sive Catalogus librorum in omnibus fere scientiis et artibus* [...] *quos sibi collegerat* [...] *Emo Lucius Vriemoet*. Leeuwarden: Apud Abrahamum Ferwerda.

Bibliotheca Wasteau = Linden, Johann vander. 1719. *Splendidissima et excellentissima bibliotheca, continens omni in lingua libros exquisitissimos* [...] *Abrahami Wasteau*. Leiden: Apud J. vander Linden sen.

Bibliothecae Duncanniana = Boucquet, Cornelis, and Gaillard, Mattheus. 1754. *Pars bibliothecae Duncannianae sive Catalogus librorum* [...] *quibus* [...] *usus fuit* [...] *Joan Duncan*. The Hague: By Cornelis Boucquet, en Mattheus Gaillard.

Blijenbergh, Willem van. 1671. *De kennisse Godts en Godts-Dienst veweert tegen d'uytvluchten der atheisten*. Leiden-Amsterdam: By Daniel, Abraham en Adriaen van Gaesbeeck. First edition 1663.

Blijenbergh, Willem van. 1674. *De waerheyt van de Christelijcke Godts-Dienst en de authoriteit der H. Schriften of een Wederlegginge van dat Godt-lasterlijcke boeck, genoemt Tractatus theologico-politicus*. Leiden: By Daniel van Gaesbeeck.

Blijenbergh, Willem van. 1682. *Wederlegging van de Ethica of Zede-kunst, van Benedictus de Spinosa.* [...] *Voornamentlijk ontrent het Wesen ende de Natuur van God en van onse Ziel*. Dordrecht: Wed. J. en D. Goris.

Boerhaave 1690 = Boerhaave, Hermann. 18 December 1690. *Disputatio philosophica inauguralis de distinctione mentis a corpore*. Leiden: Apud Abrahamum Elzevier.

Boerhaave 1703 = Boerhaave, Hermann. 1703. *De usu ratiocinii mechanici in medicina oratio*. Leiden: Apud Johannem Verbessel.

Boerhaave, Hermann. 1710. *Index plantarum: quae in Horto Academico Lugduno Batavo reperiuntur*. Leiden: Apud Cornelium Boutestein.

Boerhaave 1721 = Boerhaave, Hermann. 1721. *Oratio qua repurgatae medicinae facilis asseritur simplicitas, editio altera*. Leiden: Apud Joh. Arnold Langerak. First edition 1709.

Boerhaave 1983 = Boerhaave, Hermann. 1983. *Boerhaave's Orations*, ed. and trans. Elze Kegel-Brinkgreve, and Antonie Maria Luyendijk-Elshout. Leiden: Brill.

Bolk, Bartholomaeus. 1719. *Animadversiones philosophicae in decantatam Spinozae propositionem, quae res nihil commune inter se habent, earum una alterius causa esse non potest*. Amsterdam: Excudit Jacobus Borstius.

Borelli 1667 = Borelli, Giovanni Alfonso. 1667. *De vi percussionis liber*. Bologna: Ex typographia Iacobi Montii.

Borelli 1670 = Borelli, Giovanni Alfonso. 1670. *De motionibus naturalibus a gravitate pendentibus*. Reggio di Calabria: In Officina Dominici Ferri.

Borelli, Giovanni Alfonso. 1680–1681. *De motu animalium*. Rome: Ex typographia Angeli Bernabò.

Borelli 2015 = Borelli, Giovanni Alfonso. 2015. *Borelli's On the Movement of Animals – On the Natural Motions Resulting from Gravity*, trans. Paul Maquet. Cham-Heidelberg-New York-Dordrecht-London: Springer.

Borri 1576 = Borri, Girolamo. 1576. *De motu gravium et levium*. Florence: In officina Georgii Marescotti. First edition 1575.

Boyle 1660 = Boyle, Robert. 1660. *New Experiments Physico-Mechanicall, Touching the Spring of the Air and its Effects*. Oxford: Printed by H. Hall.

Boyle 1662a = Boyle, Robert. 1662. *A Defence of the Doctrine Touching the Spring and Weight of the Air*. London: Printed by F. G.

Boyle 1662b = Boyle, Robert. 1662. *An Examen of Mr. T. Hobbes his Dialogus Physicus de Natura Aeris*. London: Printed by J. G.

Boyle 1666 = Boyle, Robert. 1666. *Hydrostatical Paradoxes, Made Out by New Experiments (for the Most Part Physical and Easie)*. Oxford: Printed by William Hall.

Boyle, Robert. 1666. *The Origine of Formes and Qualities, (According to the Corpuscular Philosophy,) Illustrated by Considerations and Experiments*. Oxford: Printed by H. Hall.

Boyle 1669a = Boyle, Robert. 1669. *Certain Physiological Essays and Other Tracts* [...], *Increased by the Addition of a Discourse about the Absolute Rest in Bodies*. London: Printed for Henry Herringman. First edition 1661.

Boyle 1669b = Boyle, Robert. 1669. *A Continuation of New Experiments Physico-Mechanical, Touching the Spring and Weight of the Air and their Effects*. Oxford: Printed by Henry Hall.

Boyle 1669c = Boyle, Robert. 1669. *Paradoxa hydrostatica: novis experimentis (maximam partem physicis ac facilibus) evicta*. Oxford: Oxonii Typis Henrici Hall.

Boyle 1670 = Boyle, Robert. 1670. *Paradoxa hydrostatica novis experimentis (maximam partem physicis ac facilibus) evicta*. Rotterdam: Ex officina Arnoldi Leers.

Boyle 1671 = Boyle, Robert. 1671. *De absoluta quiete in corporibus*. Amsterdam: Apud Johannem Janssonium a Waesberge. Hamburg: Apud Gothofredum Schultzen.

Boyle, Robert. 1672. *Tracts* [...] *Containing New Experiments, Touching the Relation Betwixt Flame and Air: And about Explosions. An Hydrostatical Discourse Occasion'd by Some Objections of Dr. Henry More* [...] *to Which is Annex't, An Hydrostatical Letter* [...] *New Experiments, of the Positive or Relative Levity of Bodies Under Water; of the Air's Spring on Bodies Under Water; about the Differing Pressure of Heavy Solids and Fluids*. London: Printed for Richard Davis.

Boyle, Robert. 1680. *The Aerial Noctiluca, or, Some New Phenomena and a Process of a Factitious Self-Shining Substance: Imparted in a Letter to a Friend Living in the Country*. London: Printed by Tho. Snowden.

Boyle 1999–2000 = Boyle, Robert. 1999–2000. *The Works of Robert Boyle*, ed. Michael Hunter and Edward B. Davis. London: Pickering & Chatto.

Braun, Johannes. 1688. *Doctrina foederum sive Systema theologiae didacticae et elencticae*. Amsterdam: Apud Abrahamum van Someren.

Braun, Johannes. 1702. *Futilis Spinosismi Depulsionis* [...] *depulsio necessaria*. Groningen: Typis Bernhardi Vriemoet.

Bredenburg, Johannes. 1675. *Enervatio Tractatus theologico-politici, una cum demonstratione, geometrico ordine disposita, naturam non esse Deum: cuius effati contrario praedictus Tractatus unice innititur*. Rotterdam: Apud Isaacum Naeranum.

Bredenburg, Johannes. 1684. *Noodige verantwoording op de ongegronde beschuldiging van Abraham Lemmerman*. Rotterdam: By Isaak Naeranus.

Bredenburg, Johannes. 1684. *Wiskunstige demonstratie, dat alle verstandelijke werking noodzakelijk is*. Amsterdam: By Daniel Bakkamude.

Bredenburg, Johannes. 1686. *Korte aanmerkingen op De Brieven van Den Hr. Philippus van Limborch* [...] *aan Pieter Smout en N.N. Rakende het geschil van 't gebruik der Reden in de Religie tussen den Hr. P. van. Limborch en J. Bredenbnrg* [*sic*]. *Waar by komen eenige stukken tot het zelve geschil behoorende*. Rotterdam. Gedruckt by Barent Bos.

De Bruyn and Posahazi 1654 = Bruyn, Johannes de (*praeses*), and Posahazi, János (*author et respondens*). 1654. *Disputatio physica de corpore fluido*. Utrecht: Typis Gisberti a Zyll, et Theodori ab Ackersdijck.

De Bruyn 1663 = Bruyn, Johannes de. 1663. *Epistola ad* [...] *Isaacum Vossium, ubi iudicium fertur super ipsius libro De natura et proprietate lucis, et simul Cartesii doctrina defenditur*. Amsterdam: Apud Ludovicum et Danielem Elzevirios.

Bruyn, Johannes de (*praeses*). 1665–1668. *Disputationum philosophicarum de naturali Dei cognitione prima*[*–decima sexta*]. Utrecht: Ex Officina Meinardi a Dreune.

Bruyn, Johannes de. 1670. *Defensio doctrinae Cartesianae, de dubitatione et dubitandi modo, ut et de idea Dei in nobis, deque existentiae Eius demonstratione ex ea idea. Adversus obiectiones Reineri Vogelsangii, insertas Indignationi iustae et c*. Amsterdam: Apud Danielem Elzevirium.

Buonamici 1591 = Buonamici, Francesco. 1591. *De motu libri X, quibus generalia naturalis philosophiae principia summo studio collecta continentur*. Florence: Apud Bartholomaeum Sermatellium.

Burgersdijk 1626 = Burgersdijk, Franco. 1626. *Institutionum logicarum libri duo ex Aristotelis, Keckermanni, aliorum praecipuorum logicorum praeceptis recensitis*. Leiden: apud Abraham Commelinum.

Burgersdijk 1652 = Burgersdijk, Franco. 1652. *Idea philosophiae naturalis, sive Methodus definitionum et controversiarum physicarum. Editio novissima*. Leiden: Ex officina Elzeviriorum.

Burman 1671–1672 = Burman, Frans. 1671–1672. *Synopsis theologiae et speciatim oeconomiae Foederum Dei, ab initio saeculorum usque ad consummationem eorum*. Utrecht: Apud Cornelium Jacobi Noenardum.

Burman 1700 = Burman, Frans Jr. 1700. *Burmannorum pietas, gratissimae beati parentis memoriae communi nomine exhibita, a Francisco Burmanno, f. Adiiciuntur mutuae Cl. Limburgii et Fr. Burmanni epistolae*. Utrecht: Ex Officina Guilielmi vande Water.

Burnet, Thomas. 1681. *Telluris theoria sacra: orbis nostri originem et mutationes generales, quas aut iam subiit, aut olim subiturus est, complectens*. London: Typis R. N.

Burnet, Thomas. 1684. *Theory of the Earth Containing an Account of the Original of the Earth, and of All the General Changes Which It Hath Already Undergone, or is to Undergo, Till the Consummation of All Things*. London: Printed by R. Norton.

Burnet, Thomas. 1689. *Telluris theoria sacra: orbis nostri originem et mutationes generales, quas aut iam subiit, aut olim subiturus est, complectens. Libri duo priores de diluvio et Paradiso. Editio secunda*. London: Typis R. N.

Bylaert 1743 = Bylaert, Jan Jacob. Circa 1743. *Gezigt op het huijs in de Nonnensteeg agter de Academie voor de Prof. in de kruijdkunde, waar naast zig vertoont het huijs waar in de Prof. in de Phisica publieke lessen houdt*. Stadsarchief Leiden, Erfgoed Leiden en omstreken, signature: PV13541.1–2. https://www.erfgoedleiden.nl/collecties/beeldmateriaal/zoeken-in-beeldmateriaal/detail/197775d6-26bd-11e3-a46c-3cd92befe4f8/media/2d8cadce-1094-d816-d508-8895abc78292. Accessed 18 February 2019.

Cardano, Girolamo. 1550. *De subtilitate libri XXI*. Nürnberg: Apud Ioh. Petreium.

Castelli 1628 = Castelli, Benedetto. 1628. *Della misura dell'acque correnti*. Rome: Nella Stamparia Camerale.

Catalogue De Mey = Cleef, Isaac van, and Scheurleer, Bernardus. 1796. *Catalogue d'une superbe collection de livres* [...] *delaissée par feu Monsieur* [...] *P. de Mey*. The Hague: J. Van Cleef & B. Scheurleer.

Catalogus [...] *Henninius* = Reets, Anton. 1705. *Catalogus selectissimorum et variorum in omni facultate librorum, quos* [...] *collegit Henricus Christianus Henninius*. Utrecht: Ex officina Guilielmi vande Water.

Catalogus De Witt (1696) = Willegaerts, Cornelius. 1696. *Catalogus exquisitissimorum, raroque occurrentium librorum, in omni studiorum genere, facultate, et lingua [...] collecta a Viro Amplissimo Jano Albino*. Dordrecht: Apud Cornelium Willegaerts.

Catalogus De Witt (1701) = Graevius, Johannes Georgius. 1701. *Catalogus bibliothecae luculentissimae, et exquisitissimis ac rarissimis in omni disciplinarum et linguarum genere libris, magno studio, dilectu et sumptu quaesitis, instructissimae, a Joanne de Witt*. Dordrecht: Apud Theodorum Goris et Joannem van Braam.

Catalogus J. A. C. = Os, Pieter van. 1764. *Catalogus van een keurlyke bibliotheek, in alle faculteyten en taalen [...]. Alle by een Verzameld door den Wel Edele Heer J. A. C.* The Hague: Pieter van Os.

Charleton 1654 = Charleton, Walter. 1654. *Physiologia Epicuro-Gassendo-Charltoniana, or, A Fabrick of Science Natural, upon the Hypothesis of Atoms*. London: Printed by Tho. Newcomb.

Chauffepié 1756 = Chauffepié, Jacques George de. 1756. Volder (Burcher de). In *Nouveau dictionnaire historique et critique: pour servir de supplement ou de continuation au Dictionnaire historique et critique de Pierre Bayle*, ed. Jacques George de Chauffepié, volume 4, 589–593. Amsterdam: Chez Z. Chatelain et al.

Clauberg, Johannes. 1654. *Logica vetus et nova, quadripartita, modum inveniendae ac tradendae veritatis, in genesi simul et analysi, facili methodo exhibens*. Amsterdam: Apud Ludovicum Elzevirium.

Clauberg, Johannes. 1658. *Logica vetus et nova modum inveniendae ac tradendae veritatis, in genesi simul et analysi, facili methodo exhibens, editio secunda*. Amsterdam: Ex officina Elzeviriana.

Clauberg 1691 = Clauberg, Johannes. 1691. *Opera omnia philosophica*. Amsterdam: Ex typographia P. et T. Blaev.

Conimbricenses 1592 = Conimbricenses. 1592. *Commentarii Collegii Conimbricensis [...] in octo libros Physicorum Aristotelis*. Coimbra: Typis et expensis Antonii a Mariz.

Conimbricenses 1603 = Conimbricenses. 1603. *Commentarii Collegii Conimbricensis [...] in quatuor libros de Caelo Aristotelis*. Cologne: Impensis Lazari Zetzneri. First edition 1594.

Conimbricenses 1604 = Conimbricenses. 1604. *Commentarii Collegi Conimbricensis [...] in Aristotelis Logicam*. Venice: Apud Rubertum Meiettum.

Copernicus, Nicolaus. 1543. *De revolutionibus orbium coelestium libri VI*. Nüremberg: Apud Ioh. Petreium.

Cordemoy, Géraud de. 1668. *Copie d'une lettre ecrite à un sçavant religieux de la Compagnie de Jesus, pour montrer, I. Que le systeme de M. Descartes, son opinion touchant les bestes, n'ont rien de dangereux. II. Et que tout ce qu'il en a écrit semble estre tiré du premier chapitre de la Genese*. S.l.: s.n.

Cornelio, Tommaso. 1648. *Timaei Locrensis Epistola, qua motuum illorum, qui vulgo ob fugam vacui fieri dicuntur, vera causa per circumpulsionem ad mentem Platonis explicatur*. Rome: Typis Manelphi Manelphii.

Cornelio 1663 = Cornelio, Tommaso. 1663. *Progymnasmata physica*. Venice: Typis Haeredum Franci Baba.

Craanen 1685 = Craanen, Theodoor. 1685. *Oeconomia animalis ad circulationem sanguinis breviter delineata. In duas partes distributa. Item generatio hominis ex legibus mechanicis*. Gouda: Ex officina Guilhelmi vander Hoeve.

Craanen, Theodoor. 1686. *Lumen rationale medicum, hoc est Praxis medica reformata sive Annotationes in Praxin Henrici Regii*. Middelburg: Apud Johannem de Reede.

Craanen 1689 = Craanen, Theodoor. 1689. *Tractatus physico-medicus de homine, in quo status eius tam naturali, quam praeternaturalis, quoad theoriam rationalem mechanice demonstratur*. Leiden: Apud Petrum vander Aa.

Dalrymple 1686 = Dalrymple, James. 1686. *Physiologia nova experimentalis: in qua, generales notiones Aristotelis, Epicuri, et Cartesii supplentur: errores deteguntur et emendantur*. Leiden: Apud Cornelium Boutesteyn.

Dati 1663 = Dati, Carlo. 1663. *Lettera a Filaleti di Timauro Antiate della vera storia della cicloide e della famosissima esperienza dell'argento vivo*. Florence: All'insegna della Stella.

De Frankenau 1696 = Frankenau, Georg Friedrich Franck de. 1696. *Disquisitio epistolaris succi nutricii per nervos transitum eiusque effectus in corpore humano expendens*. Leipzig: Impensis Joh. Melchior. Liebe.

De Monconys 1695 = Monconys, Balthasar de. 1695. *Journal des voyages* [...] *où les sçavants trouveront un nombre infini de nouveautez, en machines de mathématique, expériences physiques, raisonnemens de la belle philosophie, curiositez de chymie, et conversations des illustres de ce siècle*. Paris: Chez P. Delaulne. First edition 1665–1666.

Della Porta, Giovanni Battista. 1589. *Magiae naturalis libri XX*. Naples: Apud Horatium Salvianum. First edition 1558.

Denzinger and Schönmetzer 1991 = Denzinger, Heinrich, and Schönmetzer, Adolfus. 1991. *Enchiridion symbolorum definitionum et declarationum de rebus fidei et morum*. Freiburg-Basel-Rome-Vienna: Herder 1991. First edition 1854.

Descartes, René. 1637. *Discours de la Méthode pour bien conduire sa raison, et chercher la vérité dans les sciences. Plus la Dioptrique, les Météores et la Géométrie qui sont des Essais de cette Méthode*. Leiden: Imprimerie de Ian Maire.

Descartes 1644a = Descartes, René. 1644. *Principia philosophiae*. Amsterdam: Apud Ludovicum Elzevirium.

Descartes 1644b = Descartes, René. 1644. *Specimina philosophiae seu Dissertatio de methodo, Dioptrice et Meteora*. Amsterdam: Apud Ludovicum Elzevirium.

Descartes, René. 1647. *Les principes de la philosophie*, trans. Claude Picot. Paris: De l'Imprimerie de Pierre Des-Hayes.

Descartes, René. 1651. *Les principes de la philosophie*, trans. Claude Picot. Paris: Chez Henry Le Gras et Edmé Pepingué.

Descartes and Clerselier 1657–1667 = Descartes, René, and Clerselier, Claude. 1657–1667. *Lettres de Mr Descartes*. Paris: Chez Charles Angot.

Descartes, René et al. 1659. *Geometria*, ed. Frans van Schooten. Amsterdam: Apud Ludovicum et Danielem Elzevirios.

Descartes, René. 1662. *De homine figuris et latinitate donatus*, trans. Florentius Schuyl. Leiden: Apud Petrum Leffen et Franciscum Moyardum.

Descartes 1664 = Descartes, René. 1664. *Le monde* [...] *ou le Traité de la lumière, et des autres principaux objets des sens*. Paris: Chez Theodore Girard.

Descartes, René. 1664. *L'homme* [...] *et un Traité de la formation du foetus*, ed. Claude Clerselier. Paris: Chez Charles Angot.

Descartes, René. 1668. *Traité de la méchanique, plus l'Abrégé de la musique*. Paris: Chez Charles Angot.

Descartes, René. 1670–1672. *Opera philosophica. Editio quinta, nunc demum hac editione diligenter recognita, et mendis expurgata*. Amsterdam: Apud Danielem Elzevirium.

Descartes, René. 1682. *De verhandeling van den mensch en de makinge van de vrugt in 's moeders lichaam*, trans. Jacob Copper. Middelburg: By de Wed. van Remigius Scrijver, en Adolphus Rammazein.

AT = Descartes, René. 1897–1913. *Oeuvres*, ed. Charles Adam and Paul Tannery. Paris: L. Cerf.

Descartes 1982 = Descartes, René. 1982. *Principles of Philosophy*, ed. and trans. Valentine Rodger Miller and Reese P. Miller. Dordrecht: Springer.

Descartes 1985 = Descartes, René. 1985. *The Philosophical Writings of Descartes*, ed. and trans. John Cottingham, Robert Stoothoff and Dugald Murdoch. Cambridge: Cambridge University Press.

Descartes 1993 = Descartes, René. 1993. *Rene Descartes' Meditations on First Philosophy in Focus*, ed. and trans. Elizabeth S. Haldane and G. R. T. Ross. New York: Routledge.

Descartes 1996a = Descartes, René. 1996. *Discourse on Method and Meditations on First Philosophy*, ed. and trans. William T. Bluhm and David Weissman. Yale: Yale University Press.

Descartes 1996b = Descartes, René. 1996. *Meditations on First Philosophy: With Selections from the Objections and Replies*, ed. and trans. John Cottingham. Cambridge: Cambridge University Press.

Descartes 2004 = Descartes, René. 2004. *The World and Other Writings*, ed. and trans. Stephen Gaukroger. Cambridge: Cambridge University Press. First edition 1998.

Desmarets, Samuel (praeses). 1667. *Disputationes theologicae refutatoriae Libelli de philosophia interprete Scripturae*. Groningen: s.n. (Reported in Nauta 1935, 39. Now lost).

Dolaeus 1703 = Dolaeus, Johannes. 1703. *Opera omnia, exhibentia non modo encyclopaediam medicam dogmaticam, in qua affectus humani corporis interni, encyclopaediam chirurgicam rationalem, in qua idem affectus externi* [...] *pertractantur*. Frankfurt: Sumptibus Friderici Knochii.

Doppelmayr, Johann Gabriel. *Briefwechsel Johann Gabriel Doppelmayr*. Nürnberger Astronomische Gesellschaft. https://www.astronomie-nuernberg.de/index.php?category=doppelmayr&page=1701-04-00-Allgoewer. Accessed 6 February 2019.

Drelincourt 1681 = Drelincourt, Charles. 1681. *Experimenta anatomica: ex vivorum sectionibus petita. Edita per Ernestum Gottfried Heyseum*. Leiden: Apud Cornelium Boutesteyn.

Driessen, Antonius. 1718. *Epistola ad Virum plurimum Reverendum et Celeberrimum D. T. H. van den Honert, S. Th. D. et Professorem Leidensem qua petitur solutio quarumdam difficultatum motarum adversus V. Cl. J. Wittichii Disputationem de natura Dei, a Cl. Honertio laudatam, nec non Animadversiones in Cl. Wittichii Responsiones ad Scrupulos quosdam illi propositos*. Groningen: Ex Officina Georgii Spandaw.

Driessen 1718d = Driessen, Antonius. 1718. *Dissertatio Cl. J. Wittichii Disputationi de natura Dei opposita, qua difficultates adversus Disputationem motae vindicantur et urgentur. Porro adseritur Dei potentia, qua materiam produxit ex nihilo, atque omnimoda falsitas axiomatis Spinosistici: res quae nihil inter se commune habent una alterius causa esse nequit, ostenditur. Praeambula responsionum Wittichianarum refutationi. Praemittitur Repraesentationis Honertianae excussio*. Groningen: Ex Officina Georgii Spandaw.

Driessen 1718a = Driessen, Antonius. 1718. *Responsionum Wittichianarum refutatio. Dissertationi nuper emissae adnectenda praemittitur Alloquium ad ipsum Cl. Wittichium*. Groningen: Ex Officina Georgii Spandaw.

Driessen 1718c = Driessen, Antonius. 1718. *Alloquium alterum ad lectorem orthodoxum rei gestae narrationem vindicans adversus Repraesentationem C. Honerti alteram*. Groningen: Ex Officina Georgii Spandaw.

Driessen, Antonius. 1718. *Responsio ad Vindicationem Gr. V.* Groningen: Ex Officina Joh. a Velsen.

Driessen, Antonius. 1718. *Responsio ad Jacobi Wittichii Alloquium ad lectorem philosophum*. Ex Officina Georgii Spandaw.

Driessen 1718b = Driessen, Antonius. 1718. *Alloquium ad Ecclesiam Belgicam in brevi tabella Disputationem J. Wittichii de natura Dei exhibens et paragraphorum singulorum mentem indicans*. Groningen: Ex Officina Georgii Spandaw.

Driessen, Antonius. 1718. *Aanspraak aan de Kerk van Nederlant, vertoonende het dispuit van J. Wittichius over de natuur van Godt, en de meining van yder stelling aanwyzende*. S.l.: s.n.

Driessen, Antonius. 1719. *Tegen-berigt van Antonius Driessen Doctor en Professor der H. Godgeleerdheid ter verydeling van enige verdigte vertellingen, door welke de HHren van den Honert, Wittichius, en hunne aanhangeren zyn gedrag omtrent het dispuut van Wittichius de natura Dei tragten verdagt te maken*. Groningen: By Jurjen Spandaw.

Driessen, Antonius. 1719. *Beter berigt van Antonius Driessen Doctor en Professor der H. Theologie, en Akademi-prediker te Groningen gegeven aan den Nederlandzen lezer en dien weledelen gestrengen heer den welken de heer T. H. van den Honert Professor der H. Theologie te Leiden door zynen gedrukten brief misleid heeft*. Groningen: By Jurjen Spandaw.

Du Hamel, Jean-Baptiste. 1663. *De consensu veteris et novae philosophiae libri duo*. Paris: Apud Carolum Savreux.

Du Hamel 1678 = Du Hamel, Jean-Baptiste. 1678. *Philosophia vetus et nova: ad usum scholae accommodata In Regia Burgundia novissima hoc biennio pertractata*. Paris: Apud Stephanum Michallet.

Du Hamel 1682 = Du Hamel, Jean-Baptiste. 1682. *Philosophia vetus et nova ad usum scholae accomodata: in Regia Burgundia olim pertractata*. Nuremberg: Typis recusa a Christophoro Gerhardo.

Duchesne 1648 = Duchesne, Jaques. 1648. *Quercetanus redivivus, hoc est, ars medica dogmatico-hermetica*. Frankfurt: Sumptibus Joannis Beyer.

Ens 1636 = Ens, Gaspar. 1636. *Thaumaturgus mathematicus, id est admirabilium effectorum e mathematicarum disciplinarum fontibus profluentium sylloge*. Cologne: Apud Constantinum Münch.

Ewyck, Gerard van. 1719. *Difficultas maxima, orta ex concatenationibus idearum, quas Vir Clarissimus Jacobus Wittichius concinnavit in Disputatione sua de natura Dei, cui difficultati ut publice satisfiat, quum et plurium mentes non parum ea torqueri possint, publicitus illam quoque proponere, nec non simul axioma Spinosisticum primarium penitus subvertere voluit*. Leiden: Apud Johannem vander Linden.

Exquemelin, Alexandre Olivier. 1678. *De Americaensche zee-roovers: Behelsende een pertiente en vaerachtige beschrijving van alle de voornaemste roveryen, en onmenschelijcke wreedheden, die de Engelse en Franse rovers, tegens de Spanjaerden in America, gepleeght hebbe* [...]. *Hier achter is bygevoeght, een korte verhandeling van de macht en rijkdommen, die de koninck van Spanje, Karel de Tweede, in America heeft, nevens des selfs inkomsten en regering aldaer. Als mede een kort begrijp van alle de voornaemste plaetsen in het selve gewest, onder Christen potentaten behoorende*. Amsterdam: By Jan ten Hoorn.

Exquemelin, Alexandre Olivier. 1681. *Piratas de la America, y luz à la defensa de las costas de Indias Occidentales*, trans. Alonso de Buena Maison. Cologne: En casa de Lorenzo Struickman. (False place and name).

Fatio de Duillier, Nicolas. 1687. Reflexions sur une méthode de trouver les tangentes de certaines lignes courbes, laquelle vient d'être publiée dans un livre intitulé Medicina mentis. *Bibliothèque universelle et historique* 5: 25–33.

Fontenelle, Bernard le Bovier de. 1686. *Entretiens sur la pluralité des mondes*. Paris: Chez la veuve C. Blageart.

Gadroys, Claude. 1675. *Le système du monde selon les trois hypothèses, où conformement aux loix de lamechanique l'on explique dans la supposition du mouvement de la terre*. Paris: Chez Guillaume Desprez.

Galilei, Galileo. 1632. *Dialogo* [...] *dove ne i congressi di quattro giornate si discorre sopra i due massimi sistemi del mondo Tolemaico, e Copernicano: proponendo indeterminatamente le ragioni filosofiche, e naturali tanto per l'una, quanto per l'altra parte*. Florence: Per Gio. Batista Landini.

Galilei, Galileo. 1638. *Discorsi e dimostrazioni matematiche, intorno à due nuove scienze attenenti alla mecanica & i movimenti locali*. Leiden: Appresso gli Elsevirii.

Galileo 1655 = Galilei, Galileo. 1655. *Discorso al Serenissimo Don Cosimo II, Gran Duca di Toscana: intorno alle cose, che stanno sù l'acqua, ò che in quella si muouono*. Bologna: Per gli HH. del Dozza. First edition 1612.

Galilei, Galileo. 1655–1656. *Opere*. Bologna: Per gli HH. del Dozza.

Galileo 1699 = Galilei, Galileo. 1699. *Systema cosmicum: in quo dialogis IV. de duobus maximis mundi systematibus. Ptolemaico et Copernicano, rationibus utrinque propositis indefinite ac solide disseritur: accessit altera hac editione praeter conciliationem locorum S. Scripturae cum Terrae mobilitate, eiusdem tractatus De motu*. Leiden: Apud Fredericum Haaring et Davidem Severinum.

Galilei, Galileo. 1718. *Opere*. Florence: Nella Stamp. di S. A. R. Per Gio. Gaetano Tartini.

Galileo 1744 = Galilei, Galileo. 1744. *Opere* [...] *in questa nuova edizione accresciute di molte cose inedite*. Padua: Nella stamperia del seminario.

Galilei, Galileo. 1842–1856. *Le opere di Galileo Galilei: Prima edizione completa condotta sugli autentici manoscritti Palatini*, ed. Eugenio Alberi. Florence: Società Editrice Fiorentina.

Galileo 1890–1909 = Galilei, Galileo. 1890–1909. *Le opere di Galileo Galilei. Edizione Nazionale*, ed. Antonio Favaro. Florence: Tipografia Barbèra.

Galluzzi and Torrini 1975 = Galluzzi, Paolo, and Torrini, Maurizio (Eds.). 1975. *Le opere dei discepoli di Galileo Galilei: Carteggio, 1642–1648*. Florence: Giunti-Barbèra.

Gassendi 1644 = Gassendi, Pierre. 1644. *Disquisitio metaphysica. Seu Dubitationes et Instantiae: adversus Renati Descartes Metaphysicam et Responsa*. Amsterdam: Apud Johannem Blaeu.

Gassendi 1649 = Gassendi, Pierre. 1649. *Animadversiones in decimum librum Diogenis Laertii, qui est De vita, moribus, placitisque Epicuri. Continent autem placita, quas ille treis statuit philosophiae partes; I. Canonicam nempe, habitam Dialecticae loco. II. Physicam, ac imprimis nobilem illius partem Meteorologiam. III. Ethicam, cuius gratia ille excoluit caeteras*. Lyon: Apud Guillelmum Barbier.

Gassendi 1658 = Gassendi, Pierre. 1658. *Opera omnia in sex tomos divisa*. Lyon: Sumptibus Laurentii Anisson, et Joan. Bapt. Devenet.

Geulincx, Arnold. 1653. *Questiones quodlibeticae in utramque partem disputatae*. Antwerp: Bij de We Cnobbaert.

Geulincx, Arnold. 1663. *Methodus inveniendi argumenta, quae solertia quibusdam dicitur*. Leiden: Apud Isaacum De Waal Verbiest.

Geulincx, Arnold. 1665. *Saturnalia, seu (ut passim vocantur) Quaestiones quodlibeticae in utramque partem disputatae. Editio secunda ab auctore recognita et aucta*. Leiden: Ex officina Henrici Verbiest.

Geulincx, Arnold. 1675. *Γνῶθι σεαυτόν, sive Ethica*. Leiden: Apud Adrianum Severini.

Geulincx, Arnold. 1691. *Metaphysica vera et ad mentem peripateticam*. Amsterdam: Apud Joannem Wolters.

Geulincx 1891–1893 = Geulincx, Arnold. 1891–1893. *Opera philosophica*, ed. Jan Pieter Nicolaas Land. The Hague: Martinus Nijhoff.

Gorlaeus 1620 = Gorlaeus, David. 1620. *Exercitationes philosophicae post mortem auctoris editae, quibus universa fere discutitur philosophia theoretica, et plurima ac praecipua Peripateticorum dogmata evertuntur*. Leiden: In biblipolio Commeliniano.

Gr. V. 1718a = Gr. V. [pseudonym]. 1718. *Vindicatio nominis atque orthodoxae Wittichianae a calumnis atque iniuriis Driessenianus, ex ipsa Cl. Wittichii Disputatione de natura Dei eruta*. Leiden: Apud Samuelem Luchtmans.

Gr. V. 1718b = Gr. V. [pseudonym]. 1718. *Altera vindicatio nominis atque orthodoxiae Wittichianae a calumniis atque iniuriis Driessianis*. Leiden: Apud Samuelem Luchtmans.

Graaf 1689 = Graaf, Lieuwe Willemsz. 1689. *Eenvoudig en onvervalscht verhaal van 't voorgevallene in 's Gravenhage ter vergaaderinge van de H. M. Heeren, de Staaten Generaal der Vereenigde Nederlanden over 't aanwijsen van 't vinden der lengte van Oost en West, en de waare tijdreekeninghe*. S.l: s.n.

Graaf, Lieuwe Willemsz. 1691. *D'eerste opening en gewis onderwys van de waere vinding der lengte van Oost en West*. Leeuwarden: By Hendrik Rintjes.

Gregory, David. 1695. *Catoptricae et dioptricae sphaericae elementa*. Oxford: E theatro Sheldoniano.

Groenig, Johann. 1701. *Historia cycloeidis: qua genesis et proprietates lineae cycloeidalis praecipuae, secundum eius infantiam, adolescentiam et iuventutem ordine chronologico recensentur*. Hamburg: Apud Apud Gotfr. Liebereit.

Gronovius 1709 = Gronovius, Jakob. 1709. *Burcheri de Volder laudatio*. Leiden: Apud Cornelium Boutestein.

Grotius, Hugo. 1627. *Sensus librorum sex, quos pro veritate religionis Christianae batavice*. Leiden: Ex Off. Joh. Maire.

Grotius, Hugo. 1653. *Van de waerheyt der Christelijke Godts-dienst wt het Latijn vertaalt, door I. D. Volder*. Amsterdam: Voor Ian Rieuwerts.

Hagen 1670 = Hagen, Christiaan. Circa 1670. *De Academie. Kruythoff*. Stadsarchief Leiden, Erfgoed Leiden en omstreken, signature: PV12512. https://www.erfgoedleiden.nl/collecties/beeldmateriaal/zoeken-in-beeldmateriaal/detail/aa4cfdb6-26bc-11e3-b6dc-3cd92befe4f8/media/230f31fd-707d-f78e-3a9e-709a65c67323. Accessed 18 February 2019.

Harvey 1651 = Harvey, William. 1651. *Exercitationes de generatione animalium*. London: Typis Du-Gardianis.

Heereboord 1654 = Heereboord, Adriaan. 1654. *Meletemata philosophica, maximam partem, metaphysica*. Leiden: Ex officina Francisci Moyardi.

Heidanus, Abraham. 1645. *De causa Dei. Dat is: de sake Godts verdedight tegen den mensche. Ofte Wederlefginge van de Antwoorde van S. Episcopius. Mitsgaders een digressu tegen het: Klaer ende volkomen schriftuerlick bericht van J. Batelier*. Leiden: Gedruckt by Paulus Aertz van Ravesteyn.

Hellingwerf 1718 = Hellingwerf, Pieter. 1718. *Wiskonstige oeffening, behelsende eene verhandeling over veele voorname van de mathesis*. Amsterdam: By Joannes Loots.

Helmont, Jean Baptiste van. 1648. *Ortus medicinae, id est, Initia physicae inaudita: Progressus medicinae novus in morborum ultionem ad vitam longam*. Amsterdam: Apud Ludovicum Elzevirium.

Van Helmont 1652 = Helmont, Jean Baptiste van. 1652. *Ortus medicinae, id est, Initia physicae inaudita: progressus medicinae novus, in morborum ultionem ad vitam longam*. Amsterdam: Apud Ludovicum Elzevirium.

Van Helmont 1707 = Helmont, Jean Baptiste van. 1707. *Opera omnia*. Frankfurt: Ex bibliopolio Hafniensi Hieronymi Christiani Paulli.

Hermann, Jakob. 1704. Méthode géométrique et générale de déterminer le diamètre de l'Arc-en-ciel, quelque hypothèse de réfraction que l'on suppose dans l'eau ec. *Nouvelles de la République des Lettres*, June 1704: 658–671.

Hero and Commandino 1583 = Hero of Alexandria and Commandino, Federico. 1583. *Spiritalium Liber a Federico Commandino Urbinate, ex Graeco nuper in Latinum conversus*. Paris: Apud Aegidium Gorbinum. First edition 1575.

Hero and Aleotti 1589 = Hero of Alexandria and Aleotti, Giovanni Battista. 1589. *Gli artifitiosi et curiosi moti spiritali di Herrone*. Ferrara: Per Vittorio Baldini.

Hobbes, Thomas. 1655. *Elementorum philosophiae sectio prima De corpore*. London: Excusum sumptibus Andreae Crook.

Hobbes, Thomas. 1661. *Dialogus physicus, sive, De natura aëris: Coniectura sumpta ab experimentis nuper Londini habitis in Collegio Greshamensi, item De duplicatione cubi*. London: Typis J. B.

Hobbes, Thomas. 1667. *Leviathan: of van de stoffe, gedaente, ende magt vande kerckelycke ende wereltlycke regeeringe*, trans. Abraham van Berkel. Amsterdam: By Jacobus Wagenaar.

Hobbes 1668 = Hobbes, Thomas. 1668. *Opera philosophica quae latine scripsit omnia*. Amsterdam: Apud Ioannem Blaeu.

Hobbes, Thomas. 1670. *Leviathan, sive De materia, forma et potestate civitatis ecclesiasticae et civilis*. Amsterdam: Apud Ioannem Blaeu.

Van Hogelande 1646 = Hogelande, Cornelis van. 1646. *Cogitationes, quibus Dei existentia, item animae spiritalitas, et possibilis cum corpore unio, demonstrantur, nec non, brevis historia oeconomiae corporis animalis, proponitur, atque mechanice explicatur*. Amsterdam: Apud Ludovicum Elzevirium.

Holwarda, Johannes Phocylides. 1651. *Philosophia naturalis, seu Physica vetus-nova, ex optimis quibusque autoribus, antiquis pariter et neotericis deducta, propriisque speculationibus et inventis aucta et illustrata*. Franeker: Excudit Idzardus Alberti.

Honert, Taco Hajo van den. 1718. *Repraesentatio cur ad criminationes quibus Antonius Driessen Jacob. Wittichum ob habitam ante aliquot annos de natura Dei Disputationem a cathedra Groningana arcere conatus est nihil responderit; addita est Wittichii de natura Dei Disputatio*. Leiden: Apud Samuelem Luchtmans.

Van den Honert 1718a = Honert, Taco Hajo van den. 1718. *Repraesentatio altera, qua Driessenii rei gestae narratio ad examen revocatur, vera molitionum Driesseniarum historia traditur, ac Repraesentatio prior ab eius cavillationibus vindicatur*. Leiden: Apud Samuelem Luchtmans.

Van den Honert 1718b = Honert, Taco Hajo van den. 1718. *Briev aan een voornaam Heer, wegens het Dispuut De natura Dei van* [...] *Jacobus Wittichius: waar by gevoegt is een korte ontleding van het selve dispuut, en een antwoord op de Aanspraak van* [...] *Antonius Driessen aan de Kerk van Nederland.* Leiden: By Samuel Luchtmans.

Hooke 1679 = Hooke, Robert. 1679. P. Fran. Lana's Way of Making a Flying Chariot; with an Examination of the Grounds and Principles thereof. *Philosophical Collections* 1: 18–29.

Hooke 1705 = Hooke, Robert. 1705. *The Posthumous Works* [...] *containing his Cutlerian Lectures.* London: Printed by Sam. Smith and Benj. Walford.

Horch, Heinrich. 1719. *Invisibilia Dei in rebus ab Ipso factis facta visibilia, contra Spinozam, cum consectario pacifico inter Cll. Viros, Driessenium et Wittichium, et multiplici analogia naturae spiritualis et corporeae, tanquam genuino theologiae hieroglyphicae fundamento.* Marburg: Typis Philippi Casimiri Mülleri.

Huet, Pierre-Daniel. 1679. *Demonstratio Evangelica.* Paris: Apud Stephanum Michallet.

Huet 1689 = Huet, Pierre-Daniel. 1689. *Censura philosophiae Cartesianae.* Paris: Apud Danielem Horthemels.

Hulsius, Paulus (*praeses*), and Venhuisen, Petrus van (*defendens*). 1702. *Exercitatio theologica de unione animae cum corpore.* Groningen: Apud E. Atlas bibl.

Hulsius 1702 = Hulsius, Paulus. 1702. *Spinosismi depulsio, nominatim circa corporum peccata et poenas.* Groningen: Typis Joh. Lens.

Huygens, Christiaan. 1659. *Systema Saturnium, sive De causis mirandorum Saturni phaenomenôn, et comite eius planeta novo.* The Hague: Ex typographia Adriani Vlacq.

Huygens 1669a = Huygens, Christiaan. 1669. A Summary Account of the Laws of Motion, Communicated by Mr. Christian Hugens in a Letter to the R. Society, and since Printed in French in the Journal des Scavans of March 18. 1669. *Philosophical Transactions* 4(46): 925–928.

Huygens 1669b = Huygens, Christiaan. 1669. Extrait d'une lettre de M. Hugens à l'Auteur du Journal. *Journal des Sçavans*, 18 March 1669: 22–24.

Huygens 1669c = Huygens, Christiaan. 1669. Debat de 1669 à l'académie sur les causes de la pesanteur. Mémoire de Huygens du 28 août 1669. In Huygens, Christiaan. 1888–1950. *Oeuvres complètes*, volume 19, 631–640. The Hague: Martinus Nijhoff.

Huygens 1673 = Huygens, Christiaan. 1673. *Horologium oscillatorium: sive De motu pendulorum ad horologia aptato demonstrationes geometricae.* Paris: Apud F. Muguet.

Huygens 1690 = Huygens, Christiaan. 1690. *Traité de la lumière: avec un Discours de la cause de la pesanteur.* Leiden: Chez Pierre van der Aa.

Huygens 1698a = Huygens, Christiaan. 1698. *Κοσμοθεωρος: sive De terris coelestibus, earumque ornatu, coniecturae, ad Constantinum Hugenium fratrem.* The Hague: Apud Adrianum Moetjens.

Huygens 1698b = Huygens, Christiaan. 1698. *The Celestial World discovered, or Conjectures concerning the Inhabitants, Plants, and Productions of the Worlds in the Planets.* London: Printed for Timothy Childe.

Huygens 1703 = Huygens, Christiaan. 1703. *Opuscula postuma*, ed. Burchard de Volder and Bernard Fullenius. Leiden: Apud Cornelium Boutesteyn.

Huygens 1888–1950 = Huygens, Christiaan. 1888–1950. *Oeuvres complètes.* The Hague: Martinus Nijhoff.

Irenaeus Philalethius [pseudonym of Heidanus, Abraham]. 1656. *Bedenkingen op den Staat des geschils over de Cartesiaensche philosophie en op de Nader openinghe over eenige stucken de theologie raeckende.* Rotterdam: By Johannes Benting.

Irenaeus Philalethius [pseudonym of Heidanus, Abraham]. 1656. *De overtuigde quaetwilligheidt van Svetonius Tranquillus.* Leiden: By Adrianus Wijngaerden.

Keill, John. 1702. *Introductio ad veram physicam: seu Lectiones physicae. Habitae in schola naturalis philosophiae Academiae Oxoniensis, quibus accedunt Christiani Hugenii theoremata de vi centrifuga et motu circulari demonstrata.* Oxford: E theatro Sheldoniano.

Kepler, Johannes. 1609. *Astronomia nova αιτιολογετοσ, seu Physica coelestis, tradita commentariis de motibus stellae Martis.* S.l. (Heidelberg): s.n (Voegelin).

Kepler 1618 = Kepler, Johannes. 1618. *Epitome astronomiae Copernicanae,* [...] *quorum tres hi priores sunt de doctrina sphaerica epitome.* Linz: Excudebat Johannes Plancus.
Kepler, Johannes. 1619. *Harmonices mundi libri V.* Linz: Sumptibus Godofredi Tampachii.
Kircher 1650 = Kircher, Athanasius. 1650. *Musurgia universalis.* Rome: Ex typographia Haeredum Francisci Corbelletti.
Koecher 1738 = Koecher, Johann Christoph. 1738. *De idolatria litteraria liber singularis.* Hannover: Haerede B.N. Foersteri et Filii.
Kuyper, Frans. 1676. *Arcana atheismi revelata, philosophice et paradoxe refutata, examine Tracatus theologico-politici.* Rotterdam: Apud Isaacum Naeranum.
La Hire 1687 = La Hire, Philippe de. 1687. *Tabularum astronomicarum pars prior, de motibus Solis et Lunae.* Paris: Apud Stephanum Michallet.
Lamsweerde, Jan Baptist van. 1674. *Respirationis Swammerdammianae exspiratio, una cum Anatomia neologices Joannis de Raei.* Amsterdam: Ex officina Joannis a Someren.
Lana de Terzi 1670 = Lana de Terzi, Francesco. 1670. *Prodromo ovvero Saggio di alcune inventione nuove premesso all'arte Maestra Opera che prepara il P. Francesco Lana, Bresciano della Compagnia di Giesu. Per mostrare li più reconditi proncipij della Naturale Filosofia, riconosciuti con accurata Teorica nelle più segnalate inventioni, ed isperienze fin'hora ritrovate da gli scrittori di questa materia & altre nuove dell'autore medesimo.* Brescia: Per li Rizzardi.
Lana de Terzi, Francesco. 1684–1692. *Magisterium naturae, et artis* [...]. *In quo occultiora naturalis philosophiae principia manifestantur, et multiplici tum experimentorum, tum demonstrationum serie comprobantur.* Brescia: Apud Jo. Mariam Ricciardum; Parma: Typis Hyppoliti Rosati.
Lange 1723a = Lange, Joachim. 1723. *Causa Dei et religionis naturalis adversus atheismum.* Halle: Impensis Orphanotrophei.
Lange 1723b = Lange, Joachim. 1722. *Modesta disquisitio novi philosophiae systematis de Deo, mundo et homine.* Halle: In officina Orphanotrophei.
Lange 1727 = Lange, Joachim. 1727. *Causa Dei et religionis naturalis adversus atheismum.* Halle: Litteris et impensis Orphanotrophei, 1727.
Le Clerc, Jean. 1695. *XVIII. prima commata capitis primi Evangelii S. Joannis paraphrasi et animadversionibus illustrata a Joanne Clerico; ubi demonstratur, contra alogos, Evangelium hoc esse foetum Joannis Apostoli; et evertitur sententia Fausti Socini, de sensu primorum eius commatum.* Amsterdam: Apud Henr. Wetstenium.
Le Clerc 1699 = Le Clerc, Jean. 1699. *Parrhasiana, ou, Pensées diverses sur des matiéres de critique, d'histoire, de morale et de politique.* Amsterdam: Les Héritiers d'A. Schelte.
Le Clerc 1709 = Le Clerc, Jean. 1709. Éloge de feu Mr. De Volder Professeur en Philosophie et aux Mathematiques, dans l'Academie de Leide. *Bibliothèque choisie* 18: 346–354.
Le Clerc 1987–1997 = Le Clerc, Jean. 1987–1997. *Epistolario*, ed. Mario Sina and Maria Grazia Zaccone-Sina. Florence: Leo S. Olschki.
Le Grand 1675 = Le Grand, Antoine. 1675. *Institutio philosophia, secundum principia Renati Descartes, nova methodo adornata et explicata, ad usum iuventutis academicae.* London: Apud J. Martyn. First edition 1672.
Leibniz, Gottfried Wilhelm von. 1670. *Theoria motus abstracti; seu, Rationes motuum universales, a sensu et phaenomenis independentes.* Leipzig: s.n.
Leibniz, Gottfried Wilhelm von. 1671. *Hypothesis physica nova, qua phaenomenorum naturae plerorumque causae ab unico quodam universali motu, in globo nostro supposito, neque tychonicis, neque copernicanis aspernando, repetuntur, nec non Theoria motus abstracti.* London: Impensis J. Martyn.
Leibniz 1686 = Leibniz, Gottfried Wilhelm von. 1686. Brevis demonstratio erroris memorabilis Cartesii et aliorum circa legem naturae, secundum quam a Deo eandem semper quantitatem motus conservari: qua et in re mechanica abutuntur. *Acta Eruditorum*, March 1686: 161–163.
Leibniz, Gottfried Wilhelm von. 1689. Schediasma de resistentia medii, et motu proiectorum gravium in medio resistente. *Acta Eruditorum*, January 1689: 38–47.
Leibniz 1689 = Leibniz, Gottfried Wilhelm von. 1689. Tentamen de motuum coelestium causis. *Acta Eruditorum*, February 1689: 82–96.

Leibniz, Gottfried Wilhelm von. 1690. De causa gravitatis et defensio sententiae suae de veris naturae legibus contra Cartesianos. *Acta Eruditorum*, May 1690: 228–239.
Leibniz, Gottfried Wilhelm von. 1691. De legibus naturae et vera aestimatione virium motricium contra Cartesianos. *Acta Eruditorum*, September 1691: 439–447.
Leibniz, Gottfried Wilhelm von. 1694. De primae philosophiae emendatione, et de notione substantiae. *Acta Eruditorum*, March 1694: 110–112.
Leibniz, Gottfried Wilhelm von. 1695. Specimen dynamicum pro admirandis naturae legibus circa corporum vires et mutuas actiones detegendis, et ad suas causas revocandis. *Acta Eruditorum*, April 1695: 145–157.
Leibniz, Gottfried Wilhelm von. 1698. De ipsa natura sive De vi insita actionibusque creaturarum, pro dynamicis suis confirmandis illustrandisque. *Acta Eruditorum*, September 1698: 427–440.
Leibniz, Gottfried Wilhelm von. 1710. *Essais de theodicée sur la bonté de Dieu, la liberté de l'homme et l'origine du mal*. Amsterdam: Chez Isaac Troyel.
Leibniz, Gottfried Wilhelm von. 1720. *Lehr-Sätze über die Monadologie, ingleichen Von Gott und seiner Existentz, seinen Eigenschafften und von der Seele des Menschen &c.: wie auch Dessen letzte Vertheidigung seines Systematis Harmoniae praestabilitae wider die Einwürffe des Herrn Bayle*. Frankfurt–Leipzig: Bey Joh. Meyers sel Witbe.
GM = Leibniz, Gottfried Wilhelm von. 1849–1855. *Mathematische Schriften*, ed. Carl I. Gerhardt. Berlin: Asher; Halle: Schmidt.
GP = Leibniz, Gottfried Wilhelm von. 1875–1890. *Die philosophischen Schriften*, ed. Carl I. Gerhardt. Berlin: Weidmannsche Buchhandlung.
Leibniz 1899 = Leibniz, Gottfried Wilhelm von. 1899. *Der Briefwechsel von Gottfried Wilhelm Leibniz mit Mathematikern*, ed. Carl I. Gerhardt. Berlin: Mayer & Müller.
A = Leibniz, Gottfried Wilhelm von. 1923–present. *Sämtliche Schriften und Briefe* (*Akademie Ausgabe*). De Haan 1665 = Haan, Galenus Abrahamsz. de. 1665. *Het compromis tusschen Dr. Galenus Abramsz, nevens sijne medestanders, en Tobias Govertz van den Wyngaert*. Amsterdam: By Jan Rieuwertsen.
Lemmerman, Abraham. 1684. *Eenige bewijzen dat Johannes Bredenburg, staande zijn stellingen, geenszins kan gelooven, dat'er zulk een God is, als de H. Schrift leert*. Amsterdam: By Daniel Bakkamude.
Leupold 1712 = Leupold, Jacob. 1712. *Antlia pneumatica illustrata, oder, Deutliche Beschreibung der so genandten Lufft-Pumpe, darinnen ausführlich gezeiget wird, was solche sey, und wie sie nebst denen zugehörigen, so wohl alten als gantz neu-inventirten Machinen zu gebrauchen: alles in vielen deutlichen und accuraten Figuren entworfen, und zum andern mahl, nebst der ersten Continuation und einem Verzeichniss unterschiedl. mathematischen und physicalischen Instrumenten herausgegeben*. Leipzig: Zu finden bey dem Autore in Nosocomio und bey C. Zunkeln im kleinen Fürsten-Collegio.
Leupold 1724 = Leupold, Jacob. 1724. *Theatri Machinarum Hydraulicarum. Tomus I. Oder: Schau-Platz der Wasser-Künste, Erster Theil*. Leipzig: Druckts Christoph Zunkel.
Leydekker 1719 = Leydekker, Jacob. 1719. *De blyde Spinosist en de bedroefde Christenleeraar over de wysgeerige verhandelinge van de natuure Gods, welke Jacobus Wittichius in den jare 1711. te Duisburg uitgegeven en verdedigt, nu in Maart 1719. in 't Nederduits vertaalt en met aanmerkingen verrykt in 't ligt gebragt heeft. Hier by komt het eenparig oordeel van de philosophische Faculteit der Academie van Jenen*. Rotterdam: By Reinier van Doesburg.
Limborch, Philipp van. 1686. *Schriftelyke onderhandeling, tusschen den heer Philippus van Limborg Professor der Remonstranten, ende Johannes Breedenburg. Rakende 't gebruyk der Reden in de Religie*. Rotterdam: By Barent Bos.
Van Limborch 1686 = Limborch, Philipp van. 1686. *Theologia Christiana ad praxin pietatis ac promotionem pacis Christianae unice directa*. Amsterdam: Apud Henricum Wetstenium.
Van Limborch 1687 = Limborch, Philipp van. 1687. *De veritate religionis Christianae: amica collatio cum erudito Judaeo*. Gouda: Apud Justum ab Hoeve.
Limborch, Philipp van. 1699. *Defensio contra Joannis van der Waeyen iniquam criminationem*. Amsterdam: Apud Joannem Groenewoudt.

Limborch, Philipp van, and Hartsoeker, Christiaan (Eds.). 1704. *Praestantium ac eruditorum virorum epistolae ecclesiasticae et theologicae* [...]. *Editio tertia, nova augmento locupletata.* Amsterdam: Excudit Franciscus Halma. First edition 1660.

Van Limborch 1730 = Limborch, Philipp van. 1730. *Theologia Christiana ad praxin pietatis ac promotionem pacis Christianae unice directa. Adiuncta est relatio historica de origine et progressu controversiarum in Foederato Belgio de praedestinatione. Tractatus posthumus.* Amsterdam: Apud Balthazar Lakeman. First edition 1715.

Line, Francis. 1661. *Tractatus de corporum inseparabilitate: in quo experimenta de vacuo, tam Torricelliana, quam Magdeburgica, et Boyliana, examinantur veraque eorum causa detecta, ostenditur vacuum naturaliter darinon posse, unde et Aristotelica de rarefactione sententia.* London: Typis Thomae Roycroft.

Lipstorp, Daniel. 1653. *Specimina philosophiae Cartesianae. Quibus accedit eiusdem authoris Copernicus redivivus.* Leiden: Apud Johannem et Danielem Elzevier.

Locke 1688a = Locke, John. 1688. Philosophiae Naturalis Principia Mathematica. Auctore Js. Newton, Trin. Coll. Cantab. Soc. Matheseos Professore Lucasiano, et Societatis Regalis Sodali. 4. Londini. Prostant ap. Sam. Smith. 1687. pagg. 510. *Bibliothèque universelle et historique* 8: 436–450.

Locke 1688b = Locke, John. 1688. Extrait d'un Manuscrit Anglois, intitulé Essai Philosophique, concernant l'entendement. *Bibliothèque universelle et historique* 8: 49–142.

Locke, John. 1690. *An Essay Concerning Humane Understanding.* London: Printed for Tho. Basset.

Locke, John. 1695. *The Reasonableness of Christianity, as Delivered in the Scriptures.* London: Printed for Awnsham and John Churchil.

Locke 1976–1989 = Locke, John. 1976–1989. *The Correspondence of John Locke*, ed. Esmond Samuel De Beer. Oxford: Clarendon Press.

Lowthorp, John (Ed.). 1705. *Philosophical Transactions and Collections, abridg'd and dispos'd under general heads, to the year 1700.* London: Printed for Thomas Bennet.

Lufneu 1679 = Lufneu, Hermann. 1679. *Disputatio chemico-medica inauguralis de fermentatione.* Leiden: Apud Viduam et Heredes Johannis Elsevirii.

Lufneu 1685 = Lufneu, Hermann. 1685. Memoire communiqué [...] sur une expérience curieuse d'hydrostatique. *Nouvelles de la République des Lettres*, April 1685: 381–389.

Lufneu 1687 = Lufneu, Hermann. 1687. Réponse de Mr Lufneu à la difficulté que Mr Pujolas luy a faite dans les Nouvelles du mois de Janvier dernier. *Nouvelles de la République des Lettres*, March 1687: 239–249.

Lufneu 1759 = Lufneu, Hermann. 1759. Memoire communiqué [...] sur une expérience curieuse d'hydrostatique. *Nouveau choix de pièces tirées des anciens Mercures, et des autres journaux* 28: 193–200.

Lulofs 1764 = Lulofs, Johannes. 1764. *Grond-beginselen der wynroey- en peil-kunde, ten dienste der landgenooten.* Leiden: By de Wed. A. Honkoop en zoon.

Magalotti 1666 = Magalotti, Lorenzo. 1666. *Saggi di naturali esperienza fatte nell'Accademia del cimento sotto la protezione del serenissimo principe Leopoldo di Toscana e descritte del segretario di essa Accademia.* Florence: Per Giuseppe Cocchini all'Insegna della Stella.

Magni, Valeriano, and Roberval, Gilles Personne de. 1647. *Admiranda de vacuo, scilicet Valeriani Magni demonstratio ocularis de possibilitate vacui. Eiusdem altera pars demonstrationis ocularis. D. De Roberval Narratio de vacuo. Valeriani responsio ad D. de Roberval. Responsio eiusdem ad Peripateticum Cracoviensem.* Warsaw: Officina Petri Elert.

Maignan 1653 = Maignan, Emmanuel. 1653. *Cursus philosophicus: concinnatus ex notissimis cuiusque principiis.* Toulouse: Apud Raymundum Bosc.

Malebranche, Nicolas. 1674–1675. *De la recherche de la vérité. Où l'on traite de la nature de l'esprit de l'homme, & de l'usage qu'il doit faire pour éviter l'erreur dans les sciences.* Paris: Chez André Pralard.

Malebranche 1842 = Malebranche, Nicolas. 1842. *De la recherche de la vérité*, ed. Jules Simon. Paris: Charpentier.

Mallement de Messange, Claude. 1679. *L'ouvrage de la création, traité physique du monde, nouveau système, raisonnements différens de ceux des anciens et des nouveaux philosophes*. Paris: Vve C. Thiboust et P. Esclassan.

Van Mansvelt 1674 = Mansvelt, Regnier van. 1674. *Adversus anonymum Theologo-politicum liber singularis*. Amsterdam: Apud Abrahamum Wolfgang bibliopolam.

Marchand 1758–1759 = Marchand, Prosper. 1758–1759. *Dictionnaire historique ou Mémoires critiques et littéraires*. The Hague: Chez Pierre de Hondt.

Marckius, Johannes et al. 1723. *Iudicium Ecclesiasticum, quo opiniones quaedam Cl. Herm. Alex. Röell synodice damnatae sunt. Laudatum a Professoribus Theologiae in Academia Lugd. Batava*. Leiden: Apud Samuelem Luchtmans.

Mariotte 1673 = Mariotte, Edme. 1673. *Traité de la percussion ou chocq des corps, dans lequel les principales règles du mouvement, contraires à celles que Mr. Descartes et quelques autres modernes ont voulu établir, sont démonstrées par les véritables causes*. Paris: Chez Estienne Michallet.

Mariotte 1679 = Mariotte, Edme. 1679. *Essays de phisique, ou Mémoires pour servir à la science des choses naturelles. Second essay. De la nature de l'air*. Paris: Chez Estienne Michallet.

Mariotte, Edme. 1681. *Essays de phisique, ou Mémoires pour servir à la science des choses naturelles. Quatrieme essay. De la nature des couleurs*. Paris: Chez Estienne Michallet.

Mariotte 1686 = Mariotte, Edme. 1686. *Traité du mouvement des eaux et des autres corps fluides. Divisé en V. parties. Par feu M. Mariotte, de l'Academie royale des sciences. Mis en lumiere par les soins de M. de La Hire, lecteur & professeur du Roy pour les mathematiques, & de l'Academie royale des sciences*. Paris: Chez Estienne Michallet.

Meijer, Lodewijk. 1666. *Philosophia S. Scripturae interpres; exercitatio paradoxa, in qua, veram philosophiam infallibilem S. Literas interpretandi normam esse, apodictice demonstratur, et discrepantes ab hac sententiae expenduntur, ac refelluntur*. S.l. [Amsterdam]: s.n.

Mersenne 1636 = Mersenne, Marin. 1636. *Harmonicorum libri: in quibus agitur de sonorum natura, causis et effectibus*. Paris: Sumptibus O. Baudry.

Mersenne 1644 = Mersenne, Marin. 1644. *Cogitata physico-mathematica: in quibus tam naturae quam artis effectus admirandi certissimis demonstrationibus explicantur*. Paris: Sumptibus Antonii Bertier.

Mersenne 1647 = Mersenne, Marin. 1647. *Novae observationes physico-mathematicae*. Paris: Sumptibus Antonii Bertier.

Mersenne 1977 = Mersenne, Marin. 1977. *Correspondance du P. Marin Mersenne, Religieux Minime, XIII: 1644–1645*, ed. Cornélis de Waard and Armand Beaulieu. Paris: Editions du CNRS.

More, Henry. 1653. *Coniectura Cabbalistica. Or, A Conjectural Essay of Interpreting the Minde of Moses, according to a Threefold Cabbala: Viz. Literal, Philosophical, Mystical, or, Divinely Moral*. London: Printed by J. Flesher.

More 1671 = More, Henry. 1671. *Enchiridion metaphysicum, sive, De rebus incorporeis succincta et luculenta dissertatio*. London: Typis E. Flesher.

Morley 1684 = Morley, Christopher Love (Ed.). 1684. *Collectanea chymica Leidensia, id est, Maetsiana, Margraviana, Le Mortiana. Opus quingentis, et amplius processibus adornatam, omnibus et medicis, et chimicis, et pharmacopaeis*. Leiden: Apud Henricum Drummond.

Morley and Schacht 1680 = Morley, Christopher Love, and Schacht, Lucas. 1680. *De morbo epidemico: tam huius, quam superioris, anni, id est, 1678 et 1679, narratio*. London: Impensis Johannis Gay.

Mort, Jacob le. 1700. *Chymia ab insidiis, erroribus, et calumniis Philosophorum et Galenicorum vindicata*. Leiden: Apud Jordanum Luchtmans.

Van Musschenbroek 1700 = Musschenbroek, Johannes Joosten van. 1700. *Descriptio antliae pneumaticae et instrumentorum ad eam inprimis pertinentium*. S.l.: s.n.

Van Musschenbroek 1729 = Musschenbroek, Pieter van. 1729. *Physicae experimentales et geometricae dissertationes*. Leiden: Apud Samuelem Luchtmans.

Van Musschenbroek 1734 = Musschenbroek, Pieter van. 1734. *Elementa physicae conscripta in usus academicos*. Leiden: Apud Samuelem Luchtmans.

Van Musschenbroek 1745 = Musschenbroek, Pieter van. 1745. *Institutiones physicae: conscriptae in usus academicos*. Venice: Apud Joannem Baptistam Recurti. First edition 1734.

Newton 1687 = Newton, Isaac. 1687. *Philosophiae naturalis principia mathematica*. London: Iussu Societatis Regiae ac Typis Josephi Streater.

Newton 1803 = Newton, Isaac. 1803. *The Mathematical Principles of Natural Philosophy*, trans. Andrew Motte. London: Printed for H. D. Symonds.

Newton 1959–1977 = Newton, Isaac. 1959–1977. *The Correspondence of Isaac Newton*, ed. Herbert Westren Turnbull. Cambridge: Published for the Royal Society at the University Press.

Nierop, Dirck Rembrandtsz. van. 1653. *Nederduytsche astronomia: dat is Onderwijs van den loop des hemels, leerende het vinden der plaetsen en bewegingen der vaste sterren, Son en Maen*. Harlingen: By Jan Hessels.

Nierop, Dirck Rembrandtsz. van. Circa 1660. *Wassende graade paskaart, vertonende alle de zeekusten van Europa, de geheele Middelandsche Zee, als oock ten Noordwesten, en Noordoosten soo veer als ons tot noch toe bekent is, geteeckent door D. Rembrandtz. van Nierop Wassende graade paskaart*. Amsterdam: Bij Pieter Goos. Amsterdam University Library, signature: OTM: HB-KZL L.K. VI 4. hdl.handle.net/11245/3.1326. Accessed 7 February 2019.

Nierop, Dirck Rembrandtsz. van. 1661. *Des Aertrycks beweging en de Sonne stilstant: bewijsende dat dit geensins met de christelijke religie is strijdende*. Amsterdam: By Gerrit van Goedesbergh.

Nierop, Dirck Rembrandtsz. van. 1669. *Eenige oefeningen in Godlijcke, wiskonstige en natuerlijcke dingen*. Amsterdam: By Gerrit van Goedesbergh.

Van Nierop 1680 = Nierop, Dirck Rembrandtsz. van. 1680. *Tweede deel op de Wiskonstige rekening, welke sijn eenige aenteykeningen, dienende tot verbetering en vermeerdering des selfden*. Amsterdam: By de Weduwe Abel van der Storck.

Nieuhof, Johan. 1682. *Gedenkweerdige Brasiliaense Zee- en Lant-Reise und Zee- en Lant-Reize door verscheide Gewesten van Oostindien*. Amsterdam: By de weduwe van Iacob van Meurs.

Nieuhof, Johan. 1682. *Zee- en Lant-Reise door verscheide Gewesten van Oostindien, behelzende veele zeldzaame en wonderlijke voorvallen en geschiedenissen. Beneffens een beschrijving van lantschappen, dieren, gewassen, draghten, zeden en godsdienst der inwoonders: En inzonderheit een wijtloopig verhael der Stad Batavia*. Amsterdam: Voor de weduwe van Jacob van Meurs.

Nieuwentijt, Bernard. 1715. *Het regt gebruik der werelt beschouwingen, ter overtuiginge van ongodisten en ongelovigen aangetoont*. Amsterdam: By de Wed. J. Wolters, en J. Pauli.

Nieuwentijt, Bernard. 1718. *The Religious Philosopher, or the Right Use of Contemplating the Works of the Creator*. London: Printed for J. Senex.

Nieuwentijt 1725 = Nieuwentijt, Bernard. 1725. *L'existence de Dieu, démontrée par les merveilles de la nature: en trois parties, où l'on traite de la structure du corps de l'homme, des élémens, des astres, & de leurs divers effets: avec des figures en taille-douce*. Paris: De l'imprimerie de Jacques Vincent.

Nieuwentijt 1747 = Nieuwentijt, Bernard. 1747. *Rechter Gebrauch Der Welt-Betrachtung: Zur Erkentnis der Macht, Weisheit und Güte Gottes, Auch Ueberzeugung Der Atheisten und Ungläubigen*. Jena: Bei Christian Heinrich Cuno.

Noël 1648 = Noël, Étienne. 1648. *Gravitas comparata, seu Comparatio gravitatis aëris cum hydargyri gravitate*. Paris: Apud Sebastianum Cramoisy et Gabrielem Cramoisy.

Oldenburg, Henry. 1671. An accompt of two books. - I. Prodromo overo saggio di alcune inventioni nuove premesso all Arte Maestra di P. Francisco Lana della Campagnia di Jesu, in Brescia, 1670. in 4°. II Joh. Henr. Meibomii de cerevisiis, potibusque et ebriaminibus extra vinum aliis Commentarius, annexo libello Turnebi de vino. Helmestadii 1668. in 4°. *PhilosophicalTransactions* 6(69): 2114–2118.

Oldenburg 1965–1986 = Oldenburg, Henry. 1965–1986. *The Correspondence of Henry Oldenburg*, ed. Alfred Rupert Hall and Marie Boas Hall. Madison: University of Wisconsin Press; London: Mansel; London: Taylor & Francis.

Papin 1685a = Papin, Denis. 1685. The Description of a Siphon, performing the same things with the Sipho Wurtemburgicus; invented by Dr. Papin Fellow of the Royal Society. *Philosophical Transactions* 15(167): 847–848.

Papin 1685b = Papin, Denis. 1685. A new way of raising water. *Philosophical Transactions* 15(173): 1093–1094.

Papin 1685c = Papin, Denis. 1685. A full description, with the use, of the new contrivance for raising - water, propounded in the phil. Trans N°. 173. *Philosophical Transactions* 15(178): 1274–1278.

Papin 1689 = Papin, Denis. 1689. De gravitatis causa et proprietatibus observationes. *Acta Eruditorum*, April 1689: 183–189.

Papin, Denis. 1691. Mechanicorum de viribus motricibus sententia. *Acta Eruditorum*, January 1691: 6–13.

Parent 1701 = Parent, Antoine. 1701. Démonstration des XIII Théorèmes de la force centrifuge, proposés sans démonstration par Mr. Huygens de Zulichem, à la fin de son Traité de la Pendule, augmentés de ceux qui manquoient, & remis dans leur ordre naturel. *Journal des Sçavans* 20 (23 May 1701): 370–388.

Pascal 1663 = Pascal, Blaise. 1663. *Traitez de l'équilibre des liqueurs, et de la pesanteur de la masse de l'air*. Paris: Chez Guillaume Desprez.

Pascal 1923 = Pascal, Blaise. 1923. *Œuvres de Blaise Pascal; publiées suivant l'ordre chronologique, avec documents complémentaires, introductions et notes*, ed. Léon Brunschwig and Pierre Boutroux. Paris: Libraire Hachette. First edition 1904–1914.

Pecquet 1651 = Pecquet, Jean. 1651. *Experimenta nova anatomica* [...]. *Eiusdem Dissertatio anatomica de circulatione sanguinis, et chyli motu*. Paris: Apud Sebastianum Cramoisy et Gabrielem Cramoisy.

Pecquet 1654 = Pecquet, Jean. 1654. *Experimenta nova anatomica* [...]. *Eiusdem Dissertatio anatomica de circulatione sanguinis, et chyli motu*. Paris: Ex Officina Cramosiana.

Petit 1647 = Petit, Pierre. 1647. *Observation touchant le vuide faite pour la première fois en France, contenue en une letter écrite à Monsieur Chanut*. Paris: Chez Sebastien Cramoisy et Gabriel Cramoisy.

Pitcairne 1692 = Pitcairne, Archibald. 1692. *Oratio, qua ostenditur, medicinam ab omni philosophorum secta esse liberam*. Leiden: Apud Abrahamum Elzevier.

Plato 1892 = Plato. 1892. *The Dialogues of Plato*, ed. and trans. Benjamin Jowett. London: MacMillan and Co. First edition 1871.

Poiret, Pierre. 1677. *Cogitationum rationalium de Deo, anima, et malo libri IV. In quibus quid Cartesius eiusque sequaces, de hisce senserint, omnisque philosophiae certiora fundamenta, atque in primis tota metaphysica verior continentur*. Amsterdam: Apud Danielem Elsevirium.

Poiret, Pierre. 1687. *L'oeconomie divine ou Système universel et démontré des oeuvres et des desseins de Dieu envers les hommes*. Amsterdam: Chez Henry Wetstein.

Poiret, Pierre. 1705. *Oeconomia divina, in quibus Dei erga homines proposita, agendi rationes, atque opera demonstrantur*. Frankfurt: Fritsch.

Poiret 1715 = Poiret, Pierre. 1715. *Vera et cognita omnium prima sive De natura idearum ex origine sua repetita, asserta et adversus Cl. A. Pungelerum defensa, disquisitio theologico-philosophica, in qua Spinosismus et Socianismus toto praevertuntur* [...] *ac non paucis in theologia et philosophia momentosis difficultatibus profundius dilucidandis via aperitur*. Amsterdam: Ex Officina Wetsteniana.

Polinière, Pierre. 1709. *Expériences de physique*. Paris: Chez Jean de Laulne, Claude Jombert, Jacque Quillau.

Power 1664 = Power, Henry. 1664. *Experimental Philosophy. In Three Books. Containing New Experiments Microscopical, Mercurial, Magnetical. With some Deductions and probable Hypotheses raised from them, in Avouchment and Illustration of the now famous Atomical Hypothesis*. London: Printed by T. Roycroft.

Pujolas 1687 = Pujolas, Moïse. 1687. Extrait d'une lettre écrite de Londres à M. Silvestre, Docteur en Médecine, par M. Pujolas, touchant l'expérience curieuse d'hydrostatique communiquée par

M. M. Lufneu, et insérée dans les Nouvelles d'avril 1685, art. V. *Nouvelles de la République des Lettres*, January 1687: 20–28.
Pungeler 1712 = Pungeler, Abraham. 1712. *Dissertatio de rerum possibilium ideis in Deo Cl. Poireto opposita*. Herborn: Typis Johannis Nicolai Andreae.
Raey, Johannes de. 1651. *Oratio inauguralis de gradibus et vitiis notitiae vulgaris circa contemplationem naturae, et officio philosophi circa eandem*. Leiden: E typographeo Francisci Hackii.
Raey, Johannes de. 1651–1652. *Disputationum physicarum ad Problemata Aristotelis prima [–quinta]*. Leiden: E typographeo Francisci Hackii, et ex officina Joannis Maire.
De Raey 1654 = Raey, Johannes de. 1654. *Clavis philosophiae naturalis, seu Introductio ad naturae contemplationem, Aristotelico-Cartesiana*. Leiden: Ex officina Joannis et Danielis Elsevier.
De Raey 1666 = Raey, Johannes de (*praeses*), and Bassecour, Fabricius de la (*respondens*). 1666. *Disputatio physica de humido et sicco*. Leiden: Apud Viduam et Haeredes Johannis Elsevirii.
Raey, Johannes de (*praeses*), and Freherus, Johannes (*respondens*). 1668. *Disputatio philosophica, de constitutione logicae*. Leiden: Apud viduam et haeredes Johannis Elsevirii.
Raey, Johannes de (*praeses*), and Reus, Abrahamus de (*respondens*). 1668. *Disputatio philosophica, de constitutione physicae*. Leiden: Apud viduam et haeredes Johannis Elsevirii.
Raey, Johannes de. 1677. *Clavis philosophiae naturalis Aristotelico-Cartesiana. Editio secunda, aucta opusculis philosophicis varii argumenti*. Amsterdam: Apud Danielem Elsevirium.
De Raey 1692 = Raey, Johannes de. 1692. *Cogitata de interpretatione, quibus natura humani sermonis et illius rectus usus, tum in communi vita et disciplinis ad vitae usum sptectantibus, tum in philosophia, ab huius seculi errore et confusione vindicantur. Accedunt Notae recentes ad partem primam generalem. Cum Appendice ex olim scriptis, propter cognationem*. Amsterdam: Apud Henricum Wetstenium.
Raey, Johannes de Jr. 1680. *Dictionarium geographicum, ofte, Schat-en woordt-boeck des aerdtrycks*. Amsterdam: By Abraham Wolfganck.
Régis, Pierre-Sylvain. 1690. *Système de philosophie, contenant la logique, la métaphysique, la physique, et la morale*. Paris: De l'imprimerie de Denys Thierry.
Régis 1691 = Régis, Pierre-Sylvain. 1691. *Cours entier de philosophie, ou Systême general selon les principes de M. Descartes, contenant la logique, la metaphysique, la physique, et la morale*. Paris: Aux dépens des Huguetan.
Regius 1640 = Regius, Henricus (*praeses*), and Hayman, Johannes (*respondens*). 1640. *Disputatio medico-physiologica pro sanguinis circulatione*. Utrecht: Ex officina Aegidii Roman.
Regius 1646 = Regius, Henricus. 1646. *Fundamenta physices*. Amsterdam: Apud Lodovicum Elzevirium.
Regius 1654 = Regius, Henricus. 1654. *Philosophia naturalis*. Amsterdam: Apud Lodovicum Elzevirium.
Regius, Henricus. 1657. *Praxis medica, medicationum exemplis demonstrata*. Utrecht: Typis Theodori ab Ackersdijck, et Gisberti a Zijll.
Regius 1661 = Regius, Henricus. 1661. *Philosophia naturalis, in qua tota rerum universitas, per clara et facilia principia, explanatur*. Amsterdam: Apud Lodovicum et Danielem Elzevirios.
Regius, Henricus. 1687. *Philosophie naturelle*. Utrecht: Chez Rodolphe van Zyll.
Regius, Johannes. 1714. Kartezius Spinoza voorlichtend. In Regius, Johannes. 1714. *Principia philosophiae theoreticae. De beginselen der beschouwende filozofy*, 321–353. Rotterdam: By Reinier van Doesburg.
Regius 1719 = Regius, Johannes. 1719. *Cartesius verus Spinozismi architectus*. Franeker: Excudit Henricus Halma.
Reisel 1685 = Reisel, Salomon. 1685. Part of a letter to one of the Secr. of the R. Society, from the learned Dr. Salomon Reiselius, Chief Physician to the Duke of Wirtemberg; concerning the sipho wirtembergicus stutgardiae. 8. Oct. S. V. 1685. *Philosophical Transactions* 15(178): 1272–1273.
Reisel 1689 = Reisel, Salomon. 1689. *Sipho Wurtembergicus per maiora experimenta firmatus, in vertice effluens, correctus et detectus*. Stuttgart: Lobber.

Remus, Georgius. 1711. *Disputatio medica inauguralis de functione pulmonis*. Leiden: Apud Abrahamum Elzevier.

Reneri 1635 = Reneri, Henricus (*praeses*). 1635. *Disputationum physicarum prima[–septima]*. Utrecht: Ex Officina Abrahami ab Herwiick.

Revius 1647 = Revius, Jacobus (*praeses*), and Danburgh, Antonius (*respondens*). 1647. *Analectorum theologicorum disputatio XXIII. De cognitione Dei, tertia*. Leiden: Apud haeredes Johannis Nicolai a Dorp.

Riccioli 1651 = Riccioli, Giovanni Battista. 1651. *Almagestum novum astronomiam veterem novamque complectens observationibus aliorum, et propriis novisque theorematibus, problematibus, ac tabulis promotam*. Bologna: Ex typographia haeredis Victorii Benatii.

Riccioli 1661 = Riccioli, Giovanni Battista. 1661. *Geographiae et hydrographiae reformatae libri duodecim*. Bologna: Ex typographia haeredis Victorii Benatii.

Rittangel, Johann Stephan, and Waeyen, Johannes van der. 1698. *Libra veritatis et de Paschate tractatus, praemissa est Johannis vander Waeyen Dissertatio de Λόγῳ adversus Johannem Clericum*. Franeker: Typis et impensis Jacobi Horrei.

Rittangel, Johann Stephan, and Waeyen, Johannes van der. 1699. *Veritas religionis Christianae in articulis de Trinitate et Christo ex Scriptura, Rabbinis et Cabbala probata. Praefixa est Johannis vander Waeyen Limborgianae responsionis discussio*. Franeker: Apud Wibium Bleck.

Roberval 1923 = Roberval, Gilles Personne de. 1923. De vacuo narratio ad Nobilem Virum Dominum des Noyers, Serenissimae Reginae Poloniae a consiliis et secretis. In Pascal, Blaise. 1923. *Œuvres de Blaise Pascal; publiées suivant l'ordre chronologique, avec documents complémentaires, introductions et notes*, ed. Léon Brunschwig and Pierre Boutroux, volume 2, 310–410. Paris: Libraire Hachette. First edition 1904–1914.

Roëll 1723 = Roëll, Johannes Alexander. 1723. *Larva detracta Jacobo Wittichio* [...] *Jacobus Wittichius ontmaskert*. Utrecht: By Willem vande Water.

Rohault 1671 = Rohault, Jacques. 1671. *Traité de physique*. Paris: Chez la veuve de Charles Savieux.

Rohault and Le Grand 1691 = Rohault, Jaques, and Le Grand, Antoine. 1691. *Tractatus physicus cum animadversionibus Antonii le Grand*. Amsterdam: Apud J. Pauli. First edition 1682.

Rohault and Clarke 1710 = Rohault, Jaques, and Clarke, Samuel. 1710. *Physica. Latine vertit, recensuit, et adnotationibus ex* [...] *Isaaci Newtoni Philosophia maximam partem haustis, amplificavit et ornavit Samuel Clarke* [...] *Editio tertia*. London: Impensis Jacobi Knapton. First edition 1697.

Rosenboom 1697 = Rosenboom, Hub. 1697. *Stukken met betrekking tot de onenigheden over het curatorschap van Hub. Rosenboom aan de Universiteit van Leiden*. S.l.: s.n.

Rump 1721 = Rump, Justus Wessel. 1721. *Institutiones theologicae in tres partes distributae, in quibus fidei dogmata et controversiae fere omnes, etiam recentissimae exhibentur*. Soest: Impensis Josephi Wolschendorf; Leipzig: Apud Heredes Friderici Lanckisii.

Ruysch 1717 = Ruysch, Frederik. 1717. *Adversariorum anatomico-medico-chirurgicorum decas tertia*. Amsterdam: Apud Janssonio-Waesbergios.

Ruysch 1725 = Ruysch, Frederik. 1725. *Responsio, ad dissertationem epistolicam Jo. Christoph. Bohlii quam de usu novarum cavae propaginum in systemate chylopoeo, nec non de cortice cerebri, conscripsit*. Amsterdam: Apud Janssonio-Waesbergios.

Saint Julien-Pottier, Antonius de. 1698. *Circuli quadraturae nodum per helicem enodatum*. Utrecht: Ex Officina Francisci Halma.

Scaliger, Joseph Justus. 1594. *Cyclometrica elementa duo*. Leiden: Ex officina Plantiniana.

Schelhammer 1703 = Schelhammer, Günther Christoph. 1703. *Exercitationum* [...] *duodecima, continens theses selectas de partibus generationi dicatis earumque usu*. Kilonum: Literis Bartholdi Reutheri.

Schoock 1643 = Schoock, Martin. 1643. *Admiranda methodus novae philosophiae Renati des Cartes*. Utrecht: Ex officina Joannis van Waesberge.

Schott 1657 = Schott, Gaspar. 1657. *Mechanica hydraulico-pneumatica*. Würzburg: Sumptu heredum Joannis Godefridi, Excudebat Henricus Pignim.

Schott 1657–1659 = Schott, Gaspar. 1657–1659. *Magiae universalis naturae et artis* [...] *opus quadripartitum*. Würzburg: Excudebat Henricus Pignim.

Seerup, Nicolaus Nicolai (*praeses*), and Ivarides, Andreas (*defendens*). 1681. *Disputationum philosophicarum sive Cogitationibus rationalium de Disputationibus philosophicis sive Cogitationibus, uti inscribuntur, rationalibus Viri Clarissimi et Doctissimi Dn. Burcheri de Volder de rerum naturalium principiis prima qua contra eundem solidis, ut spes est, argumentis ac rationibus, aqua solum materiale omnium corporum principium adstruitur, et simul eadem assertio ab animadversionibus clarissimorum aliquot virorum vindicatur*. Copenhagen: Literis Christiani Wilingii Acad. Typogr.

Senguerd 1681 = Senguerd, Wolferd. 1681. *Philosophia naturalis: quatuor partibus primarias corporum species affectiones, differentias, productiones mutationes et interitus exhibens*. Leiden: Apud Danielem a Gaesbeeck.

Senguerd 1685 = Senguerd, Wolferd. 1685. *Philosophia naturalis: quatuor partibus primarias corporum species affectiones, differentias, productiones mutationes et interitus exhibens. Editio secunda, priore auctior*. Leiden: Apud Danielem a Gaesbeeck.

Senguerd, Wolferd (*praeses*), and Westhovius, Fredericus Gulielmus (*respondens*). 1698. *Exercitium experimentale quintum decimum, quod est de aëris gravitate, secundum*. Leiden: Apud Abrahamum Elzevier.

Senguerd, Wolferd (*praeses*), and Voorhoff, Seger (*respondens*). 1698. *Exercitium experimentale septimum decimum, quod est de aëreae elasticitatis et resistentiae passivae effectis*. Leiden: Apud Abrahamum Elzevier.

Senguerd, Wolferd. 1714. Annotationes circumstantiarum singularium circa cohaerentiam hemispheriorum concavorum et cylindrorum solidorum. *Acta Eruditorum*, February 1714: 82–85.

Senguerd 1715 = Senguerd, Wolferd. 1715. *Rationis atque experientiae connubium continens experimentorum physicorum, mechanicorum, hydrostaticourm, barometricorum, thermometricorum, aliorumque compendiosam narrationem*. Rotterdam: Apud Bernardum Bos.

Sennert 1629 = Sennert, Daniel. 1629. *De chymicorum cum Aristotelicis et Galenicis consensu ac dissensu liber*. Wittemberg: Sumtibus viduae et haered. Zachariae Schüreri Senioris. First edition 1619.

Sennert, Daniel. 1644. *Epitome institutionum medicinae et Liber de febribus*. Amsterdam: Apud Jodocum Jan Sonium.

Sennert 1651 = Sennert, Daniel. 1651. *Epitome naturalis scientiae*. Amsterdam: Apud Joannem Ravesteyn. First edition (as series of disputations) 1600.

Sennert 1667 = Sennert, Daniel. 1667. *Institutionum medicinae libri quinque*. Wittenberg: Sumptibus haered. Tobiae Mevii. First edition 1611.

Serrurier, Joseph. 1700. *Physicae experimentis innixae compendiosa tractatio, in usum scholarum domesticarum*. Herborn: Ex Officina Johannis Nicolai.

Sinclair 1669 = Sinclair, George. 1669. *Ars nova et magna gravitatis et levitatis*. Rotterdam: Ex officina Arnoldi Leers.

Snel, Willebrord. 1617. *Eratosthenes Batavus, de terrae ambitus vera quantitate*. Leiden: Apud Iodocum a Colster.

Spinoza 1663 = Spinoza, Baruch de. 1663. *Renati des Cartes Principiorum philosophiae pars 1., et 2., more geometrico demonstratae* [...]. *Accesserunt* [...] *Cogitata metaphysica*. Amsterdam: Apud Johannem Rieuwertsz.

Spinoza 1677 = Spinoza, Baruch de. 1677. *Opera posthuma, quorum series post praefationem exhibetur*. S.l. (Amsterdam): s.n. (Jan Rieuwertsz).

Spinoza 1912 = Spinoza, Baruch de. 1912. *The Chief Works*, ed. and trans. Robert Harvey Monro Elwes. London: George Bell and Sons. First edition 1891.

Spinoza 1974 = Spinoza, Baruch de. 1974. *The Principles of Descartes' Philosophy*, ed. and trans. Halbert Britan. LaSalle: Open Court.

Spinoza 1995 = Spinoza, Baruch de. 1995. *The Letters*, ed., intr. and trans. Samuel Shirley, Steven Barbone, Jacob Adler and Lee Rice. Indianapolis: Hackett.

Spinoza 1996 = Spinoza, Baruch de. 1996. *Ethics*, ed. and trans. Edwin Curley. London: Penguin Books.

Sprat 1667 = Sprat, Thomas. 1667. *The History of the Royal Society of London, for the Improving of Natural Knowledge*. London: Printed by T. R.
Stevin, Simon. 1586. *De beghinselen des waterwichts*. Leiden: By Françoys van Raphelingen.
Stevin 1605 = Stevin, Simon. 1605. *Hypomnemata mathematica.* [...] *Tomus quartus* [...] *de statica*. Leiden: Ex officina Ioannis Patii.
Stevin, Simon. 1634. *Les oeuvres mathematiques* [...]. *Ou sont inserées les memoires mathematiques, esquelles s'est exercé le tres-haut & tres-illustre prince Maurice de Nassau* [...]. *Le tout reveu, corrigé, & augmenté par Albert Girard.* Leiden: Chez Bonaventure & Abraham Elsevier.
Stolle 1728 = Stolle, Gottlieb. 1728. *Introductio in historiam litterariam in gratiam cultorum elegantiorum litterarum et philosophiae*. Jena: Apud vid. Joann. Meyeri.
Stouppe, Jean Baptiste. 1673. *La religion des Hollandois*. Paris: Chez François Clousier.
Strada, Famiano. 1632–1647. *De bello Belgico, decas prima*[*–secunda*]. Rome: Typis Francisci Corbelletti.
Strada, Famiano. 1682. *Primera*[*–segunda*] *Decada de las Guerras de Flandes*, trans. Melchor de Novar, ed. Alonso de Buena Maison. Cologne: s.n. (False place of publication; actually: Amsterdam).
Sturm 1676 = Sturm, Johann Christoph. 1676. *Collegium experimentale, sive curiosum*. Nuremberg: Sumptibus Wolfgangi Mauritii Endteri, et Johannis Andreae Endteri Haeredum.
Sturm 1685 = Sturm, Johann Christoph. 1685. *Collegii experimentalis sive curiosi, pars secunda in qua porro praesentis aevi experimenta et inventa physico-mathematica compluria,* [...] *Epistola ad* [...] *Henricum Morum*. Nuremberg: Sumtibus Wolfangi Mauritii Endteri.
Svetonius Tranquillus 1656a = Svetonius Tranquillus [pseudonym]. 1656. *Staat des geschils, over de Cartesiaansche philosophie*. Utrecht: By Johannes van Waesberge.
Svetonius Tranquillus 1656b = Svetonius Tranquillus [pseudonym]. 1656. *Nader openinge van eenighe stucken in de Cartesiaensche philosophie raeckende de* [...] *theologie*. Leiden: Voor Cornelis Banheining.
Svetonius Tranquillus 1656c = Svetonius Tranquillus [pseudonym]. 1656. *Den overtuyghden cartesiaen ofte Claere aenwysinge uyt de Bedenckingen van Irenaeus Philalethius*. Leiden: Gedruckt door Cornelis Banheinning.
Swammerdam 1667 = Swammerdam, Jan. 1667. *Tractatus physico-anatomico-medicus de respiratione usuque pulmonum*. Apud Danielem, Abraham, et Adrian, a Gaasbeeck.
Swammerdam 1737–1738 = Swammerdam, Jan. 1737–1738. *Bybel der natuure of Historie der insecten. Biblia naturae, sive Historia insectorum*. Leiden: By Isaak Severinus, Boudewyn vander Aa, Pieter vander Aa.
Sylvius, Franciscus. 1679. *Opera medica, tam hactenus inedita, quam variis locis et formis edita, nunc vero certo ordine disposita, et in unum volumen redacta*. Amsterdam: Apud Danielem Elsevirium, et Abrahamum Wolfgang.
Tinelis de Castelet 1677 = Tinelis de Castelet, Alexandre. 1677. Lettre à M. l'abbé Bourdelot, dans laquelle il démontre que les raisons que M. Descartes a données du flux et du reflux de la mer sont fausses. Paris: Chez François Muguet.
Torricelli 1644 = Torricelli, Evangelista. 1644. *Opera geometrica*. Florence: Typis Amatoris Masse et Laurentii de Landis.
Trigland 1647 = Trigland, Jacobus. 1647. *Disputationum theologicarum in confessionem et apologiam remonstrantium quinquagesima*. Leiden: Ex officina Bonaventurae et Abrahami Elsevir.
Tschirnhaus, Ehrenfried Walther von. 1687. *Medicina mentis; sive, Tentamen genuinae logicae, in qua disseritur de methodo detegendi incognitas veritates*. Apud Albertum Magnum et Joannem Rieuwerts Juniorem.
Uffenbach 1720 = Uffenbach, Zacharias Conrad von. 1720. *Bibliotheca Uffenbachiana manuscripta, seu, Catalogus et recensio msstorum codicum qui in bibliotheca Zachariae Conradi ab Uffenbach Traiecti ad Moenum adservantur et in varias classes distinguuntur*. Halle: Impensis Novi Bibliopolii.
Uffenbach 1729–1731 = Uffenbach, Zacharias Conrad von. 1729–1731. *Bibliotheca Uffenbachiana universalis, sive Catalogus librorum tam typis quam manu exaratatorum, quot* [...] *collegit*

Zach. Conradus ab Uffenbach, nunc vero [...] *venales prostant*. Frankfurt: Apud Jo. Benj. Andreae et Henr. Hort.

Uffenbach 1753–1754 = Uffenbach, Zacharias Conrad von. 1753–1754. *Merkwürdige Reisen durch Niedersachsen, Holland und Engelland*. Frankfurt-Leipzig: s.n.

V. Gr. [or Gr. V.]. 1719. *De aanspraak van Antonius Driessen aan de Nederlantse Kerk wegens 't Dispuut van J. Wittichius de natura Dei tegen de vernuftelose Sophisticatien van Gr. V. agter den welken schuilt de Heer T. H. van den Honert, verdedigt door V. Gr. met eene Vorrede van Anton. Driessen*. S.l: s.n.

Valentini 1709 = Valentini, Michael Bernhard. 1709. *Armamentarium naturae systematicum, seu Introductio ad philosophiam modernorum naturalem per formam institutionum aphoristice olim tradita, nunc vero perpetuo commentario et fig. aeneis illustrata. Accedit historia literaria S. R. I. academiae naturae curiosorum, antehac seorsim continuata*. Giessen: Typis et impensis Henningi Mülleri.

Valentini 1714 = Valentini, Michael Bernhard. 1714. *Neu-auffgerichtetes Rüst- und Zeughauß Der Natur: Worinnen Die so wundersame, curiöse, auch sehr nützliche Machinen und Instrumenten, deren sich die heutige Naturkündiger in Erforschung der natürlichen Ursachen bedienen, zu sehen und zu finden sind*. Frankfurt: Verlegt von Johann David Zunners sel. Erben und Johann Adam Jungen.

Varignon 1692 = Varignon, Pierre. 1692. De actione aquae erga fundum vasis in inferiori parte latioris quam in superiori. *Acta Eruditorum*, August 1692: 365–370.

Velthuysen, Lambert van. 1670. *Tractaet van de afgoderye en superstitie*. Amsterdam: Gebdruckt by Gabriël Hendricksz.

Velthuysen, Lambert van. 1680. *Opera omnia. Ante quidem separatim, tam belgice quam latine, nunc vero coniunctim latine edita. Quibus accessere duo tractatus novi, hactenus inediti: Prior est De articulis fidei fundamentalibus. Alter De cultu naturali, oppositus Tractatui theologico-politico et Operi posthumo Benedicti de Spinoza*. Rotterdam: Typis Reineri Leers.

Van Velthuysen 2013 = Velthuysen, Lambert van. 2013. *A Letter on the Principles of Justness and Decency, Containing a Defence of the Treatise De Cive of the Learned Mr Hobbes*, ed., intr. and trans. Malcolm de Mowbray and Catherine Secretan. Leiden-Boston: Brill.

Venhuisen, Petrus van. 1702. *Apologia Christiana contra theses apologeticas, quas praeses Johannes Bernoulli, et respondens, Phaebus Themmen* [...] *contra me* [...] *defendendas susceperint*. Groningen: s.n.

Versé, Noël Aubert de. 1684. *L'impie convaincu, ou Dissertation contre Spinoza*. Amsterdam: Chez Jean Crelle.

De Versé 2015 = Versé, Noël Aubert de. 2015. *L'impie convaincu, ou Dissertation contre Spinoza*, ed. Fiormichele Benigni. Rome: Edizioni di Storia e Letteratura.

Verwer, Adriaen. 1683. *'t Mom-aensicht der atheistery afgerukt door een verhandeling van den aengeboren stand der menschen*. Amsterdam: By Wilhelmus Goeree.

Visscher, Nicolaes. Circa 1689. *Africae accurata tabula*. Amsterdam: s.n.

Visscher, Nicolaes. Circa 1698. *Germaniae Inferioris XVII Provinciarum accuratissima tabula*. Amsterdam: s.n. Vrije Universiteit van Amsterdam. http://imagebase.ubvu.vu.nl/getobj.php?ppn=330047426. Accessed 3 December 2018.

Voet 1643 = Voet, Gijsbert. 1643 (held on 23 and 24 December 1641). Appendix ad Corollaria theologico philosophica nuperae disputationi de iubilaeo romano, de rerum naturis et formis substantialibus. In Voet, Gijsbert et al. 1643. *Testimonium academiae Ultraiectinae, et Narratio historica qua defensae, qua exterminatae novae philosophiae*, 36–51. Utrecht: Ex typographia Wilhelmi Strickii.

Voet 1661 = Voet, Daniel. 1661. *Physiologia, adiectis aliquot eiusdem argumenti disputationibus* [...] *Opusculum posthumum, ab eius fratre* [Paul Voet] *editum*. Amsterdam: Ex officina Joh. a Waesberge.

Voet, Daniel. 1661. *Meletemata philosophica, in quibus antiquae et receptae philosophiae fulcra magis stabiliuntur, novatorum philosophandi methodus evertitur. Adiectis selectis quibusdam disputationibus ante hac publice ab eodem dum viveret, habitis*. Utrecht: Ex officina Henrici Versteegh.

Voet and De Vries 1678 = Voet, Daniel, and Vries, Gerard de. 1678. *Physiologia. Iterata hac editione recognita, ac notis illustrata per Gerardum de Vries* [...]. *Subiecta est huius dissertatio gemina brevis, altera de lumine, altera de lunicolis.* Utrecht: Apud Guilielmum Clerck.
Vogelsang, Reinier. 1669. *Contra libellum, cui titulus, Philosophia S. Scripturae interpres, exercitatio paradoxa, indignatio iusta, continens varias digressiones: de dubitatione Cartesiana, de ethica Cartesii* [...]. *Accessit Specimen conflictus eundem inter auctorem et* [...] *Johannem de Bruin, super dubitatione Cartesiana. Nec non diatribe de idea Dei secundum Cartesium. Item Corollaria irenica de Sabbatho.* Utrecht: Ex officina Henrici Versteegh.
Von Guericke 1672 = Guericke, Otto von. 1672. *Experimenta Nova (ut vocantur) Magdeburgica De Vacuo Spatio.* Amsterdam: Apud Joannem Janssonium a Waesberge.
Von Guericke 1993 = Guericke, Otto von. 1993. *The New (So-Called) Magdeburg Experiments of Otto von Guericke*, trans. Margaret Glover Foley Ames. Dordrecht-Boston: Kluwer Academic.
Voogt, Nicolaus Jansz. 1680. *Astrolabium catholicum, ofte een grondige onderwijsing, aangaande d'oorspronk, 't maakzel en 't veelvuldigh gebruyk des algemeeen starre-thoneels.* Amsterdam: by Jacobus Robyn.
Vossius 1662 = Vossius, Isaac. 1662. *De lucis natura et proprietate.* Amsterdam: Apud Ludovicum et Danielem Elzevirios.
De Vries 1679 = Vries, Gerard de. 1679. *Narrator confutatus, sive Animadversiones in narrationem de controversiis nuperius in Academia Ultraiectina motis.* Utrecht: Ex officina G. Clerck.
Van der Waeyen 1698 = Waeyen, Johannes van der. 1698. Dissertatio de Λόγῳ adversus Johannem Clericum. In Rittangel, Johann Stephan, and Waeyen, Johannes vander. 1698. *Libra veritatis et de Paschate tractatus, praemissa est Johannis vander Waeyen Dissertatio de Λόγῳ adversus Johannem Clericum.* Franeker: Typis et impensis Jacobi Horrei.
Wallis 1665 = Wallis, John. 1665. A relation concerning the late earthquake neer Oxford; together with some observations of the sealed weatherglass, and the barometer both upon that phaenomenon, and in general. *Philosophical Transactions* 1(10): 166–171.
Wallis 1668 = Wallis, John. 1668. A summary account given by Dr. John Wallis, of the general laws of motion, by way of letter written by him to the Publisher, and communicated to the R. Society, Novemb. 26. 1668. *Philosophical Transactions* 3(43): 864–866.
Wallis 1670–1671 = Wallis, John. 1670–1671. *Mechanica, sive De motu, tractatus geometricus.* London: Typis Gulielmi Godbid.
Weismann 1722 = Weismann, Christian Eberhard. 1722. *Commentatio theologica de praeiudicio.* Tübingen: Apud Jo. Georgium Cottam.
Whiston 1710 = Whiston, William. 1710. *Praelectiones physico-mathematicae Cantabrigiae in scholis publicis habitae: quibus philosophia illustrissimi Newtoni mathematica explicatius traditur, et facilius demonstratur: cometographia etiam Halleiana commentariolo illustratur.* Cambridge: Typis Academicis.
Wit, Frederick de. Circa 1667. *Nieuw Kaertboeck van de XVII Nederlandse Provinciën.* Amsterdam: Bij Frederick de Wit.
De Wit 1688 = Wit, Frederick de. Circa 1688. *Daniae, Frisiae, Groningae et Orientalis Frisiae Littora.* Amsterdam: Apud F. de Wit. Madrid, Museo Naval, signature: A-10117_9. http://bvpb.mcu.es/es/consulta/registro.cmd?id=477833. Accessed 3 December 2018.
Witsen 1671 = Witsen, Nicolaas. 1671. *Aeloude en hedendaegsche scheeps-bouw en bestier: waer in wijtloopigh wert verhandelt, de wijze van scheeps-timmeren, by Grieken en Romeynen: scheeps-oeffeningen, strijden, tucht, straffe, wetten en gewoonten* [...] *doorgaens verciert met vele kopere platen.* Amsterdam: By Casparus Commelijn; Broer en Jan Appelaer.
Witsen, Nicolaas. 1687. *Nieuwe lantkaarte van het noorder en ooster deel van Asia en Europe strekkende van Nova Zemla tot China.* Amsterdam: s.n.
Witsen, Nicolaas, and Wit, Frederick de. Circa 1690. *Imperii Russici sive Moscoviae, status generalis, in eius regna, ducatus, provincias, populosq subiacentes divisus, et ex tabula spectat. mi domini N. Witsen* [...] *pro maiori parte excerptus, per F. de Witt.* Amsterdam: Ex Officina P. Mortier; London: By Christopher Browne.
Witsen, Nicolaas. 1692. *Noord en oost Tartarye.* Amsterdam: By François Halma.

Witte van Schooten, Johannes Franciscus de. 1712. *Dissertatio philosophica inauguralis de solido, eiusque partium, nec non hemisphaeriorum concavorum, et cylindrorum solidorum cohaerentia.* Leiden: Apud viduam Corn. Boutesteyn.

Wittich, Christoph. 1653. *Dissertationes duae, quarum Prior de S. Scripturae in rebus philosophicis abusus examinat* [...]; *altera dispositionem et ordinem totius universi* [...] *tradit.* Amsterdam: Apud Ludovicum Elzevirium.

Wittich, Christoph. 1659. *Consensus veritatis in Scriptura divina et infallibili revelatae cum veritate philosophica a Renato des Cartes detecta.* Nijmegen: Apud Adrianum Wyngaerden.

Wittich 1688 = Wittich, Christoph. 1688. *Annotationes ad Renati des Cartes Meditationes.* Dordrecht: Ex officina V. Casperi et T. Goris.

Wittich 1690 = Wittich, Christoph. 1690. *Anti-Spinoza, sive Examen Ethices Benedicti de Spinoza, et Commentarius de Deo et eius attributis.* Amsterdam: Apud Joannem Wolters.

Wittich, Jacob (*praeses*), and Brouwer, Gerhard Nicolaus (*respondens*). 1711. *Disputatio philosophica de natura Dei.* Duisburg: typis J. Sas.

Wittich 1718a = Wittich, Jacob. 1718. De natura Dei Disputatio. In Honert, Taco Hajo van den. 1718. *Repraesentatio cur ad criminationes quibus Antonius Driessen Jacob. Wittichum ob habitam ante aliquot annos de natura Dei Disputationem a cathedra Groningana arcere conatus est nihil responderit; addita est Wittichii de natura Dei Disputatio.* Leiden: Apud Samuelem Luchtmans.

Wittich, Jacob. 1718. *Abstersio calumniarum, quibus eius Disputatio de natura Dei ante aliquot iam annos habita, non ita pridem inquinata fuit a Cl. Antonio Driessen, S.S. Th. Professore Groningano.* Leiden: Apud Samuelem Luchtmans.

Wittich, Jacob. 1718. *Wysgerige twistredenering over de natuur van God, welke onder des Heeren bystand onder de bescherming van Jacobus Wittichius. Hier is tot een Toegift bygevoegt het Oordell van den Heere Ruardus Andala over het Dispuit van den Heere Jak. Wittichius, alles uit Latyn vertaalt.* S.l.: s.n.

Wittich 1718b = Wittich, Jacob. 1718. *Oratio inauguralis de evidentia et certitudine, in quibus philosophiae partibus illa reperiatur.* Leiden: Apud Samuelem Luchtmans.

Wittich, Jacob. 1719. *Wijsgerige verhandeling van de nature Gods sijnde deselve nu door den aucteur in het Nederduyts vertaalt, met aanmerkingen, tot oplossing van des Heeren Driessens beschuldigingen.* Leiden: By Samuel Luchtmans.

Wittich and Van den Honert 1718 = Wittich, Jacob (*praeses*), and Honert, Johann van den (*defensurus auctor*). 1718. *Disputatio philosophica de essentia et existentia Dei.* Leiden: Apud Petrum vander Aa.

Wittich, Jacob. 1720. *Disputatio philosophica de natura Dei. Accessit J. Fr. Buddei epistola ad theologos quosdam Roterodamenses et iudicium theologicae et philosophicae facultatis Jenensis de isthac Cl. Wittichii disputatione.* The Hague: s.n.

Wittich 1723 = Wittich, Jacob. 1723. *Zeedig antwoord op het lasterschrift tegen hem gemaekt door Johannes Alexander Roëll.* Leiden: By Andries Dyckhhuysen.

Wolff 1713 = Wolff, Christian von. 1713. *Elementa matheseos universae.* Halle: Prostat in Officina Libraria Rengeriana.

Wolff 1733 = Wolff, Christian von. 1733. *Elementa matheseos universae.* Halle: Prostat in Officina Libraria Rengeriana.

Wolff 1741–1756 = Wolff, Christian von. 1741–1756. *Elementa matheseos universae.* Halle: Prostat in Officina Libraria Rengeriana.

Wren 1668 = Wren, Christopher. 1668. Dr Christopher Wrens theory concerning the same subject; imparted to the Royal Society Decemb. 17 last, though entertain'd by the author divers years ago, and verified by many experiments, made by himself and that other excellent mathematician M. Rook before the said Society. *Philosophical Transactions* 3(43): 867–868.

Zabarella 1590 = Zabarella, Jacopo. 1590. *De rebus naturalibus libri XXX.* Venice: Apud Paulum Meietum.

Zabarella 1617 = Zabarella, Jacopo. 1617. *De rebus naturalibus libri XXX. Quibus quaestiones, quae ab Aristotelis interpretibus hodie tractari solent, accurate discutiuntur.* Frankfurt: Sumptibus Lazari Zetzneri Bibliop.

Zabarella 2016 = Zabarella, Jacopo. 2016. *De rebus naturalibus libri XXX*, ed. José M. García Valverde. Leiden-Boston: Brill.

Zoutmann, Johannes. 1702. *Disputatio medica inauguralis physico-anatomica de spirituum animalium vera existentia et operatione*. Leiden: Apud Abrahamum Elzevier.

Zucchi 1649 = Zucchi, Niccolò. 1649. *Nova de machinis philosophia, in qua, paralogismis antiquae detectis, explicantur machinarum vires unico principio, singulis immediato* [...] *Accessit exclusio vacui contra nova experimenta, contra vires machinarum. Promotio philosophiae magneticae ex ea novum argumentum contra systema Pythagoricum. Initium operis proposita distinctius exhibet, errata, suis locis asterisco notata, corrigit*. Rome: Typis Haeredum Manelphii.

Other Handwritten and Archival Sources

AC = Archieven Universiteit Leiden: Curatoren, 1574–1815, Leiden University Library.

ASF = Archieven Universiteit Leiden: Senaat en Faculteiten, 1575–1877, Leiden University Library.

Académie Royale des Sciences. 1669. *Procès-verbaux T6 (3 avril–18 décembre 1669, Registre de physique)*. Bibliothèque Nationale de France. Microfilm. Reference number: FRBNF37572027. ark:/12148/bpt6k556821. Accessed 18 February 2019.

Amsterdam, Stadsarchief, Begraafregistrer, 15-10-1681, Wester Kerk, DTB 1102.

Amsterdam, Stadsarchief, Lidmaten Doopsgezinden, 18-02-1680, Archief 1120, inv. nr. 213.

Anonymous. 1691. *Journal of my Travels (24 June – 20 August 1691)*. British Library, Ms. Add. 20705, ff. 1–46.

Craanen, Theodoor (?) (author), and Morley, Christopher Love (copist/writer). Undated. *Dictata, ut videtur, in Theodori Craanen tractatum physico-medicum de homine*. British Library, mss. Sloane 1274–1275.

Craanen, Theodoor (author), and Morley, Christopher Love (copist/writer). Undated. *Praxis medica reformata, sive Annotationes in Praxin Henrici Regii*. British Library, ms. Sloane 1276.

Gregory, David. 1693. *Memorandum*. London, Royal Society, Ms. 247, f. 81.

Hudde, Johannes et al. 1682–1697. *Stukken betreffende de uitvinding van het kompas en zeeklok door Ch. Huygensen soort gelijke uitvindingen door prof. Anthony de Sanguilon en St. Julien Pottier, met adviesen van prof. De Volder, over de jaren 1682–1697*. The Hague, Algemeen Rijksarchief, Inventaris van het archief van J. Hudde, signature: 1.10.48 no. 44.

Huet, Pierre-Daniel. Undated. *Papiers divers de Huet*. Bibliothèque Nationale de France, Ms. F. Lat. 11451.

Huygens, Christiaan. 1656–. *De motu corporum ex percussione et alia*. Leiden University Library, ms. HUG 26 A.

Huygens, Christiaan. 1680–1688. *Liber "F": varia de rebus mathematicis, astronomicis, geometricis*. Leiden University Library, ms. HUG 1.

Leiden, Inventaris van het stadsarchief: 1313–1343, Registers van te Leiden begravenen, register 1327.

Leiden, Inventaris van het stadsarchief: 1313–1343, Registers van te Leiden begravenen, register 1329.

Leiden, Inventaris van het stadsarchief: Memoriaalboek, inventarisnummer 210+6.

Morley, Christopher Love. 1682. A *letter from Dr. Buena Alaison (?) to Christopher Love Morley concerning the publishing of his work intitled Collectanea chymica Leydensia, and respecting a friend of his own, who was about with his family to emigrate to Barbadoes. Dat. Amsterdam, 24 Jul. 1682*. British Library, ms. Sloane 1270, f. 218.

Morley, Christopher Love (copist/writer). Undated. Ms. Sloane 1292, Chartaceus, in 8vo. ff. 141. sec. XVII. British Library.

Morley, Christopher Love. Undated. *Commonplace-book*. British Library, ms. Sloane 1281.

Musschenbroek, Jan van. 1744–1746. *Brie[ven]van Jan van Musschenbroek (1687–1748) aan Arnout Vosmaer (1720–1799)* [4 letters]. Leiden University Library, ms. BPL 246.

Musschenbroek, Jan Willem van. Circa 1760. *Geslachtlyst van Van Musschenbroek.* The Hague, Van Musschenbroek Foundation.

Musschenbroek, Johannes Joosten van et al. 1693–1763. *Briefe von Musschenbrock.* Marburg, Hessisches Staatsarchiv, file UniA 305a Nr 5947.

Musschenbroek, Pieter van et al. Circa 1700–1800. Van Musschenbroek family documents. Leiden University Library, mss. BOERH g a 138 d and BOERH a 251.

Musschenbroek, Pieter van. Circa 1700–1800. *Petri van Musschenbroek Collectanea.* Leiden University Library, ms. BPL 240.

Raey, Johannes de. Undated. *Dictatorum ad Scunerti epitomen libri quinque.* British Library, ms. Sloane 1733, ff. 225–306.

Senguerd, Wolferd. 1690. *Dictata Clarissimi Doctissimi Viri D. D. Wolferdii Senguerdii Philosophiae Professoris in Renati Des-Cartes Principia Anno 1690.* The Hague, Koninklijke Bibliotheek, ms. 72 A 8.

Senguerd, Wolferd. Undated. *Nobillissimi, Doctissimi Wolferdi Senguerdi notae ad Cartesii principia Philosophiae ad dua priores partes seu libros.* The Hague, Koninklijke Bibliotheek, ms. 133 M 71.

Stolle, Gottlieb, and Hallmann, Georg Theophil. 1703–1704. *Reisejournal.* Warsaw University Library, ms. R 766 and Cod. IV oct. 49. Saint Petersburg, National Library of Russia, ms. Nem. IV F 158.

Secondary Sources

Aalderink 2009 = Aalderink, Mark. 2009. *Philosophy, Scientific Knowledge, and Concept Formation in Geulincx and Descartes.* Utrecht: Utrecht University, Publications of the Department of Philosophy.

Abou-Nemeh 2012 = Abou-Nemeh, Catherine. 2012. *Nicolas Hartsoeker's* Système *of Nature: Physics by Conjecture and Optics by Design in Early Modern Europe.* Princeton: Princeton University. Doctoral dissertation.

Adams 1994 = Adams, Robert Merrihew. 1994. *Leibniz: Determinist, Theist, Idealist.* New York-Oxford: Oxford University Press.

Adomaitis 2019 = Adomaitis, Laurynas. 2019. Cause and Effect in Leibniz's *Brevis demonstratio. HOPOS: The Journal of the International Society for the History of Philosophy of Science.* DOI: doi:10.1086/701044. Accessed 15 February 2019.

Ahnert 2002 = Ahnert, Thomas. 2002. *The Culture of Experimentalism in the Holy Roman Empire: Johann Christoph Sturm (1635–1703) and the* Collegium Experimentale. http://sammelpunkt.philo.at:8080/308/. Accessed 28 January 2019.

Ait-Touati 2011 = Ait-Touati, Frédérique. 2011. *Fictions of the Cosmos: Science and Literature in the Seventeenth Century.* Chicago: University of Chicago Press.

Aiton 1955a = Aiton, Eric J. 1955. Descartes's Theory of the Tides. *Annals of Science* 11(4): 337–348.

Aiton 1955b = Aiton, Eric J. 1955. The Contributions of Newton, Bernoulli and Euler to the Theory of the Tides. *Annals of Science* 11(3): 206–223.

Aiton 1957 = Aiton, Eric J. 1957. The Vortex Theory of the Planetary Motions—I. *Annals of Science* 13(4): 249–264.

Aiton 1958a = Aiton, Eric J. 1958. The Vortex Theory of the Planetary Motions—II. *Annals of Science* 14(2): 132–147.

Aiton 1958b = Aiton, Eric J. 1958. The Vortex Theory of the Planetary Motions—III. *Annals of Science* 14(3): 157–172.

Aiton 1959 = Aiton, Eric J. 1959. The Cartesian Theory of Gravity. *Annals of Science* 15(1): 27–49.
Aiton 1972 = Aiton, Eric J. 1972. *The Vortex Theory of Planetary Motions*. London: MacDonald.
Aiton 1989 = Aiton, Eric J. 1989. The Contributions of Isaac Newton, Johann Bernoulli and Jakob Hermann to the Inverse Problem of Central Forces. In *Der Ausbau des Calculus durch Leibniz und die Brüder Bernoulli* (*Studia Leibnitiana, Sonderhefte*), ed. Heinz-Jürgen Hess and Fritz Nagel, 48–58. Stuttgart: Steiner-Verlag.
Alanen 2003 = Alanen, Lilli. 2003. *Descartes's Concept of Mind*. Cambridge-London: Harvard University Press.
Albrecht 1999 = Albrecht, Michael. 1999. Einengung und Befreiung als Wirkungen des Cartesianismus am Beispiel Lodewijk Meyers. In *Johannes Clauberg (1622–1665) and Cartesian Philosophy in the Seventeenth Century*, ed. Theo Verbeek, 161–180. Dordrecht-Boston-London: Kluwer Academic Publishers.
Album promotorum qui inde ab anno MDCXXXVI usque ad annum MDCCCXV in Academia Rheno-Trajectina gradum doctoratus adepti sunt = Ketner, Frans. 1936. *Album promotorum qui inde ab anno MDCXXXVI usque ad annum MDCCCXV in Academia Rheno-Trajectina gradum doctoratus adepti sunt*. Utrecht: Verlag Broekhoff.
Album studiosorum Academiae Lugduno-Batavae MDLXXV–MDCCCLXXV = Du Rieu, Willem Nikolaas. 1875. *Album studiosorum Academiae Lugduno-Batavae MDLXXV–MDCCCLXXV. Accedunt nomina Curatorum et Professorum per eadem secula*. The Hague: Apud Martinum Nijhoff.
Album studiosorum Academiae Rheno-Traiectinae MDCXXXVI–MDCCCLXXXVI = Anonymous. 1886. *Album studiosorum Academiae Rheno-Traiectinae MDCXXXVI–MDCCCLXXXVI. Accedunt nomina Curatorum et Professorum per eadem secula*. Utrecht: Apud J. L. Beijers et J. van Boekhoven.
Allaby 2014 = Allaby, Michael. 2014. *Dangerous Weather: Blizzards. Revised Edition*. New York: Facts On File.
Almeida 2014 = Almeida, Joseph. 2014. Simplicius on *Categories* 1a16–17 and 1b25–27: An Examination of the Interests of Ancient and Modern Commentary on the *Categories* 4. *Quaestiones Disputatae* 4: 73–99.
Althaus 1914 = Althaus, Paul. 1914. *Die Prinzipien der deutschen reformierten Dogmatik im Zeitalter der aristotelischen Scholastik*. Leipzig: Deichert.
Andrault and Laerke 2018 = Andrault, Raphaele, and Laerke, Mogens (Eds.). 2018. *Steno and the Philosophers*. Leiden-Boston: Brill.
Andriesse 2005 = Andriesse, Cornelis Dirk. 2005. *Huygens: The Man Behind the Principle*. Cambridge: Cambridge University Press.
Angyal 1941 = Angyal, Andreas. 1941. Kleiner Beitrag zur Literaturgeschichte des deutschen Cartesianismus. *Germanisch-romanische Monatsschrift* 29: 69–73.
Anonymous. Undated. *Luchtpompen* [*sic*] *van Burchardus de Volder*. Museum Boerhaave, Leiden, Nummer archiefinventaris 2.24.10.02. http://proxy.handle.net/10648/af5524ce-d0b4-102d-bcf8-003048976d84. Accessed 18 February 2019.
Anstey 2000 = Anstey, Peter. 2000. Descartes' Cardiology and Its Reception in English Physiology. In *Descartes' Natural Philosophy*, ed. Stephen Gaukroger, John Schuster, and John Sutton, 420–444. London-New York: Routledge.
Anstey 2005 = Anstey, Peter. 2005. Experimental versus Speculative Natural Philosophy. In *The Science of Nature in the Seventeenth Century: Patterns of Change in Early Modern Natural Philosophy*, ed. Peter Anstey and John Schuster, 215–242. Dordrecht: Kluwer-Springer.
Anstey 2011 = Anstey, Peter. 2011. *John Locke and Natural Philosophy*. Oxford: Oxford University Press.
Anstey 2018 = Anstey, Peter. 2018. Locke and Cartesian Cosmology. In *Locke and Cartesian Philosophy*, ed. Philippe Hamou and Martine Pécharman, 33–48. Oxford: Oxford University Press.
Anstey and Jalobeanu 2011 = Anstey, Peter, and Jalobeanu, Dana (Eds.). 2011. *Vanishing Matter and the Laws of Motion: Descartes and Beyond*. London: Routledge.

Anstey and Vanzo 2016 = Anstey, Peter, and Vanzo, Alberto. 2016. Early Modern Experimental Philosophy. In *A Companion to Experimental Philosophy*, ed. Justin Sytsma and Wesley Buckwalter, 87–102. Oxford: Wiley Blackwell.

Antognazza 2008 = Antognazza, Maria Rosa. 2008. *Leibniz: An Intellectual Biography*. Cambridge: Cambridge University Press.

Antoine-Mahut and Gaukroger 2016 = Antoine-Mahut, Delphine, and Gaukroger, Stephen (Eds.). 2016. *Descartes' Treatise on Man and Its Reception*. Dordrecht: Springer.

Antoine-Mahut and Roux 2018 = Antoine-Mahut, Delphine, and Roux, Sophie (Eds.). 2018. *Physics and Metaphysics in Descartes and in His Reception*. New York: Routledge.

Ariew 1999 = Ariew, Roger. 1999. *Descartes and the Last Scholastics*. Ithaca: Cornell University Press.

Ariew 2011a = Ariew, Roger. 2011. *Descartes Among the Scholastics*. Leiden-Boston: Brill.

Ariew 2011b = Ariew, Roger. 2011. The New Matter Theory and Its Epistemology: Descartes (and Late Scholastics) on Hypotheses and Moral Certainty. In *Vanishing Matter and the Laws of Motion: Descartes and Beyond*, ed. Peter Anstey and Dana Jalobeanu, 31–46. London: Routledge.

Ariew 2012 = Ariew, Roger. 2012. Descartes and His Critics on Matter and Form: Atomism and Individuation. In *Matter and Form in Early Modern Science and Philosophy*, ed. Gideon Manning, 33–75. Leiden: Brill.

Ariew 2013 = Ariew, Roger. 2013. Censorship, Condemnations, and the Spread of Cartesianism. In *Cartesian Empiricisms*, ed. Mihnea Dobre and Tammy Nyden, 25–46. Dordrecht-Heidelberg-New York-London: Springer.

Ariew 2014 = Ariew, Roger. 2014. *Descartes and the First Cartesians*. Oxford: Oxford University Press.

Armitage 1950 = Armitage, Angus. 1950. René Descartes (1596–1650) and the Early Royal Society. *Notes and Records of the Royal Society of London* 8: 1–19.

Armogathe 2005 = Armogathe, Jean-Robert. 2005. Cartesian Physics and the Eucharist in the Documents of the Holy Office and the Roman Index (1671–6). In *Receptions of Descartes. Cartesianism and Anti-Cartesianism in Early Modern Europe*, ed. Tad Schmaltz, 157–182. London-New York: Routledge.

Arthur 1994 = Arthur, Richard. 1994. Space and Relativity in Newton and Leibniz. *British Journal for the Philosophy of Science* 45(1): 219–240.

Arthur 2007 = Arthur, Richard. 2007. Beeckman, Descartes and the Force of Motion. *Journal of the History of Philosophy* 45(1): 1–28.

Arthur 2013 = Arthur, Richard. 2013. Leibniz's Theory of Space. *Foundations of Science* 18(3): 499–528.

Ast 1807 = Ast, Friedrich. 1807. *Grundriß einer Geschichte der Philosophie*. Landshut: J. Thomann.

Atherton 2005 = Atherton, Margaret. 2005. Descartes Among the British: The Case of the Theory of Vision. In *Receptions of Descartes. Cartesianism and Anti-Cartesianism in Early Modern Europe*, ed. Tad Schmaltz, 218–234. London-New York: Routledge.

Balázs et al. 2016 = Balázs, Mihály, Font, Zsuzsa, and Kovács, András. 2016. *Köleséri Sámuel és az európai korai felvilágosodás*. Kolozsvár: Erdélyi Múzeum-Egyesület.

Baldini 1976 = Baldini, Ugo. 1976. La struttura della materia nel pensiero di Galileo. *De Homine* 77: 91–164.

Baneke 2010 = Baneke, David. 2010. *Leiden Observatory Archives, E. F. van de Sande Bakhuyzen Archive (EB)*. https://www.strw.leidenuniv.nl/outreach/strwarchief/EB_inventory.pdf. Accessed 28 January 2019.

Barbour 1989 = Barbour, Julian. 1989. *Absolute or Relative Motion? A Study from Machian Point of View of the Discovery and the Structure of Dynamical Theories*. Cambridge: Cambridge University Press.

Bardout 2002 = Bardout, Jean-Christophe. 2002. Occasionalism: La Forge, Cordemoy, Geulincx. In *A Companion to Early Modern Philosophy*, ed. Steven Nadler, 140–151. Boston: Blackwell.

Baumann 1949 = Baumann, Evert D. 1949. *Francois de le Boe Sylvius*. Leiden: Brill.

Bazerman 1993 = Bazerman, Charles. 1993. Forums of Validation and Forms of Knowledge: The Magical Rhetoric of Otto von Guericke's Sulfer Globe. *Configurations* 1: 201–228.
Beck 1972–1991 = Beck, Hans-Ulrich. 1972–1991. *Jan van Goyen, 1596–1656*. Amsterdam: Davaco.
Beck 2001 = Beck, Andreas J. 2001. Gisbertus Voetius (1589–1676. Basic Features of His Doctrine of God. In *Reformation and Scholasticism: An Ecumenical Enterprise*, ed. Willem J. van Asselt and Eef Dekker, 205–226. Grand Rapids: Baker Academic.
Beck 2007 = Beck, Andreas J. 2007. *Gisbertus Voetius (1589–1676). Sein Theologieverständnis und seine Gotteslehre*. Göttingen: Vandenhoeck, and Ruprecht.
Belgioioso 1985 = Belgioioso, Giulia (Ed.). 1985. *Paolo Mattia Doria fra rinnovamento e tradizione*. Galatina: Congredo.
Belgioioso 1992 = Belgioioso, Giulia. 1992. *Cultura a Napoli e cartesianismo. Scritti su G. Gimma, P. M. Doria, C. Cominale*. Galatina: Congredo.
Belgioioso 1999 = Belgioioso, Giulia. 1999. *La variata immagine di Descartes. Percorsi della metafisica da Parigi a Napoli*. Lecce: Milella.
Belgioioso 2005 = Belgioioso, Giulia. 2005. Images of Descartes in Italy. In *Receptions of Descartes. Cartesianism and Anti-Cartesianism in Early Modern Europe*, ed. Tad Schmaltz, 171–196. London-New York: Routledge.
Bell 1986 = Bell, Eric Temple. 1986. *Men of Mathematics: The Lives and Achievements of the Great Mathematicians from Zeno to Poincaré*. New York: Simon and Schuster.
Bellis 2013 = Bellis, Delphine. 2013. Empiricism without Metaphysics: Regius' Cartesian Natural Philosophy. In *Cartesian Empiricisms*, ed. Mihnea Dobre and Tammy Nyden, 169–172. Dordrecht-Heidelberg-New York-London: Springer.
Benigni 2017 = Benigni, Fiormichele. 2017. Socinianismo e crisi del cartesianismo in Aubert de Versé. *Philosophica* 3: 135–151.
Beningi 2018 = Benigni, Fiormichele. 2018. *Itinerari dell'antispinozismo. Spinoza e le metafisiche cartesiane in Francia (1684–1718)*. Florence: Le Lettere.
Beretta et al. 2009 = Beretta, Marco, Clericuzio, Antonio, and Principe, Lawrence (Eds.). 2009. *The Accademia del Cimento and its European Context*. Sagamore Beach: Watson History Publishing.
Bergh 2002 = Bergh, Govaert Carolus Joannes Joseph. 2002. *Die holländische Schule. Ein Beitrag zur Geschichte von Humanismus und Rechtswissenschaft in den Niederlanden 1500–1800*. Frankfurt: Vittorio Klostermann.
Bernstein 1981 = Bernstein, Howard R. 1981. Passivity and Inertia in Leibniz's 'Dynamics'. *Studia Leibnitiana* 13: 97–113.
Bertoloni Meli 1990 = Bertoloni Meli, Domenico. 1990. The Relativization of Centrifugal Force. *Isis: A Journal of the History of Science* 81(1): 23–43.
Bertoloni Meli 1993 = Bertoloni Meli, Domenico. 1993. *Equivalence and Priority: Newton versus Leibniz*. Oxford: Oxford University Press.
Bertoloni Meli 2006 = Bertoloni Meli, Domenico. 2006. *Thinking with Objects: The Transformation of Mechanics in the Seventeenth Century*. Baltimore: Johns Hopkins University Press.
Bertoloni Meli 2017 = Bertoloni Meli, Domenico. 2017. Gerardus Blasius and the Illustrated Amsterdam *Observationes* from Nicolaas Tulp to Frederik Ruysch. In *Professors, Physicians and Practices in the History of Medicine: Essays in Honor of Nancy Siraisi*, ed. Gideon Manning and Cynthia Klestinec, 255–297. New York: Springer.
Bertrand 2015 = Bertrand, Ester. 2015. *Johannes Swartenhengst (1644–1711): A Dutch Cartesian in the Heat of Battle*. Edinburgh: University of Edinburgh; Brussels: Free University of Brussels. Doctoral dissertation.
Beukers 1980 = Beukers, Harm. 1980. Het Laboratorium van Sylvius. *Tijdschrift voor de geschiedenis der geneeskunde, natuurwetenschappen, wiskunde en techniek* 3: 28–36.
Beukers 1982 = Beukers, Harm. 1982. Mechanistische principes bij Franciscus dele Boë, Sylvius. *Tijdschrift voor de geschiedenis der geneeskunde, natuurwetenschappen, wiskunde en techniek* 5: 6–15.

Beukers 1999 = Beukers, Harm. 1999. Acid Spirits and Alkaline Salts. The Iatro-chemistry of Franciscus Dele Böe, Sylvius. *Sartoniana* 12: 39–58.

Bierens de Haan 1960 = Bierens de Haan, David. 1960. *Bibliographie néerlandaise historique-scientifique des ouvrages importants dont les auteurs sont nés aux 16e, 17e et 18e siècles, sur les sciences mathematiques et physiques*. Nieuwkoop: B. de Graaf.

Bitbol-Hespériès 1993 = Bitbol-Hespériès, Annie. 1993. Descartes et Regius: leur pensée médicale. In *Descartes et Regius. Autour de l'Explication de l'esprit humain*, ed. Theo Verbeek, 47–68. Amsterdam: Rodopi.

Blackwell 1966 = Blackwell, Richard J. 1966. Descartes' Laws of Motion. *Isis: A Journal of the History of Science* 57(2): 220–234.

Blackwell 1977 = Blackwell, Richard J. 1977. Christiaan Huygens' The Motion of Colliding Bodies. *Isis: A Journal of the History of Science* 68(4): 574–597.

Blair 2000 = Blair, Ann. 2000. *Mosaic Physics and the Search for a Pious Natural Philosophy in the Late Renaissance*. Chicago: University of Chicago Press.

Blum 1998 = Blum, Paul Richard. 1998. *Philosophenphilosophie und Schulphilosophie. Typen des Philosophierens in der Neuzeit*. Stuttgart: Franz Steiner.

Boantza 2011 = Boantza, Victor. 2011. From Cohesion to Pesanteur: The Origins of the 1669 Debate on the Causes of Gravity. In *Controversies Within the Scientific Revolution*, ed. Marcelo Dascal and Victor D. Boantza, 77–100. Amsterdam-Philadelphia: John Benjamins.

Boas 1949 = Boas, Marie. 1949. Hero's *Pneumatica*: A Study of Its Transmission and Influence. *Isis: A Journal of the History of Science* 40(1): 38–48.

Bodeüs 1991 = Bodeüs, Rirchard. 1991. Leibniz, Jean de Raey et la Physique Reformée. *Studia Leibnitiana* 33: 103–110.

Bohatec 1912 = Bohatec, Josef. 1912. *Die cartesianische Scholastik in der Philosophie und reformierten Dogmatik des 17. Jahrhunderts*. Leipzig: A. Deichert.

Booth 2006 = Booth, Emily. 2006. *A Subtle and Mysterious Machine: The Medical World of Walter Charleton (1619–1707)*. Dordrecht: Springer.

Bordoli 1997 = Bordoli, Roberto. 1997. *Ragione e Scrittura tra Descartes e Spinoza. Saggio sulla 'Philosophia S. Scripturae Interpres' di Lodewijk Meyer e sulla sua recezione*. Milan: Franco Angeli.

Bordoli 2009 = Bordoli, Roberto. 2009. *Dio ragione verità. Le polemiche su Descartes e su Spinoza presso l'Università di Franeker*. Macerata: Quodlibet.

Borghero 1998 = Borghero, Carlo. 1998. Discussioni sullo scetticismo di Descartes (1650–1712). *Giornale critico della filosofia italiana* 77(1): 1–25.

Borghero 2011 = Borghero, Carlo. 2011. *Les Cartésiens face à Newton. Philosophie, science et religion dans la première moitié du XVIIIe siècle*. Turnhout: Brepols.

Borghero and Del Prete 2011 = Borghero, Carlo, and Del Prete, Antonella (Eds.). 2011. *Immagini filosofiche e interpretazioni storiografiche del cartesianismo*. Florence: Le Lettere.

Bos 2002 = Bos, Erik-Jan. 2002. *The Correspondence between Descartes and Henricus Regius*. Utrecht: Utrecht University, Publications of the Department of Philosophy.

Bos 2013 = Bos, Erik-Jan. 2013. Henricus Regius et les limites de la philosophie cartésienne. In *Qu'est-ce qu'être cartésien?*, ed. Delphine Kolesnik-Antoine, 53–68. Paris: Editions de l'ENS de Lyon.

Bos and Krop 1993 = Bos, Egbert P., and Krop, Henri (Eds.). 1993. *Franco Burgersdijk (1590–1635): Neo-Aristotelianism in Leiden*. Amsterdam: Rodopi.

Boschiero 2007 = Boschiero, Luciano. 2007. *Experiment and Natural Philosophy in Seventeenth-Century Tuscany. The History of the Accademia del Cimento*. Dordrecht: Springer.

Bosl 1983 = Bosl, Erika. 1983. Sturm, Johann Christoph, Philosoph, Physiker und Mathematiker. In *Bosls bayerische Biographie*, ed. Karl Bosl, 766. Regensburg: Pustet.

Bouillier 1854 = Bouillier, Francisque. 1854. *Histoire de la Philosophie Cartésienne*. Paris: Durand, Lyon: Brun.

Boutroux 1914 = Boutroux, Pierre. 1914. La signification historique de la «géométrie» de Descartes. *Revue de Métaphysique et de Morale* 22(6): 814–827.

Briels 1997 = Briels, Jan. 1997. *Peintres flamands au berceau du Siècle d'or hollandais, 1585–1630: avec biographies en annexe*. Antwerp: Fonds Mercator.
Brigaglia 1995 = Brigaglia, Aldo. 1995. La riscoperta dell'analisi e i problemi apolloniani. In *Geometria, flussioni, differenziali: tradizione e innovazione nella matematica del Seicento*, ed. Marco Panza and Clara Silvia Roero, 221–269. Naples: La Città del Sole.
Brockliss 1981 = Brockliss, Laurence. 1981. Aristotle, Descartes and the New Science: Natural Philosophy at the University of Paris, 1600–1740. *Annals of Science* 38(1): 33–69.
Brockliss 1987 = Brockliss, Laurence. 1987. *French Higher Education in the Seventeenth and Eighteenth Centuries: A Cultural History*. Oxford: Clarendon Press.
Brockliss 1995 = Brockliss, Laurence. 1995. Descartes, Gassendi, and the Reception of the Mechanical Philosophy at the French Collèges de Plein Exercice, 1640–1730. *Perspectives on Science* 3(4): 450–479.
Brown 1934 = Brown, Harcourt. 1934. *Scientific Organizations in Seventeenth Century France (1620–1680)*. Baltimore: Williams and Wilkins.
Brulez 1926 = Brulez, Lucien. 1926. *Holländische Philosophie*. Breslau: Hirt.
Brunet 1926 = Brunet, Pierre. 1926. *Les physiciens hollandais et la méthode expérimentale en France au XVIIIe siècle*. Paris: Albert Blanchard.
Brykman 1987–1988 = Brykman, Geneviève. 1987–1988. Bayle's Case for Spinoza. *Proceedings of the Aristotelian Society* 88: 259–270.
Buning 2013 = Buning, Robin. 2013. *Henricus Reneri (1593–1639), Descartes' Quartermaster in Aristotelian Territory*. Utrecht: Utrecht University, Publications of the Department of Philosophy.
Buys 2011 = Buys, Ruben. 2011. Between Actor and Spectator: Arnout Geulincx and the Stoics. *British Journal for the History of Philosophy* 18(5): 741–761.
Calero 2008 = Calero, Julián Simón. 2008. *The Genesis of Fluid Mechanics 1640–1780*. Dordrecht: Springer.
Capadose 1828 = Capadose, Abraham. 1828. *Francisci Burmanni* [...] *itineris anglicani acta diurna*. Amsterdam: Apud J. H. den Ouden.
Capecchi 2012 = Capecchi, Danilo. 2012. *History of Virtual Work Laws: A History of Mechanics Prospective*. Milan: Springer.
Capecchi 2017 = Capecchi, Danilo. 2017. *The Path to Post-Galilean Epistemology: Reinterpreting the Birth of Modern Science*. Cham: Springer.
Caroti 2014 = Caroti, Stefano. 2014. Ruardus Andala e la nuova filosofia. In *Filosofia e religione. Studi in onore di Fabio Rossi*, ed. Stefano Caroti and Alberto Siclari, 236–257. Parma: E-Theca On Line Open Access Edizioni. http://www.didaschein.net/ojs/index.php/QuadernidiNoctua/article/viewFile/125/102. Accessed 15 February 2019.
Carpenter 2011 = Carpenter, Audrey T. 2011. *John Theophilus Desaguliers: A Natural Philosopher, Engineer and Freemason in Newtonian England*. New York: Continuum Publishing.
Carriero 2016 = Carriero, John. 2016. *Descartes and the Autonomy of the Human Understanding*. London: Routledge. First edition 1990.
Catalogue of the Special Loan Collection of Scientific Apparatus at the South Kensington Museum = Anonymous. 1876. *Catalogue of the Special Loan Collection of Scientific Apparatus at the South Kensington Museum: MDCCLXXVI*. London: George E. Eyre and William Spottiswoode.
Cerny 1987 = Cerny, Gerald (Ed.). 1987. *Theology, Politics and Letters at the Crossroads of European Civilization. Jacques Basnage and the Baylean Huguenot Refugees in the Dutch Republic*. Dordrecht: Kluwer.
Chabbert 1970 = Chabbert, Pierre. 1970. Jacques Borelly: Membre de l'Académie Royale des Sciences. *Revue d'histoire des sciences* 23: 203–227.
Chacón Rebollo and Lewandowski 2014 = Chacón Rebollo, Tomás, and Lewandowski, Roger (Eds.). 2014. *Mathematical and Numerical Foundations of Turbulence Models and Applications*. New York: Springer.
Chalmers 2009 = Chalmers, Alan. 2009. *The Scientist's Atom and the Philosopher's Stone – How Science Succeeded and Philosophy Failed to Gain Knowledge of Atoms*. Dordrecht: Springer.

Chalmers 2017 = Chalmers, Alan. 2017. *One Hundred Years of Pressure – Hydrostatics from Stevin to Newton*. Dordrecht: Springer.

Chareix 1996 = Chareix, Fabien. 1996. La pésanteur dans l'univers méchanique de Christiaan Huygens. *De Zeventiende Eeuw* 12(1): 244–252.

Chareix 2004 = Chareix, Fabien. 2004. Huygens and Mechanics. In *Proceedings of the International Conference TITAN, From Discovery to Encounter, 13–17 April 2004*, ed. Karen Fletcher, 55–65. Noordwijk: ESTEC.

Clair 1978 = Clair, Pierre. 1978. *Jacques Rohault (1618–1672), bio-biographie avec l'édition critique des* Entretiens sur la philosophie. Paris: CNRS.

Clarke 1982 = Clarke, Desmond. 1982. *Descartes' Philosophy of Science*. Manchester: Manchester University Press.

Clarke 1989 = Clarke, Desmond. 1989. *Occult Powers and Hypotheses. Cartesian Natural Philosophy under Louis XIV*. Oxford: Clarendon.

Clarke 2005 = Clarke, Desmond. 2005. *Descartes's Theory of Mind*. Oxford: Clarendon Press.

Clarke 2010 = Clarke, Desmond. 2010. The Physics and Metaphysics of the Mind: Descartes and Regius. In *Mind, Method, and Morality: Essays in Honour of Anthony Kenny*, ed. John Cottingham and Peter Hacker, 187–207. Oxford: Oxford University Press.

Clatterbaugh 1999 = Clatterbaugh, Kenneth. 1999. *The Causation Debate in Modern Philosophy, 1637–1739*. New York: Routledge.

Clericuzio 1998 = Clericuzio, Antonio. 1998. The Mechanical Philosophy and the Spring of Air. New Light on Robert Boyle and Robert Hooke. *Nuncius* 13: 69–75.

Clericuzio 2000 = Clericuzio, Antonio. 2000. *Elements, Principles and Corpuscles: A Study of Atomism and Chemistry in the Seventeenth Century*. Dordrecht: Kluwer.

Cobb 2000 = Cobb, Matthew. 2000. Reading and Writing The Book of Nature: Jan Swammerdam (1637–1680). *Endeavor* 24: 122–138.

Cobb 2006 = Cobb, Matthew. 2006. *Generation: The Seventeenth-Century Scientists Who Unraveled the Secrets of Sex, Life, and Growth*. New York: Bloomsbury.

Cohen 1997 = Cohen, I. Bernard. 1997. *Science and the Founding Fathers: Science in the Political Thought of Thomas Jefferson, Benjamin Franklin, John Adams, and James Madison*. New York: Norton. First edition 1995.

Colie 1957 = Colie, Rosalie L. 1957. *Light and Enlightenment: A Study of the Cambridge Platonists and the Dutch Arminians*. New York: Cambridge University Press.

Conlon 2011 = Conlon, Thomas E. 2011. *Thinking about Nothing: Otto Von Guericke and the Magdeburg Experiments on the Vacuum*. San Francisco: The Saint Austin Press.

Conti 1997 = Conti, Lino. 1997. Galileo and the Ancient Dispute about the Weight of Air. In*Die Schwere der Luft in der Diskussion des 17. Jahrhunderts*, ed. Wim Klever, 9–30. Wiesbaden: Harrassowitz.

Cook 1986 = Cook, Ian. 1986. Newton's 'Experimental' Law of Impacts. *The Mathematical Gazette* 70: 107–114.

Cook 1993 = Cook, Harold John. 1993. The Cutting Edge of a Revolution? Medicine and Natural History near the Shores of the North Sea. In *Renaissance and Revolution: Humanists, Scholars, Craftsmen and Natural Philosophers in Early Modern Europe*, ed. Judith Veronica Field and Frank A. J. L. James, 45–61. Cambridge-New York-Melbourne: Cambridge University Press.

Costabel 1973 = Costabel, Pierre. 1973. *Leibniz and Dynamics*. Ithaca: Cornell University Press.

Cottingham 1997 = Cottingham, John. 1997. Air, Gravity and Cartesian Physics. In *Die Schwere der Luft in der Diskussion des 17. Jahrhunderts*, ed. Wim Klever, 31–45. Wiesbaden: Harrassowitz.

Cousin 1866 = Cousin, Victor. 1866. *Fragments Philosophiques pour servir à l'histoire de la philosophie*. Paris: Didier.

Cover et al. 1999 = Cover, Jan A., and O'Leary-Hawthorne, John. 1999. *Substance and Individuation in Leibniz*. Cambridge: Cambridge University Press.

Crombie 1996 = Crombie, Alistair Cameron. 1996. *Science, Art and Nature in Medieval and Modern Thought*. London-Rio Grande: Bloomsbury Academic.

Crommelin 1929 = Crommelin, Claude August. 1929. *Het lenzen slijpen in de 17e eeuw*. Amsterdam: H. J. Paris.

Crowley 2008a = Crowley, Timothy J. 2008. Aristotle's "So-Called Elements". *Phronesis* 53: 223–42

Crowley 2008b = Crowley, Timothy J. 2009. *Aristotle on the Matter of the Elements*. Oxford: University of Oxford. Doctoral dissertation.

Cunning 2003 = Cunning, David. 2003. Descartes on the Immutability of the Divine Will. *Religious Studies* 39(1): 79–92.

Cunning 2014 = Cunning, David. 2014. The First Meditation: Divine Omnipotence, Necessary Truths and the Possibility of Radical Deception. In *The Cambridge Companion to Descartes' Meditations*, ed. David Cunning, 68–87. Cambridge: Cambridge University Press.

Cunningham 2010 = Cunningham, Andrew. 2010. *The Anatomist Anatomis'd: An Experimental Discipline in Enlightenment Europe*. Aldershot: Ashgate.

Curley 1977 = Curley, Edwin. 1977. Spinoza as an Expositor of Descartes. In *Speculum Spinozanum*, ed. Siegfried Hessing, 133–142. London: Routledge & Kegan Paul.

Curley 1993 = Curley, Edwin. 1993. Certainty: Psychological, Logical and Metaphysical. In *Essays on the Philosophy and Science of René Descartes*, ed. Stephen Voss, 11–30. New York: Oxford University Press.

Damiron 1846 = Damiron, Jean Philibert. 1846. *Essai sur l'histoire de la philosophie en France, au XVIIe siècle*. Paris: Hachette.

Daugirdas 2017 = Daugirdas, Kęstutis. 2017. The Biblical Hermeneutics of Philip van Limborch (1633–1712) and Its Intellectual Challenges. In *Scriptural Authority and Biblical Criticism in the Dutch Golden Age*, ed. Dirk van Miert, Henk Nellen, Piet Steenbakkers, and Jetze Touber, chapter 10. Oxford: Oxford University Press.

Daumas 1972 = Daumas, Maurice. 1972. *Scientific Instruments of the Seventeenth and Eighteenth Centuries and Their Makers*. Trans. by Mary Holbrook. London: B. T. Batsford. First edition 1953, *Les instruments scientifiques aux XVIIe et XVIIIe siècle*. Paris: Presses Universitaires de France.

Davidse, Ad. *Christiaan Huygens in het Nederlands*. https://adcs.home.xs4all.nl/Huygens/. Accessed 28 January 2019.

De Clercq 1988 = De Clercq, Peter. 1988. The 's Gravesande Collection in the Museum Boerhaave, Leiden. *Nuncius* 3(1): 127–137.

De Clercq 1989 = De Clercq, Peter. 1989. *The Leiden Cabinet of Physics*. Leiden: Museum Boerhaave.

De Clercq 1991 = De Clercq, Peter. 1991. Exporting Scientific Instruments around 1700: The Musschenbroek Documents in Marburg. *Tractrix* 3: 79–120.

De Clercq 1997a = De Clercq, Peter. 1997. *At the Sign of the Oriental Lamp. The Musschenbroek Workshop in Leiden, 1660–1750*. Rotterdam: Erasmus Publishing.

De Clercq 1997b = De Clercq, Peter. 1997. *The Leiden Cabinet of Physics. A Descriptive Catalogue*. Leiden: Museum Boerhaave.

De Clercq 2016 = De Clercq, Peter. 2016. The Travel Journals of Balthasar de Monconys (1608–1665) and Zacharias Conrad von Uffenbach (1683–1734). *Bulletin of the Scientific Instrument Society* 128: 2–14.

De Grijs 2015 = De Grijs, Richard. 2015. *The search for longitude: Preliminary insights from a 17th Century Dutch perspective*. ArXiv:1501.00627. Accessed 30 January 2019.

De Hoog 1974 = De Hoog, Adriaan Cornelis. 1974. *Some Currents of Thought in Dutch Natural Philosophy: 1675–1720*. Oxford: Oxford University. Doctoral dissertation.

De Jong and Zuidervaart 2018 = De Jong, Theo, and Zuidervaart, Huib J. 2018. 'Philosophus et Mathematicus Incomparabilis': Johannes Hudde (1628–1704), de Verlichte burgemeester van Amsterdam. *Studium* 11(1): 26–54.

De Jongste and Veenendaal 2002 = De Jongste, Jan A. F., and Veenendaal, Augustus J. Jr. (Eds.). 2002. *Anthonie Heinsius and the Dutch Republic, 1688–1720: Politics, War, and Finance*. The Hague: Institute of Netherlands History.

De Pater 1975 = De Pater, Cornelis. 1975. Experimental Physics. In *Leiden University in the Seventeenth Century: An Exchange of Learning*, ed. Theodoor Herman Lunsingh Scheurleer and Guillaume Henri Marie Posthumus Meyjes, 309–327. Leiden: Brill.

De Risi 2007 = De Risi, Vincenzo. 2007. *Geometry and Monadology: Leibniz's Analysis Situs and Philosophy of Space*. Basel: Birkhäuser.

De Sitter 1933 = De Sitter, Willem. 1933. *Short History of the Observatory of the University at Leiden 1633–1933*. Haarlem: Joh. Enschede en Zonen.

De Waard 1911 = De Waard, Cornélis. 1911. Hudde, Johannes. In *Nieuw Nederlandsch biografisch woordenboek*, ed. Philip Christiaan Molhuysen, Petrus Johannes Blok, and Laurentius Knappert, volume 1, columns 1171–1176. Leiden: A. W. Sijthoff's uitgevers- maatschappij.

Débarbat 1986 = Débarbat, Suzanne. 1986. Newton, Halley et l'Observatoire de Paris. *Revue d'histoire des sciences et de leurs applications* 39: 127–154.

Debus 2001 = Debus, Allen G. 2001. *Chemistry and Medical Debate. Van Helmont to Boerhaave*. Canton: Science History Publications.

Debus 2002 = Debus, Allen G. 2002. *The French Paracelsians: The Chemical Challenge to Medical and Scientific Tradition in Early Modern France*. Cambridge: Cambridge University Press. First edition 1991.

Del Prete 2018 = Del Prete, Antonella. 2018. *Duplex Intellectus Et Sermo Duplex*: Method and the Separation of Disciplines in Johannes De Raey. In *Physics and Metaphysics in Descartes and in his Reception*, ed. Delphine Antoine-Mahut and Sophie Roux, 161–174. New York: Routledge.

Della Rocca 1996 = Della Rocca, Michael. 1996. *Representation and the Mind-Body Problem in Spinoza*. New York: Oxford University Press.

Della Rocca 2005 = Della Rocca, Michael. 2005. Descartes, the Cartesian Circle, and Epistemology without God. *Philosophy and Phenomenological Research* 70(1): 1–33.

Della Rocca 2008 = Della Rocca, Michael. 2008. *Spinoza*. New York: Routledge.

Den Boer 2008 = Den Boer, Harem. 2008. Amsterdam as "Locus" of Iberian Printing in the Seventeenth and Eighteenth Centuries. In *The Dutch Intersection: The Jews and the Netherlands in Modern History*, ed. Yosef Kaplan, 87–110. Leiden-Boston: Brill.

Des Amorie van der Hoeven 1843 = Des Amorie van der Hoeven, Abraham. 1843. *De Joanne Clerico et Philippo a Limborch Dissertationes duae*. Amsterdam: Apud Fredericum Muller.

Des Chene 1996 = Des Chene, Dennis. 1996. *Physiologia: Natural Philosophy in Late Aristotelian and Cartesian Thought*. Ithaca-London: Cornell University Press.

Des Chene 2002 = Des Chene, Dennis. 2002. Cartesian Science: Régis and Rohault. In *A Companion to Early Modern Philosophy*, ed. Steven Nadler, 183–196. Malden: Blackwell.

Des Chene 2005 = Des Chene, Dennis. 2005. Mechanisms of Life in the Seventeenth Century: Borelli, Perrault, Régis. *Studies in History and Philosophy of Biological and Biomedical Sciences* 36(2): 245–60.

Dessì and Lotti 2011 = Dessì, Paola, and Lotti, Brunello. 2011. *Eredità cartesiane nella cultura britannica*. Florence: Le Lettere.

Dew 2013 = Dew, Nicholas. 2013. The Hive and the Pendulum: Universal Metrology and Baroque Science. In *Science in the Age of Baroque*, ed. Ofer Gal and Raz Chen-Morris, 239–256. Dordrecht-New York: Springer.

Di Biase 2014 = Di Biase, Giuliana. 2014. John Locke on Monotheism. A Dispute with Johannes Hudde. *Archivio di filosofia* 82(1): 317–329.

Dibon 1954 = Dibon, Paul. 1954. *La philosophie néerlandaise au siècle d'or. Tome I. L'enseignement philosophique dans les universités à l'époque précartésienne (1575–1650)*. Paris-Amsterdam-London-New York: Elsevier.

Dibon 1990 = Dibon, Paul. 1990. *Regards sur la Hollande du siècle d'or*. Naples: Vivarium.

Dieckhöfer 1970 = Dieckhöfer, Klemens. 1970. *Der niederländische Artzt Daniel Voet. Ein Neoaristoteliker des 17. Jahrhunderts*. Münster: Institut für Geschichte der Medizin der Universität Münster.

Dieckhöfer 1978 = Dieckhöfer, Klemens. 1978. Bemerkungen zur "Physiologia" des niederländers Daniel Voet. *Janus*, 65, 283–296.

Dijksterhuis 1950 = Dijksterhuis, Eduard Jan (Ed.). 1950. *Descartes et le cartésianisme hollandais – Etudes et documents*. Paris: Presses Universitaires de France.

Dijksterhuis 1970 = Dijksterhuis, Eduard Jan. 1970. *Simon Stevin: Science in the Netherlands around 1600*. The Hague: M. Nijhoff.
Dijksterhuis 1987 = Dijksterhuis, Eduard Jan. 1987. *Archimedes*. Princeton: Princeton University Press. First edition 1938.
Dijksterhuis 2004 = Dijksterhuis, Fokko Jan. 2004. *Lenses and Waves. Christiaan Huygens and the Mathematical Science of Optics in the Seventeenth Century*. Dordrecht: Kluwer Academic Publishers.
Dijksterhuis 2007 = Dijksterhuis, Fokko Jan. 2007. Constructive Thinking: A Case for Dioptrics. In *The Mindful Hand: Inquiry and Invention from the Late Renaissance to Early Industrialisation*, ed. Lissa Roberts, Simon Schaffer, and Peter Dear, 59–82. Amsterdam: Koninklijke Nederlandse Akademie van Wetenschappen.
Dijkstra 2007 = Dijkstra, Arjen. 2007. *Het vinden van oost en west in het Friesland van de zeventiende eeuw. De polemiek tussen Lieuwe Willemsz. Graaf (1652–1704) en Bernhardus Fullenius Jr. (1640–1707)*. Groningen: University of Groningen. Doctoral dissertation.
Dijkstra 2012 = Dijkstra, Arjen. 2012. *Between Academics and Idiots. A Cultural History of Mathematics in the Dutch Province of Friesland (1600–1700)*. Enschede: University of Twente. Doctoral dissertation.
Dobre 2010 = Dobre, Mihnea. 2010. *Metaphysics and Physics in Cartesian Natural Philosophy: Descartes and Early French Cartesians on the Foundation of Natural Philosophy*. Nijmegen: Radboud University. Doctoral dissertation.
Dobre 2013a = Dobre, Mihnea. 2013. Rohault's Cartesian Physics. In *Cartesian Empiricisms*, ed. Mihnea Dobre and Tammy Nyden, 203–226. Dordrecht-Heidelberg-New York-London: Springer.
Dobre 2013b = Dobre, Mihnea. 2013. On Glass-Drops: A Case Study of the Interplay between Experimentation and Explanation in Seventeenth-Century Natural Philosophy. *Journal of Early Modern Studies* 2: 105–124.
Dobre 2013c = Dobre, Mihnea. 2013. Knowledge and Certainty in the Foundation of Cartesian Natural Philosophy. *Revue Roumaine de philosophie* 57(1): 95–110.
Dobre 2017 = Dobre, Mihnea. 2017. *Descartes and Early French Cartesianism: Between Metaphysics and Physics*. Bucharest: Zeta Books.
Dobre and Nyden 2013 = Dobre, Mihnea, and Nyden, Tammy (Eds.). 2013. *Cartesian Empiricisms*. Dordrecht-Heidelberg-New York-London: Springer.
Dohmen 1982 = Dohmen, J. 1982. Wetenschappelijke erediensten voor publiek. De anatomische lessen van Frederik Ruysch. *Folia Civitatis* 35(19): 17–21.
Douglas 2013 = Douglas, Alexander. 2013. Spinoza and the Dutch Cartesians on Philosophy and Theology. *Journal of the History of Philosophy* 51(4): 567–588.
Douglas 2015 = Douglas, Alexander. 2015. *Spinoza & Dutch Cartesianism: Philosophy and Theology*. Oxford: Oxford University Press.
Ducheyne 2008 = Ducheyne, Steffen. 2008. Galileo and Huygens on Free Fall: Mathematical and Methodological Differences. *Dynamis* 28: 243–274.
Ducheyne 2011 = Ducheyne, Steffen. 2011. *The Main Business of Natural Philosophy: Isaac Newton's Natural-Philosophical Methodology*. Dordrecht: Springer.
Duker 1897–1915 = Duker, Arnoldus C. 1897–1915. *Gisbertus Voetius*. Leiden: Brill.
Dunham 1987 = Dunham, William. 1987. The Bernoullis and the Harmonic Series. *The College Mathematics Journal* 18(1): 18–23.
Dunkel 1755 = Dunkel, Johann Gottlob Wilhelm. 1755. Volder (Burchard van). In Dunkel, Johann Gottlob Wilhelm. 1755. *Historisch-kritische Nachrichten*, volume 2, part 1, 200. Cöthen-Dessau: In der Cörnerischen Buchhandlung.
Earman 1989 = Earman, John. 1989. Remarks on Relational Theories of Motion. *Canadian Journal of Philosophy* 19: 83–87.
Eaton 2005 = Eaton, William R. 2005. *Boyle on Fire: The Mechanical Revolution in Scientific Explanation*. London: Continuum.

Eloy 1778 = Eloy, Nicolas François Joseph. 1778. Volder (Burcher De). In Eloy, Nicolas François Joseph. 1778. *Dictionnaire historique de la médecine ancienne et moderne*, volume 4, 551–552. Mons: Chez H. Hoyoisi.

Elzinga 1971 = Elzinga, Aant. 1971. Huygens' Theory of Research and Descartes' Theory of Knowledge. *Zeitschrift für allgemeine Wissenschaftstheorie* 2: 174–194.

Erlichson 1994 = Erlichson, Herman. 1994. Huygens and Newton on the Problem of Circular Motion. *Centaurus* 37: 210–229.

Favaretti Camposampiero et al. 2018 = Favaretti Camposampiero, Matteo, Priarolo, Mariangela, and Scribano, Emanuela (Eds.). 2018. *Occasionalism: From Metaphysics to Science*. Turnhout: Brepols Publishers.

Favino 2008 = Favino, Federica. 2008. Beyond the «Moderns»? The Accademia Fisico-matematica of Rome (1677–1698) and the vacuum. In *Institutions of Knowledge, Circles of Knowledge in Early Modern Europe,* ed. Sven Duprè and Sachiko Kusukawa, special issue of *History of Universities* 23(2): 120–158

Fechner 1980 = Fechner, H. 1980. Ein neues Bild von Joost de Volder. *Oud Holland* 94: 284–286.

Fichant 1998 = Fichant, Michel. 1998. *Science et mètaphysique dans Descartes et Leibniz*. Paris: Presses Universitaires de France.

Fix 1990 = Fix, Andrew Cooper. 1990. Mennonites and Collegiants in Holland 1630–1700. *Mennonite Quarterly Review* 64: 160–177.

Fix 1991 = Fix, Andrew Cooper. 1991. *Prophecy and Reason: The Dutch Collegiants in the Early Enlightenment*. Princeton: Princeton University Press.

Fix 1999 = Fix, Andrew Cooper. 1999. *Fallen Angels: Balthasar Bekker, Spirit Belief, and Confessionalism in the Seventeenth Century Dutch Republic*. Dordrecht: Springer.

Fleckenstein 1977 = Fleckenstein, Joachim Otto. 1977. *Johann und Jakob Bernoulli*. Basel: Birkhäuser. First edition 1949.

Font 2014 = Font, Zsuzsa. 2014. Az erdélyi felvilágosodás kezdeteinek kulcsembere, ifj ú Köleséri Sámuel. *Keresztény Magvető* 120(3–4): 252–263.

Fouke 1997 = Fouke, Daniel Clifford. 1997. *The Enthusiastical Concerns of Dr. Henry More: Religious Meaning and the Psychology of Delusion*. Leiden-New York-Cologne: Brill.

Fouke 2003 = Fouke, Daniel Clifford. 2003. Pascal's Physics. In *The Cambridge Companion to Pascal*, ed. Nicholas Hammond, 75–101. Cambridge, Cambridge University Press.

Fox 1970 = Fox, Michael. 1970. Leibniz's Metaphysics of Space and Time. *Studia Leibnitiana* 2: 29–55.

Frampton 2007 = Frampton, Travis L. 2007. *Spinoza and the Rise of Historical Criticism of the Bible*. New York-London: T&T Clark.

Fredette 2001 = Fredette, Raymond. 2001. *Galileo's* De Motu Antiquiora*: Notes for a Reappraisal*. Berlin: Max Planck Institute for the History of Science, Preprint.

Freedman 1984 = Freedman, Joseph S. 1984. *Deutsche Schulphilosophie im Reformationszeitalter (1500–1650): Ein Handbuch für den Hochschulunterricht*. Münster: MAKS.

Freudenthal 1899 = Freudenthal, Jacob. 1899. *Die Lebensgeschichte Spinoza's in Quellenschriften, Urkunden und Nichtamtlichen Nachrichten*. Leipzig: Verlag Von Veit.

Freudenthal 2002 = Freudenthal, Gideon. 2002. *Perpetuum mobile*: The Leibniz-Papin Controversy. *Studies in History and Philosophy of Science Part* A 33: 573–637.

Frijhoff 2016 = Frijhoff, Willem. 2008. Multilingualism in the Dutch Golden Age: An Exploration. In *Multilingualism, Nationhood, and Cultural Identity: Northern Europe, 16th–19th Centuries*, ed. Willem Frijhoff, Marie-Christine Kok-Escalle, Karene Sanchez-Summerer, Bettina Brandt, and Mary Robitaille-Ibbett, 17–48. Amsterdam: Amsterdam University Press.

Frijhoff and Spies 2004 = Frijhoff, Willem, and Spies, Marijke. 2004. *Dutch Culture in a European Perspective: 1650, Hard-Won Unity*. Assen: Van Gorcum.

Fuchs 2001 = Fuchs, Thomas. 2001. *The Mechanization of the Heart. Harvey and Descartes*. Trans. by Marjorie Grene. Rochester: The University of Rochester Press. First edition 1992, *Die Mechanisierung des Herzens – Harvey und Descartes – Der vitale und der mechanische Aspekt des Kreislaufs*. Frankfurt: Suhrkamp.

Furley 1989 = Furley, David (Ed.). 1989. *Cosmic Problems: Essays on Greek and Roman Philosophy of Nature*. Cambridge: Cambridge University Press.
Futch 2008 = Futch, Michael. 2008. *Leibniz's Metaphysics of Time and Space*. Berlin: Springer.
Gaab et al. 2004 = Gaab, Hans, Leich, Pierre, and Löffladt, Günter (Eds.). 2004. *Johann Christoph Sturm (1635–1703)*. Frankfurt: Harri Deutsch Verlag.
Gabbey 1971 = Gabbey, Alan. 1971. Force and Inertia in Seventeenth-Century Dynamics. *Studies in History and Philosophy of Science Part A* 2(1): 1–68.
Gabbey 1980 = Gabbey, Alan. 1980. Force and Inertia in the Seventeenth Century: Descartes and Newton. In *Descartes: Philosophy, Mathematics and Physics*, ed. Stephen Gaukroger, 230–320. Brighton: Harvester Press.
Gabbey 1982 = Gabbey, Alan. 1982. *Philosophia Cartesiana Triumphata*: Henry More (1646–1671). In *Problems of Cartesianism*, ed. Thomas M. Lennon, John M. Nicholas, and John W. Davis, 171–249. Kingston-Montreal: McGill-Queens University Press.
Gabbey 1986 = Gabbey, Alan. 1986. Mariotte et Roberval, son collaborateur involontaire. In *Mariotte savant et philosophe, (†1684). Analyse d'une renommée*, ed. Pierre Costabel, 204–244. Paris: Vrin.
Gal 2005 = Gal, Ofer. 2005. The Invention of Celestial Mechanics. *Early Science and Medicine* 10(4): 529–534.
Galama 1954 = Galama, Sybrand. 1954. Het wijsgerig onderwijs aan de Hogeschool te Franeker 1585–1811. Franeker: T. Wever.
Gale 1973 = Gale, George. 1973. Leibniz's Dynamical Metaphysics and the Origins of the *Vis Viva* Controversy. *Systematics* 11: 184–207.
Gale 1988 = Gale, George. 1988. The Concept of 'Force' and Its Role in the Genesis of Leibniz's Dynamical Viewpoint. *Journal of the History of Philosophy* 26: 45–67.
Galluzzi 1979 = Galluzzi, Paolo. 1979. *Momento: studi galileiani*. Rome: Edizioni dell'Ateneo e Bizzarri.
Garber 1992 = Garber, Daniel. 1992. *Descartes' Metaphysical Physics*. Chicago: The University of Chicago Press.
Garber 2001 = Garber, Daniel. 2001. *Descartes Embodied. Reading Cartesian Philosophy Through Cartesian Science*. Cambridge: Cambridge University Press.
Garber 2002 = Garber, Daniel. 2002. Descartes, Mechanics, and the Mechanical Philosophy. *Midwest Studies in Philosophy* 26: 185–204.
Garber 2006 = Garber, Daniel. 2006. Physics and Foundations. In *Cambridge History of Science. Vol. 3: Early Modern Science*, ed. Katharine Park, 21–69. New York: Cambridge University Press.
Garber 2009 = Garber, Daniel. 2009. *Leibniz: Body, Substance, Monad*. New York: Oxford University Press.
Garber and Roux 2013 = Garber, Daniel, and Roux, Sophie (Eds.). 2013. *The Mechanization of Natural Philosophy*. New York: Springer.
Garber and Tho 2018 = Garber, Daniel, and Tho, Tzuchien. 2018. Force and Dynamics. In *The Oxford Handbook of Leibniz*, ed. Maria Rosa Antognazza, 304–331. New York: Oxford University Press.
Garber et al. 1998 = Garber, Daniel, Henry, John, Joy, Lynn, and Gabbey, Alan. 1998. New Doctrines of Body and Its Powers, Place, and Space. In *The Cambridge History of Seventeenth-Century Philosophy*, ed. Daniel Garber and Michael Ayers, volume 1, 553–623. Cambridge: Cambridge University Press.
Gariepy 1990 = Gariepy, Thomas. 1990. *Mechanism without Metaphysics. Henricus Regius and the Establishment of Cartesian Medicine*. New Haven: Yale University. Doctoral dissertation.
Garrett 2018 = Garrett, Don. 2018. *Nature and Necessity in Spinoza's Philosophy*. New York: Oxford University Press.
Gaukroger 1980 = Gaukroger, Stephen (Ed.). 1980. *Descartes: Philosophy, Mathematics and Physics*. Brighton: The Harvester Press; Totowa: Barnes and Noble.
Gaukroger 1989 = Gaukroger, Stephen. 1989. *Cartesian Logic. An Essay on Descartes's Conception of Inference*. Oxford, Oxford University Press.

Gaukroger 2000a = Gaukroger, Stephen. 2000. The Foundational Role of Hydrostatics and Statics in Descartes' Natural Philosophy. In *Descartes' Natural Philosophy*, ed. Stephen Gaukroger, John Schuster, and John Sutton, 60–80. London: Routledge.
Gaukroger 2000b = Gaukroger, Stephen. 2000. The Role of Matter Theory in Baconian and Cartesian Cosmologies. *Perspectives on Science* 8: 201–222.
Gaukroger 2002 = Gaukroger, Stephen (Ed.). 2002. *Descartes' System of Natural Philosophy*. Cambridge: Cambridge University Press.
Gaukroger 2008 = Gaukroger, Stephen. 2008. Knowledge, Evidence and Method. In *The Cambridge Companion to Early Modern Philosophy*, ed. Donald Rutherford, 39–66. Cambridge, Cambridge University Press.
Gaukroger and Schuster 2002 = Gaukroger, Stephen, and Schuster, John. 2002. The Hydrostatic Paradox and the Origins of Cartesian Dynamics. *Studies in History and Philosophy of Science Part A* 33: 535–572.
Gaukroger et al. 2000 = Gaukroger, Stephen, Schuster, John, and Sutton, John (Eds.). 2000. *Descartes' Natural Philosophy*. London: Routledge.
Gemelli 1996 = Gemelli, Benedino. 1996. *Aspetti dell'atomismo classico nella filosofia di Francis Bacon e nel Seicento*. Florence: Leo S. Olschki.
Gemelli 2002 = Gemelli, Benedino. 2002. *Isaac Beeckman atomista e lettore critico di Lucrezio*. Florence: Leo S. Olschki.
Georgescu 2017 = Georgescu, Laura. 2017. *Devising Magnetism: Concepts and Investigative Practices*. Ghent: University of Ghent. Doctoral dissertation.
Giglioni 2000 = Giglioni, Guido. 2000. *Immaginazione e malattia: saggio su Jan Baptiste van Helmont*. Milan: Franco Angeli.
Gill 2009 = Gill, Mary Louise. 2009. The Theory of the Elements in *De caelo* 3 and 4. In *New Perspectives on Aristotle's* De caelo, ed. Alan C. Bowen and Christian Wildberg, 119–161. Leiden-Boston: Brill.
Gliozzi 1930 = Gliozzi, Mario. 1930. I metodi di Galileo per determinare il peso dell'aria. *Archeion* 2: 171–173.
Goldstein 2008 = Goldstein, Catherine. 2008. Écrire l'expérience mathématique au XVIIe siècle; la méthode selon Bernard Frenicle de Bessy. In *Réduire en art. La technologie de la Renaissance aux Lumières*, ed. Pascal Dubourg Glatigny and Hélène Vérin, 213–224. Paris: Éditions de la Maison des Sciences de l'Homme.
Gootjes 2019 = Gootjes, Albert. 2019. The Collegie der Sçavanten: A Seventeenth-Century Cartesian Scholarly Society in Utrecht. In *Enlightened Religion. From Confessional Churches to Polite Piety in the Dutch Republic*, ed. Joke Spaans and Jetze Touber, 156–182. Leiden-Boston: Brill.
Gordon 2004 = Gordon, Alexander. 2004. John Gale. In *The Oxford Dictionary of National Biography*, ed. Henry Colin Gray Matthew, volume 24, 247. London: Oxford University Press.
Gori 1972 = Gori, Giambattista. 1972. *La fondazione dell'esperienza in 's Gravesande*. Florence: La Nuova Italia.
Gorman 1994 = Gorman, Michael John. 1994. Jesuit Explorations of the Torricellian Space: Carp-Bladders and Sulphurous Fumes. *Melange de L'Ecole Française de Rome* 106: 7–32.
Goudriaan 1999a = Goudriaan, Aza. 1999. Die Erkennbarkeit Gottes in der Philosophie Claubergs. In *Johannes Clauberg (1622–1665) and Cartesian Philosophy in the Seventeenth Century*, ed. Theo Verbeek, 39–59. Dordrecht: Kluwer.
Goudriaan 1999b = Goudriaan, Aza. 1999. *Philosophische Gotteserkenntnis bei Suarez und Descartes im Zusammenhang mit der niederländischen reformierten Theologie und Philosophie des 17. Jahrhunderts*. Leiden: Brill.
Goudriaan 2003 = Goudriaan, Aza. 2003. Anthonius Driessen *contra* Jacob Wittich: over God, de schepping en causaliteit. In *Spinoza en de scholastiek*, ed. Gunther Coppens, 53–68. Leuven-Leusden: Acco.
Goudriaan 2006 = Goudriaan, Aza. 2006. *Reformed Orthodoxy and Philosophy, 1625–1750. Gisbertus Voetius, Petrus van Mastricht, and Anthonius Driessen*. Leiden: Brill.

Goudriaan 2013 = Goudriaan, Aza. 2013. Theology and Philosophy. In *A Companion to Reformed Orthodoxy*, ed. Herman Selderhuis, 27–63. Leiden: Brill.
Goudriaan 2016 = Goudriaan, Aza. 2016. Descartes, Cartesianism, and Early Modern Theology. In *The Oxford Handbook of Early Modern Theology 1600–1800*, ed. Ulrich L. Lehner, Richard A. Muller, and A. G. Roeber, 533–547. Oxford: Oxford University Press.
Grabow 1985 = Grabow, Rolf. 1985. *Simon Stevin*. Leipzig: BSB B. G. Teubner Verlagsgesellschaft.
Grafton 1988 = Grafton, Anthony. 1988. Civic Humanism and Scientific Scholarship at Leiden. In *The University and the City. From Medieval Origins to the Present*, ed. Thomas Bender, 59–78. New York-Oxford: Oxford University Press.
Grant 1981 = Grant, Edward. 1981. *Much Ado About Nothing: Theories of Space and Vacuum from the Middle Ages to the Scientific Revolution*. Cambridge: Cambridge University Press.
Graupe 2012 = Graupe, Katharina. 2012. Oratio historica – *Reden über Geschichte: Untersuchungen zur praktischen Rhetorik während des spanisch-niederländischen Konfliktes im 16. und 17. Jahrhundert*. Berlin: Walter de Gruyter.
Greene 2010 = Greene, Robert A. 2010. The Origin, Definition, Assimilation and Endurance of instinctu naturae in Natural Law Parlance: From Isidore to Hobbes and Locke. *History of European Ideas* 36(4): 361–374.
Greene 1992 = Grene, Marjorie. 1992. The Heart and the Blood: Descartes, Plemp and Harvey. In *Essays on the Philosophy and Science of René Descartes*, ed. Stephen Voss, 324–336. Oxford: Oxford University Press.
Greene 1993 = Grene, Marjorie. 1993. Aristotelico-Cartesian Themes in Natural Philosophy: Some Seventeenth Century Cases. *Perspectives on Science* 1: 66–87
Grieser 1969 = Grieser, Rudolf. 1969. Korrespondenten von G. W. Leibniz: Johannes Teyler. *Studia Leibnitiana* 1: 208–227.
Grootendorst and Van Maanen 1982 = Grootendorst, Albertus Wilhelmus, and Van Maanen, Jan A. 1982. Van Heureat's Letter (1659) on the Rectification of Curves. Text, Translation (English-Dutch), Commentary. *Nieuw Archief voor Wiskunde* 30(3): 95–113.
Grosholz 1994 = Grosholz, Emily. 1994. Descartes and the Individuation of Physical Objects. In *Individuation and Identity in Early Modern Philosophy*, ed. Jorge J. E. Gracia and Kenneth F. Barber, 41–58. Albany: State University of New York Press.
Grosslight 2013 = Grosslight, Justin. 2013. Small Skills, Big Networks: Marin Mersenne as Mathematical Intelligencer. *History of Science* 41: 337–374.
Guéroult 1934 = Guéroult, Martial. 1934. *Dynamique et métaphysique leibniziennes, suivi d'une Note sur le principe de la moindre action chez Maupertuis*. Paris: Les Belles Lettres.
Guéroult 1980 = Guéroult, Martial. 1980. The Metaphysics and Physics of Force in Descartes. In *Descartes: Philosophy, Mathematics and Physics*, ed. Stephen Gaukroger, 196–229. Brighton: Harvester Press.
Guerrini 1987 = Guerrini, Anita. 1987. Archibald Pitcairne and Newtonian Medicine. *Medical History* 31(1): 70–83.
Guhrauer 1847 = Guhrauer, Gottschalk Eduard. 1847. Beiträge zur Kenntnis des 17. und 18. Jahrhunderts aus den handschriftlichen Aufzeichnungen. *Allgemeine Zeitschrift für Geschichte*, 7, 385–436, 481–531.
Guicciardini 1995 = Guicciardini, Niccolò. 1995. Johann Bernoulli, John Keill and the Inverse Problem of Central Forces. *Annals of Science* 52: 537–575.
Guicciardini 1999 = Guicciardini, Niccolò. 1999. *Reading the* Principia*: The Debate on Newton's Mathematical Methods for Natural Philosophy from 1687 to 1736*. Cambridge: Cambridge University Press.
Guicciardini 2005 = Guicciardini, Niccolò. 2005. Reconsidering the Hooke-Newton debate on Gravitation: Recent Results. *Early Science and Medicine* 10: 511–517.
Guicciardini 2017 = Guicciardini, Niccolò. 2017. The Newton-Leibniz Calculus Controversy, 1708–1730. In *The Oxford Handbook of Newton*, ed. Eric Schliesser and Chris Smeenk. New York: Oxford University Press. DOI: 10.1093/oxfordhb/9780199930418.013.9. Accessed 3 February 2019.

Guillemin 1868 = Guillemin, Amédée. 1868. *Les phénomènes de la physique*. Paris: L. Hachette.
Hacking 1980 = Hacking, Ian. 1980. Proof and Eternal Truths: Descartes and Leibniz. In *Descartes: Philosophy, Mathematics and Physics*, ed. Stephen Gaukroger, 169–179. Brighton: Harvester Press.
Hackmann 1979 = Hackmann, Willem. 1979. The Relationship Between Concept and Instrument Design in Eighteenth-Century Experimental Science. *Annals of Science* 36(3): 205–224.
Hahn 1971 = Hahn, Roger. 1971. *The Anatomy of a Scientific Institution. The Paris Academy of Sciences, 1666–1803*. Berkeley-Los Angeles-London: University of California Press.
Hall 1982 = Hall, A. Rupert. 1982. Further Newton Correspondence. *Notes and Records of the Royal Society of London* 37(1): 7–34.
Hall 1983 = Hall, A. Rupert. 1983. *The Revolution in Science 1500–1750*. London: Longman.
Hall 1990 = Hall, A. Rupert. 1990. *Henry More. Magic, Religion and Experiment*. Oxford: Blackwell.
Hansen 1996 = Hansen, Julie V. 1996. Resurrecting Death: Anatomical Art in the Cabinet of Dr. Frederik Ruysch. *The Art Bulletin* 78: 663–679.
Harrison 2000 = Harrison, Peter. 2000. The Influence of Cartesian Cosmology in England. In *Descartes' Natural Philosophy*, ed. Stephen Gaukroger, John Schuster, and John Sutton, 168–194. London-New York: Routledge.
Harsch 2007 = Harsch, Viktor. 2007. Otto von Guericke (1602–1686) and His Pioneering Vacuum Experiments. *Aviation, Space, and Environmental Medicine* 78(11): 1075–1077.
Hartog 1876 = Hartog, Jan. 1876. Het collegie der Scavanten in Utrecht. *De Gids* 40: 77–114.
Hartz 2006 = Hartz, Glenn A. 2006. *Leibniz's Final System: Monads, Matter and Animals*. London-New York: Routledge.
Hartz and Cover 1988 = Hartz, Glenn A., and Cover, Jan A. 1988. Space and Time in the Leibnizian Metaphysic. *Noûs* 22: 493–519.
Hatfield 1979 = Hatfield, Gary. 1979. Force (God) in Descartes' Physics. *Studies in History and Philosophy of Science Part A* 10: 113–140.
Hatfield 1990 = Hatfield, Gary. 1990. Metaphysics and the New Science. In *Reappraisals of the Scientific Revolution*, ed. David C. Lindberg, Robert S. Westman, 93–166. Cambridge: Cambridge University Press.
Hatfield 2013 = Hatfield, Gary. 2013. The Cartesian Psychology of Antoine Le Grand. In *Cartesian Empiricisms*, ed. Mihnea Dobre and Tammy Nyden, 251–274. Dordrecht-Heidelberg-New York-London: Springer.
Hattab 2007 = Hattab, Helen. 2007. Concurrence or Divergence? Reconciling Descartes's Metaphysics with His Physics. *Journal of the History of Philosophy* 45: 49–78.
Hattab 2009 = Hattab, Helen. 2009. *Descartes on Forms and Mechanisms*. Cambridge: Cambridge University Press.
Haven 2006 = Haven, Kendall F. 2006. *100 Greatest Science Inventions of All Time*. Westport: Libraries Unlimited.
Hedesan 2016 = Hedesan, Georgiana D. 2016. *An Alchemical Quest for Universal Knowledge: The 'Christian Philosophy' of Jan Baptist Van Helmont (1579–1644)*. New York: Routledge.
Heering 1994 = Heering, Jan Paul. 1994. Hugo Grotius' *De veritate Religionis Christianae*. In *Hugo Grotius Theologian: Essays in Honour of G. H. M. Posthumus Meyjes*, ed. Henk J. M. Nellen, and Edwin Rabbie, 41–52. Leiden: Brill.
Heimann 1977 = Heimann, Peter. 1977. 'Geometry and nature': Leibniz and Johann Bernoulli's Theory of Motion. *Centaurus* 21: 1–26.
Henry 2013 = Henry, John. 2013. The Reception of Cartesianism. In *The Oxford Handbook of British Philosophy in the Seventeenth Century*, ed. Peter Anstey, 116–143. Oxford: Oxford University Press.
Herrmann and Platz 2003 = Herrmann, Volker, and Platz, Kai Thomas (Eds). 2003. *Der Wahrheit auf der Spur: Johann Christoph Sturm (1635–1703): Mathematiker, Physiker, Astronom*. Hilpoltstein: Verlag Dr. Faustus.

Heyd 1983 = Heyd, Michael. 1983. *Between Orthodoxy and the Enlightenment. Jean-Robert Chouet and the Introduction of Cartesian Science in the Academy of Geneva*. The Hague: Martinus Nijhoff.
Heyd 1985 = Heyd, Michael. 1985. *"Be Sober and Reasonable": The Critique of Enthusiasm in the Seventeenth and Early Eighteenth Centuries*. Leiden: Brill.
Hill 1985 = Hill, Kenneth A. 1985. Hartsoeker's *Homunculus*: A Corrective Note. *Journal of the History of the Behavioral Sciences* 21(2): 178–179.
Hintikka 1962 = Hintikka, Jaakko. 1962. *Cogito, ergo sum*: Inference or Performance? *The Philosophical Review* 71: 3–32.
Hirai 2001 = Hirai, Hiro. 2001. Paracelsisme, néoplatonisme et médecine hermétique dans la théorie de la matière de Joseph Du Chesne à travers son *Ad veritatem hermeticae medicinae*. *Archives internationales d'histoire des sciences* 51: 9–37.
Hirschfield 1981 = Hirschfield, John Milton. 1981. *The Académie Royale des Sciences. 1666–1683*. New York: Arno Press.
Hogendijk 2010 = Hogendijk, Jan P. 2010. The Scholar and the Fencing Master: The Exchanges between Joseph Justus Scaliger and Ludolph van Ceulen on the Circle Quadrature (1594–1596). *Historia Mathematica* 37(3): 345–375.
Home 1968 = Home, Roderick. 1968. The Third Law in Newton's Mechanics. *The British Journal for the History of Science* 4: 39–51.
Hooijmaijers and Maas 2013 = Hooijmaijers, Hans, and Maas, Ad. 2013. Entrepreneurs in Experiments: The Leiden Cabinet of Physics and the Motives of its Founders (1675–1742). In *Cabinets of Experimental Philosophy in Eighteenth-Century Europe*, ed. Jim Bennett and Sofia Talas, 27–47. Leiden: Brill.
Horstmanshoff et al. 2012 = Horstmanshoff, Manfred, King, Helen, and Zittel, Claus (Eds.). 2012. *Blood, Sweat and Tears. The Changing Concepts of Physiology from Antiquity into Early Modern Europe*. Leiden-Boston: Brill.
Hoskin 1961 = Hoskin, Michael A. 1961. Mining All within: Clarke's Notes to Rohault's *Traité de Physique*. *The Thomist* 24(2): 253–263.
Howard 2008 = Howard, Nicole. 2008. Marketing Longitude: Clocks, Kings, Courtiers, and Christiaan Huygens. *Book History* 11: 59–88.
Howarth 2007 = Howarth, Richard. 2007. Gravity Surveying in Early Geophysics. I. From Time-Keeping to Figure of the Earth. *Earth Sciences History* 26(2): 201–228.
Howse 1989 = Howse, Derek. 1989. *Nevil Maskelyne: the Seaman's Astronomer*. Cambridge: Cambridge University Press.
Hübner 2018 = Hübner, Karolina. 2018. Spinoza's Unorthodox Metaphysics of the Will. In *The Oxford Handbook of Spinoza*, ed. Michael Della Rocca. Oxford: Oxford University Press. DOI: 10.1093/oxfordhb/9780195335828.013.015. Accessed 2 February 2018.
Hutton 1990 = Hutton, Sarah (Ed.). 1990. *Henry More (1614–1687). Tercentenary Studies*. Dordrecht: Kluwer.
Hyslop 2015 = Hyslop, Scott J. 2015. *The Mathematics of Collision and the Collision of Mathematics in the 17th Century*. Bloomington: Indiana University. Doctoral dissertation.
Iltis 1971 = Iltis, Carolyn. 1971. Leibniz and the *Vis Viva* Controversy. *Isis: A Journal of the History of Science* 62: 21–35.
Iltis 1974 = Iltis, Carolyn. 1974. Leibniz' Concept of Force: Physics and Metaphysics. *Studia Leibnitiana Supplementa* 13(2): 143–149.
Israel 1996 = Israel, Jonathan. 1996. The Banning of Spinoza's Works in the Dutch Republic (1670–1678). In *Disguised and Overt Spinozism around 1700*, ed. Wiep van Bunge, 3–14. Leiden-New York-Cologne: Brill.
Israel 2001 = Israel, Jonathan. 2001. *Radical Enlightenment: Philosophy and the Making of Modernity 1650–1750*. Oxford: Oxford University Press.
Israel 2010 = Israel, Jonathan. 2010. The Early Dutch and German Reaction to the *Tractatus Theologico-Politicus*: Foreshadowing the Enlightenment's More General Spinoza Reception? In *Spinoza's Theological-Political Treatise: A Critical Guide*, ed. Michael A. Rosenthal and Yitzhak Y. Melamed, 72–100. Cambridge: Cambridge University Press.

Jalobeanu 2002 = Jalobeanu, Dana. 2002. The Two Cosmologies of René Descartes. In *Esprits modernes*, ed. Dana Jalobeanu and Vlad Alexandrescu, 75–94. Bucharest-Arad: Editura Universitatii Bucuresti.

Jalobeanu 2011 = Jalobeanu, Dana. 2011. The Cartesians of the Royal Society. The Debate Over Collisions and the Nature of Body (1668–1670). In *Vanishing Matter and the Laws of Motion: Descartes and Beyond*, ed. Peter Anstey and Dana Jalobeanu, 103–129. London: Routledge.

Jauernig 2008 = Jauernig, Anja. 2008. Leibniz on Motion and the Equivalence of Hypotheses. *The Leibniz Review* 18: 1–40.

Jesseph 2005 = Jesseph, Douglas. 2005. Mechanism, Skepticism, and Witchcraft: More and Glanvill on the Failures of the Cartesian Philosophy. In *Receptions of Descartes. Cartesianism and Anti-Cartesianism in Early Modern Europe*, ed. Tad Schmaltz, 183–199. London, New York: Routledge.

Jöcher 1751 = Jöcher, Christian Gottlieb. 1751. De Volder (Burchard). In Jöcher, Christian Gottlieb. 1751. *Allgemeines Gelehrten-Lexicon*, volume 4, 1704. Leipzig: Verlag Johann Friedrich Gleditsch.

Johnson 2018 = Johnson, Ryan. 2018. The Cartesian Eye Without Organs: The Shaping of Subjectivity in Descartes's Optics. *Philosophy & Rhetoric* 51(1): 73–90.

Jones 2001 = Jones, Matthew L. 2001. Writing and sentiment: Blaise Pascal, the vacuum, and the *Pensées*. *Studies In History and Philosophy of Science Part A* 32(1): 139–181.

Jorink 1999 = Jorink, Erik. 1999. *Wetenschap en wereldbeeld in de Gouden Eeuw*. Hilversum: Verloren.

Jorink 2003 = Jorink, Erik. 2003. 'Outside God there is nothing'. Swammerdam, Spinoza and the Janus-face of the Early Dutch Enlightenment. In *The Early Enlightenment in the Dutch Republic, 1650–1750*, ed. Wiep van Bunge, 81–107. Leiden-Boston: Brill.

Jorink 2010 = Jorink, Erik. 2010. *Reading the Book of Nature in the Dutch Golden Age, 1575–1715*. Leiden: Brill.

Jorink 2016a = Jorink, Erik. 2016. Correspondence of Swammerdam (1664–1679). *Research Data Journal for the Humanities and Social Sciences*, 1(1). DOI: 10.1163/24523666-01000003. Accessed 3 February 2019.

Jorink 2016b = Jorink, Erik. 2016. Swammerdam, hoveling? Enige kanttekeningen bij de reputatie van een wetenschappelijk onderzoeker. *Studium*, 8(4): 173–197.

Jorink 2018 = Jorink, Erik. 2018. *Modus politicus vivendi*: Nicolaus Steno and the Dutch (Swammerdam, Spinoza and Other Friends), 1660–1664. In *Steno and the Philosophers*, ed. Raphaele Andrault and Mogens Laerke, 13–44. Leiden-Boston: Brill.

Jorink and Maas 2012 = Jorink, Erik, and Maas, Ad (Eds.). 2012. *Newton and the Netherlands. How Isaac Newton was Fashioned in the Dutch Republic*. Leiden: Leiden University Press.

Jourdan 1825 = Jourdan, Antoine Jacques Louis. 1825. Volder (Burchard de). In Jourdan, Antoine Jacques Louis. 1825. *Dictionnaire des sciences médicales: biographie médicale*, volume 7, 452. Paris: C.L.F. Panckoucke.

Joy 1987 = Joy, Lynn. 1987. *Gassendi the Atomist: Advocate of History in an Age of Science*. Cambridge: Cambridge University Press.

Juillard 1999 = Juillard, Alain. 1999. Bernard, Jacques. In *Dictionnaire des journalistes 1600–1789*, ed. Jean Sgard, volume 1, 77–80. Oxford: Voltaire Foundation.

Kaplan 1982 = Kaplan, Yosef. 1982. *From Christianity to Judaism. The Story of Isaac Orobio De Castro*. Jerusalem: Magnes Press.

Kardel 1994 = Kardel, Troels. 1994. *Steno, Life, Science, Philosophy*. Copenhagen: National Library of Science.

Khamara 1988 = Khamara, Edward J. 1988. Indiscernibles and the Absolute Theory of Space and Time. *Studia Leibnitiana* 20: 140–159.

Khamara 1993 = Khamara, Edward J. 1993. Leibniz's Theory of Space: A Reconstruction. *Philosophical Quarterly* 43: 472–488.

King 1970 = King, Lester Snow. 1970. *The Road to Medical Enlightenment*. New York: American Elsevier.

Klever 1988 = Klever, Wim. 1988. Burchardus de Volder (1643–1709): A Crypto-Spinozist on a Leiden Cathedra. *Lias. Journal of Early Modern Intellectual Culture and its Sources* 15: 191–241.

Klever 1989 = Klever, Wim. 1989. Hudde's Question on God's Uniqueness; A Reconstruction on the Basis of Van Limborch's Correspondence with John Locke. *Studia Spinozana* 5: 327–358.

Klever 1991 = Klever, Wim. 1991. Verba et sententiae Spinozae, *or, Lambertus van Velthuysen (1622–1685) on Benedictus de Spinoza*. Amsterdam-Maarssen: APA, Holland University Press.

Klever 1993 = Klever, Wim. 1993. Blijenbergh Struggle with Evil and Spinoza Response. *Tijdschrift voor Filosofie* 55(2): 307–329.

Klever 1997 = Klever, Wim. 1997. *Mannen rond Spinoza, 1650–1700: presentatie van een emanciperende generatie*. Hilversum: Verloren.

Klever 2000 = Klever, Wim. 2000. Spinozisme in het geding tussen Hulsius, Braunius en Bernoullius. Een Gronings debat anno 1702. *De Zeventiende Eeuw* 16: 118–135.

Knobloch 2006 = Knobloch, Eberhard. 2006. Beyond Cartesian Limits: Leibniz's Passage from Algebraic to "Transcendental" Mathematics. *Historia Mathematica* 33: 113–131.

Knoeff 2002 = Knoeff, Rina. 2002. *Herman Boerhaave (1668–1738): Calvinist Chemist and Physician*. Amsterdam: Koninklijke Nederlandse Akademie van Wetenschappen.

Knoeff 2015 = Knoeff, Rina. 2015. Touching Anatomy: On the Handling of Preparations in the Anatomical Cabinets of Frederik Ruysch (1638–1731). *Studies in History and Philosophy of Science Part C* 49: 32–44.

Knowles Middleton 1963 = Knowles Middleton, William Edgar. 1963. The Place of Torricelli in the History of the Barometer. *Isis: A Journal of the History of Science* 54: 11–28.

Knowles Middleton 1964 = Knowles Middleton, William Edgar. 1964. *The History of the Barometer*. Baltimore: Johns Hopkins Press.

Knowles Middleton 1971 = Knowles Middleton, William Edgar. 1971. *The Experimenters. A Study of the Accademia del Cimento*. Baltimore: The Johns Hopkins Press.

Kobus and De Rivecourt 1870 = Kobus, Jan Christiaan, and De Rivecourt, Willem George Hendrik. 1870. Volder (Burchardus de). In Kobus, Jan Christiaan, and De Rivecourt, Willem George Hendrik. 1870. *Biographisch woordenboek van Nederland*, volume 3, 249. Zutphen: Verlag A. E. C. van Someren.

Kochiras 2017 = Kochiras, Hylarie. 2017. *Newton's General Scholium and the Mechanical Philosophy*. https://hal.archives-ouvertes.fr/hal-01442121v2. Accessed 3 February 2019.

Kok and Fokke 1793 = Fokke, Jan, and Kok, Jacobus. 1793. Volder (Burchard de). In Fokke, Jan, and Kok, Jacobus. 1793. *Vaderlandsch woordenboek*, volume 29, 306–311. Amsterdam: Johannes Allart.

Kolesnik-Antoine 2013a = Kolesnik-Antoine, Delphine (Ed.) 2013. *Qu'est-ce qu'être cartésien?* Paris: Editions de l'ENS de Lyon.

Kolesnik-Antoine 2013b = Kolesnik-Antoine, Delphine. 2013. Le rôle des expériences dans la physiologie d'Henricus Regius: les 'pierres lydiennes' du cartésianisme. *Journal of Early Modern Studies* 2(1): 125–145.

Kooijmans 2004 = Kooijmans, Luuc. 2004. *De doodskunstenaar, de anatomische lessen van Frederik Ruysch*. Amsterdam: Bert Bakker.

Kooijmans 2011 = Kooijmans, Luuc. 2011. *Death Defied: The Anatomy Lessons of Frederik Ruysch*. Leiden: Brill.

Koyanagi 1989 = Koyanagi, Kimiyo. 1989. Les expériences du vide dans le vide. *Courrier du Centre international Blaise Pascal* 11: 3–23.

Koyré 1973 = Koyré, Alexander. 1973. *The Astronomical Revolution: Copernicus, Kepler, Borelli*. Ithaca: Cornell University Press.

Krop 1999 = Krop, Henri. 1999. Spinoza and the Calvinistic Cartesianism of Lambertus van Velthuysen. *Studia Spinozana* 15: 107–132.

Krop 2003a = Krop, Henri. 2003. Leiden, University of. In *The Dictionary of Seventeenth and Eighteenth-Century Dutch Philosophers*, ed. Wiep van Bunge, Henri Krop, Bart Leeuwenburgh,

Han van Ruler, Paul Schuurman, and Michiel Wielema, volume 2, 603–614. Bristol: Thoemmes Press.
Krop 2003b = Krop, Henri. 2003. Bornius, Henricus (1617–1675). In *The Dictionary of Seventeenth and Eighteenth-Century Dutch Philosophers*, ed. Wiep van Bunge, Henri Krop, Bart Leeuwenburgh, Han van Ruler, Paul Schuurman, and Michiel Wielema, volume 1, 203–209. Bristol: Thoemmes Press.
Krop 2003c = Krop, Henri. 2003. Burgersdijk, Franco (1590–1635). In *The Dictionary of Seventeenth and Eighteenth-Century Dutch Philosophers*, ed. Wiep van Bunge, Henri Krop, Bart Leeuwenburgh, Han van Ruler, Paul Schuurman, and Michiel Wielema, volume 1, 181–190. Bristol: Thoemmes Press.
Krop 2003d = Krop, Henri. 2003. Medicine and Philosophy in Leiden around 1700: Continuity or Rupture? In *The Early Enlightenment in the Dutch Republic, 1650–1750*, ed. Wiep van Bunge, 173–196. Leiden-Boston: Brill.
Krop 2004 = Krop, Henri. 2004. Spinoza en het calvinistisch cartesianisme van Lambertus Van Velthuysen (1622–1685). In *Spinoza en het Nederlands cartesianisme*, ed. Gunther Coppens, 61–78. Leuven: Acco.
Kuiper 1958 = Kuiper, E.J. 1958. *De Hollandse "Schoolordre" van 1625: Een studie over het onderwijs op de Latijnse scholen in Nederland in de 17de en 18de eeuw*. Groningen: J. B. Wolters.
Laerke 2013 = Laerke, Mogens. 2013. *Ignorantia inflat*. Leibniz, Huet and the Critique of the Cartesian Spirit. *The Leibniz Review* 23: 13–42.
Laerke 2016 = Laerke, Mogens. 2016. La méthode géométrique dans la *Demonstratio evangelica* de Pierre-Daniel Huet. *Éthique, politique, religions* 1(8): 49–67.
Laerke et al. 2013 = Laerke Mogens, Smith, Justin E. H., and Schliesser, Eric (Eds.). 2013. *Philosophy and Its History. Aims and Methods in the Study of Early Modern Philosophy*. New York: Oxford University Press.
Lambour 2012 = Lambour, Ruud. 2012. Het doopsgezind milieu van Michiel van Musscher (1645–1705) en van andere schilders in zeventiende-eeuws Amsterdam: een revisie en ontdekking. *Oud Holland* 125(4): 193–214.
Land 1887 = Land, Jan Pieter Nicolaas. 1887. Arnold Geulincx te Leiden (1658–1669). *Verslagen en Mededeelingen der Koninklijke Akademie van Wetenschappen. Afdeeling Letterkunde* 3: 277–327.
Land 1891= Land, Jan Pieter Nicolaas. 1891. Arnold Geulincx and His Works. *Mind* 16: 223–242.
Land 1895 = Land, Jan Pieter Nicolaas. 1895. *Arnold Geulincx und seine Philosophie*. The Hague: Martinus Nijhoff.
Landucci 2015 = Landucci, Sergio. 2015. Insinuare un'impostura di Mosè nel 1687. *Rivista di storia della filosofia* 2: 557–569.
Lariviere 1989 = Lariviere, Barbara. 1989. Leibnizian Relationalism and the Problem of Inertia. *Canadian Journal of Philosophy* 17: 437–448.
Le Grand 1978 = Le Grand, Homer E. 1978. Galileo's Matter Theory. In *New Perspectives on Galileo*, ed. Robert E. Butts and Joseph C. Pitt, 197–208. Dordrecht: Reidel.
Lech 1993 = Lech, Hermann. 1993. *Franciscus Sylvius' Lehre von den Schärfen*. Tübingen: University of Tübingen. Doctoral dissertation.
Lenau 1993 = Lenau, Nikolaus. 1993. *Werke und Briefe. Band 7: Aufzeichnungen. Vermischte Schriften*. Ed. by Norbert O. Eke. Vienna: Deuticke und Klett-Cotta Verlag.
Lennon 1993 = Lennon, Thomas M. 1993. *The Battle of the Gods and Giants: The Legacies of Descartes and Gassendi 1655–1715*. Princeton: Princeton University Press.
Lennon 2003 = Lennon, Thomas M. (Ed.). 2003. *Cartesian Views: Papers Presented to Richard A. Watson*. Leiden: Brill.
Lennon 2007 = Lennon, Thomas M. 2007. The Eleatic Descartes. *Journal of the History of Philosophy* 45: 29–47.
Lennon 2008 = Lennon, Thomas M. 2008. *The Plain Truth. Descartes, Huet, and Skepticism*. Leiden-Boston: Brill.

Lennon et al. 1982 = Lennon, Thomas M., Nicholas, John M., and Davis, John W. (Eds.). 1982. *Problems of Cartesianism*. Kingston: McGill-Queens University Press.
Lenz and Waldow 2013 = Lenz, Martin, and Waldow, Anik (Eds.). 2013. *Contemporary Perspectives on Early Modern Philosophy*. Boston: Springer.
Lindeboom 1975 = Lindeboom, Gerrit Arie. 1975. *The Letters of Jan Swammerdam to Melchisédec Thévenot*. Amsterdam: Taylor & Francis.
Lindner 2011 = Lindner, Nicolas. 2011. *Ein Vergleich des "Cogito" bei Augustinus und Descartes – Übereinstimmungen und Unterschiede*. München: GRIN Verlag.
Liu 2014 = Liu, Chun-Fa. 2014. *Die metaphysische Grundlage der Kontroverse um den Kraftbegriff zwischen Descartes und Leibniz*. Munich: University of Munich. Doctoral dissertation.
Lo 2017 = Lo, Melissa. 2017. The Picture Multiple: Figuring, Thinking, and Knowing in Descartes's *Essais* (1637). *Journal of the History of Ideas* 78(3): 369–399.
Lodge 1998 = Lodge, Paul. 1998. The Failure of Leibniz's Correspondence with De Volder. *Leibniz Society Review* 8: 47–67.
Lodge 2001 = Lodge, Paul. 2001. The Debate over Extended Substance in Leibniz's Correspondence with De Volder. *International Studies in the Philosophy of Science* 15: 155–166.
Lodge 2003 = Lodge, Paul. 2003. Leibniz on Relativity and the Motion of Bodies. *Philosophical Topics* 31(1–2): 277–308.
Lodge 2004 = Lodge, Paul. 2004. Leibniz's Close Encounter with Cartesianism in the Correspondence with De Volder. In *Leibniz and His Correspondents*, ed. Paul Lodge, 162–192. Cambridge: Cambridge University Press.
Lodge 2005 = Lodge, Paul. 2005. Burchard de Volder: Crypto-Spinozist or Disenchanted Cartesian? In *Receptions of Descartes. Cartesianism and Anti-Cartesianism in Early Modern Europe*, ed. Tad Schmaltz, 128–145. London-New York: Routledge.
Lodge 2013 = Lodge, Paul (Ed.). 2013. *The Leibniz-De Volder Correspondence: With Selections from the Correspondence Between Leibniz and Johann Bernoulli*. New Haven: Yale University Press.
LoLordo 2005 = LoLordo, Antonia. 2005. 'Descartes's One Rule of Logic': Gassendi's Critique of the Doctrine of Clear and Distinct Perception. *British Journal for the History of Philosophy* 13(1): 51–72.
LoLordo 2007 = LoLordo, Antonia. 2007. *Pierre Gassendi and the Birth of Early Modern Philosophy*. Cambridge-New York: Cambridge University Press.
LoLordo 2018 = LoLordo, Antonia. 2018. Gassendi on Skepticism. In *Skepticism: From Antiquity to the Present*, ed. Diego Machuca and Baron Reed, 295–305. London-New York: Bloomsbury Publishing.
Long 2012 = Long, Tony. 2012. June 11, 1644: The Barometer Gets Its First Practical Airing. *Wired*. https://www.wired.com/2012/06/june-11-1644-the-barometer-gets-its-first-practical-airing/. Accessed 13 January 2019.
Look 1999 = Look, Brandon. 1999. *Leibniz and the "Vinculum Substantiale"*. Stuttgart: Franz Steiner Verlag.
Lunsingh Scheurleer and Posthumus Meyjes 1975 = Lunsingh Scheurleer, Theodoor Herman, and Posthumus Meyjes, Guillaume Henri Marie (Eds.). 1975. *Leiden University in the Seventeenth Century. An Exchange of Learning*. Leiden: Brill.
Lüthy 1997 = Lüthy, Christoph. 1997. Thoughts and Circumstances of Sébastien Basson. Analysis, Micro-History, Questions. *Early Science and Medicine* 2: 1–73.
Lüthy 2001 = Lüthy, Christoph. 2001. An Aristotelian Watchdog as Avant-Garde Physicist: Julius Caesar Scaliger. *The Monist* 84: 542–561.
Lüthy 2005 = Lüthy, Christoph. 2005. Sennert's Slow Conversion from Hylemorphism to Atomism. *Graduate Faculty Philosophy Journal* 26: 99–121.
Lüthy 2006 = Lüthy, Christoph. 2006. Where Logical Necessity Becomes Visual Persuasion: Descartes's Clear and Distinct Illustrations. In *Transmitting Knowledge: Words, Images, and Instruments in Early Modern Europe*, ed. Ian Maclean and Sachiko Kusukawa, 97–133. Oxford: Oxford University Press.

Lüthy 2012 = Lüthy, Christoph. 2012. *David Gorlæus (1591–1612): An Enigmatic Figure in the History of Philosophy and Science*. Amsterdam: Amsterdam University Press.

Luyendijk-Elshout 1975 = Luyendijk-Elshout, Antonie M. 1975. *Oeconomia Animalis*, Pores and Particles: The Rise and Fall of the Mechanical Philosophical School of Theodoor Craanen (1621–1690). In *Leiden University in the Seventeenth Century: An Exchange of Learning*, ed. Theodoor Herman Lunsingh Scheurleer and Guillaume Henri Marie Posthumus Meyjes, 294–307. Leiden: Brill.

Lynes 1982 = Lynes, John W. 1982. Descartes' Theory of Elements: From *Le Monde* to the *Principes*. *Journal of the History of Ideas* 43(1): 55–72.

Mac Lean 1971 = Mac Lean, Johannes. 1971. De nagelaten papieren van Johannes Hudde. *Scientiarum Historia* 13: 144–162.

MacDonald Ross 1987 = MacDonald Ross, George. 1987. Leibniz and de Volder on the Infinitely Small in Metaphysics. *Studia Leibnitiana* 14: 169–177.

Mackay 1873 = Mackay, Aeneas James George. 1873. *Memoir of Sir James Dalrymple, First Viscount Stair*. Edinburgh: Edmonston and Douglas.

Maffioli 2011 = Maffioli, Cesare S. 2011. *La ragione del vacuo*: Why and How Galileo Measured the Resistance of Vacuum. *Galilæana. Studies in Renaissance and Early Modern Science* 8: 73–104.

Mahoney 1980 = Mahoney, Michael S. 1980. The Measurement of Time and of Longitude at Sea. In *Studies on Christiaan Huygens*, ed. Henk J. M. Bos, 234–270. Lisse: Swets.

Mahoney 1993 = Mahoney, Michael S. 1993. Algebraic vs. Geometric Techniques in Newton's Determination of Planetary Orbits. In *Action and Reaction: Proceedings of a Symposium to Commemorate the Tercentenary of Newton's* Principia, ed. Paul Harold Theerman, Adele F. Seeff, and Dolors Altaba-Artal, 183–205. Newark: University of Delaware Press.

Mahoney 1995 = Mahoney, Michael S. 1995. *Christiaan Huygens, On the motion of bodies resulting from impact*. http://www.princeton.edu/~hos/Mahoney/texts/huygens/impact/huyimpct.html. Accessed 3 February 2019.

Manchak 2009 = Manchak, John. 2009. On Force in Cartesian Physics. *Philosophy of Science* 76: 295–306.

Manning 2012 = Manning, Gideon (Ed.). 2012. *Matter and Form in Early Modern Science and Philosophy*. Leiden-Boston: Brill.

Manning 2013 = Manning, Gideon. 2013. Descartes' Healthy Machines and the Human Exception. In *The Mechanization of Natural Philosophy*, ed. Daniel Garber and Sophie Roux, 237–262. Boston: Springer.

Manzo 2003 = Manzo, Silvia. 2003. The Arguments on Void in the Seventeenth Century: The Case of Francis Bacon. *The British Journal for the History of Science* 36(1): 43–61.

Markie 1992 = Markie, Peter. 1992. The *Cogito* and Its Importance. In *The Cambridge Companion to Descartes*, ed. John Cottingham, 140–173. Cambridge: Cambridge University Press.

Marshall 2006 = Marshall, John. 2006. *John Locke, Toleration and Early Enlightenment Culture*. Cambridge: Cambridge University Press.

Matthews 2000 = Matthews, Michael. 2000. *Time for Science Education: How Teaching the History and Philosophy of Pendulum Motion Can Contribute to Science Literacy*. New York: Kluwer.

McClaughlin 1977 = McClaughlin, Trevor. 1977. Le concept de science chez Jacques Rohault. *Revue d'histoire des sciences* 30(3): 225–240.

McClaughlin 1979 = McClaughlin, Trevor. 1979. Censorship and Defenders of the Cartesian Faith in Mid-Seventeenth Century France. *Journal of the History of Ideas* 40(4): 563–581.

McClaughlin 1996 = McClaughlin, Trevor. 1996. Was There an Empirical Movement in Mid-Seventeenth Century France? Experiments in Jacques Rohault's *Traité de physique*/Y avait-il un mouvement empirique dans la France du milieu du XVIIe siècle? Les expériences dans le *Traité de physique* de Jacques Rohault. *Revue d'histoire des sciences* 49(4): 459–481.

McClaughlin 2000 = McClaughlin, Trevor. 2000. Descartes, Experiments, and a First Generation Cartesian: Jacques Rohault. In *Descartes' Natural Philosophy*, ed. Stephen Gaukroger, John Schuster, and John Sutton, 330–345. London: Routledge.

McConnell 2008 = McConnell, Donald R. 2008. The Nature in Natural Law. *Liberty University Law Review* 2(3): article 8. https://digitalcommons.liberty.edu/lu_law_review/vol2/iss3/8. Accessed 3 February 2019.

McGahagan 1976 = McGahagan, Thomas Arthur. 1976. *Cartesianism in the Netherlands. 1639–1676. The New Science and the Calvinist Counter-Reformation.* Philadelphia: University of Pennsylvania. Doctoral dissertation.

McLaughlin 2000 = McLaughlin, Peter. 2000. Force, Determination and Impact. In *Descartes' Natural Philosophy*, ed. Stephen Gaukroger, John Schuster, and John Sutton, 81–112. London-New York: Routledge.

Meihuizen 1952 = Meihuizen, Hendrik Wiebes. 1952. *Galenus Abrahamsz, 1622–1706, Strijder voor een onbeperkte verdraagzaamheid en verdediger van het Doperse Spiritualisme*. Haarlem: H. D. Tjeenk Willink & Zoon.

Menn 1990 = Menn, Stephen. 1990. Descartes and Some Predecessors on the Divine Conservation of Motion. *Synthese* 83(2): 215–238.

Menn 1998 = Menn, Stephen. 1998. *Descartes and Augustine*. Cambridge: Cambridge University Press.

Mercer 1999 = Mercer, Christia. 1999. Johann Clauberg, Corporeal Substance, and the German Response. In *Johannes Clauberg (1622–1665) and Cartesian Philosophy in the Seventeenth Century*, ed. Theo Verbeek, 147–159. Dordrecht-Boston-London: Kluwer Academic Publishers.

Mercer 2001 = Mercer, Christia. 2001. *Leibniz's Metaphysics: Its Origins and Development.* Cambridge: Cambridge University Press.

Michael 1997 = Michael, Emily. 1997. Daniel Sennert on Matter and Form: At the Juncture of the Old and the New. *Early Science and Medicine* 2: 272–299;

Michael 2001 = Michael, Emily. 2001. Sennert's Sea Change: Atoms and Causes. In *Late Medieval and Early Modern Corpuscular Matter Theories*, ed. Christoph Lüthy, William R. Newman, and John Emery Murdoch, 331–362. Leiden: Brill.

Mignini 1984 = Mignini, Filippo. 1984. Sur la genèse du *Court Traité*: l'hypothèse d'une dictée originaire est-elle fondée? *Cahiers Spinoza* 5: 147–165.

Mignini 2015 = Mignini, Filippo. 2015. *Introduzione a Spinoza*. Bari: Laterza. First edition 2006.

Mijers 2012 = Mijers, Esther. 2012. *'News from the Republick of Letters': Scottish Students, Charles Mackie and the United Provinces, 1650–1750.* Leiden-Boston: Brill.

Milani 2010 = Milani, Nausicaa Elena. Il *Système* di Régis e le sue immagini tra nuova filosofia e censura. *Nuncius* 25(2): 241–297.

Milani 2013 = Milani, Nausicaa Elena. 2013. *Tra cartesianesimo ed empirismo: il Système di Pierre-Sylvain Régis*. Parma: Università degli Studi di Parma. Doctoral dissertation.

Milani 2014a = Milani, Nausicaa Elena. 2014. The Empirical Interpretation of French Cartesianism: the Académie des Sciences, the *Journal des Sçavans* and the Relationship with the Royal Society. *Noctua* 1(1): 312–480.

Milani 2014b = Milani, Nausicaa Elena. 2014. Régis' Interpretation of the Nature of God and His *Refutation de l'opinion de Spinoza*. In *Filosofia e religione. Studi in onore di Fabio Rossi*, ed. Stefano Caroti and Alberto Siclari, 188–235. Parma: E-Theca On Line Open Access Edizioni. http://www.didaschein.net/ojs/index.php/QuadernidiNoctua/article/viewFile/128/105. Accessed 15 February 2019.

Milani 2015 = Milani, Nausicaa Elena. 2015. Motion and God in 17th Century Cartesian Manuals: Rohault, Régis and Gadroys. *Noctua* 2(1–2): 481–516.

Miniati 1987 = Miniati, Mara (Ed.). 1987. *L'età di Galileo. Il secolo d'oro della scienza in Toscana.* Florence: Istituto e Museo di Storia della Scienza.

Miniati 2009 = Miniati, Stefano. 2009. *Nicholas Steno's Challenge for Truth. Reconciling Science and Faith.* Milan: Franco Angeli.

Miranda 2012 = Miranda, Giacomo. 2012. *Il primato dell'interpretazione: lineamenti di una "teologia cartesiana" in Ruardus Andala (1665–1727).* Parma: Università degli Studi di Parma. Doctoral dissertation.

Mojet 2016 = Mojet, Emma. 2016. *Early Dutch Interest in Newtonian Mathematics: Adriaen Verwer (1654–1717) and Newton's Usefulness.* Utrecht: University of Utrecht. Master's dissertation.

Mojet 2017 = Mojet, Emma. 2017. Early Modern Mathematics in a Letter: Adriaen Verwer to David Gregory on Mathematics and Natural Philosophy. *Lias. Journal of Early Modern Intellectual Culture and its Sources* 44(2): 117–142.

Molhuysen 1913–1924 = Molhuysen, Philip Christiaan (Ed.). 1913–1924. *Bronnen tot de Geschiednis der Leidsche Universiteit 1574–1811*. The Hague: Martinus Nijhoff.

Monchamp 1886 = Monchamp, Georges. 1886. *Histoire du Cartésianisme en Belgique*. Brussels: F. Hayez.

Montuori 1983 = Montuori, Maurizio. 1983. *John Locke on Toleration and the Unity of God*. Amsterdam: J. C. Gieben.

Mormino 1990a = Mormino, Gianfranco. 1990. La relatività del movimento negli scritti sull'urto di Christiaan Huygens. In De motu. *Studi di storia del pensiero su Galileo, Hegel, Huygens e Gilbert*, ed. Enrico Rambaldi, 107–138. Milan: Cisalpino-Goliardica.

Mormino 1990b = Mormino, Gianfranco. 1990. *'Penetralia mortus'. Studio sulla nozione di movimento nell'opera di Christiaan Huygens, con l'edizione del 'Codex Hugenianus 7 A'*. Milan: Università degli Studi di Milano. Doctoral dissertation.

Morrow 2017 = Morrow, Jeffrey L. 2017. The Acid of History: La Peyrère, Hobbes, Spinoza, and the Separation of Faith and Reason in Modern Biblical Studies. *Heythrop Journal* 57(6): 169–180.

Mouy 1934 = Mouy, Paul. 1934. *Le développement de la physique cartésienne: 1646–1712*. Paris: Vrin.

Mugnai 1990 = Mugnai, Massimo. 1990. "Necessità ex hypothesi" e analisi infinita in Leibniz. In *L'infinito in Leibniz. Problemi e terminologia. Das Unendliche bei Leibniz. Problem und Terminologie*, ed. Antonio Lamarra, 143–155. Rome: Edizioni l'Ateneo.

Mugnai 1992 = Mugnai, Massimo. 1992. *Leibniz' Theory of Relations*. Stuttgart: Steiner Verlag.

Mulsow 2002 = Mulsow, Martin. 2002. *Moderne aus dem Untergrund. Radikale Frühaufklärung in Deutschland1680–1720*. Hamburg: Meiner.

Mulsow 2009 = Mulsow, Martin (Ed.). 2009. *Spätrenaissance-Philosophie in Deutschland 1570–1650: Entwürfe zwischen Humanismus und Konfessionalisierung, okkulten Traditionen und Schulmetaphysik*. Tübingen: Niemeyer.

Mulsow 2015 = Mulsow, Martin. 2015. *Enlightenment Underground: Radical Germany, 1680–1720*. Charlottesville: University of Virginia Press.

Mulsow, Martin, and Simons, Olaf (Eds.). Forthcoming. *Gottlieb Stolle: Eine Reise durch die Gelehrtenrepublik. Reisejournal 1703/4*. Stuttgart: Steiner.

Murray et al. 2011 = Murray, Gemma, Harper, William, and Wilson, Curtis. 2011. Huygens, Wren, Wallis, and Newton on Rules of Impact and Reflection. In *Vanishing Matter and the Laws of Motion: Descartes and Beyond*, ed. Peter Anstey and Dana Jalobeanu, 153–191. London: Routledge.

Nadler 1993 = Nadler, Steven (Ed.). 1993. *Causation in Early Modern Philosophy Cartesianism, Occasionalism, and Preestablished Harmony*. University Park: Pennsylvania State University Press.

Nadler 1998 = Nadler, Steven. 1998. Louis de La Forge and the Development of Occasionalism: Continuous Creation and the Activity of the Soul. *Journal of the History of Philosophy* 36(2): 215–231.

Nadler 1999 = Nadler, Steven. 1999. Knowledge, Volitional Agency and Causation in Malebranche and Geulincx. *British Journal for the History of Philosophy* 7: 263–274.

Nadler 2000 = Nadler, Steven. 2000. Malebranche on Causation. In *The Cambridge Companion to Malebranche*, ed. Steven Nadler, 112–138. Cambridge: Cambridge University Press.

Nadler 2005 = Nadler, Steven. 2005. Cordemoy and Occasionalism. *Journal of the History of Philosophy* 43: 37–54.

Nadler 2011a = Nadler, Steven. 2011. *A Book Forged in Hell Spinoza's Scandalous Treatise and the Birth of the Secular Age*. Princeton: Princeton University Press.

Nadler 2011b = Nadler, Steven. 2011. *Occasionalism: Causation among the Cartesians*. Oxford-New York: Oxford University Press.

Nauenberg 2005 = Nauenberg, Michael. 2005. Hooke's and Newton's Contributions to the Early Development of Orbital Mechanics and Universal Gravitation. *Early Science and Medicine* 10: 518–528.
Nauta 1935 = Nauta, Doede. 1935. *Samuel Maresius*. Amsterdam: H. J. Paris.
Naylor 1980 = Naylor, Ronald H. 1980. Galileo's Theory of Projectile Motion. *Isis: A Journal of the History of Science* 71(4): 550–570.
Nejeschleba 2017 = Nejeschleba, Tomáš. 2017. The Role of Senses and Sense Perception in Valeriano Magni's Philosophy. *Filosofický časopis* 65(2): 149–163.
Newman 2006 = Newman, William. 2006. *Atoms and Alchemy: Chymistry and the Experimental Origins of the Scientific Revolution*. Chicago-London: University of Chicago Press.
Newman and Principe 2002 = Newman, William, and Principe, Lawrence. 2002. *Alchemy Tried in the Fire: Starkey, Boyle, and the Fate of Helmontian Chymistry*. Chicago: University of Chicago Press.
Newton Project website. http://www.newtonproject.ox.ac.uk/his-library/books-in-newtons-library. Accessed 13 January 2019.
Niceron 1733 = Niceron, Jean-Pierre. 1733. Burcher de Volder. In Niceron, Jean-Pierre. 1733. *Mémoires pour servir á l'histoire des hommes illustres dans la République des Lettres. Avec un catalogue raisonné de leurs ouvrages*, volume 22, 48–57. Paris: Briasson.
Nijhoff 1862 = Nijhoff, Martinus. 1862. *Catalogue d'une collection remarquable de manuscrits, d'autographes et de quelques imprimés concernant l'histoire et le droit public des Pays-Bas*. The Hague: Martinus Nijhoff.
Nollet 1743–1748 = Nollet, Jean Antoine. 1743–1748. *Leçons de physique expérimentale*. Paris: H.-L. Guerin et L. F. Delatour.
Nonnoi 1988 = Nonnoi, Giancarlo. 1988. *Il pèlago d'aria: Galileo, Baliani, Beeckman*. Rome: Bulzoni Editore.
Nonnoi 1994 = Nonnoi, Giancarlo. 1994. Against Emptiness: Descartes's Physics and Metaphysics of Plenitude. *Studies in History and Philosophy of Science Part A* 25: 81–96.
Nordström 1955 = Nordström, Johan Nils. 1995. Swammerdamiana: Excerpts from the Travel Journal of Olaus Borrichius and Two Letters from Swammerdam to Thevenot. *Lychnos* 15: 21–65.
Normore 2008 = Normore, Calvin. 2008. Descartes and the Metaphysics of Extension. In *A Companion to Descartes*, ed. Janet Broughton and John Carriero, 271–287. Oxford: Blackwell.
Nuchelmans 1973 = Nuchelmans, Gabriël. 1973. *Theories of the Proposition: Ancient and Medieval Conceptions of the Bearers of Truth and Falsity*. Amsterdam: North-Holland.
Nuchelmans 1988 = Nuchelmans, Gabriël. 1988. *Geulincx' Containment Theory of Logic*. Amsterdam: Koninklijke nederlandse Akademie van Wetenschappen.
Nuchelmans 1998 = Nuchelmans, Gabriël. 1998. Deductive Reasoning. In *The Cambridge History of Seventeenth-Century Philosophy*, ed. Daniel Garber and Michael Ayers, volume 1, 132–146. Cambridge: Cambridge University Press.
Nyden 2007 = Nyden, Tammy. 2007. *Spinoza's Radical Cartesian Mind*. New York: Continuum.
Nyden 2013 = Nyden, Tammy. 2013. De Volder's Cartesian Physics and Experimental Pedagogy. In *Cartesian Empiricisms*, ed. Mihnea Dobre and Tammy Nyden, 227–249. Dordrecht-Heidelberg-New York-London: Springer.
Nyden 2014 = Nyden, Tammy. 2014. Living Force at Leiden: De Volder, 's Gravesande and the Reception of Newtonianism. In *Newton and Newtonianism*, ed. Zvi Biener and Eric Schliesser, 207–222. Oxford: Oxford University Press.
Oertel 1842 = Oertel, Eucharius Ferdinand Christian. 1842. *Auswahl der schönsten Denk- u. Sittensprüche, Sprüchwörter, Räthsel, Buchstabenwechsel, Grabschriften etc. aus Latein. Dichtern u. Prosaikern gezogen*. Nürnberg: Campe.
Olmsted 1942 = Olmsted, John W. 1942. The Scientific Expedition of Jean Richer to Cayenne (1672–1673). *Isis: A Journal of the History of Science* 34: 117–128.
Omodeo 2017 = Omodeo, Pietro Daniel. 2017. Lodewijk de Bils' and Tobias Andreae's Cartesian Bodies: Embalmment Experiments, Medical Controversies and Mechanical Philosophy. *Early Science and Medicine* 22(4): 301–332.

Oostwoud 1754–1769 = Oostwoud, Jacob. 1754–1769. *Mathematische Liefhebberye met het nieuws der fransche en duytsche schoolen in Nederland*. Purmerende: By P. Jordaan.

Osler 2007 = Osler, Margaret J. 2008. Descartes's Optics: Light, the Eye, and Visual Perception. In *A Companion to Descartes*, ed. Janet Broughton and John Carriero, 124–141. Oxford: Blackwell.

Otterspeer 2001 = Otterspeer, Willem. 2001. The University of Leiden: An Eclectic Institution. *Early Science and Medicine* 6(4): 324–333.

Pagel 2002 = Pagel, Walter. 2002. *Joan Baptista van Helmont: Reformer of Science and Medicine*. Cambridge: Cambridge University Press.

Palaia 2012 = Palaia, Roberto. 1989. Appunti sui concetti di natura e di forza nell'epistolario fra Leibniz e Sturm. *Lexicon Philosophicum. Quaderni di terminologia filosofica e di storia delle idee* 4: 95–100.

Palladini 2011 = Palladini, Fiammetta. 2011. *Die Berliner Hugenotten und der Fall Barbeyrac. Orthodoxe und 'Sozinianer' im Refuge (1685–1720)*. Leiden: Brill.

Palmerino 2000 = Palmerino, Carla Rita. 2000. Una nuova scienza della materia per la Scienza nova del moto. La discussione dei paradossi dell'infinito nella Prima Giornata dei *Discorsi* galileiani. In *Atomismo e continuo nel XVII secolo*, ed. Egidio Festa and Romano Gatto, 276–319. Naples: Vivarium.

Palmerino 2001 = Palmerino, Carla Rita. 2001. Galileo's and Gassendi's Solutions to the *Rota Aristotelis*' Paradox: A Bridge between Matter and Motion Theories. In *Late Medieval and Early Modern Corpuscular Matter Theories*, ed. Christoph Lüthy, William R. Newman and John Emery Murdoch, 381–422. Leiden: Brill.

Palmerino 2007 = Palmerino, Carla Rita. 2007. Bodies in Water Like Planets in the Skies: Uses and Abuses of Analogical Reasoning in the Study of Planetary Motion. In *Mechanics and Cosmology in the Medieval and Early Modern Period*, ed. Massimo Bucciantini, Michele Camerota and Sophie Roux, 145–168. Florence: Leo S. Olschki.

Palmerino 2013 = Palmerino, Carla Rita. 2013. The Composition of Space, Time and Matter according to Isaac Newton and John Keill. In *The Mechanization of Natural Philosophy*, ed. Daniel Garber and Sophie Roux, 117–142. Boston: Springer.

Palmieri 2005 = Palmieri, Paolo. 2005. The Cognitive Development of Galileo's Theory of Buoyancy. *Archive for History of Exact Sciences* 59: 189–222.

Paquot 1765 = Paquot, Jean-Noël. 1765. Burcher de Volder. In Paquot, Jean-Noël. 1765. *Mémoires pour servir à l'histoire littéraire des dix-sept provinces des Pays*, volume 1, 84–86. Leuven: De l'Imprimerie Academique.

Paradís et al. 2004 = Paradís, Jaume, Pla, Josep, and Viader, Pelegrí. 2004. Fermat and the Quadrature of the Folium of Descartes. *The American Mathematical Monthly* 111(3): 216–229.

Pasnau 2011 = Pasnau, Robert. 2011. *Metaphysical Themes 1274–1671*. Oxford: Oxford University Press.

Patterson 2008 = Patterson, Sarah. 2008. Clear and Distinct Perception. In *A Companion to Descartes*, ed. Janet Broughton and John Carriero, 216–234. Oxford: Blackwell.

Pavelich 1997 = Pavelich, Andrew. 1997. Descartes's Eternal Truths and Laws of Motion. *Southern Journal of Philosophy* 35(4): 517–537.

Payne 1894 = Payne, John Bruce. 1894. Morley, Christopher Love. In Payne, John Bruce. 1894. *Dictionary of National Biography*, volume 39, 73–74. New York: MacMillan and Co.

Perler and Rudolph 2000 = Perler, Dominik, and Rudolph, Ulrich. 2000. *Occasionalismus: Theorien der Kausalität im arabisch-islamischen und im europäischen Denken*. Göttingen: Vandenhoeck & Ruprecht.

Petrescu 2013 = Petrescu, Lucian. 2013. Descartes on the Heartbeat: The Leuven Affair. *Perspectives on Science* 21(4): 397–428.

Phemister 2005 = Phemister, Pauline. 2005. *Leibniz and the Natural World: Activity, Passivity, and Corporeal Substances in Leibniz's Philosophy*. Dordrecht: Springer.

Pisano and Capecchi 2010 = Pisano, Raffaele, and Capecchi, Danilo. 2010. On Archimedean Roots in Torricelli's Mechanics. In *The Genius of Archimedes – 23 Centuries of Influence on Mathematics, Science and Engineering*, ed. Stephanos A. Paipetis and Marco Ceccarelli, 17–27. Dordrecht-New York: Springer.

Pope 2005 = Pope, Malcolm H. 2005. Giovanni Alfonso Borelli – The Father of Biomechanics. *Spine* 30: 2350–2355.

Pourciau 1991 = Pourciau, Bruce. 1991. On Newton's Proof that Inverse-Square Orbits Must be Conics. *Annals of Science* 48: 159–172.

Prager 1981 = Prager, Frank D. 1981. Berti's Devices and Torricelli's Barometer from 1641 to 1643. *Annali dell'Istituto e Museo di Storia della Scienza* 2: 35–53.

Ragland 2008 = Ragland, Evan. 2008. Experimenting with Chymical Bodies: Reinier de Graaf's Investigations of the Pancreas. *Early Science and Medicine* 13(6): 615–664.

Ragland 2012 = Ragland, Evan. 2012. Chymistry and Taste in the Seventeenth Century: Franciscus dele Boë Sylvius as a Chymical Physician between Galenism and Cartesianism. *Ambix* 59(1): 1–21.

Ragland 2016 = Ragland, Evan. 2016. Mechanism, the Senses, and Reason: Franciscus Sylvius and Leiden Debates Over Anatomical Knowledge After Harvey and Descartes. In *Early Modern Medicine and Natural Philosophy*, ed. Peter Distelzweig, Benjamin Goldberg, and Evan Ragland 173–205. Dordrecht: Springer.

Rank 1984 = Rank, Ludwig. 1984. *Die Theorie des Segelns in ihrer Entwicklung: Geschichte eines Problems der nautischen Mechanik.* Berlin: Dietrich Reimer.

Rapetti 1999 = Rapetti, Elena. 1999. *Pierre-Daniel Huet: erudizione, filosofia, apologetica.* Milan: Vita e Pensiero.

Rapetti 2003 = Rapetti, Elena. 2003. *Percorsi anticartesiani nelle lettere a Pierre-Daniel Huet.* Florence: Leo S. Olschki.

Rapetti 2018 = Rapetti, Elena. 2018. *La stanza degli specchi. Descartes e Spinoza nella corrispondenza di Pierre-Daniel Huet.* Milan: Edizioni Universitas Studiorum.

Redgrove and Redgrove 1922 = Redgrove I. M., and Redgrove, H. Stanley. 1922. *Joannes Baptista van Helmont: Alchemist, Physician and Philosopher.* London: William Rider and Son.

Redondi 1983 = Redondi, Pietro. 1983. *Galileo eretico.* Turin: Einaudi.

Redondi 1985 = Redondi, Pietro. 1985. Atomi, indivisibili e dogma. *Quaderni storici* 20: 529–571.

Reid 2014 = Reid, Jasper. 2014. Descartes and the Individuation of Bodies. *Archiv für Geschichte der Philosophie* 96(1): 38–70.

Réka 2009 = Réka, Bozzay. 2009. *Die Peregrination ungarländischer Studenten an der Universität Leiden (1595–1796).* Budapest: MTA Egyetemtörténeti Albizottsága, ELTE Levéltára.

Renckens 1966 = Renckens, B. J. A. 1966. Joost de Volder. *Oud Holland* 81: 58–59.

Renn 1992 = Renn, Jürgen. 1992. Proofs and Paradoxes: Free Fall and Projectile Motion in Galileo's Physics. In *Exploring the Limits of Preclassical Mechanics*, ed. Peter Damerow, Gideon Freudenthal, and Peter McLaughlin, 127–268. New York-Berlin: Springer.

Rey 2009a = Rey, Anne-Lise. 2009. L'ambivalence de la notion d'action dans la Dynamique de Leibniz. La correspondance entre Leibniz et De Volder (Ière Partie). *Studia Leibnitiana* 41(1): 47–66.

Rey 2009b = Rey, Anne-Lise. 2009. L'ambivalence de la notion d'action dans la Dynamique de Leibniz. La correspondance entre Leibniz et De Volder (IIe Partie). *Studia Leibnitiana* 41(2): 157–182.

Rey 2016 = Rey, Anne-Lise (Ed.). 2016. *Correspondance – Gottfried Wilhelm Leibniz, Burcher De Volder.* Paris: Vrin.

Ribe 1997 = Ribe, Neil M. 1997. Cartesian Optics and the Mastery of Nature. *Isis: A Journal of the History of Science* 88(1): 42–61.

Rijks 2012 = Rijks, Marlise. 2012. *The Correspondence of Dirck Rembrantsz van Nierop (1610–1682).* The Hague: Dutch History of Science Web Centre – Huygens ING.

Roberts 2003 = Roberts, John. 2003. Leibniz on Force and Absolute Motion. *Philosophy of Science* 70(3): 553–573.

Rosenfeld 1969 = Rosenfeld, Leon. 1969. Newton's Views on Aether and Gravitation. *Archive for History of Exact Sciences* 6: 29–37.

Rossi 1968 = Rossi, Paolo. 1968. *Francis Bacon: From Magic to Science*. Chicago: University of Chicago Press. First edition 1957.

Rothschuh 1953 = Rothschuh, Karl E. 1953. *Geschichte der Physiologie*. Berlin-Göttingen-Heidelberg: Springer.

Rothschuh 1968 = Rothschuh, Karl E. 1968. Henricus Regius und Descartes. Neue Einblicke in die frühe Physiologie (1640–1641) des Regius. *Archives internationales d'histoire des sciences* 21: 39–66.

Rousset 1999 = Rousset, Bernard. 1999. *Geulincx entre Descartes et Spinoza*. Paris: Vrin.

Roux 1998 = Roux, Sophie. 1998. Le scepticisme et les hypothèses de la physique. *Revue de synthèse* 119(2–3): 211–255.

Roux 2006 = Roux, Sophie. 2006. La philosophie naturelle à l'époque de Le Nôtre. Remarques sur la philosophie mécanique et sur le cartésianisme. In *André Le Nôtre, fragments d'un paysage culturel. Institutions, arts, sciences et techniques*, ed. Georges Farhat, 98–111. Sceaux: Musée de l'Île-de-France.

Roux 2012a = Roux, Sophie. 2012. *L'*Essai de logique *de Mariotte. Archéologie des idées d'un savant ordinaire*. Paris: Classiques Garnier.

Roux 2012b = Roux, Sophie. 2012. Quelles Machines pour quels animaux? Jacques Rohault, Claude Perrault, Giovanni Alfonso Borelli. In *L'Automate: Machine, Modèle, Merveille*, ed. Aurélia Gaillard, Jean-Yves Goffi, Bernard Roukhomovsky, and Sophie Roux 69–113. Bordeaux: Presses Universitaires de Bordeaux.

Roux 2013a = Roux, Sophie. 2013. Was There a Cartesian Experimentalism in 1660s France? In *Cartesian Empiricisms*, ed. Mihnea Dobre and Tammy Nyden, 47–88. Dordrecht-Heidelberg-New York-London: Springer.

Roux 2013b = Roux, Sophie. 2013. An Empire Divided: French Natural Philosophy (1670–1690). In *The Mechanization of Natural Philosophy*, ed. Daniel Garber and Sophie Roux, 55–95. Boston: Springer.

Ruestow 1973 = Ruestow, Edward G. 1973. *Physics at Seventeenth and Eighteenth-Century Leiden: Philosophy and the New Science in the University*. The Hague: Martinus Nijhoff.

Ruestow 1996 = Ruestow, Edward G. 1996. *The Microscope in the Dutch Republic: The Shaping of Discovery*. New York: Cambridge University Press.

Russell 1927–1928 = Russell, Leonard James. 1927–1928. The Correspondence between Leibniz and De Volder. *Proceedings of the Aristotelian Society*, 28, 155–176.

Russo 2013 = Russo, Lucio. 2013. *The Forgotten Revolution: How Science Was Born in 300 BC and Why it Had to Be Reborn*. Berlin: Springer. First edition 1996.

Rutherford 1995 = Rutherford, Donald. 1995. *Leibniz and the Rational Order of Nature*. Cambridge: Cambridge University Press.

Sabra 1981 = Sabra, Abdelhamid. 1981. *Theories of Light: From Descartes to Newton*. Cambridge: Cambridge University Press.

Sarkar 2003 = Sarkar, Husain. 2003. *Descartes'* Cogito*: Saved from the Great Shipwreck*. Cambridge: Cambridge University Press.

Sasaki 2003 = Sasaki, Chikara. 2003. *Descartes's Mathematical Thought*. Dordrecht: Springer.

Sassen 1970 = Sassen, Saskia. 1970. The Intellectual Climate in Leiden in Boerhaave's Time. In *Boerhaave and his Time*, ed. Gerrit Arie Lindeboom, 1–16. Leiden: Brill.

Savini 2011 = Savini, Massimiliano. 2011. *Johannes Clauberg:* Methodus cartesiana *et ontologie*. Paris: Vrin.

Sayre-McCord 1984 = Sayre-McCord, Geoffrey. 1984. Leibniz, Materialism, and the Relational Account of Space and Time. *Studia Leibnitiana*, 16, 204–211.

Schliesser 1997 = Schliesser, Eric. 1997. Van Nierops paskaart van Europa in een rapport van Christiaan Huygens. *Caert-Tresoor* 16: 93–96.

Schliesser 2000 = Schliesser, Eric. 2000. Een kaart van Visscher en de grote globe van Blaeu in de bewijsvoering van een proef met slingeruurwerken van Huygens. *Caert-Tresoor* 19(2): 51–55.

Schliesser and Smith 1996 = Schliesser, Eric, and Smith, George E. 1996. Huygens's 1688 Report to the Directors of the Dutch East India Company. *De Zeventiende Eeuw* 12: 198–214.

Schliesser and Smith 2000 = Schliesser, Eric, and Smith, George E. 2000. *Huygens's 1688 Report to the Directors of the Dutch East India Company on the Measurement of Longitude at Sea and the Evidence it Offered Against Universal Gravity*. PhilSci archive. http://philsci-archive.pitt.edu/5510/1/gseric.pdf. Accessed 4 February 2019.

Schmaltz 2002 = Schmaltz, Tad. 2002. *Radical Cartesianism: The French Reception of Descartes*. Cambridge: Cambridge University Press.

Schmaltz 2005a = Schmaltz, Tad (Ed.). 2005. *Receptions of Descartes. Cartesianism and Anti-Cartesianism in Early Modern Europe*. London-New York: Routledge.

Schmaltz 2005b = Schmaltz, T. 2005. French Cartesianism in Context: The Paris Formulary and Regis's *Usage*. In *Receptions of Descartes. Cartesianism and Anti-Cartesianism in Early Modern Europe*, ed. Tad Schmaltz, 73–89. London-New York: Routledge.

Schmaltz 2008 = Schmaltz, Tad. 2008. *Descartes on Causation*. Oxford: Oxford University Press.

Schmaltz 2015 = Schmaltz, Tad. 2015. The Metaphysics of Rest in Descartes and Malebranche. *Res Philosophica* 92(1): 21–40.

Schmaltz 2016 = Schmaltz, Tad. 2016. *Early Modern Cartesianisms: Dutch and French Constructions*. New York: Oxford University Press.

Schneider 2013 = Schneider, Daniel. 2013. *Otto von Guericke – Ein Leben für die Alte Stadt Magdeburg*. Stuttgart-Leipzig: B. G. Teubner.

Schneider 2016 = Schneider, Daniel. 2016. Spinoza's Epistemological Methodism. *Journal of the History of Philosophy* 54(4): 573–599.

Schoneveld 1983 = Schoneveld, Cornelis W. 1983. *Intertraffic of the Mind. Studies in Seventeenth-Century Anglo-Dutch Translation*. Leiden: Brill.

Schouls 2000 = Schouls, Peter A. 2000. *Descartes and the Possibility of Science*. Ithaca: Cornell University Press.

Schüller 2001 = Schüller, Volkmar. 2001. Samuel Clarke's Annotations in Jacques Rohault's *Traité de Physique*, and How They Contributed to Popularising Newton's Physics. In *Between Leibniz, Newton, and Kant: Philosophy and Science in the Eighteenth Century*, ed. Wolfgang Lefèvre, 95–110. Dordrecht: Springer.

Schuster 1993 = Schuster, John. 1993. Whatever Should We Do with Cartesian Method? Reclaiming Descartes for the History of Science. In *Essays on the Philosophy and Science of René Descartes*, ed. Stephen Voss, 195–223. New York: Oxford University Press.

Schuster 2013 = Schuster, John. 2013. *Descartes-Agonistes. Physico-mathematics, Method & Corpuscular-Mechanism 1618–33*. Dordrecht-Heidelberg-New York-London: Springer.

Schuurman 2004 = Schuurman, Paul. 2004. *Ideas, Mental Faculties and Method: The Logic of Ideas of Descartes and Locke and its Reception in the Dutch Republic, 1630–1750*. Leiden-Boston: Brill.

Scribano 1988 = Scribano, Emanuela. 1988. *Da Descartes a Spinoza. Percorsi della teologia razionale nel Seicento*. Milan: Franco Angeli.

Sdzuj 2005 = Sdzuj, Reimund. 2005. 'Adamus in filiis lucis non peccavit.' Die ersten Reaktionen der reformierten Orthodoxie auf Lodewijk Meyers Programmschrift *Philosophia sacrae scripturae interpres* (1666). In *Geschichte der Hermeneutik und die Methodik der textinterpretierenden Disziplinen*, ed. Jorg Schonert, Friedrich Vollhardt, 157–186. Berlin: Walter de Gruyter.

Secundant 2016 = Secundant, Sergei. 2016. Johann Christoph Sturm: Eclecticism as a Philosophical Ideology and Methodological Program. *Epistemology & Philosophy of Science* 48(2): 104–120.

Shafer 1996 = Shafer, Glenn. 1996. The Significance of Jacob Bernoulli's *Ars Conjectandi* for the Philosophy of Probability Today. *Journal of Econometrics* 57(1): 15–32.

Shank 2008 = Shank, John Bennett. 2008. *The Newton Wars and the Beginning of the French Enlightenment*. Chicago: University of Chicago Press.

Shank 2012 = Shank, John Bennett. 2012. What Exactly Was Torricelli's "Barometer?" In Ofer Gal, and Raz Chen-Morris (Eds.), *Science in the Age of Baroque*, 161–196. Dordrecht-New York: Springer.

Shapin and Schaffer 1985 = Shapin, Steven, and Schaffer, Simon. 1985. *Leviathan and the Air-Pump: Hobbes, Boyle, and the Experimental Life*. Princeton: Princeton University Press.

Shea 1989 = Shea, William R. 1989. Galileo's Atomic Hypothesis. *Ambix* 17: 13–27.
Shea 2003 = Shea, William R. 2003. *Designing Experiments & Games of Chance: The Unconventional Science of Blaise Pascal*. Canton, MA: Science History Publications.
Shea 2004 = Shea, William R. 2004. The "Rational" Descartes and the "Empirical" Galileo. In *The Reception of the Galilean Science of Motion in Seventeenth-Century Europe*, ed. Carla Rita Palmerino and J. M. M. Hans Thijssen, 67–82. Dordrecht: Kluwer.
Shelford 2002 = Shelford, April. 2002. Thinking Geometrically in Pierre-Daniel Huet's "Demonstratio evangelica" (1679). *Journal of the History of Ideas* 63: 599–618.
Shelford 2007 = Shelford, April. 2007. *Transforming the Republic of Letters: Pierre-Daniel Huet and European Intellectual Life, 1650–1720*. Rochester: University Rochester Press.
Sierksma 1992 = Sierksma, Gerard. 1992. Johann Bernoulli (1667–1748): His Ten Turbulent Years in Groningen. *The Mathematical Intelligencer* 14(4): 22–31.
Sierksma and Sierksma 1999 = Sierksma, Gerard, and Sierksma, Wybe. 1999. The Great Leap to the Infinitely Small. Johann Bernoulli: Mathematician and Philosopher. *Annals of Science* 56(4): 433–449.
Simonutti 1984 = Simonutti, Luisa (Ed.). 1984. *Arminianesimo e tolleranza nel Seicento olandese. Il carteggio Ph. van Limborch – J. Le Clerc*. Florence: Leo S. Olschki.
Simonutti 1990 = Simonutti, Luisa. 1990. Reason and Toleration: Henry More and Philip van Limborch. In *Henry More (1614–1687). Tercentenary Studies*, ed. Sarah Hutton, 201–218. Dordrecht: Kluwer.
Simonutti 2002 = Simonutti, Luisa. 2016. Between History and Politics: Philipp van Limborch's History of the Inquisition. In *Histories of Heresy in Early Modern Europe. For, Against, and Beyond Persecution and Toleration*, ed. John Christian Laursen, 101–117. New York-Basingstoke: Palgrave.
Slaby 1907 = Slaby, Adolf (Ed.). 1907. *Otto von Guericke. Festvortrag aus Anlass der Grundsteinlegung des Deutschen Museums zu München gehalten im Wittelsbach-Palais am 13. November 1906*. Berlin-Heidelberg: Springer.
Slowik 1996 = Slowik, Edward. 1996. Perfect Solidity: Natural Laws and the Problem of Matter in Descartes' Universe. *History of Philosophy Quarterly* 13: 187–204.
Slowik 1997 = Slowik, Edward. 1997. Huygens' Center-of-Mass Space-Time Reference Frame: Constructing a Cartesian Dynamics in the Wake of Newton's "De Gravitatione" Argument. *Synthese* 112: 247–269.
Slowik 2001 = Slowik, Edward. 2001. Descartes and Individual Corporeal Substance. *British Journal for the History of Philosophy* 9: 1–15.
Slowik 2002 = Slowik, Edward. 2002. *Cartesian Spacetime. Descartes' Physics and the Relational Theory of Space and Motion*. Dordrecht: Kluwer Academic Publishers.
Smith 1987 = Smith, A. Mark. 1987. Descartes's Theory of Light and Refraction: A Discourse on Method. *Transactions of the American Philosophical Society* 77(3): 1–92.
Smith 2010 = Smith, Kurt. 2010. *Matter Matters: Metaphysics and Methodology in the Early Modern Period*. Oxford: Oxford University Press.
Smith 2013 = Smith, Justin E. H. 2013. Heat, Action, Perception: Models of Living Beings in German Medical Cartesianism. In *Cartesian Empiricisms*, ed. Mihnea Dobre and Tammy Nyden, 105–123. Dordrecht-Heidelberg-New York-London: Springer.
Smith and Nachtomy 2011 = Smith, Justin E. H., and Nachtomy, Ohad. 2011. *Machines of Nature and Corporeal Substances in Leibniz*. Dordrecht: Springer.
Snelders 1989 = Snelders, Harry A. 1989. Christiaan Huygens and Newton's Theory of Gravitation. *Notes and Records of the Royal Society of London* 43: 209–222.
Snow 1924 = Snow, Adolph Judah. 1924. Newton's Objections to Descartes's Astronomy. *The Monist* 34(4): 543–557.
Sorell 1993 = Sorell, Tom (Ed.). 1993. *The Rise of Modern Philosophy: The Tension between the New and Traditional Philosophies from Machiavelli to Leibniz*. Oxford: Clarendon.
Sorell 2005 = Sorell, Tom. 2005. *Descartes Reinvented*. Cambridge: Cambridge University Press.

Sorell 2010 = Sorell, Tom. 2010. *Scientia* and Sciences in Descartes. In Scientia *in Early Modern Philosophy. Seventeenth-Century Thinkers on Demonstrative Knowledge from First Principles*, ed. Tom Sorell, John G. A. Rogers, and Jill Kraye, 71–82. Dordrecht-Heidelberg-London-New York: Springer.

Spink 2018 = Spink, Aaron. 2018. The experimental physics of Jacques Rohault. *British Journal for the History of Philosophy* 26(5), 850–870.

Stanglin 2016 = Stanglin, Keith. 2016. Arminian, Remonstrant, and Early Methodist Theologies. In *The Oxford Handbook of Early Modern Theology, 1600–1800*, ed. Ulrich L. Lehner, Richard A. Muller, and A. G. Roeber, 387–401. New York: Oxford University Press.

State 1991 = State, Stephen A. 1991. *Thomas Hobbes and the Debate over Natural Law and Religion*. London: Routledge.

Stechow 1966 = Stechow, Wolfgang. 1966. *Dutch Landscape Painting of the Seventeenth Century*. London: Phaidon.

Steenbakkers 1997 = Steenbakkers, Piet. 1997. Johannes Braun (1628–1708), Cartesiaan in Groningen. *Nederlands Archief voor Kerkgeschiedenis/Dutch Review of Church History* 77(2): 196–210.

Strazzoni 2011 = Strazzoni, Andrea. 2011. La filosofia aristotelico-cartesiana di Johannes De Raey. *Giornale critico della filosofia italiana* 7(1): 107–132.

Strazzoni 2012 = Strazzoni, Andrea. 2012. The Dutch Fates of Bacon's Philosophy: *Libertas Philosophandi*, Cartesian Logic and Newtonianism. *Annali della Scuola Normale Superiore di Pisa – Classe di Lettere e Filosofia* 4(1): 251–281.

Strazzoni 2014 = Strazzoni, Andrea. 2014. The Crypto-Dualism of Henricus Regius. In *Individuazione, individualità, identità personale*, ed. Stefano Caroti and Mariafranca Spallanzani, 133–151. Florence: Le Lettere.

Strazzoni 2015 = Strazzoni, Andrea. 2015. The Cartesian Philosophy of Language of Johannes de Raey. *Lias. Journal of Early Modern Intellectual Culture and its Sources* 42(2): 89–120.

Strazzoni 2018a = Strazzoni, Andrea. 2018. *Dutch Cartesianism and the Birth of Philosophy of Science: From Regius to 's Gravesande*. Berlin-Boston: Walter de Gruyter.

Strazzoni 2018b = Strazzoni, Andrea. 2018. How Did Regius Become Regius? The Early Doctrinal Evolution of a Heterodox Cartesian. *Early Science and Medicine* 23(4): 362–412.

Strazzoni 2018c = Strazzoni, Andrea. 2018. The Medical Cartesianism of Henricus Regius. Disciplinary Partitions, Mechanical Reductionism and Methodological Aspects. *Galilæana. Studies in Renaissance and Early Modern Science* 15: 181–220.

Strazzoni 2019 = Strazzoni, Andrea. 2019. The Letters of Burchard de Volder to Philipp van Limborch. *Noctua* 5(2): 268–298.

Suitner 2016 = Suitner, Riccarda. 2016. *Die philosophischen Totengespräche der Frühaufklärung*. Hamburg: Meiner.

Taliaferro 1986 = Taliaferro, Charles. 1986. Dualism and the Problem of Individuation. *Religious Studies* 22: 263–276.

Taton 1963 = Taton, René. 1963. L'annonce de l'experience barometrique en France. *Revue d'histoire des sciences et de leurs applications* 16: 77–83.

Theis et al. 2009 = Theis, Robert, Ferrari, Jean, Guenancia, Pierre, Ruffing, Margit, and Vollet, Matthias (Eds.). 2009. *Descartes und Deutschland – Descartes et l'Allemagne*. Hildesheim: Olms.

Thiel 2011 = Thiel, Udo. 2011. *The Early Modern Subject. Self-Consciousness and Personal Identity from Descartes to Hume*. New York: Cambridge University Press.

Thijssen 1992 = Thijssen, J. M. M. Hans 1992. David Hume and John Keill and the Structure of Continua. *Journal of the History of Ideas* 53: 271–286.

Thijssen-Schoute 1954 = Thijssen-Schoute, C. Louise. 1954. *Nederlands cartesianisme, avec sommaire et table des matières en français*. Amsterdam: Noord-Hollandsche Uitg. Mij.

Thijssen-Schoute 1960 = Thijssen-Schoute, C. Louise. 1960. Hermanus Lufneu, Stadarts te Rotterdam. *Rotterdams Jaarboekje* 8: 180–227.

Thirion 1907–1909 = Thirion, Jean. 1907–1909. Pascal, l'horreur du vide et la pression atmosphérique. *Revue des questions scientifiques* 12–14.

Tho 2017 = Tho, Tzuchien. 2017. Vis Vim Vi*: Declinations of Force in Leibniz's Dynamics*. Cham: Springer.

Thomas 2015 = Thomas, Emily. 2015. In Defence of Real Cartesian Motion: A Reply to Lennon. *Journal of the History of Philosophy* 53: 747–762.

Thurston 1999 = Thurston, Alan J. 1999. Giovanni Borelli and the Study of Human Movement: An Historical Review. *Australian and New Zealand Journal of Surgery* 69: 276–288.

Torero-Ibad 2011 = Torero-Ibad, Alexandra. 2011. Fictions of the World in Descartes' and Gassendi's Physics. *Society and Politics* 5(2): 75–87.

Torrini 1977 = Torrini, Maurizio. 1977. *Tommaso Cornelio e la ricostruzione della scienza*. Naples: Guida.

Touber 2018 = Touber, Jetze. 2018. *Spinoza and Biblical Philology in the Dutch Republic, 1660–1710*. Oxford: Oxford University Press.

Trevisani 1992 = Trevisani, Francesco. 1992. *Descartes in Germania: La ricezione del cartesianesimo nella Facoltà filosofica e medica di Duisburg (1652–1703)*. Milan: Franco Angeli.

Ueberweg 1866 = Ueberweg, Friedrich. 1866. *Grundriss der Geschichte der Philosophie*. Berlin: Mittler.

Underwood 1972 = Underwood, Edgar A. 1972. Franciscus Sylvius and His Iatrochemical School. *Endeavour* 31(113): 73–76.

Valleriani 2010 = Valleriani, Matteo. 2010. *Galileo Engineer*. Dordrecht: Springer.

Valleriani 2014 = Valleriani, Matteo. 2014. Sixteenth-Century Hydraulic Engineers and the Emergence of Empiricism. In *Conflicting Values of Inquiry: Ideologies of Epistemology in Early Modern Europe*, ed. Claus Zittel, Tamás Demeter, and Kathryn Murphy, 39–68. Leiden: Brill.

Van Asselt and Dekker 1995 = Van Asselt, Willem J., and Dekker, Eef (Eds.). *De scholastieke Voetius: Een luisteroefening aan de hand van Voetius'* Disputationes selectae. Zoetermeer: Boekencentrum.

Van Berkel 1999a = Van Berkel, Klaas. 1999. Burchardus de Volder 1643–1709. In *A History of Science in the Netherlands: Surveys, Themes and Reference*, ed. Klaas van Berkel, Albert van Helden, and Lodewijk Palm, 589–591. Leiden: Brill.

Van Berkel 1999b = Van Berkel, Klaas. 1999. Johannes Hudde 1628–1704. In *A History of Science in the Netherlands: Surveys, Themes and Reference*, ed. Klaas van Berkel, Albert van Helden, and Lodewijk Palm, 476–478. Leiden: Brill.

Van Berkel 2013 = Van Berkel, Klaas. 2013. *Isaac Beeckman on Matter and Motion: Mechanical Philosophy in the Making*. Baltimore: Johns Hopkins University Press.

Van Berkel et al. 1999 = Van Berkel, Klaas, Van Helden, Albert, and Palm, Lodewijk (Eds.). 1999. *A History of Science in the Netherlands: Surveys, Themes and Reference*. Leiden: Brill.

Van Besouw 2016 = Van Besouw, Jip. 2016. The Impeccable Credentials of an Untrained Philosopher: Willem Jacob 's Gravesande's Career before his Leiden Professorship, 1688–1717. *Notes and Records: The Royal Society Journal of the History of Science* 70(3): 231–249.

Van Bunge 1989 = Van Bunge, Wiep. 1989. On the Early Dutch Reception of the *Tractatus Theologico-Politicus*. *Studia Spinozana* 5: 225–251

Van Bunge 1990 = Van Bunge, Wiep. 1990. *Johannes Bredenburg (1643–1691). Een Rotterdamse collegiant in de ban van Spinoza*. Rotterdam: Erasmus University Rotterdam. Doctoral dissertation.

Van Bunge 1995 = Van Bunge, Wiep. 1995. Van Velthuysen, Batelier and Bredenburg on Spinoza's Interpretation of the Scriptures. In *The Spinozistic Heresy. The Debate on the Tractatus Theologico-Politicus, 1670–1677*, ed. Paolo Cristofolini, 49–65. Amsterdam-Maarssen: APA, Holland University Press.

Van Bunge 2001 = Van Bunge, Wiep. 2001. *From Stevin to Spinoza. An Essay on Philosophy in the Seventeenth-Century Dutch Republic*. Leiden-Boston-Cologne: Brill.

Van Bunge 2012 = Van Bunge, Wiep. 2012. *Spinoza Past and Present: Essays on Spinoza, Spinozism, and Spinoza Scholarship*. Leiden: Brill.
Van Bunge 2013 = Van Bunge, Wiep. 2013. Dutch Cartesian Empiricism and the Advent of Newtonianism. In *Cartesian Empiricisms*, ed. Mihnea Dobre and Tammy Nyden, 89–104. Dordrecht-Heidelberg-New York-London: Springer.
Van Bunge 2017 = Van Bunge, Wiep. 2017. The Waning of the Radical Enlightenment in the Dutch Republic. In *Reassessing the Radical Enlightenment*, ed. Steffen Ducheyne, 178–193. London: Routledge.
Van Bunge 2018 = Van Bunge, Wiep. 2018. *From Bayle to the Batavian Revolution: Essays on Philosophy in the Eighteenth- Century Dutch Republic*. Leiden: Brill.
Van der Aa 1876 = Van der Aa, Abraham Jacob. 1876. Volder, Burchardus de. In Van der Aa, Abraham Jacob. 1876. *Biographisch woordenboek der Nederlanden*, volume 19, 316. Haarlem: J. J. Van Brederode.
Van der Schoot 2011 = Van der Schoot, Joas. 2011. *The Necessity of the Universe. Christiaan Huygens's* Cosmotheoros *(1698) and Its Philosophical and Literary Backgrounds: An Intellectual Testament*. Utrecht: University of Utrecht. Master's dissertation.
Van der Schoot 2014 = Van der Schoot, Joas. 2014. Interpreting the *Kosmotheoros* (1698). A Historiographical Essay on Theology and Philosophy in the Work of Christiaan Huygens. *De Zeventiende Eeuw* 30(1): 20–39.
Van der Vijver 1844 = Van der Vijver, Cornelis. 1844. Burchard de Volder. In Van der Vijver, Cornelis. 1844. *Geschiedkundige beschrijving der stad Amsterdam*, volume 1, 385. Amsterdam: Gebrüder Dietrichs.
Van der Wall 1996 = Van der Wall, Ernestine. 1996. Cartesianism and Cocceianism: A Natural Alliance? In *De l'humanisme aux Lumierès, Bayle et le protestantisme*, ed. Michelle Magdelaine, 445–455. Paris-Oxford: Voltaire Foundation.
Van der Willigen 1870 = Van der Willigen, Adriaan. 1870. *Les artistes de Harlem: Notices historiques avec un precis sur la gilde de St. Luc*. Haarlem: Les Heritiers F. Bohn. The Hague: Martinus Nijhoff.
Van Eijnatten 2003 = Van Eijnatten, Joris. 2003. The Huguenot Clerisy in the United Provinces: Aspects of Huguenot Influence on Dutch Intellectual Life After the Revocation. In *The Berlin Refuge 1680–1780. Learning and Science in European Context*, ed. Sandra Pott, Martin Mulsow and Lutz Danneberg, 207–235. Leiden-Cologne-New York: Brill.
Van Helden 1991 = Van Helden, Albert. 1991. The Age of the Air-Pump. *Tractrix* 3: 149–172.
Van Hoogstraten and Schuer 1733 = Van Hoogstraten, David, and Schuer, Jan Lodewyk. 1733. Volder (Burchard van). In Van Hoogstraten, David, and Schuer, Jan Lodewyk. 1733. *Groot algemeen woorden-boek, zo historisch, geografisch, genealogisch en Oordeelkundig Woordenboek*, volume 10, 117–118. Amsterdam-Utrecht-Den Haag: Brunel & Wetsteins.
Van Maanen 1993 = Van Maanen, Jan A. 1993. Johann Bernoulli, Man of Contrasts. *Nieuw archief voor wiskunde* 4(11): 241–246.
Van Miert 2009 = Van Miert, Dirk. 2009. *Humanism in an Age of Science: The Amsterdam Athenaeum in the Golden Age, 1632–1704*. Leiden-Boston: Brill.
Van Miert 2018a = Van Miert, Dirk. 2018. Education. In *The Cambridge Companion to the Dutch Golden Age*, ed. Helmer J. Helmers and Geert H. Janssen, 333–349. Cambridge: Cambridge University Press.
Van Miert 2018b = Van Miert, Dirk. 2018. *The Emancipation of Biblical Philology in the Dutch Republic, 1590–1670*. Oxford: Oxford University Press.
Van Miert et al. 2017 = Van Miert, Dirk, Nellen, Henk, Steenbakkers, Piet, and Touber, Jetze (Eds.). 2017. *Scriptural Authority and Biblical Criticism in the Dutch Golden Age*. Oxford: Oxford University Press.
Van Rooden and Wesselius 1987 = Van Rooden, Peter, and Wesselius, Jan Wim. 1987. The Early Enlightenment and Judaism: The Civil Dispute between Philippus van Limborch and Isaac Orobio de Castro (1687). *Studia Rosenthaliana* 21: 140–153.

Van Ruler 1995 = Van Ruler, Han. 1995. *The Crisis of Causality. Voetius and Descartes on God, Nature and Change*. Leiden-New York-Cologne: Brill.
Van Ruler 1999 = Van Ruler, Han. 1999. "Something, I know not what". The Concept of Substance in Early Modern Thought. In *Between Imagination and Demonstration. Essays in the History of Science and Philosophy Presented to John D. North*, ed. Lodi Nauta and Ario Vanderjagt, 365–393. Leiden: Brill.
Van Ruler 2000 = Van Ruler, Han. 2000. Minds, Forms, and Spirits: The Nature of Cartesian Disenchantment. *Journal of the History of Ideas* 61: 381–395.
Van Ruler 2003 = Van Ruler, Han. 2003. The Shipwreck of Belief and Eternal Bliss: Philosophy and Religion in Later Dutch Cartesianism. In *The Early Enlightenment in the Dutch Republic, 1650–1750*, ed. Wiep van Bunge, 109–136. Leiden-Boston: Brill.
Vanpaemel 1984 = Vanpaemel, Geert. 1984. Rohault's *Traité de physique* and the Teaching of the Cartesian Physics. *Janus: Revue Internationale de l'Histoire des Sciences, de la Médecine et de la Technique* 71: 31–40.
Vanpaemel 1985 = Vanpaemel, Geert. 1984. *De mechanistische natuurwetenschap aan de Leuvense Artesfakulteit (1650–1797)*. Leuven: Katholieke Universiteit.
Verbeek 1989 = Verbeek, Theo. 1989. Les passions et la fièvre. L'idée de la maladie chez Descartes et quelques cartésiens néerlandais. *Tractrix* 1: 45–61.
Verbeek 1992a = Verbeek, Theo. 1992. *Descartes and the Dutch: Early Reactions to Cartesian Philosophy, 1637–1650*. Carbondale-Edwardsville: Southern Illinois University Press.
Verbeek 1992b = Verbeek, Theo. 1992. "Ens per accidens": Le origini della *Querelle* di Utrecht. *Giornale critico della filosofia italiana* 71: 276–288.
Verbeek 2000 = Verbeek, Theo. 2000. The Invention of Nature. Descartes and Regius. In *Descartes' Natural Philosophy*, ed. Stephen Gaukroger, John Schuster, and John Sutton, 149–167. London: Routledge.
Verbeek 2001 = Verbeek, Theo. 2001. Notes on Ramism in the Netherlands. In *The Influence of Petrus Ramus*, ed. Mordechai Feingold, Joseph S. Freedman, and Wolfgang Rother, 38–53. Basel: Schwabe.
Verhoeven 1973 = Verhoeven, Cornelis. 1973. *Het axioma van Geulincx*. Bilthoven: Ambo.
Vermij 1991 = Vermij, Rienk H. 1991. *Secularisering en natuurwetenschap in de zeventiende en achttiende eeuw: Bernard Nieuwentijt*. Amsterdam: Rodopi.
Vermij 2002 = Vermij, Rienk H. 2002. *The Calvinist Copernicans. The Reception of the New Astronomy in the Dutch Republic, 1575–1750*. Amsterdam: Royal Netherlands Academy of Arts and Sciences.
Vermij 2003 = Vermij, Rienk H. 2003. The Formation of the Newtonian Philosophy: The Case of the Amsterdam Mathematical Amateurs. *The British Journal for the History of Science* 36(2): 183–200.
Vermij and Atzema 1995 = Vermij, Rienk H., and Atzema, Eisso. 1995. *Specilla Circularia*: An Unknown Work by Johannes Hudde. *Studia Leibnitiana* 27(1): 104–121.
Vongsuravatana 1992 = Vongsuravatana, Raphaël. 1992. *Un jésuite à la Cour de Siam*. Paris: Éditions France-Empire.
Voss 1993 = Voss, Stephen (Ed.). 1993. *Essays on the Philosophy and Science of René Descartes*. New York: Oxford University Press.
Wagenaar 1767 = Wagenaar, Jan. 1767. Volder (Burchard de). In Wagenaar, Jan. 1767. *Amsterdam in zyne Opkomst*, volume 11, 336. Amsterdam: Yntema en Tieboel.
Wagenaar 1790 = Wagenaar, Jan. 1790. Volder (Burchard de). In Wagenaar, Jan. 1790. *Beschrijving van Amsterdam*, volume 3, 237–238. Amsterdam: By J. W. Elwe.
Walsh 2018 = Walsh, Julie. 2018. Locke's Last Word on Freedom: Correspondence with Limborch. *Res Philosophica* 95(4): 637–661.
Water 1802 = Water, Jona Willem te. 1802. *Narratio de rebus Academiae Lugduno Batavae, seculo octavo et decimo, prosperis et adversis. Accedunt B. Vulcanii consilium de studio medicinae. Auctarium legati Papenbroekiani. Series curatorum et professorum Academiae Lugduno Batavae, seculo XVIII*. Leiden: Apud S. et I. Luchtmans.

Watson 1966 = Watson, Richard A. 1966. *The Downfall of Cartesianism, 1673–1712: A Study of Epistemological Issues in Late 17th Century Cartesianism*. The Hague: Martinus Nijhoff.

Watson 1998 = Watson, Richard A. 1998. *The Breakdown of Cartesian Metaphysics*. Atlantic Islands: Humanities Press International.

Webster 1965 = Webster, Charles. 1965. The Discovery of Boyle's Law, and the Concept of the Elasticity of Air in the Seventeenth Century. *Archive for the History of the Exact Sciences* 2: 441–502.

Webster 1969 = Webster, Charles. 1969. Henry More and Descartes: Some New Sources. *The British Journal for the History of Science* 4: 359–377.

Weinstock 1992 = Weinstock, Robert. 1992. Newton's *Principia* and Inverse-Square Orbits: the Flaw Reexamined. *Historia Mathematica* 19(1): 60–70.

Weller 2009 = Weller, Dennis. 2009. *Seventeenth-Century Dutch and Flemish Paintings*. Raleigh: North Carolina Museum of Art.

West 1998 = West, John B. 1998. *High Life: A History of High-Altitude Physiology and Medicine*. New York: Oxford University Press.

West 2013 = West, John B. 2013. Torricelli and the Ocean of Air: The First Measurement of Barometric Pressure. *Physiology* 28(2): 66–73.

Westfall 1971 = Westfall, Richard. 1971. *Force in Newton's Physics. The Science of Dynamics in the Seventeenth Century*. London: Macdonald; New York: Elsevier.

Westfall 1981 = Westfall, Richard. 1981. *Never at Rest: A Biography of Isaac Newton*. Cambridge: Cambridge University Press.

Wiesenfeldt 2000 = Wiesenfeldt, Gerhard. 2000. Die 'Freiheit beim Philosophieren' und das *Theatrum Physicum* in Leiden: über die Integration experimenteller Naturlehre in die Universitätskultur des 17. Jahrhunderts. In *Instrument – Experiment: historische Studien*, ed. Christoph Meinel, 269–277. Berlin: GNT-Verlag.

Wiesenfeldt 2002 = Wiesenfeldt, Gerhard. 2002. *Leerer Raum in Minervas Haus. Experimentelle Naturlehre an der Universität Leiden, 1675–1715*. Amsterdam: Koninklijke Nederlandse Akademie van Wetenschappen.

Wiesenfeldt 2003 = Wiesenfeldt, Gerhard. 2003. Volder, Burchard de (1643–1709). In *The Dictionary of Seventeenth and Eighteenth-Century Dutch Philosophers*, ed. Wiep van Bunge, Henri Krop, Bart Leeuwenburgh, Han van Ruler, Paul Schuurman, and Michiel Wielema, volume 2, 1041–1044. Bristol: Thoemmes Press.

Wiesenfeldt 2004 = Wiesenfeldt, Gerhard. 2004. Das *Collegium experimentale sive curiosum* und die Anfänge experimenteller Naturlehre in Deutschland. *Acta Historica Astronomiae* 22: 184–202.

Wiesenfeldt 2008 = Wiesenfeldt, Gerhard. 2008. The Order of Knowledge, of Instruments, and of Leiden University, ca. 1700. In *Instruments in Art and Science: On the Architectonics of Cultural Boundaries in the 17th Century*, ed. Helmar Schramm, Ludger Schwarte, and Jan Lazardzig, 222–234. Berlin-Boston: Walter de Gruyter.

Wiesenfeldt 2016 = Wiesenfeldt, Gerhard. 2016. Academic Writings and the Rituals of Early Modern Universities. *Intellectual History Review* 26(4): 1–14.

Wilson 1994 = Wilson, Curtis. 1994. Newton's Orbit Problem, A Historian's Response. *College Mathematics Journal* 25(3): 193–200.

Wilson 2000 = Wilson, Catherine. 2000. Descartes and the Corporeal Mind: Some Implications of the Regius Affair. In *Descartes' Natural Philosophy*, ed. Stephen Gaukroger, John Schuster, and John Sutton, 659–679. London-New York: Routledge.

Wilson 2009 = Wilson, David B. 2009. *Seeking Nature's Logic: Natural Philosophy in the Scottish Enlightenment*. University Park: Pennsylvania State University Press.

Wohlwill 2001 = Wohlwill, Emil. 2001. The Discovery of the Parabolic Shape of the Projectile Trajectory. In *Galileo in Context*, ed. Jürgen Renn, 375–410. Cambridge: Cambridge University Press.

Wolf-Devine 1993 = Wolf-Devine, Celia. 1993. *Descartes on Seeing: Epistemology and Visual Perception*. Carbondale: Southern Illinois University Press.

Wollgast 1988 = Wollgast, Siegfried. 1988. *Philosophie in Deutschland zwischen Reformation und Aufklärung 1550–1650*. Berlin: Walter de Gruyter.

Woo 2013 = Woo, B. Hoon. 2013. The Understanding of Gisbertus Voetius and René Descartes on the Relationship of Faith and Reason, and Theology and Philosophy. *Westminster Theological Journal* 75(1): 45–63.

Woolhouse 1994 = Woolhouse, Roger S. 1994. Descartes and the Nature of Body. *British Journal of the History of Philosophy*, 2, 19–33.

Van der Woude, Sape. Undated. *Apparaat-Van der Woude. Bibliografie van Nederlandse dissertaties en oraties verschenen vóór 1800, op fiches. Samengesteld door prof. dr. S. van der Woude*. Special catalogue, Amsterdam University Library.

Wundt 1939 = Wundt, Max. 1939. *Die deutsche Schulmetaphysik des 17. Jahrhunderts*. Tübingen: Olms.

Yoder 1988 = Yoder, Joella. 1988. *Unrolling Time: Christiaan Huygens and the Mathematization of Nature*. Cambridge-New York: Cambridge University Press.

Yoder 1998 = Yoder, Joella. 1988. The Archives of Christiaan Huygens and his Editors. In *The Formation and Exchange of Ideas in Seventeenth-Century Europe*, ed. Michael Hunter, 91–107. Woolbridge: Boydell & Brewer.

Yoder 2013 = Yoder, Joella. 2013. *A Catalogue of the Manuscripts of Christiaan Huygens Including a Concordance with his Oeuvres Complètes*. Leiden: Brill.

Zagorin 2007 = Zagorin, Perez. 2007. Hobbes as a Theorist of Natural Law. *International History Review* 17: 239–255.

Zagorin 2009 = Zagorin, Perez. 2009. *Hobbes and the Law of Nature*. Princeton: Princeton University Press.

Zedler 1746 = Zedler, Johann Heinrich. 1754. Volder, Burchard von. In Zedler, Johann Heinrich. 1754. *Grosses vollständiges Universal-Lexicon Aller Wissenschafften und Künste*, volume 50, 409. Leipzig: Spalte.

Zepeda 2013 = Zepeda, Joseph. 2013. Descartes on Physical Vacuum: Rationalism in Natural-Philosophical Debate. *Society and Politics* 7(2): 126–141.

Zepeda 2014 = Zepeda, Joseph. 2014. The Concept of Space and the Metaphysics of Extended Substance in Descartes. *History of Philosophy Quarterly* 31(1): 21–39.

Zouckermann 1981 = Zouckermann, Raymond. 1981. Air Weight and Atmospheric Pressure from Galileo to Torricelli. *Fundamenta Scientiae* 2(2): 185–204.

Zuidervaart 2007 = Zuidervaart, Huib J. 2007. *Telescopes from Leiden Observatory and Other Collections 1656–1859. A Descriptive Catalogue*. Leiden: Rijksmuseum Boerhaave.

Zuidervaart 2012a = Zuidervaart, Huib J. 2012. Jacob Ouseel ['Oiselius'] (1631–1686). A Versatile Book Collector, Telescope Maker and Dubious Scholar from the Netherlands. In *Weiter sehen/ Seeing further. Beiträge zur Frühgeschichte des Fernrohrs und zur Wissenschaftsgeschichte Augsburgs/Essays on the Early History of the Telescope and History of Science in Augsburg*, ed. Jürgen Hamel and Michael Korey, 161–190. Frankfurt: Harri Deutsch.

Zuidervaart 2012b = Zuidervaart, Huib J. 2012. The 'Invisible Technician' Made Visible: Telescope Making in the Seventeenth and Early Eighteenth-Century Dutch Republic. In *From Earth-Bound to Satellite – Telescopes, Skills and Networks*, ed. Alison D. Morrison-Low, Sven Dupré, Stephen Johnston, and Giorgio Strano, 41–102. Leiden: Brill.

Index of Manuscript

A. Strazzoni, *Burchard de Volder and the Age of the Scientific Revolution*, Studies in History and Philosophy of Science 51,
https://doi.org/10.1007/978-3-030-19878-7

Name Index

A. Strazzoni, *Burchard de Volder and the Age of the Scientific Revolution*, Studies in History and Philosophy of Science 51,
https://doi.org/10.1007/978-3-030-19878-7

M

N

T

Y

Z

MIX
Papier aus verantwortungsvollen Quellen
Paper from responsible sources
FSC® C105338

If you have any concerns about our products,
you can contact us on
ProductSafety@springernature.com

In case Publisher is established outside the EU,
the EU authorized representative is:
Springer Nature Customer Service Center GmbH
Europaplatz 3, 69115 Heidelberg, Germany

Printed by Libri Plureos GmbH
in Hamburg, Germany